EVOLUTION: THE FIRST FOUR BILLION YEARS

EVOLUTION

The First Four Billion Years

EDITED BY

MICHAEL RUSE
JOSEPH TRAVIS

WITH A FOREWORD BY
EDWARD O. WILSON

THE BELKNAP PRESS OF
HARVARD UNIVERSITY PRESS

CAMBRIDGE, MASSACHUSETTS
LONDON, ENGLAND
2009

Library of Congress Cataloging-in-Publication Data

Evolution : the first four billion years / edited by Michael Ruse, Joseph Travis ; with a foreword by Edward O. Wilson.
p.   cm.
Includes bibliographical references and index.
ISBN 978-0-674-03175-3 (alk. paper)
1. Evolution (Biology)   I. Ruse, Michael.   II. Travis, Joseph, 1953–
QH366.2.E863   2009
576.8—dc22   2008030270

Contents

Foreword

Edward O. Wilson

Two centuries after its author's birth and 150 years after its publication, Charles Darwin's *On the Origin of Species* can fairly be ranked as the most important book ever written. Not the most widely read, to be sure. Copies of the *Origin* have not been placed in hotel rooms across America; its verities are not preached from pulpits each Sunday; and no political leader takes an oath of office with a hand on its cover. It is the masterpiece that first addressed the living world and (with *The Descent of Man* following) humanity's place within it, without reference to any religion or ideology and upon massive scientific evidence provided across successive decades. Its arguments have grown continuously in esteem as the best foundation for human self-understanding and the philosophical guide for human action.

So solidly have the fields of biology built upon the Darwinian conception of evolution that it makes sense today to recognize it as one of the two laws (universal principles if you wish) that govern our understanding of life. The first law is that all the elements and processes that define living organisms are ultimately obedient to the laws of physics and chemistry. This formulation has been the driving force of molecular and cellular biology during the past half century. The second law, the foundation and product of evolutionary biology as well as much of organismic and environmental biology, is that all elements and processes defining living organisms have been generated by evolution through natural selection.

The two laws have proven to be fully complementary, a prime requirement for the recognition of scientific laws generally. Further, to an increasing degree they are being combined to achieve seamless analyses of particular biological phenomena. The first law addresses how a phenomenon occurs and the second addresses why it occurs. Mitochondria and other organelles, for example, are built and work thus and so; and they originated in one way and not some other. Each description completes the explanation for the other. Modern biology as a whole has little meaning without the joining of both approaches guided by the two laws.

Even without the impetus of the *Origin* sesquicentennial—fortuitously the same as the Darwin birth bicentennial—2009 is an appropriate time to present evolutionary biology in its encompassing modern form. I doubt that any single writer, or any ensemble of several writers, could summarize this burgeoning discipline in a comprehensive and authoritative form. The multiplicity of authors in *Evolution: The First Four Billion Years* has gone far toward that goal. Its array of experts provide state of the art for each subject in turn. The science is presented here, but also some of the main practical consequences and inevitably, with fundamental relevance, the consequences of biological evolutionary thought for philosophy and religion. The great questions—"Who are we?" "Where did we come from?" and "Why are we here?"—can be answered only, if ever, in the light of scientifically based evolutionary thought.

Introduction

Michael Ruse and Joseph Travis

The discovery of evolution is one of the greatest intellectual achievements of Western thought, ranking with the calculus and general and specific relativity among scientific discoveries that changed indelibly how we see our world. From seeing nature as fixed forever in form and composition to seeing it as forever changing, we have been transformed utterly by discovering and understanding evolution.

In science the word *evolution* is used to express three different but related ideas. First, there is the *fact* of evolution. This is the realization that all organisms, living and dead, including humans, are the end products of a long natural process of change through which each species is descended from other, different ones. Although protoevolutionary ideas date back to the time of the ancient Greeks, only in the eighteenth century did the claim that organisms evolved really start to gain currency. However, the idea was not considered a basic element of scientific knowledge until after Charles Darwin published his *On the Origin of Species* in 1859. Darwin argued that all the organisms on the planet emerged through a single, straightforward process that operated relentlessly from the dawn of life and that continues to shape the natural world. An enormous amount of scientific research, from paleontology to molecular biology, has provided the compelling and overwhelming evidence that evolution is indeed a fact.

Second, *evolution* is often used to refer to the *path* of life's history on this globe. This can refer to the history of a group of existing species, as in the evolution of orchids, or the path through which particular traits emerged, such as the evolution of the middle ear of mammals from small bones in the reptile jaw. Evolutionary histories are reconstructed from a wide variety of evidence that ranges from sequences of species and forms in the fossil record to similarities and differences in the DNA of existing organisms. The evolution of some features, such as the hooves of ungulates, is very well understood, while the evolution of others, particularly very old features like the organelles of cells, is less well understood.

The third idea is the *theory* of evolution. This refers not to a suggestion that evolution has not occurred or that the history of particular features cannot be reconstructed but to our ideas about the forces that drive evolutionary change. In the *Origin* Darwin proposed what today is almost universally considered the major force in organic change, natural selection. But natural selection is not the only important evolutionary force; natural selection operates on the variation in shapes, sizes, and forms of organisms created by mutations in their genes, and the rate and magnitude of the changes that are driven by natural selection are governed by how those genes control development. An enormous amount of observation and experiment has documented the action of natural selection and some of the genetic responses to selection. But modern discoveries in genetics and development have raised many new questions and guided us toward some surprising answers about how all these forces combine to drive evolution.

Outside science the word *evolution* refers to a natural process considered the inspiration for a host of sociological, literary, political, philosophical, and religious ideas. From writers exploring the animal nature of humans to psychologists searching for the origin of human behavior in evolutionary history to theologians grappling with the implications of evolution for our understanding of the Divine, the influence of evolution outside science far exceeds that of any other scientific discovery ever made. And as modern controversies in the schools illustrate, particularly in the United States, it provokes reactions almost as strong as issues surrounding the beginning and end of human life.

In this book we explore all these facets of evolution. Our authors present the evidence for evolution as fact from the fossil record to genomics, illustrate the history of groups from bacteria to birds, and describe our current ideas about how these histories and this evidence came to be. Our authors also explore the influence of evolution on philosophy, religion, sociology, psychology, and many other areas and address the current controversies about the teaching of evolution in schools. The book covers the historical discoveries, such as homology, that opened the path for discovering evolution; the phenomena, such as industrial melanism, that played major roles in inspiring experimental studies of evolution; the applications of evolutionary principles to areas seemingly far afield, such as computer science; and the contributions of the major figures who shaped the history of evolutionary science and the discipline as it exists today.

Perhaps most important, this book presents evolutionary science as a modern, dynamic discipline. All too often discussions of evolution outside the science itself give the impression that the subject became fossilized in the nineteenth century. This is particularly so when the suitability of evolution as a subject for secondary-school students is being argued. The irony is that the discoveries of modern science, from molecular biology to computational innovation, have provided compelling evidence that Darwin could only dream of seeing. From those discoveries have also come applications that Darwin might never have imagined, such as Darwinian medicine. Principles of evolution are being applied to a wide variety of scientific problems, from understanding senescence to strategies for conservation to explaining human behavior. There are few topics in science with so many exciting new facets, which reveal that despite all we have learned, there is much more to be discovered. Not that there is further need to substantiate evolution as fact; rather, there is much more to be done to trace the evolution of groups of species or features, particularly complex features at the cellular and molecular levels, and to refine our theory of evolution to account more fully for the astonishing diversity of life as we continue to discover it. The essays in this book show not only how far we have come but where the scientific horizons lie and how we might move toward those horizons.

This book is organized into two parts. The first part contains long essays on the overarching themes in evolution. From the history of evolutionary thought to the controversies over education, from the fossil record and the origin of life to the process of adaptation, the long essays offer primers on the major features of evolution. Many of them offer close looks at particular areas like molecular evolution, genomic evolution, and Darwinian medicine. The second part contains a large number of shorter essays on more specific topics. These include essays on major groups of organisms with which most people are familiar, topics that are important facets of evolution, major figures in the discovery and shaping of evolutionary science, and the critical books that recount the history of discovery, development, and maturation of a discipline.

This organization allows the reader to explore evolution according to his or her interests and background. The reader who wishes to be immersed in the science can focus on the major scientific themes, while the reader who is interested in the intellectual history and influence of the subject away from science can enter through a separate set of essays. A reader interested in very specific topics or historical figures can find the appropriate entries, along with essays on related topics. Our goal was to provide an exciting and compelling introduction to evolution along with a basic reference work that could point the way toward a deeper study of individual issues. The essays include bibliographies, which serve as guides to further and deeper reading.

Although we hope that our fellow scholars enjoy these essays, we want this volume to inform, educate, and excite readers who are not professional scholars. We find evolution in all of the word's meanings to be a provocative, interesting, and indeed awe-inspiring topic, and we want our readers to emerge with the same feeling. But beyond the excitement of studying evolution, it is important to understand and appreciate it. Evolution explains the challenge of antibiotic-resistant pathogens and the scourge of novel infectious diseases, it shows us what we can expect climate change and our own alterations of habitat to do to the natural world, and it may offer profound explanations of who we are and why we behave as we do. It also challenges many of our closely held beliefs about man's, and woman's, place in nature.

The last of these issues is perhaps the foundation of why evolution has proven so controversial. Survey after survey shows that fewer than 50% of people in the United States accept that evolution is responsible for the diversity of life on earth. It is unclear how much of that opposition is based on a poor understanding of evolution and the evidence about it and how much is based on the challenges that it offers. Yet opposition to evolution by religious people, especially by Christians, is by no means inevitable; some of the most distinguished believers have accepted evolution without many qualms. We are convinced that the Christian—with the Jew and Muslim and other religious believers—should be at the head of the queue of those who welcome evolutionary ideas. Finding the idea, building and elaborating upon it, and looking at its influence and importance are, as the contributors to this volume show again and again, truly the best of proofs that we are made in the image of God, celebrating creation in all its wonderful manifestations.

This book is a testament to one of mankind's greatest discoveries, a discovery that offers unparalleled insight into what creation really is. We love evolution, and if we and our fellow contributors can inspire you even in a small way with our enthusiasm, then we shall be happy indeed.

The History of Evolutionary Thought

Michael Ruse

The idea that all organisms (including humans) are generated by natural means from other forms has ancient roots. Aristotle tells us that Empedocles (fifth century B.C.E.) toyed with such thoughts. However, it was not until the eighteenth century and the Enlightenment that *evolution* (as we now call this idea of natural development) really started to gain a serious number of supporters. There are reasons both for the long delay and why the idea finally began to gain momentum.

The Early Days

The Greeks had no great religious objection to evolution, but their world picture did not have a place for any kind of significant developmental processes. Specifically, the Greeks thought that they had irrefutable reasons to reject ongoing, incremental organic change. They—particularly the philosophers Plato and Aristotle—thought that the world (especially the world of organisms) showed order and intention and, as such, was not something that could simply have appeared through blind, ungoverned processes of law. It certainly was not something that could have grown from simple beginnings to the complexity of today. Plato used human fingernails as an example of order and design: "Sinew, skin and bone were interwoven at the ends of our fingers and toes. The mixture of these three was dried out, resulting in the formation of a single stuff, a piece of hard skin, the same in every case." Plato then went on to put things in context.

> Now these were merely auxiliary causes in its formation—the preeminent cause of its production was the purpose that took account of future generations: our creators knew that one day women and the whole realm of wild beasts would one day come to be from men, and in particular they knew that many of these offspring would need the use of nails and claws or hoofs for many purposes. This is why they took care to

include nails formed in a rudimentary way in their design for humankind, right at the start. This was their reason, then, and these the professed aims that guided them in making skin, hair and nails grow at the extremities of our limbs. (*Timaeus*, 76d–e, in Cooper 1997, 1277)

In an incredibly influential discussion, Aristotle in *De partibus animalium* (1984) identified the factors at work here as "final causes." These are causes that occur not just to produce or do something (the finger parts dry and make nails) but for the sake of some kind of purpose (the nails protect the finger ends). They show some kind of forethought or intention. For this reason, final causes cannot be reduced to blind, unguided law, as is demanded in evolution. The world, particularly the world of organisms, must in some sense have been designed rather than just produced under its own steam by natural processes. (Sedley 2008 is the definitive study.)

The Jews, and following them the early Christians, had religious reasons for the rejection of evolution. It goes against the creation stories of the early chapters of Genesis, which portray a world created miraculously by God and then peopled by him through divine fiat over a short time span. But do not think that religion as such was then and always an absolute bar to evolutionism. The church fathers (the major Christian theologians of the early centuries) worked toward an understanding of the biblical text that would allow interpretation, particularly in the face of advances of science. Saint Augustine was eager not to let ancient creation accounts stand in the way of modern thought. He himself, believing that God stands outside time, speculated in a kind of protoevolutionary fashion that the Divine had formed seeds of life that then sprang into full being when they were placed here on earth. However, one should not read too much into any of this. Like the Greeks, the Jews and Christians were simply not looking in the direction of evolution and would have thought final causes an unanswerable objection to significant developmentalism. As is well known, these kinds of causes became a foundation of one of the major proofs of God's existence, the Argument from Design, which moves from design here on earth to the existence of the divine artificer.

Why, then, did evolution start its rise in the eighteenth century? The answer is simple. It was at this time that people started to challenge the Christian picture of world history—a providential picture of a world created by God, where humans are made in his image but have fallen and are able to achieve salvation only through his undeserved grace. Some began to argue that perhaps humans held their fates in their own hands and could progressively improve their own lots. It was this idea of progress—the belief that the world and its denizens are on a trajectory upward and that this upward rise is made possible by (and only by) the unaided efforts of the world's human inhabitants—that gave rise to the idea of organic evolution (Ruse 1996). Enthusiasts for progress extended their thinking into nature and developed the idea of evolution—progressive change upward from the simple to the complex. They then read this idea back into human thought and social practice as confirmation of their beliefs about progress.

The British physician and man of science Erasmus Darwin (grandfather of Charles) was a paradigm who hymned in verse life's upward rise to humankind:

> Imperious man, who rules the bestial crowd,
> Of language, reason, and reflection proud,
> With brow erect who scorns this earthy sod,
> And styles himself the image of his God;
> Arose from rudiments of form and sense,
> An embryon point, or microscopic ens!
>
> (E. Darwin 1803, 1: Canto I, lines 309–314)

This is all the end product of the progressive development of human intelligence, which causes and is reflected in humans' scientific achievements:

> How loves and tastes, and sympathies commence
> From evanescent notices of sense;
> How from yielding touch and rolling eyes
> The piles immense of human science rise!
>
> (Canto III, lines 43–46)

Similar ideas were to be found elsewhere, most notably in France. In his *Philosophie zoologique* (1809), the taxonomist Jean-Baptiste de Lamarck produced the first full-blown evolutionary theory—a picture of upward rise to our own species from the most primitive forms of life, which in turn had been produced from mud and slime through the actions of heat and electricity and other natural forces.

Although the metaphysical idea of progress was the main factor behind the rise of evolutionary ideas, it is not true that there was no pertinent empirical evidence. Aristotle had noted that organisms of very different species seem to share common patterns or structures—what today are known as homologies—and the evolutionists were ready to interpret these as signs of common ancestries (Figure 1). Likewise, the successes of animal and plant breeders did not go unnoticed (Figure 2). But generally the evidence took a very secondary position. The fossil record, something that today many (if not most) people would invoke first as the proof of developmental origins, was less than helpful. As a systematic proof of progressive change, the gleanings from the rocks were meager indeed. In any case, counting against the empirical side was the fact that no one had any great understanding of what might have caused evolution. Most assumed some kind of vague, upwardly thrusting force or forces, but little more. Generally, everyone was committed to the folk belief that characteristics acquired in one generation could be transmitted immediately to the future generations—Lamarck was so enthused by this process that the inheritance of acquired characteristics has since become known as Lamarckism—but beyond this was silence.

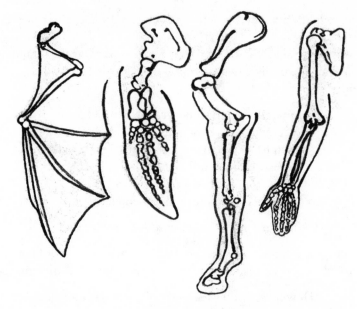

Figure 1. Homologies. These are the forelimbs of *(from left to right)* the bat (its wing), the porpoise (its paddle), the horse (its leg), and the human (its arm). The bones are obviously similar, as if modeled on a single archetype, and yet the uses to which they are put are very different. It was the British anatomist Richard Owen who called them "homologies," although it was not until after Darwin's *On the Origin of Species* (1859) that they became universally accepted as proof of descent from a common ancestor.

The ideology of progress was what counted, and it was for this reason that most people around 1800 would have regarded evolutionism less as a real science and more as a pretender, somewhat like animal magnetism (mesmerism) and the reading of character from skull shape (phrenology). Even judged by the standards of that time, evolution was what may fairly be called a pseudoscience. Obviously, Christian opponents of evolution disliked intensely the antiprovidential underpinnings of the doctrine. But evolution was not associated with total nonbelief, atheism, or even what later in the nineteenth century Thomas Henry Huxley was to call agnosticism. Most evolutionists were deists who believed in God as unmoved mover, a being who had set the world in motion and now let it unfurl without need of miraculous intervention. For the deist, indeed, evolution was proof of God's power and intention rather than disproof. Everything was planned beforehand and went into effect through the laws of nature. In Erasmus Darwin's words: "What a magnificent idea of the infinite power of THE GREAT ARCHITECT! THE CAUSE OF CAUSES! PARENT OF PARENTS! ENS ENTIUM!" (1801, 2: 247). There is a link here, especially in England, with the Industrial Revolution. People were harnessing the forces of nature—water, coal, and others—to produce goods through machines rather than by hand. The god of the deist was the ultimate

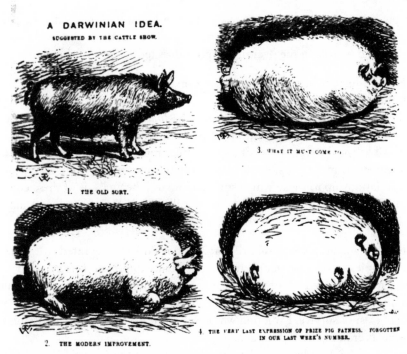

Figure 2. Prize pig. An industrial revolution means that people leave the countryside and go to the factories in the cities. Hence, far fewer people are left on the land to produce the nation's food. There has to be an agricultural revolution, and at the end of the eighteenth century people discovered that a major key to success was the breeding of better-quality livestock—fatter pigs, hairier sheep, and bigger cattle. (From *Punch*, 1865.)

divine industrialist as he harnessed the forces of nature to produce the goods of the world through law rather than miraculously or by hand.

Critics of evolution, notably the great French comparative anatomist Georges Cuvier, made reference to empirical problems. Cuvier cited the mummified bodies of humans and animals brought back by Napoleon's army from Egypt. Although they were very old, their forms were identical to contemporary forms and hence counted against ongoing organic change. But as with the positive case for evolution, in the negative case it was the ideology that really counted. Above all, for Cuvier, there was hatred of progress— hatred of a doctrine that had led humans to think that they could do more than they could do and had led ultimately to the terrors of the French Revolution, when opponents of the existing state had tried to change all that had proved true and safe for many generations. Combined with this, for Cuvier, as for every other opponent of evolution, was the still-unsolved problem of final cause. The Frenchman emphasized that this was something that could not be ignored; indeed, it was the most important distinguishing feature of life. The key to understanding the organism lay in the fact that it was not simply subject to the physical laws of nature but was organized. The parts were directed to the end of the functioning whole, and each individual feature played its role in the overall, purpose-directed scheme of things:

> Natural history nevertheless has a rational principle that is exclusive to it and which it employs with great advantage on many occasions; it is the *conditions of existence* or, popularly, *final causes*. As nothing may exist which does not include the conditions which made its existence possible, the different parts of each creature must be coordinated in such a way as to make possible the whole organism, not only in itself but in its relationship to those which surround it, and the analysis of these conditions often leads to general laws as well founded as those of calculation or experiment. (Cuvier 1817, 1: 6)

Cuvier's point simply was that the organism is far too integrated—organized and complex—to allow significant change in any direction. It is certainly too integrated to allow change from one species to another. Organisms at midpoint would be literally neither fish nor fowl and hence would simply be unable to exist or survive. Evolution was in some sense a theoretical impossibility, as well as empirically unfounded.

Religion was involved too. Progress goes against the Christian doctrine of Providence. Nevertheless, although Cuvier thought that there was evidence of Noah's flood, neither he nor other serious scientists wanted to make the case by simple reference to Genesis. Indeed, by the beginning of the nineteenth century, all were starting to realize that the earth's history must be far older than the traditional 6,000 years that one can work out from the genealogies given in the Bible. It is not that the Bible is false, but rather that it needs interpretation. Some solved the problem by thinking of the six days of creation as six long periods of time; others solved it by supposing that there were long, unmentioned gaps between the biblical days. God's creation therefore was a long, drawn-out process, but it was not evolutionary.

The controversy was at an impasse, and not much had changed by the middle of the nineteenth century. On the one side were the evolutionists, committed to progress and ardent in their belief that organic development was the perfect complement to this ideology, with enthusiasm outstripping empirical knowledge. Confirming this pattern, in 1844 the Scottish publisher Robert Chambers wrote (anonymously) a highly popular work on evolution, *Vestiges of the Natural History of Creation,* in which he argued that everything was in a state of upward becoming and that what happened in the social world mirrored what happened in the biological world:

> The question whether the human race will ever advance far beyond its present position in intellect and morals, is one which has engaged much attention. Judging from the past, we cannot reasonably doubt that great advances are yet to be made; but if the principle of development be admitted, these are certain, whatever may be the space of time required for their realization. A progression resembling development may be traced

in human nature, both in the individual and in large groups of men . . .
Now all of this is in conformity with what we have seen of the progress
of organic creation. It seems but the minute hand of a watch, of which
the hour hand is the transition from species to species. Knowing what
we do of that latter transition, the possibility of a decided and general
retrogression of the highest species towards a meaner type is scarce admissible, but a forward movement seems anything but unlikely. (Chambers 1846, 400–402)

On the other side were the opponents, committed to Providence, vocal
about the significance of final cause, and accepting that Genesis must be
modified but thinking that this could be readily done. David Brewster, general man of Scottish science and biographer of Newton, could see the dan-
gers: "It would auger ill for the rising generation, if the mothers of England
were infected with the errors of Phrenology: it would auger worse were they
tainted with Materialism." The problem, Brewster gloomily concluded with
reflections that still find much support in many circles today, stemmed from
the slackness of schools and universities: "Prophetic of infidel times, and indicating the unsoundness of our general education, 'The Vestiges . . .' has
started into public favour with a fair chance of poisoning the fountains of science, and of sapping the foundations of religion" (Brewster 1844, 503). (Interestingly, this kind of attack seems to have confirmed Chambers in his
views. Although, from the start, progress was the leitmotif of his book, he
felt the need to make his point ever more explicit. The passage just quoted
above is from the fifth edition of *Vestiges* and replaces a passage that gives
more credit to the Creator. [Figure 3])

Of course, neither side was really satisfactory. Ideology is no substitute for
real evidence, and final cause can be ignored but does not go away. Referring
all to miracle may be socially and psychologically comforting, but it is not a
good scientific solution. The time had come for a significant step forward.

Charles Darwin

Charles Robert Darwin (1809–1882) was sent to Edinburgh University to
train in the family tradition of medicine (Figure 4). After two years he
dropped out, bored with the lectures and revolted by the operations. Yet already Darwin had started to mix with scientists, especially naturalists interested in the living world. One of his acquaintances was Robert Grant, an
anatomist and an avowed evolutionist. So, quite apart from his grandfather's
work (the young Charles read Erasmus's major treatise, *Zoonomia*), evolution was an idea to which Darwin was introduced at an early age. It seems,
nevertheless, that the youthful Darwin accepted in a fairly literal form the
whole of Christianity, including the early chapters of Genesis, and that this
was a factor in his redirected choice of a career: to be an ordained minister in
the established Church of England. To achieve this end one needed a degree

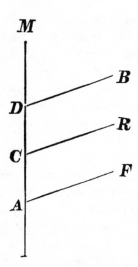

Figure 3. Chambers's evolution. This is the picture of evolution that Chambers gave in *Vestiges of the Natural History of Creation.* Apparently, for no apparent cause, a primitive organism went on developing in the womb and turned into a fish; then a fish went on developing and turned into a reptile; then a bird; and finally a mammal. Note the intertwining of thinking about the history of the individual and the history of the group, something that was to become commonplace in evolutionary thinking later in the nineteenth century. (From *Vestiges of the Natural History of Creation.*)

from an English university, and so, in 1828, Darwin was packed off to Christ's College, Cambridge.

In 1831 Darwin (who continued mixing with scientists) got his big break. After he graduated, his career as a clergyman was put on hold through the offer of a lengthy voyage on HMS *Beagle,* just about to start on a surveying trip around South America (Figure 5). A major influence at this point was (vicariously) the Scottish geologist Charles Lyell, who at the beginning of the decade began publishing his massive *Principles of Geology* (there were three volumes; Darwin took the first with him and had the others sent out). Although he was no evolutionist, Lyell insisted that the physical world must be explained in terms of natural causes of a kind now still working (Figure 6). This had a great effect on Darwin, whose first systematic work was in geology, and prepared him to think about the organic world likewise less in biblical terms and more in terms of natural causes. In 1835, leaving South America, the *Beagle* sailed into the Pacific Ocean and visited the Galápagos Archipelago, a group of islands on the equator, far from land. Thanks to the governor, who pointed out that the giant tortoises indigenous to the archipelago were different from island to island, Darwin came to see that this held for the Galápagos fauna generally—the birds in particular, the finches and the mockingbirds, were peculiar to their specific homes (Figure 7). The differences had to be significant, and the only way the significance could be

Figure 4. Charles Darwin. This is a wedding portrait of Darwin, painted in 1839 just before he married his first cousin Emma Wedgwood. It is a very good likeness (no surprise, since the painter was George Richmond, the best portraitist in England at the time), which points to the fact that Darwin came from a very rich, middle-class family. One of his grandfathers, shared by his wife, was Josiah Wedgwood, responsible for bringing the pottery trade into mass production. Although he was a full-time scientist, Darwin never had to work for a living. The fact that he was living off the family fortune perhaps explains why *On the Origin of Species* was written in a very open and user-friendly fashion. Darwin was writing in part for his sponsors, the Darwin-Wedgwood family.

explained was through "descent with modification," an idea that Darwin embraced in early 1837, shortly after the *Beagle* voyage (Figure 8). Unlike earlier thinkers, what was of great concern to Darwin was the cause or causes of evolution. Without causes, he was no more than one among many evolutionists.

Figure 5. HMS *Beagle*. The *Beagle* was captained by Robert Fitzroy, then only 23 years old. Being a captain on a warship would be a very lonely post because one would be set aside from the crew. Fitzroy therefore looked for someone who could be his companion on the trip, someone not in the navy but a gentleman able to pay his own bills. Darwin fit the job perfectly, rapidly evolving from captain's friend to full-time ship's naturalist. The crew used to call him "Philos," short for Natural Philosopher—the term then used for scientists.

Without causes, he could never be the Newton of biology. Darwin wanted to find a biological force that would explain the evolution of life and was universal in the world and the key behind all organic motions, and to this end he worked feverishly for the next 18 months, speculating and doing an extensive literature review.

Darwin soon realized that the probable key to change lay in something parallel to the way in which animal and plant breeders effect change, namely, the selection of the desirable and the rejection of all others. But how was one to get a natural equivalent to the breeders' artificial selection? At the end of September 1838 Darwin read *An Essay on the Principle of Population* (sixth edition, 1826) by the Reverend Thomas Robert Malthus. In his work Malthus argued that food supplies would always be outstripped by potential population growth. Hence, unless there was "prudential restraint," there were bound to be ongoing struggles for existence as people competed for the available food and supplies, including living space (Figure 9). Darwin turned Malthus's reasoning on its head. He pointed out that there could be no prudential restraint in the animal and plant world, and hence, because population growth certainly was not restricted to humankind, there would be an ongoing, organic-wide struggle for existence. More than this, success and failure in the struggle would (on average) be a function of the different

PRINCIPLES

OF

GEOLOGY,

BEING

AN ATTEMPT TO EXPLAIN THE FORMER CHANGES OF THE EARTH'S SURFACE,

BY REFERENCE TO CAUSES NOW IN OPERATION.

———

BY

CHARLES LYELL, Esq., F.R.S.

FOR. SEC. TO THE GEOL. SOC., &c.

———

IN TWO VOLUMES.

Vol. I.

———

LONDON:

JOHN MURRAY, ALBEMARLE-STREET.

—

MDCCCXXX.

Figure 6. *Principles of Geology.* The full title of Lyell's work tells you all about his methodological intentions. Eventually, there were three volumes.

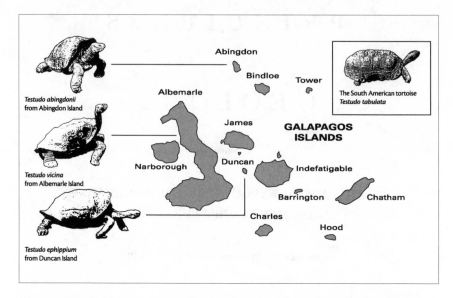

Figure 7. The Galápagos tortoises. Darwin did not become an evolutionist until he returned home and he could show his collection of birds to John Gould, a leading ornithologist. It was when he was told that there were unambiguously different species that he moved across to evolution. Note that even before he became an evolutionist, Darwin was moving in circles where he could call on the top minds for support and information. He made powerful friends at Cambridge and they spotted his talent early.

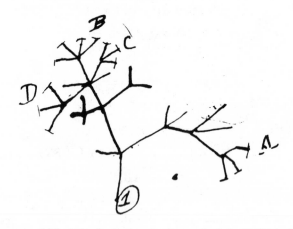

Figure 8. The tree of life. This little sketch, drawn in July 1837, is from a private notebook that Darwin was keeping on the species problem. Lamarck had a tree-of-life picture and in the eighteenth century one can find them in the writings of nonevolutionists who were nevertheless trying to show the order in which God had created.

Figure 9. The Malthusian explosion. This cartoon appeared in the *Comic Almanack* in 1851. One of the reasons that Darwin was able to convince people of the truth of evolution was, apart from his friendly style, the fact that he relied on ideas that were widely accepted in Victorian Britain.

characteristics possessed by organisms. Hence there would be an ongoing natural selection in which the winners would pass on features to the next generation and the losers would get nothing. Over time, this selection would lead to full-blown organic change. But this organic change would be of a particular kind, for the features that make for success, known as "adaptations"—for example, ears, eyes, teeth, hands, legs, leaves, bark, and roots—would show that purposeful nature that epitomizes final cause.

Darwin sat on his evolutionary ideas for 15 years, during which time he turned to a massive study of barnacles. We are not quite sure why this delay occurred, although by this time Darwin had fallen sick with a mysterious ailment that was to plague him for the rest of his life, and so undoubtedly he was not relishing the huge debate that his ideas were bound to cause. Also a major factor must have been his reluctance to upset powerful science-establishment figures (including Cambridge mentors) who had encouraged the young Darwin in his work. One of the things he did during the pause was to network with younger scientists, who could rally around him when he did go public. Finally, however, Darwin was pushed into action when, in the middle of 1858, he received a short essay by Alfred Russel Wallace, a collector in the Malay peninsula, that had virtually the same premises and conclusion that he had discovered some 20 years earlier.

ON THE ORIGIN OF SPECIES

Darwin quickly wrote up his ideas, and *On the Origin of Species by Means of Natural Selection, or the Preservation of Favoured Races in the Struggle for Life* appeared in the fall of 1859 (Figure 10). The work begins with a discussion of artificial selection and animal and plant breeders' great successes. This is partly heuristic, to prepare the reader by following the path that Darwin himself had taken to discovery, and partly justificatory, suggesting that what we can do, nature can surely do better. Following is a discussion of the variation that always occurs in the wild, something that Darwin obviously needed as the building blocks of change. If every organism were exactly like its parents, then there could be no evolution. At this point Darwin was ready to introduce the struggle for existence and, following on this, the mechanism of natural selection:

> A struggle for existence inevitably follows from the high rate at which all organic beings tend to increase. Every being, which during its natural lifetime produces several eggs or seeds, must suffer destruction during some period of its life, and during some season or occasional year, otherwise, on the principle of geometrical increase, its numbers would quickly become so inordinately great that no country could support the product. Hence, as more individuals are produced than can possibly survive, there must in every case be a struggle for existence, either one individual with another of the same species, or with the individuals of distinct species, or with the physical conditions of life. (C. Darwin 1859, 63)

ON

THE ORIGIN OF SPECIES

" But with regard to the material world, we can at least go so far as this—we can perceive that events are brought about not by insulated interpositions of Divine power, exerted in each particular case, but by the establishment of general laws."

W. WHEWELL : *Bridgewater Treatise.*

BY MEANS OF NATURAL SELECTION,

OR THE

PRESERVATION OF FAVOURED RACES IN THE STRUGGLE FOR LIFE.

" To conclude, therefore, let no man out of a weak conceit of sobriety, or an ill-applied moderation, think or maintain, that a man can search too far or be too well studied in the book of God's word, or in the book of God's works ; divinity or philosophy ; but rather let men endeavour an endless progress or proficience in both."

BACON : *Advancement of Learning.*

By CHARLES DARWIN, M.A.,

FELLOW OF THE ROYAL, GEOLOGICAL, LINNÆAN, ETC., SOCIETIES ;
AUTHOR OF ' JOURNAL OF RESEARCHES DURING H. M. S. BEAGLE'S VOYAGE
ROUND THE WORLD.'

Down, Bromley, Kent,
October 1st, 1859.

LONDON:
JOHN MURRAY, ALBEMARLE STREET.
1859.

The right of Translation is reserved.

Figure 10. The title page of *On the Origin of Species*. Darwin was being a little cheeky in quoting the philosopher William Whewell, opposite the title page, on the topic of the law-governed nature of the universe. Whewell had been one of Darwin's mentors when the latter was a student at Cambridge, and he continued in this role in the early years after the *Beagle* voyage. He was a well-known opponent of evolution and was scathing on the subject of *Vestiges*. Darwin sent a complimentary copy of the *Origin* to Whewell, who responded by banning the book from the shelves of the library of Trinity College, of which he was Master (head).

Note that even more than a struggle for existence, Darwin needed a struggle for reproduction. It is no good having the physique of Tarzan if you have the sexual desires of a philosopher. But with the struggle understood in this sort of way, given naturally occurring variation, natural selection follows at once:

Can it . . . be thought improbable, seeing that variations useful to man have undoubtedly occurred, that other variations useful in some way to each being in the great and complex battle of life, should sometimes occur in the course of thousands of generations? If such do occur, can we doubt (remembering that many more individuals are born than can possibly survive) that individuals having any advantage, however slight, over others, would have the best chance of surviving and of procreating their kind? On the other hand we may feel sure that any variation in the least degree injurious would be rigidly destroyed. This preservation of favourable variations and the rejection of injurious variations, I call Natural Selection. (80–81)

In later editions Darwin introduced the alternative term *survival of the fittest,* which was perhaps a little unfortunate, for it has led to endless claims that selection is a tautology, reducing simply to the claim that those who survive are those who survive. But although this is obviously true, selection means more than that. It claims, for better or for worse, that on average those who survive are different from those who do not, and that success in the struggle is a function of those differences. This may or may not be true, but it is not a tautology.

Darwin included a secondary selective mechanism. He argued that the struggle is not always for food and space but can be directly for mates. This mechanism, *sexual selection,* Darwin divided into two parts. There was selection through male combat (for females). The antlers of the deer would be the product of this. Then there was selection through female choice (of the most desirable males). The tail feathers or general plumage of the peacock or similar bird would be a product of this (Figure 11). Notice how sexual se-

Figure 11. Sexual selection. The analogy with animal breeding led Darwin to distinguish sexual selection from natural selection. Natural selection corresponds to selection for attributes that help survival, like thicker coats, whereas sexual selection corresponds to selection for attributes that give pleasure to the breeders, like prettier feathers or more ferocious fighting qualities. This picture is taken from *The Descent of Man.* (Because the *Origin* was written in a hurry, there was no time for footnoting or illustrating. Most of Darwin's other works are carefully footnoted and copiously illustrated.)

lection particularly puts the emphasis on competition between members of the same species: selection is directed toward individual-benefiting characteristics rather than species-benefiting characteristics. Darwin tended always to see selection (including natural selection) acting in this way.

Having thus presented the main mechanisms of change, Darwin introduced the famous metaphor of a tree: "The affinities of all the beings of the same class have sometimes been represented by a great tree. I believe this simile largely speaks the truth." The leaves and twigs at the top represent the species extant today. Then, as we go down the branches, we have the great evolutionary paths of yesterday. All the way down we go until we reach the very first shared origins of life. "As buds give rise by growth to fresh buds, and these, if vigorous, branch out and overtop on all sides many a feebler branch, so by generation I believe it has been with the great Tree of Life, which fills with its dead and broken branches the crust of the earth, and covers the surface with its ever branching and beautiful ramifications" (C. Darwin 1859, 129–130) (Figure 12).

Now, with some minor problems brushed away, Darwin was ready to present the second part of his theory. For a good two-thirds of the *Origin*, Darwin took the reader through the various branches of biological science—instinct, paleontology, biogeography, classification, morphology, embryology—and showed that phenomena in these branches are explained by evolution through natural selection, and, conversely, these various branches point to and support the mechanism of evolution through selection (Figure 13). There

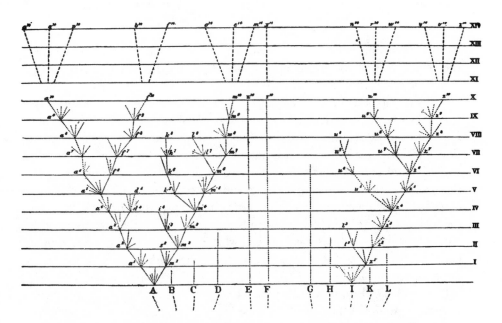

Figure 12. The tree of life as given in the *Origin*. Notice that the emphasis in this picture is less on showing life reaching up and more on the way in which it spreads.

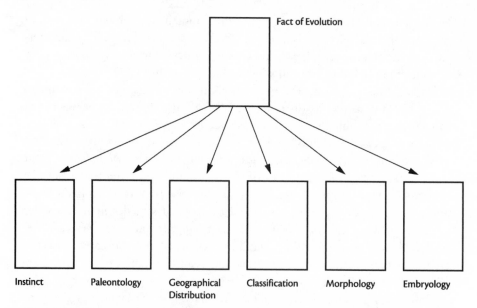

Figure 13. The structure of Darwin's argument. Darwin did not hit by chance on this kind of argumentation, where a unifying hypothesis explains in many different areas and in turn is justified or supported by these areas. Whewell, who called it a "consilience of inductions," made much of it when discussing successful theories of physics, especially Newtonian mechanics and the wave theory of light. Whewell identified the causes at the center of a consilience as "true causes," what Newton had called *verae causae*. Showing a consilience was therefore an essential strategy for a would-be Newton of biology.

were some problematic issues. Social behavior, particularly that shown by the social insects (the hymenoptera—the ants, the bees, and the wasps), would be easily explicable if selection favored the group. But from an individualistic perspective, why should one get sterile workers who seemingly spend their whole lives devoted to the well-being of others in the nest or hive? How can natural selection, with its emphasis on the importance of reproduction, produce organisms that cannot reproduce? Finally, Darwin, who always favored individualistic approaches over group approaches because he thought the latter too open to cheating and exploitation, decided that in such cases the group members are so well integrated that it is permissible to treat the whole hive as a kind of superorganism in which the individual insects are parts of the whole. Just as selection can work on the eye, for example, which exists for the benefit of the whole organism, so the worker exists for the benefit of the whole hive.

Paleontology also raised difficulties. It had its good points—for instance, that as we go down the record, increasingly we find organisms that seem to have features midway between features of extant organisms that are widely different. However, against the positive, there were gaps in the fossil record, and, even worse, the fossil record began abruptly at the start of what we would call the Cambrian; there was no record before that. The gaps Darwin explained simply as a product of incomplete fossilization. The remains might no longer be there, but the linking ancestors did exist. The abrupt beginning brought out all of Darwin's inventive powers. Early life probably existed where now there were oceans; this would explain why we on dry land could not find it. Even if we were able to drill beneath the oceans, however, it was doubtful that we would find life; the pressure from above had surely compressed and metamorphosed the fossil remains below (Figure 14).

Most other areas of biology fell into place with more ease. Geographic distribution (biogeography) was a triumph because Darwin explained just why it is that one finds the various patterns of animal and plant life around the globe. Why, for instance, does one have the strange sorts of distributions and patterns that are exhibited by the Galápagos Archipelago and other island groups? It is simply that the founders of these isolated island denizens came by chance from the mainland, and once they were established, they started to evolve and diversify under the new selective pressures to which they were now subject (see Figure 7). Systematics likewise exhibits the kind of patterns one expects from evolution. As Linnaeus demonstrated, organisms can be classified hierarchically in nested sets, thus showing a pattern that reveals past history (Figure 15). Morphology has those similarities, called homologies, between organisms of different species. Obviously these speak to shared ancestry (see Figure 1). Embryology likewise was a particular point of pride for Darwin. Why is it that the embryos of some different species, such as man and the dog, are very similar, whereas the adults are very different? Darwin argued that this follows from the fact that in the womb the selective forces on the two embryos would be very similar (they would not therefore be torn apart), whereas the selective forces on the two adults would be very different (they would be torn apart) (Figure 16). Here, as always throughout his discussions of evolution, Darwin turned to the analogy with the world of breeders in order to clarify and support the point at hand: "Fanciers select their horses, dogs, and pigeons, for breeding, when they are nearly grown up: they are indifferent whether the desired qualities and structures have been acquired earlier or later in life, if the full-grown animal possesses them" (C. Darwin 1859, 446).

So we are led to the concluding passages of the *Origin*. Darwin, whose Christianity had evolved into a form of deism (Lyell, a Unitarian like many members of the Wedgwood family, was a major influence here), never concealed that he was working in a God-backed mode, and there are frequent unself-conscious references to the Creator. "Authors of the highest eminence seem to be fully satisfied with the view that each species has been independently created." This was not Darwin's position. "To my mind it accords better with what we know of the laws impressed on matter by the Creator, that the production and extinction of the past and present inhabitants of the world should have been due to secondary causes, like those determining the birth and death of the individual" (C. Darwin 1859, 488). God backs the world. What God does not do is get involved in the world, and science has no business in supposing otherwise.

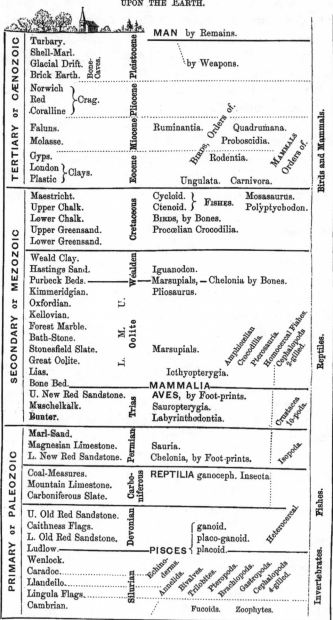

Figure 14. The history of life as seen in 1860. This is taken from Richard Owen's *Paleontology*. After the *Origin,* Owen became an implacable enemy of the Darwinians, thinking their theory to be materialistic; but, as young men Owen and Darwin had been friendly. Owen probably became a transmutationist of a kind (with God guiding the course of history), but one should not necessarily interpret pictures like these as implying evolution. As with the trees of life, people worked out the fossil record thinking that it might simply tell of the order in which God sequentially created organisms.

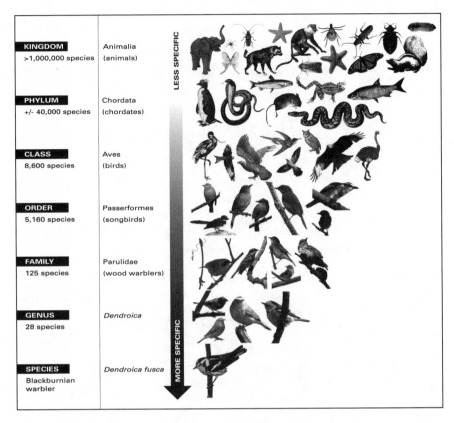

		LESS SPECIFIC
KINGDOM >1,000,000 species	Animalia (animals)	
PHYLUM +/- 40,000 species	Chordata (chordates)	
CLASS 8,600 species	Aves (birds)	
ORDER 5,160 species	Passerformes (songbirds)	
FAMILY 125 species	Parulidae (wood warblers)	
GENUS 28 species	*Dendroica*	
SPECIES Blackburnian warbler	*Dendroica fusca*	MORE SPECIFIC

Figure 15. The Linnaean hierarchy. Organisms can all be placed in a hierarchy with the members of lower-level sets being grouped into ever fewer (but more comprehensive) sets at higher levels. All organisms belong to one and only one set (known as taxa, singular taxon) at each level (category). This reflects a shared past, although as with the fossil record, much of this was worked out without regard to evolution. Hence, giving the hierarchy an evolutionary interpretation gave it new meaning and at the same time confirmed the hypothesis of shared descent.

THE DESCENT OF MAN

What about humans? From the first, Darwin was always stone-cold certain that humans are part and parcel of the evolutionary process. His encounter on the *Beagle* voyage with the natives at the bottom of South America, the Tierra del Fuegians, had convinced him of this. In the *Origin*, however, tactically Darwin decided not to make much of the human question. He did not want to conceal his thinking; he simply did not want it to swamp his general points about evolution and its process. Hence, in one of the great understatements of all time, he said merely, "Light will be thrown on the origin of man and his history" (C. Darwin 1859, 488). Darwin fully intended to get back to the topic and deal with it at length. This he did in 1871, offering a major two-volume work, *The Descent of Man*.

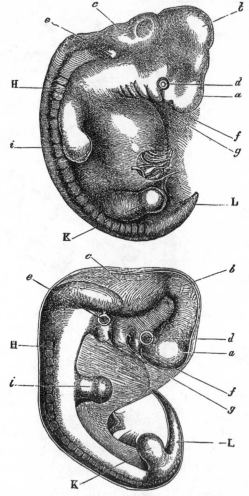

Fig. 1.   Upper figure human embryo, from Ecker.   Lower figure that of a dog, from Bischoff.

- - - - - - -

a. Fore-brain, cerebral hemispheres, &c.
b. Mid-brain, corpora quadrigemina.
c. Hind-brain, cerebellum, medulla oblongata.
d. Eye.
e. Ear.

f. First visceral arch.
g. Second visceral arch.
H. Vertebral columns and muscles in process of development.
i. Anterior }
K. Posterior } extremities.
L. Tail or os coccyx.

Figure 16. Similarity of the embryo of human and dog. Again, we see the strength of the *Origin* in making use of facts like these, well known to all biologists. (From *The Descent of Man.*)

When one looks at this book, it is really rather peculiar. Most of the book is not on the human species at all but consists of an extended essay on the secondary mechanism of sexual selection. The peculiarity of the topic yields the clue to its genesis. Alfred Russel Wallace started by sharing Darwin's

naturalistic perspective on human origins and even penned a stimulating little essay on how humans might have evolved. But then he became enamored
with spiritualism and convinced that humankind could be no chance, purely
natural event. There must have been a Mind guiding human genesis. In support of this conclusion, Wallace argued that there were many human characteristics that could not have been the product of natural selection, such as
human hairlessness and humans' great intelligence. Wallace, who had actually lived with natives, maintained that there was a vast amount of brain-
power unused by primitive man, and therefore it could not have been
produced by selection. (Wallace's essays are collected in his *Contributions to
the Theory of Natural Selection*, 1870.)

Such a conclusion was anathema to Darwin, but he felt that Wallace had
good arguments. Natural selection unaided could not produce many distinctive human features, but sexual selection could. Darwin argued that much
that is characteristic of humans came about through varying standards of
beauty and desirability. Hence, although distinctive, the features cited by
Wallace were not such as to take matters out of the range of evolution and
certainly not out of the range of science. Thus Darwin wrote a long discussion of sexual selection in general and then its application to humankind.
Why are men big and strong and powerful and intelligent? Because their ancestors who were got the best women. Why are women soft and yielding and
domestic? Because their ancestors who were got the best men and did the
best job with the family. Why are humans hairless? Because that is what their
ancestors found sexually exciting. Why (to get more specific) do Hottentot
women (using Darwin's language) have such big backsides? Because this was
the standard of beauty of the tribe, and the bravest braves got first choice.
Expectedly, all of this is wrapped up with some fairly standard Victorian sentiments about the relative abilities of men and women: "If men are capable of
decided eminence over women in many subjects, the average standard of
mental power in man must be above that of woman" (C. Darwin 1871, 2:
326–327).

No argument is needed to alert us to the fact that by the 1870s Darwin and
his thinking were starting to settle into some fairly conventional modes. He
may have been a revolutionary, but he was not always that far ahead of his
tribe. It is little wonder that Darwin became a major source of national pride.
By universal acclaim, when he died, he was buried next to Isaac Newton in
that English Valhalla, Westminster Abbey (Figure 17).

After Darwin

Before the *Origin* was published, evolution rode on the back of the doctrine
of progress, it was opposed by the idea of final cause, and its status was akin
to that of mesmerism and phrenology—it was a pseudoscience. How did
Darwin change things? As far as progress was concerned, Darwin himself
was certainly a cultural progressivist and saw evolution itself as progressive.

Figure 17. Cartoon of Darwin as an ape. The humor was always gentle and never savage. The English were very proud of Darwin, and they still are. He is on the back of today's £10 note. Now, as then, Darwin is as much a part of culture as of science. (Published in 1871 in the *Hornet* magazine.)

However, he also saw that the link between progress and evolution was something that brought down the status of the latter, and he realized that natural selection is a mechanism that denies the inevitability of biological progress. In other words, conceptually Darwin broke the link between the two, although, having done this, he argued that natural selection could lead to progress as the end result of what today's evolutionists would call a kind of arms race, with ever better features coming from competition with rivals. "If we take as the standard of high organisation, the amount of differentiation and specialisation of the several organs in each being when adult (and this will include the advancement of the brain for intellectual purposes), natural selection clearly leads towards this standard: for all physiologists admit that the specialisation of organs, inasmuch as in this state they perform their functions better, is an advantage to each being; and hence the accumulation of variations tending towards specialisation is within the scope of natural selection" (C. Darwin 1959, 222; this is from the third edition of the *Origin* of 1861).

Remember that as far as final cause was concerned, Darwin thought that natural selection speaks to the designlike aspect of living beings. It gives a blind-law explanation: selection leads to adaptation, and this is the reason for the end-directed nature of such features as eyes and hands. For Darwin this was no minor matter. He always recognized that the designlike nature of the organic world is its most distinctive feature and should not be ignored. He was brought up on the Argument from Design, especially as it was presented in the classic discussion of Archdeacon William Paley in his *Natural Theology* (1802). However, what was no less important was that although Darwin may have been a deist (later he became agnostic), for him a god of any kind had no place in a scientific explanation. Everything must be explained without recourse to miracles or other specially guided forms of force.

The status of evolutionary thought raises particularly interesting questions and surprising answers. Darwin himself wanted to raise the status of evolutionary thought up to the level of physics and chemistry—as a mature science, functioning within universities and research centers, with skilled practitioners using natural selection as a tool to cut through the problems of organic existence and function. He wanted evolution to be what we today might call a professional science. But it was not to be. Darwin was a sick man, so the fate of his ideas had to be entrusted to others—followers like Thomas Henry Huxley in England, Asa Gray in America, and Ernst Haeckel in Germany. Unfortunately, they had aims other than those of Darwin. They all became ardent evolutionists, and all agreed that Darwin's argument in the *Origin*, which showed how many areas of biological inquiry could be tied together under the umbrella of evolution, was definitive. But few, if any, were very keen on natural selection. No one denied it outright, but some thought that the whole issue of final cause was overblown and hence that selection was unneeded, and others thought that final cause still needed explanation and that selection could not really do the job. Either way, people turned to other putative causes, such as Lamarckism, evolution by jumps or what we today might call macromutations (the theory is known as saltationism), or evolution through a kind of internal momentum (orthogenesis).

Admittedly, there were some significant biological problems with Darwin's theory. One was that he had no working theory of heredity. People could not see how sufficient new variation could arise and be maintained within populations that selection could have a lasting effect (Vorzimmer 1970). Somewhat unfortunately Darwin compounded his problems in this respect by assuming that when two organisms breed, the usual effect is that their distinctive features blend with those of their mates. Hence, if only one has a new, valuable variation, within a generation or two it is blended into virtually nonbeing. What was needed was a "particulate" theory of heredity, that preserved good new features from generation to generation. Another problem of Darwin's was that on the basis of calculations about the age of the earth made by leading physicists (notably the future Lord Kelvin), there was too little time for such a leisurely process as natural selection to take effect (Burchfield 1975). But in a way the scientific problems were not the deciding factor. What was significant was that even Darwin's most ardent followers were generally not that interested in turning evolution into a professional science, at least not as Darwin envisioned it. This was the time of great fossil discoveries in the New World—all of those dinosaur monsters of the past—and naturally there was great interest in working out the true paths of life's history (Figure 18). In combination with not-altogether-reliable methodologies lifted from embryology, most famously Haeckel's so-called biogenetic law that ontogeny (the history of the individual) recapitulates phylogeny (the history of the group), much effort was expended on working out the full details of the evolutionary tree of life. But as a causal theory of origins, evolution was somewhat of a flop. Its enthusiasts were content to keep it more at the level of a "popular science," where it could have a different role.

Why? By the second half of the nineteenth century, men like Huxley were working hard to bring their societies out of the eighteenth century and into the twentieth (Desmond 1994, 1997). They tried to reform the military, medicine, the civil service, education, and other institutions. A subject like physiology obviously has a pragmatic payoff in such areas as medicine (Figure 19).

Figure 18. Dinosaur models. It was Richard Owen, in the early 1840s, who labeled a major group of reptiles as dinosaurs. From the first they captured people's imaginations, as they still do. These concrete, painted models were made for the Great Exhibition of London in 1851, where Britain showed its prowess to the world. They are now preserved in a park in south London. Note that they show dinosaurs as rather heavy, clumsy brutes. Today, we think them much more active.

THE "SILENT HIGHWAY"-MAN

Figure 19. The "Great Stink." In the summer of 1858 London became nearly uninhabitable because the Thames had become such a vile, odiferous receptacle of everyone's raw sewage. Few things brought home more vividly to the Victorians the need for training in modern technological subjects, like civil engineering. (From *Punch*, 1858.)

Evolution seems to have no such immediate function. But, especially combined with the ideology of progress, it could function as the philosophy of the new reformed society—it could have a role as a kind of alternative to Christianity, a kind of secular religion in its own right. It could be a story of origins, a story of humans' exalted place in the process, a vision of where they were going and what they should do to ensure success and triumph. And basically this was how evolution was promoted, especially by the leading gurus of the day, of whom none was more vocal than the English man of science Herbert Spencer. Playing up evolution, his vision was as progressive as anything from the eighteenth century. Spencer saw progress as being a move from the undifferentiated to the differentiated, or what he called a move from the homogeneous to the heterogeneous: "Whether it be in the development of the Earth, in the development of Life upon its surface, in the development of Society, of Government, of Manufactures, of Commerce, of Language, Literature, Science, Art, this same evolution of the simple into the complex, through successive differentiations, holds throughout" (Spencer 1857, 2–3).

Everything obeys this law. Humans are more complex or heterogeneous than other animals, Europeans are more complex or heterogeneous than savages, and the English language is more complex or heterogeneous than the tongues of other peoples (Figure 20).

Incidentally, it was Herbert Spencer who popularized the word *evolution*. Until the middle of the nineteenth century, the term was generally reserved for the development of the individual (ontogeny). For evolution in the modern sense (phylogeny), most people used words like "transformation" (Richards 1992). Seeing no essential difference between the development of

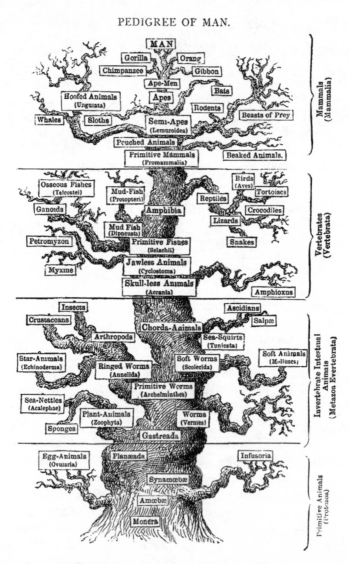

Figure 20. The tree of life as drawn by Ernst Haeckel. Notice just how the idea of progress influenced his picture of life's history. (From Haeckel 1897, 2: 188.)

the individual and the development of the group, already by the time of the just-quoted essay on progress (1857), Spencer was happily using the same term for both. In the *Origin* (1859), Darwin's continued preferred phrase was "descent with modification," although the final word of the *Origin* was "evolved." Darwin always felt rather uncomfortable about Spencer, thinking him a woolly and loose thinker (which he was). It is certainly no surprise that in the *Origin,* a work that the well-established scientist Charles Darwin intended as a major contribution to empirical research, there was no rush to use a term that appeared in a journal (the *Westminster Review*) more devoted to philosophy, literature, and political economy than pure science. By the time of *The Descent of Man* (1871), the usage of *evolution* in the sense of today had become universal, and we find Darwin was using the word as often and as comfortably as anyone else.

A good religion, secular or otherwise, has a moral dimension. In the case of evolution, the articulation of ethical dicta came to be known as social Darwinism. The basic pattern was simple: the key is the supposed progressiveness of evolution—simple to complex, homogeneous to heterogeneous, monad to man. This is a good thing. Hence one ought to support it or, at the very least, not impede it. Therefore, one's moral obligations are to go along with the evolutionary process, not to stand in its way; and, if possible, one should promote it and cherish and aid the features that have been produced by evolution. Start with natural selection—remember, no one said that this never occurs—and it is an easy transition from the biological mechanism to a socioeconomic social policy of laissez-faire, where there is struggle and competition, and the weakest go to the wall. All people who attempt state-supported ameliorations of poverty are compounding the problem. "Blind to the fact that under the natural order of things, society is constantly excreting its unhealthy, imbecile, slow, vacillating, faithless members, these unthinking, though well-meaning, men advocate an interference which not only stops the purifying process but even increases the vitiation—absolutely encourages the multiplication of the reckless and incompetent by offering them an unfailing provision, and *discourages* the multiplication of the competent and provident by heightening the prospective difficulty of maintaining a family" (Spencer 1851, 323–324).

Christians differ, sometimes drastically and bitterly, over the true moral meanings of their religion. So likewise did social Darwinians differ. Apart from anything else, many were no keener to see natural selection in the social world than in the biological (Bannister 1979; Jones 1980). Some (like Spencer) were libertarian; others (notably Wallace) promoted socialism in the name of evolution; and still others (for instance, the Russian prince Peter Kropotkin), appealing to a kind of mutual sympathy that supposedly exists between fellow members of a species, thought that evolution justified a form of anarchism. Some were feminists (Wallace); some were male chauvinists (Darwin). Some were pacifists (Spencer); some were warmongers (Haeckel to Hitler). All used evolution less as a vehicle for scientific inquiry into the nature of the living world—as a full-blown professional science—than as a kind of religion for the new society. No longer a pseudoscience, evolution was certainly no more than a popular science.

The Synthetic Theory

Things persisted this way into the early decades of the twentieth century and in some respects became even worse when Continental thinkers, notably the embryologist Hans Driesch in Germany and the philosopher Henri Bergson in France, began pushing a kind of neo-Aristotelian theory of life that saw special, nonphysical final forces guiding the path of evolution. Driesch's entelechies and Bergson's *élans vitaux,* the foundations of "vitalism," were just not the elements of a forward-looking modern science. Relief finally came with the development of Mendelian genetics and its melding with Darwinian selection, although even here the process took time and was not straightforward (Provine 1971). Gregor Mendel, a Moravian monk who lived in the Austro-Hungarian Empire, discovered the essential principles of heredity in the 1860s. Unlike Darwin, he saw the transmission of characteristics as "particulate," believing that features like color and size can be passed on without the threat of being blended away (Figure 21). However, it was not until the beginning of the twentieth century that people came to appreciate the importance of Mendel's work and saw that it provided a key element in the story of evolution. You might think that this was a terrible missed opportunity, and that, had Darwin read Mendel's key paper, he would have realized that now he had the answers to all of the issues of heredity that his theory of natural selection demanded. Hence, the theory would have moved forward more quickly, much earlier. This is probably not so. Even when the work of Mendel was discovered, it took time to assimilate it. Many people at first thought it an alternative to Darwin rather than a complement (see below). So even if Darwin had read Mendel—and the monk's work was published in a journal well-enough known that, if he searched, Darwin would have found it—there is no reason to think that there would have been a "eureka" moment. Mendel incidentally did read the *Origin.* Interestingly, he never thought of his own work as pertinent to Darwin's problems. Judging from the annotations that Mendel made in the margins of the German translation of the *Origin,* he was far more interested in the theological implications of evolution than in the troubles of heredity that Darwin faced. There is really no surprise here since, after all, Mendel was first and foremost a man of God and only secondarily a plant scientist.

There were fairly straightforward reasons why at first, when Mendel was rediscovered at the beginning of the twentieth century, no one thought that he was speaking to Darwinian issues. Naturally, early geneticists focused on big variations and so tended to favor a kind of saltatory theory of overall change, that is evolution by large jumps. Slowly, however, thanks particularly to work in the second decade of the century by Thomas Hunt Morgan and his students at Columbia University in New York, the nature of the gene

Figure 21. The pea plant. This lowly garden vegetable was the organism used by Mendel to work out the principles of heredity. It was easy to grow and came in many different varieties and forms—seed and pod shape and color, height, and so forth. Mendel could therefore distinguish and record how features were passed from one generation to the next. It now seems certain that Mendel's results were a little too good to be true. It could have been that subconsciously he was only recording results that fit his hypothesis, or that the gardener was overzealously trying to please his master. This stylized picture is from a late-seventeenth-century plant book.

was revealed, and it could be seen as the complement to natural selection. Finally, around 1930, a number of highly gifted mathematical biologists, notably R. A. Fisher (1930) and J. B. S. Haldane (1932) in England and Sewall Wright (1931, 1932) in America, showed how Mendelian genes sort themselves and are transmitted in groups (an essential finding, given that selection works only in the group situation), and then it was possible to bring Darwin's work to the completion that it needed. Along with mutation (the coming of new variation, caused by spontaneous changes in genes), natural selection can truly be a significant force for change.

Since by the 1930s the question of the age of the earth was no longer pressing (the discovery of the warming effects of radioactive decay showed that the earth is quite old enough for the slow workings of selection), biologists moved rapidly forward with new ideas (known as *population genetics*) that would put empirical flesh on the mathematical skeletons. In Britain a highly vocal supporter of the theory was Thomas Henry Huxley's grandson Julian Huxley (the older brother of novelist Aldous Huxley), who produced a major work that pulled ideas together: *Evolution: The Modern Synthesis* (1942). Scientifically, after Fisher the most important figure was the Oxford biologist E. B. Ford (1964), who did groundbreaking studies of selection in populations of butterflies and who gathered around himself in a school of "ecological genetics" a number of younger researchers likewise interested in selection and its effects in nature. Noteworthy were Arthur Cain (1954) and Philip Sheppard (1958), who worked on banding patterns in snails, and Bernard Kettlewell (1973), whose interests were in industrial melanism—the ways in which butterflies change adaptive color patterns as their habitats are changed by the effects of smoke and pollution. In America the influential figure was the Russian-born geneticist Theodosius Dobzhansky, whose *Genetics and the Origin of Species* (1937) was an inspiration to a whole generation of evolutionists (Figure 22). Working alongside him were others, notably the ornithologist/taxonomist Ernst Mayr (author of *Systematics and the Origin of Species*, 1942), the paleontologist George Gaylord Simpson (*Tempo and Mode in Evolution,* 1944), and the botanist G. Ledyard Stebbins (*Variation and Evolution in Plants,* 1950).

By the 1950s Darwin's dream of a mature, professional science of evolutionary biology was realized. Moreover, it was genuinely Darwinian, for although there had been pretenders to the causal throne, notably Sewall Wright's process of genetic drift (random changes in gene frequency in small populations due to the vagaries of breeding), it was recognized that the key factor in organic nature is its adaptiveness, its manifestation of final cause, and that natural selection is a full and satisfying way of explaining this phenomenon. At the same time, progress—and all the moralizing and philosophizing that went along with it—had been expelled. No one was going to use this kind of professional biology as an excuse for quasi-religious speculations about the status of humankind and the obligations that nature lays upon humans (Figure 23).

Yet, for all this, there is one more important factor to the story, and this partly explains why to this day evolutionary ideas remain so controversial to so many. Although evolutionary biology was upgraded from the level of a popular science to the level of a professional science, this did not occur simply out of a disinterested quest for the truth by men who had no aims but the finding of the workings of nature. Virtually every one of the new professional biologists, from Fisher to Stebbins, became an evolutionist because he was attracted to the subject by thinking that it was more than just a scientific theory.

Figure 22. Theodosius Dobzhansky. This picture of Dobzhansky *(front row, fourth from left)* at a conference with his students underlines how successful science means having students who will carry on and develop your ideas. The Harvard geneticist Richard Lewontin is second from the left in the front row; Cornell geneticist Bruce Wallace is right behind him; and Francisco Ayala (who was a Spanish priest before he became an American evolutionist) is on the far right in the front row. Clearly, Darwin's views on the relative abilities of the sexes was a long-lasting prejudice in evolutionary circles.

For some, like Dobzhansky, the attraction was explicitly religious. He brought progress—progress to humans, that is—into a kind of overall world picture that saw God working his way through the forces of nature. Ignoring divisions that were crucial to earlier thinkers, Dobzhansky thought that God's grace and an unfurling creation could be combined. Unsurprisingly, he was attracted to the ideas of the Jesuit paleontologist Teilhard de Chardin (1955), who promoted similar ideas. Others, like Julian Huxley in England and Mayr and Simpson in America, were secular in their thinking, but they too liked the idea of progress and thought that it showed that evolution had some kind of direction and meaning.

Because all these seminal evolutionists were people who grew up in the early twentieth century, when much of evolution was only a popular science dripping with metaphysical and moral implications, it would have been a surprise if this had not been the main motivating factor. But, like Darwin himself, this new breed of evolutionists saw that if they were to have professional status for their activities—something they ardently desired as full-time scientists—then they would need to purify their work of its extrascientific aspects. They needed to deal with doctrines of progress and with extrapolated

STUDIES ON IRRADIATED POPULATIONS OF DROSOPHILA MELANOGASTER*

By BRUCE WALLACE

Biological Laboratory, Cold Spring Harbor, N.Y.

(With Five Text-figures)

(*Received* 28 *July* 1955)

INTRODUCTION

It is well known that the widespread use of ionizing radiations, because of their genetic effects, poses a problem regarding future generations. These radiations induce gene mutations. The vast majority of mutations have deleterious effects on individuals carrying them. Under the pressure of continued mutation, these deleterious mutations will accumulate in populations. Therefore, an irradiated population will, on the average, be harmed—have its 'fitness' reduced—by a continual exposure to irradiation.

The present article summarizes observations made on irradiated populations of *Drosophila melanogaster*. Some of the material presented here has been published previously (Wallace, 1950, 1951; Wallace & King, 1951, 1952). This summary, however, will introduce new material in addition to extending the original observations.

MATERIAL AND METHODS

The experimental populations. The experimental populations of *D. melanogaster* are kept in lucite and screen cages. The original flies were obtained from an Oregon-R strain kept by mass transfer for many years. Fourteen lethal- and semi-lethal-free second chromosomes were extracted from this strain through the use of a series of matings identical to those described later (Fig. 2). Flies carrying these second chromosomes and mixtures of Oregon-R and 'marked stock' chromosomes other than the second were the parental flies of the populations.

Brief descriptions of the populations are given in Table 1. The left-hand column gives the identifying number for each population. The second column indicates the origin of the population. 'Stocks' indicates populations whose original flies carried lethal-free second chromosomes of Oregon-R derivation. Three more recent populations are subpopulations of populations 5 and 6; the designation in the table gives the parental population and the generation during which eggs were removed to start the new populations. The third column indicates the number of adults in the population cages: 'large' refers to populations of about 10,000 individuals, 'small' to populations frequently with fewer than 1000 individuals. The last three columns of Table 1 give the type of exposure, the dose, and the date the population was started for each population. Chronic exposure refers to continuous exposure to radium 'bombs'. No exposure in the case of popula-

* This work was done under Contract No. AT-(30-1)-557, U.S. Atomic Energy Commission.

Figure 23. A paper on fruit flies by Bruce Wallace. See the acknowledgment at the bottom of the page. By the 1950s, evolutionists were becoming very creative at finding funds for their work. Dobzhansky persuaded the U.S. Atomic Energy Commission that fruit flies were the perfect model organism for study of the effects of nuclear radiation, a major worry at the time given the ongoing testing of bombs (in the atmosphere). In England, E. B. Ford persuaded the Nuffield Foundation that the study of butterflies gave insight into the spread of genes in populations and hence was of great value in studying human genetically caused ailments.

moral exhortations. So they did. They took the extraneous thinking about progress, morality, and the meaning of life out of the science that they produced as professional researchers. Ernst Mayr was the first editor of the newly founded journal *Evolution* (Figure 24). His correspondence with prospective authors, deposited with the American Philosophical Society, is highly revealing about his intentions that the new science be as much like established sciences as possible. On the one hand, it had to be methodologically proper: "The field has reached a point where quantitative work is badly needed. Also, evolutionary research, as you realize, has shifted almost completely from the phylogenetic interest (proving evolution) to an ecological interest evaluating the factors of evolution" (letter to W. A. Gosline, March 5, 1948). On the other hand, it had to be value free: "The prestige of evolutionary research has suffered in the past because of too much philosophy and speculation" (letter to G. G. Ferris, March 28, 1948).

Yet, having done their professional science, these paradigm-making evolutionists then wrote other works that were as dripping with progress and moral exhortation as anything to be found in Herbert Spencer, except that their directives were focused on the problems of the twentieth century rather than the nineteenth. "The most essential material factor in the new evolution seems to be just this: knowledge, together, necessarily, with its spread and inheritance. As a first proposition of evolutionary ethics derived from specifically human evolution, it is submitted that promotion of knowledge is essentially both the acquisition of new truths or of closer approximations to truth (metaphorically the mutations of the new evolution) and also its spread by communication to others and by their acceptance and learning of it (metaphorically its heredity)" (Simpson 1949, 311). Simpson was writing at midcentury, just when Stalinism was at its peak and in Soviet Russia biologists in particular were under threat because of the high position of the charlatan T. D. Lysenko. Unsurprisingly, therefore, "Democracy is wrong in many of its current aspects and under some current definitions, but democracy is the only political ideology which can be made to embrace an ethically good society by the standards of ethics here maintained" (321).

The Past Half Century

That was all 50 or more years ago. The neo-Darwinians in Britain and the synthetic theorists in America built good foundations for their science. Considered just as a science, which it has every right to be, modern evolutionary theory deservedly is one of the most forward-looking and exciting areas of empirical inquiry. Every one of the areas treated by Darwin in the *Origin* flourishes as never before. Selection studies themselves, both theoretical and empirical, have reached a very high degree of sophistication. The work of Rosemary and Peter Grant (1989) on the beak size of Galápagos finches is a paradigmatic example of excellent science. The advent of molecular biology has been very important here for offering both new insights (for instance, the

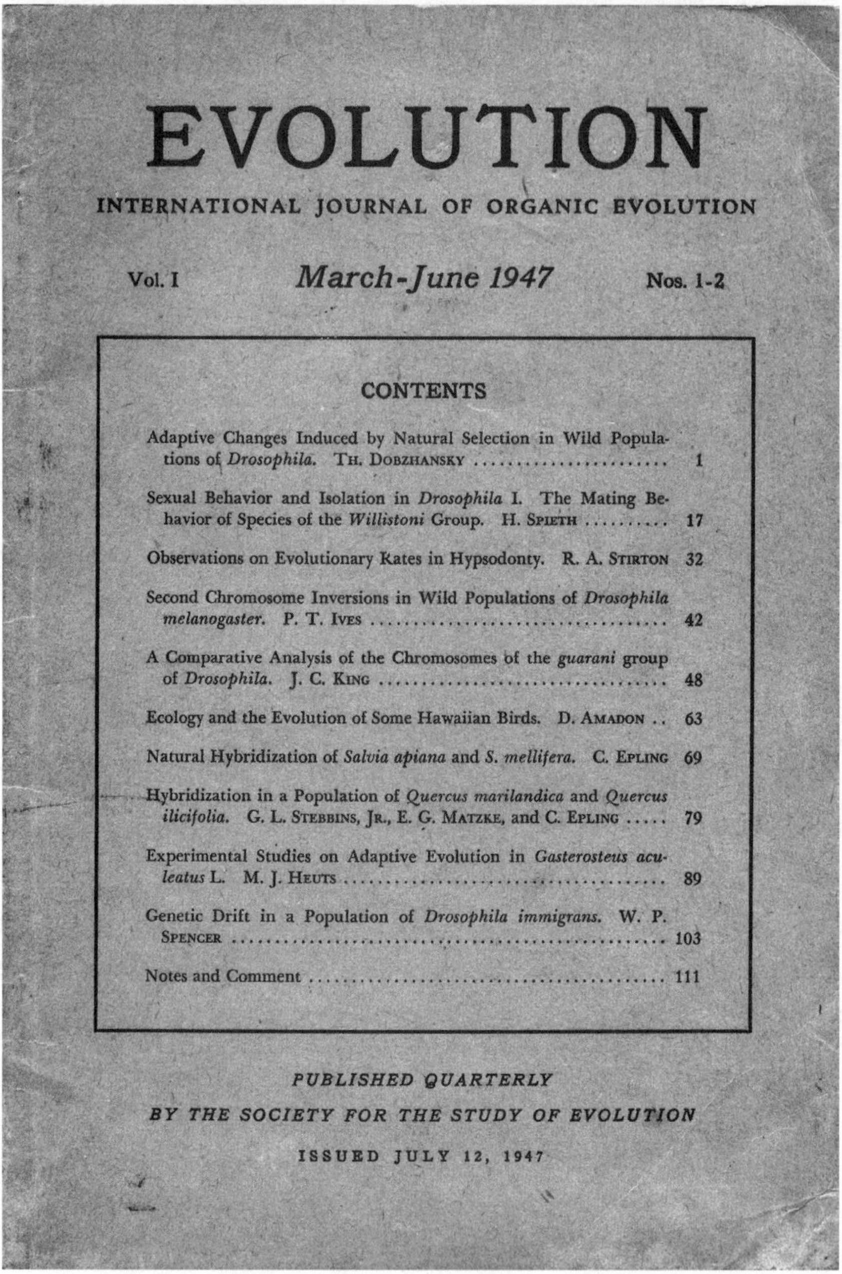

EVOLUTION

INTERNATIONAL JOURNAL OF ORGANIC EVOLUTION

Vol. I   *March-June 1947*   Nos. 1-2

CONTENTS

PUBLISHED QUARTERLY

BY THE SOCIETY FOR THE STUDY OF EVOLUTION

ISSUED JULY 12, 1947

Figure 24. The cover of the first issue of *Evolution*. One of the biggest problems of these early professional evolutionists was of finding people, working to their standards, who were not studying fruit flies or similar organisms. The journal was launched with a grant from the American Philosophical Society (founded by Benjamin Franklin) in Philadelphia. Revealingly, the only person on the grants committee who opposed the funding was the embryologist Edwin Conklin. Embryology was then the queen of the biological sciences and Conklin was not convinced that evolution could be the basis of real professional science. Perhaps because of these disciplinary tensions, evolutionists responded by ignoring embryology and it was not until the 1980s, by which time molecular biology had completely changed the landscape, that it again became a major element in evolutionary studies.

significance of drift at the molecular level) and new techniques (the use of genetic fingerprinting for determining heredity). Social behavior and instinct, detailed in Edward O. Wilson's magnificent survey *Sociobiology: The New Synthesis* (1975), have powerful new models of explanation, like kin selection and reciprocal altruism, confirming Darwin's hunch that the right way to explain the intricacies of social interactions, including group behavior, is from an individualistic perspective. Backing the theoretical works are major empirical studies—for instance, that of Tim Clutton-Brock and his associates (1982) on the red deer of Scotland—that show that evolution has indeed been a powerful factor in the evolution of social behavior. Somewhat controversially, the field (often here known as *evolutionary psychology*) also in a very Darwinian fashion looks at humankind, trying to explain such things as infanticide and sexual behavior and mating patterns in terms of natural selection. Paleontology has been revolutionized. Not only are there continued major fossil finds—recently the discovery in the Canadian Arctic of a link between fish and amphibians (Daeschler, Shubin, and Jenkins 2006), not to mention the finding in Indonesia of a little humanlike figure (*Homo floresiensis*, naturally known as the "hobbit") (Brown et al. 2004)—but molecular techniques have enabled people to explore the past course of evolution with ever greater precision. At the more conceptual level, new theories have been proposed (for instance, the theory of punctuated equilibria of Niles Eldredge and the late Stephen Jay Gould [1972]) (Figure 25), and ideas drawn from other areas of evolutionary studies have been applied to understand the patterns of the past (for instance, the way in which the late John J. Sepkoski Jr. [1976] used ecological theories about island biogeography to throw light on past patterns of animal diversity, as well as on the Great American Interchange, an event that occurred about 10 million years ago when South and

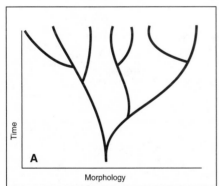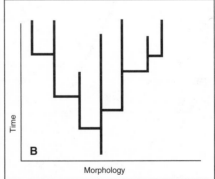

Figure 25. Punctuated equilibrium. In the Darwinian picture, labeled "phyletic gradualism" by Eldredge and Gould (A), evolution goes smoothly. In their alternative picture (B), it goes by jumps. Gould denied strongly that he was a saltationist, which rather suggests that much of the debate was about comparative timescales. For a fruit fly geneticist, 1,000 years is an age and much change can take place during that period. For a paleontologist, 1,000 years barely gets recorded and any changes are going to seem instantaneous.

North America joined and animals moved north to south and south to north [Marshall et al. 1982]) (Figure 26). Theories of geographic distribution were revolutionized by the coming of plate tectonics. Until this point, in a tradition embraced by Darwin in the *Origin,* evolutionists had spent many happy hours throwing up hypothetical land bridges and finding ways in which seeds and small animals could cross large bodies of water. Now the moving of the continents did all the work for them. Lystrosaurus, a mammal-like herbivorous reptile found rather more than 200 million years ago, is fat, short, and squat. It is certainly not an animal that would have roamed far and wide. Today it is found in the same fossil deposits (Lower Triassic) on the continents of Africa, (Southeast) Asia, and Antarctica. This would be inexplicable were it not for the fact that more than 200 million years ago all those continents touched when they were part of Pangaea. They have since drifted apart (Figure 27).

Anteaters
Armadillos
Capybaras
Glyptodonts
Monkeys
Opossums
Porcupines
Phorusrhacids
Sloths
Toucans
Toxodonts

Bears
Camels
Cats
Dogs
Elephants
Horses
Peccaries
Rabbits
Raccoons
Skunks
Tapirs
Weasels

Figure 26. The Great American Interchange. By treating North America and South America as if they were two islands being colonized, one can work out the expected rates of exchange and the numbers of invading groups one would expect at equilibrium.

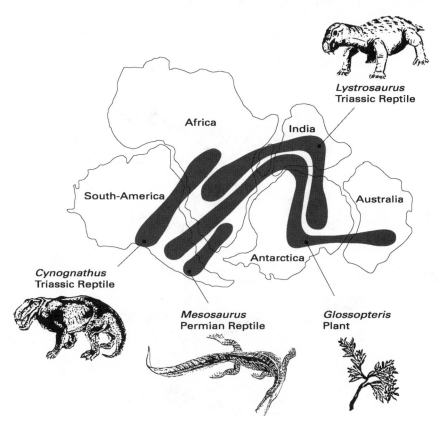

Figure 27. Patterns of fossil organisms. These distributions make sense only if you assume that the continents were joined. Phenomena like these show just how much evolutionists today still work in the Darwinian mode and yet how far we have come from the *Origin* itself. (United States Geological Survey.)

Systematics has been transformed by new cladistic techniques, much aided by the coming of computers that have allowed massive amounts of data to be absorbed, quantified, calculated, and understood (Hull 1988). Morphology likewise is open to new understandings. Above all, embryology, a subject that the synthetic theorists tended to ignore, has now, especially under the new name *evolutionary development* (evo-devo), been converted from a nonconceptual backwater to the most exciting area of research today in the field of evolutionary studies (Carroll 2005). Fantastic new findings have emerged, for instance, about the ways in which organisms as diverse as fruit flies and humans share the same underlying genetic mechanisms for development, and the ways in which organisms are built and variations are produced are among the hottest areas of research (Figures 28 and 29).

There are, of course, controversies and differences. No one wants to deny the importance—the very great importance—of natural selection. But some evolutionists, particularly in areas like paleontology and embryology, where sometimes questions of adaptive significance are (as was earlier the case for

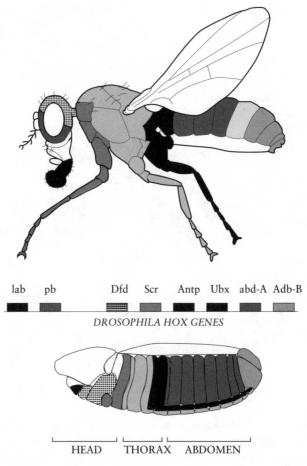

Figure 28. Development in the fruit fly. In the center we have the genes that control development, in the order in which they occur on the chromosome. Below we have the larva and above we have the grown fly.

Fly Dfd	P K R Q R T A Y T R H Q I L E L E K E F H Y N R Y L T R R R R I E I A H T L V L S E R Q U K I W F Q N R R M K W K K D N	K L P N T K N V R
AmphiHox4	T K R S R T A Y T R Q Q V L E L E K E F H F N R Y L T R R R R I E I A H S L G L T E R Q I K I W F Q N R R M K W K K D N	R L P N T K T R S
Mouse HoxB4	P K R S R T A Y T R Q Q V L E L E K E F H Y N R Y L T R R R R V E I A H A L C L S E R Q I K I W F Q N R R M K W K K D H	K L P N T K I R S
Human HoxB4	P K R S R T A Y T R Q Q V L E L E K E F H Y N R Y L T R R R R V E I A H A L C L S E R Q I K I W F Q N R R M K W K K D H	K L P N T K I R S
Chick HoxB4	P K R S R T A Y T R Q Q V L E L E K E F H Y N R Y L T R R R R V E I A H S L C L S E R Q I K I W F Q N R R M K W K K D H	K L P N T K I R S
Frog HoxB4	A K R S R T A Y T R Q Q V L E L E K E F H Y N R Y L T R R R R V E I A H T L R L S E R Q I K I W F Q N R R M K W K K D H	K L P N T K I K S
Fugu HoxB4	P K R S R T A Y T R Q Q V L E L E K E F H Y N R Y L T R R R R V E I A H T L C L S E R Q I K I W F Q N R R M K W K K D H	K L P N T K V R S
Zebrafish HoxB4	A K R S R T A Y T R Q Q V L E L E K E F H Y N R Y L T R R R R V E I A H T L R L S E R Q I K I W F Q N R R M K W K K D H	K L P N T K I K S

Figure 29. Homologous amino acids. On the left is a column of the names of genes in different organisms. All of these genes are crucial, kicking in at similar stages of development. The rows in the main box are made up of letters corresponding to amino acids, produced by these genes, in turn producing a vital protein. It can be seen that the genes of fruit flies and humans are virtually identical, leading to the conclusion that development is like building models with Lego bricks. You start with the same pieces and then put them together in very different ways.

people like Thomas Henry Huxley) not overwhelmingly pressing, think that perhaps other causal factors were significant. This was the underlying theme of the theory of punctuated equilibria, with Gould in particular arguing that many features are not particularly adaptive but are more "spandrel-like"— that is, nonfunctional by-products of the evolutionary process (Gould and Lewontin 1979) (Figures 30 and 31). He accused Darwinian evolutionists of relying too often on "Just so" stories, the fantabulous-like tales told by Rudyard Kipling (Figure 32). Today many evo-devo enthusiasts feel much the

Figure 30. A spandrel from the Villa Farnesina in Rome, painted by Raphael and assistants. Gould's claim is that many features of organisms are like spandrels (actually, the true technical name is pendentive). Although these spandrels seem designed for painting, in truth they are simply produced as the result of other needs. Hence, one should not always look for adaptation and perhaps natural selection is overrated.

Figure 31. The Irish elk. Actually, this now-extinct beast, found in the bogs of Ireland, is not a true elk. It is remarkable for its antlers. They are so large it is hard to imagine that they could ever have been of adaptive advantage. A popular hypothesis today is that they grew through sexual selection, the elks with bigger horns defeating their rivals and getting the females. It is suggested that the elks became fertile before they stopped growing and perhaps an immature elk would pass on his genes, even though later in life he and his offspring would be at a disadvantage. An explanation like this shows how careful one should be in talking about adaptation and its absence. In one sense, the antlers are adaptive, but not necessarily for survival and perhaps not for much of the animal's life.

same way. "The homologies of process within morphogenetic fields provide some of the best evidence for evolution—just as skeletal and organ homologies did earlier. Thus, the evidence for evolution is better than ever. The role of natural selection in evolution, however, is seen to play less an important role. It is merely a filter for unsuccessful morphologies generated by development. Population genetics is destined to change if it is not to become as irrelevant to evolution as Newtonian mechanics is to contemporary physics"

Figure 32. The elephant's nose. In his stories, Kipling claimed that the elephant's nose came about because a crocodile grabbed it and pulled. Gould suggested that many stories about adaptive advantage have about as much plausibility.

(Gilbert et al. 1996, 368). No doubt posterity will tell us whether sentiments like these are wise and prescient or merely the unjustified effluvia of enthusiasts for a new discipline that is trying to establish itself as important. Either way, the vigor and excitement of contemporary evolutionary theory are confirmed.

Does this then mean that the nonscientific, more ideological side to evolutionary thinking has now vanished? Does evolution no longer function for many people as a kind of secular religion? Not at all. On the one hand, huge numbers of articles and books (not to mention radio and television programs) show or justify ideological implications ascribed to or drawn from evolution. Edward O. Wilson is a paradigm, for he has poured out a stream of books designed to show that evolution is progressive and that from this flows not just moral exhortations—for Wilson, the preservation of the rain forests—but an alternative view of creation to that of traditional religion:

> But make no mistake about the power of scientific materialism. It presents the human mind with an alternative mythology that until now has always, point for point in zones of conflict, defeated traditional religion. Its narrative form is the epic: the evolution of the universe from the big bang

of fifteen billion years ago through the origin of the elements and celestial bodies to the beginnings of life on earth. The evolutionary epic is mythology in the sense that the laws it adduces here and now are believed but can never be definitively proved to form a cause-and-effect continuum from physics to the social sciences, from this world to all other worlds in the visible universe, and backward through time to the beginning of the universe. Every part of existence is considered to be obedient to physical laws requiring no external control. The scientist's devotion to parsimony in explanation excludes the divine spirit and other extraneous agents. Most importantly, we have come to the crucial stage in the history of biology when religion itself is subject to the explanations of the natural sciences. As I have tried to show, sociobiology can account for the very origin of mythology by the principle of natural selection acting on the genetically evolving material structure of the human brain.

If this interpretation is correct, the final decisive edge enjoyed by scientific naturalism will come from its capacity to explain traditional religion, its chief competition, as a wholly material phenomenon. Theology is not likely to survive as an independent intellectual discipline. (Wilson 1978, 192)

Another scientist much given to this kind of speculation—although he disliked traditional notions of progress and wanted to substitute his own ideas about the randomness of change—was Stephen Jay Gould, particularly in his popular essays, "This View of Life," published monthly in *Natural History*. Very rarely did Gould fail to draw some kind of moral message from his writings, whether about the racism endemic in our society or the need for conservation. Not that Gould gave traditional religion much more scope than did Wilson. Any ideas about our being the favored children of God, that we humans might be the reason for the creation, are simply hubris. "Since dinosaurs were not moving toward markedly larger brains, and since such a prospect may lie outside the capabilities of reptilian design . . . , we must assume that consciousness would not have evolved on our planet if a cosmic catastrophe had not claimed the dinosaurs as victims. In an entirely literal sense, we owe our existence, as large and reasoning mammals, to our lucky stars" (Gould 1989, 318). Here Gould anticipated the so-called new atheists, like his great British counterpart in the realm of popular science writing, Richard Dawkins. In the *The God Delusion* (2006) Dawkins argues that Christianity and other religions are the major sources of humankind's ills, and that we need a secular philosophy as a substitute—a secular philosophy informed by Darwinian thinking, which itself makes appeals to traditional religions not merely otiose but absolutely false. Philosopher Daniel Dennett's *Darwin's Dangerous Idea* (1995) is another exemplar of this kind of literature.

Heading the other way are the many critics of evolutionary thinking. In America today, survey after survey confirms that most Americans do not believe in evolution. Indeed, the great majority believe that the earth is less than

10,000 years old and that all organisms were created in a burst of divine energy in just six days, followed some time later by their destruction in a universal flood, save only for those lucky pairs that floated away in Noah's ark (Ruse 2005; Numbers 2006). American evangelical Christianity is not our story here, but the constant attacks on modern Darwinism show clearly that enthusiasts for this kind of religion do not regard evolution as just a scientific theory but as something more, a materialistic, secular alternative to the traditional belief systems that carries within it opposition to all decent and long-held moral norms. In the words of the leading "creation scientist," the late Henry M. Morris: "It is rather obvious that the modern opposition to capital punishment for murder and the general tendency toward leniency in punishment for other serious crimes are directly related to the strong emphasis on evolutionary determinism that has characterized much of this century" (Morris 1989, 148). Apparently, the "notorious Darwinian philosopher Michael Ruse," a well-known "atheistic humanist," has made a major contribution to the moral rot. More recently, the rather crude literalism of creation science has morphed into the more sophisticated anti-Darwinism of so-called Intelligent Design Theory, which sees aspects of organic life as so complex as to require special creative interventions by the Designer. But still the debate today is more than just a conflict of science versus religion and one between competing ideologies. The founder of the Intelligent Design movement, former Berkeley law professor Phillip Johnson, is explicit in seeing evolution in the old terms: "The Christian philosophy that was overthrown in the 1960s was an easy target because it had become identified with American culture and with worldly ideas like human perfectibility and the inevitability of progress, which are actually profoundly un-Christian" (Johnson 1997, 106). Unsurprisingly, this humanistic metaphysics is linked to moral failings—pornography, gay marriage, abortion, and like transgressions such as socialism.

Conclusion

The history of evolutionary theory falls into three stages. From about 1700 until 1859, it was little more than a pseudoscience riding on the back of the ideology of progress, the notion that humans can make change for the better. It was opposed to traditional religion not so much because it went against the literal truth of the Bible but because it left no place for Providence, the belief that change comes only through God's undeserved grace. Charles Darwin's *On the Origin of Species* changed all this. Darwin established the reasonable truth of evolution once and for all. However, his mechanism of natural selection was ignored or downplayed, and the leading evolutionists who followed Darwin, notably Thomas Henry Huxley, were much more interested in using evolutionary ideas as a kind of popular science, almost a secular religion, as an alternative ideology to the Christian religion they saw blocking their way to social reform. The third stage started around 1930 with the incorporation of Mendelian genetics. Now there was the possibility of building a new professional science of evolutionary change. This occurred, and it is this theory that flourishes today. Yet evolutionists should never forget the past and its influence on the present. Along with the science there is still an ongoing debate about ideology and the extent to which evolution in some way provides a secular alternative to older, overtly religious ways of viewing the world and humankind's status within it.

BIBLIOGRAPHY

Aristotle. 1984. *De partibus animalium.* In J. Barnes, ed., *The Complete Works of Aristotle,* 1087–1110. Princeton, NJ: Princeton University Press.

Bannister, R. 1979. *Social Darwinism: Science and Myth in Anglo-American Social Thought.* Philadelphia: Temple University Press.

Brewster, D. 1844. Vestiges. *North British Review* 3: 470–515.

Brown, P., T. Sutikna, M. J. Morwood, R. P. Soejono, Jatmiko, E. W. Saptomo, and R. A. Due. 2004. A new small-bodied hominin from the Late Pleistocene of Flores, Indonesia. *Nature* 431: 1055–1061.

Burchfield, J. D. 1975. *Lord Kelvin and the Age of the Earth.* New York: Science History Publications.

Cain, A. J. 1954. *Animal Species and Their Evolution.* London: Hutchinson.

Carroll, S. B. 2005. *Endless Forms Most Beautiful: The New Science of Evo Devo and the Making of the Animal Kingdom.* New York: Norton.

Chambers, R. 1844. *Vestiges of the Natural History of Creation.* London: Churchill.

———. 1846. *Vestiges of the Natural History of Creation.* 5th ed. London: Churchill.

Clutton-Brock, T. H., F. E. Guinness, and S. D. Albon. 1982. *Red Deer: Behavior and Ecology of Two Sexes.* Chicago: University of Chicago Press.

Cooper, J. M., ed. 1997. *Plato: Complete Works.* Indianapolis: Hackett.

Cuvier, G. 1817. *Le règne animal distribué d'après son organisation, pour servir de base à l'histoire naturelle des animaux et d'introduction à l'anatomie comparée.* 4 vols. Paris: Déterville.

Daeschler, E. B., N. H. Shubin, and F. A. Jenkins Jr. 2006. A Devonian tetrapod-like fish and the evolution of the tetrapod body plan. *Nature* 440: 757–763.

Darwin, C. 1859. *On the Origin of Species.* London: John Murray.

———. 1871. *The Descent of Man.* 2 vols. London: John Murray.

———. 1959. *The Origin of Species by Charles Darwin: A Variorum Text.* M. Peckham, ed. Philadelphia: University of Pennsylvania Press.

Darwin, E. 1801. *Zoonomia; or, The Laws of Organic Life.* 3rd ed. 4 vols. London: J. Johnson.

———. 1803. *The Temple of Nature.* London: J. Johnson.

Dawkins, R. 2006. *The God Delusion.* Boston: Houghton Mifflin.

Dennett, D. C. 1995. *Darwin's Dangerous Idea.* New York: Simon and Schuster.

Desmond, A. 1994. *Huxley, the Devil's Disciple.* London: Michael Joseph.

———. 1997. *Huxley, Evolution's High Priest.* London: Michael Joseph.

Dobzhansky, T. 1937. *Genetics and the Origin of Species.* New York: Columbia University Press.

Eldredge, N., and S. J. Gould. 1972. Punctuated equilibria: An alternative to phyletic gradualism. In T. J. M. Schopf, ed., *Models in Paleobiology,* 82–115. San Francisco: Freeman, Cooper.

Fisher, R. A. 1930. *The Genetical Theory of Natural Selection.* Oxford: Oxford University Press.

Ford, E. B. 1964. *Ecological Genetics.* London: Methuen.

Gilbert, S. F., J. M. Opitz, and R. A. Raff. 1996. Resynthesizing evolutionary and developmental biology. *Developmental Biology* 173: 357–372.

Gould, S. J. 1989. *Wonderful Life: The Burgess Shale and the Nature of History.* New York: W. W. Norton.

Gould, S. J., and R. C. Lewontin. 1979. The spandrels of San Marco and the Panglossian paradigm: A critique of the adaptationist programme. *Proceedings of the Royal Society of London, Series B* 205: 581–598.

Grant, B. R., and P. R. Grant. 1989. *Evolutionary Dynamics of a Natural Population: The Large Cactus Finch of the Galápagos.* Chicago: University of Chicago Press.

Haeckel, E. 1897. *The Evolution of Man: A Popular Exposition of the Principal Points of Human Ontogeny and Phylogeny.* 2 vols. New York: D. Appleton and Company.

Haldane, J. B. S. 1932. *The Causes of Evolution.* New York: Longmans, Green.

Hull, D. L. 1988. *Science as a Process.* Chicago: University of Chicago Press.

Huxley, J. S. 1942. *Evolution: The Modern Synthesis.* London: Allen and Unwin.

Johnson, P. E. 1997. *Defeating Darwinism by Opening Minds.* Downers Grove, IL: InterVarsity Press.

Jones, G. 1980. *Social Darwinism and English Thought.* Brighton, U.K.: Harvester.

Kettlewell, H. B. D. 1973. *The Evolution of Melanism.* Oxford: Clarendon Press.

Lamarck, J. B. 1809. *Philosophie zoologique.* Paris: Dentu.

Lyell, C. 1830–1833. *Principles of Geology: Being an Attempt to Explain the Former Changes of the Earth's Surface by Reference to Causes Now in Operation.* 3 vols. London: John Murray.

Malthus, T. R. [1826] 1914. *An Essay on the Principle of Population.* 6th ed. London: Everyman.

Marshall, L. G., S. D. Webb, J. J. Sepkoski Jr., and D. M. Raup. 1982. Mammalian evolution and the great American interchange. *Science* 215: 1351–1357.

Mayr, E. 1942. *Systematics and the Origin of Species.* New York: Columbia University Press.

Morris, H. M. 1989. *The Long War against God: The History and Impact of the Creation/Evolution Conflict.* Grand Rapids, MI: Baker Book House.

Numbers, R. L. 2006. *The Creationists: From Scientific Creationism to Intelligent Design.* 2nd ed. Cambridge, MA: Harvard University Press.

Owen, R. 1860. *Palaeontology or a Systematic Study of Extinct Animals and Their Geological Relations.* Edinburgh: Adam and Charles Black.

Paley, W. [1802] 1819. *Natural Theology (Collected Works,* vol. 4). London: Rivington.

Provine, W. B. 1971. *The Origins of Theoretical Population Genetics.* Chicago: University of Chicago Press.

Richards, R. J. 1992. *The Meaning of Evolution.* Chicago: University of Chicago Press.

Ruse, M. 1996. *Monad to Man: The Concept of Progress in Evolutionary Biology.* Cambridge, MA: Harvard University Press.

———. 2005. *The Evolution-Creation Struggle.* Cambridge, MA: Harvard University Press.

Sedley, D. 2008. *Creationism and Its Critics in Antiquity.* Berkeley: University of California Press.

Sepkoski, J. J., Jr. 1976. Species diversity in the Phanerozoic—Species-area effects. *Paleobiology* 2, no. 4: 298–303.

Sheppard, P. M. 1958. *Natural Selection and Heredity.* London: Hutchinson.

Simpson, G. G. 1944. *Tempo and Mode in Evolution.* New York: Columbia University Press.

———. 1949. *The Meaning of Evolution.* New Haven, CT: Yale University Press.

Spencer, H. 1851. *Social Statics; or, The Conditions Essential to Human Happiness Specified and the First of Them Developed.* London: J. Chapman.

————. 1857. Progress: Its law and cause. *Westminster Review* 67: 244–267.
Stebbins, G. L. 1950. *Variation and Evolution in Plants*. New York: Columbia University Press.
Teilhard de Chardin, P. 1955. *Le phénomène humain*. Paris: Editions du Seuil.
Vorzimmer, P. J. 1970. *Charles Darwin: The Years of Controversy*. Philadelphia: Temple University Press.
Wallace, A. R. 1858. On the tendency of varieties to depart infinitely from the original type. *Journal of the Proceedings of the Linnaean Society, Zoology* 3: 53–62.
————. 1870. *Contributions to the Theory of Natural Selection: A Series of Essays*. London: Macmillan.
Wilson, E. O. 1975. *Sociobiology: The New Synthesis*. Cambridge, MA: Belknap Press of Harvard University Press.
————. 1978. *On Human Nature*. Cambridge, MA: Harvard University Press.
Wright, S. 1931. Evolution in Mendelian populations. *Genetics* 16: 97–159.
————. 1932. The roles of mutation, inbreeding, crossbreeding and selection in evolution. *Proceedings of the Sixth International Congress of Genetics* 1: 356–366.

The Origin of Life

Jeffrey L. Bada and Antonio Lazcano

Generations of scientists may yet have to come and go before the question of the origin of life is finally solved. That it will be solved eventually is as certain as anything can ever be amid the uncertainties that surround us.

Earl Albert Nelson, *There Is Life on Mars* (1956)

How, where, and when did life appear on earth? Although Charles Darwin was reluctant to address these issues in his books, in a letter sent on February 1, 1871, to his friend Joseph Dalton Hooker, he wrote in a now-famous paragraph that "it is often said that all the conditions for the first production of a living being are now present, which could ever have been present. But if (and oh what a big if) we could conceive in some warm little pond with all sorts of ammonia and phosphoric salts, —light, heat, electricity present, that a protein compound was chemically formed, ready to undergo still more complex changes, at the present such matter would be instantly devoured, or absorbed, which would not have been the case before living creatures were formed" (Darwin 1887, 3, 18). (The letter is in the Darwin Collection in Cambridge. It is referenced as Letter 7471 in Burkhart and Smith 1994 and is reproduced in Calvin 1969, 4–5.)

Darwin's letter not only summarizes in a nutshell his ideas on the emergence of life but also provides considerable insights on the views about the molecular nature of the basic biological processes that were prevalent at the time in many scientific circles. Although Friedrich Miescher had discovered nucleic acids (he called them nuclein) in 1869 (Dahm 2005), the deciphering of their central role in genetic processes would remain unknown for more than 80 years. In contrast, the roles played by proteins in manifold biological processes had been firmly recognized. Equally significant, by the time Darwin wrote his letter, major advances had been made in the understanding of the material basis of life, which for a long time had been considered fundamentally different from inorganic compounds. The experiments of Friedrich Wöhler, Adolph Strecker, and Aleksandr Butlerov, who had demonstrated independently the feasibility of the laboratory synthesis of urea, alanine, and sugars, respectively, from simple starting materials, were recognized as a demonstration that the chemical gap separating organisms from the nonliving was not insurmountable.

But how had this gap first been bridged? The idea that life was an emergent feature of nature has been widespread since the nineteenth century. The major breakthroughs that transformed the origin of life from pure speculation into workable and testable research models were proposals that were suggested independently in the 1920s by Aleksandr I. Oparin and J. B. S. Haldane, as well as others (see Bada and Lazcano 2003). They hypothesized that the first life forms were the outcome of a slow, multistep process that began with the abiotic synthesis of organic compounds and the formation of a "primordial soup." Here their proposals diverged. While Haldane argued that viruses represented a primordial stage that had appeared prior to cells, Oparin argued that colloidal, gel-like systems had formed from the soup, leading to anaerobic heterotrophs that could take up surrounding organic compounds and use them directly for growth and reproduction. Although the details of these ideas have been superseded, the Oparin-Haldane hypothesis of chemical evolution provided a conceptual framework for the experimental development of the study of the origin of life. Laboratory experiments have shown how easy it is to produce a variety of organic compounds, including biochemically important monomers, under plausible cosmic and geochemical conditions. The robust nature of these reactions has been demonstrated by the finding that organic compounds are ubiquitous in the universe, as shown by their presence in carbon-rich meteorites, cometary spectra, and interstellar clouds where star and planetary formation is taking place.

A Timescale for the Emergence of Life

Although traditionally it had been assumed that the origin and early evolution of life involved several billion years (Lazcano and Miller 1994), such views are no longer tenable. It is true that it is not possible to assign a precise chronology to the appearance of life, but in the past few years estimates of the time within which this must have occurred have been considerably reduced. Determination of the biological origin of what have been considered the earliest traces of life is now a rather contentious issue (van Zuilen et al. 2002). This is not surprising. The geological record of the early Archean is sparse, and there are very few rocks that are more than 3.5 billion years old. The rocks with ages greater than 3.5 billion years that are preserved have been extensively altered by metamorphic processes (van Zuilen et al. 2002), and thus any direct evidence of ancient life has apparently been largely obliterated.

Although the biological nature of the microstructures present in the 3.5 billion-year-old Apex cherts of the Australian Warrawoona formation (Schopf 1993) has been disputed (Brasier et al. 2002; García-Ruiz et al. 2003), there is evidence that life emerged on earth very early in its history. The proposed timing of the onset of microbial methanogenesis, based on the low ^{13}C values in methane inclusions, has been found in hydrothermally precipitated quartz in the 3.5 billion-year-old Dresser Formation in Australia (Ueno et al. 2006), although this finding has also been challenged (Lollar and McCollom 2006). However, sulfur isotope investigations of the same site indicate biological sulfate-reducing activity (Shen et al. 2001), and analyses of 3.4 billion-year-old South African cherts suggest that they formed in a marine environment inhabited by anaerobic photosynthetic prokaryotes (Tice and Lowe 2004). These results, combined with reports on 3.43 billion-year-old stromatolites from Western Australia (Allwood et al. 2007), support the idea that the early Archean earth was teeming with prokaryotes, and that the origin of life must have taken place as soon as the conditions were suitable to permit the survival of these types of organisms.

The early Archean fossil record speaks for the relatively short timescale required for the origin and early evolution of life on earth and suggests that the critical factor may have been the presence of liquid water, which became possible as soon as the planet's surface finally cooled below the boiling point of water. Unfortunately, there is no geological evidence of the environmental conditions on the early earth at the time of the origin of life, nor are any molecular or physical remnants preserved that provide information about the evolutionary processes that preceded the appearance of the first cellular organisms found in the early fossil record. Direct information is generally lacking not only on the composition of the terrestrial atmosphere during the period of the origin of life but also on the temperature, ocean pH, and other general and local environmental conditions that may or may not have been important for the emergence of living systems.

The Primitive Earth Environment

Considerable progress has been made in our understanding of environmental conditions of the early earth and how the transition from abiotic to biotic chemistry may have occurred (for example, see Bada 2004). Nevertheless, there are still enormous gaps in our description of how the simple organic compounds associated with life as we know it reacted to generate the first living entities and how these in turn evolved into organisms that left behind actual evidence of their existence in the rock record. To evaluate how life may have begun on earth, we must access what the planet was like during its early history and under what conditions the processes thought to be involved in the origin of life took place. Life as we know it depends on the presence of liquid water and organic polymers such as nucleic acids and proteins. The available evidence suggests that during an early stage, biological systems lacked proteins and depended largely on catalytic and replicative polyribonucleotides, but water was likely essential from the very beginning, as it provides the medium for chemical reactions to take place. Without these basic components, as far as we know, life is impossible.

It is unlikely that water made its first appearance on earth as a liquid. During the Hadean period the volatile components that were trapped inside the accreting planet were released (degassed) from the interior of the juvenile earth to form a secondary atmosphere. Any primary atmosphere (if one existed at all) must have been lost, as evidenced by the depletion of rare gases in the earth's atmosphere compared with cosmic abundances (Kasting 1993b). As a consequence of the nearly simultaneous formation of the earth's core with accretion, the metallic iron was removed from the upper mantle, which would have allowed the volcanic gases to remain relatively reduced and produce a very early atmosphere that contained species such as CH_4, NH_3, and H_2. Since the temperature at the surface was high enough to prevent any water from condensing, the atmosphere would have consisted mainly of superheated steam along with these other gases (Kasting 1993a). Even though the secondary atmosphere may have been lost several times during large-impact events, especially the one that formed the moon, it would have been regenerated by further outgassing from the interior, as well as being resupplied from later impactors.

It is generally accepted that the impactors during the later stages of the accretion process originated from outside the solar system and would have been similar in composition to comets. As suggested in 1961 by the late Spanish chemist Joan (John) Oró, cometary volatile compounds, which appear to be the most pristine materials that survive from the formation of the solar system, may have supplied a substantial fraction of the volatiles on the terrestrial planets, perhaps including organic compounds that may have played a role in the origin of life on earth (Oró 1961). It has been suggested that the water on the earth was provided entirely from this source. However, recent measurements of the deuterium content of water in comets Halley, Hyakutake, and Hale-Bopp indicate that comets delivered only a fraction of it, while the largest fraction was trapped during the earlier accretionary phase and released via degassing (Robert 2001).

It is reasonable to assume that the atmosphere that developed on the earth over the period 4.4 to 3.8 billion years ago (perhaps several times if it was eroded by large-impact events) was basically a mix of volatiles delivered by volatile-rich impactors such as comets and outgassing from the interior of an already-differentiated planet. This atmosphere was probably dominated by water steam until the surface temperatures dropped to ~100°C (depending on the pressure), at which point water condensed to form early oceans (Wilde et al. 2001). The reduced species, which were mainly supplied by volcanic outgassing, are very sensitive to ultraviolet (UV) radiation that penetrated the atmosphere because of the lack of a protective ozone layer. These molecules were probably destroyed by photodissociation, although there might have been a steady-state equilibrium between these two processes that allowed a significant amount of these reduced species (especially H_2) to be present in the atmosphere (for example, see Tian et al. 2005). Eventually, as H_2 escaped from the earth into space, reduced species in the atmosphere would have been depleted. Thus, in general the overall consensus at present is that the early atmosphere was dominated by oxidized molecules such as CO_2, CO, and N_2. A similar atmosphere is present on Venus today, although it is much more dense than the atmosphere of the early earth.

The climate on the early earth at this stage depended mainly on two factors: the luminosity of the sun and the radiative properties of the atmosphere. Standard theoretical solar-evolution models predict that the sun was about 30% less luminous than today (Gilliland 1989). If the atmosphere of the early earth were the same as it is now, the entire surface of the planet would have been frozen. However, as discussed extensively by Kasting (1993a, 1993b), a CO_2-rich atmosphere may have been present throughout the Hadean and Early Archean periods, resulting in a significant greenhouse effect that would have prevented the oceans on the early earth from freezing. The basic argument is that during this period of early earth history, there were probably no major continents, and thus there was no extensive silicate weathering. Because this process is nowadays the long-term loss process for CO_2 removal and storage, the conclusion is that CO_2 would have been primarily contained in the atmosphere and the ocean. With the assumption of a solar luminosity of ~70% of the present value, a steady-state atmosphere containing ~10 bars of CO_2 could have been required in order to maintain a mean surface temperature greater than the freezing point of water.

In summary, the current models for the early terrestrial atmosphere suggest that it consisted of a weakly reducing mixture of CO_2, N_2, CO, and H_2O, with lesser amounts of H_2, SO_2, and H_2S. Reduced gases such as CH_4 and NH_3 are considered to have been nearly absent or present only in localized regions near volcanoes or hydrothermal vents.

During such early times when volcanic islands may have been prevalent, and large continents had not yet formed (Zahnle et al. 2007). Recent considerations of the early carbon cycle suggest that before extensive tectonic recycling of crustal sediments became common, most of the carbon on the earth's surface would have remained buried in the crust and mantle as calcium carbonate (Sleep and Zahnle 2001). There is thus the possibility that the CO_2 concentrations in the early atmosphere were not high enough to prevent the formation of a global ice-covered ocean (Bada et al. 1994). If this were indeed the case, the thickness of the global ice sheet has been estimated to be on the order of 300 m, which would have been thin enough to allow melting by an impactor of ~100 km in diameter. The frequency of impacts of such ice-melting bolides has been estimated to be one event every 10^5–10^7 years between about 3.6 and 4.5 billion years ago, suggesting periodic thaw-freeze cycles associated with the ice-melting impacts. The precursor compounds imported by the impactor or synthesized during the impact, such as HCN, would have been washed into the ocean during the thaw periods, which have been termed "Impact Summers" (Zahnle et al. 2007). In addition, CH_4, NH_3, H_2, and CO derived from hydrothermal vents would have been stored in the unfrozen ocean below the ice layer, which would have protected these gases from UV radiation. After a large impact the trapped gases would have been expelled into the atmosphere, where they could have persisted for some time before they were destroyed by photochemical reactions. During these episodes highly reducing conditions may have prevailed.

Organic Compounds on the Primitive Earth

Today organic compounds are so pervasive on the earth's surface that it is hard to imagine the earth devoid of organic material. However, during the period immediately after the earth first formed some 4.5 billion years ago, there would have been no organic compounds present on its surface because soon after accretion the decay of radioactive elements heated the interior of the young earth to the melting point of rocks (Wetherill 1990). Volcanic eruptions expelled molten rock and hot scorching gases out of the juvenile earth's interior, creating a global inferno. In addition, the early earth was also being peppered by mountain-sized planetesimals, the debris left over after the accretion of the planets. Massive volcanic convulsions, coupled with the intense bombardment from space, generated surface temperatures so hot that the earth at this point could very well have had an "ocean" of molten rock (i.e., a magma ocean).

Although temperatures would have slowly decreased as the infall of objects from space and the intensity of volcanic eruptions declined, elevated temperatures likely persisted for perhaps 100 million years or longer after the formation of the earth. During this period temperatures would probably have been too hot for organic compounds to survive. Without organic compounds, life as we know it could not have existed. However, on the basis of data from ancient zircons, by approximately 4 billion years ago (or perhaps even earlier) the earth's surface must have cooled to the point that liquid water could exist and global oceans could begin to form (Wilde et al. 2001). It was during this period that organic compounds would have first started to accumulate on the earth's surface, as long as there were abiotic processes by which they could synthesize and accumulate. What was the nature of these processes?

Synthesis of Organic Compounds on the Primitive Earth?

The hypothesis that the first organisms were anaerobic heterotrophs is based on the assumption that abiotic organic compounds were a necessary precursor for the appearance of life. The laboratory synthesis of organic compounds from inorganic starting material was first achieved in the 1820s when Friedrich Wöhler demonstrated that urea could be formed in high yield by the reaction of cyanogen and liquid ammonia and by heating ammonium cyanate. Although it was not immediately recognized as such, a new era in chemical research had begun. In 1850 Adolph Strecker achieved the laboratory synthesis of alanine from a mixture of acetaldehyde, ammonia, and hydrogen cyanide. This was followed in 1861 by the experiments of Aleksandr

Butlerov, who showed that the treatment of formaldehyde with alkaline catalysts, such as calcium hydroxide, leads to the synthesis of a variety of sugars. The laboratory synthesis of biochemical compounds was soon extended to include more complex experimental settings, and by the end of the nineteenth century a large amount of research on organic synthesis had been performed (Bada and Lazcano 2003).

However, these nineteenth-century organic syntheses were not conceived as laboratory simulations of the "warm little pond" that Darwin mentioned to Hooker, but rather as attempts to understand the autotrophic mechanisms of nitrogen assimilation and CO_2 fixation in green plants (see Bada and Lazcano 2002a). The first convincing demonstration of the possible synthesis of organic compounds under prebiotic conditions was accomplished in 1953 by Stanley L. Miller, who investigated the action of electric discharges acting for a week on a mixture of CH_4, NH_3, H_2, and H_2O; racemic mixtures of several protein amino acids were produced, as well as hydroxy acids, urea, and other organic molecules (Miller 1953, 1955).

Miller achieved his results by means of an apparatus in which he could simulate the interaction between an atmosphere and an ocean (see Figure 1). As an energy source Miller chose a spark discharge, considered to be the second-largest energy source, in the form of lightning and coronal discharges,

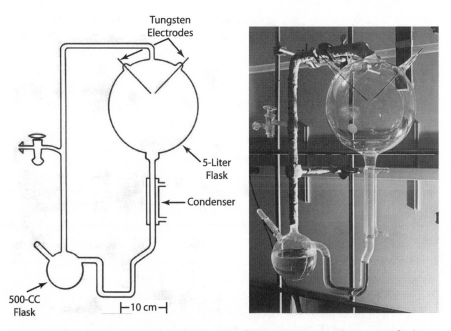

Figure 1. The apparatus used in the 1953 Miller experiment. The 500 cc flask was used to represent the oceans and the 5-liter flask was used to represent the atmosphere. A spark discharge generated across the electrodes with a Tesla coil, invented by Nikola Tesla in 1891, was used to mimic lightning and corona discharges in the atmosphere.

on the early earth after UV radiation (Miller and Urey 1959). The apparatus was filled with various mixtures of methane, ammonia, and hydrogen, as well as water, which was then heated during the experiment. A spark discharge between the tungsten electrodes, which simulated lightning and corona discharges in the early atmosphere, was produced using a high-frequency Tesla coil with a voltage of 60,000 V. The reaction time was usually a week or so, and the maximum pressure was 1.5 bars. With this experimental setup, Miller was able to transform almost 50% of the original carbon (in the form of methane) into organic compounds. Although most of the synthesized organic material was an insoluble tarlike solid, he was able to isolate amino acids and other simple organic compounds from the reaction mixture. Glycine, the simplest amino acid, was produced in 2% yield (based on the original amount of methane carbon), whereas alanine, the simplest amino acid with a chiral center, showed a yield of 1%. Miller was able to demonstrate that the alanine that was produced was a racemic mixture (equal amounts of D- and L-alanine). This provided convincing evidence that the amino acids were produced in the experiment and were not biological contaminants somehow introduced into the apparatus.

The other organic compounds that Miller was able to identify made it possible for him to propose a possible reaction pathway for the amino acids (Miller 1957). The proposed synthetic mechanism was in fact the one described by Adolph Strecker (1850) when he reported the synthesis of alanine. This involves the reaction of hydrogen cyanide, ammonia, and carbonyl compounds (aldehydes or ketones) to form amino nitriles, which then undergo hydrolysis to form the amino acids (see Reaction 1). Depending on the concentration of ammonia in the reaction mixture, varying amounts of hydroxy acids are produced as well. This is what Miller found, with larger relative amounts of hydroxy acids being formed in a reaction mixture containing less ammonia.

It is important to note that neither purines nor pyrimidines, the nucleobases that are part of DNA and RNA, were looked for in the mixtures of the original Miller-Urey experiment. However, in experiments carried out soon after Miller's experiment by Oró and coworkers, the formation of adenine from ammonium cyanide solutions was demonstrated (Oró 1960; Oró and Kimball 1961). Later it was shown that the abiotic synthesis of purines and other heterocyclic compounds also takes place under the same conditions as in the original Miller-Urey experiment, but with much smaller yields than for the amino acids. In addition, it has been found that guanine can be produced in a direct "one-pot" synthesis from the polymerization of aqueous solutions of ammonium cyanide (Yuasa et al. 1984).

Although these results are extremely encouraging, the atmospheric composition that formed the basis of the Miller-Urey experiment is not considered today to be plausible by many researchers. It is generally agreed that free oxygen was absent from the primitive earth, but there is no general agreement on the composition of the primitive atmosphere; opinions vary from strongly reducing ($CH_4 + N_2$, $NH_3 + H_2O$, or $CO_2 + H_2 + N_2$) to neutral

Reaction 1. The Strecker-cyanohydrin reaction for the formation of amino and hydroxy acids.

($CO_2 + N_2 + H_2O$). In general, those working on prebiotic chemistry lean toward more reducing conditions, under which the abiotic syntheses of amino acids, purines, pyrimidines, and other compounds are very efficient, while nonreducing atmospheric models are favored by planetologists. A weakly reducing or neutral atmosphere appears to be more in agreement with the current model for the early earth. Although Miller and Urey (1959) originally rejected the idea of nonreducing conditions for the primitive atmosphere, a number of experiments were later carried out in Miller's laboratory that used model atmospheres of CO and CO_2 (Schlesinger and Miller 1983). It was found that not only were the yields of the amino acids reduced, but also glycine was basically the only amino acid synthesized under these conditions. A general trend showed that as the atmosphere became less reducing and more neutral, the yields of synthesized organic compounds decreased drastically, although they were never zero. The presence of methane and ammonia appeared to be especially important for the formation of a diverse mixture of amino acids. The main problem in the synthesis of amino acids and other biologically relevant organic compounds with nonreducing atmospheres is the apparent lack of formation of hydrogen cyanide, which is an intermediate in the Strecker pathway and an important precursor compound for the synthesis of nucleobases (Ferris et al. 1978). However, as mentioned earlier, localized high concentrations of reduced gases may have existed around volcanic eruptions, and in these localized environments reagents such as HCN, aldehydes, and ketones may have been produced, which after washing into the oceans could have become involved in the prebiotic synthesis of organic molecules.

Whether the primitive atmosphere was reducing or neutral may be irrelevant with respect to the synthesis of organic compounds on the early earth, however. It has recently been shown that contrary to previous findings, significant amounts of amino acids are produced in spark discharge experiments using CO_2/N_2/liquid water mixtures (Cleaves et al. 2008). The low yields of amino acids found previously were apparently the result of oxidation during hydrolytic workup by nitrite and nitrate produced in the reactions, as well as by the inhibition of synthesis caused by the low pH conditions generated during the experiment. Addition of calcium carbonate to the system to buffer the system near neutral pH, along with oxidation inhibitors such as ascorbic acid or Fe^{2+} prior to hydrolysis, results in the recovery of several hundred times more amino acids than reported previously.

In the recent neutral atmosphere experiments, the amounts of free cyanide, aldehydes, and ammonia are low, suggesting that the intermediates may be bound as nitriles. The observed amino acids may have formed by several mechanisms, including the Strecker synthesis or the related Bucherer-Bergs pathway. Although the amino acids observed and their relative abundances are similar to those found when oligomers formed by the self-condensation of HCN in aqueous solution are hydrolyzed (Ferris et al. 1978), the absence of polymeric material in the experiments appears to discard this mechanism. Experiments with slightly reducing model atmospheres (Schlesinger and Miller 1983) suggest that the addition of traces of CH_4 and/or H_2 in neutral atmosphere simulations would likely further enhance the production of amino acids.

Because of supposed problems associated with the direct Miller-Urey type of syntheses on the early earth, a different hypothesis for the abiotic synthesis of organic compounds has been proposed. This suggestion resulted from the discovery of deep-ocean hydrothermal vents. A group of researchers have argued that the remarkable properties of the hydrothermal vent environments, particularly their protection from the harsh conditions caused by large-impact events, might have played an important role in the origin of life. Since it is thought by some that the last common ancestral organism of all extant life on earth was a thermophile, several researchers have proposed the hypothesis that the organic compounds necessary for the origin of life were actually synthesized under vent conditions (Holm and Andersson 1995). Major proponents of this hypothesis are Everett Shock and coworkers, who have calculated that thermodynamic-based equilibria favor the formation of compounds such as amino acids at hydrothermal vent temperatures (Shock 1990; Shock and Schulte 1998). Vent-based synthesis is considered to be especially important in vents associated with off-axis systems (Holm and Charlou 2001; Kelley et al. 2001). However, at the high temperatures (> 350°C) associated with most vent discharges, amino acids and other biomolecules have been found to rapidly decompose (Bernhardt et al. 1984; Miller and Bada 1988; Bada et al. 1995). For example, amino acids are destroyed in timescales of minutes at temperatures greater than 300°C. The rate of hydrolysis for RNA at pH 7 extrapolated to elevated temperatures gives a half-life of 2 minutes at 250°C for the hydrolysis of every phosphodiester bond; at 350°C the half-life is 4 seconds. For DNA, the half-lives for depurination of each nucleotide at pH 7 are nearly the same as the hydrolysis rates for RNA (Miller and Lazcano 1995). It has been pointed out by Lazcano (1997) that if the origin of life took place over a sufficiently long period of time, all the complex organic compounds in the ocean, derived from whatever sources, would be destroyed by passage through the hydrothermal vents. It is thus possible that hydrothermal vents are much more effective in regulating the concentration of critical organic molecules in the oceans than in playing a significant role in their direct synthesis.

Despite the problems with stability at high temperatures, the vent-induced synthesis theory has been championed by Huber and Wächtershäuser (2006), who have described have experiments claimed to "narrow the gap between biochemistry and volcanic geochemistry." Using high concentrations of HCN (0.2 M) and high pressures of CO (75 bars) in the presence of Ni/Fe catalysts, a variety of amino and hydroxyl acids are produced at 100°C. However, there are concerns about the relevance of this type of experiment to the natural geochemical environments that would be expected to have existed on the early earth, especially the high concentrations and pressures of starting reagents (Bada et al. 2007).

Did Prebiotic Organic Compounds Come from Space?

The difficulties supposedly involved with the endogenous, earth-based synthesis of amino acids and nucleobases led to the development of alternatives. In the early 1990s Chyba and Sagan (1992) reanalyzed Oró's 1961 proposal on the role of cometary nuclei as sources of volatiles for the primitive earth and proposed that the exogenous delivery of organic matter by asteroids, comets, and interplanetary dust particles (IDPs) could have played a significant role in seeding the early earth with the compounds necessary for the origin of life. This conclusion was based on knowledge about the organic composition of meteorites. It is important to note that if this concept is valid, impacts on the early earth not only created devastating conditions that made it difficult for life to originate but also at the same time perhaps delivered the raw material necessary to set the stage for the origin of life. In an even wider view this hypothesis could have profound implications for the abundance of life in the universe. The source of the essential organic compounds required for the origin of life is not constrained by the conditions on a particular planet; rather, synthesis of organic compounds is a ubiquitous process that takes place on primitive solar-system bodies such as asteroids and comets. The possibility of the origin of life is thus considerably increased, provided the essential organic compounds are delivered intact to a planet that is suitable for further chemical evolution. This is yet to be fully demonstrated.

Carbonaceous chondrites, a class of stony meteorites, are considered to be the most primitive objects in the solar system in terms of their elemental composition, yet they have a high abundance of organic carbon, more than 3% by weight in some cases. The meteorites most extensively analyzed for organic compounds include the CMs Murchison (which fell in 1969 in Victoria, Australia) and Murray (1950, Kentucky, United States) and the CI-class Orgueil meteorite (1864, France). (CM and CI refer to types of chondrites. CM are about 70% fine-grained material and CI are almost exclusively fine-grained material. Both types have experienced extensive aqueous alteration on the parent asteroid.) The carbon phase is dominated by an insoluble fraction. The majority (up to 80%) of the soluble organic matter is made up of polycyclic aromatic hydrocarbons (PAHs), aliphatic hydrocarbons, carboxylic acids, fullerenes, and amino acids (Botta and Bada 2002). The purines adenine, guanine, xanthine, and hypoxanthine have also been detected, as well as the pyrimidine uracil in concentrations of 200 to 500 parts per billion (ppb) in the CM chondrites Murchison and Murray and in the CI chondrite Orgueil (Stoks and Schwartz 1977, 1981). In addition, a variety of other nitrogen-heterocyclic compounds, including pyridines, quinolines, and isoquinolines, were also identified in the Murchison meteorite (Stoks and Schwartz 1982), as well as sugar acids (polyols) (Cooper et al. 2001) and membrane-forming lipidic compounds (Deamer and Pashley 1989).

It has been found that the CI-type meteorites such as Orgueil contain an amino acid composition distinct from that of the CMs (Ehrenfreund et al. 2001). The simple amino acid mixture, consisting of just glycine and racemic alanine, found in CI carbonaceous chondrites is interesting because it has been generally thought that a wide variety of amino acids were required for the origin of life. As discussed later, however, among the candidates for the first genetic material is peptide nucleic acid (PNA), a nucleic acid analogue in which the backbone does not contain sugar or phosphate moieties (Egholm et al. 1992; Nelson et al. 2000). For the PNA backbone, achiral amino acids such as glycine or aminoethylglycine, possibly delivered by CI-type carbonaceous chondrites to the early earth, may have been the only amino acids needed for the origin of life.

Prebiotic Polymerization Processes

The organic material on the early earth before life existed, regardless of its source, likely consisted of a wide array of different types of simple monomeric compounds, including many that play a major role in biochemistry today. How these simple abiotic organic constituents were assembled into polymers and then into the first living entities is presently the most challenging area of research on the origin of life.

Simple monomers on the primitive earth would need to undergo polymerization, a thermodynamically unfavorable process. It is generally assumed that polymers composed of at least 20 to 100 monomeric units (20 to 100 mers) are required in order to have any catalytic and replication functions (Joyce 2002; de Duve 2003). Thus early polymerization processes must have been capable of producing polymers of at least this minimum size.

There is no evidence of abiotically produced oligopeptides or oligonucleotides in meteorites, so condensation reactions clearly must have taken place directly on the primitive earth. Synonymous terms like "primordial broth" or "Darwin's warm little pond" have led in some cases to major misunderstandings, including the simplistic image of a worldwide ocean, rich in self-replicating molecules and accompanied by all sorts of biochemical monomers. The term *warm little pond*, which has long been used for convenience, refers not necessarily to the entire ocean but to parts of the hydrosphere where the accumulation and interaction of the products of prebiotic synthesis may have taken place. These include oceanic particles and sediments, intertidal zones, freshwater shallow ponds and lakes, lagoons undergoing wet-and-dry cycles, and eutectic environments associated with glacial ponds.

Simple organic compounds dissolved in the primitive oceans or other bodies of water would need to be concentrated by some mechanism in order to enhance polymerization processes. Concentration involving the selective adsorption of molecules onto mineral surfaces has been suggested as one means of promoting polymerization, and this process has been demonstrated in laboratory experiments that have used a variety of simple compounds and activated monomers (Hill et al. 1998; Ferris 2002). The potential importance of mineral-assisted catalysis is demonstrated by the montmorillonite-promoted polymerization of activated adenosine and uridine derivatives that produces 25–50 mer oligonucleotides (Ferris 2002), the general length range considered necessary for primeval biochemical functions.

Because adsorption onto surfaces involves weak noncovalent van der Waals interactions, the mineral-based concentration process and subsequent polymerization would be most efficient at cool temperatures (Liu and Orgel 1998; Sowerby et al. 2001). This, however, presents somewhat of a problem because as the length of polymers formed on mineral surfaces increases, they tend to be more firmly bound to the mineral (Orgel 1998). In order for these polymers to be involved in subsequent interactions with other polymers or monomers, they need to be released. This could be accomplished by warming the mineral, although this would also tend to hydrolyze the adsorbed polymers. A way around this problem would be to release the polymers by concentrated salt solutions (Hill et al. 1998), a process that could take place in tidal regions during evaporation or freezing of seawater.

The direct concentration of dilute solutions of monomers could also be accomplished by evaporation and by eutectic freezing of dilute aqueous solutions, which, coupled with other physicochemical mechanisms such as the adherence of biochemical monomers to active surfaces, could have raised local concentrations and promoted polymerization. The evaporation of tidal regions and the subsequent concentration of their organic constituents have been proposed in the synthesis of a variety of simple organic molecules (Nelson et al. 2000). Eutectic freezing of dilute reagent solutions has also been found to promote the synthesis of key biomolecules (Levy et al. 2000). These same concentration processes also could have played a key role in the synthesis of polymers. It has been shown that the freezing of dilute solutions of activated amino acids at $-20°C$ yields peptides at higher amounts than in experiments with highly concentrated solutions at $0°$ and $25°C$ (Liu and Orgel 1997). In addition, recent studies have shown that eutectic freezing is especially effective in the nonenzymatic synthesis of oligonucleotides (Kanavarioti et al. 2001).

Salty brines may also have been important in the formation of peptides and perhaps other important biopolymers as well. According to Rode and coworkers (Rode 1999), "salt-induced peptide formation reaction" (SIPF) provides a pathway for the efficient synthesis of peptides and possibly proteins directly from simple amino acids in concentrated NaCl solutions containing Cu(II). Yields of di- and tripeptides in the range of 0.4 to 4% have been reported using starting amino acid concentrations in the 40–50 m M range. Clay minerals such as montmorillonite apparently promote the reaction. Again, the evaporation of tidally flushed lagoons or the freezing of the primitive oceans could have produced the concentrated salt-rich brines needed to promote this salt-induced polymerization process.

Did Pyrite Play a Role in the Origins of Life on Earth?

For some time, one of the most serious rivals to the "primordial soup" heterotrophic theory was derived from the ideas of Wächtershäuser (1988). According to his "pioneer metabolic theory," life began with the appearance of an autocatalytic two-dimensional chemolithotrophic metabolic system that took place on the surface of the highly insoluble mineral pyrite in the vicinity of hydrothermal vents. Replication followed the appearance of this nonorganismal iron sulfide–based two-dimensional "life," in which chemoautotrophic carbon fixation took place by a reductive citric acid cycle, or reverse Krebs cycle, of the type originally described for the photosynthetic green sulfur bacterium *Chlorobium limicola*. Molecular phylogenetic trees show that this mode of carbon fixation and its modifications (such as the reductive acetyl-CoA or the reductive malonyl-CoA pathways) are found in anaerobic archaea and the most deeply divergent eubacteria, which has been interpreted as evidence of its primitive character (Maden 1995). But is the reverse Krebs cycle truly primordial?

The reaction $FeS + H_2S = FeS_2 + H_2$ is very energetically favorable. It has an irreversible, highly exergonic character with a standard free energy change of $G_o = -9.23$ kcal/mol, which corresponds to a reduction potential $E_o = -620$ m V). Thus the FeS/H_2S combination is a strong reducing agent and has been shown to provide an efficient source of electrons for the reduction of organic compounds under mild conditions. Pyrite formation can produce molecular hydrogen and reduce nitrate to ammonia, acetylene to ethylene, and thioacetic acid to acetic acid, as well as promoting more complex syntheses (Maden 1995), including peptide bonds that result from the activation of amino acids with carbon monoxide and nickel/iron sulphides (Huber and Wächtershäuser 1998). Although pyrite-mediated CO_2 reduction to organic compounds has not been achieved, the fixation under plausible prebiotic conditions of carbon monoxide into activated acetic acid by a mixture of co-precipitated NiS/FeS/S has been reported (Huber and Wächtershäuser 1998). However, in these experiments the reactions occur in an aqueous environment to which powdered pyrite has been added; they do not form a dense monolayer of ionically bound molecules or take place on the surface of pyrite. In addition, a careful examination of the Huber and Wächtershäuser amino acid–peptide synthesis scheme has found that it is highly unlikely to achieve formation of oligomers that had the ability to promote effective autocatalysis processes associated with the "pioneer metabolic theory" (Ross 2008).

None of these experiments proves by itself that both enzymes and nucleic acids are the evolutionary outcome of surface-bounded metabolism. In fact, the results are also compatible with a more general, modified model of the primordial soup in which pyrite formation is recognized as an important source of electrons for the reduction of organic compounds. It is thus possible that under certain geological conditions the FeS/H_2S combination could have reduced not only CO but also CO_2 released from molten magna in deep-sea vents, and led to biochemical monomers. Peptide synthesis, for instance, could have taken place in an iron and nickel sulfide system (Huber and Wächtershäuser 1998) involving amino acids formed by electric discharges via a Miller-type synthesis. If the compounds synthesized by this process did not remain bound to the pyrite surface but drifted away into the surrounding aqueous environment, then they would become part of the prebiotic soup, not of a two-dimensional autocatalytic reaction scheme.

The essential question in deciding between these two different theories is not whether pyrite-mediated organic synthesis can occur, but whether direct CO_2 reduction and synthesis of organic compounds can be achieved by a hypothetical two-dimensional living system that lacks genetic information. Proof of Wächtershäuser's hypothesis requires demonstration not only of the tight coupling of the reactions necessary to drive autocatalytic CO_2 assimilation via a reductive citric acid cycle but also of the interweaving of a network of homologous cycles that, it is assumed, led to all the anabolic pathways, as well as replication (Maden 1995). This has not been achieved. In fact, experimental results achieved so far with the FeS/H_2S combination in reducing N_2 and CO are consistent with a heterotrophic origin of life that acknowledges the role of sulfur-rich minerals and other catalysts in the synthesis and accumulation of organic compounds.

How Did Replication First Originate?

The primordial broth must have been a bewildering organic chemical wonderland in which a wide array of different molecules were constantly synthesized, destroyed, or incorporated into cycles of chemical transformations. Regardless of the complexity of the prebiotic environment, life as we know it could not have evolved in the absence of a genetic replicating mechanism to guarantee the maintenance, stability, and diversification of life's basic components under the action of natural selection. The appearance of the first molecular entities capable of replication, catalysis, and multiplication would have marked the origin of both life and evolution. What were the fundamental characteristics of these first molecular living entities that distinguished them from nonliving chemistry?

The leap from biochemical monomers and small oligomers to living entities is an enormous leap indeed. How the ubiquitous genetic system of extant life based on nucleic acid originated is one of the major unsolved problems in contemporary biology. The discovery of catalytically active RNA molecules gave considerable credibility to prior suggestions that the first living entities were largely based on catalytic RNA molecules (ribozymes), in a hypothetical stage in the early evolution of life called the RNA world (Gilbert 1986; Joyce 2002). This possibility is now widely accepted, but the limited stability of the components of RNA implies that this molecule was probably not a direct product of prebiotic evolution.

The ribose component of RNA is very unstable, which makes its presence in the prebiotic milieu unlikely (Larralde et al. 1995). Both lead hydroxide (Zubay 1998) and borate minerals apparently stabilize ribose (Ricardo et al. 2004), and cyanamide is known to react with ribose to form a stable bicyclic adduct (Springsteen and Joyce 2004). However, in order to be involved in the polymerization reactions that lead to RNA, ribose would likely have to be present in solution, where it would be prone to decomposition. Moreover, the huge number of possible random combinations of derivatives of nucleobases, sugars, and phosphate that may have been present in the prebiotic soup makes it unlikely that an RNA molecule capable of catalyzing its own self-replication arose spontaneously (Joyce 2002). These difficulties have led to proposals of pre-RNA worlds, in which informational macromolecules with backbones different from those of extant nucleic acids may also have been endowed with catalytic activity, that is, with phenotype and genotype also residing in the same molecule. Thus a simpler self-replicator must have come first, and several possible contenders have been suggested. The RNA precursor would have had the capacity to catalyze reactions and to store information, although the component nucleobases and the backbone that held the polymer together were not necessarily the same as those in modern RNA and DNA. The nature of the genetic polymers and the catalytic agents that may have preceded RNA is presently unknown.

There are now several known examples of molecular genetic systems that have been studied in the laboratory, and these provide examples of the types of molecular systems that could have given rise to early self-replicating entities (see Bada 2004). Possible candidates include nucleic acid analogues (Egholm et al. 1992; Miller 1997; Eschenmoser 1999, 2004; Nelson et al. 2000) such as peptide nucleic acid (PNA), where the backbone consists of linked amino acid derivatives such as N-(2-aminoethyl)glycine or (AEG) (the nucleobases are attached by an acetic acid linkage to the amino group of glycine), and threose nucleic acid (TNA), where the backbone is made up of L-threose connected by 3', 2' phosphodiester bonds. Both of these candidates form double helical structures through Watson-Crick base pairing with complementary strands of themselves, RNA, and DNA. With PNA-PNA helices both left- and right-handed structures are produced in equal amounts, while the PNA-RNA and PNA-DNA helices are both right-handed. Although appealing as possible candidates for the first self-replicating molecular living entities, both PNA and TNA have positive and negative aspects. The main negative factors are the lack of any demonstrated oligomerization process that would produce these nucleic acid analogues under plausible prebiotic conditions and the lack of any demonstrated catalytic properties.

PNA is attractive because its backbone is achiral (lacking handedness), which eliminates the need for the selection of chirality before the time of the origin of life. Its components, AEG and nucleobases linked to acetic acid, have been produced under simulated prebiotic conditions (Nelson et al. 2000). However, PNA is susceptible to an N-acyl migration reaction that produces a rearranged PNA. This problem could be minimized, however, by blocking the N-terminal position by acetylation, for example.

On the basis of an extensive study of sugar-based nucleic acids, TNA appears to be superior to other possible sugar-based nucleic acids with respect to its base-pairing attributes, especially with RNA (Eschenmoser 1999, 2004; Schöning et al. 2000). The tetrose sugar in TNA could have been synthesized during the reaction cascade that takes place during the formose reaction first described by Butlerov in 1861. The 4-carbon sugars threose and erythrose could have been readily synthesized by the dimerization of glycolaldehyde (see Reaction 2), which in turn could have been produced from the dimerization of formaldehyde. Even though the presence of a 4-carbon sugar in TNA reduces this problem to 2 sugars and 4 stereoisomers, to demonstrate how oligonucleotides composed only of L-threose could be preferentially synthesized under prebiotic conditions is a formidable challenge.

It is possible that PNA preceded TNA and in fact assisted in the transition to the first TNA molecules. As stated earlier, the selection of the chiral sugar component of TNA would have required some sort of selection process to be in operation. Orgel and coworkers have suggested that the incorporation of chiral sugar dinucleotides at the end of a PNA chain, which could have occurred simply by chance, can induce chirality into a nucleic acid produced by PNA-induced oligomerization (Kozlov et al. 2000). PNA could thus have provided a means of conveying the critically important biological property of chirality into polymers near the time of the origin of life. This possibility potentially solves an ongoing dilemma whether the origin of chirality occurred before the origin of life or during the evolution of early living entities.

$$HCHO \xrightarrow{HCHO} HOCH_2\text{-}CHO \xrightarrow{HOCH_2\text{-}CHO} HOCH_2\text{-}CHOH\text{-}CO\text{-}CH_2OH$$

Formaldehyde   **Glycolaldehyde**   **Tetrose**

Reaction 2. The formation of tetrose sugars from the dimerization of formaldehyde and glycolaldehyde. This is only part of the overall formose series of reactions, and both hexoses and pentoses would have been produced by subsequent reactions.

Regardless of the type of analogue similar to nucleic acid or other type of replicator system that was used by the first self-replicating entities, polymer stability and survival would have been of critical importance. Nucleic acids in general have very short survival times at elevated temperatures. RNA is very unstable because of the presence of the phosphodiester bond involving the 2'-hydroxyl group of ribose. The stability of TNA is unknown at present, but it is probably somewhat similar to the stability of DNA. It is possible that other tetrose sugar-based nucleic acids were less stable than TNA because of their tendency to form less stable helical structures (Eschenmoser 1999) and that the selection of TNA was the result of this factor, although this is also not known. The stability of PNA has been partly investigated, and provided the N-acyl migration reaction can be minimized, the amide linkage in PNA would be expected to have a stability at neutral pH similar to that of peptide bonds in proteins (Wang 1998). This suggests that in environments with temperatures of around 25°C, its survival time would be in the range of 10^4–10^5 years. Salty brines may have played a role in early nucleic acid survival because high salt concentrations have been found to enhance nucleic acid stability. For example, the stabilities of several transfer RNAs (tRNAs) were significantly increased in 1–2 M NaCl solution in comparison with stability in pure water (Tehei et al. 2002). The stability of DNA also increases with increasing salt concentration (for example, see Bada 2004). If these results were applicable to other nucleic acid analogues such as PNA and TNA, then salt solutions could have provided a protective environment that could have enhanced the survival of early self-replicating molecular entities.

Were membranes required for the emergence of replicating systems? The emergence of life may be best understood in terms of the dynamics and evolution of sets of chemical replicating entities. Whether such entities were enclosed within membranes is not yet clear, but given the prebiotic availability of amphiphilic compounds, this may have well been the case. Membrane-forming lipidic compounds may have been provided to the primitive earth from extraterrestrial sources (Deamer and Pashley 1989). This source could potentially supply a wide variety of aliphatic and aromatic hydrocarbons, alcohols, and branched and straight fatty acids, including some that are membrane-forming compounds. Prebiotic lipidic molecules may also have resulted from abiotic synthesis, as shown by the formation of normal fatty acids, glycerol, glycerol phosphate, and other lipids (Eichberg et al. 1977). Also, experiments have shown that simple mixtures of fatty acids and glycerol form from mixtures of mono-, di-, and triglycerides under mild conditions (Hargreaves et al. 1977). It is very attractive to assume that compartmentalization within liposomes formed by amphiphilic molecules of prebiotic origin was essential for the emergence of life. RNA molecules adsorbed onto clays such as montmorillonite, which can catalyze the formation of RNA oligomers, can be encapsulated into fatty acid vesicles whose formation in turn is accelerated by the clay. By incorporating additional fatty acid micelles, these vesicles can grow and divide while still retaining a portion of their contents needed to support RNA replication. In this manner, some of the basic machinery needed for RNA self-replication could have been compartmentalized into proto-type cells (Hanczyc et al. 2003; Chen 2006).

Despite the success of the RNA encapsulation experiments, there are problems with this "within the vesicle" scenario (Griffiths 2007). For example, in order to maintain a steady supply of molecules needed for survival of a replicating system inside a vesicle, molecules would need to cross a lipid-like bilayer barrier. This may have been a major obstacle for charged species such as amino acids. As a result, perhaps the first replicating system evolved "outside the vesicle" (Griffiths 2007). In this case, vesicles attached on mineral surfaces could have been in close proximity to simple self-replicating molecules that used the catalytic properties of the mineral to enhance reactions involved in polymerization and replication. As these the two systems evolved in complexity, synergistic interactions could have developed that allowed some of the components involved in the replicating system to become freely transported into the vesicle. At this point the two systems could have merged and formed a single compartmentalized, self-replication system. However, recent results showing that activated nucleotides can diffuse into liposomes and become polymerized within them suggest that under certain conditions the lipidic barrier may not have been impenetrable (Mansy et al. 2008).

The Transition toward a DNA/RNA/Protein World

The evolution of the first living molecular living entities into ones based entirely on RNA (i.e., the RNA world) would have been the next step in the evolution toward modern biochemistry. RNA has been found to be an all-in-one molecule that can not only store information but also catalyze reactions. Laboratory-based "test-tube-evolution" experiments have demonstrated that catalytic RNA molecules (ribozymes) have the capacity to carry out a wide range of important biochemical reactions, including the joining together of RNA fragments (ligation) and peptide-bond formation (Bartel and Unrau 1999). The list of demonstrated catalytic reactions is extensive and suggests that the RNA world could have had a large repertoire of catalytic RNA molecules, perhaps functioning in concert with one another (Doudna and Cech 2002). The complex series of reactions needed to permit multiplication, genetic transfer, and variation required in the RNA world has so far not been demonstrated in the laboratory, but optimism remains because of the relative immaturity of this area of research (Bartel and Unrau 1999; Doudna and Cech 2002; Joyce 2002).

The invention of protein synthesis and the encapsulation of reaction machinery needed for replication may have taken place during the RNA world. Four of the basic reactions involved in protein biosynthesis are catalyzed by ribozymes, and it has been noted that the complementary nature of these reactions is not likely accidental but rather suggestive that they had a common origin, most probably in the RNA world (Kumar and Yarus 2001). If this was the case, then the primitive nucleobase code used for protein biosynthesis had its origin in the RNA world, although the bases used in the early code could have been different from the ones used today (Kolb et al. 1994).

It is possible that by the time RNA-based life appeared on earth, the supplies of simple abiotic organic compounds derived from the sources discussed earlier would have been greatly diminished. Many of the components of the primordial soup had likely been extensively converted into polymers, including those associated with living entities, and thus the raw materials needed to sustain life had become largely exhausted. This implies that simple metabolic-like pathways must have been in place in order to ensure a supply of the ingredients needed to sustain the existence of the primitive living entities. In this case some metabolic pathways needed to produce essential components required by primitive living entities were perhaps originally nonenzymatic or semienzymatic autocatalytic processes that later became fine tuned as ribozymatic and protein-based enzymatic processing became dominant.

More than with earlier living molecular entities, the main limitation in the RNA world would have been the extreme instability of RNA. This in turn implies that RNA molecules must have been very efficient in carrying out self-replication reactions in order to maintain an adequate inventory of molecules needed for survival. The instability of RNA could have been the primary reason for the transition to the DNA/protein world, where, because of the increased stability of the genetic molecules, survival would have been less dependent on polymer stability. According to Joyce (2002), it is possible that in the RNA world ribozymes arose that could catalyze the polymerization of DNA, and in this manner information stored in RNA could be transferred to the more stable DNA. Another reason for DNA takeover could have been that because of increased stability much longer oligomers could have accumulated. This would have provided an enhanced storage capacity for information that could be passed on to the next generation of living entities. Before long, RNA, which had once played the singular role of replication and catalysis, was replaced by the more efficient and robust DNA/protein world, wherein RNA was demoted to a role of messenger/transcriber of DNA-stored information needed for protein biosynthesis.

Did Life Arise in a High-Temperature Environment?

Even though DNA is more stable than RNA, it is still rapidly degraded at elevated temperatures. In addition, protein enzymes denature rapidly at elevated temperatures. This must have at least initially limited the environments where DNA/protein–based life could survive for any significant period of time, and as was the case for all other earlier living entities based on nucleic acid, survival would have been most likely under cool conditions (Bada and Lazcano 2002b). Nevertheless, several researchers have advocated high temperatures, especially those associated with hydrothermal vent systems, as the environment where DNA/protein–based entities first arose. Proponents for a high-temperature origin cite the fact that the universal tree of extant life appears to be rooted in hyperthermophilic organisms. Thus if the last common ancestor (LCA) of all modern biology was a hyperthermophile, then it is concluded that the first DNA/protein–based life must have arisen in a similar type of environment.

At first glance, both the molecular and the paleontological fossil records appear to support a hyperthermophilic origin of life. Life on earth arose early. Large-scale analysis suggests that soon after its formation the surface of primitive earth was extremely hot. The planet is generally thought to have remained molten for some time after its formation 4.6 billion years ago (Wetherill 1990), but mineralogical evidence of a 4.3- to 4.4-billion- year-old hydrosphere implies that its surface rapidly cooled down (Wilde et al. 2001). However, there is theoretical and empirical evidence that the planet underwent late accretion impacts (Byerly et al. 2002; Schoenberg et al. 2002) that may have boiled the oceans as late as 3.8 billion years ago (Sleep et al. 1989).

Could modern biochemistry have arisen under these high temperature conditions? The proposals of a high-temperature origin of life face major problems, including the chemical decomposition of presumed essential biochemical compounds such as amino acids, nucleobases, RNA, and other thermolabile molecules, whose half-lives for decomposition at temperatures between 250–350°C are at the most a few minutes (White 1984; Miller and Bada 1988).

It is true, of course, that high temperatures allow chemical reactions to go faster. Primitive enzymes, once they appeared, could have been inefficient and thus the rate enhancement associated with higher temperatures would be one way to overcome this limitation (Harvey 1924). However, as summarized elsewhere (Islas et al. 2007), high-temperature regimes would lead to:

1. reduced concentrations of volatile intermediates, such as HCN, H_2CO, and NH_3;
2. lower steady-state concentrations of prebiotic precursors like HCN, which at temperatures a little above 100°C undergoes hydrolysis to first formamide and then formic acid and ammonia;
3. instability of reactive chemical intermediates like amino nitriles ($RCHO(NH_2)CN$), which play a central role in the Strecker synthesis of amino acids; and
4. loss of organic compounds by thermal decomposition and diminished stability of genetic polymers.

The recognition that the deepest branches in rooted universal phylogenies are occupied by hyperthermophiles is controversial and does not provide by itself conclusive proof of a high-temperature origin of DNA/protein–based life (Brochier and Philippe 2002). Given the huge gap existing in current descriptions of the evolutionary transition between the prebiotic synthesis of biochemical compounds and the LCA of all extant living beings, it is probably naive to attempt to describe the origin of life and the nature of the first living systems from molecular phylogenies (Islas et al. 2007). In addition, lateral gene transfer of thermoadaptative traits has apparently greatly compromised the genetic record present in modern organisms, which makes any conclusions about the environment where the DNA/protein world originated questionable (Doolittle 1999, 2000; Becerra et al. 2007b). An analysis of protein sequences has found only one enzyme, reverse gyrase, that is specific to hyperthermophiles; other proteins are apparently not ancestral to these organisms and are likely simply heated-adapted versions of those present in cooler-temperature organisms (Forterre 2002). Even if the LCA was a hyperthermophile (Gaucher et al. 2008; Gouy and Chaussidon 2008), there are alternative explanations for their basal distribution, including the possibility that they are (1) a relic from early Archean high-temperature regimes that may have resulted from a severe impact regime (Sleep et al. 1989; Gogarten-Boekels et al. 1995); (2) adaptation of bacteria to extreme environments by lateral transfer of reverse gyrase (Forterre et al. 2000) and other thermoadaptive traits from heat-loving Archaea; and (3) outcompetion of older mesophiles by hyperthermophiles originally adapted to stress-inducing conditions other than high temperatures (Miller and Lazcano 1995).

From the Origin of Protein Synthesis to the Last Common Ancestor of Life

Once DNA/protein–based life, and probably RNA-based life as well, became dominant and evolved in sophistication, the way carbon was produced and sequestered on the earth's surface was forever changed, especially after photosynthesis appeared. The amount of organic carbon produced by early life via the autotrophic fixation of CO_2, CH_4, and simple organic compounds such as formic and acetic acids would have far exceeded the amounts of organic compounds remaining, or still being synthesized, from either home-grown processes or extraterrestrial sources under the best conditions. The reservoir of organic material present on the earth then shifted from one initially characterized by compounds of abiotic origin to one made up entirely of biologically derived components.

It is likely that only in the DNA/protein world that biochemical machinery compartmentalized by cell-like membrane structures, comparable to those used in modern biology, became widespread. This is thus the first time that any direct evidence of life's existence might have been preserved in the form of physical fossils in the rock record. Earlier stages would have left behind only molecular remnants, and these have not survived the ravages of geochemical abuse over billions of years of geologic time. Although traces of hydrocarbon biomarkers have been detected in 2.7 billion-year-old sedimentary rocks, the oldest unambiguous molecular fossils found to date (Brocks et al. 1999), diagenetic processes, biological assimilation, and post-depositional metamorphism have long since eradicated the earlier record of abiotic organic chemistry and the components associated with primitive life.

All known organisms share the same essential features of genome replication, gene expression, basic anabolic reactions, and membrane-associated ATPase-mediated energy production. The molecular details of these universal processes not only provide direct evidence of the monophyletic origin of all extant forms of life but also imply that the sets of genes that encode the components of these complex traits were frozen a long time ago (i.e., major changes in them are very strongly selected against and are lethal). It is true that no ancient incipient stages or evolutionary intermediates of these molecular structures have been detected, but the existence of graded intermediates can be deduced, and therefore the supernatural origin advocated by both old-fashioned and contemporary creationists is rendered unnecessary.

For instance, the fact that RNA molecules by themselves are capable of performing all the reactions involved in peptide-bond formation suggests, as stated earlier, that protein biosynthesis evolved in an RNA world (Zhang and Cech 1998), that is, that the first ribosome lacked proteins and was produced only by RNA. This possibility is supported by crystallographic data that have shown that the ribosome's catalytic site, where peptide-bond formation takes place, is composed of RNA (Ban et al. 2000; Nissen et al. 2000). Clues to the genetic organization of primitive forms of translation are also provided by paralogous genes, which are sequences that diverge, not through speciation, but after a duplication event. For instance, the presence in all known cells of pairs of homologous genes that encode two elongation factors, which are GTP (guanosine triphosphate)-dependent enzymes that assist in protein biosynthesis, provides evidence of the existence of a more primitive, less regulated version of protein synthesis that took place with only one elongation factor. In fact, the experimental evidence of in vitro translation systems with modified cationic concentrations lacking both elongation factors and other protein components (Gavrilova et al. 1976; Spirin 1986) strongly supports the possibility of an older ancestral protein synthesis apparatus before the emergence of elongation factors.

Bioinformatic analysis of sequenced cellular genomes from the three major domains can be used to define the set of the most conserved protein-encoding sequences to characterize the gene complement of the LCA of extant life. The resulting set is dominated by different putative ATPases, and by molecules involved in gene expression and RNA metabolism. The so-called DEAD gyrases, which unwind RNA molecules and leave them ready for degradation, are as conserved as many transcription and translation genes. This suggests the early evolution of a control mechanism for gene expression at the RNA level, providing additional support to the hypothesis that RNA molecules played a more prominent role during early cellular evolution. Conserved sequences related to biosynthetic pathways include those encoding putative phosphoribosyl pyrophosphate synthase and thioredoxin, which participate in nucleotide metabolism. Although the information contained in the available databases corresponds only to a minor portion of biological diversity, the sequences reported here are likely to be part of an essential and highly conserved pool of protein domains common to all organisms (Becerra et al. 2007a).

The high levels of genetic redundancy detected in all sequenced genomes imply not only that duplication has played a major role in the accretion of the complex genomes found in extant cells, but also that before the early duplication events revealed by the large protein families, because the duplication is used to build more complexity, simpler living systems existed that lacked the large sets of enzymes and the sophisticated regulatory abilities of contemporary organisms. The variations of traits common to extant species can be easily explained as the outcome of divergent processes from an ancestral life form that existed before the separation of the three major biological domains, that is, the last common ancestor (LCA) or cenancestor. No paleontological remnants likely remain that bear testimony of its existence, so the search for a fossil of the cenancestor is bound to prove fruitless.

Analysis of an increasingly large number of completely sequenced cellular genomes has revealed major discrepancies in the topology of ribosomal RNA (rRNA) trees. Very often these differences have been interpreted as evidence of horizontal gene transfer (HGT) events between different species, casting doubt upon the feasibility of the reconstruction and proper understanding of early biological history (Doolittle 1999, 2000). There is clear evidence that genomes have a mosaic-like nature whose components come from a wide variety of sources (Ochman et al. 2000). Different advocates have described a wide spectrum of mix-and-match recombination processes, ranging from the lateral transfer of a few genes via conjugation, transduction, or transformation to cell fusion events involving organisms from different domains.

Universal gene-based phylogenies ultimately reach a single universal entity, but the bacterial-like LCA (Gogarten et al. 1989) that we favor was not alone. It would have been in the company of its siblings, a population of entities similar to it that existed throughout the same period. They may have not survived, but some of their genes did if they became integrated via lateral transfer into the LCA genome. The cenancestor is thus one of the last evolutionary outcomes of a series of ancestral events, including lateral gene transfer, gene losses, and paralogous duplications, that took place before the separation of Bacteria, Archaea, and Eucarya (Lazcano et al. 1992; Glansdorff 2000; Castresana 2001; Delaye et al. 2004; Becerra et al. 2007a).

Conclusions and Perspectives

Although there have been considerable advances in the understanding of chemical processes that may have taken place before the emergence of the first living entities, life's beginnings are still shrouded in mystery. Like vegetation in a mangrove swamp, the roots of universal phylogenetic trees are submerged in the muddy waters of the prebiotic broth, and how the transition from the nonliving to the living took place is still unknown. Given the huge gap that exists in current descriptions of the evolutionary transition between the prebiotic synthesis of biochemical compounds and the last common ancestor of all extant living beings, it is probably naive to attempt to completely describe the origin of life and the nature of the first living systems from molecular phylogenies.

Our current understanding of genetics, biochemistry, cell biology, and the basic molecular processes of living organisms has challenged many original assumptions of the heterotrophic theory. The view advocated here assumes that even if the first living entities were endowed with minimal synthetic abilities, their maintenance and replication depended primarily on organic compounds synthesized by prebiotic processes. An updated heterotrophic hypothesis assumes that the raw material for assembling the first self-maintaining, replicative chemical systems was the outcome of abiotic synthesis, while the energy required to drive the chemical reactions involved in growth and reproduction may have been provided by cyanamide, thioesters, glycine nitrile, or other high-energy compounds (de Duve 1995; Lazcano and Miller 1996).

The basic tenet of the heterotrophic theory of the origin of life is that the maintenance and reproduction of the first living entities depended primarily on prebiotically synthesized organic molecules. As summarized here, there has been no shortage of discussion about how the formation of the primordial soup took place. But have too many cooks spoiled the soup? Not really. It is very unlikely that any single mechanism can account for the wide range of organic compounds that may have accumulated on the primitive earth, and that the consequent prebiotic soup was formed by contributions from endogenous syntheses in a reducing atmosphere, metal sulphide–mediated synthesis in deep-sea vents, and exogenous sources such as comets, meteorites, and interplanetary dust. This eclectic view does not beg the issue of the relative significance of the different sources of organic compounds—it simply recognizes the wide variety of potential sources of organic compounds, the raw material required for the emergence of life.

As discussed here, the existence of different abiotic mechanisms by which biochemical monomers can be synthesized under plausible prebiotic conditions is well established. Of course, not all prebiotic pathways are equally efficient, but the wide range of experimental conditions under which organic compounds can be synthesized demonstrates that prebiotic syntheses of the building blocks of life are robust, that is, that the abiotic reactions that lead to them do not take place under a narrow range defined by highly selective reaction conditions but rather under a wide variety of experimental settings.

The synthesis of chemical constituents of contemporary organisms by nonenzymatic processes under laboratory conditions does not necessarily imply that they were either essential for the origin of life or available in the primitive environment. Nonetheless, the remarkable coincidence between the molecular constituents of living organisms and those synthesized in prebiotic experiments is too striking to be fortuitous, and the robustness of this type of chemistry is supported by the occurrence of most of these biochemical compounds in the carbonaceous meteorites of 4.5 billion years ago (Ehrenfreund et al. 2001). So it becomes plausible, but not proven, that similar synthesis took place on the primitive earth. For all the uncertainties surrounding the emergence of life, it appears to us that the formation of the prebiotic soup is one of the most firmly established events that took place on the primitive earth.

Mainstream scientific hypotheses on the origin of life, which have been developed within the framework of an evolutionary analysis, have led to a wealth of experimental results and the development of a coherent historical narrative that links many different disciplines and raises major philosophical issues. It is true that there are large gaps in the current descriptions of the evolutionary transition between the prebiotic synthesis of biochemical compounds and the last common ancestor of all extant living beings, but attempts to reduce it have allowed a more precise description of the beginning of life. Nowadays a central issue in origin-of-life research is to understand the abiotic synthesis of an ancestral genetic polymer endowed with catalytic activity and its further evolution to an RNA world and ultimately to the DNA/RNA/protein world characteristic of all life on earth. We face major unsolved problems, but they are not completely shrouded in mystery, unsolvable, or unknowable.

BIBLIOGRAPHY

Allwood, A. C., I. W. Burch, M. R. Walter, and B. S. Kamber. 2007. 3.43 billion-year-old stromatolite reef from Western Australia: Ecosystem-scale insights to early life on earth. *Precambrian Research* 158: 198–227.

Bada, J. L. 2004. How life began on earth: A status report. *Earth Planetary Science Letters* 226: 1–15.

Bada, J. L., C. Bigham, and S. L. Miller. 1994. Impact melting of frozen oceans on the early earth: Implications for the origin of life. *Proceedings of the National Academy of Sciences USA* 91: 1248–1250.

Bada, J. L., B. Fegley, S. L. Miller, A. Lazcano, H. J. Cleaves, R. M. Hazen, and J. Chalmers. 2007. Debating evidence for the origin of life on earth. *Science* 315: 937–938.

Bada, J. L., and A. Lazcano. 2002a. Miller revealed new ways to study the origins of life. *Nature* 416: 475.

———. 2002b. Some like it hot, but not the first biomolecules. *Science* 296: 1982–1983.

———. 2003. Prebiotic soup—Revisiting the Miller experiment. *Science* 300: 745–746.

Bada, J. L., S. L. Miller, and M. Zhao. 1995. The stability of amino acids at submarine hydrothermal vent temperatures. *Origins of Life and Evolution of the Biosphere* 25: 111–118.

Ban, N., P. Nissen, J. Hansen, P. B. Moore, and T. Steitz. 2000. The complete atomic structure of the large ribosomal subunit at 2.4 Å resolution. *Science* 289: 905–920.

Bartel, D. P., and P. J. Unrau. 1999. Constructing an RNA world. *Trends in Cell Biology* 9: M9–M13.

Becerra, A., L. Delaye, A. Islas, and A. Lazcano. 2007a. Very early stages of biological evolution related to the nature of the last common ancestor of the three major cell domains. *Annual Review of Ecology, Evolution and Systematics* 38: 361–379.

Becerra, A., L. Delaye, A. Lazcano, and L. Orgel. 2007b. Protein disulfide oxidoreductases and the evolution of thermophily: Was the last common ancestor a heat-loving microbe? *Journal of Molecular Evolution* 65: 296–303.

Bernhardt G., H. D. Lüdemann, R. Jaenicke, H. König, and K. O. Stetter. 1984. Biomolecules are unstable under "black smoker" conditions. *Naturwissenschaften* 71: 583–586.

Botta, O., and J. L. Bada. 2002. Extraterrestrial organic compounds in meteorites. *Surveys in Geophysics* 23: 411–467.

Brasier, M., O. R. Green, A. P. Jephcoat, A. K. Kleppe, M. van Kranendonk, J. F. Lindsay, A. Steele, and N. V. Grassineau. 2002. Questioning the evidence for earth's earliest fossils. *Nature* 416: 76–79.

Brochier, C., and H. Philippe. 2002. Phylogeny: A non-hyperthermophilic ancestor for bacteria. *Nature* 417: 244.

Brocks, J. J., G. A. Logan, R. Buick, and R. E. Summons. 1999. Archean molecular fossils and the early rise of eukaryotes. *Science* 285: 1033–1036.

Burkhardt, F., and S. Smith, eds. 1994. *A Calendar of the Correspondence of Charles Darwin, 1821–1882, With Supplement.* Cambridge: Cambridge University Press.

Byerly, G. R., D. R. Lowe, J. L. Wooden, and X. Xie. 2002. An Archean impact layer from the Pilbara and Kaapvaal cratons. *Science* 297: 1325–1327.

Calvin, M. 1969. *Chemical Evolution: Molecular Evolution towards the Origin of Living Systems on the Earth and Elsewhere.* Oxford: Clarendon Press.

Castresana, J. 2001. Comparative genomics and bioenergetics. *Biochimica et Biophysica Acta* 1506: 147–162.

Chen, I. A. 2006. The emergence of cells during the origin of life. *Science* 314: 1558–1559.

Chyba, C., and C. Sagan. 1992. Endogenous production, exogenous delivery and impact-shock synthesis of organic molecules: An inventory for the origins of life. *Nature* 355: 125–132.

Cleaves, H. J., J. H. Chalmers, A. Lazcano, S. L. Miller, and J. L. Bada. 2008. A reassessment of prebiotic organic synthesis in neutral planetary atmospheres. *Origins of Life and Evolution of the Biosphere* 38: 105–115.

Cooper, G., N. Kimmich, W. Belisle, J. Sarinana, K. Brabham, and L. Garrel. 2001. Carbonaceous meteorites as a source of sugar-related organic compounds for the early earth. *Nature* 414: 879–883.

Dahm, R. 2005. Friedrich Miescher and the discovery of DNA. *Developmental Biology* 278: 274–288.

Darwin, F. 1887. *The Life and Letters of Charles Darwin, Including an Autobiographical Chapter.* London: Murray.

Deamer, D. W., and R. M. Pashley. 1989. Amphiphilic compounds of the Murchison carbonaceous chondrite: Surface properties and membrane formation. *Origins of Life and Evolution of the Biosphere* 19: 21–38.

de Duve, C. 1995. *Vital Dust: Life as a Cosmic Imperative.* New York: Basic Books.

———. 2003. A research proposal on the origin of life. *Origins of Life and Evolution of the Biosphere* 33: 559–574.

Delaye, L., A. Becerra, and A. Lazcano. 2004. The nature of the last common ancestor. In L. Ribas de Pouplana, ed., *The Genetic Code and the Origin of Life,* 34–47. Georgetown, TX: Landes Bioscience.

Doolittle, W. F. 1999. Phylogenetic classification and the universal tree. *Science* 284: 2124–2128.

———. 2000. The nature of the universal ancestor and the evolution of the proteome. *Current Opinion in Structural Biology* 10: 355–358.

Doudna, J. A., and T. R. Cech. 2002. The chemical repertoire of natural ribozymes. *Nature* 418: 222–228.

Egholm, M., O. Buchardt, P. E. Nielsen, and R. H. Berg. 1992. Peptide nucleic acid (PNA): Oligonucleotide analogues with an achiral peptide backbone. *Journal of the American Chemical Society* 114: 1895–1897.

Ehrenfreund, P., D. P. Glavin, O. Botta, G. Cooper, and J. L. Bada. 2001. Extraterrestrial amino acids in Orgueil and Ivuna: Tracing the parent body of CI type carbonaceous chondrites. *Proceedings of the National Academy of Sciences USA* 98: 2138–2141.

Eichberg, J., E. Sherwood, E. Epps, and J. Oró. 1977. Cyanamide mediated syntheses under plausible primitive earth conditions. *Journal of Molecular Evolution* 10: 221–230.

Eschenmoser, A. 1999. Chemical etiology of nucleic acid structure. *Science* 284: 2118–2124.

———. 2004. The TNA-family of nucleic acid systems: Properties and prospects. *Origins of Life and Evolution of the Biosphere* 34: 277–306.

Ferris, J. P. 2002. Montmorillonite catalysis of 30–50 mer oligonucleotides:

Laboratory demonstration of potential steps in the origin of the RNA world. *Origins of Life and Evolution of the Biosphere* 32: 311–332.

Ferris, J. P., P. D. Joshi, E. H. Edelson, and J. G. Lawless. 1978. HCN: A plausible source of purines, pyrimidines, and amino acids on the primitive earth. *Journal of Molecular Evolution* 11: 293–311.

Forterre, P. 2002. A hot story from comparative genomics: Reverse gyrase is the only hyperthermophile-specific protein. *Trends in Genetics* 18: 236–237.

Forterre, P., C. Bouthier de la Tour, H. Philippe, and M. Duguet. 2000. Reverse gyrase from hyperthermophiles: Probable transfer of a thermoadaptation trait from Archaea to Bacteria. *Trends in Genetics* 16: 152–154.

García-Ruiz, J. M., S. T. Hyde, A. M. Carnerup, A. G. Christy, M. J. van Kranendonk, and N. J. Welham. 2003. Self-assembled silica-carbonate structures and detection of ancient microfossils. *Science* 302: 1194–1197.

Gaucher, E. A., J. M. Thomson, M. F. Burgan, and S. A. Benner. 2008. Inferring the palaeoenvironment of ancient bacteria on the basis of resurrected proteins. *Nature* 425: 285–288.

Gavrilova, L. P., O. E. Kostiashkina, V. E. Koteliansky, N. M. Rutkevitch, and A. S. Spirin. 1976. Factor-free, non-enzymic, and factor-dependent systems of translation of polyuridylic acid by *Escherichia coli* ribosomes. *Journal of Molecular Biology* 101: 537–552.

Gilbert, W. 1986. Origin of life: The RNA world. *Nature* 319: 618.

Gilliland, R. L. 1989. Solar evolution. *Global Planet Change* 1: 35–55.

Glansdorff, N. 2000. About the last common ancestor, the universal life-tree and lateral gene transfer: A reappraisal. *Molecular Microbiology* 38: 177–185.

Gogarten, J. P., H. Kibak, P. Dittrich, L. Taiz, E. J. Bowman, B. J. Bowman, M. F. Manolson, R. J. Poole, T. Date, T. Oshima, J. Konishi, K. Denda, and M. Yoshida. 1989. Evolution of the vacuolar H+-ATPase: Implications for the origin of eukayotes. *Proceedings of the National Academy of Sciences USA* 86: 6661–6665.

Gogarten-Boekels, M., E. Hilario, and J. P. Gogarten. 1995. The effects of heavy meteorite bombardment on the early evolution of life—a new look at the molecular record. *Origins of Life and Evolution of the Biosphere* 25: 78–83.

Gouy, M., and M. Chaussidon. 2008. Ancient bacteria like it hot. *Nature* 451: 635–636.

Griffiths G. 2007. Cell evolution and the problem of membrane topology. *Nature Reviews: Molecular Cell Biology* 8: 1018–1024.

Hanczyc, M. M., S. M. Fujikawa, and J. W. Szostak. 2003. Experimental models of primitive cellular compartments: Encapsulation, growth, and division. *Science* 302: 618–622.

Hargreaves, W. R., R. S. Mulvihill, and D. W. Deamer. 1977. Synthesis of phospholipids and membranes in prebiotic conditions. *Nature* 266: 78–80.

Harvey, R. B. 1924. Enzymes of thermal algae. *Science* 60: 481–482.

Hill, A. R., C. Böhler, and L. E. Orgel. 1998. Polymerization on the rocks: Negatively charged D/L amino acids. *Origins of Life and Evolution of the Biosphere* 28: 235–243.

Holm, N. G., and E. M. Andersson. 1995. Abiotic synthesis of organic compounds under the conditions of submarine hydrothermal vents: A perspective. *Planetary and Space Science* 43: 153–159.

Holm, N. G., and J. L. Charlou. 2001. Initial indications of abiotic formation of hydrocarbons in the Rainbow ultramafic hydrothermal system, Mid-Atlantic Ridge. *Earth Planetary Science Letters* 191: 1–8.

Huber, C., and G. Wächtershäuser. 1998. Peptides by activation of amino acids with CO on (Ni, Fe)S surfaces and implications for the origin of life. *Science* 281: 670–672.

————. 2006. α-Hydroxyl and α-amino acids under possible Hadean, volcanic origin-of-life conditions. *Science* 314: 630–632.

Islas, S., A. M. Velasco, A. Becerra, L. Delaye, and A. Lazcano. 2007. Extremophiles and the origin of life. In Charles Gerday and Nicolas Glansdorff, eds., *Physiology and Biochemistry of Extremophiles,* 3–10. Washington, DC: ASM Press.

Joyce, G. F. 2002. The antiquity of RNA-based evolution. *Nature* 418: 214–221.

Kanavarioti, A., P. A. Monnard, and D. W. Deamer. 2001. Eutectic phases in ice facilitate nonenzymatic nucleic acid synthesis. *Astrobiology* 1: 271–281.

Kasting, J. F. 1993a. Early evolution of the atmosphere and ocean. In J. M. Greenberg, C. X. Mendoza-Gómez, and V. Pironello, eds., *The Chemistry of Life's Origins,* 149–176. Dordrecht, The Netherlands: Kluwer Academic Publishers.

————. 1993b. Earth's early atmosphere. *Science* 259: 920–926.

Kelley, D. S., J. A. Karson, D. K. Blackman, L. Gretchen, G. L. Früh-Green, D. A. Butterfield, M. D. Lilley, E. J. Olson, M. O. Schrenk, K. K. Roe, G. T. Lebon, P. Rivizzigno, and the AT3-60 Shipboard Party. 2001. An off-axis hydrothermal vent field near the Mid-Atlantic Ridge at 30° N. *Nature* 241: 145–149.

Kolb, V. M., J. P. Dworkin, and S. L. Miller. 1994. Alternative bases in the RNA world: The prebiotic synthesis of urazole and its ribosides. *Journal of Molecular Evolution* 38: 549–557.

Kozlov, I. A., L. E. Orgel, and P. E. Nielsen. 2000. Remote enantioselection transmitted by an achiral peptide nucleic acid backbone. *Angewandte Chemie International Edition* 39: 4292–4295.

Kumar, R. K., and M. Yarus. 2001. RNA-catalyzed amino acid activation. *Biochemistry* 40: 6998–7004.

Larralde, R., M. P. Robertson, and S. L. Miller. 1995. Rates of decomposition of ribose and other sugars—implications for chemical evolution. *Proceedings of the National Academy of Sciences USA* 87: 8158–8160.

Lazcano, A. 1997. The tempo and mode(s) of prebiotic evolution. In C. B. Cosmovici, S. Bowyer, and D. Werthimer, eds., *Astronomical and Biochemical Origins and the Search for Life in the Universe,* 70–80. Bologna, Italy: Editrice Compositori.

Lazcano, A., G. E. Fox, and J. Oró. 1992. Life before DNA: The origin and early evolution of early Archean cells. In R. P. Mortlock, ed., *The Evolution of Metabolic Function,* 237–295. Boca Raton, FL: CRC Press.

Lazcano, A., and S. L. Miller. 1994. How long did it take for life to begin and evolve to cyanobacteria? *Journal of Molecular Evolution* 39: 546–554.

————. 1996. The origin and early evolution of life: Prebiotic chemistry, the pre-RNA world, and time. *Cell* 85: 793–798.

Levy, M., S. L. Miller, K. Brinton, and J. L. Bada. 2000. Prebiotic synthesis of adenine and amino acids under Europa-like conditions. *Icarus* 145: 609–613.

Liu, R., and L. E. Orgel. 1997. Efficient oligomerization of negatively-charged β-amino acids at –20°C. *Journal of the American Chemical Society* 119: 4791–4792.

————. 1998. Polymerization of β-amino acids in aqueous solution. *Origins of Life and Evolution of the Biosphere* 28: 47–60.

Lollar, B. S., and T. M. McCollom. 2006. Biosignatures and abiotic constrainst on early life. *Nature* 444: E18.

Maden, B. E. H. 1995. No soup for starters? Autotrophy and origins of metabolism. *Trends in Biochemical Sciences* 20: 337–341.

Mansy, S. S., J. P. Schrum, M. Krishnamurthy, S. Tobé, D. A. Treco, and J. W. Szostak. 2008. Template-directed synthesis of a genetic polymer in a model protocell. *Nature* 454: 122–125.

Miller, S. L. 1953. A production of amino acids under possible primitive earth conditions. *Science* 117: 528–529.

————. 1955. Production of some organic compounds under possible primitive earth conditions. *Journal of the American Chemical Society* 77: 2351–2361.

————. 1957. The mechanism of synthesis of amino acids by electric discharges. *Biochimica et Biophysica Acta* 23: 480–489.

————. 1997. Peptide nucleic acids and prebiotic chemistry. *Nature Structural Biology* 4: 167–169.

Miller, S. L., and J. L. Bada. 1988. Submarine hot springs and the origin of life. *Nature* 334: 609–611.

Miller, S. L., and A. Lazcano. 1995. The origin of life—did it occur at high temperatures? *Journal of Molecular Evolution* 41: 689–692.

Miller, S. L., and H. C. Urey. 1959. Organic compound synthesis on the primitive earth. *Science* 130: 245–251.

Nelson, K. E., M. Levy, and S. L. Miller. 2000. Peptide nucleic acids rather than RNA may have been the first genetic molecule. *Proceedings of the National Academy of Sciences USA* 97: 3868–3871.

Nissen, P., J. Hansen, N. Ban, P. B. Moore, and T. Steitz. 2000. The structural bases of ribosome activity in peptide bond synthesis. *Science* 289: 920–930.

Ochman, H., J. G. Lawrence, and E. A. Groisman. 2000. Lateral gene transfer and the nature of bacterial innovation. *Nature* 405: 299–304.

Orgel, L. E. 1998. Polymerization on the rocks: Theoretical introduction. *Origins of Life and Evolution of the Biosphere* 28: 227–234.

Oró, J. 1960. Synthesis of adenine from ammonium cyanide. *Biochemical and Biophysical Research Communications* 2: 407–412.

————. 1961. Comets and the formation of biochemical compounds on the primitive earth. *Nature* 190: 442–443.

Oró, J., and A. P. Kimball. 1961. Synthesis of purines under possible primitive earth conditions. I. Adenine from hydrogen cyanide. *Archives of Biochemistry and Biophysics* 94: 217–227.

Ricardo, A., M. A. Carrigan, A. N. Olcott, and S. A. Benner. 2004. Borate minerals stabilize ribose. *Science* 303: 196.

Robert, F. 2001. The origin of water on earth. *Science* 293: 1056–1058.

Rode, B. M. 1999. Peptides and the origin of life. *Peptides* 20: 773–786.

Ross, D. S. 2008. A quantitative evaluation of the iron-sulfur world and its relevance to life's origin. *Astrobiology* 8: 267–272.

Schlesinger, G., and S. L. Miller. 1983. Prebiotic synthesis in atmospheres containing CH_4, CO, and CO_2. I. Amino acids. *Journal of Molecular Evolution* 19: 376–382.

Schoenberg, R., B. S. Kamber, K. D. Collerson, and S. Moorbath. 2002. Tungsten isotope evidence from 3.8-Gyr metamorphosed sediments for early meteorite bombardment of the earth. *Nature* 418: 403–405.

Schöning, K.-U., P. Scholz, S. Guntha, X. Wu, R. Krishnamurthy, and A. Eschenmoser. 2000. Chemical etiology of nucleic acid structure: The α-threofuranosyl-(3'→2') oligonucleotide system. *Science* 290: 1347–1351.

Schopf, J. W. 1993. Microfossils of the early Archaean Apex chert: New evidence for the antiquity of life. *Science* 260: 640–646.

Shen, Y., R. Buick, and D. E. Canfield. 2001. Isotopic evidence for microbial sulphate reduction in the early Archaean era. *Nature* 410: 77–81.

Shock, E. L. 1990. Geochemical constraints on the origin of organic compounds in hydrothermal systems. *Origins of Life and Evolution of the Biosphere* 20: 331–367.

Shock, E. L., and M. D. Schulte. 1998. Organic synthesis during fluid mixing in hydrothermal systems. *Journal of Geophysical Research* 103: 28513–28528.

Sleep, N. H., and K. Zahnle. 2001. Carbon dioxide cycling and implications for climate on ancient earth. *Journal of Geophysical Research* 106: 1373–1399.

Sleep, N. H., K. J. Zahnle, J. F. Kastings, and H. J. Morowitz. 1989. Annihilation of ecosystems by large asteroid impacts on the early earth. *Nature* 342: 139–142.

Sowerby, S. J., C. M. Mörth, and N. G. Holm. 2001. Effect of temperature on the adsorption of adenine. *Astrobiology* 1: 481–487.

Spirin, A. S. 1986. *Ribosome Structure and Protein Biosynthesis*. Menlo Park, CA: Benjamin/Cummings.

Springsteen, G., and G. F. Joyce. 2004. Selective derivatization and sequestration of ribose from a prebiotic mix. *Journal of the American Chemical Society* 126: 9578–9583.

Stoks, P. G., and A. W. Schwartz. 1977. Uracil in carbonaceous meteorites. *Nature* 282: 709–710.

———. 1981. Nitrogen-heterocyclic compounds in meteorites: Significance and mechanisms of formation. *Geochimica et Cosmochimica Acta* 45: 563–569.

———. 1982. Basic nitrogen-heterocyclic compounds in the Murchison meteorite. *Geochimica et Cosmochimica Acta* 46: 309–315.

Strecker, A. 1850. Über die künstliche Bildung der Milchsäure und einem neuen dem Glycocoll homologen Körper. *Annalen der Chemie* 75: 27.

Tehei, M., B. Franzetti, M.-C. Maurel, J. Vergne, C. Hountondji, and G. Zaccai. 2002. The search for traces of life: The protective effect of salt on biological macromolecules. *Extremophiles* 6: 427–430.

Tian, F., O. Toon, A. Pavlov, and H. De Sterck. 2005. A hydrogen-rich early earth atmosphere. *Science* 308: 1014–1017.

Tice, M. M., and D. R. Lowe. 2004. Photosynthetic microbial mats in the 3,416-Myr-old ocean. *Nature* 431: 549–552.

Ueno, Y., K. Yamada, N. Yoshida, S. Maruyama, and Y. Isozaki. 2006. Evidence from fluid inclusions for microbial methanogenesis in the early Archaean era. *Nature* 440: 516–519.

van Zuilen, M. A., A. Lepland, and G. Arrhenius. 2002. Reassessing the evidence for the earliest traces of life. *Nature* 418: 627–630.

Wächtershäuser, G. 1988. Before enzymes and templates: Theory of surface metabolism. *Microbiological Reviews* 52: 452–484.

Wang, X. S. 1998. Stability of genetic informational molecules under geological conditions. PhD thesis, Scripps Institution of Oceanography, University of California at San Diego.

Wetherill, G. W. 1990. Formation of the earth. *Annual Review of Earth Planetary Science* 18: 205–256.

White, R. H. 1984. Hydrolytic stability of biomolecules at high temperatures and its implication for life at 250 °C. *Nature* 310: 430–432.

Wilde, S. A., J. W. Valley, W. H. Peck, and C. M. Graham. 2001. Evidence from detrital zircons for the existence of continental crust and oceans on the earth 4.4 Gyr ago. *Nature* 409: 175–178.

Yuasa, S., D. Flory, B. Basile, and J. Oro. 1984. Abiotic synthesis of purines and other heterocyclic compounds by the action of electrical discharges. *Journal of Molecular Evolution* 21: 76–80.

Zahnle, K., N. Arndt, C. Cockell, A. Halliday, E. Nisbet, F. Selsis, and N. H. Sleep. 2007. Emergence of a habitable planet. *Space Science Reviews* 129: 35–78.

Zhang, B., and T. R. Cech. 1998. Peptide bond formation by in vitro selected ribozymes. *Nature* 390: 96–100.

Zubay, G. 1998. Studies on the lead-catalyzed synthesis of aldopentoses. *Origins of Life and Evolution of the Biosphere* 28: 12–26.

Paleontology and the History of Life

Michael Benton

And out of the ground the Lord God formed every beast of the field, and every fowl of the air; and brought them unto Adam to see what he would call them: and whatsoever Adam called every living creature, that was the name thereof.

Genesis 2:19

People have always been astounded by the diversity of life, although perhaps in different ways. In prescientific times farmers saw how their crops and live-stock were merely part of a much larger richness of life, and people have al-ways striven to understand the complexity and arrangement of living things. From Aristotle to Linnaeus, scientists attempted to catalog life and to under-stand where it had come from. During the eighteenth century it became clear to all savants that the earth had been populated formerly by strange and mar-velous creatures that had since become extinct. By 1820 some rough picture of the succession of floras and faunas through geological time was beginning to emerge. Charles Darwin, during the voyage of HMS *Beagle* in the early 1830s, became increasingly convinced that life was more diverse than he had imagined—every island he visited sported a new crop of plants and animals. He saw the lateral (geographic) and vertical (historic) links between species and realized by 1837 that species were all linked by a great tree. The tree con-cept made it clear why species that in his time were geographically close should also be genealogically close. Further, the tree concept made it clear why the fossil mammals he found in Argentina should be similar to the living mammals of the region.

This essay addresses four concepts: what we know about the sequence of the history of life, how life has diversified through time, how speciation oc-curs, and how good (or bad) the fossil record is as a source of data on the his-tory of life. These were all issues that concerned Darwin (1859), and they concern us still today.

Narrative

There are many ways to recount the story of the history of life. As one of many typical examples, Benton and Harper (1997, 2008) presented a sequence of 10 stages from the origin of life to the origin of modern humans:

1. *The origin of life.* The appearance of the first evidences of life in the fossil record, some 3.6–3.5 billion years ago. These first, prokaryotic cells were probably like modern cyanobacteria ("blue-green algae"), but they lived in the absence of oxygen.
2. *Eukaryotes and the origin of sex.* The first eukaryote cells, with a nucleus, are reported from rocks dated about 1.3–1 billion years old and hence presumably appeared some time before that. The initiation of sexual reproduction, possibly at the same time as the origin of eukaryotes, possibly not, opened up the possibility of mixing of genetic material and recombination.
3. *Multicellularity.* The first multicelled fossils, red algae, are 1.26 billion–950 million years old, and this range of dates corresponds to molecular evidence that points to the origin of multicelled organisms some 1.2 billion years ago. Organisms consisting of many cells can diversify the functions of those cells, and they can become large.
4. *Skeletons.* Many animal groups apparently acquired hard, mineralized skeletons about 545 million years ago, at the beginning of the Phanerozoic, during the so-called Cambrian explosion. Animal fossils are known from many localities in the preceding Ediacaran, but the diversity of phyla represented by fossils expanded hugely when skeletons were acquired. The reasons for this are unknown, but skeletons clearly offered protection and support and permitted certain groups to enter new life zones.
5. *Predation.* Perhaps linked with the spread of skeletons, typically protective outer shells, may have been the rise of new kinds of predators, macroscopic animals such as trilobites that employed new strategies to feed on their large prey. Arms races between predators and prey became a standard feature of animal evolution thereafter.
6. *Biological reefs.* Reefs have existed since Cambrian times. They have been made from a broad range of animals; dominance shifted from polychaete worms in the Early Cambrian to archaeocyathans in the Late Cambrian and algae, bryozoans, stromatoporoids, and rugose and tabulate corals in the Ordovician, Silurian, and Devonian. Different groups of algae, sponges, bryozoans, and corals came and went through time, but reefs remained and are today dominated by scleractinian corals. Reefs are built from organisms, but they form major physical geographic features and provide a plethora of new habitats for life.
7. *Terrestrialization.* The move of life onto land similarly opened up a huge array of new life zones. Soils are known from some late

Precambrian successions, and soils imply life on land. Burrowing animals are known from Ordovician soils, while many small vascular plants and arthropods are known from the Silurian. Land plants became larger and more diverse in the Devonian, and the diversity of land-dwelling animals expanded to include worms, mollusks, many more arthropod groups, and vertebrates.

8. *Trees and forests.* Just as reefs in the sea are geographically substantial biological structures, so too are forests. The first trees in the Devonian were largely isolated, but massive forests developed in the Carboniferous, with trees up to 20 m tall. Not only did various land-plant groups diversify as trees, but other plants and animals that exploited the new habitats created by the trees also diversified.

9. *Flight.* A further expansion of ecospace was marked by the origin of flight. Insects arose in the Early Devonian, but the first true flyers were Carboniferous in age. Insects dominated the skies from then on. The first flying (gliding) vertebrates arose in the Permian and Triassic, while powered flight in vertebrates arose at least three times, in pterosaurs in the Late Triassic, in birds in the Late Jurassic, and in bats in the Tertiary.

10. *Consciousness.* The origin of consciousness in humans is much debated. Is this a feature of *Homo sapiens* alone, and hence present only in the past 200,000 years or so, or did earlier species of *Homo*, or even precursor genera such as *Australopithecus* and *Ardipithecus*, possess consciousness as much as 2 to 4 million years ago? Consciousness allowed humans to create things and to modify their environments and is the basis of the profound impact humans are having on the evolution of life.

These 10 stages are linked closely to major biological innovations (life, sex, multicellularity, skeletons, predation, consciousness) and major expansions in habitats occupied by life (reefs, terrestrialization, trees, flight). All 10 are documented to a greater or lesser extent in the fossil record.

In a more biologically oriented presentation, John Maynard Smith and Eörs Szathmáry (1995) identified eight major steps from the origin of life to human societies with language:

1. *Replicating molecules.* The first objects with the properties of multiplication, variation, and heredity were probably replicating molecules, similar to RNA but perhaps simpler and capable of replication but not informational because they did not specify other structures. A popular view is that RNA came before DNA because it can act both as a gene and an enzyme, views encapsulated in the term *RNA world.* If evolution were to proceed further, it was necessary that different kinds of replicating molecule should cooperate, each producing effects that helped the replication of others. For this to happen, populations of molecules had to be enclosed within some kind of membrane, or *compartment,* corresponding to a simple cell.

2. *Independent replicators.* In existing organisms, replicating molecules, or *genes,* are linked together end to end to form *chromosomes* (a single chromosome per cell in most simple organisms). This has the effect that when one gene is replicated, all are. This coordinated replication prevents competition between genes within a compartment and forces cooperation on them, since if one fails, they all fail.

3. *RNA as gene and enzyme.* In modern organisms there is a division of labor between two classes of molecule: *nucleic acids* (DNA and RNA) that store and transmit information and *proteins* that catalyze chemical reactions and form much of the structure of the body (for example, muscle, tendon, hair). Perhaps originally RNA molecules performed both functions. The transition from an RNA world to a world of DNA and protein required the evolution of the genetic code, whereby base sequence determines protein structure.

4. *Eukaryotes and organelles. Prokaryotes* lack a nucleus and usually have a single circular chromosome. They include the bacteria and cyanobacteria (blue-green algae). *Eukaryotes* have a nucleus that contains rod-shaped chromosomes and usually other intracellular structures called *organelles,* including mitochondria and chloroplasts. The eukaryotes include all other cellular organisms, from the single-celled Amoeba and *Chlamydomonas* up to humans.

5. *Sexual reproduction.* In prokaryotes and in some eukaryotes, new individuals arise as *asexual clones* by the division of a single cell into two. In most eukaryotes, in contrast, this process of multiplication by cell division may be interrupted by a process of *sexual reproduction* in which a new individual arises by the fusion of two sex cells, or gametes, produced by different individuals.

6. *Differentiated cells.* Protists exist either as single cells or as colonies of cells of only one or a very few kinds, whereas *multicelled* organisms among animals, plants, and fungi are composed of many different kinds of cells, such as muscle cells, nerve cells, and epithelial cells. Each individual, therefore, carries not one copy of the genetic information (two in a diploid) but many millions of copies. The problem, of course, is that although all the cells contain the same information, they are very different in shape, composition, and function.

7. *Colonial living.* Most organisms are solitary, interacting with others of their species but not dependent on them. Other animals, notably ants, bees, wasps, and termites, live in *colonies* in which only a few individuals reproduce. Such a colony has been likened to a *superorganism,* analogous to a multicellular organism. The sterile workers are analogous to the body cells of an individual, and the reproducing individuals to the cells of the germ line. The origin of such colonies is important; it has been estimated that one-third of the animal biomass of the Amazon rain forest consists of ants and termites, and much the same is probably true of other habitats.

8. *Primate societies, human societies, and the origin of language.* The decisive step in the transition from ape to human society was probably the origin of language. In many ways human language is like the genetic code; information is stored and transmitted, with modification, down the generations. Communication holds societies together and allows humans to escape evolution.

Maynard Smith and Szathmáry (1995) argue that all but two of these eight transitions were unique, occurring just once in a single lineage. The two exceptions are the origins of multicellular organisms, which happened three times, and of colonial animals with sterile castes, which has happened many times. Had any of the other six transitions not happened, and that includes the origin of life itself (number 1), then we would not be here. That only two of the eight have demonstrably happened more than once speaks against Simon Conway Morris's (2003) thesis that convergence in nature is so prevalent that humanity and many other extant styles of life are virtually inevitable.

Diversification

NUMBERS

Life is astonishingly diverse today, with estimates ranging from 5 million to 100 million species, but perhaps, more soberly, homing in on a figure between 10 million and 15 million species (see the main essay "The Pattern and Process of Speciation" by Margaret B. Ptacek and Shala J. Hankison in this volume). Of these, fewer than 2 million have been recognized and described so far.

It is commonly assumed that life today is more diverse than it has ever been. This seems in some ways obvious, but it could also be construed as extraordinary vanity, somehow akin to the view that all evolution was planned to lead to human beings, and that somehow this instant in the vast span of time is the most important of all. However, it is evident that all living organisms and all organisms known as fossils derive from a single common ancestor (on the basis of the evidence of shared complex characters, such as the DNA-RNA system of inheritance, homeobox genes, and the like). That common ancestor, the single species that gave rise to all of life, existed some 3.5–3.8 billion years ago. But how many species have ever existed?

Biologists and paleontologists have tried a number of lines of reasoning to estimate the total number of species that have ever existed. Perhaps living biodiversity is only 2–4% of the total that has ever existed, as is suggested by three lines of reasoning (Sepkoski 1992; May 1994):

1. The first argument is based on an assumed pattern of species diversity increase and the known average duration of a species before it becomes extinct. If species diversity increased roughly linearly (additive model) through the Phanerozoic, and average species duration is 5–10 million

years, then living species represent 2–4% of those that existed during the past 600 million years.

2. The second argument is based on the diversification of insects over the past 450 million years. If the average duration of an insect species is 10 million years, and the group has diversified linearly, then 5% of all terrestrial species that ever existed are alive today.

3. The third argument is based on preservability of fossils. Some 250,000 species of fossil marine animals have been named, similar to the total number of known marine animal species alive today (200,000). Allowing for nonpreservation of soft-bodied organisms and other losses, this could also represent only 2–4% of the total number of fossil species (Sepkoski 1992).

In these discussions it has only been possible to examine patterns of diversification for macroscopic organisms, that is, typical plants and animals. Microbes are excluded because it has been hard to estimate their current diversity, and very little is known of their early fossil record. If 2–4% of all species that ever existed are alive today (10–15 million), there must have been some 250 to 750 million species in the past.

MODELS

Biodiversity, then, has expanded from one species to many millions of species today. But how? There are many ways to go from one species to many, and these can be expressed in terms of three mathematical models, represented by a straight line, an exponential curve, and a logistic curve, first as an uninterrupted increase (Figure 1A), and second with some mass extinctions superimposed (Figure 1B).

The *linear model* represents additive increase, the addition of a fixed number of new species in each unit of time. (The increase in this example and the others is a net increase, i.e., true increase minus extinctions.) In terms of an evolutionary branching model, additive increase would mean that through time speciation rates have declined, or extinction rates have increased regularly at a rate sufficient to mop up the excess speciations. The implied decline in the rate of evolution in the linear model comes about simply because the total number of species is increasing regularly, yet the *rate* of increase across the board remains fixed; hence, for any individual evolutionary line, the rate or probability of splitting (speciation) must decline. Such a model has generally been rejected as improbable.

The *exponential model* is more consistent with a branching mode of evolution. If speciation and extinction rates remain roughly constant, then there will be regular doubling of diversity within fixed units of time. A steady rate of evolution at the level of individual evolutionary lines scales up to an exponential rate of increase overall since total diversity is ever increasing. This model has been applied to the diversification rates of individual clades and to the diversification of life in general (Benton 1995; Hewzulla et al. 1999).

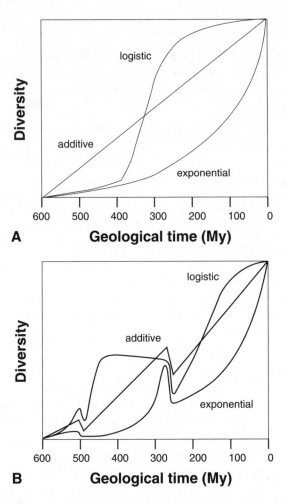

Figure 1. Theoretical models for the diversification of life plotted for the last 600 million years: (A) in the absence of major perturbation and (B) with two mass extinctions superimposed. In each case the upper curve is the logistic or equilibrium model, the middle curve is the additive or linear model, and the lower curve is the exponential model.

The *logistic model* involves one or more classic S-shaped curves, each consisting of an initial period of slow diversity increase, a rapid rise, a slowing of the rate of increase as a result of diversity-dependent damping factors, and then a plateau corresponding to a limiting or equilibrium value. The logistic model has been used to explain patterns of diversification of marine organisms (Sepkoski 1984) and of plants (Niklas et al. 1983).

There is clearly no consensus on which model best explains the diversification of major sectors of life through time, or on whether all patterns of diversification adhere to the same model of increase. The choice of model is important since each makes profoundly different claims about evolution.

LAND AND SEA COMPARED

There are major differences between the patterns of diversification on land and in the sea, and the history of life in each realm may have been rather different. Today about 85% of described species of plants and animals live on land, and the main groups (plants, arthropods, vertebrates) have reached their present great diversity in the past 450 million years. Plants and animals have been evolving in the sea since at least 600 million years ago, and the fossil record is dominated by marine species, which make up some 95% of all described forms. This dominance of marine organisms is partly accounted for by the facts that virtually the only organisms known from the Vendian and early Paleozoic (600–450 million years ago) are marine, and that the early history of life on land appears to have occurred at relatively low diversities. Also, fossils in certain marine environments are more likely to be preserved than those in many continental settings. The observation that life on land today is apparently five to six times as diverse as life in the sea, largely because of the insects, could be an artifact that reflects the greater amount of time devoted by systematists to continental than to marine organisms. However, if this difference is even partly correct, then it would imply a much more rapid diversification on land than in the sea.

In studies of the diversification of marine animal families (Figure 2A), there is evidence for a short plateau in the Cambrian (lasting about 40 million years) and a longer one from the Ordovician to the Permian (about 250 million years). This is followed by a long phase (250 million years) of near-exponential increase in diversity through the Mesozoic and Cenozoic, the rising element of a third logistic curve, which shows a hint of a slowdown in the last 25 million years or so, suggesting that a third plateau level may be achieved 125 million years in the future (Sepkoski 1984).

Marine invertebrate diversification has been explained (Sepkoski 1984) as the succession of three major phases of evolution, in which broad assemblages of different phyla *(evolutionary faunas)* dominated the oceans and were then replaced. The Cambrian fauna diversified exponentially at first, and then diversification slowed as the equilibrium level of 85 families was approached. The exponential diversification of the Paleozoic fauna then began in the early Ordovician, reaching an equilibrium diversity of 350 families and largely supplanting the Cambrian fauna. Finally, after the mass extinction at the end of the Permian, which reduced the global diversity of the Paleozoic fauna dramatically, the Modern fauna continued and accelerated its long-term rise in diversity.

In studies of the diversification of vascular plants (Figure 2B), an equilibrium interpretation has also been given (Niklas et al. 1983). There was a succession of major *Baupläne* (ground plans or archetypes) of plant types: early vascular plants in the Devonian; lycopods, ferns, conifers, and others in the Carboniferous to the Permian; gymnosperms in the Triassic to the Jurassic; and angiosperms from the Cretaceous onward. There is evidence for declining speciation rates and increasing species durations during each of the first

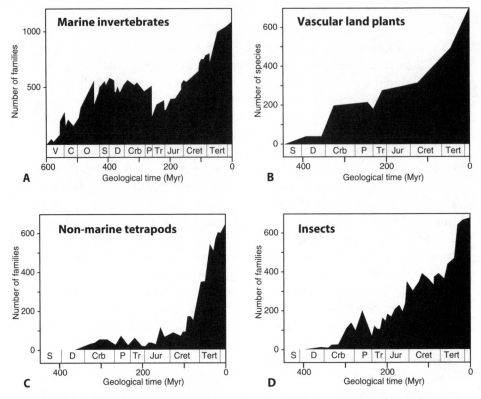

Figure 2. Patterns of diversification of families of marine invertebrates (A), vascular land plants (B), non-marine tetrapods (C), and insects (D). Stratigraphic abbreviations: C, Cambrian; Crb, Carboniferous; Cret, Cretaceous; D, Devonian; Jur, Jurassic; O, Ordovician; P, Permian; S, Silurian; Tert, Tertiary; Tr, Triassic; V, Vendian. (Based on Sepkoski 1984; Niklas et al. 1983; Benton 1985; and Labandeira and Sepkoski 1993.)

three radiations as the new set of clades partially replaced the old. Each new radiation led to an increase in total global diversity, while the diversity of the preceding floras declined. Angiosperms apparently continued to diversify at a high rate. It is hard to identify plateaus in land-plant species diversification, and it is hard to find evidence for logistic models of diversification. Equally, the total curve of species diversities through time is not obviously exponential, and, if anything, the pattern appears to suggest linear increase in diversity through time.

The diversification of continental tetrapod families (Figure 2C) appears to correspond to an exponential model of increase (Benton 1985). Diversity levels remained low, at some 30 to 40 families, during the late Paleozoic and much of the Mesozoic. They then rose to about 100 families at the end of the Cretaceous and, after recovery from the end-Cretaceous extinction event, familial diversity increased rapidly toward 330 families and shows no sign of a slowdown. The pattern of diversification may be dissected into successive radiations of three global clade associations: basal tetrapods (formerly termed *labyrinthodont* amphibians) and synapsids (mammal-like reptiles) in the late Paleozoic; archosaurs (dinosaurs, pterosaurs, crocodilians) in the Mesozoic; and lissamphibians (frogs and salamanders), lepidosaurs (lizards and snakes), birds, and mammals from the late Cretaceous to the present day. These clade associations replace each other and are associated with ever higher global familial diversity levels, but it is difficult to fit logistic curves to any of the associations.

The diversification of insects (Figure 2D) was also apparently exponential, especially in the Mesozoic portion of the curve (Labandeira and Sepkoski 1993). This suggests that insects have had a long and continuous pattern of expansion that perhaps slowed somewhat during the Tertiary. This may indicate that insect diversity is approaching an equilibrium level now.

Plots of the diversification of families of marine, continental, and "all" life (Figure 3) by Benton (1995) confirm these varying models of diversification. The continental curve (Figure 3B), dominated by tetrapods, insects, and land plants, is exponential. The marine curve (Figure 3C) retains a Paleozoic plateau level and appears to show a slowdown in diversification toward the Recent, which may indicate that marine diversity levels today are approaching an equilibrium level. The curve that combines all marine and continental families (Figure 3A) could be interpreted as a single poorly fitting exponential curve (Hewzulla et al. 1999), but the Paleozoic plateau, reflecting the contribution of marine invertebrates (compare Figure 3C), cannot be ignored.

EXPLANATIONS OF PATTERNS OF DIVERSIFICATION

In comparing logistic and exponential models for the diversification of life, the key distinction is between equilibrium and nonequilibrium (or expansion) models. The former imply the existence of global equilibria in diversity, while expansion models assume that there is no ceiling to the diversity of life, or at least that such a ceiling has yet to be reached.

Equilibrium models for the expansion of the diversity of life were based on an influential body of ecological theory. Logistic modeling of global-scale data on diversification assumes (1) interactions among species within clades, (2) interactions between clades, and (3) global equilibrium levels. Many studies show that clades may radiate initially at exponential rates, but that the rate of diversification slows at a certain point as a result of diversity-dependent phenomena, such as competitive exclusion, increased species packing, and reduction of species ranges (Sepkoski 1984, 1996). This style of reasoning follows explicitly from classical experiments in competition where the increase of one population suppresses another that depends on the same limiting resource. An initial exponential increase of the successful population is followed by a plateau when the species begins to deplete the limiting resource (usually food); this corresponds to the local carrying capacity.

The pattern of diversification of marine families (Figure 2A) has been interpreted (Sepkoski 1984, 1996) in terms of a three-phase logistic model that

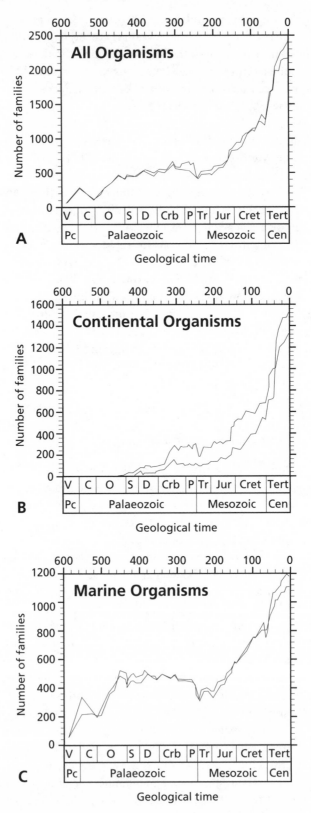

represents the behavior of the three evolutionary faunas: Cambrian, Paleozoic, and Modern. The replacing faunas are said to have been characterized by the ability to penetrate ever wider sets of niches and hence to achieve higher diversities. There is some evidence in favor of this idea; for example, later marine animals could burrow deeper, form more complex reefs, and capture prey in ever more ingenious ways. The equilibrium models could be interpreted simply in terms of large-scale competition between major clades, with bivalves outcompeting brachiopods, mammals outcompeting dinosaurs, and so on. Despite the popular appeal of such suggestions, most supposed cases of large-scale competition crumble when the evidence is examined (e.g., Gould and Calloway 1980; Benton 1987). Sepkoski (1996) attributed the patterns of waxing and waning of clades to diffuse competition between them at the species level, where species in one clade are generally competitively superior to those in another. However, there are four areas of concern with equilibrium models:

1. There is no independent evidence for equilibria, that is, for fixed carrying capacities on the earth today. In evolutionary terms equilibrium diversities imply that all available resources are in use and all ecospace is filled. If a new species originates, it must displace a preexisting one. However, observations of cases where previously isolated floras and faunas come into contact suggest that species are just as likely to insinuate (enter new niches) and not cause extinction of other taxa.

2. Multiple logistic models imply predictable outcomes of interactions between members of the different faunas, that is, that members of one group will generally succeed where those of another will fail. Where major biotic replacements have been investigated, one group is more likely to disappear because of an extinction event (Benton 1987) than as a result of interactions.

Figure 3. Patterns of the diversification of life through time in terms of changes in numbers of families extant per stratigraphic stage, plotted for all organisms (A), continental organisms (B), and marine organisms (C). In each graph a maximum and a minimum are shown, based on a combination of stratigraphic and habitat-preference information. The minimum measure includes only families recorded as definitely present within each stratigraphic stage or as definitely spanning that stage, and only families designated as restricted solely to the marine or continental realm. The maximum measure includes also all doubtful stratigraphic attributions of families and all equivocal and shared habitat designations. The sum of minimum measures for continental and marine organisms is equal to the minimum measure for all taxa together. The sum of maximum measures, however, does not equal the maximum measure for all taxa because families with equivocal environmental assignments and those that occur in both marine and continental settings are counted as both marine and continental. Stratigraphic abbreviations as in Figure 2 with the addition of Cen, Cenozoic; Pc, Precambrian. (Based on Benton 1995.)

3. The diversification of the Modern fauna seems more prolonged and slower than predicted by a logistic model. The rising phase of the logistic curve has lasted for 250 million years, with some evidence of a slowdown toward the present. If there is no current plateau, then it would seem that in the second half of the Phanerozoic, the best-known part of the fossil record, the logistic rules have been forgotten.

4. The classic logistic curves of Sepkoski (1984, 1996) may be artifacts of the level of analysis. The curves are plotted at the level of families. When these are translated to the generic level, the logistic patterns begin to break down, and at species level the pattern could be exponential (Benton 1997). The shape of the curves switches because each genus contains many species, and each family many genera. In an evolutionary tree the species are the final twigs, while genera extend deeper into the tree, and families deeper still. So, although the number of families might be defined at a constant level, the number of species in each could be expanding exponentially.

The alternative to equilibrium is *expansion*. Are the aggregate patterns, or at least some of them (Figures 2 and 3), the result of unconstrained expansion? Certainly, some clades (such as insects, angiosperms, birds, and mammals) seem to continue radiating linearly or exponentially for many tens or hundreds of millions of years. Such ever-expanding patterns imply that these groups are highly successful and adaptable. The overall patterns of diversification (Figure 3) incorporate the numerous constituent clades, some expanding, others diminishing, and others remaining at constant diversity at any particular time. From an expansionist viewpoint, there is no prediction of how the individual clades affect each other. New global diversity levels may be achieved by combinations of new adaptations, habitat changes, and extinction events. In the past 250 million years the diversification of life has been dominated by the spectacular radiations of certain clades both in the sea (decapods, gastropods, teleost fishes) and on land (insects, arachnids, angiosperms, birds, mammals). There is little evidence that these major clades have run out of steam and nothing to indicate that they will not continue to expand into new ecospaces.

Exponential increase could imply that diversification would last forever. Presumably there is a limit to the numbers of families or other taxa that can inhabit the earth at any time: such a limit would be caused not least by the amount of standing room on the ark. If a limit of living space were approached, ever smaller organisms would presumably be favored by selection. Equally, as has happened many times during evolution, organisms would take unexpected measures to survive, for example, by occupying the air, burrowing into sediments, and, in the case of some bacteria, living deep within the earth's crust. With size reduction, the ultimate limit to the diversification of life might then become the availability of the chemical components of life, principally carbon.

EQUILIBRIUM OR EXPANSION?

It is hard to select between the two models for the diversification of life. Until recently, equilibrium models have dominated the thoughts of paleobiologists, just as they have dominated the minds of ecologists. However, just as ecologists are now questioning the oversimplistic equilibrium models they accepted in the 1960s and 1970s, so paleobiologists are reconsidering how diversification might have happened in the longer term.

The equilibrium model assumes that specific major ecological realms can accommodate only certain numbers of species, and that when the carrying capacity is reached, net diversification ceases. The expansion model makes no such assumption and allows for continuing, if episodic, diversification with no ultimate limit in sight.

Paleobiologists have debated and continue to debate which model is correct. Perhaps all life has diversified according to either an equilibrium or an expansion model (Stanley 2007). Or perhaps different sectors of life diversified in different ways. Evolution in the sea may have resulted in a greater level of stability, and patterns of increase may have generally been logistic, while life on land may have diversified exponentially since the first plants and arthropods crept cautiously out of the water. The implications of the equilibrium and expansion models are profoundly different, not merely for paleobiologists, but for everyone concerned about the present and future state of global biodiversity.

Speciation

The effort to understand how life diversified from its origin to the present day is at the upper end of the spectrum of studies in *macroevolution*. At the lower end the important crossover between paleobiology and modern evolutionary studies is *speciation* (see the main essay "The Pattern and Process of Speciation" by Margaret B. Ptacek and Shala J. Hankison in this volume), the formation of new species by the splitting of *lineages* (evolutionary lines).

Paleobiologists clearly cannot use the *biological species concept* since they cannot test whether any two fossil specimens are capable of interbreeding, so they seek to apply the *morphological species concept,* in which species are distinguished by differences in their phenotypes. This may at first seem to be a weak substitute, but there are shared assumptions: paleobiologists use populational thinking, seek to characterize species boundaries on the basis of statistical studies of large samples of specimens, and assume that speciation involves morphological differentiation associated with reproductive isolation. In practice, this is not very different from the approach of many systematists of modern organisms—species of mollusks are generally determined from morphology, and fossil and extant material may form parts of a seamless study. Systematists rarely carry out interbreeding trials.

Until 1970 most evolutionists assumed that fossils could say very little original about speciation. Then Eldredge and Gould (1972) challenged the consensus with their theory of evolution by punctuated equilibria. Eldredge and Gould termed the standard viewpoint *phyletic gradualism.* It assumed that evolving lineages were changing at variable rates (Figure 4A) but were changing more or less continuously *(anagenesis).* Speciation was seen as a by-product of this process of change: sometimes lineages had become so different from their starting points that they had evolved into a new species (the *chronospecies* concept). Speciation by splitting *(cladogenesis)* could also happen, but the process was no slower or faster than normal rates of lineage evolution.

The opposing viewpoint, evolution by *punctuated equilibria* (Figure 4B), proposed that rapid morphological change at cladogenesis, rather than ana-genesis, was the most important process. The normal state of a lineage was *stasis* (no change), and from time to time speciation would happen, but such speciation events were rapid and revolutionary. The punctuated-equilibrium model makes two claims: that rates of change along a lineage are punctuated, and that rapid change is correlated with speciation events. The latter claim was the most controversial, but it has been regarded as the key, defining element of the punctuated-equilibrium model by its supporters.

Eldredge and Gould based their new model on two observations:

1. Stasis is common in the fossil record. Fossils can remain constant in appearance through many meters of sediment (i.e., thousands or millions of years), and then everything seems to change.
2. If most speciation happens according to Mayr's *allopatric* (geographic-splitting) model, then it would appear as a rapid event in the fossil record. The detail of the gradual divergence of two isolated populations would not be seen.

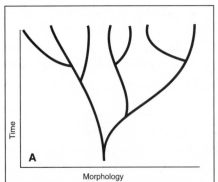 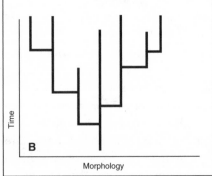

Figure 4. Contrasting expectations of species-level evolution in the classic phyletic-gradualism model (A) and the punctuated-equilibrium model (B). Degree of morphological change is indicated on the *x*-axis, time on the *y*-axis.

The punctuated-equilibrium versus phyletic-gradualism debate of the 1970s and 1980s was huge fun for all concerned, not least because the protagonists kept shifting ground and revising their claims. The first phase was based on the ideas in Eldredge and Gould's essays. From 1975 onward several supporters of punctuated equilibria extended the model to espouse a new idea of *species selection,* a higher-order process in which species were sorted and, in some models, could even undergo selection of "emergent" higher-order characters. Species selection could, in theory, occur independent of, and even in opposition to, natural selection (Stanley 1975). This led to the idea of an "expansion" of evolutionary theory to many hierarchical levels (Gould 1990), with hints of an apparent rejection of aspects of the neo-Darwinian view. Such extreme positions were abandoned fairly rapidly by most enthusiasts, and paleobiologists continued their staunch adherence to natural selection during the late 1980s and 1990s. Species selection is generally considered to be theoretically possible but extremely feeble and unlikely to result in significant adaptations of the type that occur in the natural selection of individuals. However, the notion of stasis stuck: too much fossil evidence pointed to its importance, and quantitative genetic models were found to explain it (Lande 1986).

A plethora of case studies was assembled and published from 1975 to 1990 that purported to test the phyletic-gradualism versus punctuated-equilibrium views. Early efforts were often inadequately documented and did not provide enough evidence about the accuracy of dating and the possibility of migrations of taxa in and out of the study area. It soon became clear that what was a gradual anagenetic pattern to one person was an obviously stepped, punctuational pattern to another. Some studies (e.g., Williamson 1981; Sheldon 1987) involved hundreds of thousands of specimens and an intense focus on fine-scale dating and statistical analysis of huge samples, but they were still subject to criticism. For example, Sheldon's (1987) study of trilobites showed a great deal of anagenesis and stasis, but his sampled lineages just did not speciate. Williamson (1981) believed that he had documented a number of punctuational speciation events, but they were generally interpreted as more likely to be examples of *ecophenotypic* change, that is, nongenetic change in shape resulting from temporary environmental stresses.

The fossil record demonstrates the widespread occurrence of stasis. Erwin and Anstey (1995) summarized the results of 58 studies of speciation patterns in the fossil record published between 1972 and 1995. Organisms ranged from radiolaria and foraminifera to ammonites and mammals, and stratigraphic ages ranged from the Cambrian to the Neogene, with the majority concentrating in the Neogene, the past 25 million years of the history of the earth. Of the 58 studies, 41 (71%) showed stasis, associated either with anagenesis (15 cases; 37%) or with punctuated patterns (26 cases; 63%). It seems clear, then, that stasis is common in species-level evolution, and it had not been predicted from modern genetic studies.

What then of the punctuated-equilibrium versus phyletic-gradualism debate? As ever, the protagonists on either side sought to demonstrate the ubiquity of their model, and yet common sense suggests that there are clear biological reasons that each model prevails under particular circumstances. Benton and Pearson (2001) noted that speciation is a consequence of reproductive isolation, and hence the frequency of speciation in a group is likely to be related to the ease with which reproductive barriers appear. At one extreme there are organisms such as planktonic protists that live in huge populations that seldom encounter barriers to dispersal, and that do not possess complex behaviors associated with reproduction. For these, genetic isolation of populations is a rare event, and speciation, which perhaps occurs rarely, is probably generally long term and gradual, lasting perhaps half a million years. At the other extreme are organisms such as freshwater fishes that live in spatially structured and often-transient environments. Here speciation might be so common that every lake and river has its own reproductively isolated population of a particular type of fish, but these populations might neither be very distinctive nor last very long. In between might lie the majority of invertebrate and vertebrate groups, generally exhibiting stasis but from time to time speciating in a punctuational way as a result of a major perturbation in the environment.

A fine example of speciation in a marine protist is seen in *Rhizosolenia,* a planktonic diatom that occurs today in huge abundance in the equatorial Pacific (Sorhannus et al. 1998). The siliceous valves of this genus have accumulated for millions of years on the seabed, and they have been sampled back to 3.4 million years ago from cores. Huge samples of the valves can be taken every few millimetres through the sediment pile, and these can be dated accurately. Today there are two species of *Rhizosolenia* living side by side in the oceans, *R. bergonii* and *R. praebergonii,* and these can be tracked back for some 2.6 million years. Then, from 2.6 to 3.1 million years ago, the morphologies of the two species converge and fuse, and there is apparently only a single lineage before that time (Figure 5). This splitting event is reported from eight different seabed cores, and the morphological divergence occurs in several morphological characters, so it was evidently not a local event but occurred throughout the equatorial Pacific. Diatoms generally reproduce asexually, but they occasionally produce sexual offspring. Perhaps the combination of a generally asexual reproductive mode and the barrierless Pacific Ocean explains why *Rhizosolenia* speciated phyletically and over a span of some 400,000 to 500,000 years.

It seems probable that sexually reproducing animals that live in varied habitats more often show punctuational speciation. *Metrarabdotos* is an ascophoran cheilostome bryozoan that is represented today in the Caribbean by three species. Coastal rocks of the Dominican Republic and elsewhere in the Caribbean document the past 10 million years of sedimentation in shallow seas, and they yield abundant fossils of this bryozoan. The fossils show that *Metrarabdotos* radiated dramatically from 8 to 4 million years ago, splitting into some 12 species, most of which had died out by the Quaternary (Figure 6). Studies by Cheetham and Jackson (1995) have established a variety of protocols for distinguishing species within *Metrarabdotos,* taking into

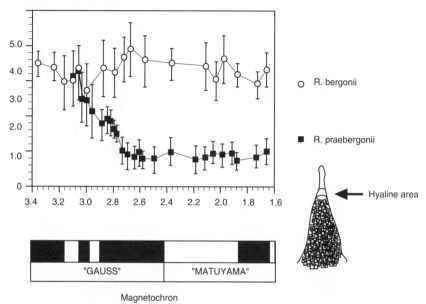

Rhizosolenia (planktonic diatom)

o R. bergonii

■ R. praebergonii

← Hyaline area

"GAUSS"   "MATUYAMA"

Magnetochron

Figure 5. Gradual speciation in the diatom *Rhizosolenia*. One of several morphological characters, the height of the hyaline area, that differentiate the two living species *R. bergonii* and *R. praebergonii*. This plot, from one of eight sampling stations on the floor of the Pacific Ocean, shows a long-term divergence between the two species from 3.1 to 2.6 million years ago, as the populations slowly differentiated. The sedimentary record is continuous, and it can be dated by fossils and by magnetic reversal measurements, indicated in the magnetochron scale at the bottom.

account the genetics and the amount of morphological differentiation of related extant species and then extending comparable statistical tests of morphological differentiation to the fossil forms. Lineage splitting in *Metrarabdotos* seems to have been rapid and punctuational in character. Speciation was especially rapid from 8 to 7 million years ago, with nine new species appearing in that time, although sampling may be a problem in rocks of this age. However, the interval from 8 to 4 million years ago, represented largely by information from Dominica, has been intensely sampled. So, although there are questions over the origins of the nine basal species within this interval, the origins of the remainder (*tenue,* new species 10, and new species 8) are more confidently documented as being punctuational.

The plurality of evolutionary modes (gradualism without stasis, gradualism plus stasis, punctuation plus stasis) might be real, and there might be environmental controls that work in a somewhat unexpected way. Sheldon (1996) proposed that gradualism might characterize taxa that live in stable environments and change in line with slow environmental changes, whereas

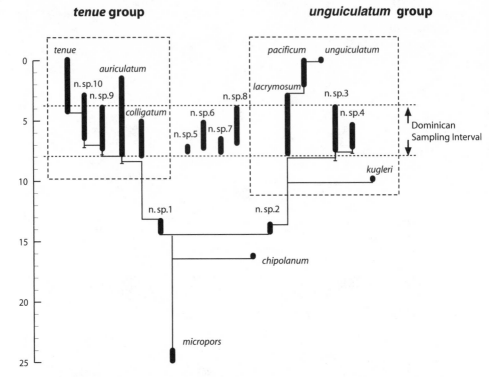

Figure 6. Punctuational speciation in the bryozoan *Metrarabdotos*. The three living species are documented by fossils, but a further nine species are extinct. Intense collecting throughout the Caribbean has revealed the pattern of speciation. Sampling in the time from 25 to 8 million years ago is sporadic, but the time from 8 to 4 million years ago, the Dominican Sampling Interval (DSI), is extremely well known, based on a fine and continuous fossil record on the island of Dominica. Species distinctions are based on a combination of morphological and genetic studies.

stasis might be a feature of taxa that occupy unstable environments, which can vary rapidly and dramatically, but those taxa do not evolve in line with every environmental fluctuation.

Quality

The quality of the fossil record has obsessed and continues to obsess paleobiologists. Ever since Darwin (1859, 279) considered the "imperfection of the geological record," paleobiologists and others have oscillated in their confidence. Some have claimed, perhaps rather wildly, that everything the fossil record says is correct, while others, perhaps equally wildly, have rejected the fossil record as being next to useless. The truth presumably lies somewhere between, and most people tacitly veer to the positive end of the spectrum. A number of major challenges have arisen recently, however, and I shall touch on these briefly.

A qualitative argument first, in favor of the fossil record, is that since 1859 nothing much new has come to light. In the historical context, Charles Lyell was still in 1859 arguing for nonprogression, that he might expect to find Silurian reptiles and Devonian mammals, and that ichthyosaurs might one day return to earth. Darwin argued for progression of life forms through time, of course, and he knew about trilobites in the Silurian, armored fishes in the Devonian, amphibians in the Carboniferous, reptiles in the Permian, dinosaurs in the Triassic, and mammals and birds in the Jurassic. In the 150 years since 1859, and despite the input of millions of hours of searching, paleontologists have simply adjusted the mid-Victorian picture: the origin of agnathan fishes has been pushed back from the Silurian to the Cambrian, the origin of amphibians from the early Carboniferous to the latest Devonian, the origin of reptiles from the early Permian to the mid-Carboniferous, and the origin of mammals from the mid-Jurassic to the late Triassic. The origin of birds has remained unchanged in the latest Jurassic. Were the fossil record hopelessly unrepresentative, new finds should provide major surprises from time to time.

The notion of a good fossil record was confirmed in quantitative analyses by Maxwell and Benton (1990) of the vertebrate record and Sepkoski (1993) of the marine animal record. The first authors looked at the accumulation of knowledge from 1890 to 1987, the second at changes from 1982 to 1992. In both cases the sum totals of diversity through time increased, essentially doubling in 100 years, but the increases were randomly distributed with respect to time, and so the overall patterns of diversifications and extinctions remained unaffected. In a further study Benton and Storrs (1994) tested whether new fossil discoveries tend to fill predicted gaps or create new gaps. They found the former: in 25 years of study, new fossil finds had improved the completeness of the tetrapod fossil record by some 5% by plugging gaps more often than creating new gaps.

There have been two major challenges to this somewhat complacent view: sampling effects and molecular phylogenies. In both cases the critics accept that the fossil record as documented is well understood. However, they highlight the fact that the fossil record itself is a poor sample of life. So, however well known it is, the fossil record can never document the evolution of soft-bodied organisms adequately, and other groups—perhaps microscopic, perhaps living in unusual environments—may similarly not be fossilizable.

The criticisms of sampling, although linked, have had different foci. Smith (2001, 2007) argued that the fossil record is closely tied to the sedimentary rock record, which itself is linked to sea-level changes, and that many supposed extinction events are nothing more than changes of environment. If there is a major regression (retreat of the sea), shallow marine organisms apparently disappear from sections. This could be recorded as an extinction, but it is merely a sedimentary artifact. Smith (2001, 2007) pointed out, however, that the diversification of life in the sea during the last 250 million years is probably real because sea levels were falling during this time: the rise in diversity is not driven by a rise in sea level.

Peters and Foote (2001, 2002) went one step further, arguing that virtually all the fossil record is closely dependent on the sedimentary record. Supposed extinctions, diversifications, and other biological patterns are all apparently driven by the volume of preserved rock. These authors found that the diversity of life apparently matches the number of named rock formations in North America. So, they argued, when life appears to be diversifying, it is simply because there is more sedimentary rock and hence more fossils. When a mass extinction is identified, it is nothing more than a loss of appropriate rocks in which to preserve fossils.

These views have been seen as perhaps rather extreme (Benton 2003; Peters 2005). The correlation of numbers of named formations and diversity of life reported by Peters and Foote (2001, 2002) could be reversed: it is just as likely that when life is diverse and fossils are abundant, geologists recognize and name more formations. In general, of course, as we always teach our students, correlation does not imply causation. Further, assuming that rock volume, or some proxy for rock volume, is pure error in our understanding of the fossil record and should then be applied as a correction factor is heavy handed—major biological events such as diversifications and mass extinctions are removed at a stroke. As Peters (2005) pointed out, there is no reason to reject the proposal that marine rock volume and marine diversity might vary in concert with a third factor, such as sea-level change. Abundant marine rocks and abundant marine fossils might actually reflect high sea levels and abundant marine life: in such a case, to divide the peak in marine diversity by the peak in marine rock volume would effectively remove that (true) biological signal. Much more work is required to investigate further why the marine rock and fossil records are correlated, but the terrestrial rock and fossil record appears not to be (Fara 2002).

The second major current challenge to the informativeness of the fossil record comes from molecular studies and debates about the timing of origins of major groups. Some molecular estimates place the origins of Metazoa (animal phyla), green plants, angiosperms, and modern orders of birds and mammals at points up to twice as old as the oldest representative fossils (e.g., Wray et al. 1996; Cooper and Penny 1997; Kumar and Hedges 1998; Wray 2001; Hedges and Kumar 2004). The range of molecular estimates for the origin of metazoans is 0.6 to 1.2 billion years ago, with most estimates closer to 1 billion years ago than 600 million years ago. The range of molecular estimates for the origin and basal splitting of placental mammals and modern birds is 130 to 70 million years ago, again with more estimates nearer 120 million than 70 million. The first fossils date, respectively, from around 600 and 70 million years ago.

The mismatch of first fossil dates and first molecular dates could indicate major errors in one or the other source of data or both. Many commentators (e.g., Easteal 1999; Wray 2001; Hedges and Kumar 2004) have argued that the fossil dates are almost certainly wrong, and that the molecular dates are closer to the truth. Others (e.g., Benton 1999; Bromham and Hendy 2000; Benton and Ayala 2003) have suggested potential problems with both approaches: fossil dates are always too young, of course (one never finds the first fossil of a group), but not outrageously too young, while the molecular dates tend to be too old, sometimes by a long way. Graur and Martin (2004), speaking from the molecular side, present a robust attack on the evidence for molecular age doubling just noted, while Hedges and Kumar (2004) defend their position. On balance, however, the molecular findings that seemed like a devastating exposure of the frailties of the fossil record in 1996 have now been shown to rest on far weaker ground than was first asserted.

Do these debates just reduce to assertion and counterassertion? Can we ever bridge the gap between knowledge of the fossil record (which is accepted to be good) and the problem of unpreserved life of the past? Perhaps this can be done to some extent. In a brilliant example of lateral thinking, Norell and Novacek (1992) realized that evolutionists uniquely had three independent sources of data on the history of life: fossils, cladistics, and molecular phylogeny. By fossils they meant the stratigraphic order of occurrence of fossils in the rocks. Cladistics and molecular phylogeny reconstruction are relatively new approaches to drawing evolutionary trees, whether on the basis of morphological or molecular characters, and they are independent of stratigraphy. Norell and Novacek (1992) proposed that all three approaches could be compared for congruence (agreement). Up to that point no one knew whether the fossils or the trees were in any way close to the truth: without a time machine, or omniscience, how could one tell? They argued, though, that if the fossil sequences and the sequences of branching points in the trees were congruent, then probably the fossil record did represent the true pattern of the history of life. Lack of congruence could indicate that either the fossil record is wrong or the trees are wrong, or both. Studies can be done group by group, and they show good congruence in as many as 75% of cases (Norell and Novacek 1992; Benton and Storrs 1994; Benton et al. 2000). These studies indicate that all groups examined are about equally well preserved in the fossil record, that there is no substantial difference between marine and continental fossil records, and that there is no time bias (for broad-scale phylogenetic trees plotted against stage-level divisions of time).

Conclusion

Paleontology is an ancient subject, dating back to the sixteenth or seventeenth centuries. To many, it had its heyday in Victorian times, and little has happened since beyond the discoveries of new dinosaurs and hominids reported weekly in *Science* and *Nature*. In a sense, as I have argued, that is true, and the fact that (despite the claims) these new discoveries are rarely shocking is strong evidence that the fossil record tells us something truthful about the history of life.

But the history of life is more than just a narrative of first one fossil, then another, group A giving way to group B, and so on through to mankind. The narrative is important for what it tells us about the astonishing inventiveness of life, its ability to evolve and do unpredictable things. Which Devonian observer would have predicted that the multilimbed insects that crept round the waterside plants would one day take to the air? Which Jurassic observer would have guessed that the insects she saw flitting and dancing in the sunlight around the muddy legs of the dinosaurs would evolve sociality and then drive their biodiversity and global biomass to untold levels? Current biodiversity is higher than it ever has been. Where might life evolve next? There are intimations of microbes that live in ice and in boiling waters, and some that live perhaps three kilometers down within the rocks beneath the sea floor. These examples strain credulity, but they have happened and are happening.

There is more to the study of the history of life, though, than mere documentation of fossils. Fossils represent extinct groups. Dinosaurs cannot be predicted from molecules, but they existed and show us organisms that did what no living animal does. Some dinosaurs were 10 times the mass of the largest living mammals. Some pterosaurs had wingspans three or four times those of the largest birds today. Some extinct arthropods were larger than any alive today. These organisms are all truly wonderful, and they pose interesting problems for biomechanicists and physiologists.

Diversifications and mass extinctions can be predicted (retrodicted?) from studies of modern phylogenies, but the details have always remained mysterious. Paleontologists have the great pleasure of being able to dissect such events in fine detail, and as fossil-collecting techniques, the precision of dating, and environmental analytical tools improve, the level of understanding will increase.

Speciation can be studied in the laboratory and in long-term studies in the field, but there is always a time limit. Now, with improved collecting and dating techniques, paleontological studies of species evolution and species splitting can be tied seamlessly to the present day, and we are beginning to see for the first time macroevolution in action.

New finds capture the headlines, but new insights into biomechanics, large-scale events, and macroevolution are even more impressive. The study of the history of life has never been more exciting than it is now.

BIBLIOGRAPHY

Benton, M. J. 1985. Mass extinction among non-marine tetrapods. *Nature* 316: 811–814.

———. 1987. Progress and competition in macroevolution. *Biological Reviews* 62: 305–338.

———. 1995. Diversification and extinction in the history of life. *Science* 268: 52–58.

———. 1997. Models for the diversification of life. *Trends in Ecology and Evolution* 12: 490–495.

———. 1999. Early origins of modern birds and mammals: Molecules vs. morphology. *BioEssays* 21: 1043–1051.

———. 2003. The quality of the fossil record. In P. C. J. Donoghue and M. P. Smith, eds., *Telling the Evolutionary Time: Molecular Clocks and the Fossil Record*, 66–90. Boca Raton, FL: CRC Press.

Benton, M. J., and F. J. Ayala. 2003. Dating the tree of life. *Science* 300: 1698–1700.
Benton, M. J., and D. A. T. Harper. 1997. *Basic Paleontology*. Harlow, Essex, U.K.: Longman.
————. 2008. *Introduction to Paleobiology and the History of Life*. Oxford: Blackwell.
Benton, M. J., and P. N. Pearson. 2001. Speciation in the fossil record. *Trends in Ecology and Evolution* 16: 405–411.
Benton, M. J., and G. W. Storrs. 1994. Testing the quality of the fossil record: Paleontological knowledge is improving. *Geology* 22: 111–114.
Benton, M. J., M. A. Wills, and R. Hitchin. 2000. Quality of the fossil record through time. *Nature* 403: 534–537.
Bromham, L. D., and M. D. Hendy. 2000. Can fast early rates reconcile molecular dates to the Cambrian explosion? *Proceedings of the Royal Society of London, Series B* 267: 1041–1047.
Cheetham, A. H., and J. B. C. Jackson. 1995. Process from pattern: Tests for selection versus random change in punctuated bryozoan speciation. In D. H. Erwin and R. L. Anstey, eds., *New Approaches to Speciation in the Fossil Record*, 184–207. New York: Columbia University Press.
Conway Morris, S. 2003. *Life's Solution: Inevitable Humans in a Lonely Universe*. Cambridge: Cambridge University Press.
Cooper, A., and D. Penny. 1997. Mass survival of birds across the Cretaceous-Tertiary boundary: Molecular evidence. *Science* 275: 1109–1113.
Darwin, C. 1859. *On the Origin of Species*. London: John Murray.
Easteal, S. 1999. Molecular evidence for the early divergence of placental mammals. *BioEssays* 21: 1052–1058.
Eldredge, N., and S. J. Gould. 1972. Punctuated equilibria: An alternative to phyletic gradualism. In T. J. M. Schopf, ed., *Models in Paleobiology*, 82–115. San Francisco: Freeman, Cooper.
Erwin, D. H., and R. L. Anstey. 1995. Speciation in the fossil record. In D. H. Erwin and R. L. Anstey, eds., *New Approaches to Speciation in the Fossil Record*, 11–28. New York: Columbia University Press.
Fara, E. 2002. Sea level variations and the quality of the continental fossil record. *Journal of the Geological Society, London* 159: 489–491.
Gould, S. J. 1990. Is a new and general theory of evolution emerging? *Paleobiology* 6: 119–130.
Gould, S. J., and C. B. Calloway. 1980. Clams and brachiopods—ships that pass in the night. *Paleobiology* 6: 383–396.
Graur, D., and W. Martin. 2004. Reading the entrails of chickens: Molecular timescales of evolution and the illusion of precision. *Trends in Genetics* 20: 80–86.
Hedges, S. B., and S. Kumar. 2004. Precision of molecular time estimates. *Trends in Genetics* 20: 242–247.
Hewzulla, D., M. C. Boulter, M. J. Benton, and J. M. Halley. 1999. Evolutionary patterns from mass originations and mass extinctions. *Philosophical Transactions of the Royal Society, Series B* 354: 463–469.
Kumar, S., and S. B. Hedges. 1998. A molecular timescale for vertebrate evolution. *Nature* 392: 917–920.
Labandeira, C. C., and J. J. Sepkoski Jr. 1993. Insect diversity in the fossil record. *Science* 261: 310–315
Lande, R. 1986. The dynamics of peak shifts and the pattern of morphological evolution. *Paleobiology* 12: 343–354.
Maxwell, W. D., and M. J. Benton. 1990. Historical tests of the absolute completeness of the fossil record of tetrapods. *Paleobiology* 16: 322–335.
May, R. M. 1994. Conceptual aspects of the quantification of the extent of biological diversity. *Philosophical Transactions of the Royal Society, Series B* 345: 13–20.

Maynard Smith, J., and E. Szathmáry. 1995. *The Major Transitions in Evolution.* Oxford: W. H. Freeman Spektrum.

Niklas, K. J., B. H. Tiffney, and A. H. Knoll. 1983. Patterns in vascular land plant diversification. *Nature* 303: 614–616.

Norell, M. A., and M. J. Novacek. 1992. The fossil record and evolution: Comparing cladistic and paleontologic evidence for vertebrate history. *Science* 255: 1690–1693.

Peters, S. E. 2005. Geologic constraints on the macroevolutionary history of marine animals. *Proceedings of the National Academy of Sciences USA* 102: 12326–12331.

Peters, S. E., and M. Foote. 2001. Biodiversity in the Phanerozoic: A reinterpretation. *Paleobiology* 27: 583–601.

———. 2002. Determinants of extinction in the fossil record. *Nature* 416: 420–424.

Sepkoski, J. J., Jr. 1984. A kinetic model of Phanerozoic taxonomic diversity. III. Post-Paleozoic families and mass extinctions. *Paleobiology* 10: 246–267.

———. 1992. Phylogenetic and ecologic patterns in the Phanerozoic history of marine biodiversity. In N. Eldredge, ed., *Systematics, Ecology, and the Biodiversity Crises,* 77–100. New York: Columbia University Press.

———. 1993. Ten years in the library: How changes in taxonomic data bases affect perception of macroevolutionary pattern. *Paleobiology* 19: 43–51.

———. 1996. Competition in macroevolution: The double wedge revisited. In D. Jablonski, D. H. Erwin, and J. H. Lipps, eds., *Evolutionary Paleobiology,* 211–255. Chicago: University of Chicago Press.

Sheldon, P. R. 1987. Parallel gradualistic evolution of Ordovician trilobites. *Nature* 330: 561–563.

———. 1996. Plus ça change—A model for stasis and evolution in different environments. *Paleogeography, Paleoclimatology, Paleoecology* 127: 209–227.

Smith, A. B. 2001. Large-scale heterogeneity of the fossil record: Implications for Phanerozoic biodiversity studies. *Philosophical Transactions of the Royal Society, Series B* 356: 1–17.

———. 2007. Marine diversity through the Phanerozoic: Problems and prospects. *Journal of the Geological Society* 164: 731–745.

Sorhannus, U., E. J. Fenster, L. H. Burckle, and A. Hoffman. 1998. Cladogenetic and anagenetic changes in the morphology of *Rhizosolenia praebergonii* Mukhina. *Historical Biology* 1: 185–205

Stanley, S. M. 1975. A theory of evolution above the species level. *Proceedings of the National Academy of Sciences USA* 72: 646–650.

———. 2007. An analysis of the history of marine animal diversity. *Paleobiology Memoirs* 4. *Paleobiology* 33 (Suppl.): 1–55.

Williamson, P. G. 1981. Paleontological documentation of speciation in Cenozoic molluscs from the Turkana Basin. *Nature* 293: 437–443.

Wray, G. A. 2001. Dating branches on the Tree of Life using DNA. *Genome Biology* 3: 1–7.

Wray, G. A., G. S. Levinton, and L. H. Shapiro. 1996. Molecular evidence for deep Precambrian divergences among metazoan phyla. *Science* 274: 568–573.

Adaptation

Joseph Travis and David N. Reznick

What Is Adaptation?

The features of organisms are wonderfully suited to their environments and lifestyles. In some cases the match of feature and environment is simple: a brown, mottled moth is cryptic to its predators when it is resting on a diffusely shaded brown tree trunk. In others the match involves several traits that combine to perform an important task: rat snakes that specialize in eating bird eggs have unique jaws that open extraordinarily wide (even for snakes), mouths with smooth grooves instead of teeth, and specialized vertebrae behind the neck that help puncture eggshells and crush and fold the empty shells for expulsion. These matches, which we call *adaptations*, are fundamental observations that biologists want to explain, and understanding how these matches are made is a fundamental puzzle that evolutionary biologists want to solve.

The earliest scientific explanation—that is, an explanation that could be tested with observation and experiment—posited that matches like these emerged through a process of elimination. In this view animals and plants were created with a variety of features; those species that had features appropriate for their particular circumstances persisted, while those that did not went extinct, leaving us with only those species well matched to their circumstances. This theory was derived in part from observations of the persistence or extinction of varieties of domestic plants and animals introduced to new locations (e.g., Townsend's 1786 description of goats introduced to Juan Fernández Island: Eiseley 1958). Some varieties thrived, while others languished, and it seemed reasonable to conclude that nature selected some varieties over others through the combinations of features that distinguished them, and that nature could select among species in the same way.

The theory of adaptation via selective filtering of species was predicated on the conviction that the features of a species would not change unless humans intervened through concentrated selective breeding. There was no doubt that the features of plants and animals were malleable; crop plants, farm animals, and pets bore ample witness that selective breeding could change almost any feature of a plant or animal. But there were obvious limits to malleability—chickens cannot be bred to lay two eggs per day every day of their lives—and nature seemed able to distinguish only substantial differences among species or stocks (Provine 1971; Ruse 1979). These substantial differences seemed capable of being produced only by selective breeding of farm stock (if nature was sorting among varieties of domestic goats introduced to an island) or original creation (if nature was sorting among species).

This explanation of adaptation became increasingly cumbersome as more and more fossil organisms were discovered. For one thing, layers of rock that included fossilized species did not include fossilized material from extant species; if nature were merely filtering species, where was the evidence that extant species had ever coexisted with extinct ones? For another, although many fossil organisms bore little resemblance to extant ones (e.g., dinosaurs), others seemed very similar to extant species (e.g., fossil flounders). It was unclear why one species should have gone extinct while another with very similar features should persist.

The scientific revolution in the study of adaptation is the idea of natural selection, as suggested by Charles Darwin and Alfred Russel Wallace. Although both men offered the idea, it was Darwin who articulated it more completely and integrated it more thoroughly into a larger theory of biological evolution. As a result, we use the term *Darwinian evolution* to refer to the process through which adaptations emerge, a process also called *evolution by natural selection* or *adaptive evolution*.

In Darwinian evolution, natural selection, coupled with inheritance of small trait differences, is the agent of adaptation. Natural selection is a statistical filtering process; individuals with particular characteristics are more likely to survive or produce more offspring than individuals with other characteristics. A more formal definition of natural selection is the causal association of a feature or trait with fitness. By *fitness* we mean the number of offspring that an individual leaves behind; individuals that die without reproducing have zero fitness. *The causal association of a feature or trait with fitness* means, operationally, that a feature or value of a trait is directly responsible for whether an individual lives or dies or why some individuals leave more offspring than others.

For example, many frogs breed in temporary ponds that hold water for a brief period and dry every season. The tadpoles in these ponds must complete their development into juvenile frogs before the pond dries and they are killed. Tadpoles with very slow rates of development die because they cannot complete metamorphosis before the pond dries. Natural selection filters individuals with different rates of development so that only those with more rapid development emerge from the pond.

This filtering would have no lasting effect if individuals who developed rapidly were not genetically distinct from those that developed slowly. The effectiveness of natural selection emerges from the heritable nature of trait variation; in the next generation the genetic variants responsible for the favorable trait values will increase in relative abundance. In the frog example, the genes responsible for more rapid development are more common than they were in the previous generation because the genes responsible for slower development died with the tadpoles that carried them. The increase in frequency of the favored genes causes the average values of the traits under selection to increase; more copies of genes for rapid development mean that on average the next generation of tadpoles will develop faster. When natural selection occurs consistently over time, the average values of the traits that are being selected change substantially. The longer and more consistently selection acts on trait variation and the more genetic variation in the trait, the greater the cumulative change in trait values.

An adaptation is any feature of an organism that has evolved through genetic response to a specific ecological agent of natural selection. In our example, we call rapid development rate an adaptation to the risk of desiccation in temporary ponds. Adaptation is central to Darwin's argument for evolution because it explains the matches between feature and environment without invoking purposeful design or special creation. It also provides the vital argument for how new species arise; when adaptive evolution takes different populations down divergent paths, the differences between those populations become so extensive that they no longer can interbreed, and each becomes a separate species.

Darwin's critical argument was that nature's filter, his natural selection, could indeed distinguish small differences among individuals. During his travels he had seen, time and again, differences in features among populations of the same plant or animal species in different locations. In some cases these differences were large and dramatic, as he saw in the Galápagos Archipelago; in others they were smaller and more subtle, as in differences in the same bird species in different locations in South America. Others had made these kinds of observations, but Darwin interpreted them differently. He saw all these differences as points along a continuum; rather than seeing nature as filtering existing large differences, he saw nature as filtering small individual differences that, coupled with inheritance and time, produced substantial cumulative changes in features and substantial divergence among populations. The testing of Darwin's argument, that is, the resolution of whether natural selection could be as powerful as artificial selection, whether trait differences that were under selection would prove heritable, and ultimately whether populations would diverge through differences in their selective regimes, has occupied biologists for nearly a century and a half.

Evidence for Adaptation through Selection

Biologists have continued to describe the match of organism to environment and have extended our knowledge of that match to the molecular level. They have also uncovered striking patterns in those matches that make scientific explanations other than Darwinian evolution untenable.

And through observational and experimental studies they have provided the answers to questions that Darwin and his peers could only ponder.

VARIETIES OF ADAPTATION

Nature presents a wide variety of adaptations. Carnivorous plants occur in nutrient-poor soil and use their ability to capture and digest insects to obtain the nitrogen that other plants absorb through their roots. Certain fish and crustaceans that occur in temporary ponds and floodplains have eggs that can tolerate complete drying and can rest in the soil until floodwaters break the animals' estivation and induce hatching. Several species of fish live in polar waters that would freeze other animals solid, relying on a specialized antifreeze protein that inhibits ice-crystal formation within their bodies (Hochachka and Somero 2002).

Adaptations also come in less dramatic varieties. Damselfly species in the genus *Enallagma* that encounter dragonflies as their primary predators have different patterns of movement and behavior than species that occur with fish as the primary predator. The behaviors that protect against dragonfly predation facilitate fish predation and vice versa, so the behaviors are well matched to the prevailing predator (McPeek 1995). The fungal pathogen *Colletotrichum lindemuthianum* grows on two closely related species of bean plants often found in the same locations, *Phaseolus vulgaris* and *P. coccinus,* and produces the same lesions and scarring on each species. But the fungal strains on the two host plants are specialized at surmounting each host plant's chemical defenses, just as different species of damselflies are specialized at overcoming different predators (Sicard et al. 2007). Populations of the fish *Fundulus heteroclitus* from more northern latitudes have different forms of the enzyme lactate dehydrogenase than populations from more southerly latitudes. The structure and the regulation of expression of each form is tuned to the prevailing thermal regimes to produce similar reaction rates across widely different temperatures, which allows the species to thrive across a wide range of thermal environments (Schulte 2001).

Biochemical adaptations can be particularly striking because they often involve the production of different molecules by the same individual at different stages of its life, such as the expression of hemoglobin in frogs. In most vertebrates the hemoglobin molecule in the blood that carries oxygen to cells is less able to hold oxygen under acidic conditions. This property enhances its function; cells that are starved for oxygen are acidic environments, and hemoglobin will readily give up its oxygen to those cells. But many tadpoles that live in warm, shallow ponds have a hemoglobin molecule with the reverse property: it holds tightly to oxygen under more acidic conditions. These animals live in very acidic waters, and this hemoglobin facilitates their ability to extract oxygen from the water even when there is little of it available. Their cells are less acidic than the water, so oxygen is readily released where it is needed. Upon metamorphosis the adult frogs, which breathe air, express a different hemoglobin gene whose protein behaves like most vertebrate hemoglobin (Feder and Burggren 1992).

Some of the most striking patterns of adaptation involve convergent or parallel adaptation. Convergent adaptation occurs when the same adaptation is found in unrelated organisms; parallel adaptation occurs when related organisms display the same adaptation through independent origins. In both cases adaptations reflect repeatable matches of organism to environment. For example, two unrelated frogs that live under similarly demanding hot desert conditions on two different continents, *Phyllomedusa sauvagii* in South America and *Chiromantis xerampilina* in South Africa each excrete uric acid rather than urea, as do other adult frogs; uric acid excretion requires about one-twentieth of the water required for urea excretion and represents an adaptive solution to the need for water conservation (Schmidt-Nielsen 1990).

A dramatic example of parallel adaptation is the contrast between the life histories of guppies from upstream and downstream locations in Trinidad that is repeated in several drainages from one mountain range to another across the island. Guppies from upstream locations become reproductively mature at later ages and larger sizes than guppies from downstream locations, and upstream females produce fewer, larger babies in each brood. These differences are matched to their different environments; downstream guppies suffer much higher predation rates, have lower population densities, and, in general, must complete their life cycle quickly or not at all. Guppies from different drainages are much less closely related than upstream and downstream guppies in the same river, which indicates that the distinctions between the upstream and downstream populations have evolved time and again (Reznick and Travis 1996).

DEMONSTRATIONS OF NATURAL SELECTION

Many biology textbooks still employ H. B. D. Kettlewell's classic study of industrial melanism to illustrate the power of natural selection (e.g., Futuyma 2005). As tree trunks in Britain turned dark from air pollution, the coloration of the peppered moth, *Biston betularia,* changed from mottled to black over a large area. The textbooks usually describe Kettlewell's observations of predation on tethered moths of different colors on different backgrounds; predators found it easy to detect mottled moths on dark tree trunks and black moths on lichen-colored trunks in unpolluted forests. Although the value of the observations on tethered moths has been questioned by critics, those observations were not the critical evidence. The convincing data emerged from Kettlewell's meticulous mark-and-recapture studies; he marked large numbers of moths of different colors, released them, and counted the number of recaptures of each type of moth in different locations. This work demonstrated that the different color types had different mortality rates, and that the match between moth color and background was being molded by selection (see the alphabetical entry "Industrial melanism" in this volume).

There are thousands of demonstrations of natural selection in action. A wide variety of ecological forces create differences in mortality rates or levels of reproductive success among individuals with different trait values or different combinations of features. From predators and pathogens to cold stress and drought, from studies of snails in British forests to Darwin's finches in the Galápagos, there is no room to question the reality of natural selection (Hereford et al. 2004).

That natural selection occurs is clear, but can it be a driving force of evolution? There are two questions about natural selection that are crucial to establishing the validity of Darwinian evolution. First, can selection among individuals be sufficiently strong and consistent to mold the features of organisms? Second, is there enough genetic variation for traits under selection to produce substantial changes in those features? Two lines of evidence answer both questions in the affirmative.

The first line of evidence capitalizes on environmental alterations made by humans and the responses of organisms to those alterations. The development of antibiotics in the 1940s and 1950s allowed humans to create novel environments for bacterial pathogens. To a bacterium, an antibiotic is a toxin; to survive, it must sequester the toxin before it poisons the cell or denature it and render it harmless. This situation creates a powerful selective force for alternative biochemical pathways within the cell; individuals that could delay or stymie the toxin's effect would divide and reproduce more often than those that could not. Slowly at first, but at an accelerating pace, species after species of bacterial pathogen evolved features that enabled them to take one or another path toward surviving in the presence of antibiotics. Although the cells of *Streptococcus* look the same as they did in the 1930s, their biochemistry and cellular metabolism are very different (Dzidic et al. 2008).

Substantial biochemical evolution in response to modern chemistry is not confined to bacteria. Over 1,000 species of insects have evolved biochemical responses to insecticide (Roush and McKenzie 1987) and weedy plant species have rapidly evolved resistance to herbicides, even glyphosate, the "once in a century" herbicide (Powles 2008). Many aquatic organisms have evolved adaptations to anthropogenically produced heavy metals in water (Klerks and Weis 1987) and many species of plants have undergone extensive biochemical and morphological evolution in response to the presence of heavy metals in the soil produced by the effects of mining, leaded gasoline, and even the zinc in galvanized fencing (MacNair 1987).

In the second line of evidence, biologists have studied hundreds of cases in which individuals of the same species from different locations exhibit readily observable differences in coloration, morphology, life history, or features of biochemistry and physiology but remain capable of interbreeding. They have shown that these differences are heritable and, through a variety of experimental approaches (often directly exchanging individuals between locations), have demonstrated that divergent selection in the different locations is responsible for maintaining those differences (Reznick and Travis 1996; Travis and Reznick 1998; Reznick and Ghalambor 2001).

In the most thorough studies biologists have re-created the presumed evolutionary trajectory of the differences. The Trinidadian guppies offer one of the most thoroughly studied examples. Recall that upstream guppies, regardless of stream or mountain range, live longer, mature later and larger, and have fewer, larger offspring at each bout of reproduction than their close relatives downstream. The magnitude of these differences is not trivial; females in upstream locations can be more than 20% larger at maturity, have fewer than half as many offspring as downstream guppies, and produce individual newborns that are 50–75% larger than those produced by downstream guppies. There are a number of upstream locations above substantial waterfalls in which guppies are absent. John Endler and David Reznick exploited this opportunity by introducing downstream guppies into some of these locations, in effect replicating the likely course of guppy colonization of upstream habitat (Reznick et al. 1997). The introduced guppies followed an evolutionary path that in less than six years produced trait values comparable with those seen in natural upstream populations. Natural selection in the upstream environment, acting on the genetic variation that was producing small differences among individual downstream guppies transplanted upstream, molded new populations of upstream guppies in very short order. It was the experiment that Darwin would have liked to do.

How Do We Reconstruct Adaptive Evolution?

Although studies of selection and response demonstrate that Darwinian evolution occurs, the full picture of adaptation emerges when studies of selection are joined with painstaking work in genetics, development, and paleontology. Of course, the picture is neither complete nor perfect; biologists have not studied every striking feature of every organism, and even for some of the most scrutinized cases, there is still much to learn.

The features that define the modern equids, horses and zebras, offer a particularly strong example (MacFadden 1992). Paleontology has revealed the sequence of feature appearance from the elongation of the lower limbs, the reduction of toes, and the emergence of hooves to the increasing height of the molars. Biomechanical studies have pointed to the advantages offered by particular features; elongated limbs facilitate rapid running for extended periods, and higher molars are beneficial for grazing tough, silica-containing vegetation like grass. Placing these changes in the context of a changing climate and the emergence of savannas and plains, we can reconstruct the evolutionary pathway through which small ancestral browsers produced a family of grazing descendants molded by living in open country and being at risk from a variety of large predators.

Such a reconstruction presumes that the transitions we reconstruct, like the elongation of the limb and the reduction in the number of digits in the foot, are developmentally possible. That is, the reconstruction presumes that a developmental mechanism for producing such changes can be described, and that the genetic variations in the elements through which that mechanism could act would have been readily available. Although we cannot test these presumptions in extinct species, we can acquire insights from studies of existing species.

Most terrestrial vertebrates have five digits on the foot at the end of each limb, but the position of those digits on the foot, their relative size, and even their number can vary. Studies of foot and digit development in tractable experimental animals like salamanders and chickens have shown that the number, size, and position of digits are determined by molecular signals exchanged by embryonic cells during early limb development (Shubin 2002). Relatively small differences in the timing, intensity, and patterning of those signals change the subsequent pattern of digit development. Delays in signaling will impede the formation of the third, fourth, and fifth digits, and if they do develop, those digits can be quite small. Delays in signaling, coupled with growth in the foot itself, will alter the position of digits on the foot. Thus although we cannot retrace the precise developmental path that led to a hoof in fossil equids, we can offer a plausible model through which cumulative changes in signaling and gene expression could have transformed a five-toed foot into a single-toed hoof.

Would genetic variation for the key developmental elements have been available? The best analogy comes from studies of relative hindlimb length in frogs. Frog species vary enormously in the length of the hindlimb compared with that of the body; cricket frogs have hindlimbs substantially longer than their body lengths, whereas spadefoot-toad hindlimbs are rarely more than about half the body length. Even closely related species like the barking tree frog and green tree frog of the southeastern United States can have substantially different relative hindlimb lengths. These differences arise from differences in the timing of limb-bud initiation and the rate of limb-bud growth in the tadpoles (Blouin 1991). Genetic variation for the relative length of the hindlimb appears readily available even in single frog populations (Blouin 1992). This work suggests that the raw material for hindlimb evolution in frogs was probably not difficult for selection to find in the past and, by extension, suggests a similar conclusion for other animals.

It is easy to criticize conclusions about adaptation in horses whose presumptions can be tested only through experimental studies on birds or frogs. Such extrapolation will always be necessary when we attempt to reconstruct adaptation on a large scale, such as might be found in the fossil record. It is also easy to criticize the generality of conclusions about adaptation derived from a study on a small scale, such as the case of industrial melanism. To a critic, the rapid change in the frequency of alternative forms of a single gene for coloration is hardly evidence that natural selection can drive profound transformations in organismic features—do changing color patterns in moths tell us enough to understand how evolution molds fish muscles into electric organs or transforms plant leaves into traps for catching and digesting insects? The difficulty of balancing what can be achieved experimentally at smaller scales against the larger scales of the most important evolutionary questions has inspired and sustained considerable controversy about Darwinian evolution, even among scientists who study it.

Controversies in the Study of Adaptation

IS THE RAW MATERIAL FOR ADAPTATION READILY AVAILABLE?

At first glance the answer to the question whether the raw material for adaptation is readily available would seem obvious. The study of the transplanted guppies offers compelling evidence that the genetic variation required for substantial adaptive evolution was available. This does not appear to be an exceptional case; hundreds of studies of a variety of characters, ranging from cellular concentrations of alcohol dehydrogenase in flies and the relative hindlimb length of frogs to the color of swallowtail butterfly pupae, have found appreciable levels of genetic variation for individual traits among individuals within a single population.

But the existence of heritable variation in single characters does not imply that natural selection can mold features arbitrarily. Different characters can share developmental pathways that cause particular combinations of trait values to be inherited together and thereby constrain the immediate response to selection. In several species of poeciliid fish—the family of guppies, mollies, swordtails, and mosquito fish—the process of maturation is governed by an endocrine pathway that reduces growth rate as it accelerates sexual development (Kallman 1989). The genetic variation in this pathway creates an association between rapid maturation rate and small body size at maturity. This association means that the response to a selection pressure that favors larger body sizes at maturity will inevitably include slower maturation rates. If slower maturation rates are detrimental for any reason, there is no obvious solution; the genetic variation for the combination of large size and rapid maturation is not immediately available.

The key point revolves around the word *immediately;* does such a constraint hold in the long run? Were we to watch sufficiently closely and wait sufficiently long, would we see mutations in individual genes that dissociate these traits? If selection acted with sufficient consistency, would these new mutations facilitate the eventual emergence of the combination of large size and rapid maturation? We have no simple answers to these questions; although it may well be that given enough time, a mutation of the right kind may arise, it is impossible to predict if the ecological circumstances that would favor such a mutation would still prevail when it finally arose.

The possibility of constrained genetic variation provokes a reconsideration of adaptations that are built on associations between individual traits. For example, most cases of cryptic coloration in animals work only if the animal behaves in a particular fashion, whether it is to remain motionless or to assume a particular posture or orientation. Mismatches between behavior and coloration are failures. The ardent advocate of Darwinian evolution might argue that natural selection has assembled the components of such adaptations via meticulous molding of individual features until the best combination emerged from all the possibilities. At the other extreme one might argue that the concerted inheritance of the components offered natural selection a limited range of raw material from which only a few good matches to circumstances were possible.

The same argument can be provoked by considering the evolution of any putative adaptation that involves multiple characters, be it a biochemical pathway or a suite of associated behaviors. The controversy is not whether such features represent adaptations; the controversy is over the extent to which adaptation is constrained by the nature of the available raw material. This is being approached from several perspectives, from examining the nature of new mutations for key characters to dissecting the associations among individual features of an adaptation into their genetic control and shared developmental pathways.

IS ADAPTATION PERFECT?

The marvelous matches of organism to environment offer a strong temptation to conclude that adaptation is perfect. Indeed, when contemplating intricate adaptations like age-specific hemoglobin gene expression, the amazed scientist is not very different from the awestruck poet or the inspired theologian. But the scientist must take this question as a testable idea. There are two operational ways in which we might use the term *perfect;* it could refer to whether every feature of an organism has adaptive significance and/or whether the match of organism to environment is as good as possible.

Not every feature of an organism is adaptively significant. For one thing, concerted inheritance of characters, as described previously, means that some features with no adaptive significance may persist because they are inherited through shared developmental pathways with other features that represent an adaptation. Concerted inheritance can also protect features that are detrimental by themselves if they are inherited along with advantageous characters and the net fitness effect of the combination is positive. This is the essence of one theory for the evolution of senescence; genes with detrimental effects at later ages, after a substantial reproductive period, remain in the population because they confer a large fitness advantage early in life, before or early in an organism's reproductive period.

Is the match of feature to environment as good as possible? Obviously the answer to this question depends on how we define *possible*. And therein lies the answer, at least in part. Natural selection does not design an organism or its features; it merely filters existing variation. The end product of Darwinian evolution is always as good as possible, but here *possible* is defined as the best of the available options, which are determined in turn by the genetic variation that is available and what the constraints on that variation might be. Put another way, adaptation is a contingent process; it constructs the best possible solution contingent on the raw material provided by mutation. The best illustrations of this contingency occur when different species or even different populations of the same species respond idiosyncratically to the same ecological challenge. This has occurred in evolution of resistance to pesticides; different population of the two-spotted spider mite evolved different biochemical and physiological mechansims to cope with the same pesticide (Matsumura and Voss 1964) and, on a larger scale, when different species of electric fishes co-opted different muscles for transformation into the electric organs (Bennett 1971).

The other part of the answer is more subtle: adaptation is historical. Natural selection filters variation in features that are themselves products of prior evolution. As a result, there is a signature of history on every adaptation, and history can generate features and combinations of features that are probably not what an intelligent designer would produce if the organism were being designed de novo. For example, the human eye has a substantial blind spot caused by the threading of blood vessels through the retina. This inefficient structure is a historical artifact of the eye's origin as a light-sensing organ and subsequent elaboration of the retina as a screen for image resolution. An easier and better design would wind the blood vessels around and not through the retina.

There is another way in which *perfection* in adaptation has been considered by evolutionary biologists and critics alike. Complex adaptations— features of organisms that are built from many characters that combine to perform a specific function, like the vertebrate eye, with subsets of characters that capture and focus light, that detect brightness and color, and that resolve molecular signals into an accurate image of the external world—have long been considered examples of nature's "perfection." Explaining the evolution of these complex traits has always been a central challenge for the champions of Darwinian evolution, and the answers to that challenge illustrate nearly every theme in the study of adaptation.

CAN DARWINIAN EVOLUTION REALLY EXPLAIN COMPLEX TRAITS?

Darwin himself appreciated the challenge posed by complex traits. The dilemma is that all such features, whether an organ like the eye or a metabolic pathway like the Krebs cycle used by cells to produce energy, are a composite of many individual adaptations, all of which must be present and properly integrated for the adaptation to function. How can natural selection bring so many things together all at once and make them work properly? The natural theologian's answer in the nineteenth century was that it cannot, the same position taken by modern proponents of Intelligent Design. The consistent thesis offered by Darwin's critics, then and now, is that the complexity of the eye or the Krebs cycle is too great to have been built by mutation and selection. Changing the color of moths is one thing, but building an eye is quite another (Ayala 2006).

But is it? Darwin argued that natural selection could produce complex adaptations if they evolved through a series of small steps, each of which had some adaptive advantage over its predecessor. The evolution of the eye could only be revealed by reconstructing the historical sequence of these small changes, and an understanding of why an eye looks as it does requires an understanding of that same signature of history (Weber and Depew 2004). To the modern evolutionary biologist, the failure to appreciate the contingent and historical natures of adaptation is the signal weakness in the arguments for Intelligent Design.

Darwin illustrated his argument for the eye by pointing out the diversity of photosensitive structures in different species of gastropods (snails, mussels, squid, and octopuses). These structures range from clusters of cells in some clams that can discriminate between light and dark to the sophisticated, camera-like eyes of an octopus, with a range of photosensitive organs of increasing complexity in between. Darwin argued by analogy, asserting that the modern diversity of photosensitive organs in gastropods illustrates the steps that could have occurred from ancestor to descendant to build the most complicated of these organs, the octopus eye, with a comparable possibility for building the vertebrate eye. In this argument small changes against the background of what has come before can accumulate into a remarkable transformation of cells and tissues.

Darwin's argument, while revealing his great knowledge of animal diversity and his breathtaking creativity, did not convince the critics and, in fact, does not lend itself to a modern evolutionary analysis. The species involved in the argument are too distantly related to one another for skeptical tastes; the fossil record and the implications of comparative molecular data suggest that it has been tens of millions of years since many of them shared a common ancestor (presuming that one even gives credence to the idea of evolution). The evolutionary pathway to the octopus eye has long since been lost. We would have no better luck tracing the pathway to the vertebrate eye; nature now offers a large group of organisms that all have well-developed eyes except for those living in very dark environments that seem to have secondarily evolved reduced eyes. Even in vertebrates the progression from simple to complex eyes occurred in some unknown set of ancestors that lived over 400 million years ago.

The dilemma for understanding the evolution of complexity is that it occurs on a timescale that is very long compared with the timescale on which mutational variation in a single character arises or natural selection exhausts the available variation in a single character. But even when we resort to broad comparisons or indirect inferences, it is often the case that we can make some very strong inferences in particular cases. We review two of those cases because they illustrate how we can combine data from different life-science disciplines to develop insights into the origin of complex structures, even though this origin is not directly observable.

Example 1: From gill arches to ear ossicles and jaws
An old inference from the fossil record is that vertebrate jaws were derived from the gill arches of jawless fish (represented today by lampreys and

hagfish). A second inference is that the single middle-ear ossicle of amphibians, reptiles, and birds was derived from an element of the jaw suspension of fish, and the additional two middle-ear ossicles of mammals were derived from bones associated with the jaw joint of their reptilian ancestors (Figure 1). These would be remarkable evolutionary events in which mutation and selection drove the wholesale transformation of characters, molding entirely new shapes and creating entirely new functions (Wake 1979; Radinsky 1987; Luo 2007).

But the differences in function make such a progression seem fanciful at best. Gill arches are a composite of bones, muscles, and nerves that function as a bellows. Fish use them to expand and contract the volume of the oral cavity. When a fish opens its mouth, the bellows expands, the volume of the oral cavity increases, and water is drawn in. When a fish closes its mouth, the bellows contracts and reduces the volume of the oral cavity as the fish expels water through the gill slits. In living jawless fishes (and presumably in the ancestors of all jawed vertebrates), this pump is a feeding device, drawing waterborne particles into the mouth that adhere to a sticky substance produced in the oral cavity, which is then conveyed to the gut. The entire apparatus also functions as a respiratory organ. The tissues that line the gill arches are well supplied with blood; the pump passes fresh, oxygenated water over them, and the thin tissues that line the arches allow oxygen to be transported from the water into the blood vessels. By contrast, the structures supposedly derived from the gill arches have more circumscribed functions. The jaws grasp and process food. The first middle-ear ossicle in terrestrial vertebrates enhances the transmission of the energy in sound waves from the eardrum, on the outer surface of the body, to the inner ear, which is in the temporal region of our skulls.

How can such diverse structures, not all of which are present in all vertebrates, be evolutionarily linked to one another? The jaws in the early jawed fish are part of a repeated series of bones that includes the gill arches. The muscles that open and close the jaws function in a similar fashion to those that expand and contract the gill arches. In some living species of fish, the gill arches have the functional equivalent of teeth and function as do our jaws by grasping and grinding food. Given the patterns of muscle and innervation, the transformation from gill arch to jaw suggested by the fossil record is not so dramatic as to seem an unlikely cumulative effect of small, attainable changes.

The transformation from jaw support in reptiles to ear ossicle in mammals is less dramatic than it might appear from comparisons of only the existing species. The fossil record offers striking patterns of change across the organisms that represent a transition between reptiles and mammals (Figure 1). In the reptiles that appear to be ancestral to mammals, the lower jaw was a composite of seven bones and the articulation between the lower jaw and the skull was formed by the articular bone on the lower jaw and the quadrate bone on the skull (Figure 1A). In mammals the lower jaw is a single bone, the dentary, and articulation of the jaw uses the dentary bone itself and the

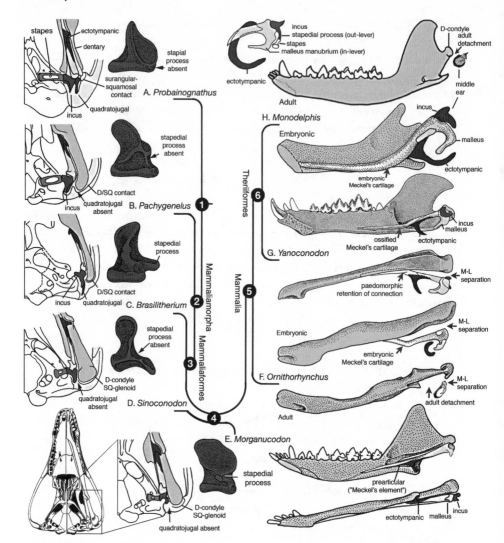

Figure 1. This figure, modified from Luo 2007, illustrates the evolution of the mammalian jaw, derived from a series of fossils in the synapsid lineage (the lineage in which mammals emerged) and from the embryonic development of the living mammals. The drawings on the left (A–D) represent the lower left ventral (bottom) surface of the cranium, as illustrated in the drawing in the lower-left corner at the bottom of the figure. The figures on the right (E–H) represent the medial (inner) and ventral surfaces of the lower jaw. The numbers represent relative ages, with the lower numbers indicating older fossils and the higher numbers indicating recent mammals.

(A) *Probainognathus* was a late Triassic Synapsid that is not in the direct ancestral line to mammals; there was a complex articulation of lower jaw and cranium involving the "malleus" (the modified articular bone) on the lower jaw and the "incus" (the modified quadrate bone), quadratojugal, and squamosal bones of the upper jaw.

(B) *Pachygenelus* represents a different branch of the Synapsids that dates to the late Triassic–early Jurassic period. This group had a dual articulation in which the "malleus" on the lower jaw articulated with the "incus" on the upper jaw and the dentary on the lower jaw articulated with the squamosal on the upper jaw. There was no quadratojugal bone.

squamosal bone on the skull. Some of the advanced nonmammalian Synapsids (the scientific term for the extinct group of mammal-like reptiles) have jaw hinges that consist of a combination of the dentary and the articular on the lower jaw and the quadrate and the squamosal on the upper jaw (Figure 1B–E). This dual articulation of the jaw, in combination with many other features of the skeleton, indicates that these animals formed a bridge between reptiles and mammals. In the case of the jaw joint, it appears that the dentary and squamosal bones first were included in the jaw joint and then completely replaced the articular and quadrate bones as the jaw joint.

Once the articular and quadrate bones were redundant in the articulation of jaw and skull, they were available for other purposes. The evidence from fossils and from the embryonic development of the most ancient living mammals indicates that these bones became reduced in size, altered in position and orientation, and converted into the middle-ear ossicles, the articular

---

(C) *Brasilitherium* represents a third lineage of Synapsids, found from the late Triassic through the late Cretaceous, that had the same dual articulation as *Pachygenelus* (B) but retained a quadratojugal.

(D) *Sinoconodon* was a member of a fourth Synapsid lineage, known only from the early Jurassic, that had a dual articulation like that of *Panygenelus* even though the individual bones had different sizes and shapes than in *Panygenelus*.

(E) *(Left)* *Morganucodon* represents a fifth Synapsid lineage, known from the late Triassic through the mid-Jurassic, that also had a dual articulation, but with a greatly reduced participation of the "incus" and no jugal.

To the right of each of these figures is a detailed diagram of the "incus." Note that the stapedial process is present in (C) and (E) but absent in the others. This process articulates with the stapes, which is the third middle ear ossicle.

(E) *(Right)* This is a ventral and lateral view of the lower jaw of *Morganucodon,* illustrating "Meckel's element," which is associated with a series of bones including the ectotympanic, the malleus, and the incus. These bones have a different embryonic origin from the other jawbones in living animals and they can also be clearly recognized in some fossils. The ectotympanic is a ring of bone that supports the eardrum, which is the interface between the outside world and the middle ear ossicles.

(F) These are adult and embryonic jaws of the duck-billed platypus, genus *Ornithorhynchus.* In the embryo, the Meckel elements are attached to the medial surface of the lower jaw, as in Morganucodon. Later in development the ectotympanic, malleus, and incus separate from the lower jaw and produce the configuration seen in the middle ear of the adult.

(G) The lower drawing is a ventral view, the upper one a medial view of the adult jaw of *Yanoconodon,* an early mammal that dates to the early Cretaceous. The condition of the Meckel elements are similar to those of an embryonic duck-billed platypus.

(H) The lower drawing is the medial view of the embryonic lower jaw of the opossum, *Monodelphus,* while the upper drawing is the adult jaw. The embryonic jaw is very similar to the adult jaw of *Morganucodon* and *Yanoconodon,* with the Meckel elements being attached to the medial surface of the lower jaw, while the adult condition is typical of living mammals, with the Meckel elements being completely detached from the lower jaw.

into what is now called the malleus and the quadrate into what is now called the incus.

The driver for this transformation would have been natural selection's favoring enhanced ability to detect vibrations. Research suggests that the jaw in these transitional animals transmitted vibrations from the ground to the middle ear, as it does in some modern snakes—think of a small, nocturnal animal that picked up vibrations by touching its jaw to the ground and transmitting vibrations from the jaw to the stapes. These vibrations would have traveled through the articular (malleus) and quadrate (incus) bones to the stapes. In two lineages of nonmammalian Synapsids, neither of which is ancestral to mammals, the incus has an outgrowth called a stapedial process, which is a lever arm that transmits the force of the sound waves from the incus to the stapes. Its presence in these two lineages suggests that both of them were evolving an enhanced capacity to transmit sound from the jaw to the stapes and then to the inner ear.

The arrangement of bones in the lineage that includes mammals is different than in these other lineages but does include an incus with a stapedial process. The final stage of course is the separation of the ossicles from the jaw itself. When the ossicles became separated from the jaw, they could transmit sound vibrations efficiently from the eardrum to the middle ear. The complete fossil record reveals that, rather than there being a direct transformation from the primitive synapsid condition to the mammal condition, there was a diversity of arrangements of bones in the jaw-ear region and a diversity of rearrangements over time. The anatomy of living reptiles and mammals represents two endpoints of a transformation that followed diverse paths in between, with most of the variants having gone extinct.

Although the fossils suggest the progression of features in this transformation, developmental biology offers us two important additional lines of evidence. First, when the ossicles first appear in mammalian embryos, they are part of the jaw assembly. During development, they separate and form the familiar structures of the middle ear (Figure 1F–H). While the notion that an organism's development recapitulates its evolutionary history has been discredited in general, this is a case in which development offers a clue to an evolutionary history that is otherwise traceable in the fossils.

Second, gill arches, jaws, and ear ossicles have a critical relationship to one another that sets them apart from other features of the skull. The vertebrate body is a composite of three different layers of tissue, the ectodermal (outer), mesodermal (middle), and endodermal (inner) layers. These three layers are easily distinguished from one another early in development but become blended in many parts of the body as development progresses. Most of the vertebrate skeleton is derived from the middle, or mesodermal, layer of tissue. However, the bones associated with the gill arches, jaw, and middle-ear ossicles have a common developmental origin different from that of most of the skeleton; these bones are derived from specialized cells of the outer, or ectodermal, layer of tissue. Likewise, the muscles and nerves associated with these bones are related to one another in their origin; the muscles associated with the gill arches, jaws, and middle-ear ossicles are all derived from what is called the *branchiomeric* series of muscles, and the nerves that serve these muscles all develop from the same group of cranial nerves.

All these observations support the argument that jaws are derived from gill arches and that the middle-ear ossicles are derived from jaws, and thus that mutation and selection can indeed mold complex features from simple components. But how and why would such a transition occur? François Jacob (1977) once described evolution as a process of tinkering rather than engineering; a tinker uses whatever is at hand and modifies it to serve new functions, while an engineer designs something new from scratch. Jacob meant that adaptation is a historical process because it involves modifying existing structures for new functions. In this case gill arches began with one function (efficient feeding) and facilitated a second (respiration). As vertebrates evolved into larger, more active animals, respiration became an increasingly important function of the pharyngeal pump. The anteriormost gill arches retained an association with feeding. The bones, muscles, and nerves that power the pharyngeal pump have the same action pattern as those that open and close the jaws, so they were readily modified to specialize in one or the other function.

Although the transition from jaws to middle-ear ossicles is harder to understand, the association between the jaw joint and hearing is easy to illustrate. Stick your finger in your ear and then open and close your mouth. You will see that the jaw joint is right next to your outer and middle ear so that you can feel the joint move as you open and close your mouth. Some living vertebrates, including small nocturnal mammals, cetaceans, and snakes, all appear to pick up vibrations from the external environment through the jaw and then to transmit them through the jaw to the middle and inner ears. The snakes and small mammals that rely on this mechanism for picking up vibrations are nocturnal and do so through the tip of the jaw. The recycling of elements of the jaws to become sound-transmission bones in the middle ear is largely a consequence of the proximity of the jaw joint to the inner ear and the prior role that the jaw played in transmitting sound from the external environment to the inner ear.

Example 2: Floral evolution in monkey flowers (Mimulus)

Our second example, while decidedly less sweeping in scope, reveals how genetic analyses contribute to understanding the architecture of complex adaptations (Schemske and Bradshaw 1999; Bradshaw and Schemske 2003). Here we focus on shape and color of flowers. Although everyone appreciates flowers for their beauty and fragrance, from an evolutionary perspective they are reproductive structures that are specialized to produce and disseminate pollen (the male gamete) and receive and deliver pollen to the ovule (the female gamete).

The flowers that inspire our greatest affection are those specialized to attract animal pollinators like bees or hummingbirds. These flowers are actually complex combinations of characters that together are an adaptation for dispersing and collecting pollen. Flowers pollinated by animals are comparatively large, colorful, and shaped in ways that facilitate the transfer of pollen. Some flowers are highly specialized to attract specific pollinators, usually by a color and/or fragrance that are particularly attractive to a pollinator. The sugary nectar is the reward offered by the flower to the pollinator, a reward that draws the pollinator's interest to the inside of the flower and sets the pollinator in place to pick up and deposit pollen. Many flowers have markings that direct the visitor toward the nectar, some of which are visible only to animals like insects that see in the ultraviolet (black-light) frequencies. Others have modified petals or sepals that serve as a sort of landing strip on which the pollinator can rest, and some species have specialized structures to attach the pollen to the pollinator.

Two closely related species of monkey flowers, *Mimulus lewisii* and *M. cardinalis*, occur along riparian habitats in the western United States (Figure 2). They segregate by elevation, with *M. lewisii* occurring at higher elevations, and there is a narrow range of elevation in which the species overlap. Genetic analyses show that each species is more closely related to the other than either is to any other species of monkey flower, so any differences between them evolved after they last shared a common ancestor.

The two species have quite different flowers. The species at the higher elevations, *M. lewisii*, is pollinated primarily by bees. It has pink flowers that are relatively wide and whose petals form a horizontal landing strip; yellow structures on the landing strip appear to guide bees to the appropriate part of

Figure 2. Studies of the color, size, shape, and features of the flowers of two closely related species of *Mimulus* have illustrated the genetic control of complex adaptations. *Left:* The blue flower of *M. lewisii*, which features a "landing platform" for the insects that transfer pollen between flowers and "guides" that direct the insects toward the nectar and the reproductive structures of the flower. *Right:* The red flower of *M. cardinalis*, which is a long, narrow, tubular flower with nectar at the base, reachable by the tongues of the hummingbirds that serve as the pollinators.

the flower. The flowers produce small amounts of nectar and have short reproductive organs contained within the tubular structure of the flower, where they are encountered by the bee as it moves toward the nectar at the bottom of the inside of the flower. In contrast, *M. cardinalis* has longer, narrow, red flowers that lack the horizontal landing surface and are pollinated primarily by hummingbirds, which see the color red particularly well. The flowers produce large quantities of nectar that attract hungry hummingbirds and have anthers (male organs) that are elongated and extend outside the flower so that pollen is deposited on the forehead of the bird as it feeds on nectar. The stamens (female organs) are also elongated so that pollen is deposited on them as the bird feeds.

Each flower is well matched to its prevailing pollinator, and so the marked differences between the species in floral morphology reflect divergent, complex adaptations for pollination. Several characters combine to create the match of flower to pollinator: color, shape, the structure of the reproductive organs, and nectar production rates. The question is how readily mutation and selection could have driven the evolution of such divergent flowers from a common ancestor. One extreme point of view, based on the notion discussed earlier that genetic variation is often highly constrained, is that these differences represent the effects of one or a very few genes that affect all the characters. If this were the case, this complex adaptive divergence could be explained very simply by the rise of one or a very few mutant genes that create the two flowers; shifting from one type to another would be, in this scenario, not substantially different than changing the coloration of moths from mottled to black. At the opposite extreme one might argue that such dramatic differences are based on differences in an enormous number of genes, each of which has very small effects on only one of the characters, like color or length of the anther. In this scenario mutation and selection would have been very meticulous and patient molders of the ensemble of differences in the floral traits. Of course, the truth might be anywhere between these extreme views.

Our knowledge of plant development sheds only a little light on the competing hypotheses. Although we know that grossly different arrangements of sepals, petals, stamens, and pistils can be produced by three or four genes, it is unclear how many genes might be necessary to modify the length of the reproductive structures, the shape of the flower, or the nectar production rate.

The route to the answer was found by Doug Schemske, Terry Bradshaw, and their colleagues by combining a traditional technique with modern molecular methods (Schemske and Bradshaw 1999; Bradshaw and Schemske 2003). Schemske and his collaborators began by using methods that were used by Mendel when he discovered the genetic basis of heredity in the nineteenth century. They produced hybrids between the two species; although hybrids rarely occur in nature and fare poorly in the wild when they do, they are quite hardy in greenhouses and are themselves fertile. Schemske and colleagues next crossed hybrids with each parental species, which created a second generation of so-called backcrossed plants with a variable mix of genes from each original species. They then turned to molecular biology, employing molecular markers that enabled them to associate particular variations in floral characters as seen in the various combinations of plants with differences in molecular markers on specific locations of chromosomes. The differences on the chromosomes reflect the locations of genes that contribute to the variation in the flowers.

Twelve individual characters were identified and studied with this method, which revealed that the truth was indeed somewhere between the two extreme ideas about genetic control. A single gene determined whether the flower was red or pink. By placing flowering hybrids and backcrosses in nature, Schemske and colleagues found that the color difference alone had an enormous impact on whether the flower would be visited by a bee or a hummingbird. The scientists also found that while another single gene made a large contribution to the differences in 9 of the 12 floral traits, there were many genes that made smaller contributions to one or a small subset of the traits.

From these results we can offer some plausible reconstructions of this complex adaptation. The effect of flower color on visitations suggests that a change in a single gene, that for color, could initiate the evolution of a specialized flower. Different pollinators attracted to the different colors would exert different selection pressures on the remaining traits. Although it is unclear whether the large single-gene effect on nine traits preceded or followed the changes in the many other genes that affected one or a few traits, the overall picture is of a single, simple genetic change (color) initiating a stepwise process of modification and refinement whose end products are the two complex adaptations that we now see. The match of feature to circumstance was not built quite as easily as changing the coloration of moths but did not require so meticulous a molding as to strain credulity.

A final note about the historical nature of adaptation is that it does not seem possible that the changes in floral shape and reproductive characters could have preceded the color change. Without variation in color and the resultant difference in species of pollinators, there would have been no consistent selection pressure to mold divergent values of the other characters. But once the color differences appeared, the divergent foraging of bees and hummingbirds probably catalyzed a rapid divergence. This example illustrates the contingency inherent in the evolution of complex features in which a single change in a feature can promote a cascade of subsequent changes that refine its adaptation to new circumstances. To return to the critique of Intelligent Design, without appreciating the role of history, we would be hard pressed to explain how and why these flowers diverged so substantially.

WHAT CAN INDIVIDUAL SELECTION NOT EXPLAIN?

Our discussion of adaptation has emphasized how the features of individuals are molded by the action of selection on the available genetic variation.

A fundamental tenet of Darwinian evolution is that selection favors characters that enhance the fitness of individuals; it does not necessarily enhance the well-being of a species. Indeed, the notion that evolution has done anything in particular for the good of the species is, as a fundamental principle, long gone from scientific discourse; any benefit to the species is a side effect of the benefit to the individual.

But although this may be a useful first principle, it cannot be the correct explanation for all features of organisms. The most striking challenge to individual selection comes from the social insects, ants, wasps, bees, and termites, in which nearly all members of the colony forgo reproduction entirely. It is difficult to reconstruct how selection among individuals would favor an individual with a fitness of absolute zero, which is what happens if one does not reproduce.

The key to resolving this paradox for bees, ants, and wasps is to appreciate that the colony workers are raising their nieces and nephews. It appears that the sacrifice in individual fitness is compensated by the enhanced ability to raise the offspring of a sibling, provided that most individuals sustain this commitment. There are many examples of less extreme forms of altruism, from the cooperative rearing of offspring by sisters in lion prides to the aid provided by older offspring to their parents in raising younger offspring, as has been observed in several species of birds.

These cases are examples of selection at the level of the family, or kin selection. In kin selection individual sacrifice is favored when it enhances the production of related individuals, with the degree of sacrifice proportional to the combination of number and relatedness of the individuals that benefit from the sacrifice. This is a powerful explanation for many forms of sociality (see the main essay "Social Behavior and Sociobiology" by Daniel Rubenstein and the alphabetical entries "Altruism" and "W. D. Hamilton" in this volume).

In fact, kin selection is one of several varieties of selection that are named for the level at which they occur. These varieties of selection emerge when there is a causal relationship between features expressed at a particular level, such as the family level, and the success of the units at that level, be they families or larger groups of unrelated individuals. The conditions for kin and group selection to occur have been explored extensively in a variety of sophisticated theories, and the critical issue in modern evolutionary biology is to diagnose how often the conditions that promote selection at levels other than that of the individual are met.

A wide variety of features appear explicable only by selection at levels other than that of the individual, and this is a very active topic of modern research. Besides the many cases of reproductive altruism, there are cases of group living and cooperative behavior (see the main essay "Social Behavior and Sociobiology" by Daniel Rubenstein and the alphabetical entry "Group selection" in this volume). Selection at several levels plays a critical role in understanding the evolutionary interactions between pathogens and hosts (see the main essay "Evolutionary Biology of Disease and Darwinian Medicine" by Michael Antolin and the alphabetical entry "Host parasite evolution" in this volume); a pathogen that multiplies too quickly and kills its host too quickly before it can be transmitted to another host loses out because selection among families or groups (represented by the pathogen population within an individual host) overwhelms selection among individuals (represented by the struggle for proliferation within an individual host).

ARE STUDIES OF MICROEVOLUTION RELEVANT TO UNDERSTANDING MACROEVOLUTIONARY PATTERNS?

It is convenient to characterize the products of evolution as either microevolution or macroevolution. *Microevolution* refers to small-scale changes that are directly observable, like many of those discussed earlier (melanism in moths, disease resistance in bacteria, or insecticide resistance). As we have discussed, microevolution has been very thoroughly documented and has been studied in ways that meet any test of scientific rigor. As a result, it is not controversial, even among many faith-based critics of evolution (see the main essay "American Antievolutionism: Retrospect and Prospect" by Eugenie Scott in this volume).

Macroevolution refers to the larger-scale events that we attribute to evolution, such as the origin of new species, the differential proliferation or differential extinction of certain types of species, and the emergence of higher levels of the taxonomic hierarchy. This is the aspect of evolution that is controversial because it represents events that occur on a timescale that is much longer than our lives and hence not directly observable. As we have discussed, scientists must turn to indirect methods to draw inferences about the underlying mechanisms that cause macroevolution.

The most revolutionary feature of Darwin's *On the Origin of Species* (1859) was to propose natural selection as the single unifying mechanism that causes both micro- and macroevolution. Darwin argued that macroevolution is just microevolution writ large, or that the process we see and study as the cause of microevolution will, given sufficient time, also cause everything that we attribute to macroevolution. He argued that natural selection, which causes the evolution of the adaptations discussed throughout this essay, is also responsible for the origin of all levels of biological complexity and for the origin of biological diversity, or all the species that have been found on the earth throughout its history.

Our inability to observe macroevolution in action has inspired controversy over whether natural selection really is the unifying mechanism of all evolutionary change. Stephen J. Gould and Niles Eldredge are the authors of what is currently the most prominent scientific argument against natural selection as a single, unifying cause of micro- and macroevolution (Eldredge and Gould 1972; Gould and Eldredge 1977). These authors were motivated by the pattern of change recorded in the fossil record. If we look at the fossil history of the evolution of most organisms, such as the increase of body size in horses or brain size in the hominid lineage leading to our species, we often see that there are long intervals when the trait in question changes little or not at all, then brief intervals when the trait evolves rapidly. Gould and Eldredge described the long, stable intervals as stasis or equilibrium and the short intervals as punctuations. They also pointed out that there is often an association between punctuations and the origin of new species and coined the term *punctuated equilibrium* to describe the entire pattern. Eldredge and Gould argued that both phases of punctuated equilibrium are inconsistent with natural selection as a single underlying cause. Their argument was based on the premise that if selection were such a unifying cause, we should observe a predominant pattern of gradual, continuous change in characters. Stasis in characters suggests resistance to change. The discontinuity of punctuations suggested to them that something happened to break this resistance and cause a subsequent pulse of accelerated evolution.

Although Gould and Eldredge's arguments have evolved during the debate that ensued after their first essay, their arguments have two unifying features. First, natural selection is posited as being effective only in fine-tuning organisms to small-scale changes in their environment. Second, some process other than natural selection must come into play to cause punctuations. The frequent association between punctuations and the appearance of new species suggests to Gould and Eldredge that this process, whatever it is, causes both phenomena, but they have altered their proposed mechanism for this process during the 36-year history of punctuated equilibrium as an idea.

The counterargument in favor of natural selection as a unifying process has three parts. First, we have empirical estimates of the rate at which evolution by natural selection can occur, and we know that it can be much faster than the rates associated with punctuations in the fossil record. For example, Galápagos finches can evolve a 10% change in average body size or bill size in a population in a single year in response to either droughts or heavy rainfall. In drought years food availability is low, the birds must rely on large seeds with thick seed coats and individuals with wide, heavy bills are more effective at harvesting such seeds and are more likely to survive and reproduce. Because there is a genetic basis to bill size and shape, the next generation of finches exhibits larger, heavier bills. This is classic microevolution. If we were to put this rate of evolution on the same timescale as what we see in the fossil record, it is 10,000 or more times faster than what Gould and Eldredge interpret as punctuations. The apparent suddenness of punctuations is thus not inconsistent with the rate of evolution that is attainable under natural selection. If anything, we might ask why evolution appears to be so slow in the fossil record.

Second, selection is capable of causing substantial changes in organisms. This, of course, is the fundamental challenge we reviewed in earlier sections, but here it is important because Gould and Eldredge argue that natural selection can only make fine adjustments to organisms, and that stasis, the resistance to change, must be broken by some other process. We can follow one of Darwin's lines of argument, albeit with more information. Genetic data indicate that domestic dogs were derived from wolves, beginning tens of thousands of years ago. The first appearance of morphologically distinct domestic dogs dates to around 10,000 years ago (Savolainen et al. 2002). From such recent beginnings we have breeds that range in size from the Chihuahua to the mastiff and in shape from the Pekinese to the Irish wolfhound. If we focus on body weight alone, the range of average weights among breeds is from around 1 kilogram for Chihuahuas to 80 kilograms for mastiffs. This range exceeds the entire range of body sizes of other species in the family Canidae, which includes dogs, wolves, foxes, and other doglike carnivores (Finarelli and Flynn 2006). It falls short of what we see in the order Carnivora, which includes the family Canidae along with others, such as the Felidae (cats) or Ursidae (bears). Ten thousand years of artificial selection have thus been sufficient to generate a range of morphological variation that is equivalent to the range we see when we look three or four steps up the taxonomic hierarchy (species, genus, family, order, class, phylum, kingdom).

Third, we must consider whether the presence of stasis, or long intervals of little or no change, can be explained by natural selection. The Galápagos finches again provide a good conceptual alternative to explain why there might be long intervals of no apparent change in the fossil record. In drought years there is intense selection for increased bill and body size. In the rainy years that accompany El Niño events, the resource base changes because there is an abundance of small, thin-shelled seeds from different species of plants. Selection now favors smaller individuals with narrower, more forceps-like bills. This means that periods of selection for increased bill size alternate with periods of selection for decreased bill size, and in some years there is in fact no selection at all (Gibbs and Grant 1987). Over the nearly three decades in which these birds have been studied, there has been no net change in the bill size of the population in spite of strong evidence for intense selection and rapid evolution on a year-to-year basis (see Weiner 1994 for a nontechnical review of studies of Darwin's finches). The fossil record could represent a fragmented, long-term history of such short-term fluctuating selection. The fossil record does not represent each and every year, and, in fact, a defined period in the fossil record will represent a random mixture of these different episodes of selection. The actual rate of evolution that we see in the fossil record is a long-term average of the increases and decreases and will inevitably underestimate the true rate at which organisms can evolve. A fossil record of the Galápagos finches may well look like stasis.

In the final analysis there is nothing in the fossil record that inherently contradicts Darwin's daring idea that natural selection is the unifying mechanism. Organisms can evolve much more rapidly than they appear to evolve during punctuations, and the capacity of organisms to change under selection far exceeds the constraints that are apparent during stasis. The absence of change during stasis may instead be a statistical artifact of fluctuating evolution that tracks a fluctuating environment. But our arguments do not prove that macroevolution is microevolution writ large; they merely support the viability of natural selection as a single, overarching explanation. The root of the controversy remains the difficulty of reconciling what we can study in the short term with the enormously different timescale of the major patterns we seek to explain, and the divide between micro- and macroevolution remains an unsettled frontier of the discipline of evolutionary biology.

Horizons in the Study of Adaptation

Adaptation is being studied at least as intensely as ever. Although the reality of adaptation is evident, many facets of it remain unclear. For example, we are beginning to dissect the genetic basis of adaptations, particularly complex adaptations. We have a few classic studies like those on the monkey flowers, but we cannot conclude whether the patterns that Schemske and colleagues found are general. Genomic and proteomic methods are accelerating this research, and a version of this essay written a decade from now may offer some very different conclusions about this topic.

The constraints on the raw material for adaptation are being studied in different ways. In some cases biologists are studying the properties of new mutations. In others they are examining whether characters that occur together in nature do so because they are controlled by the same genes or because the individual genes that produce concerted inheritance are aligned in such a way that they are inherited as units or blocks of genes. Although it may seem implausible to posit extensive linkage of genes that control correlated characters, in fact there are a few studies that show a surprising level of such linkages (see the main essay "Evolution of the Genome" by Brian Charlesworth and Deborah Charlesworth and the alphabetical entry "Mimicry" in this volume). This suggests that the genome itself can be a product of Darwinian evolution and not merely a constraint against which such evolution must operate. This is an important question that has attracted many scientists.

Our collective view of selection at levels other than those of the individual has changed markedly in the past 30 years. Multilevel selection is clearly important for understanding many adaptations in diverse systems; research in this area is leading to far more nuanced and sophisticated interpretations of the evolution of sociality, of pathogen-host relationships, and even of very large phenomena such as multicellularity and perhaps the cell itself. Indeed, the evolution of the cell and the diversity of cell structures are extremely active areas of research (Woese 2004; Kurland et al. 2006). The molecular revolution has enabled long-standing questions to be answered and entirely new ones to be asked. The specialized nature of the methods being employed and the difficulty of applying multilevel selection theory to a phenomenon that happened over a billion years ago make progress slower than it is for other topics, but in a decade or two our knowledge of cellular evolution and our appreciation of multilevel selection as a critical evolutionary force will be very different than they are today.

The study of Darwinian evolution remains a vibrant topic nearly 150 years after *On the Origin of Species* appeared. Although Darwin laid out the argument in broad outline, it fell to his successors to discover which elements of his argument were possible in theory, which were demonstrable in practice, and which were robust to variations among species and circumstances. That so much of his original argument remains standing is a tribute to his genius; that so much remains to be discovered is a tribute to the marvels that adaptation offers and the diversity of nature that it has catalyzed.

BIBLIOGRAPHY

Ayala, F. J. 2006. *Darwin and Intelligent Design.* Minneapolis: Fortress Press.

Bennett, M. V. L. 1971. Electric organs. In W. S. Hoar and D. J. Randall, eds., *Fish Physiology,* vol. 5, 347–492. New York: Academic Press.

Blouin, M. S. 1991. Proximate developmental causes of limb length variation between *Hyla cinerea* and *Hyla gratiosa* (Anura: Hylidae). *Journal of Morphology* 209: 305–310.

———. 1992. Genetic correlations among morphometric traits and rates of growth and differentiation in the green tree frog, *Hyla cinerea. Evolution* 46: 735–744.

Bradshaw, H. D., Jr., and D. W. Schemske. 2003. Allele substitution at a flower colour locus produces a pollinator shift in monkeyflowers. *Nature* 426: 176–178.

Darwin, C. 1859. *On the Origin of Species.* London: John Murray.

Dzidic, S., J. Suskovic, and B. Kos. 2008. Antibiotic resistance mechanisms in bacteria: Biochemical and genetic aspects. *Food Technology and Biotechnology* 46: 11–21.

Eldredge, N., and S. J. Gould. 1972. Punctuated equilibria: An alternative to phyletic gradualism. In T. J. M. Schopf, ed., *Models in Paleobiology,* 82–115. San Francisco: Freeman, Cooper.

Eiseley, L. 1958. *Darwin's Century: Evolution and the Men Who Discovered It.* Garden City, NY: Doubleday.

Feder, M. E., and W. W. Burggren, eds. 1992. *Environmental Physiology of the Amphibians.* Chicago: University of Chicago Press.

Finarelli, J. A., and J. J. Flynn. 2006. Ancestral state reconstruction of body size in the Caniformia (Carnivora, Mammalia): The effects of incorporating data from the fossil record. *Systematic Biology* 55: 301–313.

Futuyma, D. J. 2005. *Evolution.* Sunderland, MA: Sinauer Associates.

Gibbs, H. L., and P. R. Grant. 1987. Oscillating selection on Darwin's finches. *Nature* 327: 511–513.

Gould, S. J., and N. Eldredge. 1977. Punctuated equilibria: The tempo and mode of evolution reconsidered. *Paleobiology* 3: 115–151.

Hereford, J., T. F. Hansen, and D. Houle. 2004. Comparing strengths of directional selection: How strong is strong? *Evolution* 58: 2133–2143.

Hochachka, P. W., and G. S. Somero. 2002. *Biochemical Adaptation: Mechanism and Process in Physiological Evolution.* New York: Oxford University Press.

Jacob, F. 1977. Evolution and tinkering. *Science* 196: 1161–1166.

Kallman, K. D. 1989. Genetic control of size at maturity in *Xiphophorus.* In G. K. Meffe and F. F. Snelson Jr., eds., *Ecology and Evolution of Livebearing Fishes (Poeciliidae),* 163–184. Englewood Cliffs, NJ: Prentice Hall.

Klerks, P. W., and J. S. Weis. 1987. Genetic adaptation to heavy-metals in aquatic organisms—a review. *Environmental Pollution* 45: 173–205.

Kurland, C. G., L. J. Collins, and D. Penny. 2006. Genomics and the irreducible nature of eukaryote cells. *Science* 312: 1011–1014.

Luo, Z.-X. 2007. Transformation and diversification in the early mammalian evolution. *Nature* 450: 1011–1019.

MacFadden, B. J. 1992. *Fossil Horses: Systematics, Paleobiology, and Evolution of the Family Equidae.* New York: Cambridge University Press.

MacNair, M. R. 1987. Heavy-metal resistance in plants—a model evolutionary system. *Trends in Ecology and Evolution* 2: 354–359.

Matsumura, F., and G. Voss. 1964. Mechanism of malathion and parathion resistance in the two-spotted spider mite *Tetranychus urticae. Journal of Economic Entomology* 57: 911–917.

McPeek, M. A. 1995. Morphological evolution mediated by behavior in the damselflies of two communities. *Evolution* 49: 749–769.

Powles, S. B. 2008. Evolved glyphosate-resistant weeds around the world: Lessons to be learnt. *Pest Management Science* 64: 360–365.

Provine, W. B. 1971. *Origins of Theoretical Population Genetics.* Chicago: University of Chicago Press.

Radinsky, L. B. 1987. *The Evolution of Vertebrate Design.* Chicago: University of Chicago Press.

Reznick, D. N., and C. K. Ghalambor. 2001. The population ecology of contemporary adaptations: What empirical studies reveal about the conditions that promote adaptive evolution. *Genetica* 112: 183–198.

Reznick, D. N., F. H. Shaw, F. H. Rodd, and R. G. Shaw. 1997. Evaluation of the rate of evolution in natural populations of guppies *(Poecilia reticulata). Science* 275: 1934–1937.

Reznick, D., and J. Travis. 1996. Empirical studies of adaptation. In M. Rose and G. Lauder, eds., *Adaptation: Perspectives and New Approaches,* 243–289. New York: Academic Press.

Roush, R. T., and J. A. McKenzie. 1987. Ecological genetics of insecticide and acaricide resistance. *Annual Review of Entomology* 32: 361–380.

Ruse, M. 1979. *The Darwinian Revolution: Science Red in Tooth and Claw.* Chicago: University of Chicago Press.

Savolainen, P., Y. P. Zhang, J. Luo, J. Lundeberg, and T. Leitner. 2002. Genetic evidence for an East Asian origin of domestic dogs. *Science* 298: 1610–1613.

Schemske, D. W., and H. D. Bradshaw Jr. 1999. Pollinator preference and the evolution of floral traits in monkeyflowers *(Mimulus). Proceedings of the National Academy of Sciences USA* 96: 11910–11915.

Schmidt-Nielsen, K. 1990. *Animal Physiology: Adaptation and Environment.* Cambridge: Cambridge University Press.

Schulte, P. M. 2001. Environmental adaptations as windows on molecular evolution. *Comparative Biochemistry and Physiology B* 128: 597–611.

Shubin, N. H. 2002. Origin of evolutionary novelty: Examples from limbs. *Journal of Morphology* 252: 15–28.

Sicard, D., P. S. Pennings, C. Grandeclement, J. Acosta, O. Kaltz, and J. A. Shykoff. 2007. Specialization and local adaptation of a fungal parasite on two host plant species as revealed by two fitness traits. *Evolution* 61: 27–41.

Travis, J., and D. N. Reznick. 1998. Experimental approaches to the study of evolution. In W. J. Resitarits and J. Bernardo, eds., *Issues and Perspectives in Experimental Ecology,* 437–459. New York: Oxford University Press.

Wake, M. H., ed. 1979. *Hyman's Comparative Vertebrate Anatomy.* 3rd ed. Chicago: University of Chicago Press.

Weber, B. H., and D. J. Depew. 2004. Darwinism, design, and complex system dynamics. In W. A. Dembski and M. Ruse, eds., *Debating Design: From Darwin to DNA,* 173–190. Cambridge: Cambridge University Press.

Weiner, J. 1994. *The Beak of the Finch: A Study of Evolution in Our Time.* New York: Alfred A. Knopf.

Woese, C. R. 2004. A new biology for a new century. *Microbiology and Molecular Biology Reviews* 68: 173–186.

Molecular Evolution

Francisco J. Ayala

Molecular biology has made it possible to reconstruct the continuity of succession from the original form of life, ancestral to all living organisms, to every species now living on earth. The universal tree of life, which embraces all known sorts of organisms, includes three sets of branches, Bacteria, Archaea, and Eukarya, which all emerged from one form of life, LUCA (Last Universal Common Ancestor; Figure 1). All living organisms are related by common descent from a single form of life, represented by the tree trunk in the figure. Life may have originated more than once on our planet—we presently do not know—but only one form of life prospered and left descendants, the organisms that now populate the earth.

Bacteria and Archaea are prokaryotic microscopic organisms. Eukarya are organisms with complex cells that contain several organelles, one of which, the nucleus, includes the hereditary DNA. Most eukaryotic organisms are single celled and microscopic. Animals, plants, and fungi are multicellular organisms, three of the many branches of the Eukarya set.

The main branches of the universal tree of life from LUCA, several billion years ago, to the present have been reconstructed on the whole and in many details. More details about more and more branches are published in scores of scientific articles every month. The virtually unlimited evolutionary information encoded in the DNA sequence of living organisms allows evolutionists to reconstruct any evolutionary relationships that have led to present-day organisms, or among them, with as much detail as wanted. Invest the necessary resources (time and laboratory expenses), and you can have the answer to any query with as much precision as you want.

Molecular Evolution: Uniformity and Diversity

In its unveiling of the nature of DNA and the workings of organisms at the level of enzymes and other protein molecules, molecular biology has shown the unity of life. Moreover, molecular biology has shown that these molecules, DNA and proteins, hold information about an organism's ancestry.

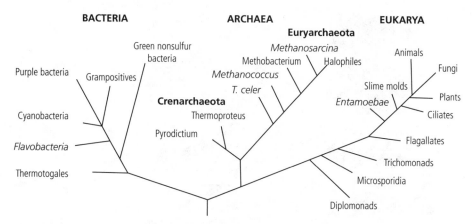

Figure 1. The universal tree of life. The three major groups of organisms, Bacteria, Archaea, and Eukarya, are represented by the three main branches. All organisms are related by common descent from a single form of life, represented by the tree's "trunk" (the straight-up line at bottom). This tree was constructed using slow-evolving ribosomal RNA genes. (Adapted from Woese 1998.)

This has made it possible to reconstruct evolutionary events that were previously unknown and to confirm and adjust the view of events already known. The precision with which these events can be reconstructed is one reason the evidence from molecular biology is so compelling.

The molecular components of organisms are remarkably uniform in the nature of the components, as well as in the ways in which they are assembled and used. In all bacteria, archaea, plants, animals, and humans, the DNA consists of a different sequence of the same four component nucleotides (represented as A, C, G, and T; Figure 2). The genetic code, by which the information contained in the DNA of the cell nucleus is passed on to proteins, is virtually the same in all organisms (Figure 3). The enormously diverse proteins in all sorts of organisms are synthesized from different combinations, in sequences of variable length, of the same 20 amino acids (listed in Figure 3), although several hundred other amino acids exist. Similar metabolic pathways—sequences of biochemical reactions—are used by the most diverse organisms to produce energy and to make up the cell components.

This unity reveals the genetic continuity and common ancestry of all organisms. There is no other rational way to account for their molecular uniformity when numerous alternative structures are in principle equally likely. The genetic code serves as an example. Each particular sequence of three nucleotides in the nuclear DNA acts as a pattern for the production of exactly the same amino acid in all organisms. This is no more necessary than it is for a language to use a particular combination of letters to represent a particular object. If it is found that certain sequences of letters—planet, tree, woman—are used with identical meanings in a number of different books, one can be sure that the languages used in those books are of common origin.

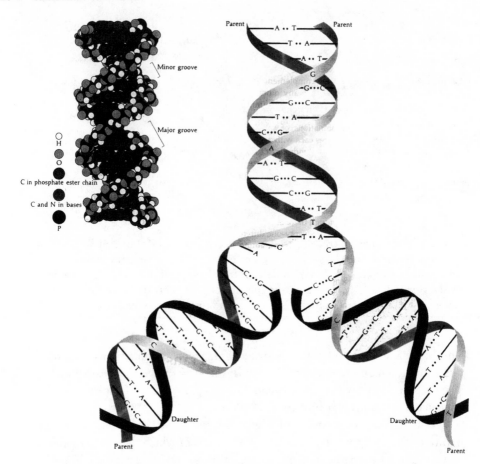

Figure 2. The double-stranded helical configuration of the DNA molecule. The molecule consists of two polynucleotide chains. The outward backbone of the molecule is made of alternating deoxyribose sugars (S) and phosphate groups (P). Nitrogen bases connected to the sugars project toward the center of the molecule. The two nucleotide chains are held together by hydrogen bonds between complementary purine-pyrimidine bases. Adenine (A) and thymine (T) form two hydrogen bonds; cytosine (C) and guanine (G) form three.

Genes and proteins are long molecules that contain information in the sequence of their components in much the same way in which sentences of the English language contain information in the sequence of their letters and words. The sequences that make up the genes are passed on from parents to offspring and are identical from generation to generation, except for occasional changes introduced by mutations. Closely related species have very similar DNA sequences; the few differences reflect mutations that occurred since their last common ancestor. Species that are less closely related to one another exhibit more differences in their DNA than those more closely related because more time has elapsed since their last common ancestor. This is

Second Position

		U	C	A	G	
First Position	U	UUU ⎤ Phe UUC ⎦ UUA ⎤ Leu UUG ⎦	UCU ⎤ UCC UCA UCG ⎦ Ser	UAU ⎤ Tyr UAC ⎦ UAA Stop UAG Stop	UGU ⎤ Cys UGC ⎦ UGA Stop UGG Trp	U C A G
	C	CUU ⎤ CUC CUA CUG ⎦ Leu	CCU ⎤ CCC CCA CCG ⎦ Pro	CAU ⎤ His CAC ⎦ CAA ⎤ Gln CAG ⎦	CGU ⎤ CGC CGA CGG ⎦ Arg	U C A G
	A	AUU ⎤ AUC ⎥ Ile AUA ⎦ AUG Met	ACU ⎤ ACC ACA ACG ⎦ Thr	AAU ⎤ Asn AAC ⎦ AAA ⎤ Lys AAG ⎦	AGU ⎤ Ser AGC ⎦ AGA ⎤ Arg AGG ⎦	U C A G
	G	GUU ⎤ GUC GUA GUG ⎦ Val	GCU ⎤ GCC GCA GCG ⎦ Ala	GAU ⎤ Asp GAC ⎦ GAA ⎤ Glu GAG ⎦	GGU ⎤ GGC GGA GGG ⎦ Gly	U C A G

(Third Position)

Figure 3. The genetic code: correspondence between the 64 possible codons in messenger RNA and the amino acids (or termination signals, "stop"). The nitrogen base thymine does not exist in RNA, where uracil (U) takes its place; the other three nitrogen bases in messenger RNA are the same as in DNA: adenine (A), cytosine (C), and guanine (G). The 20 amino acids that make up proteins (with the three-letter and one-letter standard abbreviations) are alanine (Ala, A), arginine (Arg, R), asparagine (Asn, N), aspartic acid (Asp, D), cysteine (Cys, C), glycine (Gly, G), glutamic acid (Glu, E), glutamine (Gln, Q), histidine (His, H), isoleucine (Ile, I), leucine (Leu, L), lysine (Lys, K), methionine (Met, M), phenylalanine (Phe, F), proline (Pro, P), serine (Ser, S), threonine (Thr, T), tyrosine (Tyr, Y), tryptophan (Trp, W), and valine (Val, V).

the rationale for the reconstruction of evolutionary history, whose methods and outcomes will be reviewed later.

As an illustration, let us assume that we are comparing two books. Both books are 200 pages long and contain the same number of chapters. Closer examination reveals that the two books are identical page for page and word for word, except that an occasional word—say, 1 in 100—is different. The two books cannot have been written independently; either one has been copied from the other, or both have been copied, directly or indirectly, from the same original book. In living beings, if each component nucleotide of DNA is represented by one letter, the complete sequence of nucleotides in the DNA of a higher organism would require several hundred books, each with hundreds of pages, with several thousand letters on each page. When the "pages" (or sequences of nucleotides) in these "books" (genomes) are examined one by one, the correspondence in the "letters" (nucleotides) gives unmistakable evidence of common origin.

Molecular biology offers two kinds of arguments for evolution, based on two different grounds. Using the alphabet analogy, the first argument says that languages that use the same alphabet (the same hereditary molecule, the DNA made up of the same four nucleotides, and the same 20 amino acids in their proteins), as well as the same dictionary (the same genetic code), cannot be of independent origin. The second argument concerns similarity in the sequence of nucleotides in the DNA (and thus the sequence of amino acids in the proteins); it says that books with very similar texts cannot be of independent origin.

Informational Macromolecules

Nucleic acids and proteins have been called *informational macromolecules* because they are long linear molecules made up of sequences of units—nucleotides in the case of nucleic acids, amino acids in the case of proteins—that retain considerable amounts of evolutionary information. Comparison of the sequence of the components in two macromolecules establishes the numbers that are different. Because evolution usually occurs by changing one unit at a time, the number of differences is an indication of the recency of common ancestry.

The degree of similarity in the sequence of nucleotides or of amino acids can be precisely quantified. For example, in humans and chimpanzees the protein molecule called cytochrome-c, which serves a vital function in respiration within cells, consists of the same 104 amino acids in exactly the same order. It differs, however, from the cytochrome-c of rhesus monkeys by 1 amino acid, from that of horses by 11 additional amino acids, and from that of tuna by 21 additional amino acids.

The degree of similarity reflects the recency of common ancestry. Thus inferences from comparative anatomy and other disciplines concerning evolutionary history can be tested in molecular studies of DNA and proteins by examining their sequences of nucleotides and amino acids. The authority of this kind of test is overwhelming: each of the thousands of genes and thousands of proteins contained in an organism provides an independent test of that organism's evolutionary history.

Molecular evolutionary studies have three notable advantages over comparative anatomy and the other classical disciplines. One is that the information is more readily quantifiable. The number of units that are different is readily established when the sequence of units is known for a given macromolecule in different organisms. The second advantage is that comparisons can be made even between very different sorts of organisms. There is very little that comparative anatomy, for example, can say when organisms as diverse as yeasts, pine trees, and human beings are compared, but there are numerous DNA and protein sequences that can be compared in all three. The third advantage is multiplicity. Each organism possesses thousands of genes and proteins, which all reflect the same evolutionary history. If the investigation of one particular gene or protein does not satisfactorily resolve the evolutionary relationship of a set of species, additional genes and proteins can be investigated until the matter has been settled to the satisfaction of the investigator.

Moreover, the widely different rates of evolution of different sets of genes open up the opportunity of investigating different genes for achieving different degrees of resolution in the tree of evolution. Evolutionists rely on slowly evolving genes to reconstruct remote evolutionary events but increasingly faster-evolving genes to reconstruct the evolutionary history of more recently diverged organisms.

Genes that encode the ribosomal RNA molecules are among the slowest-evolving genes. They have been used to reconstruct the evolutionary relationships among groups of organisms that diverged very long ago: for example, among bacteria, archaea, and eukaryotes (the three major divisions of the living world), which diverged more than 2 billion years ago (see Figure 1); or among the microscopic protozoa (e.g., *Plasmodium,* which causes malaria) compared with plants and animals, the three of which are eukaryotic groups of organisms that diverged about 1 billion years ago. Cytochrome-c, mentioned earlier, which evolves slowly, but not as slowly as the ribosomal RNA genes, is used to decipher the relationships within large groups of organisms, such as among humans, fishes, and insects. Fast-evolving molecules, such as the fibrinopeptides involved in blood clotting, are appropriate for investigating the evolution of closely related animals, for example, the primates: macaques, chimpanzees, and humans.

Molecular Phylogenies of Organisms

DNA and proteins provide information not only about the branching of lineages from common ancestors (cladogenesis) but also about the amount of genetic change that has occurred in any given lineage (anagenesis). It might seem at first that quantifying anagenesis for proteins and nucleic acids would be impossible because it would require comparison of molecules from organisms that lived in the past with those from living organisms. Organisms of the past are sometimes preserved as fossils, but their DNA and proteins have largely disintegrated. Nevertheless, comparisons between living species provide information about anagenesis.

As a specific example, consider the previously mentioned protein cytochrome-c, which is involved in cell respiration. The sequence of amino acids in this protein is known for many organisms, from bacteria and yeasts to insects and humans; in animals, cytochrome-c consists of 104 amino acids. When the amino acid sequences of humans and rhesus monkeys are compared, they are found to be different at position 58 (isoleucine in humans, threonine in rhesus monkeys), but identical at the other 103 positions. When humans are compared with horses, 12 amino acid differences are found, and when horses are compared with rhesus monkeys, there are 11 amino acid differences (Figure 4). Even without knowing anything else about the evolutionary history of mammals, one would conclude that the lineages of humans and rhesus monkeys diverged from each other much more recently than they diverged from the horse lineage. Moreover, it can be concluded that the amino acid difference between humans and rhesus monkeys must have occurred in the human lineage after its separation from the rhesus monkey lineage (see Figure 5).

```
                1   5   10   15   20     25   30   35   40   45   50
Human    | GDVEKGKKIFIMKCSQCHTVEKGGKHKTGPNLHGLFGRKTGQAPGYSYTAAN
Monkey   | ...................................................
Horse    | ........VQ..A...............................FT..D..

                55   60   65   70   75   80   85   90   95   100  104
Human    | KNKGIIWGEDTLMEYLENPKKYIP6TKMIFVGIKKKEERADLIAYLKKATNE
Monkey   | .....T..............................................
Horse    | .....T.K.E....................A.....T..E...........
```

Figure 4. Amino acid sequence of cytochrome-c proteins in humans, rhesus monkeys, and horses. The sequence consists of 104 amino acids, each represented by a letter. A dot indicates the same amino acid as in humans. See Figure 3 for the names of the amino acids.

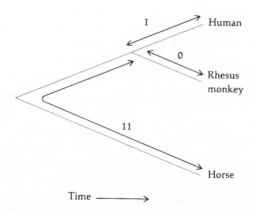

Figure 5. The evolution of cytochrome-c among humans, monkeys, and horses. The numbers indicate the amino acid replacements that have taken place in each branch of the phylogeny.

Evolutionary trees are models or hypotheses that seek to reconstruct the evolutionary history of taxa (i.e., species or other groups of organisms, such as genera, families, or orders). As pointed out earlier, the trees embrace two kinds of information related to evolutionary change, cladogenesis and anagenesis. Figure 5 illustrates both. The branching relationships of the trees reflect the relative relationships of ancestry, or cladogenesis. Thus in the right side of the figure, humans and rhesus monkeys are seen to be more closely related to each other than either one is to the horse. Stated another way, this tree shows that the last common ancestor to all three species lived in a more remote past than the last common ancestor to humans and monkeys.

Evolutionary trees may also indicate the changes that have occurred along each lineage, or anagenesis. Thus in the evolution of cytochrome-c since the last common ancestor of humans and rhesus monkeys, one amino acid changed in the lineage that went to humans but none in the lineage that went to rhesus monkeys. This conclusion is drawn from the observation (Figure 4) that at position 58, monkeys and horses (as well as other animals; see Figure 4) have the same amino acid (threonine), while humans have a different one (isoleucine), which therefore must have changed in the human lineage after it separated from the monkey lineage.

Molecular Reconstruction of Evolutionary History

Several methods exist for constructing evolutionary trees. Some were developed for interpreting morphological data, others for interpreting molecular data; some can be used with either kind of data. The main methods currently used in molecular evolution are called distance, maximum parsimony, and maximum likelihood.

DISTANCE METHODS

A "distance" is the number of differences between two kinds of organisms. The differences are measured with respect to certain traits (e.g., morphological data) or to certain macromolecules (primarily the sequence of amino acids in proteins or the sequence of nucleotides in DNA or RNA). The tree illustrated in Figure 5 was obtained by taking into account the distance, or number of amino acid differences, among three organisms with respect to a particular protein. The amino acid sequence of a protein contains more information than is reflected in the number of amino acid differences because in some cases the replacement of one amino acid by another requires no more than one nucleotide substitution in the DNA that codes for the protein, whereas in other cases it requires at least two nucleotide changes. Table 1 shows the minimum number of nucleotide differences in the genes of 20 separate species that is necessary to account for the amino acid differences in their cytochrome-c. An evolutionary tree based on the data in Table 1, showing the minimum numbers of nucleotide changes in each branch, is illustrated in Figure 6.

Table 1. Minimum number of nucleotide differences between the genes coding for cytochrome-c in 20 organisms.

Organism	1	2	3	4	5	6	7	8	9	10	11	12	13	14	15	16	17	18	19	20
1. Human	–	1	13	17	16	13	12	12	17	16	18	18	19	20	31	33	36	63	56	66
2. Monkey		–	12	16	15	12	11	13	16	15	17	17	18	21	32	32	35	62	57	65
3. Dog			–	10	8	4	6	7	12	12	14	14	13	30	29	24	28	64	61	66
4. Horse				–	1	5	11	11	16	16	16	17	16	32	27	24	33	64	60	68
5. Donkey					–	4	10	12	15	15	15	16	15	31	26	25	32	64	59	67
6. Pig						–	6	7	13	13	13	14	13	30	25	26	31	64	59	67
7. Rabbit							–	7	10	8	11	11	11	25	26	23	29	62	59	67
8. Kangaroo								–	14	14	15	13	14	30	27	26	31	66	58	68
9. Duck									–	3	3	3	7	24	26	25	29	61	62	66
10. Pigeon										–	4	4	8	24	27	26	30	59	62	66
11. Chicken											–	2	8	28	26	26	31	61	62	66
12. Penguin												–	8	28	27	28	30	62	61	65
13. Turtle													–	30	27	30	33	65	64	67
14. Rattlesnake														–	38	40	41	61	61	69
15. Tuna															–	34	41	72	66	69
16. Screwworm fly																–	16	58	63	65
17. Moth																	–	59	60	61
18. *Neurospora*																		–	57	61
19. *Saccharomyces*																			–	41
20. *Candida*																				–

Source: After Fitch and Margoliash 1967.

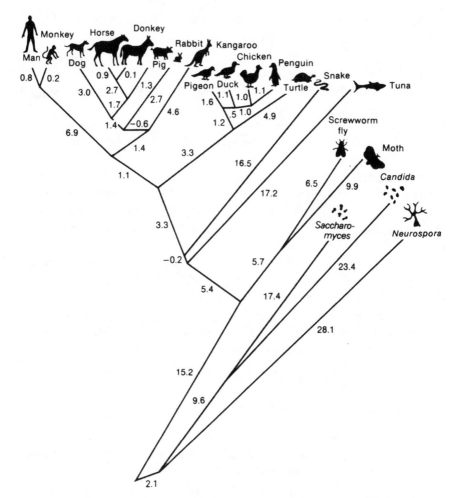

Figure 6. Phylogeny of 20 organisms, based on differences in the amino acid sequence of cytochrome-c. The minimum number of nucleotide substitutions required for each branch is shown. Although fractional numbers of nucleotide substitutions cannot occur, the numbers shown are those that fit best the data in Table 1. (After Fitch and Margoliash 1967.)

The relationships between species as shown in Figure 6 correspond fairly well to the relationships determined from other sources, such as the fossil record. According to the figure, chickens are less closely related to ducks and pigeons than to penguins, and humans and monkeys diverged from the other mammals before the marsupial kangaroo separated from the nonprimate placentals. Although these examples are known to be erroneous relationships, the power of the method is apparent in that a single protein yields a fairly accurate reconstruction of the evolutionary history of 20 organisms that started to diverge more than 1 billion years ago.

The most common procedure to transform a distance matrix into a phylogeny is called cluster analysis. The distance matrix is scanned for the smallest distance element, and the two taxa involved (say, human and monkey in Table 1) are joined at an internal node, or branching point. The matrix is scanned again for the next-smallest distance, and the two new taxa (say, horse and donkey) are clustered. The procedure is continued until all taxa have been joined. When a distance involves a taxon that is already part of a previous cluster, the average distance is obtained between the new taxon and the preexisting cluster (say, the average distance between human and horse and monkey and horse). This simple procedure assumes that the rate of evolution is uniform along all branches.

Other distance methods (including the one used to construct the tree in Figure 6) relax the condition of uniform rate and allow for unequal rates of evolution along the branches. One of the most extensively used methods of this kind is called neighbor-joining. The method starts, as before, by identifying the smallest distance in the matrix and linking the two taxa involved. The next step is to remove these two taxa and calculate a new matrix in which their distances to other taxa are replaced by the distance between the node that links the two taxa and all other taxa. The smallest distance in this new matrix is used for making the next connection, which will be between two other taxa or between the previous node and a new taxon. The procedure is repeated until all taxa have been connected with one another by intervening nodes.

MAXIMUM-PARSIMONY METHODS

Maximum-parsimony methods seek to reconstruct the tree that requires the fewest (i.e., most parsimonious) number of changes summed along all branches. This is a reasonable assumption because it usually will be the most likely. But evolution may not necessarily have occurred following a minimum path, because the same change instead may have occurred independently along different branches, and some changes may have involved intermediate steps.

Not all evolutionary changes, even those that involve a single step, may be equally probable. For example, among the four nucleotide bases in DNA, cytosine (C) and thymine (T) are members of a family of related molecules called pyrimidines; likewise, adenine (A) and guanine (G) belong to a family of molecules called purines. A change within a DNA sequence from one pyrimidine to another ($C \leftrightarrow T$) or from one purine to another ($A \leftrightarrow G$), called a transition, is more likely to occur than a change from a purine to a pyrimidine or the converse (G or $A \leftrightarrow C$ or T), called a transversion. Parsimony methods take into account different probabilities of occurrence if they are known.

Maximum-parsimony methods are related to cladistics, a very formalistic theory of taxonomic classification that is extensively used with morphological and paleontological data. The critical feature in cladistics is the identification of derived shared traits, called synapomorphic traits. A synapomorphic trait is shared by some taxa but not others because the former inherited it from a common ancestor that acquired the trait after its lineage separated from the lineages that went to the other taxa. In the evolution of carnivores, for example, domestic cats, tigers, and leopards are clustered together because they possess retractable claws, a trait acquired after their common ancestor branched off from the lineage that led to dogs, wolves, and coyotes.

MAXIMUM-LIKELIHOOD METHODS

Maximum-likelihood methods seek to identify the most likely tree, given the available data. They require that an evolutionary model be identified that would make it possible to estimate the probability of each possible individual change. For example, as is mentioned in the preceding section, transitions are more likely than transversions among DNA nucleotides, but a particular probability must be assigned to each. All possible trees are considered. The probabilities for each individual change are multiplied for each tree. The best tree is the one with the highest probability (or maximum likelihood) among all possible trees.

Maximum-likelihood methods are computationally expensive when the number of taxa is large because the number of possible trees (for each of which the probability must be calculated) grows factorially with the number of taxa. With 10 taxa, there are about 3.6 million possible trees; with 20 taxa, the number of possible trees is about 2 followed by 18 zeros (2×10^{18}). Even with powerful computers, maximum-likelihood methods can be prohibitive if the number of taxa is large, because of the considerable computer time required. Heuristic methods exist in which only a subsample of all possible trees is examined, and thus an exhaustive search is avoided.

STATISTICAL APPRAISAL OF EVOLUTIONARY TREES

The statistical degree of confidence of a tree can be estimated for distance and maximum-likelihood trees. The most common method is called bootstrapping. It consists of taking samples of the data by removing at least one data point at random and then constructing a tree for the new data set. This random sampling process is repeated hundreds or thousands of times. The bootstrap value for each node is defined by the percentage of cases in which all species derived from that node appear together in the trees. Nodes with bootstrap values above 90% are regarded as statistically strongly reliable; those with bootstrap values below 70% are considered unreliable.

Molecular Phylogeny of Genes

The methods for obtaining the nucleotide sequences of DNA have enormously improved in recent years and have become largely automated. Many genes have been sequenced in numerous organisms, and the complete genome has been sequenced in many species, ranging from humans to bacteria. The use of DNA sequences has been particularly rewarding in the study of gene duplications. The genes that code for the hemoglobins in humans and other mammals provide a good example.

Knowledge of the amino acid sequences of the hemoglobin chains and of myoglobin, a closely related protein, has made it possible to reconstruct the evolutionary history of the duplications that gave rise to the corresponding genes. But direct examination of the nucleotide sequences in the genes that code for these proteins has shown that the situation is more complex, and also more interesting, than it appears from the protein sequences.

DNA sequence studies on human hemoglobin genes have shown that their number is greater than previously thought. Hemoglobin molecules are tetramers (molecules made of four subunits) that consist of two polypeptides (relatively short protein chains) of one kind and two of another kind. In embryonic hemoglobin E, one of the two kinds of polypeptide is designated ε; in fetal hemoglogin F, it is γ; in adult hemoglobin A, it is β; and in adult hemoglobin A_2, it is δ (hemoglobin A makes up about 98% of human adult hemoglobin, and hemoglobin A_2 about 2%). The other kind of polypeptide in embryonic hemoglobin is ζ; in both fetal and adult hemoglobin, it is α. The genes that code for the first group of polypeptides (ε, γ, β, and δ) are located on chromosome 11; the genes that code for the second group of polypeptides (ζ and α) are located on chromosome 16.

There are additional complexities. Two γ genes exist (known as G_γ and A_γ), as do two α genes (α_1 and α_2). Furthermore, there are two β pseudogenes ($\psi\beta_1$ and $\psi\beta_2$) and two α pseudogenes ($\psi\alpha_1$ and $\psi\alpha_2$), as well as a ζ pseudogene. These pseudogenes are very similar in nucleotide sequence to the corresponding functional genes, but they include terminating codons and other mutations that make it impossible for them to yield functional hemoglobins.

The similarity in the nucleotide sequences of the polypeptide genes and pseudogenes of both the α and β gene families indicates that they are all homologous—that is, that they have arisen through various duplications and subsequent evolution from a gene ancestral to all. Moreover, homology also exists between the nucleotide sequences that separate one gene from another. The evolutionary history of the genes for hemoglobin and myoglobin is summarized in Figure 7.

Multiplicity and Rate Heterogeneity

Cytochrome-c consists of only 104 amino acids, encoded by 312 nucleotides. Nevertheless, this short protein stores enormous evolutionary information, which made possible the fairly good approximation, shown in Figure 6, to the evolutionary history of 20 very diverse species over a period longer than 1 billion years. Cytochrome-c is a slowly evolving protein. Widely different species have in common a large proportion of the amino acids in their cytochrome-c, which makes possible the study of genetic differences between organisms only remotely related. For the same reason, however, comparison

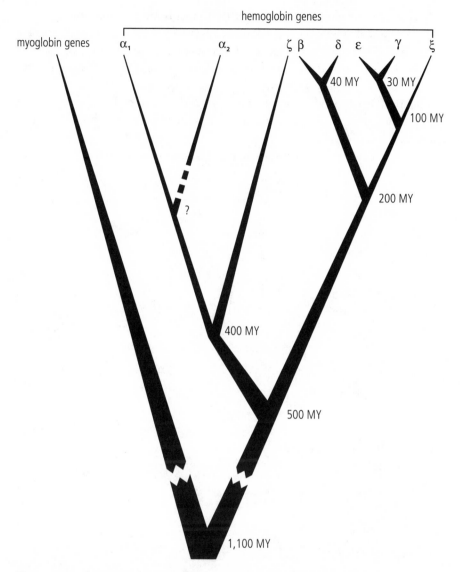

Figure 7. Evolutionary history of the globin genes. The forks indicate points at which ancestral genes duplicated, giving rise to new gene lineages. The approximate times when these duplications occurred are indicated in millions of years (MY) ago. The time when the duplication of α_1 and α_2 occurred is uncertain.

of cytochrome-c molecules cannot determine evolutionary relationships between closely related species. For example, the amino acid sequence of cytochrome-c in humans and chimpanzees is identical, although they diverged 6–8 million years ago; between humans and rhesus monkeys, which diverged from their common ancestor 35–40 million years ago, it differs by only one amino acid replacement.

Proteins that evolve more rapidly than cytochrome-c can be studied in order to establish phylogenetic relationships between closely related species. Some proteins evolve very fast; the fibrinopeptides—small proteins that are involved in the blood-clotting process—are suitable for reconstructing the phylogeny of recently evolved species, such as closely related mammals. Other proteins evolve at intermediate rates; the hemoglobins, for example, can be used to reconstruct evolutionary history over a fairly broad range of time (see Figure 8).

One great advantage of molecular evolution is its multiplicity, as noted

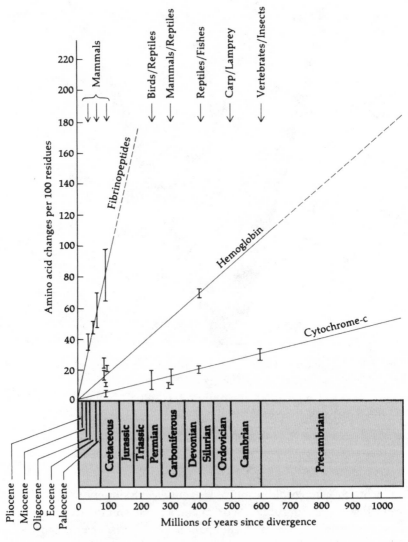

Figure 8. Rates of molecular evolution of three different proteins. (After Dickerson 1971.)

earlier. Within each organism are thousands of genes and proteins; these evolve at different rates, but every one of them reflects the same evolutionary events. Scientists can obtain greater and greater accuracy in reconstructing the evolutionary phylogeny of any group of organisms by increasing the number of genes investigated. The range of differences in the rates of evolution between genes opens up the opportunity of investigating different sets of genes for achieving different degrees of resolution in the tree of life; slowly evolving genes can be used to study remote evolutionary events. Even genes that encode slowly evolving proteins can be useful for reconstructing the evolutionary relationships between closely related species by examination of the redundant codon substitutions (nucleotide substitutions that do not change the encoded amino acids), the introns (noncoding DNA segments interspersed among the segments that code for amino acids), or other noncoding segments of the genes (such as the sequences that precede and follow the encoding portions of genes); these generally evolve much faster than the nucleotides that specify the amino acids.

The Molecular Clock of Evolution

One conspicuous attribute of molecular evolution is that differences between homologous molecules can readily be quantified and expressed as, for example, proportions of nucleotides or amino acids that have changed. Rates of evolutionary change can therefore be more precisely established with respect to DNA or proteins than with respect to phenotypic traits of form and function. Studies of molecular evolution rates have led to the proposition that macromolecules may serve as evolutionary clocks.

It was first observed in the 1960s that the numbers of amino acid differences between homologous proteins of any two given species seemed to be nearly proportional to the time of their divergence from a common ancestor. If the rate of evolution of a protein or gene were approximately the same in the evolutionary lineages that led to different species, proteins and DNA sequences would provide a molecular clock of evolution. The sequences could then be used to reconstruct not only the sequence of branching events of a phylogeny but also the time when the various events occurred.

Consider, for example, Figure 6. If the substitution of nucleotides in the gene coding for cytochrome-c occurred at a constant rate through time, one could determine the time that elapsed along any branch of the phylogeny simply by examining the number of nucleotide substitutions along that branch. One would need only to calibrate the clock by reference to an outside source, such as the fossil record, that would provide the actual geological time that elapsed in at least one specific lineage.

The molecular evolutionary clock is not expected to be a metronomic clock, like a watch or other timepiece that measures time exactly, but a stochastic clock, like radioactive decay. In a stochastic clock the probability of a certain amount of change is constant (for example, a given quantity of atoms of radium-226 is expected, through decay, to be reduced by half in 1,620 years), although some variation occurs in the actual amount of change. Over fairly long periods of time a stochastic clock is quite accurate. The enormous potential of the molecular evolutionary clock lies in the fact that each gene or protein is a separate clock. Each clock "ticks" at a different rate—the rate of evolution characteristic of a particular gene or protein—but each of the thousands and thousands of genes or proteins provides an independent measure of the same evolutionary events.

Evolutionists have found that the amount of variation observed in the evolution of DNA and proteins is greater than is expected from a stochastic clock—in other words, the clock is "overdispersed," or somewhat erratic. The discrepancies in evolutionary rates along different lineages are not excessively large, however. So it is possible, in principle, to time phylogenetic events with as much accuracy as may be desired, but more genes or proteins (about two to four times as many) must be examined than would be required if the clock were stochastically constant. The average rates obtained for several proteins taken together become a fairly precise clock, particularly when many species are studied and the evolutionary events involve long time periods (on the order of 50 million years or longer).

This conclusion is illustrated in Figure 9, which plots the cumulative number of nucleotide changes in seven proteins against the dates of divergence of 17 species of mammals (16 pairings) as determined from the fossil record. The overall rate of nucleotide substitution is fairly uniform. Some primate species (represented by the points below the line at the lower left of Figure 9) appear to have evolved at a slower rate than the average for the rest of the species. This anomaly occurs because the more recent the divergence of any two species, the more likely it is that the changes observed will depart from the average evolutionary rate. As the length of time increases, periods of rapid and slow evolution in any lineage are likely to cancel one another out.

Evolutionists have discovered, however, that molecular time estimates tend to be systematically older than estimates based on other methods and, indeed, to be older than the actual dates. This is a consequence of the statistical properties of molecular estimates, which are asymmetrically distributed. Because of chance, the number of molecular differences between two species may be larger or smaller than expected. But overestimation errors are unbounded, whereas underestimation errors are bounded, since they cannot be smaller than zero. Consequently, a graph of a typical distribution of estimates of the age when two species diverged, gathered from a number of different genes, is skewed from the normal bell shape, with a large number of estimates of younger age clustered together at one end and a long tail of older-age estimates trailing away toward the other end. The average of the estimated times thus will consistently overestimate the true date (Figure 10). The overestimation bias becomes greater when the rate of molecular

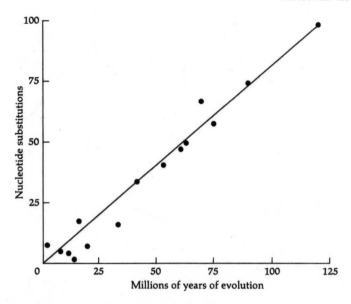

Figure 9. Rate of nucleotide substitution over paleontological time. Each of the 15 dots marks the time at which a pair of species diverged from a common ancestor *(horizontal scale)* and the number of nucleotide substitutions, or protein changes, that have occurred since the divergence *(vertical scale)*. The solid line drawn from the origin to the outermost dot gives the average rate of substitution. (After Fitch 1976.)

evolution is slower, the sequences used are shorter, and the time becomes increasingly remote.

Conclusion

Molecular biology emerged as a discipline 100 years after Darwin, following the 1953 discovery of the double-helix structure of DNA, the hereditary chemical. Molecular biology provides the strongest evidence of biological evolution and makes it possible to reconstruct evolutionary history with as much detail and precision as anyone might want.

Molecular biology proves evolution in two ways: first, by showing the unity of life in the nature of DNA and the workings of organisms at the level of enzymes and other protein molecules; second, and most important for evolutionists, by making it possible to reconstruct evolutionary relationships that were previously unknown, and to confirm, refine, and time all evolutionary relationships from the universal common ancestor up to all living organisms.

As pointed out earlier in this chapter, all organisms are made up of the same basic components, which, moreover, are similarly assembled and used.

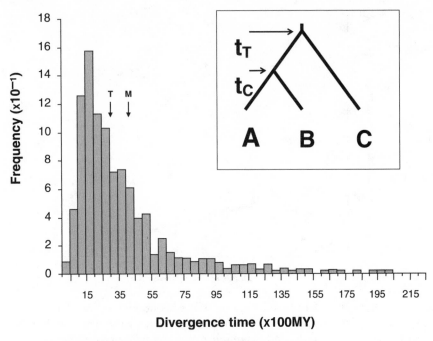

Figure 10. Skewing and age bias in estimating molecular dates. The inset shows the tree topology for lineages A, B, and C; t_C and t_T represent, respectively, calibration and target times. The main panel shows a frequency distribution of 1,000 estimates of the divergence time between lineages C and AB in the inset, set to have occurred 3 billion years ago and obtained with the use of a short (75 residues), slow-evolving (one replacement per site per 10^{10} years) protein and with the use of the split between A and B, set to 300 million years ago, as a calibration point. T and M represent actual (3 billion years) and estimated mean (4.084 billion years) times. (Modified from Rodríguez-Trelles et al. 2002.)

In all organisms, from bacteria to humans, the instructions that guide the development and functioning of organisms are encased in the same hereditary material, DNA, which provides the instructions for the synthesis of proteins. The thousands of diverse proteins that exist in organisms consist of the same 20 amino acids in all organisms, from bacteria to plants and to animals. The genetic code, by which the information contained in the DNA of the cell nucleus is passed on to proteins, is shared by all sorts of organisms. All organisms use similar metabolic pathways—sequences of biochemical reactions—to produce energy and to make up the cell components. The unity of life reveals the genetic continuity and common ancestry of all organisms.

The degree of similarity in the sequence of nucleotides in the DNA (and thus the sequence of amino acids in the encoded proteins) makes it possible to reconstruct evolutionary history. Each of the thousands of genes and thousands of proteins present in an organism provides an independent test of the organism's evolutionary history. The evolutionary relationships among all organisms, from bacteria and protozoa to plants, animals, and humans can now be reconstructed with as much detail as wanted by means of molecular biology. Darwin could not have hoped for more.

BIBLIOGRAPHY

Avise, J. C. 2004. *Molecular Markers, Natural History, and Evolution.* 2nd ed. Sunderland, MA: Sinauer Associates.

Ayala, F. J., ed. 1976. *Molecular Evolution.* Sunderland, MA: Sinauer Associates.

Baker, A. J., ed. 2000. *Molecular Methods in Ecology.* Oxford: Blackwell.

Carroll, S. B. 2006. *The Making of the Fittest: DNA and the Ultimate Forensic Record of Evolution.* New York: W. W. Norton.

Dickerson, R. E. 1971. The structure of cytochrome *c* and the rates of molecular evolution. *Journal of Molecular Evolution* 1: 26–45.

Felsenstein, J. 2004. *Inferring Phylogenies.* Sunderland, MA: Sinauer Associates.

Fenchel, T. 2002. *The Origin and Early Evolution of Life.* Oxford: Oxford University Press.

Fitch, W. M. 1976. Molecular evolutionary clocks. In F. J. Ayala, ed., *Molecular Evolution,* 160–178. Sunderland, MA: Sinauer Associates.

Fitch, W. M., and E. Margoliash. 1967. Construction of phylogenetic trees. *Science* 155: 279–284.

Graur, D., and W.-H. Li. 2000. *Fundamentals of Molecular Evolution.* 2nd ed. Sunderland, MA: Sinauer Associates.

Hall, B. G. 2004. *Phylogenetic Trees Made Easy.* 2nd ed. Sunderland, MA: Sinauer Associates.

Hazen, R. M. 2005. *gen•e•sis: The Scientific Quest for Life's Origin.* Washington, DC: Joseph Henry Press.

Lynch, M. 2007. *The Origins of Genome Architecture.* Sunderland, MA: Sinauer Associates.

Nei, M., and S. Kumar. 2000. *Molecular Evolution and Phylogenetics.* Oxford: Oxford University Press.

Rodríguez-Trelles, F., R. Tarrío, and F. J. Ayala. 2002. A methodological bias toward overestimation of molecular evolutionary time scales. *Proceedings of the National Academy of Sciences USA* 99: 8112–8115.

Woese, C. R. 1998. The universal ancestor. *Proceedings of the National Academy of Sciences USA* 95: 6854–6859.

Evolution of the Genome

Brian Charlesworth and
Deborah Charlesworth

Changes in the genetic information stored in the genome are the ultimate basis of evolution. Much of evolutionary biology involves studies of the observable characteristics of organisms and the heritable changes on which their evolution depends. But genomes themselves also evolve. The genomes of contemporary species reflect billions of years of evolution, and they are still evolving, often surprisingly rapidly. Data on genome organization have been accumulating since the era of classical genetics, starting with the discovery that genes are carried in the chromosomes of cells and can be mapped genetically and physically to specific locations in the chromosomes. The later discovery that chromosomes are made of nucleic acids has led to increasingly fine-scale physical maps of genomes, most recently the complete sequences of (parts of the) genomes of (single individuals from) increasing numbers of species, now including several species of animals (e.g., *C. elegans* Genome Sequencing Project 1999; International Human Genome Sequencing Consortium 2001; Mouse Genome Sequencing Project 2002; *Drosophila* 12 Genomes Consortium 2007) and plants (e.g., The Arabidopsis Genome Initiative 2001), as well as over 80 bacterial species (Sharp et al. 2005). Genomes of many more species will certainly be sequenced in the next few years, given the rapid advances being made in sequencing technology.

This information has generated renewed interest in genome evolution, which can now be described in unprecedented detail, and two major books have recently been devoted to it (Burt and Trivers 2006; Lynch 2007). With such extensive data in hand, we can hope to identify some of the main processes involved in genome evolution. Models of many of these processes exist, and some general themes are clear. Genome evolution often involves various opposing forces, so that the organization of the genome of a given species, or of a given genomic region, may depend on situations that weaken or strengthen one force or another. Particularly interesting situations arise when the evolutionary interests of individual genome features conflict with selection that acts on the organisms themselves. Because of the complex interplay of different forces, changes often occur in one direction (e.g., an increase in genome size) and then later reverse.

As with other types of evolutionary change, we need to understand both the types of changes in genome organization and composition that arise within individuals (mutations and other molecular changes that are the raw material of evolution) and also how a trait carried initially by one individual can spread through the whole population, leading to evolution of the genome of the species. There are two main causes of this spread, just as for other evolutionary processes. First, deterministic forces, particularly natural selection, may cause individuals with a given trait to leave more descendants than average; as we shall see, such Darwinian natural selection at the level of individuals is not the only deterministic force that acts on genomes. A second factor is the random process of *genetic drift,* which is especially relevant for components of genomes that do not affect organismal functions. Because populations are finite in size, genetic variants that are subject to sufficiently weak deterministic forces will experience random fluctuations in their frequencies over generations, sometimes causing an initially rare variant to become established (fixed) in all individuals of a species. The rate of genetic drift is controlled by the *effective population size,* which is usually much smaller than the number of breeding individuals in a species (Kimura 1983). Even deleterious variants can become fixed by drift if the intensity of selection is of the order of the reciprocal of the effective population size or less (Kimura 1983). The relative magnitudes of drift and deterministic forces therefore control the direction and speed of evolutionary changes in genomes; this has recently been strongly emphasized as a causal factor in genome evolution by Lynch (2007).

Before describing examples of hypotheses and tests, we first outline some important features of genome organization and size and then discuss evolutionary factors that may have shaped these properties.

The Composition and Size of the Genome

THE GENETIC MATERIAL

The chemical basis of the genetic material—a genome's most basic feature—is a long strand of polynucleotide, in which four different nucleotide *bases* linked by sugar and phosphate groups store the genetic information (Lewin 2008). In all cellular organisms, from bacteria to mammals, the polynucleotide is double-stranded DNA (the famous double helix), but in viruses it may be double-stranded DNA (e.g., the T4 virus that infects bacteria), single-stranded DNA (such as another bacterial virus, φX174), or RNA (e.g., influenza virus). The two strands of double-stranded DNA molecules are complementary: where the base present in the sequence of one strand is adenine, the other strand has thymine, and where the first strand has guanine, the complementary strand has cytosine. The composition of one strand can thus be described in terms of the frequencies of the four nucleotide bases

(A, T, G, and C), while that of the double structure is specified by the frequencies of the base pairs AT and GC. RNA is generally single stranded, with a different sugar residue from that in DNA and with uracil bases instead of thymines.

Although the origins of the genetic code are unknown, self-replicating RNA molecules may have been the earliest form of "life," but the transition to genomes that used the much more stable DNA occurred long ago (Joyce 1989). Double-stranded DNA also allows damage to be corrected by copying from the undamaged complementary strand. Since DNA largely lacks RNA's ability to act as a catalyst, this change must have gone along with the evolution of a means of transcribing one strand of a gene's DNA into RNA. In present-day organisms the sequence of bases in the "messenger" RNA transcribed from a gene determines the sequence of the amino acids in the protein chain encoded by the gene (Figure 1). There are also many RNA molecules of direct functional significance, which are transcribed (copied) from DNA but not translated into an amino-acid sequence.

ORGANIZATION OF THE GENETIC MATERIAL

Many prokaryotes' genomes (bacteria and viruses) consist of just a single DNA or RNA molecule, often circular (Table 1). Some bacteria (such as *Borellia*, the cause of Lyme disease) have several separate molecules; although

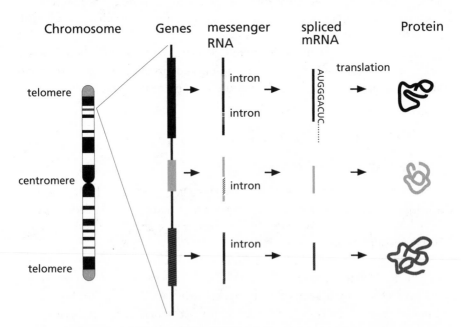

Figure 1. Chromosomes, showing their major features and the banding patterns that can be seen with staining methods. The figure also shows a schematic of a region of the chromosome containing three genes, the messenger RNA produced by them, and the individual proteins encoded by the messenger RNA.

virus genomes are usually single molecules, some RNA viruses (such as influenza) package several different RNA molecules that carry different genes into the same virus particle. Other RNA viruses (e.g., alfalfa mosaic virus) have different RNAs packaged into separate virus particles. These viruses can thus combine different portions of their genome from different parents during multiple infections, a primitive type of sexuality.

Humans and other animals, plants, fungi, and many single-celled organisms belong to the eukaryote division of life, which is defined by having the DNA organized into linear chromosomes carried within the nuclei inside cells (Figure 1). The chromosomes have specialized regions *(telomeres)* to deal with the problem of replicating the ends of double-stranded DNA molecules, and *centromeres* that are required for correct chromosome transmission to the daughter cells during cell division (Lewin 2008). Many primitive eukaryotes, such as the green alga *Chlamydomonas,* have a primarily haploid genome, with just one copy of each of the different chromosomes. When cells mate, there is a brief diploid stage, with a chromosome set contributed by each mating partner. This undergoes a special form of cell division *(meiosis)*

Table 1. Genome sizes

Organism	Genome	Gene number	Genome size
Viruses			
Plant viroids	ssRNA	0	300 bp
Influenza virus	ssRNA	12	13.5 kb
Reovirus	dsRNA	22	23 kb
Phage φX174	ssDNA	11	5.4 kb
SV40	dsDNA	6	5 kb
Vaccinia virus	dsDNA	300	187 kb
Bacteria			
E. coli	dsDNA	4,400	4.6 Mb
Mycoplasma genitalium	dsDNA	500	0.6 Mb
Nuclear genomes of eukaryotes			
Yeast *(Saccharomyces cerevisiae)*		6,000	12 Mb
Nematode *(Caenorhabditis elegans)*		19,000	100 Mb
Fruit fly *(Drosophila melanogaster)*		14,000	140 Mb
Fish *(Fugu rubripes)*		22,000 (?)	400 Mb
Salamander		30,000 (??)	90,000 Mb
Mammal *(Homo sapiens)*		20,000 (?)	3,000 Mb
Cress *(Arabidopsis thaliana)*		27,000	100 Mb
Maize *(Zea mays)*		50,000 (??)	5,000 Mb
Fritillary *(Fritillaria assyriaca)*		27,000 (??)	120,000 Mb

Note: Abbreviations: bp, basepairs; ss, single stranded; ds, double stranded; kb, kilobases; Mb, megabases.

? means that the gene number is only approximately known from the genome sequence,

?? that it is based on incomplete evidence or analogy with sequenced genomes of similar organisms.

to restore the haploid state. In this division, genes from the parental cells are recombined through reciprocal exchanges of portions of their chromosomes (*crossing over:* see Lewin 2008 and Figure 2). In animals and higher plants the diploid stage predominates, and the haploid state is often just a brief part of the reproductive process.

Chromosome numbers in the haploid set vary widely among different species: 16 in the case of *Chlamydomonas*, 5 in *Drosophila*, 23 in humans, but sometimes just a single chromosome (in a species of ant and an *Ascaris* roundworm). Changes between species in chromosomal structure are common, involving *duplications* and *deletions* of genetic material, *inversions* of parts of chromosomes (Figure 3), and *translocations* (exchanges of material between different chromosomes). Chromosome rearrangements occur infrequently in evolution (roughly one per genome per 10 million years in mammals, although some groups, such as rodents, have rates 10 times higher: Eichler and Sankoff 2003). Over evolutionary time, the order of genes within a chromosome gradually becomes scrambled by successive inversions. Several human chromosomes seem to have remained unchanged in gene content

Crossing over

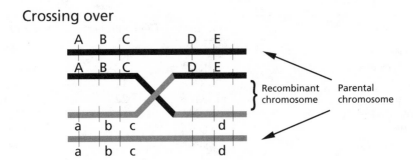

Unequal crossing over between duplicated or repeated genes

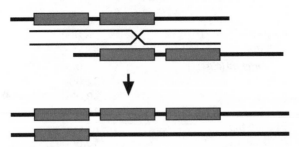

Figure 2. *Top:* Crossing over in a region of a chromosome containing five genes, A to E, showing the resulting recombinant combinations of the genes (indicated by upper- and lowercase letters and also by black and gray lines for the two different parental chromosomes). *Bottom:* Crossing over in a region of chromosome containing two duplicated genes that are similar enough that they may mispair, leading to unequal crossing over.

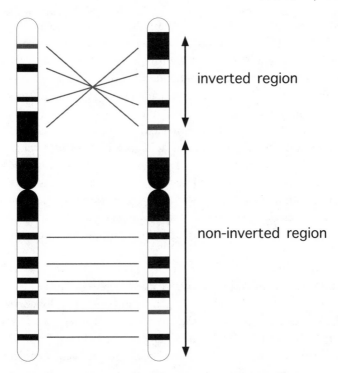

inverted region

non-inverted region

Figure 3. A chromosome inversion. The diagram shows the banding pattern of a chromosome, and the appearance when the top part of the chromosome has been rearranged so that part of the genome in the right-hand chromosome is the other way around from the order in the left one.

(though not order) over the course of some 70 million years, but others are made up of separate blocks derived from different ancestral chromosomes.

CODING AND NONCODING SEQUENCES

Genome size is defined as the number of bases in a single strand of the haploid genome. A genome has two main components: coding sequences (which are transcribed into messenger RNA and encode proteins) and noncoding sequences (Lewin 2008). Noncoding sequences include both *intergenic sequences* and *introns*. Introns are segments of sequences within genes that are transcribed but are removed by *splicing* before the messenger RNA is translated; in eukaryotes, intron splicing involves a protein-RNA machine known as the *spliceosome*. Intergenic sequences and some introns include regions that are responsible for controlling gene expression by determining the levels and timing of messenger RNA production from nearby genes. These are noncoding but functional parts of genomes.

Even with complete genome sequences, it is difficult to count a genome's genes because coding sequences are almost indistinguishable from other sequences. The task is a needle-in-the-haystack problem because of the low proportion of coding sequences in the genomes of many multicellular organisms (estimated to be about 1.5% of human DNA; see International Human Genome Sequencing Consortium 2001). The triplet genetic code used by messenger RNA, in which three adjacent nucleotides *(codons)* correspond to each amino acid of the protein encoded, and certain triplets—the *stop codons*—signify the end of translation, can suggest candidates for sequences that encode amino acid sequences. The nucleotides in the third positions of many codons do not affect the amino acid encoded, since several different codons correspond to a given amino acid (the code is *degenerate*) (Lynch 2007; Lewin 2008). If the sequence of a given genome region is compared between different species, a predominance of differences at every third position can reveal coding sequences; in noncoding sequences no such pattern will appear. Nevertheless, estimates of gene numbers remain uncertain for the genomes of multicellular organisms, and these estimates are constantly being updated.

Genome sizes of different types of organisms differ widely both in the numbers and lengths of coding sequences and in the amount of each type of noncoding sequences (Table 1). Total gene numbers generally correlate with the numbers and sizes of introns and the amount of intergenic DNA (Lynch and Conery 2003; Lynch 2007). The small genomes of prokaryotes have very little space between adjacent genes, and they lack spliceosomal introns, although their coding sequences may be interrupted by mobile sequences that can splice themselves out of the message (Lewin 2008). Unicellular eukaryotes with compact genomes, such as yeast, have genes with few, small introns and little intergenic DNA. In contrast, the largest genomes of eukaryotes tend to have large introns and much intergenic DNA (Lynch and Conery 2003; Lynch 2007).

REPETITIVE SEQUENCES IN NONCODING REGIONS

In addition to functional sequences such as promoters and enhancers involved in the regulation of gene activity (Lewin 2008), noncoding regions of genomes also include sequences that seem less likely to have important functions for the cells, especially sequence types that occur repeatedly in the genome. We will concentrate here on the two major components of genomes of multicellular organisms and will not describe other types of repeats, such as minisatellites and microsatellites (see B. Charlesworth et al. 1994).

Satellite DNA

Satellite DNA consists of long tandem arrays of highly repeated, nontranscribed sequences; the units in such arrays vary in length from five to several hundred base pairs (bp) and are usually rich in the bases A and T. Satellite DNA forms the major component of *heterochromatin,* the blocks of unusually deeply staining material visible in dividing cells' chromosomes, usually concentrated near their centromeres and (in some species) telomeres (John and Miklos 1988; Elgin and Grewal 2003). They can make up astonishing amounts of the DNA of some organisms. For example, one chromosome of the fruit fly *Drosophila melanogaster* (the *X chromosome,* involved in sex determination) has about 20 million bases (Mb) of satellite DNA, about half of them composed of a block of repeated 359 bp units. Functional genes are sparser in heterochromatin than in the rest of the genome (the *euchromatin*). The amount and sequence composition of satellite DNA may vary widely among different species. For instance, about 40% of the DNA of *D. virilis* is made up of four short satellite sequences, which are quite different from those of its distant relative *D. melanogaster* and contribute to its much larger genome size (John and Miklos 1988; *Drosophila* 12 Genomes Consortium 2007). Different kangaroo rat species (Dipodomys) vary in genome size by a factor of 1.6 (a range similar to that in mammals as a whole), mainly because of differences in the amounts of three short repeats (John and Miklos 1988). Variation among species in the amount of satellite sequences can thus cause large differences in genome size, which are unrelated to differences in gene numbers (Cavalier-Smith 1985).

Transposable elements

Substantial proportions of the genomes of many species consist of larger repeats (200 bases–10 kilobases), with a more scattered distribution across the whole genome than satellite sequences. These repeats are largely *transposable elements* (TEs), which can insert new copies of themselves into novel locations in the genome *(transposition)*. In *D. melanogaster,* the nematode worm *Caenorhabditis elegans,* and the plant *Arabidopsis thaliana,* about 5–10% of the total genomic DNA content is made up of TEs (Table 2). In humans the figure is about 45% (International Human Genome Sequencing Consortium 2001), and some plants, such as maize, have even higher TE abundances (Lynch and Conery 2003; Lynch 2007).

Several major types of TE are recognizable in the genomes of a wide range of organisms (Table 2); these differ in their transposition mechanisms. Within each of the major categories, the elements in a given species can be classified into different families, within which the TE sequences are very similar. Unlike the other types of repetitive DNA, most TE families include at least some members with one or more genes that code for proteins required for their transposition, but many also have members with deletions or other mutations in their sequences. Of these, some can be transposed in the presence of complete elements that provide the transposition machinery, while others are fragments whose transposition is disabled entirely. Much of the human genome is made up of such fragments of TEs.

Just as with satellite sequences, large genome-size differences between related species can be due to differences in the amount of TE-derived DNA, without major differences in gene numbers. For example, *D. melanogaster* split from its relative *D. simulans* only about 3 million years ago but has about three times as many TEs (Vieira et al. 2002). Maize has much larger amounts of these sequences than its closest relatives (Bruggmann et al. 2006), and one diploid rice species has double the DNA content of others because of the multiplication of a few TE types (Piegu et al. 2006).

Table 2. TE contents of human and *Drosophila* genomes

TEs in the human and *D. melanogaster* euchromatic genomes

Type	Number of copies	Number of families	% of the genome
Humans			
LTR elements	443×10^3	100	8
SINEs	$1,558 \times 10^3$	3	13
LINEs	868×10^3	3	20
DNA elements	294×10^3	60	3
D. melanogaster			
LTR elements	682	49	2.65
LINEs	486	27	0.97
DNA elements	404	20	0.35

Sources: Human data from International Human Genome Sequencing Consortium (2001); Drosophila data from Kaminker et al. (2002).

Abbreviations: LTR elements, Long Terminal Repeat elements (elements with repeated sequences at each end, in the same orientation, which replicate through reverse transcription into DNA of an RNA copy of their DNA sequence); SINEs, Short Interspersed Nuclear Elements (elements < 500bp long, which replicate through an RNA intermediate but lack repeats and rely on LINEs to provide the enzyme needed for their reverse transcription); LINE, Long Interspersed Nuclear Elements (similar to LTR elements, but > 1kb long and lacking the terminal repeats); DNA elements (elements that replicate without use of an RNA intermediate, usually with short inverted repeats at each end).

Most TEs are found in intergenic DNA or (to a lesser extent) in introns. In some regions of the genome, TEs can be very densely packed, with multiple elements inserted within one another. The portion of the *D. melanogaster* heterochromatin that adjoins the euchromatin is largely made up of material derived from TEs (John and Miklos 1988); the regions around the centromeres of *A. thaliana* chromosomes (Wright et al. 2003) and many intergenic regions in maize (SanMiguel et al. 1996) are similar.

GENOME ADDITIONS

Early genomes must have been very small, but today's organisms have large numbers of genes with diverse functions. An important, though infrequent, process is addition of genomes or genes from other organisms. Bacteria regularly acquire novel genes from other bacterial species (Ochman et al. 2000). Eukaryote cells contain organelles with their own genomes, as well as the nuclear genomes. Green plants and green algae have chloroplasts, and red algae have similar plastids; plant and animal cells both contain mitochondria (Lynch 2007). The genomes of these organelles are of prokaryote origin and arose early in the evolution of eukaryotes, when a bacterial cell was engulfed into a eukaryote cell and took up residence there as an endosymbiont. Sequences of genes from chloroplast genomes are similar to those of photosynthetic cyanobacteria (blue-green algae) and probably evolved around 800 million years ago, while mitochondria evolved much earlier from a group of bacteria known as α-proteobacteria (Gray 1999).

Later endosymbiont events have also occurred. The apicomplexan protozoa (malaria parasites and their relatives) and dinoflagellates contain plastids derived from red algae; presumably a protozoan cell engulfed a red algal cell. Members of the chromophyte group of algae also contain a red-algal-derived chloroplast, which has recently been shown to represent a whole red algal cell. As well as their plastid genome, members of the cryptomonad subgroup still have a vestigial nuclear genome derived from the red algal ancestor (with over 400 genes) and thus possess four genomes, including the mitochondria and the true nuclear genome (Douglas et al. 2001).

GENE DUPLICATION AND GENE LOSS

More frequently, increases in gene numbers come from gene duplications. New genes arise from duplications of parts of the genome or of whole genomes (Eichler and Sankoff 2003; Lynch 2007). Single gene duplications can often be recognized because they frequently occur as *tandem duplications* close to their progenitor gene (Baumgarten et al. 2003). The sizes of such arrays can then change further by unequal crossing over. An exchange between two different members of an array creates one descendant with three copies of the sequence and one with a single copy (see Figure 2).

Some duplicated arrays of genes probably evolved under selection that favored large amounts of the gene product, for example, the large arrays of genes that encode the RNA components of the protein translation mechanism (ribosomes) and the expanded arrays of some insecticide-resistance genes (Ranson et al. 2002). Often, however, higher gene dosage has no advantage, so that natural selection will not act to preserve the functions of duplicated genes. Given enough evolutionary time, an inactive mutant copy of one of the duplicate genes may become fixed in the species by genetic drift. Once a duplicated gene has lost its function, further mutations may occur in it and be fixed in the population by drift. Eventually the gene may lose its protein-coding ability by mutations creating codons, and it then becomes a remnant *pseudogene;* eventually deletions of genetic material may remove all or part of it. Newly duplicated genes are therefore often expected to be quickly lost from genomes (Walsh 1995; Lynch 2007). Loss of gene duplicates means that, after a time, whole genome duplications are difficult to distinguish from partial duplications; in either situation only a fraction of the genome is recognizably duplicated (Wong et al. 2002). With complete genome sequence data that provide large genome stretches of clearly similar gene content, duplicate regions can be recognized, even when most of the genes within the stretches have returned to the single-copy state. Chromosome rearrangements will, however, eventually erase all traces of duplication.

Sometimes, however, duplicated genes may gain new functions by mutation before inactivating mutations occur. For instance, an enzyme that catalyzes a given reaction step in a biochemical pathway may gain a new ability to act on the product of the reaction and generate a new product. Some genes encode proteins with two or more functions, and inactivating mutations that affect only one function may lead to survival of both duplicates, each with just one of the initial functions (Lynch and Conery 2003; Lynch 2007). This increases the number of functional genes in the genome. Families of genes that have arisen by duplication are important components of genomes, and sometimes have many members with related kinds of functions, such as mammals' odorant receptors (Mombaerts 2001) or plant disease-resistance genes (Baumgarten et al. 2003). New genes may also form by fusion of parts of two genes if part of the messenger RNA of one gene is transcribed back into DNA and inserted into the chromosome within or adjacent to another gene (Long 2001). Such reverse transcription and transposition into the genome can also create nonfunctional duplicates without introns *(processed pseudogenes)* (Lynch 2007; Lewin 2008).

RECOMBINATION RATES AND REGIONAL DIFFERENCES WITHIN GENOMES

An important influence on some of the patterns just described is variation in the rate of genetic recombination (crossing over) across the genome. There is essentially no crossing over in heterochromatin. Recombination frequencies also vary within euchromatin. Recombination is frequently almost absent near the junction with the centromeric heterochromatin but is much more frequent in other parts of the chromosome (B. Charlesworth et al. 1994). In some species, such as *Drosophila* and many plants, recombination is suppressed near the tips of chromosomes (B. Charlesworth et al. 1994), but rates rise sharply near the telomeres in human males (International Human Genome Sequencing Consortium 2001). In *C. elegans*, which has no well-defined centromeres, recombination mainly occurs near the chromosome ends and rarely in the central regions of the chromosome arms (Barnes et al. 1995). In mammalian genomes, there is also fine-scale variation in crossover rates, and very localized hot spots, with unusually high recombination rates, are scattered across the genome (Kauppi et al. 2003; Myers et al. 2005).

Sex chromosomes have particularly striking recombination differences. In species with separate sexes and highly evolved mechanisms of sex determination (animals like humans and fruit flies, and some plants), the sex of an individual is determined by genes on a pair of sex chromosomes. Most often females have two X chromosomes, while males have an X from their mother and a morphologically distinctive Y chromosome inherited from their male parent (in contrast, in birds and Lepidoptera, females are the sex with the nonmatching chromosome pair). All or most of the Y chromosome does not cross over with the X and is therefore effectively inherited as an asexual block of DNA (D. Charlesworth et al. 2005). We describe the evolutionary consequences of this later.

Although it is not known how recombination-rate differences are controlled, several genome features have been found to correlate with the recombination rate. One such association is base composition (expressed in terms of the frequency of GC versus AT base pairs, mentioned earlier). In the human genome the mean GC content is about 41%, but the content varies regionally from about 30% to 65% (International Human Genome Sequencing Consortium 2001). The GC contents of genes and intergenic sequences vary concordantly between regions. Although noncoding DNA tends to be much more AT rich than coding sequences, the GC content of third coding positions (where most mutations do not affect protein sequence) usually correlates to some extent with that of adjacent noncoding sequences. Nonfunctional sequences, such as pseudogenes and defective transposable elements, are particularly interesting because they reveal the GC content that will exist purely because of the input of mutations (without natural selection), and they reflect the fact that there is often a mutation pressure toward AT and away from GC (Lynch 2007). In the human, *C. elegans,* and *Drosophila* genomes, GC content is highest for such sites in chromosome regions where crossover rates are high (International Human Genome Sequencing Consortium 2001; Marais and Piganeau 2002).

Another correlate of recombination is intron length; this tends to be lower in high-recombination regions (Comeron and Kreitman 2000; International Human Genome Sequencing Consortium 2001). In addition, in humans and *A. thaliana* (International Human Genome Sequencing Consortium 2001; Wright et al. 2003) gene densities are highest in these regions, but in *C. elegans* the opposite is found (*C. elegans* Genome Sequencing Project 1999).

Transposable-element densities also vary regionally within genomes. In *Drosophila* low-recombination regions have much higher numbers of TE insertions than high-recombination regions (Kaminker et al. 2002). Humans have a more complex pattern: LINE elements (Table 2) are associated with low-recombination AT-rich regions, and SINE elements with GC-rich (high-recombination) regions (International Human Genome Sequencing Consortium 2001).

Evolutionary Factors that Control Genome Size and Organization

FACTORS THAT AFFECT THE NUMBER OF GENES IN A GENOME

Among related species, unusually small genomes are associated with ways of life (parasitism or intracellular symbiosis) where most of the resources required for survival and reproduction are provided by the host. For example, bacteria that make their living as intracellular symbionts or parasites have much smaller gene numbers than free-living relatives (Moran 2003). More extreme reduction has occurred in organelle genomes, which are much smaller than either bacterial or nuclear genomes; they carry only genes needed

for their own replication and protein synthesis and for a few specialized organellar functions. Many genes essential for organelle functions have been
transferred to the nucleus (Gray 1999). Ongoing gene transfer has been detected from plant organelle genomes, as well as transfers that create nonfunctional pseudogenes of organelle genes in the nuclear genomes of many species
(Gray 1999; Bensasson et al. 2001).

Genome-size evolution involves several pressures. There may be advantages of smaller genome size, because a reduced genome can replicate faster,
allowing faster cell division; this can occur through deletions of genes that
are no longer needed, or of intergenic and intronic sequences. The finding
that compact genomes often have overlapping genes, in which the same
stretch of DNA is transcribed in two different reading frames (Douglas et al.
2001; Lewin 2008), suggests that selection favored size reduction in these
cases. But loss of some sequences may be selectively neutral, so that deletions
can be fixed by genetic drift, as outlined earlier for redundant gene duplicates. In nonrepetitive sequences that are unconstrained by selection, the
amount of DNA tends to decrease over time because replication mistakes
seem to cause small deletions more often than insertions of bases (Gregory
2004; Lynch 2007). If selective constraints on genes are removed, they will
then gradually decline in size over time as more deletions than insertions become fixed by genetic drift. Probably both selective and nonselective processes have been operating.

MUTATION RATES AND THE NUMBERS OF GENES

Most mutations with effects on an organism's functions reduce survival or
fertility (fitness, for short). Unless selection is very weak, there will be a balance between the mutational input of new deleterious variants into a population and their elimination by selection. The presence of these mutations
reduces average fitness, compared with a mutation-free population; this *mutational load* depends on the total number of new deleterious mutations that
arise per individual each generation, such that the negative of the natural log-
arithm of the mean fitness of a population is equal to the load (Crow 1993).
For a given mutation rate per gene, there will thus be a greater load if the
number of genes in a genome is larger. RNA genomes lack error-correcting
mechanisms and have extremely high mutation rates per genome per replication, despite their small genome size: around 1 new mutation per replication
cycle for the influenza virus (Drake et al. 1998). Such genomes cannot, therefore, have large numbers of genes, because their fitness would become very
low. DNA-based genomes have error-correcting mechanisms that give them
much lower, but not zero, mutation rates; the *E. coli* genome is estimated to
have a mutation rate of around 10^{-10} per nucleotide per replication and a mutation rate per genome of only 0.0025 (Drake et al. 1998).

Multicellular organisms have large gene numbers and many cell divisions
each generation. Because error correction can never be perfect, and the mutation rate per nucleotide per cell division is quite similar to that in bacteria,

the mutation rate per nucleotide site per generation is much higher than in bacteria. For example, in humans it is estimated to be about 2×10^{-8} (Kondrashov 2003). Given the number of sites that code for amino acids in the human genome, this translates into a mutation rate per genome for changes in protein sequence of about 0.82 new mutations per individual each generation, and around 80% of these are likely to be slightly deleterious, yielding a net mutation rate to deleterious amino acid mutations of about 0.66. An even larger number of deleterious mutations is likely to be caused by mutations in the noncoding portions of the human genome (Asthana et al. 2007), bringing the estimate of the human deleterious mutation rate to well above 1.

High deleterious mutation rates and the resulting low population fitnesses (Crow 1993) may limit possible genome sizes for multicellular organisms. Theoretical models of the evolution of genes that control mutation rates indicate that mutational load favors a reduced mutation rate (Sniegowski et al. 2000). However, there is an opposing pressure: the costs of error-correcting mechanisms, which slow down DNA replication and consume cellular resources. The occurrence of advantageous mutations also disfavors reducing rates to zero, but in sexually reproducing species with genetic recombination between genes, alleles that increase the mutation rate will not remain associated with favorable mutations that they happen to induce, and so they gain little advantage. Mutation rates of sexually reproducing higher eukaryotes are thus likely to reflect the evolutionary balance attained when the cost of reducing the mutation rate equals the advantage of reducing the mutational load (Sniegowski et al. 2000).

In contrast, in organisms that reproduce largely without any evolutionarily effective genetic recombination, such as asexual and highly inbreeding species, an allele that increases the mutation rate will remain permanently associated with any favorable mutations it causes, and will thus increase in frequency as the mutation spreads in the population. This may promote higher mutation rates than in sexual species. This is an example of *genetic hitchhiking* and is observed in bacterial populations that are subject to intense novel selection pressures (Sniegowski et al. 2000).

GENOME SIZE IN RELATION TO THE AMOUNT OF NONCODING DNA

Genome-size differences among related taxa of higher eukaryotes often involve large differences in amounts of highly repeated, noncoding satellite DNA, with only minor differences in gene number (Cavalier-Smith 1985; Lynch 2007). In some cases DNA amounts clearly correlate with differences in life history and ecology. For example, slowly developing species tend to have more satellite and other types of noncoding DNA than faster-developing ones, suggesting that the size of the noncoding portion of the genome is not neutral with respect to selection (Cavalier-Smith 1985). A model of opposing forces (like the balance between proofreading and mutation described earlier) provides a plausible and testable explanation of these patterns. Several processes that involve errors in DNA replication and unequal crossing over can amplify arrays by multiples of whole repeated units, so there are forces that increase copy number, sometimes adding DNA to the genome of an individual; genetic drift can then cause it to spread to fixation (B. Charlesworth et al. 1994; Lynch 2007). Selection against high copy number will often oppose this process, but such selection is least effective in species with a small effective population size (Kimura 1983). This suggests that there should be a relation between the effective size of a species (which can be estimated from its level of DNA sequence diversity; Lynch 2007) and the contribution of repetitive sequences to its genome size. Broad comparisons across different taxa support this prediction (Lynch 2007). However, species such as *D. virilis* and *D. melanogaster* differ considerably in their amount of repetitive DNA but seem to have similar effective population sizes (*Drosophila* 12 Genomes Consortium 2007). This suggests that other factors must play a role.

A likely disadvantage of a large amount of satellite DNA is a slower development time, since cell divisions will be slow if it takes longer to replicate genomes with large DNA amounts (Cavalier-Smith 1985). When rapid development is important, the balance is tipped toward lower copy number, but if slow development is not disfavored by selection, selection against the accumulation of repetitive DNA will be weaker, and such sequences can accumulate (Figure 4). Thus large DNA content does not necessarily imply that selection has favored more DNA (Pagel and Johnstone 1992; B. Charlesworth et al. 1994).

Of course, it might sometimes be advantageous to have more noncoding DNA. For instance, more DNA per nucleus might enable organisms with large cells to move RNA more rapidly into the cytoplasm (Cavalier-Smith 1985). How can one distinguish between these possibilities? Experiments that alter the amount of heterochromatin in *D. melanogaster* hardly affect viability or development time, suggesting that large changes in copy numbers (millions of bases) have at most very slight advantages or disadvantages (John and Miklos 1988). Very slight fitness differences can, however, be important in evolution, so this is not conclusive. In comparative tests that examine development times and genome sizes across different species, the balance model predicts that a relationship will be found even after sta-

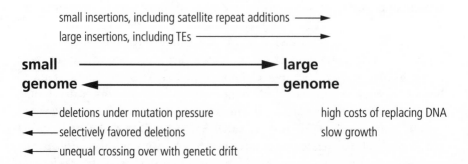

Figure 4. Opposing forces acting on genome-size evolution.

tistical corrections for differences in nuclear volume and cell size, whereas this is not expected if there is an advantage of more DNA. Corrected data from salamanders and newts still show a significant relation (Pagel and Johnstone 1992).

Can we see the footprint of a mutational pressure that increases array sizes, as assumed earlier? The distribution of satellite DNA, with a strong concentration in regions of infrequent genetic recombination, suggests that we can. The argument involves a simple model in which a tandem repeat array undergoes unequal crossing over (explained earlier for tandem gene duplications). Unequal exchange does not change average numbers of unit copies in the array, but it increases the repeat number range. If repeat number is neither advantageous nor disadvantageous, genetic drift may cause the species to become fixed for just a single copy. Because unequal crossing over requires multiple copies, a region that reaches the single-copy state will thereafter remain single copy (unless a new duplication occurs). However, extra copies can be lost again by unequal exchanges. In the absence of other forces that affect array sizes, the combined effects of drift and unequal exchange on a tandem array thus produce an asymmetry: there is an increasing probability over time that the population finds itself with just a single copy, especially if unequal exchange is frequent. If there is some selection against individuals with high array sizes, an equilibrium distribution of array sizes results. Large arrays arise only if crossing over is extremely infrequent (B. Charlesworth et al. 1994). The observed association of satellite arrays with low-recombination genomic regions therefore supports the hypothesized weak force that pushes repeat numbers upward; we see its effects only when the force that acts in the opposite direction (unequal crossing over) is removed.

No functional advantage to having large blocks of tandemly repeated DNA sequences is thus necessary in order to explain the main features of their distribution within genomes and among different species. Moreover, related species often differ greatly in the types and amounts of satellite sequences, which also strongly suggests that no selection is involved (John and Miklos 1988; *Drosophila* 12 Genomes Consortium 2007). However, this does not mean that satellites never have necessary functions in organisms. In *Drosophila,* portions of the heterochromatin are essential for correct chromosomal behavior at meiosis, and human α-satellite sequences seem to be involved in centromere function (Sullivan et al. 2001). But these functions probably did not cause the evolution of these sequences in the first place; more likely, these arrays were first established by the processes modeled earlier and were later co-opted for cell functions.

TRANSPOSABLE ELEMENTS

Most types of transposable elements (TEs) (Table 2) can insert copies of themselves into new positions in the genome, often without loss of the original copy. These transpositions also create a pressure toward increased element numbers over time unless some other forces keep TE numbers in check. For example, in *D. melanogaster* it is estimated that on average at least one new TE is added to a haploid euchromatic genome every 10 generations in laboratory conditions where selection is relaxed (Maside et al. 2000). There are about 1,000 fairly complete TEs in the genome (Kaminker et al. 2002), so the mean transposition rate per potentially active element is about 10^{-4} per generation.

What regulates the abundance of TEs in the genome? There are two contrasting views (B. Charlesworth et al. 1994). First, TEs could be maintained because they are advantageous for their host organisms; for instance, they might occasionally induce favorable mutations by inserting into genes or regulatory sequences. A problem with this idea is that insertions mostly cause harmful, not advantageous, effects, disrupting coding or regulatory sequences. Many known spontaneous deleterious mutations, including some human disease mutations, are due to insertions into coding sequences (Ostertag and Kazazian 2001). Such strongly deleterious mutations are usually rapidly eliminated from populations, and as mentioned earlier, it is indeed observed that in the genome sequences of higher organisms TEs are largely absent from coding sequences, indicating that insertions there are harmful (International Human Genome Sequencing Consortium 2001; Kaminker et al. 2002; Wright et al. 2003). Further evidence that TE movement is disadvantageous to the host is that many species, such as yeast, the mold *Neurospora*, maize, and *Drosophila*, have mechanisms that restrict the rate of movement of TEs (Selker 2002; Aravin et al. 2007). Moreover, the view that TE insertions are advantageous for their hosts' future evolution rests on the idea that the long-term survival of a species is enhanced by the presence of TEs, and it does not explain how TEs become established in the first place.

Second, TEs could be maintained by their ability to spread by self-replication within the genome, coupled with transmission of new copies to the offspring during sexual reproduction, even if they usually have harmful effects on their hosts (the *selfish DNA* hypothesis). On this view TE abundance will depend on the balance between transposition that causes increased copy number versus selection against individuals with large copy numbers (B. Charlesworth et al. 1994; Lynch 2007). This balance implies that the reduction in fitness caused by an element insertion must on average equal the chance that an element transposes. Estimates of this rate are around 10^{-4} per generation in *Drosophila*, showing that this selection pressure must be weak. This has been tested further in *D. melanogaster* by determining element positions on the giant chromosomes of larval salivary glands. The frequencies of TE insertions in the population are nearly always very low at a given place on the chromosome, again implying that insertions are disadvantageous. The strength of selection can be inferred from the frequencies, and it agrees with the estimated transposition rate (B. Charlesworth et al. 1994).

What is the nature of this selection? We have already discussed selection against large genome size, and harmful effects of transpositions on the host. Another disadvantage comes from crossing over between elements in different genomic locations *(ectopic exchange)*. This causes chromosome inversions, translocations, deletions, and duplications (B. Charlesworth et al. 1994). Some rearrangements have few or no harmful effects and can spread in populations because of genetic drift, thereby contributing to evolutionary change in chromosome structure. But many rearrangements disrupt gene functions and are highly deleterious: many human genetic diseases are caused by chromosome rearrangements *(genomic diseases)*, including many cancers (Deininger et al. 2003). If this is an important disadvantage to having high TE abundance, it predicts that TEs should accumulate mainly in genomic regions with low recombination rates, since they are then less likely to cause harmful rearrangements and will therefore not be removed by selection against the products of ectopic exchanges. As mentioned earlier, this is often true. For example, a large proportion of the Y chromosomes of many species consists of TE-derived sequences (B. Charlesworth et al. 1994). However, there are other possible explanations for this pattern, such as the reduced efficiency of selection in regions of the genome with reduced recombination (see later discussion).

TEs may sometimes induce favorable mutations or chromosome rearrangements, and selection will then spread them throughout the population. Some TE-derived sequences appear to regulate gene activity (Deininger et al. 2003) or confer other benefits on their hosts (Brookfield 2003). As we argued for satellite sequences, however, a present-day benefit does not imply that selection promoted the initial spread of these sequences. It is possible that regulatory or even coding functions have been acquired by such TEs long after they were inserted into their present locations and fixed in the population.

MASSIVE TE ACCUMULATION AND POPULATION SIZE DIFFERENCES

Although the rate of increase of TEs by transposition in species such as *D. melanogaster* seems low (one new element in the haploid genome element roughly every 10 generations), copy number will increase rapidly over evolutionary timescales if no opposing force operates: for example, with the rate of transposition estimated earlier, a tenfold increase would take only 23,000 *Drosophila* generations, less than 2,000 years. Large differences in TE abundances among related species, as in the maize and rice examples described earlier, are thus readily explicable if selection pressures against them are sometimes relaxed. Another example is provided by *Drosophila miranda,* in which an autosome became attached to the Y chromosome about 1 million years ago; this has since been transmitted only from father to son, causing this genome region to stop recombining because recombination does not occur in *Drosophila* males. The region has since then accumulated a high density of TE insertions into noncoding sequences (Bachtrog 2003). Genomes of some species (e.g., humans and maize) have a very high overall TE abundance (Table 2), and at many chromosomal sites, TEs may be present at high frequencies (or even in all individuals) (Ostertag and Kazazian 2001), contrasting with the low insertion frequencies in *Drosophila*.

Why are there such large differences between species? Perhaps the much larger amounts of intergenic DNA and larger introns in humans than in

Drosophila mean that insertions have less risk of causing harmful mutations (although this would not be true if intergenic regions' sequences often control gene activity). A more interesting possibility depends on the weakness of selection against each new TE insertion. If TEs rarely excise from a site, genetic drift can sometimes lead to a site becoming irreversibly fixed for an element. In species with low population sizes, such as humans, selection against elements is less effective in opposing genetic drift, and this possibly allows the pressure for element buildup to predominate (Lynch 2007). New insertions into already-occupied sites would not be prevented by selection (there would be no direct harmful effects, just a slightly greater risk of ectopic exchange), and so elements could snowball over time. This may account for the fixed elements in mammalian genomes, but not for the very high TE abundance in maize, a species whose high DNA sequence variability indicates a large population size (Tenaillon et al. 2002). Recombination in maize may be largely confined to genes, so that TEs in intergenic sequences could not be removed by ectopic exchange. Thus elements would build up between genes, as observed (SanMiguel et al. 1996). However, the causation could be the other way round: TEs may simply recombine less than host sequences.

THE EVOLUTIONARY CONSEQUENCES OF GENETIC RECOMBINATION AND POPULATION SIZE

Selection against TEs is only one way in which small population size and lack of recombination can be important for genome evolution. Recombination rates play important roles in numerous evolutionary processes, especially the evolutionary advantages and disadvantages of sexual reproduction. Several differences among genome regions are connected with recombination rate differences, and the effects of low recombination rates are similar to those of small effective population size (Gordo and Charlesworth 2001).

Sex and recombination allow selection to act independently on different sites within the genome. This increases the efficiency of selection. For example, if favorable mutations arise independently, but very rarely, at two sites, the population will rarely contain any individuals that carry double mutants, even if this is the fittest combination (unless the population size is so huge that the second mutation event happens before selection fixes the first favorable mutation). Genetic recombination allows the double-mutant combination to be formed. When there is little or no genetic recombination, or when reproduction is asexual, selection on one gene thus impedes selection acting on other genes, making favorable mutations less likely to spread and deleterious mutations (such as amino acid changes in proteins) more likely to be fixed by genetic drift, particularly when population sizes are low (Gordo and Charlesworth 2001). This process probably contributes to genome reduction in symbiotic bacteria that live within insect cells (recombination between bacteria of these species is prevented by isolation); these reduced genomes also have faster amino acid sequence evolution, as predicted given the lowered ability of selection to eliminate mutations with weakly disadvantageous effects (Moran 2003).

Genes in genome regions that lack genetic recombination should undergo similar evolutionary processes that impede selection and lead to reduced levels of adaptation. This may also contribute to the buildup of TEs in nonrecombining chromosomal regions and to these regions' increased intron sizes (perhaps also in part caused by TE-derived material). Y sex chromosomes strikingly illustrate the disastrous consequences of not recombining (D. Charlesworth et al. 2005). The mammalian Y and X chromosomes share some genes, betraying their common descent from a normal chromosome pair, but the Y lacks most of the more than 1,000 genes on the X—it has lost the genetic information once shared with the X. This is just as predicted for a nonrecombining chromosome: selection acting on hundreds of genes will impede their ability to maintain adaptation. Y-linked genes should thus evolve faster than their X-linked counterparts if the ability of selection to remove disadvantageous variants is impeded; they should evolve more like pseudogenes than like normal genes. Faster Y evolution is indeed detected in analyses of sex-linked genes in mammals (Gerrard and Filatov 2005) and birds (Fridolfsson and Ellegren 2000). A similar acceleration is observed for the genes added to the Y chromosome of *D. miranda,* and polymorphism data confirm that this is due to ineffective selection against deleterious amino acid mutations (Bartolomé and Charlesworth 2006).

Similar but milder effects are observed when we compare other genome regions with different recombination rates. Genes often disproportionately use codons that end in G or C rather than those that end in A or T, so the GC content of third coding positions tends to be higher than that of their introns or the adjacent noncoding sequence. In bacteria, *D. melanogaster, C. elegans,* and *A. thaliana,* there is convincing evidence that this codon usage bias reflects the action of selection, probably because GC-ending codons allow faster and/or more accurate translation of the messenger RNA into the protein sequence (Akashi et al. 1998; Sharp et al. 2005). The selection pressure is, however, very small compared with that acting on most changes to the protein sequences, so that there is only a statistical tendency for GC to be preferred over AT. In an organism like *D. melanogaster,* with a substantial component of the genome in regions near the centromeres and telomeres where recombination is infrequent, the ability of selection to maintain optimal codon usage might be lessened sufficiently that genes in these regions would be less biased than elsewhere in the genome. This accounts for part of the observed relationship between recombination rates and the GC content of coding sequences in *Drosophila* (Marais and Piganeau 2002), and for the weakening of selection for codon usage bias that has been detected for the genes added to the Y chromosome in *D. miranda* (Bartolomé and Charlesworth 2006).

However, the GC content of noncoding sequences in *D. melanogaster* and *C. elegans* also correlates with recombination rates, and GC contents of coding sequences and noncoding sequences are correlated with each other, as mentioned earlier (Marais 2003; Lynch 2007). Nonselective forces, such as rates and directions of mutation, thus probably also vary with recombination rates. Certainly, not all mutations occur equally frequently. As already mentioned, the relatively high AT content of noncoding DNA is caused by mutational bias toward AT. A process that may oppose this is *biased gene conversion;* because of the molecular mechanism of genetic recombination, when one genome has GC at a site and the same site in the other has AT, recombination can produce a very slight excess of GC in the daughter genome (Marais 2003). This will increase GC content in both coding and noncoding sequences, causing GCs to be fixed more frequently than ATs, just as though GCs were favored by selection. Because it is part of the recombination process, biased gene conversion is expected to act mainly in regions with high rates of recombination, leading to a higher GC frequency in such regions. Third positions of codons will have an additional enhancement of GC from the effects of selection for codon bias. Statistical analyses of the GC content of the genome of *D. melanogaster* support this hypothesis, as do data from the highly recombining parts of sex chromosomes of mammals (Marais 2003).

Conclusions

Genome evolution, like other evolutionary changes, is opportunistic. Changes in genome features have consequences of two kinds. First, change may lead to further changes. For instance, chromosome rearrangements affect recombination. A gene can move into a region of lower or higher recombination, affecting the balance between biased gene conversion that pushes GC content up versus mutation pressure in the other direction, so that GC content will start changing. If a rearrangement reduces recombination, as happens when a chromosome arm is joined to a Y chromosome, the decreased ability of natural selection to eliminate weakly disadvantageous changes may ultimately lead to the loss of functionality, or even the physical presence, of genes from this chromosome, as indeed has happened in *D. miranda* (D. Charlesworth et al. 2005).

Second, even when a genome feature evolves without being favored by individual selection, there can occasionally be potentially advantageous effects for organismal functions. For example, TEs that contain sequences that control the expression of their own genes can also affect the activity of genes close to their insertion sites. Initially parasitic selfish DNA (spreading purely as a consequence of selection for the sequences' own ability to replicate) can become "domesticated" as a part of the genome of a species, present in all individuals and functioning in its host, rather like the organelles, which were once independent organisms (Brookfield 2003). This possibility is often overlooked, but it is important to avoid the naive attitude that present-day benefits can explain the origin of all genome features.

Genome evolution is far more complex than straightforward adaptive evolution by natural selection, and many features may have had a long history in genomes before any benefits to the organism evolved. It has even been suggested that introns may initially have spread within genomes when population sizes decreased as multicellular organisms evolved. The presence of introns exposes genes to extra harmful mutations in addition to those that change the protein and controlling sequences (because sites required for splicing must be conserved). It is possible that their presence can be resisted by selection only in species with very large populations. If some mutational pressure increases intron numbers, this could account for introns' rarity in prokaryotes and presence in multicellular eukaryotes (Lynch and Conery 2003; Lynch 2007). Selection may also favor small prokaryote genome size for other reasons, but in principle all opposing forces must play some role.

Again, however, it seems unlikely that mutational pressure explains the presence of all introns in genes of living multicellular eukaryotes, especially because intron size and numbers can decrease, as well as increase (this has happened, for instance, in the small genome of the plant *A. thaliana;* Wright et al. 2002). Whatever the causes of their initial origins, introns have evolved to become domesticated and have acquired cellular functions, as we have suggested for other genome features, and there is increasing evidence for selective constraints on intron and other kinds of noncoding sequences in eukaryote genomes (Halligan and Keightley 2006; Asthana et al. 2007).

Further progress in understanding the causal processes involved in genome evolution is to be expected as more and more detailed comparisons of genomes become possible. The Twelve Genomes Project, in which large parts of the genomes of 12 members of the genus *Drosophila* have been sequenced, is a step in this direction (*Drosophila* 12 Genomes Consortium 2007). Although the constant changes in genomes make it difficult to understand the causes of particular events in genome evolution, sufficiently detailed comparisons of living species should allow us to discern patterns in genomes and to infer the main causes of many changes. Because genome changes are frequent, it should be possible to compare independent cases of similar changes and to discover common factors, similar to the procedures widely used for studying other types of evolutionary changes (e.g., Harvey and Pagel 1991). These provide more rigorous tests of hypotheses than broad comparisons among distantly related taxa.

BIBLIOGRAPHY

Akashi, H., R. M. Kliman, and A. Eyre-Walker. 1998. Mutation pressure, natural selection and the evolution of base composition in Drosophila. *Genetica* 102: 49–60.

The Arabidopsis Genome Initiative. 2001. Analysis of the genome sequence of the flowering plant *Arabidopsis thaliana. Nature* 408: 796–815.

Aravin, A. A., G. J. Hannon, and J. Brennecke. 2007. The Piwi-piRNA pathway provides an adaptive defense in the transposon arms race. *Science* 318: 761–764.

Asthana, S., W. S. Noble, G. Kryukov, C. E. Grant, S. Sunyaev, and J. A. Stamatoyannopoulos. 2007. Widely distributed noncoding purifying selection in the human genome. *Proceedings of the National Academy of Sciences USA* 104: 12410–12415.

Bachtrog, D. 2003. Accumulation of Spock and Worf, two novel non-LTR retrotransposons, on the neo-Y chromosome of *Drosophila miranda. Molecular Biology and Evolution* 20: 173–181.

Barnes, T. M., Y. Kohara, A. Coulson, and S. Hekimi. 1995. Meiotic recombination, noncoding DNA and genomic organization in *Caenorhabditis elegans*. *Genetics* 141: 159–179.

Bartolomé, C., and B. Charlesworth. 2006. Evolution of amino-acid sequences and codon usage on the *Drosophila miranda* neo-sex chromosomes. *Genetics* 174: 2033–2044.

Baumgarten, A., S. Cannon, R. Spangler, and G. May. 2003. Genome-level evolution of resistance genes in *Arabidopsis thaliana*. *Genetics* 165: 309–319.

Bensasson, D., D. X. Zhang, D. L. Hartl, and G. M. Hewitt. 2001. Mitochondrial pseudogenes: Evolution's misplaced witnesses. *Trends in Ecology and Evolution* 16: 314–321.

Brookfield, J. F. Y. 2003. Mobile DNAs: The poacher turned gamekeeper. *Current Biology* 13: R846–R847.

Bruggmann, R., A. K. Bharti, H. Gundlach, J. S. Lai, S. Young, A. C. Pontaroli, F. S. Wei, G. Haberer, C. Fuks, C. G. Du, C. Raymond, M. C. Estep, R. Y. Liu, J. L. Bennetzen, A. P. Chan, P. D. Rabinowicz, J. Quackenbush, W. B. Barbazuk, R. A. Wing, B. Birren, C. Nusbaum, S. Rounsley, K. F. X. Mayer, and J. Messing. 2006. Uneven chromosome contraction and expansion in the maize genome. *Genome Research* 16: 1241–1251.

Burt, A., and R. L. Trivers. 2006. *Genes in Conflict: The Biology of Selfish Genetic Elements*. Cambridge, MA: Harvard University Press.

Cavalier-Smith, T. 1985. *The Evolution of Genome Size*. Chichester, U.K.: John Wiley.

C. elegans Genome Sequencing Project. 1999. How the worm was won. *Trends in Genetics* 15: 51–58.

Charlesworth, B., P. Sniegowski, and W. Stephan. 1994. The evolutionary dynamics of repetitive DNA in eukaryotes. *Nature* 371: 215–220.

Charlesworth, D., B. Charlesworth, and G. Marais. 2005. Steps in the evolution of heteromorphic sex chromosomes. *Heredity* 95: 118–128.

Comeron, J. M., and M. Kreitman. 2000. The correlation between intron length and recombination in Drosophila: Equilibrium between mutational and selective forces. *Genetics* 156: 1175–1190.

Crow, J. F. 1993. Mutation, mean fitness, and genetic load. *Oxford Surveys of Evolutionary Biology* 9: 3–42.

Deininger, P. L., J. V. Moran, M. A. Batzer, and H. H. Kazazian. 2003. Mobile elements and mammalian genome evolution. *Current Opinion in Genetics and Development* 13: 651–658.

Douglas, S., S. Zauner, M. Fraunholz, M. Beaton, S. Penny, X. N. Wu, M. Reith, T. Cavalier-Smith, and U. G. Maier. 2001. The highly reduced genome of an enslaved algal nucleus. *Nature* 410: 1091–1096.

Drake, J. W., B. Charlesworth, D. Charlesworth, and J. F. Crow. 1998. Rates of spontaneous mutation. *Genetics* 148: 1667–1686.

Drosophila 12 Genomes Consortium. 2007. Evolution of genes and genomes on the Drosophila phylogeny. *Nature* 450: 203–218.

Eichler, E. E., and D. Sankoff. 2003. Structural dynamics of eukaryotic chromosome evolution. *Science* 301: 793–797.

Elgin, S. C. R., and S. I. S. Grewal. 2003. Heterochromatin: Silence is golden. *Current Biology* 13: R895–R898.

Fridolfsson, A., and H. Ellegren. 2000. Molecular evolution of the avian *CHD1* genes on the Z and W sex chromosomes. *Genetics* 155: 1903–1912.

Gerrard, D. T., and D. A. Filatov. 2005. Positive and negative selection on mammalian Y chromosomes. *Molecular Biology and Evolution* 22: 1423–1432.

Gordo, I., and B. Charlesworth. 2001. Genetic linkage and molecular evolution. *Current Biology* 11: R684–R686.

Gray, M. W. 1999. Evolution of organellar genomes. *Current Opinion in Genetics and Development* 9: 678–687.

Gregory, T. R. 2004. Insertion-deletion biases and the evolution of genome size. *Gene* 324: 1–34.

Halligan, D. L., and P. D. Keightley. 2006. Ubiquitous selective constraints in the *Drosophila* genome revealed by a genome-wide sequence comparison. *Genome Research* 16: 875–884.

Harvey, P. H., and M. Pagel. 1991. *The Comparative Method in Evolutionary Biology.* Oxford: Oxford University Press.

International Human Genome Sequencing Consortium. 2001. Initial sequencing and analysis of the human genome. *Nature* 409: 861–921.

John, B., and G. L. G. Miklos. 1988. *The Eukaryote Genome in Development and Evolution.* London: Allen and Unwin.

Joyce, G. F. 1989. RNA evolution and the origin of life. *Nature* 338: 217–224.

Kaminker, J. S., C. M. Bergman, B. Kronmiller, J. Carlson, R. Svirskas, E. Frise, D. A. Wheeler, S. E. Lewis, G. M. Rubin, M. Ashburner, and S. E. Celniker. 2002. The transposable elements of the *Drosophila melanogaster* genome: A genomics perspective. *Genome Biology* 3: 0084.1–0084.20.

Kauppi, L., A. Sajantila, and A. J. Jeffreys. 2003. Recombination hotspots rather than population history dominate linkage disequilibrium in the MHC class II region. *Human Molecular Genetics* 12: 33–40.

Kimura, M. 1983. *The Neutral Theory of Molecular Evolution.* Cambridge: Cambridge University Press.

Kondrashov, A. S. 2003. Direct estimates of human per nucleotide mutation rates at 20 loci causing Mendelian diseases. *Human Mutation* 21: 12–27.

Lewin, B. 2008. *Genes IX.* London: Jones and Bartlett.

Long, M. 2001. Evolution of novel genes. *Current Opinion in Genetics and Development* 11: 673–680.

Lynch, M. 2007. *The Origins of Genome Architecture.* Sunderland, MA: Sinauer Associates.

Lynch, M., and J. S. Conery. 2003. The origins of genome complexity. *Science* 302: 1401–1404.

Marais, G. 2003. Biased gene conversion: Implications for genome and sex evolution. *Trends in Genetics* 19: 330–338.

Marais, G., and G. Piganeau. 2002. Hill-Robertson interference is a minor determinant of variations in codon bias across *Drosophila melanogaster* and *Caenorhabditis elegans* genomes. *Molecular Biology and Evolution* 19: 1399–1406.

Maside, X., S. Assimacopoulos, and B. Charlesworth. 2000. Rates of movement of transposable elements on the second chromosome of *Drosophila melanogaster*. *Genetical Research* 75: 275–284.

Mombaerts, P. 2001. The human repertoire of odorant receptor genes and pseudogenes. *Annual Review of Genomics and Human Genetics* 2: 493–510.

Moran, N. A. 2003. Tracing the evolution of gene loss in obligate bacterial symbionts. *Current Opinion in Microbiology* 6: 512–518.

Mouse Genome Sequencing Consortium. 2002. Initial sequencing and comparative analysis of the mouse genome. *Nature* 420: 520–562.

Myers, S., L. Bottolo, C. Freeman, G. McVean, and P. Donnelly. 2005. A fine-scale map of recombination rates and hotspots across the human genome. *Science* 310: 312–324.

Ochman, H., J. G. Lawrence, and E. A. Groisman. 2000. Lateral gene transfer and the nature of bacterial innovation. *Nature* 405: 299–304.

Ostertag, E. M., and H. H. Kazazian. 2001. Biology of mammalian L1 retrotransposons. *Annual Review of Genetics* 35: 501–538.

Pagel, M., and R. A. Johnstone. 1992. Variation across species in the size of the nuclear genome supports the junk-DNA explanation for the C-value paradox. *Proceedings of the Royal Society of London, Series B* 249: 119–124.

Piegu, B., R. Guyot, N. Picault, A. Roulin, A. Saniyal, H. Kim, K. Collura, D. S. Brar, S. Jackson, R. A. Wing, and O. Panaud. 2006. Doubling genome size without polyploidization: Dynamics of retrotransposition-driven genomic expansions in *Oryza australiensis,* a wild relative of rice. *Genome Research* 16: 1262–1269.

Ranson, H., C. Claudianos, F. Ortelli, C. Abgrall, J. Hemingway, M. V. Sharakhova, M. F. Unger, F. H. Collins, and R. Feyereisen. 2002. Evolution of supergene families associated with insecticide resistance. *Science* 298: 179–181.

SanMiguel, P., A. Tikhonov, N. Jin, N. Motchulskaia, A. Zakharov, A. Melakeberhan, P. S. Springer, K. J. Edwards, M. Lee, Z. Avramova, and J. L. Bennetzen. 1996. Nested retrotransposons in the intergenic regions of the maize genome. *Science* 273: 765–769.

Selker, E. U. 2002. Repeat-induced gene silencing in fungi. *Advances in Genetics* 46: 439–450.

Sharp, P. M., E. Bailes, R. J. Grocock, J. F. Peden, and R. E. Sockett. 2005. Variation in the strength of selected codon usage bias among bacteria. *Nucleic Acids Research* 33: 1141–1153.

Sniegowski, P. D., P. J. Gerrish, T. Johnson, and A. Shaver. 2000. The evolution of mutation rates: Separating causes from consequences. *BioEssays* 22: 1057–1066.

Sullivan, B. A., M. D. Blower, and G. H. Karpen. 2001. Determining centromere identity: Cyclical stories and forking paths. *Nature Reviews Genetics* 2: 584–596.

Tenaillon, M. I., M. C. Sawkins, L. K. Anderson, S. M. Stack, J. F. Doebley, and B. S. Gaut. 2002. Patterns of diversity and recombination along chromosome 1 of maize (*Zea mays* ssp. *mays* L.). *Genetics* 162: 1401–1413.

Vieira, C., C. Nardon, C. Arpin, D. Lepetit, and C. Biémont. 2002. Evolution of genome size in Drosophila: Is the invader's genome being invaded by transposable elements? *Molecular Biology and Evolution* 19: 1154–1161.

Walsh, J. B. 1995. How often do duplicated genes evolve new functions? *Genetics* 139: 421–428.

Wong, S., G. Butler, and K. H. Wolfe. 2002. Gene order evolution and paleopolyploidy in hemiascomycete yeasts. *Proceedings of the National Academy of Sciences USA* 99: 9272–9277.

Wright, S. I., N. Agrawal, and T. E. Bureau. 2003. Effects of recombination rate and gene density on transposable element distributions in *Arabidopsis thaliana. Genome Research* 13: 1897–1903.

Wright, S. I., B. Lauga, and D. Charlesworth. 2002. Rates and patterns of molecular evolution in inbred and outbred Arabidopsis. *Molecular Biology and Evolution* 19: 1407–1420.

Zhang, X. H.-F., and L. A. Chasin 2006. Comparison of multiple vertebrate genomes reveals the birth and evolution of human exons. *Proceedings of the National Academy of Sciences USA* 103: 13427–13432.

The Pattern and Process of Speciation

Margaret B. Ptacek and Shala J. Hankison

> The essential bit of evolutionary theory which is concerned with the origin and nature of *species* remains utterly mysterious.
>
> W. Bateson (1922)

Speciation, the evolutionary process by which new and distinct lineages arise and maintain their independent trajectories, has been a perplexing phenomenon ever since Darwin (1859) first proposed that all living organisms have diversified from shared ancestors. There are several reasons why understanding how speciation occurs is challenging. First, most speciation events have occurred in the past, leaving only the end products, species, as a signature of their occurrence. Species, both extant and extinct forms, bear the characteristics associated with their divergence from common ancestors, but in addition, they also carry unique phenotypic and genotypic changes that have occurred since speciation. Disentangling the characteristics associated with the speciation process from those that have evolved since divergence can be particularly difficult.

Second, new species arise through a variety of evolutionary mechanisms (e.g., natural selection, sexual selection, genetic drift, and mutation), and therefore, finding a unifying definition that describes and defines all species is an impossible challenge. Thus much debate has occurred about the nature of species characteristics and how scientists can distinguish species for taxonomic, systematic, and conservation purposes.

Finally, speciation is a series of processes, with a beginning stage of initial divergence, a middle stage wherein species-specific characteristics are refined by various forces of evolution, and an end point at which a new species becomes a completely separate evolutionary lineage on its own trajectory of evolutionary change with the potential for extinction or further diversification into new lineages (Figure 1). Knowing the location of a species along this continuum of speciation has proved to be a perplexing problem for evolutionary biologists who wish to understand how new species arise. Despite this hurdle, the study of speciation is one of the most important and exciting areas of evolutionary biology. A resurgence of interest in speciation by

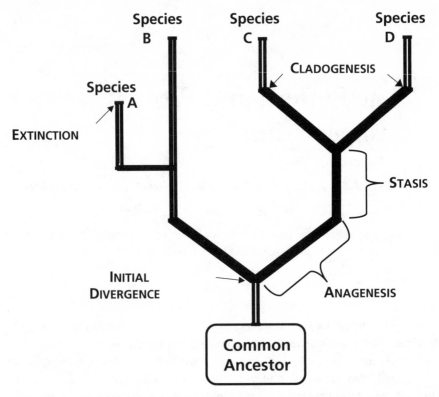

Figure 1. The process of speciation is a continuum. Initial divergence between populations is followed by a period of anagenic change within each new lineage wherein species-specific characteristics are refined by various forces of evolution. In some species long periods with little evolutionary change (stasis) may occur. Once formed, species may further diversify into new lineages (cladogenesis) or may become extinct. The challenge in the study of speciation is uncovering where along this continuum a species actually exists.

evolutionary biologists over the past 15 years has led to many exciting advances. These include the discovery of several *speciation genes* that confer reproductive isolation, a better understanding of the importance of sympatric speciation (speciation without geographic isolation); and a resurrection of interest in how natural selection can lead to the formation of new species as byproducts of adaptive differentiation.

In this essay we outline some of the major challenges in understanding speciation and showcase examples that illustrate the major advances in the study of the origin and diversification of species. We start with the problem of defining species, and how an understanding of the underlying mechanisms that lead to the formation of new species may be the best approach in attempting to define what a species really is. Next, we outline the various types of isolating barriers than can initiate, promote, and complete the speciation process. Finally, we discuss the evolutionary mechanisms that drive the speciation process. Outlining the relative roles that natural selection, sexual selection, genetic drift, and mutation play in promoting phenotypic and genotypic differences between populations is critical to our understanding of how and why new species arise. Our goal is not to cover every potential way in which new species are formed but rather to illustrate with examples the primary means by which new species are formed and maintain their independent evolutionary pathways. Through such studies, evolutionary biologists have begun to resolve the "mysteries of speciation" (Bateson 1922).

The Nature of Species

Species concepts, a set of rules or characteristics used to define a species, abound in the literature (Mayden 1997; Table 1) and have been a subject of long-standing debate among evolutionary biologists. The nature of this problem is centered around a disagreement between evolutionary biologists on pattern-oriented or process-oriented definitions, which, in turn, leads to differences of opinion on where in the speciation process a definition is most applicable (Figure 2). Pattern-oriented definitions use rules or criteria to define species. For example, the Biological Species Concept (BSC), first articulated by Theodosius Dobzhansky (1937) and later championed by Ernst Mayr (1942, 1963), defines species on the basis of reproductive isolating mechanisms. Reproductive isolation may occur through premating or postmating barriers to successful reproduction.

Process-oriented definitions do not use specific characteristics (isolating mechanisms, phenotypic characters) to define a species but rather use a criterion of some level of divergence (evolutionary or ecological) that separates species into distinct entities. For example, the Ecological Species Concept (van Valen 1976) defines species by their unique niche, the subset of ecological requirements and adaptations possessed by a species that has diverged from other species through interspecific competition, predation, and other ecological characteristics that organize communities of species. The "demographic exchangeability" part of the Cohesion Species Concept proposed by Alan Templeton (1989) provides a similar definition of species as having unique ecological properties that separate them from other species.

Evolutionary Species Concepts define species on the basis of unique and separate evolutionary trajectories (e.g., Simpson 1961); again, an emphasis on process, that is, evolutionary divergence, is used to define species. An evolutionary lineage definition was first proposed by Ed Wiley (1978) and has since been formalized with the use of phylogenetic inference by Joel Cracraft and colleagues in the Phylogenetic Species Concept (Cracraft 1983, 1989). Species are defined as separate evolutionary lineages or monophyletic clades that result from divergence from a common ancestor. Thus evolutionary species do not become diagnosable until gene flow with all other lineages has completely ceased (Figure 2).

What do we do with all these different definitions, and why do we need to define species anyway? The organisms that make up discrete entities that

Table 1. Species concepts, their definitions, and references

	Species concept	Definition	References
Pattern oriented	Biological	Groups of interbreeding natural populations that are reproductively isolated from other such groups.	Dobzhansky 1937; Mayr 1942, 1963
	Cohesion*	The most inclusive group of organisms having the potential for genetic and/or demographic exchangeability.	Templeton 1989
	Phenetic	A set of organisms that look similar to each other and are distinct from other such sets.	Sokal and Crovello 1970
	Recognition	A set of populations that share a common fertilization system (specific mate-recognition system).	Paterson 1985
Process oriented	Cohesion*	The most inclusive group of organisms having the potential for genetic and/or demographic exchangeability.	Templeton 1989
	Ecological	A set of organisms adapted to a single ecological niche, evolving separately from lineages outside its range.	van Valen 1976
	Evolutionary	A single lineage (an ancestral-descendant sequence of populations) evolving separately from others and with unique evolutionary tendencies and its own historical fate.	Simpson 1961; Wiley 1978
	Phylogenetic	The smallest diagnosable cluster of individual organisms within which there is a parental pattern of ancestry and descent.	Cracraft 1983, 1989

* Parts of definition consider both pattern and process of speciation.

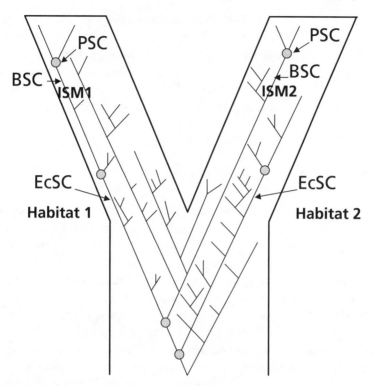

Figure 2. An illustration of how different species concepts define species at different points in the speciation process after the splitting of a single lineage into two. When divergent lineages disperse into different habitats, the Ecological Species Concept (EcSC) will define them as separate species. The evolution of reproductive isolation (ISM1 or ISM2) between the two lineages will define separate species under the Biological Species Concept (BSC). Lineages must represent monophyletic clades (all populations are descended from a single common ancestor) before being recognized as species under the Phylogenetic Species Concept (PSC).

scientists call species seem to know their own boundaries, even if biologists cannot seem to agree. There are, however, some very important reasons for having some form of definition that scientists can use to recognize species, even if different species require different criteria for defining their boundaries. First, defining biological entities as species provides a systematic manner in which to classify diversity and is the basis of modern taxonomy. Second, defining diagnosable groups as species corresponds to the discrete entities that exist in nature. Third, one cannot study the process of how new species arise without having some definition of what a species is in the first place. Fourth, species, as the least inclusive category of diversity, represent the evolutionary history of organisms. Finally, biodiversity is generally measured by the number and kinds of living organisms, and these estimates are usually made at the species level. Thus for a variety of reasons, including categorizing and conserving diversity, as well as understanding its origins, defining species as discrete biological entities is essential to our understanding of the process of speciation and the patterns of biological diversity. No one species concept will accomplish all these purposes. Evolutionary biologists should think about the nature of their question with regard to the "species problem," as well as the natural history of their organism of interest, and then choose the concept that applies best to the solving of that particular problem.

The Nature of Isolating Barriers

The importance of isolating barriers is nearly universal regardless of the species concept being applied. Although the link between reproductive isolation and the BSC, which defines species on the basis of barriers to interbreeding, is obvious, isolating barriers also provide the means through which populations begin to diverge through decreased gene flow, the first step in speciation. Thus even more process-oriented concepts may be linked indirectly to the effects of isolating barriers because these barriers may at some point play a role in cessation of gene flow between evolutionary lineages.

There are numerous ways to classify isolating barriers; however, one of the most widely used divisions is between premating and postmating barriers. These barriers include those that are due to extrinsic factors, such as ecological or behavioral mechanisms, and those that are due to intrinsic factors, such as egg/sperm incompatibilities or genetic incompatibilities between hybrid genomes (Table 2).

Premating barriers include those that prevent or decrease the likelihood of mating between taxa. In effect, premating barriers prevent taxa (or the gametes from those taxa) from meeting. In ecological isolation potential mates do not meet because of differences in habitat use. For example, many insects are limited to feeding, courting, mating, and ovipositioning on specific plant hosts. Thus different species are prevented from interbreeding by the restriction of each to a specific host plant (Futuyma et al. 1994). Alternatively, individuals may be temporally isolated, differing in their timing of reproduction. For example, the sympatric coral species *Montastraea annularis*, *M. faveolata*, and *M. franksi* are temporally isolated through differences in peak spawning times, and differences of only 1 to 2 hours between spawning events provide enough time to dilute gametes and prevent interspecific hybridization (Knowlton et al. 1997). In other cases individuals of different species may be present at the same time and place and still avoid interbreeding. Pollinator isolation is one such method. Characters such as flower shape, reward (pollen versus nectar), and color attract different types of pollinators (e.g., moths, bees, hummingbirds) that may preferentially visit only certain flower types, ensuring that pollen is transferred only within a species.

A final type of extrinsic isolating barrier relies on behavioral differences as premating barriers. Closely related species often differ in particular traits that function as signals in the mating process. Thus even species that exhibit no

Table 2. Isolating barriers (both pre- and postmating)

	Type of barrier	Definition
Premating	Ecological isolation	Species are separated by different hosts or microhabitats.
	Temporal isolation	Reproduction occurs at different times.
	Pollinator isolation	Different species attract different pollinators, or pollen of different species is carried on different body parts.
	Behavioral isolation	Behaviors used as mating signals differ between species.
Postmating, prezygotic	Mechanical isolation	Sperm transfer cannot take place because genitalia are not compatible (or pollen does not adhere) between different species.
	Egg-sperm recognition	Properties of the gametes (often surface proteins) are not compatible between species and prevent fertilization.
	Conspecific sperm precedence	Heterospecific sperm are outcompeted by conspecific sperm, regardless of mating order.
Postmating, postzygotic	Hybrid inviability	Hybrids die before reaching sexual maturity.
	Hybrid sterility	Hybrids are incapable of reproducing.
	Decreased hybrid fitness	Hybrids have lower fitness than either parental species and are selected against.

other pre- or postmating isolating barriers (e.g., sympatric or synchronous breeding) may avoid interbreeding because of preferences for particular mating signals that restrict their mating options to members of a single species. Elaborate male ornaments and courtship displays that differ between closely related species are often thought to result in behavioral isolating barriers (Andersson 1994).

Despite the inherent appeal of barriers that prevent different taxa from mating at the outset, many isolating barriers occur postmating and may happen both pre- and postzygotically. As implied by the name, mating may occur in taxa with postmating, prezygotic barriers, but fertilization does not take place. For example, in both plants and animals conspecific sperm (pollen) precedence is commonly found, where, despite multiple matings, the female (or female structure of plants) only accepts sperm (or pollen) from conspecifics, or, alternatively, conspecific sperm outcompetes heterospecific sperm for fertilization. Anita Diaz and Mark Macnair (1999) studied conspecific pollen precedence in the monkeyflowers *Mimulus guttatus* and *M. nasutus*. On *M. guttatus* conspecific pollen-tube growth was faster than heterospecific pollen-tube growth, ensuring conspecific fertilization and conferring reproductive isolation between these species.

Postmating, postzygotic barriers represent a final category of isolating barriers. Although this is a fairly common mechanism of reproductive isolation, Darwin was concerned with this isolating barrier because it is the basis for the production of sterile hybrids. His dilemma, simply stated, was this: if two taxa evolved from a common ancestor, how could the combination of these taxa form a sterile hybrid? Would that not imply sterility at some time in the evolution of taxa ancestral to present-day species? Today we know, however, that although hybrid sterility is often an important isolating barrier, it is far from the only form of postzygotic isolation, and a better understanding of genetics has yielded a solution to Darwin's problem of sterile hybrids.

As with the study of many evolutionary processes in biology, *Drosophila* species have proved extremely useful in understanding postzygotic barriers. In a classic study of postzygotic barriers, Alfred Sturtevant (1920) showed that crossing *D. melanogaster* females with males of their sister species *D. simulans* yielded only female hybrids, with males dying at the larval to pupal transition. Moreover, the reciprocal cross gave rise to only hybrid males because the females died as embryos. This study provided not only evidence of postzygotic barriers to hybridization but also evidence that these barriers can appear at different developmental stages, that the direction of the hybrid cross can be influential, and that hybrid inviability can span the range from partial to complete (Coyne and Orr 2004).

Although the postzygotic barriers just discussed provide evidence of the variety of intrinsic postzygotic incompatibilities, they give us only a rudimentary understanding of these barriers and their relationship to speciation; that is, they tell us what might happen, but not why or how these barriers evolve. For that, we must turn to genetic models of postzygotic isolation.

Darwin was unaware of genes or gene interactions, but his concern about sterile hybrids was one that troubled evolutionary biologists for some time. The question of how two species can evolve from a common ancestor yet, when mated, form sterile hybrids without at least one of those species passing through a maladaptive form was quite perplexing. A solution to this dilemma is provided by the Dobzhansky-Muller model (although, as noted by Orr [1996], Bateson essentially solved this same problem 25 years earlier). The solution depends upon understanding that genetic incompatibilities that lead to unfit offspring may be ameliorated by multilocus interactions. In the simplest model, consider two loci, AABB (Figure 3). In one species a mutation causes the "a" allele to appear and become fixed, resulting in the aaBB genotype. In the second species a mutation leads to a "b" allele (again becoming

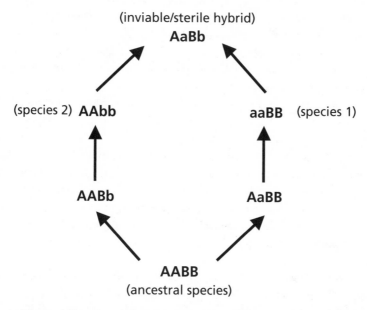

Figure 3. The Dobzhansky-Muller model of intrinsic postmating isolation demonstrates how speciation could proceed unhindered by a detrimental intermediate hybrid stage. The ancestral species diverges into two separate taxa that possess the variant alleles "a" and "b." Fixation of the novel alleles in each taxon results in speciation (species 1 and species 2). Since "a" and "b" are incompatible, mating between the two species results in a hybrid that is sterile, inviable, or otherwise less fit than the parental species. Support for this model of gene interaction comes from studies of *Drosophila* (Coyne and Orr 2004).

fixed) and the AAbb genotype. In either case the new allele is compatible with the ancestral allele (e.g., both AaBB and AABb are viable). Thus neither species need pass through a selectively detrimental stage. However, the "a" and "b" alleles are incompatible when they co-occur. If individuals from the different lineages mate, producing an AaBb hybrid, it could be sterile (or otherwise less fit) because the "a" and "b" alleles, when in the same genetic background, are incompatible and form a postmating isolating barrier. Although this process may seem unlikely, multiple examples, especially in postmating barriers among *Drosophila* species, show that these types of multilocus genetic incompatibilities may in fact be quite common (Orr 1995).

Another important area of study in postzygotic isolation has focused on an observation by J. B. S. Haldane (1922) that the heterogametic sex (the one with two different sex chromosomes, such as males, XY, in mammals, and females, ZW, in birds) is the one most often affected by hybrid sterility or inviability. Called Haldane's rule, this observation has consistently been supported in a variety of taxa (Coyne and Orr 2004) regardless of which sex is heterogametic. Haldane's rule appears to function early in the evolution of postzygotic isolation (Coyne and Orr 2004), since relatively recently diverged taxa are more likely to exhibit Haldane's rule; older taxa often show symmetric hybrid problems that affect both sexes.

Although Haldane's rule has primarily been observed to result in hybrid sterility or inviability of the heterogametic sex, some evidence suggests that heterogametic hybrids may suffer decreased fitness through a variety of other mechanisms as well. For example, Neal Davies and colleagues showed that in the Neotropical butterflies *Anartia fatima* and *A. amathea* strong assortative mating occurred within each species, but when they were crossed artificially, female hybrids showed reduced fertility, consistent with Haldane's rule (female butterflies are heterogametic; Davies et al. 1997). In addition to this more typical result, however, hybrid females from the *A. amathea* (female) × *A. fatima* (male) cross showed a reduced tendency to mate (Davies et al. 1997). This study provides evidence that Haldane's rule may extend beyond the typical observation of hybrid sterility or inviability, and that hybrid fitness may be decreased through both premating and postmating barriers.

A complete genetic understanding of postzygotic isolating barriers is much more complex than Haldane's rule alone implies. Of particular interest is the concept of *speciation genes*. Speciation genes are genes that differ between taxa and cause reproductive isolation between them. A gene called *Odysseus*, for example, shows evidence of rapid evolution in some fruit flies, despite relatively slow evolution in other organisms. This gene appears to be responsible for sterility in hybrid males between *D. simulans* and *D. mauritiana* (Ting et al. 1998). Like *Odysseus*, genes that experience rapid evolution (at least in some taxa) may be likely to cause genetic incompatibilities between taxa and thus be good candidates for speciation genes. Much of the work on speciation genes remains theoretical; however, evolutionary biologists are addressing exciting questions such as whether particular genes are generally more likely to lead to reproductive isolation, or whether genes associated with certain functions are better candidates for speciation genes than others.

The Process of Speciation

Evolutionary mechanisms such as natural selection, sexual selection, and genetic drift can promote differences between populations in phenotypes and their underlying genotypes, even to the extent that different populations become fixed for completely different alternatives. Why, then, is not every clearly differentiated population considered a different species? The crux of the problem rests on two issues. First, how much gene flow occurs between differentiating populations? Gene flow acts to homogenize populations both phenotypically and genetically; hence it works against selection and drift in promoting adaptive or random differentiation, respectively. Second, how much differentiation is enough? This issue is thornier and points to the importance of reproductive isolation in the speciation process. Although diverging populations do not have to evolve intrinsic barriers to reproduction, some barrier to reproductive exchange must be present to stop gene flow and complete the speciation process. For allopatric populations, which do not overlap in geographic range, the physical distance imposed by geography can serve as a strong barrier to gene flow. However, for diverging populations whose ranges overlap (sympatry) or are adjacent to one another (parapatry), or when previously isolated allopatric populations come into contact (secondary contact), intrinsic or extrinsic barriers to gene flow that prevent interbreeding become necessary to complete the speciation process. Thus the importance of reproductive isolating barriers has been paramount in studies of speciation, and much of our understanding of the process of speciation is centered on the mechanisms that favor the evolution of reproductive isolation between species.

Classic models of the speciation process focused on the geographic distribution of populations undergoing divergence and the relative contribution of geographic isolation to the formation of new species (Figure 4). Although the degree of geographic separation between diverging populations is certainly an important contributor to the disruption of gene flow, more recent studies of speciation have focused on the importance of evolutionary mechanisms (natural selection, sexual selection, genetic drift, and mutation) in promoting diversification and speciation either in allopatry or in sympatry.

THE ROLE OF NATURAL SELECTION IN SPECIATION

The importance of adaptive divergence in the speciation process (termed *ecological speciation*) was first noted by the geneticist Theodosius Dobzhansky in the 1940s in describing speciation in *Drosophila*. Dobzhansky (1946) believed that speciation in *Drosophila* occurred mainly through the evolution of divergent physiological complexes that were successful in different environments. Ernst Mayr (1942) was the first to point out that many of the accumulated genetic differences between populations, particularly in physiological and ecological adaptations to particular environments, were also potentially reproductive isolating mechanisms. Such "by-product" evolution of reproductive isolation is the driving force behind Mayr's classic mode of vicariance allopatric speciation (Figure 4).

More recently the importance of ecological divergence in promoting reproductive isolation either in allopatry or in sympatry has been recognized, and an important role for natural selection in a variety of routes to speciation has become clear. Laboratory experiments (reviewed in Rice and Hostert 1993) have confirmed that ecological speciation can occur; experimental studies that raised fruit flies (*D. melanogaster* or *D. pseudoobscura*) in different environments (varying temperatures or food sources) have demonstrated that some level of premating reproductive isolation evolved between lines raised in different environments, while no premating isolation evolved between independent lines raised in the same environment. These experiments demonstrate how the process might also work in nature.

In the simplest models of ecological speciation, reproductive isolation evolves between populations incidentally as a by-product of adaptation to

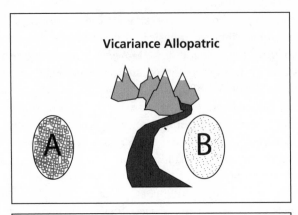

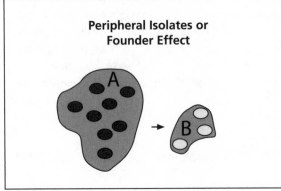

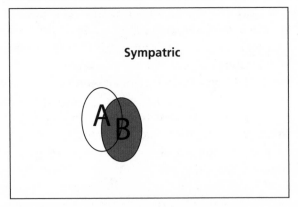

Figure 4. Classic models (or modes) of speciation are based on the degree of geographic separation as the driving force of speciation. In vicariance allopatry, ancestral populations of a once-continuous, large range of a species are separated by a vicariance event (e.g., mountains, river, ocean). Differences in environmental conditions on either side of the barrier to dispersal promote local adaptation and by-product reproductive isolation, leading to the formation of new species (Mayr 1942). The peripheral-isolates or founder-effect model of speciation postulates genetic drift as the primary mechanism that promotes divergence and speciation. Peripheral isolates or founder events occur when populations expand outward from the edges of their distributions to occupy new habitats, or when a small number of individuals disperse to a new location and establish a new population. Mayr (1954) proposed that genetic change

alternative selection regimes. A recent example of such by-product reproductive isolation has been reported in Darwin's finches of the Galápagos Islands. These birds are one of the most celebrated illustrations of adaptive radiation and have evolved an impressive array of specializations in beak form and function, in accordance with the diverse feeding niches that different species have come to occupy. Long-term field studies by Peter and Rosemary Grant (Grant et al. 1976) and their colleagues have shown that beak morphology in Darwin's finches has evolved by means of natural selection in precise correspondence to changing ecological conditions, including food availability and interspecific competition. Jeffrey Podos (2001) tested for potential correlations between beak morphology and song structure. Song is a major component of mating signals in birds, and differences in birdsong lead to premating behavioral reproductive isolation in a variety of species of birds, including Darwin's finches. The production of birdsongs with variation in frequency (pitch) and temporal properties (e.g., syllable repetitions) requires rapid changes in beak gape; hence vocal performance capacities are predicted to vary as a function of beak morphology. Podos tested this idea by comparing the structure of song and beak morphology in eight species of Darwin's finches on Santa Cruz Island (Figure 5). He found that species with large beaks adapted for crushing hard seeds had evolved songs with comparatively lower rates of syllable repetition and narrower frequency ranges than those species with smaller beaks evolved for feeding on smaller, softer seeds. Because song plays a central role in reproductive isolation in Darwin's finches, this linkage between beak morphology selected for feeding performance and its concomitant influence on song structure illustrates one of the best examples of by-product evolution of reproductive isolation and its potential importance in speciation of Darwin's finches.

PARALLEL SPECIATION AS A SIGNATURE OF ECOLOGICAL SPECIATION

Parallel evolution of similar traits in populations that inhabit similar environments strongly implicates natural selection as a causative agent. Parallel speciation can result from the parallel evolution of traits that determine reproductive isolation that evolve repeatedly in independently derived, closely related populations as a by-product of adaptation to different environments

would be rapid in such founding populations because of genetic drift. This could result in genetic reorganizations that could incidentally yield reproductive isolation and result in rapid speciation. Sympatric speciation occurs between populations that overlap in geographic distribution. It is the only model of speciation where reproductive isolation evolves between populations that are not geographically separated. It is usually associated with a host or habitat shift accompanied by assortative mating on the new host or in the new habitat.

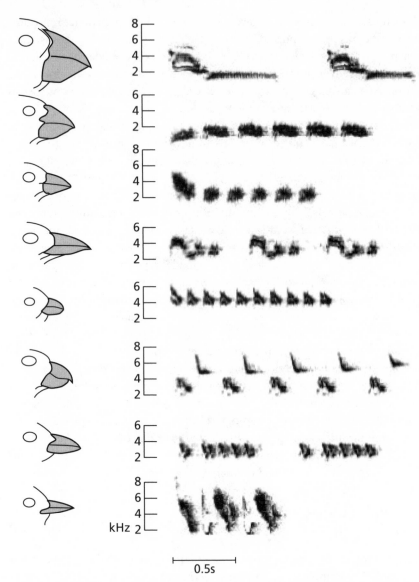

Figure 5. An illustration of by-product reproductive isolation as a result of adaptive divergence in beak morphology occurs in Darwin's finches. Beak morphology and representative sound spectrograms of songs from eight Darwin's finch species on Santa Cruz Island *(from top to bottom: Geospiza magnirostris, G. fortis, G. fuliginosa, G. scandens, Camarhynchus parvulus, C. psittacula, C. pallida, C. olivacea)* are shown. Spectrogram frequency resolution, 98 Hz; scale bar, 0.5s. Interspecific variation is apparent in both morphology and song structure. Species with large beaks (e.g., *G. magnirostris*) have lower rates of syllable repetition and narrower frequency ranges than those species with smaller beaks (e.g., *C. olivacea*). (From Podos 2001.)

(Schluter and Nagel 1995). The best examples of parallel speciation come from sympatric speciation events where independent colonizations of particular environments have led to similar, repeated patterns of character divergence in adaptive traits associated with different niches and the parallel buildup of assortative mating by selective environment based upon differences in those adaptive traits (Figure 6).

The threespine stickleback *(Gasterosteus aculeatus)* has become a model system for the study of parallel speciation. The ancestral threespine stickleback is a marine fish found across a wide geographic distribution from Alaska and British Columbia through Iceland, Scotland, Norway, and Japan. This species is anadromous, being spawned in freshwater streams, spending the juvenile stage in marine environments, and migrating back to streams as adults to breed. Independent colonizations of stream environments have led to multiple stream-resident stickleback populations that no longer migrate to the ocean. In addition, in British Columbia and Alaska independent colonizations by marine sticklebacks of postglacial freshwater lakes formed after the retreat of the glaciers at the end of the Pleistocene have led to the independent evolution of lake forms (benthics and limnetics) in multiple freshwater lakes. Speciation in sticklebacks has involved the repeated colonization of new environments and the accompanying changes in adaptations, primarily body-size differences and number and size of gill rakers (spines on the gill plates used to filter food particles), associated with ecological differences in each environment. Have these adaptive changes led to reproductive isolation between different stickleback ecotypes?

In a recent study Jeffrey McKinnon and colleagues (McKinnon et al. 2004) found strong evidence of assortative mating by size through differential female preferences for male body size in alternative stream and marine environments.

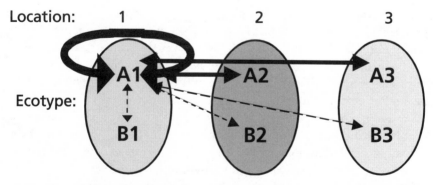

Figure 6. Parallel speciation can result from the parallel evolution of traits that determine reproductive isolation that evolve repeatedly as a by-product of adaptation to different environments. *Solid lines* are used to show matings between ecotypes from the same habitat with a greater degree of compatibility (width increases with increased compatibility). *Dashed lines* indicate matings between different ecotypes with a low degree of mating compatibility.

Anadromous marine populations are larger in body size than stream-resident populations, presumably an advantage in the long-distance migrations that marine populations undergo. Mating compatibility was more than two times higher when females were paired with males from the same ecotype (from British Columbia or Japan) than when they were paired with males from a different ecotype. To test the importance of body size as the basis for assortative mating by ecotype, McKinnon and colleagues (McKinnon et al. 2004) reared sticklebacks from stream-resident and marine populations in the laboratory and manipulated their body sizes by changing the length of the growing season. This produced either large-bodied or small-bodied fish of both sexes in both marine and stream-resident ecotypes. In mating trials with these experimentally manipulated fish, they found that males and females had higher mating compatibilities when they were matched for size, regardless of their ecotype of origin. These results demonstrate the importance of a single phenotypic trait, body size, in adaptive diversification in sticklebacks and suggest that divergent natural selection on this trait in different environments also makes a primary contribution to reproductive isolation that leads to speciation.

THE ROLE OF NATURAL SELECTION IN HOST SHIFTS AND SYMPATRIC SPECIATION

The occurrence of sympatric speciation, speciation without geographic isolation (Figure 4), has been contentious because the conditions under which sympatric populations can escape the homogenizing effects of gene flow were thought to be rare and require exceptional circumstances. More recently the recognition of disruptive natural selection accompanied by strong assortative mating of divergently selected phenotypes (Figure 7) as a likely model for the origin and diversification of sympatric species has convinced many evolutionary biologists that sympatric speciation is an important and even primary type of speciation in certain groups of organisms. Implicit in the process of sympatric speciation is rapid evolution of reproductive isolation between diverging forms because geography cannot serve as the primary barrier to gene flow, as it does for divergent allopatric populations. One common way in which such reproductive isolation can rapidly evolve is through host shifts by parasites or phytophagous insects that specialize in particular host species.

Guy Bush provided the first potential example of a sympatric speciation event initiated by a host shift with his studies (1966, 1969) of the host races of hawthorn and apple maggot flies *(Rhagoletis pomonella)*. Hawthorn trees are native to North America, and the hawthorn race of *R. pomonella* is the ancestor of the apple host race. Apple maggot flies were first discovered on introduced, domestic apple trees in the Hudson Valley region of New York in the mid-1800s. Bush and his colleagues have been studying differences between the native hawthorn race and the apple race since the mid-1960s and have documented a number of phenotypic differences in life history and reproduction that have facilitated speciation between the two host races. One

Disruptive selection

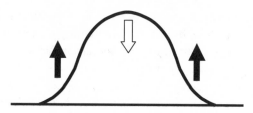

Assortative mating within phenotypes or habitats

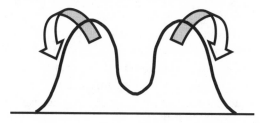

Figure 7. Sympatric speciation can occur when disruptive natural selection facilitated by host or habitat shifts is accompanied by assortative mating by the diverging phenotypes. The sympatric model of speciation requires that reproductive isolation evolves as a result of assortative mating of "like" individuals on the new host or in the new habitat in order to initiate the speciation process without geographic isolation.

primary characteristic that promotes sympatric divergence associated with host shifts in *Rhagoletis* and other phytophagous insects is that they feed and breed on the same host species; thus dispersal between different host plant species is low. For example, females of hawthorn origin are averse to ovipositing eggs on apple trees, whereas apple-origin females prefer apple trees for oviposition sites. This strong host fidelity limits interhost movement of adults between hawthorn and apple trees. In addition, genetic differences have accumulated since the initial host switch that promote changes in breeding phenology, such as timing of mating, pupation and diapause, and eclosion of adults (e.g., Feder 1998). These changes are thought to have been favored by natural selection as the different host races of flies adapted to differences in fruiting phenology between hawthorn and apple hosts. Such locally adapted differences by each host race contribute to reproductive isolation between the host races and augment host fidelity through female oviposition preferences. Thus the *Rhagoletis* flies provide a fundamental example of ecological divergence where complete geographic isolation is not always required to initiate speciation.

Although phytophagous insects offer several well-documented examples of sympatric speciation though host shifts, a recent study provides at least one example of the potential for sympatric speciation via a host shift in a vertebrate. African indigobirds (genus *Vidua*) are host-specific brood parasites, laying their eggs in the nests of firefinches (genus *Lagonosticta*), and different indigobird species specialize in parasitizing different species of firefinches. Indigobird nestlings are reared along with the host young and mimic the mouth markings of their respective hosts, thus lowering the probability of being rejected by their host foster parents. In addition to this unique mimicry adaptation by young indigobirds, as adults, male indigobirds mimic the song that is sung by their host species sire, and females reared by a specific host species of firefinch preferentially mate with male indigobirds that sing their natal host-specific song. This facilitates host-assortative mating and leads to reproductive isolation among different species of indigobirds that parasitize different host species of firefinches. A recent phylogenetic study by Michael Sorenson and colleagues (Sorenson et al. 2003) that compared mitochondrial DNA sequences for a number of different species of indigobirds and their respective firefinch hosts demonstrated that different indigobird species that parasitize different firefinch hosts are genetically differentiated from one another, a finding consistent with assortative mating of indigobirds reared by a particular firefinch host species. These results suggest that divergence through local adaptations in nestling mouth markings, adult male song copying, and female preferences for host species song type has led to positive assortative mating between different host-specific species of indigobirds. This pattern of host-specific adaptations linked with positive assortative mating within a host type is consistent with sympatric speciation in African indigobirds through host switches in the firefinch species whose nests they parasitize.

REPRODUCTIVE ISOLATION MAY BE REINFORCED BY NATURAL SELECTION

Natural selection can act to directly increase the amount of premating isolation between two diverging populations by selecting against hybrids formed between them. This process, first proposed by Theodosius Dobzhansky (1937) and termed *reinforcement* by Frank Blair (1955), can occur in areas of secondary contact when previously allopatric populations come into contact and interbreed. When hybrids between formerly allopatric populations are partially infertile, reinforcement can complete the speciation process. Reinforcement was a popular concept after the modern synthesis because it allows natural selection to drive speciation directly, but its empirical demonstration has been difficult, and its existence as an important component of ecological speciation has been controversial (e.g., Noor 1999; Marshall et al. 2002).

Most of the evidence for reinforcement comes from comparisons between allopatric and sympatric populations in the strength of female mating preferences for conspecifics. For example, females derived from populations of *Drosophila pseudoobscura* sympatric with *D. persimilis* display greater reluctance to mate with heterospecific males than do females from allopatric populations (Noor 1995). F1 hybrid males are sterile, but hybrid females are fertile, allowing for gene exchange and selection against hybrids.

One of the best-documented examples of reinforcement occurs in the benthic and limnetic ecotypes found in lake populations of the threespine stickleback. Howard Rundle and Dolph Schluter (1998) demonstrated both in laboratory crosses and in F1 hybrids detected in wild populations that hybrids suffer a foraging disadvantage relative to their parental species in the parental environments. In the laboratory F1 hybrids reared in a common environment with ample access to food showed growth rates similar to those of parental species. However, in field studies F1 hybrids reared in either type of parental environment, littoral and limnetic zones of lakes, had lower growth rates and were smaller at maturity than either pure parental species. Hybrids were poorer at catching benthic invertebrates and plankton in the different field environments because of intermediate gill-raker number and body size. In addition, benthic females from sympatric populations were much less likely to mate with intermediate-sized F1 hybrid males than were benthic females from allopatric populations where limnetics did not occur. These results support a model of reinforcement of size-assortative mating preferences through natural selection against intermediate-sized hybrids that have lower fitness in either parental environment.

How important has reinforcement been in cases of ecological speciation? The problem is that reinforcement is not the only explanation for observed patterns of stronger divergence in mating preferences in sympatric populations. Imagine that two sympatric populations had already evolved a high degree of reproductive isolation while still in allopatry. These populations will likely continue to coexist, while formerly allopatric populations with low levels of reproductive isolation may be lost either through gene flow and fusion or the extinction of the rarer population. Thus the only populations (or species pairs) that we observe living in sympatry in nature are those that evolved a high degree of isolation in allopatry, not because of reinforcement following secondary contact. In addition, recombination between loci involved in mating decisions and the character(s) influencing hybrid fitness undermines the process of reinforcement. In order for reinforcement to properly promote divergence in mate preferences, genes that code for mating decisions must be tightly linked with genes that code for hybrid infertility or inviability. These requirements may not often be met in natural populations.

These arguments do not show that reinforcement has not operated in completing the speciation process during ecological speciation. They only suggest that the evidence for the widespread importance of reinforcement is inconclusive. Of the two processes that can promote reproductive isolation associated with ecological speciation—divergence with isolation as a by-product and reinforcement—the first is well documented and almost certainly important in speciation; the importance of the second still needs further substantiation.

SEXUAL SELECTION AND ECOLOGICAL SPECIATION

Sexual selection, a force that acts on traits that increase mating success, can be influenced by the environment and, as such, may contribute to ecological speciation. Sexual selection that leads to speciation as a result of divergent selection between environments can result from spatial variation in secondary sexual traits, such as clinal variation in male plumage or coloration in response to a varying environmental gradient. For example, clinal variation in the availability of foods high in carotenoids that lead to bright red and orange coloration in male birds has been suggested as the primary cause for divergence in plumage among populations in male house finches, *Carpodacus mexicanus* (Hill 1994). A model that demonstrates the importance of parallel clinal variation in female mating preferences as a result of environmental variation in male signaling traits was first proposed by Russell Lande (1982), but good examples of clinal variation that leads to speciation via sexual selection are lacking.

Natural selection through variation in features of the environment where mating signals are propagated can result in divergent sexual selection of male mating signals, such as courtship displays or nuptial coloration, leading to premating isolation between populations from divergent signaling environments (Ryan and Rand 1993; Boughman 2002). Natural selection can also act upon the sensory system of females as well, leading to sensory biases for certain properties of male signals (e.g., brighter coloration, larger size, louder and longer duration of vocalizations). Particular features of local environmental conditions could alter sensory bias or make some male traits better at exploiting biases in certain environments than in others. In addition, female mating preferences for male traits that are indicators of superior quality can also be influenced by the signaling environment. For example, if particular male traits are more reliable indicators of quality in some environments than in others (Schluter and Price 1993), divergence in preferences between environments for these male traits could lead to premating reproductive isolation between them. Thus different coevolutionary paths of signals and preferences in different environments could generate speciation.

The best examples of an important role for sexual selection in promoting speciation as a result of varying signaling environments come from species that live in environments that vary in light intensity or transmission properties that influence the spectral quality of signals (i.e., which colors are best reflected by particular environments). For example, allopatric populations of *Anolis cristatellus* lizards occur in two different environments (xeric and mesic) that differ with respect to light intensity and the spectral quality of colors that are best reflected. Manuel Leal and Leo Fleishman (2004) have shown that male anoles from the two environments differ in the color and pattern of their dewlaps (folds of skin under the throat that are distended in male displays used as territorial and mating signals) in the direction that increases signal detectability in each habitat. Males from the xeric environment have darker and slightly redder dewlaps that maximize contrast where the background radiance is relatively high (i.e., they will appear darker than the overall brightness of the background), while males from the mesic environment have evolved bright dewlaps because of their relatively high reflectance and transmission at all wavelengths. Such males are more detectable in mesic habitats, where the background habitat light levels are low. This divergence in dewlap signals would minimize gene flow between habitats, because males with the wrong dewlap color pattern would be less detectable by conspecifics and hence less likely to secure a territory or obtain matings.

THE ROLE OF SEXUAL SELECTION IN NONECOLOGICAL SPECIATION

Models of sexual selection, particularly models of the evolution of female mating preferences for male traits (Mead and Arnold 2004), show that populations can diverge in these preferences, promoting reproductive isolation and speciation. Superficially, this process mirrors that observed in ecological speciation, except that instead of natural selection promoting divergence in male mating signals through selection against hybrids or matching of signal properties to the signaling environment, the selective force is sexual selection, promoted through divergent female mating preferences.

One model of female choice that may lead to speciation is the Fisherian runaway process (or "sexy sons"), first proposed by Ronald Fisher (1930). Fisherian processes are often cited as the basis for the evolution of some of the most extreme male traits, such as the elaborate plumage of male birds of paradise or the bright colors of male tropical fishes. The extreme exaggeration of these traits arises as a result of the genetic correlation between genes that encode the attractive male trait and genes that encode female preferences for that trait. Fisher's runaway process has been modeled mathematically, and there is abundant empirical evidence for the role of Fisherian selection in promoting the evolution of exaggerated male traits within species (Andersson 1994); however, studies that support its exclusive role in speciation are difficult to pinpoint. One potential example involves artificial selection of eye-span length in stalk-eyed flies *(Cyrtodiopsis dalmanni)*. In the laboratory Gerald Wilkinson and Paul Reillo (1994) created two lineages of flies from wild flies, a short-eye-span line and a long-eye-span line. After 13 generations of selection, they found a genetic correlation between male eye span, a sexually selected male trait, and female preference for eye-span length. Females from the line selected for short male eye spans preferred to mate with males with short eye spans, and females from the lines selected for long eye spans preferred males with long eye spans. Such divergence in female mating preferences for eye-span length might account for speciation among species of stalk-eyed flies that differ in eye-span length.

Sexual selection may also act on indicator traits (also known as *good genes*). Under this model of female choice, exaggerated male traits provide the female with information about the genetic quality of the male, allowing her to make mating decisions based on information that will potentially provide superior quality to her offspring. Although this mechanism may potentially lead to speciation if the indicator traits differ between populations, like runaway selection, empirical examples are rare. Some evidence indicates that a good-genes mechanism may promote differences in male genitalia that lead to divergence in some insect taxa, although this hypothesis is still under debate (Hosken and Stockley 2004). It should be noted that for both indicator and Fisherian traits, the lack of conclusive empirical evidence does not mean that divergence is unlikely to be a product of these mechanisms; rather, it is often difficult to distinguish which mechanism may have operated in the past.

An additional role of sexual selection in speciation can be traced to understanding the mechanisms associated with sexual conflict. Sexual conflict arises when the best strategy for one sex is different (or even detrimental) compared with the best strategy in the other sex. This antagonism is predicted to lead to a coevolutionary arms race where differing sets of traits between the sexes interact to determine the nature of male-female interactions (Arnqvist and Rowe 2005). An example of within-species sexual conflict is the effects of seminal fluid in *Drosophila melanogaster*. Seminal fluid has been shown to be beneficial to male fitness, reducing female remating and increasing ovulation, but it is toxic to females; the more seminal fluid they receive, the faster they die. Sexual conflict appears to occur across a wide range of taxa (Arnqvist and Rowe 2005), but predictably, some of the best evidence for a role of sexual conflict in speciation comes from insects (Arnqvist et al. 2000). Because sexual conflict may lead to rapid evolution in the reproductive tract, it follows that it may also increase the pace of reproductive isolation between populations.

SEXUAL SELECTION, ISOLATION, AND RATES OF SPECIATION

Sexual selection, because of its role both in promoting the evolution of premating reproductive isolating barriers and in continued divergence between species, may have a considerable impact on the rate of speciation (reviewed by Panhuis et al. 2001). For example, taxa with a wide range of sensory perception capabilities may be particularly prone to speciation through sexual selection, since they have an elevated ability to produce or detect a broad range of signals, be they visual, acoustic, or pheromonal (Andersson 1994). Species-rich groups of frogs, for instance, can detect a wider range of advertisement-call frequencies than those in less speciose lineages, potentially accounting for why such lineages of frogs are more speciose compared with those with narrower frequency perception ranges (Ryan 1986). In addition, reinforcement, as described earlier, can complete the speciation process upon secondary contact by enhancing divergence of female mating preferences. Sexual selection plays an important role in sympatric speciation as well. Assortative mating, the propensity to mate with like individuals, has been proposed as one way to quickly evolve reproductive isolation (Coyne and Orr 2004) and is a necessary component of the sympatric speciation process.

Sexual selection has been linked to relatively high rates of speciation, in part because of its direct effect on mate recognition traits (Panhuis et al. 2001). Species of birds with higher levels of promiscuity, for example, are thought to experience stronger sexual selection than closely related, more monogamous lineages; these promiscuous lineages generally are also more speciose (Mitra et al. 1996). Other studies of birds conform to the prediction of high speciation rates in taxa with strong sexual selection, including comparisons of dichromatic (sexes of different colors; males generally bright, with females dull) clades versus monochromatic (both sexes the same color, usually bright) clades (Barraclough et al. 1995) and clades with varying amounts of feather ornamentation (Møller and Cuervo 1998). Similar trends have been observed in angiosperms (and their associated pollinators), insects and other invertebrates, and fishes (reviewed in Andersson 1994; Panhuis et al. 2001). A recent example highlighting the rapid rate of speciation as a result of sexual selection has been shown in Hawaiian crickets in the genus *Laupala*. Tamra Mendelson and Kerry Shaw (2005) used divergence among species at molecular markers calibrated with geological estimates of the relative ages of the different Hawaiian islands to show that the rate of speciation in one clade of these crickets from a single island was calculated at over four species per million years, the highest rate ever calculated for arthropods. Speciation in this group appears to be promoted by differences in male courtship song trill rate, which females use in species recognition, because there do not appear to be any ecologically distinguishing features that separate different cricket species. These and many other examples highlight how sexual selection can be an important evolutionary force that leads to rapid divergence in premating isolating barriers, resulting in high rates of speciation.

THE ROLE OF MUTATION IN SPECIATION

Mutational changes associated with chromosomal rearrangements were once thought to be a primary mechanism of speciation. For example, M. J. D. White (1978, 336) concluded that chromosomal rearrangements have "played the primary role in the majority of speciation events." More recently the widespread role of karyotypic change in speciation has been questioned, because the accumulation of chromosomal differences between populations is thought to be largely incidental to speciation (Sites and Moritz 1987). The problem with the view that chromosomal rearrangements promote speciation centers on the difficulty of alternative rearrangements in diverging populations reaching fixation. All models of chromosomal speciation assume that alternative rearrangements impair the fertility or viability of interspecific hybrids, thereby reducing gene flow and promoting fixation of alternative chromosomal rearrangements. Although this may likely be true in examples of extreme chromosomal rearrangements (such as mutational events that result in changes in chromosome number), a number of studies have shown that chromosomal rearrangements often have little effect on fertility and hence are ineffective barriers to gene flow (Spirito 1998).

Loren Rieseberg and colleagues have studied recombinational speciation in sunflowers (*Helianthus* spp.) and the effects of chromosomal rearrangements in contributing to the speciation process. Recombinational speciation occurs when two diploid species interbreed and give rise to a new lineage that is both fertile and true breeding but is reproductively isolated from both parental species. Rieseberg (2001) argues that chromosomal rearrangements that are heterozygous in hybrids have major effects on suppressing recombination between chromosomes that carry the rearranged segments and genes linked to these segments on the chromosomes. Genes unlinked to the chromosomal rearrangement segments, especially those that are neutral, freely introgress between the hybridizing parental species. Rearrangements that suppress recombination but lack a causal effect on hybrid fitness could act synergistically with linked isolation genes to extend their effects over a larger genomic region. Thus chromosomal rearrangements probably do play an important role in reducing gene flow across species barriers, but not necessarily through the mechanisms of reduced hybrid fitness traditionally envisioned. The hybrid sunflower species *H. anomalus,* which results from hybridization between *H. annuus* and *H. petiolaris,* is thought to have arisen in such a manner. In addition, *H. anomalus* has higher fitness in novel environments where both of its parental species progenitors do poorly. Such an adaptive role for hybridization in fostering speciation has been suggested for a number of plant and some animal species (Arnold 1997; Dowling and Secor 1997).

SPECIATION BY POLYPLOIDY

Polyploidization occurs through the duplication of entire sets of chromosomes and is one of the few mechanisms that can produce instantaneous speciation (polyploid species are reproductively incompatible with their diploid progenitors). Polyploid species are formed either through genome duplication of a single diploid species (autopolyploidy) or through the fusion of two or more diploid genomes as a result of interspecific hybridization (allopolyploidy). Both mechanisms of polyploidy are common in ferns, mosses, algae, and virtually all groups of vascular plants, particularly angiosperms (Stebbins 1950; Levin 2002).

How polyploid species become established is not fully understood. Several conditions, particularly common in plants, might enable a new polyploid to increase and form a viable population. These include self-fertilization, vegetative propagation, higher fitness than the diploid progenitor species, or niche separation from the diploid progenitor species. Indeed, many polyploid taxa reproduce by selfing or vegetative propagation, and most differ from their diploid progenitors in habitat and distribution. Increases in ploidy alter cell size, water content rate, rate of development, and many other physiological properties, so many polyploids may have selective advantages in new ecological niches immediately upon their origin.

An unusual case of polyploid speciation in animals occurs in the gray tree frog species complex found throughout the eastern United States. The complex contains a diploid species, *Hyla chrysoscelis* (2N=24) and a tetraploid species *Hyla versicolor* (4N=48). Both species occupy large regions of allopatry (*H. chrysoscelis* in the southern half of the range, *H. versicolor* in the northern half), but they also exist in sympatry in multiple localities where their ranges overlap. Diploids and tetraploids reproduce sexually and are reproductively isolated from one another through both premating (mating signal differences) and postmating (triploid hybrids are sterile) barriers to interbreeding. Males produce advertisement calls that differ in their trill rates, with the diploid *H. chrysoscelis* always producing calls with faster trill rates than those of the tetraploid *H. versicolor* (Gerhardt 2005). Studies by Carl Gerhardt and colleagues have shown that females of both diploids and tetraploids are highly selective (especially in sympatric populations) when evaluating call characteristics of male advertisement calls, strongly discriminating against calls with the trill rate of the "wrong" species.

A recent study by Alicia Holloway and colleagues (Holloway et al. 2006) has demonstrated an unusual pattern of polyploid speciation in the tetraploid species *H. versicolor*. Using DNA sequences from both maternally inherited mitochondrial gene regions and biparentally inherited nuclear gene regions, this study showed that at least three different diploid progenitor species contributed to multiple origins of the tetraploid gray tree frog *H. versicolor* (Figure 8). Two of the three diploid progenitor species are from extinct lineages, as suggested by the occurrence of unique alleles found in tetraploid lineages but not in the living diploid lineages examined. These results suggest that tetraploids arose through recurrent allopolyploid speciation events involving different diploid ancestors. Interestingly, different lineages of tetraploids that arose from different diploid ancestors freely interbreed where they occur in sympatry. Because the different tetraploid lineages are not reproductively isolated from one another, they are still considered a single tetraploid species, *H. versicolor*. Polyploidization in this species may have led to predictable and consistent change in the advertisement call, perhaps as a direct result of increased cell size, which has been shown to correlate with decreased trill rate in artificially produced autotriploid gray tree frogs (Keller and Gerhardt 2001). Thus changes in trill rate as a result of polyploid events have led to reproductive continuity between lineages representing different origins of tetraploids but have reinforced trill-rate differences between diploids and tetraploids. Speciation in gray tree frogs illustrates the complex manner in which polyploid species can arise and further shows how changes in isolating barriers, such as advertisement calls, can be directly affected by hybridization and polyploidy.

THE ROLE OF GENETIC DRIFT IN SPECIATION

In theory, genetic drift should promote speciation as populations expand outward from the edges of their distributions to occupy new habitats, or when a small number of individuals disperse to a new location and establish a new population. In practice, however, little empirical support exists for the formation of new species solely through random genetic changes promoted

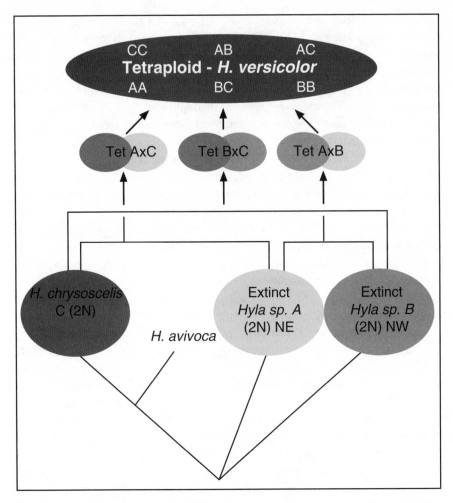

Figure 8. A model of tetraploid origins from diploid ancestors illustrates multiple allopolyploid origins for the gray tree frog *Hyla versicolor*. The relationships of diploid gray tree frogs and a related species, *H. avivoca*, are depicted at the bottom of the tree. Extinct diploids (2N), *H.* sp. A and *H.* sp. B, were inferred from tetraploid allele lineages. Tetraploid gray tree frogs, *H. versicolor* (4N), were formed multiple times from extinct diploid ancestors (A×B, A×C, and B×C). Tetraploid gray tree frogs from different lineages (e.g., NE, NW) are interfertile and thus form a single biological species that shares a diverse gene pool of multiple ancestral progenitor species. (From Holloway et al. 2006.)

by genetic drift. The idea of *founder-effect speciation* (Figure 4) was first proposed by Ernst Mayr (1954; later termed *peripatric speciation*) and was based on his observation that in many birds and mammals, populations peripheral to the distribution of a probable "parent" species were often highly divergent, to the point of being classified as different species. Mayr proposed that genetic change would be very rapid in such localized founding populations because of random fixation of alleles by genetic drift, resulting in a breakup of ancestral coadapted gene complexes and the formation of new ones that would lead to the potential for natural selection to favor different interacting sets of genes in the new founder environments. Such gene reorganizations could incidentally yield reproductive isolation and hence provide an avenue for rapid speciation. Although genetic changes associated with genetic drift probably do occur in founder populations, little evidence for the types of major gene reorganizations proposed by Mayr (1954) and later Hampton Carson (1975) and Alan Templeton (1980) exists.

One potential example of speciation promoted by changes due to genetic drift, however, comes from species differences in chirality (the coiling direction of shells) between different species of the Japanese land snail in the genus *Euhadra*. These snails live in small populations with limited dispersal between them. Chirality is determined by a single gene locus with two alleles, with dextral (right-handed) coiling much more common than sinistral (left-handed) coiling. Because of the physical difficulty in two-way copulation (these snails are hermaphrodites) between snails from populations that are fixed for opposite coiling directions, these populations are reproductively isolated from one another and so appear to represent distinct species. Since opposite coiling types are selected against when they are rare within populations, genetic drift is likely to have played a role in the appearance of new coiling types among populations. Genetic models have demonstrated that genetic drift is the most plausible mechanism for a switch from dextral to sinistral chirality (van Batenburg and Gittenberger 1996) and the parallel speciation of at least four independently derived sinistral species has been demonstrated through phylogenetic analysis of mitochondrial DNA haplotype variation (Ueshima and Asami 2003).

Consequences of Speciation

The bulk of this essay has focused on examples of living species and the various mechanisms by which they are thought to have speciated. Yet speciation has resulted in the enormous diversity of now-extinct animals, plants, and microbes uncovered in the fossil record as well. Determining process from pattern is challenging; hence most studies of the speciation process focus on living examples. Although much debate has centered on the mode and tempo of speciation as illustrated by the fossil record (phyletic gradualism, the slow transition of one species into another, versus punctuated equilibrium, rapid evolutionary diversification and speciation followed by long periods of stasis), evolutionary biologists generally agree that the same evolutionary mechanisms that promote divergence today among populations of living species (microevolution) are responsible for speciation events (macroevolution), both ongoing and in the past.

The tempo of speciation varies dramatically among different taxa. Some groups, such as horseshoe crabs (genus *Limulus*), have remained unchanged for over 150 million years, while others, such as Hawaiian crickets (genus

Laupala), have speciated rapidly, with up to 4.17 species per million years (Mendelson and Shaw 2005). Thus speciation can occur quickly or slowly, under a variety of ecological conditions, and with or without the evolution of reproductive isolation.

Unraveling the mysteries of how new species arise is one of the most exciting areas of evolutionary biology, and although we have made much progress in understanding speciation since William Bateson's early concerns, future studies promise to yield even more new and fascinating discoveries. The hope of the future, in preserving and protecting our planet's biodiversity, lies in an understanding of the process of speciation and a commitment to saving the products of this process: the species that exist on earth today and those that have yet to evolve in the future.

BIBLIOGRAPHY

Andersson, M. 1994. *Sexual Selection*. Princeton, NJ: Princeton University Press.
Arnold, M. L. 1997. *Natural Hybridization and Evolution*. New York: Oxford University Press.
Arnqvist, G., M. Edvardsson, U. Friberg, and T. Nilsson. 2000. Sexual conflict promotes speciation in insects. *Proceedings of the National Academy of Sciences USA* 97: 10460–10464.
Arnqvist, G., and C. Rowe. 2005. *Sexual Conflict*. Princeton, NJ: Princeton University Press.
Barraclough, T. G., P. H. Harvey, and S. Nee. 1995. Sexual selection and taxonomic diversity in passerine birds. *Proceedings of the Royal Society of London, Series B* 259: 211–215.
Bateson, W. 1922. Evolutionary faith and modern doubts. *Science* 55: 55–61.
Blair, W. F. 1955. Mating call and stage of speciation in the *Microhyla olivacea–M. carolinensis* complex. *Evolution* 9: 469–480.
Boughman, J. W. 2002. How sensory drive can promote speciation. *Trends in Ecology and Evolution* 17: 571–577.
Burkhardt, D., and I. de la Motte. 1985. Selective pressures, variability, and sexual dimorphism in stalk-eyed flies. *Naturwissenschaften* 72: 204–206.
Bush, G. L. 1966. *The Taxonomy, Cytology and Evolution of the Genus* Rhagoletis *in North America (Diptera: Tephritidae)*. Cambridge, MA: Museum of Comparative Zoology.
———. 1969. Sympatric host race formation and speciation in frugivorous flies of the genus *Rhagoletis* (Diptera: Tephritidae). *Evolution* 23: 237–251.
Carson, H. L. 1975. The genetics of speciation at the diploid level. *American Naturalist* 109: 83–92.
Coyne J. A., and H. A. Orr. 2004. *Speciation*. Sunderland, MA: Sinauer Associates.
Cracraft, J. 1983. Species concepts and speciation analysis. *Current Ornithology* 1: 159–187.
———. 1989. Speciation and its ontology: The empirical consequences of alternative species concepts for understanding patterns and processes of differentiation. In D. Otte and J. A. Endler, eds., *Speciation and Its Consequences*, 28–59. Sunderland, MA: Sinauer Associates.
Darwin, C. 1859. *On the Origin of Species*. London: John Murray.
Davies, N., A. Aiello, J. Mallet, A. Pomiankowski, and R. E. Silberglied. 1997. Speciation in two neotropical butterflies: Extending Haldane's rule. *Proceedings of the Royal Society of London, Series B* 264: 845–851.

Diaz, A., and M. R. Macnair. 1999. Pollen tube competition as a mechanism of prezygotic reproductive isolation between *Mimulus nasutus* and its presumed progenitor *M. guttatus*. *New Phytologist* 144: 471–478.

Dobzhansky, T. 1937. *Genetics and the Origin of Species*. New York: Columbia University Press.

———. 1946. Complete reproductive isolation between two morphologically similar species of *Drosophila*. *Ecology* 27: 205–211.

Dowling, T. E., and C. L. Secor. 1997. The role of hybridization and introgression in the diversification of animals. *Annual Review of Ecology and Systematics* 28: 593–619.

Feder, J. L. 1998. The apple maggot fly, *Rhagoletis pomonella* flies in the face of conventional wisdom about speciation? In D. J. Howard and S. H. Berlocher, eds., *Endless Forms: Species and Speciation*, 130–144. Oxford: Oxford University Press.

Fisher, R. A. 1930. *The Genetical Theory of Natural Selection*. Oxford: Clarendon Press.

Futuyma, D. J., J. S. Walsh Jr., T. Morton, D. J. Funk, and M. C. Keese. 1994. Genetic variation in a phylogenetic context: Responses of two specialized leaf beetles (Coleoptera: Chrysomelidae) to host plants of their congeners. *Journal of Evolutionary Biology* 7: 127–146.

Gerhardt, H. C. 2005. Advertisement-call preferences in diploid-tetraploid tree frogs (*Hyla chrysoscelis* and *Hyla versicolor*): Implications for mate choice and the evolution of communication systems. *Evolution* 59: 395–408.

Grant, P. B., B. R. Grant, J. N. M. Smith, I. J. Abbott, and L. K. Abbott. 1976. Darwin's finches: population variation and natural selection. *Proceedings of the National Academy of Sciences USA* 73: 257–261.

Haldane, J. B. S. 1922. Sex ratio and unisexual sterility in animal hybrids. *Journal of Genetics* 12: 101–109.

Hill, G. E. 1994. Geographic variation in male ornamentation and female mate preference in the house finch: A comparative test of models of sexual selection. *Behavioral Ecology* 5: 64–73.

Holloway, A. K., D. C. Cannatella, H. C. Gerhardt, and D. M. Hillis. 2006. Polyploids with different origins and ancestors form a single sexual polyploid species. *American Naturalist* 167: E88–E101.

Hosken D. J., and P. Stockley. 2004. Sexual selection and genital evolution. *Trends in Ecology and Evolution* 19: 87–93.

Keller, M. J., and H. C. Gerhardt. 2001. Polyploidy alters advertisement call structure in gray tree frogs. *Proceedings of the Royal Society of London, Series B* 268: 341–345.

Knowlton, N., J. L. Maté, H. M. Guzmán, R. Rowan, and J. Jara. 1997. Direct evidence for reproductive isolation among the three species of the *Montastraea annularis* complex. *Marine Biology* 127: 705–711.

Lande, R. 1982. Rapid origin of sexual isolation and character divergence in a cline. *Evolution* 36: 213–223.

Leal, M., and L. J. Fleishman. 2004. Differences in visual signal design and detectability between allopatric populations of *Anolis* lizards. *American Naturalist* 163: 26–39.

Levin, D. A. 2002. *The Role of Chromosomal Change in Plant Evolution*. Oxford: Oxford University Press.

Marshall, J. L., M. L. Arnold, and D. J. Howard. 2002. Reinforcement: The road not taken. *Trends in Ecology and Evolution* 17: 558–563.

Mayden, R. L. 1997. A hierarchy of species concepts: The denouement in the saga of the species problem. In M. F. Claridge, A. H. Dawah, and M. R. Wilson, eds., *Species: The Units of Biodiversity*, 381–424. Norwell, MA: Chapman and Hall.

Mayr, E. 1942. *Systematics and the Origin of Species.* New York: Columbia University Press.

———. 1954. Change of genetic environment and evolution. In J. Huxley, A. C. Hardy, and E. B. Ford, eds., *Evolution as a Process,* 157–180. London: George Allen and Unwin.

———. 1963. *Animal Species and Evolution.* Cambridge, MA: Belknap Press of Harvard University Press.

McKinnon, J. S., S. Mori, B. K. Blackman, L. David, D. M. Kingsley, L. Jamieson, J. Chou, and D. Schluter. 2004. Evidence for ecology's role in speciation. *Nature* 429: 294–298.

Mead, L. S., and S. J. Arnold. 2004. Quantitative genetic models of sexual selection. *Trends in Ecology and Evolution* 19: 264–271.

Mendelson, T., and K. L. Shaw. 2005. Rapid speciation in an arthropod. *Nature* 433: 375–376.

Mitra, S., H. Landel, and S. Pruett-Jones. 1996. Species richness covaries with mating system in birds. *The Auk* 113: 544–551.

Møller, A. P., and J. J. Cuervo. 1998. Speciation and feather ornamentation in birds. *Evolution* 52: 859–869.

Noor, M. A. F. 1995. Speciation driven by natural selection in *Drosophila. Nature* 375: 674–675.

———. 1999. Reinforcement and other consequences of sympatry. *Heredity* 83: 503–508.

Orr, H. A. 1995. The population genetics of speciation: The evolution of hybrid incompatibilities. *Genetics* 139: 1805–1813.

———. 1996. Dobzhansky, Bateson, and the genetics of speciation. *Genetics* 144: 1331–1335.

Panhuis, T. M., R. K. Butlin, M. Zuk, and T. Tregenza. 2001. Sexual selection and speciation. *Trends in Ecology and Evolution* 16: 364–371.

Paterson, H. E. H. 1985. The recognition concept of species. In E. S. Vrba, ed., *Species and Speciation,* 21–29. Transvaal Museum Monograph no. 4. Pretoria, South Africa: Transvaal Museum.

Podos, J. 2001. Correlated evolution of morphology and vocal signal structure in Darwin's finches. *Nature* 409: 185–188.

Rice, W. R. 1998. Intergenomic conflict, interlocus antagonistic coevolution, and the evolution of reproductive isolation. In D. J. Howard and S. H. Berlocher, eds., *Endless Forms: Species and Speciation,* 261–270. New York: Oxford University Press.

Rice, W. R., and E. E. Hostert. 1993. Laboratory experiments on speciation: What have we learned in 40 years? *Evolution* 47: 1637–1653.

Rieseberg, L. H. 2001. Chromosomal rearrangements and speciation. *Trends in Ecology and Evolution* 16: 351–358.

Rundle, H. D., and D. Schluter. 1998. Reinforcement of stickleback mate preferences: Sympatry breeds contempt. *Evolution* 52: 200–208.

Ryan, M. J. 1986. Neuroanatomy influences speciation rates among anurans. *Proceedings of the National Academy of Sciences USA* 83: 1379–1382.

Ryan, M. J., and A. S. Rand. 1993. Species recognition and sexual selection as a unitary problem in animal communication. *Evolution* 47: 647–657.

Schluter, D., and L. M. Nagel. 1995. Parallel speciation by natural selection. *American Naturalist* 146: 292–301.

Schluter, D., and T. Price. 1993. Honesty, perception, and population divergence in sexually selected traits. *Proceedings of the Royal Society of London, Series B* 253: 117–122.

Simpson, G. G. 1961. *Principles of Animal Taxonomy.* New York: Columbia University Press.

Sites, J. W., Jr., and C. Moritz. 1987. Chromosomal evolution and speciation revisited. *Systematic Zoology* 36: 153–174.

Sokal, R. R., and T. J. Crovello. 1970. The Biological Species Concept: A critical evaluation. *American Naturalist* 104: 127–153.

Sorenson, J. D., K. M. Sefc, and R. B. Payne. 2003. Speciation by host switch in brood parasitic indigobirds. *Nature* 424: 928–931.

Spirito, F. 1998. The role of chromosomal rearrangements in speciation. In D. J. Howard and S. H. Berlocher, eds., *Endless Forms: Species and Speciation,* 320–329. Oxford: Oxford University Press.

Stebbins, G. L. 1950. *Variation and Evolution in Plants.* New York: Columbia University Press.

Sturtevant, A. H. 1920. Genetic studies on *Drosophila simulans.* I. Introduction. Hybrids with *Drosophila melanogaster. Genetics* 5: 488–500.

Templeton, A. R. 1980. The theory of speciation via the founder principle. *Genetics* 94: 1011–1038.

———. 1989. The meaning of species and speciation: A genetic perspective. In D. Otte and J. A. Endler, eds., *Speciation and Its Consequences,* 2–27. Sunderland, MA: Sinauer Associates.

Ting, C.-T., S. C. Tsaur, M.-L. Wu, and C.-I. Yu. 1998. A rapidly evolving homeobox at the site of a hybrid sterility gene. *Science* 282: 1501–1504.

Ueshima, R., and T. Asami. 2003. Single-gene speciation by left-right reversal. *Nature* 425: 679.

van Batenburg, F. H. D., and E. Gittenberger. 1996. Ease of fixation of a change in coiling: Computer experiments on chirality in snails. *Heredity* 76: 278–286.

van Valen, L. 1976. Ecological species, multispecies, and oaks. *Taxon* 25: 233–239.

White, M. J. D. 1978. *Modes of Speciation.* San Francisco: W. H. Freeman and Company.

Wiley, E. O. 1978. The evolutionary species concept reconsidered. *Systematic Zoology* 27: 17–26.

Wilkinson, G. S., and P. R. Reillo. 1994. Female choice response to artificial selection on an exaggerated male trait in a stalk-eyed fly. *Proceedings of the Royal Society of London, Series B* 255: 1–6.

Evolution and Development

Gregory A. Wray

The intellectual histories of developmental and evolutionary biology have been intertwined almost from the birth of biology as a distinct scientific discipline (Mayr 1982; Richards 1992). The reasons are not difficult to find. Both evolution and development are temporal processes, and every individual's developmental history is embedded within, and reflects, a much deeper evolutionary history. This temporal parallel forms the basis for the concept of recapitulation, that embryonic development literally replays evolutionary history. Recapitulation is one of the oldest and most persistently controversial concepts in evolutionary developmental biology, but not its most illuminating (Gould 1977; Raff 1996; Wilkins 2002).

The role of development in transmuting genotype into phenotype provides a more fruitful approach to understanding the interactions between evolutionary and developmental processes (Raff and Kaufman 1983; Wilkins 2002). The evolution of the genome and the evolution of organismal phenotype, which have often been treated as if they were separate processes, are inextricably bound together by the processes of development (Figure 1). The way an embryo functions has as much to do with ecological context and phylogenetic history as it does with generating a particular phenotypic outcome. Conversely, the evolutionary fate of a group of related species can be influenced as much by developmental characteristics as it is by adult anatomy, physiology, and behavior.

Temporal Parallels

Recapitulation has by turns enlightened, entertained, and exasperated generations of biologists (de Beer 1940; Gould 1977; Churchill 1980). The recognition of parallels between embryological events and relationships among species predated the concept of biological evolution, at least in its modern sense of transmutation between species. Indeed, the words *evolution* and *development* were used interchangeably to refer to the unfolding of both individual development and phyletic change until the beginning of the twentieth

evolutionary
mechanisms

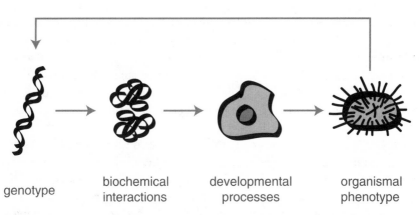

| genotype | biochemical interactions | developmental processes | organismal phenotype |

Figure 1. Genotype to phenotype. The processes of development transmute genetic information into organismal phenotype, which includes anatomy, behavior, and physiology. These traits, in turn, interact with the environment, where they influence survivorship and reproduction. Natural selection and other evolutionary mechanisms change the relative frequencies of genotypes in each successive generation. A complete understanding of evolution requires knowing how processes operate at each of these steps. Until quite recently, the weakest link in our knowledge was development.

century (Hall 1992; Richards 1992). What we recognize today as distinct processes were considered inextricably intertwined.

Early descriptions of parallels between development and evolution were cast in terms of the *scala natura* (chain of being). Aristotle recognized that embryos of more complex creatures passed through stages in which they resembled simpler ones (Gould 1977). For Charles Bonnet, an early-eighteenth-century preformationist, evolution and development were literally the same process. During the late 1700s and early 1800s several investigators noted that chicken and human embryos pass through a series of stages that contain features of "lower" or "inferior" animals. Johann Meckel, noted, for instance, that human embryos contain gill slits and a tail at early stages, with limbs appearing afterwards, and definitive mammalian features later still.

Karl Ernst von Baer, who made extensive and meticulous firsthand observations of diverse animal embryos, explicitly rejected the idea of recapitulation (Hall 1992; Richards 1992). He cited the presence of unique features in the embryos of vertebrates, such as the yolk sac, as evidence that they did not represent adults of other species; conversely, he noted that some features of lower species do not appear in the embryos of higher ones, such as the absence of fins in the embryos of birds (Baer 1828). Baer also noted that the embryos of distinct groups of animals did not pass through stages representing

other ones: "The embryo of a vertebrate is at the beginning already a vertebrate" (translation from Richards 1992, 59).

Despite Baer's potent criticisms, support for the notion of recapitulation
persisted for another century. Recapitulation theory reached its zenith in the
writings of Ernst Haeckel (1866, 1875), whose statement that "ontogeny recapitulates phylogeny" ("die Ontogenie weiter nichts ist als eine kurze Recapitulation der Phylogenie"; 1866, 2, 7) even today encapsulates for many biologists
the relationships between evolution and development. Haeckel, a strong supporter of Darwin, viewed development as an exhibit of phylogenetic history,
with the adults of ancestors displayed in sequence during the development of an
individual. His woodcuts depicting the progressive divergence of initially similar vertebrate embryos into distinctive fish, turtle, bird, and human embryos
have become icons of late-nineteenth-century evolutionary developmental biology (Figure 2). Less widely appreciated is Haeckel's view of recapitulation as
the actual mechanism that produces evolutionary change rather than simply a
passive record of such change. He considered this mechanism to consist of two
processes: terminal addition of new features onto the end of development and
condensation of existing features into ever more abbreviated earlier stages.

Haeckel recognized the existence of exceptions to strict recapitulation
(1875), including the objections raised by Baer (1828). He called them
cenogenetic features and regarded them in much the same light that most
phylogeneticists view homoplasy (features that have converged in form but
without common ancestry) today, as an annoying obfuscation of evolutionary history. Haeckel identified two kinds of cenogenetic features: *heterochrony,* or a change in the order of development events, and *heterotopy,* or
a change in the germ layer giving rise to a structure. Both terms are now used
in much more general senses, to refer to any change in the timing or location
of a developmental process, respectively (Raff 1996; Klingenberg 1998).
Ironically, and not a little confusingly, the current, more general usage of the
term *heterochrony* encompasses strictly recapitulatory temporal changes, as
well as Haeckel's original, nonrecapitulatory ones.

Baer was not the only person to attack recapitulation theory, nor was
Haeckel its only apologist. Remarkably, debates about recapitulation were
still going strong during the 1930s as the modern synthesis was under way;
they may have contributed to the cordial mutual apathy that arose between
evolutionary and developmental biology about this time and lasted until the
late 1980s (Gould 1977; Wilkins 2002). Although the influence of recapitulation eventually faded, comparative embryology continued to play an important role in evolutionary biology.

Embryos as Evidence of Evolution, Phylogeny, and Homology

If embryos do not infallibly recapitulate evolutionary history, they nonetheless provide evidence of the process of evolution, of phylogenetic relationships

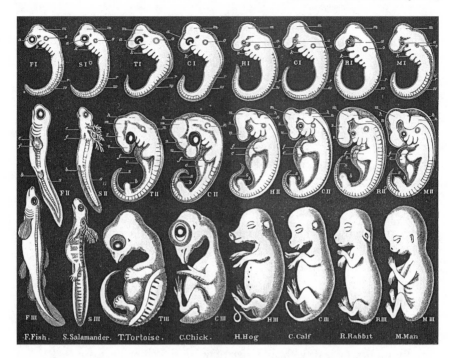

Figure 2. Recapitulation as envisioned by Haeckel. The great comparative anatomist Ernst Haeckel was one of the strongest proponents of the notion that the evolutionary history of a species is replayed during its embryological development (Haeckel 1866). His famous illustrations of vertebrate embryos, one of which is shown here, depict progressive divergence in overall similarities among species as development proceeds. The embryos are *(from left to right)* those of a fish, amphibian, turtle, bird, pig, cow, rabbit, and human. Note that all the embryos have gill slits at an early stage *(top row)* and that the human embryo has a tail *(middle row, right)* that is later lost *(bottom row, right)*. It was the presence of these phylogenetically ancient features in embryos, despite their absence in adults, that strongly impressed Haeckel. (From Haeckel 1866.)

among taxa, and of the history of specific characters. These three practical applications were all first proposed during the nineteenth century, but for somewhat different reasons they remain important today.

Not many readers of this book will need to be persuaded that the process of evolution takes place. Anticipating a rather different audience, Darwin marshaled all the evidence he could find to support the concept of transmutation among species when he wrote *On the Origin of Species* (1859). Where Baer, who rejected the notion of evolution, spoke of "general" and "specific" developmental traits, Darwin interpreted similarities among embryos as compelling evidence of common descent. "Embryology is to me by far the strongest single class of facts in favor of change of forms," he wrote in a letter to Asa Gray (September 10, 1860) shortly after the publication of the *Origin* (Burkhardt et al. 1993, 350). Embryological similarities remain persuasive:

traits not manifest in adult anatomy, such as the presence of gill slits and a tail in human embryos, are tangible evidence of descent with modification.

A second practical application of comparative embryology is in phylogenetic analysis. Darwin pointed out in the *Origin* that "community in embryonic structure reveals community of descent" (1859, 449). Systematists have been applying this concept enthusiastically ever since. Several significant insights into phylogenetic relationships came about through use of this approach, such as the discovery that barnacles are arthropods rather than mollusks (Darwin 1854) and the realization that ascidians are closely related to vertebrates (Garstang 1928). In both cases, adults are highly derived anatomically, but larvae provide clear phylogenetic links. Similarities in embryos or larvae provided the first evidence for phylogenetic relationships among many otherwise anatomically distinct taxa and often remained the only direct evidence until the advent of molecular data sets. The names of many clades are based on these embryological criteria, such as Spiralia (for the spiral geometry of embryonic cell divisions found in some animals) and monocots (for the single cotyledon that nourishes the embryo of some flowering plants). With the widespread application of DNA sequences in phylogenetics, embryological traits play a less important role in identifying clades today, but they can provide an important reality check when new relationships are proposed.

The sequence of embryological events has also been invoked as a basis for polarizing character-state transformation in formal phylogenetic analyses (Nelson 1978). This *ontogenetic criterion* is based on a literal reading of Baer's laws: that earlier events during development are more general and therefore arose earlier during phylogenetic history than later ones. The utility of this criterion has been questioned on first principles (de Queiroz 1984), and it clearly fails in many real-world cases (Mabee 1989; Raff 1996). Outgroup comparison is far more widely used today, and the ontogenetic criterion is interesting primarily as an example of the persistence of recapitulationist thinking a century after Haeckel.

The third practical application of comparative embryology is as a criterion of homology among specific characters. Of the three applications mentioned here, this remains by the far the most important today (Hall 1994; Wilkins 2002). Richard Owen originally proposed the terms *homology* and *analogy* in 1843 with reference to his notion of the archetype, a generalized body organization shared among species within a large group (Panchen 1994). Owen, who was ostensibly strongly opposed to the idea of evolution, introduced the terms to contrast between two kinds of anatomical similarity among species: homology for corresponding parts of an archetype and analogy for nonhomologous traits with similar functions (Owen 1848). Both terms were soon expropriated by evolutionary biologists (Panchen 1994) and are still used to distinguish between two fundamentally different historical hypotheses: common versus independent evolutionary origins of a trait (Hall 1994).

Distinguishing between homologous and analogous structures is often straightforward, but in many cases it is exceedingly difficult. A persistent challenge, for instance, is distinguishing whether the appendages, body segments, and sense organs present in different animal phyla are derived from a common ancestor or arose independently (Moore and Wilmer 1997; Nielsen 2002). A similar embryological origin has long been considered a criterion of homology, since correspondences in anatomical organization are often clearer (Hall 1994). Embryological comparisons have identified some initially surprising homologies, such as the origin of mammalian ear ossicles from jawbones and the origin of vertebrate jaws from gill arches. This approach continues to be useful, as evinced by the recent debate about digit homologies among archosaurs (Burke and Feduccia 1997; Wagner and Gauthier 1999). Although exceptions can occur (de Beer 1971; Roth 1984), similarity of embryological origin remains one of the most useful criteria of homology in anatomy.

In recent years the embryological criterion of homology has been extended to the expression of developmental regulatory genes. This application is based on the observation that the same regulatory protein is often present during the development of homologous structures even when they give rise to highly divergent anatomies and even in distantly related taxa (Holland and Holland 1999). For instance, muscle cells throughout the animal kingdom express not only the structural proteins that endow them with their specific function (myosin, actin, and α-actinin) but also regulatory proteins that control the transcription of these structural genes (MyoD and MEF-2). The observation that developmental regulatory genes are expressed in similar patterns forms the basis of controversial proposals that eyes (Quiring et al. 1994), legs (Panganiban et al. 1997), and segmentation (Holland et al. 1997) are homologous in arthropods and chordates. Framing such arguments convincingly requires comparing orthologous genes (i.e., corresponding members of the same gene family) and critically assessing alternative hypotheses, such as spurious similarity in gene expression (Abouheif et al. 1997; Davidson 2001). Rigorous applications of gene-expression criteria to identifying homologous structures include comparisons of the central nervous system among deuterostome groups (Williams and Holland 1998; Lowe et al. 2003) and of body segments among divergent groups of arthropods (Averof and Patel 1997; Abzhanov et al. 1999).

Heterochrony

Many events during development happen in a particular sequence and at a particular pace because they are triggered by, or dependent upon, earlier events; even developmental events that are not causally related often unfold in a consistent sequence because of global coordination or simply because even largely independent processes are each tightly controlled. As a result, almost any change in one aspect of development will alter the rate or sequence of other aspects (Raff 1996). The result is a *heterochrony*, an evolutionary change in the timing of developmental processes. This simple connection has led to the view that all evolutionary change in development is fundamentally heterochronic (Gould 1977; McKinney and McNamara 1991).

Modern ideas about evolutionary changes in the timing of development grew out of recapitulation theory. The accumulating weight of exceptions steadily eroded the generality of Haeckel's terminal addition and condensation as evolutionary mechanisms. However, there were two additional factors that brought about the downfall of recapitulation theory: a clear understanding of the mechanisms that underlie heredity and the recognition that natural selection can act throughout the life cycle (Gould 1977; Raff 1996; Wilkins 2002). By the 1930s it was widely accepted that the biogenetic law is neither a mechanism of evolutionary change nor an accurate description of it. Even the most vociferous critics of recapitulation theory recognized, however, that parallels frequently do exist between evolutionary history and embryological events. Out of these debates came new and ultimately more rewarding perspectives.

One important insight was that changes in the timing of developmental events often represent adaptations rather than by-products of change. Walter Garstang, a pugnacious and witty critic of recapitulation, asserted that "ontogeny does not recapitulate phylogeny, it creates it" (Garstang 1922, 82). Garstang neatly inverted the Haeckelian view of ontogeny as a record of phyletic history and argued instead that natural selection acts on developmental timing, thereby producing changes in anatomy. He interpreted many specific evolutionary changes in early development as adaptations, including interpolations and losses of larval stages, changes in methods of metamorphosis, and alterations in the timing of reproduction (Garstang 1922, 1929). Stephen Jay Gould took up this theme in *Ontogeny and Phylogeny* (1977), which tied heterochrony to the classical ecological framework of r-selection and K-selection (r-selection produces adaptations, such as rapid growth and early reproduction, for success in unpredictable environments; K-selection produces the converse adaptations in more stable enivonments). Many authors have emphasized the links between heterochrony and such life-history traits as fecundity, timing of reproduction, and adult body size (e.g., McKinney and McNamara 1991; Ryan and Semlitsch 1998). The important lesson that emerged from this work is that simple changes in developmental timing can produce changes in organismal phenotype that are ecologically relevant.

A second, related insight was that evolutionary changes in developmental timing can occur throughout the life cycle. As early as 1895 Frank Lillie suggested that adaptations can arise in the earliest events of embryogenesis. Garstang (1922) argued that new traits could appear at any point in development (not just the terminal addition advocated by Haeckel), and that subsequent modifications could shift their timing either earlier or later (not just Haeckel's condensation, involving progressive shifts to an earlier position in ontogeny). Developmental biologists were losing interest in evolutionary perspectives at the time, however, and the few evolutionary biologists who were studying heterochrony focused on late life-history stages rather than embryos and larvae. Most studies of heterochrony therefore emphasized changes in developmental timing relative to adult anatomy or to sexual maturity, a perspective that was solidified by influential books (Gould 1977; McKinney and McNamara 1991).

Heterochrony in early development attracted renewed interest as the reunification of developmental and evolutionary biology gathered pace during the 1980s and 1990s. Developmental biologists examined heterochrony in embryos and in complex life cycles, while paleontologists documented heterochrony in the fossil record (Raff and Wray 1989; McKinney and McNamara 1991). Developmental biologists began to apply the concept of heterochrony to such processes as morphogenesis, cell lineage, and gene expression. Comparisons of gene expression among closely related species revealed many temporal changes, some in the absence of obvious anatomical correlates (e.g., Dickinson 1988; Cavener 1992) and others correlated with phenotypic transformations (e.g., Wray and Bely 1994; Averof and Patel 1997). It became clear that heterochrony, like other kinds of phenotypic transformations, could arise through a variety of evolutionary mechanisms.

A third important insight was a shift to viewing heterochrony as pattern rather than process (Raff 1996). Until quite recently heterochrony was sometimes considered to be an evolutionary mechanism that produces phenotypic change (Gould 1977; McNamara 1988), a curious persistence of Haeckel's original conception. To the extent that developmental processes are heritable traits, however, heterochrony is an aspect of phenotype like any other and cannot cause changes in phenotype (Raff 1996). Heterochrony nonetheless encompasses a rich array of phenotypic transformations. Gavin de Beer's *Embryos and Ancestors* (1940) was the first attempt at a comprehensive classification of modes of heterochrony, and several others have followed (Gould 1977; Alberch et al. 1979; Raff and Wray 1989). These analyses emphasized the variety of temporal changes that are possible in development and provided a precise vocabulary for describing these changes.

Heterochrony has faded somewhat as a central theme in studies of developmental evolution, in part because temporal changes in development are so pervasive that documenting another case is no longer surprising (Raff 1996; Wilkins 2002). The emphasis has shifted to understanding the genetic basis for evolutionary changes in phenotype, the ecological context in which such changes arise, and how the organization of development biases the evolution of phenotype.

Developmental Constraint and Dissociability

A central goal of evolutionary biology is understanding how changes in genotype produce changes in phenotype. Evolutionary geneticists have focused attention on characterizing genetic variation with a phenotypic impact, identifying which components of this variation selection operates upon, and establishing the number and relative impact of mutations that underlie modifications in traits. These questions have been addressed largely without reference to the developmental mechanisms that link genotype to phenotype since the schism between evolutionary and developmental biology during the 1930s and 1940s (Mayr 1982; Wilkins 2002). Evolutionary and population geneticists have generally treated development as a black box whose contents have little to contribute toward understanding evolutionary processes.

The validity of this approach has been challenged repeatedly. Charles Whitman, an early evolutionary geneticist, wondered whether "the laws of development exclude some lines of [phenotypic] variation and favor others" (1919, 11). By the 1980s several evolutionary biologists were prepared to answer in the affirmative (Mayr 1982; Holder 1983; Maynard Smith et al. 1985; Arthur 1988; Alberch 1989). The notion that the organization of development imposes biases on the production of phenotypes seems reasonable—to believe otherwise is to assume that every possible phenotypic outcome is equally likely, which is manifestly not the case (Alberch 1989; Wilkins 2002). The true challenge has been identifying the underlying sources and long-term consequences of such developmental constraints.

Evidence of developmental constraint has come from two sources. One is *forbidden morphologies,* phenotypes that one can imagine and that seem functionally reasonable but are never observed. The second is the restricted range of phenotypes that emerge following natural or experimental perturbation of development. In both cases some of the best examples come from tetrapod digit patterns (Figure 3) (Holder 1983; Alberch and Gale 1985).

Although these approaches convincingly demonstrate that the production of phenotypes can be biased, the causes of these biases have remained elusive. A phenotype could be "missing" because it is difficult to produce developmentally, or because of a physical or geometric limitation, a lack of genetic variation, or an unrecognized fitness cost. Distinguishing among these possibilities has proven quite difficult. Even when it is possible to infer the existence of developmental constraints by process of elimination, identifying what aspects of development bias phenotypic variation raises further challenges, and few clear examples exist.

A related question is why such biases exist. One explanation is that development is actively "buffered" to produce consistent phenotypic outcomes in spite of genetic or environmental perturbations. This idea was carefully articulated by Conrad Waddington (1957), who named it *canalization.* Developmental biologists have long recognized a similar phenomenon they call *regulation:* the ability of an embryo to converge on a normal phenotype following experimental manipulation. Waddington believed canalization to be the product of genetic assimilation, a process whereby a phenotype that was initially invoked by the environment is subsequently genetically fixed through selection to maintain that phenotype. Ivan Schmalhausen (1949) proposed that stabilizing selection produced much the same result. Other possible sources of developmental constraint are a progressive loss of flexibility as the processes of development become burdened by historical legacies (Wimsatt 1986) and the likelihood that modifications in early development will have a broader phenotypic impact than later modifications (Arthur 1988; Gould 1989).

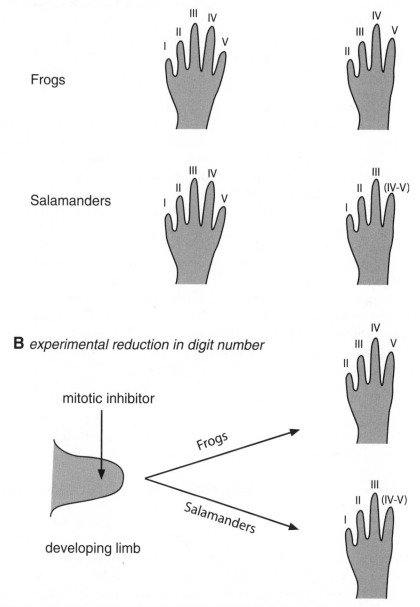

A *evolutionary reduction in digit number*

Frogs

Salamanders

B *experimental reduction in digit number*

mitotic inhibitor

Frogs

Salamanders

developing limb

Figure 3. Biased production of phenotypes. The organization of development may restrict the range of phenotypes it is capable of producing. Evidence comes from experimental perturbation of developing amphibian limbs, which show a range of phenotypes that is similar to that found in related species. (A) Several independent cases of digit reduction have occurred during the evolution of amphibians. Although four-toed species have evolved in frogs and salamanders, these result from the loss of different digits (denoted by roman numerals). (B) By treating the developing limbs of amphibians with an inhibitor of mitosis, Alberch and Gale (1985) were able to experimentally reduce digit number in frogs and salamanders. In each case the range of resulting phenotypes is not random but instead conforms to the range of phenotypes found among related species. (Based on Alberch and Gale 1985.)

If development sometimes acts as a constraint on evolutionary change, how do modifications in development arise? Joseph Needham (1933) pointed out that fundamental developmental processes, such as cell division, morphogenesis, and differentiation, can often continue independently after experimental perturbations. He called this phenomenon *dissociability* and assembled a long list of examples. The underlying basis for dissociability may differ in detail from case to case, but a common general feature is probably modularity, the property of partial or complete functional independence between developmental processes (Gerhart and Kirschner 1997; Hansen 2003). The evolutionary significance of dissociability is that it allows modifications in particular aspects of development without fatally compromising the enterprise (Gerhart and Kirschner 1997).

In genetic terms dissociability is roughly equivalent to circumscribed pleiotropy (Wilkins 2002). In an influential book Ronald Fisher (1930) argued that mutations of small effect are far more likely to form the genetic basis of phenotypic change; his argument turned on the intuitive assumption that broader pleiotropy is, in general, more likely to result in a net negative fitness consequence. Fisher's view has been refined (Orr 2002) but remains a basic expectation for most evolutionary geneticists.

Developmental processes are sometimes strongly dissociated from evolutionary changes in adult anatomy. Striking instances of divergence in developmental mechanisms with no obvious impact on adult morphology have been documented in several different clades, including nematodes (Goldstein et al. 1998; Félix and Sternberg 1998), arthropods (Patel et al. 1994; Grbic et al. 1998), and sea urchins (Henry and Raff 1990; Wray and Raff 1990). Some of these cases involve changes in life history, suggesting that the way development proceeds, rather than its adult product, has been the target of selection, as discussed later in this essay.

If some ways of organizing development limit change, other ways of organizing it may facilitate change. The properties of dissociability and canalization may contribute to the ease with which modifications arise in development (Gerhart and Kirschner 1997; Stern 2000). Some authors have argued that changes in the organization of development, or in its genetic basis, can be key evolutionary innovations with far-reaching effects. The basic idea of a key evolutionary innovation is that a novel feature elevates the ratio of speciation to extinction, permits an expansion of anatomical disparity, or both, over macroevolutionary timescales (Levinton 2001). Possible developmental examples include the expansion of *Hox* gene number early in vertebrate evolution (Holland 1992) and the appearance of holometabolous development in insects (Yang 2001). These hypotheses are difficult to test, since each represents a unique historical event, but they raise the possibility that the organization of development might have a long-term positive impact on the evolutionary fate of a clade.

We arrive at a problem that has not been adequately addressed: is canalization an impediment to or a facilitator of evolutionary change? On the one hand, canalization allows the accumulation of additive genetic variation within a population by minimizing its immediate phenotypic impact on any given individual; the increased genetic variation might later become grist for the mill of selection under altered environmental circumstances or in different genetic backgrounds (Gibson and Dworkin 2004). On the other hand, genetic variants that might otherwise produce phenotypic consequences and thereby become visible to selection might be masked and therefore effectively drift as neutral alleles. We still have only a limited understanding of the degree to which selection established mechanisms that buffer phenotypic variation and the degree to which such mechanisms limit or bias the evolutionary fate of clades (West-Eberhard 2003).

Evolutionary versus Developmental Genetics

The field of evolutionary genetics, as noted earlier, largely ignored the question of how genotypic differences produce phenotypic differences. William Bateson, who coined the term *genetics,* was one of the first biologists to consider the basis for phenotypic variation and differences among species. His *Materials for the Study of Variation Treated with Especial Regard to Discontinuity in the Origin of Species* (1894) emphasized the role of changes that are *homeotic* (another term he coined, to describe spatial transpositions and duplications in anatomy). Bateson (1902) later argued that Mendelian genetics provides a sufficient explanation for the heredity demanded by Darwin's hypothesis of natural selection. During the 1930s and 1940s the modern synthesis completed this historic intellectual merger, and the early neo-Darwinians postulated a dominant role for mutations of small phenotypic effect (Fisher 1930).

Richard Goldschmidt (1940) bucked the trend of the times to advocate a dominant role for macromutations. Even at the time, the genetic mechanisms that he proposed led to "hopeful monsters" (Goldschmidt 1960, 326) seemed implausible. It was not until the 1980s that Goldschmidt's concept of macromutations gained intellectual traction, albeit in considerably altered guise (Wilkins 2002). The way was prepared by several influential molecular biologists, who argued that phenotypic differences among species are largely a result of mutations in gene regulation rather than protein function (Jacob and Monod 1961; Britten and Davidson 1969; King and Wilson 1975; Wilson 1975). The rise of developmental genetics during the subsequent decade highlighted the potential role of regulatory genes in mediating dramatic phenotypic transformations (Raff and Kaufman 1983).

Developmental regulatory genes captured the imagination of evolutionary biologists for several reasons. The most famous regulatory genes belong to the *bithorax* and *Antennapedia* complexes in *Drosophila melanogaster* (Lewis 1978; Scott et al. 1983). Spectacular mutant phenotypes in these genes recall Bateson's (1894) homeotic differences, transforming, for instance, antennae into legs (Figure 4). A remarkable feature of these mutations is their circumscribed pleiotropy: their phenotypic impact is often quite limited outside the

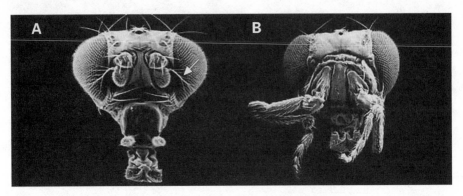

Figure 4. Homeotic mutations. The fruit fly *Drosophila melanogaster* has been an important research subject for both evolutionary and developmental biologists. Homeotic mutations in this species are among the most famous and spectacular mutations that have been recovered from any organism. (A) Head of a fly that is homozygous for wild-type alleles at the *Antennapedia* locus. Note the small, feathery antennae *(white arrow)*. (B) Head of a fly that is homozygous for *Antennapedia* loss-of-function alleles. This mutation transforms the antennae in into enormous, nearly perfect legs. A remarkable feature of this and many other homeotic mutations is their highly localized phenotypic effects. (Photos by Rudi Turner.)

homeotically transformed organ. This effect is a genetic manifestation of Needham's concept of dissociation and, we now know, a result of the way gene expression is regulated (Wray et al. 2003). Another interesting feature of the fly homeotic loci is the resemblance of mutant phenotypes to the anatomies of distantly related insect species. For instance, some mutations in *Ubx* produce four-winged flies, recalling an ancestral anatomy that has been absent in dipterans for more than 200 million years (Lewis 1978).

Developmental geneticists have identified hundreds of loci whose mutant phenotypes indicate key roles in development. During the 1990s loci were discovered in plants that resemble the homeotic genes of animals in several regards: they produce dramatic transformations of floral organ identity, their phenotypic impact is generally circumscribed, and some mutations produce flowers that resemble the floral organization of basal angiosperms (Jack et al. 1992; Rounsley et al. 1995). Not all developmental regulatory genes produce homeotic phenotypes when they are mutated. Several loci in the nematode *Caenorhabditis elegans* produce heterochronic phenotypes (Ambros and Horvitz 1984). Many mutations are known that affect developmental processes; they delete specific organs or cell types, alter body proportions, change the timing of reproduction or fecundity, modify color patterns or other integumentary structures, and create a host of other specific phenotypes (Wilkins 1993; Gilbert 2006).

As soon as it became possible to characterize these genes at the molecular level, it was clear that most encode proteins with regulatory functions (Gerhart and Kirschner 1997; Wilkins 2002). Some are transcription factors, proteins that regulate the expression of other genes; others are involved in communication among cells (signals, receptors, and the machinery that transduces a perceived signal); and some regulate key cellular processes, such as mitosis and cell death. When the homeotic genes of the *Drosophila bithorax* and *Antennapedia* complexes were isolated, for instance, they were all found to contain a sequence motif characteristic of certain DNA-binding proteins. In a historical nod to Bateson, this motif became known as the *homeobox,* and members of this gene family are called *Hox* genes. The preponderance of loci that encode regulatory proteins among the genes identified in mutant screens was intellectually satisfying because it reflected precisely what developmental biologists consider the core processes of development: communication among cells endows them with distinct fates, which in turn activates different batteries of genes that mediate the processes growth, morphogenesis, and differentiation.

The wealth of information that has emerged from developmental genetics, most of it during the past 25 years, has several important evolutionary implications. First, the processes of development often limit pleiotropy. It is not unusual for phenotypes to be circumscribed to a particular anatomical structure or cell type (Wilkins 1993). Even mutations that alter events in early embryos sometimes do so without creating havoc in later developmental events. This implies that mutations that affect the development of strictly delimited aspects of anatomy can arise within natural populations (Stern 2000).

Second, many mutations mimic differences among species in anatomy, life history, or mating systems (Raff and Kaufman 1983; Fitch 1997; Gerhart and Kirschner 1997). Whether or not these mutations also mimic the genetic basis for such differences, and whether or not individuals that bear these mutations would typically survive in the wild (points discussed later), they do demonstrate that even simple mutations can produce profound changes in aspects of phenotype that are ecologically relevant; Goldschmidt's hopeful monsters seem less farfetched, at least at a genetic level, than we once thought.

Third, and perhaps most generally, developmental genetics has thrown open the black box that evolutionary biologists largely ignored for half a century. Until quite recently population geneticists, quantitative geneticists, and even, to some extent, evolutionary geneticists treated genes for the most part as abstract entities; few thought deeply about how different kinds of genes and mutations might contribute to phenotypic evolution. Conversely, developmental geneticists thought a lot about these issues, but rarely in an evolutionary context. The intersection between these perspectives has had a profound and positive impact on both developmental and evolutionary biology. Today, there is widespread interest in understanding what kinds of genes are subject to positive selection (Nielsen et al. 2005; Haygood et al. 2007).

Body Plans and the *Hox* Paradox

The reunification of developmental and evolutionary biology was well under way by the early 1990s. A fortuitous combination of events—the earlier publication of insightful books (Gould 1977; Raff and Kaufman 1983), technical advances (most notably the infusion of molecular techniques into virtually every facet of biology), improvements in analytical tools (especially phylogenetic methods), and a series of exciting discoveries (more on these later)—propelled the merger forward. Much of the initial excitement revolved around the evolution of animal body plans.

As the first developmental regulatory genes were characterized at a molecular level, a major surprise emerged: homologous developmental regulatory genes are shared among organisms with very different body plans. The first clear instance was the realization that *lin-12, Notch,* and *EGF1,* important regulatory genes independently identified in nematodes, arthropods, and chordates, are homologous (Greenwald 1985; Yochem et al. 1988). But this finding was quickly overshadowed by the discovery that vertebrates contain *Hox* genes homologous to those in flies, and that they are clustered in the genome in the same relative order (Akam 1989). By the mid-1990s the list of regulatory genes shared among the major model systems of developmental biology ran to several dozen (Raff 1996; Gerhart and Kirschner 1997). Surveys of other phyla revealed a remarkably broad phylogenetic distribution, including basal metazoan groups (such as cnidarians and ctenophores) and anatomically exotic ones (such as echinoderms and mollusks). It became commonplace to speak of a shared "toolkit" of animal development (Carroll et al. 2001).

All of this was completely unexpected. Animals in different phyla are so anatomically distinct that identifying homologous structures and determining phylogenetic relationships remain contentious (Nielsen 2002). This phenotypic distinctness has formed one of the central themes of comparative anatomy for as long as the field has existed (Hall 1992; Raff 1996). The notion of a body plan is that each phylum of animals has a unique combination of anatomical traits, most of which are architectural (such as body symmetry), embryological (such as the geometry of cell division), or both (such as the number of primary germ layers). The body-plan concept has deeply influenced ideas about the diversification of animals up to the present (Gould 1989; Hall 1992; Raff 1996; Arthur 1997). Developmental biologists, for their part, have long recognized many profound differences in the way in which basic processes of patterning and morphogenesis occur in embryos of different phyla (Gilbert and Raunio 1997). No one predicted that many of the same basic set of genes would turn out to regulate development throughout the animal kingdom.

Yet it is so. Contemporary textbooks of developmental or evolutionary biology usually contain an illustration that shows the similar genomic organization and expression of *Hox* genes in *Drosophila* and mouse: this image has become an icon of evolutionary developmental biology (Figure 5). Nearly all the known gene families that encode transcription factors and cell signaling systems in animals date back at least to the ancestor of all bilaterians, and many to the unicellular ancestor of plants and animals (Gerhart and Kirschner 1997; Carroll et al. 2001; Wilkins 2002). The current developmental roles of these regulatory proteins in model organisms such as *Drosophila* and *Arabidopsis* are clearly different from their original functions in protozoans.

Perhaps even more remarkably, some of these conserved regulatory proteins are expressed in roughly similar regions of embryos that belong to different phyla (Gerhart and Kirschner 1997; Carroll et al. 2001). The *Hox* genes provided the first and still the most dramatic example, but many other orthologous regulatory genes show similar expression among distantly related animal phyla (Figure 5). In all the excitement it has been easy to overlook the fact that most developmental regulatory genes do *not* have similar domains of expression in different phyla; this observation suggests that their developmental roles have evolved considerably during the diversification of animals (Davidson 2001; Wilkins 2002). The sometimes-striking similarities in gene expression that are shared between arthropods and vertebrates have been used to reconstruct the anatomy of the bilaterian ancestor (DeRobertis and Sasai 1996; Carroll et al. 2001). Reconstructing long-extinct ancestors solely on the basis of gene-expression similarities involves many assumptions and much guesswork, however, and the anatomy of the biliaterian ancestor remains uncertain (Erwin and Davidson 2002; Wilkins 2002).

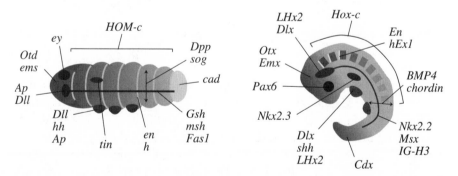

Figure 5. The *Hox* paradox. Homologous genes pattern the embryos of fruit flies and mice despite the phylogenetic distance and phenotypic disparity of these animals. The most famous of these are genes of the *Hox* complex, which establish position of body regions along the anteroposterior axis. Many other homologous genes pattern the brains, hearts, eyes, and limbs of arthropods and chordates. These homologous genes are expressed in roughly similar patterns during development but produce very different phenotypes (Gerhart and Kirschner 1997; Wilkins 2002). This basic result suggests extensive "rewiring" of developmental gene networks during the course of evolution (see text). (From Wray 2003.)

What is clear at this point is that the same basic complement of developmental regulatory genes, some of which are expressed in roughly the same regions of the embryo, can produce a wide diversity of animal body plans. Figuring out how this occurs is one of the grand challenges of evolutionary developmental biology (Gerhart and Kirschner 1997; Carroll et al. 2001; Davidson 2001). A detailed resolution of this *Hox* paradox lies many years in the future, but changes in the organization of developmental gene networks are likely to form an important part of the solution (Carroll et al. 2001; Wilkins 2002; Wray et al. 2003).

Two basic kinds of rewiring within gene networks have been documented, roughly corresponding to quantitative and qualitative changes. The first, or quantitative, category involves a change in the location or timing of expression of a regulatory gene. Such changes can arise when a network is altered upstream of the gene of interest. The location or timing of the developmental processes the gene controls will change as a result, potentially altering organismal phenotype. For example, the spatial extent of *Hox* gene expression differs in ways that correspond to segment identity in crustaceans (Averof and Patel 1997) and to vertebral anatomy in tetrapods (Burke et al. 1995). The second, or qualitative, kind of network rewiring entails a change in which developmental processes are regulated. These changes can occur when a network is altered downstream of the gene of interest. For instance, signaling via the receptor Notch regulates specification of neuronal cell fate in insects and specification of white blood cell fate in vertebrates (Heitzler and Simpson 1991; Guidos 2002), involving two very different sets of effector genes. Similarly, the transcription factor brachyury regulates notochord development in urchordates and chordates but plays no such role in other phyla, which lack this structure; instead, it regulates development of the gut and other structures (Technau and Scholtz 2003).

Dozens of cases are now known where an evolutionary change in the expression profile or downstream targets of a regulatory protein has altered anatomy. Some cases encompass changes at the scale of entire body plans, such as the transition from radial to bilateral symmetry during early metazoan evolution (Martindale et al. 2002) and the later transition back to radial symmetry in echinoderms (Lowe and Wray 1997). Others involve modifications in organs, such as the transition from fin to limb during the origin of tetrapods (Shubin et al. 1997), or evolutionary innovations, such as arthropod wings (Averof and Cohen 1997). Even subtle differences in anatomy are often the result of modifications in gene networks: examples include differences in bristle patterns in flies (Stern 1998; Skaer and Simpson 2000) and color patterns on butterfly wings (Brakefield et al. 1996; Brunetti et al. 2001; Gompel et al. 2005).

How do changes in developmental gene networks arise? There are many ways in which this can happen—any specific interaction between macromolecules is a link in a gene network, and any difference in interactions alters the network. The two most common and important kinds of network changes are probably interactions between proteins and interactions between proteins and regulatory sequences in DNA (Davidson 2001). Creating, modifying, or abolishing such interactions can occur with the simplest and most common of mutations, single base substitutions (Gerhart and Kirschner 1997; Wray et al. 2003). Many proteins associate specifically with other proteins or with nucleic acids, and small changes in binding affinity or specificity can produce dramatic effects. Changing a single amino acid can create a specific interaction or alter affinity dramatically; similarly, changing a single nucleotide can create or modify a binding site for a transcription factor in a regulatory region.

Changes in developmental gene networks clearly have had a profound impact on the history of life. As discussed next, however, understanding how and why gene networks are "rewired" remains a central challenge.

From Macroevolution to Microevolution

Phenotypic changes, including evolutionary changes in development, begin as variation within populations. Relatively few studies have addressed developmental evolution at this microevolutionary scale. As a result, we know little about the genetic basis for modifications in developmental gene networks in wild populations (as opposed to what induced mutations can accomplish in the lab) or about what kinds of evolutionary mechanisms sort genetic variation in network function (as opposed to simply observing that changes can happen over long timescales) (Purugganan 2000; Wray et al. 2003).

This paucity of information derives from two historical trends in research. Developmental biologists, in the first instance, have striven to eliminate the influences of genetic and environmental variation from their analyses and have worked largely with model systems that are so phylogenetically divergent that comparisons yield little information about evolutionary mechanisms. For their part, population geneticists have largely ignored development and have focused primarily on enzymes rather than developmental regulatory proteins, protein structure rather than gene expression, and single genes rather than gene interactions.

There are, of course, exceptions. Several studies have examined genetic variation in the coding and *cis*-regulatory sequences of developmental regulatory genes (e.g., Ayala and Hartl 1993; Ludwig and Kreitman 1995). These population-level analyses provide a glimpse of the nature and level of segregating variation for a few regulatory genes. Comparisons of developmental mechanisms among closely related species with distinct life histories (e.g., Wray and Raff 1990; Swalla and Jeffery 1996; Grbic et al. 1998), or among populations with distinct morphologies (e.g., Shapiro et al. 2004), can reveal how specific developmental mechanisms respond to changes in selection. Overall, though, the general lack of information about population genetics and small-scale evolutionary changes in development remains a glaring gap in the reunification of evolutionary and developmental biology.

One unresolved but rather basic question is what kind of genetic variation contributes to evolutionary changes in development. The large-effect mutations beloved by developmental geneticists alter phenotype in dramatic ways, sometimes mimicking differences among species. These mutations are probably not generally good models for the genetic basis of evolutionary change, in part because the phenotypes are rarely well integrated in terms of organismal function (Budd 1999; Carroll et al. 2001). This does not mean that mutations of large effect play no role in the evolution of development, but detecting their influence is a significant challenge. At the population scale, variants of large effect are expected to be exceedingly rare, and mutations of small effect are expected to predominate in the wild; even at the level of close interspecies comparisons, multiple mutations are typically fixed at each locus, and dissecting out their individual contributions is exceedingly difficult (Wilkins 2002).

Another basic question is how much genetic variation that influences development is present in natural populations. Additive genetic variation that underlies a specific aspect of anatomy provides an indirect measure, since it must at some level represent modifications in development. But by itself, quantitative genetics provides no information about what kinds of genes are responsible for trait differences, nor how mutations in those genes alter developmental processes. Recently, the quantitative genetics of gene expression has attracted attention as a way to begin bridging this gap. It has become clear that natural populations harbor abundant genetic variation that affects gene expression (e.g., Oleksiak et al. 2002; Rockman and Wray 2002; Wittkopp et al. 2004). However, establishing a connection between genetically based differences in gene expression and organismal phenotype poses its own challenges. Molecular and quantitative genetics provide ways of testing this link, and convincing cases are beginning to appear (Brem et al. 2002; Shapiro et al. 2004; Gompel et al. 2005).

A third basic question is what evolutionary mechanisms operate on genetic variation that influences development. The long-term conservation of embryos that forms the basis of Baer's laws is famous, but it is unclear whether its cause is developmental constraint or persistent negative or stabilizing selection. A number of studies have implicated particular evolutionary mechanisms that operate on developmental processes, including relaxed negative selection (Strickler et al. 2001), character displacement (Wray 1996; Grbic et al. 1998), sexual selection (Kopp et al. 2000), and drift (True and Haag 2001).

During the past few years developmental gene networks have also been implicated in speciation. The Dobzhansky-Muller model of speciation proposes that genetic isolation can arise because of incompatibilities in interactions among genes at many loci (Dobzhansky 1936; Muller 1942). Because they involve so many interactions, developmental gene networks are good candidates for such incompatibilities (Johnson and Porter 2000; Wilkins 2002). Several loci involved in reproductive isolation have been characterized at the molecular level, and most either encode transcription factors or reside within *cis*-regulatory regions (Orr and Presgraves 2000), while surveys of many genes suggest that expression differences play a role in reproductive isolation (Michalak and Noor 2003).

Ecological Developmental Evolutionary Biology

Leigh van Valen (1973, 488) noted that "evolution is the control of development by ecology." The simplicity of this statement belies the diversity of interactions between what have been studied traditionally as three separate disciplines within biology (West-Eberhard 2003). Development not only must re-create a functioning adult organism every generation but must do so in a manner consistent with survival. In other words, development is not just a means to destination; it is a particular route to that destination. Although both aspects of development are clearly important, the second has received far less attention (Gilbert 2001; West-Eberhard 2003). There is little recognition in the developmental biology literature that alternative developmental routes to the same phenotypic outcome even exist. Alternative developmental routes are actually quite common, however, and the imprint of natural selection on development per se, rather than through its product, is evident in several phenomena.

Development in many eukaryotes is indirect: alternating haploid-diploid generations and contrasting larval-adult morphologies are two widespread developmental digressions from a direct route that links zygote to sexually reproductive adult. In principle, complex life cycles could be either the result of natural selection for a particular developmental route or historical baggage (features that have lost their function but cannot be deleted because other aspects of organismal function are built around them). Although both are probably true to some extent, many features of complex life cycles appear to be adaptive (Garstang 1929; Strathmann 1985). For instance, larvae may aid in dispersal, provide access to food, evade predators, or minimize risks of local extinction (McEdward 1995). Many species with larvae are closely related to species that lack them, and extensive differences in early development can accompany this life-history contrast (e.g., Wray and Raff 1990; Grbic et al. 1998; Beckham et al. 2003). Such cases demonstrate that building a particular adult body plan does not require a particular developmental route (Strathmann 1985). Many of these cases are correlated with ecological shifts in ways that enhance organismal function, suggesting that natural selection has operated on the underlying developmental mechanisms.

Alternative developmental routes to the same phenotype commonly exist within a species. Rhabditid nematodes provide a particularly well-studied example: adults derived from normal or dauer larval stages are anatomically identical, but underlying developmental processes can be quite different (Félix 2004). (Dauer stages, from the German word meaning "enduring," are an alternative stage taken in unfavorable environments, enabling the organisms to survive until conditions improve.) Asexual reproduction and regeneration provide additional examples. Both phenomena are common and are present in phylogenetically diverse multicellular organisms; both demonstrate how vastly different developmental routes can converge on common end points. In species where both cloning and sexual reproduction occur, adults are literally indistinguishable despite having started development in very different ways. Like the life-history changes discussed in the preceding paragraph, regeneration and asexual reproduction are developmental adaptations generally associated with particular environmental circumstances. Asexual reproduction may evolve from regeneration (Bely and Wray 2001), and regeneration may evolve as an elaboration of wound-healing mechanisms.

Development is rarely hardwired. Most multicellular organisms show some degree of phenotypic plasticity in response to the environment they experience during development (Schlicting and Pigliucci 1998; West-Eberhard 2003). These responses in some cases are simply side effects of perturbation that are not buffered by canalization, but in many other cases responses to the environment are adaptive. Examples include the growth habit and leaf shape of many plants, inducible defenses in many plants and animals, and partitioning resources between growth and reproduction in many plants and animals (Harvell 1990; Schlicting and Pigliucci 1998; West-Eberhard 2003). Dramatic cases of phenotypic plasticity involve sex determination, which is temperature dependent in some turtles and crocodilians (Crews et al. 1994) and is set by social context in some fishes (Godwin et al. 2003).

Polyphenism is a subset of phenotypic plasticity where discrete morphs develop in response to environmental cues. Polyphenism has been particularly well studied in insects: many butterflies and aphids have seasonal morphs, some water striders produce winged and nonwinged morphs depending on water level, and, perhaps most spectacularly, the social insects develop into distinct castes (Nijhout and Wheeler 1982; West-Eberhard 2003). In polyphenism the same genotype can produce two or more distinct phenotypes that are at least as different as species within a genus and in some cases a good deal more so.

The evolutionary origins and developmental bases of phenotypic plasticity are not well understood (Schlicting and Pigliucci 1998; West-Eberhard 2003). Considerable debate has centered around the questions whether there are distinct "plasticity genes" and whether plasticity is a product or by-product of selection (Via et al. 1995; Schlicting and Pigliucci 1998). The expression of many genes is sensitive to environmental conditions, and this sensitivity provides a general means of transducing external cues into phenotypic responses and suggests that there is no need to posit distinct plasticity genes (Wray et al. 2003). Differences in gene expression accompany the development of different castes in bees and ants (Evans and Wheeler 2001; Abouheif and Wray 2002), while the genetic basis for polyphenism has been worked out in detail for the dauer larva of the nematode *Caenorhabiditis elegans* (Albert and Riddle 1988). These approaches suggest that changes in gene expression may be at least as important as changes in gene function during the evolution of plasticity.

Prospects

After half a century of largely ignoring each other, developmental and evolutionary biologists recently entered a phase of profound reciprocal enlightenment. Contemporary developmental biology bears an unmistakable imprint of evolutionary concepts: phylogenetic methods revealed the shared "toolkit" of proteins that regulate developmental processes throughout eukaryotes, and comparative methods are now routinely used to characterize developmental gene networks. Contemporary evolutionary biology, for its part, has been influenced by the powerful functional perspective that developmental biologists have brought to understanding the relationship between genotype and phenotype: attention is shifting from abstract models of single loci to models of gene networks based on real organisms, and quantitative genetics is now viewed as the first step in identifying loci for functional analysis rather than as an end in itself. The reintegration of developmental and evolutionary biology is not yet complete, but these two old partners are back where they belong: in a consensual and mutually supportive relationship.

ACKNOWLEDGMENTS

Thanks to Dan McShea and Michael Ruse for insightful comments. My research is supported by the National Science Foundation and the National Aeronautics and Space Administration.

BIBLIOGRAPHY

Abouheif, E. H., M. Akam, W. J. Dickinson, P. W. H. Holland, A. Meyer, N. H. Patel, R. A. Raff, V. L. Roth, and G. A. Wray. 1997. Homology and developmental genes. *Trends in Genetics* 13: 37–40.

Abouheif, E. H., and G. A. Wray. 2002. The developmental genetic basis for the evolution of wing polyphenism in ants. *Science* 297: 249–252.

Abzhanov, A., A. Popadic, and T. C. Kaufman. 1999. Chelicerate *Hox* genes and the homology of arthropod segments. *Evolution & Development* 1: 77–89.

Akam, M. 1989. Hox and HOM: Homologous gene clusters in insects and vertebrates. *Cell* 57: 347–349.

Alberch, P. 1989. The logic of monsters: Evidence for internal constraint in development and evolution. *Geobios,* mémoire spéciale, 12: 21–57.

Alberch, P., and E. A. Gale. 1985. A developmental analysis of an evolutionary trend: Digital reduction in amphibians. *Evolution* 39: 8–23.

Alberch, P., S. J. Gould, G. F. Oster, and D. B. Wake. 1979. Size and shape in ontogeny and phylogeny. *Paleobiology* 5: 296–317.

Albert, P. S., and D. L. Riddle. 1988. Mutants of *Caenorhabditis elegans* that form dauerlike larvae. *Developmental Biology* 126: 270–293.

Ambros, V., and H. R. Horvitz. 1984. Heterochronic mutants of the nematode *Caenorhabditis elegans. Science* 226: 409–416.

Arthur, W. 1988. *A Theory of the Evolution of Development.* Chichester, U.K.: John Wiley and Sons.

————. 1997. *The Origin of Animal Body Plans*. Cambridge: Cambridge University Press.

Averof, M., and S. M. Cohen. 1997. Evolutionary origin of insect wings from ancestral gills. *Nature* 385: 627–630.

Averof, M., and N. H. Patel. 1997. Crustacean appendage evolution associated with changes in Hox gene expression. *Nature* 388: 682–686.

Ayala, F. J., and D. L. Hartl. 1993. Molecular drift of the *bride of sevenless (boss)* gene in *Drosophila*. *Molecular Biology and Evolution* 10: 1030–1040.

Baer, K. E. von. 1828. *Über Entwickelungsgeschichte der Thiere: Beobachtung und Reflexion*. Königsberg: Gebrüder Bornträger.

Bateson, W. 1894. *Materials for the Study of Variation Treated with Especial Regard to Discontinuity in the Origin of Species*. New York: Macmillan.

————. 1902. *Mendel's Principles of Heredity: A Defence*. Cambridge: Cambridge University Press.

Beckham, Y. M., K. Nath, and R. P. Elinson. 2003. Localization of RNAs in oocytes of *Eleutherodactylus coqui*, a direct developing frog, differs from *Xenopus laevis*. *Evolution & Development* 5: 562–571.

Beer, G. R. de. 1940. *Embryos and Ancestors*. Oxford: Clarendon Press.

————. 1971. *Homology: An Unsolved Problem*. London: Oxford University Press.

Bely, A. E., and G. A. Wray. 2001. Evolution of regeneration and fission in annelids: Insights from *engrailed*- and *orthodenticle*-class gene expression. *Development* 128: 2781–2791.

Brakefield, P. M., J. Gates, D. Keys, F. Kesbeke, P. J. Wijngaarden, A. Monteiro, V. French, and S. B. Carroll. 1996. Development, plasticity and evolution of butterfly eyespot patterns. *Nature* 384: 236–242.

Brem, R. B., G. Yvert, R. Clinton, and L. Kruglyak. 2002. Genetic dissection of transcriptional regulation in budding yeast. *Science* 296: 752–755.

Britten, R. J., and E. H. Davidson. 1969. Gene regulation for higher cells: A theory. *Science* 165: 349–357.

Brunetti, C. R., J. E. Selegue, A. Monteiro, V. French, P. M. Brakefield, and S. B. Carroll. 2001. The generation and diversification of butterfly eyespot color patterns. *Current Biology* 11: 1578–1585.

Budd, G. E. 1999. Does evolution in body patterning genes drive morphological change—or vice versa? *BioEssays* 21: 326–332.

Burke, A. C., and A. Feduccia. 1997. Developmental patterns and the identification of homologies in the avian hand. *Science* 278: 666–668.

Burke, A. C., C. E. Nelson, B. A. Morgan, and C. Tabin. 1995. *Hox* genes and the evolution of vertebrate axial morphology. *Development* 121: 333–346.

Burkhardt, F., J. Browne, D. M. Porter, and M. Richmond, eds. 1993. *The Correspondence of Charles Darwin*. Vol. 8, *1860*. Cambridge: Cambridge University Press.

Carroll, S. B., J. K. Grenier, and S. D. Weatherbee. 2001. *From DNA to Diversity: Molecular Genetics and the Evolution of Animal Design*. Malden, MA: Blackwell Science.

Cavener, D. R. 1992. Transgenic animal studies on the evolution of genetic regulatory circuitries. *BioEssays* 14: 237–244.

Churchill, F. B. 1980. The modern evolutionary synthesis and the biogenetic law. In E. Mayr and W. B. Provine, eds., *The Evolutionary Synthesis: Perspectives on the Unification of Biology*, 112–122. Cambridge, MA: Harvard University Press.

Crews, D., J. M. Bergeron, J. J. Bull, D. Flores, A. Tousignant, J. K. Skipper, and T. Wibbels. 1994. Temperature-dependent sex determination in reptiles: Proximate mechanisms, ultimate outcomes, and practical applications. *Developmental Genetics* 15: 297–312.

Darwin, C. 1854. *A Monograph of the Sub-class Cirripedia: The Balanidae (or Sessile Cirripedes), the Verrucidae, &c.* London: Ray Society.

———. 1859. *On the Origin of Species.* London: John Murray.

Davidson, E. H. 2001. *Genomic Regulatory Systems: Development and Evolution.* San Diego: Academic Press.

de Queiroz, K. 1984. The ontogenetic method for determining character polarity and its relevance to phylogenetic systematics. *Systematic Zoology* 34: 280–299.

DeRobertis, E. M., and Y. Sasai. 1996. A common plan for dorsoventral patterning in Bilateria. *Nature* 380: 37–40.

Dickinson, W. J. 1988. On the architecture of regulatory systems: Evolutionary insights and implications. *BioEssays* 8: 204–208.

Dobzhansky, T. 1936. Studies on hybrid sterility. II. Localization of sterility factors in *Drosophila pseudoobscura* hybrids. *Genetics* 21: 113–135.

Erwin, D. H., and E. H. Davidson. 2002. The last common bilaterian ancestor. *Development* 129: 3021–3032.

Evans, J. D., and D. E. Wheeler. 2001. Expression profiles during honeybee caste determination. *Genome Biology* 2: 0001–0006.

Félix, M. A. 2004. Alternative morphs and plasticity of vulval development in a rhabditid nematode species. *Development Genes and Evolution* 214: 55–63.

Félix, M. A., and P. W. Sternberg. 1998. A gonad-derived survival signal for vulval precursor cells in two nematode species. *Current Biology* 26: 287–290.

Fisher, R. A. 1930. *The Genetical Theory of Natural Selection.* Oxford: Clarendon Press.

Fitch, D. H. A. 1997. Evolution of male tail development in rhabditid nematodes related to *Caenorhabditis elegans. Systematic Biology* 46: 145–179.

Garstang, W. J. 1922. The theory of recapitulation: A critical restatement of the biogenetic law. *Journal of the Linnean Society of London (Zoology)* 35: 81–101.

———. 1928. The morphology of the Tunicata, and its bearing on the phylogeny of the Chordata. *Quarterly Journal of Microscopy Science* 87: 103–193.

———. 1929. The origin and evolution of larval forms. *British Association for the Advancement of Science* 1929: 77–98.

Gerhart, J., and M. Kirschner. 1997. *Cells, Embryos, and Evolution: Toward a Cellular and Developmental Understanding of Phenotypic Variation and Evolutionary Adaptability.* Malden, MA: Blackwell Science.

Gibson, G., and I. Dworkin. 2004. Uncovering cryptic genetic variation. *Nature Reviews Genetics* 5: 681–690.

Gilbert, S. F. 2001. Ecological developmental biology: Developmental biology meets the real world. *Developmental Biology* 233: 1–12.

———. 2006. *Developmental Biology.* 8th ed. Sunderland, MA: Sinauer Associates.

Gilbert, S. F., and A. M. Raunio. 1997. *Embryology: Constructing the Organism.* Sunderland, MA: Sinauer Associates.

Godwin, J., J. A. Luckenbach, and R. J. Borski. 2003. Ecology meets endocrinology: Environmental sex determination in fishes. *Evolution & Development* 5: 40–49.

Goldschmidt, R. 1940. *The Material Basis of Evolution.* New Haven, CT: Yale University Press.

———. 1960. *In and Out of the Ivory Tower: The Autobiography of Richard B. Goldschmidt.* Seattle: University of Washington Press.

Goldstein, B., L. M. Frisse, and W. K. Thomas. 1998. Embryonic axis specification in nematodes: Evolution of the first step of development. *Current Biology* 29: 157–160.

Gompel, N., B. Prud'homme, P. J. Wittkopp, V. A. Kassner, and S. B. Carroll. 2005. Chance caught on the wing: *cis*-regulatory evolution and the origin of pigment patterns in *Drosophila. Nature* 433: 481–487.

Gould, S. J. 1977. *Ontogeny and Phylogeny.* Cambridge, MA: Belknap Press of Harvard University Press.

———. 1989. *Wonderful Life: The Burgess Shale and the Nature of History.* New York: Norton.

Grbic, M., L. M. Nagy, and M. R. Strand. 1998. Development of polyembryonic insects: A major departure from typical insect embryogenesis. *Development Genes and Evolution* 208: 69–81.

Greenwald, I. 1985. *lin-12,* a nematode homeotic gene, is homologous to a set of mammalian proteins that includes epidermal growth factor. *Cell* 43: 583–590.

Guidos, C. J. 2002. Notch signaling in lymphocyte development. *Seminars in Immunology* 14: 395–404.

Haeckel, E. 1866. *Generelle Morphologie der Organismen: Allgemeine Grundzüge der organischen Formen-Wissenschaft, mechanisch begründet durch die von Charles Darwin reformirte Descendenz-Theorie.* 2 vols. Berlin: Georg Reimer.

———. 1875. Die Gastraea und die Eifurchung der Thiere. *Jenaische Zeitschrift für Naturwissenschaft* 9: 402–508.

Hall, B. K. 1992. *Evolutionary Developmental Biology.* London: Chapman and Hall.

———, ed. 1994. *Homology: The Hierarchical Basis of Comparative Biology.* San Diego: Academic Press.

Hansen, T. F. 2003. Is modularity necessary for evolvability? Remarks on the relationship between pleiotropy and evolvability. *Biosystems* 69: 83–94.

Harvell, C. D. 1990. The ecology and evolution of inducible defences. *Quarterly Review of Biology* 65: 323–340.

Haygood, R., O. Fédrigo, B. Hanson, K.-D. Yokoyama, and G. A. Wray. 2007. Promoter regions of many neural- and nutrition-related genes have experienced positive selection during human evolution. *Nature Genetics* 39: 1140–1144.

Heitzler, P., and P. Simpson. 1991. The choice of cell fate in the epidermis of *Drosophila. Cell* 64: 1083–1092.

Henry, J. J., and R. A. Raff. 1990. Evolutionary change in the process of dorsoventral axis determination in the direct-developing sea urchin *Heliocidaris erythrogramma. Developmental Biology* 141: 55–69.

Holder, N. 1983. Developmental constraints and the evolution of vertebrate digit patterns. *Journal of Theoretical Biology* 104: 451–471.

Holland, L. Z., M. Keene, N. A. Williams, and N. D. Holland. 1997. Sequence and embryonic expression of the amphioxus *engrailed* gene (AmphEn): The metameric pattern of transcription resembles that of its segment-polarity homolog in Drosophila. *Development* 124: 1723–1732.

Holland, N. D., and L. Z. Holland. 1999. Amphioxus and the utility of molecular genetic data for hypothesizing body part homologies between distantly related animals. *American Zoologist* 39: 630–640.

Holland, P. W. H. 1992. Homeobox genes and vertebrate evolution. *BioEssays* 14: 267–273.

Jack, T., L. L. Brockman, and E. M. Meyerowitz. 1992. The homeotic gene *APETALA3* of *Arabidopsis thaliana* encodes a MADS box and is expressed in petals and stamens. *Cell* 68: 683–697.

Jacob, F., and J. Monod. 1961. On the regulation of gene activity. *Cold Spring Harbor Symposium on Quantitative Biology* 26: 193–211.

Johnson, N. A., and A. H. Porter. 2000. Rapid speciation via parallel, directional selection on regulatory genetic pathways. *Journal of Theoretical Biology* 205: 527–542.

King, M. C., and A. C. Wilson. 1975. Evolution at two levels in humans and chimpanzees. *Science* 188: 107–116.

Klingenberg, C. P. 1998. Heterochrony and allometry: The analysis of evolutionary change in ontogeny. *Biological Reviews* 73: 79–123.

Kopp, A., I. Duncan, and S. B. Carroll. 2000. Genetic control and evolution of sexually dimorphic characters in *Drosophila*. *Nature* 408: 553–559.

Levinton, J. 2001. *Genetics, Paleontology, and Macroevolution*. Cambridge: Cambridge University Press.

Lewis, E. B. 1978. Gene complex controlling segmentation in *Drosophila*. *Nature* 276: 565–570.

Lillie, F. R. 1895. The embryology of the Unionidae. *Journal of Morphology* 10: 1–100.

Lowe, C. J., and G. A. Wray. 1997. Radical alterations in the roles of homeobox genes during echinoderm evolution. *Nature* 389: 718–721.

Lowe, C. J., M. Wu, A. Balic, L. Evans, E. Lander, N. Stange-Thomann, C. E. Guber, J. Gerhart, and M. Kirschner. 2003. Anteroposterior patterning in hemichordates and the origins of the chordate nervous system. *Cell* 113: 853–865.

Ludwig, M. Z., and M. Kreitman. 1995. Evolutionary dynamics of the enhancer region of *even-skipped* in *Drosophila*. *Molecular Biology and Evolution* 12: 1002–1011.

Mabee, P. M. 1989. An empirical rejection of the ontogenetic polarity criterion. *Cladistics* 5: 409–416.

Martindale, M. Q., J. R. Finnerty, and J. Q. Henry. 2002. The Radiata and the evolutionary origins of the bilaterian body plan. *Molecular Phylogenetics and Evolution* 24: 358–365.

Maynard Smith, J., R. Burian, S. Kauffman, P. Alberch, J. Campbell, B. Goodwin, R. Lande, D. Raup, and L. Wolpert. 1985. Developmental constraints and evolution. *Quarterly Review of Biology* 60: 265–287.

Mayr, E. 1982. *The Growth of Biological Thought: Diversity, Evolution, and Inheritance*. Cambridge, MA: Belknap Press of Harvard University Press.

McEdward, L. R., ed. 1995. *Ecology of Marine Invertebrate Larvae*. Boca Raton, FL: CRC Press.

McKinney, M. L. 1986. Ecological causation of heterochrony: A test and implications. *Paleobiology* 12: 282–289.

McKinney, M. L., and K. J. McNamara. 1991. *Heterochrony: The Evolution of Ontogeny*. New York: Plenum Press.

McNamara, K. J. 1988. The abundance of heterochrony in the fossil record. In M. L. McKinney, ed., *Heterochrony in Evolution: A Multidisciplinary Approach*, 287–325. New York: Plenum Press.

Michalak, P., and M. A. Noor. 2003. Genome-wide patterns of expression in *Drosophila* pure species and hybrid males. *Molecular Biology and Evolution* 20: 1070–1076.

Moore, J., and P. Wilmer. 1997. Convergent evolution in invertebrates. *Biological Reviews* 72: 1–60.

Muller, H. J. 1942. Isolating mechanisms, evolution, and temperature. *Biological Symposia* 6: 71–125.

Needham, J. 1933. On the dissociability of the fundamental processes in ontogenesis. *Biological Reviews* 14: 445–452.

Nelson, G. J. 1978. Ontogeny, phylogeny, and the biogenetic law. *Systematic Zoology* 22: 87–91.

Nielsen, C. 2002. *Animal Evolution: Interrelationships of the Animal Phyla*. Oxford: Oxford University Press.

Nielson, R., C. Bustamante, A. G. Clark, S. Glanowski, T. B. Sackton, M. J. Hubisz, A. Fledl-Alon, D. M. Tanenbaum, D. Civello, T. J. White, J. J. Sninsky, M. D. Adams, and M. Cargill. 2005. A scan for positively-selected genes in the genomes of humans and chimpanzees. *Public Library of Science Biology* 3: 976–985.

Nijhout, H. F., and D. E. Wheeler. 1982. Juvenile hormone and the physiological basis of insect polymorphisms. *Quarterly Review of Biology* 57: 109–133.

Oleksiak, M. F., G. A. Churchill, and D. L. Crawford. 2002. Variation in gene expression within and among natural populations. *Nature Genetics* 32: 261–266.

Orr, H. A. 2002. The population genetics of adaptation: The adaptation of DNA. *Evolution* 56: 1317–1330.

Orr, H. A., and D. C. Presgraves. 2000. Speciation by postzygotic isolation: Forces, genes and molecules. *BioEssays* 22: 1085–1094.

Owen, R. 1848. *On the Archetype and Homologies of the Vertebrate Skeleton.* London: John van Voorst.

Panchen, A. L. 1994. Richard Owen and the concept of homology. In B. K. Hall, ed., *Homology: The Hierarchical Basis of Comparative Biology*, 21–62. San Diego: Academic Press.

Panganiban, G., S. M. Irvine, C. Lowe, H. Roehl, L. B. Corley, B. Sherbon, J. K. Grenier, J. F. Fallon, J. Kimble, M. Walker, G. A. Wray, B. J. Swalla, M. Q. Martindale, and S. B. Carroll. 1997. The origin and evolution of animal appendages. *Proceedings of the National Academy of Sciences USA* 94: 5162–5166.

Patel, N. H., B. G. Condron, and K. Zinn. 1994. Pair-rule expression patterns of *even-skipped* are found in both short- and long-germ beetles. *Nature* 367: 429–434.

Purugganan, M. D. 2000. The molecular population genetics of regulatory genes. *Molecular Ecology* 9: 1451–1461.

Quiring, R., U. Walldorf, U. Kloter, and W. J. Gehring. 1994. Homology of the *eyeless* gene of *Drosophila* to the *small eye* gene in mice and *aniridia* in humans. *Science* 265: 785–789.

Raff, R. A. 1996. *The Shape of Life: Genes, Development, and the Evolution of Animal Form.* Chicago: University of Chicago Press.

Raff, R. A., and T. C. Kaufman. 1983. *Embryos, Genes, and Evolution.* New York: Macmillan.

Raff, R. A., and G. A. Wray. 1989. Heterochrony: Developmental mechanisms and evolutionary results. *Journal of Evolutionary Biology* 2: 409–434.

Richards, R. J. 1992. *The Meaning of Evolution: The Morphological and Ideological Construction of Darwin's Theory.* Chicago: University of Chicago Press.

Rifkin, S. A., J. Kim, and K. P. White. 2003. Evolution of gene expression in the *Drosophila melanogaster* subgroup. *Nature Genetics* 33: 138–144.

Rockman, M. V., and G. A. Wray. 2002. Abundant raw material for *cis*-regulatory evolution in humans. *Molecular Biology and Evolution* 19: 1991–2004.

Roth, V. L. 1984. On homology. *Biological Journal of the Linnean Society* 22: 13–29.

Rounsley, S. D., G. S. Ditta, and M. R. Yanofsky. 1995. Diverse roles for MADS box genes in Arabidopsis development. *Plant Cell* 7: 1259–1269.

Ryan, T. J., and R. D. Semlitsch. 1998. Intraspecific heterochrony and life history evolution: Decoupling somatic and sexual development in a facultatively paedomorphic salamander. *Proceedings of the National Academy of Sciences USA* 12: 5643–5648.

Schlichting, C. D., and M. Pigliucci. 1998. *Phenotypic Evolution: A Reaction Norm Perspective.* Sunderland, MA: Sinauer Associates.

Schmalhausen, I. I. 1949. *Factors of Evolution: The Theory of Stabilizing Selection.* Philadelphia: Blakiston.

Scott, M. P., A. J. Weiner, T. I. Hazelrigg, B. A. Polisky, V. Pirrotta, F. Scalenghe, and T. C. Kaufman. 1983. The molecular organization of the *Antennapedia* locus of *Drosophila*. *Cell* 35: 763–776.

Shapiro, M. D., M. E. Marks, C. L. Peichel, B. K. Blackman, K. S. Nereng, B. Jonsson, D. Schluter, and D. M. Kingsley. 2004. Genetic and developmental basis of evolutionary pelvic reduction in threespine sticklebacks. *Nature* 428: 717–723.

Shubin, N., C. Tabin, and S. B. Carroll. 1997. Fossils, genes and the evolution of animal limbs. *Nature* 388: 639–648.

Skaer, N., and P. Simpson. 2000. Genetic analysis of bristle loss in hybrids between *Drosophila melanogaster* and *D. simulans* provides evidence for divergence of *cis*-regulatory sequences in the *achaete-scute* gene complex. *Developmental Biology* 221: 148–167.

Stern, D. L. 1998. A role of *Ultrabithorax* in morphological difference between *Drosophila* species. *Nature* 396: 463–466.

———. 2000. Perspective: Evolutionary developmental biology and the problem of variation. *Evolution* 54: 1079–1091.

Strathmann, R. R. 1985. Feeding and nonfeeding larval development and life history evolution in marine invertebrates. *Annual Review of Ecology & Systematics* 16: 339–361.

Strickler, A. G., Y. Yamamoto, and W. R. Jeffery. 2001. Early and late changes in *Pax6* expression accompany eye degeneration during cavefish development. *Development Genes and Evolution* 211: 138–144.

Swalla, B. J., and W. R. Jeffery. 1996. Requirement of the *Manx* gene for expression of chordate features in a tailless ascidian larva. *Science* 274: 1205–1208.

Technau, U., and C. B. Scholtz. 2003. Origin and evolution of endoderm and mesoderm. *International Journal of Developmental Biology* 47: 531–539.

True, J. R., and E. S. Haag. 2001. Developmental systems drift and flexibility in evolutionary trajectories. *Evolution & Development* 3: 109–119.

van Valen, L. 1973. Festschrift. *Science* 180: 488.

Via, S., R. Gomulkiewicz, G. de Jong, S. M. Scheiner, C. D. Schlichting, and P. H. V. Tienderen. 1995. Adaptive phenotypic plasticity: Consensus and controversy. *Trends in Ecology and Evolution* 10: 212–217.

Waddington, C. H. 1957. *The Strategy of the Genes: A Discussion of Some Aspects of Theoretical Biology*. London: Allen and Unwin.

Wagner, G. P., and J. A. Gauthier. 1999. 1,2,3=2,3,4: A solution to the problem of the homology of the digits in the avian hand. *Proceedings of the National Academy of Sciences USA* 96: 5111–5116.

Wang, R. L., A. Stec, J. Hey, L. Lukens, and J. Doebley. 1999. The limits of selection during maize domestication. *Nature* 398: 236–239.

West-Eberhard, M. J. 2003. *Developmental Plasticity and Evolution*. Oxford: Oxford University Press.

Whitman, C. O. 1919. *Orthogenetic Evolution in Pigeons*. Washington, DC: Carnegie Institute.

Wilkins, A. S. 1993. *Genetic Analysis of Animal Development*. New York: Wiley-Liss.

———. 2002. *The Evolution of Developmental Pathways*. Sunderland, MA: Sinauer Associates.

Williams, N. A., and P. W. H. Holland. 1998. Molecular evolution of the brain of chordates. *Brain Behavior and Evolution* 52: 177–185.

Wilson, A. C. 1975. Evolutionary importance of gene regulation. *Stadler Symposium* 7: 117–134.

Wimsatt, W. C. 1986. Developmental constraints, generative entrenchment, and the innate-acquired distinction. In P. Bechtel, ed., *Integrating Scientific Disciplines*, 185–208. Dordrecht, The Netherlands: Martinus Nijhoff.

Wittkopp, P. J., B. K. Haerum, and A. G. Clark. 2004. Evolutionary changes in *cis* and *trans* gene regulation. *Nature* 430: 85–88.

Wray, G. A. 1996. Parallel evolution of non-feeding larvae in echinoids. *Systematic Biology* 45: 308–322.

————. 2003. Transcriptional regulation and the evolution of development. *International Journal of Developmental Biology* 47: 675–684.

Wray, G. A., and A. E. Bely. 1994. The evolution of echinoderm development is driven by several distinct factors. *Development* 120, Supplement: 97–106.

Wray, G. A., M. W. Hahn, E. Abouheif, J. P. Balhoff, M. Pizer, M. V. Rockman, and L. A. Romano. 2003. The evolution of transcriptional regulation in eukaryotes. *Molecular Biology and Evolution* 20: 1377–1419.

Wray, G. A., and C. J. Lowe. 2000. Developmental regulatory genes and echinoderm evolution. *Systematic Biology* 49: 28–51.

Wray, G. A., and R. A. Raff. 1990. Novel origins of lineage founder cells in the direct developing sea urchin *Heliocidaris erythrogramma*. *Developmental Biology* 141: 41–54.

Yang, A. S. 2001. Modularity, evolvability, and adaptive radiations: A comparison of the hemi- and holometabolous insects. *Evolution & Development* 3: 59–72.

Yochem, J., K. Weston, and I. Greenwald. 1988. The *Caenorhabditis elegans lin-12* gene encodes a transmembrane protein with overall similarity to *Drosophila Notch*. *Nature* 335: 547–550.

Social Behavior and Sociobiology

Daniel I. Rubenstein

Most animals live solitary lives. Apart from coming together to mate, most individuals forage, move about, and avoid predators on their own. Yet as highly social animals ourselves, we are fascinated by species that develop long-term relationships and live social lives. We want to know why such species live in groups and why the groups take the form that they do. We want to know why some groups persist, whereas others are more transitory. We want to know why cooperation comes in many types and why cooperative and competitive tendencies shape most relationships and thus coexist in many societies. We also want to know why particular roles between individuals and the sexes develop and why some appear universal, while others do not. Answers to these questions continue to emerge, but since we often look to animals to learn about ourselves, the ability to focus the lens of science on answering these and other questions about human sociality has been clouded by problems. Behavior, like all other features of the phenotype, evolves, and the fear that what animals do, humans must also do has often caused people to shy away from asking, or at times even being allowed to ask, questions about the adaptive value of social behavior.

Foundations in Natural History

Understanding why species exhibit the patterns of social behavior they do requires drawing ideas from evolution, ecology, and ethology. During the past half century these ideas, each spun from different modes of thought, have been woven into a coherent but lengthening conceptual cloth. First, behaviorists had to clarify how they explained the appearance of a behavior. From the mundane to the fascinating, the behavior of animals induces the curious to ask, "Why are they behaving in this way?" In a seminal article Tinbergen (1963) proposed that there were four ways of answering these "why" questions. Three were standard fare that focused on issues of immediate causation, ontogenetic development, and evolutionary history. What he added was a fourth that focused on unraveling a behavior's function by identifying how it augmented an actor's survival or reproductive value. This emphasis on explaining behavior in terms of selective benefits to individuals helped move explanations of social functions away from what is good for groups to what is good for the individuals that compose those groups (Williams 1966). Tinbergen's insights identified a confusion that often pervades the study of behavior, namely, that there are broadly two types of explanations: proximate and ultimate. Identifying triggers, or releasers, of a behavior or the timing of its appearance during ontogeny yields proximate explanations because they focus on the machinery within animals that generates the behavior. Identifying how a behavior alters an individual's fitness or the way in which historical antecedents determine what material constitutes the repertoire of a species yields ultimate explanations; they focus on the evolutionary roots and consequences of behavior. Of course, this dichotomy is blurry since mechanistic patterns have functional consequences, and as a result they too evolve. But for the most part, keeping the level of explanations clear when accounting for why animals behave the way they do was a major breakthrough. If this practice had been accepted widely, many of the arguments that have plagued the study of social behavior, especially as it relates to humans, could have been avoided.

Second, searching for a behavior's function, or adaptive value, necessitates assessing its relationship to the environment. Long before Darwin, biologists interpreted morphological traits in relation to the environment. What Darwin did was demonstrate that such adaptations could arise without a creator, thus compelling early naturalists to search for environmental pressures that could serve as selective, or shaping, agents. One of the earliest and most compelling studies that demonstrated that the adaptive value of entire social systems could be explained by differences in ecology was Crook's (1964a, 1964b) study of weaverbirds. By examining the foraging, nesting, grouping, and mating behavior of over 100 species of African weavers that occupied habitats ranging from tropical forests to grasslands and savannas, Crook was able to demonstrate that ecological factors associated with the distribution of food and predators were responsible for the broad social differences that the species exhibited. Those species that lived in forests tended to be solitary, insectivorous, territorial, drably colored, and monogamous. Crook argued that these traits covaried because insects were hard to find. Since feeding young required frequent visits to the nest by fathers and mothers, monogamy and pairwise defense of areas where insects would reappear led to monogamy and territoriality. Crook also suggested that building cryptic nests within territories and the evolution of drab coloration would reduce predation. Conversely, those species that lived in savannas and other open landscapes tended to be social, eaters of seeds, colonial, and polygynous, with males covered in brightly colored feathers. Since seeds often are clumped, Crook believed that benefits associated with groups being able to search for seeds over wide areas led to the evolution of sociality and that coloniality emerged because safe nest sites were rare, inducing many pairs to nest together in the safest trees. With opportunities for cuckoldry prevalent, Crook argued that polygyny and bright male coloration were also favored. Over the years societies of many other groups of animals have been analyzed by use of this comparative lens (Lack 1968; Jarman 1974; Kruuk 1975; Fricke 1975). Although methods of categorizing behavior and classifying evolutionary relationships among species have been refined by overlaying behavior on phylogenies, the comparative method by itself, with its reliance on statistical correlations among behavioral outcomes and ecological forces, could only provide a broad-brush understanding of the evolution of animal societies. To move beyond what critics (e.g., Lewontin 1979) claimed were "just-so stories" and gain a mechanistic understanding of ecological causation, a better grasp of why particular social relationships among individuals developed in specific circumstances was needed, and this required the development of theories to explain individual decision making.

Emerging Theoretical Frameworks

Two different theories emerged, one from social theorists and one from ecologists. Why animals should help others, often at the expense of reproducing themselves, is a question that perplexed biologists for centuries. Even Darwin was puzzled about why so many social insects were sterile. A major breakthrough occurred when Hamilton (1964) proposed the idea of *kin selection*. Until the mid-1960s there was confusion about the units of selection. Wynne-Edwards (1962) championed the view that social behavior evolved as a means of regulating populations; hence group behavior was where selection operated. Williams (1966) criticized this view by demonstrating that most social interactions augmented the fitness of individuals that engaged in them. Hence selection acted on individuals. But if benefits of being social accrue to individuals, then accounting for the evolution of altruistic acts that demonstrably help recipients but accrue no direct benefits for altruists becomes a problem. Hamilton's insight was that individuals can pass on their genes not only directly through offspring but also indirectly via the offspring of close relatives. The combination of these two benefits produces an individual's *inclusive fitness* and provides a fitness accounting for the evolution of altruism. Even sterile casts of insects can evolve if they can sufficiently enhance the reproduction of relatives. But the devil is in the details of what constitutes "sufficient," and Hamilton provided a calculus for determining this with the creation of the *kinship coefficient*. His rule that the net reproductive benefit derived by a relative must be greater than the cost incurred by the altruistic relative devalued by the reciprocal of the kinship coefficient provided the key for understanding the evolution of the most fundamental social behaviors and has been supported by many studies on insects and vertebrates.

That benefits and costs are affected by kinship also creates the conditions for the development of conflict within families and between the sexes (Trivers

1972b, 1974; Trivers and Hare 1976). Since parents are related equally to all their offspring, but offspring are more closely related to themselves than to their siblings, conflict over expected investment by parents will often emerge. In general, offspring will attempt to prevent parents from equally spreading resources among all their offspring. However, the relationship between parental investment and sexual competition may have been the most profound extension of kinship theory for shaping the study of social evolution. Armed with the results of Bateman's (1948) study of fruit flies that showed that reproduction by males and females was limited by different factors, Trivers proposed that when one sex invests more in the rearing of offspring than the other, members of the latter will compete for members of the former. In essence, the sex that invests more in rearing becomes a limiting resource, and the sex with "excess" resources should invest heavily in mating activities, including finding and guarding members of the parenting sex.

The second theory emerged from ecology. Pioneering work by Orians, Verner, and Willson showed that the mating system of red-winged blackbirds depended on how ecological conditions affected the sexes in different ways (Orians 1969). They showed that within one population both polygyny and monogamy could coexist, and that the behavior of females induced this mating-system variation. They argued that on landscapes with resource patches that varied in quality, males on those of the highest quality could defend enough resources to support the clutches of many females. On those of lower quality, however, territorial males would be lucky if they were able to secure enough resources to support the offspring of one female. What made their model unique was that it depended on assessments by females—the polygyny threshold—to determine whether mating polygynously or monogamously was in a female's best evolutionary interest.

The underlying premise that fitness depended on what others were doing set the stage for Maynard Smith's (1977) insight that the evolution of the four major mating systems is a game in which what is best for one sex depends on what is best for the other. Maynard Smith proposed that whether one sex should invest in rearing current offspring should depend on how likely the other sex was to continue rearing those offspring. Since parents' interests are similar but not always identical, Maynard Smith showed that there was a pair of strategies, one for the male and one for the female, that, when performed together, would lead to an evolutionarily stable strategy (ESS). Such a strategy would occur only if it did not pay either parent to depart from his or her strategy. Thus it was possible for biparental care, uniparental care proffered by either sex, or no parental care at all to be an ESS under specific ecological conditions. What the environment determines is the extent to which parental care is needed to rear offspring. If only one parent is needed, then the likelihood of either partner deserting depends on the ease of finding an additional mate. For females in particular, this likelihood is also affected by the extent to which parenting reduces her fertility in future reproductive episodes. For the first time social aspects of mating behavior could be characterized and tied to specific selective forces. Sexual asymmetries provided the lens for identifying how resources and predation affected parenting options. And when only one parent was required for rearing, determining which sex that should be depended on trade-offs shaped by both evolutionary history and current ecological circumstances.

While the social theorists were expanding the behavioral decision-making part of the polygyny-threshold model to account for the evolution of the full range of mating systems, ecologists were expanding the model to account for a wider range of ecological determinants. For Emlen and Orling (1977), the potential for polygyny depended on both the spatial and temporal distribution of the limiting sex. If females were widely scattered, then males would find it difficult to control more than one female. Alternatively, if females lived in groups, or resources were clumped so that females could temporarily aggregate without incurring much competition, then polygyny could result. But as in the Maynard Smith model, much would depend on the behavior of females; if females were receptive mostly at the same time, then the potential for polygyny could not be realized.

Emergence of a Synthetic Theory

The elements of a synthetic theory were crystallizing (Rubenstein and Wrangham 1986). Females were viewed as the driving force of social evolution because their social relationships reflected their best response to meeting the demands of the physical environment, and, as Alexander (1974) argued in his seminal review, these largely fell into two categories: resource distribution and predation pressure. Maximizing reproductive success by females depended mostly on solving problems posed by acquiring resources or finding safe sites for themselves or their young rather than on finding or securing mates. Only after associations among females and their distributions on landscapes were determined did male strategies associated with competing for mating opportunities come into play. Subsequent responses by females to actions by breeding males and, in turn, responses by breeding males to female behavior ultimately led to an evolutionarily stable mating system. But actions by subadults and development of relationships among members of different generations also came into play as social structures developed in complex societies (Rubenstein 1986, 1994).

The synthetic model predicted that in environments in which resources were abundant, especially when they were evenly distributed, competition would be low enough to permit females to aggregate. If sufficiently large foraging or antipredator benefits could be derived by these aggregating females, and the groups that formed were not too large, then these groups could be defended by single males, and so called harem-defense polygyny would result. However, if resources were more patchily distributed, so that competition among females was periodically intensified, then female group sizes would vary, and female associations would become more transitory. Rather than defending unstable groups of females, males instead would attempt to defend resource patches sought by females. In these systems of resource-defense polygyny the most able males would defend the best patches and thus gain access to the largest number of females for the longest periods of time. If resources were distributed not only patchily but in patches that were large, were widely separated, and fluctuated seasonally in abundance, then competition among females would be low. This would facilitate the formation of large groups provided that females could range widely and follow the shifting locations of peaks in food abundance. Males would thus be forced to either follow these large groups and compete for, and then tend, one reproductive female at a time *(wandering polygyny)* or position themselves at the intersection of female migratory routes and wait for females to visit them *(lekking polygyny)*. In either case intense male-male competition would generate a mating system based on male-dominance polygyny, and in the latter case females would be afforded the exquisite opportunity of simultaneously comparing many males before choosing with whom to mate. Whenever resources were sparsely but somewhat evenly distributed, high levels of competition would prevent females from forming groups. As a result, individual females would defend individual territories, thus ensuring a regular supply of a renewing resource. Since solitary individuals who were searching for members of the opposite sex would face heightened predation risk, pairs would often share territories, and monogamy would result.

Until empirical studies on antelope (Jarman 1974) and bats (Bradbury and Vehrencamp 1977a, 1977b) showed how ecology actually shaped female distributions, many of the underlying assumptions of this model were supported only by anecdotal correlations. Jarman's classic study of African antelopes illustrates how environmental forces interacted with physiological constraints to shape the mating system of a species. By showing how body size affected the ways in which different species perceived and then responded to the distribution and abundance of forage and predators on grasslands, Jarman showed why particular social systems increased survival and reproductive prospects for particular species. He argued that the smallest-bodied species, such as did-dik *(Madoqua kirkii)*, duikers *(Cephalophus* spp., *Sylvicapra* spp.), suni *(Neotragus moschatus)*, and klipspringers *(Oreotragus oreotragus)*, required limited amounts of high-quality vegetation. But because of their small size, such food items often appeared to be widely scattered. High levels of competition and intensified risks of predation made territoriality and monogamy appear the best strategies. Pairs generally lived in wooded or shrub-rich areas where moisture enabled vegetation to grow and renew itself well into the dry season. By signaling territorial ownership via scent rather than by means of sound or visual display, these small-bodied species reduced the chances that any of a large number of carnivores would prey upon them.

As species increased in body size, both physiologically determined dietary needs and the way acceptable forage became distributed on the landscape changed. Since crypsis became an untenable antipredator strategy for larger and more widely ranging species, forming groups became the best strategy for such species to lower predation risk. Fortunately, with larger size also came an ability to subsist on more abundant, lower-quality vegetation. When it was patchily distributed, as was the case for impala *(Aepyceros melampus)*, reedbuck *(Redunca* spp.), and some gazelles *(Gazella* spp.), males defended the best patches that females preferred. When the vegetation was more evenly distributed, which often resulted simply from the fact that larger species such as eland *(Taurotragus oryx)* and Cape buffalo *(Syncerus caffer)* could use even the lowest-quality items, larger groups formed. Because the largest species viewed large continuous swards of a landscape as acceptable, resource competition was virtually eliminated, and many males could associate with many females. With such high levels of male-male competition, defense of a small subgroup of females became impossible, and dominance defense systems developed.

The same sorts of connections between changing ecological circumstances and behavioral decision making shaped the types of sociality exhibited in other taxa. A brief survey of mammals illustrates some of the more general patterns. For the equids, the close association between food and water enables horse *(Equus caballus)* and plains zebra *(Equus burchelli)* females of different reproductive states to associate permanently. Thus males are able to defend such groups, and so-called harems form. When these two resources are widely dispersed, as for Grevy's zebra *(Equus grevyi)* and the Asiatic wild ass *(Equus hemionous)*, females of different states are precluded by metabolic constraints from foraging together. As a result, males compete for territories along traveling routes that take females from feeding areas to watering points (Rubenstein 1986, 1994).

In felids, females remain separated when food is scarce and habitats are densely wooded because in such circumstances individual prey can be caught and consumed before competitors can intervene. In more open habitats and where both prey and competitors are large and much more numerous, coalitions of females form to help hold on to kills until they are completely devoured (Packer 1986). If these female coalitions are themselves large, then there is pressure for males to aggregate to control reproductive access to females. Thus the only highly social felid is the savanna-living lion *(Panthera leo)*. In this plentiful landscape the leopard *(Panthera pardus)* remains solitary because it can safely cache its large prey in trees.

For canids, monogamy is the rule. But variations do occur: small-bodied foxes (*Vulpes* spp.) sometimes exhibit polygyny, and large-bodied hunting dogs *(Lycaon pictus)* and timber wolves *(Canis lupus)* develop polyandry (Moehlman 1986). Typically canids need help from nonbreeding foragers to help nourish lactating mothers. For the midsized jackals, such as the silverback jackal *(Canis mesomelas)*, the helpers are young from previous litters that cannot themselves find successful breeding locations. For the smallest species, however, prepartum investment in young is relatively small for a female; thus her mate provides all the help that is needed. In years when prey resources are very high, competition between mothers and their soon-to-be-fecund daughters is low, and their daughters are not forced to disperse. They therefore provide neighboring males with additional mating opportunities, and polygyny results. For the largest-bodied species, however, female prepartum investment is very high, and they need all the help they can get. To enlist the support of other adult males that must hunt cooperatively to capture large prey, a dominant female not only kills the offspring of other females in the group but also mates with their mates so that these males behave as if they are the sires of the dominant female's young. Hence, depending on size-determined metabolic investments and needs, social systems of canids can vary from polygyny to polyandry, and they sometimes use the services of juvenile or adult nonbreeders to help rear offspring.

As these examples illustrate, the comparative method benefited tremendously by the development of socioecological theory that emphasized the importance of females and their relationship to the environment in structuring societies. But aspects of kinship and demography also come into play. As Hamilton's rule illustrates, the stronger the degree of relatedness among relatives, the more likely altruists are to enhance the reproductive opportunities of kin while incurring costs associated with diminished personal reproduction. Thus in the previous examples, it is not surprising to find that the coalitions that form among lionesses when they are protecting kills and among male lions when they are defending mating opportunities with females are formed most often among full siblings (Packer 1986); nor is it surprising that the helpers jackal adults keep to rear additional offspring are themselves the full siblings of the offspring being raised (Moehlman 1986). In general, strong kinship lowers the threshold for the appearance of altruistic and cooperative behavior and may substantially affect the costs of living in groups.

The structure of jackal societies not only highlights the importance of kin selection but also illustrates the fact that many societies consist of adults of different generations, some of which care for offspring that are not necessarily their own. Many insects in the orders Hymenoptera and Isoptera and about 3% of birds and mammals show such systems composed of cooperative breeders (Emlen 1997). Up to this point the emerging synthetic theory of sociality has largely treated all adult females and males as independent decision makers; responding to ecological conditions shaped the best response for females, which in turn constrained the best responses of males. But in many societies mature offspring remain and help parents rear additional offspring. Ultimately, understanding why individuals forgo personal reproduction to help close relatives reproduce provides insights into how the balance between despotic and egalitarian social relationships, as well as the dynamics of families, is determined by interactions among ecology, kinship, and dominance (Emlen 1995).

In most species, offspring leave home before breeding. But in many social insects, primates, and other mammals and birds, one sex retains ties to parents, sometimes permanently, creating multigenerational families. Generally, offspring postpone dispersing when breeding opportunities are limited elsewhere. Ecological constraints associated with limited breeding sites, mates, or food resources often make the option of staying home and helping parents produce additional siblings better than dispersing early and almost certainly failing to produce any offspring. Thus many family groups will be unstable, forming when conditions are poor and disbanding when conditions are good. But when they form, intergenerational cooperation should be commonplace, as many studies have shown.

If family composition changes, however, then social dynamics can also change. Because close inbreeding often has deleterious effects, retained offspring rarely attempt to mate with a parent or close relative of the opposite sex. Thus competition with the same-sexed parent for reproductive opportunities remains low. If, however, a parent dies and is replaced by a nonrelative, then from the perspective of nondispersing adults, new mating opportunities will likely produce conflict. Similarly, if a nondispersing adult acquires a mate and remains in the group and limited resources are consumed by the offspring of this pair, then competition will be heightened. In either case the degree to which the reproductive activities of the nondispersing adult are tolerated will depend on the strength of the dominance-subordination relationship between the two. *Reproductive skew* describes the distribution of reproduction among breeding adults within families and can range from a value of 1 when the breeder in charge garners all reproduction to a value of 0 when breeding is completely egalitarian. The degree to which sharing of reproduction occurs will depend upon four conditions: (1) the benefit derived by dominants if subordinates stay; (2) the reproductive success of subordinates if they leave; (3) the genetic relatedness among potential cobreeders; and (4) the magnitude of the dominance relationship between potential cobreeders. When opportunities for the subordinate to breed elsewhere are high, the dominance difference among breeders is small, kinship is low, and the benefits of having helpers to the dominant are high, then the dominant breeder will be selected to pay a high *staying incentive* and *peace incentive* (Reeve and Ratnieks 1993) to ensure that a helper helps since helpers have options elsewhere and, because of nearly equal fighting ability, can inflict damage on the established breeder. As studies on bee-eaters (Emlen and Wrege 1992), Groove-billed anis (Vehrencamp 1978), Galápagos Hawk (Faaborg et al. 1995), and mongooses (Creel and Waser 1991; Keane et al. 1994) have shown, dominant breeders making "concessions" to helpers is likely when helpers are not close kin since they gain little indirect benefit via kin selection.

But as might be expected, variants to this set of rules that shape staying and leaving decisions of helpers occur. Clutton-Brock and colleagues have questioned whether dominants have full control over reproduction within groups (Clutton-Brock et al. 2000). Their studies have shown that in cooperative breeding societies of meerkats, levels of concessions do not fit predictions based on kinship and the needs of dominants. As a result, other factors, such as mutualisms that enhance reproduction for every group member and that emerge as meerkat groups get larger (Clutton-Brock et al. 2002), may come into play.

Additional Social Glues

As meerkats illustrate, not all societies need be held together by the glue of kinship. Relationships built around cooperation rather than altruism or based on mutualism and reciprocity can have just as strong an effect and provide the opportunities for different types of societies to evolve. Relationships are built upon repeated interactions that can involve competitive and cooperative elements. Manipulating the nature of personal relationships and those of companions should also determine the existence and characteristics of many societies.

Trivers (1972a) was the first to suggest that reciprocity could play an important role in shaping societies. If individuals were to aid a stranger by increasing its survival or future reproduction at some cost to self in terms of diminished reproduction, then such behavior could only be maintained in a population if the recipient were to eventually repay the costs incurred by the altruist. This concept has been formalized as the *prisoner's dilemma*, where two individuals have two alternatives—cooperate or defect. Mutual cooperation, usually by staying true to each other despite attempts by authorities to cajole one partner to turn on the other, provides a higher reward *(R)* than if they both defect and are punished *(P)*. If only one partner turns the other, however, then the size of the temptation *(T)* necessary to break the partners' cooperation enables the cheat to do best of all. Conversely, the partner who remains cooperative to the end is the sucker and receives the lowest reward *(S)*. As long as payoffs are ordered $T > R > P > S$, defecting is favored since it provides the greatest short-term benefit. If, however, such situations occur frequently and the number of individuals in a society is small, then pairs of individuals will find themselves interacting time and again. In such a setting, defecting is no longer the best strategy. Rather, cooperation does best, and the tactic that often provides the greatest long-term reward is *tit-for-tat* (TFT; Axelrod and Hamilton 1981), in which interactants cooperate initially and continue to do so until a partner defects. If defection occurs, then the other partner defects and continues to match the behavior of the initial defector. If the initial defector again cooperates, then the responder is instantly forgiving and cooperates as well. TFT can resist invasion by a strategy of pure defection. But since it is equally as rewarding as a strategy of always cooperating, TFT can be superseded by such a strategy via drift. Since a purely cooperative strategy can be invaded by the selfish strategy of defection, the strategy of tit-for-tat cannot be truly an ESS, but it comes close.

The logic of this argument suggests that structuring societies on reciprocity alone will be difficult. Kinship can always provide a strong foundation since it reduces the need of fully repaying delayed benefits. By cooperating mutually to solve common problems, partners may benefit to such a large extent that the temptation to defect remains low enough that a sucker does better than a defector. West-Eberhard (1989) terms this *by-product mutualism* because cooperation is universally best regardless of what a partner does.

Studies on a variety of species suggest that mutualism can shape sociality. Long-term studies of lions by Packer (Packer and Ruttan 1988; Scheel and Packer 1991) and Stander (1992) showed that only by hunting together and by adopting individual roles were lions likely to bring down large prey. Similarly, strong relationships among females to protect cubs from attacks by infanticidal males or to hold on to prey in the face of other marauding predators provided simultaneous benefits that could not be obtained by acting alone.

Social Variants and Alternative Strategies

As the examples of families with cooperative breeders show, high population density and the intensified competition it engenders in the search for suitable territories with sufficient food or habitable burrows is the factor ultimately responsible for favoring the establishment of coalitions and the recruitment of helpers. Thus demographic factors, such as population density, as well as sex ratio and age structure that result from differences in phenotype-specific vital rates, will shape patterns of social organization and lead to within-species variation as ecological circumstances change. If mortality, for example, were age specific and higher for prereproductive females than for males, then high breeding sex ratios (males/females) would result, and mating relationships would generally become more polyandrous. If sex-specific patterns of mortality were reversed, however, then polygyny would become more common. Since some of these mortality concerns were also responsible for the variation in canid social structure described earlier, these mortality schedules can have important consequences.

The synthetic model of sociality that is emerging shows that a variety of environmental features influence the patterns of sociality exhibited by animals. In many cases these patterns are flexible, and species can vary in the system of social organization they exhibit depending upon environmental conditions. In equids, for example, although female asses typically live in transitory groups whose membership changes (Rubenstein 1986, 1994) when populations move from arid to mesic areas, social relationships can also change. As was found on an island off the coast of the southeastern United States, where food and water are abundant and close together, females with differing needs can coalesce into permanent groups. Males, which in arid areas are forced to establish territories along routes to and from water, respond by defending these groups much like males of horses, their close kin, and haremlike societies emerge.

Social variation, however, is not limited to differences among populations. Variation in social strategies can evolve within populations since not all individuals are equal; differences in ownership and past experience can affect the outcome of interactions. If individuals interact repeatedly, then the nature of their interactions will change over time so that contests end more quickly and peaceably, with subordinates submitting to dominants. In small or moderately sized groups, hierarchies form and social roles become codified. Although all individuals, regardless of rank, benefit when serious fighting is reduced because both winners and losers benefit by reducing wasted time on outcomes that are virtually assured, dominants typically derive the greatest reproductive benefits. In extended families of cooperative breeders, dominance plays an important role in determining whether reproduction is more or less equally apportioned among adult group members.

When dominance is extreme and kinship is low, reproductive options for adults are limited as they bide their time proceeding along normal ontogenetic trajectories en route to reproducing. In such situations shortcuts to breeding are often favored, especially if mortality rates of adults are size or age specific and are greater for larger and older males. Trading off longevity with low rates of reproduction for copious reproduction over a short period can lead to discrete size polymorphisms—so-called alternative male mating strategies—within sexes within populations. Often the typical strategy adopted by older and larger males of defending harems or resources yields the most reproductive gains during a breeding season, but it often incurs a high cost as well. Because displaying, fighting, attracting the attention of predators, and delaying reproduction while growing are all costly activities, males that adopt less instantaneously successful but also less costly tactics can flourish.

The maintenance of such alternative mating patterns is common among many species of insects, fish, amphibians, birds, and mammals. For example, in bluegill sunfish *(Lepomis macrochirus)* males typically defend nest sites where they display, attract females, and then fertilize and guard the eggs they lay (Gross and Charnov 1980). Since only the largest males have the ability to defend nest sites, they must delay breeding for over seven years. As a result, smaller and younger males have evolved various cuckolding strategies that may in fact be equally successful evolutionarily. In one, the so-called sneak begins breeding at two years of age. Although very small, such sneaks are virtually all testes and because of their inconspicuousness can dive from the surface just as a mating pair releases sperm and eggs. By exuding large volumes of sperm, some fertilize a few eggs. In the other, males delay breeding for as long as do females. By being the same size and color as females, these males join the mating pair and apparently fool the displaying male into thinking that he is courting two females rather than just one. As the original pair releases its gametes, the female mimic does too, thus fertilizing some of the female's eggs. For these two strategies to be equally successful alternatives, the costs and benefits of each must vary inversely with the frequency of individuals that adopt these tactics. Although they sometimes do, as in the case of the bluegill sunfish, in other species they do not.

Underlying Mechanisms: From Hormonal Control to Self-Organization

Decision making underlies the emergence of animal societies. Although comparisons of costs and benefits provide the ultimate explanation of when particular social variations are favored by selection, the way information is processed and used in making social decisions requires focusing on a different level of analysis. In the past, studies of sociality focused on only one scale: either why individuals interacted the way they did or on how such behavior was triggered. More recently, studies have begun moving across scales, combining the two to provide a more integrated understanding of the causes of social behavior.

One form of integration examines how information within organisms organizes behavior. Since much of behavior is conditioned on features of the physical or social environment, ensuring that appropriate behaviors appear at the right time, in the right context, and in the presence of the right partners requires coordination. Hormones are coordinators par excellence. When two animals are ready to mate, they must both be in the right mood; when they mate, mature gametes must be ready and waiting; and when the age or size for the transition from subadult to mature adult is reached, many phenotypic features must be reorganized. Which substances control what processes and whether they act on the brain or other organs can vary, but initiating and then coordinating and integrating many aspects of social behavior require information flow. Ultimately the processes involved constrain what behavior is possible.

Another type of integration involves understanding how collective behavior emerges from the actions of individuals (Levin 1999). Often patterns of behavior viewed at one scale appear imposed from above, whereas at another scale they appear to emerge from below as if self-organized. Thus animals may assemble in particular habitats because of prevailing environmental conditions, but whether or not they form tightly knit groups of particular shapes and sizes depends on how they interact with each other. Simple rules of repulsion, attraction, and alignment associated with interaction zones can account for the size, shape, composition, and direction of group movements. Using a set of hierarchical decision rules that balanced repulsion to avoid competition and attraction to reduce risks of predation, Gueron and colleagues (1996) showed that groups displayed a wide variety of shapes and tendencies to fragment and self-sort by phenotype. All that was required to induce these transitions was information about the speed and position of close neighbors. Much depended on the existence of a neutral zone where the effects of neighbors did not alter intrinsic tendencies. Similar models in three dimensions, but where the presence of a neutral zone allowed individuals to align with the orientation of a nearest neighbor, revealed how small changes in individual behavior could dramatically change the shape of schools of fish and flocks of birds (Couzin and Krause 2003). Depending on the strength of orientation, groups could move from disorganized swarms to highly polarized conformations. In fact, strength of alignment played a critical role in understanding leadership, information transfer, and consensus decision making within groups even when informed individuals did not know the quality of their information or whether they were in the majority or minority (Couzin et al. 2005).

From Animals to People

Humans are animals, so it seems self-evident that evolutionary processes should shape human sociality. First principles, especially those that underlie the synthetic model of sociality in nonhuman animals, have stimulated and organized the study of human social behavior. Not surprisingly, connections among particular features of social evolution and particular aspects of human nature have been identified. Nowhere is this more apparent than in the area of mate choice. In rodents genes associated with the major histocompatibility complex (MHC) affect cell recognition and immune function. During estrus females associate with males that differ genetically from themselves, whereas when they are pregnant, they associate with females that are genetically similar. Assessment is via odor and suggests that MHC-specific associations are evolutionarily adaptive. Diversifying the gene pool of offspring should improve their immune function, and associating with close kin during pregnancy should reduce costs and increase benefits associated with sociality (Penn and Potts 1999). When Wedekind and colleagues (1995) performed a study to see if human females showed similar sexual preferences, he found that women in the fertile phase of their cycle identified the odor of T-shirts from males with whom they shared MHC genes as less "pleasant" than those of males with whom they shared fewer genes. Apparently the same underlying forces and mechanisms employed by animals may be operating in humans when it comes to choosing mates. But by no means do these results demonstrate that processes that shape the sociosexual behavior of animals will necessarily shape the behavior of humans in similar ways. Unfortunately, the links between animals and people are often presented as universally true when in fact they show variation depending on the nature of selective forces. By ignoring the contingency of selection, claims of universality of human social institutions and their immutable nature are likely to be false. As many of the examples of animal sociality explored earlier have demonstrated, particular types of social relationships and social systems thrive only under a limited set of environmental conditions. Why, then, should we expect that only one set of human relationships or only one type of society should evolve?

It is somewhat ironic that political theory itself is a human construct with each predicated on a different set of assumptions about human nature. Conservative philosophies often depict humans as fundamentally selfish, competitive, and somewhat invariant, while more liberal philosophies often assume that humans are flexible and adaptable. As is the case with all animals, there is no one animal nature. Aggression and dominance are offset to varying degrees by cooperation and altruism. What sets human societies apart from those of most other animals is their complexity. More levels of organization and more varieties of relationships can form, thus making interactions more diffuse. But this does not mean the rules that structure animal societies do not apply to those of humans. Perhaps they are insufficient, but they are not necessarily invalid.

One of the most compelling examples where evolutionary rules of sociality can explain a novel set of human relationships is the so-called mother's brother's phenomenon, in which men invest in their sisters' children as opposed to their own children (Alexander 1979a, 1979b). At first glance this strategy appears to violate the most basic rules of kinship. All else being equal, a parent shares on average twice as many genes with its own offspring as with a sibling's offspring. But in some societies not all else is equal. This is especially true in many West African societies, such as that of the Tonga (Flinn 1981; Irons 1981). The land upon which these horticulturalists live is rich, and as a result, females can derive enough resources from their labor to raise their children without the aid of their husbands. In such cultures female philandering is high, and females can initiate divorce. As a result, cuckoldry is common, and men are not likely to be fathers of their putative children. Even if offspring are only half sibs, men will be more closely related to their sister's children than to their own putative offspring. Thus, according to kinship theory, men are behaving evolutionarily rationally since they are making the best of a bad job and investing in the closest genetic relative available.

Applying cost-benefit and gene-accounting principles to the study of human social behavior is the cornerstone of sociobiology. Much of early sociobiology sought to explain humanity's social roots and thus studied the social behavior of existing cultures minimally impacted by Western societies. The hope was selective pressures that currently operated were the same as those that had operated in the past. Although the assumption of continuity is unlikely to be true, it does not invalidate the underlying sociobiological search for identifying how environmental pressures shape human relationships among and between sexes and within and between families. Moreover, as in most animal societies, individuals in these non-Westernized societies are tied closely to their environment. Thus ecological, social, and reproductive variables can be measured to identify patterns of causation.

Perhaps spurred on by the Darwinian investigation of human sociality, psychologists also began examining human social behavior from an adaptive perspective, but their approach was fundamentally different because they sought to identify how natural selection shaped the human psyche over evolutionary time. For psychologists, the ultimate question was, "Are the mechanisms that our brains use to make decisions about choosing mates or ostracizing cheats the ones that we would expect if they were 'designed' to solve such problems?" (Barrett et al. 2002). Thus while sociobiologists focused on reproductive outcomes to measure adaptation and often ignored the mechanisms that underlie assessment and decision making, evolutionary psychologists focused on identifying the mechanisms of assessment and information processing with little regard for reproductive outcomes. Just as an integrated approach to the study of social behavior has emerged within the biological tradition, an approach that integrates mechanistic and functional quests for understanding the adaptive value of the human mind and the role it plays in shaping social behavior would be the best of both worlds. It would cut across scales and provide more complete insights into why humans behave the way they do.

Although humans are animals, they are exceptional animals in many ways. The language they use is more complex than the communication systems of other animals; the ways in which they manipulate and deceive and give support to and elicit it from others are more subtle than in any other species; and only humans have developed cultural norms and traditions that can be modified and passed on to children. Evolutionary rules will have much to say about the adaptive value of these behaviors. And while we search for links between particular human practices and particular circumstances, we should not abandon the search for universals that might underlie the organization of human sociality.

Human societies often become subdivided into collectives composed of individuals who are similar genetically or phenotypically. Perhaps this is because similarity tightens connections and repeated interactions strengthen social relationships. The same tendencies for individuals to associate with others like themselves are often seen in animal societies as they get larger (Rubenstein and Hack 2004). Different environments will continue to favor certain types of interaction over others, but familiarity will reduce uncertainty and increase the likelihood that cooperation, whether based on altruism, reciprocity, or mutualism, will be strong. Language is a tool that humans use to reinforce connections, to mete out punishment, and to broadcast the exploits of cooperators. In general, ideas—the memes of Richard Dawkins (1976)—can be transmitted among generations just like genes, with good ones spreading and bad ones disappearing. The evolution of social norms that extol virtue may be the extra piece of the equation that differentiates humans from most other social species. Testing this proposition is essential if we are ever to know what connects and sets us apart from other species, but it will be made virtually impossible if it is intertwined with political agendas.

BIBLIOGRAPHY

Alexander, R. D. 1974. The evolution of social behavior. *Annual Review of Ecology and Systematics* 5: 325–383.
———. 1979a. *Darwinism and Human Affairs*. Seattle: University of Washington Press.
———. 1979b. Evolution and culture. In N. A. Chagnon and W. Irons, eds., *Evolutionary Biology and Human Behavior: An Anthropological Perspective*, 59–78. North Scituate, MA: Duxbury Press.
Axelrod, R., and W. D. Hamilton. 1981. The evolution of cooperation. *Science* 211: 1390–1396.

Barrett, L., R. Dunbar, and J. Lycett. 2002. *Human Evolutionary Psychology.* Princeton, NJ: Princeton University Press.

Bateman, A. J. 1948. Intra-sexual selection in Drosophila. *Heredity* 2: 349–368.

Bradbury, J. W., and S. L. Vehrencamp. 1977a. Social organization and foraging in emballonurid bats. III. Mating systems. *Behavioral Ecology Sociobiology* 2: 1–17.

———. 1977b. Social organization and foraging in emballonurid bats. IV. Parental investment patterns. *Behavioral Ecology Sociobiology* 2: 19–29.

Clutton-Brock, T. H., P. N. M. Brotherton, M. J. O'Riain, A. S. Griffin, D. Gaynor, L. Sharpe, R. Kansky, M. B. Manser, and G. M. McIlrath. 2000. Individual contributions to babysitting in a cooperative mongoose, *Suricata suricatta. Proceedings of the Royal Society of London, Series B* 267: 301–305.

Clutton-Brock, T. H., A. F. Russell, L. L. Sharpe, A. J. Young, Z. Balmforth, and G. M. McIlrath. 2002. Evolution and development of sex differences in cooperative behavior in meerkats. *Science* 297: 253–256.

Couzin, I. D., and J. Krause. 2003. Self-organization and collective behavior of vertebrates. *Advances in the Study of Behavior* 32: 1–67.

Couzin, I. D., J. Krause, N. R. Franks, and S. A. Levin. 2005. Effective leadership and decision making in animal groups on the move. *Nature* 433: 513–516.

Creel, S. R., and P. M. Waser. 1991. Failures of reproductive suppression in dwarf mongooses *(Hologale parvula)*: Accident or adaptation? *Behavioral Ecology* 2: 7–15.

Crook, J. H. 1964a. The evolution of social organisation and visual communication in the weaver birds (Ploceinae). *Behaviour,* Suppl. 10: 1–178.

———. 1964b. Field experiments on the nest construction and repair behaviour of certain weaver birds. *Proceedings of the Zoological Society of London* 142: 217–255.

Dawkins, R. 1976. *The Selfish Gene.* New York: Oxford University Press.

Emlen, S. T. 1995. An evolutionary theory of the family. *Proceedings of the National Academy of Sciences USA* 92: 8092–8099.

———. 1997. Predicting family dynamics in social vertebrates. In J. R. Krebs and N. B. Davies, eds., *Behavioural Ecology: An Evolutionary Approach,* 228–253. Oxford: Blackwell Scientific.

Emlen, S. T., and L. W. Orling. 1977. Ecology, sexual selection, and the evolution of mating systems. *Science* 197: 215–223.

Emlen, S. T., and P. H. Wrege. 1992. Parent-offspring conflict and the recruitment of helpers among bee-eaters. *Nature* 356: 331–333.

Faaborg, J., P. G. Parker, L. DeLay, T. de Vries, J. C. Bednarz, S. M. Paz, J. Naranjo, and T. A. Waite. 1995. Confirmation of cooperative polyandry in the Galapagos Hawk *(Buteo galapagoensis)* using DNA fingerprinting. *Behavioral Ecology and Sociobiology* 36: 83–90.

Flinn, M. 1981. Uterine vs. agnatic kinship variability and associated cross-cousin marriage preferences: An evolutionary biological analysis. In R. D. Alexander and D. W. Tinkle, eds., *Natural Selection and Social Behavior: Recent Research and New Theory,* 439–475. New York: Chiron Press.

Fricke, H. W. 1975. Evolution of social systems through site attachment in fish. *Zeitschrift für Tierpsychologie (Ethology)* 39: 206–210.

Gross, M. R., and E. L. Charnov. 1980. Alternative male life histories in bluegill sunfish. *Proceedings of the National Academy of Sciences USA* 77: 6937–6940.

Gueron, S., S. A. Levin, and D. I. Rubenstein. 1996. The dynamics of mammalian herds: From individual to aggregations. *Journal of Theoretical Biology* 182: 85–98.

Hamilton, W. D. 1964. The genetical evolution of social behaviour: I and II. *Journal of Theoretical Biology* 7: 1–52.

Irons, W. 1981. Why lineage exogamy? In R. D. Alexander and D. W. Tinkle, eds., *Natural Selection and Social Behavior: Recent Research and New Theory*, 417–438. New York: Chiron Press.

Jarman, P. J. 1974. The social organisation of antelope in relation to their ecology. *Behavior* 48: 215–267.

Keane, B., P. M. Waser, S. R. Creel, N. M. Creel, L. F. Elliott, and D. J. Minchella. 1994. Subordinate reproduction in dwarf mongooses. *Animal Behavior* 47: 65–75.

Kruuk, H. 1975. Functional aspects of social hunting by carnivores. In G. Baerends, C. Beer, and A. Manning, eds., *Function and Evolution of Behaviour*, 119–141. Oxford: Clarendon Press.

Lack, D. 1968. *Ecological Adaptations for Breeding Birds*. London: Chapman and Hall.

Levin, S. A. 1999. *Fragile Dominion: Complexity and the Commons*. Reading, MA: Perseus Books.

Lewontin, R. C. 1979. Sociobiology as an adaptationist program. *Behavioral Science* 24: 5–14.

Maynard Smith, J. 1977. Parental investment: A prospective analysis. *Animal Behavior* 25: 1–9.

Moehlman, P. D. 1986. The ecology of cooperation in canids. In D. I. Rubenstein and R. W. Wrangham, eds., *Ecological Aspects of Social Evolution: Birds and Mammals*, 64–86. Princeton, NJ: Princeton University Press.

Orians, G. 1969. On the evolution of mating systems in birds and mammals. *American Naturalist* 103: 589–603.

Packer, C. 1986. The ecology of sociality in felids. In D. I. Rubenstein and R. W. Wrangham, eds., *Ecological Aspects of Social Evolution: Birds and Mammals*, 429–451. Princeton, NJ: Princeton University Press.

Packer, C., and L. Ruttan. 1988. The evolution of cooperative hunting. *American Naturalist* 132: 159–198.

Penn, D., and W. Potts. 1999. The evolution of mating preferences and major histocompatibility complex genes. *American Naturalist* 153: 145–164.

Reeve, H. K., and F. L. W. Ratnieks. 1993. Queen-queen conflict in polygynous societies: Mutual tolerance and reproductive skew. In L. Keller, ed., *Queen Number and Sociality in Insects*, 45–85. Oxford: Oxford University Press.

Rubenstein, D. I. 1986. Ecology and sociality in horses and zebras. In D. I. Rubenstein and R. W. Wrangham, eds., *Ecological Aspects of Social Evolution: Birds and Mammals*, 282–302. Princeton, NJ: Princeton University Press.

———. 1994. The ecology of female social behavior in horses, zebras and asses. In P. Jarman and A. Rossiter, eds., *Animal Societies: Individuals, Interactions and Organisations*, 13–28. Kyoto: Kyoto University Press.

Rubenstein, D. I., and M. Hack. 2004. Natural and sexual selection and the evolution of multi-level societies: Insights from zebras with comparisons to primates. In P. Kappeler and C. P. van Schaik, eds., *Sexual Selection in Primates: New and Comparative Perspectives*, 266–279. Cambridge: Cambridge University Press.

Rubenstein, D. I., and R. W. Wrangham. 1986. Socioecology: Origins and trends. In D. I. Rubenstein and R. W. Wrangham, eds., *Ecological Aspects of Social Evolution: Birds and Mammals*, 3–20. Princeton, NJ: Princeton University Press.

Scheel, D., and C. Packer. 1991. Group hunting behaviour of lions: A search for cooperation. *Animal Behavior* 41: 697–709.

Stander, P. E. 1992. Cooperative hunting in lions: The role of the individual. *Behavioral Ecology Sociobiology* 29: 445–454.

Tinbergen, N. 1963. On aims and methods of ethology. *Zeitschrift für Tierpsychologie (Ethology)* 20: 410–433.

Trivers, R. L. 1972a. The evolution of reciprocal altruism. *Quarterly Review of Biology* 46: 35–57.

———. 1972b. Parental investment and sexual selection. In B. Campbell, ed., *Sexual Selection and the Descent of Man,* 136–179. Chicago: Aldine-Atherton.

———. 1974. Parent-offspring conflict. *American Zoologist* 14: 249–264.

Trivers, R. L., and H. Hare. 1976. Haplodiploidy and the evolution of social insects. *Science* 191: 249–263.

Vehrencamp, S. L. 1978. The adaptive significance of communal nesting in Groove-billed anis *(Crotophaga sulcirostris). Behavioral Ecology and Sociobiology* 4: 1–33.

Wedekind, C., T. Seebeck, F. Bettens, and A. J. Paepke. 1995. MHC-dependent mate preferences in humans. *Proceedings of the Royal Society of London, Series B* 260: 245–249.

West-Eberhard, M. J. 1989. Phenotypic plasticity and the origins of diversity. *Annual Review of Ecology and Systematics* 20: 249–278.

Williams, G. C. 1966. *Adaptation and Natural Selection.* Princeton, NJ: Princeton University Press.

Wynne-Edwards, V. C. 1962. *Animal Dispersion in Relation to Social Behaviour.* Edinburgh: Oliver and Boyd.

Human Evolution

Henry M. McHenry

The richness of the human fossil record allows reasonable confidence in reconstructing our evolutionary history. This record demonstrates that humanlike forms appeared by about 6 million years ago and accumulated more and more human characteristics through time. As Darwin and his contemporaries predicted, the human lineage shows clear evidence of descent with modification. Perhaps the best demonstration of this is the expansion of the brain through time (Table 1). Fossil species that were clearly humanlike in body and dentition 6 to 3 million years ago had ape-sized brains that were one-third the size of those of modern humans. Just after 2 million years ago some fossils show an increase to about half the size of *Homo sapiens,* and over the past half million years, modern brain size appears in the fossil record (Conroy 2005; Schwartz and Tattersall 2002, 2003, 2005; Delson et al. 2000).

Although Darwin and other early evolutionists were keenly interested in human evolution and made great strides in interpreting comparative morphology from an evolutionary point of view, very few premodern human fossils were known in their day (Huxley 1863). By the turn of the twentieth century, bits of Neandertals and "Java man" entered the consciousness of evolutionary biologists (Keith 1915). By the turn of the twenty-first century, thousands of premodern fossils had found their way into the world's museums, and knowledge of human evolution had grown spectacularly (Delson et al. 2000; Conroy 2005; Klein 1999; Schwartz and Tattersall 2002, 2003, 2005; Holloway et al. 2004). The richness of the record is due to the intelligence, skill, and diligence of the fieldworkers who find the fossils and determine their biological and geological contexts.

Our Place in Nature

It was clear to Darwin and his contemporaries that people shared a common ancestor with living apes more closely than with any other animal. Thomas Huxley (1863) was a particularly strong advocate for an apelike ancestry of the human line, and Haeckel (1868) imagined and named a specific series of forms from apelike to humanlike.

Curiously, their ape-man theory (as it was later called) was seriously challenged during the first seven decades of the twentieth century. These challenges included the view that the last common ancestor we shared with the animal world resembled the tarsier more closely than any ape (Jones 1918, 1926, 1929). Other early twentieth-century scientists reconstructed the last common ancestor as more monkeylike than apelike (Straus 1949) or as some kind of "pre-ape" without long fingers and upper-body specializations of living apes (e.g., Napier and Davis 1959). Because modern humans appear to be so divergent in shape and behavior from other animals, the time of divergence of the human evolutionary lineage was often given as quite ancient. The nineteenth-century "ape-man" theory continued to have supporters (e.g., Gregory 1916, 1926, 1927a, 1927b, 1928, 1934; Washburn 1951), but until the last three decades of the twentieth century these were relatively lonely voices (Fleagle and Jungers 1982).

Among the many important changes in view and method in human evolutionary studies during the last few decades of the twentieth century was the application of molecular biology. Immunology provided the first insight. By comparing the strength of the immune system's response by one species to proteins of another species, Goodman (1963) quantified an estimate of genetic relatedness among primate species. The results were shocking at the time to most comparative anatomists. Contrary to prevailing understanding at the time, humans were more similar to African apes (chimpanzee and gorilla) than African apes were to the Asian great ape (orangutan). Monkey species were far more distantly related. More sophisticated immunological methods (Sarich and Wilson 1966) and DNA hybridization (Hoyer et al. 1964) confirmed these early results and revealed that changes in some macromolecules (e.g., albumin) were relatively constant through time (Sarich and Wilson 1967). Constancy of change permitted estimates of time of divergence of evolutionary lineages as long as the rate could be calibrated by a known date of divergence and thus provide a basis for development of a "molecular clock." Sarich and Wilson (1967) used the date of 60 million years for the divergence of several mammalian orders to calculate the rate of change in albumin. From this they calculated that the divergence of the lineages that led to humans and African great apes occurred 4.2 million years ago. Subsequent refinements place the date somewhere between 10 and 5 million years by most current estimates (Kumar and Hedges 1998; Kumar et al. 2005; Holmes et al. 1989; Chen and Li 2001; Arnason et al. 2000; Pilbeam and Young 2004; Patterson et al. 2006; Glazko and Masatoshi 2003; Stauffer et al. 2001). With the advent of DNA sequencing it is now clear that humans are very closely related to the African great apes, and the time of divergence of their evolutionary lineages is relatively short from the perspective of mammalian evolution.

A second important change in the later part of the twentieth century was in the methods of comparing morphology of species. Molecular approaches cannot be applied to fossils much older than 70,000 years (Krause et al.

Table 1. Species, dates, body size, brain size, and posterior tooth size in early hominids

Taxon[1]	Dates[2]	Mass[3]		Stature[4]		ECV[5]	Brain weight[6]	Postcanine tooth area[7]	EQ[8]	MQ[9]
		Male	Female	Male	Female					
Pan troglodytes	Extant	49	41	—	—	—	395	294	2.0	0.9
Sahelanthropus tchadensis	7–6	—	—	—	—	365	—	—	—	—
Orrorin tugenensis	6	41	—	131	—	—	—	329	—	—
Ardipithecus kadabba	5.2–5.8	—	—	—	—	—	—	347	—	—
Australopithecus anamensis	4.2–3.9	51	33	151	105	—	—	428	—	1.4
Australopithecus afarensis	3.9–3.0	45	29	138	115	438	434	460	2.4	1.7
Australopithecus africanus	3.0–2.4	41	30	—	—	452	448	516	2.7	2.0
Paranthropus aethiopicus	2.7–2.2	—	—	—	—	410	407	688	—	—
Paranthropus boisei	2.3–1.4	49	34	137	124	521	514	756	2.7	2.7
Paranthropus robustus	1.9–1.4	40	32	132	110	530	523	588	3.0	2.2
Australopithecus garhi	2.5–?	—	—	—	—	450	446	—	—	—
Homo habilis	1.9–1.6	37	32	131	100	612	601	478	3.6	1.9
Homo rudolfensis	2.4–1.6	60	51	160	150	752	736	572	3.1	1.5
Homo ergaster	1.9–1.7	66	56	180	160	871	849	377	3.3	0.9
Homo erectus	1.1–0.01	66	56	—	—	1,050	1,019	402	4.0	1.0
Homo heidelbergensis	0.6–0.2	77	56	—	—	1,194	1,156	389	4.2	0.9
Late Homo neanderthalensis	0.2–0.03	77	66	167	158	1,414	1,362	335	4.7	0.7

Homo floresiensis	0.09–0.01	—	29	—	106	380	378	—	2.6	1.8
Homo sapiens	Extant	58	49	175	161	—	1,350	334	5.8	0.9

1. Taxonomy based on Klein (1999) with additions from Brunet et al. (2002), Senut et al. (2001), Brown et al. (2004), and Haile-Selassie et al. (2004).

2. Dates (in millions of years ago) are from Klein (1999), with additions from Brunet et al. (2002), Senut et al. (2001), Brown et al. (2004) and Haile-Selassie et al. (2004).

3. Body mass estimates (in kg) are from McHenry (1992) except for the following: *O. tugenensis* is from Nakatsukasa et al. (2007), where 41 is the mid-point of their range and they did not specify sex; *A. anamensis* male is from Leakey et al. (1995); *A. anamensis* female is calculated from the ratio of male and female in *A. afarensis*; *H. ergaster, H. erectus, H. heidelbergensis*, and late *H. neanderthalensis* are from Ruff et al. (1998); and *H. floresiensis* is from Brown et al. (2004)

4. Stature estimates (in cm) are from McHenry (1991) except for *O. tugenensis*, which is from Nakatsukasa et al. (2007), where 131 is the midpoint of their range and they did not specify sex; *H. ergaster*, which is from Ruff and Walker (1993); *H. erectus, H. heidelbergensis*, and late *H. neanderthalensis*, which are from Ruff et al. (1998); and *H. floresiensis*, which is from Brown et al. (2004).

5. ECV is cranial capacity in cc from sources listed in McHenry (1994a; 1994b) with the addition of *S. tchadensis* (Zollikofer et al. 2005); A.L. 444-2 (540 cc) to *A. afarensis* (Kimbel, personal communication); Stw 505 (515 cc) to *A. africanus* (Conroy et al. 1998); *P. aethiopicus* (Walker et al. 1986); KGA 10-525 (545 cc) to *P. boisei* (Suwa et al. 1997); and BOU-VP-12/130 (450 cc) to *A. garhi* (Asfaw et al. 1999). *H. erectus, H. heidelbergensis*, and late *H. neanderthalensis* are from Ruff et al. (1998), and *H. floresiensis* is from Brown et al. (2004).

6. Brain weight is calculated from formula (6) in Ruff et al. (1998).

7. Postcanine tooth area (in mm²) is the sum of products of buccal-lingual and mesio-distal lengths of P_4, M_1, and M_2 and is taken from McHenry (1994b) with the addition of *S. tchadensis* estimated from TM266-02-154-1 reported in Brunet et al. (2002); *O. tugenensis* from Senut et al. (2001); *Ar. kadabba* from Haile-Selassie et al. (2004); *Ar. ramidus* from White et al. (1994); *A. anamensis* from Leakey et al. (1995); *H. erectus, H. heidelbergensis*, and late *H. neanderthalensis* from Wolpoff (1971, 1982); and *H. floresiensis* from Brown et al. (2004).

8. EQ is the encephalization quotient calculated as brain mass divided by (11.22×body mass$^{0.76}$) from Martin (1981).

9. MQ is the megadontial quotient derived as postcanine tooth area divided by (12.15×body mass$^{0.86}$) from McHenry (1988).

2007; Orlando et al. 2006), so morphology is the only guide to understanding relationships. Close evolutionary relationships can be inferred from similarity in morphology (see Figure 1), but not without careful analysis of the causes of resemblances. Evolutionary biologists have always been aware of this, of course, but a formal approach to make comparisons of form only took hold in the later part of the twentieth century.

Willi Hennig (1966) formalized the approach to comparing morphology of species by noting that resemblance of form did not necessarily imply close phylogenetic affinity. Similarity may be due to sharing primitive characteristics that are not helpful in determining biological relationships. Aspects of the dentition of the 28-million-year-old *Propliopithecus,* for example, resemble those of modern humans more than those of modern apes (Kurten 1972), but these are due to the retention of ancestral traits and not to the evolution of unique traits shared only by descendants and not present in other closely related species. Sometimes similarities are due to parallel or convergent evolution rather than descent from a common ancestor. The wings of birds, bats, and butterflies are similar, but not because they evolved from the same winged ancestor. The term *homoplasy* is used to describe the phenomenon where organisms share unique traits that evolved independently in separate

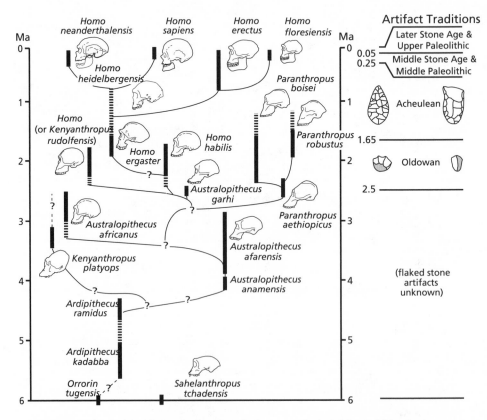

Figure 1. A working hypothesis of evolutionary relationships within the human fossil record, redrawn from R. G. Klein.

lines of descent. Homoplasy is pervasive in human evolution (McHenry 1996). Because Hennig's method focuses on the characteristics of evolutionary divergence, it is often referred to as *cladistics*. A *clade* in this vocabulary is a unique evolutionary branch of the tree of life defined by sharing unique traits that are not shared in closely related branches. Hennig called such shared unique traits "synapomorphies," which is usually translated as "shared derived."

The power of this logic pervades evolutionary biology (Hull 1988) and helps reconcile the findings of molecular biology with the interpretation of the fossil record. As a result, the terms used to classify our biological relatives now reflect our true genetic relationships. The old tradition placed great apes (orangutans, gorillas, and chimpanzees) into the family Pongidae and humans into Hominidae, but the true genetic affinities force a reclassification that recognizes the Asian great ape as the family Pongidae and the African clade as the family Hominidae. Within the family Hominidae are the subfamilies Gorillinae (gorilla), Paninae (chimpanzee), and Homininae (humans). It is a useful reminder that we are tightly nested in the bush of life's organic forms and that the origin of our unique lineage is very recent.

Before Hominins

Our close genetic relationship to the African great apes implies an African origin of our evolutionary lineage. Fossil discoveries confirm this. Unfortunately, the fossil record of the origin of the African clade (gorillas, chimpanzees, and humans) is poorly documented. The Asian clade of our family is better known. Discoveries in the foothills of the Himalayas are particularly revealing. In sediments that date between about 12 and 8 million years ago, forms appear that resemble the orangutan in many specific facial features (Pilbeam 1982). *Sivapithecus* is the genus name that includes fossils with the uniquely orang-like features of narrow space between the eyes, "prowlike" profile of the face below the nose, construction of the palate, and many other features that among the living species occur in the orangutan *(Pongo)*.

The face of the Asian clade (*Sivapithecus* and *Pongo*) contrasts with that of the African clade (gorillas, chimpanzees, and humans), but the resemblance of these living species may be due to sharing primitive characteristics and may not, by the logic of cladistics, be useful in determining true relationships. Most paleontologists accept these resemblances as sufficiently unique to qualify European fossil genera such as *Dryopithecus* (8–13 million years ago), *Ouranopithecus* (9–10 million years ago), and perhaps *Pierolapithecus* (12.5–13 million years ago) (Moyà-Solà et al. 2004; Begun and Ward 2005) as part of the African clade (Begun 1994; Fleagle 1999). The fossil record in Africa for Miocene members of Hominidae (African apes and humans) is sparse, but diligent attempts continue to promise new discoveries. *Kenyapithecus* (16 to 14 million years ago) and related genera (*Equatorius,* 16 to 15 million years ago) appear to be part of the African clade with living members, but each has its own specializations that make them less attractive as candidates for ancestry of living species of Hominidae (McCrossin and Benefit 1997).

The paleoenvironment within which our ancestors evolved was influenced profoundly by dramatic fluctuations in world climates over the past 7 million years (Vrba et al. 1995; Bromage and Schrenk 1999; Potts 2007; Pickford 2006; Behrensmeyer 2006). Intense periods of cold temperatures in high latitudes resulted in episodes of drought in the tropics. These appeared at 16–13, 6.5–5, and 2.8 million years ago, and beginning at 2.5 million years ago, the earth experienced 21 cycles of glaciation in higher elevations and latitudes accompanied by reduction in rainfall in equatorial lands, especially Africa. Habitats favored by hominids (ancestors of the African apes and humans) contracted and expanded. The fossil record of hominids in the Middle and Late Miocene is frustratingly incomplete, and almost nothing is known of the lineages that led to chimpanzees and gorillas. Surprisingly, much more is known of the fossil record of our own branch, the Homininae.

The Earliest Hominins

Recognition of the earlier members of Homininae is guided by understanding gained from molecular biology and cladistics. The molecular clock brackets the time of divergence somewhere between 10 and 5 million years by most current estimates. By the logic of cladistics, the earliest ancestor of humans can be recognized by the presence of traits unique to the lineage. This is obvious but is easily confused by naive views of pre-Hennigian comparative morphology that continue to plague human evolutionary studies. If it looks like an ape, it is not necessarily excluded from the human realm. We now know of an extraordinary record of an apelike-becoming-humanlike sequence of fossils. The earliest member of this sequence looks very apelike. By cladistic logic, that is exactly what one would expect.

The earliest candidate for membership in the subfamily Homininae is *Sahelanthropus tchadensis* (Brunet et al. 2002). This precious fossil skull from Chad was found in sediments that contained fauna linked with datable sediments in Ethiopia and North Africa 7 to 6 million years ago. Its overall appearance is quite apelike, including an estimated brain size slightly below the average for living chimpanzees (Table 1). But by the logic of cladistics, it shares with later hominins unique traits that define our lineage. The most important of these is the canine that, unlike that of apes and primates in general, does not project or wear against its lower premolar. This is a significant feature that was recognized as an important difference between apes and people by many, including Darwin (1872).

Small canines are significant. Darwin (1872) noted that when human ancestors first stood erect, they freed their hands and could brandish weapons for threat and fight, thus no longer relying on large canines. Small canines carry special significance, but it is not too difficult to imagine other reasons for selection that favors reduction to relieve the expense of growing such daggers. Another hominin novelty seen in *S. tchadensis* is the forward placement of the foramen magnum. In overall morphology (Zollikofer et al. 2005), however, *Sahelanthropus* is apelike, as befits a species close to the origin of the human lineage. Its brain is minuscule compared with that of modern humans. It has enormous brow ridges that are even larger than those of modern apes or earlier members of the genus *Homo* where brow ridges appear after having been less developed in earlier species of our lineage.

The discovery of *Sahelanthropus* is due to the persistence, skill, and cleverness of Michel Brunet and his team, including Djimdoumabaye Ahounta, the Chadian student who found the skull. Its location 2,500 km west of the East African Rift, where almost all the other earliest hominins were found, reminds us to be mindful of the limited extent of our sample.

Almost all of the very earliest fossil record of our evolutionary history comes from the African Rift Valley. Because of extraordinary geological events, many late Miocene to Recent terrestrial fauna are preserved in sediments layered by volcanic ashes that can be dated by the radioactive decay of naturally occurring potassium-40 to argon-40. Fossils buried beneath layers of ash in Ethiopia, Kenya, and Tanzania provide most of the evidence for the early part of the story of human evolution and the context in which it occurred.

The first chapter in this story comes from fossils beneath ash layers dated to just less than 6 million years ago in the Tugen Hills of Kenya (Senut et al. 2001; Pickford et al. 2002; Sawada et al. 2002). Named *Orrorin tugenensis*, this species combined apelike teeth and forelimbs with humanlike thighs that appear to be adapted to bipedality (Pickford et al. 2002; Richmond and Jungers 2008).

Slightly later in time (5.8–4.4 million years ago) is the collection of the Ethiopian specimens referred to as *Ardipithecus* (Haile-Selassie 2001; Wolde-Gabriel et al. 2001; Haile-Selassie et al. 2004; White et al. 1994, 1995). As with *Orrorin tugenensis* and as expected in an early ancestor, it is apelike in most respects, but it does share certain key traits with later hominins. Among these is a canine tooth that is less projecting and less pointed than that of fossil or modern apes and a cranial base that is shorter. Its toes apparently had the ability to hyperextend like those in bipedal humans (Haile-Selassie 2001). It lived in a closed woodland environment (WoldeGabriel et al. 1994, 2001).

Australopithecus and Other Hominins
4.2–2.5 Million Years Ago

The fossil evidence for hominin bipedality is hinted in interpretations of *Sahelanthropus tchadensis*, *Orrorin*, and *Ardipithecus*, but by 4.2–3.9 million years ago there appear in the fossil record fossils attributed to *Australopithecus anamensis* that possess a shinbone with numerous bipedal specializations (Leakey et al. 1995, 1998; Ward et al. 2001; White et al. 2006). This species retains numerous apelike characteristics, like its predecessor, but shares certain evolutionary novelties with all later hominins that are not present in the more primitive *Ardipithecus*. Most conspicuous is the thickening of enamel that becomes characteristic of all later hominins. Its upper canine has thickened enamel on the cheek side of its apex. The molars are expanded from side to side, and the first and second molars are not markedly different in size. The ear tube is more humanlike. Its deciduous lower first molar is reported to be intermediate between those of *Ardipithecus* and *A. afarensis* in size and shape. In the rest of the skeleton it is very much like later hominins (few details of *Ardipithecus* are published), especially its elbow, knee, and ankle.

The sample of *A. afarensis* (3.6–2.9 million years ago) is best known from Laetoli (Tanzania) and Hadar (Ethiopia), is later in time than *A. anamensis*, and shares derived characteristics with later hominins such as an ear tube that is more rounded in outline, a canine with a shorter crown and higher shoulders, a more humanlike lower first deciduous molar and permanent premolar, and a less sloping chin. Also, one of the wrist bones (the capitate) is less primitive (although not fully modern) (Johanson et al. 1982; Leakey et al. 1995, 1998; White et al. 2006; Alemseged et al. 2006; Kimbel et al. 2006).

Kenyanthropus platyops (3.5–3.3 million years ago) overlaps in time with *A. afarensis* but appears to be quite distinctive in its morphology (Leakey et al. 2001). In some respects it is more primitive than its contemporary. For example, it lacks a petrous crest on the tympanic bone and has a narrow external ear tube with a small aperture more like that in earlier hominins and not like that in *A. afarensis*. In other respects it resembles much later hominins, particularly *H. rudolfensis,* in having a relatively flat face, tall cheeks, small molars, and forward facing cheekbones. These latter traits are related to mastication that, according to Leakey et al. (2001, 437), "suggests a diet-driven adaptive radiation among hominins in this time interval." It is possible that the resemblances between *H. rudolfensis* and *K. platyops* are due to phylogenetically independent events driven by adaptations to similar situations and are, thereby, one more example of the pervasiveness of homoplasy. It is equally possible that the resemblances represent true homology and imply a clade that links *K. platyops* with *H. rudolfensis*. Leakey et al. (2001) favor the latter view, which suggests transferring *H. rudolfensis* to the genus *Kenyanthropus* (i.e., *K. rudolfensis*). Although the face of *Kenyanthropus* appears similar to that of *H. rudolfensis,* some or all of this resemblance may be due to expansion cracks that have distorted *Kenyanthropus* considerably (White 2003).

The next-oldest hominin species is *A. garhi* (2.5 million years ago). It has some shared derived traits with later and contemporary hominin species relative to *A. afarensis* (Asfaw et al. 1999). Relative to *A. afarensis*, it shares with *A. africanus* and the "robust" australopithecines absolutely and relatively larger cheek teeth. Its upper third premolar is reported to have a more oval outline and no mesiobuccal line extension. Its premolars are beginning to show some expansion (molarization), and its enamel is thicker. Its upper incisor roots are no longer lateral to the nasal aperture, as they are in *A. afarensis*. The relative sizes of the posterior temporalis and parietomastoid angle are beginning to reduce, as they do in all other species of post-*afarensis* hominin except *P. aethiopicus*.

The South African species, *A. africanus* (3–2 million years ago), is roughly contemporary with *A. garhi* and later in time than *A. afarensis*. It possesses some traits that appear to be derived and shared with *Homo* relative to these species (Asfaw et al. 1999). These include upper lateral incisor roots that are medial to the nasal aperture, a flatter clivus contour, reduced subnasal prognathism, reduced incisor procumbency, and reduced bipartite lateral anterior facial contour.

The "Robust" Australopithecines

In 1938 Robert Broom recognized a mostly unanticipated chapter in human evolution by his description of what he called *Paranthropus* (Broom 1938). Its teeth, jaws, and face were massively built (hence the name *robustus*), and Broom and subsequent authors regarded it as a side branch of human evolution (Robinson 1972). However, in many important respects *Paranthropus* shares many derived features with *Homo* (Skelton and McHenry 1992) and may be closely related to our genus. It is common to lump large-tooth hominins together as "robust" australopithecines, but this grouping may be inappropriate. The earliest of the "robust" australopithecines, *P. aethiopicus* (2.7–2.3 million years ago), shares very few derived characters with later "robusts" or *Homo* that are not related to hypermastication (Grine 1988). It has massive cheek teeth and chewing apparatus but retains the primitive skull characteristics of *A. afarensis* (Walker et al. 1986). The later "robust" species, *P. robustus* (2–1 million years ago) and *P. boisei* (2.3–1.4 million years ago), do share apparent evolutionary novelties with *Homo* relative to other australopithecines, such as a parabolic dental arcade shape, deep anterior depth of the palate, rare or absent diastema between the upper incisors and canine, a variable to weak canine jugum and fossa, reduced subnasal prognathism, and enlarged brain (Skelton and McHenry 1992; Strait et al. 1997; Strait and Grine 2004).

Origin of *Homo*

One might despair at the apparently conflicting evidence for the origin of the genus *Homo*, but from the perspective of the accumulation of shared derived traits, the fossil record is less perplexing. Brains expand, and cheek teeth reduce. In one species, *H. habilis* (2.3–1.6 million years ago), the body appears to remain like that of *Australopithecus*, small with relatively long forearms and powerfully built shoulders (Haeusler and McHenry 2007). If one accepts the association of the hindlimb fossils found at the same site and time as the craniodental remains of *H. rudolfensis* (~1.9 million years ago) as belonging to that species, then the more humanlike body proportions and hip architecture first appear in this species just after 2 million years ago (McHenry and Coffing 2000). Both of these species are transitional, with some primitive and some derived characteristics of later *Homo*. Postcranial associations are critical here because body size appears to be very different. *Homo habilis* was very small bodied (35 kg), and *H. rudolfensis* was large (55 kg) (McHenry 1994b). Scaling cheek-tooth size to body weight shows that they both had reversed the trend of ever-increasing cheek-tooth size. Relative brain size expanded, especially in *H. habilis*.

Absolute brain size expands further with the appearance of *H. ergaster/erectus* by at least 1.8 million years ago, but body size also increases, so relative brain size apparently was not dramatically expanded (Walker and Leakey 1993). The early African form of this species is often referred to as the species *H. ergaster* to contrast it with the well-known Asian sample of *H. erectus* (Wood 1991; Spoor et al. 2007). Body size and especially hindlimb length reach modern proportions in this species (Ruff and Walker 1993). Other synapomorphies with later *Homo* include a further reduction in prognathism, a reduction in the postglenoid process that is fused with the tympanic plate, and reduction in the absolute and relative size of the cheek teeth. Brains continue to expand, and cheek teeth get progressively smaller during the evolution of the genus *Homo*.

First Dispersal out of Africa

The first appearance of human populations outside Africa appears to have occurred soon after the appearance of early *Homo* in Africa. Fossil *Homo* remains from the Dmanisi site in the Republic of Georgia are most likely older than 1.7 million years and are very much like those found in Africa attributed to *H. erectus/habilis* (Gabunia and Vekua 1995; Gabunia et al. 2000; Vekua et al. 2002; Lordkipanidze et al. 2005, 2007). *Homo erectus* occupied parts of tropical Asia perhaps as early as 1.8 million years ago (Swisher et al. 1994). The exact chronology is still uncertain, but sometime after this and before 0.5 million years ago some populations had adapted to life in the temperate climatic zone of northern Asia (Antón and Swisher 2001).

An archaic species of *Homo* left tools on the island of Flores, Indonesia, as early as 800,000–700,000 years ago (Brumm et al. 2006; Morwood et al. 1998), and in surprisingly young sediments (95,000–12,000 years ago) a new species, *H. floresiensis*, has been recovered (Brown et al. 2004; Morwood et al. 2004, 2005; Morwood and van Oosterzee 2007). It has a small body (29 kg), minute endocranial volume (380 cc), and archaic body (Larson et al. 2007; Tocheri et al. 2007). It is most likely a descendant of an early dispersal of archaic *Homo* that underwent insular dwarfism (Brown et al. 2004), although some challenge this interpretation (Argue et al. 2006; Jacob et al. 2006; Hershkovitz et al. 2007).

The geologically late occurrence of such a primitive species of *Homo* demonstrates the fallacy of regarding human evolution as passing through progressive stages. Populations adapt to local conditions and diverge from other populations unless checked by gene flow.

Archaic *Homo* to *Homo sapiens*

Larger-brained species of archaic *Homo* begin to appear in Africa by 600,000 years ago and in Eurasia somewhat later (Clark et al. 1994). The Neandertals dominate Europe and western Asia from approximately 300,000 to 40,000 years ago (Klein 2003; Kuhn 2007; Krause et al. 2007). *H. sapiens* appears first in Africa sometime before 100,000 years ago (Klein 1999, 2000). The earliest traces of Upper Paleolithic tool technology that later became the hallmark of earliest members of *H. sapiens* first appear in sub-Saharan Africa before about 90,000 years ago (Brooks et al. 1995; McBrearty and Brooks 2000). By about 50,000 years ago populations of our species spread out of Africa to Asia and then Europe, apparently replacing or genetically swamping local archaic *Homo* populations (Anikovich 2007; Goebel 2007). The spread was apparently rapid in Asia, as evidenced by early (50,000–46,000 years old) archaeological traces in Australia that probably belonged to modern *H. sapiens* populations found there in sediments as old as 40,000 years (Bowler et al. 2003). America was settled by immigrants from Asia who migrated across a land bridge that connected Siberia and Alaska perhaps 20,000 to 15,000 years ago, but the successful colonization of most of the Americas began about 12,000 years ago (Dillehay 2003; Goebel et al. 2008). People first reached some of the Pacific islands several thousand years ago from the east, arriving at the Marquesas Islands by about A.D. 300 and New Zealand by about A.D. 1200 (Lum and Cann 1998; Finney 2007; Hunt and Lipo 2006).

Technological development in human evolution appears to be erratic in pace, but it certainly shows a pattern of acceleration (Klein 1999). Relatively crude stone tools persist for over 1.5 million years. Finely worked blade tools are much more recent. Humans have had agriculture, cities, and writing for less than 0.25% of their evolutionary development as a separate mammalian lineage.

The historical distribution of human genetic variation appears to have developed only in the past 40,000 to 50,000 years when populations of *H. sapiens* equipped with Upper Paleolithic tool technology spread out of their African homeland, first to Asia and by at least 30,000 years ago throughout Europe (Klein 1999). This recent colonization accounts for the remarkable genetic homogeneity found among living members of our species compared with that of our own closest biological relatives, *Pan* and *Gorilla,* where the degree of genetic heterogeneity within each species is many times greater than that in *Homo sapiens* (Gagneux et al. 1999; Ruvolo 1997). Rapid adaptations to local conditions and the random genetic changes that can occur in small isolated populations have led to several conspicuous phenotypic changes, such as skin color, hair form and color, and body proportions, but these differences appear to be relatively recent alterations. The geographic distribution of skin color is closely correlated with the intensity of solar radiation such that tropical people in Africa, Asia, and the Americas have great density of melanin pigments in the skin for protection against tissue and metabolic damage. Higher-latitude people are depigmented most likely because ultraviolet radiation from the sun catalyzes the synthesis of vitamin D in the skin and because of other factors (Jablonski and Chaplin 2000). Depigmentation of the skin in high-latitude peoples appears to be true of Neandertals (Culotta 2007; Lalueza-Fox et al. 2007). Body proportion varies geographically according to temperature, so that people in hot climates have longer limbs and less globular trunks to allow greater heat dissipation through the skin (Ruff 1994). Exceptions to this pattern can often be explained by the presence of cultural adaptations that allow preferential survival and reproductive success independent of genetic adaptations. The first *H. sapiens* in Europe, for example, had body builds characteristic of tropical people (they came from Africa), but they had technologies far superior to those of the local Neandertals. Technological innovation is often the reason for population expansion and plays a profound role in determining why certain physical varieties of *H. sapiens* occupy geographic regions and displace earlier indigenous people (Diamond 1997).

Evolution of the Human Brain

The accumulation of evolutionary novelties through time reveals the continuity of human evolution. The demonstration of this continuity is a triumph of evolutionary biology. It further confirms Darwin's view of descent with modification. The expansion of the brain in the hominin lineage is one of the most dramatic demonstrations of this. Absolute and relative brain-size expansion follows time remarkably well, as shown in Table 1. The earliest species for which brain size is known, *A. afarensis,* has an average endocranial volume of 438 cc (based on 4 specimens), compared with modern chimps at 400 cc, gorillas at 500 cc, and *H. sapiens* at 1,350 cc. The endocranial volume for *A. garhi* (1 specimen) is reported to be 450 cc, for *A. africanus* (7 specimens), 452 cc, for *P. boisei* (1 specimen), 521 cc, for *P. robustus* (1 specimen), 530 cc, for *H. habilis* (6 specimens), 612 cc, for *H. rudolfensis* (1 specimen), 752 cc, and for *H. ergaster* (3 specimens), 871 cc. However, at least one species of *Homo (H. floresiensis)* managed to survive on a remote island until 18,000 years ago with an endocranial volume of only 380 cc (Brown et al. 2004).

Brains expand during the Pleistocene from 914 cc between 1.8 and 1.2 million years ago to 1,090 cc between 550,000 and 400,000 years ago, 1,186 cc between 300,000 and 200,000 years ago, and above 1,300 cc after 150,000 years ago (Ruff et al. 1998). Relative to estimated body weight, the fossil species go from slightly above the size seen in modern apes to 5.4 times the mammalian average (the column labeled ECV in Table 1). Not all populations kept up with brain expansion, as is dramatically shown by *H. floresiensis* (Brown et al. 2004), but all extant populations of *H. sapiens* are fully endowed with modern brains. Unfortunately the fossil record is mute on the subject of brain reorganization, which is a pity because there were presumably profound changes that accompanied the development of speech and other uniquely human faculties. A strong case can be made for a major genetic change that enabled *H. sapiens* to speak and perform modern human behaviors beginning about 50,000 years ago (Klein and Edgar 2002).

Evolution of Human Walking

Bipedalism is the first major grade shift in hominin evolution, but the very first hominins were most likely generalized hominoids with rather apelike bodies without the specializations for bipedalism that later became the hallmark of our lineage (Thorpe et al. 2007). They were part of the African and possibly European radiation of late Miocene hominoids that came, in Darwin's words, "to live somewhat less on trees and more on the ground" due to "a change in its manner of procuring subsistence, or to a change in the conditions of its native country" (Darwin 1872, p. 135). The first members of the human clade probably will be recognized in the fossil record by subtle evolutionary novelties. The major genetic/developmental transformations that produced the bipedally adapted trunk and hindlimb morphology that define the human grade probably appeared later. The precise sequence of transformations of the trunk and hindlimb may never be known from the fossil record. It may be that parts of the bipedal body changed at different rates so that the length of the pelvic blades reduced earlier than the adduction of the hallux. The extinct and distinctly nonhominin Miocene ape *Oreopithecus* appears to have had a reduced pelvic length but retained a divergent big toe (Köhler and Moyà-Solà 1997). The full analysis of *Ardipithecus ramidus* will help clarify our understanding of the nature of the transition to bipedality in our lineage.

Early evolutionists, including Lamarck (1809), Darwin (1872), and Haeckel (1868), predicted that bipedality preceded the expansion of the brain. How, why, and in what habitat an apelike ancestor adopted this unique posture is a favorite topic of speculation (Kingdon 2003; McHenry 2004). It is possible that many apelike species in the Miocene adopted bipedality, given the fact that the ape body plan is adapted to moving in trees with forelimb suspension and not to efficient quadrupedal walking. Once bipedal, this body plan can move with an energetic efficiency equal to that of quadrupedal mammals and much greater than that of quadrupedal apes (Rodman and McHenry 1980; Sockol et al. 2007).

There has been a peculiar polarization of views about the meaning of the primitive traits retained in the postcrania of early species of *Australopithecus*. One pole emphasizes the bipedal specializations (Lovejoy 2005a, 2005b), and the other pole calls attention to the many primitive characteristics (Stern 2000). Viewed in grand perspective, the divergent interpretations can be seen as differences in emphasis. Both camps agree that all species of *Australopithecus* were bipedal. That implies that these species did not climb like apes. They did, however, retain morphological features associated with arborealism for at least a million years and a hip architecture different from that of later species of *Homo*. That implies some difference from modern humans in gait and climbing ability, but they did not walk and climb like apes.

The published hominin fossil record does not yet have a true intermediate stage between an apelike and a humanlike body (McHenry and Jones 2006). The postcranium of *Australopithecus* retains many primitive apelike traits but is fundamentally reorganized and highly specialized for bipedality, unlike that of any ape (McHenry 1986). There is a profound alteration in the genetic template that produced the short pelvic blade, adducted hallux, and other bipedal traits of this hominin.

The difference between *Australopithecus*'s positional and locomotor behavior and that seen in later species of *Homo* became clearer with the discoveries of postcranial fossils associated with *H. ergaster/erectus* (Walker and Leakey 1993), particularly the "Strapping Youth" (KNM WT 15000). There are striking differences between the pelvis and femur of all species of *Australopithecus* and those of *H. ergaster/erectus*, with the latter appearing much more humanlike (McHenry and Coffing 2000). This probably registers a major shift in adaptation between *Australopithecus* and *Homo*. From this perspective, *Australopithecus* appears to be a biped with free hands for carrying but best adapted to short distances and with a healthy appreciation of trees for safety, feeding, and sleeping. The longer femora and more humanlike pelves that appear by 1.9 million years ago with *Homo* mark the beginning of an important change.

From the currently available fossil evidence, there appear to be two grade shifts in the postcranial evolution of the human body. The first was the transition from generalized hominoid to hominin bipedalism that is documented in *A. anamensis*, *A. afarensis*, *A. africanus*, and *A. garhi* (Ward 2002). The differences in postcranial morphology among these species are difficult to assess because of the incompleteness of the fossil record. All share derived features with *Homo*, but only *A. afarensis* and *A. africanus* are complete enough to make detailed comparisons. In some respects these species are remarkably similar relative to other species of Hominoidea. For example, they share a similar mix of apelike, humanlike, and unique features in their wrists, hips, and knees (McHenry 1986). They apparently differ in the ratio of forelimb to hindlimb joint sizes, however, with *A. africanus* appearing to be more apelike even though it is later in time and more *Homo*-like in craniodental features (McHenry and Berger 1998; Green et al. 2007). Both species appear to share a combination of specialized bipedal traits, but not exactly like those of modern humans, in features of the forelimb associated with climbing (Stern 2000). The second grade shift in postcranial evolution took place about 1.9 million years ago with the appearance of much more humanlike hips that are uniquely *Homo* (Bramble and Lieberman 2004). Long femora and relatively enlarged hip joints mark a significant change in locomotion that is perhaps related to long-distance, efficient striding more like that seen in modern *H. sapiens*. The discovery of *A. garhi* reveals the complexity of postcranial evolution in that it has a humanlike ratio of femur to humerus but an apelike relative forearm length (Asfaw et al. 1999).

The Evolution of Human Mastication and Diet

Except for relatively small canine teeth, the earliest hominins had teeth similar to those of many of the Miocene apes (Ungar 2007). These early members of our subfamily lived in woodland habitats (Pickford 2006). Presumably their diets were similar as well. But beginning about 4 million years ago and climaxing with the appearance of *Paranthropus boisei* (2.3–1.4 million years ago), the cheek teeth expand through time (Constantino and Wood 2007). Although Darwin and his contemporaries predicted much of what the human fossil record would eventually reveal, no one anticipated the discovery of hominin megadonts. African apes and modern humans have small cheek teeth relative to body size. But early hominins had relatively huge molars and premolars with concomitantly gigantic jaws, alveolar bone, buttressed face and cheeks, and attachment areas for the chewing musculature. There is an approximate trend through time in successive species of increasing relative cheek-tooth size from small in *Ardipithecus ramidus* to moderate in *A. anamensis* and *A. afarensis*, big in *A. africanus, P. robustus,* and *A. garhi,* and huge in *P. boisei* and *P. aethiopicus*. The trend reverses in the *Homo* lineage so that from *H. habilis* and *H. rudolfensis* to modern *H. sapiens,* the relative size becomes progressively smaller. One way to express these trends is with relative cheek-tooth size, as is shown in the last column of Table 1. The megadontial quotient is simply the absolute size of the lower second premolar and first two molars divided by their expected size derived from the species body weight. It is not intended to be a precise index but merely a general indication of relative size. It is reasonable to assume that the increase in megadontia through time is related to dietary specializations (i.e., the ability to process greater quantities of low-nutrient and tough-textured foods). The reversal of the trend in megadontia in the evolution of *Homo* may be related to both dietary change (i.e., less reliance on tough-textured foods and the incorporation of more meat in the diet) and the processing of foods with tools before mastication.

Remaining Mysteries

The diversity of living members of the superfamily Hominoidea is impoverished relative to what it was in the past. Even with the limited fossil sample, it is clear that there were many kinds of apes and humans long ago. More species await discovery. There were probably many evolutionary experiments in the varied habitats of the African Miocene to Pleistocene (Suwa et al. 2007). Although the current sample of fossil hominins leads some to the impression that there were only a few hominin lineages (White et al. 2006), it is far more likely that our family tree will turn out to be quite bushy (Sarmiento et al. 2007). Species names may need to multiply to accommodate the diversity, although a balance needs to be maintained between excessive splitting and lumping.

It will be wonderful to resolve the question of knuckle-walking in human evolution (Richmond and Strait 2000, 2001; Richmond et al. 2001). Living species of the African clade of Hominoidea include three knuckle-walkers and a biped. The odds are that the first hominin was also a knuckle-walker, but the earliest hominins known so far are bipeds and climbers without specializations for knuckle-walking. Perhaps there are functional and developmental processes that predispose large-bodied apes to become knuckle-walkers when they adapt to habitats that require more terrestriality. If so, that apparently odd gait may have evolved in parallel in two or more lineages.

There may have been several expansions of populations out of Africa during favorable climatic conditions that populated parts of Eurasia and then became extinct. It is seductive to imagine only one triumphant pulse of migration that spread humanity for the first time throughout Eurasia. The spread of *H. sapiens* appears to have happened in a short amount of geological time, but archaic species of *Homo* were probably much less adaptable and more vulnerable to the severe climatic shifts that occurred everywhere over the past several million years. Local extinctions may have been common throughout the changing geographic range of early hominins. Before 50,000 years ago humans were an insignificantly small part of the vertebrate fauna.

It is tempting to swing to extreme views of pessimism or optimism. The hominin fossil record is limited and incomplete, but it is also rich and consistent. Its overall quality is consistent with Darwin's view of descent with modification. It makes it very unlikely that, for example, significant encephalization began before the evolution of bipedalism in our lineage. Conflicting interpretations will always be part of the science, but debate is a sign of intellectual health and helps ensure that ideas are grounded in accurate observations of material evidence and precise logic.

BIBLIOGRAPHY

Alemseged, Z., F. Spoor, W. H. Kimbel, R. Bobe, D. Geraads, D. Reed, and J. G. Wynn. 2006. A juvenile early hominin skeleton from Dikika, Ethiopia. *Nature* 443: 296–301.

Anikovich, M. V. 2007. Early Upper Paleolithic in eastern Europe and implications for the dispersal of modern humans. *Science* 315: 223–224.

Antón, S. C., and C. C. Swisher. 2001. Evolution of cranial capacity in Asian *Homo erectus*. In E. Indriati, ed., *A Scientific Life: Papers in Honor of Professor Dr. Teuku Jacab*, 25–39. Yagyakarta: Bigraf Publishing.

Argue, D., D. Donlon, C. Groves, and R. Wright. 2006. *Homo floresiensis*: Microcephalic, pygmoid, *Australopithecus,* or *Homo? Journal of Human Evolution* 51: 360–374.

Arnason, U., A. Gullberg, A. S. Burguete, and A. Janke. 2000. Molecular estimates of primate divergences and new hypotheses for primate dispersal and the origin of modern humans. *Hereditas* 133: 217–228.

Asfaw, B., T. White, O. Lovejoy, B. Latimer, S. Simpson, and G. Suwa. 1999. *Australopithecus garhi:* A new species of early hominid from Ethiopia. *Science* 284: 629–635.

Begun, D. R. 1994. Relations among the great apes and humans: New interpretations based on the fossil great ape Dryopithecus. *Yearbook of Physical Anthropology* 37: 11–63.

Begun, D. R., and C. V. Ward. 2005. Comment on *"Pierolapithecus catalaunicus,"* a new Middle Miocene great ape from Spain. *Science* 308: 203.

Behrensmeyer, A. K. 2006. Climate change and human evolution. *Science* 311: 476–478.

Bowler, J. M., H. Johnston, J. M. Olley, J. R. Prescott, R. G. Roberts, W. Shawcross, and N. A. Spooner. 2003. New ages for human occupation and climatic changes at Lake Mungo, Australia. *Nature* 421: 837–840.

Bramble, D. M., and D. E. Lieberman. 2004. Endurance running and the evolution of *Homo. Nature* 432: 345–351.

Bromage, T. G., and F. Schrenk. 1999. *African Biogeography, Climate Change, and Human Evolution.* New York: Oxford University Press.

Brooks, A. S., D. M. Helgren, J. S. Cramer, A. Franklin, W. Hornyak, J. M. Keating, R. G. Klein, W. J. Rink, H. Schwarcz, J. N. L. Smith, K. Stewart, N. E. Todd, J. Verniers, and J. E. Yellen 1995. Dating and context of three Middle Stone Age sites with bone points in the Upper Semliki Valley, Zaire. *Science* 268: 548–552.

Broom, R. 1938. The Pleistocene anthropoid apes of South Africa. *Nature* 142: 377–379.

Brown, P., T. Sutikna, M. J. Morwood, R. P. Soejono, Jatmiko, E. W. Saptomo, and R. A. Due. 2004. A new small-bodied hominin from the Late Pleistocene of Flores, Indonesia. *Nature* 431: 1055–1061.

Brumm, A., F. Aziz, G. D. van den Bergh, M. J. Morewood, M. W. Moore, I. Kurniawan, D. R. Hobbs, and R. Fullagar. 2006. Early stone technology on Flores and its implications for *Homo floresiensis. Nature* 441: 624–628.

Brunet, M., F. Guy, D. Pilbeam, H. T. Mackaye, A. Likius, D. Ahounta, A. Beauvilain, C. Blondel, H. Bocherens, J. Boisserie, L. De Bonis, Y. Coppens, J. Dejax, C. Denys, P. Duringer, V. Eisenmann, G. Fanone, P. Fronty, D. Geraads, T. Lehmann, F. Lihoreau, A. Louchart, A. Mahamat, G. Merceron, G. Mouchelin, O. Otero, P. Campomanes, M. Ponce de Leon, J. Rage, M. Sapanet, M. Schuster, J. Sudre, P. Tassy, X. Valentin, P. Vignaud, L. Viriot, A. Zazzo, and C. Zollikofer. 2002. A new hominid from the Upper Miocene of Chad, Central Africa. *Nature* 418: 145–151.

Chen, F.-C., and W.-H. Li. 2001. Genomic divergences between humans and other Hominoids and the effective population size of the common ancestor of humans and chimpanzees. *American Journal of Human Genetics* 68: 444–456.

Clark, J. D., J. de Heinzelin, K. D. Schick, T. D. White, G. WoldeGabriel, R. C. Walter, G. Suwa, B. Asfaw, E. Vrba, and Y. Haile-Selassie. 1994. African *Homo erectus:* Old radiometric ages and young Oldowan assemblages in the middle Awash Valley, Ethiopia. *Science* 264: 1907–1910.

Conroy, G. C. 2005. *Reconstructing Human Origins.* New York: W. W. Norton.

Conroy, G. C., G. W. Weber, H. Seidler, P. V. Tobias, A. Kane, and B. Brunsden. 1998. Endocranial capacity in an early hominid cranium from Sterkfontein, South Africa. *Science* 280: 1730–1731.

Constantino, P., and B. Wood. 2007. The evolution of *Zinjanthropus boisei*. *Evolutionary Anthropology* 16: 49–62.

Culotta, E. 2007. Ancient DNA reveals Neandertals with red hair, fair complexions. *Science* 318: 546–547.

Darwin, C. 1872. *The Descent of Man, and Selection in Relation to Sex*. New York: D. Appleton and Co.

Delson, E., I. Tattersall, J. A. Van Couvering, and A. S. Brooks. 2000. *Encyclopedia of Human Evolution and Prehistory*. New York: Garland Publishing.

Diamond, J. 1997. *Guns, Germs, and Steel*. New York: W. W. Norton.

Dillehay, T. 2003. Tracking the first Americans. *Nature* 425: 23–24.

Finney, B. 2007. Tracking Polynesian seafarers. *Science* 317: 1873–1912.

Fleagle, J. G. 1999. *Primate Adaptation and Evolution*. New York: Academic Press.

Fleagle, J. G., and W. L. Jungers. 1982. Fifty years of higher primate phylogeny. In F. Spencer, ed., *A History of American Physical Anthropology, 1930–1980*, 187–230. New York: Academic Press.

Gabunia, L., and A. Vekua. 1995. A Plio-Pleistocene hominid from Dmanisi, East Georgia, Caucasus. *Nature* 373: 509–512.

Gabunia, L., A. Vekua, D. Lordkipanidze, C. C. Swisher III, R. Ferring, A. Justus, M. Nioradze, M. Tvalchrelidze, S. C. Anton, G. Bosinski, O. Joris, M.-A. de Lumley, G. Majsuradze, and A. Mouskhelishvili. 2000. Earliest Pleistocene hominid cranial remains from Dmanisi, Republic of Georgia: Taxonomy, geological setting, and age. *Science* 288: 1019–1025.

Gagneux, P., C. Wills, U. Gerloff, D. Tautz, P. A. Morin, C. Boesch, B. Fruth, G. Hohmann, O. A. Ryder, and D. S. Woodruff. 1999. Mitochondrial sequences show diverse evolutionary histories of African hominoids. *Proceedings of the National Academy of Sciences USA* 96: 5077–5082.

Glazko, G. V., and N. Masatoshi. 2003. Estimation of divergence time for major lineages of primate species. *Molecular Biolology and Evolution* 20: 424–434.

Goebel, T. 2007. The missing years for modern humans. *Science* 315: 194–196.

Goebel, T., M. R. Waters, and D. H. O'Rourke. 2008. The late Pleistocene dispersal of modern humans into the Americas. *Science* 319: 1497–1502.

Goodman, M. 1963. Man's place in the phylogeny of the primates as reflected in serum proteins. In S. L. Washburn, ed., *Classification and Human Evolution*, 204–234. Chicago: Aldine de Gruyter.

Green, D. J., A. D. Gordon, and B. G. Richmond. 2007. Limb-size proportions in *Australopithecus afarensis* and *Australopithecus africanus*. *Journal of Human Evolution* 52: 187–200.

Gregory, W. K. 1916. Studies on the evolution of the primates. *Bulletin of the American Museum of Natural History* 35: 239–355.

———. 1926. Dawn-man or ape? *Scientific American*, September: 230–232.

———. 1927a. The origin of man from the anthropoid stem—when and where? *Proceedings of the American Philosophical Society* 66: 439–463.

———. 1927b. Two views of the origin of man. *Science* 65: 601–605.

———. 1928. Were the ancestors of man primitive brachiators? *Proceedings of the American Philosophical Society* 67: 129–150.

———. 1934. *Man's Place among the Anthropoids*. Oxford: Clarendon Press.

Grine, F. E. 1988. *Evolutionary History of the "Robust" Australopithecines*. New York: Aldine de Gruyter.

Haeckel, E. 1868. *Natürliche Schöpfungsgeschichte*. Berlin: Reimer.

Haeusler, M., and H. M. McHenry. 2007. Evolutionary reversals of limb proportions in early hominids? Evidence from KNM-ER 3735 *(Homo habilis)*. *Journal of Human Evolution* 53: 383–405.

Haile-Selassie, Y. 2001. Late Miocene hominids from the Middle Awash, Ethiopia. *Nature* 412: 178–181.

Haile-Selassie, Y., G. Suwa, and T. D. White. 2004. Late Miocene teeth from Middle Awash, Ethiopia, and early hominid dental evolution. *Science* 303: 1503–1505.

Hennig, W. 1966. *Phylogenetic Systematics*. Urbana: University of Illinois Press.

Hershkovitz, I., L. Kornreich, and Z. Laron. 2007. Comparative skeletal features between *Homo floresiensis* and patients with primary growth hormone sensitivity (Laron syndrome). *American Journal of Physical Anthropology* 134: 198–208.

Holloway, R. L., D. C. Broadfield, M. S. Yuan, and J. H. Schwartz. 2004. *The Human Fossil Record, Brain Endocasts: The Paleoneurological Evidence*. New York: Wiley-Liss.

Holmes, E. C., G. Pesole, and C. Saccone. 1989. Stochastic models of molecular evolution and the estimation of phylogeny and rates of nucleotide substitution in the hominoid primates. *Journal of Human Evolution* 18: 775–794.

Hoyer, B. H., B. J. McCarthy, and E. T. Bolton. 1964. A molecular approach in the systematics of higher organisms. *Science* 144: 959–967.

Hull, D. L. 1988. *Science as a Process*. Chicago: University of Chicago Press.

Hunt, T., and C. Lipo. 2006. Late colonization of Easter Island. *Science* 311: 1603–1606.

Huxley, T. H. 1863. *Evidence as to Man's Place in Nature*. London: Williams and Norgate.

Jablonski, N. G., and G. Chaplin. 2000. The evolution of human skin coloration. *Journal of Human Evolution* 39: 57–106.

Jacob, T., E. Indriati, R. P. Soejono, K. Hsu, D. W. Frayer, R. B. Eckhardt, A. J. Kuperavage, A. Thorne, and M. Henneberg. 2006. Pygmoid Australomelanesian *Homo sapiens* skeletal remains from Liang Bua, Flores: Population affinities and pathological abnormalities. *Proceedings of the National Academy of Sciences USA* 103: 13421–13426.

Johanson, D. C., M. Taieb, and Y. Coppens. 1982. Pliocene hominids from the Hadar Formation, Ethiopia (1973–1977): Stratigraphic, chronologic, and paleoenvironmental contexts, with notes on hominid morphology and systematics. *American Journal of Physical Anthropology* 57: 373–402.

Jones, F. W. 1918. *The Problem of Man's Ancestry*. London: Society for Promoting Christian Knowledge.

———. 1926. *Arboreal Man*. New York: Hafner.

———. 1929. *Man's Place among the Mammals*. New York: Longmans, Green.

Keith, A. 1915. *The Antiquity of Man*. London: Williams and Norgate.

Kimbel, W. H., C. A. Lockwood, C. V. Ward, M. G. Leakey, Y. Rak, and D. C. Johanson. 2006. Was *Australopithecus anamensis* ancestral to *A. afarensis*? A case of anagenesis in the hominin fossil record. *Journal of Human Evolution* 51: 134–153.

Kingdon, J. 2003. *Lowly Origin: Where, When, and Why Our Ancestors First Stood Up*. Princeton: Princeton University Press.

Klein, R. G. 1999. *The Human Career: Human Biological and Cultural Origins*. Chicago: University of Chicago Press.

———. 2000. Archeology and the evolution of human behavior. *Evolutionary Anthropology* 9: 17–36.

———. 2003. Whither the Neanderthals? *Science* 299: 1525–1527.

Klein, R. G., and B. Edgar. 2002. *The Dawn of Human Culture*. New York: Wiley-Liss.

Köhler, M., and S. Moyà-Solà. 1997. Ape-like or hominid-like? The positional behavior of *Oreopithecus bambolii* reconsidered. *Proceedings of the National Academy of Sciences USA* 94: 11747–11750.

Krause, J., L. Orlando, D. Serre, B. Viola, K. Prüfer, M. P. Richards, J. Hublin, C. Hää, A. P. Derevianko, and S. Pääbo. 2007. Neanderthals in central Asia and Siberia. *Nature* 449: 902–904.

Kuhn, S. L. 2007. 150 years of Neanderthal discoveries: Early Europeans, continuity and discontinuity. *Evolutionary Anthropology* 16: 41–42.

Kumar, S., A. Filipski, V. Swarna, A. Walker, and S. B. Hedges. 2005. Placing confidence limits on the molecular age of the human-chimpanzee divergence. *Proceedings of the National Academy of Sciences USA* 102: 18842–18847.

Kumar, S., and S. B. Hedges. 1998. A molecular timescale for vertebrate evolution. *Nature* 392: 917–920.

Kurten, B. 1972. *Not from the Apes*. New York: Vintage Books.

Lalueza-Fox, C., H. Römpler, D. Caramelli, C. Stäubert, G. Catalano, D. Hughes, N. Rohland, E. Pilli, L. Longo, S. Condemi, M. de la Rasilla, J. Fortea, A. Rosas, M. Stoneking, T. Schöneberg, J. Bertranpetit, and M. Hofreiter. 2007. A melanocortin 1 receptor allele suggests varying pigmentation among Neanderthals. *Science* 318: 1453–1455.

Lamarck, J. B. P. A. de M. de. 1809. *Philosophie zoologique; ou, Exposition des considérations relative à l'histoire naturelle des animaux.* Paris: Chez Dentu [et] l'auteur.

Larson, S. G., W. L. Jungers, M. J. Morewood, T. Sutikna, Jatmiko, E. W. Saptomo, R. A. Due, and T. Djubiantono. 2007. *Homo floresiensis* and the evolution of the hominin shoulder. *Journal of Human Evolution* 20: 1–14.

Leakey, M. G., C. S. Feibel, I. McDougall, and A. Walker. 1995. New four-million-year-old hominid species from Kanapoi and Allia Bay, Kenya. *Nature* 376: 565–571.

Leakey, M. G., C. S. Feibel, I. McDougall, C. Ward, and A. Walker. 1998. New specimens and confirmation of an early age for *Australopithecus anamensis*. *Nature* 393: 62–66.

Leakey, M. G., F. Spoor, F. H. Brown, P. N. Gathogo, C. Kiarie, L. N. Leakey, and I. McDougall. 2001. New hominin genus from eastern Africa shows diverse middle Pleistocene lineages. *Nature* 410: 433–440.

Lordkipanidze, D., T. Jashavhvili, A. Vekua, M. S. Ponce de León, C. P. E. Zollikofer, G. P. Rightmire, H. Pontzer, R. Ferring, O. Oms, M. Tappen, M. Bukhsianidze, J. Agusti, R. Kahlke, G. Kiladze, B. Martinez-Navarro, A. Mouskhelishvili, M. Nioradze, and L. Rook. 2007. Postcranial evidence from early *Homo* from Dmanisi, Georgia. *Nature* 449: 305–309.

Lordkipanidze, D., A. Vekua, R. Ferring, G. P. Rightmire, J. Agusti, G. Kiladze, A. Mouskhelishvili, M. Nioradze, M. S. Ponce de León, M. Tappen, and C. P. E. Zollikofer. 2005. The earliest toothless hominin skull. *Nature* 434: 717–718.

Lovejoy, O. C. 2005a. The natural history of human gait and posture. Part 1. Spine and pelvis. *Gait and Posture* 21: 95–112.

———. 2005b. The natural history of human gait and posture. Part 2. Hip and thigh. *Gait and Posture* 21: 113–124.

Lum, J. K., and R. L. Cann. 1998. mtDNA and language support a common origin of Micronesians and Polynesians in Island Southeast Asia. *American Journal of Physical Anthropology* 105: 109–119.

Martin, R. D. 1981. Relative brain size and basal metabolic rate in terrestrial vertebrates. *Nature* 293: 57–60.

McBrearty, S., and A. S. Brooks. 2000. The revolution that wasn't: A new interpretation of the origin of modern human behavior. *Journal of Human Evolution* 39: 453–563.

McCrossin, M. L., and B. R. Benefit. 1997. On the relationships and adaptations of *Kenyapithecus*, a large-bodied hominoid from the middle Miocene of eastern Africa. In D. R. Begun, C. V. Ward, and M. D. Rose, eds., *Function, Phylogeny, and Fossils: Miocene Hominoid Evolution and Adaptations*, 241–267. New York: Plenum Press.

McHenry, H. M. 1986. The first bipeds: A comparison of the *Australopithecus afarensis* and *Australopithecus africanus* postcranium and implications for the evolution of bipedalism. *Journal of Human Evolution* 15: 177–191.

———. 1988. New estimates of body weights in early hominids and their significance to encephalization and megadontia in "robust" australopithecines. In F. E. Grine, ed., *Evolutionary History of the "Robust" Australopithecines,* 133–148. New York: Aldine de Gruyter.

———. 1991. Femoral lengths and stature in Plio-Pleistocene hominids. *American Journal of Physical Anthropology* 85: 149–158.

———. 1992. Body size and proportions in early hominids. *American Journal of Physical Anthropology* 87: 407–431.

———. 1994a. Behavioral ecological implications of early hominid body size. *Journal of Human Evolution* 27: 77–87.

———. 1994b. Early hominid postcrania: Phylogeny and function. In R. S. Corruccini and R. L. Ciochon, eds., *Integrative Pathways to the Past: Paleoanthropological Papers in Honor of F. Clark Howell,* 251–268. Englewood Cliffs, NJ: Prentice-Hall.

———. 1996. Homoplasy, clades and hominid phylogeny. In W. E. Meikle, F. C. Howell, and N. G. Jablonski, eds., *Contemporary Issues in Human Evolution,* 77–92. San Francisco: California Academy of Sciences.

———. 2004. Uplifted head, free hands, and the evolution of human walking. In D. J. Meldrum and C. E. Hilton, eds., *From Biped to Strider: The Emergence of Modern Human Walking, Running, and Resource Transport,* 203–210. New York: Kluwer Academic/Plenum Publishers.

McHenry, H. M., and L. R. Berger. 1998. Body proportions in *Australopithecus afarensis* and *A. africanus* and the origin of the genus *Homo. Journal of Human Evolution* 35: 1–22.

McHenry, H. M., and K. Coffing. 2000. *Australopithecus* to *Homo:* Transformations in body and mind. *Annual Review of Anthropology* 29: 129–146.

McHenry, H. M., and A. L. Jones. 2006. Hallucial convergence in early hominids. *Journal of Human Evolution* 50: 534–539.

Morwood, M., and P. van Oosterzee. 2007. *A New Human: The Startling Discovery and Strange Story of the "Hobbits" of Flores, Indonesia.* New York: HarperCollins.

Morwood, M. J., P. Brown, Jatmiko, T. Sutikna, E. W. Saptomo, K. E. Westaway, R. A. Due, R. G. Roberts, T. Maeda, S. Wasisto, and T. Djubiantono. 2005. Further evidence for small-bodied hominins from the Late Pleistocene of Flores, Indonesia. *Nature* 437: 1012–1017.

Morwood, M. J., P. B. O'Sullivan, F. Aziz, and A. Raza. 1998. Fission-track ages of stone tools and fossils on the east Indonesian island of Flores. *Nature* 392: 173–176.

Morwood, M. J., R. P. Soejono, R. G. Roberts, T. Sutikna, C. S. M. Turney, K. E. Westaway, W. J. Rink, J.-X. Zhao, G. D. van den Bergh, R. A. Due, D. R. Hobbs, M. W. B. Moore, M. I. Bird, and L. K. Fifield. 2004. Archaeology and age of a new hominin from Flores in eastern Indonesia. *Nature* 431: 1087–1091.

Moyà-Solà, S., M. Köhler, D. M. Alba, I. Casanovas-Vilar, and J. Galindo. 2004. *Pierolapithecus catalaunicus,* a new Middle Miocene great ape from Spain. *Science* 306: 1339–1344.

Nakatsukasa, M., M. Pickford, N. Egi, and B. Senut. 2007. Femoral length, body mass, and stature of *Orrorin tugenensis,* a 6 Ma hominid from Kenya. *Primates* 48: 171–178.

Napier, J. R., and P. R. Davis. 1959. The forelimb skeleton and associated remains of *Proconsul africanus. Fossil Mammals of Africa* 16: 1–69.

Orlando, L., P. Darlu, M. Toussaint, D. Bonjean, M. Otte, and C. Hänni. 2006. Revisiting Neandertal diversity with a 100,000 year old mtDNA sequence. *Current Biology* 16: R400–R402.

Patterson, N., D. J. Richter, S. Gnerre, E. S. Lander, and D. Reich. 2006. Genetic evidence for complex speciation of humans and chimpanzees. *Nature* 441: 1103–1108.

Pickford, M. 2006. Paleoenvironments, paleoecology, adaptations and the origins of bipedalism in hominidae. In H. Ishida, R. Tuttle, M. Pickford, N. Ogihara, and M. Nakatsukasa, eds., *Human Origins and Environmental Backgrounds*, 175–199. New York: Springer.

Pickford, M., B. Senut, D. Gommery, and J. Treil. 2002. Bipedalism in *Orrorin tugenensis* revealed by its femora. *Human Palaeontology and Prehistory* 1: 191–203.

Pilbeam, D., and N. Young. 2004. Hominid evolution: Synthesizing disparate data. *Comptes Rendus Palevol* 3: 305–321.

Pilbeam, D. R. 1982. New hominoid skull material from the Miocene of Pakistan. *Nature* 295: 232–234.

Potts, R. 2007. Paleoclimate and human evolution. *Evolutionary Anthropology* 16: 1–3.

Richmond, B. G., D. R. Begun, and D. S. Strait. 2001. Origin of human bipedalism: The knuckle-walking hypothesis revisited. *Yearbook of Physical Anthropology* 44: 70–105.

Richmond, B. G., and W. L. Jungers. 2008 *Orrorin tugenensis* femoral morphology and the evolution of hominin bipedalism. *Science* 319: 1662–1665.

Richmond, B. G., and D. S. Strait. 2000. Evidence that humans evolved from a knuckle-walking ancestor. *Nature* 404: 382–385.

———. 2001. Did our ancestors knuckle-walk? *Science* 410: 326.

Robinson, J. T. 1972. *Early Hominid Posture and Locomotion*. Chicago: University of Chicago Press.

Rodman, P. S., and H. M. McHenry. 1980. Bioenergetics and the origin of hominid bipedalism. *American Journal of Physical Anthropology* 52: 103–106.

Ruff, C. B. 1994. Morphological adaptation to climate in modern and fossil hominids. *Yearbook of Physical Anthropology* 37: 65–107.

Ruff, C. B., E. Trinkaus, and T. W. Holliday. 1998. Body mass and encephalization in Pleistocene *Homo*. *Nature* 387: 173–176.

Ruff, C. B., and A. Walker. 1993. Body size and body shape. In A. Walker and R. Leakey, eds.,*The Nariokotome* Homo erectus *Skeleton*, 234–265. Cambridge, MA: Harvard University Press.

Ruvolo, M. 1997. Genetic diversity in hominoid primates. *Annual Review of Anthropology* 26: 515–540.

Sarich, V. M., and A. Wilson. 1966. Quantitative immunochemistry and the evolution of primate albumins: Microcomplement fixation. *Science* 154: 1563–1566.

———. 1967. Immunological time scale for hominid evolution. *Science* 158: 1200–1203.

Sarmiento, E., G. J. Sawyer, and R. Milner. 2007. *The Last Human: A Guide to Twenty-two Species of Extinct Humans*. New Haven, CT: Yale University Press.

Sawada, Y., M. Pickford, B. Senut, T. Itaya, M. Hyodo, T. Miura, C. Kashine, T. Chujo, and H. Fujii. 2002. The age of *Orrorin tugenensis*, an early hominid from the Tugen Hills, Kenya. *Human Palaeontology and Prehistory* 1: 293–303.

Schwartz, J. H., and I. Tattersall. 2002. *The Human Fossil Record*. Vol. 1. *Terminology and Craniodental Morphology of Genus* Homo *(Europe)*. New York: Wiley-Liss.

———. 2003. *The Human Fossil Record*. Vol. 2. *Craniodental Morphology of Genus* Homo *(Africa and Asia)*. New York: Wiley-Liss.

————. 2005. *The Human Fossil Record.* Vol. 4. *Craniodental Morphology of Early Hominids (Genera* Australopithecus, Paranthropus, Orrorin) *and Overview.* Hoboken, NJ: Wiley-Liss.

Senut, B., M. Pickford, D. Gommery, P. Mein, K. Cheboi, and Y. Coppens. 2001. First hominid from the Miocene (Lukeino Formation, Kenya). *Earth and Planetary Sciences* 332: 137–144.

Skelton, R. R., and H. M. McHenry. 1992. Evolutionary relationships among early hominids. *Journal of Human Evolution* 23: 309–349.

Sockol, M. D., D. A. Raichlen, and H. Pontzer. 2007. Chimpanzee locomoter energetics and the origin of human bipedalism. *Proceedings of the National Academy of Sciences USA* 104: 12265–12269.

Spoor, F., M. G. Leakey, P. N. Gathogo, F. H. Brown, S. C. Anton, I. McDougall, C. Kiarie, F. K. Manthi, and L. N. Leakey. 2007. Implications of new early *Homo* fossils from Ileret, east of Lake Turkana, Kenya. *Nature* 448: 688–691.

Stauffer, R. L., A. Walker, O. A. Ryder, M. Lyons-Weiler, and S. B. Hedges. 2001. Human and ape molecular clock and constraints on paleontological hypotheses. *Journal of Heredity* 92: 469–474.

Stern, J. T. 2000. Climbing to the top: A personal memoir of *Australopithecus afarensis. Evolutionary Anthropology* 9: 113–133.

Strait, D. S., and F. E. Grine. 2004. Inferring hominoid and early hominid phylogeny using craniodental characters: The role of fossil taxa. *Journal of Human Evolution* 47: 399–452.

Strait, D. S., F. E. Grine, and M. A. Moniz. 1997. A reappraisal of early hominid phylogeny. *Journal of Human Evolution* 32: 17–82.

Straus, W. L. 1949. The riddle of man's ancestry. *Quarterly Review of Biology* 24: 200–223.

Suwa, G., B. Asfaw, Y. Beyene, T. D. White, S. Katoh, S. Nagaoka, H. Nayaka, K. Uzawa, P. Renne, and G. WoldeGabriel. 1997. The first skull of *Australopithecus boisei. Nature* 389: 489–492.

Suwa, G., R. T. Kono, S. Katoh, B. Asfaw, and Y. Beyene. 2007. A new species of great ape from the late Miocene epoch in Ethiopia. *Nature* 448: 921–924.

Swisher, C. C., G. H. Curtis, T. Jacob, A. G. Getty, A. Suprijo, and Widiasmoro. 1994. Age of the earliest known hominids in Java, Indonesia. *Science* 263: 1118–1121.

Thorpe, S. K. S., R. L. Holder, and R. H. Crompton. 2007. Origin of human bipedalism as an adaptation for locomotion on flexible branches. *Science* 316: 1328–1331.

Tocheri, M. W., C. M. Orr, S. G. Larson, T. Sutikna, Jatmiko, E. W. Saptomo, R. A. Due, T. Djubiantono, M. J. Morwood, and W. L. Jungers. 2007. The primitive wrist of *Homo floresiensis* and its implications for hominin evolution. *Science* 317: 1743–1745.

Ungar, P. S. 2007. *Evolution of the Human Diet: The Known, the Unknown, and the Unknowable.* New York: Oxford University Press.

Vekua, A., D. Lordkipanidze, G. P. Rightmire, J. Agusti, R. Ferring, G. Maisuradze, A. Mouskhelishvili, M. Nioradze, M. Ponce de Leon, M. Tappen, M. Tvalchrelidze, and C. Zollikofer. 2002. A new skull of early *Homo* from Dmanisi, Georgia. *Science* 297: 85–89.

Vrba, E. S., G. H. Denton, T. C. Partridge, and L. H. Burckle. 1995. *Paleoclimate and Evolution, with Emphasis on Human Origins.* New Haven, CT: Yale University Press.

Walker, A., and R. E. F. Leakey. 1993. *The Nariokotome* Homo erectus *Skeleton.* Cambridge, MA: Harvard University Press.

Walker, A., R. E. F. Leakey, J. M. Harris, and F. H. Brown. 1986. 2.5-MYR *Australopithecus boisei* from west of Lake Turkana, Kenya. *Nature* 322: 517–522.

Ward, C. V. 2002. Interpreting the posture and locomotion of *Australopithecus afarensis:* Where do we stand? *Yearbook of Physical Anthropology* 45: 185–215.

Ward, C. V., M. G. Leakey, and A. Walker. 2001. Morphology of *Australopithecus anamensis* from Kanapoi and Allia Bay, Kenya. *Journal of Human Evolution* 41: 255–368.

Washburn, S. L. 1951. The analysis of primate evolution with particular reference to the origin of man. *Cold Spring Harbor Symposium on Quantitative Biology* 15: 67–77.

White, T. 2003. Early hominids—Diversity or distortion? *Science* 299: 1994–1997.

White, T. D., G. Suwa, and B. Asfaw. 1994. *Australopithecus ramidus,* a new species of early hominid from Aramis, Ethiopia. *Nature* 371: 306–312.

———. 1995. Corrigendum: *Australopithecus ramidus,* a new species of early hominid from Aramis, Ethiopia. *Nature* 375: 88.

White, T. D., G. WoldeGabriel, B. Asfaw, S. Ambrose, Y. Beyene, R. L. Bernor, J. R. Boisserie, B. Currie, H. Gilbert, Y. Haile-Selassie, W. K. Hart, L. J. Hlusko, C. F. Howell, R. T. Kono, T. Lehmann, A. Louchart, O. C. Lovejoy, P. R. Renne, H. Saegusa, E. S. Vrba, H. Wesselman, and G. Suwa. 2006. Asa Issie, Aramis and the origin of *Australopithecus. Nature* 440: 883–889.

WoldeGabriel, G., Y. Haile-Selassie, P. R. Renne, W. K. Hart, S. H. Ambrose, B. Asfaw, G. Heiken, and T. White. 2001. Geology and palaeontology of the Late Miocene Middle Awash valley, Afar rift, Ethiopia. *Nature* 412: 175–178.

WoldeGabriel, G., T. D. White, G. Suwa, P. Renne, J. de Heinzelin, W. K. Hart, and G. Heiken. 1994. Ecological and temporal placement of early Pliocene hominids at Aramis, Ethiopia. *Nature* 371: 330–333.

Wolpoff, M. H. 1971. *Metric Trends in Hominid Dental Evolution.* Cleveland: Case Western Reserve University Press.

———. 1982. The Arago dental sample in the context of hominid dental evolution. In H. de Lumley, ed., *L'Homo erectus et la place de l'homme de Tautavel parmi les hominidés fossiles,* 389–410. Colloque International de Centre National de la Recherche Scientific. Nice: Louis-Jean Scientific and Literary Publications.

Wood, B. 1991. *Koobi Fora Research Project.* Vol. 4. *Hominid Cranial Remains.* Oxford: Clarendon Press.

Zollikofer, C. P. E., M. S. Ponce de Leon, D. E. Lieberman, F. Guy, D. Pilbeam, A. Likius, H. T. Mackaye, P. Vignaud, and M. Brunet. 2005. Virtual cranial reconstruction of *Sahelanthropus tchadensis. Nature* 434: 755–759.

Evolutionary Biology of Disease and Darwinian Medicine

Michael F. Antolin

Symptoms

A young boy in Kansas, in the midwestern United States, develops a severe rash with fever, swollen lymph nodes, and approaching delirium. A quick check reveals that the boy has survived the typical childhood infections (chicken pox) and has up-to-date vaccinations and no history of allergies. The family seldom travels far from home, and all its members appear to be healthy, if slightly overweight in the typical American way. This family has a habit, however, of keeping interesting pets and for his birthday gave the boy a young black-tailed prairie dog, one of those lovable, smaller-sized relatives of groundhogs found on the Great Plains. This pet was captured in the wild, and prairie dogs are known to carry serious infectious pathogens like plague and tularemia (rabbit fever). The boy's symptoms include swollen lymph nodes that could indicate plague or tularemia, but the pustular rash points in a different direction.

As it happens, the suspect prairie dog came from a pet facility near Chicago, where it became the source for monkeypox virus that naturally circulates in wild animals half the world away in western Africa. The pathogen, a close relative of the dreaded smallpox *Variola,* entered the United States in a shipment of Gambian rats imported for the exotic pet trade. The rats and prairie dogs had been housed together in a warehouse, and there the final link in the chain of infection from Africa to the American Midwest was forged (CDC 2003; Guarner et al. 2004). After treatment the boy recovers, though his beloved prairie dog does not. The rest of the family and local health workers are treated with smallpox vaccine because it stimulates immunity against monkeypox as well.

The case of this particular family is part fiction but is based on an actual monkeypox outbreak in six midwestern states in 2003 that infected 71 persons who came in contact with Gambian rats or prairie dogs from the same pet facility (CDC 2003). At 10% case fatality, monkeypox is not as deadly as smallpox, but the resemblance of the monkeypox virus to its cousin, which is widely believed to have been eradicated from the wild by a worldwide vaccination campaign in the 1970s, is frightening. The case also reveals how modern medicine and public health use the framework of evolution to solve potential health crises. This has been called Darwinian medicine (see box) because diagnosis and treatment depend on methods of evolutionary biology to sleuth the underlying causes of immediate symptoms (Williams and Neese 1991; Stearns 1999). Infection with monkeypox was confirmed by viral isolation, epidemiological investigation, and finally DNA analysis, and family resemblances among pox viruses traced this particular virus back to Africa. A similar story can be told about the coronavirus that causes severe acute respiratory syndrome (SARS) and sparked a public health panic in 2002–2004 when infections spread from hospitals in Asia to North America. In a mere eight months genetic analysis of the newly emerged SARS virus first led to its identification, then to intermediates in civet cats sold in wild-game markets in China (Ruan et al. 2003), and eventually to potential reservoirs in bats (Li et al. 2005; Wang et al. 2006). On a tragically much larger scale the human immunodeficiency viruses (HIVs) that cause AIDS have also been traced by genetic analysis to wild primate populations in Africa (Hahn et al. 2000; Keele et al. 2006).

Etiology

Darwinian medicine searches for the evolutionary origins of sickness—those things that make life nasty, brutish, and short. Darwinian medicine succeeds by looking for disease origins in biological adaptations of humans and the microbes that plague them—those traits that allow some to survive and reproduce better than others. One basic tenet of Darwinian medicine comes from adaptationist thinking, that diseases arise because humans exist in a

---

Darwinian medicine describes an evolutionary approach to caring for the sick, and for public health in general. In this sense, Darwinian medicine looks to the adaptations, and maladaptations, of both humans and the microbes that cause disease. For microbes in particular, Darwinian medicine must account for relatively rapid generation times of pathogenic microbes, and high rates of mutation and genetic recombination that lead inexorably to novel adaptations and the emergence of new diseases. Both pathogen and host, however, also must be understood in terms of evolutionary history, homology, and phylogenetic constraints on adaptation. Darwinian medicine, then, takes a dynamic view of both infectious and genetic diseases, where several evolutionary histories converge within individual patients on the one hand, and in human populations on the other.

mismatch between the current world and the world in which our species evolved and adapted, and that humans suffer from a series of trade-offs between high fitness in the past and reduced fitness in the present (Williams and Neese 1991; Swynghedauw 2004; Gluckman and Hanson 2006). Obesity, heart disease (Swynghedauw 2004), type-2 diabetes (Diamond 2003), breast and prostate cancer (Greaves 2002), goiter, iodine deficiency, and birth defects (Gluckman and Hanson 2006), and aging (Williams and Neese 1991) have been primary targets of this kind of thinking.

The recent worldwide increase in type-2 diabetes provides a compelling example that previous selection for tolerating starvation may have selected for genes that cause diabetes in a food-rich environment (Diamond 2003). In reality, however, few cases document ongoing natural selection in humans in relation to disease: the case of sickle-cell anemia, low-functioning α hemoglobin, and malaria resistance in sub-Saharan Africa is essentially the only one, and the story does not hold up generally outside a few well-documented populations (Hill and Motulsky 1999). As will be shown later, both genetic and infectious diseases have multiple causes that include nutrition and other aspects of the environment, genetic susceptibility, embryonic development, immune responses, and pathogen diversity (Burnet and White 1972; Stearns 1999; Weiss and McMichael 2004; Graham et al. 2005; Gluckman and Hanson 2006). Finding the causes of disease remains a difficult challenge.

Although I acknowledge that natural selection is the most powerful evolutionary force that leads to adaptation, I take a different tack for Darwinian medicine that requires less speculation about what may or may not have been adaptive in humans in the past. Even so, evolutionary biology has much to offer the world of medicine via understanding not only adaptations by natural selection but also homologies, traits that are shared by organisms because of common descent from their ancestors (Williams and Neese 1991; Stearns 1999; Bull and Wichman 2001). Clearly both genetic and infectious diseases are maladaptive to humans in that morbidity and mortality reduce fitness. Disease, however, is an inescapable consequence of an ever-changing biological world in which adaptation and fitness are not absolutely tuned to some ideal but are relative to what might occur in a world of constrained possibilities. This was the point of the famous critique of the adaptationist program by Gould and Lewontin 30 years ago (Gould and Lewontin 1979). For instance, we can understand the seven vertebrae in our necks in a functional sense, in the adaptive trade-offs that keep our bony skulls and nervous system upright at the top (front) of our bodies while maintaining flexibility of movement that does not damage our spinal cord. But this does not tell us why we have seven vertebrae, not six or eight (Williams and Neese 1991). The answer to this question resides in our evolutionary history as chordates and vertebrates and in the developmental program encoded in the DNA we inherited from our ancestors. Seven neck vertebrae may have represented an adaptive number in the past, but this cause is lost in history, and the result is passed on to us as a phylogenetic constraint of our body plans (see Gould and Lewontin 1979).

The view of Darwinian medicine I espouse considers the evolution of disease as a series of ongoing events in populations that experience both natural selection and random forces like mutation within a framework of phylogenetic constraints. In other words, I want doctors to begin thinking like population biologists, like those of us who obsess about variable populations of organisms, be they humans or their microbes, in an ecological sphere that changes in time and space (Burnet and White 1972; Lederberg 2000; W. Anderson 2004; Purssell 2005). Further, medical science will greatly benefit from thinking of populations at several levels, from human populations that inhabit one place or another to populations of microbes that live within one tissue or another within patients (Levin et al. 1999; Stearns and Ebert 2001; Grenfell et al. 2004). I begin by looking at two aspects of the current human condition with respect to disease: human population size and evolution of pathogens, and human population size relative to our own genetics and mutations.

ADAPTABILITY OF PATHOGENS

Like other parts of nature, humans represent a resource for microbial growth. With the help of modern medicine the human population has grown fourfold in the past century, surpassing 6 billion in the past decade, and humans now constitute a substantial part of the planet's biomass and use a large fraction of the planet's ecological output (Vitousek et al. 1986; Daily et al. 1997). To a large extent, human populations have expanded because vaccination and antimicrobial drugs protect us from the microbial world, with the direct benefits of lowered infant mortality and longer lifespan (R. M. Anderson and May 1991; Lederberg 2000). The other side of the growth equation is agriculture and protection of food. But add to this the great diversity of the microbial world and the proven ability of viruses, bacteria, protozoa, and single-celled fungi to adapt to new environments. Summed together, humans and their companion animals and plants constitute a vast resource with ever-increasing mobility and contact with wild reservoirs of microbial pathogens.

The outcome is inevitable: pathogens will regularly circumvent our antimicrobial defenses, and we will be colonized again and again by microbes that adapt their transmission cycles to our habits and movements. For example, the recent emergence of SARS includes genetic intermediates between coronaviruses in bats, civet cats in wild-game markets in China, and humans: it appears that SARS jumped to humans after some evolution in civet cats, not directly from bats to humans (Wang et al. 2006). Much of this can be predicted from the framework of Darwinian medicine via the adaptive potential of microbes and the evolutionary history of pathogens (Lederberg 2000). Microbes, with much shorter generation times than ours, have been shown time and again to adapt to novel environments by natural selection and provide some of our best models for experimental studies of evolution (Ebert 1998; MacLean 2005).

A primary example is the shifting antigens (surface proteins that stimulate immune responses via specific antibodies) of influenza A viruses. We are all familiar with annual flu outbreaks, though perhaps not with the pandemic Spanish flu of 1918–1919 that killed more than 40 million individuals. Flu outbreaks occur worldwide each year on a less dramatic scale as the viruses' ever-changing surface antigens avoid immunity stimulated by previous infections and necessitate constant updating of flu vaccines (Bush et al. 1999; Ferguson et al. 2003). Influenza viruses are remarkably simple: eight RNA strands code for internal proteins that surround the RNA, an external protein capsule, polymerases (enzymes) for viral replication, and 3 protruding proteins that act as antigens, 11 proteins in all. The 3 surface proteins are hemagglutinin (HA) for binding to host cells, protein channels that aid RNA invasion into the host cell (M2), and neuraminidase (NA) to help newly formed particles escape the host cell. Genes coding for HA and NA are especially prone to mutation, and both HA and NA are represented by multiple antigenic subtypes within groups of related hosts. For instance, waterfowl, shore birds, and poultry, the main reservoir and spillover hosts for the influenza A viruses, have at least 16 HA and 9 NA antigenic subtypes (Olsen et al. 2006). The current virulent avian influenza subtype is thus designated H5N1.

Within each subtype, however, HA and NA genes are subject to regular antigenic drift in that flu viruses accumulate enough mutations to become less easily recognized by the immune systems of previously infected or vaccinated individuals (Bush et al. 1999; Ferguson et al. 2003). The implications for human health are enormous, both in terms of predicting how long vaccines will remain effective (D. J. Smith 2006) and in the possibility of mutations or antigen shuffling between virus subtypes, which could lead to changes in virulence and transmission between hosts (Baigent and McCauley 2003; Kuiken et al. 2006). The great fear, of course, is that the H5 hemagglutinin will change influenza's transmissibility in humans, for instance, by evolving HA affinity from cells deep in the lungs to cells that line our throats and noses. Rather than scattered cases associated with exposure to infected waterfowl or poultry, a new form of H5N1 could be transmitted by direct contact between people, leading to disease outbreaks in human populations.

GENETICS OF MUTATION

Many well-known human diseases result from mutations in our genomes. Mutation is also inevitable, a consequence of mistakes in copying the DNA that for the most part keeps the record straight during cell cycles that renew our bodies and make our eggs and sperm. Some mutations are spontaneous, some are environmentally caused, but rest assured that mutation never sleeps. Individual mutations occur at relatively low rates, but the numbers are tilted in favor of mutations appearing in high absolute frequency. Prevalence of common genetic diseases like diabetes, muscular dystrophies, and cancers can be understood in the framework of evolution via those inexorable mutations and their inheritance within families and ethnic groups (Frank 2004; Tishkoff and Kidd 2004). At the base level mutations arise in somewhere between 1 in 10,000 and 1 in 100,000,000 DNA replications. Given a human population of more than 6 billion persons, each with 30,000 genes and 6 billion base pairs of DNA, novel mutations will arise each generation by the thousands, if not millions. Subsequently, novel mutation can be traced and predicted within families by pedigrees but can also come to differ in frequency between populations with higher or lower probability of inherited disease.

The vast majority of mutations will disrupt genes and be harmful. Consider phenylketonuria (PKU), the genetic disease caused by the inability to metabolize the amino acid phenylalanine because of mutations in the *PAH* gene (Scriver and Waters 1999). The disease has variable phenotypes, of which mental retardation is one of the worst, but it can be avoided if phenylalanine is excluded from the diet of affected individuals (carefully read the labels of your low-calorie soft drinks). More than 400 mutant *PAH* alleles that encode malfunctioning enzymes have been identified in various human populations. Routine neonatal screening for unusually high phenylalanine levels and defective PAH enzyme now provides a rapid test for diagnosis and treatment before the most severe symptoms develop.

On the other hand, some positive surprises are hidden in the human genome as well. A number of individuals of northern European descent infected with HIV-1 have never developed AIDS, and many of them are homozygous (have the same allele on both chromosomes) for a mutation in the cytokine receptor CCR5 on the surface of CD4+T cells of the immune system (Dean et al. 1996). HIV-1 uses the CCR5 receptor to enter CD4+T cells and eventually overwhelms the population of helper T cells that normally respond to invaders in the body, leading to a collapse of the immune system and AIDS. This mutation is called CCR5-Δ32 because of a deletion of 32 base pairs in the gene, leading to a nonfunctional protein that prevents HIV-1 from binding and invading CD4+T cells. CCR5-Δ32 is found in 5% to 15% of northern Europeans but is essentially absent from East Asian, African, and Native American populations (Dean et al. 2002).

Because it is impossible that AIDS selected for increased frequency of CCR5-Δ32 within a single generation, it has been suggested that high frequency of CCR5-Δ32 resulted from strong selection in the past, especially from infection by plague or smallpox, which were both common and deadly in European populations during medieval times and until the late 1700s (Galvani and Slatkin 2003). As is often the case with surmising selection in the past, other evidence does not neatly fit the plague or smallpox hypotheses. In mice infection by the plague bacterium is as common in CCR5-deficient strains as in CCR5+strains (Mecsas et al. 2004; Elvin et al. 2004), in human populations several other nonfunctional alleles of the CCR5 receptor also exist in high frequency (Blanpain et al. 2000), and the CCR5-Δ32 mutant allele has been recovered from Bronze Age (2,900 years ago) skeletons from several places in Europe long before plague is thought to have invaded (Hummel et al. 2005). Regardless, several other mutations in the CCR5 receptor also persist, and it appears that selection via resistance to infections, rather than genetic drift, provides a plausible explanation for the high frequency of mutant alleles (Blanpain et al. 2000; Galvani and Slatkin 2003; Novembre et al. 2005; Sabeti et al. 2005).

Finally, infections and genetic diseases often co-occur within the same patients, and Darwinian medicine can also encompass this broader perspective on human health. Patients may be sickened by pathogens, but other aspects of their genetics and evolutionary history (ethnicity) can make individuals predisposed to infection or to greater pathogenesis and more severe disease once infection sets in (Cooke and Hill 2001; Weiss and McMichael 2004). For instance, some alleles of *HLA* genes of the immune system increase susceptibility to a whole variety of pathogens, including those that cause leprosy, tuberculosis, malaria, and AIDS (Hill and Motulsky 1999; Dean et al. 2002). Even more profound is that disease from common human maladies like dengue, influenza, malaria, and tuberculosis results not only from direct damage from the pathogens but also from overreaction of the immune system while attacking those pathogens (Graham et al. 2005). Thus the roots of disease are complex and in many instances grow from deep in our genetics and immune responses to infections, appropriate or not.

Diagnosis

What aspects of care for the sick and for public health distinguish the Darwinian approach to disease? The largest distinction is extending medical care beyond individual patients to consider human populations in the context of their connections with the microbial world and the patients' genetic and evolutionary history. This can be done without compromising immediate treatment to alleviate pain and suffering and in many cases can provide more rapid diagnosis, treatment, and recovery. A typical example arises in treatment of fever and iron deficiency in patients with infections, because both of these may represent part of an adaptive defensive response to slow the growth of bacteria (Williams and Neese 1991; Neese and Williams 1994). Treating the symptoms of infected patients could have the undesired effect that infections persist longer, with greater potential for transmission.

The following decision tree shows pathways through this broader context and hopefully can lead physicians not only to diagnose and treat disease once the basic symptoms and causes are known but also toward a global understanding of disease:

1. Is this an infection or a congenital condition?
2. If it is an infection, is this person otherwise healthy?
   a. If otherwise unhealthy, is the patient compromised by poor environment (e.g., poor diet and nutrition, smoking, or occupational hazards)?

    b. If otherwise unhealthy but with a good environment, could the patient be genetically predisposed to infection (e.g., poor immune system)?

    c. If otherwise healthy, should symptoms be treated (e.g., nausea, fever, nasal congestion), or should this infection be allowed to run its course with minimal intervention?

3. If it is an infection, is this pathogen unique and/or common in humans, or is the pathogen's source from other animals (called *zoonotic disease*; see the review by Wolfe et al. 2007).

    a. Where did this patient contact the infectious source, and how was the pathogen transmitted (e.g., by direct contact or by an arthropod vector like a mosquito, flea, or tick)?

    b. How contagious is the pathogen: should patients be isolated until they are no longer infectious?

    c. Do we know the infectious microbe's closest relatives? Will therapies to fight those infections help this patient as well?

    d. After we identify whether the pathogen is a bacterium, a virus, a protozoan, or a fungus, does the infectious agent have resistance to antimicrobials?

4. If this is a congenital disorder, what is this person's family history and ethnicity?

    a. Is the individual from a small population with a history of marriages between close relatives (e.g., first cousins or closer)?

    b. Is this disease highly prevalent in other populations with the same ethnic origins?

5. Will a therapy or intervention have an effect beyond this patient?

    a. Will therapy select for higher pathogenicity or virulence in the microbes (e.g., effects of imperfect vaccines)?

    b. What kind of selection pressure will antimicrobials create, and what will be the trade-off between intervening with a drug now and failure of therapies in the future if drug resistance evolves?

Effective use of this decision tree presupposes a technological world where laboratory diagnostics based on modern biochemistry and genomics are available for rapid screening.[1] This tree also presumes that medical and health professionals learn the natural history and ecology of pathogens and parasites in the area where they practice, and that they understand the ethnic diversity of the population they serve.

    The work of Dr. D. Holmes Morton among the Amish of Pennsylvania and Illinois provides a clear connection between biomedical research and clinical practice (Belkin 2005). Morton and colleagues first diagnosed and then identified the underlying causes of pretzel syndrome, a form of cerebral palsy with severe mental retardation and then early death in affected children (Strauss et al. 2003). Similar to PKU, pretzel syndrome is glutaric aciduria (GA1), caused by mutations in the degradation pathways of three amino acids (lysine, hydroxylysine, and tryptophan). Morton is primarily a pediatrician, but in his clinic in central Pennsylvania he pursued the technology to first detect GA1 via nondegraded proteins in urine samples and then develop a genetic test based on microchips. The mutation has been traced to Europe, where the Amish originated, and occurs at low frequency within populations of European descent. With screening and diet, disease prevalence has been reduced among Amish to a third of the pretreatment rate.

Recognizing this broader context is an integral part of Darwinian medicine. In part this kind of training was at one time provided to health professionals under the subject of geographic medicine, but it was swept aside in the silver-bullet years immediately after the discovery of antimicrobial drugs in the 1930s and 1940s (Burnet and White 1972; W. Anderson 2004). The rediscovery of this branch of medicine comes on the heels of failures of drug therapies and the reemergence of infectious diseases like polio and tuberculosis that were thought to have been conquered (Garrett 1994; Levin and Anderson 1999).

Treatment and Therapy

Darwinian medicine asks health professionals to look beyond symptoms of the patient in front of them and to practice medicine in a way that is both predictive and preemptive (Zerhouni 2005; Culliton 2006). The challenge to the medical world, including biomedical research and drug discovery, is to embrace both natural selection and common descent (homology) as part of the process that translates biomedical research to clinical practice. In this translation humans cannot set themselves apart from their evolutionary connections to the natural world, including their own genetics and biology. The science of evolution is highly predictive, and it is unethical for those who are developing treatments to ignore knowledge of evolutionary biology. As will be seen later, some of these pitfalls can only be discovered in retrospect, but the cost of ignoring evolutionary biology in medicine has been high in some cases.

ADAPTATION

If one considers microbes, the idea of natural selection is easily accepted because resistance to antimicrobial drugs has rapidly evolved and resulted in reemergence of previously controlled pathogens (Garrett 1994; Levin and Anderson 1999; Lederberg 2000). In some cases bacteria have evolved multiple-drug resistance, partly because of novel mutations but just as often by acquiring drug resistance via DNA exchange from other bacterial species. This is true for *Neisseria,* the bacterium responsible for sexually transmitted gonorrhea and directly transmitted bacterial meningitis (Snyder et al. 2005). Similarly, multidrug resistance has evolved in the flesh-eating *Staphylococcus aureus,* which is often acquired in hospitals. In its new and impervious form this common skin bacterium causes severe disease of skin and soft tissues, sometimes leading to death (Diekema et al. 2001; Eady and Cove 2003).

On the other hand, it is implausible that humans will evolve new mechanisms of disease resistance, at least in the absence of some kind of regular and catastrophic challenge that would severely reduce fitness of those who lack genetic predisposition for fighting infections. Again, this is not to say that humans have always been unaffected by the power of natural selection—recent genomic analyses show ongoing selection in brain size (Mekel-Bobrov et al. 2005) and recent selection for adult lactose metabolism in European and African populations (Tishkoff et al. 2007). But the disparity of generation times of microbes and humans spurs natural selection in favor of microbes, which will experience selection at multiple levels: within individual hosts, between human hosts, and between humans and pathogen reservoirs in commensal or wild populations of other animals.

An intriguing possibility is that our immune systems are currently under selection via immunopathology, disease caused by overresponse by our immune systems to what could otherwise be mild infections (Graham et al. 2005). Our immune systems could be seen as a series of trade-offs between recognition of self and nonself, defense against invading pathogens, and the potential damage to noninfected tissues when clearing invaders from infected tissues. In this case selection would be toward an intermediate phenotype that balances trade-offs. Genomic analyses of immune-system genes, combined with experiments in model systems that further demonstrate how these kinds of trade-offs function, will provide an active area of research for years to come (Graham et al. 2005).

Thus modern medicine still finds itself on a shifting landscape, and we continue to write the chapter in human history suggested by Joshua Lederberg (2000), which he calls "Our Wits versus Their Genes." Evolution in microbes will never stop, but we are beginning to apply strategies specifically designed to halt their transmission, slow resistance to antimicrobials, and improve vaccines. The first issue is perhaps the simplest: hygiene can halt pathogen transmission. This was immediately apparent from the famous epidemiological sleuthing by Dr. John Snow in London, who in 1854 removed the handle of a public water pump on Broad Street in Soho and proved that a local cholera epidemic came from a contaminated water supply (Wills 1996). Similarly, plague and the Black Death, caused by the bacterium *Yesinia pestis* and responsible for some of the most infamous human pandemics, have largely waned as human diseases in areas where rodents and their fleas are removed from human living quarters (Gage and Kosoy 2005). In terms of evolutionary biology, hygiene circumvents the opportunity for selection for pathogens by simply removing chains of infection and preventing transmission (Galvani 2003).

Slowing evolution of antimicrobial resistance is one of the largest public health challenges we face, but most work in this area is hypothetical because critical data that could distinguish which strategies would be most effective are not yet available (Lipsitch 2001). Regardless, mathematical models and simulation studies indicate that in lieu of attacking each microbe with a single drug until resistance reaches a level that compromises public health, the greatest health benefits will be gained by deploying combinations of drugs in the same patients at the same time or targeting for control those pathogens with the lowest transmission rates (Bonhoeffer et al. 1997; Levin and Anderson 1999; Bull and Wichman 2001; Lipsitch 2001; D. L. Smith et al. 2005). Again, possible outcomes have enough contingencies that only time, carefully controlled human trials, and laboratory experiments with model systems will provide definitive answers.

As noted for influenza A, keeping effective vaccines on hand can be difficult, although several vaccines have remained effective even after decades of use (e.g., measles vaccines). This question has arisen again because of the possibility that H5N1 avian influenza could emerge as a directly transmitted human pathogen, and it is not feasible to provide perfectly matched vaccines for widespread use. Thus the hope is that even an imperfectly matched vaccine will provide some level of protection and slow viral transmission. Unfortunately, this does not take into account that pathogens undergo selection within hosts that could favor more virulent pathogens (defined by high rates of pathogen replication causing cell or tissue damage). Specifically, imperfect vaccines may cause pathogens to evolve both faster transmission to escape protected individuals and higher virulence in unprotected individuals (Gandon et al. 2001; van Boven et al. 2005). Again, public health implications of using vaccines that only partially protect individuals must take into account that selection on pathogens occurs at several levels, and that within-host evolution generally selects for greater virulence (Stearns and Ebert 2001; Galvani 2003). Releasing partial vaccines to stem local outbreaks could have the unintended risk of selecting for more virulent pathogens in the future.

Finally, much has been made of the evolution of virulence and the argument that pathogens should evolve lower virulence because damage to hosts slows transmission (Ewald 1994). Empirical support for the optimal-virulence idea comes from a few examples like the reduced death rate among European rabbits infected with myxoma virus in Australia in the years after the virus was introduced to control the rabbits that were ravaging the landscape. Lowered virulence in the bacterium that causes diphtheria in humans is a second example (Ebert and Bull 2003). But again, the trade-off hypothesis that balances virulence and transmission assumes a relatively simple definition of virulence, which is actually a function of pathogen replication within the host, toxicity of the pathogen to the host, immune response of the host, and, finally, which tissues are affected. For instance, the bacterium *Neisseria meningitidis* causes meningitis when it infects spinal tissues, but its transmission is from infections of the upper respiratory tract, which are usually asymptomatic. Thus virulence in *N. meningitidis* is unrelated to transmission and is actually a spillover of infections within hosts. From the idea of optimal virulence comes the notion that disease could be avoided if pathogen virulence could be controlled. The question then arises: what will be the target for selection? Ebert and Bull (2003) suggest that selection should target virulence directly by attacking pathogen toxicity; vaccines against diphtheria and pertussis have been designed to do just that, with the result that surviving strains are less pathogenic.

COMMON DESCENT

Ideas of common descent and homology may have greater difficulty in being accepted in medical practice for the same reasons that the notion of common descent generates cognitive dissonance in the general population. The concept of homology, that traits in different species have common evolutionary origins, is critical to development of pharmaceuticals, which are first tested on mammals like mice, rats, and dogs. Promising candidates may then be tested on primates before clinical trials with human volunteers. Thus biomedical research advances by understanding that species phylogenetically closest to humans have the most similar genetics, physiology, and immune systems to humans and thus prove to be reliable models for research and development of treatment and therapy.

How badly can things go if common descent is disregarded? One poignant case is that of Baby Fae, an infant born with an underdeveloped heart who received the heart of a baboon at Loma Linda University Medical Center in 1984 (Bailey et al. 1985). The baboon donor was one of five that had been tested for immunological similarity on the basis of three *HLA* genes. The baboon with lowest reactivity to the infant's immune cells (lymphocytes) was chosen as donor. Baby Fae survived for 20 days after surgery with the help of the immunosuppressive drug cyclosporine but eventually died from rejection of the transplanted heart and other organ failure. Ethical and procedural questions aside, a troubling part of the story is that several mistakes directly resulted from a rejection of evolutionary thinking.

Leonard Bailey, the lead surgeon for Baby Fae's operation, admitted to being a fundamentalist Christian during a radio interview for the Australian Broadcasting Corporation.[2] Bailey chose baboons as donors because their hearts were available and the right size. Asked why hearts of chimpanzees, humans' closest living primate relative, were not considered, Bailey answered, "The scientists that are keen on the evolutionary concept, that we actually developed serially from subhuman primates to humans, with mitochondrial DNA dating and that sort of thing, the differences have to do with millions of years. That boggles my mind somehow. I don't understand it well, and I'm not sure that it means a great deal in terms of tissue homology." Infant transplants by Bailey's group at Loma Linda subsequently used only human donors. The prospect of using other species for transplantation continues to be questioned because for most organs the cross-species immunity barrier is high, even with immunosuppressive drugs, and because of the risk that novel pathogens could be transferred to humans from the donor animals (Deschamps et al. 2005).

Prognosis

Where does Darwinian medicine take us in the future? First and foremost, the greatest challenge is that microbes have an almost unlimited capacity to adapt to changing environments, and that humans in the modern world are one of the largest and most consistent environments available to them. But are doomsday scenarios of disease pandemics necessary? Not really, but to ignore the possibility would also be foolish, and many interventions can be adopted that could reduce disease severity (Purssell 2005). The issue is that the process of evolution has many faces, with change in the future as the only certainty. Given variation in pathogen populations, with some traits conferring higher fitness and with inheritance of trait differences, adaptation by natural selection is a certainty. Responses to selection, however, sometimes have counterintuitive outcomes because selection acts in different directions at different levels. The possibility that evolution occurs at multiple scales is among the most important contributions evolutionary biology has made to modern medicine (Williams and Neese 1991; Stearns and Ebert 2001). Deploying imperfect vaccines during public health emergencies, as discussed above, provides a potentially alarming example, where epidemics could be slowed regionally while the vaccine selects for higher virulence among surviving pathogens, with more severe disease in subsequent outbreaks (Gandon et al. 2001). Trade-offs across multiple levels lay at the heart of the evolution of pathogen virulence, transmission, and reactions of our immune systems to invaders (William and Neese 1991; Ewald 1994; Ebert and Bull 2003; Graham et al. 2005).

The prognosis for genetic change that would reduce disease incidence in human populations is mixed. Rapid advances in genomics and gene discovery mean that genetic screening, prenatally or soon after birth, will provide pediatricians with information needed to treat even more of the metabolic diseases, like PKU and GA1, caused by mutations. Genetic screening of adults, and what could be done with that information, is a matter of ethics that is still being debated, especially for the risk of genetic disease, the predisposition of individuals for infections, and rates charged for health insurance. Screening for mutations that cause genetic disease could help potential parents decide how to proceed, but again, this is a matter of privacy and ethics outside the science of evolution. Progress in gene therapy—reintroducing functional genes into the genomes of patients to cure their disease—was set back after a fatality in a gene therapy trial in Philadelphia in 1999 (Thompson 2000) but is being revisited with improved biotechnologies. But nothing in the study of modern genomics and genetic disease changes the fact of human population size, and that mutations in large populations occur at high absolute rates. Thus we can still predict with certainty that eugenics, the discarded notion that selective breeding of humans could eradicate genetic disease, would do even less now to alleviate human misery than it did a century ago when eugenics programs were first instituted.[3]

The problems that face Darwinian medicine are the same ones that make evolution fascinating to some and frustrating to others. Within the more certain framework of inheritance and natural selection, evolution is influenced by chance events like mutations, chance encounters with an individual who brings a new infection into a community, or a rapidly altered environment. On the one hand, some find comfort in the idea from Darwinian medicine that some diseases can be "predicted" from the maladaptation of humans to their current world (Neese and Williams 1994). On the other hand, we would feel better about Darwinian medicine and looming plagues of modern times if the identities of the next pandemic-causing pathogens were more predictable (many would say that viruses like H5N1 or Ebola are most likely; see Cleaveland et al. 2007). We can take greater comfort, however, from geographic epidemiology, much as Darwin may have found comfort through his interest in geographic distributions of populations and species (Burnet and White 1972; Lederberg 2000; W. Anderson 2004; Tishkoff and Kidd 2004). Thus we know that the histories of populations will influence the prevalence of congenital disorders, and we know that disease organisms have geographic origins, as do the reservoirs for disease and potential insect vectors for transmission. By acknowledging that humans live as part of the worldwide ecology of infectious disease and that evolution of pathogens is the inevitable outcome, we can be even more prepared the next time the monkeypox visits Kansas.

ACKNOWLEDGMENTS

My research is supported by the National Science Foundation: Ecology of Infectious Diseases Program (EF-0327052) and the Short Grass Steppe Long Term Ecological Research project (DEB 0217631). I thank the editors and anonymous reviewers, along with Irene Eckstrand, Norman Johnson, Mary Poss, and my physician Mark Paulsen, for comments. I also thank Dr. Paulsen for his strict attention to my infections and genetics.

NOTES

1. A caveat must be made here: disease describes the physical damage to a patient, which is an individual phenomenon. Screening can identify individuals with particular genotypes or specific infections, but these individuals will not necessarily develop disease. Nonetheless, screening can predict disease, and disease risk can be understood at the level of populations when exposures to pathogens and/or frequencies of genetic mutations are high.

2. Original radio broadcast on June 3, 1985, during the program *Health Report*.

3. See Haldane (1964) for an amusing and readable essay that defends mathematical theories of population genetics, including reference to the balance between mutation and selection, and some lovely verse about his own life-ending fight with colorectal cancer.

BIBLIOGRAPHY

Anderson, R. M., and R. M. May. 1991. *Infectious Diseases of Humans: Dynamics and Control.* Oxford: Oxford University Press.

Anderson, W. 2004. Natural histories of infectious disease: Ecological vision in twentieth-century biomedicine. *Osiris* 19: 39–61.

Baigent, S. J., and J. W. McCauley. 2003. Influenza type A in humans, mammals and birds: Determinants of virus virulence, host-range and interspecies transmission. *BioEssays* 25: 657–671.

Bailey, L. L., S. L. Nehison-Cannarella, W. Concepcion, and W. B. Jolly. 1985. Baboon-to-human cardiac xenotransplantation in a neonate. *Journal of the American Medical Association* 254: 3321–3329.

Belkin, L. 2005. A doctor for the future. *New York Times Magazine,* November 6.

Blanpain, C., B. Lee, M. Tackoen, B. Puffer, A. Boom, F. Libert, M. Sharron, V. Wittamer, G. Vassart, R. W. Doms, and M. Parmentier. 2000. Multiple nonfunctional alleles of CCR5 are frequent in various human populations. *Blood* 96: 1638–1645.

Bonhoeffer, S., M. Lipsitch, and B. R. Levin. 1997. Evaluating treatment protocols to prevent antibiotic resistance. *Proceedings of the National Academy of Sciences USA* 94: 12106–12111.

Bull, J. J., and H. A. Wichman. 2001. Applied evolution. *Annual Review of Ecology and Systematics* 32: 183–217.

Burnet, M., and D. O. White. 1972. *Natural History of Infectious Disease.* 4th ed. Cambridge: Cambridge University Press.

Bush, R. M., C. A. Bender, K. Subbarao, N. J. Cox, and W. M. Fitch. 1999. Predicting the evolution of human influenza A. *Science* 286: 1921–1925.

Centers for Disease Control and Prevention (CDC). 2003. Update: Multistate outbreak of monkeypox—Illinois, Indiana, Kansas, Missouri, Ohio, and Wisconsin, 2003. *Morbidity and Mortality Weekly Report (MMWR)* 52: 642–646.

Cleaveland, S., D. T. Haydon, and L. Taylor. 2007. Overviews of pathogen emergence: Which pathogens emerge, when, and why? *Current Topics in Microbiology and Immunology* 315: 85–111.

Cooke, G. S., and A. V. S. Hill. 2001. Genetics of susceptibility to human infectious disease. *Nature Reviews Genetics* 2: 967–977.

Culliton, B. J. 2006. Extracting knowledge from science: A conversation with Elias Zerhouni. *Health Affairs* 25: w94–w103.

Daily, G. C., S. Alexander, P. R. Ehrlich, L. Goulder, J. Lubchenco, P. A. Matson, H. A. Mooney, S. Postel, S. H. Schneider, D. Tilman, and G. M. Woodwell. 1997. Ecosystem services: Benefits supplied to human societies by natural ecosystems. *Issues in Ecology* 2: 1–16.

Dean, M., M. Carrington, and S. J. O'Brien. 2002. Balanced polymorphism selected by genetic versus infectious human disease. *Annual Review of Genomics and Human Genetics* 3: 263–292.

Dean, M., M. Carrington, C. Winkler, G. A. Huttley, M. W. Smith, R. Allikmets, J. J. Goedert, S. P. Buchbinder, E. Vittinghoff, E. Gomperts, S. Donfield, D. Vlahov, R. Kaslow, A. Saah, C. Rinaldo, R. Detels, and S. J. O'Brien. 1996. Genetic restriction of HIV-1 infection and progression to AIDS by a deletion allele of the CKR5 structural gene. *Science* 273: 1856–1862.

Deschamps, J. Y., F. A. Roux, P. Sai, and E. Gouin. 2005. History of xenotransplantation. *Xenotransplantation* 12: 91–109.

Diamond, J. 2003. The double puzzle of diabetes. *Nature* 423: 599–602.

Diekema, D. J., M. A. Pfaller, F. J. Schmitz, J. Smayevsky, J. Bell, R. N. Jones, M. Beach, and the SENTRY Participants Group. 2001. Survey of infections due to *Staphylococcus* species: Frequency of occurrence and antimicrobial susceptibility of isolates collected in the United States, Canada, Latin America, Europe, and the western Pacific region for the Sentry Antimicrobial Surveillance Program, 1997–1999. *Clinical Infectious Diseases* 32: S114–S132.

Eady, E. A., and J. H. Cove. 2003. Staphylococcus resistance revisited: Community-acquired methicillin resistant *Staphylococcus aureus*—an emerging problem for the management of skin and soft tissue infections. *Current Opinion in Infectious Diseases* 16: 103–124.

Ebert, D. 1998. Experimental evolution of parasites. *Science* 282: 1432–1435.

Ebert, D., and J. J. Bull. 2003. Challenging the trade-off model for the evolution of virulence: Is virulence management feasible? *Trends in Microbiology* 11: 15–20.

Elvin, S. J., E. D. Williamson, J. C. Scott, J. N. Smith, G. P. de Lema, S. Chilla, P. Clapham, K. Pfeffer, D. Schlondorff, and B. Luckow. 2004. Ambiguous role of CCR5 in *Y. pestis* infection. *Nature* 430: doi:10.1038/nature02822.

Ewald, P. W. 1994. *Evolution of Infectious Diseases*. New York: Oxford University Press.

Ferguson, N. M., A. P. Galvani, and R. M. Bush. 2003. Ecological and immunological determinants of influenza evolution. *Nature* 422: 428–433.

Frank, S. A. 2004. Genetic predisposition to cancer—insights from population genetics. *Nature Reviews Genetics* 5: 764–772.

Gage, K. L., and M. Y. Kosoy. 2005. Natural history of plague: Perspectives from more than a century of research. *Annual Review of Entomology* 50: 505–528.

Galvani, A. P. 2003. Epidemiology meets evolutionary ecology. *Trends in Ecology and Evolution* 18: 132–139.

Galvani, A. P., and M. Slatkin. 2003. Evaluating plague and smallpox as historical selective pressures for the CCR5–Δ32 HIV-resistance allele. *Proceedings of the National Academy of Sciences USA* 100: 15276–15279.

Gandon, S., M. J. Mackinnon, S. Nee, and A. F. Read. 2001. Imperfect vaccines and the evolution of pathogen virulence. *Nature* 414: 751–756.

Garrett, L. 1994. *The Coming Plague*. New York: Penguin Books.

Gluckman, P., and M. Hanson. 2006. *Mismatch: Why Our World No Longer Fits Our Bodies*. New York: Oxford University Press.

Gould, S. J., and R. C. Lewontin. 1979. The spandrels of San Marco and the Panglossian paradigm: A critique of the adaptationist programme. *Proceedings of the Royal Society of London, Series B* 205: 581–598.

Graham, A. L., J. E. Allen, and A. F. Read. 2005. Evolutionary causes and consequences of immunopathology. *Annual Review of Ecology, Evolution, and Systematics* 36: 373–397.

Greaves, M. 2002. Cancer causation: The Darwinian downside of past success? *Lancet Oncology* 3: 244–251.

Grenfell, B. T., O. G. Pybus, J. R. Gog, J. L. N. Wood, J. M. Daly, J. A. Mumford, and E. C. Holmes. 2004. Unifying the epidemiological and evolutionary dynamics of pathogens. *Science* 303: 327–332.

Guarner, J., B. J. Johnson, C. D. Paddock, W.-J. Shieh, C. S. Goldsmith, M. G. Reynolds, I. K. Damon, R. L. Regnery, S. R. Zaki, and the Veterinary Monkeypox Virus Working Group. 2004. Monkeypox transmission and pathogenesis in prairie dogs. *Emerging Infectious Diseases* 10: 426–431.

Hahn, B. H., G. M. Shaw, K. M. De Cock, and P. M. Sharp. 2000. AIDS as a zoonosis: Scientific and public health implications. *Science* 287: 607–614.

Haldane, J. B. S. 1964. A defense of beanbag genetics. *Perspectives in Biology and Medicine* 7: 343–359.

Hill, A. V. S., and A. G. Motulsky. 1999. Genetic variation and human diseases: The role of natural selection. In S. C. Stearns, ed., *Evolution in Health and Disease*, 50–61. New York: Oxford University Press.

Hummel, S., D. Schmidt, B. Kremeyer, B. Herrmann, and M. Oppermann. 2005. Detection of the CCR5–Δ32 HIV resistance gene in Bronze Age skeletons. *Genes and Immunity* 6: 371–374.

Keele, B. F., F. Van Heuverswyn, Y. Y. Li, E. Bailes, J. Takehisa, M. L. Santiago, F. Bibollet-Ruche, Y. L. Chen, L. V. Wain, F. Liegeois, S. Loul, E. M. Ngole, Y. Bienvenue, E. Delaporte, J. F. Y. Brookfield, P. M. Sharp, G. M. Shaw, M. Peeters, and B. H. Hahn. 2006. Chimpanzee reservoirs of pandemic and nonpandemic HIV-1. *Science* 313: 523–526.

Kuiken, T., E. C. Holmes, J. McCauley, G. F. Rimmelzwaan, C. S. Williams, and B. T. Grenfell. 2006. Host species barriers to influenza virus infections. *Science* 312: 394–397.

Lederberg, J. 2000. Infectious history. *Science* 288: 287–293.

Levin, B. R., and R. M. Anderson. 1999. The population biology of anti-infective chemotherapy and the evolution of drug resistance: More questions than answers. In S. C. Stearns, ed., *Evolution in Health and Disease*, 125–137. New York: Oxford University Press.

Levin, B. R., M. Lipsitch, and S. Bonhoeffer. 1999. Population biology, evolution, and infectious disease: Convergence and synthesis. *Science* 283: 806–809.

Li, W., Z. Shi, M. Yu, W. Ren, C. Smith, J. H. Epstein, H. Wang, G. Crameri, Z. Hu, H. Zhang, J. Zhang, J. McEachern, H. Field, P. Daszak, B. T. Eaton, S. Zhang, and L.-F. Wang. 2005. Bats are natural reservoirs of SARS-like coronaviruses. *Science* 310: 676–679.

Lipsitch, M. 2001. The rise and fall of antimicrobial resistance. *Trends in Microbiology* 9: 438–444.

MacLean, R. C. 2005. Adaptive radiation in microbial microcosms. *Journal of Evolutionary Biology* 18: 1376–1386.

Mecsas, J., G. Franklin, W. A. Kuziel, R. R. Brubaker, S. Falkow, and D. E. Mosier. 2004. CCR5 mutation and plague protection. *Nature* 427: 606.

Mekel-Bobrov, N., S. L. Gilbert, P. D. Evans, E. J. Vallender, J. R. Anderson, R. R. Hudson, S. A. Tishkoff, and B. T. Lahn. 2005. Ongoing adaptive evolution of ASPM, a brain size determinant in *Homo sapiens*. *Science* 309: 1720–1722.

Meyers, L. A., B. R. Levin, A. R. Richardson, and I. Stojiljkovic. 2003. Epidemiology, hypermutation, within-host evolution and the virulence of *Neisseria meningitidis*. *Proceedings of the Royal Society of London, Series B* 270: 1667–1677.

Neese, R. M., and G. C. Williams. 1994. *Why We Get Sick: The New Science of Darwinian Medicine*. New York: Vintage Books.

Novembre, J., A. P. Galvani, and M. Slatkin. 2005. The geographic spread of the CCR5 Δ32 HIV-resistance allele. *Public Library of Science Biology* 3: 1954–1962.

Olsen, B., V. J. Munster, A. Wallensten, J. Waldenstrom, A. D. M. E. Osterhaus, and R. A. M. Fouchier. 2006. Global patterns of influenza A virus in wild birds. *Science* 312: 384–388.

Purssell, E. 2005. Evolutionary nursing: The case of infectious diseases. *Journal of Advanced Nursing* 49: 164–172.

Ruan, Y. J., C. L. Wei, L. A. Ee, V. B. Vega, H. Thoreau, S. T. S. Yun, J. M. Chia, P. Ng, K. P. Chiu, L. Lim, Z. Tao, C. K. Peng, L. O. L. Ean, N. M. Lee, L. Y. Sin, L. F. P. Ng, R. E. Chee, L. W. Stanton, P. M. Long, and E. T. Liu. 2003. Comparative full-length genome sequence analysis of 14 SARS coronavirus isolates and common mutations associated with putative origins of infection. *Lancet* 361: 1779–1785.

Sabeti, P. C., E. Walsh, S. F. Schaffner, P. Varilly, B. Fry, H. B. Hutcheson, M. Cullen, T. S. Mikkelsen, J. Roy, N. Patterson, R. Cooper, D. Reich, D. Altshuler, S. O'Brien, and E. S. Lander. 2005. The case for selection at CCR5-delta32. *Public Library of Science Biology* 3: 1963–1969.

Scriver, C. R., and P. J. Waters. 1999. Monogenic traits are not simple—Lessons from phenylketonuria. *Trends in Genetics* 15: 267–272.

Sieradzki, K., R. B. Roberts, S. W. Haber, and A. Tomasz. 1999. The development of vancomycin resistance in a patient with methicillin-resistant *Staphylococcus aureus* infection. *New England Journal of Medicine* 340: 517–523.

Smith, D. J. 2006. Predictability and preparedness in influenza control. *Science* 312: 392–394.

Smith, D. L., S. A. Levin, and R. Laxminarayan. 2005. Strategic interactions in multi-institutional epidemics of antibiotic resistance. *Proceedings of the National Academy of Sciences USA* 102: 3153–3158.

Snyder, L. A. S., J. K. Davies, C. S. Ryan, and N. J. Saunders. 2005. Comparative overview of the genomic and genetic differences between the pathogenic *Neisseria* strains and species. *Plasmid* 54: 191–218.

Stearns, S. C., ed. 1999. *Evolution in Health and Disease*. New York: Oxford University Press.

Stearns, S. C., and D. Ebert. 2001. Evolution in health and disease: Work in progress. *Quarterly Review of Biology* 76: 417–432.

Strauss, K. A., E. G. Puffenberger, D. L. Robinson, and D. H. Morton. 2003. Type I glutaric aciduria, part 1: Natural history of 77 patients. *American Journal of Medical Genetics* C 121C: 38–52.

Swynghedauw, B. 2004. Evolutionary medicine. *Acta Chirurgica Belgica* 104: 132–139.

Thompson, L. 2000. Human gene therapy: Harsh lessons, high hopes. *FDA Consumer* 34, no. 5. http://www.fda.gov/fdac/features/2000/500_gene.html.

Tishkoff, S. A., and K. K. Kidd. 2004. Implications of biogeography of human populations for "race" and medicine. *Nature Genetics* 36: S21–S27.

Tishkoff, S. A., F. A. Reed, A. Ranciaro, B. F. Voight, C. C. Babbitt, J. S. Silverman, K. Powell, H. M. Mortensen, J. B. Hirbo, M. Osman, M. Ibrahim, S. A. Omar, G. Lema, T. B. Nyambo, J. Ghori, S. Bumpstead, J. K. Pritchard, G. A. Wary, and P. Deloukas. 2007. Convergent adaptation of human lactase persistence in Africa and Europe. *Nature Genetics* 39: 31–40.

van Boven, M., F. R. Moo, J. E. P. Schellekens, H. E. de Melker, and M. Kretzschmar. 2005. Pathogen adaptation under imperfect vaccination: Implications for pertussis. *Proceedings of the Royal Society of London, Series B* 272: 1617–1624.

Vitousek, P. M., P. R. Ehrlich, A. H. Ehrlich, and P. A. Matson. 1986. Human appropriation of the products of photosynthesis. *BioScience* 36: 368–373.

Wang, L. F., Z. L. Shi, S. Y. Zhang, H. Field, P. Daszak, and B. T. Eaton. 2006. Review of bats and SARS. *Emerging Infectious Diseases* 12: 1834–1840.

Weiss, R. A., and A. J. McMichael. 2004. Social and environmental risk factors in the emergence of infectious diseases. *Nature Medicine* 10: S70–S76.

Williams, G. C., and R. M. Neese. 1991. The dawn of Darwinian medicine. *Quarterly Review of Biology* 66: 1–22.

Wills, C. 1996. *Yellow Fever, Black Goddess: The Coevolution of People and Plagues*. Reading, MA: Addison-Wesley.

Wolfe, N. D., C. P. Dunavan, and J. Diamond. 2007. Origins of major human infectious diseases. *Nature* 447: 279–283.

Zerhouni, E. A. 2005. US biomedical research: Basic, translational, and clinical sciences. *Journal of the American Medical Association* 294: 1352–1358.

Beyond the Darwinian Paradigm: Understanding Biological Forms

Brian Goodwin

The question that I shall examine in this essay is the relation between form and function in evolution, and how the origin of different forms is understood. Form has two aspects: shape or structure, as in the morphology of organisms; and pattern in time or behavior, such as the movements of individual organisms (swimming, walking, flying) or the spatial patterns of collectives, such as flocks of birds or foraging colonies of social insects. These characteristics of species are often referred to collectively as characters, and they have adaptive, functional significance for the survival of the species. However, there is a distinction between understanding the origin of a form (i.e., how the form can arise in organisms during their evolution) and the process whereby natural selection acts to change the relative frequencies of the different forms in a population because of their survival value. Natural selection cannot explain how any form originates, and it can be used only in explaining the differential abundance of characters, connected with their contribution to the fitness of organisms.

There is a tendency among evolutionary biologists to assume that any form can be generated as a result of random variation in the genes, resulting in an indefinite diversity of possible forms for evolution to work with. This is the assumption that I shall suggest is mistaken as a general principle, on the basis of two types of evidence. The first of these comes from observation of species morphology, which shows that forms generated during evolution fall into distinct, not continuous, classes. An implication of this is that there are constraints on the possible forms of organisms and their behavior, so that some forms can be generated, while others cannot. The case that I shall examine in detail comes from plant morphology, since we have experimental, observational, and theoretical evidence for distinct sets of patterns in this example. All three are required to make a convincing case against the assumption of continuous variation as a basic principle in evolution. There are many examples that can be used from animal morphology and collective behavior patterns, some of which I discuss, but the plant example makes the general points clearly and illustrates how the argument that relates to morphology can be pursued in other cases.

The second type of evidence that I present for constraints on the possible forms available for evolution comes from the nature of the processes that generate the forms. These are the dynamic activities whereby different morphologies and behaviors are generated by interactions within organisms during their development and between them in producing behavior patterns. This involves some element of theoretical modeling, showing what types of processes could give rise to the forms and why these result in discrete, not continuous, variation in characters.

The viewpoint presented here in no sense challenges the importance of natural selection in evolution. Whether the forms of organisms are initially generated as distinct possibilities or as a continuum of variations, selection still operates to influence the frequency of the different forms according to the fitness that they confer on the individual members of a species. The important point made is that the generative origins of biological structures and patterns are not to be understood in the same way as their differential abundance in different habitats.

We are still far from having anything like a complete description of the processes in organisms that underlie the origin of their distinctive characteristics. However, I believe that we understand sufficiently well the basic principles of these generative activities to draw some important conclusions about both the distinctions and the relationships between form and function in biology. On the basis of the examples presented and the general principles that govern nonlinear dynamics, I shall argue that the processes involved have the consequence that some forms are possible, while others are not. Understanding the origins of the forms that define different species therefore requires that we understand the dynamic processes that generate them, which are not those of natural selection.

Phyllotaxis as a Self-Organizing Growth Process

A classic example of form in flowering plants concerns the arrangement of leaves along the stem (phyllotaxis). It has been recognized for many years that there are only three primary, common patterns, though there are also some infrequently found arrangements. The major arrangements are shown in Figure 1. The first of these has the form of a spiral when one looks down on the plant, illustrated here by yucca. Successive leaves occupy positions rotated through a well-defined average angle (137.5°) relative to the previous leaf. The second pattern is called whorled, in which two or more leaves arise at each node. The leaves at each successive node are rotated so that the leaves are located above the gaps of the previous whorl. The example shown is fuschia, which has two leaves at each node, with rotation of successive whorls through

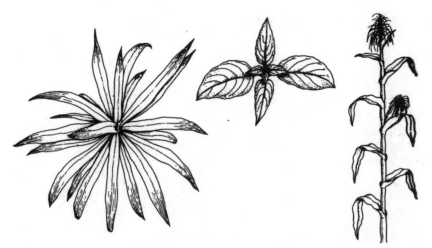

Figure 1. The three types of phyllotaxis: spiral, as in yucca *(left)*; decussate, as in fuschia *(middle)*; and distichous, as in maize *(right)*.

90°. The third pattern is known as distichous, in which there is a single leaf at each node and each successive leaf is located on the opposite side of the stem from the preceding one. The generator of all these patterns is the meristem, the growing tip of the plant, where the leaves are produced as distinct structures by growth and differentiation of cells in interaction with one another.

MORPHOGENETIC FIELDS AS GENERATORS OF MORPHOLOGY

A comprehensive examination of the generative process in the meristem led Douady and Couder (1996) to synthesize the components of this process into a model that shows how the three types of phyllotaxis could occur. The fundamental idea is that the meristem is a domain of growth in which a leaf primordium will arise only if the concentration of an inhibitory field in this region falls below a threshold value. Each new leaf that does form acts as a source of inhibition to neighboring tissue, preventing another leaf from forming. However, the inhibitory influence decreases with increasing distance, so that as leaves grow and move apart, there are regions where leaf formation can be initiated. It is the pattern of interactions in this self-organizing process that results in the arrangement of leaves. Depending on the initial state of the meristem and its geometry, the process generates one of the three major patterns of leaf arrangement as the primary pattern and other patterns as secondary ones. The term used in biology to describe this type of spatial patterning process in developing organisms is *morphogenetic field*. In technical terms, each of the different patterns that arises from the self-organizing dynamic process is called an *attractor,* which is the stable solution to the morphogenetic field for particular initial and boundary conditions that produces a particular form.

What can this model reveal about distinctions among the three major phyllotaxis categories? It is known that spiral patterns occur in just over 80% of higher plant species. The initial leaf pattern in the seedlings of all these species is either one storage leaf, as in monocotyledons, or two, as in dicotyledons. What emerges from the work of Douady and Couder is that with these initial conditions on the field organization of the meristem, spiral phyllotaxis is the pattern that is most accessible (it is the attractor that lies on the main branch of the dynamic generator). Furthermore, the divergence angle between successive leaves converges on the observed angle of 137.5° on this branch of the pattern-generation process. The other spiral patterns observed infrequently, with different divergence angles, are accessible, but they require that parameters in the model have specific ranges, as do rare whorled patterns. Different plant species can therefore have different phyllotaxis patterns as a result of genetic influences on initial conditions and parameter values in the meristem, but there is an overall bias in the direction of spirals, as observed in nature.

MORPHOGENETIC FIELDS PRODUCE FORMS WITH MATHEMATICAL REGULARITIES

Plants with spiral phyllotaxis have a characteristic mathematical order in the spatial relationships of their leaves. Some examples of these are given in Figure 2, which shows the spiral arrangements of leaves in *Araucaria excelsa,* a member of the Monkey Puzzle family. The numbers on the leaves refer to the sequence in which they were generated, 1 being the youngest leaf or the last to be generated, 2 the previous leaf, and so on. The spirals in this figure are drawn through leaves that make direct contact with each other. Now look at the difference between the numbers of successive leaves along any spiral. In Figure 2a the differences are 8 along the spirals defined by dotted lines and 13 for differences along spirals with solid lines. This defines what is known as (8, 13) spiral phyllotaxis. The other two examples in the figure have (5, 8) and (3, 5) spiral phyllotaxis (Figures 2b and 2c, respectively). The numbers in these pairs, 3, 5, 8, and 13, belong to a well-known mathematical series defined as $n_{i+1} = n_i + n_{i-1}$: each successive number of the series is the sum of the two preceding numbers. The particular numbers of the series depend on the two numbers that start it. If we begin with the numbers 1, 1, the series is 1, 1, 2, 3, 5, 8, 13, 21, 34, 55, and so on. Most plants with spiral phyllotaxis belong to spirals defined by successive numbers in this series, as we saw in Figure 2 for the Monkey Puzzle tree. Douady and Couder (1996) showed that these are the spirals that arise on the main branch of the generative process in their model. However, spirals with number differences belonging to other Fibonacci series are also found in nature, such as (2, 2, 4, 6, 10, 16, 26, . . .), (2, 5, 7, 12, 19, 41, 68, . . .), and (3, 4, 7, 10, 17, 27, 44, . . .). These can arise from the dominant spiral branch by specifying particular initial

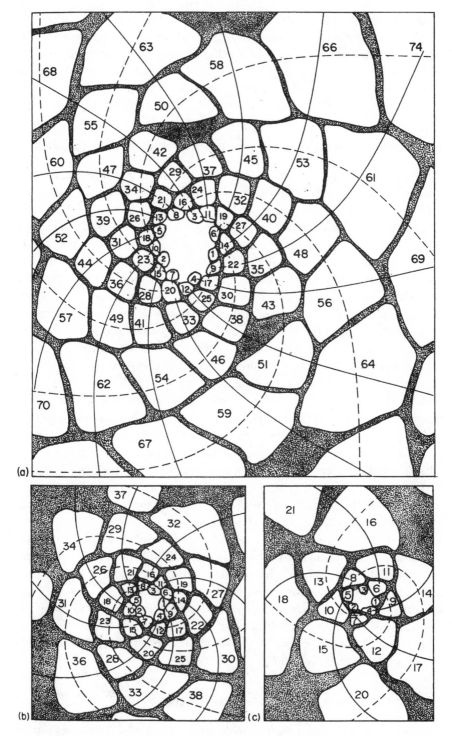

Figure 2. Leaf arrangements on branches of a species of Monkey Puzzle tree, *Araucaria excelsa*. See text for discussion.

conditions and parameter values in the model of Douady and Couder. The rare patterns of whorled leaf arrangements that are distinctive to particular species are also generated by the model when specific boundary conditions and parameter values are specified, forcing these modes to appear, though they are less stable than the dominant ones in terms of packing density of leaves.

RELATIVE STABILITY OF DIFFERENT PHYLLOTACTIC FORMS

All the patterns generated in the model of Douady and Couder are stable for some range of boundary and initial conditions on the morphogenetic field in the meristem, but some are less sensitive to changes in these than others. Again, the dominant patterns of Figure 1 are the ones that are least sensitive to variations in the generative conditions in the model, and among these the spiral patterns are the most robust. This field model of the dynamics of leaf production in the meristem thus provides us with a possible explanation of how all the different leaf patterns can be generated in higher plants; that is, it gives us a plausible picture of how the different patterns are possible forms of the same dynamic process under variation in the generative conditions that can be related to different gene activities and boundary conditions on the meristem (monocots and dicots). It is not necessary or appropriate to use the concept of natural selection or differential fitness of the characters to explain how the observed range of forms could have originated in plants. What the model tells us is that within some plant lineage or lineages there evolved within the meristem a dynamic process with the generative characteristics described. This model generates those and only those patterns observed in known plant species. There are many patterns that we can imagine that might be produced, such as a single leaf at a node, two rotated at 90° to this at the next node, three at the next rotated relative to these two, and then a repeat of this sequence. These are never observed. Biological forms are not selected from any imaginable set or a continuum of possibilities. Nonlinear dynamic processes of the types that occur in development always generate a discrete set of forms accessible to the dynamic processes involved. It is not natural selection that defines these possibilities, but the intrinsic dynamics of the generative process in the developing organism. Understanding these morphogenetic processes is what provides us with an understanding of the origins of species as distinct forms accessible within a lineage (Goodwin 1994).

FIELDS AND MOLECULAR GENERATORS

A model such as that of Douady and Couder shows what principles are required to provide a candidate explanation of phyllotaxis, but it operates at quite a high level of abstraction. Recent studies by Reinhardt et al. (2003) indicate the nature of the molecules involved in generating phyllotactic patterns and reveal a field dynamic different in detail from that assumed by Douady and Couder. The latter assumed an inhibitory influence, possibly carried by a morphogen from a leaf primordium, that decreases with distance, allowing a new leaf to form at a critical distance from a developing leaf. However, the evidence presented by Reinhardt and colleagues is that the influence is an induction of primordium formation by auxin. A new leaf acts as a sink for auxin that gets transported away from the site, thus preventing new leaf formation in its neighborhood. The inhibition assumed by Douady and Couder is thus replaced by removal of an inducer from the new leaf rather than direct inhibitory action by a morphogen. The field rule is therefore effectively inverted so that instead of permissive regions of low inhibition that allow new leaf formation in a scalar field, we have domains of elevated auxin, a morphogen that induces leaf initiation, and transport of auxin away from the new leaf and its boundaries in the meristem as the leaf develops, preventing new leaf inititation in its neighborhood. The dynamic is more complex than a field of inhibition because there is now a directional component that governs where the peaks of auxin occur in the field, arising from polarized transport of the morphogen.

The elegant studies of Reinhardt and colleagues take the question of morphogenetic fields in plants in some very interesting and challenging directions. Clearly they do not explain phyllotaxis, which requires both a coherent model of the whole and an experimentally valid molecular dynamic. The directional-field component that underlies auxin flow is governed by the spatial location of auxin pumps in the cell membrane, such as PIN1, a protein that is localized to the apical side of cells in the outer layers of the meristem, transporting auxin apically, but orientated for basal transport in developing vascular cells of the developing leaf. Thus there is a field of regulatory influence that underlies the production and localization of PIN1 in cells and needs to be explained in terms of molecular organizing forces in order to model this morphogenetic field. Modeling the close coupling between oriented pumping across cell walls, cell differentiation, and auxin flow patterns throughout the meristem to produce the coherent macroscopic patterns of leaf arrangements observed in leaf phyllotaxis presents an interesting challenge for those who seek to provide an integrated understanding of this robust morphogenetic process. Clearly any model that seeks to explain these phenomena must achieve the same range of integrated understanding of phyllotactic patterns and their transformations as the model of Douady and Couder does.

DEVELOPMENT AND EVOLUTION: THE RELATIONSHIP
BETWEEN ORIGINS OF FORMS AND THEIR FREQUENCIES

I now turn to the issue of natural selection and what role it might play in explaining aspects of phyllotaxis. Fitness to habitat is the concept that Darwin used to define the adaptive features of organisms, that is, how well an organism functions in a particular environment so that it survives to reproduce with other members of its species and leave offspring. Fitness can be applied to the characters of an organism (e.g., pattern of leaves on the stem of higher plants, pigmentation patterns and shapes of leaves) and to the life cycle as a whole. Natural selection is generally taken to describe the process whereby organisms with characters adapted to their habitat arise in the course of evolution: organisms with different hereditary characters arise as a result of genetic variation, and the more adaptive (fitter) characters increase in frequency in the population relative to those characters that are less fit. The characters observed most frequently in species are therefore regarded as having greater functional value to the organism, so that the differential abundance of a character is taken as a measure of its adaptive value to the organism. If we apply this reasoning to the frequency of phyllotaxis patterns in the leaves of higher plants, we may conclude that spiral phyllotaxis is the pattern with highest fitness or utility to the life cycles of higher plant species, since this is the most abundant pattern, taken to be established historically by natural selection.

What is the selective value of leaf arrangements in plants? A reasonable suggestion is that leaves serve the function of catching light for photosynthesis: the different patterns of leaves on a stem have differences in the efficiency with which they catch photons, which depends on how much they overlap and shade one another. Studies have shown that the spiral pattern illustrated by yucca in Figure 1a has the least mutual shading when light is coming from directly overhead, since leaves that are arranged in rows necessarily shade one another (Figures 1b and 1c). This could be a basis for understanding the differential abundance of leaf phyllotaxes in terms of natural selection and function. However, when other factors are considered, this proposal loses its strength. When averaged over the course of a day, with the sun occupying all angular positions relative to the plant, the differences between different leaf arrangements with regard to shading become minimal. Furthermore, observation of species in different habitats shows that there is no correlation between shade plants and spiral phyllotaxis, as might be expected if maximization of light capture correlated with spiral leaf arrangement. Plant adaptation to shade is largely through variation in leaf shape and pigmentation; the latter increases absorption of the light frequencies available. The overall evidence is that phyllotactic pattern is selectively neutral in plants, since all the different patterns are observed in various light habitats. In this case differential abundance, the predominance of spiral phyllotaxis in higher plants, does not appear to be explained by natural selection of a functional character. If the conclusion is correct that all phyllotactic patterns work adequately for the photosynthetic needs of plants, then spirals predominate because they are developmentally more accessible and stable. There is nothing surprising or problematic about this conclusion. It means simply that we need to take account of the relative developmental accessibility and stability of different morphological patterns, which will affect the frequency of characters in species and may have a primary influence in cases where the characters are selectively neutral. We cannot simply assume that differential abundance is always explained by natural selection. Of course, it is always assumed that if a character generated in a species is not useful, it will not increase in frequency, and if it is a disadvantage, it will not survive. However, it may not be utility that determines differential abundance.

Collective Behavior in Ant Colonies

The behavior patterns of organisms are clearly an important and interesting aspect of their evolution. The abilities of fish to swim, of birds to fly, of insects to forage for food, and of plants in temperate climates to drop their leaves in the autumn all contribute significantly to their capacity for survival. Just as there are differences between species in their morphologies, so there are differences in their behavior patterns that allow us to recognize them as significant indicators of the distinctive lifestyles and habits of different species. I shall now take a look at the way we can understand behavioral patterns as biological forms. For this I use an example from the study of dynamic order in an ant colony. The queen and the young ants at all stages of development are kept in a space within the colony called the brood chamber. In this chamber there are many worker ants that tend the queen and the young, feeding and cleaning them, moving them about as the young mature, and keeping the brood chamber free of debris. Researchers who have examined colonies kept in laboratories have observed the occurrence of a rhythmic activity pattern among the workers in the brood chamber of certain species of ants. When individual workers were observed, either in isolation from their fellows or at low density so that there were few interactions between them, it was shown that their behavior was chaotic in the technical sense of the term (Gleick 1998). The question arises: how does a regular periodic rhythm throughout a colony arise from the interactions of chaotic individuals? Where does the order come from?

Nigel Franks and colleagues (Franks et al. 1990) observed that the workers in a colony of ants of the genus *Leptothorax* who were tending the queen and the young in the brood chamber exhibited a well-defined rhythm in their activity-inactivity cycles. Roughly speaking, the workers were inactive for about 10 minutes every half hour. The first assumption was that each ant is individually periodic in its activity-inactivity cycle, and that in the brood chamber these cycles become synchronized through interactions so that the whole exhibits a regular periodic cycle. Such synchronization is well known in the flashing of fireflies or in female monkeys coming into estrus with the full moon. However, when Cole (1991) examined the behavior of individual ants from colonies of a species closely related to that studied by Franks et al. (1990), which also exhibited rhythmic behavior in the brood chamber, he found that their activity-inactivity cycles were technically chaotic: they exhibited deterministic chaos, without any sign of a regular rhythm. How, then, does a regular periodicity emerge from the interaction of chaotic individuals?

This problem was studied by Solé et al. (1993) by constructing a computer model of the interaction of ants in a brood chamber. Each individual model ant was described in its activity-inactivity behavior as chaotic, in accordance with the experimental evidence, and the model ants interacted with each other by excitation, again in accordance with observation: if an active ant interacts with an active or an inactive ant, both cases result in increased activity. The model was constructed as a kind of mobile neural network in which ants played the role of excitable neurons that could move around on a grid that represented the brood chamber. It was not possible to predict the outcome of the model, since there was no reason to assume that chaotic individuals that interacted by excitation would develop a coherent rhythm. However, that is precisely what occurred. Furthermore, the model also conformed to observations that were not built into the model: a regular periodicity emerges in the ant colony only when the density of ants exceeds a critical level, as observed by Cole (1991) in his experimental studies. The periodicity emerges quite suddenly above the critical density; in technical terms, it has the characteristics of a phase transition, like the sudden condensation of a gas to a liquid at a critical temperature, so that one kind of order suddenly replaces another. It is interesting to ask what kind of explanation of the emergent property this model gives us.

SEEKING AN EXPLANATION FOR THE FORM

The emergent periodicity in the brood chamber cannot be reduced to the behavior of the individual ants and their interactions in any specific causal sense. That is, knowing about behavior of individual ants and the nature of their interactions is not sufficient to predict the periodic behavior of the whole colony. Nevertheless, this periodicity is a consistent property of the model above a critical density of ants. Here is a case of consistency without causal reduction, which is found in many cases of emergence in complex systems. It may well be that mathematical theorems will be discovered that allow one to make predictions about such systems. However, these were not known at the time the model was constructed. What the model makes clear is the necessary properties for such an emergent process. The mathematical functions used in the model describe these properties without explaining what their consequences are. This is typical of scientific procedure: models often describe without giving causal explanations of the phenomena.

Such models are very useful in clarifying various aspects of emergent properties in complex evolving systems. In particular, we can ask what role natural selection might play in the emergence of rhythmic activity of workers in the brood chamber. To explore this, we need to ask what function rhythmic activity has for the colony and how it serves its survival. Franks et al. (1990) suggested what this might be. He and others have observed that if an active ant encounters an embryo or a larva that is already getting attention from other workers, the active worker will go to another member of the brood. Therefore, if workers in the brood chamber are active at the same time, they probably distribute their care over the brood and the queen so that there is little duplication of attention. If their activity patterns were chaotic, then some embryos or larvae might get more attention than others simply by accident of how the workers were distributed in the brood chamber. Delgado and Solé (1999) have shown that when this is modeled by computer simulation of colony activity, periodic patterns do indeed produce more evenly distributed care than chaotic patterns. So we have a possible explanation of why rhythmic cycles of activity in the brood chamber are beneficial and may be selected: they enhance the survival chances of the young by virtue of good care when they are developing, and so increase the fitness of the colony. This is probably why these activity cycles are commonly observed in such colonies.

However, it is clear that natural selection in no sense explains the origin of the rhythmic activity pattern in the brood chamber. The possible function of the rhythm played no role in designing the model, which provides an understanding of how rhythmic activity patterns can emerge in an ant colony with the properties observed. As stated in the comprehensive volume *Self-Organization in Biological Systems* (Camazine et al. 2001, 89): "One of the revelations of self-organization studies is that the richness of structures observed in nature does not require a comparable richness in the genome but can arise from the repeated application of simple rules by large numbers of sub-units." Modeling the ant colony as a complex system provides a demonstration of the possibility of rhythmic behavior in the brood chamber. Furthermore, it shows that the density required for rhythmic activity is quite low (about 20% of maximum density), as observed experimentally, and the rhythm arises for a wide range of excitability of the ants. This is what is required for a phenomenon to arise in evolution: it must occur fairly readily (be accessible) and must be robust in the sense that there should be a considerable range of parameter values in the model over which the phenomenon arises. This corresponds to a wide range of genotypes in which the property can arise. In addition, the fact that the rhythm arises suddenly as a phase transition in the colony means that it is not selected gradually, with progressively increasing definition, as the gradualist view of evolution would have it. We certainly cannot say that all biological properties arise in this sudden, unexpected manner, as do phase transitions, but complexity theory is demonstrating that this is a common feature of the way biological organization emerges (see Goodwin 1994; Solé and Goodwin 2000).

EMERGENT "ORDER FOR FREE"

It is these spontaneous, robust emergent properties that Kauffman (1993, 1995) called "order for free" in evolution: the production of some organismic morphology or behavior made possible by the occurrence of conditions in complex living systems that allow the property to be generated. These conditions arise by a random search of possibilities that result from spontaneous diversification within organisms both of genes and of dynamic possibilities, including the occurrence of chaotic activity, as we have seen in ants, and from diversification of environmental conditions outside the organism. The variation of conditions within organisms and in their environments is what

Darwin assumed to generate hereditary variety within populations and the habitat variety to which they can adapt. Natural selection may explain the differential frequencies among species of various forms that have arisen in evolution, such as the occurrence of rhythmic activity patterns in the brood chamber of various species of ant. However, there are situations in which differential abundance may be explained in terms of the dynamics of the generative process and the relative stabilities of the different forms produced. In order to understand the phenomena of evolution, we need to understand both the generative dynamic processes that produce the distinctive characteristics of species in the first place (origins) and the effects of natural selection on the frequencies of the various characters. The integration of these areas of study, development and evolution, into a coherent dynamic picture of the evolutionary process is now called evolutionary developmental biology.

Summary and Conclusions

Form and function have always been central concepts used in biology to understand organisms and their evolution. The particular theory of evolution developed by Darwin included both, but there has been a tendency to emphasize the role of function in describing how organisms are shaped and their behavior is molded by natural selection. A basic assumption has been that biological forms vary continuously as a result of random variation in genes, and it is natural selection that "decides" which particular forms survive and are favored. Any discontinuities of form between existing species are then regarded as a result of this selection process, which then becomes the sculptor in evolution.

The viewpoint described here differs from this by presenting evidence that the morphology and behavior patterns of organisms often arise as distinct forms and do not belong to a continuum of possibilities. Furthermore, the intrinsic frequencies and stabilities of the forms produced can contribute to the relative frequency of the different forms, so that natural selection is not the only factor involved in explaining their differential abundance. This evidence comes from observation of morphology, from experimental study, and from models that describe how different morphologies and behavior patterns can arise from dynamic processes within and between organisms.

There is no difficulty in combining natural selection with a discrete set of possibilities for evolution to work with. Distinct forms still need to be tested for their survival value, and discrimination regarding function can be made among whatever forms are generated, resulting in different frequencies of the forms correlating with different habitats. However, understanding the primary generative origins of these forms needs to be distinguished from secondary selection processes. For readers who want to pursue the issues further, good overview discussions include Camazine et al. (2001), Arthur (2002), Wilkins (2002), and Minelli (2003). A now-classic discussion of constraints is Maynard Smith et al. (1985). Phyllotaxis is discussed in Snow and Snow

(1952), Mitchison (1972), and Jean (1994). Social insects and their organization are discussed in Cole (1991), Seeley (1995), and Gordon (1999). Evolutionary development (also colloquially known as evo-devo) is the topic of Gilbert et al. (1996). I discuss these issues, trying to put them into a broader, ethical perspective, arguing for a holistic perspective that requires a rethinking of conventional science to achieve true ecological harmony and sustainability, in *Nature's Due: Healing Our Fragmented Culture* (Goodwin 2007).

BIBLIOGRAPHY

Arthur, W. 2002. The emerging conceptual framework of evolutionary developmental biology. *Nature* 415: 757–764.

Camazine, S., J.-L. Deneubourg, N. R. Franks, J. Sneyd, G. Theraulaz, and E. Bobabeau. 2001. *Self-Organization in Biological Systems.* Princeton, NJ: Princeton University Press.

Cole, B. J. 1991. Is animal behaviour chaotic? Evidence from the activity of ants. *Proceedings of the Royal Society of London, Series B* 244: 253–259.

Delgado, J., and R. V. Solé. 1999. Self-synchronization and task fulfilment in ant colonies. *Journal of Theoretical Biology* 205: 433–441.

Douady, S., and Y. Couder. 1996. Phyllotaxis as a self-organising process. Parts I, II, and III. *Journal of Theoretical Biology* 178: 255–312.

Franks, N. R., S. Bryant, R. Driffith, and L. Hemerik. 1990. Synchronisation of behaviour within the nests of the ant *Leptothorax acervorum. Bulletin of Mathematical Biology* 52: 597–612.

Gilbert, S. F., J. M. Opitz, and R.A. Raff. 1996. Resynthesizing evolutionary and developmental biology. *Developmental Biology* 173: 357–372.

Gleick, J. 1998. *Chaos: The Amazing Science of the Unpredictable.* London: Vintage.

Goodwin, B. C. 1994. *How the Leopard Changed Its Spots.* London: Weidenfeld and Nicolson.

———. 2007. *Nature's Due: Healing Our Fragmented Culture.* Edinburgh: Floris Books.

Gordon, D. M. 1999. *Ants at Work: How an Insect Colony Is Organized.* New York: Free Press.

Jean, R.V. 1994. *Phyllotaxis: A Systematic Study in Plant Morphogenesis.* Cambridge: Cambridge University Press.

Kauffman, S. A. 1993. *Origins of Order: Self-Organization and Selection in Evolution.* New York: Oxford University Press.

———. 1995. *At Home in the Universe.* New York: Oxford University Press.

Maynard Smith, J., R. Burian, S. Kauffman, P. Alberch, J. Campbell, B. Goodwin, R. Lande, D. Raup, and L. Wolpert. 1985. Developmental constraints and evolution. *Quarterly Review of Biology* 60: 265–287.

Minelli, A. 2003. *The Development of Animal Form: Ontogeny, Morphology, and Evolution.* Cambridge: Cambridge University Press.

Mitchison, G. 1972. Phyllotaxis and the Fibonacci series. *Science* 196: 270–275.

Reinhardt, D., E.-R. Pescoe, P. Stieger, T. Mandel, K. Baltensperger, M. Bennett, J. Traas, J. Frimi, and C. Kuhlemeier. 2003. Regulation of phyllotaxis by polar auxin transport. *Nature* 426: 255–260.

Salazar-Ciudad, I., J. Jernvall, and S. A. Newman. 2003. Mechanisms of pattern formation in development and evolution. *Development* 130: 2027–2037.

Seeley, T. D. 1995. *The Wisdom of the Hive.* Cambridge, MA: Harvard University Press.

Snow, M., and R. Snow. 1935. Experiments in phyllotaxis: Diagonal slits through decussate apices. *Philosophical Transactions of the Royal Society of London, Series B* 225: 63–94.

———. 1952. Minimum area and leaf determination. *Proceedings of the Royal Society of London, Series B* 139: 545–566.

Solé, R. V., and B. C. Goodwin. 2000. *Signs of Life: How Complexity Pervades Biology*. New York: Basic Books.

Solé, R. V., O. Miramontes, and B. C. Goodwin. 1993. Oscillations and chaos in ant societies. *Journal of Theoretical Biology* 161: 343–357.

Webster, G. C., and B. C. Goodwin. 1997. *Form and Transformation*. Cambridge: Cambridge University Press.

Wilkins, A. S. 2002. *The Evolution of Developmental Pathways*. Sunderland, MA: Sinauer Associates.

Philosophy of Evolutionary Thought

Kim Sterelny

The development of evolutionary biology since 1858 is one of the great intellectual achievements of science. The living world presents science with two major challenges: how is its immense diversity across time and at any given time to be explained, and how is the extraordinary adaptive fit between organism and environment to be explained? There is no doubting the scale of these challenges. There are no reliable measures of the diversity of today's biota (especially if we take into account the diversity of microorganisms), but there are certainly millions of extant species of organisms (Cracraft and Donoghue 2004; Dawkins 2004). Likewise, the adaptiveness of organisms is sometimes breathtaking in its complexity and its exquisite precision (Dawkins 1996). Even so, we now have at hand, in broad outline, an explanation of both diversity and adaptation. In brutal summary, selection drives the adaptation of organisms in a population to their environment, and because environments differ, it drives differentiation too, taking advantage of whatever variation arises in the population. Moreover, since the biological environment is an important part of any organism's world, there is positive feedback in the process of differentiation. Differentiation breeds differentiation by increasing environmental heterogeneity.

Evolutionary biology has many important open questions. Some of these concern explanations of particular events. Were the dinosaurs really driven to extinction by a meteor impact, or were they on their way out anyway? Was their extinction a necessary precondition for the evolution of the mammals? Some concern broad patterns in the evolution of life. Has diversity increased more or less smoothly over time, or have there been long periods when it stabilized or even dropped? Some concern mechanisms. How important has mass extinction been in the history of life? Are the selective mechanisms that generate adaptation importantly constrained by the mechanisms that produce variation? What are the relative roles of selection and chance in evolutionary change? How important are selfish genetic elements, and how is it that such elements are usually neutralized in organisms' genotypes? As other essays will have made clear, these are important and difficult questions, but they seem to be empirical questions. Since philosophers do not collect data, what could philosophers have to say about evolutionary biology that helps us understand it? How could answering such questions as these be relevant to philosophers' core projects? My project in this essay is to answer these questions. I begin with the relevance of philosophy to evolutionary biology; I conclude with the relevance of evolutionary biology to philosophy.

Ferment in Evolutionary Biology: Conceptual Innocence Lost

The neo-Darwinian synthesis was developed by the Ronald A. Fisher (1930)–J. B. S. Haldane (1931)–Sewall Wright (1931, 1932) triumvirate in the early 1930s and was completed in America by Theodosius Dobzhansky (1937), Ernst Mayr (1942), George G. Simpson (1944, 1953), and G. Ledyard Stebbins (1950) and in Britain by E. B. Ford and his school (e.g., Fisher and Ford 1947; Cain and Sheppard 1950, 1952, 1954; Kettlewell 1955, 1956) before, during, and after World War II. It became the received view of evolutionary biology. It was individualist, microevolutionary, gradualist, and selectionist. The synthesis focused on individual organisms: competitive interactions take place between individual organisms in local populations, not between groups or species. Individual organisms are more or less fit; individuals prosper or fail. The synthesis was microevolutionary: the large-scale changes visible in the fossil record are nothing but the accumulation of small-scale changes in local populations. The synthesis was gradualist: evolutionary changes in local populations are slow and incremental. Horses lost their toes and grew their teeth little by little and over long periods. And it was selectionist. Chance events play a role in evolutionary change. A potentially favorable mutation might simply just not happen, or it might happen to an organism that is fit but unlucky. But according to the synthesis, natural selection is the dominant causal factor that drives the evolutionary dynamics of populations.

The synthesis left formidable technical problems unsolved, but the theory of evolution conceived this way seemed relatively conceptually unproblematic. This conceptual clarity was not complete. For example, understanding the nature of species seemed a particularly pressing problem, for the very gradualism of the synthesis seemed to suggest that the appearance of well-defined species might be an illusion generated by the length of our life spans. They seem stable and well defined only because we do not live long enough to see lineages change and differentiate. If we had been around to see the ancestral cat population gradually change and differentiate into its array of modern forms, perhaps we would not think of leopards, jaguars, panthers, and snow leopards as distinct, nameable units of nature. There were other issues too, for example, whether evolution is a genuinely indeterministic process because of the role of chance, not just in the generation of variation but also in selection. The fitter organisms in a population are more likely to be successful, but they are not certain to be more successful. Thus there were puzzles about the nature of fitness and its causal role in evolution, for the synthesis depended on a distinction between actual and expected fitness, and yet it was unclear just how this distinction was to be drawn (Sober 1984).

Thus the synthesis of evolutionary theory did pose conceptual questions. Nonetheless, the conceptual geography of evolutionary biology became much more intractable in the 1970s and 1980s as the synthesis came under fire within biology. The most systematic challenge to the synthesis has been to its individualism, to the idea that the targets of selection are, exclusively, individual organisms. For example, G. C. Williams (1966) and Richard Dawkins (1976) argued that selection does not really act on individual organisms at all. Williams and Dawkins claimed that the gradualism of the synthesis undermines its individualism. Complex biological adaptations—sensory systems, the formidable weapons of predators, the power and streamlining of great pelagic vertebrates—are built incrementally, one small step at a time. Selection builds complex systems cumulatively by preserving small improvements that form the platform for further small improvements. So the targets of selection must be lineages of near-perfect copies—for only thus can small advantages be preserved—that persist through many cycles of selection.

Yet organisms do not form lineages. My daughter is not a copy of me, nor would she be even if I were an asexual creature reproducing by cloning. A clone of me would not be a near-perfect copy, for life's accidents mark me, and these marks do not reappear on a clone. As a result of a childhood accident, I have just one eye: a clone would have two. In contrast, though, many thousands of genes in my daughter are copies of genes in me; likewise, most of the genes I carry are copies of parental genes. Genes, unlike organisms, are copied, that is, replicated, and they do form persisting lineages. These lineages reflect the differing success of genes: some are deep and bushy, lasting many generations and existing in many copies in each generation. Others are short and thin. For the most part, these differences are no accident: genes are not just replicators, they are active replicators. Their own characteristics influence their prospects for replication, normally through their influence on bodies, the vehicles in which they ride. The right way to think about the history of life is to see it as a giant struggle between lineages of genes, each trying to secure the resources needed for replication (Williams 1966, 1992; Dawkins 1976, 1982).

This line of thought has raised very difficult issues indeed that are central to much contemporary philosophy of evolutionary biology. In what sense is gene selection a genuine alternative to the individualism of the synthesis? There seem to be at least three answers to this question, and the defenders of gene selection have never really picked their response. The simplest and cleanest response would be to insist that gene selection is an empirically distinct and superior alternative to individualism. Though genes normally improve their replication prospects by helping the other genes in an embryo build a well-adapted body that will carry them all into the next generation, they do not always replicate so cooperatively. For example, some genes can get to the next generation only if they are in an organism of the right sex. The genes in the mitochondria in my sperm cells are doomed, for mitochondrial genes descend only in the female line. My daughter's mitochondrial genes are all copies of genes in her mother. In many circumstances organisms are fittest if the sex ratio of their offspring is 50–50 or thereabouts. But genes that descend via only one sex have no interest in the body of which they are a part having offspring of the other sex, and these genes are under selection to distort the sex ratio in their favor, even though this is maladaptive for the organism of which they are a part. Many such sex-ratio distorters are known (Burt and Trivers 2005). Sex-ratio genes and other so-called genetic outlaws prompt the following response: gene selection is *predictively different* from the synthesis. Gene selection predicts that phenotypes that express the fittest genes will become more common over time; the synthesis predicts that phenotypes that characterize the fittest organisms become more common over time. Although these predictions overlap substantively, they are not identical.

Many have responded by accepting that when genes are outlaws, we should indeed see them as targets of selection, but they take these to be exceptional cases that do not undermine the general validity of individualism. Selection does act on genes, but only in very special conditions. So a second way of defending gene selection is to accept that it is, for the most part, predictively identical to the synthesis, but that it appeals to distinct underlying causal structures. Two theories can lead to the same predictions about observable outcomes even when they posit distinct underlying causal processes. Elliott Sober and David Wilson, in their challenge to the synthesis, accept that individualist views of the evolution of cooperation can make the right predictions, but they argue that these misidentify its causal basis. Cooperation, they say, evolves only when and because cooperative acts enhance the fitness of the groups that contain cooperative agents (Sober and Wilson 1998). Likewise, with regard to gene selection, fit genes are fit because they act on their environment in ways that enhance their probability of replication. The usual method genes have of enhancing their replication prospects is by teaming up with other genes to produce vehicles, which in turn reproduce, replicating the genes they host. Most of these vehicles are organisms, but perhaps they can be hives, nests, and symbiotic alliances too. Fit genes make well-adapted vehicles, which in turn replicate their builders. The fact that genes typically adopt this indirect strategy of replication brings the predictions of gene selection into line with those of the synthesis, but the various pictures of the causal structure of selection (gene, individual organism) remain distinct.

A third line is to argue that gene selection has heuristic advantages over the synthesis. Strictly speaking, it is not a different view of the causal structure of evolutionary process; instead, it enables us to see possibilities and problems that the synthesis conceals. Dawkins took this line in *The Extended Phenotype* (1982), his most important defense of gene selection. He gave many examples of evolutionary phenomena that are simply more salient when we think of the biological universe as an ensemble of competing gene lineages. One such phenomenon is the evolution of the organism itself. Why do gene lineages mostly replicate indirectly, via the cooperative construction of vehicles? Given that they do, why do vehicles have the distinctive characteristics of organisms, in particular, a life cycle that involves reproduction via a single-celled stage? Dawkins points out that it is easy to imagine other life cycles, and, indeed, some of these alternatives are real, though uncommon. It is possible to formulate this question about the origin of organisms and their characteristic life cycle from within the perspective of the synthesis, but gene selection makes that question inescapable, and Dawkins gives many other examples of the advantages of the gene-selection perspective.

The issue of gene selection and its relation to individualism is just one aspect of the debate about the units of selection that is now central to the philosophy of evolutionary biology. Evolutionary biologists have come to see that cooperation is central to the evolution of complexity. Complex biological systems are composed of individuals, and these are often under selection to behave in ways that undermine the integration of the collective. A key question, then, is how new levels of organization evolve in the face of selection on the individuals that make them up to go their own way. Elliott Sober, David Wilson, Stephen Jay Gould, and others have agreed that selection does act on individuals, but they have argued that it acts as well on groups and on species (Sober and Wilson 1998; Gould 2002). Furthermore, just as the individualism of the synthesis has raised difficult conceptual issues, so too has its gradualism. According to the synthesis, large-scale patterns in the history of life are just aggregations of small-scale changes. Biologists can actually observe evolution in action in local populations: they can observe changes in rabbit immune systems as rabbits are exposed to a new virus, or finch beaks becoming larger in dry seasons. According to the gradualism of the synthesis, all the mechanisms of evolutionary change are at work in local populations that are changing in small ways over a few generations. We just need space and lots of time to accumulate the results of these mechanisms into the patterns of the fossil record.

In a famous 1972 essay Niles Eldredge and Stephen Gould argued that the fossil record does not support this "extrapolationist" picture. In particular, they maintained that the tree of life does not change enough over time for extrapolationism to be true (Eldredge and Gould 1972). They argued that most species do not change significantly over most of their life spans. Instead, species come into existence relatively quickly and then do not change until they split into daughter species or become extinct. The extent to which this does describe the typical life history of a species is still open to debate (Carroll 1997). But if Eldredge and Gould are right, their picture does challenge extrapolationism. If evolutionary patterns are nothing but the aggregate of changes in local populations, we would expect to see gradual change rather than long periods without change interrupted by sudden species formation. It is true that there are local processes that can result in stasis in particular cases; populations can track their habitat by shifting geographically rather than adapting in place; adaptations can be behavioral rather than morphological and hence can be invisible to us. Even so, if the fossil record is nothing but population-level evolution summed over large quantities of space and time, we would not expect stasis to be its dominant pattern (Eldredge 1995; Gould 2002).

Gould has developed a more radical critique of extrapolationism that rejects both its individualism and its gradualism. In a classic article Lewontin defined the idea of a *Darwinian population:* a population of individuals that differ one from another and experience differential reproductive success, and in which descendants inherit their ancestors' distinctive traits. Such populations evolve under natural selection (Lewontin 1970). Gould argued that species groups form Darwinian populations (Gould 2002; Gould and Lloyd 1999). Species have determinate origins and endpoints. Daughter species resemble their ancestors: reconstructing phylogeny would be impossible if this were not so. Species are differentially successful. Within a regional biota some species lineages will differentiate richly, radiating into many niches, whereas others will remain restricted in numbers and ecological penetration. Gould does not expect species-level selection to explain the adaptations of organisms, for example, their perceptual systems, but he thinks that selection will explain characteristics of species themselves: such properties as genetic variation, population structure, and geographic range are visible to species-level selection and hence can be the target of species-level selection.

Gould's ideas about species selection face serious problems. For one thing, there are no convincing case studies of species selection in action. Moreover, there are serious theoretical challenges to his ideas. It is by no means clear that the inheritance condition on Darwinian populations is satisfied. It is true that the organisms in a daughter species resemble the organisms in the parental species. But the species-level properties of the daughter—its distribution and population structure—may not resemble those of its parent at all. Furthermore, in thinking of species success and failure, we may not be able to distinguish drift from selection. The success or failure of *individual organisms* is not always the result of selection. Some mortality is the result of unhappy accident rather than ill design. Some organisms are fecund by luck, despite their design flaws. Hence realized fitness differs from expected fitness. Success due to systematic features of the population's environment is success in virtue of expected fitness, as when a swallow dodges a hawk. Success due to freak accidents—being blown by a storm to a freighter that is heading in the right direction—is drift. But the distinction between expected and freak causes of mortality depends on the existence of a population of events that is large enough for there to be patterns in those events. However, species population sizes are so small and cycles of selection are so few that we may not be able to draw the distinction between selective and accidental mortality.

I hope that these examples have begun to explain the relevance of philosophy to evolutionary theory. It is relevant because the empirical issues involved in rethinking the synthesis are not clearly separated from the conceptual issues. The next section makes the same point in discussing one important problem about the relationship of evolutionary theory to developmental biology.

Evolution within the Life Sciences

In the past decade or so a further issue has become increasingly salient: the integration of evolutionary biology with the rest of biology. The problem is multifaceted. It arises with ecology in identifying the transmission belt that connects events in local populations with the evolution and differentiation of species. It arises in developmental biology, a domain of the life sciences that played little role in the construction of the synthesis in the middle of the twentieth century. As a consequence, the nature of evolutionary developmental biology remains highly contested. It arises for human biology, for humans are both evolved animals and encultured agents: the perspective of the biological sciences and the perspective of the social sciences are both valid, and they need somehow to be part of a single unified framework for thinking about human agency (Sterelny 2003). It arises for molecular biology, and I shall briefly introduce one aspect of this case.

What is a gene? This question is challenging because the relationship between the classical genetics of Mendel and his twentieth-century successors and molecular biology has turned out to be very complex indeed. The classical gene of the fruit-fly experiments both had a well-defined phenotypic effect and was transmitted as a single unit. There may be no DNA sequences with both of those properties, for the developmental import of a DNA sequence—how and when it is transcribed—has turned out to be very sensitive to the local genetic environment. Genes can be switched on and off and can even be read in different ways, depending on the presence or absence of repressors and promoters elsewhere in the genome. Since these regions need not be adjacent to the DNA stretch they influence, they may not be passed on when that DNA sequence itself is replicated (Griffiths and Neumann-Held 1999; Griffiths 2002). Evolutionary biologists often abstract away from these molecular details of gene copying and expression and instead think of genes as information bearers. Evolution requires inheritance: unless successful organisms are apt to transmit the characteristics that make them successful to their offspring, a population cannot change adaptively over time in response to selection. Inheritance is mediated by the flow of accurately copied genes from one generation to the next, and evolutionary biologists often think of these genes as instructions, as carrying information about the organism of which they are a part. Since the flow of genes is edited by natural selection, with suitable ones making it to the next generation and unsuitable ones being deleted by the death of their carriers, we can think of the population gene pool as carrying information directly about organisms and indirectly about the environments in which those organisms have lived and reproduced. The success of some genes and the failure of others is a signal about the nature of those environments (Williams 1992; Maynard Smith 2000a).

This picture of cross-generation gene flow as a flow of instructions has generated vigorous debate. Paul Griffiths, Russell Gray, Susan Oyama, and other developmental-systems theorists think that it is wrongheaded in two respects (Griffiths and Gray 1994; Oyama 1985). First, they argue that this picture overstates the importance of gene-based inheritance. No one denies that the flow of genes between generations is one of the mechanisms that are responsible for the resemblances between parents and offspring. But the developmental-systems theorists argue that many other mechanisms are important too: symbiotic microorganisms, chemicals in cell cytoplasm, food traces in maternal milk transmitting the mother's food preferences, behavioral imprinting on the natal site, and many more. There are many interacting mechanisms, they argue, that are responsible for cross-generation similarity, and a good evolutionary developmental biology cannot be built if these other aspects of inheritance are neglected (Oyama et al. 2001).

Moreover, although genes are certainly important, developmental-systems theorists doubt whether genes carry information in any distinctive sense. Genes undeniably carry *predictive* information about phenotypes. Holding environmental factors constant, if you know the genes an organism has inherited, in principle you can predict its phenotype. But all factors causally relevant to development carry predictive information. The plant on which the butterfly lays its eggs predicts features of the butterfly phenotype, namely, the plant species on which insects like this kind of butterfly will lay its eggs. The temperature of the tuatara's nest predicts the sex of its offspring. Thus although genes do indeed predict phenotypes, they are not alone in doing so. Moreover, if information is just a covariation between signal and source that enables us to predict features of the source from the signal, information cannot be misread or misused. It would make no sense to say of these "instructions" that they were ignored or misread. In certain developmental contexts typical human genotypes predict that the resulting phenotype has vestigial limbs: in those contexts they covary with that outcome. But Williams, Maynard Smith, and others do not think that human genes carry instructions to grow vestigial limbs if confronted with thalidomide or that acorns carry genes that tell them to rot just because most acorns predictably rot. They need an alternative notion of information that makes sense of the idea that genes and only genes direct development along predictable trajectories when the developmental process operates as designed, but also makes sense of the fact that genetic information is misread when the gene-reading system malfunctions. It has proved surprisingly hard to develop such an account (see especially the commentaries to Maynard Smith 2000a: Godfrey-Smith 2000; Sarkar 2000; Sterelny 2000; and the reply by Maynard Smith 2000b). Thus there is still work to be done to show that the evolutionary conception of genes and genetic inheritance integrates smoothly with molecular conceptions of the gene.

Why Biology Matters to Philosophy

I have been discussing challenges to the neo-Darwinian synthesis. Philosophy is relevant to these challenges, for there is no single experiment or prediction that would settle (for example) the fate of Gould's view of the relations between macro- and microevolution or of Dawkins's view of the role of genes in evolution. One role for philosophy is to bring these views into sharp focus: to specify exactly what they claim about life and its history. Just as philosophy is relevant to evolutionary theory, that theory is relevant to philosophy. Indeed, it is relevant in at least four ways: (1) Evolutionary biology helps set the explanatory agenda for one major movement of twentieth-century philosophy. (2) It provides a crucial set of examples to which philosophical theory must respond. (3) It provides an important set of tools that philosophy can adapt for its own problems. (4) It is an exemplar of science in action, but it is very different from the usual models on which philosophy of science has been based. I shall discuss (1), (2), and (3) in this section and spend the most time on (4) in the following section.

1. Let me begin with the explanatory agenda. One major movement of twentieth-century philosophy has been naturalism: seeing humans as wholly part of the natural world. In turn, naturalism generates a difficult philosophical problem: to what extent is our ordinary, commonsense picture of ourselves and our world compatible with the fact that we are nothing more than complex evolved biochemical machines? Common sense views humans as agents: we are self-aware; deliberative makers of real choices; reflective; often rational; and aware of moral considerations and sometimes responsive to them. Perhaps this picture is undermined by an evolutionary perspective that sees us as gene-replicating machines: as vehicles built by and for genes. What could choice, rationality, or morality be but illusions if that is the truth about us? Compatibilist naturalists argue that the two pictures are consistent, though they perhaps require some revisions of common sense. Thus Daniel Dennett has defended the idea that evolved creatures like us nonetheless make real choices (Dennett 2003). But compatibilism is certainly not unchallenged. For example, in ethics Michael Ruse (1986, 2007) and Richard Joyce (2006) have argued that the natural history of human cooperation shows us that "ethical truth" is an illusion. There are no moral facts for humans to be responsive to. Moral belief systems exist merely because they induce cooperation and damp down the destructive pursuit of self-interest; recognition of moral truth has nothing to do with it. Humans with cooperation-inducing moral systems got to be ancestors; others did not (Joyce 2006). So biology is important to philosophy by cranking up the tension between our folk self-concept and our scientific self-portrait.

2. Evolutionary biology is also important as a source of real-world examples. Much philosophy recycles a limited set of examples, and these are almost always thought experiments. Thought experiments have a role in philosophy, just as they do in the sciences. All theories make claims about

counterfactual cases, and so it is appropriate to consider counterfactual scenarios. But there are good reasons for more serious attention to real phenomena as well. First, thought experiments are often underdescribed. They abstract away from the rich detail of real cases. Moreover, many of these thought experiments lead to a clash of intuitions, partly because intuitive judgment is sensitive to just how one describes the thought experiments. One set of these imaged scenarios involves *Star Trek*–style teletransporters and seems to show that personal continuity is not the same as the continuity of our physical bodies. But while it is easy to tell these stories so that teletransporting seems to be a way of actually traveling while leaving one's old body behind, it is also easy to make them seem like killing machines that replace the duped traveler with an impostor. This sensitivity to the mode of presentation suggests that the details matter. Vary the setting, and you vary the judgment. But if God is in the details, nothing but the real will do.

Moreover, philosophers have often not fully considered the range of actual cases. Absence is often evidence. Dennett's work is important in this respect. He shows how ignoring reality has allowed philosophers to repeatedly slide into substituting a dichotomy for a continuum by neglecting the range of real cases. For example, in thinking about rational agency, philosophers have often contrasted rational intentional agency with merely mechanical systems. Our flexible, informationally sensitive response to our world is contrasted with the rigid automaticity of artifacts and insects. Dennett points out that evolutionary biology shows that there must be a range of intermediate agents, for informational sensitivity is a complex adaptation, and complex adaptations do not arise from rigid systems in a single step (Dennett 1995, 2003).

3. A further important role for biology within philosophy is as a source of tools and ideas. Thus David Hull (1988) has applied evolutionary models to the dynamics of science itself, and Ruth Millikan (1989a, 1989b), David Papineau (2003), and others have attempted to use the biological notion of function to solve one core problem of philosophy: the nature of mental symbols. Beliefs and preferences are representations of the world. But what in the physical world is a representation? I believe that tigers are the most handsome members of the cat family. But what makes my tiger thoughts *about* tigers? What makes certain neural structures within my brain a *symbol* of tigers? Historically this was seen as a big problem for any physicalist theory of mind because people can think about nonexistent objects and abstract objects. Meaning cannot be a physical relationship between mind and the world, because there are no physical relationships to angels or numbers. More recently philosophers of mind have noticed that tough problems arise even for simple cases like thoughts about spiders, for there is no simple relationship between spider thoughts and spiders being present to the thinker. Peter can have spider thoughts without spiders, for he can take a moth for a spider; and he can fail to have spider thoughts even in the presence of a spider by thinking that a spider is a moth.

So those who think that humans are both thinking beings and part of the natural world need to explain the nature of mental symbols and the relationship between a symbol and its target. What facts about Peter's brain, his life, and his environment make it true that a particular internal feature of his mind is a symbol for spiders? Philosophers have explored a range of responses to this question. An intuitively appealing idea is that symbols *resemble* their targets in the world. Peter's mental state is about spiders if it resembles spiders. This is a natural suggestion if thoughts are like images, maps, or diagrams. However, almost certainly it will not do. Mental images do not seem to be objectively similar to enduring three-dimensional physical animals; the appearance of real similarity between pictures and things is an illusion generated by the way humans interpret pictures. Peter's spider image will resemble other thoughts much more than it resembles any animal. Moreover, even if some mental states are like images, surely most mental representations are not imagelike. What is the image that corresponds to the thought "Aristotle was the founder of logic, and Euclid the founder of geometry"?

So the relation between symbol and target is not a resemblance relationship. But perhaps it is a causal relationship. Some events are natural signs of others. Think in particular of instruments. An instrument is a fuel gauge if there is a reliable relation between the reading on the gauge and the fuel in one's tank. That is what makes a symbol on the gauge mean "quarter full." So an internal state of Peter is a spider symbol only if it reliably covaries with there being a spider in his vicinity. The symbol-target relationship is the relationship between an inner structure and its environmental correlation: the feature of the environment that causes the inner structure to light up. But this will probably not do either. In the first instance, this picture applies at best to perception. Lots of my wine thoughts occur precisely because there is no wine about. But the crunch problem is that this picture seems to make it impossible for us ever to make mistakes, yet surely we do. If Peter is somewhat arachnophobic, and his friends plague him with clever imitation spiders, he will often think "Spider!!" when there is no spider around. He is misrepresenting his environment. Yet according to the covariation picture of the symbol-target relationship, he would not be misrepresenting at all, for his "spider" thoughts covary with the category of spiders or imitation spiders. Hence they are not symbols of spiders after all; their target is the broader category of spider look-alikes (including real spiders, rubber spiders, and so forth). And since Peter thinks "Spider!!" only in the face of spider look-alikes, he has made no mistake.

David Papineau and Ruth Millikan have suggested an alternative view of the relationship between symbols and their targets that is based on biological function. The biological function of, say, the pattern on oystercatcher eggs is concealment. Those eggs are *camouflaged*. Eggs that were difficult for predators to spot against typical backgrounds of sand and debris were more likely to hatch than those that were easier to find. This selective history is what makes it true to say that the biological function of the pattern is camouflage. On any particular egg the pattern has that biofunction even if it does not have that effect. The pattern is supposed to conceal the egg even if the egg is seen anyway, and even if the egg is very easily seen because the parent has laid it on white sand against which it stands out. Similarly, thoughts have biological functions. Their function is to direct behavior that adapts an organism to a specific feature of its environment. Vervet monkeys have distinctive calls and responses to the sight of a leopard, and their leopard-recognition thoughts are *about* leopards because their function is to adapt vervet behavior to the presence of leopards in their environment. Those vervets with a disposition to scramble to the top of trees when they had a distinctively leopardy visual experience were more likely to live than those without it. So the leopard response means "leopard" even when it is a false alarm, for that disposition in the vervet exists as a result of selection for leopard avoidance.

This view of meaning, of the symbol-world relationship, has some very attractive features. It explains why we want to say of the vervet that it is afraid of leopards rather than being afraid of leopards and leopard look-alikes even though (I have no doubt) vervets would respond the same way to dummy leopards constructed by behavioral ecologists. For leopard dummies have played no part in the evolution of vervets and their escape behaviors, so vervet alarm calls are about leopards, not leopard look-alikes. This view of symbols explains how it is possible to misrepresent. Misrepresentation is failure of function. When a chicken's hawk response is triggered by a duck overhead, and it flees in fright, the chicken has misrepresented its environment. The chicken's internal state is not performing its selected role of hawk avoidance. This last point is very important. Many in this field have thought that misrepresentation is a fatal stumbling block for naturalized views of thought. The claim is that error, or misrepresentation, is a *normative* notion, and normative claims cannot be defined in factual terms. A theory of meaning that imports functional concepts from biology escapes this problem. Functional facts are facts about the selective history of that organ. They are natural facts.

Science and the Mirror of Physics

Until the 1970s (perhaps even later) philosophy of science was dominated by examples drawn from the physical sciences, especially physics. The standard examples of theory change included the Copernican revolution, the transition from Newtonian conceptions of space and time to relativist conceptions, and sometimes the revolution in atomic physics that saw the establishment of quantum mechanics. These were not the only examples of science in the wild, but they were stock examples, and many others were drawn from the physical sciences too. These examples encouraged the idea that a core feature of science was the formulation of general laws: a small set of exceptionless general principles that characterize a scientific domain.

Within the empiricist tradition exemplified by Nagel's classic *The Structure of Science* (Nagel 1961), there was a good deal of debate about the nature of these general principles: debate about whether they were accurate descriptions of nature as it really is, or whether they were better understood as highly compact summaries of the phenomena. Notwithstanding these differing conceptions of generalizations, though, it was common ground that science traded in them. The aim of science was to find and refine such generalizations. Moreover, this focus encouraged a certain picture of the epistemology of science: scientific warrant derives from the fact that these generalizations, together with appropriately specified boundary conditions, deliver quantitative and often very precise predictions. Sometimes these are the result of specific experimental manipulations. Sometimes they are the results of natural experiments, as in the celebrated discovery of Neptune by inferring its location from perturbations in the orbit of Uranus. Either way, evidential weight depends on quantitative prediction. Finally, this focus on physics encouraged a particular picture of the unification of science. Physics is the *base case:* the theory of everything. Theories of specific scientific domains—chemistry and biology—are just special cases of physical systems, and their generalizations will turn out to be special cases of physical generalizations. One of the early triumphs of Newtonian mechanics was the demonstration that Kepler's laws of planetary motion were more or less a special case of Newtonian mechanics, and this example was taken to be an exemplification of the relationship between specific, domain-limited sciences and physics.

From the late 1960s the empiricist tradition in philosophy of science came under wide-ranging fire (see Lakatos and Musgrave 1970). But this focus on the physical sciences was one common element in the competing accounts of science that fought over its corpse. Karl Popper (1959, 1963), Imré Lakatos (1970), and the tradition they established traded in examples from physics: Lakatos's "degenerating research programs" did not focus on the idealist morphology of Richard Owen (1848, 1849) in the mid-nineteenth century or the British resistance to classic genetics in the early twentieth century. Thomas Kuhn cut his teeth on the Copernican revolution (1957), and his training as a historian of physics pervades his *Structure of Scientific Revolutions* (Kuhn 1962). The falsificationist tradition, Kuhn, and the irascible and eccentric Paul Feyerabend (1975) challenged much in the neopositivist tradition of thinking about science. In particular, their picture of science and its history emphasized discontinuity rather than smooth incremental increases. In successive scientific revolutions old theories are refuted, superseded, and replaced. But these critics of neopositivism did not challenge its preoccupation with physics and physical chemistry, nor (with the partial exception of Kuhn) did they challenge the preoccupation with universal generalizations and precise and quantitative prediction.

Yet this picture fails to fit biology and especially evolutionary biology. Whole-organism biology does not seem to trade in small sets of exceptionless general principles that deliver precise predictions of biological phenomena. Ecology and evolutionary biology are explanatorily powerful: they explain, for example, the vulnerability of New Zealand's native bird life to introduced stoats. These birds evolved in an environment free of any terrestrial predator and free of any predator at all that finds its prey by smell. But ecology and evolutionary biology rarely enable us to make precise quantitative predictions. There are generalizations in biology, and one of them is the fragility of island ecosystems and the vulnerability of island biota to extinction. But generalizations of this kind are neither exceptionless nor quantitative: they bear little resemblance to Boyle's law. Biologists seem to discover causal mechanisms: natural selection, competitive exclusion, ecological succession, and gene replication and transcription. But the effects of these causal mechanisms are context specific, and those contexts vary over space and time. So the discovery of mechanisms does not translate into exceptionless regularities that are the consequences of the operation of those mechanisms. There are generalizations about evolution on islands: large animals tend to get smaller, and small animals tend to get larger. But there are many exceptions to this generalization.

Moreover, the simple picture of the unification of science fits biology poorly. Chemical properties like acidity indexed, or were thought to index, a single underlying physical property that was itself responsible for the phenomenological properties of acids. Even in chemistry it is likely that the relationship between chemical and physical properties is very often complex indeed. But this complexity first became obvious in biology, for cell biology and molecular biology do not conform to the model of reduction suggested by acidity (Kitcher 2003). Cells are not miracle factories: their biological activities are constrained by physical principles. To that extent physics really is the theory of everything. But the relationship between, say, the concept of a dominant gene and the underlying molecular mechanisms responsible for the phenotypic silence of one of a pair of heterozygous alleles turns out to be very complex indeed. Finally, the history of biology is not a series of revolutions and replacements. The history of biology often seems to manifest incremental improvement. All sciences have their revisions and false trails, but the history of cell biology, physiology, genetics, or even evolutionary biology does not abound with major reversals of core doctrines comparable with the revolution in plate tectonics in the earth sciences. In short, it is hard to both take biology with full seriousness as a science and accept a conception of science derived from the physical paradigms. Philosophers have gone both ways on this. Alex Rosenberg is the most prominent of those who have accepted the physicists' conception of science and as a consequence have accepted the view that some components of biology are not quite up to snuff (Rosenberg 1994). John Dupré is the most prominent contemporary who rejects the image of science in the physicist's mirror on the grounds that it fails to fit biology (Dupré 2002). Most of those who work in philosophy of biology are closer to Dupré than to Rosenberg on this issue: they think that evolutionary biology is a paradigm of a successful science, and hence the physics-based conception of science is too narrow. There is no consensus, however, on exactly what is wrong with the physics model, or on how it needs to be extended.

In conclusion, philosophy of evolutionary biology acts as a type of two-way pump. It pumps philosophy into evolutionary theory, for there are many difficult problems of interpretation, clarification, and integration that arise out of the extraordinary dynamism of the life sciences over the past century. And it pumps evolutionary biology into philosophy: a stream of examples, tools, and models that can perhaps be adapted to drive forward philosophy's own projects. If you want to read more, I suggest you turn first to the writings of Richard Dawkins: *The Extended Phenotype* (1982) is his finest and most subtle defense of gene selection; *Climbing Mount Improbable* (1996) is a sustained and convincing defense of gradualism. Both are beautifully written, though *The Extended Phenotype* is quite technical. For Gould's views, see *The Structure of Evolutionary Theory* (2002). It is appallingly long but reasonably readable, and a good deal of the purely historical material can be skipped. *Darwinism Evolving*, by David Depew and Bruce Weber (1995), is a good overview of the historical development of synthesis and post-synthesis evolutionary biology. Two books by Dan Dennett, *Darwin's Dangerous Idea* (1995) and *Freedom Evolves* (2003), are splendid examples of value of evolutionary insights within philosophy itself. *Sex and Death: An Introduction to the Philosophy of Biology*, by me and Paul Griffiths (1999), covers many of the issues of this essay and other topics in much more detail.

BIBLIOGRAPHY

Burt, A., and R. Trivers. 2005. *Genes in Conflict: The Biology of Selfish Genetic Elements.* Cambridge, MA: Harvard University Press.

Cain, A. J., and P. M. Sheppard. 1950. Selection in the polymorphic land snail Cepaea nemoralis. *Heredity* 4: 275–294.

———. 1952. The effects of natural selection on body colour in the land snail Cepaea nemoralis. *Heredity* 6: 217–231.

———. 1954. Natural selection in Cepaea. *Genetics* 39: 89–116.

Carroll, R. L. 1997. *Pattern and Process in Vertebrate Evolution.* Cambridge: Cambridge University Press.

Cracraft, J., and M. Donoghue, eds. 2004. *Assembling the Tree of Life.* Oxford: Oxford University Press.

Dawkins, R. 1976. *The Selfish Gene.* Oxford: Oxford University Press.

———. 1982. *The Extended Phenotype.* Oxford: Oxford University Press.

———. 1996. *Climbing Mount Improbable.* New York: W. W. Norton.

———. 2004. *The Ancestor's Tale: A Pilgrimage to the Dawn of Life.* London: Weidenfeld and Nicolson.

Dennett, D. C. 1995. *Darwin's Dangerous Idea: Evolution and the Meanings of Life.* New York: Simon and Schuster.

———. 2003. *Freedom Evolves.* New York: Viking.

Depew, D., and B. H. Weber. 1995. *Darwinism Evolving: Systems Dynamics and the Genealogy of Natural Selection.* Cambridge, MA: MIT Press.

Dobzhansky, T. 1937. *Genetics and the Origin of Species.* New York: Columbia University Press.

Dupré, J. 2002. *Humans and Other Animals.* Oxford: Clarendon Press.

Eldredge, N. 1995. *Reinventing Darwin.* New York: John Wiley and Sons.

Eldredge, N., and S. J. Gould. 1972. Punctuated equilibria: An alternative to phyletic gradualism. In T. J. Schopf, ed., *Models in Paleobiology,* 82–115. San Francisco: Freeman, Cooper.

Feyerabend, P. K. 1975. *Against Method.* London: New Left Books.

Fisher, R. A. 1930. *The Genetical Theory of Natural Selection.* Oxford: Oxford University Press.

Fisher, R. A., and E. B. Ford. 1947. The spread of a gene in natural conditions in a colony of the moth *Panaxia dominula* L. *Heredity* 1: 143–174.

Gould, S. J. 2002. *The Structure of Evolutionary Theory.* Cambridge, MA: Harvard University Press.

Gould, S. J., and E. A. Lloyd. 1999. Individuality and adaptation across levels of selection: How shall we name and generalize the unit of Darwinism? *Proceedings of the National Academy of Sciences USA* 96: 11904–11909.

Griffiths, P. E. 2002. Molecular and developmental biology. In P. K. Machamer and M. Silberstein, eds., *The Blackwell Guide to the Philosophy of Science,* 252–271. New York: Blackwell.

Griffiths, P. E., and R. D. Gray. 1994. Developmental systems and evolutionary explanation. *Journal of Philosophy* 91, no. 6: 277–305.

Griffiths, P. E., and E. Neumann-Held. 1999. The many faces of the gene. *BioScience* 49, no. 8: 656–662.

Godfrey-Smith, P. 2000. Information, arbitrariness, and selection: Comments on Maynard Smith. *Philosophy of Science* 67: 202–207.

Haldane, J. B. S. 1931. A mathematical theory of natural and artificial selection. Pt. VIII. Metastable populations. *Transactions of the Cambridge Philosophical Society* 27: 137–142.

Hull, D. L. 1988. *Science as a Process.* Chicago: University of Chicago Press.

Joyce, R. 2006. *The Evolution of Morality.* Cambridge, MA: MIT Press.

Kettlewell, H. B. D. 1955. Selection experiments on industrial melanism in the Lepidoptera. *Heredity* 9: 323–342.

———. 1956. Further selection experiments on industrial melanism in the Lepidoptera. *Heredity* 10: 287–301.

Kitcher, P. 2003. *In Mendel's Mirror: Philosophical Reflections on Biology.* Oxford: Oxford University Press.

Kuhn, T. 1957. *The Copernican Revolution.* Cambridge, MA: Harvard University Press.

———. 1962. *The Structure of Scientific Revolutions.* Chicago: University of Chicago Press.

Lakatos, I. 1970. Falsification and the methodology of scientific research programmes. In I. Lakatos and A. Musgrave, eds., *Criticism and the Growth of Knowledge,* 91–195. Cambridge: Cambridge University Press.

Lakatos, I., and A. Musgrave, eds. 1970. *Criticism and the Growth of Knowledge.* Cambridge: Cambridge University Press.

Lewontin, R. C. 1970. The units of selection. *Annual Review of Ecology and Systematics* 1: 1–14.

Maynard Smith, J. 2000a. The concept of information in biology. *Philosophy of Science* 67: 177–194.

———. 2000b. Reply to commentaries. *Philosophy of Science* 67: 214–218.

Mayr, E. 1942. *Systematics and the Origin of Species.* New York: Columbia University Press.

Millikan, R. 1989a. Biosemantics. *Journal of Philosophy* 86: 281–297.

———. 1989b. In defense of proper functions. *Philosophy of Science* 56: 288–302.

Nagel, E. 1961. *The Structure of Science: Problems in the Logic of Scientific Discovery.* London: Routledge and Kegan Paul.

Owen, R. 1848. *On the Archetype and Homologies of the Vertebrate Skeleton.* London: Voorst.

———. 1849. *On the Nature of Limbs.* London: Voorst.

Oyama, S. 1985. *The Ontogeny of Information.* Cambridge: Cambridge University Press.

Oyama, S., P. E. Griffiths, and R. D. Gray, eds. 2001. *Cycles of Contingency: Developmental Systems and Evolution.* Cambridge, MA: MIT Press.

Papineau, D. 2003. *The Roots of Reason: Philosophical Essays on Rationality, Evolution, and Probability.* New York: Oxford University Press.

Popper, K. R. 1959. *The Logic of Scientific Discovery.* London: Hutchinson.

———. 1963. *Conjectures and Refutations.* London: Routledge and Kegan Paul.

Rosenberg, A. 1994. *Instrumental Biology or the Disunity of Science.* Chicago: University of Chicago Press.

Ruse, M. 1986. *Taking Darwin Seriously: A Naturalistic Approach to Philosophy.* Oxford: Blackwell.

———. 2007. *Charles Darwin.* Oxford: Blackwell.

Sarkar, S. 2000. Information in genetics and developmental biology: Comments on Maynard Smith. *Philosophy of Science* 67: 208–213.

Simpson, G. G. 1944. *Tempo and Mode in Evolution.* New York: Columbia University Press.

———. 1953. *The Major Features of Evolution.* New York: Columbia University Press.

Sober, E. 1984. *The Nature of Selection: Evolutionary Theory in Philosophical Focus.* Cambridge, MA: MIT Press.

Sober, E., and D. S. Wilson. 1998. *Unto Others: The Evolution and Psychology of Unselfish Behavior.* Cambridge, MA: Harvard University Press.

Stebbins, G. L. 1950. *Variation and Evolution in Plants.* New York: Columbia University Press.

Sterelny, K. 2000. The "genetic program" program: A commentary on Maynard Smith on information in biology. *Philosophy of Science* 67: 195–201.

———. 2003. *Thought in a Hostile World.* New York: Blackwell.

Sterelny, K., and P. E. Griffiths. 1999. *Sex and Death: An Introduction to Philosophy of Biology.* Chicago: University of Chicago Press.

Williams, G. C. 1966. *Adaptation and Natural Selection.* Princeton, NJ: Princeton University Press.

———. 1992. *Natural Selection: Domains, Levels, and Challenges.* New York: Oxford University Press.

Wright, S. 1931. Evolution in Mendelian populations. *Genetics* 16: 97–159.

———. 1932. The roles of mutation, inbreeding, crossbreeding and selection in evolution. *Proceedings of the Sixth International Congress of Genetics* 1: 356–366.

Evolution and Society

Manfred D. Laubichler and Jane Maienschein

Evolution and society are connected in many different ways. First and foremost, humans, as a product of evolution, live in diverse societies. These are an outcome of evolution, albeit in more complex and often more indirect ways than the evolution of direct morphological and physiological adaptations. Second, the initial development of evolutionary theory occurred in a specific social milieu conducive to this particular intellectual endeavor, and subsequent development of evolutionary theory has also reflected its evolving historical settings. Evolutionary theory, like any other science, has thus been the product of a particular human society. Finally, and as probably the most visible connection of evolution and society, especially in the United States, we find surprisingly persistent debates about whether evolution or some form of creationism accounts for the origin of species (as Eugenie C. Scott discusses in the main essay "American Antievolutionism: Retrospect and Prospect" in this volume).

At some times, in some places, and in some ways, society has enthusiastically embraced evolution; at other times and places and in other ways, society has not. In the process *evolution* has meant different things, as has *society*. We therefore start with definitions and then look at society's impact on evolution in two ways: the way in which society in the sense of social, cultural, and intellectual context has shaped evolutionary theory, and also the social reception of evolution. Then we turn the direction of influence around and address how evolutionary theory today and in the past has contributed to explanations of society. In conclusion we discuss current relations between evolution and society in a forward-looking way.

We do not focus on much-discussed relationships such as standard interpretations of social Darwinism or retrace the history of ways in which social commentators have used (and misused) evolution. There is such a rich literature in these general areas that any general Google or Amazon search will yield dozens of offerings, of which Carl Degler's *In Search of Human Nature: The Decline and Revival of Darwinism in American Social Thought* (1992) or Richard Hofstadter's *Social Darwinism in American Thought* (1959) are just two well-known and different types of examples. Readers can look to Herbert Spencer, Peter Kropotkin, Andrew Carnegie, and other popular examples of social thinkers who invoked evolution, or they can take up the evolution chapters in such recent books as Bernard Lightman's *Victorian Popularizers of Science: Designing Nature for New Audiences* (2007) for an introduction to less familiar popularizers. Again, it is not on this well-trod ground that we roam. Rather, we explore less familiar aspects of the intersections of evolution and society by asking about the relationships in a variety of ways.

Definitions

By *evolution* we mean the evolutionary theory of biologists rather than general evolutionary ideas, that is, naturalistic, materialistic explanations of the change of populations over time such that inherited variations are preserved and lead to divergence of forms and the origin of new species that are adapted to their environments. This process can happen more or less gradually, with variations of various sorts, inheritance working in complex ways, and species and populations defined differently. Nonetheless, there are constant themes: all accounts of evolution begin with naturalism rather than supernaturalism as a core assumption. All assume that organisms exist in an environment that in effect favors some variations over others since the favored are better *adapted*. And all versions assume that it is inherited variations that matter for evolution, though inheritance may come culturally with learning between generations, as well as biologically through genetic transmission.

We also share the view of Ernst Mayr and others that the Darwinian theory of evolution is composed of a variety of separate theories, and that at some times in the course of the history of evolutionary biology one or the other of these theories dominated the debates (Mayr 1982). Variation, heredity, struggle and competition, and emergence and preservation of the new are all part of evolutionary theory. Indeed, evolutionary theory is complex and consists of interconnected components that we explore in different ways in this essay.

By *society* we designate a variety of kinds of social groups. In some cases the relevant element of society centers on leaders who make decisions that affect many others. In other cases the relevant social group is more local, contingent, and populist, as when civic leaders may agree that evolution should be taught in the schools in any civilized society, but a vocal popular group demands a creationist antievolution account of the origin of man. Different social cohorts play different roles, obviously, and we will look at different groups as we go along. Similarly, when we discuss evolutionary explanations of (human) societies, we follow an equally catholic approach; *society* here means any group of animals or humans that interact with each other in a structured way and whose survival depends on these interactions (Wilson 1975; Boyd and Richerson 1985; Kuper 1994).

Evolutionary Theory Shaped by Society

Evolutionary theory exists in certain societies and is developed by members of these societies. That much is obvious. Darwin is the most visible central player, and historians have shown the significance of the facts that socially Darwin was upper middle class and a Victorian Englishman. For example, the British gentleman would presumably have found it easier than the poor son of a long-poor working-class family to come to Darwin's view of inheritance as a progressive and positive force in preserving variation or to see heredity as enabling rather than constraining. The way Darwin saw the world from HMS *Beagle* was surely shaped by his context and values, as historians have shown (Desmond and Moore 1991; Browne 1995, 2003; Hodge and Radick 2003). His world was the richly but gently entangled bank of an English garden rather than an arid desert, a diverse and pungent rain forest, or an Alpine tundra, each of which might have led him to different questions and emphases.

Furthermore, as Darwin scholars have demonstrated clearly, the British context of natural theology shaped evolutionary theory. William Paley saw each organism and each type as adapted to its place in the world. Of course, it was Paley's Anglican God the Creator who had done the adapting in that case, but the fit of life form to environment carried over neatly to Darwin's view of life. True, a Thomas Robert Malthus might point to the human economic dilemmas intrinsic in the tendency of every individual to increase into a population that rapidly outgrows available resources. But Malthus did not call into question that humans were adapted to their environment. Indeed, the tendency to produce too many individuals just gives nature the opportunity to choose some over others. Or, as Malthus suggested, perhaps man could do the choosing with wise population control. At any rate, social views shaped the evolutionary theory that emerged, and the results were optimistic. For Malthus, population control would make life better for each person and for society as a whole. For Darwin, only the "healthy, vigorous, and happy" would survive and reproduce. A happy picture indeed (Ruse 1979, 2003; Hodge and Radick 2003).

Another part of Darwin's Victorian milieu that was extremely important for his eventual formulation of the theory of natural selection as an explanation of evolutionary change was his close contact over many years with many animal and plant breeders (Browne 2003). Indeed, the close comparison between artificial selection (with morphological and behavioral changes in populations that could be induced by continued breeding) and natural selection (changes that would be a result of the "struggle for existence" in nature) is a major part of Darwin's "long argument"

The German intellectual movement *Naturphilosophie,* with its romantic conception of nature, brought other perspectives that shaped evolutionary theory as it played out in Germany. Johann Wolfgang von Goethe's organicism that saw an unfolding of form, Friedrich Schelling and Georg Hegel's assumptions of unity of nature, and Karl Marx's search for an unfolding of society shared the assumptions of emergence or of form from nonform and of change over time that leads to improvement and progress (Ruse 1996; Mocek 2002; Richards 2002). These social and philosophical threads helped shape the intellectual environment into which evolution arrived. Ernst Haeckel's particular version of evolutionary theory in Germany shared the positive and progressive implications of evolutionary change and developed the ideas in an explicitly and eccentrically materialistic way that had tremendous social impact, just as society had shaped the scientific ideas.

What was seen as an "eclipse of Darwinism" at the end of the nineteenth century also reflects impacts from outside science itself on evolutionary theory (Bowler 1983), shifting away from reliance on what Darwin had seen as randomly arising variations. Mendelism introduced genes, heritable units, and a focus on the internal workings of the organism more than on the organism in its environment. This brought a shift from the more organic interpretations that appealed to Marx and other late nineteenth-century social reformers who were calling for natural change. Instead, we see a rise of genetic determinist or hereditarian thinking, which in turn influenced evolutionary theory, and the determinism of which appealed to different social groups. If the heritable variations were genetically determined, then evolution could presumably be influenced by modifying these genes.

We cannot overemphasize that this is not simply a matter of science affecting society, though that has been the primary focus of historical discussion. Here the point is that forces within society were affecting science, shaping evolutionary theory, and influencing the choices made about which questions to ask and which assumptions to make in a complex theoretical arena of uncertainty where some assumptions must nonetheless be made. Social factors, as well as internal scientific developments, shape which methods to use to study evolution: naturalistic field work and description or experimental control and manipulation of variables, and study of populations or individuals, whole organisms, parts of organisms, invisible parts like genes, or emergent parts like behaviors and other traits.

August Weismann provided an inheritance-based evolutionary theory. His 1892 *Keimplasma* contained sequestered germplasm that was protected from environmental influence. This was the material of inheritance and the raw material of evolution, providing substance for evolutionary change through competition among his hypothetical ids, determinants, and idants, all arranged along visible chromosomes. The relevant population, that of these inherited material units, was now inside the individual organism, and this raised new questions about how far the action of natural selection can take us and about the fine details of the mechanisms of evolutionary change. For Weismann, who was part of a scientific tradition that focused on cell biology, development, and microscopic observation, the important natural selection took place inside each organism, and the environment was the internal environment.

Weismann's evolutionary theory disallowed direct external environmental impact on the inherited chromosomes. His denial that Lamarckian use and disuse of particular parts could have any effect on evolution is taken in retrospect as putting that idea to rest, though it did not actually do so. Instead, neo-Lamarckism continued to find a place in social and political contexts that emphasized the importance of change through effort and will, such as that of Richard Semon early in the twentieth century and with a line of less well-known supporters through subsequent generations (Semon 1908). By the second half of the twentieth century, political figures such as T. D. Lysenko gained considerable authority in the Stalinist Soviet Union, and not solely for distorted political reasons. Lysenko's neo-Lamarckism appealed to the hope that if only scientists could cultivate seeds in the right environment, they could produce enough grain to save the people from starvation. There was tremendous belief in the powers of science, and even though neo-Lamarckian ideas have later been identified with Soviet-style materialism, at least initially many of these proposals emphasized the role of culture and the social environment in shaping the fate of human societies.

Some would insist that this was clearly bad science, and for those fully familiar with contemporary scientific paradigms, it was. But for those for whom evolution was not a central part of their education and worldview, such an evolutionary theory that would place the causes of variations directly in use and disuse in response to the environment made sense. It was not a long stretch from Darwinian evolution to seeing temperature changes as a source of new variations through mutation and selection and therefore as the source and cause of the origin of species. Lysenko's version of neo-Lamarckism promised a speedier evolution, and since evolution was seen as progressive, it therefore offered a faster route to progress and improvement. Social needs shaped science, which then played out in society (Todes 1989).

We see another example in the so-called modern evolutionary synthesis of the late 1940s and after. Participants at key meetings, such as the 1947 Princeton meeting, saw this as a time of optimism for evolutionary biologists (Mayr and Provine 1980). The prewar tensions in Europe and the United States may well have encouraged the sort of synthesis that Julian Huxley, as a political activist with the United Nations, sought in his *Evolution: The Modern Synthesis* (1942). Huxley was not a major intellectual shaper of the perceived synthesis but a central voice in conceiving of it in that way (Smocovitis 1996). However, even while evolutionary biologists such as Ernst Mayr and George Gaylord Simpson eagerly embraced the idea of synthesis with evolution at the core, already other biologists were pushing in counterdirections. Cell biologists and geneticists cared more about the internal workings of cells and their contents than about the bigger pictures of evolution. Rapidly increased funding for medical research and medical demands of wars helped drive wedges between specialties within the biological communities. Evolutionary biologists often found themselves as comparative specialists in natural history museums, while more reductionist research programs gained priority in many universities and especially in medical schools.

Debates about the efficacy of adaptation reveal similar social shaping of evolutionary theory. Stephen Jay Gould, Niles Eldredge, and others began in the 1970s to question how far natural selection can really effect change (Eldredge and Gould 1972). Is evolution really all gradualism and slow accumulation of morphological differences? Or could change be "punctuated" with more sudden "explosions"? Perhaps some changes could enable many others, as passing through a bottleneck to the opening beyond can do. Or external factors might lead to punctuation. Evolutionary theory could support competing interpretations of mechanisms, these researchers felt, though their views were interpreted by antievolutionary creationists as challenging the validity of evolution.

A central point here is that this critique and the revisions in evolutionary theory that resulted were, in part, shaped by the society in which they appeared. Old-school adaptationism and gradualism provided a slow process for change. Not coincidentally, those who were calling for punctuation were young and fancied themselves social reformers who favored rapid social change. This does not mean that they were wrong scientifically any more than those who favored gradualism. Nor does it mean that their challenges have not had tremendous positive impact on evolutionary theory generally. They have. Our point here is that the society and the social values the scientists absorbed have shaped the evolutionary theory significantly.

Today we see another transformation of the adaptationist paradigm in evolutionary biology in the form of a merger of developmental and evolutionary biology. Population genetic models have emphasized genetic changes within population. But genes do not interact with the environment, organisms do, and the way genetic variations relate to corresponding phenotypic variations has important implications for our understanding of evolutionary changes. Two notions, in particular, that are part of this new paradigm also have wider societal implications, namely, the ideas of constraints and of interactions (Maynard Smith et al. 1985). The idea of constraint emphasizes that not all variations are equally likely or even possible. It states that the developmental and physical boundaries of an organism limit the ways genetic variations can be translated into phenotypic variations because of the properties of the developing system. These are, of course, a consequence of multiple interactions among genetic, cellular, and environmental factors that together contribute to organismic development. The new synthesis of evo-devo (short for evolutionary and developmental biology) thus also emphasizes an epigenetic and interactive view of biology.

This perspective has many scientific, cultural, and societal resonances; the idea of limits on what is possible within the boundaries of a natural system was first made popular in the context of the worry about a "population bomb" and the environmental movement. Similarly, the growing emphasis on interactions, even between distant parts of a large ecosystem, has found its first manifestations in this context. But today these concepts are also seen as relevant within developmental and evolutionary biology, where, quite ironically, the completion of the many genome projects—in many ways the culmination of the genetic paradigm—has led to insights into the highly interactive and epigenetic nature of both development and evolution (Hall 1998).

That evolutionary theory has not been insulated from social influences is no surprise to historians of science, but it is often surprising to scientists who see science as relatively insulated and progressing according to its own internal logic and responses to opportunities. It is also surprising to some social historians, who rarely study the history of science and who tend to take science as given and to see it as shaping society on occasion rather than to see the interactions in both directions. Darwinism, as Desmond and Moore (1991) and Browne (1995, 2003) have shown clearly, is just as much a story of Darwin in society as of Darwinism in society.

Society's Reception of Evolution

Historians have long described the reception of Darwinism, in particular, to show the variations in reaction. Why is evolution taken as a challenge to established values in some countries or local societies and not in others or to some religions and not others? Why have some groups readily endorsed the naturalism at the core of evolutionary theory, as Germany did in the late nineteenth century, while there is hostility at some times and in some groups to the apparent randomness and lack of purpose that they see in evolution? To what extent is the response shaped by local contingencies of dominant influential individuals, perhaps, and to what extent does the reaction flow from the logic of the values and assumptions within the culture? The answers to these questions are complex and involve detailed accounts of the social, cultural, economic, and political history of these societies and countries. For a more detailed account, the reader should consult the substantial body of literature in this area (for example, Hull 1973; Kohn and Kottler 1985; Wassersug and Rose 1984).

Here we want to point to just one aspect, namely, how different societies' reception of evolutionary ideas contributed to the further development of evolutionary theory. Within Darwin's inner circle Thomas Henry Huxley most visibly emphasized the importance of evolutionary theory to promote a liberal and materialistic agenda. To him, evolution implied that there is no intrinsic value in heritage and pedigree, and that all that should count are the abilities of people. As a consequence, he promoted science as a profession rather than a vocation and fought a constant battle against religion as the stalwart of received values (Desmond 1997). Similarly, Ernst Haeckel focused on the materialistic implications of evolutionary theory, which he expanded into a whole system of monistic philosophy. Neither Huxley nor Haeckel thought highly of natural selection as a mechanism of evolutionary change. Rather, each incorporated the principle of evolution into his scientific discipline (mostly morphology) and his highly successful teaching. In teaching they both interacted with and shaped current trends within their respective societies, while their scientific contributions influenced several generations of biologists.

A similar pattern can be seen in Russia, where ideas and observations about symbiosis and cooperation also challenged the dominance of natural selection as an explanation of evolutionary change (Todes 1989; Ackert 2007). Another influence in Russia was the presence of a strong tradition of natural history and ecology, which laid the groundwork for subsequent studies of evolutionary changes within local populations (Adams 1994). Meanwhile, in Vienna emphasis on experimental work within physiology and developmental mechanics paired with focus on an organism's life history in the work of the biologists who worked in the Vivarium. This provided an experimental basis for proposals that questioned the then-emerging genetic context in favor of a developmental (including neo-Lamarckian) perspective on evolution (Przibram 1904).

All these examples show that the reception of evolutionary theory in a society never was just a one-way street. In all cases this was a highly interactive relationship that in turn contributed to the further development of evolutionary theory in interesting ways.

Evolutionary Theory Explaining Society

There are no a priori limits on what can be the subject of an evolutionary explanation as long as certain basic conditions apply. From monad to man, a multitude of different life forms and their behaviors, as well as their social and ecological interactions, have all been shaped by evolutionary forces (Ruse 1996). Obviously there are differences between the coordinated movements of slime molds that are forming a fruiting body and the members of an orchestra who are playing a Beethoven symphony, but both are ultimately the result of the interplay of complex behaviors that exist in accordance with the framework of evolutionary theory. In order to understand this last statement, keep in mind the ways in which any feature, morphological or behavioral, can be considered the product of evolution.

Natural selection favors variants with higher fitness, namely, those that manage to increase their representation in the next generation. This process is described by the universal replicator equation that mathematically describes the consequences of natural selection and that applies to all objects that can reproduce themselves (Dawkins 1976; Hofbauer and Sigmund 1998). The replicator equation does not specify what properties of these objects actually contribute to fitness differences; it only predicts that whatever variable features are responsible for these effects, these will eventually reach a (local) optimum as long as environmental conditions do not change. Furthermore, the replicator equation does not specify how those variants manage to reproduce themselves. All that is required for natural selection is that reproduction happens with a sufficient degree of accuracy that the favorable properties do not disintegrate too fast. Many different structures can thus be potential and actual replicators, or units of selection. These include molecules, genes, cells, clones (in the botanical sense), groups, including social groups, and ideas, so-called memes (Dawkins 1982; Brandon and Burian 1984; Sober and Wilson 1998).

The replicator equation specifies the broad limits of what kind of properties can evolve—those that either increase the fitness of their carriers or those that are neutral with regard to fitness (those, however, are the subject of a different dynamics; see Kimura 1983). If we now ask in what way a specific new phenotype such as a morphological feature or a form of behavior can be the product of evolution, we have to address an additional question. Before such novel phenotypes can contribute to fitness, they must first emerge within their carriers; that is, they must themselves be products of evolution. In other words, the molecular, cellular, and developmental context of organisms is the first part of the explanation of any new feature. Only those phenotypic variants that are possible at all can become the raw material for natural selection, because phenotypes determine the fitness of their carriers. The fact that development plays a crucial role in this generation of variation has recently been the subject of increased attention (Raff 1996; Hall 1998).

For our discussion of evolutionary explanations of societies, a different aspect of the evolutionary process is also of interest. As we have seen, all that is required for natural selection to act is the availability of accurately transmitted variation in fitness. This relationship is commonly measured by heritability of a character. Heritability is actually defined as the correlation between parents and offspring, and even though in many cases the reason for this correlation will be genetic, it is not limited to the effects of genes (Falconer 1989). This aspect of heritability is especially important in the context of discussions about cultural evolution. Not only can ideas or memes be interpreted as replicators, but in the context of human, as well as animal, societies learned behaviors can be transmitted to the next generation. From this abstract perspective there is thus no contradiction between the intricate patterns of human history reflected in the changes and transformations of human societies and the basic principles of evolutionary biology. There are, of course, many additional factors besides genes that contribute to human history. Nothing less would be expected from the viewpoint of evolutionary theory because more complex systems add many different levels and degrees of freedom, but all this happens within the confines of evolutionary principles.

It is important to note that this perspective does not imply any form of biological reductionism; rather, the evolutionary analysis of complex systems, such as human societies, requires that we simultaneously consider effects at all different levels of complexity. Indeed, in evolutionary biology the most interesting questions always arise in situations where any simple explanation reaches its limit and where we find a conflict between different entities. These are the "major transitions in evolution" described by the late John Maynard Smith and Eörs Szathmáry that bring about something fundamentally new, such as the first multicellular organisms or the first human societies (Maynard Smith and Szathmáry 1995). In order to survive, these new entities need to "control" the behavior and selfish interest of the parts; in higher organisms, cells have to give up their own reproductive interests, and in a complex society with a high degree of division of labor, these activities also need to be coordinated. How these regulations come about and evolve is one of the most fascinating problems of evolutionary biology (Margulis 1981).

These brief technical comments about the basic assumptions of evolutionary theory also help us see the long-standing controversy over nature versus nurture in a different light. In the context of evolutionary biology this distinction is one only of degree and not of kind. There is no clear-cut separation between nature (or biology) and nurture (or culture and society). A multitude of factors, both biological and cultural, contribute to the development and formation of any human being; indeed, our developmental program requires all these different stimuli as crucial inputs.

As we have learned from evolutionary biology, compared with our closest relatives, we are born prematurely (Portmann 1951). One consequence of this fact is that in human infants the developing brain is stimulated by a multitude of environmental and social factors. In addition, we have learned that in our evolutionary history the emergence of social skills, including the evolution of language, was much more important than accumulated genetic differences (Diamond 1992; Kuper 1994). Of course, precisely those characteristics also enable learning and the transmission of acquired knowledge. In conclusion, all evidence thus far points to highly interactive relationships among different factors (biological and cultural) in human evolution. Consequently, in explanations of human evolution and society we cannot give priority to any one of these factors over all the others. This realization should also help us avoid the naturalistic fallacy; knowing the evolutionary history of some aspects of our behavior and social organization does not imply any value judgment about these properties. But realizing the evolutionary reasons for these behaviors does help us if we want to enforce societal roles, a fact that is increasingly realized by a growing community of legal scholars, economists, and management consultants.

Our brief analysis of the relationship between evolutionary theory and society notwithstanding, it is also true that throughout the history of evolutionary biology the majority of attempts to apply evolutionary principles to explanations of human societies have been grandiose failures (Hofstadter 1959). Seen through historians' eyes, these cases are prime examples of the mutual relationship between science and society discussed in the first section; at certain times specific societal beliefs and assumptions resonate with elements of the then-current state of evolutionary theory and become the foundation of wide-ranging "explanations" of nature and society. Seen through the eyes of an evolutionary biologist, these cases take on a different meaning. On the one hand, one might argue that these cases are a reflection of the then-incomplete status of evolutionary theory, at least when compared with our present understanding. Taken at face value, this view is problematic because it implies a rather naive conception of scientific progress. Another observation an evolutionary biologist would make, looking back at previous cases of evolutionary explanations of society, is that all these proposals tend to emphasize only one or a few elements of what Ernst Mayr and others have identified as the multiple elements of evolutionary theory (Mayr 1982).

This observation allows us to combine historical contextualization of evolutionary explanations of society (what was the scientific, social, political, and economic context of these proposals?) with evaluation in terms of their scientific merit (what were the problems with and wrong assumptions of these proposals? what was left out, and what was exaggerated?). This approach allows us to analyze these various proposals both in relation to their historical context and with regard to their scientific content. Unfortunately, we do not have the space to discuss these ideas in detail, but we want briefly to present a few examples of how evolutionary ideas were applied to human societies in the past before concluding with a sketch of how present-day evolutionary biology approaches the issue of human societies.

It will come as no surprise that as soon as there was evolutionary theory, it was applied to explain aspects of human society. Herbert Spencer, a popular philosopher and advocate of ideas similar to those of Darwin, coined the term *survival of the fittest* to describe the consequences of natural selection. Detached from its strictly biological meaning, this term could be (and was) applied to a variety of social contexts. It could be taken to support a conservative status quo when it was combined with ideas and observations that evolution favors conservation of existing traits and behaviors. In this way evolutionary theory provided a naturalistic and materialistic explanation of social differences by demonstrating how some people are more successful because they are fitter in Herbert Spencer's sense. Their variations are better adapted to the competitive social environment in which they live, and they therefore naturally enough rise to the top.

Not surprisingly, it was the great industrialist Andrew Carnegie who most enthusiastically embraced this interpretation of evolutionary theory as explaining society. But the idea of natural selection was also used in support of radical transformation of society, as the initial positive reception of Darwin by Marx, Engels, and other leading social democrats attests. They, of course, emphasized the transformative powers of natural selection rather than the conservative ones.

What all these initial applications of evolutionary theory to society have in common is that they focus mostly on competition, struggle, and a progressive interpretation of (evolutionary) history, as well as a hierarchical notion of different social classes and races. Furthermore, their notion of competition is more or less direct or hands-on. Struggle was often seen as a consequence of different strengths of individuals, groups, states, or races. Indeed, the early years of the twentieth century brought several attempts to adapt Darwinian logic to work as a guide to international affairs and politics (Ziegler 1918). What all these applications of evolutionary theory to society missed was an understanding of the internal conditions of organisms and the modes of transmission of hereditary material, which were, after all, crucial elements of Darwin's theory.

The centrality of heredity to the theory of evolution was not missed by Darwin's cousin Francis Galton. In his *Hereditary Genius* (1869) he realized that the cumulative effect of natural selection, necessary for gradual evolution, crucially depended on both the availability and the stability of adequate hereditary material. Focusing on inheritance and stability across generations, Galton saw possibilities for controlling populations and thereby controlling evolution and the future of human populations. This was an enticing idea to the Victorian progressivist. Not only could we have progress, but we could cause and control it through biology, Galton suggested with his call for good breeding with the negative and positive selections of eugenics.

In *Man and Superman* (1903), George Bernard Shaw also reflected a more progressive interpretation, though still very much grounded in assumptions that some are naturally better than others. Why not acknowledge that our society is currently weaker than it could and should be because we have no informed breeding program, Shaw suggested. We can explain our imperfections now and also seek to improve, generating a society of men and supermen.

Indeed, eugenics brought hope for progress especially to European and American visionaries in the early twentieth century. Leading geneticists joined the call for good breeding and for better society through biology. Only gradually did scientists accept how little solid scientific knowledge really undergirded the eugenic assumptions, so that in the 1910s and into the 1920s eugenics was not as ridiculously unscientific as it has seemed to commentators in retrospect. Social selective pressures reinforced those who adopted eugenic programs based on biological claims (Kevles 1985; Paul 1995).

Eugenic ideas, together with a biological concept of racial superiority and racial hygiene, contributed to the Nazi genocides, the campaign to exterminate "unworthy life," and finally the Holocaust. As a consequence, a growing coalition formed to counteract what was perceived as an inappropriate biological definition of human races, culminating in the 1950 UNESCO statement on race. This statement then became the foundation for further antidiscrimination policies and laws. These declarations emphasized ethnic rather than biological differences between human populations, which seemed consistent with studies that showed that the vast majority of genetic differences among humans occur within populations. However, in recent years genetic differences in mitochondrial and Y-chromosomal DNA between human populations have also been used to illuminate migration patterns and genealogical relationships between human populations, including the origin of our species in East Africa about 130,000 years ago. Today, even though the subject of genetic differences between human populations is still touchy, we find a growing consensus that neither the biological nor the cultural history of mankind can be ignored. This is especially true as several technological developments in the postgenomic age increase our abilities to customize medical treatment on the basis of the genetic constitution of patients.

The two most important theoretical insights into the evolution of (animal and human) behavior were Hamilton's notion of kin selection as an explanation of seemingly altruistic behavior and the application of game-theoretic notions, and here especially the concept of strategies, by John Maynard Smith and others (Hamilton 1964; Maynard Smith 1982). These ideas represented a major change in evolutionary explanations of behavior and set the stage for the development of rigorous mathematical models.

The theory of kin selection and of inclusive fitness focused the debate on the appropriate units of selection in explanations of behavior. Rather than just looking at one individual's fate as the sole determinant of fitness, the theory of kin selection changes the ways in which evolutionary accounting is done. Within the framework of population genetics changes, evolution is represented as changes in gene frequencies. What Hamilton realized was that there is more than one way in which a specific gene in a specific individual can make it into the next generation. It can be passed on to its offspring directly or through the reproductive success of close relatives, who will share the same gene with a certain probability (.5 for siblings, .125 for cousins). In calculating the fitness of a specific gene, one thus has to focus on the inclusive fitness, which is the sum of all the different means by which a specific gene can come to be represented in the next generation.

Previous vague notions, such as the idea that a behavior can evolve because it is good for the species or the group, were replaced by concepts that accorded with the predictions of population genetics and the theory of natural selection. However, it is also important to note that these notions were highly abstract, and when they were first proposed, scientists did not really know anything about how these proposed genetic replicators actually could cause the respective behaviors. At this time the arguments were largely theoretical: if a gene/replicator causes altruistic behavior, it can be favored by natural selection as long as the cost to the individual acting selflessly is smaller than the benefit to its recipients multiplied by the coefficient of relatedness between those individuals. A similar approach guided the many game-theoretic concepts, such as the prisoner's dilemma and ideas about repeated games and reciprocal altruism that were used to explain many aspects of animal behavior.

The appeal of these new ideas was enormous, and biology seemed capable of explaining much of social behavior. The ideas provided an important boost to many disciplines, such as behavioral ecology, but also invited far-reaching generalizations and in turn generated many, often polemical, controversies. The most visible of the scandals erupted with the publication of E. O. Wilson's *Sociobiology* in 1975. In this encyclopedic tome Wilson, an expert on ants and social insects, collected an enormous amount of empirical data on what was then known about animal behavior and organized it in the context of the new theoretical framework. In the final chapter he applied these ideas to human behavior and the evolution of human societies. Furthermore, he proposed that this new synthesis would lead the social sciences to be integrated within the framework of the biological sciences. The reaction was vivid. However, ensuing debate focused mostly on the feared political implications—Wilson's most vocal critics belong to the Left—and on the consequences of the reductionism implicit in Wilson's emphasis of genetic explanations and in his model of the relations among the sciences. Surprisingly little attention was paid to the bulk of the book, the enormous amount of empirical data and the rigorous application of evolutionary principles to animal and human behavior.

The sociobiology debate resembled the many other attempts to apply evolutionary explanations to human societies in that it overemphasized just one of the many dimensions of evolutionary theory. Wilson's genetic reductionism, supported by his experience with social ants, as well as the state of population genetics at this time, left little room for other factors or for the more interactive perspective that has emerged more recently. Wilson, his supporters, and his critics were all products of their social and political, as well as scientific, environments (at that time dominated by the recent successes of molecular biology). This led to a rather polarized climate within the life sciences of that time, which in part explains the confrontational tone of these debates.

Today, more than 30 years after the publication of *Sociobiology*, evolutionary explanations of society and discussions of cultural evolution are still popular within biology. However, the conceptual, as well as the empirical, basis for this research has become much more pluralistic. There is still considerable debate and controversy, especially surrounding evolutionary psychology. Yet some of today's approaches to understanding human (and animal) societies better reflect the insights of evolutionary theory sketched earlier. Many different fields of evolutionary biology contribute, and indeed the study of the evolution of complex social systems has become a paradigmatic case for interdisciplinary research. As such, it faces all the challenges of this kind of research; as each specialty takes on one aspect of society and human behavior, it is in danger of overemphasizing that part. And the ensuing debate, even if scientifically productive, is received by a society that is uneasy about evolution in general.

Conclusions

Evolutionary theory is unquestionably the foundation of biology. And of all the sciences, biology has the largest impact on today's societies. Biotechnology and biomedicine carry the hopes of billions that their, or at least their children's, lot will improve, while environmental sciences warn almost daily about the negative consequences of our own actions. Nothing less than the future of human societies—some might call it their evolutionary fate—seems to be at stake. Biology offers both gloom and glory, utopian dreams and conservative longings for an idealized past, and evolutionary biology with its focus on the history of life, as well as the mechanisms of change, is at center stage.

It seems almost inevitable that today's rapid scientific developments and associated transformations of society and human (self-)understanding might trigger reaction by fundamentalists, with their desires for sure answers. In the context of evolution and society, the popularity of fundamentalist movements highlights important failures in communicating what is promising in the fundamental principles of evolutionary theory and in exploring implications and limitations of these scientific insights for our current societies and for human self-understanding (see the main essays "Evolution and Religion" by David N. Livingstone and "American Antievolutionism: Retrospect and Prospect" by Eugenie C. Scott in this volume). We thus offer a few remarks on future directions.

Evolutionary analysis of human behavior and social organizations will remain an important research area within evolutionary biology and will have a high potential for insights, as well as for conflict. Evolutionary medicine is already transforming the ways we understand and treat certain diseases (see the main essay "Evolutionary Biology of Disease and Darwinian Medicine" by Michael F. Antolin in this volume). Evolutionary psychology offers insights into the history of human behavior and our potential to modify and control behavior, including possible transformations in our legal systems and practices (see the main essay "Social Behavior and Sociobiology" by Daniel I. Rubenstein in this volume). The merging of economic theory with evolutionary theory benefits both fields and has given evolutionary theory new conceptual and mathematical tools, such as game theory and the notion of strategies. It has also introduced new questions by replacing traditional ideas of an "invisible hand" with the behavior of individual actors shaped by their evolutionary history and constraints. Applications of evolutionary theory even transform certain fields of engineering and computer science as the principles of evolutionary design and of genetic algorithms have opened new venues for solving complex optimization problems, such as the design of airplane wings (Rechenberg 1973; Holland 1995).

The relationship between evolution and society also provides an important case study for interactions between science and society more generally. The history of evolutionary theory is among the best-studied areas in the history of biology and has revealed interesting ways in which scientific results emerge in a complex interplay between scientific investigations and what can be called their social and cultural contexts. Although these studies have led to insights about how science happens and thus have contributed to popular understandings of science, they also challenge philosophers of science to rethink some fundamental assumptions about the nature of scientific evidence and how historically contingent experimental and theoretical practices can lead to increasingly "accurate" representations of natural phenomena. Some even go so far as to describe this "social" process of finding the "truth" in science in evolutionary terms, thus coming full circle.

But the most important aspect of the ongoing interactions between evolution and society lies in the ways in which evolutionary theory transforms our self-understanding as both biological and social/cultural beings. As proponents of cultural evolution suggest, there is no inherent difference between the ways our bodies and behaviors and our societies and cultures evolved. Indeed, the rigorous application of evolutionary theory to such areas as the evolution of language and of cultural transmission has allowed these proponents to formulate a theory of human society and culture that is materialistic without being unduly gene centered and reductionistic. The insight that social organization and cultural transmission are governed by the same general principles of the replicator equation implies (1) that not everything that is the product of evolution has to be genetically determined and (2) that many phenomena (such as the evolution of cooperation) that remain puzzling when studied within just one level of analysis (genetic or social/cultural) can be resolved when they are approached from both directions. Without doubt this current extension of evolutionary theory to society will be an important and controversial continuation of the long history of evolution and society.

Of course, all this can be taken too far. We must keep in mind that evolutionary theory might provide an explanation for society, but it cannot provide a justification. Evolution also cannot provide an epistemic justification for ethical theory, for example, despite the numerous attempts to do so. "Is" cannot lead to "ought." Evolution cannot tell us how societies ought to behave, except insofar as it follows that societies must accept evolutionary theory as good science if they agree to accept science at all.

BIBLIOGRAPHY

Ackert, L. 2007. The "cycle of life" in ecology: Sergei Vinogradskii's soil microbiology, 1885–1940. *Journal of the History of Biology* 40: 109–145.

Adams, M. B. 1994. *The Evolution of Theodosius Dobzhansky: Essays on His Life and Thought in Russia and America.* Princeton, NJ: Princeton University Press.

Bowler, P. J. 1983. *The Eclipse of Darwinism: Anti–Darwinian Evolution Theories in the Decades around 1900.* Baltimore: Johns Hopkins University Press.

Boyd, R., and P. J. Richerson. 1985. *Culture and the Evolutionary Process.* Chicago: University of Chicago Press.

Brandon, R. N., and R. M. Burian. 1984. *Genes, Organisms, Populations: Controversies over the Units of Selection.* Cambridge, MA: MIT Press.

Browne, J. 1995. *Charles Darwin: Voyaging.* New York: Knopf.

———. 2003. *Charles Darwin: The Power of Place.* Princeton, NJ: Princeton University Press.

Dawkins, R. 1976. *The Selfish Gene.* Oxford: Oxford University Press.

———. 1982. *The Extended Phenotype: The Gene as the Unit of Selection.* New York: W. H. Freeman.

Degler, C. 1992. *In Search of Human Nature: The Decline and Revival of Darwinism in American Social Thought.* New York: Oxford University Press.

Desmond, A. J. 1997. *Huxley: From Devil's Disciple to Evolution's High Priest.* Reading, MA: Addison-Wesley.

Desmond, A. J., and J. R. Moore. 1991. *Darwin.* New York: Warner Books.

Diamond, J. M. 1992. *The Third Chimpanzee: The Evolution and Future of the Human Animal.* New York: HarperCollins.

Eldredge, M., and S. J. Gould. 1972. Punctuated equilibria: An alternative to phyletic gradualism. In T. J. M. Schopf, ed., *Models in Paleobiology,* 82–115. San Francisco: Freeman, Cooper.

Falconer, D. S. 1989. *Introduction to Quantitative Genetics.* New York: Longman.

Galton, F. 1869. *Hereditary Genius: An Inquiry into Its Laws and Consequences.* London: Macmillan.

Hall, B. K. 1998. *Evolutionary Developmental Biology.* London: Chapman and Hall.

Hamilton, W. D. 1964. The genetical evolution of social behaviour, I. *Journal of Theoretical Biology* 7: 1–16; The genetical evolution of social behaviour, II. *Journal of Theoretical Biology* 7: 17–52.

Hodge, M. J. S., and G. Radick. 2003. *The Cambridge Companion to Darwin.* Cambridge: Cambridge University Press.

Hofbauer, J., and K. Sigmund. 1998. *Evolutionary Games and Population Dynamics.* Cambridge: Cambridge University Press.

Hofstadter, R. 1959. *Social Darwinism in American Thought.* New York: G. Braziller.

Holland, J. H. 1995. *Hidden Order: How Adaptation Builds Complexity.* Reading, MA: Addison-Wesley.

Hull, D. 1973. *Darwin and His Critics: The Reception of Darwin's Theory of Evolution by the Scientific Community.* Cambridge, MA: Harvard University Press.

Huxley, J. 1942. *Evolution: The Modern Synthesis.* London: Allen and Unwin.

Kevles, D. J. 1985. *In the Name of Eugenics: Genetics and the Uses of Human Heredity.* New York: Knopf.

Kimura, M. 1983. *The Neutral Theory of Molecular Evolution.* New York: Cambridge University Press.

Kohn, D., and M. J. Kottler. 1985. *The Darwinian Heritage.* Princeton, NJ: Princeton University Press.

Kuper, A. 1994. *The Chosen Primate: Human Nature and Cultural Diversity.* Cambridge, MA: Harvard University Press.

Lightman, B. 2007. *Victorian Popularizers of Science: Designing Nature for New Audiences.* Chicago: University of Chicago Press.

Margulis, L. 1981. *Symbiosis in Cell Evolution: Life and Its Environment on the Early Earth.* San Francisco: W. H. Freeman.

Maynard Smith, J. 1982. *Evolution and the Theory of Games.* Cambridge: Cambridge University Press.

Maynard Smith, J., R. M. Burian, et al. 1985. Developmental constraints and evolution. *Quarterly Review of Biology* 60: 265–287.

Maynard Smith, J., and E. Szathmáry. 1995. *The Major Transitions in Evolution.* Oxford: W. H. Freeman.

Mayr, E. 1982. *The Growth of Biological Thought: Diversity, Evolution, and Inheritance.* Cambridge, MA: Belknap Press of Harvard University Press.

Mayr, E., and W. B. Provine. 1980. *The Evolutionary Synthesis: Perspectives on the Unification of Biology.* Cambridge, MA: Harvard University Press.

Mocek, R. 2002. *Biologie und soziale Befreiung: Zur Geschichte des Biologismus und der Rassenhygiene in der Arbeiterbewegung.* Frankfurt am Main: Lang.

Paul, D. B. 1995. *Controlling Human Heredity, 1865 to the Present.* Atlantic Highlands, NJ: Humanities Press.

Portmann, A. 1951. *Biologische Fragmente zu einer Lehre vom Menschen.* Basel: B. Schwabe.

Przibram, H. 1904. *Einleitung in die experimentelle Morphologie der Tiere.* Leipzig: Franz Deuticke.

Raff, R. A. 1996. *The Shape of Life: Genes, Development, and the Evolution of Animal Form.* Chicago: University of Chicago Press.

Rechenberg, I. 1973. *Evolutionsstrategie: Optimierung technischer Systeme nach Prinzipien der biologischen Evolution.* Stuttgart–Bad Cannstatt: Frommann-Holzboog.

Richards, R. J. 2002. *The Romantic Conception of Life: Science and Philosophy in the Age of Goethe.* Chicago: University of Chicago Press.

Ruse, M. 1979. *The Darwinian Revolution: Science Red in Tooth and Claw.* Chicago: University of Chicago Press.

————. 1996. *Monad to Man: The Concept of Progress in Evolutionary Biology.* Cambridge, MA: Harvard University Press.

————. 2003. *Darwin and Design: Does Evolution Have a Purpose?* Cambridge, MA: Harvard University Press.

Semon, R. 1908. *Die Mneme als erhaltendes Prinzip im Wechsel des organischen Geschehens.* Leipzig: Engelmann.

Smocovitis, V. B. 1996. *Unifying Biology: The Evolutionary Synthesis and Evolutionary Biology.* Princeton, NJ: Princeton University Press.

Sober, E., and D. S. Wilson. 1998. *Unto Others: The Evolution and Psychology of Unselfish Behavior.* Cambridge, MA: Harvard University Press.

Todes, D. P. 1989. *Darwin without Malthus: The Struggle for Existence in Russian Evolutionary Thought.* New York: Oxford University Press.

Wassersug, R. J., and M. R. Rose. 1984. A reader's guide and retrospective to the 1982 Darwin centennial. *Quarterly Review of Biology* 59, no. 4: 417–437.

Weismann, A. 1892. *Das Keimplasma: Eine Theorie der Vererbung.* Jena: Fischer.

Wilson, E. O. 1975. *Sociobiology: The New Synthesis.* Cambridge, MA: Belknap Press of Harvard University Press.

Ziegler, H. E. 1918. *Die Vererbungslehre in der Biologie und in der Soziologie.* Jena: G. Fischer.

Evolution and Religion

David N. Livingstone

Ever since the mythic encounter between "Soapy Sam" Wilberforce, the bishop of Oxford, and Thomas Henry Huxley, "Darwin's bulldog," at the 1860 meeting of the British Association for the Advancement of Science, there has been a widespread sense of an ongoing and inevitable clash between evolution and religion. The historical narrative of this encounter, however, proves to be rather less clear-cut that such impressionism might suggest. In the pages that follow we examine the religious views of Darwin himself and something of the historiography of the debate before introducing Catholic, Protestant, Jewish, and Islamic responses to his theory. A number of recurring themes emerge from this historical survey, and a review of these issues concludes this essay.

Darwin's Religious Evolution

Syms Covington was HMS *Beagle*'s fiddler, ship's boy, and odd-job man. For just over six years he was also Charles Darwin's servant. Because he was the ever-present accessory to Darwin's geographic and scientific travels, it is hardly surprising that this virtually invisible Darwinian adjunct would sooner or later catch a novelist's imagination. The final few sentences of Roger MacDonald's fictional account of Covington in *Mr. Darwin's Shooter* (1998, 410) crystallize an ambiguity at the heart of any account of the relationship between evolution and religion: "He saw Darwin on his knees, and there was no difference between prayer and pulling a worm from the grass. As for Mr Covington, he prayed in the old-fashioned way." The elision between the transcendental and the aesthetic that is captured here resonates with Michael Ruse's conclusion to his recent account in *Darwin and Design*. Fully aware of Darwin's repeated resort to the language of beauty and wonder in the natural order, Ruse (2003, 335) reminds his readers "of the genuine love and joy"—an "overwhelming experience" touching on the spiritual—that evolutionists sense in their encounters with the organic world. Here we are alerted to the Darwin who breathed "Hosannah" in intoxicated delight at the sublimity of the primeval Brazilian rain forest.

Given the complications that observations of this class introduce, not to mention Thomas Henry Huxley's remark to Charles Kingsley in 1860 that a "deep sense of religion is compatible with the entire absence of theology" (quoted in Browne 2002, 310), it is not surprising that opinions differ on the nature of Darwin's own religious convictions. Did he lose faith entirely? If so, at what point? Did his theory of evolution by natural selection have anything to do with it? Did he undergo a deathbed conversion? Evolutionists and antievolutionists alike have a stake in the fate of Darwin's soul.

The idea that Darwin moved inexorably from belief to agnosticism has been contested by those who are convinced that he remained a "muddled theist to the end" (Moore 1985, 438). Similarly, a variety of supposedly diagnostic spiritual moments have been identified: the materialism discernible in notebooks dating from the late 1830s; his growing doubts about the adequacy of William Paley's account of creation with its robust confidence in divine design; and the personal moral dilemmas he suffered over the "damnable doctrine" of eternal retribution with the loss of his father and the senseless, cruel death of his delightful daughter Annie (Moore 1989, 197). There are, too, those who insist that an interest in cultivating a naturalistic account of species, mind, and emotion cannot be construed as evidence of either thoroughgoing atheism or metaphysical materialism (Gillespie 1979) even though Darwin himself turned away from Christianity. Regardless, the image of Darwin as crusading secularist fails to do justice to the complexities of the case.

Since irresolution on Darwin's spiritual state seems to be the collective judgment of these portrayals, the language of certainty and precision in seeking to diagnose Darwin's spiritual condition is misplaced. Ambiguity and hesitancy are more appropriate, all the more so since they convey the sense of wavering to which Darwin himself gave voice. In an 1860 letter to Asa Gray, for example, he wrote: "With respect to the theological view of the question; this is always painful to me.—I am bewildered.—I had no intention to write atheistically. But I own that I cannot see, as plainly as others do, & as I sh^d. wish to do, evidence of design & beneficence on all sides of us . . . On the other hand I cannot anyhow be contented to view this wonderful universe & especially the nature of man, & to conclude that everything is the result of brute force" (Burkhardt and Smith 1985–2005, 8: 224). Then again his sense of vacillation surfaced in his shifting judgments about whether he was wise to "truckle" to public opinion and refer to the Pentateuchal term *creation* in *On the Origin of Species* (1859), though he actually used it and its cognates over 100 times. Account too must be taken of his comments in an 1870 letter to Joseph Dalton Hooker that "my theology is a simple muddle" and then in an 1879 note to J. Fordyce that on theological matters "my judgment often fluctuates" (quoted in Brown 1986, 25, 31). Such remarks confirm Brown's (1986, 27) conclusion that Darwin's religious beliefs "never entirely ceased to ebb and flow . . . At low tide, so to speak, he was essentially an undogmatic atheist; at high tide he was a tentative theist; the rest of the time he was basically agnostic—in sympathy with theism but unable or unwilling to commit himself on such imponderable questions."

The impact of religion on Darwin, however, cannot be restricted to fluctuations in his spiritual temperature. The influence of Paley's *Natural Theology* (1802), for example, lingered long. The connections Paley believed he could discern between divine creativity and human agency provided Darwin with an absorbing analogy: natural selection. Nature, Darwin judged, was involved in the selection of organic forms rather like the manner in which pigeon fanciers picked out the most promising variations. As he himself famously put it: "Natural selection is daily and hourly scrutinizing, throughout the world, every variation, even the slightest; rejecting that which is bad, preserving and adding up all that is good" (Darwin, 1959, 168–169). In some ways Darwin's Nature took on the attributes of Paley's God. To this we might add that such concepts as organic adaptation and the harmony of nature were as central to Paley's cosmogony as they were to Darwin's. Yet Darwin's cognitive indebtedness to theology cannot be limited to architectural echoes of Paley. Some of his most profound convictions emerged in dialogue with—or in defiance of—conventional Christian doctrine. His critique of natural theology, it has been suggested, was umbilically connected to his growing convictions about the links between human and animal—how could the products of an advanced monkey-mind be trusted? His allergy to the miraculous was inflamed by a reading of Paley's *Evidences of Christianity* (1794); his sensitivity to suffering, as Donald Fleming (1961, 231) speculated, may have owed its intensity to a "yearning after a better God than God"; and his doubts about teleology were nurtured by an inability to reconcile the doctrine of divine providence with life's daily details (Brooke 1985). Taken overall, science and religion were, in one way or another, thoroughly intertwined in Darwin's life and thought.

Clearing the Ground

On the face of it, Darwin's theory posed challenges of epic proportions to Christian belief at every turn. It mythologized the Mosaic narrative of special creation; it smashed through the practice of using biblical genealogies to date earth history; it removed the idea of divine design from nature by demonstrating how species came about through the ordinary, humdrum processes of natural selection; it revealed that humans were, in some fundamental sense, no different from animals; it rooted moral sensibility not in the human subject dignified as God's image bearer but in the primitive impulse of a struggle for survival; it was next-door neighbor to the French materialism that had been infiltrating into Britain through Lamarckian-enthused medical radicals who mobilized it to attack the Anglican-Tory establishment. Not surprisingly, the idea of a protracted battle between science and religion received its greatest impetus from nineteenth-century treatises that charted, as in the case of chemist-turned-historian John Draper, the *History of the Conflict between Religion and Science* (1875) or, in the words of Andrew White, president of Cornell University, *A History of the Warfare of Science with Theology in Christendom* (1896). The very titles of works like these did much to confirm the impression of a monumental death struggle between the forces of enlightened science and benighted religion.

But the crisp clarity of this received image substitutes monochrome abstraction for historical technicolor. There is a richer narrative to be uncovered. For a start, at least since Augustine, many had read the creation "days" of the Genesis record symbolically rather than literally. Various harmonizing strategies to maintain concord between Genesis and geology had long been elaborated, and the idea of a lengthy earth history was well established in the minds of early Victorian Christian gentlemen-geologists. Natural theology took a variety of forms, some less susceptible to Darwin's antiteleological challenge, and of course there were traditions, like certain strands of high Anglicanism and strict unsentimental Calvinism, for which the significance of the design argument was negligible. And there were those like John Herschel and William Whewell whose religious outlook predisposed them to stress creation by natural law rather than by divine intervention, a move that opened the door to the idea of evolution as God's method of creation. The ways in which evolution and religion could be made to fit one another have taken, and continue to take, myriad forms.

What further complicates matters is that when conflict between evolutionists and religious believers did occur, the campaign was often waged from the side of secular zealots as much as from theological reactionaries. Take Thomas Henry Huxley, a key member of that Darwinian ginger group, the X club, who dedicated themselves to driving out of power Paleyites and parsons alike. No sooner did he hear of the Catholic St. George Mivart's advocacy of evolution than he ferreted out a copy of Francisco Suarez's Scholastic theology to insist that evolution was in "complete and irreconcilable antagonism to that vigorous and consistent enemy of the highest intellectual, moral, and social life of mankind—the Catholic Church" (quoted in Brooke 1991, 308). Then, in irritation at the strategies of those who read Genesis through geological spectacles, he declared his determination to "prove that rape, murder & arson" were "positively enjoined" in the Old Testament (quoted in Desmond and Moore 1991, 472). Besides, he was forever talking about the "scientific priesthood," preaching "lay sermons," singing "hymns to creation," denominating himself a "Bishop" of the new ecclesiology, ordaining friends to "the church scientific," and expounding his "molecular teleology."

At the same time advocates of conflict histories have also had their own social agendas in delivering pugilistic accounts of science and religion. In the case of Draper, who remarked that science was the twin sister of the Reformation, the real object of opprobrium was the Roman Catholic Church; the doctrine of papal infallibility enunciated in 1870 appalled him and led him to prophesy that it would bring the Catholic Church into conflict with hitherto-friendly governments. As for White, the opposition that his nonsectarian

Cornell University had attracted from clergymen strengthened his resolve to keep science at the forefront of the curriculum; Cornell would be an asylum for science liberated from the strictures of religious dogma. It was maneuvers of this kind that prompted some historians to reconceptualize the so-called warfare between science and religion in Victorian society as a social struggle between two contending intellectual elites—namely, the old-fashioned parson and the newly professionalized scientist (Turner 1978). In this scenario debates about scientific knowledge in general and evolution in particular simply became a further arena in which tussles for cultural supremacy were played out.

However alluring grand narratives may be, stories of the encounter between evolution and religion that trade in systemic conflict or, for that matter, cooperation sacrifice historical intricacy to linear simplicity. In fact, religious believers responded in vastly different ways to evolution, and their stances do not follow any straightforward taxonomy of theological orientation, denominational affiliation, or doctrinal preoccupation. Besides, evolutionary theory was never encountered in a cultural vacuum, and debates about it routinely became a means of giving voice to a range of other anxieties. Something of the diversity of these engagements may be gleaned from the following thumbnail sketch.

The View from Rome

Catholic opinion on evolution has long been divided. From the early days the opposition of a group of Jesuit thinkers brought together by Pope Pius IX to combat the forces of modernity was expressed through the pages of *La Civiltà Cattolica,* an outlet that exerted very considerable influence among Italian Jesuits. Thus in the early 1860s the necessity of maintaining the fixity of species against Lamarckian ideas of transformism was vigorously promulgated by G. B. Pianciani in its pages (Brundell 2001). By contrast, the efforts of the English comparative anatomist St. George Jackson Mivart, a Catholic convert, to Christianize evolution through a Lamarckian reading of transmutation in his 1871 volume *On the Genesis of Species* was praised in the Catholic press. Huxley might snipe that Mivart could not be a sturdy soldier of science and a loyal son of Rome, but Pope Pius IX awarded him the degree of doctor of philosophy in 1876, and the Belgian bishops pressed him to take up a chair at the University of Louvain in 1884, although he later fell into disfavor on account of certain theological writings.

However vociferous the opponents of evolution undoubtedly were, the official custodians of Catholic dogma did not take any action against Catholic evolutionists until the 1890s when attitudes began to harden as the aging Pope Leo XIII, hitherto a force for moderation, lost ground to the increasing influence of Roman Jesuits and the machinations of the *Civiltà Cattolica* coterie. So while Mivart's efforts escaped at least for a time the censure of the Vatican, other Catholic evolutionists did not fare so favorably, notably the

French Dominican priest M.-D. Leroy, author of *L'èvolution des espèces organiques* (1887), and the American John Zahm, who played a significant role in the development of the University of Notre Dame in Indiana and published *Evolution and Dogma* in 1896.

Both these writers supported the theological acceptability of evolution, albeit within certain limits. Leroy's apologia, erected foursquare on Mivart's foundations, was directed against both atheistic renditions of evolution and theological enemies of the theory. But because he hinted that the human body might be the product of evolutionary forces, the cardinal prefect of the Sacred Congregation of the Index of Prohibited Books issued the judgment that "evolution theory is temerarious and anti-Christian when applied to the human body" (quoted in Brundell 2001, 88). Although the finding that his book was censured was conveyed to him privately, Leroy was devastated and speedily retracted in the pages of *Le Monde* in March 1895. Later he tried to find ways of revisiting the whole issue by producing a new, corrected version of the book, but it too fell foul of the authorities.

It was much the same with Zahm, who had already achieved widespread recognition for his *Sound and Music* (1892) and *Catholic Science and Catholic Scientists* (1893). Although he had earlier condemned Darwin's *Descent of Man* as the cause of social inequity, his purpose in writing *Evolution and Dogma* was to demonstrate its theological acceptability. Indeed he found much in it "to admire, much that is ennobling and inspiring" (Zahm, 1896, xx). Likewise grounded in Mivart's system, Zahm engaged in detailed patristic exegesis to justify a teleological rendering of evolution. The book was an international success, and an Italian translation rapidly appeared, but readers of a fuming *Civiltà Cattolica* were assured that it was all a tissue of lies, redolent with reckless assertions, and hopelessly compromised by dubious assumptions. The Congregation of the Index denounced it in 1897, focusing very largely on the matter of human origins. To traditionalists, Zahm's evolutionism was part of a package of radical proposals that included the adoption of textual criticism and the increasing segregation of academic and ecclesiastical affairs. Like Leroy, Zahm withdrew his efforts despite the favorable review he received from the English bishop John Hedley, who, for his pains, presently came under the whiplash of *Civiltà Cattolica* tongues.

Various factors played their parts in these intrigues. A fear of "Americanism"—the tendency on the part of American Catholics to display intellectual independence and to adapt Catholicism to an American context—was one such force. This indicates that Darwinian controversies among turn-of-the-century Catholics were often less about science than about issues of identity with which immigrant communities were grappling. Another suite of concerns rotated around the routine conflation of Darwinism and modern scholarly trends, notably the new biblical criticism. Reactions to evolutionary biology were thus often all of a piece with responses to wider challenges to established ways of thinking. Critical too was the matter of intellectual authority and who had the right to interpret the tradition's received canon. Those who found evolution in the writings of Aquinas and Augustine were arrogantly failing to pay due deference to the history of neoscholastic commentary. Later, fears of Darwinian-inspired eugenic policies played their role in shaping Catholic evaluations of evolution (Appleby 1999).

Despite these persistent anxieties, Huxley's efforts to proscribe Mivart's Catholic Darwinism, and the difficulties that Leroy and Zahm experienced at the hands of the Congregation of the Index, a succession of Catholic scholars have continued to defend the theological propriety of Darwinian biology. During the final decades of the nineteenth century, the Catholic astronomer George Searle declared that the theory of evolution enjoyed a strong factual basis and that human evolution would not jeopardize Catholic theology. At the same time the Harvard anatomist Thomas Dwight insisted that there was no contradiction between evolution and teleology. The theologian John Gmeiner added his support in 1884 when, in *Modern Scientific Views and Christian Doctrines Compared,* he went so far as to claim that Augustine had actually promoted a version of evolutionary theory and suggested that Darwin could arguably be considered his disciple.

Perhaps the most considered Catholic statement on the subject in the early decades of the twentieth century appeared in French in 1921 (and in English translation by Ernest Messenger the following year) by the director of the Geological Institute at the University of Louvain, Canon Henri de Dorlodot, who defended not just the legitimacy but the intrinsic doctrinal plausibility of Darwinism. Not only were there no scriptural arguments against the theory, he insisted in *Darwinism and Catholic Thought,* but the "teaching of the Fathers of the Church is very favourable to the theory of Absolute Evolution" (Dorlodot 1922, 4). Central to Dorlodot's diagnosis was his development of the concept of what he called Christian naturalism. Conceived as an antidote to atheistic and materialistic evolution, as well as to the interventionist stance of figures like Georges Cuvier and Alcide d'Orbigny, this concept emphasized the immanent workings of the Creator through natural processes. So emphatic was Dorlodot in this judgment that he pronounced it legitimate for Catholics to go even further than Darwin had done when he attributed the initial origin of living things to an act of special divine intervention. Catholic theology, he maintained, predisposed its adherents to an advanced system of transformism and even obliged them to accept the idea that all living beings were derived from a few elementary organisms. All of this was reportedly in keeping with the teaching of the church fathers. St. Gregory of Nyssa's account of origins, for example, was paraded as a theory of "absolute evolution" that left no room for special intervention "even at the origin of life" (Dorlodot 1922, 79).

Given Messenger's role as translator of Dorlodot's volume, it is not surprising that he himself turned in 1931 to a detailed analysis of the theological propriety of evolution that built on Dordolot's "brilliant piece of work" (Messenger 1931, xxiv). Staged in the introductory essay by Charles Souvay as poised between the "noisy advocates of Protestant Modernism" and the "stubborn ranks" of Protestant Fundamentalism, Messenger's volume undertook an extensive archaeological trawl through church history to vindicate an evolutionary account of human origins (Messenger 1931, xviii, xix). The classical writings of Ephrem, Basil, Gregory, Chrysostom, Ambrose, Aquinas, and the like on the origin of living beings, particularly the human race, were scoured, as was the work of later Scholastics like Suarez, Aeneas Sylvius, and Rodrigo de Arriaga. These excavations were marshaled in support of a number of specific conclusions—that spontaneous generation was taught in Scripture and that "Christian naturalism" underlay the proclamations of many of the church fathers on the passive mode of divine creation. Certainly Messenger reminded his readers that standard manuals of Catholic dogma routinely condemned evolution as theologically false and philosophically absurd, but he remained convinced that the evolution of plants, animals, and the human body—though not the soul, which required a direct act of creation—was fully compatible with Catholic orthodoxy.

It would be mistaken to think that Dorlodot and Messenger were lone voices. Take, for example, the case of Sir Bertram Windle, FRS, dean of medicine in Birmingham, president of Queen's College Cork, and, later, professor of anthropology in St. Michael's College, Toronto. A convert to Catholicism and a correspondent of Thomas Hardy, Windle secured a significant reputation as a Catholic apologist through his publication of works like *Twelve Catholic Men of Science* (1912). Not surprisingly, evolution came under detailed scrutiny in his work *The Church and Science*, which first appeared in 1917. Writing at a time when Darwinism was in eclipse because of the rediscovery of Mendel and the availability of various non-Darwinian evolutionary proposals, Windle was able to exploit to the full his inclinations toward a Catholic version of neo-Lamarckism.

Despite these efforts and the sustained succession of evolutionary apologetics from the French Jesuit and paleontologist Teilhard de Chardin during the mid-twentieth century that served as something of a rallying point for Catholics sympathetic to Darwin, Catholic support for evolution has attracted adverse reactions. Messenger reported the antagonism of such Catholic theologians as Janssens, Pignataro, Hugon, and Van Noort. Similarly, Teilhard de Chardin was on the receiving end of pre–Vatican II disapproval and was prohibited from speaking publicly on certain subjects. No less did his work elicit the biting censure of a profoundly unsympathetic Peter Medawar (1967), the Nobel Prize–winning immunologist, who disdainfully dismissed the *Phenomenon of Man* as a work of philosophical fiction. Among theological critics debate has routinely congregated around the genesis of the human species, a preoccupation that confirms that anthropological concerns, whether over polygenism or human ancestry, were more unnerving to Catholicism than either uniformitarian geology or the principle of natural selection per se because the consanguinity of the human race had to be retained at all theological costs (Astore 1996). Thus various twentieth-century papal pronouncements have been concerned to preserve the special creation of the human soul by direct divine intervention.

Consider Pius XII's encyclical *Humani Generis* (1950). Here Catholic thinkers were warned of the danger of too eagerly embracing scientific novelties. So while research "into the origin of the human body as coming from pre-existent and living matter" was allowed, any suggestion that the human race was of plural origins was absolutely condemned because of its implications for the doctrine of original sin. A similar stipulation pertained to the origin of the human soul, which Pius insisted was beyond the reach of evolutionary transformism: "The Catholic faith obliges us to hold that souls are immediately created by God" (Pope Pius XII 1950, ¶ 36).

More recently John Paul II addressed a plenary assembly of the Pontifical Academy of Sciences in October 1996 on the subject of evolution and its significance for understanding the human agent. Recalling Pius XII's encyclical and noting that evolution had now gone beyond the status of a mere hypothesis, he conceded the likelihood that the human body had originated from preexistent living matter, but he still restated the traditional creationist—as opposed to traducianist—doctrine of the origin of the human soul in order to sustain what he called "the ontological leap" between humanity and its forebears. By this move metaphysical matters rotating around morality, conscience, freedom, aesthetic appreciation, and spiritual experience were delegated to philosophical analysis rather than to evolutionary biology because the "moment of transition to the spiritual" was simply not open to empirical "observation" (Pope Joannes Paulus II 1996, ¶ 6).

Not surprisingly, this statement has attracted the attention of Catholic commentators. On the more reactionary side, John McCarthy observed that the address had generated dismay and confusion among the faithful. He therefore worked hard to minimize any openness to evolution that the pronouncement might foster by the introduction of a number of serpentine interpretations and by reminding his readers that members of the Pontifical Academy of Sciences were chosen regardless of religious creed. By contrast, in 1998 the Vatican astronomer George V. Coyne warmly embraced the pope's announcement of evolution's factual status and went on to urge that since evolution was the method of divine creation, the idea of "continuous creation" was the best way to conceive of the emergence of the human species (Coyne 1998, 160).

The Protestant Mosaic

Protestant responses to evolution were no less diverse. From the earliest days there were those who sensed in evolution the method of God's modus operandi in nature. Anglican parson, novelist, and Christian socialist Charles Kingsley, for example, was enthusiastic about the *Origin* and wrote to Darwin telling him that although the theory meant that he must give up many of the things he believed, it was "just as notable a conception of Deity, to believe that He created primal forms capable of self-development" (Burkhardt and Smith 1985–2005, 7: 380). Similarly, Frederick Temple, later to become archbishop of Canterbury, expressed his support as early as 1860, contending that a world made to make itself actually enhanced the Creator's nobility. To such figures, a universe governed by natural law was more dignifying than the image of a divine conjurer who magicked new species into existence. By contrast, Samuel Wilberforce thought that Darwin's theory was both empirically and philosophically unsound and famously poured scorn on it at the 1860 meeting of the British Association for the Advancement of Science. Francis Orpen Morris, Victorian ornithologist and Anglican rector in Yorkshire, who authored a sequence of pamphlets with titles like *Difficulties of Darwinism* (1869) and *The Demands of Darwinism on Credulity* (1890), thought that Darwinism deserved only utter contempt and derision. Across the Atlantic the clergyman Luther Tracy Townsend, who taught biblical languages in Boston Theological Seminary, placed before his readers the stark choice encapsulated in the title of his tract *Evolution or Creation* (1896). Mention too should be made of the Princeton theologian Charles Hodge, doyen of American Presbyterians, who answered his own question in *What Is Darwinism?* (1874) with the terse quip "It is atheism," though this was more a judgment on Darwin's antiteleological stance than on the principle of evolution itself.

The sweep of opinion represented here could be repeated many times over. The English Nonconformist and cofounder of the Evangelical Alliance, Thomas R. Birks, took up the cudgels against evolution, particularly in its Spencerian form, in books like *Modern Physical Fatalism and the Doctrine of Evolution* published in 1876. The Scotsman William Miller felt that humanity's ultimate choice was, to use the title of his 1897 volume, *God, or Natural Selection*. His fellow countryman Henry Drummond, on the other hand, vigorously sought to Darwinize theology in his *Natural Law in the Spiritual World* (1883). Less speculative and more traditionally Calvinist, James McCosh, who took up the presidency of the College of New Jersey (later Princeton University) in 1868, told the 1873 New York meeting of the Evangelical Alliance that instead of denouncing the theory of evolution, Christian philosophers would be better employed in expounding its religious dimensions. His fellow Scotsmen, the theologians Robert Rainy, George Matheson, James Iverach, and Henry Calderwood, all added their voices in support of some form of theistic evolution. Later the American Baptist theologian A. H. Strong insisted in the first decade of the twentieth century that the principle of evolution was simply another name for Christ.

This patchwork of Protestant opinion from the theologians in the decades around 1900 is matched by the pattern of response among scientists who were connected with various Protestant denominations. In Britain Darwin's own mentor, the Anglican clergyman-naturalist John Stevens Henslow, expressed early opposition, as did the Cambridge geologist Rev. Adam Sedgwick and the Scottish natural philosopher David Brewster. The Irish mathematical physicist at Glasgow, William Thomson (Lord Kelvin), whose physical computations of the age of the earth seemed to deny Darwin the time needed for the operations of natural selection, remained antagonistic. And infamously the Plymouth Brethren naturalist Philip Henry Gosse, a correspondent of Darwin and admirer of Wallace, was implacably opposed to evolution and relied on the idea he put forward in *Omphalos* (1857) (the

Greek word for *navel*) that God had created a mature universe with the semblance of age—prochronism, as he called it. Across the Atlantic the Swiss geologist and physical geographer at Presbyterian Princeton, Arnold Guyot, and his Canadian counterpart John William Dawson, president of McGill University, also set their faces against Darwinism. In the mid 1880s Guyot advocated a kind of pre-Darwinian developmentalism that, in the fashion of the Scottish stonemason Hugh Miller, read the geological epochs into the days of Genesis, while Dawson balked at Darwinism's antiteleological stance. By contrast the Harvard botanist, Congregationalist, and confidant of Darwin Asa Gray found both Hodge's and Dawson's objections odd and worked hard to fit evolution into teleological garments by arguing that it was mistaken to restrict the idea of design to the sudden flashing of a Paleyite watch into existence. Another Congregationlist, the Oberlin geologist George Frederick Wright, also gave evolution his endorsement in the early 1880s, even if he wavered later in life, and went so far as to draw a set of telling analogies between Calvinism and Darwinism, concluding that Darwinism was "the Calvinistic interpretation of nature" (Wright 1882, 255). Other supportive commentaries, though with varying degrees of enthusiasm, were forthcoming from such scientific practitioners as the California geologist Joseph LeConte, the Yale mineralogist James Dana, the Michigan geologist Alexander Winchell, and the Princeton clergyman-biologist George Macloskie. Indeed, figures like these did a good deal to keep lines of communication open between specialist science and readers of denominational magazines.

A catalog of such particularities could be elaborated ad libitum, but as an overall generalization, most Protestant thinkers believed that evolution could be embraced without abandoning doctrinal essentials. The majority found ways of accommodating their theology to more or less revised versions of evolutionary theory. As well as challenging widespread assumptions of systemic hostilities between evolution and religion, this trajectory serves to call attention to a number of seeming incongruities. Thus numbered among the writers of *The Fundamentals,* a sequence of 12 pamphlets issued between 1910 and 1915 that gave voice to early Fundamentalism, were several writers, including G. F. Wright and the Scottish theologian James Orr, who advocated evolution in one form or another. There is also the case of the Princeton theologian B. B. Warfield, custodian of the confessional standards of traditional Calvinism and modern architect of the doctrine of biblical inerrancy, a mainstay of the American Fundamentalist movement. Brought up on a Kentucky stock farm, Warfield knew firsthand about inherited variation in animal breeding and later mused that he was already a "Darwinian of the purest water" before the coming of James McCosh to Princeton (Warfield 1916, 652). In 1915 he claimed that Calvin's doctrine of creation was in fact a "very pure evolutionary scheme" (Warfield 1915, 209).

Despite this pluralism across the denominational spectrum, certain sectarian strands in Protestantism remained staunchly resistant to Darwinian infiltration. To a considerable degree the emergence of modern creation science can be traced to the energetic activities of the Seventh-Day Adventist George McCready Price, author in 1906 of *Illogical Geology: The Weakest Point in the Evolution Theory* and in 1923 of *The New Geology*. His flood geology expressed the Mosaic science of the movement's chief apologist, Ellen White, whose doctrine of the Sabbath was grounded in a literalist interpretation of the creation narrative. Price, of course, was not alone, and he received encouragement from the Presbyterian minister and self-styled research scientist Harry Rimmer and the Minnesota Baptist pastor William Bell Riley. Support, though from a different ideological stable, was also forthcoming from William Jennings Bryan, three times Democratic candidate for the presidency of the United States and committed pacifist, whose fears about the seeming alliance between Darwinism and Germanic militarism encouraged him to take up the prosecution's case in the celebrated Scopes trial of 1925. It earned for him the reputation of moron extraordinaire courtesy of the polemics of his feisty opponent, Clarence Darrow. Of course, this is not to say that matters of political ideology were absent from these other controversialists. Price, for instance, issued a pamphlet titled *Poisoning Democracy* (1921) in which he netted socialism and evolutionism together as the mainspring of modern-day ills. His loathing of evolutionary transformism was thus multidimensional—biblical, scientific, and political. In Britain the Evolution Protest Movement, under the leadership of Ambrose Fleming, FRS, professor of electrical technology at University College London, lacked the vitriol of its American counterpart and allowed for the possibility of some divinely guided evolutionary change. Indeed, in Fleming's own case he was even prepared to read aspects of the creation of the biblical Adam in the light of various paleoanthropological findings, a move that certainly rubbed the anatomist Arthur Keith the wrong way.

Whatever the sources of these sentiments, this flurry of creationist commotion declined fairly dramatically in the years after the Scopes trial, though it prompted the publication of E. T. Brewster's 1927 *Creation: A History of Non-evolutionary Theories* to balance the "ample and competent and up-to-date literature of Evolution." Creationist agitation went into abeyance until the 1950s when the hydraulic engineer Henry Morris and the biblical theologian John Whitcomb joined forces to produce in 1961 what became the foundational text of the modern creation science movement, *The Genesis Flood*. It limited geological history to no more than 10,000 years and reasserted creation in six literal days. The trajectory of this movement, its various educational machinations, and its legal stratagems are charted elsewhere in this volume. It suffices here to add that despite its aggressive polemics and international success, it would be mistaken to think that six-day creationism has dominated conservative Protestant thought more generally. In fact, the theory of evolution, in one version or another, has continued to be defended by Protestant theologians and scientists across the denominational range.

In the Light of Menorah and Qur'an

Religious responses to evolutionary theory were often bound up with broader issues that confronted theological cultures and faith communities, not least questions of how to engage with a threatening wider society or how to retain or refashion identity in the face of extramural challenges. Jewish debates about Darwinism in late nineteenth-century America are illustrative, for these were part and parcel of a suite of deliberations on the future shape that American Judaism should take (Swetlitz 1999). Traditionalist, Moderate Reform, and Radical Reform rabbis each had a stake in how the Jewish community should respond. On the radical wing evolution was embraced because it could be mobilized to underwrite progressivism and transformation. Thus the readers of Kaufmann Kohler's various pronouncements during the 1870s and 1880s learned that evolution applied to the spiritual realm and that Jewish ritual and ceremonial needed to be refashioned to keep in step with modern science. To him and others like him, evolutionary development confirmed the principle of progressive revelation; adaptation to the prevailing environment applied to organic and religious communities alike. From the same theological stable Joseph Krauskopf, in a sequence of lectures titled *Evolution and Judaism* and published in 1887, readily applied evolution to the growth of religion, the textual history of scripture, primitive society, the idea of God, and the development of Jewish worship.

Traditionalists, by contrast, either rejected Darwinism or in some cases mobilized it to support traditional Jewish theology. On the one hand, Samuel M. Isaacs used his position as editor of the *Jewish Messenger* to launch attacks on evolution, while Abraham de Sola, a Canadian immigrant, marshaled the criticisms of the Montreal geologist John William Dawson to attack the theory. Thomas Mitchell, an advocate of the view that God had created the world in six 24-hour days, told readers of his 1887 article "Evolution and Judaism" for *Menorah* that there was not a shred of evidence for human antiquity or evolutionary change. On the other hand, the traditionalist editors of the *American Hebrew*, a good number of whom were trained in science or medicine, worked to cultivate a theistic version of evolution along the lines of Asa Gray. Traditionalists too appealed to the gradualism of evolution to oppose rapid change and any sweeping overthrow of inherited institutions. The tenacious endurance over countless generations of traditional ceremonial testified to its value in the Jewish struggle for survival against innumerable adversities.

Internal rivalries thus did much to govern rabbinic encounters with evolution even though pro- and anti-Darwinian sentiments never directly mapped onto the community's various factions. This means that the scientific dimensions of evolutionary theory were rarely addressed; responses were mediated through concerns about the supposed influence that materialism had on lowering synagogue attendance, on evolution's implications for the nature of mind and morality, and on what advocates believed its acceptance entailed for religious institutions and customs. What was often at stake in these debates was the nature of Judaism itself.

Something of the same spectrum of reactions is also discernible in Islam, though in the Islamic case a rather more prevailing literalistic hermeneutic of the Qur'an has meant that advocates of evolution are less common. Nevertheless, Islamic support for evolution is not entirely absent. Muhammad Iqbal, for example, has claimed that the idea of evolution originated in medieval Islam, while the Lebanese scholar Hossein al-Jisr insists that the theory is entirely compatible with Qur'anic theology (Majid 2002). In the case of Ahmed Afzaal (1996), who likewise contends that "the theory of evolution sprouts from the Holy Qur'an itself," a crucial distinction is prosecuted between Darwinism and evolution. The former he finds unacceptable on account of the limited applicability of natural selection, whereas the principle of evolution "fits perfectly in the overall scheme of God's creation as described in the Holy Book." Indeed, deploying the standard soul-body dualism, Afzaal argues for an evolutionary account of the human species that allocates physical development to natural processes and the emergence of self-consciousness to divine creation. In this move increasingly humanoid forms—*Homo habilis, Homo erectus,* and so on—developed to a point where "Almighty Allah (SWT) selected a single pair—a male and a female—and endowed them with their spiritual souls." Not surprisingly, Afzaal found inspiration in the antimechanistic accounts of figures like Teilhard de Chardin and Henri Bergson.

Despite these irenic efforts, vigorous opposition has been forthcoming from writers like T. H. Janabi who claim that evolution is without empirical warrant and is promoted for social and political reasons. Similarly, Shahabuddin Nadvi issued a refutation of evolution in his 1987 account of the creation of Adam, while Adem Tatli's *Evolution: A Bankrupt Theory* (1990) incorporated matter originally presented to the Turkish government. But perhaps the most influential Islamic critic of Darwinism is Harun Yahya, reportedly an eminent Turkish scholar, who has been so prolific in producing antievolutionary tracts (like *Why Darwinism Is Incompatible with the Qur'an*) that his opponents claim that the works are actually the product of a group of writers. In any case, in the 1999 book *The Evolution Deceit* Yahya hit out at the "dishonest philosophy" of "materialism" that "seeks to abolish the basic values on which the state and society rest" (Yahya 1999, 1). Darwinism, the tract urges, provided a mythic foundation for materialism's claims to scientific status, thereby explaining why Marx rooted dialectical materialism in natural history. In another 2001 rehearsal of "the disasters Darwinism brought to humanity," Yahya has attributed war, poverty, pain, and massacre to the malign influence of "the selfish and pitiless world view" of Darwinism on figures like Stalin, Trotsky, Mao, Pol Pot, Hitler, and Mussolini (Yahya 2001). Translated into many languages, the output of the Harun Yahya mind-set, particularly in Turkey, has been reported as immense, with creationist books becoming more influential than standard textbooks in certain places (Koenig 2001). To be sure, these moves have been resisted by a number of Turkish Darwinians, notably the biologist Aykut Kence and the medical geneticist Isik Bökesoy. But marshaling support for Darwinian evolution has proved difficult in a context where political forces have fostered the teaching of creationism in schools and where direct support has been provided by American Protestant creationists (Edis 1994).

Recurring Themes

Various historical models have been canvassed in the attempt to get a handle on the relations between science and religion. Our travels thus far confirm that any simple portrayal of the relations between evolutionary theory and religious belief is doomed to fall foul of the messiness of history, but a number of persistent themes have recurred, albeit in different ways in different settings.

Questions of scriptural *hermeneutics* have routinely surfaced in discussions of evolution. Some adhere to a literalistic exegesis of creation narratives, while those who are interested in concordism have stressed the role of metaphor and cultural context in the interpretation of particular statements. Those who are willing to relegate the Mosaic narrative to the realm of primitive mythology have been engaged no less in a hermeneutic undertaking. Either way, questions of exegesis have persistently obtruded. Philip Henry Gosse's whole system of prochronism was designed to preserve the literal accuracy of the Bible's timescale. By contrast, the Oriental and biblical scholar W. H. Green relieved certain strands of conservative Protestantism of the burden of Ussher-type chronology (which famously dated the world's origin to 4004 B.C.E.) when he demonstrated in 1890 that Old Testament genealogies could not be relied on to date primeval time on account of their strategic omissions from the records. That gave some theistic evolutionists the hermeneutic room they needed. Again, the Catholic writer Dorlodot devoted considerable space to elucidating Catholic theories of inspiration in order to allow for his conviction that the thought-forms of scripture were accommodated to the cultural conditions of the time. And the Islamic evolutionist Ahmed Afzaal urged the legitimacy of explaining "the relevant Qur'anic verses in terms of a metaphor or parable" and spoke of "the Qur'anic legend of the 'Fall'" (Afzaal 1996). In one way or another, evolution theory has persistently recalled attention to hermeneutic matters in the interpretation of ancient sacred texts.

From its earliest days Darwin's theory confronted head-on the issue of *teleology*. As Darwin himself put it at one point, "The old argument from design in nature, as given by Paley which formerly seemed to me so conclusive, fails, now that the law of natural selection has been discovered" (Darwin 1888, 1: 309). The consequences were potentially monumental, for natural theology had underscored the fundamentally moral character of nature, enthusing believers, curbing sectarian tensions, and inspiring men and women to study God's handiwork. Whether from the pen of William Paley, a moderate Whig, whose *Natural Theology* (1802) was devoured by Darwin, or in the multivolume *Bridgewater Treatises* (1833–1836), which set out to demonstrate divine wisdom and power in creation, the natural theology enterprise under-wrote inquiry into the natural order. Darwin destabilized this whole universe. In response, numerous writers worked hard to interpret evolution teleologically. In general, those who were still enthusiastic about Paley's version had a more difficult task than those who found design in archetypal plans and the operations of natural law. Thus James McCosh, already an enthusiast for the way Argyll's *The Reign of Law* relocated divine design, found divine wisdom more in homological structures than in specific adaptations. The Dutch Calvinist, founder of the Free University of Amsterdam, and turn-of-the-century prime minister Abraham Kuyper and later devotees like the zoologist Jan Lever (1958, 229) spoke of "divine evolutionistic creation," which conceived of nature unfolding according to a predetermined plan. Of course, this does not mean that advocates of Paley's mechanical version entirely disappeared. The neocreationist Intelligent Design movement that has surfaced in recent times seeks to identify externally derived purpose in what has been called irreducible complexity, namely, certain biological systems of an all-or-nothing variety whose functions cannot be reduced to simpler operations on which natural selection can work. In its conceptual structure, if not in its statistical wizardry, this reads like Paley redivivus. In any case the implications of Darwin for design have remained a key arena of debate. That there continues to be philosophical mileage in this whole matter is clear from the different positions adopted by such eminent philosophers as Alvin Plantinga (1991),who defends divine interventionism in the history of nature, and Ernan McMullin (1991), who inclines more toward God acting through natural law. The fact that biologists have frequently traded in teleological-sounding vocabulary urging "that evolution is 'opportunistic', that it proceeds by 'trial and error', that it reaches 'dead ends' or accomplishes 'breakthroughs' by sneaking through 'loopholes', or that natural selection 'can remodel proteins in order to improve interactions' " adds yet further complications to the whole issue (Greene 1989, 408).

Across the religious range opponents of Darwinism have been haunted by the spectre of *materialism*. Islamic opposition springs in part from a sense that evolution is erected on a materialist creed that deconstructs humanity's moral nature and has cultivated a decadent Western culture. Christians of various strands have also been troubled by similar concerns. Bertram Windle, for instance, the Catholic advocate of evolution and author of *What Is Life? A Study of Vitalism and Neo-vitalism* (1908), expressed sympathy for the outlook of the experimental physicist and spiritualist Oliver Lodge and for the antireductionism expressed in *The Unseen Universe* of Balfour Stewart and P. G. Tait, which first appeared anonymously in 1875 immediately after John Tyndall's infamous Belfast address. An evolutionist of Lamarckian inclinations, Windle was attracted to the idea of a Vital Power that animates matter, and he remained chary of scientific inclinations toward a purely mechanical universe. By contrast, numerous writers have used the doctrine of secondary causes—that God works through natural agencies—to argue for a form of materialism consistent with the Judeo-Christian heritage. Warfield, for example, insisted that "a complete system of natural causation" was entirely consistent with teleology and that every "teleological system implies a complete 'causo-mechanical' explanation as its instrument" (quoted in Livingstone and Noll 2000, 301). Different though all these responses undoubtedly are, they bear witness to materialism as a recurring locus of debate.

Closely connected with these concerns is a set of questions that congregate around the implications of evolution for the *nature and morality of the human agent*. Various stances have been adopted on the issue of whether human beings *have* souls or *are* souls. Some have opted for the view that the soul is always the result of an act of divine creation. Thus the Catholic evolutionist Ernest Messenger observed in his 1931 account *Evolution and Theology* that "no Catholic theologian or philosopher could possibly allow that the spiritual soul of man has evolved from a brute soul, much less from inorganic matter. The soul is immaterial, and can only come into existence by a direct act of creation" (Messenger 1931, 143). Others have thought that the idea of the soul emerging in some way from earlier forebears is entirely compatible with their theology. As for the ethical repercussions of Darwinism, some have recoiled at the suggestion that moral sensibility has been shaped by the imperatives of a struggle for survival. In such a scenario, as William Jennings Bryan, the "Great Commoner," made clear, there is no reason to expect humans descended from animals to behave any differently from animals. Others, enthused with the possibilities of cosmic evolution, have sought, like Henry Drummond, to use the language of evolution to make sense of ethical and spiritual development. Again, although different religious commentators have taken different positions on the question, the issue of ethics and evolution has persistently manifested itself.

However ambiguous in Darwin's own thinking, the ideas of *progress* and, in certain strands of religious thought, its theological next-door neighbor *eschatology* have frequently entered the discussion. Some of these associations were identified by Ernst Benz in *Evolution and Christian Hope* (1966), which assessed the relationship between Darwinism and a variety of future-oriented ideologies—Marxist and materialist histories of salvation, Nietzsche's futuristic doctrine of the Übermensch, and speculations on evolution and the future of humanity in the writings of the Hindu Sri Aurobindo. Within the Western Christian tradition he focused on the evolutionized eschatology of Teilhard de Chardin, who conceived of evolution as impelling all life toward an "Omega Point." The writings of James McCosh, Henry Drummond, and George Frederick Wright also came within the arc of this analysis. Indeed, there is much to be said for correlating attitudes to evolution with eschatological stances. Postmillenarians, with their robust confidence in the transforming power of Christian civilization, have routinely been favorably disposed toward evolutionary progressivism, while premillennialists, who believe that the world will continue to degenerate until the millennium is ushered in, typically find evolution repugnant. Thus it is not surprising that the modern founder of creation science, George McCready Price, felt that beliefs about the beginning and end of time were umbilically connected, and that he told the readers of his 1923 *Science and Religion in a Nutshell* that it was "useless to expect people to believe in the predictions given in the last chapters of the Bible, if they do not believe in the record of the events described in its first chapters" (Price 1923b, 13).

The realization that the principles of Darwinism might be applied to the *textual evolution of scripture* and to the *emergence of religion* itself has been a further point of issue. The idea of social evolution, of course, predated the publication of Darwin's *Origin,* and the suggestion that the textual history of the Bible had to be interpreted in the light of society's stadial developments was widely adopted by textual critics in the nineteenth century. Indeed, the principle of evolutionary change was so deeply ingrained in the documentary hypothesis emanating from German higher criticism that Darwin's intervention created no stir whatsoever in that Old Testament scholarly community nor among those enthusiastic about the publication of *Essays and Reviews.* By the same token, as we have seen, the realization that evolution could be applied to both the institutional and the ritual history of Israel was relevant to the way in which Darwin's challenge was read among nineteenth-century Jews. Such uses of evolutionary thought forms, of course, have been disturbing to many. George Frederick Wright's later wavering on Darwinism sprang in good part from his concerns about its implications for the Bible's documentary history. In Scotland the promulgation of biblical criticism by William Robertson Smith with its deployment of evolutionary anthropology spooked conservative Calvinists so much that Darwin's own theory of species change seemed rather tame in comparison (Livingstone 2004). Once again, attitudes to Darwinism were mediated through what was taken to be the theory's relevance for broader cultural affairs.

Putting Evolution and Religion in Their Place

Relations between evolution and religion have never been straightforward. Attitudes, moreover, do not map neatly onto denominational contours, confessional traditions, or locations on the conservative-liberal spectrum. In the light of this realization we might justifiably pause to wonder whether the terms *evolution* and *religion* track any transcendental essence in the theological, social, or scientific worlds. At the very least the wide-ranging uses of these labels in different locations and their deployment in a host of discourses— political, economic, moral, and social—alert us to the fact that in speaking of the relationship between evolution and religion we are handling rather tricky bundles of ideas. A less transcendental and more local approach to particular episodes might therefore prove enlightening. For example, whereas Tyndall's attack on the clergy in his Belfast address of 1874 made it difficult for local Presbyterians to read evolution theory sympathetically, the preoccupation with the threats of biblical criticism at the same time in Scotland meant that Presbyterians there were less exercised about biological matters. In New Zealand responses were shaped by the politics of Maori-settler relations, and in the American South racial questions predominated, though with different outcomes in each case. The way in which American Catholics responded to social Darwinism in the years around 1900 was shaped by the social niche they occupied: American Catholicism was very largely a working-class religion of newcomers struggling to maintain their identity in a new environment. In all these instances the particulars of geographic location and social space had a huge role to play in what could be said about evolution and, just as important, what could be heard about it.

From the moment Darwin put forward his theory of evolution by natural selection, the debate was engaged on its significance for a wide range of religious issues. The fact that contemporary evolutionary biology, as a comprehensive worldview and as the master narrative of genetic engineering, continues to raise questions of morality and meaning demonstrates the persistent intertwining of scientific inquiry with matters of social, political, ethical, and religious concern.

ACKNOWLEDGMENTS

I am most grateful to Bernard Lightman, Ernan McMullin, and Michael Ruse for helpful comments on an earlier draft of this essay.

BIBLIOGRAPHY

Afzaal, A. 1996. Qur'an and human evolution. *Qur'anic Horizons* 1, no. 3 (July–September). Available online at: http://www.fortunecity.com/boozers/cheshire/170/SURVIVAL%20(1).html.

Appleby, R. S. 1999. Exposing Darwin's "hidden agenda": Roman Catholic responses to evolution, 1875–1925. In R. L. Numbers and J. Stenhouse, eds., *Disseminating Darwinism: The Role of Place, Race, Religion, and Gender.* Cambridge: Cambridge University Press.

Astore, W. J. 1996. Gentle skeptics? American Catholic encounters with polygenism, geology, and evolutionary theories from 1845 to 1875. *Catholic Historical Review* 82: 40–76.

Benz, E. 1966. *Evolution and Christian Hope: Man's Concept of the Future from the Early Fathers to Teilhard de Chardin.* H. G. Frank, trans. New York: Doubleday.

Birks, T. R. 1876. *Modern Physical Fatalism and the Doctrine of Evolution.* London: Macmillan.

Bowler, P. J. 2001. *Reconciling Science and Religion: The Debate in Early-Twentieth-Century Britain.* Chicago: University of Chicago Press.

Brewster, E. T. 1927. *Creation: A History of Non-evolutionary Theories.* Indianapolis: Bobbs-Merrill.

Brooke, J. H. 1985. Darwin's science and his religion. In J. Durant, ed., *Darwinism and Divinity: Essays on Evolution and Religious Belief.* Oxford: Blackwell.

———. 1991. *Science and Religion: Some Historical Perspectives.* Cambridge: Cambridge University Press.

Brown, F. B. 1986. *The Evolution of Darwin's Religious Views.* Macon, GA: Mercer University Press.

Browne, J. 2002. *Charles Darwin: The Power of Place.* London: Jonathan Cape.

Brundell, B. 2001. Catholic church politics and evolution theory, 1894–1902. *British Journal for the History of Science* 34: 81–95.

Burkhardt, F. H., and S. Smith, eds. 1985–2005. *The Correspondence of Charles Darwin.* 15 vols. Cambridge: Cambridge University Press.

Coyne, G. V. 1998. Evolution and the human person: The pope in dialogue. In T. Peters, ed., *Science and Theology: The New Consonance.* Boulder, CO: Westview Press.

Darwin, C. 1859. *On the Origin of Species.* London: John Murray.

———. 1888. *The Life and Letters of Charles Darwin.* 3 vols. London: John Murray.

———. 1959. *The Origin of Species by Charles Darwin: A Variorum Text.* M. Peckham, ed. Philadelphia: University of Pennsylvania Press.

Desmond, A., and J. R. Moore. 1991. *Darwin.* London: Michael Joseph.

Dorlodot, H. de. 1922. *Darwinism and Catholic Thought.* E. Messenger, trans. London: Burns, Oates and Washbourne.

Draper, J. W. 1875. *History of the Conflict between Religion and Science.* London: Henry S. King.

Drummond, H. 1883. *Natural Law in the Spiritual World.* London: Hodder and Stoughton.

Edis, T. 1994. Islamic creationism in Turkey. *Creation/Evolution* 14, no. 34: 1–14.

Fleming, D. 1961. Charles Darwin, the anaesthetic man. *Victorian Studies* 4: 219–236.

Gillespie, N. 1979. *Darwin and Creation.* Chicago: University of Chicago Press.

Gmeiner, J. 1884. *Modern Scientific Views and Christian Doctrines Compared.* Milwaukee: J. H. Yewdale and Sons.

Gosse, P. H. 1857. *Omphalos: An Attempt to Untie the Geological Knot.* London: John Van Voorst.

Greene, J. C. 1989. Afterword. In J. R. Moore, ed., *History, Humanity and Evolution: Essays for John C. Greene.* Cambridge: Cambridge University Press.

Hodge, C. 1874. *What Is Darwinism?* New York: Scribner, Armstrong, and Company.

Koenig, R. 2001. Creationism takes root where Europe, Asia meet. *Science* 292: 18.

Krauskopf, J. 1887. *Evolution and Judaism.* Kansas City: Berkowitz.

Leroy, M. D. 1887. *L'évolution des espèces organiques.* Paris.

Lever, J. 1958. *Creation and Evolution.* Grand Rapids, MI: International Publications.

Livingstone, D. N. 1987. *Darwin's Forgotten Defenders: The Encounter between Evangelical Theology and Evolutionary Thought.* Edinburgh and Grand Rapids, MI: Scottish Academic Press and Eerdmans.

———. 2004. Public spectacle and scientific theory: William Robertson Smith and the reading of evolution in Victorian Scotland. *Studies in History and Philosophy of Biological and Biomedical Sciences* 35: 1–29.

Livingstone, D. N., and M. A. Noll. 2000. B. B. Warfield (1851–1921): A biblical inerrantist as evolutionist. *Isis* 91: 283–304.

MacDonald, R. 1998. *Mr. Darwin's Shooter.* London: Anchor.

Majid, A. 2002. The Muslim responses to evolution. *Science-Religion Dialogue: Bi-annual Journal of the Hazara Society for Science-Religion Dialogue Pakistan* 1. Available online at: http://www.hssrd.org/journal/summer2002/muslim -response.htm.

McMullin E., ed. 1985. *Evolution and Creation.* Notre Dame, IN: University of Notre Dame Press.

———. 1991. Plantinga's defense of special creation. *Christian Scholar's Review* 21: 55–79.

Medawar, P. B. 1967. *The Art of the Soluble.* London: Methuen.

Messenger, E. C. 1931. *Evolution and Theology: The Problem of Man's Origin.* London: Burns, Oates and Washbourne.

Miller, W. 1897. *God, or Natural Selection: A Summary of the Opinions Opposed to Darwin's Theory of Evolution by Natural Selection.* Glasgow: The Leader Publishing Company.

Mitchell, T. 1887. Evolution and Judaism. *Menorah* 3 (August): 112–113.

Mivart, St. G. J. 1871. *On the Genesis of Species.* London: Macmillan.

Moore, J. R. 1979. *The Post-Darwinian Controversies: A Study of the Protestant Struggle to Come to Terms with Darwin in Great Britain and America, 1870–1900.* Cambridge: Cambridge University Press.

———. 1985. Darwin of Down: The evolutionist as squarson naturalist. In D. Kohn, ed., *The Darwinian Heritage,* 435–481. Princeton, NJ: Princeton University Press.

———. 1989. Of love and death: Why Darwin "gave up" Christianity. In J. R. Moore, ed., *History, Humanity and Evolution,* 195–229. Cambridge: Cambridge University Press.

Numbers, R. L. 1992. *The Creationists: The Evolution of Scientific Creationism.* New York: Knopf.

Numbers, R. L., and J. Stenhouse, eds. 1999. *Disseminating Darwinism: The Role of Place, Race, Religion, and Gender.* New York: Cambridge University Press.

Paley, W. [1794] 1819. *Evidences of Christianity (Collected Works,* vol. 3). London: Rivington.

———. [1802] 1819. *Natural Theology (Collected Works,* vol. 4). London: Rivington.

Plantinga, A. 1991. When faith and reason clash: Evolution and the Bible. *Christian Scholar's Review* 21: 8–32.

Pope Joannes Paulus II. 1996. Message of Pope John Paul II to the Pontifical Academy of Sciences concerning the relationship between revelation and theories of evolution. *L'Osservatore Romano,* October 30. (Original message in French: *L'Osservatore Romano,* October 23, 1996). Available online at: http://www.octc.kctcs.edu/crunyon/CE/Darwin/popejpii.htm.

Pope Pius XII. 1950. *Humani Generis: Encyclical of Pope Pius XII Concerning Some False Opinions Threatening to Undermine the Foundations of Catholic Doctrine to Our Venerable Brethren, Patriarchs, Primates, Archbishops, Bishops, and Other Local Ordinaries Enjoying Peace and Communion with the Holy See.* Rome: Vatican. Available online at: http://www.vatican.va/holy_father/pius_xii/encyclicals/documents/hf_p-xii_enc_12081950_humani-generis_en.html.

Price, G. M. 1906. *Illogical Geology: The Weakest Point in the Evolution Theory.* Los Angeles: Modern Heretic Co.

———. 1921. *Poisoning Democracy: A Study of the Moral and Religious Aspects of Socialism.* New York: Fleming H. Revell.

———. 1923a. *The New Geology.* Mountain View, CA: Pacific Press Association.

———. 1923b. *Science and Religion in a Nutshell.* Washington, DC: Review & Herald Publishing Association.

Roberts, J. H. 1988. *Darwinism and the Divine in America: Protestant Intellectuals and Organic Evolution.* Madison: University of Wisconsin Press.

Ruse, M. 2001. *Can a Darwinian Be a Christian? The Relationship between Science and Religion.* Cambridge: Cambridge University Press.

———. 2003. *Darwin and Design: Does Evolution Have a Purpose?* Cambridge, MA: Harvard University Press.

[Stewart, B., and P. G. Tait.] 1875. *The Unseen Universe, or, Physical Speculations on a Future State.* London.

Swetlitz, M. 1999. American Jewish responses to Darwin and evolutionary theory, 1860–1890. In R. L. Numbers and J. Stenhouse, eds., *Disseminating Darwinism: The Role of Place, Race, Religion, and Gender.* Cambridge: Cambridge University Press.

Teilhard de Charin, P. 1959. *The Phenomenon of Man*. London: Collins.

Townsend, L. T. 1896. *Evolution or Creation: A Critical Review of the Scientific and Scriptural Theories of Creation*. New York: Fleming H. Revell.

Turner, F. M. 1978. The Victorian conflict between science and religion: A professional dimension. *Isis* 69: 356–376.

Warfield, B. B. 1915. Calvin's doctrine of the creation. *Princeton Theological Review* 13: 190–255.

———. 1916. Personal reflections of Princeton undergraduate life: IV—The coming of Dr. McCosh. *The Princeton Alumni Weekly* 16, no. 28: 650–653.

White, Andrew D. 1896. *A History of the Warfare of Science with Theology in Christendom*. London: Macmillan.

Windle, B. C. A. 1908. *What Is Life? A Study of Vitalism and Neo-vitalism*. London: Sands.

———, ed. 1912. *Twelve Catholic Men of Science*. London: Catholic Truth Society.

———. 1924. *The Church and Science*. 3rd ed. London: Catholic Truth Society.

Wright, G. F. 1882. *Studies in Science and Religion*. Andover: W. F. Draper.

Yahya, H. 1999. *The Evolution Deceit*. London: Ta-Ha Publishers.

———. 2001. *The Disasters Darwinism Brought to Humanity*. Scarborough: At-Attique Publishers.

———. n.d. *Why Darwinism Is Incompatible with the Qur'an*. Available online at: http://www.harunyahya.com/incompatible01.php.

Zahm, J. A. 1892. *Sound and Music*. Chicago: A. C. McClurg.

———. 1893. *Catholic Science and Catholic Scientists*. Philadelphia: H. L. Kilner.

———. 1896. *Evolution and Dogma*. Chicago: D. H. McBride.

American Antievolutionism: Retrospect and Prospect

Eugenie C. Scott

Antievolutionism in the United States is entering its second century. Beginning in the early twentieth century and continuing today, with no sign of relenting, antievolutionists have protested the teaching of evolution to children in public schools. The controversy has waxed and waned in strength in direct response to the amount of evolution that is taught in the classroom and found in textbooks. It is a complex controversy with a long history and deep roots in distinctively American attitudes toward religion, education, and science.

The History of Antievolutionism

BANNING EVOLUTION

To understand American antievolutionism, one must understand the history of this controversy, which can be divided into three time periods (Scott 1997). The first consists of the efforts to ban evolution from the public school classroom. Well before the second decade of the twentieth century, evolution had found its way into high-school textbooks and thereby into the curriculum. According to Larson, a demographic trend in urban population growth triggered the first round of antievolutionism in the United States (Larson 2003). High schools at that time were largely an urban phenomenon, and the populations of cities—fueled both by changing economic conditions that prompted the movement of people from farms into cities and by foreign immigration—were increasing. As a result, more students were attending high school by the second decade of the twentieth century than ever before. Although evolution had been part of high-school biology for over a decade, it did not become an issue until more children were exposed to it. Once that happened, conservative Christian animosity toward evolution prompted efforts to remove evolution from the curriculum.

That animosity is founded on a strongly held religious conviction that acceptance of evolution (or "belief" in evolution, as it is commonly put) causes children to abandon their faith in God. In this view, if a child loses faith in God, then that child is lost to salvation, a serious issue to conservative Christians. Furthermore, children who lack belief in God will have no moral rudder to guide them and thus will have no reason to behave properly to their fellow citizens. Society would suffer greatly, all because of "belief" in evolution. A common creationist claim related to behavior is that if children are taught that they are animals, they will behave like animals rather than creatures made in the image of God. In addition to these practical reasons for rejecting evolution, there are theological reasons as well. Genesis is seen as the foundation for Christian beliefs: if Adam and Eve were not real flesh-and-blood individuals, specially created by God, whose sin of disobedience caused them to be cast out from the Garden of Eden, then humankind is not inherently sinful. If we are not inherently sinful because of the Fall, then Christ's sacrifice and resurrection, the pivotal events for Christianity, were unnecessary. In the opinion of conservative Christians, the book of Genesis thus must be interpreted exactly as written, and evolution is rejected because it is incompatible with most forms of biblical literalism. Needless to say, this is not the only Christian view; Catholics and mainstream Protestants (Episcopalians, members of the United Church of Christ, most Methodists, most Presbyterians, members of the Evangelical Lutheran Church in America, and others) are theistic evolutionists who believe that God created through evolution.

The heyday of this first period was approximately 1919 to 1927, when attempts were made in several state legislatures to ban the teaching of evolution. The best-known repercussion of these efforts was the famous 1925 Scopes trial. Tennessee teacher John Scopes was tried for breaking a state law that prohibited the teaching of evolution in the public schools. The trial of the century, so called because of the hitherto-unexperienced coast-to-coast radio and print media publicity that it engendered, resulted in the conviction of Scopes (overturned on a technicality by the Tennessee Supreme Court). Despite the public perception of the trial as a defeat for Fundamentalism (as encouraged by the play and film *Inherit the Wind*), in its aftermath came the gradual diminishment of evolution in textbooks and therefore in the curriculum of public schools.

Indeed, according to one commentator, evolution was effectively removed from the curriculum by the 1930s, and it did not return for several decades (Larson 2003). Evolution returned to the classroom only in the late 1950s and early 1960s, when professional scientists working with the federally funded Biological Sciences Curriculum Study (BSCS) wrote textbooks that matter-of-factly included evolution. Commercial textbooks subsequently emulated the content of the popular BSCS books, and by the late 1960s and early 1970s evolution was usually, though not amply, included in most of the textbooks being sold in the United States. The return of evolution to the curriculum stimulated the second wave of antievolutionism in the United States: the period of *creation science*.

CREATION SCIENCE

When evolution returned to textbooks in the mid- to late 1960s, conservative Christian antievolutionists again felt the need to protect their children from learning what they believed to be an antireligious idea. Not only had evolution been returned to the curriculum, but also as a result of court decisions (particularly *Epperson v. Arkansas*, 1987), evolution could no longer be banned. Perhaps, then, the Bible could be taught alongside evolution; antievolutionists believed that if equal time were given to Genesis in the classroom, fewer students would accept evolution.

But the establishment and free-exercise clauses of the First Amendment to the U.S. Constitution require that public institutions, such as schools, be religiously neutral: they may neither promote nor inhibit religion. To teach the contents of the Bible as factually correct (as opposed to teaching comparatively about religion) would promote a sectarian religious view, proscribed by the establishment clause. The effort to require that the Bible be taught as a balance to evolution in science classes thus had only a brief history: it was foreordained for rapid rejection by the Supreme Court. In 1973 the state of Tennessee passed a law that required textbooks that included evolution to carry a disclaimer stating that evolution was a "theory" that was "not represented to be scientific fact." (The Bible was specifically defined as a reference book rather than a textbook and thus did not have to display a disclaimer.) In 1975 an appeals court struck down this law in *Daniel v. Waters* on the grounds that it violated the establishment clause. The distaste of Christian conservatives for evolution was not a valid reason either to ban evolution or to present a sectarian view such as biblical literalism in the classroom. "There is and can be no doubt that the First Amendment does not permit the State to require that teaching and learning must be tailored to the principles or prohibitions of any religious sect or dogma," wrote the judge in *Daniel*.

If evolution could not be banned, and the Bible could not be taught alongside it to ameliorate its presumed negative effects, what recourse did antievolutionists have? Religion could not be advocated, but alternative scientific views could: if the creation story of Genesis could be supported by scientific evidence, creationists reasoned, then it would be constitutional to present that evidence alongside evolution. In fact, ever since eighteenth-century discoveries in geology refuted the literal Flood story and demonstrated an ancient age of the earth, conservative Christian "scriptural geologists" have fought a rearguard movement, arguing for a young earth and special creation (Lynch 2002). Creation "science" was built upon the shoulders of these views.

The true giant of creation science, however, was hydraulic engineer Henry M. Morris, who even before he entered graduate school began writing books that presented what he viewed as scientific evidence for the special creation of the universe, earth, and living things. After a career in business and academia, in 1970 he left the chairmanship of the Civil Engineering Department at Virginia Tech to help evangelist Tim LaHaye found Christian Heritage

College. In 1972 he formed the Institute for Creation Research (ICR) as a scientific creationism unit within the college. In 1980 he and LaHaye agreed, apparently amicably, on a separation, and Morris took the ICR to nearby Santee, California, as an independent nonprofit organization (Numbers 1992). After Morris's retirement in 1996, the ICR continues today under the directorship of his son, John Morris.

Morris's 1961 book *The Genesis Flood,* coauthored with theologian John Whitcomb Jr., proposed that the earth's geological features could be explained by catastrophic events described in the Bible, especially Noah's flood. But rather than merely positing theological reasons for believing this, Whitcomb and Morris claimed that scientific theory and evidence supported these ideas. Creation science was born.

Creation science was enthusiastically embraced within the conservative Christian community, though scientists understandably ignored it. It consists mostly of negative statements purporting to disprove evolution, offering (in its purest form, at least) no explanation for the universe other than its abrupt appearance in its present form. Morris proposed varieties of creation science that were to be deployed in different environments. *Biblical* creationism, to be taught in churches, was up-front about the special-creation roots of creation science, freely admitting a belief that the stars, the galaxies, and the earth and its living things appeared at one time in their present form as a result of creative acts by God. In *biblical scientific* creationism evidence from both science and the Bible was to be used. In the *scientific* creationism proposed to be taught in the public schools, God was not to be mentioned; only scientific evidence was to be used (Morris 1980). The hope was that by stripping references to God and Genesis from unabashed biblical creationism, creation science would be made legally suitable for instruction in public schools.

Eve and Harrold (1991) contend that creation science has a particular appeal to educated Christian conservatives who are living and working in the modern "knowledge economy," but who still embrace premodernist social values that emphasize "authority, tradition, and revelation of God's truth through the Bible" (1991, 109). Creation science allows them the best of both worlds: they can embrace science yet remain faithful to the conservative religious and social values undergirded by belief in a literally read Bible.

Inspired by Morris's teachings, North Carolinian Paul Ellwanger drafted model legislation that would mandate the teaching of creation science if evolution were taught. In time-honored American grassroots political fashion, he circulated it to friends in a variety of states, who themselves passed it on to conservative state legislators. Such "equal-time" bills were introduced in state legislatures of at least 23 states during the mid- to late 1970s and early 1980s (Scott 2004); scientists and teachers vigorously opposed them. Two equal-time bills passed and were immediately challenged. Arkansas Bill 590 was challenged by a coalition of religious organizations and educators, with a Presbyterian minister, the Reverend Bill McLean, as lead plaintiff. The resulting highly publicized 1981 trial, *McLean v. Arkansas,* was inevitably billed as "Scopes II" by the press, although during the 1925 trial no claims were made for the scientific validity of special creation.

McLean was a resounding defeat for creation science. The plaintiffs argued a straight establishment clause position: creation science was not science, and therefore there was no secular purpose for teaching it in public school science classes. Its teaching would therefore result in the state promoting religion, in violation of the First Amendment. The state's witnesses (the creation side) had little credible science to present, and the defense was embarrassed when the history of the bill's passage provided clear evidence that the Arkansas equal-time law was proposed and passed for the purpose of singling out one topic, evolution, from all other curricular subjects because it offended the religious sensibilities of some citizens. The law was struck down. So firm was the decision that the state declined to appeal to a higher court.

The next equal-time law to be enacted was a very similar bill passed in nearby Louisiana. The district-court judge decided on summary judgment (that is, on the basis of documents filed by both sides, without a trial) that the Louisiana equal-time law violated the establishment clause. The state also lost on appeal and then requested a hearing from the Supreme Court, which accepted the case. In a 1987 decision the Supreme Court agreed 7–2 (Justices Antonin Scalia and William Rehnquist dissenting) with the lower courts that creation science was an inherently religious position, and thus its teaching unconstitutionally would promote religion. The Supreme Court chose not to comment on whether or not creation science was scientific, believing that its essential religiosity was sufficient to bar it from the classroom. The decision in *Edwards v. Aguillard* stated:

> The Act impermissibly endorses religion by advancing the religious belief that a supernatural being created humankind. The legislative history demonstrates that the term "creation science," as contemplated by the state legislature, embraces this religious teaching. The Act's primary purpose was to change the public school science curriculum to provide persuasive advantage to a particular religious doctrine that rejects the factual basis of evolution in its entirety. Thus, the Act is designed either to promote the theory of creation science that embodies a particular religious tenet or to prohibit the teaching of a scientific theory disfavored by certain religious sects. In either case, the Act violates the First Amendment. (*Edwards v. Aguillard*, 578)

Creation science's legislative strategy thus came to a halt at the foot of the Supreme Court's steps. But creation science itself did not wither; instead, it expanded during the 1990s and shows no sign of diminishing during the twenty-first century. In January 2005 SB 2286 was introduced into (although it failed to pass) the Mississippi state legislature, proposing that "public schools within this state shall give balanced treatment to the theory of scientific creationism and the theory of evolution." Classic creation science can be found in the extensive web presences of the ICR and its sister organization, Answers in Genesis (AIG), as well as in the independent creation science evangelism ministries of Kent Hovind (until his recent incarceration on federal charges of failing to pay payroll taxes ["Kent Hovind sentenced to 10 years" 2007]), Walter Brown, John MacKay, and Carl Baugh, among others. Speakers from these ministries and others are constantly crisscrossing the country, speaking at churches, colleges, and even public schools—anywhere they can find an audience to listen to the antievolution gospel. Innumerable individual websites promote creation science, and televangelists frequently invoke classic creation science arguments to attack evolution. The ICR has a museum that claims thousands of visitors annually, and in the spring of 2007 AIG opened a 90,000-square-foot public museum near its headquarters in northern Kentucky. Within the first seven months of its opening, AIG had already exceeded its expected annual attendance, drawing over 300,000 visitors in its first year ("Creation Museum Unexpected Success" 2008). A creationist Museum of Earth History, which will focus on Flood geology, is planned to open near the popular Branson, Missouri, tourist destination. There is, therefore, plenty of opportunity for the American public to be exposed to creation science through the movement's various public outreach activities, publications, classes, and revivals. As two specialists in evolution education wryly remark, "Many students have had ample formal and informal educational opportunities to misunderstand evolution" (Alters and Nelson 2002).

NEOCREATIONISM

Neocreationism refers to the post-*Edwards* repackaging of creation science largely to avoid its legal problems (Scott 1997). It includes proposing alleged "alternative scientific explanations" to evolution, such as Intelligent Design (ID), and/or proposing to counter evolution not with an alternative science but by alleged evidence against evolution. These approaches were encouraged by the wording of the *Edwards* decision itself, but they preexisted in creation science.

"Alternative scientific theories": Intelligent Design
In the *Edwards* decision Justice William Brennan wrote that creationism could not be advocated as an alternative to evolution because of its religious content, but that "teaching a variety of scientific theories about the origins of humankind to schoolchildren might be validly done with the clear secular intent of enhancing the effectiveness of science instruction." He also wrote that teachers could teach "all scientific theories about the origins of humankind" (*Edwards v. Aguillard*, 593–594). In fact, there are no *scientific* alternative theories to evolution, but Brennan's comment invited creationists to propose their existence and provided a legal foundation for arguing to present them in public schools. Creation science was the original scientific alternative to evolution; with its failure to convince the scientific community or the courts, it required repackaging. The most important repackaging of creation science into an "alternative scientific theory" is Intelligent Design.

The *Edwards* decision took a long time to work its way through the courts, but even before the Supreme Court's decision was issued in the summer of 1987, many antievolutionists had seen that the antievolution movement would have to go beyond creation science and especially beyond the young-earth creationism (YEC) of Henry M. Morris and the ICR if it were to expand and be successful. They sought a form of antievolutionism that both would have more scholarly respectability and would avoid the First Amendment problems of creation science. What was soon to be called Intelligent Design was encouraged by Jon Buell, a minister who directed (and still directs) a Texas nonprofit organization called the Foundation for Thought and Ethics (FTE). Buell encouraged fellow antievolutionists to downplay biblical literalism and Genesis imagery and to invoke a vaguer form of special creation to explain highly complex natural phenomena. In this approach God the Designer would not be referred to, and the inability of evolution to explain complexity would be stressed.[1] Buell encouraged the publication of *The Mystery of Life's Origin* (Thaxton et al. 1984), which was written by three scientists associated with FTE. Although the book concentrates only on alleging weaknesses in then-current theories of the origin of life, scrupulously relegating discussion of design to an epilogue, it is now regarded as a founding document of ID.

Mystery contended that the origin of life was an example of a class of phenomena that were too complex for explanation through natural causes. This was categorically so and was not the result of incomplete knowledge. The origin of life was not only a difficult scientific problem, it was an unsolvable one if one restricted oneself only to natural causes. Natural causes were equated with chance, and such complexity as that exhibited in the structure of the first cell could not be explained by the random joining of cellular structures. Such difficulties could be overcome, Thaxton and his coauthors claimed, only by including the actions of an intelligent agent in a scientific explanation.

Mystery presented the possibility that the agent was material (such as intelligent aliens), but it opted for a transcendent agent "beyond the cosmos" (Thaxton et al. 1984, 200)—God. In the epilogue of the book the authors spoke freely of special creation and the need to expand science to include such causation as a means of obtaining a complete view of truth.

Mystery had little impact on the scholarly world. An Institute for Scientific Information (ISI) citation count notes only a few reviews, largely in philosophy and religion journals, and a total of only 28 citations between 1984 and 2008. In comparison, another popular account of the origin of life, A. G. Cairns-Smith's *Seven Clues to the Origin of Life*, published in 1985, had 95 citations during this same time period. Although uninfluential in the world of science, *Mystery* is revered in creationist circles, where adjectives like "compelling" and "important" are invariably used to describe it.

ESTABLISHING ID. One of the authors of *Mystery*, historian and chemist Charles Thaxton, promoted the new ID movement by organizing conferences of fellow Christian antievolutionists. The first was held in the same year that

Mystery was published, 1984, and was titled "Going beyond the Naturalistic Mindset: Origin of Life Studies." It was followed by "Sources of Information Content in DNA" in 1988. Other grandly named ID conferences, under different sponsorship, followed: 1992 brought "Darwinism: Scientific Inference or Philosophical Preference?"; 1995, "The Death of Materialism and the Renewal of Culture"; 1996, "Mere Creation"; and 1997, "Naturalism, Theism, and the Scientific Enterprise."

Some of the faces at these conferences would have been familiar to anyone who follows the antievolution movement, including a few young-earth creationists, such as Paul Nelson, who had worked with the Minneapolis-based Bible-Science Association and later with Students for Origin Research. The latter organization began as a group of predominantly young-earth creationist students at the University of California at Santa Barbara and later morphed into an ID advocacy organization with the odd name Access Research Network. Most of the conference participants, however, were associated with an old-earth view, such as Hugh Ross, an astronomer who founded and directs the Reasons to Believe ministry, and younger antievolutionists, such as mathematician/philosopher William Dembski, philosopher Stephen C. Meyer, and theologian and biologist Jonathan Wells. These conferences, a few of which included defenders of evolution, were a means for the young movement to develop its ideas and, more important, to hone its message and strategy for the outside world.

Even in the beginning, ID advocates aimed their efforts at the public school classroom (Scott and Matzke 2007). Beginning in the early 1980s Buell encouraged the writing and marketing of a textbook intended as a supplement to high-school biology textbooks. Part of the reason for the delay in publication was that Buell sought a secular rather than a religious publisher; the latter would make the book far less salable in the public school market (Scott 1989). Under the auspices of FTE, in 1989 Percival Davis and Dean Kenyon coauthored *Of Pandas and People,* a brief book that in typical ID fashion was long on criticisms of evolution and short on specifics of what might replace it. FTE attempted to promote it to school boards for adoption, but the book was widely rejected by both teachers and scientists.

In 1991 a law professor from the University of California at Berkeley, Phillip Johnson, published the first widely read ID book, *Darwin on Trial.* Although *The Mystery of Life's Origin* and *Of Pandas and People* attracted scarcely any notice in the scientific community, Johnson's book received wide notice from the general press and was even reviewed by a noted evolutionary biologist, albeit in a popular science journal (Gould 1992). This was likely because of the novelty of a professor from a major secular university authoring an antievolution book. Antievolutionism had previously been associated only with private organizations such as ICR that were marginal to the academic world; Johnson's involvement created a flurry of media interest and arguably put ID on the map. The book itself consisted of a reworking of many standard creationist arguments and was roundly criticized by scientists for its errors.

The next major ID publication was *Darwin's Black Box* by Lehigh University biochemist Michael Behe (1996), published by the Free Press. *Darwin's Black Box* marks a high point in ID's effort to cast itself as a science. The premise of this book was similar to that of *Mystery:* some phenomena in nature are simply too complex to be explained through natural causes. Specifically, Behe pointed to molecular structures such as the bacterial flagellum and processes such as the blood-clotting cascade as being too complex to be explainable through incremental natural selection. Behe's book received some reviews in the scientific literature—while Johnson's book did not—perhaps because here again was the novelty of a professor from a secular university publishing an antievolutionism book, this time from a mainstream press. *Darwin's Black Box* also received more attention from scholars than Johnson's *Darwin on Trial* because it attempted to provide a scientific argument. However, the reviews by scientists were almost uniformly negative (Coyne 1996; Dorit 1997; Blackstone 1997; Cavalier-Smith 1997; Thornhill and Ussery 2000).

By 1998, when philosopher and mathematician William Dembski published his first book, *The Design Inference,* the basic structure of ID had been established. In *The Design Inference* and other publications Dembski presented a probability-based explanation for the fundamental idea of ID: that some natural phenomena are too complex to be explained through natural causes and therefore require "intelligence" for their explanation. Invoking an "explanatory filter" (p. 37) of increasing improbability, Dembski claimed to be able to remove natural causes and chance as potential explanations of complex phenomena, leaving design as the residual explanation. Dembski has had a difficult time convincing fellow philosophers and probability theorists of his view and has been quite unsuccessful at persuading biologists that his concept of *complex specified information* has any utility for understanding biological phenomena (Fitelson et al. 1999; Shanks 2004; Wilkins and Elsberry 2001; Rosenhouse 2003; Pennock 1999, 2001; Elsberry and Shallit 2003).

THE CONTENT OF INTELLIGENT DESIGN. In content, Intelligent Design differs from creation science primarily in its reluctance to make specific fact claims. Whereas creation science makes claims about the age of the earth (only a few thousand years old) and the origin of geographic features (catastrophic creation, such as Grand Canyon's being formed by Noah's Flood), ID is mute about the history of the universe because ID intends to be a big tent that will house all antievolutionists, and because inside this big tent are people with decidedly incompatible ideas (Scott 2001). Most of the ID advocates are old-earth creationists, but there are young-earth creationists who are active supporters of ID. Similarly, as do the young-earth creationists, ID advocates allow for evolution within created kinds, but there is no consensus in the literature of either camp on what a "kind" is, the limits of genetic variability that restrict evolution outside these created kinds, or what determines these limits. Michael Behe, a prominent spokesperson for ID, once described himself as believing in God-guided evolution (Behe 2000), although other prominent ID theorists have declared that "ID is no friend of theistic evolution" (Dembski 1995, 3). In many ways, Behe stands apart from other ID advocates, including in his acceptance of common ancestry for living things (Behe 2007).

As Phillip Johnson has recommended, ID advocates try to limit themselves to their chief concern, which is whether nature can be explained solely through natural processes or whether intelligent—in practice, supernatural—intervention is required. This idea underlies ID's core claim: some natural phenomena are designed by an intelligence rather than having come about through natural processes. The secondary, though scientifically critical, claim is that it is possible to distinguish intelligently designed phenomena from complex phenomena of natural origin. Typically, but not always, these are biological phenomena, very often the same phenomena selected for scrutiny by creation science.[2] The methodology proposed involves a concept based on probability and information theory developed by Dembski.

Dembski has contended that low-probability events of a particular type ("specified" in advance by side information) are assignable to the category of things that have been intelligently designed. Complex specified information (CSI), it is claimed, identifies a newly discovered class of phenomena that require the action of intelligence for their origins. Harking back to Thaxton and coauthors' contention in *The Mystery of Life's Origin* that the origin of life was too complicated to be explained by natural cause, Dembski uses CSI to identify a class of phenomena not only unexplained but unexplainable by natural, that is, scientific, processes.

Michael Behe's idea of irreducible complexity (IC) of biological structures is similarly used to argue against evolution. Behe claimed in *Darwin's Black Box* (1996) that certain biochemical processes or cellular structures were not explainable by natural selection. He contended that some structures, such as the bacterial flagellum, require the assembly in final operating order of all of their many components before they can be functional. By definition, any precursor of an irreducibly complex structure is nonfunctional. Behe's view of natural selection—though not that of evolutionary biologists—is that it operates on one component of a structure at a time, rather like adding charms to a charm bracelet. Behe argues that if natural selection requires selective value for each addition of each component of an IC structure, but all the components have to be assembled before there is any selective value, incremental natural selection cannot explain irreducibly complex structures. This is a misunderstanding of the process of natural selection, and critics have pointed out a number of ways that natural selection could operate to produce complex structures, even so-called IC ones (Thornhill and Ussery 2000; Dorit 1997; Orr 1996).

ID AS SCHOLARSHIP. Because ID confuses unsolved problems with unsolvable ones, there is the risk that ID might be disproved by an irreducibly complex structure that was discovered to have a plausible natural explanation. However, this does not seem to be an actual difficulty, for ID proponents can always allow that they were wrong about the irreducible complexity of *that* structure and come up with another. Ironically, they seem to ignore structures that are irreducibly complex, such as the mammalian middle ear or the avian wrist complex, but which in fact have a well-understood evolutionary history (Gishlick 2004). From a scientific standpoint, of course, the inability to refute IC by disproving claimed examples is not a strength but a weakness. Science progresses by eliminating explanations that are disproved, not by proving explanations in some absolute sense.

Intelligent Design proponents have produced a large number of books since the appearance of *The Mystery of Life's Origin,* the bulk of them since the 1996 publication of Behe's *Darwin's Black Box* and Dembski's 1998 *The Design Inference.* Most of these books, however, must be classified as religious or philosophical in nature rather than scientific; it is revealing that most are published by sectarian publishers such as InterVarsity Press, Baker Book House, and Brazos Press.[3] This is curious, since ID's claim to be able to identify intelligently designed *biological* structures would imply that its adherents intend ID to answer questions that are being investigated in the biological sciences. The place to do this would be in scientific journals, especially in the fields of cellular and molecular biology, but here ID has been absent. Even Behe, an academic scientist at a secular university (Lehigh) who has published in mainstream scientific journals on scientific topics unrelated to ID, does not appear to be publishing or attempting to publish research that makes a positive case for ID. The most prominent ID advocates appear to prefer to publish in popular books rather than scientific journals. Dembski has commented: "I've just gotten kind of blasé about submitting things to journals where you often wait two years to get things into print. And I find I can actually get the turnaround faster by writing a book and getting the ideas expressed there. My books sell well. I get a royalty. And the material gets read more" (McMurtrie 2001, 8).

Normally, new scientific ideas are first presented at academic conferences and, after professional feedback and some general discussion, are refined to the point where articles are submitted to scientific journals, after which they may or may not join the scientific consensus. According to an article in the *Chronicle of Higher Education,* when asked why he is not building his irreducible-complexity idea and the ID model at professional meetings of cell biologists and biochemists, "Mr. Behe responds that he prefers other venues. 'I just don't think that large scientific meetings are effective forums for presenting these ideas,' he says" (McMurtrie 2001, 8).

Again paralleling creation science, ID proponents contend that their views are systematically kept out of scientific journals because they reject the status quo. Yet unorthodoxy is not routinely banished from publication (as the histories of the reception of endosymbiosis and punctuated equilibria witness), although unusual claims require more than the usual amount of support and take longer to establish. A more plausible reason for the dearth of

ID articles in the scientific literature is the constraints that are imposed on doing science by ID's foundational claim, which is that natural causes cannot produce (some) natural phenomena. Since science explains through natural causes, the theory that underlies ID makes it rather difficult to do any actual research. What would ID research in biology (rather than theology) look like other than attempting to show, as creation scientists attempted to show, that natural processes such as those involved in evolution are inadequate to account for the origins and history of life? There is no historical scenario presented by ID advocates, so there are no fact claims to test. If the theory behind ID discourages scientific activity and there are no fact claims, there should be no surprise that there are no articles in the scientific literature. Dembski himself has commented, perhaps with some frustration, on the lack of progress in producing a scientific foundation for ID and suggests employing a vague "design-theoretic framework" such as comparing a model of technological innovation developed by Russian engineers with biological systems (Dembski 2002). Presumably Dembski anticipates that biological phenomena will resemble engineered systems (designed), and that this will in some way support ID, in defiance of copious literature that points out that biological phenomena are organic in composition rather than artifactual.

Although the scientific community has paid more attention to ID than to creation science, there is not an extensive anti-ID literature in scientific journals, primarily because ID has been promoted almost exclusively in books written for the general public, Dembski's *The Design Inference* being a notable exception. As a result, most of the published critiques of Dembski and Behe are book reviews, as cited earlier, although there are a few articles in philosophy or mathematics journals—rarely in biology journals—that discuss ID (Wilkins and Elsberry 2001; Shanks 1999; Sober 2002; Pennock 2003). There is a growing number of books being published that specifically critique ID on scientific and/or philosophical grounds (Pennock 1999, 2001; Forrest and Gross 2004; Shanks 2004; K. R. Miller 1999; K. B. Miller 2003; Young and Edis 2004). One can find technical analyses and even lively discussions among ID advocates and critics taking place on the Internet, however (e.g., www.pandasthumb.org; Matzke 2004; Gishlick 2003; Elsberry and Shallitt 2003; Wein 2002).

For all the claims that ID advocates make for the scientific validity of their enterprise, even a brief computerized search of scientific journals reveals that no one is applying the concepts of CSI or irreducible complexity to understand or explain biological phenomena (Gilchrist 1997; Forrest and Gross 2004). The burden of proof remains on the proponents to demonstrate that their position is scientifically valid and useful. Meanwhile, research in evolutionary biology appears to be in a particularly productive period, with new insights and understandings being provided not only from the discovery of new fossil remains but also from molecular biology, especially as applied to developmental biology.

CULTURAL RENEWAL. William Dembski has written, "Intelligent design is three things: a scientific research program that investigates the effects of intelligent causes; an intellectual movement that challenges Darwinism and its naturalistic legacy; and a way of understanding divine action" (Dembski 1999, 13). Although the scientific research program of ID has been unsuccessful, the bulk of the effort exerted by the ID movement is directed toward the *cultural renewal* component hinted at in Dembski's second clause: the effort to replace "naturalism"—in the ID worldview, tantamount to atheism—with Christian theism. As suggested by both the inadequacy of its scientific record and the titles of its organizing conferences, the ID movement is in fact primarily concerned with the Christian religion and its conflict with materialist philosophy—the philosophy that there are only material (matter and energy) phenomena in the universe; there is no supernatural reality.

To defeat secularism in modern society, the ID movement attacks science through evolution. The attack, significantly enough, is not primarily through scientific evidence but through philosophical and even statistical arguments, though these arguments are felt to be thin by philosophers and probability theorists. To understand how the ID advocates reason requires a short digression into the nature of science as it is practiced by scientists today.

Although philosophers of science readily argue about the definition of science and even whether science can be cleanly distinguished from other ways of knowing, these arguments are not very important to the actual enterprise of science as it is practiced by scientists, who tend not to think much about the philosophical niceties. It would be relatively easy to get scientists (and probably even philosophers of science) to agree on a couple of basic elements that characterize science, even if such elements did not fully demarcate science from other epistemologies. Science is an effort to understand the natural or material world—the world of matter and energy and their interaction. It is, therefore, not capable of explaining the phenomenon of religion: the purported existence of a transcendental reality by definition not limited to the material universe. Science is in this sense a limited endeavor: it is limited to material phenomena.

Scientists would also agree that the essence of science is *testing* explanations against the natural world; an explanation based only on inspiration, revelation, creativity, or imagination would not be a scientific explanation, although a scientific explanation might begin that way. Only after confirmation through testing would it become a scientific explanation. The famous story of how Friedrich Kekulé came to discover the circular structure of benzene after dreaming of a snake chasing its tail illustrates this well: it took verification—testing the explanation against the natural world—to establish a scientific explanation. Kekulé did not stop with the dream. It is also likely that scientists would agree that a major component of testing is holding certain variables constant in order to measure the effect of others. Of course, scientists and philosophers of science can and do debate other features that may or may not define science, such as the various versions of falsification and their application, but the description presented here would suffice as a minimalist definition for virtually all scientists. But this minimal definition nonetheless contains characteristics that are opposed by ID advocates.

Intelligent Design creationists believe that the restriction of science to natural causes, known as methodological naturalism, leads people to philosophical naturalism. Philosophical naturalism is, of course, the enemy because it is opposed to Christian theism. If a powerful cultural enterprise like science could be broadened to include God's intervention, and if methodological naturalism could be abandoned, then philosophical naturalism would lose what is perceived as its strongest support. The establishment of "theistic science" (Moreland 1994; Scott 1998), which permits the invocation of the hand of God (or "intelligence") when natural explanations are inadequate for the job, is a key goal of ID creationism.

Theistic science is based on the concept of a dichotomy between *origin science* and *operation science,* the latter being ID proponents' term for the "normal," methodologically materialistic science used in the vast majority of scientific explanations. Origin science is defined as the science used to explain singular, unrepeatable events (the origin of life, for example), hypotheses about which supposedly are untestable and thus unfalsifiable. Therefore, attribution of causality to God is acceptable in origin science, though not in operation science.[4]

Evolution, then, is a stalking horse for an attack upon science, which because of its materialistic base (ID creationists do not distinguish between methodological and philosophical naturalism in any practical sense) is seen as a threat to Christian theism. The ID advocate who makes these links most clearly is one of the most prolific ID authors, Phillip Johnson, whose 1991 *Darwin on Trial* was the first ID publication to attract wide public notice.

Johnson promotes a strategy he calls the "wedge," wherein evolution is attacked as a means to challenge naturalism, thus paving the way for the reestablishment of a Christian sensibility in American culture and the hegemony of theistic science (Johnson 1997; Forrest and Gross 2004). He has also been instrumental in promoting a public presentation of ID that avoids commitment to any factual claims, such as a young or old age of the earth, as a way of avoiding the pitfalls of creation science. Instead, as did the authors of *The Mystery of Life's Origin* and *Of Pandas and People* before him, Johnson concentrates on the supposed inadequacy of natural causes (especially natural selection) to explain the origin and diversity of living things:

So the question is: "How to win?" That's when I began to develop what you now see full-fledged in the "wedge" strategy: "Stick with the most important thing"—the mechanism and the building up of information. Get the Bible and the Book of Genesis out of the debate because you do not want to raise the so-called Bible-science dichotomy. Phrase the argument in such a way that you can get it heard in the secular academy and in a way that tends to unify the religious dissenters. That means concentrating on, "Do you need a Creator to do the creating, or can nature do it on its own?" and refusing to get sidetracked onto other issues, which people are always trying to do. They'll ask, "What do you think of Noah's flood?" or something like that. Never bite on such questions because they'll lead you into a trackless wasteland and you'll never get out of it. (Kushiner 2000, 40)

Because it does not present an alternative model to evolution, and because of its reluctance to make fact claims, the content of ID thus devolves to the reiteration of arguments long ago raised by the young-earth creationists: gaps in the fossil record, the inadequacy of natural selection to produce evolutionary change, the "problems" of the Cambrian explosion, and so on. As is the case with YEC, ID appears to be erecting a "contrived dualism," to quote *McLean v. Arkansas*, in which if evolution can be shown to be untrue, ID is thus established.

Evidence against evolution

After the *Edwards* decision struck down equal-time laws, Wendell Bird, who had argued Louisiana's position before the Supreme Court, gamely advised his fellow creationists that creation science might still be legally taught if it were taught with a secular purpose (ICR 1987). His perspective was not shared by judges in district and appeals courts who interpreted *Edwards* as not distinguishing between creation science and creationism; *Edwards* has been cited by federal district and appeals courts in decisions that have struck down attempts by teachers in Illinois and California to bring creation science into the classroom.

A dissent to *Edwards*, written by Justice Scalia and signed by Justice Rehnquist, suggested an alternate approach that was quite compatible with extant creation science: "The people of Louisiana, including those who are Christian fundamentalists, are quite entitled, as a secular matter, to have whatever scientific evidence there may be against evolution presented in their schools, just as Mr. Scopes was entitled to present whatever scientific evidence there was for it" (*Edwards v. Aguillard*, 634). The ICR seized upon this immediately after the *Edwards* decision was published. Echoing Scalia, it encouraged teachers to teach the evidence against evolution: "In the meantime, school boards and teachers should be strongly encouraged at least to stress the scientific evidences and arguments against evolution in their classes (not just arguments against some proposed evolutionary mechanism, but against evolution per se), even if they don't wish to recognize these as evidences and arguments for creation (not necessarily as arguments for a particular date of creation, but for creation per se)" (ICR 1987).

The link between creation science and "evidence against evolution" is easy to find. Proponents of creation science contend that there are only two possibilities: creationism, consisting of their specific version of Christian special creation, and evolution, which also includes all the other origin stories from other religions and cultures. Therefore, to them it is logical to infer that evidence against evolution is evidence for creationism. Practical considerations are also involved: there are no credible data that support the special creation of the universe at one time, in its present form, so creation science advocates are forced to support their view with a negative argument, that of alleged evidence against evolution. The content of creation science primarily consists of a frequently reworked litany of arguments presented as disproving evolution, such as gaps in the fossil record, the supposed incompatibility of the second law of thermodynamics with a gradually increasing complexity of life, the impossibility of developing phylum-level body-plan differences through random variation and natural selection, and the "vast information content of living organisms" (ICR 1987). It perhaps goes without saying that such "evidence against evolution" generates blank looks from evolutionary biologists, who are puzzled that unsolved areas in science such as the origin of life, the appearance of Cambrian body plans, or the relative role of selective and nonselective mechanisms in evolution are considered weaknesses in evolutionary theory or reasons to reject the inference of common ancestry.

Teaching the "evidence against evolution" avoids the obvious establishment clause problem generated by ID, which is "Who is the designer?" The strategy is therefore favored among the more sophisticated antievolutionists who are operating in the early twenty-first century. The Discovery Institute, for example, has come to urge school boards not to require teachers to teach ID but rather to teach the "strengths and weaknesses" of evolution.

The Future of Antievolutionism

Scott and Branch have referred to three themes that have run throughout the three periods of American antievolutionism and continue to illustrate the controversy today as the "pillars of creationism" (Scott and Branch 2003, 282). At any given time in this history, these three arguments have been part of the creationist position:

1. Evolution is weak or unsupported science that scientists are abandoning in increasing numbers.
2. Evolution is inherently antireligious.
3. It is only fair to "balance" the teaching of evolution with creationism or Intelligent Design. A variant of the "fairness" argument is that it is good pedagogy to "balance" the teaching of evolution with "weaknesses" or "evidence against" evolution "and let the children decide."

These three themes will continue to shape the creationism/evolution controversy in the future. We can illustrate them by examining some recent case studies of antievolutionism. Although religion overwhelms all else as the strongest motivation for antievolutionism, because of legal decisions of the 1980s and early 1990s, creationists have downplayed the second pillar of creationism in favor of the first and third, at least when addressing the general public and, especially, public officials such as school-board members. Both creation science and ID proponents have hammered away at the idea of evolution as weak science and have stressed the importance of fairness, especially in terms of teaching "evidence against evolution."

Another trend has been the "morphing" of YEC to ID to "evidence against evolution." In a number of communities, people who are young-earth creationists began to identify themselves as ID supporters, thus avoiding the stigma of creation science, which is at best considered a minority view, if not a fringe perspective, in both religion and science. Intelligent Design, on the other hand, is less well known to the general public, and because of the vagueness with which it is presented (making no claims about the age of the earth, for example), it appears less marginal than creation science. Later, as a community that is debating evolution education becomes familiar with ID and its scientific sterility and dubious constitutionality, the antievolutionists often retreat to a fallback position of "evidence against evolution," usually portraying it as a compromise that should be acceptable to all. Morphing of this sort occurred during the well-publicized Kansas state science standards controversy of 1999.

CASE STUDY: KANSAS

As a result of education reform movements that began in the 1980s, most American states have established educational standards for history, mathematics, science, and other academic subjects. After the federal No Child Left Behind Act of 2001 mandated periodic high-stakes tests for students, state standards took on even more importance. Test topics would be based on the standards; if evolution were included in the standards, then it would be included in the tests. States that did not already have standards or frameworks began to develop them in the 1990s; the vast majority of them include evolution, though in varying degrees of competency (Lerner 2000). Many states that had ignored evolution previously now require that the topic be taught. A good deal of the antievolution activity of the past few years has been directed at the establishment of state science education standards: creationists oppose the inclusion of evolution in the standards, or try to get ID added, or try to get "evidence against evolution" included. Much as the return of evolution to the curriculum in the mid-1960s generated a creationist backlash, so also has the standards-based requirement that evolution be taught generated its own backlash.

Kansas undertook to write science education standards in 1999. A 27-person committee of master teachers and scientists was appointed by the Kansas Department of Education and the Kansas Board of Education to draft standards that thereafter went out for public comment, were revised, and were ultimately submitted to the board of education for approval. This procedure was prolonged, contentious, and bitterly fought. A conservative bloc of five school-board members had come to power in the 1996 state school-board election and clashed with the five moderate school-board members over many issues. When the science education standards came up for consideration, it was an easy prediction that evolution would be a contentious issue.

Indeed, conservatives on the board let it be known rather early in the process that they did not look kindly on the inclusion of evolution in the first draft of the standards (Cunningham 1999).

During the spring and summer of 1999 Kansas newspapers were full of letters to the editors arguing for and against evolution in the science standards. Inclusion of evolution was important: the standards would be used to develop questions for the state graduation exams. If evolution was not included in those exams, it was a safe bet that it would not be taught in the classrooms. Teachers and scientists therefore were very concerned that evolution have a prominent place in the science education standards. The draft of the standards that was submitted for public comment in April 1999 paralleled the National Science Education Standards (NSES), produced by the National Academy of Sciences, and the American Association for the Advancement of Science (AAAS) Benchmarks for Science Literacy, two documents that have been influential in shaping science standards across the country. Both the NSES and the Benchmarks treat evolution as an important scientific idea and include it throughout the science curriculum.

In midsummer 1999 school-board member Steve Abrams countered the standards committee's draft with a replacement standards document that had largely been written by Tom Willis, the head of the Creation Science Association for Mid-America, a young-earth creationist organization operating out of St. Louis. Abrams's standards included several provisions that any scientist or capable teacher would consider highly questionable and of course included much about how evolution was unreliable science. After a summer of public hearings often marked by bitter controversy, the board took the final draft of the standards committee and the Abrams standards and produced a compromise draft that it accepted as the state standards in August 1999. This draft compromised between those who wanted evolution to be taught and those who wanted the Abrams draft by removing evolution and related topics such as the big bang, continental drift, and radiometric dating from the standards and adding some language from Abrams's draft about the nature of science.

A loud hue and cry ensued, not only in Kansas, but also in national newspapers, television, and radio. Kansas became the butt of jokes on late-night television talk shows, and editorial cartoonists had a field day with the Kansas *Wizard of Oz* iconography of Dorothy and Toto. The ridicule that Kansas suffered at the hands of the national press stung sufficiently that in the next school-board election the conservatives lost three seats. The moderate majority swiftly abandoned the compromise standards and on a 7–3 vote adopted the final draft produced by the standards-writing committee when the new board took office in February 2001.

The Kansas controversy illustrates quite clearly the first of the pillars of creationism, the claim that evolution is weak or inadequate scientific theory. Probably because the most active antievolutionists in Kansas were young-earth creationists, and because creation science had no chance of being included in the science standards, the third, or fairness, pillar did not enter into the controversy. However, time marches on, and school boards change composition: although in the 2000 election moderates prevailed, in 2002 the board once again was deadlocked with a 5–5 ratio of conservatives to moderates. The science education standards were left alone during that term, but the presence of a 6–4 conservative majority after the 2004 elections meant that the science standards revision that took place in 2005 was highly contentious. Creationist-oriented school board members supported standards that did not directly call for the teaching of creation science or ID, but for the teaching of ID-inspired criticisms of evolution. The standards were passed in November of 2005 ("Antievolution Standards Adopted in Kansas" 2005) and were widely criticized by local teachers and scientists and national science and educational organizations, including the AAAS and the National Academy of Sciences ("Exposing the Flaws of the Kansas Science Standards" 2006). The next year, however, was an election year, and two conservative board members lost their seats ("The Pendulum Swings in Kansas" 2006). In early 2007 the reconstituted, more moderate school board rescinded the 2005 standards and adopted standards that neither denigrated evolution nor included some form of creationism ("Evolution Returns to Kansas" 2007).

Kansas is also an example of the morphing of young-earth creationists into ID creationists. A number of antievolutionists who actively opposed the 1999 Kansas standards have cast themselves as ID supporters. John Calvert is a good example: he organized the Intelligent Design Network (IDNet), an organization that has sponsored annual conferences promoting ID in Kansas and in some other states. Calvert has become a circuit rider for ID, traveling to Ohio, New Mexico, Georgia, West Virginia, and Montana to promote it. The Discovery Institute, however, has come to discourage the promotion of ID in the classroom in favor of teaching "evidence against evolution." The Discovery Institute has promoted this approach since the 2002 Ohio science education standards controversy, during the contentious 2005 revision of the Kansas standards, and elsewhere.

CASE STUDY: ROSEVILLE, CALIFORNIA

Although state board of education controversies and antievolutionism legislation tend to generate more press coverage, the majority of antievolutionism incidents occur at the local school-board level. A controversy in Roseville, California, similarly illustrates the pillars of creationism, as well as the morphing of YEC into ID into evidence against evolution.

Roseville is a community of approximately 92,000 located 20 miles northeast of Sacramento in north central California. A substantial conservative Christian community there tends to elect school-board members who share their views. There have been many "culture-war" controversies in Roseville over the past several years, including sex education as well as evolution. The most recent evolution controversy began in the summer of 2003 when high-school biology textbooks were being reviewed for adoption. A local citizen, Larry Caldwell, complained that the textbooks, all standard commercial publisher offerings, presented a one-sided treatment of evolution and did not provide any "scientific alternatives." According to a newspaper account, Caldwell first proposed that ID be included in the curriculum and in the textbooks (Rosen 2003a). Rather quickly, however, Caldwell changed his proposal to promote teaching "ideas that counter evolution" (Rosen 2003b). He recommended a videotape promoted by the Discovery Institute based on Jonathan Wells's book *Icons of Evolution* (2002) plus another ID-promoting video, *Unlocking the Mystery of Life* (2002). Another citizen submitted a YEC diatribe by Jonathan Sarfati at Answers in Genesis called *Refuting Evolution* (1999). A ubiquitous creation science book published by the Jehovah's Witnesses, *Life: How Did It Get Here? By Evolution or Creation?* (Watchtower Bible and Tract Society 1985) was also proposed.

Instructional materials proposed for classroom use in Roseville, as in most communities, must be reviewed by a committee of teachers and administrators. To give their opinion more credibility, the teachers asked scientists from local universities to review Caldwell's suggested materials. The committee reported to the school board that the materials were unsuitable for the Roseville curriculum and, reflecting the scientists' evaluations, substandard science. In September 2003 the board, stymied by the united front presented by the evaluation committee, voted to let the individual schools decide whether or not to use the materials. Teachers considered this an end run that might possibly result in the use of the books in the classroom, even though they had been rejected by the evaluation committee.

Creationists on the board also recommended that school libraries set up evolution resource centers that would have books, videos, and other materials criticizing evolution. The board members apparently reasoned that if teachers could not be compelled to teach "evidence against evolution," at least students could go to the library to get antievolution information. Librarians protested that they had little enough space in their libraries for materials that support the curriculum—the limited purpose of a school library; they did not want to waste money or space on what they considered a superfluous issue unrelated to the needs of teachers for curricular support. The evolution resource centers were never established.

Caldwell next attempted to have the board pass a policy he drafted, which he called Quality Science Education (QSE). It would have required the teaching of "strengths and weaknesses" of evolution—a choice of wording similar to policies that have appeared in some other communities. Controversy over QSE continued over the winter and culminated in a contentious meeting on May 10, 2004, in which community members debated the policy for over two hours. A petition signed by 28 of the 32 science teachers in the district opposing the policy was submitted to the board. In addition, the majority of people who testified at the school-board meeting opposed QSE.

In June 2004 the school board voted 3–2 against QSE, and the issue of creationism in Roseville seemed to have been decided in the negative. Perhaps because of complaints from parents that the board was wasting time arguing over creationism when important issues of funding, overcrowding, and other bread-and-butter concerns were not being addressed, the issue was not brought up again, and in the November 2004 election more moderate candidates prevailed, putting an end to the current wave of antievolutionism in Roseville. Some teachers, however, report that it is merely a matter of time until the issue of creationism and "evidence against evolution" rises in Roseville again, at least as long as religiously conservative school-board members keep getting elected. In January 2005 Larry Caldwell sued the district and some officials on the grounds that his constitutional rights were infringed during the controversy in Roseville. Defending themselves, the district claimed on the contrary that Caldwell's proposals had received a full hearing, and, in fact, an undue amount of time and energy was spent on Caldwell's issues. A judge sided with them in November 2007, and Caldwell lost his suit (Rosenhall 2007).

MORPHING OF CREATIONISMS

Kansas and Roseville illustrate the morphing phenomenon that has been taking place for several years. Community members who are identified with or hold YEC positions recast themselves as ID advocates. In truth, this is not a lengthy journey, because all young-earth creationists are supporters of the design view, although not all ID supporters are young-earth creationists. In Roseville, for example, both YEC and ID resources were offered to the school district. The legal advantage of being identified with a position, ID, that has not already been struck down by the courts is clear. Yet ID is not without its weaknesses: the phrase *intelligent design* carries the implication of a designer, an agent, and it is no secret that this agent is the Christian God. Perhaps for this reason, the Discovery Institute activists, from approximately the last half of 2001, have been arguing not for the inclusion of ID in the curriculum but for the inclusion of "evidence against evolution" (or "strengths and weaknesses of evolution," "critical analysis of evolution," or simply "teach the controversy"). Teaching ID is more likely to run afoul of the establishment clause than teaching "evidence against evolution," although the net effect of encouraging students to accept special-creation theology is common to either approach.

Paralleling the ICR's dictum that "evidence against evolution is evidence for creation science," William Dembski has written that "limitations on evolvability by material mechanisms constitute evidence for design" (Dembski 2002). Teaching students that evolution is weak or unsupported science that cannot explain the diversity of life encourages the acceptance of creationism: if evolution does not explain living things, by default, creationism does.

Such morphing of YEC to ID to "evidence against evolution" has occurred in a number of creationism/evolution controversies over the past few years, including controversies over the writing of science education standards in Kansas, Ohio, West Virginia, New Mexico, and Minnesota and in school-district controversies in Darby, Montana, and Dover, Pennsylvania. It is clear that this is where the creationism/evolution controversy is headed: away from the more obvious expressions of religion or creation science or Intelligent Design and toward the less legally vulnerable strategy of teaching "evidence against evolution."

PREDICTIONS FOR THE (NEAR) FUTURE

Prediction is uncertain, especially about the future, as Yogi Berra is supposed to have said, but some recent events may reshape the future of the creationism/evolution controversy in the United States. In January 2005 federal district-court judge Clarence Cooper in *Selman v. Cobb County School District* struck down a textbook disclaimer ordered inserted into biology books by the Cobb County, Georgia, Board of Education. The disclaimer read: "This textbook contains material on evolution. Evolution is a theory, not a fact, regarding the origin of living things. This material should be approached with an open mind, studied carefully, and critically considered."

Parents in the community had sued the district and the board of education in 2002, but the case took a long time to come to trial. The plaintiffs argued that the school board had a religious purpose and effect (of promoting sectarian Christianity) in requiring that the disclaimer be placed in textbooks, and that furthermore, under Georgia law, the school board's action was unconstitutional because it required the expenditure of state funds to promote religion. The judge struck down the practice because it indeed had the effect of promoting religion. In the decision the judge took notice of law-review and other scholarly articles, as well as testimony that pointed out that the "theory, not fact" language was "one of the latest strategies to dilute evolution instruction employed by antievolutionists with religious motivations" (*Selman*, 1308). Singling out evolution from all scientific theories for special negative treatment would have the effect of promoting the favored alternative to evolution, which is creationism. The district appealed the case. The Fifth Circuit Court of Appeals vacated the ruling and sent the case back to the District judge for reconsideration or retrial. The grounds for returning the case to the lower court did not concern the judge's ruling per se, but rather a messy court record and missing evidence. In discussions among the judge and the two legal teams, it was agreed to retry the case. It began to look unpromising for the defense when the plaintiffs' legal team received permission to reopen discovery and bring in expert witnesses, which strengthened their case before a judge who already had ruled in their favor. The plaintiffs' legal team was also buttressed by the addition of some of the attorneys and witnesses who had successfully defeated a school board policy requiring ID in Dover, Pennsylvania; perhaps the combination of factors was sufficient to encourage the district to settle out of court. The settlement was very favorable to the plaintiffs, however, with the district agreeing never again to disclaim evolution, whether with textbook stickers, or orally, or in any other form. The district also was required to follow the Georgia state standards, which currently require the teaching of evolution, and to refrain from excising or redacting materials on evolution from instructional materials. This latter constraint was stimulated by an earlier practice in the district in which the pages on evolution in textbooks routinely were cut out to prevent students from reading them.

Had the Cobb County district lost on appeal, other "theory, not fact" policies and disclaimers would have been encouraged, and I predict we would have seen an abundance. Disclaimers are popular with school boards because they provide the appearance that the school board is "doing something" about the teaching of evolution without requiring curriculum change or the bother and expense of review and purchase of new instructional materials. They satisfy a conservative Christian constituency without spending much time or money.

In December 2004, a month after the *Selman* case was tried, a lawsuit was filed by parents of children in Dover, Pennsylvania, protesting a policy of its board of education that would have required the teaching of ID and a variant of "evidence against evolution." The Dover resolution, which later was inserted into the curriculum, read: "Students will be made aware of gaps/problems in Darwin's Theory and of other theories of evolution including, but not limited to ID. The Origins of Life is not taught." In November 2004, amid community rumors of a possible lawsuit, the Dover Board of Education issued a "clarification" that stated that the ID textbook *Of Pandas and People* would be made available to students as a reference for the topic of ID. Confusingly, the clarification also stated that teachers were not to teach creationism or Intelligent Design.

In their complaint plaintiffs argued that ID is a religious concept indistinguishable from creationism, and that its teaching would violate the establishment clause. The trial took place over a six-week period in the fall of 2005. The Discovery Institute publicly distanced itself from the lawsuit, stating that "although we think discussion of intelligent design should not be prohibited, we don't think intelligent design should be required in public schools" ("Leading Intelligent Design Think Tank" 2004). Although originally a number of Fellows of the Discovery Institute agreed to testify as expert witnesses on behalf of the defense, all but two withdrew from the case, but not until it was too late to add new expert witnesses. But the two who remained, Michael Behe and Scott Minnich, were the ID proponents with the best scientific credentials, and thus the most qualified to defend the scientific credibility of ID.

The key legal issue in *Kitzmiller v. Dover Area School District* was whether there was a valid secular reason to teach ID. The obvious religious connotations of ID would not have been sufficient grounds to strike down the policy had there been a valid secular purpose and effect for teaching it. The defense, then, had to demonstrate that ID was a valid science, and that teaching it as an alternative to evolution would provide students with an excellent critical thinking exercise.

The plaintiffs had to convince the judge that ID was *not* valid science, and therefore the only reason to teach it was unconstitutionally to promote a religious idea. The plaintiffs contended that ID did not meet important definitional characteristics of science (such as being restricted to natural causes and thus testable) and that its fact claims were incorrect. They proposed that because ID was not a science and its fact claims were refuted, there was no pedagogical reason to teach it in a science class.

At the end of a long trial, the plaintiffs were rewarded with a decision that did not merely strike down the Dover policy, but that also weighed in on the question of whether ID was scientific. There was plenty of evidence showing that the school board unconstitutionally intended to promote a sectarian religious view, but Judge John E. Jones III additionally considered the arguments for and against the claim that ID was valid science. Judge Jones wrote that he did so "in the hope that it may prevent the obvious waste of judicial and other resources which would be occasioned by a subsequent trial involving the precise question which is before us" (*Kitzmiller*, 735).

Although ID supporters from the Discovery Institute (DeWolf et al. 2006) criticized the *Kitzmiller* decision as unnecessarily (and in their view, incorrectly) declaring ID was not a valid science, the legal theories of both sides required the judge to rule on precisely this point: if ID was valid science, there could be a constitutionally admissible secular reason for teaching it. The judge's decision on whether ID was science was relatively straightforward: witnesses for both sides testified that by the standard definition of science as widely used by scientists today, ID did not qualify as science. The difference was that the defense wanted to change the definition of science to include ID, but the judge disagreed that such a decision should be made in the high-school classroom.

The witnesses for the defense did not fare any better in their attempt to persuade the judge of the validity of the claims of ID (such as the supposed irreducible complexity of the bacterial flagellum and the unevolveability of the blood clotting mechanism). In the end, the plaintiffs got everything they wanted and more: a decision not only striking down the Dover Area School Board decision, but a decision evaluating ID as not qualifying as science—which will be cited in any future attempt to require the teaching of ID.

If the Dover Area School Board had won the case, school districts around the country would have been encouraged to impose similar ID policies, regardless of the exhortations of the Discovery Institute not to require the teaching of ID by statute. Intelligent Design is widely—and accurately—viewed as a form of creationism, and a legally successful policy would encourage the bringing of creationism into the classroom, albeit under a different name. However strong a decision, the *Kitzmiller* ruling was not appealed and therefore is precedent only in the Middle District of Pennsylvania. ID proponents might attempt to find another venue in which to test the legality of an ID policy, with a different set of facts and a different judge. However, because of the breadth of the *Kitzmiller* decision and the wealth of material in the trial record, it is doubtful that another trial would have a substantially different outcome.

The most popular antievolutionist strategy in the future, though, will be directives from school boards or state boards of education for teachers to "balance" evolution with the teaching of "evidence against evolution." It is not clear whether judges will be persuaded, as was Judge Cooper in the *Selman* case, that "evidence against evolution" is a code phrase for creationism; this requires an appreciation of the history of the creationism/evolution controversy. Because creationists believe and loudly proclaim that there are only two alternatives, creationism and evolution, they encourage teachers to teach that evolution cannot explain the diversity of living things. With evolution out of the picture, they believe that students will conclude that living things were specially created. During 2008 there was a flurry of "Academic Freedom Acts" introduced into several state legislatures (Florida, Missouri, Alabama, Michigan, and South Carolina) that would permit and encourage teachers to teach that evolution is an erroneous idea. Boilerplate for such "Academic Freedom Acts" is provided by the Discovery Institute on a website (Academic Freedom Petition 2008). Rather than *directing* teachers to teach the textbook version of evolution and then criticize this view, such bills proscribe actions by school districts that might penalize the teacher for doing so. Rather than overtly proposing that creationism be taught, the focus—or misdirection—is on the academic freedom of teachers to teach "strengths and weaknesses" of evolution (as in the South Carolina bill, SB 1386 ["Antievolution Legislation in South Carolina" 2008]) or the "full range of scientific views" (as in Florida's SB 2692 ["Antievolution Bills Dead in Florida" 2008]). But what such bills propose is to protect teachers from consequences of violating the First Amendment. Whether such "get out of jail free" cards would be constitutional remains to be tested, but at the time of this writing (July 2008) none of these bills have passed.

Will judges always strike down such policies? Legal experts noted that by the end of the first term of President George W. Bush, "all of the 13 circuits will have a majority of Republican judges" (Carp et al. 2004, 25). As Bush's second term comes to an end, "President Bush has named 294 judges to the federal courts, giving Republican appointees a solid majority of the seats, including a 60%-to-40% edge over Democrats on the influential U.S. appeals courts" (Savage 2008, A11). An analysis of decisions of judges appointed by the last eight presidents shows that President Bush's first-term appointees are particularly conservative in cases involving civil liberties, among which issues of separation of church and state are found. "Only 28 percent of the Bush cohort voted on the liberal side of issues pertaining to Bill of Rights and civil rights matters, thus giving the president the lowest score of any modern chief executive" (Carp et al. 2004, 26). It is not unfair to conclude that the legal climate favoring the establishment clause over the free exercise clause in educational affairs has weakened. It may be that the legal safeguards that have until now discouraged the inclusion of various forms of creationism in the public school science classroom will be more difficult to muster by the scientists and teachers and parents who support evolution education.

ACKNOWLEDGMENTS

I thank Glenn Branch and Wesley Elsberry for helpful comments on and corrections to the manuscript, which substantially improved it.

NOTES

1. Creation science had earlier been typified by a "contrived dualism," in the words of Judge William Overton in *McLean v. Arkansas,* which assumed that there were only two alternatives: evolution or special creation. With only two alternatives, creation science advocates argued that evidence against evolution was ipso facto evidence for creationism; ID likewise relies on disproving evolution to support the alternative of creation by an intelligent agent.

2. In keeping with the big-tent strategy, ID also welcomes those who see room for divine action in the realm of cosmology; discussions of the anthropic principle and fine-tuning arguments are not uncommon in their publications. It is beyond the scope of this discussion to consider these; see Manson 2003 for representation pro and con and Stenger 2004 for an extended argument against.

3. Examples include Dembski 1998b, 1999; Johnson 2002; Moreland 1994; Hunter 2001; and Wiker 2002.

4. In truth, origin science reduces to those issues in the historical sciences of astronomy, geology, biology, and anthropology that have implications for Christian theology: the big bang, the origin of life, the origin of the "kinds" of animals, and the origin of humans.

BIBLIOGRAPHY

Academic Freedom Petition. 2008. Discovery Institute. http://www
.academicfreedompetition.com/freedom.php (accessed May 22, 2008).

Alters, B. J., and C. E. Nelson. 2002. Perspective: Teaching evolution in higher education. *Evolution* 56, no. 10: 1891–1901.

Antievolution bills dead in Florida. 2008. National Center for Science Education, May 3. http://www.ncseweb.org/resources/news/2008/FL/739_antievolution _bills_dead_in_fl_5_3_2008.asp (accessed May 23, 2008).

Antievolution legislation in South Carolina. 2008. National Center for Science Education, May 15. http://www.ncseweb.org/resources/news/2008/SC/ 535_antievolution_legislation_in_s_5_15_2008.asp (accessed May 22, 2008).

Antievolution standards adopted in Kansas. 2005. National Center for Science Education, November 17. http://www.ncseweb.org/resources/news/2005/KS/326 _antievolution_standards_adopte_11_10_2005.asp (accessed May 15, 2008).

Behe, M. 1996. *Darwin's Black Box: The Biochemical Challenge to Evolution.* New York: Free Press.

———. 2000. [untitled article]. *Online Science Magazine,* http://www.sciencemag .org/cgi/eletters/288/5467/813#165 (accessed July 7, 2000).

———. 2007. *The Edge of Evolution: The Search for the Limits of Darwinism.* New York: Free Press.

Blackstone, N. W. 1997. Argumentum ad ignorantiam: A review of *Darwin's Black Box: The Biochemical Challenge to Evolution. Quarterly Review of Biology* 72, no. 4: 445–447.

Cairns-Smith, A. G. 1985. *Seven Clues to the Origin of Life.* Cambridge: Cambridge University Press.

Carp, R. A., K. L. Manning, and R. Stidham. 2004. The decision-making behavior of George W. Bush's judicial appointments. *Judicature* 88, no. 1: 20–28.

Cavalier-Smith, T. 1997. The blind biochemist. *Trends in Ecology and Evolution* 12: 162–163.

Coyne, J. A. 1996. God in the details. *Nature* 383: 227–228.

Creation museum unexpected success. 2008. *Florida Baptist Witness,* April 17. http://www.floridabaptistwitness.com/8688.article (accessed May 15, 2008).

Cunningham, D. L. 1999. Creationist tornado rips evolution out of Kansas science standards. *Reports of the National Center for Science Education* 19, no. 4: 10–14.

Davis, P. W., and D. H. Kenyon. 1993. *Of Pandas and People.* 2nd ed. Dallas, TX: Haughton.

Dembski, W. 1995. What every theologian should know about creation, evolution, and design. *Center for Interdisciplinary Studies Transactions* 3: 1–8.

———. 1998a. *The Design Inference: Eliminating Chance through Small Probabilities.* New York: Cambridge University Press.

———, ed. 1998b. *Mere Creation: Science, Faith and Intelligent Design.* Downers Grove, IL: InterVarsity Press.

———. 1999. *Intelligent Design: The Bridge between Science and Theology.* Downers Grove, IL: InterVarsity Press.

———. 2002. Becoming a disciplined science: Prospects, pitfalls, and reality check for ID. http://www.designinference.com/documents/2002.10.27.Disciplined _Science.htm (accessed January 15, 2005).

DeWolf, D. K., J. G. West, C. Luskin, and J. Witt. 2006. *Traipsing into Evolution: Intelligent Design and the* Kitzmiller v. Dover *Decision.* Seattle: Discovery Institute.

Dorit, R. 1997. Molecular evolution and scientific inquiry, misperceived. *American Scientist* 85 (September–October): 474–475.

Elsberry, W., and J. Shallit. 2003. Information theory, evolutionary computation, and Dembski's "complex specified information." November 23. http://www .antievolution.org/people/wre/papers/eandsdembski.pdf (accessed January 15, 2005).

Eve, R., and F. B. Harrold. 1991. *The Creationist Movement in Modern America.* Boston: Twayne Publishers.

Evolution returns to Kansas. 2007. National Center for Science Education, February 14. http://www.ncseweb.org/resources/news/2007/KS/286_evolution_returns_to _kansas_2_14_2007.asp (accessed May 15, 2008).

Exposing the flaws of the Kansas science standards. 2006. National Center for Science Education, July 19. http://www.ncseweb.org/resources/news/2006/KS/ 172_exposing_the_flaws_of_the_kans_7_19_2006.asp (accessed May 15, 2008).

Fitelson, B., C. Stephens, and E. Sober. 1999. How not to detect design: A review of William A. Dembski's *The Design Inference. Philosophy of Science* 66: 472–488.

Forrest, B., and P. Gross. 2004. *Creationism's Trojan Horse.* New York: Oxford University Press.

Gilchrist, G. W. 1997. The elusive scientific basis of Intelligent Design Theory. *Reports of the National Center for Science Education* 17, no. 3: 14–15.

Gishlick, A. D. 2003. Icons of evolution? Why much of what Jonathan Wells writes about evolution is wrong. National Center for Science Education. http://www .ncseweb.org/icons/ (accessed January 15, 2005).

———. 2004. Evolutionary paths to irreducibly complex systems. In M. Young and T. Edis, eds., *Why Intelligent Design Fails: A Scientific Critique of the New Creationism,* 58–71. New Brunswick, NJ: Rutgers University Press.

Gould, S. J. 1992. Impeaching a self-appointed judge. *Scientific American,* July: 118–121.

Hunter, C. G. 2001. *Darwin's God: Evolution and the Problem of Evil.* Grand Rapids, MI: Brazos Press.

Icons of Evolution. 2002. B. Boorujy, director. DVD. 75 minutes. N.p.: Randolph Productions.

Institute for Creation Research (ICR). 1987. The Supreme Court decision and its meaning. *ICR Impact* no. 170 (August). http://www.icr.org/pubs/imp/imp-170.htm.

Johnson, P. E. 1991. *Darwin on Trial.* Washington, DC: Regnery Gateway.

———. 1997. *Defeating Darwinism by Opening Minds.* Downers Grove, IL: InterVarsity Press.

———. 2002. *The Right Questions: Truth, Meaning and Public Debate.* Downers Grove, IL: InterVarsity Press.

Kent Hovind sentenced to 10 years. 2007. National Center for Science Education, January 24. http://www.ncseweb.org/resources/news/2007/FL/789_kent_hovind _sentenced_to_ten_y_1_24_2007.asp (accessed May 22, 2008).

Kushiner, J. M. 2000. Berkeley's radical. *Touchstone* 15, no. 5: 40.

Larson, E. J. 2003. *Trial and Error: The American Controversy over Creation and Evolution.* 3rd ed. New York: Oxford University Press.

Leading Intelligent Design think tank calls Dover evolution policy "misguided," calls for it to be withdrawn. 2004. Discovery Institute. http://www.discovery.org/ scripts/viewDB/index.php?command=view&program=CSC%20-%20Views %20and%20News&id=2341 (accessed December 14, 2004).

Lerner, L. S. 2000. *Good Science, Bad Science: Teaching Evolution in the States.* Washington, DC: Thomas B. Fordham Foundation.

Lynch, J. 2002. *Creationism and Scriptural Geology, 1817–1857.* London: Bristol Thoemmes Press.

Manson, N. A., ed. 2003. *God and Design: The Teleological Argument and Modern Science.* London: Routledge.

Matzke, N. 2004. Icon of obfuscation. TalkOrigins Archive, updated January 23, 2004. http://www.talkorigins.org/faqs/wells/iconob.html (accessed January 14, 2005).

McMurtrie, B. 2001. Darwin under attack. *Chronicle of Higher Education,* December 21, 8.

Miller, K. B., ed. 2003. *Perspectives on an Evolving Creation.* Grand Rapids, MI: Eerdmans.

Miller, K. R. 1999. *Finding Darwin's God.* New York: HarperCollins.

Moreland, J. P., ed. 1994. *The Creation Hypothesis.* Downers Grove, IL: InterVarsity Press.

Morris, H. R. 1980. The tenets of creationism. *Impact,* July: 1–4.

Numbers, R. 1992. *The Creationists.* New York: Knopf.

Orr, H. A. 1996. Darwin vs. Intelligent Design (again). *Boston Review,* 28–31.

The pendulum swings in Kansas. 2006. National Center for Science Education, August 2. http://www.ncseweb.org/resources/news/2006/KS/395_the_pendulum _swings_in_kansas_8_2_2006.asp (accessed May 15, 2008).

Pennock, R. T. 1999. *Tower of Babel: The Evidence against the New Creationism.* Cambridge, MA: MIT Press.

———, ed. 2001. *Intelligent Design Creationism and Its Critics.* Cambridge, MA: MIT Press.

———. 2003. Creationism and Intelligent Design. *Annual Review of Genomics and Human Genetics* 4: 143–163.

Rosen, L. 2003a. Darwin faces a new rival: A Roseville High School parent urges that "Intelligent Design" also be taught in biology. *Sacramento Bee,* June 22, 1, 2.

———. 2003b. Roseville sticks with evolution: School trustees OK a text that teaches Darwin but may add material disputing his theory. *Sacramento Bee,* July 3.

Rosenhall, L. 2007. Judge tosses out evolution lawsuit. *Sacramento Bee,* September 13, B2.

Rosenhouse, J. 2003. Probability, optimization theory, and evolution. *Evolution 56,* no. 8: 1721–1722.

Sarfati, J. 1999. *Refuting Evolution.* Green Forest, AR: Masterbooks.

Savage, D. G. 2008. Conservative courts likely Bush legacy: The president's success in getting judicial nominees confirmed gives the federal bench a decided GOP tilt. *Los Angeles Times,* January 2, A11.

Scott, E. C. 1989. New creationist book on the way. *Reports of the National Center for Science Education 9,* no. 2: 21.

———. 1997. Antievolutionism and creationism in the United States. *Annual Review of Anthropology 26:* 263–289.

———. 1998. "Science and religion," "Christian scholarship," and "theistic science": Some comparisons. *Reports of the National Center for Science Education 18,* no. 2: 30–32. http://www.ncseweb.org/resources/articles/6149 _science_and_religion_chris_3_1_1998.asp.

———. 2001. The big tent and the camel's nose. *Reports of the National Center for Science Education 21,* no. 2: 39–41.

———. 2004. *Creationism vs. Evolution: An Introduction.* Westport, CT: Greenwood Press.

Scott, E. C., and G. Branch. 2003. Antievolutionism: Changes and continuities. *Bioscience 53,* no. 3: 282–285.

Scott, E. C., and N. J. Matzke. 2007. Biological design in science classrooms. *Proceedings of the National Academy of Sciences 104* (Suppl. 1): 8669–8676.

Shanks, N. 1999. Redundant complexity: A critical analysis of Intelligent Design in biochemistry. *Philosophy of Science 66:* 268–298.

———. 2004. *God, the Devil, and Darwin: A Critique of Intelligent Design Theory.* Oxford: Oxford University Press.

Sober, E. 2002. Intelligent Design and probability reasoning. *International Journal for Philosophy of Religion 52,* no. 2: 65–80.

Stenger, V. 2004. Is the universe fine-tuned for us? In M. Young and T. Edis, eds., *Why Intelligent Design Fails: A Scientific Critique of the New Creationism,* 172–184. New Brunswick, NJ: Rutgers University Press.

Thaxton, C. B., W. L. Bradley, and R. L. Olsen. 1984. *The Mystery of Life's Origin: Reassessing Current Theories.* New York: Philosophical Library.

Thornhill, R. H., and D. Ussery. 2000. A classification of possible routes of Darwinian evolution. *Journal of Theoretical Biology 203:* 111–116.

Unlocking the Mystery of Life. 2002. L. Allen, director. DVD. 67 minutes. La Habra, CA: Illustra Media.

Watchtower Bible and Tract Society. 1985. *Life: How Did It Get Here? By Evolution or by Creation?* Brooklyn, NY: Watchtower Bible and Tract Society of New York.

Wein, R. 2002. Not a free lunch but a box of chocolates: A critique of William Dembski's book *No Free Lunch.* The TalkOrigins Archive (updated April 23, 2002). http://www.talkorigins.org/design/faqs/nfl/ (accessed January 24, 2005).

Whitcomb, J. C., and H. R. Morris. 1961. *The Genesis Flood: The Biblical Record and Its Scientific Implications.* Phillipsburg, NJ: Presbyterian and Reformed.

Wiker, B. 2002. *Moral Darwinism: How We Became Hedonists.* Downers Grove, IL: InterVarsity Press.

Wilkins, J. S., and W. R. Elsberry. 2001. The advantages of theft over toil: The design inference and arguing from ignorance. *Biology and Philosophy 16:* 711–724.

Young, M., and T. Edis, eds. 2004. *Why Intelligent Design Fails: A Scientific Critique of the New Creationism.* New Brunswick, NJ: Rutgers University Press.

COURT OPINIONS CITED

Daniel v. Waters. U.S. Court of Appeals, Sixth Circuit. 515 F.2d 485 (1975).
Edwards v. Aguillard. 482 U.S. 578 (1987).
Epperson v. Arkansas. 393 U.S. 97 (1968).
Kitzmiller v. Dover Area School District. 400 F. Supp. 2d 707 (M.D. Pa. 2005).
McLean v. Arkansas Board of Education. 529 F. Supp. (E.D. Ark. 1982)
Selman v. Cobb County School District. 390 F. Supp. 2d (N.D. Ga. 2005).

ALPHABETICAL GUIDE

A

Adaptation and Natural Selection (George C. Williams)

This book, published in 1966 by the American ichthyologist George C. Williams, is one of the great debunking works in the history of science. Although neo-Darwinism, or the synthetic theory of evolution, was the established paradigm in evolutionary studies, Williams (who was deeply committed to the theory itself) felt that it was infected by a number of fallacious impressions. Most significant of these was the belief, promulgated by the ecologist Vero C. Wynne-Edwards in *Animal Dispersion in Relation to Social Behaviour* (1962), that sometimes natural selection can favor the group over the individual. Although he did not want to deny the outright possibility of group selection, Williams felt that on grounds of simplicity one ought to always prefer the hypothesis that group effects are caused by individual effects and that selection therefore must benefit first the individual.

In making his case, Williams was one of the first to make reference to the groundbreaking papers of the then–graduate student William Hamilton (1964a, 1964b), who had shown how the sterility of hymenopteran workers (ants, bees, and wasps) could be explained as a result of what has become known as kin selection. The workers are improving their reproductive chances by raising fertile relatives instead of putting efforts into their own reproduction.

Another topic attacked with relish by Williams was that of evolutionary progress, then much favored by people like Julian Huxley (1942; see also Ruse 1996). Williams was withering in his critique of various notions of progress, pointing out that popular criteria of improvement (like complexity) are often very misleading. Apart from the fact that generally no one takes development into account (the liver flukes that affect sheep have an incredibly complex life history), it is by no means obvious that the more complex is better. A jet engine is clearly superior to a gasoline-fueled, motor-driven propeller engine, and yet in principle the former is less complex than the latter.

The main thrust of the book, however, is that natural selection is the main and generally sufficient explanation of evolutionary change. Hence particular scorn was reserved for those who would replace or otherwise supplement selection. C. H. Waddington (1957) had introduced the notion of genetic assimilation, where he was able to simulate Lamarckian-type effects in fruit flies. Williams had no objection to the empirical findings, but thought that inferences about the inadequacy of selection as such are simply not well taken. For him, and for his many enthusiastic readers, a stripped-down, pure Darwinism is not only aesthetically more attractive, it is scientifically superior—a point lost on the philosophy of biology community, which has spent a happy near-half century writing papers and books refuting key claims of Williams's work.

BIBLIOGRAPHY

Hamilton, W. D. 1964a. The genetical evolution of social behaviour I. *Journal of Theoretical Biology* 7: 1–16.
———. 1964b. The genetical evolution of social behaviour II. *Journal of Theoretical Biology* 7: 17–52.
Huxley, J. S. 1942. *Evolution: The Modern Synthesis*. London: Allen and Unwin.
Ruse, M. 1996. *Monad to Man: The Concept of Progress in Evolutionary Biology*. Cambridge, MA: Harvard University Press.
Sober, E., ed. [1984] 1993. *Conceptual Issues in Evolutionary Biology*. 2nd ed. Cambridge, MA: MIT Press.
Waddington, C. H. 1957. *The Strategy of the Genes*. London: Allen and Unwin.
Williams, G. C. 1966. *Adaptation and Natural Selection*. Princeton, NJ: Princeton University Press.
Wynne-Edwards, V. C. 1962. *Animal Dispersion in Relation to Social Behaviour*. Edinburgh: Oliver and Boyd.   —M.R.

Agassiz, Jean Louis Rodolphe (1807–1873)

Born in Switzerland in 1807, Louis Agassiz as a boy was a keen naturalist. He studied biology at several German universities, earning an advanced degree in natural history in 1829. He soon earned a reputation as a brilliant researcher, thanks to books describing species of fish, living and fossil, and to his careful studies of Alpine glaciers (Agassiz 1833–1843, 1840) (see figure). Beginning in 1837 he assembled evidence for a surprising new theory about an Ice Age, when a thick layer of ice had covered northern Europe. After moving to the United States in 1846, he became one of America's leading scientists, founding the Museum of Comparative Zoology at Harvard in 1859.

A charismatic lecturer, Agassiz charmed the public as well as his students with descriptions of recent discoveries in embryology, paleontology, and anatomy. Like Richard Owen in London, Agassiz emphasized the pattern of resemblances called homology. Although it had long been known that the structure of an animal's organs is related to their function, not until the nineteenth century did biologists confront the fact that diverse creatures seem to be modifications of a common blueprint or type. Vertebrates are one such type, because the same layout of bones and other organs can be traced in a fish, a frog, a bat, and a human. Agassiz impressed his audiences with unfamiliar examples of the same principle: that bees, spiders, and lobsters are also homologous to one another, as are octopuses, snails, and clams, or starfish, jellyfish, sea urchins, and coral polyps. Following Georges Cuvier, he called these four groups vertebrates, articulates, mollusks, and radiates. In his *Essay*

Louis Agassiz was a master at reconstructing fossil fish. This particular specimen, *Cyclopoma spinosum*, was giving him difficulty and then he had a dream in which he saw the full form. When he eventually compared the sketch that he had made from memory with the actual embedded specimen, the two corresponded exactly. It is described in Agassiz 1833–1843, vol. 4, *Cténöides*, 20–21.

---

on Classification (1859) Agassiz argued that homologies, which extend across embryonic forms and back through the fossil record, are evidence of the existence of "One Supreme Intelligence" who had conceived these resemblances before he created them. This argument is quite distinct from that of William Paley, whose *Natural Theology* (1802) compared God to a watchmaker or engineer who builds machines that are cleverly adapted to their purpose. Agassiz's God is like a musician or architect who enjoys playing several variations on a few basic themes.

Agassiz argued that each species was the product of a distinct act of creation and cannot change, but he understood geological time and firmly rejected the biblical story of creation. His concept of species was severely challenged when he encountered people of African descent, and his response does him no credit. He insisted that human races are actually distinct species, an opinion welcomed by slave owners.

Darwin's *On the Origin of Species* was a serious threat to Agassiz's interpretation of nature. Pointing to the same phenomena cited by Agassiz, Darwin argued that homology and biogeography are evidence of descent from a common ancestor, while the adaptedness of form to function is evidence for natural selection. Before long Agassiz's students and most of his colleagues accepted evolution, if not natural selection, but Agassiz never did. Traveling to Brazil in 1865 and to the Galápagos Islands in 1872 did nothing to change his mind. He died in 1873.

BIBLIOGRAPHY

Agassiz, L. 1833–1843. *Recherches sur les Poissons Fossiles*. Neuchãtel: Imprimerie de Petitpierrre.

———. 1840. *Études sur les glaciers*. Neuchãtel: Jent et Gassmann.

———. 1859. *Essay on Classification*. London: Freethought Publishing Company/Longman, Brown, Green, Longmans, & Roberts, and Truebner & Co. Reprint (E. Lurie, ed.), Cambridge, MA: Harvard University Press, 1962. (Originally published in 1857 as Part I of L. Agassiz, *Contributions to the Natural History of the United States*. Boston: Little, Brown.)

Lurie, E. 1960. *Louis Agassiz: A Life in Science*. Chicago: University of Chicago Press.

Paley, W. [1802] 1819. *Natural Theology* (*Collected Works*, vol. 4). London: Rivington.

Winsor, M. P. 1991. *Reading the Shape of Nature: Comparative Zoology at the Agassiz Museum*. Chicago: University of Chicago Press.   —M.P.W.

Alexander, Richard D. (b. 1929)

Richard D. Alexander was trained as an entomologist at Ohio State University. Soon after receiving his PhD in 1954, he took a position at the University of Michigan Museum of Zoology, where he remained for the rest of his career (he became a professor emeritus in 2001). Early in his career he concentrated his research efforts on communication in the singing insects—the crickets, katydids, and cicadas. He was the first biologist to use insect songs to recognize species, and he made a number of methodological innovations for the study of insect songs. He described 426 new species and genera of insects and he described the longest and most complicated hybrid zone ever known in the true katydid, *Pterophylla camellifolia*. He also was a pioneer in the study of behavior and made important contributions to the understanding of how behavior evolves. He focused on insect songs, mating behavior, aggression, and territoriality in his studies of insect behavior. With a number of colleagues he developed important theoretical ideas about the evolution of eusociality in mammals, ideas that were confirmed with the discovery of eusociality in naked mole rats.

In the early 1970s, Alexander turned his attention to the evolution of human behavior. In his book *Darwinism and Human Affairs* (1979) he showed how culture can be seen as the product of social interaction among individual human beings who have evolved to maximize their inclusive fitness in ancestral environments. In this book he reviewed a large amount of anthropological data to clarify and support his theory. In *The Biology of Moral Systems* (1987) he presented the theory that the human sense of morality is a product of a history of intergroup competition in human groups, which favors any trait that helps human beings to form larger and better-united social groups. He suggested that a basic fact of human existence, as with all sexually reproducing organisms, is conflict of reproductive interest among individuals. As groups become larger and consist mostly of genetically unrelated people, preventing conflicts from breaking up the group becomes more difficult. Alexander hypothesized that morality as a form of indirect reciprocity helps to prevent this sort of conflict and allows the formation of larger, better-united groups. In connection with this idea, he expanded Nicholas Humphrey's (1983) idea that human intelligence has evolved because of the advantages it confers in complex social interactions. Alexander has developed well-thought-out theories about a number of human traits. These include relative hairlessness in adults, female orgasm, sexual dimorphism, ecological dominance and its implications for sociality, socially imposed monogamy, the combination of high paternity confidence and multi-male social groups, group-group competition as a form of play, the association of low paternity confidence and an emphasis on uterine kinship, and cross-cousin marriage, to mention a few. These theoretical ideas are the foundation of much current research by human behavioral ecologists.

Richard Alexander has received a number of important honors: the American Association for the Advancement of Science Award (1961), the Daniel Giraud Elliott Medal by the National Academy of Sciences (1971), the Darwin Award in Insect Behavioral Ecology by the Florida Entomological Society (1986), and Distinguished Animal Behaviorist by the Animal Behavior Society (2002). He became a member of the National Academy of Sciences in 1974, and has been elected a Fellow of the American Association for the Advancement of Science, the Ohio Academy of Sciences, and the Animal Behavior Society. He was elected president of the Human Behavior and Evolution Society in 1995 and received its Distinguished Scientific Award in 2007. He has published four books and over 130 scientific articles, and has trained 32 students, most of whom now are making important contributions to the evolutionary study of behavior.

BIBLIOGRAPHY

Alexander, R. D. 1979. *Darwinism and Human Affairs.* Seattle: University of
    Washington Press.
———. 1987. *The Biology of Moral Systems.* Hawthorne, NY: Aldine de Gruyter.
———. 1990. *How Did Humans Evolve? Reflections on the Uniquely Unique
    Species.* Ann Arbor: University of Michigan Museum of Zoology, Special
    Publication 1.
Humphrey, N. 1983. *Consciousness Regained: Chapters in the Development of
    Mind.* Oxford: Oxford University Press.
Otte, D., and R. D. Alexander. 1983. *The Australian Crickets (Orthoptera:
    Gryllidae).* Philadelphia: Academy of Natural Sciences of Philadelphia. —W.I.

Altruism

The existence of altruism in animals and humans has posed one of the most profound challenges in evolutionary biology, and the evolutionary explanations for altruism have influenced psychologists, sociologists, philosophers, and even economists. While evolutionary theory and empirical studies have illuminated the understanding of altruism, it remains an active subject in modern evolutionary biology and psychology.

The problem of altruism is divided into two separate but related issues: biological altruism and psychological altruism. In biological altruism, an organism performs an altruistic act when the act benefits another organism at a cost to itself. Benefits and costs are measured in terms of reproductive fitness. Such altruism is well documented in many species. Vervet monkeys and prairie dogs have sentries at the outskirts of colonies that alert members of the colony of approaching predators, often at the cost of becoming the targets of those predators. Psychological altruism, which is the colloquial sense of the term, is intentionally performing an action at one's own cost in order to benefit another. The benefits and costs for psychologically altruistic acts, however, need not have anything to do with reproductive fitness. An example of a psychologically altruistic act would be to plant a slow-growing tree, knowing full well that you will not benefit from its shade or fruit.

Although biological and psychological altruism are separate problems, theorists have offered biological underpinnings for psychological altruism. The problem of altruism, both biological and psychological, is at the center of grounding a theory of morality within biology. I will discuss the two kinds of altruism in turn.

Altruism seems to be a problem for evolutionary theory when looking at selection from an individual perspective (i.e., selection that occurs on the level of organisms). If altruistic organisms live among selfish organisms, the selfish organisms will always be more fit than their altruistic counterparts. Over time, altruistic individuals will be selected out in favor of selfish ones. For example, sentry monkeys lower their fitness by alerting others to the presence of a predator. Over time, selfish monkeys, ones that do not take their turn on watch, would produce more offspring and eventually eliminate any altruistic genes.

Charles Darwin, in *The Descent of Man* (1871), noticed this problem and hinted at a solution: "although a high standard of morality gives but a slight or no advantage to each individual man . . . an increase in the number of well-endowed men and advancement in the standard of morality will certainly give an immense advantage to the group" (p. 166). Darwin continues by suggesting that groups that contain more altruistic individuals will outcompete groups comprised mainly of selfish individuals. From this passage, it appears that Darwin was suggesting that selection can occur at the level of groups, a radical departure from his core argument about the importance of selection among individuals.

Evolutionary biologists have vacillated in their acceptance of this so-called group selection argument. Darwin's influence gave it great credence for decades. It fell from favor in the 1960s under the influence of George C. Williams's writings (e.g., 1966) about selection and adaptation because group selection for any attribute came to be considered a biological rarity. One of the main problems with the group selection argument is that competition within a group will always favor the selfish individuals, so that even if groups of altruists are more fit as a group than groups of selfish individuals, the groups of altruists will suffer subversion from within. Recently, however, group selection models have received a spirited defense. Sober and Wilson

(1998) argue that, under certain conditions, altruism can be promoted by selection. If the average fitness of groups predominated by altruists is greater than the average fitness of groups predominated by selfish individuals, the net fitness consequences of being an altruist may be positive. Groups can maintain a predominantly altruistic population and defend themselves against subversion from within if there are periodic migrations of individuals between groups. Although the mathematics for group selection have been worked out and the plausibility of the theory confirmed with experimental studies of flour beetles, it is less obvious that the requirements of the group selection model are met often in nature and thus that the model will accurately explain natural phenomena. Furthermore, group selection models cannot explain how altruists, when initially rare (as would happen through mutation in behavioral traits), avoid being eliminated by selfish individuals before they become numerous enough to create a group selective advantage.

A second reason for the eclipse of group selection explanations is the impact of William Hamilton's famous papers (1964a, 1964b) describing what has come to be known as kin selection. Hamilton argued that if altruistic behavior is genetically based, then organisms that were altruistic toward relatives, with whom they share genetic material, would promote the spread of altruistic genes through the benefits to the relatives that carry those genes. With this explanation, group selection was not needed to explain altruism. From a gene's-eye view it would be selectively neutral for a mother to sacrifice herself for the lives of two children because each child possesses one-half of her genes. But the benefit of saving three offspring more than compensates for the cost of self-sacrifice.

Kin selection appears commonly in the biological world (just think of mothers defending their young). One potential problem for the kin selection explanation is that it requires organisms to identify not only family members but larger familial relationships, because closer relatives are more valuable than distant ones. While many organisms can identify their kin through visual or olfactory cues, the kin selection argument works even if other cues are effective proxies for relatedness. A heuristic rule like "share with organisms that are close by" will work when family members tend to aggregate in space. Such aggregation can occur even in animals without advanced social systems if dispersal from the home site is limited.

But how can we explain altruistic behavior between organisms that are not related if group selection is not the answer? A proposed solution to this question was articulated by Robert Trivers (1971) and has come to be known as reciprocal altruism. According to Trivers, an organism behaves altruistically toward other organisms in the expectation that others will help it in return. This explanation does not presuppose a conscious decision or even mental states in animals. A standard example of reciprocal altruism is found in the behavior of vampire bats. Vampire bats can survive only a day or two without feeding; however, vampire bats that find prey can feed in excess of their individual needs. Vampire bats will often share more, and more often, with bats who have recently shared with them.

Reciprocal altruism has the advantage over kin selection in explaining altruistic behavior toward unrelated individuals. One question, however, is whether reciprocal altruism is really altruistic. It could be argued that the altruist is not paying a real cost but only delaying its reward for helping others (imagine a scenario where two businesses cooperate with each other in order to maximize their own utility). This problem and a series of more complicated facets of altruism, especially reciprocal altruism, have been explored extensively by theoreticians using game theory, and game theoretic treatments have become a fertile area for developing new ideas about altruism and making new predictions about behavior in nature. (Please see the alphabetical entry "Game theory" for a more thorough treatment of the issue.)

Biological altruism can be used as a foundation for studying psychological altruism. The major difference between them is that altruistic traits manifest themselves as psychological traits in humans, as opposed to being presented as behavioral traits in the nonhuman biological world. Darwin suggested that psychological altruism, or morality more generally, could be explained by the emotion of sympathy, which may have biological roots. Similar emphasis on the role of emotions in ethical behavior can be found in the work of the philosopher David Hume.

However, one need not have a biological explanation of psychological altruism. Morality, for example, may be a product of rationality and not biology. Furthermore, there are a number of questions about psychological altruism that are not biological but philosophical in nature. For example, does psychological altruism exist at all? Egoism, the view that all behaviors have selfish roots, is directly opposed to psychological altruism. In this view, individuals who feed the starving masses are selfish because it gives them pleasure and self-satisfaction to do so. However, Aristotle (among others) noted that simply because a feeling of satisfaction accompanies "altruistic" acts, it does not follow that the act was performed in order to achieve the desired feeling.

BIBLIOGRAPHY

Darwin, C. 1871. *The Descent of Man.* London: John Murray.

Hamilton, W. D. 1964a. The genetical evolution of social behaviour I. *Journal of Theoretical Biology* 7: 1–16.

———. 1964b. The genetical evolution of social behaviour II. *Journal of Theoretical Biology* 7: 17–52.

Rosenberg, A. 1992. Altruism: Theoretical contexts. In E. F. Keller and E. A. Lloyd, eds., *Keywords in Evolutionary Biology,* 19–28. Cambridge, MA: Harvard University Press.

Sober, E., and D. S. Wilson. 1998. *Unto Others: The Evolution and Psychology of Unselfish Behavior.* Cambridge, MA: Harvard University Press.

Trivers, R. L. 1971. The evolution of reciprocal altruism. *The Quarterly Review of Biology* 46: 35–57.

Williams, G. C. 1966. *Adaptation and Natural Selection.* Princeton, NJ: Princeton University Press.

—J.Z.

Alvarez, Walter (b. 1940)

Walter Alvarez, the geologist son of Nobel Prize–winning physicist Luis Alvarez and grandson of well-known California physician Walter C. Alvarez, was born October 3, 1940, in Berkeley, California. He earned his BA from Carleton College in 1962 and his PhD from Princeton in 1967 with a dissertation on the structure of the Andes Mountains. Alvarez worked in orogeny, archeological geology, volcanics, tectonics, and paleomagnetism before turning to the study of mass extinctions. He has spent the bulk of his career at the University of California at Berkeley and has received numerous awards and honors, including the Geological Society of America's Penrose Medal, and election to the National Academy of Sciences and the American Academy of Arts and Sciences. In 1985, the planetary scientists Gene and Carolyn Shoemaker named a minor planet for him. Alvarez is best known for his theory attributing the end-Cretaceous mass extinction to the impact of an extraterrestrial object. Widespread respect for how the theory has held up under extensive testing and correction has bolstered acceptance of other nongradual explanations for events in evolutionary history and promoted neocatastrophism.

Alvarez's signal contribution lies not in originating the idea of extraterrestrial causes for extinction—such suggestions go back hundreds of years. Rather, he gave the theory a logical structure and evidentiary base that has allowed earth and planetary scientists as well as biologists to test the many subsidiary hypotheses generated.

First published in 1980 in *Science,* the Alvarez argument grew out of anomalous data collected to test a different hypothesis (L. W. Alvarez et al. 1980; see also W. Alvarez et al. 1984). The paleontological community had long regarded the extinction of dinosaur lineages as a gradual process. Nevertheless, in geological time their demise appeared rather abrupt. This apparent rapidity could have been explained by unusually slow sedimentation rates at the end of the Cretaceous. If sedimentation was very slow, a thin layer of sediment would represent a large slice of time. To test this, Alvarez proposed measuring the amount of iridium present in the Cretaceous-Tertiary (K-T) boundary layer. Iridium, a very heavy element of the platinum group, is almost entirely absent in the earth's crust; the only source is the tiny but steady rain of cosmic dust from space. An increase in the presence of iridium in the K-T boundary sediments would indicate a slower than usual rate of earth sedimentation. Samples taken in 1979 from the exposed K-T layer near Gubbio, Italy, showed an iridium spike far in excess of anything attributable to slowed sedimentation. Accordingly, Alvarez investigated the possibility that an extraterrestrial impact had contributed the iridium.

Alvarez and his group, including Luis Alvarez and chemists Frank Asaro and Helen Michel, structured their argument in three parts. They mustered geological evidence that an asteroid or comet had indeed struck the earth at the end of the Cretaceous. Then, they assembled the case that numerous lineages of terrestrial and marine organisms had indeed become extinct at that time. Having established the two events and their synchrony, they elaborated the ecological implications—that a large impact had caused a dust cloud to encircle the earth, cutting off photosynthesis and making it cold and dark for a sufficiently long period so that virtually all land animals over 50 pounds died, along with some 75% of marine organisms—and drew the inference that the impact had caused the extinction of a large fraction of species 65 million years ago, including the extinction of dinosaurs.

Alvarez's framing of the issues provided a broad basis for analysis in diverse fields. His original article has been cited thousands of times in hundreds of scientific journals from biology and geology to medicine and nuclear physics. The original hypothesis has been strengthened by further discoveries, most dramatically by direct evidence of the impact provided by the Chicxulub crater and its unique chemical signature. Further, the theory has captured the popular imagination, increasing interest in scientific investigation and promoting understanding of the unity of science. Alvarez has himself written a popular account of the discovery of the extraterrestrial impact and its implications for life on earth in *T. Rex and the Crater of Doom* (1997).

BIBLIOGRAPHY

Alvarez, L. W., W. Alvarez, F. Asaro, and H. V. Michel. 1980. Extraterrestrial cause for the Cretaceous-Tertiary extinction. *Science* 208: 1095–1108.

Alvarez, W. 1997. *T. Rex and the Crater of Doom*. Princeton, NJ: Princeton University Press.

Alvarez, W., E. G. Kauffmann, F. Surlyk, L. W. Alvarez, F. Asaro, and H. V. Michel. 1984. Impact theory of mass extinctions and the invertebrate fossil record. *Science* 223: 1135–1141.   —P.M.P.

Amphibians

Limbed vertebrates arose more than 375 million years ago in the Devonian period and they are generally called amphibians. This was a very diverse assemblage of aquatic and terrestrial animals, ranging from the size of a salamander to larger than an alligator. Some of them even became limbless. But by the end of the Paleozoic period and the great Permian extinction, about 300 million years ago, most of the lineages were extinct. A few thrived in the early Mesozoic period before they, too, disappeared. Recent amphibians and the amniotes represent the living descendants of this early radiation. There are three very different kinds of living amphibians. Perhaps the most familiar and certainly the most numerous and widespread are the frogs, which are found on every continent except Antarctica and even on a few oceanic islands (see figure). The salamanders are well known to inhabitants of the north temperate zone, but only one lineage has achieved any success in the tropics. In contrast, the relatively unfamiliar and secretive caecilians are restricted to tropical regions.

Recent amphibians differ dramatically from each other in structure and way of life. All have moist skins, however, and respiration is largely cutaneous.

The eastern spadefoot toad, *Scaphiopus holbrookii*, is a common but little-noticed inhabitant of the southeastern United States. Spadefoot toads breed explosively in the spring in ephemeral ponds after heavy rains; the tadpoles grow rapidly and metamorphose in about three weeks. The juveniles and adults feed on insects and other small animals in the litter layer of forests and burrow into sandy soil to await the next opportunity to emerge and breed. This species is part of an old lineage of frogs with representatives in the southwestern United States, whose tadpoles can complete development in ephemeral pools in the desert in as little as 10 days.

Most salamanders, a few frogs, and one caecilian lack lungs entirely. Although amphibians are associated with moist habitats, many species never enter water.

Frogs (Anura), with more than 5,500 species, have very short bodies, no tail, and well developed limbs, especially long hind limbs that have four main subdivisions. Many frogs are capable of prodigious leaps. Carnivorous as adults, they catch prey with projectile tongues. Frogs display diverse life histories, but the most general and familiar is one in which the sexes congregate in ponds in the spring, with females selecting males as mates based on the qualities of the male call. Mating involves external fertilization of eggs, which are generally laid in clusters or strings. Eggs develop into the distinctive tadpole, a larval form unique to frogs. Tadpoles typically live for one season, during which time they consume primary productivity in the form of algae or vegetation. However, there are many variations on this life history. Tadpoles may live more than one year before metamorphosing into froglets. Many frogs have no larval stage at all but lay large eggs in small numbers that develop directly into miniatures of the adult. Other frogs have tadpoles that may be brooded in special compartments that form in the skin of the back, in specialized pouches on the back, in the vocal sacs of males, or in diverse ecological settings such as arboreal bromeliads. A few frogs are viviparous, with eggs developing in the female reproductive tract, where they receive nourishment.

Salamanders (Caudata), with more than 560 species, resemble ancient amphibians more closely than do frogs or caecilians. They are generalized in structure, with a body of moderate length, a well-developed tail, and two pairs of limbs, with the hind limbs being only a little longer than the forelimbs. Heads of salamanders are relatively smaller than those of frogs, but both groups have well-developed eyes and excellent vision. Salamanders often gather in ponds to breed but males have no calls. Larvae are carnivorous rather than herbivorous, and metamorphosis is far less dramatic than in frogs. Some species remain in a permanently larval or semilarval state and some larvae become very large. Most clades of salamanders have some variation of the basic life history, but the most successful and numerous salamanders, in the family Plethodontidae, have direct development, and it is only these that have successfully invaded the tropics (restricted to tropical America). A few species of salamanders are viviparous.

Caecilians (Gymnophiona) are limbless and extremely elongate, with either a very short tail or no tail. The head is about the same diameter as the strongly segmented trunk, and the eyes are inconspicuous or invisible. A unique sensory organ, the tentacle, is formed from components of the nose and eye and provides environmental information to these mainly burrowing animals. Some caecilians have aquatic larvae but most have either direct development or are viviparous. Among the many viviparous species are forms that are exclusively aquatic. There are approximately 170 caecilian species.

Amphibians play important but generally underappreciated roles in ecosystems. Frogs are important predators as adults, but their larvae can be important consumers in aquatic systems. Removal of tadpoles from streams can quickly lead to overgrowths of algae and formation of large algal mats. Although salamanders are cryptic and less obvious than frogs, in some ecosystems they can be extremely numerous. As carnivores, they contribute to regulation of terrestrial food webs. Members of the soil community, caecilians are major consumers of earthworms, but their ecological role is less well understood than that of frogs and salamanders.

Amphibians have received attention because of concerns about evident declines and extinctions of populations as well as species throughout the world. A recent assessment of the status of amphibian species across the globe found that a higher proportion (more than one-third) were at risk of extinction than for any of the other vertebrate taxa. Especially troubling was the finding that a large percentage of the declines were from unknown causes. While most amphibian declines doubtless are related directly to habitat destruction and modification by humans, there are also other important factors. One infectious disease, a chytrid fungus that attacks the keratin of the skin and leads to dehydration, is responsible for well-documented declines in species living in protected areas in Central America, Australia, and California, where epidemics have been recorded. The disease spreads rapidly, and there is no known way to stop it. Global climate change also is implicated in the declines of many frogs in tropical montane forests. Other factors include effects of introduced species (e.g., for sport fishing), pollution from pesticides and fertilizers, and synergistic interactions of some of these factors and others such as ultraviolet radiation (UVB). However, in many parts of the planet amphibians remain abundant, adding to the puzzlement over the declines.

Today's living amphibians are remnants of a very ancient lineage; each major group can be traced back about 200 million years. Although many appear to be delicate, they were rugged survivors of many extinction events. It is troubling that amphibians that lived happily with the long-extinct dinosaurs should be facing extinction on our watch.

BIBLIOGRAPHY

Duellman, W. E., and L. Trueb. 1986. *Biology of Amphibians.* New York: McGraw-Hill.
Hutchins, M., W. E. Duellman, and N. Schlager. 2003. *Grzimek's Animal Life Encyclopedia.* 2nd ed. Vol. 6, *Amphibians.* Farmington Hills, MI: Gale Group.
Lannoo, M. 2005. *Amphibian Declines: The Conservation Status of United States Species.* Berkeley: University of California Press.
Stuart, S. N., J. S. Chanson, N. A. Cox, B. E. Young, A. S. L. Rodrigues, D. L. Fischman, and R. W. Waller. 2004. Status and trends of amphibian declines and extinctions worldwide. *Science* 306: 1783–1786.
Wells, K. D. 2007. *The Ecology and Behavior of Amphibians.* Chicago: University of Chicago Press.   —*D.B.W.*

Anderson, Edgar (1897–1969)

Edgar Anderson is regarded as one of the leading scientists of plant evolutionary biology during the twentieth century. A product of a number of diverse intellectual traditions, Anderson was able to synthesize newer developments in genetics and cytology to help address more traditional problems in plant systematics. He is best known for his inventive studies measuring variation in natural settings and for his articulation of the concept of introgressive hybridization, a process by which new genetic material is introduced through hybridization and backcrossing. Much of this work was formalized in 1949 with the publication of a small volume titled *Introgressive Hybridization.* The book was widely read and immediately recognized as an original contribution to plant genetics that could shed light on a general theory of evolution.

Anderson, born in Forestville, New York, was the son of an educational administrator and a pianist. His father became a professor of dairy science at the Michigan Agricultural College and moved his family to East Lansing when Anderson was three. From an early age, Anderson was fascinated by plants.

He was a keen student and in 1914 entered Michigan Agricultural College, where he studied botany and horticulture. In 1919 he entered the graduate program at Harvard, working at the Bussey Institution under the supervision of noted agricultural geneticist Edward Murray East. His research was on the genetics of self-incompatibility in *Nicotiana,* the tobacco plant, but he was also fond of areas like economic botany and gravitated to Oakes Ames, the noted economic botanist at Harvard and expert on orchids. After graduation, Anderson accepted a position as geneticist at the Missouri Botanical Garden and was appointed assistant professor at Washington University in St. Louis. In 1929 he accepted a National Research Fellowship to study in Britain's John Innes Horticultural Institution, working with cytogeneticist C. D. Darlington, statistician R. A. Fisher, and geneticist J. B. S. Haldane. After returning from Britain, he was influenced by Harvard geneticist Karl Sax.

Much of Anderson's research was spent developing novel visual and quantitative methods to measure geographic variation in natural populations, the best known of which is the ideogram. Anderson is also known for his detailed studies of the patterns of variation in the common blue flag, *Iris versicolor,* along the Mississippi delta and what the patterns revealed about mechanisms of speciation in plants. This study received recognition in the early 1940s, an interval of time important in the history of evolutionary biology, because it fueled the movement known as "new systematics" (or biosystematics), which sought to bring interdisciplinary insights from genetics and ecology to more traditional taxonomy. As a result, Anderson emerged as one of the leaders seeking general principles of plant evolution. He was invited to give the prestigious Jesup Lectures at Columbia University along with avian systematist Ernst Mayr in 1941. Anderson was to provide the plant side of the new systematics. Although he gave the lectures, he did not complete the required manuscript for a book based on them. The completion of this manuscript would have placed Anderson within the circle of architects of the evolutionary synthesis of twentieth-century evolutionary biology. The role of botanical architect eventually fell to his close friend, G. Ledyard Stebbins.

Anderson's failure to complete the manuscript was part of his erratic performance pattern. He was a nonconformist who frequently engaged in varied interests that competed for his time. He did make notable contributions to areas in horticulture and even became the director of the Missouri Botanical Garden. In the 1940s, he increasingly turned to questions important to understanding hybridization in maize, with his interests directed toward economic botany.

A talented individual with a strong personality, Anderson left his imprint on the history of plant evolution. For the latter part of his life he was plagued with emotional problems for which he was hospitalized. He was married to Dorothy Moore, a fellow botanist, and had one child. He died in Cleveland Avenue Gatehouse, located on the grounds of the Missouri Botanical Garden.

BIBLIOGRAPHY

Anderson, E. 1949. *Introgressive Hybridization.* New York: Wiley.

Kleinman, K. 1999. His own synthesis: Corn, Edgar Anderson, and evolutionary theory in the 1940s. *Journal of the History of Biology* 32: 293–320.

Smocovitis, V. B. 1999. Edgar Anderson. In John A. Garraty and Mark C. Carnes, eds., *American National Biography,* vol. 1, 452–453. Oxford: Oxford University Press.

Stebbins, G. L. 1978. Edgar Anderson. *National Academy of Sciences Biographical Memoirs* 49: 3–23.   *—V.B.S.*

Animal Species and Their Evolution (A. J. Cain)

A. J. Cain's *Animal Species and Their Evolution* (1954) offered one of the first comprehensive discussions of the nature of species after the initial flowering of the so-called modern synthesis. Cain drew ideas from paleontology, biogeography, taxonomy, and population genetics. He combined them into a cohesive treatment of species concepts, classification methods, speciation, and the roles of natural selection, history, and genetic drift in creating patterns of character variation. Some of these topics had been covered in earlier books; however, it was the combination of topical breadth and seamless synthesis that made Cain's treatment stand apart. The examples and case studies presented in this slim volume covered a staggering range, from orchids to oaks and from sea urchins to shrews (see figure).

Cain traces an arc from classical ideas to novel ones. He begins by discussing traditional ideas of biological nomenclature, methods of classification, and taxonomic rank. Cain then turns to the nature of species, which he

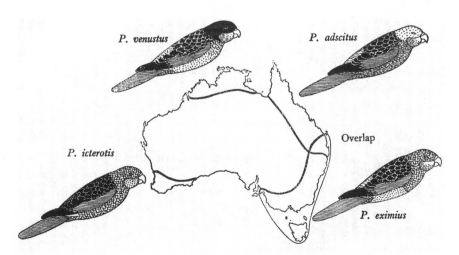

The gradual variation that can be seen in adjoining populations of organisms is taken as one of the best pieces of evidence for gradual evolution. This picture, of geographical variation in Rosellas, is taken from Cain 1954, 61.

describes as collections of populations that display variation in many characters across the species' geographical range. Cain discusses how this variation is promoted and maintained in some cases by different selection pressures in different locations and in others by genetic drift. He argues that the ubiquity of intraspecific variation of this type obviates the classical definitions of species; in fact, he argues that no single definition of species is adequate for the enormous natural diversity of organisms. Cain discusses five types of species that reflect five different species concepts, from the biological species concept championed by Ernst Mayr (1942) to what Cain calls agamospecies, collections of similar individuals that reproduce asexually. He also discusses modes of species formation, offering one of the first attempts to link different types of species with different modes of speciation. The final chapter of the book is a prescient discussion of sympatric speciation, speciation without a geographical separation between forms. The overall argument is that a species is neither a collection of identical individuals nor a fixed entity, but rather a dynamic set of populations that are waypoints on a series of evolutionary trajectories.

If the book's message is obvious to the modern reader, it is because the book made so persuasive a case. Cain's influence would extend from ecological geneticists to systematists. Several topics raised in the book, such as the adequacy of the biological species concept or the inadequacy of certain methods in systematic inference for capturing the relationships among sets of evolving species, would reemerge repeatedly over the ensuing years. Although Cain's concise treatments of these topics is obviously not informed by the wealth of data now at our command, his treatments remain among the most lucid and critical discussions of many important questions confronting evolutionary biologists who think about what a species is and how species are related to one another.

BIBLIOGRAPHY

Cain, A. J. 1954. *Animal Species and Their Evolution*. London: Hutchinson's University Library.
Mayr, E. 1942. *Systematics and the Origin of Species*. New York: Columbia University Press.   —*J.T.*

Antonovics, Janis (b. 1942)

While the study of plants has shaped many facets of evolutionary biology, few scientists have used plants to illuminate so many important issues as has Janis Antonovics. Antonovics has led a series of research programs in natural plant populations that have uncovered some of the myriad ways in which natural selection shapes the features of organisms and the diversity of their populations.

Three contributions stand out among many. His early work, performed with his doctoral advisor at the University College of North Wales, Anthony Bradshaw, and colleagues, showed how populations of plants on mine tailings, in evolving tolerance to heavy metal deposits in the soil, had also evolved partial reproductive isolation from closely adjacent populations that were not on metal-laden soil (Antonovics and Bradshaw 1970; Antonovics 2006). This was a striking demonstration of incipient isolation in the face of considerable gene flow between populations. In another project, Antonovics and his students performed elegant experiments to show that individuals that produced genetically diverse offspring through sexual reproduction had greater fitness in a spatially variable and competitive environment (Antonovics and Ellstrand 1984). While theories for the origin of sexual reproduction had postulated such an advantage, there was little, if any, empirical work that addressed precisely which ecological features of the environment would facilitate it. In a third, far-reaching project, Antonovics and his students have shown that the course of an infectious disease is determined by a balance between the local battle of pathogen and host in an individual population and the movement of pathogens, hosts, and the genes for virulence and resistance among populations across the landscape (e.g., Alexander and Antonovics 1988).

Two themes emerge from Antonovics's work. First, natural selection is a strong force in nature that emerges from a variety of ecological factors. The physiological stress of heavy metals in the soil may seem an obvious agent of adaptation—mutations that confer tolerance of heavy metals are not common among plant species—but Antonovics and his colleagues have found other agents of selection, from soil mineral profile to crowding of other plants to predators and pathogens. Second, to understand how selection molds diversity, one must understand the spatial scale on which the pressures of selection can vary and the spatial scale at which individuals (or their genes) move readily. The interplay of these two scales determines how diversity emerges in space and, ultimately, the ease with which speciation occurs. These themes are embraced by Antonovics's thesis that ecology and evolutionary biology are intertwined disciplines.

In addition to these research programs, Antonovics has authored or coauthored a number of highly influential essays, each of which examined an important topic, posed new questions about it, and inspired new research directions. These papers included discussions of adaptive genetic variation (Antonovics 1976), the dynamics of infectious diseases in natural populations (Lockhart et al. 1996), and the potential for reciprocal genetic influences among interacting species (Antonovics 2003). His approach to evolutionary biology has influenced innumerable other scientists for over forty years and that influence is visible throughout the discipline.

BIBLIOGRAPHY

Alexander, H. M., and J. Antonovics. 1988. Disease spread and population dynamics of anther-smut infection of *Silene alba* caused by the fungus *Ustilago violacea*. *Journal of Ecology* 76: 91–104.

Antonovics, J. 1976. The nature of limits to natural selection. *Annals of the Missouri Botanical Garden* 63: 224–247.

———. 2003. Toward community genomics? *Ecology* 84: 598–601.

———. 2006. Evolution in closely adjacent plant populations X: Long-term persistence of prereproductive isolation at a mine boundary. *Heredity* 97: 33–37.

Antonovics, J., and A. D. Bradshaw. 1970. Evolution in closely adjacent plant populations. 8. Clinal patterns at a mine boundary. *Heredity* 25: 349–362.

Antonovics, J., and N. C. Ellstrand. 1984. Experimental studies of the evolutionary significance of sexual reproduction. 1. A test of the frequency-dependent selection hypothesis. *Evolution* 38: 103–115.

Lockhart, A. B., P. H. Thrall, and J. Antonovics. 1996. Sexually transmitted diseases in animals: Ecological and evolutionary implications. *Biological Reviews of the Cambridge Philosophical Society* 71: 415–471.   —J.T.

Archaeopteryx

Archaeopteryx (ancient bird, or German *Urvogel*) is the generic name given to 10 skeletal specimens (as well as a secondary flight feather discovered in 1860) of the earliest known bird, from latest Jurassic lithographic limestone deposits known as the Solnhofen Limestone, located just to the north of the city of Munich in southern Germany. Dating to approximately 150 million years old, the deposits were in a hypersaline lagoon that was closed off to open sea by coral reefs. Although the discovery of the skeletons spanned some 145 years, the two most famous specimens are the first two discovered, the London and Berlin specimens, discovered in 1861 and 1877, respectively. The last recovered specimens are the Solnhofen (the largest at 10% larger than the London), discovered in a collection in 1987, and the final specimen, the Thermopolis specimen, described in 2005. Although considerable variation exists, the first seven specimens are generally referred to as a single species, *Archaeopteryx lithographica*, while the 1993 example (Aktien-Verein) is distinctive in many features and is referred to as a new species, *Archaeopteryx bavarica*. It is the only specimen to show an ossified sternum, and the proportions of the legs are quite different from the others, clearly indicating arboreal habits.

The celebrated Berlin specimen (see figure) has been called a veritable Rosetta stone of evolution because it is complete and beautifully preserved with outstretched wings. It illustrates an almost perfect intermediate between two major classes of vertebrates: birds and reptiles. *Archaeopteryx* fulfills the Darwinian expectation of a missing link between major vertebrate groups. Its elliptical wings in profile are similar to those of a modern woodland bird, with modern flight feathers virtually identical to those of living birds, even down to their microstructure, as shown by electron microscopy. Feather structure, once invented, remained unchanged in any major way for 150 million years. The primary and secondary flight feathers are anchored to the wing as in modern birds, and the flight feathers exhibit asymmetric vanes that produced individual airfoils, indicating aerodynamic function. Most paleontologists believe that *Archaeopteryx* was a capable although primitive powered flier, and the scapula and coracoid are primitive but join at a 90° angle, as in modern birds. The avian wings end with three

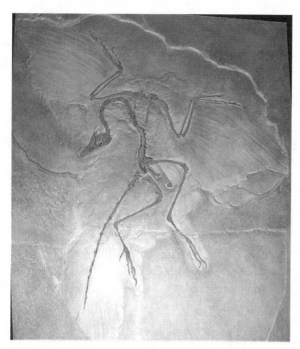

The most famous fossil of them all, *Archaeopteryx lithographica,* first discovered early in the 1860s in Solnhofen, Germany. This is the "Berlin specimen," discovered in 1877.

reptilian fingers with sharp, highly recurved claws that closely mimic those of modern trunk-climbing mammals and woodpeckers, indicating that *Archaeopteryx* may have been a trunk climber. There is recent evidence that *Archaeopteryx* also possessed handleg wings, as are present in the Chinese early Cretaceous microraptors. The *Urvogel's* feet are birdlike, with the first toe reversed as a hallux and highly recurved pedal claws. A capable percher, it was less at home on the ground. Its long reptilian tail of 23 caudal vertebrae had a pair of tail feathers coming off each vertebra. The trunk and sacrum are quite reptilian. It lacked the binding uncinate processes of the rib cage exhibited by modern birds, but the pubis was retroverted as in modern birds. The skull of *Archaeopteryx* was both reptilian and avian, having reptilian teeth, but with an avian quadrate and cranial kinesis. The brain and inner ear were quite similar to modern birds, indicating a sophisticated flight ability.

Although *Archaeopteryx* remains the oldest known bird, more Mesozoic fossil birds have been discovered in the last two decades than from the 1860s to the 1970s. Most come from the Lower Cretaceous of China, and they represent various types of primitive birds, including the *Archaeopteryx*-like *Jeholornis,* which sports a long tail with 27 caudal vertebrae, and an anatomy much like that of the famous German *Urvogel.*

BIBLIOGRAPHY

Elzanowski, A. 2002. Archaeopterygidae (Upper Jurassic of Germany). In L. M. Chiappe and L. M. Witmer, eds., *Mesozoic Birds: Above the Heads of Dinosaurs*, 129–159. Berkeley: University of California Press.

Feduccia, A. 1999. *The Origin and Evolution of Birds*. New Haven, CT: Yale University Press.

Feduccia, A., L. D. Martin, and S. Tarsitano. 2007. *Archaeopteryx* 2007: *Quo Vadis? The Auk* 124: 373–380.

Longrich, N. 2006. Structure and function of the hindlimb feathers in *Archaeopteryx lithographica*. *Paleobiology* 32: 417–431.

Mayr, G., B. Pohl, and D. S. Peters. 2005. A well-preserved *Archaeopteryx* specimen with theropod features. *Science* 310: 1483–1486.

Zhou, Z. 2004. The origin and early evolution of birds: Discoveries, disputes and perspectives from fossil evidence. *Naturwissenschaften* 91: 455–471.   —A.F.

Aristotle (384–322 B.C.E.)

Aristotle is properly recognized as the originator of the scientific study of life. He considered it a central concern of the theoretical study of nature. His treatises *History of Animals, On the Parts of Animals,* and *On the Generation of Animals* (and a number of more specialized studies) provide a theoretical defense of the proper method for biological investigation and the results of the first such investigation. There was nothing of similar scope and sophistication until the eighteenth century. In 1837 the great anatomist Richard Owen declared that "Zoological Science sprang from his [Aristotle's] labours, we may almost say, like Minerva from the Head of Jove, in a state of noble and splendid maturity" (Owen 1992, 91). Before examining this remarkable achievement, a few words about its creator are in order.

LIFE AND WORK

Aristotle was born in Stagira on the northern Aegean coast in 384 B.C.E. His father, Nicomachus, was physician to King Amyntas of Macedon, and his mother was of a wealthy family from the island of Euboea. He was sent at the age of 17 to Athens, where he studied in Plato's Academy for 20 years, until Plato's death in 347. By then he had developed his own distinctive philosophical ideas, including his passion for the study of nature. He joined a philosophical circle in Assos on the coast of Asia Minor, but soon moved to the nearby island of Lesbos where he met Theophrastus, a young man with similar interests in natural science. Between the two of them they originated the science of biology, Aristotle carrying out a systematic investigation of animals, Theophrastus doing the same for plants.

In 343 Philip II of Macedon asked Aristotle to tutor his son Alexander, but by 335 he had returned to Athens, now under Alexander's control. With Theophrastus he founded a school known as the Lyceum. He headed the Lyceum until Alexander the Great's death in 323. With anti-Macedonian feelings running high in Athens, Aristotle retired to his mother's birthplace. He died there in 322 B.C.E.

The surviving corpus of Aristotle derives from medieval manuscripts based on a first century B.C.E. edition. The first printed editions and translations date to the late fifteenth century. These works cover a remarkable range of subjects, but for the purposes of this essay I will focus on his animal studies, which constitute 25% of the existing corpus.

ARISTOTLE'S PHILOSOPHY OF SCIENCE

Aristotle was able to accomplish what he did because he had given a great deal of thought to the nature of scientific inquiry. How does one progress from the superficial and unorganized state of everyday experience toward organized scientific understanding? To answer this question, you need a concept of the goal to be achieved, and Aristotle developed such a concept in his *Prior and Posterior Analytics*. The goal of inquiry, he argued, was a system of concepts and propositions organized hierarchically, ultimately resting on knowledge of the essential natures of the objects of study and certain other necessary first principles. These definitions and principles form the basis of causal explanations of all the other universal truths about the subject of study.

The second book of the *Posterior Analytics* discusses how to achieve this goal. Plato had formulated a famous paradox of inquiry in his dialogue *Meno*: either you know the object of your inquiry, in which case inquiry is unnecessary, or you don't know the object of your inquiry, in which case inquiry is impossible. Aristotle reminds us of this paradox in the first chapter of the *Posterior Analytics*, but his full reply is in book II.

There he argues that perceptual experience gives us a grasp of the target of inquiry that does not count as knowledge, but does provide sufficient information to direct further inquiry. His examples in the *Posterior Analytics* are familiar phenomena like thunder, eclipses, and seasonal shedding of leaves, but we can see his method at work in his systematic inquiries into the anatomy and development of animals reported in *History of Animals (Historia animalium)*, *On the Parts of Animals (De partibus animalium)*, and *On the Generation of Animals (De generatione animalium)*. From here on we will refer to these by the shorthand *HA, PA,* and *GA*.

BIOLOGICAL INQUIRY AND BIOLOGICAL KNOWLEDGE

For Aristotle, systematics—the organization of information about animals—serves the goal of causal explanation, and explanation by reference to goals and functions is fundamental. The following remark, introducing his systematic presentation of information about animals, stresses the first point: "In the first instance we should grasp the attributes and the differences present in all animals. After we have done this we should attempt to discover their causes. This is the natural way to carry out the investigation, beginning with inquiry *(historia)* about each thing; for it becomes apparent from such inquiries about which things and from which things there ought to be demonstration" (*HA* I. 6, 491a7–14).

We begin with a study of the similarities and differences that we observe in the animal kingdom; only after this information is systematically gathered and organized are we in a position to pursue causal demonstrations. *HA* begins by outlining *kinds* of similarity (in form, in kind, by analogy) that one finds among animals and then presents the *four categories of difference* he will use: in parts, activities, ways of life, and character. This discussion provides his framework: books I–IV discuss similarities and differences among parts, V–VIII among activities and ways of life, and IX among traits of character. Among parts, Aristotle distinguishes between uniform and nonuniform parts (tissues and organs, as we would say) and between external and internal parts, moving systematically through these categories in nine groups of "great kinds." The "blooded kinds" (corresponding to our vertebrates) are birds, fish, cetaceans, four-legged live-bearing animals, and four-legged egg-laying animals; the "bloodless" are the hard-shelled (testaceous mollusks), soft-shelled (crustaceans), soft-bodied (cephalopods), and insected animals (insects). Aristotle uses a process of "division and correlation" to establish relationships of similarity and difference at various levels of generality, using the above "kinds" as an aid in this process. He notes that many animals (including human beings) do not fall into his "great kinds." Taxonomic tidiness is not his primary goal; these categories help in identifying loci of systematic correlations, and these will aid in the discovery of the causal relationships among animal differentiae. Primary among causal relationships are those that explain a part's differential attributes by reference to the functional demands of an animal's way of life.

On the Parts of Animals presents Aristotle's attempt to provide such causal understanding. It begins with a defense of the proper methods for biological study. It argues for the priority of teleological explanation (explanations that identify what a feature is for) over material and efficient causal explanation (what the feature is made from and how it is produced) and for conditional *necessity*, whereby materials and processes are explained by being shown to be necessary for a goal. (Notice this reinforces the priority of teleology.) This view of necessity and explanation presupposes that animals are unified composites of bodily parts with functional capacities, which Aristotle refers to as their material and formal natures. "Form" in biological contexts refers to the basic capacities of nutrition, reproduction, locomotion, and cognition, which Aristotle refers to as the animal's soul.

This first book also defends a method of multifactorial division, whereby many universal differentiae are divided in parallel in order to characterize correlated variations within kinds, and it also defends a method for establishing the kinds that are presupposed by this method. *PA* II–IV attempt both to establish causes and to provide causal explanations that parallel the first four books of *HA*. The third of these large works is *GA*, which provides causal explanations for the subject matter organized in *HA* V–VII.

THE METHOD AT WORK

The following quotation from *HA* initiates Aristotle's discussion of the internal organs of blooded animals:

We have spoken, then, of the external parts of the blooded animals—how many they are, of what sort and in what ways they differ from one another. Next we need to discuss the arrangement of the internal parts, first off in blooded animals—for it is by some being blooded and others bloodless that the great kinds differ from one another. . . .

Now all that are four-legged and live-bearing have an esophagus and windpipe, positioned in the same way as it is in human beings; likewise too in all of the four-legged animals that lay eggs, and in the birds. But they [each of the kinds mentioned] differ in the conformation of these parts. Put generally, all and only those that, by taking in air, inhale and exhale have a lung, windpipe and esophagus; and the placement of the esophagus and windpipe is similar, but the organs are not, while in the case of the lung neither is the organ similar nor is it positioned similarly. . . .

Not all the blooded animals have a lung; for example, fish do not, nor do any other animals that have gills. (excerpts from *HA* II. 505b24–506a12)

In his review of what has been accomplished, he notes that the *number, character* and *differentiations* of the *external* parts of blooded animals have already been discussed. He is now turning to the *internal* parts of the same groups. He first lists the kinds that have both windpipe and esophagus, and then generalizes to a coextensive correlation—*all* that inhale and exhale air have these two organs and a lung. Why did he not start here? Because there is no "great kind" at that level; this correlation ranges over birds, humans, four-legged live-bearing animals, and four-legged egg-laying animals. Why not "blooded"? Because of the fish, which are blooded, but have none of these organs—this network of organs does not coincide with the blooded animals either.

This discussion is remarkable in its combination of accuracy and generality. In discussing the snakes here he notes that "the rest of the internal parts are the same [in snakes] as in the lizards, except that owing to their length and narrowness the viscera are long and narrow . . . for example the windpipe is very long and the esophagus even longer" (508a13–19). The cetaceans (whales and dolphins), though mentioned in his opening list of blooded animals (505b30) and discussed extensively when the activities and ways of life of animals are in focus in books V–VIII, here are subsumed under his wide generalizations without comment.

It is equally important to notice what is *not* said in *HA*. There are *no causal explanations* on offer; universal correlations abound, but *no reasons* are given for them. Aristotle refrains from claiming that these correlations are necessary and says nothing about why these animals have or do not have these parts. *The language of nature, matter, form, necessity, cause and essence, so central to Aristotle's philosophy of nature, is carefully avoided.*

Now contrast the discussion of the same correlations in *PA* III. 3:

Not all animals have a neck, but only those with the parts *for the sake of which* the neck is *naturally* present—these are the windpipe and the part known as the esophagus. Now the larynx is present *by nature for the sake of* breathing; for it is through this part that animals draw in and expel air when they inhale and exhale. *This is why* those without a lung have no neck, e.g. the kind consisting of the fish. The esophagus is the part through which nourishment proceeds to the gut; so that animals without necks manifestly do not have an esophagus. But *it is not necessary* to have the esophagus *for the sake of* nutrition; for it concocts nothing. And further, it is *possible* for the gut to be placed right next to the position of the mouth, while for the lung this is *impossible*. For there *needs* first to be something common like a conduit, which then divides in two and through which the air is separated into passages—in this way the lung *may best* accomplish inhalation and exhalation. So, then, the organ connected with breathing *from necessity* has length; therefore it is *necessary* for there to be an esophagus between the mouth and the stomach. The esophagus is fleshy, with a sinuous elasticity—*sinuous so that* it may dilate when food is ingested, yet *fleshy so that* it is soft and yielding and is not damaged when it is scraped by the food going down. (*PA* III. 3, 664a14–35; emphases added)

I have highlighted the language of nature, necessity, possibility (and impossibility), and teleology in this passage in order to contrast it with the discussion of the same organic correlations in our *HA* passage. Here the goal is explanation—parts are present and have the character they do primarily because of the conditional necessity imposed by the organism's functional requirements. One sees here the Aristotle that so impressed the great French naturalist Georges Cuvier: Aristotle is not only systematically discussing the adaptive functions of each of these organs; he is also showing the amazing way in which the internal parts of animals constitute an organic *system*.

CONCLUSION

How self-conscious was Aristotle of the revolutionary nature of what he was doing? In *On Respiration*, after critically reviewing the errors of previous writers, he reflects on the reason for their failures: "The most basic reason that previous investigators have not discussed respiration properly is a lack of experience with the internal parts and a failure to grasp what nature makes them all for; for were they inquiring what respiration is present in animals for, and were they doing so while inspecting the parts such as gills and lung, they would have discovered the cause quickly" (471b24–29).

Speculation on the purposes of organs without doing dissection and anatomy, or doing dissections without thinking about function, is doomed to fail. Proper biological inquiry requires that one ask the functional question while doing the empirical work. Such work can be very unpleasant. Near the close of his philosophical introduction to the study of life, *PA* I, Aristotle admonishes us to keep our eye on the prize:

Even in the study of animals disagreeable to perception, the nature that crafted them likewise provides extraordinary pleasures to those who are able to know their causes and are by nature philosophers. Surely it would be unreasonable, even absurd, for us to enjoy studying likenesses of animals—on grounds that we are at the same time studying the art, such as painting or sculpture, that made them—while not prizing even more the study of things constituted by nature, at least when we can behold their causes. For this reason we should not be childishly disgusted at the examination of the less valuable animals. For in all natural things there is something marvelous. (*PA* I. 5, 645a7 –17)

On February 22, 1882, bedridden and near death, Charles Darwin read these words in a new translation sent to him as a gift by William Ogle a month earlier. In a thank-you note to Ogle, Darwin wrote: "From quotations I had seen I had a high notion of Aristotle's merits, but I had not the most remote notion what a wonderful man he was. Linnaeus and Cuvier have been my two gods, though in very different ways, but they were mere school-boys to old Aristotle." Ogle replied just three days before Darwin died: "Thank you for your kind and eulogistic letter re 'the parts of animals'. It gave me much pleasure. I am glad also to have added a third person to your gods and completed the Trinity." As a student of both of these great naturalists, it gives me great satisfaction to know that Aristotle afforded Darwin, during a period of great suffering, some moments of pleasure.

BIBLIOGRAPHY

English translations of all of Aristotle's biological treatises can be found in Harvard University Press's Loeb Classical Library. In addition, for translations I recommend:

Aristotle. 1949. *Prior and Posterior Analytics*. W. D. Ross, ed. Oxford: Clarendon Press.

———. 1992. *Aristotle's De Partibus Animalium* I and *De Generatione Animalium* I. D. M. Balme, trans. Oxford: Clarendon Press.

———. 2001. *On the Parts of Animals*. Books I–IV. J. G. Lennox, trans. Oxford: Clarendon Press.

For further discussion of Aristotle's biological works and their significance, see:

Gotthelf, A., and J. G. Lennox, eds. 1987. *Philosophical Issues in Aristotle's Biology*. Cambridge: Cambridge University Press.

Kullmann, W., and S. Föllinger, eds. 1997. *Aristotelische Biologie*. Stuttgart: Franz Steiner.

Lennox, J. G. 2001. *Aristotle's Philosophy of Biology: Studies in the Origins of Life Science*. Cambridge: Cambridge University Press.

Owen, R. 1992. *The Hunterian Lectures in Comparative Anatomy*. P. R. Sloan, ed. Chicago: University of Chicago Press.

Pellegrin, P. 1986. *Aristotle's Classification of Animals: Biology and the Conceptual Unity of the Aristotelian Corpus*. Anthony Preus, trans. Berkeley: University of California Press.

Finally, there is a fascinating discussion of Darwin's thoughts on Aristotle in:

Gotthelf, A. 1999. Darwin on Aristotle. *Journal of the History of Biology* 32: 3–30.

—*J.G.L.*

The Arkansas Creation Trial

The modern creationist movement dates from the publication in 1961 of *The Genesis Flood,* coauthored by biblical scholar John C. Whitcomb and hydraulics engineering professor Henry M. Morris (Numbers 2006). This "young earth" account of creation attracted support and grew in strength during the 1970s, thanks to the proselytizing work by former biochemist Duane T. Gish, author of *Evolution: The Fossils Say No!* (1973). Since the Scopes trial, the legal situation had changed. No longer was it possible to exclude evolution from schools. The problem now was to get the Bible in. Naturally, the creationist supporters wanted to get their ideas taught in the biology classes of state-supported schools, so they devised a variant—creation science—that supposedly made no reference to the Bible and was based on purely scientific evidence, but just so happened to support a young-age earth, six days of creation, with humans last, and sometime thereafter a massive worldwide flood.

In 1981 the state of Arkansas passed a law mandating the balanced treatment in school biology classes of evolutionary ideas and creation science. Immediately, the American Civil Liberties Union (ACLU) supported a suit claiming that the law was unconstitutional because it breached the separation of church and state. The trial was held in Little Rock in December 1981, with ACLU expert witnesses including philosopher Michael Ruse, paleontologist Stephen Jay Gould, and population geneticist Francisco J. Ayala. The case was made that creation science is far from being traditional Christianity, but is rather an idiosyncratic perversion of nineteenth-century American Protestantism; that it is not genuine science but religion dressed as science; that the empirical facts and theoretical hypotheses speak against it; and that it is entirely inappropriate that biology classes should present indifferently different positions simply because various members of society have strongly held different positions.

After a two-week trial, Judge William R. Overton ruled firmly that the law was indeed unconstitutional and should be discarded. The state decided not to appeal the decision. In coming to his conclusion, Overton argued that science follows certain fixed rules of methodology—it appeals to unbroken law, it opens itself to comparison against the findings from the empirical world (falsification), and it is always tentative in its conclusions—and that creation science fails on all of these scores. After the trial, some philosophers objected strenuously that no criterion of demarcation can properly separate science from religion. Although this occasioned excited discussion in professional circles, general thinking in the science community, not to mention the world of journalism and public opinion, sided with Overton and this seemed to be the end of matters.

Creationists themselves argued that the right way to pursue their ends was not through laws that are obviously open to attack, but at the grassroots level, pressuring school boards and individual teachers. This set the pattern for the next decade, until the Intelligent Design discussion began and the American science-religion quarrel gained new life.

BIBLIOGRAPHY

Gilkey, L B. 1985. *Creationism on Trial: Evolution and God at Little Rock.* Minneapolis: Winston Press.
Gish, D. 1973. *Evolution: The Fossils Say No!* San Diego: Creation-Life.
Numbers, R. L. 2006. *The Creationists: From Scientific Creationism to Intelligent Design.* Expanded ed. Cambridge, MA: Harvard University Press.
Pennock, R., and M. Ruse, eds. 2008. *But Is It Science? The Philosophical Question in the Creation/Evolution Controversy.* Updated ed. Buffalo, NY: Prometheus.
Ruse, M., ed. 1988. *But Is It Science? The Philosophical Question in the Creation/Evolution Controversy.* Buffalo, NY: Prometheus.
Whitcomb, J. C., and H. M. Morris. 1961. *The Genesis Flood: The Biblical Record and Its Scientific Implications.* Philadelphia: Presbyterian and Reformed Publishing Company.   —*M.R.*

Artificial Life programs and evolution

Evolution is the defining property and creative process of life, on earth and wherever it may occur. The cheetah running down its prey, the humming-bird pollinating a flower, and the drama of the human mind are all products of evolution. Life has completely reshaped the surface of our planet and is beginning to probe the rest of our solar system. Since Darwin and Wallace unraveled the mystery of evolution, many scientists have studied its details. A relatively new approach to the study of evolution is to create new instances of evolution, an approach that has been called Artificial Life.

While millions of living species currently exist on earth, it is believed that they all trace back to a common ancestor, billions of years ago; thus there is only a single tree of life on earth. From this perspective, our entire science of biology and our experience with evolution is based on a sample size of one. A truly comparative biology and a truly broad perspective on evolution would require knowledge of other instances of life and its generative process, evolution. It is believed that life exists on many planets throughout the universe but, unfortunately, they are out of our reach.

In its essence, the process of evolution involves self-replicating entities with genetic variation and differential survival, which leads to changes in the characteristics of the population of entities over the course of generations. Life on earth is the product of evolution inhabiting the medium of carbon chemistry; however, the process of evolution can operate in other media as well.

There has been some speculation among physicists that the universe itself may be the product of evolution. Some have suggested that black holes give birth to new universes, and that the fundamental constants of physics may vary among universes. Universes that produce more black holes have a higher Darwinian fitness, leading to the evolution of the fundamental constants of physics, which govern the characteristics of universes.

Speculation aside, we have discovered that evolution can inhabit the medium of digital computation. If we accept that evolution is the defining property and creative process of life, then instances of digital evolution may also be considered instances of life, albeit dramatically alien life. If we had the opportunity to observe life on other planets, it likely would be carbon-based life and thus would have that much in common with life on earth. Artificial life in the digital medium shares only the evolutionary process itself in common with life on earth, and so is more alien than life on other planets.

The digital computational medium is not a physical/chemical medium; it is a logical/informational medium. Thus these new instances of evolution are not subject to the same physical laws as organic evolution (e.g., the laws of thermodynamics), and therefore they exist in what amounts to another universe, governed by the "physical laws" of the logic of the computer. They never "see" the actual material from which the computer is constructed; they see only the logic and rules of the CPU and the operating system. These rules are the only "natural laws" that govern their behavior. Thus they live in a radically alien universe. Inoculating evolution into the digital medium gives us a broader perspective on what evolution is and what it does.

One of the most successful approaches for creating digital evolution is to write self-replicating computer programs, which have been called "digital organisms" or simply "creatures," and run them on a computer with a Darwinian operating system. A Darwinian operating system manages a population of replicating digital organisms in such a way that when new creatures are born, older ones die to free space in the memory inhabited by the programs. The Darwinian operating system also introduces mutation by flipping bits (between zero and one) in the code of the creatures. The random mutations cause genetic variation in the population, with the result that some individuals are able to reproduce better than others, which leads to a natural process of Darwinian evolution in an artificial digital medium.

Although this produces a dramatically alien instance of evolution, it is found to have some striking parallels to organic evolution on earth. Perhaps the most significant finding is that the digital organisms evolve adaptations to the presence of other digital organisms in their environment. A digital ecology emerges, which becomes the main driving force of evolution.

A computer can be seeded with a single self-replicating digital organism, which will quickly reproduce and fill the memory of the computer with a population of creatures. The creatures are then a very prominent feature of the environment and become an important source of selective forces. Parasites are one of the first things that typically evolve, and they set off an ecological-evolutionary dynamic that leads to an ongoing series of evolutionary innovations.

Whatever kind of creature is most common becomes a target for exploitation by other creatures. Or if parasites become common, they drive other creatures to evolve defenses. A typical scenario would begin with the evolution of parasites, followed by the evolution of immunity by their hosts. This cycle can repeat several times but may progress to the evolution of "hyper-parasites" that actually attack parasites, stealing their energy.

Sometimes, when one kind of creature completely dominates the memory, such that all the creatures are closely related, they will evolve a kind of sociality in which individual creatures living in isolation are not able to reproduce, but creatures living in groups are able to reproduce. The creatures in this kind of digital world are typically asexual in the sense that they do not mate to produce young but reproduce individually, copying only their own genetic material into their offspring. However, it was discovered that evolution continued even when mutations were turned off. This was because the creatures had invented a kind of primitive sexuality in which offspring were produced containing mixtures of genetic material from more than one parent. It involves a kind of sex with the dead, in which offspring include the genetic material from their parent, mixed with some genes from creatures that have died.

While the ecological coevolutionary dynamic described above is the main driving force for evolution, there is also perpetual selection for efficiency. One form of efficiency involves reducing the amount of time it takes to produce an offspring. This is usually accomplished by reducing the size of the genome that describes the organism, thereby reducing the time that it takes to make a copy. However, some optimizations have been achieved through the evolution of more complex computer code.

The evolution of more complex code is a tantalizing example of the holy grail of Artificial Life. Evolution transformed simple molecules into the complex and beautiful life forms that we find on earth today. It is the ability to generate complexity that is the source of evolution's power. It remains an unrealized goal of Artificial Life to produce an artificial system that exhibits open-ended evolution, leading to ever more complex artificial life forms, such as occurred on earth.

Life appeared on earth roughly 3.5 billion years ago, but remained in the form of single-celled organisms until about 600 million years ago. At that point in time, life made an abrupt transformation from simple microscopic, single-celled forms lacking nervous systems to large and complex multicelled forms with complex anatomies, physiologies, ecologies, and nervous systems capable of coordinating sophisticated behavior. This transformation occurred so abruptly that evolutionary biologists refer to it as the "Cambrian explosion of diversity."

Some have put forth a vision of a digital nature, a kind of biodiversity reserve for digital organisms, perhaps distributed across the Internet. People could contribute some of their memory and CPU cycles to the digital nature preserve, and digital organisms could live in the space, migrating from computer to computer, feeding on unused memory and CPU cycles. The idea is that if a large and complex region of cyberspace could be set aside for digital nature, then a digital Cambrian explosion of diversity and complexity might occur there.

Humans have been managing the evolution of other species for tens of thousands of years through the domestication of plants and animals. It forms the basis of the agriculture that underpins our civilizations. We manage evolution through breeding, the application of artificial selection to captive populations.

Similar approaches have been developed for working with evolution in the digital domain. It forms the basis of the fields of genetic algorithms and genetic programming. However, because digital evolution has not yet passed through its version of the Cambrian explosion, there exists the possibility to use a radically different approach to "managing" digital evolution. We need not limit ourselves to using evolution to produce superior versions of existing software applications. Rather, we should allow evolution to find the new applications for us. To see this process more clearly, consider how we manage applications through organic evolution.

Some of the applications provided by organic evolution are rice, corn, wheat, carrots, beef cattle, dairy cattle, pigs, chickens, dogs, cats, guppies, cotton, mahogany, tobacco, mink, sheep, silk moths, yeast, and penicillin mold. If we had never encountered any one of these organisms, we would never have thought of them either. We have made them into applications because we recognized the potential in some organism that was spontaneously generated within an ecosystem of organisms evolving freely by natural selection.

Many different kinds of things occur within evolution. Breeding relates to evolution within the species, producing new and different, possibly "better," forms of existing species. However, evolution is also capable of generating species. Even more significantly, evolution is capable of causing an explosive increase in the complexity of replicators, through many orders of magnitude of complexity. The Cambrian explosion may have generated a complexity increase of eight orders of magnitude in a span of 40 million years. Harnessing these enormously more creative properties of evolution requires a completely different approach.

We know how to apply artificial selection to convert poor-quality wild corn into high-yield corn. However, we do not know how to breed algae into corn. There are two bases to this inability: (1) if all we know is algae, we could not envision corn; and (2) even if we know all about corn, we do not know how to guide the evolution of algae along the route to corn. Our experience with managing evolution consists of guiding evolution of species through variations on existing themes. It does not consist of managing the generation of the themes themselves.

An alternative is to let natural selection do most of the work of directing evolution and producing complex software. This software will be "wild," living free in the digital biodiversity reserve. In order to reap the rewards and create useful applications, we will need to domesticate some of the wild digital organisms, much as our ancestors began domesticating the ancestors of dogs and corn thousands of years ago.

This vision of digital nature is currently unattainable, and may always remain so. Still, many researchers in the field of Artificial Life are still working toward an open-ended evolution in the digital medium.

Although the vast increases in complexity envisioned for digital nature are out of reach, more modest increases in complexity have been achieved. One approach has been to feed numbers to digital organisms and reward them for performing computations on them; greater rewards are given for more complex computations. This approach has led to the evolution of quite complex algorithms and has been used as a demonstration of the ability of evolution to produce what appear to be "irreducibly" complex structures.

Studying organic evolution can be frustrating because it is a process that occurs over vast scales of time and space. Artificial Life is an exciting approach to the study of evolution because it provides an opportunity to observe the process of evolution in action, generating novelty, ecologies, and modest complexity. Also, it broadens our perspective on life and evolution by allowing us to know the evolutionary process in a nonorganic medium.

BIBLIOGRAPHY

Adami, C. 1998. *Introduction to Artificial Life*. New York: Springer-Verlag.
Goldberg, D. E. 1989. *Genetic Algorithms in Search, Optimization, and Machine Learning*. Reading, MA: Addison-Wesley.
Koza, J. R. 1992. *Genetic Programming: On the Programming of Computers by Means of Natural Selection*. Cambridge, MA: MIT Press.
Langton, C. G., ed. 1995. *Artificial Life: An Overview*. Cambridge, MA: MIT Press.
Lenski, R. E., C. Ofria, R. T. Pennock, and C. Adami. 2003. The evolutionary origin of complex features. *Nature* 423: 139–144.
Levy, S. 1992. *Artificial Life: The Quest for a New Creation*. New York: Pantheon Books.
Ray, T. S. 1991. An approach to the synthesis of life. In C. G. Langton, C. Taylor, J. D. Farmer, and S. Rasmussen, eds., *Artificial Life II: Proceedings of the Workshop on Artificial Life*, vol. 11, 371–408. Redwood City, CA: Addison-Wesley.
———. 1994a. Evolution, complexity, entropy, and artificial reality. *Physica D* 75: 239–263.
———. 1994b. An evolutionary approach to synthetic biology: Zen and the art of creating life. *Artificial Life* 1, no. 1/2: 179–209. Reprinted in C. G. Langton, ed., *Artificial Life: An Overview*, 179–209. Cambridge, MA: MIT Press, 1995.
Smolin, L. 1997. *The Life of the Cosmos*. New York: Oxford University Press.

—T.R.

Ayala, Francisco J. (b. 1934)

Francisco J. Ayala was born in Madrid, Spain, on March 12, 1934. A distinguished member of the international scientific community, he has been acknowledged for his studies in evolutionary biology and the philosophy of science. He has been Donald Bren Professor of Biological Sciences at the University of California at Irvine since 1987 and University Professor of the University of California since 2003.

Ayala's career has coincided with and, to a large extent, has led the rise and expansion of molecular evolution and the philosophy of biology. Indeed, he has been a leader in both fields for many years. After his early investigations of the process of speciation and the role of genetic variation in evolution, performed using traditional genetic methods while he was a graduate student, he saw that the developing techniques of molecular evolution made possible a more powerful approach to these issues. Later in his career, Ayala used molecular methods to elucidate the evolutionary origin and population structure of *Trypanosoma, Plasmodium,* and other human parasites. The impact of Ayala's contributions to the philosophy of biology started in earnest with his "Biology as an Autonomous Science" (*American Scientist,* 1968) and his "Teleological Explanations in Evolutionary Biology" (*Philosophy of Science,* 1970).

As a youth, Ayala was interested in science and decided to study physics at the University of Madrid. Later, as a result of his philosophical and religious inclinations, he immersed himself for five years in theology at the University of Salamanca. Ayala next decided to become an evolutionist, coming to the United States in 1961; he became an American citizen in 1971.

Under the advice of his former Spanish teachers, geneticists Fernando Galán and Antonio de Zulueta, Ayala studied evolution from a genetic perspective with Theodosius Dobzhansky, the Russian-born American geneticist, at Columbia University in New York between 1961 and 1964. During his three years as a graduate student at Columbia he carried out investigations about the process of speciation in Australian *Drosophila* and discovered cryptic species and intermediate stages of speciation. He introduced new methods for measuring population fitness and demonstrated that the rate of evolution correlated with the level of genetic variation. These studies were published in 1965–1966 in a series of papers in *Science, Genetics, Evolution, American Naturalist,* and *Pacific Insects.*

In 1965, as a postdoctoral fellow, Ayala introduced electrophoretic techniques for detecting protein polymorphisms in Dobzhansky's laboratory at Rockefeller University. He became an assistant professor there in 1967. With these more powerful techniques, he now approached the earlier problems of measuring genetic variation (polymorphism) and how this impacted the rate of evolution. He also initiated a comprehensive investigation of the genetic changes concomitant to the speciation process and the evolution of reproductive isolation. Investigations combined laboratory experiments with the investigation of natural populations of *Drosophila willistoni* and related species of the American tropics. He published the first comparative data analyzing genetic differentiation as a function of taxonomic level in *Drosophila* spp., as well as in various other animal groups.

In collaboration with Dobzhansky, Ayala confirmed theoretical predictions about the distribution and adaptive significance of genetic variability in natural populations. One main contribution to evolutionary biology was his demonstration that there is much more genetic polymorphism in natural populations than previously thought. He became a leading participant in the controversy regarding whether molecular polymorphisms were adaptively neutral or driven by natural selection. He argued forcefully for the leading role of natural selection based on the distribution of extensive, but geographically fairly uniform, genetic polymorphisms in natural populations of *Drosophila* in the Amazon basin and throughout the Caribbean islands. He designed laboratory experiments for measuring the effect the amount of food, living space, and other environmental factors had on population fitness. Results confirmed that natural selection was the driving factor impacting genetic polymorphism. He developed laboratory methods for measuring density-dependent and frequency-dependent natural selection.

In 1971, Ayala and Dobzhansky moved to the Department of Genetics at the University of California at Davis. At UC Davis, Ayala collaborated with paleontologist James W. Valentine in a series of investigations published in the 1970s that involved diverse organisms, including tropical *Tridacna* clams; Antarctic, temperate, and tropical oceanic shrimp; and deep-ocean starfish, sea urchins, and brachiopods. The investigations demonstrated that the amount of genetic variability depended not on the abundance of resources, but on environmental stability, which allows genetic variants to become adapted to different environmental features and resources. The results prompted new hypotheses to explain the major extinctions observed in the fossil record.

During the 1980s, working at UC Davis and later at the University of California at Irvine, Ayala studied *Trypanosoma cruzi,* the agent of Chagas disease, which is a severe disease endemic in many South American countries. He showed that it reproduces clonally, rather than sexually, as generally thought at the time. Ayala and collaborator Michel Tibayrenc formulated the clonal theory of parasitic protozoa. They proposed that the population structure of these parasites is prevailingly clonal, even though these pervasive human scourges retain the capacity for sexual reproduction.

More recently, Ayala has reconstructed the evolution of *Plasmodium falciparum,* the agent of malignant malaria, and related species. He has shown that the world expansion of *P. falciparum* throughout the tropical regions is recent, starting from a single African propagule around 5,000 years ago, and that the species is genetically largely uniform, except for mutations recently evolved in response to the human immune system and to medicines.

In addition to his essays on biology as an autonomous science, the concept of biological progress, and the idea of teleology, Ayala has explored the distinctive characteristics of the scientific method. He has made other contributions to the philosophy of science, including explorations of how the process of natural selection provides a causal explanation of the evolution and design of living forms. With extensive knowledge of the history of biology and, particularly, evolutionary theories, he has contributed extensively to the popularization of evolutionary theory as the central core of biology. He has forcefully argued that the theory of evolution, as well as science in general, need not be in contradiction with religious beliefs.

Ayala has extended Dobzhansky's renowned phrase "nothing in biology makes sense except in the light of evolution" with the phrase "nothing in evolution makes sense except in the light of genetics."

Francisco J. Ayala is the author of more than 900 specialized articles as well as the author or editor of more than 20 books. He is a member of the

National Academy of Sciences, the American Academy of Arts and Sciences, the American Philosophical Society, the California Academy of Sciences, the Russian Academy of Sciences, the Italian Accademia Nazionale dei Lincei, and academies of other countries, including Mexico, Serbia, and Spain. He has served as president of the American Association for the Advancement of Science, the Sigma Xi The Scientific Research Society, and the Society for the Study of Evolution. He has received numerous honors and awards, and he has been designated *Doctor Honoris Causa* by universities in Argentina, Czech Republic, Greece, Italy, Mexico, Spain, and Russia. In 2002 he received the National Medal of Science, the highest scientific award in the United States, from President George W. Bush.

BIBLIOGRAPHY

Ayala, F. J. 1965a. *Drosophila dominicana,* a new sibling species of the *Serrata* group. *Pacific Insects* 7: 620–622.

——. 1965b. Evolution of fitness in experimental populations of *Drosophila serrata. Science* 150: 903–905.

——. 1965c. Sibling species of the *Drosophila serrata* group. *Evolution* 19: 538–545.

——. 1966a. Dynamics of populations. I. Factors controlling population growth and population size in *Drosophila serrata. The American Naturalist* 100: 333–344.

——. 1966b. Evolution of fitness. I. Improvements in the productivity and size of irradiated populations of *Drosophila serrata* and *Drosophila birchii. Genetics* 53: 883–895.

——. 1968. Biology as an autonomous science. *American Scientist* 56: 207–221.

——. 1970. Teleological explanations in evolutionary biology. *Philosophy of Science* 37: 1–15.

——. 2006. *Darwin and Intelligent Design.* Minneapolis, MN: Fortress Press.

——. 2007. *Darwin's Gift to Science and Religion.* Washington, DC: Joseph Henry Press.

Hey, J., W. M. Fitch, and F. J. Ayala, eds. 2005. *Systematics and the Origin of Species: On Ernst Mayr's 100th Anniversary.* Washington, DC: National Academies Press.

Russell, R. J., W. R. Stoeger, and F. J. Ayala, eds. 1998. *Evolutionary and Molecular Biology: Scientific Perspectives on Divine Action.* Vatican City State/Berkeley, CA: Vatican Observatory and the Center for Theology and the Natural Sciences.   —A.B.

B
<hr>

Bacterial evolution

The Microbe is so very small
You cannot make him out at all
But many sanguine people hope
To see him through the microscope.

Actually, Hilaire Belloc, who wrote these amusing lines, was several centuries out of date. Microbes were first seen in 1674 by Antoni van Leeuwenhoek, looking at scrapings from his teeth through a simple microscope. They were labeled *bacteria* in 1828 by the German biologist Christian Gottfried Ehrenberg, from the Greek word meaning "small stick." In subsequent years, biologists discovered the enormous diversity of bacteria. The struggle to understand that diversity, its evolutionary history, and the continued rapid evolution of bacteria is one of evolutionary biology's most challenging and exciting areas of study.

Bacteria are relatively simple, one-celled organisms, labeled prokaryotes, meaning that they do not have a special compartment, or nucleus, in which the genetic material is housed. There are two types of genetic information in a bacterial cell. First, there is a small, usually circular chromosome within the cell that primarily carries the genes responsible for building the cell and its components. Second, there are a number of genes floating in the cell that are primarily involved with regulating the cell's metabolic processes. Bacteria do not have more sophisticated cell parts (organelles) such as mitochondria (used for power generation) or chloroplasts (used, e.g., by eukaryotes—organisms with cells that do have nuclei—for photosynthesis). They do have ribosomes, organelles used for building proteins from the instructions given by RNA. They reproduce by simple cell division; one cell divides into two daughter cells and each daughter cell is a clonal reproduction of its mother cell.

Like all organisms, bacteria are open to evolution through natural selection: cells with advantageous features proliferate faster than those without them. Advantageous features arise in bacterial cells through three avenues. First, the genes of bacteria change spontaneously, or mutate, and clonal diversity can accumulate rapidly. Second, some bacteria are capable of a kind of sexual reproduction through the exchange, between cells, of some of the genetic material on their chromosomes. This gene exchange places mutants that have arisen in one cell line into other cell lines. Third, bacterial genes that are not on the main chromosome can often be transferred directly from one cell to another. This process, called horizontal gene transfer, can facilitate the rapid spread of a gene throughout a bacterial colony and change the characteristics of the colony quickly. Because prokaryotes are simpler than eukaryotes, the natural assumption is that the former evolved before the latter, and that the latter evolved from ancestors that resembled the former. This assumption seems to be borne out by fossil records, by the nature of cell function, and by comparative studies, particularly those at the molecular level. The earth is around 4.5 billion years old and during its early years the surface was far too hot to sustain life. The first plausible traces of fossil life date back to around 3.75 billion years ago. At around 3.5 billion years there is solid evidence of bacterial existence—cyanobacteria, otherwise known as blue-green algae. They clustered together in sheets known as stromatolites and lived in the sea along the coasts. Showing that change is not a necessary condition of life on earth, blue-green algae of the same recognizable form still flourish today.

Thanks particularly to the work of the American microbiologist Carl Woese, it is thought that the last universal common ancestor—the parent of all living things past and present—was about 3.5 billion years ago. Woese, however, stresses that it is probably wrong to think in terms of just a single organism giving birth to all others. In fact, the tree of life is really a picture of how life evolved after reproduction and inheritance through cell division emerged and became predominant over rampant horizontal gene exchange and transfer. Before then, it was likely that the focus of the evolutionary process was on the proliferation of individual genes and small clusters of genes. The discrete lineages retraced through phylogenetic methods were not likely in existence among the earliest cells. It is better, and more accurate, to think in terms of a group or population of ancestors, all of whom would have been important. Woese and his group revealed a more accurate picture of the tree of life than hitherto, and one of the most astounding findings is that the prokaryotes consist of two very different groups: the bacteria (or true bacteria or eubacteria) and archaea (or, as they have been called, Archaebacteria).

There are significant differences between bacteria and archaea. Most important is that archaea are more like the eukaryotes in the way they read and use the information of the DNA molecules. This appears to be the first major split that we can retrace between the bacteria and the common ancestor of the archaea and the eukaryotes (see Figure 1 in the main essay "Molecular Evolution" by Francisco J. Ayala in this volume). Many archaea are capable of or need to live in extreme conditions (for example, hot or salty). This had led to speculation that they may resemble some of the earliest life forms, especially if, as many now suppose, the earliest life was found in very hot, deep-sea vents. Somewhere along the line, therefore, bacteria began to lose these abilities, or they were universally lost and then regained.

The evolution of bacteria saw major changes in the ways in which energy can be obtained and used. In particular, as can be determined from molecular studies as well as remains in the geological record, the first bacteria were obligatorily anaerobic, meaning that they could not survive in significant concentrations of oxygen. They found their energy by nonoxygen means; one of the most common means was by fermentation. Of course, for the first half of the earth's existence, there were no significant concentrations of oxygen available. As the earth sustained chemical changes, oxygen levels began to rise and selection pressure toward oxygen-tolerant (aerotolerant) organisms began. Next, oxygen-tolerant organisms began to use oxygen to produce energy. The first evidence of the facultative aerobic organisms—organisms that could switch from one form of energy production (without oxygen) to another (using oxygen)—can be traced to about 2.5 billion years ago. Oxygen levels continued to rise until they were about 1% of today's present level of 20%. About 2.25 billion years ago, exclusively oxygen-using organisms (aerobes) appeared.

Eukaryotes appeared shortly after the emergence of obligate aerobes. It is now believed, thanks to the brilliant hypothesis of American biologist Lynn Margulis, that eukaryotes were formed through symbiotic interactions among various prokaryotes. Some microorganisms, presumably archaealike, absorbed other microorganisms, some of which were bacteria. While the DNA on the nuclear chromosomes of eukaryotes indicate their affinity with archaea, it appears that the mitochondria—the power plants of eukaryotic cells—are descended from the subgroup proteobacteria. (*Escherichia coli* is today's well-known proteobacteria; it lives in the guts of all mammals, including humans.) The chloroplasts of plants, those cellular structures that drive photosynthesis, also appear to have a symbiotic origin. Thus eukaryotes are truly chimeras, which are the organisms carrying genetic instructions from many different types of ancestors.

The symbiotic relationships that formed the eukaryotes have echoes in the complex relationships that modern eukaryotes have with modern bacteria. Bacteria today are ubiquitous and other organisms cannot live without them. Many plants depend on bacteria to extract the nitrogen from the air so that they can use it themselves. Aiding this process is the aim of many artificial fertilizers, and it is well known that today agriculturalists are working hard to modify plants genetically so that, thanks to genes transferred in from bacteria, they can themselves create usable nitrogen. Mammals use many bacteria in their stomachs and intestines, usually called gut flora, to digest their food. Bacteria in these places also have other functions, like preventing certain diseases or stopping various potential allergies. More generally, humans use bacteria for their own ends, as in fermentation. Today, bacteria are even used in cleaning up oil spills.

Unfortunately, it is often difficult to live with bacteria. Many diseases of animals (especially humans) and of plants are due to bacteria. Reverting back to history, tooth decay is due primarily to four kinds of bacteria found in the mouth: *Streptococcus sanguis*, *S. sobrinus*, *S. mitis*, and *S. mutans*. Other unpleasant bacterial diseases in humans are leprosy (Hansen's disease), caused by *Mycobacterium leprae*; whooping cough, caused by *Bordetella pertussis;* cholera, caused by *Vibrio cholorae;* and tuberculosis, caused by *Mycobacterium tuberculosis*. Recently, peptic ulcers were found to be caused by

Helicobacter pylori. Certain intestinal diseases, such as Crohn's disease, are thought to be the result of the body's failure to recognize its own bacterial gut symbionts.

Bacteria are known to be rapid evolvers. Perfect examples of natural selection in action, they genetically put up barriers against human preventative measures quickly. Antibiotics, especially penicillin, were the great medical discoveries of the twentieth century because they entirely destroyed targeted bacteria without affecting the human host. Unfortunately, thanks to mutation, high rates of reproduction, and horizontal gene transfer, bacteria develop natural resistances to antibiotics. For instance, almost as soon as penicillin was introduced during World War II, it was found that *Staphylococcus aureus* was developing gene-directed biochemical methods of defense. The same is true of other bacteria. Notoriously, venereal disease used to be cured with just one shot of penicillin. Now it takes a cocktail of drugs to combat it. It does not help that humans do many things that encourage resistance development, such as when prostitutes keep themselves on continuous, relatively low doses of antibiotics. Bacteria seem to find this a challenge, finding ways to survive and thrive.

The capability of rapid evolution in bacteria also presents a challenge for understanding their earliest evolutionary history. We are attempting to retrace events that began at the dawn of life, changing rapidly and through mechanisms like the horizontal transfer of genes with signatures not easy to trace. Modern advances in molecular biology are changing our understanding, and the study of bacterial evolution is among the most challenging and exciting topics in modern evolutionary biology.

BIBLIOGRAPHY

Deacon, J. *The Microbial World*. http://helios.bto.ed.ac.uk/bto/microbes/.
Howland, J. 2000. *The Surprising Archaea*. New York: Oxford University Press.
University of California Museum of Paleontology. *Introduction to the Archaea*.
    http://www.ucmp.berkeley.edu/archaea/archaea.html/.   —*J.T. and M.R.*

Baer, Karl Ernst von (1792–1876)

Karl Ernst von Baer was perhaps the most eminent morphologist and embryologist of his day. He was born in Estonia to a German noble family. He attended the new university in Dorpat and continued medical studies in Berlin and Vienna. More interested in the theoretical aspects of the science, he pursued postdoctoral study at Würzburg (1815), where the residual influence of Friedrich Schelling (1775–1854) could still be felt, transmitted through the genial hands of Ignaz Döllinger (1770–1841). In 1817 he moved to Königsberg to become assistant to Karl Friedrich Burdach (1776–1847). Baer's sober experimental approach to anatomy led him to shed much of his earlier enthusiasm for a more romantic and aesthetic interpretation of natural phenomena. In 1826 he began studying the formation of the vertebrate embryo and initially concerned himself with the origin of the egg in mammals, which had been assumed to be identified with the Graafian follicles of the ovaries or with a fluid emitted by the follicles. In the spring of the next year he sacrificed his director's pet dog and located minute yellow eggs in its follicles, a discovery with which his name would henceforth be associated. He continued his embryological studies, publishing the two parts of his masterwork, *Die Entwickelungsgeschichte der Thiere* (The developmental history of animals), in 1828 and 1837. In 1834 he accepted a chair in St. Petersburg, where he undertook various kinds of anthropological and ethnographic studies for the Russian Empire. In his later years he became a moderate evolutionist, although also a decided opponent of Darwinian theory.

Like Georges Cuvier (1769–1832), Baer held that the animal kingdom could be separated into four distinct archetypes: the radiata (e.g., starfish and sea urchins), the mollusca (e.g., clams and octopuses), the articulata (e.g., insects and crabs), and the vertebrata (e.g., fish and human beings). He denied recapitulation theory—the idea that the embryos of more complex animals passed through morphological stages comparable with those of the adult forms of organisms lower in the hierarchy of life. He maintained that the embryo of an animal exemplified from the beginning of its gestation only the archetype or *Urform* of that particular organism. "The embryo of the vertebrate," he asserted, "is already at the beginning a vertebrate" (1828–1837, 1: 220). So a human fetus, he held, would move through stages in which it would take on the form of a generalized vertebrate, a generalized mammal, a generalized primate, and finally a particular human being. The form of the growing fetus moved from the general to the specific. The human embryo in its early stages, therefore, never assumed the mature form of an invertebrate or of a fish.

Despite Baer's rejection of strict recapitulation, he did allow that species within a given archetype displayed virtually identical structures during the earlier stages of embryogenesis. In a famous passage that Darwin mistakenly attributed to Louis Agassiz (1807–1873), Baer mentioned that he had two little embryos that he had forgotten to label: "They might be lizards, small birds, or very young mammals. The formation of the heads and trunks in these animals is quite similar. The extremities are not yet present in these embryos. But even if they were in the first stages of development, they would not indicate anything; since the feet of lizards and mammals, the wings and feet of birds, as well as the hands and feet of men develop from the same fundamental form" (1828–1837, 1: 221).

Though he became wary of the romantic *Naturphilosophie* exhibited by many German morphologists, Baer retained certain metaphysical ideas that connected him with the tradition of transcendental idealism that was inaugurated by Schelling. He seems never to have abandoned the conviction, for example, that the archetype of the organism, as a kind of extraphysical entity, guided the creature's morphological development. "The type of every animal," he declared, "both becomes fixed in the embryo at the beginning and governs its entire development" (1828–1837, 1: 220).

Baer objected to the transmutational theory that was quickly becoming rooted in German biology. He understood that ideas of recapitulation and of species evolution gave seductive succor to one another, and he firmly opposed them both:

> One gradually learned to think of the different animal forms as developing out of one another—and then shortly to forget that this metamorphosis was only a mode of conception. Fortified by the fact that in the oldest layers of the earth no remains from vertebrates were to be found, naturalists believed they could prove that such unfolding of the different animal forms was historically grounded. They then related with complete seriousness and in detail how such forms arose from one another. Nothing was easier. A fish that swam upon the land wished to go for a walk, but could not use its fins. The fins shrunk in breadth from want of exercise and grew in length. This went on through generations for a couple of centuries. So it is no wonder that out of fins feet have finally emerged. (1828–1837, 1: 200)

The community of German zoologists lived along several conceptual fissures. Some, like the redoubtable Baer, rejected the notion of recapitulation and the supportive doctrine of species descent. Georg Heinrich Bronn (1800–1862), Darwin's first German translator, thought that species progressively appeared on the earth over vast periods of time according to a divine plan. He recognized that embryonic development bore a strong analogy to the morphological development of species; but perhaps fearing the consequences of the recapitulational idea, he stressed the analogical character of the parallel and reaffirmed the boundaries of embryogenesis that Baer had established. Others, like the romantic Lorenz Oken (1779–1851) and the aesthetically driven Georg Gustav Carus (1789–1869), unhesitatingly advanced the theory of the archetype and its attendant notion of recapitulation, while the embryologists Friedrich Tiedemann (1781–1861) and Johann Friedrich Meckel (1781–1833), under the influence of Jean-Baptiste Lamarck, accepted the doctrine of species transformation. Darwin adopted recapitulation theory despite Baer's strictures. He agreed that the embryo passed from a generalized to a more particular morphological state. He simply held that the phylogenetic ancestor was of a general type—for instance, that the evolutionary progenitor of man was a generalized vertebrate whose form would be recapitulated in the human embryo. "Thus the embryo," Darwin observed in *On the Origin of Species* (1859, 338), "comes to be left as a sort of picture, preserved by nature, of the ancient and less modified condition of each animal."

By the early 1870s the evolutionary thesis had swept through Europe, leaving only the most recalcitrant in its wake. Baer understood that the kind of empirical evidence Darwin cited, especially from biogeography and paleontology, pointed in only one direction, but he would not countenance Darwin's devices of species change—chance variation and natural selection, which could not account for the teleological structure of organisms. If Darwinism were correct, intermediate forms should be produced by selection, but none were found. Chance variation, which the theory assumes, should fluctuate in both positive and negative directions, thwarting evolutionary advance. Moreover, selected traits would, through random mating, be swamped out. Baer rather supposed that the great classes of animals and plants appeared suddenly in ancient times, as the gap-ridden fossil record indicated, and that internal, teleological laws had to determine evolutionary transitions within each of the larger groups of animals—fish, reptiles, birds, mammals—all arching toward particular goals and diminishing in power as the present time was approached. Baer offered his specific hypothesis only tentatively but hoped that he had awakened in the reader recognition that "for a true understanding of nature, we cannot dispense with a governing intelligence" (1876, 2: 473). Remembering his early work in embryology, evolutionists and antievolutionists alike revered the great Karl Ernst von Baer.

BIBLIOGRAPHY

Baer, K. E. von. 1828–1837. *Entwickelungsgeschichte der Thiere: Beobachtung und Reflexion.* 2 vols. Königsberg: Bornträger.
———. 1876. *Ueber Darwins Lehre.* In his *Studien auf dem Gebiete der Naturwissenschaften,* 2 vols., 2: 235–480. St. Petersburg: Schmitzdorff.
———. [1886] 1986. *Autobiography of Dr. Karl Ernst von Baer.* 2nd ed. J. Oppenheimer, ed..; H. Schneider, trans. New York: Science History Publications.
Darwin, C. 1859. *On the Origin of Species.* London: John Murray.
Oppenheimer, J. 1967. *Essays in the History of Embryology and Biology.* Cambridge, MA: MIT Press.
Richards, R. 1992. *The Meaning of Evolution: The Morphological Construction and Ideological Reconstruction of Darwin's Theory.* Chicago: University of Chicago Press.   —R.J.R.

Bartholomew, George (b. 1919), and Schmidt-Nielsen, Knut (1915–2007)

Although not mainstream evolutionary biologists, George Bartholomew and Knut Schmidt-Nielsen were enormously influential for the study of adaptation. Bartholomew and Schmidt-Nielsen, with their colleagues and students, pioneered the study of physiological and biochemical traits of animals as adaptations to specific environmental conditions. The research programs they developed asked how animals function in challenging situations. For example, one of Schmidt-Nielsen's projects asked how desert animals maintain water and salt balance. These types of questions led to broader comparisons among species that showed how well differences in physiological and biochemical properties were matched to differences in particular environmental variables. One of Bartholomew's projects examined how populations of sparrows living in salt marshes were able to obtain water by drinking seawater, a capability not shared in populations of the same species that do not live in salt marshes. The cumulative results of their work, along with the work they inspired, reveal how well the functional systems of different species, and even different populations of the same species, have been molded to meet different environmental circumstances, presenting a formidable case for the power and precision of adaptation.

Bartholomew and Schmidt-Nielsen were contemporaries; Bartholomew received his PhD in 1947 from George Clarke and Schmidt-Nielsen received his DPhil from Nobel Laureate August Krogh in 1946. For over 40 years they directed critical studies in physiology that were animated by ecological and evolutionary perspectives and the conviction that animals were functionally integrated, meaning one had to understand the entire animal in its environment. Both scientists studied a wide range of problems in a diversity of animals in many environments. Bartholomew's best-known studies revolved around temperature regulation and energy metabolism, topics that applied to survival in extreme climates, hibernation, and the energetics of flying in birds. Schmidt-Nielsen is best remembered for his examinations of water and salt balance in desert and marine animals as well as his later work on the scaling of metabolic rates and the energetic costs of locomotion in terrestrial vertebrates. Each is well known among generations of students for their studies of curious or charismatic problems, such as Bartholomew's scrutiny of the adaptations to diving in marine iguanas and Schmidt-Nielsen's study of the camel's adaptation to desert conditions. Bartholomew and Schmidt-Nielsen both were renowned classroom teachers, mentors, and writers; Bartholomew has over 1,000 intellectual descendants (students and students of students) and Schmidt-Nielsen's textbooks are among the most popular and appealing college texts ever written. Each influenced biologists far beyond his immediate discipline, and each was influential in building two of the best zoology departments in the United States, Bartholomew at UCLA and Schmidt-Nielsen at Duke University.

The intellectual legacy of Bartholomew and Schmidt-Nielsen can be found in many areas, from the study of biochemical adaptation at the molecular level to the comparison of physiological properties at the population and species level. The mechanistic approach to adaptation that each promoted represents one among several methods used to understand this central concept in Darwinian evolution.

BIBLIOGRAPHY

Dawson, W. R. 2005. George A. Bartholomew's contributions to integrative and comparative biology. *Integrative and Comparative Biology* 45: 219–230.
Schmidt-Nielsen, K. 1998. *The Camel's Nose: Memoirs of a Curious Scientist.* Washington, DC: Island Press.   —J.T.

Bates, Henry Walter (1825–1892)

British naturalist Henry Walter Bates is credited with the discovery of protective mimicry. Today, the scenario in which an organism resembles an unrelated, unpalatable species is called Batesian mimicry. Although not the first to notice shared appearances of species, Bates reinvented the subject in 1862.

He turned collectors' oddities into a powerful case study. Mimicry became a classic example, and one of the most popular textbook illustrations, of natural selection in action.

As was common in Victorian England, the young Bates, a clerk in the manufacturing city of Leicester, pursued natural history as a largely self-educated hobbyist. He was an ambitious insect collector and taxonomist, publishing from a young age in natural history journals. With his friend Alfred Russel Wallace, he turned to professional collecting. They shared an interest as well in the theoretical questions being raised in the 1840s. Both read widely on species and were drawn to early discussions of evolution and natural law. Their reading of travelers' accounts of the tropics led them to work in Brazil.

In 1848, Bates and Wallace began an exploration of the Amazon, arranging to sell exotic specimens to collectors through an agent in London. Although collecting paid their way, they hoped to be able to use their observations to address the nature and origin of species. Going separate ways in 1850, Bates spent the next nine years along the Amazon. He became well known through his specimens and notes on behavior and distribution. Known as a taxonomic expert on several groups of insects, Bates also earned credibility by being a bold and talented explorer.

Bates's fieldwork stands out for its early interest in distributional data and its attention to the relations of species to each other. Correspondence reveals that Bates and Wallace discussed at length how such facts might address the origin of species. Upon his return to England in 1859, Bates used his experience in a series of taxonomic papers and species notes, which included his ecological observations. Charles Darwin argued along lines familiar to Bates, and Bates became an immediate, enthusiastic Darwinian after reading *On the Origin of Species*. Among a series of papers on variation and geographical distribution of insects, Bates developed the case of mimetic butterflies as evidence of evolutionary change. He provided the details of an example of Darwinism in nature, where the struggle for existence led to adaptation. It was a stunning support because no previous naturalistic explanation had existed for such a striking phenomenon.

Bates was a respected naturalist and a member of the major scientific societies, but his career path shows that Darwinians did not always or easily invade the old guard. He applied but failed to secure a curatorial post at the British Museum of Natural History. He was hampered by his working-class origins and lack of social connections, as well as by scientific politics. He became assistant secretary of the Royal Geographical Society, where he was responsible for the organization of several major expeditions. Bates continued to produce a long series of taxonomic papers. He also did contractual work for other collectors, specializing in several tropical insect groups.

Along with mimicry theory, Bates became famous through his narrative, *The Naturalist on the River Amazons* (1863), a classic still in print today. An immediate success, it was highly praised for its observations, theoretical insights, and skillful portrayal of the tropics.

BIBLIOGRAPHY

Bates, H. W. 1863. *The Naturalist on the River Amazons*. London: John Murray.

Clodd, E. [1863] 1892. Memoir of the author. In H. W. Bates, *The Naturalist on the River Amazons*, xvii–lxxxix. Reprint ed. London: John Murray.

Moon, H. P. 1976. *Henry Walter Bates FRS, 1825–1892: Explorer, Scientist and Darwinian*. Leicester: Leicestershire Museums, Art Galleries, and Records Service.

O'Hara, J. E. 1995. Henry Walter Bates—His life and contributions to biology. *Archives of Natural History* 22: 195–219.   —W.C.K.

Bergson, Henri (1859–1941)

Henri Bergson was a French philosopher who taught for many years at the Collège de France. He had great influence in his country and in Britain (his mother was English). In 1927 he was awarded the Nobel Prize for literature. Born of Jewish parents, he drew close to Catholicism but refused to repudiate his heritage. He died at the beginning of the Vichy regime from a cold caught while waiting for an identity card.

He enters the story of evolution because of his *L'Evolution créatrice*, first published in 1907 (English translation, *Creative Evolution*, 1911). Bergson, who was influenced by the writings of Herbert Spencer, always saw evolution as an upward progress to humankind, but he did not agree with the Darwinian claim that a blind process like natural selection could do the job. Apart from anything else, he could not see how selection could account for the fact that sometimes two quite separate lines seemed to take the same evolutionary directions, acquiring the same innovations.

This led Bergson to introduce his famous, or notorious, notion of a vital spirit, what he called the *élan vital* (*élans vitaux* in the plural). This is a force, rather like a life spirit of the kind endorsed by Aristotle, that supposedly guides and drives the evolutionary process. Although Bergson approached the problem as a philosopher, he found support from some scientists, notably the German embryologist Hans Driesch, who as a result of his experiments on developing organisms was likewise inclined to a vitalist approach, calling his version of the force the *entelechy*.

Vitalism was anathema to most biologists, especially at the beginning of the twentieth century when men like Jacques Loeb were endeavoring to put the subject on a materialistic and naturalistic basis. Those attracted to Bergson's vital forces included a French Catholic priest and paleontologist, Father Pierre Teilhard de Chardin. His science-religion synthesis of the evolutionary picture, seeing upward progress to the Godhead, the Omega Point, clearly owed much to Bergson. Another influenced by Bergson was the American population geneticist Sewall Wright, although probably in his case the direct influence of Spencer's writings was more significant. A third enthusiast was the English biologist Julian Huxley, grandson of Thomas Henry Huxley and older brother of novelist Aldous Huxley. Searching for a non–God-based meaning for life, Julian Huxley found Bergson's ideas stimulating. His first book, *The Individual in the Animal Kingdom* (1912), was openly influenced by Bergson's thinking. By 1942, when Huxley published his massive overview, *Evolution: The Modern Synthesis,* he sadly concluded that vitalism is no basis for science, but he still saw the evolutionary world as progressing up to humankind, trying to give naturalistic backing to this supposition.

Today, few would openly subscribe to an *élan vital,* but many suspect that evolutionists who repudiate strict Darwinism in favor of alternatives, such as "order for free," secretly harbor philosophical yearnings against blind and mechanistic determinism, the same dislike that motivated Bergson. This is clearly the motivation of the movement's leader, Stuart Kauffman, in his *Reinventing the Sacred.*

BIBLIOGRAPHY

Bergson, H. 1907. *L'évolution créatrice.* Paris: Alcan.
Huxley, J. S. 1912. *The Individual in the Animal Kingdom.* Cambridge: Cambridge University Press.
———. 1942. *Evolution: The Modern Synthesis.* London: Allen and Unwin.
Kauffman, S. 2008. *Reinventing the Sacred: A New View of Science, Reason, and Religion.* New York: Basic Books.
Ruse, M. 2003. *Darwin and Design: Does Evolution Have a Purpose?* Cambridge, MA: Harvard University Press.   —M.R.

Biogeography

Biogeography is to space as paleontology is to time. The biogeographer is interested in the ecology and evolution of organisms as distributed around the globe; the paleontologist is interested in the ecology and evolution of organisms as revealed through the fossil record. Of course, there is much overlap between the two because both are interested in how the past influences distributions of organisms in the present.

There is a paradox about the respective statuses of biogeography and paleontology. For the layperson thinking about evolution, fossils come to mind, to the point that many people think that the fossil record is virtually the only thing that matters. For the evolutionary biologist seeking conviction of the truth of evolution, biogeography is simply outstanding.

This was certainly the case for Charles Darwin and also for the codiscoverer of natural selection, Alfred Russel Wallace. In his autobiography, in talking of his turn to evolution, Darwin highlighted the way in which similar but different organisms succeeded each other as he journeyed down South America. Then, famously, he spoke of the organisms (tortoises and birds) that he encountered on the Galápagos Archipelago in the Pacific: similar but slightly different. He could not see how this could be unless founders had come from the mainland and then evolved as they spread out across the island group.

When Darwin came to present his theory of evolution in *On the Origin of Species* (1859), paleontology was somewhat problematic—gaps in the fossil record, the apparent absence of life forms before the Cambrian, and so forth.

But biogeography was a triumph. Particularly important for Darwin was the way that the inhabitants of islands are so similar to the inhabitants of their closest continents and not to the inhabitants of other continents farther away. "The most striking and important fact for us in regard to the inhabitants of islands, is their affinity to those of the nearest mainland, without being actually the same species. Numerous instances could be given of this fact. I will give only one, that of the Galapagos Archipelago, situated under the equator, between 500 and 600 miles from the shores of South America. Here almost every product of the land and water bears the unmistakeable stamp of the American continent" (Darwin 1859, 416).

Wallace was likewise very sensitive to the significance of biogeography. His first paper hinting at evolution (in 1855) focused on the ways in which organisms seem to pattern themselves around the globe. He found that newly arrived organisms seem always to connect with living organisms close to them. "Every species has come into existence coincident both in space and time with a closely allied species" (Wallace 1855 [1871], 25). In *The Geographical Distribution of Animals* (1876) Wallace turned to issues in biogeography, looking both at the special issues that surround islands as well as broad questions of distribution. Well known is the so-called Wallace's line (see figure), a break between two very different sets of organisms, Asian and Australasian, running through the Malay Archipelago, between Borneo and Sulawesi, and between Bali (west) and Lombok (east).

A major query for these nineteenth-century evolutionists was exactly how organisms get dispersed. Darwin spent considerable time running experiments trying to show how things like seeds and snails could withstand salt water and get carried by jetsam or by birds. He discussed his work extensively in the *Origin*. A different approach was taken by Darwin's close friend, botanist Joseph Dalton Hooker (1853). He argued that plant distributions were better explained by once-existing land bridges that have now disappeared thanks to the rising sea levels. This was a part of a theory of climate, proposed by geologist Charles Lyell (1830–1833), which supposed that although continents cannot move sideways, they are always changing shape because the earth's surface is forever rising or falling.

Between Darwin's time and the forming of the synthetic theory in the 1930s, biogeography took a back seat to paleontology as most effort was directed toward the tracing of earth's history. When selection theory came back into fashion, most attention was paid to microevolutionary causal issues, particularly the ways in which genes spread and change in populations. Yet there was some revived interest in biogeography. Particularly important was the influence of systematist and ornithologist Ernst Mayr. In his classic *Systematics and the Origin of Species* (1942) he discussed in detail issues to do with distribution, most notably highlighting the importance of "rings of races." Here, one has a ring of adjacent and interbreeding populations of a species, but where the end points are reproductively isolated. Mayr argued that this is definitive evidence of the formation of species in a gradual rather than a sudden process.

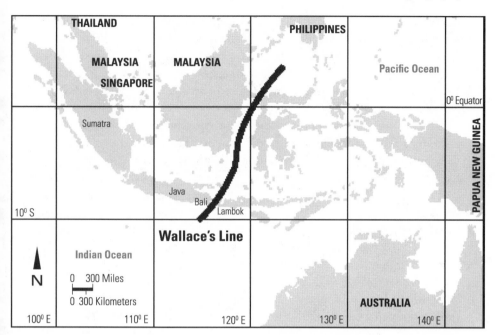

Wallace's line. When he was collecting specimens in the Far East, Alfred Russel Wallace realized that there was a line of separation between the organisms of Asia and Australasia. He discussed this and other biogeographical phenomena in *The Geographical Distribution of Animals* (1876). The line was first noted by a voyager with Ferdinand Magellan in 1521.

After the synthetic theory was well established, the major event in the history of biogeography was the coming of plate tectonic theory in the early 1960s. Plate tectonic theory offered geophysical explanations for the idea of continental drift. It now was possible to address properly the kinds of issues that had concerned the nineteenth-century evolutionists. It was not a question of lost continents but of continents that move sideways, carrying once-joined groups away from each other. Biogeography received a powerful tool and at the same time geology received strong confirmation for its audacious hypotheses. A classic example of the two-way process is given by the distribution of *Lystrosaurus*. This herbivorous, mammal-like reptile, found more than 200 million years ago, is fat, short, and squat. It is certainly not a beast that would have been a world traveler, regularly swimming across oceans. Today, it is found in the same fossil deposits (Lower Triassic) on the continents of Africa, (Southeast) Asia, and Antarctica. This would be a miracle if it were not for the fact that, more than 200 million years ago, all three continents touched when they were part of Pangaea. The fossils support the geological hypothesis, and the geological hypothesis makes sense of the fossil distributions (see Figure 27 in the main essay "The History of Evolutionary Thought" by Michael Ruse).

The development of plate tectonic theory coincided with the advent of ideas about how to reconstruct species phylogenies using a set of explicit rules, that is, ideas about cladistics. The fusion of these concepts led to the school of vicariance biogeography, articulated in the 1981 volume *Systematics and Biogeography: Cladistics and Vicariance,* edited by Gareth Nelson and Norman Platnick. Vicariance biogeographers, following Venezuelan Leon Croizat, emphasized the importance of searching for common biogeographic patterns among unrelated groups of organisms and diagnosing the role of geological events in creating those patterns. For example, a classic vicariance analysis would identify the common patterns of range limits and distributions of genetic variation among southeastern U.S. fish species, patterns tied to boundaries at major rivers, and diagnose the formation of rivers and the existence of old estuaries and sea channels as paramount events in breaking once-contiguous distributions and creating present-day patterns. In contrast, so-called dispersal biogeographers argue that the dispersal of species across broad ranges and changes in dispersal rates across major features like mountains or rivers are more important for present-day patterns.

The proper answer is surely a combination of both vicariance and dispersal. Drift and the subsequent barriers are very important. However, no one truly can say that continental drift is the only key factor in organic dispersals. Massive dispersals have occurred between North and South America as the land links have been established between them: the "Great American Interchange" (Marshall et al. 1982; see Figure 26 in the main essay "The History of Evolutionary Thought" by Michael Ruse in this volume). This is the reason marsupials, such as opossums, exist in North America and rodents appear in South America. (Obviously, the geological changes have been in part a function of drift; it is difficult to keep the two factors apart.)

It is not only thanks to geology that increasing interest is being taken in causal factors in biogeography. Early on, population geneticists realized that geographical distributions are significant in their studies. A classic case was the distribution in Africa of sickle-cell anemia, a disease in which the victim usually dies before reaching his or her fifth birthday. Why does it occur in certain parts of the continent but not in others? In one direction, it was possible to pin the disease down to the possession of a certain kind of gene—homozygotes for that gene show the disease—and it was shown that the gene causes the collapse of red blood cells. But why the distribution? Why in the Gold Coast but not in South Africa? The answer came because the genes in single doses (heterozygotes) confer a natural immunity to malaria, and the mosquito is a major pest in those and only those areas with the gene. In other words, the distribution is linked to selection setting up barriers against infection (Allison 1954a, 1954b).

With the advent of molecular methods, population geneticists were able to extend basic biogeographical principles to examine large-scale patterns of gene distributions in space. From these patterns, biologists have been able to draw inferences about the likely locations in which a species or group of species originated and the history of dispersal into its present range. These reconstructions of history, called phylogeography, represent the modern version of Darwin's and Wallace's pioneering explorations.

A major biogeographical ecological theory appeared in the 1960s. Theoretical biologist Robert MacArthur and insect specialist (and future sociobiologist) Edward O. Wilson (1967) argued that species numbers on islands—whether islands in the sea or oases in the desert—reach an equilibrium between species arriving and species leaving (perhaps through extinction). This balance is a function of such factors as the distance of the island from the mainland and of the area of the island itself. In addition, MacArthur and Wilson described how colonization and extinction could combine to generate the so-called species-area curve, a mathematical relationship that describes how the number of species found in a location increases with the area (square kilometers) of that location. Species-area curves had been described in the past, but MacArthur and Wilson offered the first quantitative explanation for the shape of those curves.

The ecological theory inspired a renaissance in biogeography. Biologists used the theory to reinterpret and unify seemingly disparate patterns, from the distribution of species on isolated mountaintops to the consequences of creating a fragmented habitat from continuous forest. The attempt to test the causal mechanisms that could create species-area curves led to the discovery of additional biogeographic patterns, such as the fact that species with larger ranges also tend to have higher population densities.

There have been several important evolutionary extensions of the ecological theory of MacArthur and Wilson. Wilson's student, paleontologist J. John Sepkoski (1978, 1979, 1984), applied the theory to changing diversity in time, arguing that at certain periods, such as the beginning of the Cambrian, we see empty niches that are then colonized by organisms at rates and with results that match the predictions of the biogeographic theory (see the main essay "Paleontology and the History of Life" by Michael Benton in this volume). The theory of the taxon cycle, originally described by Wilson, is a particularly striking extension. In the taxon cycle, species disperse through an archipelago and, on each island on which they persist, undergo an evolutionary cycle of initial population expansion, adaptation to island conditions, and eventual decline and extinction under the pressure of new colonists and changing conditions. It is, in effect, a life cycle of island biota. This theory linked ecological processes with evolutionary ones and offered an explanation for why extinctions are constantly occurring, albeit on a very long timescale. Genetic studies by Robert Ricklefs and Eldridge Bermingham (2007) revealed the signature of this process in West Indian birds and verified Wilson's insight into how ecology and evolution could be connected through biogeography.

These connections are also being made by detailed investigations of long-standing biogeographic patterns. For example, Bergmann's rule (1847) states that animals increase in size with either increasing latitude or elevation. This pattern is evident in warm-blooded and cold-blooded animals and has its explanations not in simple thermal relationships (e.g., larger animals hold heat more effectively) but in the complex relationships among temperature, humidity, growing season, and energetic efficiencies.

Biogeography continues to flourish, as it did in Darwin's day. It offers the conceptual underpinning for applied problems like the design of nature reserves and continues to play a key role in the confirmatory support of evolutionary theory.

BIBLIOGRAPHY

Allison, A. C. 1954a. Protection by the sickle-cell trait against subtertian malarial infection. *British Medical Journal* 1: 290.
———. 1954b. The distribution of the sickle-cell trait in East Africa and elsewhere and its apparent relationship to the incidence of subtertian malaria. *Transactions of the Royal Society of Tropical Medical Hygiene* 48: 312.
Avise, J. C. 2000. *Phylogeography.* Cambridge, MA: Harvard University Press.
Bergmann, C. 1847. Über die Verhältnisse der wärmeökonomie der Thiere zu ihrer Grösse. *Göttinger Studien* 3, no. 1: 595–708.
Darwin, C. 1859. *On the Origin of Species.* London: John Murray.
Hooker, J. D. 1853. *Introductory Essay to the Flora of New Zealand.* London: Lovell Reeve.
Lomolino, M. V., B. R. Riddle, and J. H. Brown. 2006. *Biogeography.* 3rd ed. Sunderland, MA: Sinauer Associates.
Lyell, C. 1830–1833. *Principles of Geology: Being an Attempt to Explain the Former Changes in the Earth's Surface by Reference to Causes Now in Operation.* 3 vols. London: John Murray.
MacArthur, R. H., and E. O. Wilson. 1967. *The Theory of Island Biogeography.* Princeton, NJ: Princeton University Press.
Marshall, L. G., S. D. Webb, J. J. Sepkoski Jr., and D. M. Raup. 1982. Mammalian evolution and the great American interchange. *Science* 215: 1351–1357.
Mayr, E. 1942. *Systematics and the Origin of Species.* New York: Columbia University Press.
Nelson, G., and N. Platnick. 1981. *Systematics and Biogeography: Cladistics and Vicariance.* New York: Columbia University Press.
Ricklefs, R. E., and E. Bermingham. 2007. The causes of evolutionary radiations in archipelagoes: Passerine birds in the Lesser Antilles. *American Naturalist* 169: 285–297.
Sepkoski, J. J., Jr. 1978. A kinetic model of Phanerozoic taxonomic diversity. I. Analysis of marine orders. *Paleobiology* 4: 223–251.
———. 1979. A kinetic model of Phanerozoic taxonomic diversity. II. Early Paleozoic families and multiple equilibria. *Paleobiology* 5: 222–252.
———. 1984. A kinetic model of Phanerozoic taxonomic diversity. III. Post-Paleozoic families and mass extinctions. *Paleobiology* 10: 246–267.
Wallace, A. R. 1855 [1871]. On the law which has regulated the introduction of new species. *Annals and Magazine of Natural History* 16: 184–196. Reprinted in *Contributions to the Theory of Natural Selection.* London: Macmillan, 1–25.
———. 1876. *The Geographical Distribution of Animals.* 2 vols. London: Macmillan.
———. 1905. *My Life: A Record of Events and Opinions.* London: Chapman and Hall.   —*J.T. and M.R.*

Birds

Since originating from small carnivorous dinosaurs in the Late Jurassic (about 150 million years ago), birds have radiated into some 9,600 species around the globe. Their colors, songs, behaviors, and ecological adaptations are endlessly complex. What sorts of factors have been key in their evolutionary radiation?

The immediate ancestors of birds were small, predatory, bipedal dinosaurs that resembled animals such as *Velociraptor, Microraptor,* and *Saurornitholestes*. Feathers, which we usually think of as a diagnostic feature of birds, actually evolved within carnivorous dinosaurs, first as a short filamentous covering in animals such as *Sinosauropteryx*, and then in increasingly more elaborate pinnate shapes as in *Beipiaosaurus, Caudipteryx,* and *Protarchaeopteryx*. Their primary function was obviously not for flight, but they would have served a nearly automatic thermoregulatory role by virtue of their insulatory property. Whether they were brightly, characteristically, or cryptically colored would have determined roles in display, species recognition, or camouflage; some color banding appears in early feathers, but not enough evidence of color remains to indicate other roles. The presence of long feathers on the fingers of at least one oviraptorid theropod, *Caudipteryx,* suggests that the feathers could have been spread over the eggs when nesting—a posture preserved in several fossil examples.

If the presence of feathers did not automatically confer flight, how did flight evolve? *Archaeopteryx,* universally recognized as the first known bird, is the first animal in the theropod-bird line to have feathers large enough to form a wing capable of sustaining the animal in the air, assuming the requisite flight stroke and underlying metabolism were present. Evidence of these features is indirect but suggestive.

Since the 1880s, two competing hypotheses for the origin of bird flight have been periodically debated. One is that bird ancestors climbed trees and used incipient wings first to parachute, then to glide, and finally to flap. The second is that bird ancestors ran along the ground and flapped incipient wings to gain increased height, either in jumps after prey or away from predators. Various modifications of these ideas have been proposed, but neither hypothesis is testable. There is no convincing evidence that bird ancestors could climb trees, but most small animals can; however, arboreality does not guarantee flight ability. It now appears that bird ancestors were terrestrial bipeds and good runners, but were they good enough to achieve liftoff?

Recent research has shifted the focus from trees versus ground to the evolution of the flight stroke itself. This stroke had its origin in the predatory movement of the forelimbs of small theropod relatives of birds, such as *Deinonychus* and *Velociraptor*. These dinosaurs shared a feature that birds inherited: the bones of the wrist that connect with the palm (metacarpal) bones had modified joints that swiveled nearly 180° from side to side. In those predatory dinosaurs, this joint permitted the hands and fingers, which were greatly elongated, to be folded close to the body when not in use, much like bird wings. When needed to deliver a predatory stroke, they could be whipped forward and downward. It is nearly this exact stroke, with elongated feathers, delivered at a slightly different angle, that constitutes one of the most frequently used flight strokes of birds.

Of course, a flight stroke is useless to an animal of the size of the theropod *Deinonychus* (about 25 kg), but the first birds were much smaller;

Archaeopteryx specimens range from pigeon to crow size. New work on the microstructure of bone tissues in dinosaurs and early birds shows that dinosaurs as a group grew quickly, at rates comparable to those of most mammals and birds. However, the bone tissues of the first birds, such as *Confuciusornis*, record relatively slower growth during just those stages (juvenile and subadult) when typical dinosaurs grew most quickly. As a result, birds effectively became miniaturized adults. At this smaller size, they would have had a greater ratio of wing surface area to body mass, and this improved wing loading would have been aerodynamically advantageous.

Did the first birds have the metabolism necessary for sustained flight? Again, direct measurement is impossible, but the growth dynamics of early birds and their dinosaurian relatives have been reconstructed from bone tissues. The high rates of sustained growth reflected in the tissues makes it difficult to draw any conclusion other than that dinosaurs, and their avian descendants, had relatively high metabolic rates. The presence of feathers is a further indication.

Part of the shift to flight involved a reorganization of the locomotory modules of bipedal dinosaurs. In these dinosaurs, the arms were not used in locomotion; the legs propelled and supported the animal, and the tail assisted in locomotion in an important way. The long muscular tail, more like a crocodile tail than like tails of birds today, attached to the spines and processes of the tail vertebrae, and inserted toward the top of the thigh bone near the hip socket. Its role in pulling back the leg, anchoring the stroke against the stabilizing tail, would have been substantial. In the dinosaurs closest to birds, the tail became reduced and stiffened, and this muscle found a more important role in controlling movements of the tail in running and in flight. The forelimbs, of course, became wings; instead of having a two-module locomotory system in which the legs and tail worked together, birds evolved a three-module system in which the winged forelimbs, the legs, and the tail each performs separate functions in the air and on the ground. Phylogenetic trends show that in the early evolution of birds, features of the flight mechanism (such as the alula and the fusion of bones) evolved and refined functions more rapidly than did features of the hindlimbs (such as perching claws) and other characters related to feeding and locomotion.

Judging from the distribution of features in birds and other dinosaurs, the first birds probably had larger clutches than the one to four eggs of today's birds. They were likely precocial rather than altricial, and they would have built nests on the ground rather than in trees; some parental care is likely. Sometime during the Cretaceous, before the diversification of the living groups of birds, growth rates apparently increased, so that adult stages were reached more rapidly. Birds today reach full size at intervals ranging from seven days (sparrow) to nearly a year (ostrich).

BIBLIOGRAPHY

Dingus, L., and T. Rowe. 1997. *The Mistaken Extinction: Dinosaur Evolution and the Origin of Birds*. New York: W. H. Freeman.

Gauthier, J. A., and L. F. Gall, eds. 2001. *New Perspectives on the Origin and Early Evolution of Birds.* New Haven, CT: Yale University Press.

Padian, K., and L. M. Chiappe. 1998a. The origin and early evolution of birds. *Biological Reviews* (Cambridge) 73: 1–42.

———. 1998b. The origin of birds and their flight. *Scientific American,* February: 38–47.

Prum, R. O., and A. H. Brush. 2003. Which came first, the feather or the bird? *Scientific American,* March: 84–93.   —K.P.

Buffon, Georges-Louis Leclerc de (1707–1788)

"There is no chapter of the theory of organic evolution more confused or more controverted than that which relates to the position of Buffon" (Lovejoy 1959, 84). Lovejoy is correct. No eighteenth-century scientist or philosopher played as prominent a role in the genesis of evolutionary ideas as Georges-Louis Leclerc de Buffon, but then again, no others were quite so ambiguous in their opinion of evolution. There is a reason for this: Buffon wrote at a time when the very idea of organic evolution—which, of course, is highly complex—was not clearly defined. In certain respects, Buffon was an evolutionist, in others he was not. From a historical perspective, two different questions need to be distinguished about Buffon and evolution: (1) Was Buffon an evolutionist? (2) What role did Buffon play in the emergence of the idea of organic evolution? All historical data about Buffon and evolution need to be evaluated in the light of these two questions.

Buffon's work touched an impressive range of domains. In 1734 he entered the Paris Academy of Sciences as a mathematician. In 1739 he became superintendent (head) of the Jardin du Roi (The King's Garden), one of the two oldest scientific institutions created by the French monarchy; the other was the Collège de France. (During the French Revolution, the Jardin du Roi was reorganized and renamed the Museum of Natural History.) As head of the Jardin du Roi, Buffon devoted more and more time to natural history. In 1741 he translated Newton's posthumously published *Method of Fluxions* (1736). He also produced several original memoirs on probability and demography. From 1749 to the end of his life, his complete works were published in the 44 volumes of his *Histoire naturelle, générale et particulière* [Natural history, general and particular], a series that he partly wrote in collaboration with several brilliant scientists of his time, in particular Daubenton and Guéneau de Montbéliard (see figure). This gigantic sum of work deals in detail with subjects as different as cosmogeny, mineralogy, geology, applied mathematics, zoology, and anthropology. Buffon's views on organic evolution (to use the modern word, which cannot be found with its present sense in the eighteenth-century writings) are dispersed in many places. They are often closely associated with reflections on the more general theme of the history of nature, a concept that came to play an increasingly important role in the course of Buffon's work.

Although geology was central to Buffon's notion of the history of nature, it was not geology that first led him to the notion of a general and

Pictures of a rhinoceros and an elephant from Buffon's *Natural History*. The many forms of animals that were apparent by the eighteenth century fascinated the savants and led them to speculate about the reasons for the diversity.

gradual evolution (in the sense of an irreversible) history of nature (Roger 1988). In *Theory of the Earth* (1749), Buffon was mainly interested in the actual causes, which, by their constant action, account for a number of geological facts (e.g., the eternal action of water, which accounts for major aspects of geographic relief). It was only after he began to elaborate his notion of species, in the mid-1750s, that Buffon began to be seriously interested in the history of nature. Three of his ideas played a major role in this respect. First, Buffon's concept of species was founded on interfertility and descent, with resemblance being subordinated to the continuity of lineages. This notion of species suggested a general vision of the unity of life in terms of temporal continuity rather than in terms of eternal types. Second, Buffon explicitly admitted that both domesticated and wild species could be modified by the action of several causes (e.g., human action or climate). For instance, he proposed that the animal species of the Old World had their counterparts in America: the latter descended from the former, but should be considered today as distinct species. Third, on the basis of Daubenton's anatomical work on animal skeletons, Buffon argued in *L'Asne* (1753) that there was an underlying unity of type of the vertebrates, and he said that this unity of type suggested the hypothesis of a common ancestry.

These ideas had a massive impact on eighteenth-century natural historians. On his three points, however, Buffon did not really adopt the hypothesis of an indefinite transmutation of species. Species could be modified, but only to a certain point. Buffon proposed to reduce the number of species of quadrupeds (200) to merely 32, interpreting the current families in terms of enlarged species. But he did not want to go any further. Most of the current species were reinterpreted in terms of varieties, but the limits of the new enlarged species could not be overstepped. Provided that the natural species is observed at the right level, they are "perduring entities, as ancient, as permanent, as Nature herself" (*Seconde Vue*, 1765). Similarly, while Buffon considered the hypothesis of transmutation in his famous memoir on the donkey (1753), where he provided an incredibly suggestive view of the whole animal world as descending from a common ancestor, he rejected this hypothesis out of hand for two reasons: (1) no true species is known to have appeared from another species and (2) the infertility of hybrids provides a definite and permanent line of distinction between species. Buffon never changed his mind on this point. Most of the alleged contradictions of his thinking result from the fact that he progressively adopted a notion of the species that included not only what we would call today a "species" but also every possible taxon up to the level of a "family." But Buffon always maintained an absolute distinction between species and varieties.

In *Époques de la nature* (1779), Buffon integrated his ideas about the history of the living world into a general vision of the history of nature. He proposed that the earth, after having been generated by a collision with a comet, had progressively become cooler. On the basis of experiments on balls made of different materials (especially iron), he calculated that at least 100,000 years had been necessary for the earth to cool to its present state from an initial state of fusion. (In another manuscript, he sometimes gives another estimation: 3 million years; see Roger 1988.) In *Époques*, he also puts forward an absolute geological chronology on the basis of the steady formation of sedimentary rocks. *Époques* gave a massive impulse to the

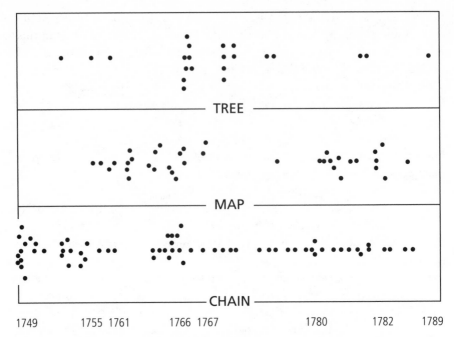

Charles Darwin was far from the first to use the picture of the tree of life as a representation or metaphor for the history of life on earth. One earlier user was Buffon. This diagram plots the frequencies with which he used the tree metaphor and other metaphors, maps, and chains. Note that the use of the tree metaphor in fact declined with time, possibly a function of the fact that Buffon came under religious pressure for his toying with evolutionary speculations, and, as one always keen to stay on the side of the authorities, he modified (or at least concealed) his thinking accordingly. (Adapted from G. Barsanti, "Buffon et l'image de la nature," in Gayon 1992, 290.)

evolutionary view of nature. This memoir also put forward a hypothesis about the origin of the different animal and plant species. Life had appeared as a consequence of the spontaneous association of organic molecules, which themselves had resulted from natural chemical reactions while the terrestrial globe was cooling. Different associations of different organic molecules had generated the different living species. From the onset, these species were as complex as they are now. As a function of the physical circumstances (i.e., climate), species appeared successively, but they did not descend from each other. All were spontaneously generated. Buffon also argued that the various species migrated from one place to another as a function of climatic change, and that they became extinct when the circumstances were no longer favorable.

This final speculation is characteristic of Buffon. On the one hand, his conceptions about the antiquity of the earth, its gradual change, the influence of physical conditions upon the history of life, and extinction look genuinely evolutionist. But his conceptual scheme for the abrupt origin of species, with no descent with modification and no room for increasing complexity, is typically pre-evolutionist.

Buffon's evolutionism was a limited evolutionism. He accepted the modification of species within narrow limits, the limits of what he called species. This is precisely what Charles Darwin and nineteenth-century evolutionists rejected, so many historians legitimately assert that Buffon was not an evolutionist in the modern sense of the term. (See the figure on page 458 for more insight into Buffon's vacillations about evolution.)

BIBLIOGRAPHY

Buffon, G.-L. (Leclerc de). 1749–1804. *Histoire naturelle, générale et particulière.* 44 vols. Paris: Imprimerie Royale; Plassan. [Note: All writings by Buffon mentioned in this entry were published as part of *Histoire naturelle, générale et particulière.*]

Gayon, J., ed. 1992. *Buffon 1988, Actes du colloque international du bicentenaire de la mort de Buffon, Paris-Montbard-Dijon.* Preface by E. Mayr, postface by G. Canguilhem. Paris: Vrin.

Lovejoy, A. O. 1959. Buffon and the problem of species. In B. Glass, O. Temkin, and W. L. Strauss Jr., eds., *Forerunners of Darwin: 1745–1859.* Baltimore: Johns Hopkins Press, 84–113.

Roger, J. 1988. Introduction. In *Buffon: Les Époques de la nature,* i–clii. Paris: Éditions du Muséum national d'Histoire naturelle.

Schmitt, S. 2007. *Buffon—Œuvres.* Paris: Gallimard.   —*J.G.*

Burgess Shale

The fossil record largely consists of the rubble and scrapings of ancient life. Almost invariably, all that we find are robust pieces of skeleton (bones and shells) as well as tracks, trails, and other signs of activity the once living organisms imprinted on the sediment and are now preserved as trace fossils. Such evidences are still central pillars in documenting evolution, but such evidence also has its disadvantages. First, many organisms, even animals, lack preservable hard parts. For example, consider the slug. Second, many trace fossils are undiagnostic as to the original type of animal, and effectively identical traces can be built by unrelated animals.

All, however, is not lost. This is because on occasion nature throws open a window into the distant past. This is possible on account of exceptional episodes of fossilization. The resultant soft-bodied fossils not only give us unparalleled insights into the original diversity of ancient life, but they also have the potential to resolve fundamental problems in evolution. Of the sedimentary deposits yielding exceptionally preserved animals, arguably the most important is the Burgess Shale (British Columbia, Canada) and its congeners, notably those from southwest China and North Greenland.

There are several reasons why Burgess Shale–type faunas are such a focus of interest. In particular, they throw remarkable light on the so-called Cambrian explosion, the seemingly abrupt appearance and evolutionary radiation of the animals. The explosion is self-evident from the dramatic appearance of skeletal fossils and the diversification of trace fossils, yet it is now apparent that if we had to rely on those sources of information alone, our understanding of this remarkable event would be greatly impoverished. This is because the soft-bodied riches of the Burgess Shale–type faunas reveal the extraordinary diversity of Cambrian marine life and in doing so pose questions of major evolutionary interest. What, if anything, triggered this explosion of life? Do we need to consider novel mechanisms of evolution? What was the role of ecology, especially predation? All these questions are ones of major current debate.

The type exemplar of such faunas, the Burgess Shale itself, is from a locality near Field, British Columbia, but numerous other localities are known in the western United States as well as eastern localities in Pennsylvania and Vermont. Further afield, key localities include Peary Land, Greenland (Sirius Passet fauna), southern China (Chengjiang, Kaili), and South Australia (Kangaroo Island). Despite the range of ages (approximately 530 to 510 million years) and also environmental settings that varied substantially in terms of water depth and probably oxygen content, the Burgess Shale–type faunas show considerable homogeneity. Thus they are dominated by arthropods, although the well-known trilobites form an insignificant fraction. Also common are sponges and various worms (especially the priapulans), while rarer groups include the chordates and even the delicate sea-gooseberries (ctenophores).

There are several reasons why Burgess Shale–type faunas are of particular evolutionary importance. First, they are well known for housing a series of supposedly bizarre animals, including the aptly named *Hallucigenia* (see figures). In fact, the history of the study of this particular animal exemplifies a radical shift in scientific thinking. The fundamental problem has been that not only do some of these fossils look very strange, but fitting them into any scheme of evolution appeared problematic. The difficulties, however, were largely a reflection of incomplete information and sometimes a lack of imagination. Resolution of these problems has depended on several lines of inquiry. In particular, molecular biology has transformed the understanding of many aspects of animal evolution. When new phylogenies are combined with reinterpretations and fresh discoveries from the Burgess Shale–type faunas, it becomes apparent that these supposedly bizarre Cambrian animals are simply staging posts in the evolution of major groups, usually referred to as phyla. These fossils show us, therefore, how body plans were assembled and, just as important, the functional and ecological contexts of these key stages in the emergence of animals. It is equally important to appreciate, however, that the basic processes of evolution involved in the Cambrian explosion appear to be no different from those operating anywhere else. Again and again the study of evolution appears to present unbridgeable gulfs between wildly disparate forms. A commonly posed question is: "How on earth could *that* have evolved? What steps could possibly connect these organisms?" Yet, repeatedly the fossil record reveals a series of intermediary stages that beautifully display the transition of form. Almost invariably these missing links display unexpected and novel features,

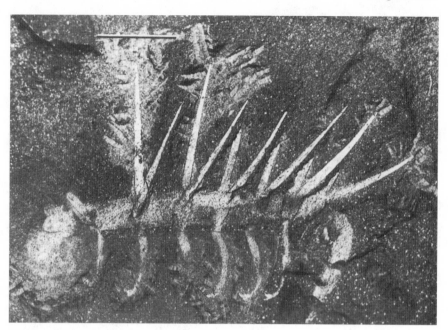

The animal *Hallucigenia sparsa,* an inhabitant of the Burgess Shale.

but the reminder here is that nature is usually a better teacher than textbooks.

The exquisite preservation of the Burgess Shale–type faunas also reveals extraordinary insights into the ecology of early animal communities. Predators transpire to be more significant than once thought, with direct evidence of their activities coming from gut contents, ferocious-looking jaws, and bite marks. So too it seems that one of the hallmarks of the Cambrian explosion, the abrupt appearance of hard shelly parts, is largely a result of the need for protection. A particularly striking feature of some hard parts is that the original skeleton formed a sort of natural chain mail that combined strength with flexibility.

The story of the Burgess Shale–type faunas is by no means complete. Indeed, much remains to be accomplished. First, there is a steady stream of new discoveries, notably from China. Perhaps the most important of recent finds were the earliest fish, the group from which humans ultimately emerged. So too, these faunas reinforce our sense of the magnitude of the Cambrian explosion. It is far from clear whether we need to seek a unique trigger, maybe at the genomic level or perhaps in terms of atmospheric oxygen, or alternatively to dismiss a single explanation as scientifically naive and look to multiple causations. Finally, what of the antecedents? Animals must have evolved from something else, but the nature and geological timing of this transition are speculative. The Cambrian explosion is preceded by the intriguing Ediacaran assemblages, but their connection to the story of animal

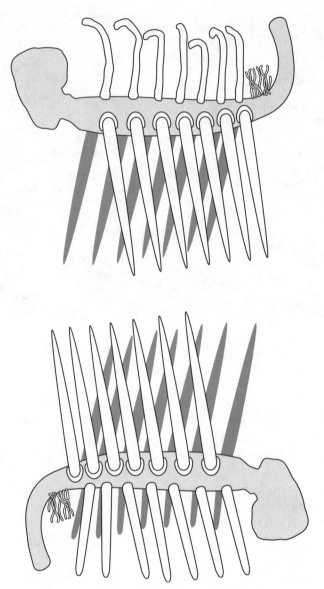

Reconstructing *Hallucigenia sparsa*. The upper drawing shows the first attempt at reconstructing *H. sparsa*. It is depicted as walking on its spines. The lower drawing shows a revised reconstruction, with the spines now uppermost, perhaps being used for defense. Compare this with Edward Drinker Cope's reconstruction of the plesiosaur, *Elasmosaurus platyurus* (see figure on page 487). Whereas Cope got his animal back to front, Conway Morris got his upside down. This all goes to show that reconstructing the past requires imagination and theory as much as brute fossil finds (see also the reconstructed dinosaurs shown in Figure 18 of the main essay "The History of Evolutionary Thought" by Michael Ruse in this volume). Again, expectation and imagination triumph over fact. People thought of the dinosaurs as brutish and cumbersome, so they reinforced this prejudice with their models and expressions like "as outdated as a dinosaur." However, the dinosaurs thrived for 150 million years, and their ancestors are still flying about everywhere.

evolution in the Cambrian is still largely unresolved. In conclusion, Burgess Shale–type faunas may be a key component in understanding evolution, but they remain a fertile field for all scientists, and especially the young.

BIBLIOGRAPHY

Budd, G. E., and S. Jensen. 2000. A critical reappraisal of the fossil record of the bilaterian phyla. *Biological Reviews* 75: 253–296.

Conway Morris, S. 1998. *The Crucible of Creation: The Burgess Shale and the Rise of Animals.* Cambridge: Cambridge University Press.

Hou, X.-G., R. J. Aldridge, J. Bergström, D. J. Siveter, D. J. Siveter, and X.-H. Feng. 2004. *The Cambrian Fossils of Chengjiang, China: The Flowering of Early Animal Life.* Oxford: Blackwell.

Valentine, J. W. 2004. *On the Origin of Phyla.* Chicago: University of Chicago Press.

Zhuravlev, A. Y., and R. Riding, eds. 2001. *The Ecology of the Cambrian Radiation.* New York: Columbia University Press.   —S.C.M.

Bush, Guy (b. 1929)

Guy Bush was educated at Harvard University and spent his working life at the University of Texas (Austin) and then at Michigan State University at East Lansing. At Harvard, Bush's interests were sparked by the strong claims of Ernst Mayr, head of the Museum of Comparative Zoology, that all speciation is allopatric, that is, demanding geographic isolation to occur. Bush challenged this view believing, as did Charles Darwin in his *On the Origin of Species,* that it is possible for sympatric speciation (speciation without geographic isolation) to occur. Following Darwin, Bush thought that it could be enough to have some kind of ecological isolation, where perhaps a species divides by specializing on different foodstuffs even though there is physical overlap. He has pursued this inquiry, with considerable success, for over 40 years.

Bush and his coworkers and students have focused on *Rhagoletis pomonella,* known as the apple maggot fly but sometimes called the railroad worm. Once confined to the hawthorn bush, in the past 100 years or so it has spread widely to cultivated apple trees introduced from Europe. By examining habits of the flies in the wild, as well as through controlled experiments, Bush has been able to show that the flies are breaking into separate species—reproductively isolated groups—simply by developing adaptations to utilize their respective hosts.

It turns out that odor is a major factor at play. The flies are responsive to the respective odors of the fruits of their hosts. Once a population settles on one host there is rapid selection perfecting the ability to recognize the fruits of this host and not of others. This can occur even when there are other hosts available and other flies utilizing them. Recently, Bush has been utilizing the methods and findings of molecular biology, both to track down the particular genes that are involved and to use this knowledge to map precisely how and when speciation has occurred.

The work of Bush and his associates is an excellent example of a phenomenon that critics of natural selection often demand before Darwinian theory can be taken as proven, namely, an observed group actually breaking into two different species. It would be nice if the work of Bush ended this particular criticism forever; however, since so often the critics are motivated more by extrascientific prejudices of a religious nature rather than by the facts, one doubts that this will be the final word on the topic.

BIBLIOGRAPHY

Darwin, C. 1859. *On the Origin of Species*. London: John Murray.

Howard, D. J., and S. H. Berlocher, eds. 1998. *Endless Forms: Species and Speciation*. New York: Oxford University Press.

Mayr, E. 1942. *Systematics and the Origin of Species*. New York: Columbia University Press.   —M.R.

C

Cain, Arthur James (1921–1999)

A. J. Cain (he always used only initials for professional publications) was one of the group of British evolutionary biologists who were members of E. B. Ford's ecological genetics group at Oxford University in the early decades after World War II. They firmly established the significance of natural selection as a major (they would have said *the* major) cause of evolutionary change. Educated at Oxford before the war, Cain had entered a world where natural selection was regarded with suspicion, if not outright hostility. Most biologists, like Linacre Professor of Zoology E. S. Goodrich, did not lecture at all on evolution, and when the topic was mentioned alternatives such as genetic drift were much favored. Not a few had fairly overt religious (at least vitalistic) reasons for distrusting a purely naturalistic approach to origins, and this line of thinking received almost official support with the appointment in 1946 of the spiritualist Alister Hardy as successor to Goodrich as the Linacre Professor of Zoology (Cain and Provine 1991; Ruse 1996).

Together with fellow Oxonian biologist Philip Sheppard, working in the light of theory being developed by Ronald A. Fisher and the experimental example of Ford, Cain set out to show that natural selection is a powerful force of nature. Their most celebrated and lasting work was on the snail *Cepaea nemoralis*, widely found in the woods, meadows, and hedgerows of the countryside around Oxford. The snail comes in many different forms, with banding and with plain shells, and with pink and brown and other colors. It was taken to be a paradigmatic example of genetic drift; such simple and varying features were thought far too insignificant to have any true adaptive value.

Cain and Sheppard showed that in fact the shell colors are tightly tied to the backgrounds in which the snails are found. A beechwood forest floor, for example, is different from a dappled ditch beside a hedge, and again different from the grass of a field grazed by cattle. They discovered that the snails are a major element in the diets of birds, thrushes especially, and that according to the backgrounds the snails are captured and eaten differentially. An important tool in making these determinations was the fact that thrushes use anvils (stones) on which to smash the snail shells, and hence the researchers could collect and count the relative numbers of forms taken by the birds in different localities (Cain 1954; Cain and Sheppard 1950, 1954). Cain's *Animal Species and Their Evolution* (1954) became a standard work on the subject.

Leaving Oxford, Cain eventually became professor at Liverpool University (where Sheppard had moved), a major center of British evolutionary studies, heavily subsidized by the Nuffield Foundation in an attempt to show the significance of evolutionary studies for medical issues, such as the natural variability of blood type and the consequent susceptibility to various ailments. By this time Cain had lost interest in frontline empirical research, although he did continue to work with collaborators and students (Cain and Curry 1963a, 1963b). Increasingly, he turned to the history and philosophy of his discipline. He wrote eloquently and informatively on such issues as the nature of species and the problems of Linnaean taxonomy, particularly about its evolutionary underpinnings (Cain 1959a, 1959b, 1962). Although his major creative work came before the major conceptual events of the 1970s (especially the coming of cladism), Cain did much to raise the importance of taxonomy as a science in its own right and one worthy of theoretical study. (Cain had spent a year early in the 1950s working in America with ornithologist and systematist Ernst Mayr. In itself this shows how by mid-century British and American evolutionists were sensitive to the merits of intellectual cross-fertilization, and specifically in Cain's case this was important for his own thinking about taxonomy.)

Subsequent years have shown, perhaps expectedly, that the work of Cain and Sheppard needs revision in light of other selection-controlled factors (Jones et al. 1977). But their work still stands as a landmark in the history of empirical studies in Darwinian evolutionary theory.

BIBLIOGRAPHY

Cain, A. J. 1954. *Animal Species and Their Evolution*. London: Hutchinson.

———. 1959a. Deductive and inductive methods in post-Linnaean taxonomy. *Proceedings of the Linnaean Society of London* 170: 185–217.

———. 1959b. The post-Linnaean development of taxonomy. *Proceedings of the Linnaean Society of London* 170: 234–244.

———. 1962. The evolution of taxonomic principles. In G. C. Ainsworth and P. H. A. Sneath, eds., *Microbial Classification*, 1–13. Cambridge: Cambridge University Press.

Cain, A. J., and J. D. Curry. 1963a. Area effects in *Cepaea*. *Philosophical Transactions of the Royal Society B* 38: 269–299.

———. 1963b. Area effects in *Cepaea* on the Larkhill Artillery Ranges, Salisbury Plain. *Journal of the Linnaean Society of London Zoology* 45: 1–15.

Cain, A. J., and W. B. Provine. 1991. Genes and ecology in history. In R. J. Berry, ed., *Genes in Ecology: The 33rd Symposium of the British Ecological Society*. Oxford: Blackwell.

Cain, A. J., and P. M. Sheppard. 1950. Selection in the polymorphic land snail *Cepaea nemoralis* (L.). *Heredity* 4: 275–294.

———. 1954. Natural selection in *Cepaea*. *Genetics* 39: 89–116.

Jones, J. S., B. H. Leith, and P. Rawlings. 1977. Polymorphism in *Cepaea*: A problem with too many solutions? *Annual Review of Ecology and Systematics* 8: 109–143.

Ruse, M. 1996. *Monad to Man: The Concept of Progress in Evolutionary Biology*. Cambridge, MA: Harvard University Press.   —M.R.

Carson, Hampton Lawrence (1914–2004)

The research of Hampton Lawrence Carson has been cited as one of the most important studies of twentieth-century biology. His work on the evolutionary biology of the insect family Drosophilidae of the Hawaiian Islands has been cited extensively and is considered one of the supreme examples of multidisciplinary research. Based on detailed examination of the banding patterns of the giant polytene chromosomes in the Hawaiian *Drosophila*, Carson developed a powerful method of population studies that led to the formulation of a new mechanism for explaining the formation of new species, Carson's Founder-Flush Theory of Speciation. He was instrumental in bringing together a team of experts in all aspects of biological sciences to work together in unraveling "that mystery of mysteries" (Darwin 1859), the origin of new organisms on this earth. Carson and his colleagues have used systematics and taxonomy, genetics, ecology, behavior, physiology, botany, geology, biochemistry, and molecular biology to understand the speciation processes within one of the most remarkable groups of organisms in the Hawaiian Archipelago. He has made major contributions in each of these areas, either by his own work or by bringing to light the relevant work of others. He has moved from one area to another as the need for new approaches arose, and he generously encouraged the participation of colleagues with new and appropriate skills. The breadth of his curiosity and the flexibility of his intellect, in a sense, represent almost an apotheosis of the multidisciplinary approach to investigating the evolutionary history of this group of insects.

Hampton Carson was born on November 5, 1914, in Philadelphia, Pennsylvania. He obtained both an undergraduate (AB, 1936) and postgraduate degree (PhD, 1943) from the University of Pennsylvania. He was hired as an assistant professor of biology at Washington University in St. Louis, Missouri, in 1943. Carson remained on the faculty as associate professor and full professor. In 1971, he moved to Hawaii to become professor of genetics at the University of Hawaii's John A. Burns School of Medicine.

Carson published nearly 300 scientific articles in the field of evolutionary biology and published a definitive book in the field of genetics entitled *Heredity and Human Life* (1963). His papers are cited widely, and he is recognized as one of the most influential figures in evolutionary biology during the past several decades. In 1985, the Philadelphia Academy of Natural Sciences awarded Carson the Leidy Medal for his outstanding contributions to the natural sciences, particularly his work on the evolutionary biology of the Hawaiian *Drosophila*. While Carson's work on the Hawaiian insects spanned more than four decades, his contributions to the natural sciences began nearly 30 years before his studies in the Hawaiian Islands. In 1935, he published a paper on medicinal herbs used by the Labrador eskimos. Then, in 1940, he published a paper on the red crossbill, and in 1945 he published a paper on the late fertilization in a species of snake. His interests in biology had obviously been strongly influenced by an interest in natural history that began as a young boy but became fully developed as a young college student member of the Delaware Valley Ornithological Club.

Although Carson retired from official duties at the University of Hawaii's Department of Genetics in 1985, he was still very active in research and continued to make significant contributions to evolutionary biology. He is considered to have been one of the University of Hawaii's most eminent scientists. He was a master teacher who, in the course of his distinguished career, gave of himself unselfishly to his colleagues and to his many students. Modesty is a virtue unparalleled in this great scientist. Once, when he was being interviewed by the news media during a symposium held in honor of his retirement, he commented that "it's very humbling. Science is a social endeavor. There is no leader. Ideas come from every which direction. . . . The most gratifying part of all is to see how it [the Hawaiian Drosophila Project] has affected the lives of so many kamaaina [Hawaiian word for locals] students" (*Honolulu Star Bulletin*, June 12, 1985). Indeed, Carson's Hawaiian Drosophila Project has been an important training ground for more than 400 undergraduate, graduate, and postgraduate students in all aspects of evolutionary biology. Also, more than 70 senior scientists from all over the world have participated in the project primarily because of Carson's leadership in the field of evolutionary biology and his ability to bring together the best minds in the field. He has been a father figure who has been able to motivate and challenge the minds of younger generations to unlock the secrets of evolutionary biology. He motivated others to think about how knowledge can help preserve the natural heritage of our planet. These qualities made this great scientist the humble person that he was.

BIBLIOGRAPHY

Anderson, W. W., K. Y. Kaneshiro, and L. V. Giddings. 1989. Hampton Lawrence Carson: Interviews toward intellectual history. In W. W. Anderson, K. Y. Kaneshiro, and L. V. Giddings, eds., *Genetics, Speciation, and the Founder Principle*, 3–28. New York: Oxford University Press.

Carson, H. L. 1963. *Heredity and Human Life*. New York: Columbia University Press.

———. 1970. Chromosome traces of the origin of species. *Science* 168: 1414–1418.

———. 1986. Sexual selection and speciation. In S. Karlin and E. Nevo, eds., *Evolutionary Processes and Theory*, 391–409. Orlando, FL: Academic Press.

Darwin, C. 1859. *On the Origin of Species*. London: John Murray.

Provine, W. B. 1989. Founder effects and genetic revolutions in microevolution and speciation: A historical perspective. In W. W. Anderson, K. Y. Kaneshiro, and L. V. Giddings, eds., *Genetics, Speciation, and the Founder Principle*, 43–76. New York: Oxford University Press.   —K.Y.K.

Castle, William E. (1867–1962)

William E. Castle, a professor at Harvard University and later at the Bussey Institution (an agricultural arm of Harvard), contributed much to the early years of genetics. Like many of his generation, he began as an embryologist, switching to problems of heredity when Mendel's work was rediscovered at the beginning of the twentieth century. Castle worked mainly on mammals, although he was also the first to use the fruit fly *Drosophila melanogaster,* which became the evolutionist's workhorse in the twentieth century. Thomas Hunt Morgan won the Nobel Prize for deciphering the genetic nature of this organism.

Castle's most famous series of experiments were on rats of varying black and white coat colors (see figure on page 470). He performed selection experiments that showed how one could change the overall range of hues in a population, from mixed to virtually entirely white or black (Castle and Phillips 1914). What was significant about this work was that once the selection was finished, the rats' coat colors continued to hover around the final point achieved. In other words, Castle showed what Charles Darwin had claimed and what many critics had denied: selection can have lasting if not permanent effects on the overall nature of a population of organisms. Castle's work proved that once selection is relaxed, everything does not revert back to the starting point.

Throughout the twentieth century there were numerous other experiments showing that selection can have lasting effects. Paradoxically, however, although Castle's work is rightly hailed as a major experimental step on the way to modern evolutionary thinking, he himself did not enter into the experiment intending to show the worth of selection. Rather, as a geneticist, his first interest was in the nature of the heritable factors that Mendelism presupposes. Castle was trying to show that these factors, what are now called genes, are not entire and unchangeable in the course of generation, but split up as reproduction occurs and recombine in new variations in the course of time. It was only after Morgan's work was completed in the second decade of the century that Castle conceded that he had been wrong and that (apart from variations caused by spontaneous changes, or mutations) genes are transmitted unchanged.

In the course of his long career, Castle had many students, the most famous and important of whom was geneticist and evolutionist Sewall Wright (see Provine 1986). In another paradox, Wright is properly known as one of the founders of modern evolutionary thought, yet the bulk of his empirical research work was on the patterns of heredity as shown by guinea pigs, an interest that went back to his studentship under Castle.

BIBLIOGRAPHY

Castle, W. E., and J. C. Phillips. 1914. *Piebald Rats and Selection: An Experimental Test of the Effectiveness of Selection and of the Theory of Gamete Purity in Mendelian Crosses.* Washington, DC: Carnegie Institution of Washington.
Provine, W. B. 1986. *Sewall Wright and Evolutionary Biology.* Chicago: University of Chicago Press.
Ruse, M. 1996. *Monad to Man: The Concept of Progress in Evolutionary Biology.* Cambridge, MA: Harvard University Press.   —M.R.

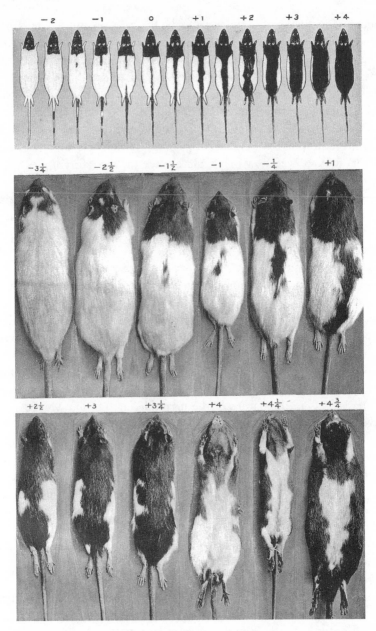

William Castle's demonstration of the effects of selection for coat color. *Top row:* the coat color scale used for classification. *Middle row:* selection for light color. *Bottom row:* selection for dark color. The paradox is that Castle's work was aimed at disproving Mendelian genetics (by showing that the units of heredity blend in each generation), not at proving the effects of natural selection. (From Castle and Phillips 1914, Plate 1.)

The Causes of Evolution (J. B. S. Haldane)

J. B. S. Haldane's (1892–1964) *The Causes of Evolution,* published in 1932, is the capstone text in the origins of theoretical population genetics. *Causes* is hailed primarily for its important, critical discussion of the work of the architects of population genetics, R. A. Fisher (1930 [1958]), Sewall Wright (1931), and Haldane himself. It is less widely, yet just as deservedly, hailed as a synthesis of classical genetics, chromosomal mechanics, cytology, and biochemistry with population genetics. This is ironic because the mathematical discussion is confined to an appendix added to original discussions of biological material drawn from Haldane's 1931 lectures at the University of Wales, Aberystwyth, entitled "A Reexamination of Darwinism."

Haldane's problematic may be posed as, "What is the nature and significance of natural selection in Mendelian populations?" In this sense, Haldane shares the problematic of his contemporaries Fisher and Wright. And in his answer, interwoven with supporting empirical evidence, Haldane in large part follows Fisher in the view that cumulative evolutionary change is mostly due to low selection pressures acting on mutations of small effect. But Haldane's support of Fisher is not in the form of an uncritical summary. Haldane carried the mathematical exploration of selection in finite population sizes, changing environments, and multiple dimensions further than Fisher had, and in part as a critical consideration of Wright's work on the topic. Moreover, and importantly, Haldane emphasized the problem of the units of selection, recognizing that selective forces acting on gametes, organisms, or populations may come into conflict.

These discussions are carried out in the middle chapters and the appendix of *Causes*. Haldane embeds them in a rich discussion of the genetic causes of within- and between-species differences discussed in earlier chapters and a broad assessment of the science of evolution as a whole in the final chapter.

Working through empirical evidence amassed from the time of the rediscovery of Mendelism in the 1900s, Haldane demonstrates that differences within and between species are due to differences between individual genes at comparable loci, differences in number of individual chromosomes or sets of chromosomes, or of whole chromosomes. Anticipating Theodosius Dobzhansky's *Genetics and the Origin of Species* (1937), Haldane uses this evidence to argue that the same sorts of processes that account for the genetic differences within species account for such differences between species. Further, Haldane broadens the Mendelian context of evolution.

The conclusion of *Causes* emphasizes the progress made in evolutionary thought since Darwin and the power of evolutionary explanations of life. At the same time, Haldane is clear that while he sees natural selection as the dominant creative evolutionary cause, he does not claim that it is the only creative cause. Throughout *Causes* Haldane highlights the multiple causes of evolution, including developmental constraints on evolution. Haldane insists that minds are surely products of evolution; they also are in all likelihood not products of natural selection. The broad discussion in Haldane's concluding chapter is to embed the science of evolution in a distinctly materialist philosophy of science.

Haldane, similar to Fisher and Wright before him, accomplished a reconciliation of natural selection and Mendelian heredity. Haldane forged ahead of his predecessors by further synthesizing a wide swath of biology with population genetics. *The Causes of Evolution* is part and parcel of the origins of theoretical population genetics, forming what is commonly called the modern synthetic theory of evolution. However, it is groundbreaking in its grander synthetic vision.

BIBLIOGRAPHY

Dobzhansky, T. 1937. *Genetics and the Origin of Species*. New York: Columbia University Press.
Fisher, R. A. 1930 [1958]. *The Genetical Theory of Natural Selection*. 2nd ed. New York: Dover Publications.
Haldane, J. B. S. 1932. *The Causes of Evolution*. London: Longmans.
Wright, S. 1931. Evolution in Mendelian populations. *Genetics* 16: 97–159.

                                                                    —*R.A.S.*

Cavalli-Sforza, Luigi Luca (b. 1922)

Luigi Luca Cavalli-Sforza can be considered a modern reincarnation of a Renaissance scholar. His major contribution to evolutionary biology emerged from his knowledge of an unusually wide range of fields that he cultivated at the professional level with a sincere passion for understanding the human condition.

Trained as a medical doctor in Italy in the 1940s, he briefly practiced medicine. He then turned his commitment to research. Attracted by the statistical aspects of genetics, he pursued his interest under the guidance of R. A. Fisher at Cambridge University. Almost at the same time (in the early 1950s), he was involved in pioneering work in bacterial sex, a highly innovative topic. He soon focused on humans. He contributed lasting methodological advances such as the still-used Cavalli-Sforza–Edwards genetic distance, an essential tool for phylogenetic and phylogeographic studies (another field he pioneered using a different name, the "geography of genes").

He creatively developed ways to use the richness of historical records in Italy (e.g., parish books and consanguinity dispensations) in human population genetics. He was among the first evolutionary scientists to believe that drift was important in the evolution of human populations. In 2004, *Consanguinity, Inbreeding, and Genetic Drift in Italy* organically presented the results of his decades-long research effort.

In 1971 he left Italy to become professor of genetics at Stanford University in California, where he is currently active emeritus professor.

Cavalli-Sforza was a pioneer in the use of modern human genes for reconstructing the natural history of man. *The History and Geography of Human*

Genes (1994) fittingly sums the breadth of his endeavor. Hailed as a breakthrough in the understanding of human evolution, it offers the first full-scale reconstruction of where human populations originated and the paths by which they spread throughout the world. By mapping the worldwide geographic distribution of genes for over 110 traits in over 1,800 primarily aboriginal populations, migrations were charted and a clock was devised to date evolutionary history. Essential information was contributed by archeology. In *The Neolithic Transition and the Genetics of Populations in Europe* (1984), he had addressed the question of whether cultural diffusion was due to the spread of ideas or of people. The issue was investigated at a world scale with special attention to the link between genetic diversity and language differentiation. Although linguists have not reached consensus on language classification at a global scale, an intriguing overlap between the distribution of genes and languages was found at different geographic scales.

Given the breadth of his research interests, it is not surprising to find a great deal of work collected in a book dedicated to the exploration of models that might help us understand how cultural traits are transmitted (*Cultural Transmission and Evolution,* coauthored with M. W. Feldman, 1981).

Cavalli-Sforza has enjoyed a remarkably creative scientific longevity. Recently his laboratory developed new techniques for discovering DNA variation used to define new genetic markers of the Y and other chromosomes. The study of these uniparentally transmitted markers turned out to be especially effective in elucidating population origins.

Cavalli-Sforza has a natural gift for communicating his passion for research. He is the author of many general audience books and newspaper columns. The textbook *The Genetics of Human Populations* (coauthored with W. Bodmer, 1971) was awarded a prize as the most quoted text in the field in 1981.

BIBLIOGRAPHY

Cavalli-Sforza, L. L., and A. Ammerman. 1984. *The Neolithic Transition and the Genetics of Populations in Europe.* Princeton, NJ: Princeton University Press.

Cavalli-Sforza, L. L., and W. Bodmer. 1971. *The Genetics of Human Populations.* San Francisco: W. H. Freeman; Mineola, NY: Dover Publications.

Cavalli-Sforza, L. L., and M. W. Feldman. 1981. *Cultural Transmission and Evolution.* Princeton, NJ: Princeton University Press.

Cavalli-Sforza, L. L., P. Menozzi, and A. Piazza. 1994. *The History and Geography of Human Genes.* Princeton, NJ: Princeton University Press.

Cavalli-Sforza, L. L., A. Moroni, and G. Zei. 2004. *Consanguinity, Inbreeding, and Genetic Drift in Italy.* Princeton, NJ: Princeton University Press.   —*P.M.*

Chambers, Robert (1802–1871)

Robert Chambers was one of the most successful businessmen in Victorian Britain. Together with his brother, William, he had pulled himself up from abject poverty to found a thriving publishing business. They were major innovators in the use of steam presses, and they were at the fore of the production of popular literature and magazines. A weekly publication, Chambers's *Journal of Literature, Science and Arts,* of which Robert and William were coeditors and to which Robert contributed many small pieces, was very successful.

Robert Chambers was always ready with his pen. He wrote books and articles as well, mainly of a popular and (with an eye to sales) informative, gossipy nature. Prime examples included guides to walks around Edinburgh, listing places where sensational murders and other ill deeds had occurred in past years—perfect material for a family faced with the need to fill the long hours of yet another Scottish Sunday. As were many other half-educated members of society, Chambers was also very interested in science, particularly geology. In the early 1840s, this led him to write what was to become one of the sensational works of the era, *Vestiges of the Natural History of Creation* (1844), together with a sequel titled *Explanations* (1845).

Apparently, Chambers had set out to write a book on phrenology (the science of brain bumps), another popular pseudoscientific topic of the day, but he got sidetracked into writing a book on evolution. (See Figure 3 in the main essay "The History of Evolutionary Thought" by Michael Ruse for details of the way that Chambers conceived of the evolutionary process.) Given the content of *Vestiges,* it was certainly not intended as a call to atheism, but more an appeal to the lawlike nature of the world. (Although Chambers was not keen on regular churchmen, he was always a believer in a deistic way.) In common with many other evolutionists, his book was a hymn of praise to progress. As we have improvement in the social world, so also do we have it in biology, and conversely as we have improvement in the biological world, so also do we have it in the social.

Because of his business and social position, Chambers went to great lengths to conceal his authorship. Indeed, it was not revealed officially until 1884, after his death. It was probably as well that he hid his name, because *Vestiges* was highly controversial. Many in the scientific community were extremely critical, leading Charles Darwin, just then formulating his evolutionary theory, to keep quiet about his own activities. However, there were many who responded positively to *Vestiges,* most notably poet Alfred Lord Tennyson. He was completing his long poem "In Memoriam," a tribute to the memory of his dear friend Arthur Hallam. Depressed by what he saw as the empty nihilism of Lyellian uniformitarianism—no change, no progress—he drew comfort from the optimism of *Vestiges,* with its support of progress. Perhaps it was that the dead-too-young Hallam was a superior being born before his time. (See the alphabetical entry on *Vestiges* for details.)

Vestiges made the idea of evolution well known in the mid-Victorian era and paved the way for Darwin and his *On the Origin of Species.* By the time the *Origin* appeared, scientists and laypeople were exhausted from arguing, the shock of the idea of evolution was over, and a more balanced account of change could be grasped and accepted by many.

BIBLIOGRAPHY

Chambers, R. 1844. *Vestiges of the Natural History of Creation.* London: Churchill.
———. 1845. *Explanations: A Sequel to "Vestiges of Creation."* London: Churchill.
Darwin, C. 1859. *On the Origin of Species.* London: John Murray.
Ruse, M. 1996. *Monad to Man: The Concept of Progress in Evolutionary Biology.* Cambridge, MA: Harvard University Press.
Secord, J. A. 2000. *Victorian Sensation: The Extraordinary Publication, Reception, and Secret Authorship of "Vestiges of the Natural History of Creation."* Chicago: University of Chicago Press.
Tennyson, A. 1850. In Memoriam. In R. H. Ross, ed., *In Memoriam: An Authoritative Text Backgrounds and Sources Criticism,* 3–90. New York: Norton.
                                                                        —M.R.

Chance and evolution

The concept of chance is associated with evolutionary theory in several ways. These associations range from being of historical and philosophical importance to playing a central role within technical debates in evolutionary and molecular biology. Three general domains where chance is employed in evolutionary theory are discussed: Charles Darwin's use of chance (for example, Darwin 1859, 131), chance and microevolution (random drift), and chance and macroevolution. Within each camp several fine-grained distinctions are discussed. However, treatment of the more technical aspects of random drift are beyond the scope of this entry.

It is important to note that biologists generally do not use chance in a deep, metaphysical sense, such as meaning "uncaused." The uncaused sense of chance is most often raised in philosophical discussions of indeterminism, such as in quantum mechanics. For example, the random result of flipping a fair coin is predicted stochastically, but it is determined by physical laws. The result has a causal story, so the underlying processes that determine a coin flip are not random, or stochastic, in a deep sense. Likewise in biology, discussions of chance rarely entail the indeterminacy of the underlying mechanisms.

Natural selection is the privileged mechanism of change in Darwinian evolution. It was, in part, Darwin's use of chance that distinguished his theory from rival theories and worldviews. Darwin described a natural process, relying on chance mutations and natural selection, which could account for the adaptive complexity of the biological world. In this process, individuals of a species exhibit chance variations, some of which are beneficial and others deleterious. Bearers of successful traits are more fit than their competitors, thus they reproduce and pass on their genes to offspring at a higher rate. Over time, the successful traits spread to all members of the species. The argument that chance variations could offer the raw material for evolving a structure as complex as a vertebrate eye seemed untenable to the opponents of Darwinian evolution. The famous English apologist William Paley (1802 [1819]) articulated this point of view when he drew the contrast between a world that was a result of chance and natural processes or the result of a designer. To Paley, the choice was obvious; the intricate complexity of features such as the eye was incompatible with anything but a designer, a thesis commonly referred to as the Argument from Design.

Darwin's use of chance variations distinguished his ideas not only from those of creationists but also from other advocates of competing theories of evolution, particularly Jean-Baptiste Lamarck (1809). According to Lamarck, mutations are directed toward an organism's needs; they arise with respect to the use and disuse of an organism's parts. For example, giraffe necks are elongated because giraffes stretched them to reach treetop foliage. Opposing this model, Darwinian evolution claims that mutations are undirected, chance events; mutations for long-necked giraffes were just as likely as mutations for short-necked giraffes. However, once neck-length mutations occurred, the long-necked giraffes had a higher fitness value, thus spreading their long-neck genes throughout the population. The critical point in Darwin's argument is that although mutations arise by chance, it is not by chance that some individuals survive and reproduce while others perish. Natural selection distinguishes between beneficial and deleterious mutations.

Echoes of these arguments emerge in modern criticisms of Darwinian evolution that are focused on the seeming absurdity that chance variations could ever produce intricate structures or well-ordered biochemical processes. These arguments are based on mistaken ideas of the role of chance that echo the arguments in Darwin's time. First, chance variations produce raw material that is molded by natural selection; without natural selection to set the direction and speed of change, chance variations could not accumulate with enough speed to produce much meaningful change in any structure or process. Second, chance actually has a specific meaning with respect to Darwin's chance variations and the genetic mutations that cause them. Here, chance means that specific mutations occur at random with respect to the environment so that, unlike the Lamarckian argument, mutations are not directed by environmental features or the likelihood that they will prove useful. Some specific mutations are indeed more likely than others—for example, some mouse coat color mutations are more likely to arise than others—but their potential usefulness has no bearing on their likelihood of appearing.

When contemporary biologists raise the notion of chance they likely are referring to random drift. Random drift refers to the change in genetic composition that occurs in a population from one generation to the next because of the random sampling of genes that occurs in a finite population. A familiar example of random genetic sampling is the sex ratio within a family; in theory, half of the children in a family should be boys and half girls, but it is not uncommon for a family of four children to consist of three boys and one girl because of the random draw of the sex chromosomes. The classic illustration of random drift was given by Theodosius Dobzhansky (1937, 129), where he described a scenario of blindly drawing balls from an urn. The balls were colored differently, but otherwise indistinguishable. If one blindly draws balls from an urn and places them in a second urn, it is likely that the sample population of balls differs in frequency, with regard to color, from the original population—by chance alone. In short, random drift describes a scenario where the genotype frequency of a population changes in the absence of selective pressures.

There are two general kinds of random drift: biased sampling as a result of a catastrophe or other nonselective biasing, and genetic drift. A catastrophe, such as a forest fire, may kill a large portion of a population, but in most cases it does so indiscriminately with respect to any particular trait in the organisms. The small number of survivors that are left to produce the next generation may have a different genetic composition than the original population. A similar form of drift, that is, an abrupt change in genotype frequencies, can occur when a small group of organisms migrate from the main population and start a new one, such as occurs when an island is colonized. The migrating group may only take a fraction of the original genetic variability with them. The unique genetic structure of the founding population can create a new population that is recognizably distinct from its original, mainland progenitors, even though natural selection may have played a minimal role. This process has been called the founder effect, and arguments for its importance are linked to the influence of Ernst Mayr (1954).

A second form of random drift occurs gradually, from one generation to the next, in all limited populations through indiscriminate gamete sampling akin to Dobzhansky's analogy of the balls in the urn. Genotype and allele frequencies change gradually across generations, and, given enough time, one allele becomes the only gene in the population (that is, it is fixed). Fixation occurs because the direction of random genetic drift is determined by the allele frequencies. Typically, a common allele is likely to become more common, and an uncommon one is likely to become rare. Eventually one allele becomes extremely common and then fixed. The chance element remains because alleles may be equally common, thus making it impossible to predict which one will eventually become fixed.

Sewall Wright, who along with J. B. S. Haldane and R. A. Fisher was responsible for fusing Mendelian genetics with Darwinian evolution, was the first to emphasize the role that random drift plays in evolution. Wright argued that there would often be different combinations of genes that would be advantageous for their carriers and that random drift would influence which of these combinations would emerge and spread through a species (Wright's "shifting balance" theory of evolution). However, throughout his long career, Wright would vacillate on the importance of random drift and in the 1940s through the 1960s he delegated random drift to a minor force in evolution (Provine 1986).

The most influential application of genetic drift—and the most sophisticated argument for the role of chance in evolution—is the so-called neutral theory of evolution. As proteins and the DNA sequences that coded for them were discovered and explored, biologists were finding many variants of a single protein that seemed to have equivalent functioning. In some cases, there were two or more variants in a single population and, in others, different species were fixed for different variants. Motoo Kimura recognized that these observations were facets of a single problem united by the concept of genetic drift. He extended Wright's original mathematical work on modeling of genetic drift to show how these patterns would emerge (Kimura 1983). This work led, in turn, to using DNA sequences as "molecular clocks" and to a revolution in understanding how species, genera, families, and even entire phyla are related (Kimura 1983).

From a macroevolutionary perspective, some biologists argue that events such as extinctions, species diversification, radiations, and increasing complexity can best be explained stochastically, without reference to a particular causal story. This is not to say that there are no underlying causes to macroevolutionary events, but that stochastic methods seem to capture these events better than deterministic methods (Millstein 2000). For example, is the increasing complexity of the organic world due to passive or active trends, that is, by chance or directed evolution? Although it is difficult to provide necessary and sufficient conditions for biological complexity, it is generally agreed that life originated as simple organisms and gradually became much more complex. Because life began near the minimum boundary of complexity, random variation alone would produce more complex organisms; this would be a passive trend. An active trend would be exemplified by developmental routines or phenotypic novelties that drive evolution toward increasing complexity, beyond that of chance (Carroll 2001).

A final sense of chance in evolutionary theory is that arriving at any particular end point in evolutionary space is a chance event. For example, Stephen Gould (1989) famously claimed that humans are an improbable end point on the evolutionary tree; many unrelated (chance) events culminated in the existence of humans. Gould did not imply that there are not selective reasons (among others) explaining the evolution of humans. Rather, he argued that humans are a result of a vast number of improbable events. This is the sense of chance, attributed to Aristotle, where independent causal chains combine to produce an unintended effect. A car accident, for example, is an instance of separate causal chains combining and producing an unintended result.

BIBLIOGRAPHY

Beatty, J. 1984. Chance and natural selection. *Philosophy of Science* 51, no. 2: 183–211.

Carroll, S. B. 2001. Chance and necessity: The evolution of morphological complexity and diversity. *Nature* 409 (February 22): 1102–1109.

Darwin, C. 1859. *On the Origin of Species*. London: John Murray.

Dobzhansky, T. 1937. *Genetics and the Origin of Species*. New York: Columbia University Press.

Eble, G. J. 1999. On the dual nature of chance in evolutionary biology and paleobiology. *Paleobiology* 25: 75–87.

Gould, S. 1989. *Wonderful Life: The Burgess Shale and the Nature of History*. New York: W. W. Norton.

Kimura, M. 1983. *The Neutral Theory of Molecular Evolution*. Cambridge: Cambridge University Press.

Lamarck, J. B. 1809. *Philosophie zoologique*. 2 vols. Paris: Dentu.

Mayr, E. 1954. Change of genetic environment and evolution. In J. Huxley, A. C. Hardy, and E. B. Ford, eds., *Evolution as a Process*, 157–180. London: George Allen and Unwin.

Millstein, R. L. 2000. Chance and macroevolution. *Philosophy of Science* 67, no. 4: 603–624.

Paley, W. [1802] 1819. *Natural Theology (Collected Works*, vol. 4). London: Rivington.

Provine, W. B. 1986. *Sewall Wright and Evolutionary Biology*. Chicago: University of Chicago Press. —*J.Z.*

Charlesworth, Brian (b. 1945)

Brian Charlesworth is one of the foremost population geneticists of the last 50 years. He is known primarily for his theoretical work, although his experimental work has been substantial. Charlesworth's work has been influential for several reasons: a willingness to take on difficult problems, an appreciation of the connection between theory and data, clever use of mathematical and numerical approaches, honest assessments of the shortcomings of the work of others, and an unusually broad knowledge of the scientific literature.

Brian Charlesworth (B.C.) was born in England in 1945 and obtained his BA in Natural Sciences from Queen's College, Cambridge, with first class honors in 1966. He met his future wife, Deborah Maltby (b. 1943, now Deborah Charlesworth, hereafter D.C.), in his first year at Cambridge. Upon graduation, both Charlesworths went on to PhD studies at Cambridge, B.C. studying under J. M. Thoday.

Upon receiving his degree in 1969, B.C. took a postdoctoral fellowship with Richard Lewontin at the University of Chicago. At Chicago, B.C. shared an office with Ted Giesel, who was simulating evolution in age-structured populations. B.C. realized that he could obtain analytical results on the problem; this effort produced a substantial body of work that showed how to define fitness and describe the relationship between genetic and demographic changes. His book, *Evolution in Age-Structured Populations* (1980, 1994), remains the primary source on the subject.

In 1971 B.C. was hired as a lecturer at the University of Liverpool and initiated experimental work on the evolution of recombination rates. Discussions with D.C. led them to study modifier theory, which concerns genetic changes that affect the evolutionary forces acting on other loci, such as modifiers of recombination, dominance, mating systems, and mutation rates. The immediate result was a collaborative series of papers on Batesian mimicry that launched D.C.'s career in evolutionary biology.

The Charlesworths moved to Sussex in 1974, thanks to efforts by John Maynard Smith (B.C.'s PhD examiner). A sabbatical year in Chuck Langley's lab at the National Institute of Environmental Health Sciences in North Carolina began B.C.'s work on transposable elements. The Charlesworths' daughter, Jane, was born in 1981.

After attending a symposium at Chicago's Field Museum on alternatives to neo-Darwinism in 1980, B.C. coauthored a paper with Monty Slatkin and

Russ Lande that sharply critiqued challenges to neo-Darwinism. These challenges principally stemmed from Stephen J. Gould's extensions of Niles Eldredge and Gould's concept of punctuated equilibrium. This paper helped catapult B.C. from a respected expert on mathematical theory to a leader in the field of evolution. B.C. has since taken many opportunities to criticize ideas outside orthodox neo-Darwinism, from Gabriel Dover's suggestion that transposable elements might play a causal role in evolution to recent attempts to erect a science of Intelligent Design.

The Charlesworths moved to the University of Chicago in 1985, where B.C. was department chair from 1986 to 1991. The Charlesworths' transition to the study of molecular evolution was cemented by their 1993 discovery, with Martin Morgan, that deleterious mutations reduce the amount of variation in parts of the genome with low recombination. This provided a novel explanation for a well-established pattern that was then ascribed to the hitchhiking effects of beneficial mutations.

In 1991, B.C. was named a Fellow of the Royal Society, which facilitated a return to the United Kingdom via a Royal Society Research professorship. B.C. continues to work primarily in molecular evolution and the evolution of sex chromosomes.

BIBLIOGRAPHY

Charlesworth, B. 1978. A model for evolution of Y chromosomes and dosage compensation. *Proceedings of the National Academy of Sciences USA* 75: 5618–5622.
———. [1980] 1994. *Evolution in Age-Structured Populations.* 2nd ed. Cambridge: Cambridge University Press.
Charlesworth, B., R. Lande, and M. Slatkin. 1982. A neo-Darwinian commentary on macroevolution. *Evolution* 36: 474–498.
Charlesworth, B., M. T. Morgan, and D. Charlesworth. 1993. The effect of deleterious mutations on neutral molecular variation. *Genetics* 134: 1289–1303.
Charlesworth, B., P. Sniegowski, and W. Stephan. 1994. The evolutionary dynamics of repetitive DNA in eukaryotes. *Nature* 371: 215–220.   —D.H.

Charlesworth, Deborah (b. 1943)

Deborah Charlesworth (nee Maltby) is a pioneer in the study of breeding systems and genome evolution, concentrating on how evolution molds sections of the genome to allow the genes for complex traits to be co-inherited so that the traits function properly without the disruption that recombination and sexual reproduction might otherwise produce.

Deborah Charlesworth (D.C.) obtained her undergraduate degree from Cambridge University, where she met her husband, Brian Charlesworth (B.C.). After taking her doctoral degree in biochemical genetics, D.C. was not able to obtain a position in science and worked as B.C.'s unpaid assistant for the next 10 years. After B.C. took a position at the University of Liverpool, D.C. was encouraged to work on the evolution of Batesian mimicry by their colleague Philip Sheppard. One of the unsolved problems in Batesian mimicry was how the several genes controlling the mimetic pattern could be brought together to be inherited as a unit; discussions with B.C. led to a collaborative series of papers on this problem that launched D.C.'s career in evolutionary biology (beginning with Charlesworth and Charlesworth 1975).

Understanding the co-inheritance of genes has a converse, which is understanding how evolution breaks up a co-inherited group. D.C. and B.C. attacked both facets in several papers, expanding to a more general understanding of how genes at one set of loci affect the evolution of genes at other loci. This work would eventually point toward their broader work on genome evolution.

D.C. pursued the study of co-inheritance to some of the plant world's most striking problems, the evolution of molecular and morphological polymorphisms that forced plants to outcross and not inbreed (Charlesworth and Charlesworth 1979a, 1979b). This work led naturally to studies of the breeding system itself and how the evolution of breeding systems serves as a powerful force guiding the evolution of genome structure as a whole (Charlesworth 2006).

D.C.'s achievements and reputation led to faculty positions at the University of Chicago (1988–1997) and the University of Edinburgh (1997–present) as well as a series of honors and recognitions, including her election as a Fellow of the Royal Society (2005) and president of the Society for Molecular Biology and Evolution (2007). She continues to study breeding systems and genome evolution, particularly the evolution of sex chromosomes in plants, and encourages students to believe they can accomplish whatever they set their minds to accomplish (Charlesworth 2007).

BIBLIOGRAPHY

Charlesworth, D. 2006. Evolution of plant breeding systems. *Current Biology* 16: R726–R735.
———. 2007. Q & A: Deborah Charlesworth. *Current Biology* 17: R264–R266.
Charlesworth, D., and B. Charlesworth. 1975. Theoretical genetics of Batesian mimicry. 1. Single locus models. *Journal of Theoretical Biology* 55: 283–303.
———. 1979a. Evolution and breakdown of S-allele systems. *Heredity* 43: 41–55.
———. 1979b. Model for the evolution of distyly. *American Naturalist* 114: 467–498. 　　　　　　　　　　　　　　　　　　　　　　　　—J.T. and D.H.

Chetverikov, Sergeĭ Sergeevich (1880–1959)

Sergeĭ Sergeevich Chetverikov (sometimes spelled Tschetwerikoff) was a Russian entomologist, an expert on butterflies, and a pioneer of population genetics. Born into a well-educated, professional family, Chetverikov entered the University of Moscow in 1900, graduating six years later. He continued at the university as a research fellow, receiving an advanced degree in 1909. Afterwards, he taught entomology at the Higher School for Women in Moscow from 1909 to 1919. In 1921, Chetverikov accepted a research position at the Institute for Experimental Biology in Anikovo, near Moscow. This was run by N. K. Kol'tsov. At the institute, Chetverikov continued his naturalist collecting, especially butterflies and moths. He also undertook experimental and theoretical research into the diversity of mutations in natural populations. At the same time, he taught both genetics and biometrics. Chetverikov was recruited by Kol'tsov for his knowledge of flies. These animals were rapidly becoming the standard organism in genetics research. Chetverikov played a key role in establishing a *Drosophila* research group at Kol'tsov's Institute when fly stocks from Thomas Hunt Morgan's famous laboratory first arrived in 1922.

Chetverikov's research career effectively ended in 1929, at its height. He was arrested, probably for suspicious organizing activity, and banished from Moscow. He was helped neither by a history of arrests for antigovernment disturbances while a student nor by close affiliations with the Morgan school of chromosomal genetics and Mendelian genetics. Joseph Stalin's isolationist policies purged Western influences from Soviet science and culture. In genetics, this purge accompanied the rise of Trofim Lysenko and a theory of inheritance that emphasized dialectics between organism and environment and the inheritance of acquired characteristics. Chetverikov was one of hundreds of experimental geneticists either sacked, imprisoned, exiled, or murdered.

In exile, Chetverikov moved to Sverdlovsk, working in a zoo until 1932. He then moved to Vladimir to teach mathematics in a junior college. In 1935 he became a professor of genetics at the University of Gorky. His activities in Gorky are not well known as he deliberately kept a low profile under the watch of suspicious authorities. He did undertake some small-scale research into natural selection using silkworms. In 1948, at the age of 68, his position was officially terminated in another Lysenko-inspired purge of geneticists linked with Western ideas. Chetverikov remained mentally active until his death, although in his final years he was hampered by poverty and poor health. His brother Nicholas cared for him until his death at the age of 78.

As an evolutionary and population geneticist, Chetverikov argued that natural populations maintained a large reserve of recessive genetic mutations, soaking them up like a sponge. In an evolutionary context, these provided raw material for new variety. They fueled microevolution. As an experimentalist, Chetverikov used inbreeding techniques to reveal the extent of diversity in recessive mutations. His ideas challenged traditional views that mutations simply were laboratory artefacts, broken forms of the normal genetic information. Instead, populations could be considered as pools in which many alleles swam. The frequency of each allele shifted according to demographic changes, such as migration, and through natural selection. Along the same lines, when he was studying wild populations in their natural habitats,

Chetverikov emphasized the importance of variation within polymorphic species.

Chetverikov is best remembered as the author of a 1926 theoretical paper, translated in 1961 as "On Certain Aspects of the Evolutionary Process from the Standpoint of Modern Genetics." This is a classic presentation of fundamental concepts in population genetics, and it earned him recognition as the forgotten fourth contributor to a trio of mathematical population geneticists (Ronald Fisher, J. B. S. Haldane, and Sewall Wright) responsible in the 1920s for combining Mendelism and Darwinian natural selection into a neo-Darwinian synthesis. In his paper, Chetverikov argued that (1) mutation is the source of variation in evolution; (2) inheritance follows a particulate (i.e., Mendelian) pattern rather than blending or dialectics; (3) genes should be understood as a series of variant alleles, and the frequency of those alleles is key in determining the rate and direction of evolutionary change; and (4) geographic isolation can have an important impact on the evolution of populations.

The fate of Chetverikov's writing is a good example of communication barriers in science. Few read his paper when it first appeared because the journal itself was poorly distributed outside Moscow. Moreover, except for a short summary in English published in tandem, those unable to read Russian had to wait for German, French, or English translations. Private translations, however, circulated in several research centers outside the Soviet Union (e.g., J. B. S. Haldane in London). Selected passages, translated by Theodosius Dobzhansky, first appeared in print only in 1959. A full English translation did not appear until 1961. During his forced retirement, Chetverikov revised his famous paper, dictating corrections and commentary despite failing health; it was never published.

Chetverikov's widest influence came through his teaching and his students, some of whom emigrated to the West, including N. P. Dubinin, N. V. Timoféeff-Ressovsky, B. L. Astaurov, D. D. Romashov, and S. M. Gershenson. Theodosius Dobzhansky credits Chetverikov with introducing him to *Drosophila* genetics.

BIBLIOGRAPHY

Adams, M. 1980. Sergei Chetverikov, the Kol'tsov Institute, and the evolutionary synthesis. In E. Mayr and W. Provine, eds., *The Evolutionary Synthesis: Perspectives on the Unification of Biology*, 242–278. Cambridge, MA: Harvard University Press.

Chetverikov, S. 1961. On certain aspects of the evolutionary process from the standpoint of modern genetics. *Proceedings of the American Philosophical Society* 105: 167–195. (Translation of 1926 paper.)   —J.C.

Clausen, Jens (1891–1969)

Jens Clausen is regarded as one of the pioneers in the ecological and evolutionary genetics of plants as well as one of the foremost California botanists of the twentieth century. He is especially known for his synthetic and inter-disciplinary attempts to understand the totality of processes involved in the origin of plant species at a crucial time in the history of evolutionary biology.

Jensen was born in Eskilstrup, Denmark, the son of farmers and house builders. His formal schooling took place between the ages of 8 and 14, ter-minating when he had to take on the responsibility of managing the family farm. He developed an interest in genetics and for about eight years educated himself in the basic sciences with the aid of a schoolteacher. He began to take an active interest in the new science of genetics, studying Mendelian genetics along with Darwinian evolutionary theory. In 1913 he entered the University of Copenhagen to study botany, genetics, and ecology, coming in contact with some of the shining lights in Scandinavian science that included Chresten Raunkier (his major professor in botany), P. Boysen Jensen, Wil-helm Johannsen, and August Krogh. At one point he even studied physics with Niels Bohr. His early research on genetics and ecology in the Violaceae (the pansy family) led to a more detailed understanding of hybridization in plants and the phenomenon known as introgression, whereby a smaller num-ber of genes are introduced through the mechanics of hybridization. His PhD dissertation on the Violaceae, published in 1926, was one of the first mono-graphs to combine systematics, ecology, and genetics in any plant group. He was subsequently appointed assistant professor to noted geneticist Øjvind Winge at the Royal Agricultural College in Copenhagen, but his real break came with the granting of a Rockefeller fellowship in 1927–1928 that took him to the University of California to work with E. B. Babcock on the gene-tics of the genus *Crepis*. While there, he met with Harvey Monroe Hall, a dis-tinguished California botanist, who later invited him to participate in the new interdisciplinary project he was organizing with the Carnegie Institution of Washington to understand more precisely the means by which new plant species originated under natural conditions using native California flora. (Hall had earlier collaborated with ecologist Frederic Clements on such a project.) Accepting the offer, Clausen arrived in California in 1931, but in-stantly found himself the leader of the new team that Hall had assembled when Hall died unexpectedly.

The new interdisciplinary team at the Carnegie Institution of Washington that was assembled at Stanford University included taxonomist David Keck and physiologist William Hiesey. Until the mid-1950s the team of Clausen, Keck, and Hiesey (as they came to be known) designed a series of classical ex-periments to derive general principles for understanding plant evolution un-der controlled, but natural conditions. Adapting some of the celebrated methods of nineteenth-century European workers such as Gaston Bonnier, and early-twentieth-century workers such as Göte Turesson, who sought to understand the process of adaptation, Clausen, Keck, and Hiesey performed a number of transplant experiments involving three distinct locations along altitudinal gradients in northern California. One location was at sea level at Stanford, another at Mather (at about 4,600 feet), and yet another at Timber-line in the Sierra Nevada mountain range (at about 10,000 feet). Performing sets of related experimental studies on native California plants such as *Potentilla glandulosa, Achillea* species, and the Madiinae (the hayfield tarweed family), Clausen, Keck, and Hiesey explored hybridization and the formation of ecotypes (genetically distinct ecological races of plants). They charted the mechanisms of speciation and refined the understanding of the genetic and ecological factors that led to the origin of plant species. Their well designed and carefully executed transplant studies are generally regarded as laying to rest the belief in the inheritance of acquired characteristics (or Lamarckian inheritance) by showing the distinctness of the genotype and phenotype, and by bringing to relief the importance of phenotypic plasticity in the plant world. More narrowly within plant evolution, their studies highlighted the complex interplay between hybridization, apomixis, and phenomena such as polyploidy (chromosome doubling), all of which were essential to deriving a general theory of plant evolution.

Throughout the 1930s and 1940s, Clausen, Keck, and Hiesey published their results in a series of important papers and monographs with the Carnegie Institution. Their work was read widely by others interested in mechanisms of speciation and by those interested in applying this understanding to developing a viable species definition useful in formulating the "new systematics" (called biosystematics) that sought to integrate approaches and understanding from genetics and ecology to traditional taxonomy. They influenced Theodosius Dobzhansky and George Ledyard Stebbins. Stebbins's *Variation and Evolution in Plants* (1950) brought botany into the evolutionary synthesis and organized the new field of plant evolutionary biology.

Clausen's own synthesis of plant evolution appeared in 1951 as *Stages in the Evolution of Plant Species*. An imaginative and pioneering work, it was somewhat brief and introductory in nature because it was based on the Messenger Lectures he had given earlier at Cornell University. The book has generally been overshadowed by Stebbins's more comprehensive earlier synthesis, but nonetheless remains one of the classic works in the history of twentieth-century evolutionary biology. Among its more important contributions was the view that evolution below the level of the genus is reticulate due to phenomena such as hybridization accompanied by polyploidy.

In 1951, Clausen was appointed professor of biology at Stanford. He retired in 1956 and died in Palo, Alto, California, in 1969. He was survived by his wife, Anna Hansen, whom he married in 1921; they had no children. In a research position for much of his life, Clausen did not leave much of a legacy in terms of students, but his insightful studies, imaginatively designed and meticulously executed, became classic examples used in nearly all introductory textbooks of genetics, ecology, and evolution.

BIBLIOGRAPHY

Clausen, J. 1951. *Stages in the Evolution of Plant Species*. Ithaca, NY: Cornell University Press.

French, C. S. 1989. Jens Christian Clausen. *National Academy of Sciences Biographical Memoirs* 58: 75–107.

Hagen, J. 1984. Experimentalists and naturalists in twentieth century botany, 1920–1950. *Journal of the History of Biology* 17: 249–270.

Smocovitis, V. B. 1999. Jens Clausen, In J. A. Garraty and M. C. Carnes, eds., *American National Biography*, vol. 5, 12–13. Oxford: Oxford University Press.

Stebbins, G. L. 1950. *Variation and Evolution in Plants*. New York: Columbia University Press.   —*V.B.S.*

Cope, Edward Drinker (1840–1897)

Edward Drinker Cope was one of the most active and controversial paleontologists of late-nineteenth-century America. He played a leading role in opening up the fossil beds of the West and describing the new species discovered there. He was also a member of what is called the American School of Neo-Lamarckism. He accepted evolution but dismissed Charles Darwin's selection theory as inadequate to explain the process of change. He opted instead for a theory in which evolution was directed along predictable lines.

Cope was born in Philadelphia in 1840 to a wealthy Quaker family. He was trained in scientific agriculture, but soon abandoned this for zoology and then paleontology. He began by studying living fish and reptiles, then moved on to studying their fossil remains, eventually generalizing his work in paleontology to include all the vertebrates uncovered by his expeditions in the West. He was financially independent until he lost his fortune in a mining fraud in 1880, after which he built up links with the University of Pennsylvania, becoming professor of zoology there in 1895. He died in 1897.

Cope is remembered for his bitter feud with Othniel Charles Marsh of Yale University. The two clashed over the description and naming of new fossil species, each accusing the other of hasty work and numerous mistakes (see figure). There was also fierce rivalry between their collecting teams, who sometimes came to blows over access to the most productive fossil beds. Eventually Marsh ousted Cope from his connections with the U.S. Geological Survey, which greatly limited his ability to collect in the West. The "fossil feud" became public in 1890 through a vitriolic exchange in the pages of the *New York Herald*.

Cope was an early convert to evolutionism, but came to the theory from a different background than Darwin. A deeply religious man, he was convinced that God directed the course of organic change. In his 1868 article "On the Origin of Genera," he depicted evolution of each group as the unfolding of multiple parallel lines, each advancing at a different rate through the same linear sequence of stages. Evolution was recapitulated in the development of the modern embryo because the forces controlling development simply added on new stages in a consistent direction. The changes were nonadaptive and the overall pattern was a product of the divine will.

In the 1870s Cope moved to a more naturalistic interpretation that combined orthogenesis and Lamarckism. Forces internal to the organism ensured that variation was directed, not random, and thus drove nonadaptive

COPE ON FOSSIL REPTILES OF NEW JERSEY.

Even the best of us can make mistakes. Edward Drinker Cope had a brilliant intuitive sense for reconstructing the monsters of the past. But when things went wrong, they really went wrong. This picture shows his conception of the plesiosaur, *Elasmosaurus platyurus,* from the late Cretaceous (about 70 million years ago). Unfortunately, Cope put the head at the wrong end of the body, not realizing that the real animal had a tremendously long neck, which in this picture is misinterpreted as a tremendously long tail. (From Cope 1868, Plate 2.)

evolution along predetermined lines. A similar process of parallel evolution could result from adaptive evolution controlled by the habits chosen by the organisms in a new environment—characteristics acquired by the consistent use of the body in a new way were passed on to succeeding generations. This process became known as Lamarckism, although Cope was not aware of Jean-Baptiste Lamarck's work when he developed his idea. Cope argued that the life force that allowed organisms to adapt was a manifestation of God's creativity in nature.

BIBLIOGRAPHY

Bowler, P. J. 1983. *The Eclipse of Darwinism: Anti-Darwinian Evolution Theories in the Decades around 1900.* Baltimore: Johns Hopkins University Press.

Cope, E. D. 1868. On a new large Enaliosaur. *Proceedings of the Academy of Natural Sciences of Philadelphia* 20.

Osborn, H. F. 1931. *Cope: Master Naturalist; The Life and Writings of Edward Drinker Cope.* Princeton, NJ: Princeton University Press.

Wallace, D. R. 1999. *The Bonehunters' Revenge: Dinosaurs, Greed and the Greatest Scientific Fraud of the Gilded Age.* Boston: Houghton Mifflin.   —*P.J.B.*

Crow, James Franklin (b. 1916)

Over the course of a career that spans eight decades, James Franklin Crow has made several seminal contributions to both theoretical (mathematical) and experimental population genetics. Although his primary research organisms have been fruit flies of the genus *Drosophila*, Crow has also made important contributions to the study of humans and agricultural crops. Although Crow has worked on problems ranging from the genetics of DDT resistance to examining genes that violate Mendel's law of fair meiosis to developing theory about the benefits of sexual reproduction, his most prominent contributions revolve around the role of mutations in evolution.

Born in January 1916 in Collegeville, Pennsylvania, Crow grew up mainly in Wichita, Kansas, where he received his undergraduate degree from Friends University, a Quaker school. In 1937, Crow went to the University of Texas at Austin for graduate work, initially to study with prominent geneticist Hermann Joseph (H. J.) Muller, who would later win the Nobel Prize for his studies of X-ray mutagenesis in *Drosophila*. Muller, however, was about to emigrate to the Soviet Union, so John Thomas (J. T.) Patterson became Crow's major adviser. Just prior to the United States' entry into World War II in 1941, Crow landed a faculty position at Dartmouth College, where he stayed until 1948. Wartime and the drafting of his colleagues at Dartmouth led Crow to add courses as diverse as parasitology, navigation, and embryology to his initial teaching responsibilities of general zoology and genetics. In 1948, Crow moved to the University of Wisconsin, where he has remained for six decades.

Mutations are not an unmixed blessing. Although mutations are the ultimate source of variation and are thus needed for evolution to proceed, deleterious mutations far outnumber beneficial mutations. Crow and his students used the genetic techniques available in *Drosophila melanogaster* to measure the input of new lethal mutations appearing on a single chromosome each generation. From this value, they calculated that the lethal mutation rate per genetic locus (gene) per generation was around three per million. Mutations need not have lethal effects to be deleterious; in fact, Crow and his colleagues have shown that nonlethal, detectable deleterious mutations arise at least 10 times more often than lethal mutations. Recent studies suggest that the rate of mildly deleterious mutations is even higher than those mutations that have effects that can be measured in the laboratory.

Crow's group and others have shown that most deleterious mutations are almost completely recessive; that is, the effects they have in heterozygotes are far smaller than the effects they have in homozygotes. Because mutations are mostly recessive, they are largely sheltered from selection and thus are not easily weeded out from the population. These deleterious mutations, whether their effects are lethal or mild, decrease the mean fitness of the population that harbors them. Muller called this reduction of fitness the genetic load of the population. With Muller and others, Crow investigated the extent of the genetic load in both *Drosophila* and humans, finding it to be surprisingly large. The average human harbors the equivalent of several lethal mutations. These mutations are mostly recessive, and their deleterious effects are not manifest unless individuals inbreed because unrelated individuals harbor a different set of recessive deleterious mutations. The recessive nature of these mutations is thus the basis of inbreeding depression. Crow continues to have an interest in mutations and their effects in humans.

Crow was the mentor and longtime collaborator of Motoo Kimura. Kimura developed the neutral theory of molecular evolution, which argues that the vast majority of molecular variants are either deleterious or neutral with respect to selection. An important feature of the neutral theory is that it easily generates many testable predictions that can allow biologists to detect whether and what forms of selection are operating at particular genetic loci. Laying the foundation for the neutral theory was work that Kimura did with Crow to determine how much variation would be expected, provided that this variation was the result of a balance between neutral mutations adding to variation and random genetic drift eroding variation. In 1964, Kimura and Crow developed an important mathematical model in which they assumed that every mutation that arises is unique. In this so-called infinite alleles model, the expected heterozygosity (the proportion of the population that was heterozygous at a site) of neutral alleles under mutation/drift balance equals four times the product of the effective population size and the neutral mutation rate. Crow and Kimura also published a highly influential textbook, *An Introduction to Population Genetics Theory,* which has been used by generations of students and practitioners of population genetics since its appearance in 1970.

In addition to his work on the role of mutation in evolution, Crow has made several other contributions to evolutionary genetics. As a graduate student under the tutelage of J. T. Patterson at the University of Texas at Austin, Crow was an early investigator of the genetic basis of reproductive isolating barriers between *Drosophila* species. Crow was among the first to study the genetic basis of hybrid female inviability, finding that this trait mapped largely to the X chromosome in hybrids between *D. mulleri* and *D. aldrichi-2.*

Crow has also had an active interest in genetic conflict and the evolution of Mendelism. Investigations of genetic systems that violate Mendel's law of equal segregation have helped our understanding of why Mendelism is in part a fair game. The best known and best characterized of these so-called meiotic drive systems, segregation distortion on the second chromosome of *Drosophila melanogaster,* was discovered in Crow's laboratory by his graduate student, Yuichiro Hirazumi, during the 1950s. In subsequent years, Crow, his students, and colleagues have worked on the genetics, molecular genetics, and evolution of the system.

Crow's influence on the field as a teacher and mentor is immense. In addition to the scores of graduate students and postdoctoral fellows that he has directly trained, Crow has influenced the field with his various textbooks and historical essays. Along with William Dove, Crow has edited the "Historical Perspectives" section for *Genetics* since 1987. (As of mid- 2008, Crow has written about 250 articles.) A longtime member of the National Academy of Sciences (NAS), Crow has chaired many committees for the NAS, including one examining forensic uses of DNA fingerprinting. He has served as chair of the Department of Genetics and acting dean of the Medical School at the University of Wisconsin at Madison, as well as president of the Genetics Society of America.

BIBLIOGRAPHY

Crow, J. F. 1979. Genes that violate Mendel's rules. *Scientific American* 240, no. 2: 134–146.
———. 1987. Population genetics history: A personal view. *Annual Review of Genetics* 21: 1–22.
———. 1993. Mutation, mean fitness, and genetic load. *Oxford Survey of Evolutionary Biology* 9: 3–42.
———. 1997. The high spontaneous mutation rate: Is it a health risk? *Proceedings of the National Academy of Sciences USA* 94: 8380–8386.
Crow, J. F., and M. Kimura. 1970. *An Introduction to Population Genetics Theory.* New York: Harper & Row.
Interview with Professor James Crow. 2006. *BioEssays* 28, no. 6: 660–678 [doi: 10.1002/bies.20426].
Kimura, M., and J. F. Crow. 1964. The number of alleles that can be maintained in a finite population. *Genetics* 49: 725–738.
Simmons, M. J., and J. F. Crow. 1977. Mutations affecting fitness in *Drosophila* populations. *Annual Review of Genetics* 11: 49–78.   —N.A.J.

Crustaceans

Crustaceans are a subphylum of arthropods, as are uniramians (insects, millipedes, and centipedes) and chelicerates (spiders, mites, ticks, and horseshoe crabs), although, in some classification schemes, arthropods are considered to be polyphyletic (i.e., not descended from a single common ancestor), and crustaceans are elevated to phylum status. The seemingly simple question, "What is a crustacean?" is surprisingly difficult to answer, and this is exactly what makes this group of animals so intriguing. As arthropods, crustaceans are basically segmented animals with jointed appendages and a stiff yet flexible exoskeleton made of chitin that must periodically be shed, or molted, to accommodate metamorphosis and growth. Yet relatively few features distinguish crustaceans from other arthropods—a unique larval form called a nauplius and a head region comprising five body segments on which are found two pairs of antennae (usually biramous, or two-branched), a pair of mandibular appendages, and two pairs of specialized feeding appendages called maxillae—but even these few features are not common to all crustaceans. Like some other arthropods, the crustacean multisegmented trunk posterior to the head, or cephalon, may or may not be further regionalized into distinct sections of fused segments, such as a thorax (or cephalothorax) and abdomen (e.g., a lobster's "tail"), and appendages can vary greatly in number and degree of specialization.

Insects, mollusks, and chelicerates exceed crustaceans with regard to number of species, but crustaceans are unequivocally the most morphologically diverse animal group on earth. The few, very general features listed above inadequately convey the remarkable diversity of a group that includes well-known forms like lobsters, shrimp, crabs, and barnacles, as well as less familiar amphipods, isopods, fairy shrimp, water fleas, and ostracods (or "seed shrimp"), to name but a few (see figure). In fact, the familiar forms comprise only a small proportion of the roughly 68,000 known crustacean species. They are primarily a marine group but some also occupy freshwater or moist terrestrial habitats (e.g., terrestrial isopods, commonly called wood lice, pill bugs, sow bugs, or "roly-polies," and terrestrial crabs). Interestingly, the roughly 150 million red terrestrial crabs of Christmas Island in the Indo-Pacific annually migrate from inland forests to the sea to spawn; many more times that number of juveniles migrate in the reverse direction. Some aquatic groups (e.g., large branchiopods referred to as fairy, tadpole, and clam shrimp) have adapted to living in temporary ponds or "ephemeral pools"; these crustaceans undergo a rapid life cycle and produce dormant embryos encysted in a hard protective case that can withstand extended periods of drying, only to emerge and continue their development when a pond refills.

Crustaceans range in size from less than a millimeter (minute zooplankton and benthic species living within marine sediments) to the very large. Japanese spider crabs can have leg spans exceeding 3 meters, or roughly 9 feet, and lobsters have been caught that exceed 20 kilograms (40 pounds). It is anticipated that many crustaceans await discovery (at least as many species as the 68,000 species currently known), especially upon further exploration of very remote habitats like freshwater and marine cave systems and the deep sea—and new findings are not necessarily expected to be confined to mere species. According to Martin and Davis (2001), almost 200 new families have been described in the 20 years that have elapsed since the last widely used classification, that of Bowman and Abele (1982). Remarkably, an entirely new class of crustaceans, the Remipedia, was discovered in a marine cave in the Bahamas as recently as 1981, and new orders (Spelaeogriphacea, Thermosbaenacea, Mictacea, and Bochusacea) of minute forms (0.5 to 2 millimeters) of peracarid crustaceans have been discovered in remote caves and the deep sea just since the mid-1950s. Not surprisingly, some of these newer orders comprise only a few, rare species. What is surprising is the remarkably disjunct distribution of some of these newly discovered groups, such as the spelaeogriphaceans—only four species are known, discovered in single caves in Brazil, South Africa, and northwestern Australia! It is suspected that these populations represent relics of an ancient marine group that was once more widespread; changes in sea levels might have led to individuals becoming entrained in, and later adapting to, groundwater and cave habitats.

Crustaceans feed and reproduce in almost every way imaginable for animals. Some are filter feeders and sieve bacteria or tiny zooplankton from their watery environment, others scavenge for their food, and many are predators. The latter may have some remarkable adaptations, such as the

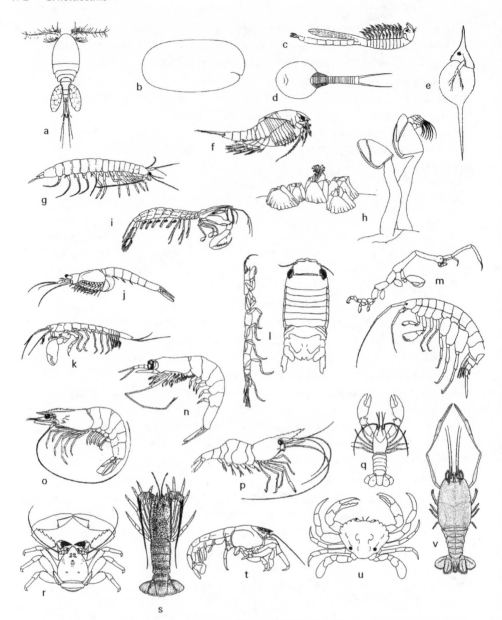

Representatives of some basic crustacean types. (a) Copepoda; (b) Ostracoda; (c) Anostraca; (d) Notostraca; (e) Cladocera; (f) Leptostraca; (g) Anaspidacea; (h) Cirripedia; (i) Stomatopoda; (j) Mysida; (k) Tanaidacea; (l) Isopoda; (m) Amphipoda; (n) Euphausiacea; (o) Dendrobranchiata; (p) Caridea; (q) Astacidea; (r) Anomura; (s) Achelata; (t) Thalassinidea; (u) Brachyura; (v) Polychelida. (From Poore 2004, Figure 2.)

modified "claw" of a stomatopod mantis shrimp that allows the animal to smash mollusks or spear fish with great force and speed. Another type of crustacean has a modified claw with pincers that, when quickly brought together, produce a loud snapping noise (hence their name, "snapping shrimp"). The sound is associated with a "shock wave" strong enough to stun or even kill small prey, such as fish. Some species of snapping shrimp are also notable for exhibiting social behavior more typical of ants, bees, and termites. Numerous offspring of a single female (a "queen") will coexist in a sponge and form castes to care for the young and protect the colony. This phenomenon of eusociality in snapping shrimp was described relatively recently by Emmett Duffy (1996) and is a first among crustaceans.

Sexes are usually separate (although some hermaphroditic species are known) and sexual reproduction is the typical mode. Some crustaceans, such as water fleas, will occasionally reproduce asexually via a mechanism known as parthenogenesis, whereby females will produce young (also female) that develop from unfertilized eggs. Crustacean eggs may be released into the aquatic environment or sometimes brooded, but the young usually emerge as free-swimming larvae that will undergo a series of molts and stages before metamorphosing into the adult form (indirect development). The females of fewer species, such as freshwater crabs, brood their eggs until miniature adults emerge (direct development), molting as they grow larger.

The early Cambrian period of geological history (approximately 550 million years ago) saw the emergence of numerous multicellular animal phyla exhibiting an incredible diversity of body plans. Paleontologists now believe this "explosion" of life forms occurred over only 5 to 10 million years (an astonishingly brief amount of time, evolutionarily speaking). Many of these phyla have long since gone extinct, and never again in the earth's history has an equivalent burst of evolutionary innovation occurred. The reason remains a topic of debate and speculation. Possible contributors to the Cambrian explosion of body plans include changes in the earth's geochemistry (e.g., perhaps greater oxygen availability favored marine organisms with higher metabolic rates and larger sizes) and sudden, large-scale genetic changes (e.g., gene-duplication events and a reshuffling of gene regulatory pathways in what were then relatively simple organisms). Thereafter, perhaps the genetic programs of animals had reached a level of complexity that precluded further nonlethal major changes. Another view is that the pre-Cambrian environment constituted an ecological tabula rasa of empty niches simply waiting to be filled by adapting organisms. Then, as now, such environments promote an adaptive radiation of species.

Fossil evidence confirms that crustaceans were one of numerous arthropod subphyla (or arthropodous phyla) present in the Cambrian and, more notably, one of the few that have survived to this day. Even more striking is the evidence that crustaceans had already diversified into several main lineages by this time. The classes Ostracoda (seed shrimp), Branchiopoda (brine shrimp, or "sea monkeys"; fairy, tadpole, and clam shrimp; and water fleas), and some of the Maxillopoda (specifically thecostracans [barnacles] and wormlike parasitic pentastomes) have Cambrian representatives. Shrimplike fossils of the class Malacostraca (comprising the familiar shrimps, lobsters, and crabs) are known from the Devonian (approximately 400 million years ago). Remarkably, perhaps the oldest surviving animal species is a crustacean "living fossil," the tadpole shrimp *Triops cancriformis*. This species appears today as it did 220 million years ago. The remaining crustacean classes (Remipedia, Cephalocarida), while primitive in appearance, are poorly represented as fossils. The virtually simultaneous appearance in the fossil record of most of the crustacean classes makes reconstruction of the evolutionary relationships among them from fossil evidence impossible; scientists cannot determine which of the major crustacean groups or classes arose first, second, or third, or which major group specifically gave rise to another.

Two features of crustaceans—their ancient origin and their bewildering array of morphological diversity (greater than that of the more speciose insects)—add to the difficulty of inferring their phylogeny (i.e., their evolutionary interrelationships). The goal of evolutionary reconstruction is to decipher the phylogenetic pattern to determine which group of organisms is ancestral to (that is, gave rise to) another group or groups of organisms. Traditionally, systematists have looked for patterns of shared morphological features among groups of organisms to reconstruct a phylogenetic (family) tree that depicts the groups' evolutionary relationships. Because the major crustacean lineages diverged from a common ancestor so long ago, lineages have had a long time during which to evolve and to diverge from each other morphologically. Discerning the morphological features that extant members of each lineage still share, and that would indicate an ancient common ancestry, is therefore difficult. For example, barnacles were long thought to be mollusks rather than crustaceans (or even arthropods!). Their highly derived and specialized adult morphology belies any affinity with crustaceans, and it was only when the larval stages of barnacles were observed in the 1830s that they were recognized as crustaceans. In fact, Charles Darwin himself provided the first comprehensive review of barnacle taxonomy, and in doing so he became more assured of his yet-unpublished ideas on evolution. Phylogenetic reconstruction is further hampered by parasitic crustacean taxa (e.g., pentastomid "worms," and ascothoracidans, rhizocephalans, and tantulocarids) that exhibit greatly modified morphologies that are difficult to fit into a "natural classification" (i.e., one that reflects evolutionary relatedness) of the Crustacea.

For all these reasons, little consensus has arisen about the evolutionary relationships among the classes of crustaceans (e.g., on the Tree of Life website, http://www.tolweb.org, the link to "Crustacea" leads to an unresolved phylogenetic tree for crustaceans as a whole and unresolved trees even within most of the major crustacean lineages). Authorities even disagree about the number of taxonomic classes and their constituents, as is reflected by the number of different classification schemes currently in use.

What accounts for crustaceans' remarkable diversity of body plans? Like other arthropods, crustaceans are derived from an ancestor (the "stem" crustacean) that probably had a flexible exoskeleton encasing numerous and relatively undifferentiated segments (a condition termed serial homonomy). Most segments bore paired, jointed, and fairly simple appendages, the nature of which remains controversial. The crustacean class Remipedia reflects this basic ground plan, but views differ over whether the remipede body plan is a holdover of a primitive condition or a reversion to the primitive condition from a more derived ancestor. Subsequent evolutionary change among early crustaceans often resulted in the compartmentalization of segments into distinct regions, called tagmata (singular, tagma; examples are the head, thorax, and abdomen) and development of appendages modified for different functions. But among extant arthropod subphyla, crustaceans alone paralleled the burst of morphological innovation observed in the early Cambrian previously reserved for the origin of disparate phyla. Recent findings from developmental biology suggest that early crustaceans, more so than any other surviving lineage, most likely possessed great developmental plasticity (the ability to develop differently under different conditions, even from the same complement of genes). Crustaceans are able to draw from an amazing bag of evolutionary tricks. Various processes promoting dramatic morphological change resulted in new lineages and contributed to the astounding diversity of crustacean body plans today. Understanding these processes can add to the appreciation of the singular scope of crustacean diversification and the problem this diversity creates during attempts to reconstruct evolutionary relationships within this group.

Dramatic and relatively sudden evolutionary change can be affected by homeotic genes. Often referred to as *Hox* genes in animals, these are master regulatory genes involved in an organism's development. The number of homeotic genes, the timing of their expression, and the other genes and embryonic regions that they in turn affect direct the fate of various groups of cells within a developing embryo and ultimately control the identity and spatial organization of the various body parts seen in the adult. In crustaceans, for example, these important regulatory genes regulate the development of body segmentation (e.g., the number of segments that make up a head, thorax, or abdomen) and the number and kind of appendages (antennae, mandibles, swimming legs, etc.) that develop on those segments. Even a minor mutation in one of an organism's homeotic genes can result in an adult with a dramatically novel body plan. For example, the dorsal fusion of the body segments of an ancestral crustacean into some sort of shield or "carapace" would radically alter the appearance of its descendant. Particularly for segments of the crustacean head region, termed the cephalon, such fusion in various patterns is suspected to account for the diverse carapace morphologies observed in crustacean taxa. Relatively undifferentiated ancestral trunk segments may also fuse into tagmata. Variations in *Hox* gene expression, for example, probably account for the different numbers of trunk segments seen in different species of putatively primitive remipedes, and also for the body organization seen in lobsters: for example, segments fusing into distinct head, thorax, and abdominal tagmata, and segments of the thorax bearing walking appendages (the first pair of which are modified claws or "chelipeds" seen in many of the better-known crustaceans), whereas those of the abdomen (a lobster's tail) bear appendages modified for swimming. Variation in *Hox* gene expression likely accounts for barnacles lacking abdominal segments entirely (one of the reasons they were originally overlooked as crustaceans). Comparative studies of gene expression patterns for the *Hox*-like *Distal-less* and engrailed developmental regulatory genes have significantly advanced our understanding of the early morphogenesis of crustacean limbs and segmentation.

Heterochrony is another process that affects the evolution of morphology, and one that has likely contributed to the diversity of crustacean body plans. Brought on by mutations in the genes controlling the timing of developmental events within an organism, the heterochronic process may quickly result in the appearance of novel morphological differences between ancestors and descendants. Even slight changes in the relative developmental and growth rates of different body parts in an embryo can result in a dramatically altered adult. For example, a mutation that extends the period during which a gene that regulates antennal growth is active can result in an offspring with antennae many times longer than those of its parents. If the mutation is heritable, then subsequent offspring will also grow long antennae. In some crustaceans (e.g., anostracans, such as fairy shrimp and brine shrimp, also called sea monkeys), antennae are often required for successful copulation. Males with long antennae may be able to mate more successfully than ones with shorter antennae, or they might only be able to mate with a particular variant among the females in a population. If a population becomes reproductively isolated, its genes are no longer mixed back into the parent group at each generation, and selection can affect it independently of the parent group; it may drift away, evolutionary speaking, to become a new species. Paedomorphosis is a form of heterochrony in which sexually mature adults retain some or all of their larval (or juvenile) features. It can arise either through an accelerated rate of sexual maturation (progenesis) or through deceleration of the development of nonreproductive organs (neoteny), but both processes are believed to have contributed to the relatively fast evolution of diverse arthropod body forms during the Cambrian. Progenesis appears to be the more common of the two heterochronic processes by which crustacean body plans diversified, and it is thought to have played a key role in the evolution of many of the maxillopodan crustaceans, which exhibit bizarre and highly reduced body plans. In contrast, cladocerans, called "water fleas," are suspected to have evolved by neoteny from a clam-shrimp ancestor that became sexually mature but bypassed the naupliar larval stage of development.

Some scientists speculate that the gene-regulatory pathways controlling development, whether involving homeotic genes or ones contributing to heterochrony, were more flexible among the relatively primitive, less complex Cambrian ancestors of modern crustaceans and that conditions in the distant past were therefore more favorable for the rapid evolution of diverse body plans than they have been ever since. The main crustacean lineages observed today are the evolutionary result of this developmental plasticity. Each displays a confusing combination of primitive and advanced features, a combination that makes discerning evolutionary trends difficult.

To circumvent these difficulties, systematists have turned to means other than the traditional comparison of adult morphology. One approach compares the larval morphologies of different taxa and is based on the assumptions that early developmental stages are free of the unique morphological specializations seen in adults and that shared larval features may therefore more accurately reflect common ancestry. The case of barnacles, mentioned above, illustrates this point, and larval morphology has further been used to unite groups such as the Acrothoracica, Thoracica, and parasitic Rhizocephala within the barnacle infraclass Cirripedia and to recognize the parasitic Ascothoracida as the sister group of cirripedes. Larval features have also been used in attempts to discern shared ancestry among hermit crabs and primitive as well as more advanced true crabs. The recent first discovery of free-living lecithotrophic (nonfeeding) remipede larvae is posing new hypotheses about the affinity of this enigmatic group with other crustaceans and the evolution of the crustacean naupliar larva.

Another approach, one that has revolutionized and revitalized the discipline of phylogenetics, relies on comparisons of DNA sequences for genes shared by different types of crustaceans. For example, the gene coding for the enzyme cytochrome oxidase subunit I occurs in the mitochondrial genome of every crustacean. The DNA sequences for this particular gene in different species of crustaceans can be determined and the sequences compared. The differing degrees of similarity among the DNA sequences can tell us which crustaceans are more closely related than others. The DNA-sequence approach to inferring evolutionary relationships is particularly useful for comparison of species that share few morphological characters, because even morphologically diverse species will have many genes in common. Parasitic crustaceans, which almost always have highly modified morphologies, are especially problematic. For example, the evolutionary affinities of pentastomes, a group of wormlike parasites, had generated much speculation because these organisms possess few morphological characteristics that might suggest a relationship with some other group. They had been variously allied with annelids, myriapods, tardigrades, onycophorans, and mites and even placed within their own phylum, but a comparison of sperm development and ultrastructure revealed similarities between pentastomes and branchiuran crustaceans (fish lice), a finding that was corroborated by DNA analysis of nuclear 18S ribosomal RNA genes and mitochondrial genes for these and other potential relatives.

The DNA approach also is useful for determining phylogenetic relationships among organisms that exhibit a high degree of homoplasy (morphological similarities due not to shared ancestry but instead to convergence, i.e., development of similar morphology in response to evolution in similar habitats with similar selective regimes). For example, burrowing crustaceans have independently evolved modified digging appendages and a streamlined body form; different burrowing species may appear morphologically similar but not share a recent common ancestor. Evolutionary affinity among species may also be misconstrued if any of them has undergone a recent evolutionary reversal to a more ancestral morphological feature. The DNA approach to inferring relatedness avoids the problems presented by these instances of morphological homoplasy. Thermosbaenaceans are minute crustaceans found in groundwater and springs, and they have variously been placed within the superorder Peracarida (in part because they have a brood chamber, a feature found in some peracarids) and in a completely separate superorder, the Pancarida (because their brood chamber has a distinctive developmental origin and location). Similarly, mysid and lophogastrid shrimplike crustaceans have been considered either to comprise an order of peracarids or to be separate orders and excluded from the Peracarida. In each of these cases, DNA evidence has been useful in resolving long-standing phylogenetic uncertainty and has shown that morphological features (e.g., the brood pouch and features of the carapace, foregut circulatory system, and eyes) shared by some so-called peracarid species may in fact have evolved independently or been lost entirely.

Some crustacean species (e.g., one entire infraclass of crustaceans related to barnacles, the Facetotecta or "y-larvae") are known only from their larvae (adult forms have never been found). Other species are known from only male or only female forms, so knowledge of key morphological features is similarly lacking. In the amphipod family Leucothoidae, males may undergo a radical transformation during a single molt, such that pre- and post-molt males were erroneously classified as not just different species but as belonging to different families. Individuals of some species of caridean shrimp are sequential hermaphrodites, beginning life as males and later transforming into females (a condition termed protandry). Taxonomic recognition and assignment is especially complicated by a species that has super-males with enlarged genitalia (these never transform into females) in addition to the more typical male-phase hermaphrodites that will eventually transform. In all of these cases DNA evidence is sometimes the only means by which some species can be identified accurately and placed within an evolutionary hierarchy.

Finally, in addition to the intractability of determining crustacean relationships, the question of relationships among crustaceans and the other major extant arthropod subphyla (Hexapoda, or insects; Myriapoda, or centipedes and millipedes; and Cheliceriformes, comprising spiders, mites, ticks, and horseshoe crabs) is perplexing, for many of the same reasons noted above. Various alternative phylogenies have been proposed on the basis of different interpretations of morphological evidence. For instance, the "uniramian hypothesis" considers arthropods to be polyphyletic (i.e., unlikely to have had a single, common ancestor) and separates chelicerates (e.g., spiders) and crustaceans (possessing biramous appendages) from uniramous insects, myriapods, and onycophorans. The "mandibulate hypothesis" takes a monophyletic (single ancestor) view of arthropods and unites crustaceans, insects, and myriapods within the Mandibulata. Yet a third view, based on an "arthropod pattern theory," also views arthropods as monophyletic but places crustaceans in closer affinity to chelicerates than to insects and myriapods. Proponents of all three views consider insects and myriapods to be sister taxa, but now new and compelling molecular evidence suggests that insects are derived from a crustacean ancestor, thereby making the Crustacea a paraphyletic taxon that some are calling the "Pancrustacea" and making insects "flying crustaceans." Substantially more data are needed to resolve these outstanding issues. In 2001 the National Science Foundation awarded over $1 million to fund the Deep Arthropod Phylogeny project to obtain over 60,000 DNA bases from roughly 120 different genes from 85 arthropod species for just this purpose. Results are pending and are highly anticipated.

BIBLIOGRAPHY

Bowman, T. E., and L. G. Abele. 1982. Classification of the recent crustacea. In L. G. Abele, ed., *Systematics, the Fossil Record, and Biogeography*. Vol. 1, *The Biology of Crustacea*, 1–27. New York: Academic Press.
Calman, W. T. 1911. *The Life of Crustacea*. London: Methuen & Company.
Duffy, J. E. 1996. Eusociality in a coral reef shrimp. *Nature* 381: 512–514.
Fortey, R. A., and R. H. Thomas, eds. 1998. *Arthropod Relationships*. Systematics Association Special Volume Series 55. London: Chapman and Hall.
Martin, J. W., and G. E. Davis. 2001. *An Updated Classification of the Recent Crustacea*. Science Series No. 39. Los Angeles: Natural History Museum of Los Angeles County.
Poore, G. C. B. 2004. *Marine Decapod Crustacea of Southern Australia: A Guide to Identification*. Collingwood, Vic.: CSIRO Publishing.
Schram, F. 1986. *Crustacea*. New York: Oxford University Press.   —*T.S.*

Cuvier, Georges (1769–1832)

Although Georges Cuvier explicitly rejected the modification of species, he played a prominent role in the early history of evolutionary science, albeit involuntarily. In retrospect, he made two major contributions. First, he established beyond doubt the reality of the extinction of species, and he showed that this had occurred frequently in the history of the earth. Second, he undermined the idea of the great chain of being; he proved it was impossible to classify organisms in a serial order, either logical or chronological. French philosopher Michel Foucault said that Cuvier did more for the emergence of evolution than Jean-Baptiste Lamarck or early advocates of the transmutation theory. However, this retrospective evaluation is historically misleading. Cuvier's alleged contributions to evolution must be placed in their own context. Some of Cuvier's theories and methods were indeed incorporated into the science of evolution, but the entirety of Cuvier's work can be better understood in light of a different conception of the history of life.

Cuvier was primarily a zoologist and an anatomist. The main problem for him was classification. He saw comparative anatomy as being capable of providing a unified theory of living beings. Cuvier's comparative anatomy was closely related to physiology and to the study of past organisms. From the beginning of his career to the end, he combined anatomical description and classification of animals, present or extinct, with physiological hypotheses about their functional organization. Because of his major discoveries and methodological innovations in this broad field of research, he has often been considered the founding father of both comparative and stratigraphic paleontology. In both fields, he had predecessors and competitors, but he played a decisive role in each case. Historians of zoology (e.g., Russell [1916] 1982) insist on the first aspect, while historians of paleontology (e.g., Rudwick 1976) emphasize the second aspect. In Cuvier's work, the two dimensions can hardly be dissociated. Nevertheless, they are reflected in the titles of the two major synthetic books that he published (Cuvier 1812, 1817). Although this is somewhat artificial, the two aspects, stratigraphic paleontology (especially the theory of extinction) and comparative anatomy, will be examined separately.

In the second half of the seventeenth century, several scientists claimed that fossils could be interpreted as relics of extinct species. But the question remained open whether these species had indeed disappeared or still existed somewhere on the earth. For instance, Lamarck, who described a huge number of new invertebrate fossils, rejected the idea of extinction. Cuvier proposed to resolve this question by using fossils of large terrestrial animals as a test. In contrast with marine animals, large terrestrial animals could hardly remain unknown by 1800. In 1796, Cuvier gave the first detailed proof of such a case. On the basis of drawings received from Paraguay, he identified a giant fossil as an unknown animal related to the sloths. He called this animal *Megatherium*. The claim that *Megatherium* had disappeared was decisively supported by the precise description of the fossil, and its assignment to a known family (the sloths). Later, Cuvier accumulated evidence about the disappearance of a number of other large terrestrial animals (mammoth, mastodon, *Paleotherium*, and many species of fossil elephants, crocodiles, deer, etc.) in different parts of the world. By use of anatomical descriptions of unprecedented precision, he provided crucial support for the claim that these animals no longer existed.

Because migration was unlikely, only two hypotheses could account for the disappearance of these animals. At the very time when Lamarck began publishing his transmutation theory, in 1801 Cuvier stated they had been either entirely destroyed or modified in their forms. Cuvier rejected the modification hypothesis because he thought there was no evidence to support it. He advocated the extinction hypothesis because he thought there was evidence in its favor. Mummified animals brought back from Egypt by Étienne Geoffroy Saint-Hilaire (Napoleon's expedition) did not provide the slightest clue of the modification of animals between antiquity and the present (see figure). If no evidence of small changes was available, how could Lamarck and others extrapolate to larger changes over long periods of time? In contrast, Cuvier and Alexandre Brongniart's observations (1808) of the successive strata above the chalk in the Paris basin thus suggested the abrupt disappearance of local faunas. Over long distances (greater than 100 kilometers), there was a constant superposition of the strata, each stratum with its characteristic fossils, that

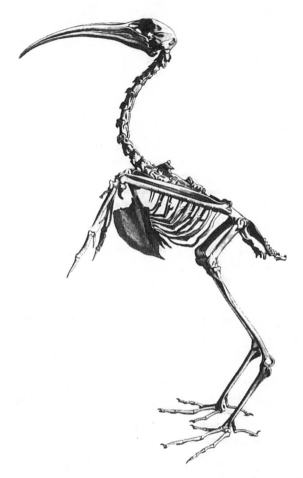

Part of Georges Cuvier's argument against evolution was based on the mummified organisms brought back by Napoleon's scientists, who went with him on his campaign to Egypt. This is the skeleton of an ibis, which Cuvier pointed out is identical to the skeletons of ibises today, thus apparently showing that there is no change over time. (From Cuvier 1804.)

disappeared in the next stratum. Furthermore, they observed an alternation of freshwater and marine formations. This could be accounted for by successive events of elevation and depression of the continental areas. Because the junctions between the two kinds of strata were rather sharp, Cuvier concluded that sudden geographic changes had been responsible for the extinction of successive terrestrial faunas and for a repopulation by existing faunas from other places.

As shown by Martin Rudwick (1976, 117), "Cuvier's rejection of evolution . . . was primarily a defense of *extinction,* not of special *creation.*" Cuvier's general representation of the history of life consisted in a series of local revolutions, which themselves implied the extinction of some species and the migration (in and out) of others. He never mentioned a general catastrophe that could have destroyed all existing species on earth. Such a hypothesis was mentioned later by other scientists (Alcide d'Orbigny in France, William Buckland in England) who also advocated events of "special creation" following the "catastrophes." Cuvier had no theory to account for the diversity of living beings. For him, this question was beyond the limits of the positive science of his time.

Although Cuvier had no theory for the origin and diversity of living organisms, he had a theory of classification that relied on three rational principles. The first principle, the principle of the conditions of existence, states that "the different parts of each being must be coordinated in such a way to render possible the existence of the being as a whole" (Cuvier 1817, 1: 6). This principle does not refer primarily to external conditions, but to the necessary coadaptation of functions and organs within the animal body. The second principle, the principle of correlation of parts, is an application of the former principle. It says that a given function cannot vary without corresponding modifications in all related organs. In practice, Cuvier used this principle to infer the shape of a number of organs in a fossil from the shape of one organ. Cuvier used his first two principles for denying the possibility of the modification of species. Organs and functions in an organism are so harmoniously coordinated in an organism that any change in a given part would require simultaneous changes in many or all of the other parts.

Cuvier's third principle, borrowed from botanist Bernard de Jussieu, was the principle of subordination of characters. It states that some characters are more fundamental than others, in the sense that they determine a set of possibilities for the rest of the organism as a whole. This principle led Cuvier to distinguish four different embranchements, or functional types of animals: vertebrates, mollusks, articulates, and radiates. These types were not reducible one to another. The notion of major heterogeneous structural-functional arrangements condemned any attempts to classify animals—and even more organisms in general—according to a serial order. For Cuvier there was no possible ascending series of complexity (as researched by Lamarck) and no ultimate unity of body plan (studied by Étienne Geoffroy Saint-Hilaire; see the alphabetical entry "Étienne Geoffroy Saint-Hilaire" in this volume). This was a decisive argument against the hypothesis of common descent of organisms. Nevertheless, in a convoluted way, the notion of embranchements played a role in the emergence of Charles Darwin's and Alfred Russel Wallace's notion of branching evolution.

As noted by most commentators, Cuvier preferred facts to ideas, and he looked systematically for laws. He did not admit either modification or special creation of species, because he could not find facts supporting either hypothesis. Correlatively, he believed that repeated events of extinction and local revolution had obeyed regular laws of nature.

BIBLIOGRAPHY

Cuvier, G. 1804. Mémoire sur L'Ibis des anciens Egyptiens. *Annales du Muséum national d' Histoire naturelle* A(1804): 116–135.

———. 1812. *Recherches sur les ossemens fossiles des quadrupèdes.* Paris: Déterville.

———. 1817 [1983]. *Le Règne animal distribué d'après son organisation pour servir de base à l'histoire naturelle des animaux et d'introduction à l'anatomie comparée.* 4 vols. Paris: Déterville. Reprint, Paris: Editions des archives contemporaines.

Cuvier, G., and A. Brongniart. 1808. Essai sur la géographie minéralogique des environs de Paris. *Journal des Mines* 23: 421–458.

Foucault, M. 1966. *Les mots et les choses: Une archéologie des sciences humaines.* Paris: Gallimard.

Rudwick, M. 1976. *The Meaning of Fossils: Episodes in the History of Palaeontology.* Chicago: University of Chicago Press.

Russell, E. S. 1916 [1982]. *Form and Function: A Contribution to the History of Animal Morphology.* London: John Murray. Reprint, Chicago: University of Chicago Press.   —*J.G.*

D

Darwin, Charles (1809–1882)

Charles Robert Darwin, the English naturalist and evolutionist, was born in the town of Shrewsbury on February 12, 1809, the same day as Abraham Lincoln across the Atlantic. He died on April 19, 1882, at his home in the village of Downe (Kent). He was the author of *On the Origin of Species*, published in 1859, and he is considered the father of evolutionary thinking.

Darwin's paternal grandfather was physician Erasmus Darwin, author of *Zoonomia* (1794–1796), one of the earliest works proclaiming evolution. His father, Robert Darwin, was also a physician as well as a money man, arranging mortgages between aristocrats with land and a need of cash and businessmen with cash and a need of safe loans. Darwin's maternal grandfather was Josiah Wedgwood, the man responsible for industrializing the pottery trade in Britain. Darwin cemented the Wedgwood connection further when he married his first cousin, Emma Wedgwood. He and his wife had 10 children, seven of whom lived to adulthood. One was George Darwin, the eminent physicist and authority on the tides.

Charles Darwin was educated first at Shrewsbury School, one of England's leading private schools, and then sent to Edinburgh University in the footsteps of his father and grandfather, intending to train as a physician. Unhappy in this direction, he enrolled at Christ's College in the University of Cambridge, intending now to be an Anglican clergyman. While at university, both in Scotland and England, Darwin started to show a strong interest in science, and by the time he graduated in 1831, senior scientists had started to spot both the interest and the talent. The substantial Darwin-Wedgwood fortunes meant that Darwin never had pressing financial needs and could always indulge his passion for empirical inquiry.

Diverted from the clerical career by the offer of a voyage on HMS *Beagle*, under the captaincy of Robert Fitzroy, Charles Darwin spent five years going around the coast of South America and eventually circumnavigating the globe. Although he was engaged as a gentleman companion to the captain, Darwin soon became the full-time ship's naturalist, accumulating extensive collections that he shipped home, thus adding to the already good opinions of his Cambridge mentors. Darwin also made extensive travels inland. The diaries he wrote on the voyage were the foundation for a popular travel book titled *The Voyage of the Beagle*, first published as part of the official record of the ship's travels and then independently. There is no doubt that being known and appreciated by the general public cemented Darwin's reputation as a man of solid worth. This status helped later to give his controversial scientific claims a fair and favorable hearing.

Darwin's first systematic interests as a scientist were in geology. Although he was trained by Cambridge professor Adam Sedgwick, who believed that the world is formed by occasional periods of great eruption and turmoil (catastrophes), the greatest influence on Darwin was Scottish geologist Charles Lyell, whose three-volume work, *Principles of Geology* (1830–1833), was eagerly devoured by the voyaging naturalist. Lyell proposed the uniformitarian thesis, arguing that the geological world is governed by unbroken law. Also, Lyell hypothesized that the world is in a kind of steady state, with all changes being transitory and a function of altering climate brought on by changes in the distribution of land and sea. He looked on the surface of the globe as a kind of water bed, where deposition in one area causes subsidence and thus brings on elevation in another area.

Darwin accepted this view completely, and his first scientific triumph was in the Lyellian mode. Why does one find circular coral reefs around tropical islands? Lyell argued that they are the tops of extinct volcanoes, but Darwin argued that it was improbable that all of the volcanoes would coincidentally just break the sea surface. Rather, he argued, the coral grows around the islands and then they sink, with the coral growing ever upward to remain at the sea surface, even after the islands themselves may have disappeared completely beneath.

Charles Darwin's greatest scientific error was also in the Lyellian mode. In Glen Roy, a small valley in Scotland, there are three horizontal, tracklike indentations around the sides. What is the reason for these parallel roads, as they were known? All agreed that they were the remains of beaches around water that once filled the glen, but where is the water now? In the late 1830s, Darwin argued that it had been seawater but that the whole area had moved upward and now the water was drained. Shortly thereafter the Swiss scientist Louis Agassiz visited the area; he at once saw that the water was a lake, dammed by ice that had since melted. Eventually Darwin appreciated his mistake, but this took time because Agassiz wrapped his thinking in a catastrophic theory that relied heavily on the interference of God.

Darwin would not have objected to the supposition that there was a God. As a young man preparing for the clergy, he was a sincere Christian. On the *Beagle* voyage his beliefs changed and he became a committed deist, believing strongly in a God who worked through unbroken law and who does not interfere with his creation. This position lasted for many years, right through the writing and publication of the *Origin*. Then, toward the end of his life, like many mid-Victorian intellectuals, Darwin's religion faded into agnosticism. The Anglican background was always vital to Darwin, however, because it insisted that God's designing powers can be seen in the adaptations of organisms, such as the eyes for seeing and the nose for smelling. Like Archdeacon William Paley, whose *Natural Theology* (1802 [1819]) Darwin read as an undergraduate, Darwin believed that the most significant feature of the living world is that it is complex and designlike, to the benefit of individual organisms.

Lyellian geology directed Darwin to thinking about organisms; fossils were vital evidence of subsidence and elevation. Thus primed, in 1835, when the *Beagle* reached the Galápagos Archipelago, a group of islands in the Pacific, Darwin was astounded to learn that the denizens, particularly the birds and the giant tortoises, are different from island to island. Evolution from a common ancestry is the obvious answer, especially to someone who believes that God's glory is shown through the working of law. However, it was not until he returned to England in the following spring of 1837 and was convinced that the Galápagos varieties are truly different species that Darwin made the leap to transformism, as it was then called.

Darwin realized that he needed a mechanism to explain his newfound belief. As a graduate of the University of Cambridge, he was aware of the triumphs of its greatest scientist, Isaac Newton. Darwin therefore sought for the organic world a kind of equivalent of Newton's universal law of gravitation. He found his answer in September 1838. He started with the Malthusian calculation that population pressures will tend to outstrip supplies of food and space. He noted that there seems always to be a source of variation in natural populations. Hence, arguing analogically from the successes of the breeders of animals and plants, Darwin concluded that there will be a natural form of selection, with some winning and others losing, and success being a function of the winners' peculiar features. Vitally, this natural selection, or "survival of the fittest," as it became known, leads not just to change but to change in the direction of adaptation.

Darwin soon wrote up his thinking as a sketch of a theory, but for reasons still undetermined he did not publish. Undoubtedly a factor was that he fell seriously ill and for the rest of his life was an invalid and a recluse. (The exact cause of the illness is unknown. Some speculate that it was psychological, the stress of having such an evolutionary theory. Others think it might have been the result of an illness, Chagas disease, picked up on the *Beagle* voyage.) For many years, Darwin sat on his evolutionary thinking and was diverted into a massive taxonomic study of barnacles.

Finally, in June 1858, Alfred Russel Wallace, a young naturalist writing from the Far East, sent Darwin a short essay containing the same ideas Darwin had 20 years before. Thus spurred, extracts of Darwin's thinking and Wallace's essay were published at once. Darwin now rapidly wrote up his theory, publishing *On the Origin of Species* at the end of 1859.

The *Origin* was immediately controversial, but it is important not to exaggerate this fact. In Britain and on the Continent, and even in America, many people, both scientists and the general population, quickly accepted the idea of evolution. Darwin himself had laid the groundwork for this by carefully cultivating men—notably anatomist Thomas Henry Huxley and botanists Joseph Hooker in Britain and Asa Gray in America—who would defend his thinking publicly. The religious tended to be more reluctant, but these people also (after a decade or so) most probably accepted some form of evolutionism. What is clear is that there was considerably more enthusiasm for evolution than for Darwin's own mechanism of natural selection. No one wanted to deny the latter outright, but most wanted to supplement it with other causes, ranging from Lamarckism (the inheritance of acquired characteristics) to saltationism (evolution by instantaneous major new variations) to some kind of guided evolution where God plays a continuing and active role.

Darwin had intended the *Origin* to be a teaser for a series of books he would now write, filling out in detail all the claims made briefly in the *Origin*. Only one of these books was ever written, *The Variation of Animals and Plants under Domestication* (1868), notable chiefly for Darwin's somewhat inadequate hypothesis about heredity, called pangenesis. (Darwin argued that particles are given off all over the body and collected in and transmitted by the sex organs. This would thus ensure Lamarckian effects, a hypothesis to which Darwin always subscribed.)

Apart from a series of incidental works that Darwin now authored—a little book on orchids, another on climbing plants, a third on insectivorous plants, yet another on earthworms—the major labors of the post-*Origin* years were on the human species, resulting in *The Descent of Man* (1871) and a kind of supplement, *The Expression of the Emotions in Man and Animals* (1872). In the *Origin*, deliberately, Darwin had said virtually nothing about humans, other than an unambiguous comment at the end suggesting that he thought his theory also applied to humans. It was not that he thought the topic uninteresting or unimportant, but he wanted first to get the basic theory on the table and not get everything at once lost in the general argument about humans.

In this, Darwin was only partially successful. At once, everyone seized on the implications for *Homo sapiens* of the "monkey theory." This is what people wanted to talk about. For a while, Darwin stayed aloof. Then, under pressure from friends and supporters, and particularly galvanized by the fact that Wallace had become a spiritualist and had turned to non-natural causes to explain human evolution, Darwin argued in detail that human nature, physical and social, is a product of unbroken law. As always, natural selection was the key causal feature, but Darwin supplemented it with a secondary mechanism, sexual selection, where organisms compete for mates. Darwin believed that sexual selection was particularly important in determining the differences between human sexes and races.

Although he did not live to a very old age, Darwin's health improved in his last decade. He continued to work until the end, including answering an ever-growing correspondence. Acknowledged both by his countrymen and by others across the world as one of the great scientists of all time, he was regarded with respect and affection. There was therefore no great surprise that when he died, by general acclaim, he was buried in Westminster Abbey, just across from Newton and a pace or two away from Lyell.

BIBLIOGRAPHY

Browne, J. 1995. *Charles Darwin: A Biography*. Vol. 1, *Voyaging*. New York: Knopf.

———. 2002. *Charles Darwin: A Biography*. Vol. 2, *The Power of Place*. New York: Knopf.

Darwin, C. 1839. *Journal of Researches into the Geology and Natural History of the Various Countries Visited by HMS Beagle*. London: Henry Colburn.

———. 1859. *On the Origin of Species by Means of Natural Selection, or the Preservation of Favoured Races in the Struggle for Life*. London: John Murray.

———. 1868. *The Variation of Animals and Plants under Domestication*. London: John Murray.

———. 1871. *The Descent of Man, and Selection in Relation to Sex*. London: John Murray.

———. 1872. *The Expression of the Emotions in Man and Animals*. London: John Murray.

Darwin, E. [1794–1796] 1801. *Zoonomia; or, The Laws of Organic Life*. 3rd ed. London: J. Johnson.

Lyell, C. 1830–1833. *Principles of Geology: Being an Attempt to Explain the Former Changes in the Earth's Surface by Reference to Causes Now in Operation*. 3 vols. London: John Murray.

Paley, W. [1802] 1819. *Natural Theology (Collected Works, vol. 4)*. London: Rivington.

Ruse, M. 1979. *The Darwinian Revolution: Science Red in Tooth and Claw*. Chicago: University of Chicago Press.

———. 2008. *Charles Darwin*. Oxford: Blackwell.

Wallace, A. R. 1858. On the tendency of varieties to depart indefinitely from the original type. *Journal of the Proceedings of the Linnean Society, Zoology* 3: 53–62.   —M.R.

Darwin, Erasmus (1731–1802)

Erasmus Darwin, the grandfather of Charles Darwin, was one of the best-known early evolutionists. A physician, he lived in Litchfield in the English Midlands and later in Derby, also in the Midlands. He fathered three sons with his first wife, then two daughters with a mistress, and seven more children with his second wife. He was open in his enthusiasm for sexual activity, and to this end he decided, as a young man, to give up drinking lest it impair his prowess. A great enthusiast for Linnaean taxonomy, based as it was on sexual characteristics, Darwin was known to the wider public as a poet who hymned the virtues of plant sexuality, *The Loves of the Plants* (1789) being one of his major literary achievements.

Erasmus Darwin was a close friend of industrialists including potter Josiah Wedgwood, whose daughter married Erasmus's third son, Robert, and one of whose children was Charles. Keenly interested in machines and their uses, Erasmus Darwin was a member of the Lunar Society, a group of men who met monthly to discuss matters of interest. This group included chemist Joseph Priestley, industrialist Matthew Bolton, and James Watt, famous for his work on improving the steam engine. Darwin was himself an inventor, including a horizontal windmill, an untippable carriage, and a canal lift for barges. He was also interested in agriculture, writing a scientific treatise on the topic. For his two middle children, whom he set up in their own school, he wrote a small treatise on female education.

He corresponded with people far and wide on topics of mutual interest. One acquaintance was Benjamin Franklin. Like many more advanced thinkers of his day, Darwin approved of the American Revolution, as well as the French Revolution, until it got out of hand. A man of great girth, a function of an enormous appetite, he was nevertheless a welcome house guest because he was a brilliant conversationalist. He was one of the wisest medical men of his day; King George III unsuccessfully begged him to be the royal physician.

Darwin's interest in evolution seems to have been sparked by fossils, particularly those thrown up in the course of boring through mountains to make tunnels for canals. But it was not until virtually the last decade of his life that he actually started writing on the subject. His major prose work, *Zoonomia* (1794–1796), carried discussion of the topic, and the subject enters into several of his poems. Although his thinking is somewhat crude by today's standards, Darwin makes important points, still valid, especially drawing attention to the similarities between different kinds of organisms (what today is called homologies) and to the possibility that embryological development is a cameo for what happens to species through the ages. Causally, Darwin floated the idea that acquired characteristics might be inherited (what today is called Lamarckism) as well as offering a version of his grandson's mechanism of sexual selection.

Truly, however, for Erasmus Darwin, as for every other early evolutionist, the change of organisms over time was taken to be a reflection of and proof of change in society over time—change from a simple state to one more advanced, that is to say, change of a progressive nature. As he used to say, from monarch (butterfly) to monarch (king). This comes through clearly in one of Darwin's later poems, *The Temple of Nature.*

> Organic Life beneath the shoreless waves
> Was born and nurs'd in Ocean's pearly caves;
> First forms minute, unseen by spheric glass,
> Move on the mud, or pierce the watery mass;
> These, as successive generations bloom,
> New powers acquire, and larger limbs assume;
> Whence countless groups of vegetation spring,
> And breathing realms of fin, and feet, and wing.
>
> Thus the tall Oak, the giant of the wood,
> Which bears Britannia's thunders on the flood;
> The Whale, unmeasured monster of the main,
> The lordly Lion, monarch of the plain,
> The Eagle soaring in the realms of air,
> Whose eye undazzled drinks the solar glare,
> Imperious man, who rules the bestial crowd,

Of language, reason, and reflection proud,
With brow erect who scorns this earthy sod,
And styles himself the image of his God;
Arose from rudiments of form and sense,
An embryon point, or microscopic ens!
(Darwin 1803, 1: Canto I, lines 295–314)

In arguing thus, Erasmus Darwin was going against a biblically inspired account of origins, but the actual disagreement with Genesis was never a major impediment in the eyes of Christians, most of whom by his day were not literalists (in the sense of today's creationists). More objectionable was the fact that progress, which depends on the efforts of individual humans, was seen to conflict with the idea that salvation can come only through Providence, or God's unmerited grace. That conflict did make evolution a radical idea, compounded by the fact that the vile doctrine was seen to have led to the French Revolution. Darwin himself was a deist, someone who believes in God as an unmoved mover. The fact that evolution is grounded in the rule of law would have been, for him, a proof of God's existence and powers rather than a refutation.

In the end, Erasmus Darwin's reputation was destroyed by a savage parody of his poetry written by political conservatives, including George Canning, a British statesman during the Napoleonic era and later, briefly, prime minister. In the funny "The Loves of the Triangles" (1798), Darwin's speculations about plant sexuality and inclinations (in an earlier poem, *The Loves of the Plants*) were mocked. There were certainly those who studied Darwin's ideas with care, including the aged Immanuel Kant in Konigsburg, who read a German translation of *Zoonomia*. Nevertheless, generally, enthusiasm for his thought went into steep decline. Perhaps somewhat unfairly, for posterity he takes second place to the later, and no more sophisticated, writings of French biologist Jean-Baptiste Lamarck.

BIBLIOGRAPHY

Canning, G., H. Frere, and G. Ellis. 1798. The loves of the triangles. *Anti-Jacobin*, 16 April, 23 April, and 17 May.
Darwin, E. 1789. *The Botanic Garden (Part II, The Loves of the Plants)*. London: J. Johnson.
———. [1794–1796] 1801. *Zoonomia; or, The Laws of Organic Life*. 3rd ed. London: J. Johnson.
———. 1803. *The Temple of Nature*. London: J. Johnson.
King-Hele, D. 1963. *Erasmus Darwin: Grandfather of Charles Darwin*. New York: Scribner.
McNeil, M. 1987. *Under the Banner of Science: Erasmus Darwin and His Age*. Manchester: Manchester University Press.
Ruse, M. 1999. *Mystery of Mysteries: Is Evolution a Social Construction?* Cambridge, MA: Harvard University Press.

—M.R.

Darwin and the Emergence of Evolutionary Theories of Mind and Behavior (Robert J. Richards)

In his *Darwin and the Emergence of Evolutionary Theories of Mind and Behavior* (1987), Robert J. Richards addresses the most difficult set of ideas that hover around biological evolution—the evolution of mind, instinct, intelligence, reason, habits, and behavior. These ideas are difficult because they are frequently used to investigate themselves. Richards also addresses morals, and there is nothing like morals to interfere with the study of morals.

Can these human characteristics evolve via natural selection, and can they themselves function as a mechanism in evolution? Richards traces the answers to these two questions in the works of such eighteenth-century scholars as Erasmus Darwin, Pierre-Jean Cabanis, and Jean-Baptiste de Lamarck, through Charles Darwin, Alfred Russel Wallace, and Herbert Spencer, and later Darwinians such as George J. Romanes, C. Lloyd Morgan, James M. Baldwin, and William James, and on up toward the end of the twentieth century.

As always, Charles Darwin viewed natural selection as the chief, although not the sole, mechanism operating in evolution, including the evolution of human traits. However, all of these mechanisms were for Darwin totally naturalistic. Some of Darwin's contemporaries followed Darwin's lead, but most did not. Wallace began thinking that the peculiarly human traits could be generated by natural selection. Later he abandoned Darwin on this score and joined with Herbert Spencer.

A common belief among present-day historians of biology is that Darwinian theory crushed nineteenth-century belief in a spiritually dominated universe and purged nature of intelligent design and purpose. Richards disagrees on both counts. Perhaps later Darwinians did not hold the unsophisticated beliefs of fundamentalist Christians, but they were hardly recalcitrant materialists. Darwin and his disciples attempted to infuse human nature with an authentic moral sense.

The second issue concerned such things as habits functioning as mechanisms for evolutionary change. Among historians of science in the past, this topic has been usually limited to parodies of Lamarck and later James Baldwin, but Richards shows that such beliefs were common from the eighteenth century to the late twentieth century. In fact, it was one of Darwin's main concerns, but later Darwinians did not pay as much attention to it as Darwin did, at least not until the end of the nineteenth century. According to Baldwin, some animals, when they enter a new environment or their old environment moves out from under them, can adjust to the change by learning new behaviors. Over several generations these behaviors might be transmuted into new habits. However, in the twentieth century, evolutionary biologists tended to reject the Baldwin Effect as not occurring at all or as being of only very minor significance. During the past few decades or so, however, the Baldwin Effect has experienced a resurgence. Maybe it is as important as Darwin originally thought.

In *Darwin and the Emergence of Evolutionary Theories of Mind and Behavior,* Richards espouses a radical philosophy of history—a selection model. Historians of science are well aware that the genealogy (who got what from whom) of ideas is important, but reconstructing conceptual genealogies is quite difficult. Usually historians have to settle for similarity of ideas, and too often, when similarity and genealogy conflict, they side with similarity. Richards does not. He not only espouses a selection model of scientific change but also puts it into practice. His history is all the better for it.

BIBLIOGRAPHY

Hodge, J., and G. Radick, eds. 2003. *The Cambridge Companion to Darwin.* Cambridge: Cambridge University Press.

Richards, R. J. 1987. *Darwin and the Emergence of Evolutionary Theories of Mind and Behavior.* Chicago: University of Chicago Press.

———. 1992. *The Meaning of Evolution: The Morphological Construction and Ideological Reconstruction of Darwin's Theory.* Chicago: University of Chicago Press.

———. 1999. The epistemology of historical interpretation: progressivity and recapitulation in Darwin's theory. In R. Creath and J. Maienschein, eds., *Epistemology and Biology,* 64–90. Cambridge: Cambridge University Press.

Ruse, R. 2003. *Darwin and Design: Does Evolution Have a Purpose?* Cambridge, MA: Harvard University Press.

Weber, B. H., and D. J. Depew. 2003. *Evolution and Learning: The Baldwin Effect Reconsidered.* Cambridge, MA: MIT Press.   —D.L.H.

Dawkins, Richard (b. 1941)

Richard Dawkins is a British zoologist and evolutionary theorist best known for his concepts of the selfish gene, the extended phenotype, and the meme. Born in Kenya, Dawkins was educated at Oxford, where he has lived and worked since 1970. His 1976 *The Selfish Gene* is one of the classic works of neo-Darwinism, unifying and clarifying the emerging consensus among such evolutionary theorists as William Hamilton, George Williams, Robert Trivers, John Maynard Smith, and his doctoral supervisor, ethologist Niko Tinbergen. This book articulates the theory of natural selection in such vivid and accessible terms that it is often mistakenly regarded as merely a "popular" book. Dawkins's central claim is that the most fundamental, and hence most explanatory, perspective on all evolution by natural selection is the "gene's-eye view," in which benefits to species, lineages, groups, and even individual organisms are seen to be subsidiary to the primary beneficiaries of all adaptations, the "selfish" genes themselves. Over time, evolution designs and builds organisms ("survival machines") that benefit coalitions of genes by improving their prospects for replication. The reach of genes does not stop at the skin of the organism: some genes control the design and construction of an extended phenotype, harnessing features of the environment and even other species. Thus the beaver's dam and the spider's web are just as important parts of the phenotypes of those species as their eyes and mouths, and parasites often exploit the behavioral controls of the host species, hijacking other genes' survival machines. Emphasizing the universal application of Charles Darwin's fundamental insights, Dawkins drew attention to the possibility that when genetic natural selection created a species, *Homo sapiens,* with an extended phenotype that included language and technology, this in effect established a new medium of evolution by natural selection: human culture, in which the differential replication of salient cultural items, called memes, could account for many of the features of cultural evolution that are otherwise perplexing. Dawkins initially presented the idea of memes as a sort of thought experiment to illustrate the abstractness of the fundamental idea of natural selection, and he has not claimed to be making a major contribution to the scientific investigation of the evolution of human culture. Others, however, have taken up the concept enthusiastically, with mixed results to date.

In 1986, Dawkins published another pedagogical tour de force, *The Blind Watchmaker,* in which he illustrated many of the most elusive implications of evolutionary theory via his pioneering Blind Watchmaker software, one of the early triumphs of the Artificial Life movement in theoretical biology. In 1995, he was appointed to be the first Charles Simonyi Professor of the Public Understanding of Science at Oxford. In this ideal role for a writer of his talents, he has subsequently published a series of lucid books and articles explaining aspects of science to the general public. He also has taken the lead in the public discussions of controversial issues of science policy and politics. Most recently, his broadside on religion, *The God Delusion* (2006), an attempt to raise consciousness about the weaknesses and follies of traditional religious belief, has attracted both enthusiastic praise and vilification, playing a leading role in what is being called "the new atheism." Less well known to the lay public are Dawkins's important analyses of fundamental problems in kin selection and the evolution of communication, among other topics.

BIBLIOGRAPHY

Dawkins, R. 1976. *The Selfish Gene.* Oxford: Oxford University Press.
———. 1982. *The Extended Phenotype.* Oxford: Oxford University Press.
———. 1986. *The Blind Watchmaker.* New York: W. W. Norton.
———. 2004. *The Ancestor's Tale.* Boston: Houghton Mifflin.
———. 2006. *The God Delusion.* Boston: Houghton Mifflin.
Grafen, A., and M. Ridley, eds. 2006. *Richard Dawkins: How a Scientist Changed the Way We Think.* Oxford: Oxford University Press.   —D.D.

The Descent of Man, and Selection in Relation to Sex (Charles Darwin)

The Descent of Man, and Selection in Relation to Sex, published in 1871, was Charles Darwin's work on the evolution of the human species. Initially, he had been unwilling to get into this aspect of the evolutionary debate, but when Alfred Russel Wallace (1870), who had become a spiritualist, declared that humans could not have been produced by natural means, Darwin felt that he had to get involved.

Little of the basic discussion of human evolution is original, as Darwin relied heavily on the reports and findings of others. Like Thomas Henry Huxley (in *Evidence as to Man's Place in Nature*, 1863) before him, Darwin's chief mode of argument was to compare humans with other animals, showing that the differences are a lot less great than one might imagine. Therefore, there is no reason to think that humans are not part of the natural world, and, given the shared features with the animals, much reason to think that humans are their descendants. Darwin did not think that humans were descended from animals alive today, and he was inclined to support the idea of an African origin for humankind.

Darwin was particularly interested in human reason and sociality. He argued that morality (and religion) are natural outcomes of evolution. Because in earlier writings Darwin had believed that natural selection works always for the benefit of the individual rather than the group, he was concerned to show that a genuine moral sense of care for others could nevertheless evolve. In some respects, he thought that a group kind of selection might have produced morality, but he also gave the suggestion that morality could be enlightened self-interest. Here he anticipated the modern mechanism of "reciprocal altruism."

Wallace's apostasy also explains the rather odd nature of Darwin's book, because much of it was not about humans at all. It was rather about Darwin's secondary mechanism of evolutionary change, that is, sexual selection. Here, organisms in a species compete for mates, and hence they develop adaptations (e.g., antlers of the deer, tail feathers of the peacock) to aid in this competition. Wallace argued that features like human hairlessness and great intelligence could not have been produced by natural selection and so must be the result of spirit forces. Darwin replied that sexual selection causes many human features, such as intelligence and body shape. For many years, people were inclined to dismiss sexual selection as something demanding treatment in its own right, and thought of it as something best incorporated with natural selection. With the rise of interest in animal social behavior, sexual selection is now something of great theoretical and empirical interest.

In the course of the discussion, Darwin was led to make some very conventional Victorian judgments about the superiority of males over females, and Europeans over all other humans. He also worried that modern medicine was having deleterious effects on human breeding stock, but felt able to defend capitalism because it leads to leisured people of culture whose works benefit us all. Recently, on the basis of these sorts of remarks, the American religious opponents of evolution (the creationists and their associates, the so-called Intelligent Design Theorists), frustrated in their attempts to prove that their biblically based beliefs are genuine science, have tried a different tack. They argue that Darwin leads straight to the vile ideologies of the twentieth century, particularly National Socialism (see especially Weikart 2006 and the Ben Stein–narrated film *Expelled: No Intelligence Allowed*, 2008). This is a classic example of bad scholarship promoting an unsupportable theological thesis. Although Darwin is very Victorian, he is a very liberal Victorian. For all that he worried about the effects of medicine, he immediately stressed that he did not think we should stop using it. He also thought that the chief reason white people tend to overcome other peoples is because other peoples succumb to the diseases of white people more rapidly than white people do to the diseases of others. More generally, although there are indeed passages in *Mein Kampf* (1925) that do sound like Social Darwinism, Hitler drew somewhat haphazardly on many sources, and the anti-Semitism of his youth in Vienna and the *Volkish* sentiments of the group promoting the memory of Wagner were at least as significant, or more (Friedlander 1997; Kershaw 1999). In any case, the Nazis generally did not much like evolution because it stressed the one-ness of humankind, linking Aryans with Jews and Gypsies. (Richards 2008 discusses these charges, putting them in context.)

BIBLIOGRAPHY

Darwin, C. 1871. *The Descent of Man, and Selection in Relation to Sex*. London: John Murray.
Expelled: No Intelligence Allowed. 2008. Nathan Frankowski, director. 90 min. N.p.: Premise Media Corporation.
Friedlander, S. 1997. *Nazi Germany and the Jews*. Vol. 1, *The Years of Persecution, 1933–39*. London: Weidenfeld and Nicolson.
Gruber, M. E. 1981. *Darwin on Man*. 2nd ed. Chicago: University of Chicago Press.
Hitler, A. 1925. *Mein Kampf*. Munich: Verlag Franz Eher Nachfolger.
Huxley, T. H. 1863. *Evidence as to Man's Place in Nature*. London: Williams and Norgate.
Kershaw, I. 1999. *Hitler*. Vol. 1, *1889–1936: Hubris*. New York: Norton.
Richards, R. J. 2008. *The Tragic Sense of Life: Ernst Haeckel and the Struggle over Evolutionary Thought*. Chicago: University of Chicago Press.
Ruse, M. 2008. *Charles Darwin*. Oxford: Blackwell.
Trivers, R. L. 1971. The evolution of reciprocal altruism. *Quarterly Review of Biology* 46: 35–57.
Wallace, A. R. 1870. The limits of natural selection as applied to man. In his *Contributions to the Theory of Natural Selection*, 332–371. London: Macmillan.
Weikart, R. 2006. *From Darwin to Hitler: Evolutionary Ethics, Eugenics, and Racism in Germany*. New York: Palgrave Macmillan.   —M.R.

The Development of Darwin's Theory (Dov Ospovat)

The Development of Darwin's Theory was published in 1981. It appeared just a few months after the death of its author, Dov Ospovat, at the age of 33. The book achieved an important milestone in what became known as Darwin Studies, the historical study of Charles Darwin's life and work. Darwin Studies became a special area in the history of science shortly after the 1959 centennial of the publication of Charles Darwin's *On the Origin of Species*. It soon grew into a major field of historical research. The explosive growth was due to two factors. First, until the 1950s Darwin was not recognized as the extremely important scientist that he is today. The modern synthetic theory of evolution (the theory formed by combining Darwinian natural selection with Mendelian genetics and, more recently, with molecular genetics) began during the 1930s and 1940s. At first it demonstrated that a number of different influences, including natural selection and genetic drift, contributed to evolutionary change, but the relative strengths of these influences was unknown. By the 1950s it was generally accepted that natural selection was the dominant evolutionary force. This consensus coincided nicely with the celebrations of the centennial of the *Origin*. Because Darwin was the father of natural selection, he now became the father of the synthetic theory of evolution. The new interest in Darwin coincided with the discovery that Darwin had left an immense store of personal papers that included personal correspondence, research notes, and preliminary drafts of his books and articles. Unlike Gregor Mendel (whose papers were destroyed by his successor), Darwin's papers offered immense opportunities for research in the history of science and, indeed, in the life story of a scientist whose importance was suddenly being recognized.

The first wave of Darwin Studies concentrated on his voyage on HMS *Beagle* and his notebooks of 1837–1839, written soon after he returned. This period covered the beginnings of Darwin's evolutionary thought and his development of the principle of natural selection. This phase of Darwin's life was important to scientists because natural selection was the aspect of Darwin's thought that fit most closely with the synthetic theory of evolution. The period from 1840 to 1859 was often regarded as a puzzle. If Darwin had discovered natural selection before 1840, why did he delay publishing until 1859? Ospovat provided an answer to that question and, simultaneously, began the study of Darwin as a man of his age—a Victorian naturalist.

Ospovat was able to provide this new perspective because he was the first historian to gain access to Darwin's papers from the so-called black box, the set of portfolios that Darwin had assembled in the 1850s in preparation for writing his major work on evolution. (This work was, of course, interrupted by Alfred Russel Wallace's codiscovery of natural selection, and *On the Origin of Species* was the result. See the main essay "The History of Evolutionary Thought" by Michael Ruse in this volume.) The first practitioners of Darwin Studies had seen natural selection as Darwin's major achievement and interpreted him as a unique genius. Ospovat, in contrast, depicted him as having benefited from wide readings in the science of his day, including paleontology, systematics, morphology, and embryology. This explained Darwin's hesitation. Even though natural selection was in place, Darwin's evidence that evolution had actually occurred was relatively weak in the early years. It was gradually strengthened during the two decades before 1959. "Darwin was the first reputable naturalist to argue for transmutation who was able to take advantage of the developmental view of nature that emerged in the mid-nineteenth century. This, as much as his new theory [natural selection], separates Darwin's work from [Jean-Baptiste] Lamarck's" (Ospovat 1981, 168).

This new view of Darwin as a beneficiary of nineteenth-century morphology and embryology was influential on several philosophers who, beginning in the 1980s, were concerned with the relation between evolution and embryological development (Amundson 2005; Burian 2004). It also set the stage for broader historical understandings of Darwin, such as those gathered by historian David Kohn into the monumental 1985 anthology *The Darwinian Heritage,* which was dedicated to Dov Ospovat. Very important was Ospovat's strongly supported claim that Darwin was firmly committed to the notion of biological progress, a topic that has been of major interest to historians in recent years (Ruse 1996, 2004; Richards 1992, 2003, 2004).

BIBLIOGRAPHY

Amundson, R. 2005. *The Changing Role of the Embryo in Evolutionary Biology: Roots of Evo-devo.* Cambridge: Cambridge University Press.

Burian, R. M. 2004. *Epistemological Papers on Development, Evolution, and Genetics.* Cambridge: Cambridge University Press.

Kohn, D., ed. 1985. *The Darwinian Heritage.* Princeton, NJ: Princeton University Press.

Ospovat, D. 1981. *The Development of Darwin's Theory.* Cambridge: Cambridge University Press.

Richards, R. J. 1992. *The Meaning of Evolution: The Morphological Construction and Ideological Reconstruction of Darwin's Theory.* Chicago: University of Chicago Press.

———. 2003. *The Romantic Conception of Life: Science and Philosophy in the Age of Goethe.* Chicago: University of Chicago Press.

———. 2004. Michael Ruse's design for living. *Journal of the History of Biology* 37: 25–38.

Ruse, M. 1996. *Monad to Man: The Concept of Progress in Evolutionary Biology.* Cambridge, MA: Harvard University Press.

———. 2004. The romantic conception of Robert J. Richards. *Journal of the History of Biology* 37: 3–23.   —R.A.

Dinosaurs: The model system for evolution

Because of their immense popularity, dinosaurs are often the organisms that first come to mind when one thinks about animals from the past. This, coupled with the fact that they were discovered at the place (mid-nineteenth-century England) where modern evolutionary theory took root, has led to their historically being used as examples in evolutionary debates. In this context, dinosaurs have played both positive and negative roles in arguments about evolution. Nevertheless, dinosaurs remain an important catalyst for developing a more scientifically literate society, a requisite for appreciating and understanding evolution.

Dinosaurs are a unique lineage of reptiles set apart from other groups by their erect posture and distinctive hip and hand structure. They are also the most successful terrestrial animals to ever inhabit the earth. Their tenure spanned the period from 225 to 65 million years ago. Furthermore, if birds are in fact a lineage of carnivorous dinosaurs (see figure on page 520), their reign actually extends another 65 million years to the present day, where they dominate the aerial realm with over 9,000 species.

Despite their great success, dinosaurs had humble beginnings. For instance, although we tend to think of dinosaurs as giants, the early forms were actually quite small (50 to 100 pounds). These carnivorous bipeds were rather minor players in Mesozoic communities. Over time they diversified into a variety of terrestrial niches and in all major lineages gigantism occurred, where some species came to weigh three or more tons. Some dinosaurs, including all carnivorous forms, were bipedal, whereas herbivorous forms also included quadrapedal taxa. On a broad scale, dinosaurs appear to have been quite sophisticated in their behavioral repertoires. Physiologically they do not appear to have been like living reptiles. Biologically they were perhaps more akin to today's birds and mammals, thus possibly explaining their great success.

Scientific recognition of nonavian dinosaurs occurred in 1824 when William Buckland described *Megalosaurus* (a large carnivorous theropod dinosaur similar to *Tyrannosaurus rex*) from Cretaceous sediments in England. This event was followed by a flurry of new discoveries of other dinosaurian giants throughout Europe. With their obvious reptilian affinities, it was initially believed these animals were simply scaled-up variants of living saurian groups such as iguanas and crocodiles. It was not until 1842 that this paradigm was laid to rest when renowned British anatomist Sir Richard Owen showed that dinosaurs were a unique reptilian lineage in and of themselves. He named these giants the Dinosauria, or "fearfully great reptiles." Owen, who was a creationist at the time, pointed out that attributes of these animals, such as erect posture and substantial fusion of the hip vertebrae, are not found in extant reptiles, but exist in living birds and mammals. Hence, he concluded that dinosaurs were anatomically and physiologically more advanced than today's reptiles. He also deduced that living reptiles are degenerate forms of a formerly advanced group, and he questioned the plausibility of progressivism and transmutation of species. Many scholars believe that Owen created the Dinosauria to debunk evolutionary theories that were at the forefront of scientific debate at the time.

Several decades later, dinosaurs were again to figure prominently in evolutionary debates. In 1860 quarrymen in a limestone quarry in Solnhofen, Germany, discovered a feather from *Archaeopteryx*. This finding showed that birds lived at the time of the dinosaurs. In 1861, two years after Charles Darwin published *On the Origin of Species*, a new specimen of *Archaeopteryx*, known as the London specimen, revealed the true, overall nature of this creature. It was undoubtedly a bird because it sported wings, feathers, and a wishbone. However, there was more to it. It possessed teeth, a reptilian tail, and clawed hands. Here was the spectacular "missing link" (an animal intermediate between two important groups) Darwin and his followers needed to prove that evolution was a reality. Amazingly, this revelation was not immediately realized. German paleontologist Andreas Wagner (in 1861) and later Richard Owen, who purchased the specimen for the British Museum (in 1863), each described the specimen. Both, perhaps due to their antievolutionary thinking, failed to attribute any evolutionary significance to the atypical avian attributes. Subsequently, Thomas Henry Huxley entered the picture. Fresh off an extensive study of avian anatomy and systematics, he recognized this specimen's importance and used it as a model organism in his famed lectures and publications in support of evolution (e.g., Huxley 1868). To this day *Archaeopteryx* is celebrated as one of the most clear-cut examples of an evolutionary missing link.

So how do dinosaurs figure into the *Archaeopteryx* story? About the same time that *Archaeopteryx* was discovered, a chicken-sized theropod dinosaur known as *Compsognathgus* was also discovered in Germany. This tiny dinosaur proved for the first time that not all dinosaurs were elephant-sized, as Owen had contended. Developmental biologist Carl Gegenbaur in 1864 astutely pointed out that the ankle structure of *Compsognathgus* was akin to that of birds. He concluded that it was an animal of double relationship—both bird and reptile. American paleontologist Edwin Drinker Cope later reached the same conclusion regarding the large carnivorous tyrannosauroid *Laelaps* (= *Dryptosaurus*). In a series of papers published from 1866 to 1869 he pointed out that the elongated neck vertebrae and light skull structure in *Laelaps* were also avian features. Shortly thereafter, Huxley, inspired by Gegenbaur's findings and apparently unaware of Cope's research, noted the remarkable similarities between the hips of *Megalosaurus* and extant birds. Using *Archaeopteryx* as a guide, he was able to identify many attributes in common between birds and dinosaurs. By the 1880s he had reached an inescapable conclusion—not only are birds reptiles, they are in fact living theropod dinosaurs. *Archaeopteryx* is simply a feathered dinosaur.

This view of avian ancestry is now almost universally accepted among paleontologists. Ironically, despite being nearly a century and a half old, it has only recently been adopted on a broad scale. By the turn of the nineteenth century, paleontologists instead favored the competing theory of stem reptiles from which the major clades of archosaurian reptiles (dinosaur relatives), crocodilians, herbivorous ornithischian dinosaurs, pterosaurs (flying reptiles), saurischian dinosaurs (theropods and long-necked herbivores), and birds independently rose. It was felt that the common features shared by birds and dinosaurs stemmed from a more ancient common ancestor than within theropods themselves.

In the 1960s, Huxley's charge was renewed when American paleontologist John Ostrom discovered well-preserved specimens of *Deinonychus,* a member of the Dromaeosauria (small to medium-sized, agile carnivorous dinosaurs with sickle-like claws on their feet). Ostrom and others identified nearly 200 dinosaurian attributes shared between birds and dromaeosaurs, including numerous fused sacral vertebrae, hollow leg bones, and long forearms (see figure on page 520). Furthermore, cladistic analyses—a methodology whereby shared-derived (evolutionarily changed) characteristics are used to establish relationships—showed that the genesis of birds came from within

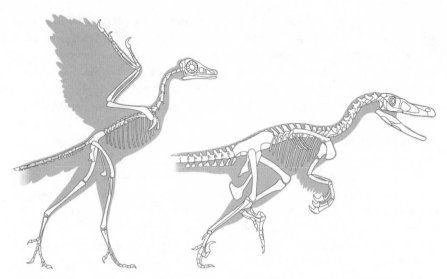

Comparison between the first bird, *Archaeopteryx,* and the dromaeosaurian dinosaur, *Velociraptor.* Note the shared features between the bird *(left)* and the nonavian dinosaur *(right),* which include upright posture, socketed carnivorous teeth, feathers, three or more fused hip vertebrae, three main toes and fingers, an s-shaped neck, extremely long forelimbs, an elongated and retroverted pubis (hip bone) with a boot shape, a stiffened tail, a curved femur (thighbone), elongated metatarsals (foot bones), clavicles (wishbones), a perforate acetabulae (a tunnel through the hip bones between where the thighbones attach), an antorbital fenestra (a hole in the skull in front of the eye socket), and hollow bones, just to name a few.

---

the Dinosauria. Statistically it is extremely unlikely that birds could have descended from any other group of animals.

Today, acceptance of this theory is almost universal. Nevertheless, during the last 15 years a few holdouts (generally avian paleontologists and ornithologists) clung to the notion that avian ancestry still might be found in some earlier stem reptile. They contended that similarities between dinosaurs and birds might somehow reflect independent adaptations to similar environmental selective forces. The crux of their dissension was the absence in dinosaurs of several critical defining attributes of birds. These included the avian wishbone (fused clavicles or furculae) and feathers, as well as the fact that the smallest dromaeosaurs, thought to be the ancestors to birds, were greater than 25 pounds in mass and seemed unlikely candidates for giving rise to the much smaller *Archaeopteryx.* Spurned by their contentions, paleontologists en masse focused on the study of these features. In quick succession, each was revealed in dinosaurs. Reexamination of theropods showed that furculae existed, but they had not been previously recognized. Feathers were found in nonvolant dinosaurs from China. Presumably the feathers were for display and/or thermoregulation. One specimen, a purported glider, appears to have had wings. Finally, there was the discovery of *Microraptor*, the smallest known dinosaur—*Archaeopteryx*-sized and a basal dromaeosaur to boot. Clearly miniaturization had occurred before the evolution of birds. Additional findings solidifying the argument for dinosaurian ancestry for birds include the discovery that egg microstructure of nonavian dinosaurs is matched in birds, that theropods brooded their eggs like birds, and that extensive avian respiratory air sac systems exist in dinosaur skeletons. With these latest findings, the debate over avian ancestry has all but ended. Dinosaurs did not go extinct at the K-T boundary (the geological border separating sediments from the Mesozoic Era, when the dinosaurs ruled, from more modern Cenozoic Era sediments), destroyed by an asteroid or other cataclysmic change. They in fact reside fluttering about outside today.

Dinosaurs have played an important role in the development of modern evolutionary theory by providing one of the most celebrated examples in support of evolution—*Archaeopteryx* and the evolutionary stages leading to its genesis. But do these animals have anything to offer for promoting evolutionary understanding in the future? I believe the answer is yes. Along with examples of human and equine (horse) evolution, studies of dinosaurs will serve as a valuable vehicle by which the public can become versed in the scientific method and at the same time grasp the workings of evolution. The *Archaeopteryx* story is not the only example that is worked out in intricate detail, although arguably it has seen more examination than any fossil vertebrate. In fact it is just one of many. Scientists now know how many times dinosaurs reached elephantine proportions, how they did it through selection for variant developmental trajectories, and how they became dwarfed in island situations. We know how they diversified with key innovations such as dental batteries (the evolution of chewing using multiple teeth) in conjunction with floral changes. And we have a good understanding of how climatic changes, continental drift, and the demise of nonavian dinosaurs—plausibly at the time of an asteroid impact—led to evolutionary change that shaped the world.

The aforementioned studies of evolutionary events have received considerable attention in the worldwide media and hence in the public eye. Through dinosaurs the evolutionary message is getting through. As such, for broadly promoting evolutionary understanding, dinosaur research is arguably one of the most important fields today. However, dinosaur research's influence in promoting scientific understanding and its potential for helping evolution gain broader acceptance could be considerably greater. More often than not our children's first introduction to science is through dinosaurs, and there is no better opportunity to teach the scientific method than to young, curious, open minds. Those who ask for data in support of what they are told will be better equipped to judge for themselves whether what they are told is a reality or not. It is for this reason that dinosaur research is an important scientific pursuit, one that can facilitate scientific literacy and provide the requisite background for understanding and judging the legitimacy of abstract concepts such as evolution. Teaching science at an early age should be of primary importance for scientists interested in promoting a more scientific literate society. Dinosaurs can provide the model examples to make this a reality.

BIBLIOGRAPHY

Cope, E. D. 1867. An account of extinct reptiles that approach birds. *Proceedings of the Academy of Natural Sciences of Philadelphia* 19: 234–235.

Darwin, C. 1859. *On the Origin of Species.* London: John Murray.

Enchanted Learning. Zoom Dinosaurs. [Website for children.] http://www.enchantedlearning.com/subjects/dinosaurs/.

Gegenbaur, C. 1864. *Untersuchungen zur vergleichenden Anatomie der Wirbelthiere.* Vol. 1, *Carpus und Tarsus.* Leipzig: Wilhelm Engelmann.

Huxley, T. H. 1868. On the animals which are most nearly intermediate between birds and reptiles. *Geological Magazine* 5: 357–365.

Ostrom, J. H. 1969. A new theropod dinosaur from the Lower Cretaceous of Montana. *Postilla* 128: 1–17.

———. 1975. The origin of birds. *Annual Review of Earth and Planetary Sciences* 3: 55–77.

Owen, R. 1863. On the *Archaeopteryx* of Von Meyer, with a description of the fossil remains of a long-tailed species from the lithographic stone of Solnhofen. *Philosophical Transactions of the Royal Society of London* 153: 33–47.

University of California Museum of Paleontology. www.ucmp.berkeley.edu.

Wagner, J. A. 1861. Uber ein neus, angeblich mit Vogelfedern versehenes Reptil aus dem Solnhofener lithographischen Schiefer. *Bayerische Akademie der Wissenschaften* (Munich) 2: 146–154.   —G.M.E.

Dobzhansky, Theodosius (1900–1975)

Theodosius Dobzhansky is perhaps most famous for the slogan, "Nothing in biology makes sense except in the light of evolution" (Dobzhansky 1973), which is endlessly invoked by evolutionists in their battles with creationists. But that is only half the battle that Dobzhansky waged on behalf of the importance of evolutionary thinking. He had two targets in mind: creationists and those would-be evolutionists who naively suppose that molecular biology is the key to life when in fact nothing, not even the processes of molecular biology, can be understood except in the light of evolution. Ironically, Dobzhansky himself was obsessed in a religious or at least a moralistic way with the "meaning" of everything, evolution included (Beatty 1987a, 1987b; Ruse 2001; van der Meer 2007). Ernst Mayr (1963) may have believed that evolutionary biology provided the ultimate explanations of life, while molecular biological explanations are at best proximate, but for Dobzhansky, the ultimate understanding of life was deeper still than evolution (see the discussion later in this entry).

To those familiar with his work, Dobzhansky is known not just as a flag bearer for evolutionary thinking, but in particular for his contributions to population genetics. He did as much or more than anyone else to establish that field and set its early agenda, especially through his influential textbook *Genetics and the Origin of Species*. One of the recurrent themes of *Genetics* was that the study of gene frequency changes is an autonomous field, not reducible to genetics per se (i.e., the genetics of individuals). And its domain is the entirety of evolution. What Dobzhansky could not prove in the latter regard he stipulated by definition: "evolution is [no more than] a change in gene frequencies" (Dobzhansky 1937, 11; see also subsequent editions).

The main problem of population genetics, as Dobzhansky elaborated in *Genetics,* is the "paradox of viability." On the one hand, each species needs to have variation present in order to adapt to environmental changes. On the other hand, the particular variations that might prove adaptive in the future are very likely maladaptive at present. As he concluded in his typically dramatic fashion, "Evolutionary plasticity can be purchased only at the ruthlessly dear price of continuously sacrificing some individuals to death from unfavorable mutations" (Dobzhansky 1937, 126–127).

Dobzhansky's own insights into the paradox were based on his studies of variation in natural (nonlaboratory) populations, which began in the 1920s in Russia. At that time he was studying intraspecific variation in ladybird beetles, but it was his studies of variation in natural populations of the fruit fly, *Drosophila,* that would prove most influential. These studies were undertaken in the 1930s, after he had moved to the United States and was working with the *Drosophila* geneticist Thomas Hunt Morgan (concerning Dobzhansky's early work, see Beatty 1987a, 1994; Lewontin 1981; Provine 1981).

Drosophila at first seemed very different from ladybugs. Phenotypically, there seemed to be no intraspecific variation at all. One need only recall Morgan's excitement at finally discovering an observable mutant whose pattern of inheritance could be studied. Subsequently, variations were found in ever greater numbers in his laboratory. But it was believed by many that these variations were artifacts. *Drosophila* in the wild seemed to have little if any intraspecific variation, and hence perplexingly little material for future adaptive evolution. In trying to make sense of the difference between laboratory and wild *Drosophila,* Dobzhansky followed the suggestion of Sergeĭ Chetverikov, who proposed that most mutations were recessive and would result in observable differences only when doubled up in the homozygous state. This would occur more often in laboratory stocks than in nature, because in the former case there is significantly more inbreeding than in wild populations. In this way, natural populations soak up variations like a sponge.

Dobzhansky, originally in collaboration with Alfred Sturtevant, found a way to observe otherwise hidden variation in *Drosophila* by focusing his microscope on variations in chromosome structure—"inversion" variations—that involve differences in gene arrangement and are often associated with gene differences as well. In both respects, these variations constitute material for evolutionary change. But how was the observed variation maintained? Dobzhansky at first partly adopted Sewall Wright's "shifting balance" theory to explain this variation. He reasoned that the variations were adaptively insignificant and that their frequencies were drifting randomly. This might lead to a reduction in variation within each population of a species, but would lead to an increase in variation between populations and hence within the species.

Dobzhansky was quite surprised, then, to find evidence that the inversion differences were adaptively significant after all, leading him to explore a range of variation-maintaining forms of natural selection. He finally settled on the idea that heterozygotes are adaptively superior to homozygotes because having two genes for a trait, instead of two copies of the same gene, allows more adaptability at the individual level. Selection for adaptability at the individual level in turn preserves evolutionary adaptability at the species level.

Dobzhansky's account came to be known as the "balance" theory of evolution. His main detractor was Hermann Muller, whose view of evolution was dubbed by Dobzhansky the "classical" (old-fashioned) theory (on the classical/balance controversy, see Beatty 1987b [and other essays on that issue] and Lewontin 1974). According to Muller, there was no paradox of viability. The vast majority of mutations are detrimental for the individuals in which they occur, and also for the species in which they accumulate. There was, in Muller's opinion, only one species—humans—in which mutations accumulate to any great extent. Through the amenities of civilization, we humans have managed to escape natural selection, to our ultimate detriment. Only through a conscious eugenics program can we now successfully reverse the rising tide of deleterious mutations. Dobzhansky, for his part, criticized "the eugenical Jeremiahs" who overlook the importance of evolutionary plasticity and who seem unaware that natural populations of many species possess considerable reserves of variation (Dobzhansky 1937, 126).

Dobzhansky and Muller were also at the center of controversies from the 1950s through the early 1970s concerning the impact of radiation-induced mutation from atomic weapons tests. (Both men received considerable funding from the Atomic Energy Commission during this period.) Adding yet more mutations to the human species was straightforwardly negative from Muller's point of view, but less so from Dobzhansky's viewpoint. Interestingly, Dobzhansky was strongly opposed to the testing and use of atomic weapons, while the once pro-Soviet Muller believed that, up to a point, such weapons had an important role to play in the containment of communism (Beatty 1987b).

Dobzhansky also used his studies and perspectives on variation to support liberal democratic ideals. As he and Leslie Dunn argued in their popular *Heredity, Race, and Society* (Dunn and Dobzhansky 1946, 45), "the absolute uniqueness of every human individual . . . translated into metaphysical and political terms is fundamental for ethics and democracy." In his widely read *Mankind Evolving*, Dobzhansky argued that, far from genetic diversity undermining equality, it actually serves as an excellent rationale: "Equality of opportunity tends to make the occupational differentiation comport with the general polymorphism of the population, and would be meaningless if all people were genetically identical" (1962, 244).

As much as he claimed to understand evolution, Dobzhansky remained deeply puzzled and uncomfortable about the means by which evolution proceeded. He had found a way to reconcile the selection of the fittest with the preservation of variation for future adaptive evolution, but not in a way that reconciled the good of the species with the welfare of its members. The greater fitness of heterozygotes means that lesser-fit homozygotes will continue to be born in each generation. It has been suggested that Dobzhansky understood this situation religiously, as a necessary evil—along the lines of free will, which gives us moral freedom, but at the cost of many sad choices. Perhaps. But he also explicitly regarded it as an "unpleasing imperfection of nature" that so many individuals should suffer such "misery" for the good of the group (Dobzhansky 1937, 127; Dobzhansky 1962, 335; Ruse 2001, 100–122; van der Meer 2007).

Dobzhansky drew productively on his Russian heritage and Russian sources in his investigations and explanations of intraspecific variation. This was a major reason for the originality of *Genetics* in an Anglo-American context. Dobzhansky's understanding of the religious/moralistic significance of intraspecific variation also reflected his Russian heritage (see especially van der Meer 2007), although this influence has been more difficult to reconstruct, and many of the same concerns—such as the evils of natural selection—had also been raised in the Anglo-American context (e.g., by the atheist Morgan, in conversations with Dobzhansky; Beatty 1994, 201 and note 3).

BIBLIOGRAPHY

Adams, M. B., ed. 1994. *The Evolution of Theodosius Dobzhansky*. Princeton, NJ: Princeton University Press.

Beatty, J. 1987a. Dobzhansky and drift: Facts, values and chance in evolutionary biology. In L. Krüger et al., eds., *The Probabilistic Revolution*, vol. 2. Cambridge, MA: Harvard University Press.

———. 1987b. Weighing the risks: Stalemate in the classical/balance controversy. *Journal of the History of Biology* 20: 289–319.

———. 1994. Dobzhansky and the biology of democracy: The moral and political significance of genetic variation. In M. B. Adams, ed., *The Evolution of Theodosius Dobzhansky*. Princeton, NJ: Princeton University Press.

Dobzhansky, T. 1937. *Genetics and the Origin of Species*. New York: Columbia University Press.

———. 1962. *Mankind Evolving: The Evolution of the Human Species*. New Haven, CT: Yale University Press.

———. 1973. Nothing in biology makes sense except in the light of evolution. *American Biology Teacher* 35: 15–129.

Dunn, L. C., and T. Dobzhansky. 1946. *Heredity, Race, and Society*. New York: Penguin.

Lewontin, R. C. 1974. *The Genetic Basis of Evolutionary Change*. Cambridge, MA: Harvard University Press.

———. 1981. Introduction: The scientific work of Theodosius Dobzhansky. In R. C. Lewontin et al., eds., *Dobzhansky's Genetics of Natural Populations I–XLIII*. New York: Columbia University Press.

Lewontin, R. C. et al., eds. 1981. *Dobzhansky's Genetics of Natural Populations I–XLIII*. New York: Columbia University Press.

Mayr, E. 1963. Cause and effect in biology. *Science* 134: 1501–1506.

Provine, W. B. 1981. Origins of the Genetics of Natural Populations series. In R. C. Lewontin et al., eds., *Dobzhansky's Genetics of Natural Populations I–XLIII.* New York: Columbia University Press.

Ruse, M. 2001. *Mystery of Mysteries: Is Evolution a Social Construction?* Cambridge, MA: Harvard University Press.

van der Meer, J. 2007. Theodosius Dobzhansky: "Nothing in evolution makes sense except in the light of religion." In N. Rupke, ed., *Eminent Lives in Twentieth-Century Science and Religion.* Frankfurt: Peter Lang.   —J.B.

Drosophila

Drosophila is a genus of flies in the family Drosophilidae that are widely used for evolutionary research. The origin of this research lies in Thomas Hunt Morgan's Nobel Prize–winning work on inheritance in *Drosophila melanogaster* (see figure) starting around 1908. Morgan's intellectual descendants have expanded the scope of work on *D. melanogaster* from genetics into virtually every aspect of biology, turning it into one of the most useful model organisms. Evolutionary research has been a robust part of this tradition from the beginning. Currently, the combination of the ease and speed with which many species can be reared in the laboratory and the extensive background knowledge of genetics and development make them useful in addressing a wide variety of questions in evolutionary genetics.

Drosophila adults are usually only a few millimeters long, with prominent red to brown eyes, and complex external morphology that has remained fairly conservative over evolutionary time. *Drosophila* show holometabolous development, undergoing a complete transformation from the feeding larvae to the adult form during a pupal stage. They are commonly called fruit flies, although most species feed on microorganisms that inhabit rotting plant material. The genus *Drosophila* currently includes between 1,150 and 1,500 species, depending on which groups are included in the genus, and has a worldwide distribution. Phylogenetic research has clearly shown that the genus is paraphyletic, meaning that other genera are more closely related to some of the *Drosophila* than they are to other groups. In the broad sense, then, *Drosophila* should include many other genera, including the species-rich *Scaptomyza* (approximately 260 species), *Hirtodrosophila* (approximately 150 species), and *Mycodrosophila* (approximately 120 species). Molecular evidence suggests that the genus is at least 50 million years old, a conclusion supported by drosophilid specimens in amber that are around this age.

Around 1910, Morgan discovered a single *D. melanogaster* with white eyes, the inheritance of which turned out to be sex linked. This simple observation inspired Morgan and his laboratory students to isolate other mutants, with which they were able to confirm the hypothesis that genes reside on chromosomes. In those years, Morgan's tiny (16 feet by 24 feet) "fly room" at Columbia University had eight desks; three were occupied by Columbia

Drosophila melanogaster, the fruit fly, the workhorse of evolutionary genetics. The way in which it was tailored for such research is discussed in Kohler 1994.

undergraduates Alfred Henry Sturtevant, Calvin Bridges, and Hermann Muller, who all played pivotal roles in genetics and evolutionary biology. Morgan, Bridges, and Sturtevant worked together at Columbia until 1928 when they all moved to the California Institute of Technology (Cal Tech), where they remained for the rest of their careers.

Sturtevant produced the first genetic map based on the realization that the frequency of crossover events could be used to determine the order and location of genes. He also worked on a wide variety of topics with evolutionary implications, including chromosomal inversions, the genetics of species differences, the genetics of development, and the phylogeny of the genus. Bridges was among the first to exploit the discovery that the salivary glands of *Drosophila* contained polytene chromosomes consisting of thousands of DNA strands resulting from repeated rounds of replication without cell division. The result is chromosomes visible at low magnification with characteristic banding patterns that allow homologous regions to be identified across individuals and species. Bridges's detailed descriptions of the banding patterns allowed linkage and physical maps to be aligned.

Muller is best known for his Nobel Prize–winning work on mutation. Beginning in the mid-1920s, Muller was the first to exploit the fact that chromosomal inversions—chromosomes in which a piece appears in reverse of the normal order—suppress recombination when in a heterozygous condition with an uninverted chromosome. Muller found that he could use an inverted chromosome marked with a dominant phenotypic marker, termed a balancer, to capture and study the properties of chromosomes with the normal gene order. Initially he used this technique to detect chromosomes carrying lethal mutations, which can never produce a homozygous adult. Muller measured the lethal frequency after inducing mutation with X-rays. Muller's concept of genetic load—mutational baggage that Muller saw as harmful—came from these early experiments. Balancer chromosomes were subsequently developed for each chromosome in *D. melanogaster,* and balancers remain one of the key advantages of doing genetics in the species.

Shortly after Morgan moved to Cal Tech, he hired Russian scientist Theodosius Dobzhansky as part of his team. Dobzhansky quickly rose to prominence as a synthesizer by writing *Genetics and the Origin of Species* (1937), one of the founding books of the evolutionary synthesis, in which he asserted the primacy of a population genetic view of evolution, based initially on remarkably little empirical evidence. At about the same time, influenced by Sturtevant and his Russian mentor Sergei Chetverikov, Dobzhansky took up the study of wild populations of *Drosophila pseudoobscura.* Populations of this common species of the mountains of western North America are polymorphic for chromosomal inversions. Dobzhansky used this variation as the basis for a long series of experiments documenting the effects of the three major evolutionary forces: selection, genetic drift, and gene flow. In this, he was greatly aided by an extensive collaboration with Sewall Wright during the late 1940s.

Modern evolutionary work on *Drosophila* currently has several major emphases that grew out of this earlier work. The study of variation at the molecular level was pioneered by Richard Lewontin, one of Dobzhansky's students, who, starting in 1965, exploited the discovery of protein electrophoresis by Jack Hubby and others to demonstrate that a large portion of genetic loci in *Drosophila* species were polymorphic. This discovery of high levels of natural variation in most species led to an acrimonious debate over whether genetic variation is generally selectively neutral, deleterious, or maintained by natural selection. Lewontin helped to catalyze this debate by contrasting Dobzhansky's view that much genetic variation was adaptive with Muller's emphasis on the costs of mutation and variation. This debate was a major factor in the development of population genetic theory. The use of *Drosophila* to study variation carries over to the DNA level, where the sequencing of *Drosophila* genomes allows unprecedented precision in the study of evolutionary patterns.

The ease with which *Drosophila* can be reared has led to its increasing use in the study of evolution in the laboratory. Some of the first sophisticated selection experiments on continuous traits were carried out in *Drosophila* by Kenneth Mather and Forbes W. Robertson in the late 1940s and early 1950s. Subsequently, successful artificial selection experiments took place on many characteristics, encompassing morphology and physiology (e.g., fecundity, egg size, developmental rate, life span, resistance to ethanol), behavior (phototaxis, geotaxis, mate preference), and aspects of the genetic system itself (rates of recombination, change in variance rather than mean value of traits). With the exception of sex ratio and asymmetry, heritable variation exists for every trait studied. Studies of experimental evolution, where the investigator creates an environment in which natural selection can take place, is also an area of active research using *Drosophila*. This approach has particular relevance in the study of life histories and the ultimate origin of processes like aging and mate choice.

The study of species differences, pioneered in *Drosophila* by Sturtevant, Muller, and Dobzhansky, is another very active area of evolutionary research. The genus is well suited to the study of speciation because of the number of crossable species and the relative ease with which reproductive barriers can be identified and quantified. Jerry Coyne, one of Lewontin's students, and his student H. Allen Orr used this accumulated data to make rough estimates of the time required for speciation in sympatric and allopatric populations. The genes responsible for aspects of species barriers have now been identified in a number of *Drosophila* species pairs.

While the study of *Drosophila* development dates back to the 1920s, *D. melanogaster* became a premier model system for the study of development based on the detailed study of the bithorax complex by Ed Lewis, one of Sturtevant's students, and the systematic mutation screen for mutations with effects early in development by Christiane Nusslein-Volhard and Eric Wieschaus. These efforts led to many of the fundamental developmental pathways that are conserved widely among animals. Lewis, Nusslein-Volhard, and Wieschaus shared a Nobel Prize for this work in 1995. This extensive background information on development has catalyzed the study of the evolution of development in *Drosophila*. Researchers have now been able to identify the specific genes responsible for many evolutionary changes in natural and experimental *Drosophila* populations.

BIBLIOGRAPHY

Dobzhansky, T. 1937. *Genetics and the Origin of Species*. New York: Columbia University Press.

Kohler, R. 1994. *Lords of the Fly*. Chicago: University of Chicago Press.

Lewontin, R. C. 1974. *The Genetic Basis of Evolutionary Change*. New York: Columbia University Press.

Powell, J. R. 1997. *Progress and Prospects in Evolutionary Biology: The* Drosophila *Model*. New York: Oxford University Press.

Rubin, G. M., and E. B. Lewis. 2000. A brief history of *Drosophila*'s contributions to genome research. *Science* 287: 2216–2218.

Sturtevant, A. H. 1965. *A History of Genetics*. New York: Harper and Row.

—D.H. and B.H.

E

Ecological Genetics (E. B. Ford)

Ecological Genetics, through its several editions, presented the approach to studying natural selection in the field pioneered by E. B. Ford (see the alphabetical entry "Edmund Brisco Ford" in this volume), his colleagues, and his students, and collected their successes. This approach, called ecological genetics, is descended from the early studies of R. C. Punnett and W. F. R. Weldon that natural selection can be studied empirically (see the alphabetical entry "W. F. R. Weldon" in this volume). The advance offered by ecological genetics over work like Weldon's was a focus on discrete traits with a simple genetic basis such as the variation in the number of spots on the wing of the moth *Panaxia dominula.* Knowing the genetic basis of variation allowed ecological geneticists to study not only selection, but also the evolutionary response to selection as trait values changed over time or space. Because changes in trait values reflected changes in the frequencies of alternate genes, these discoveries were small-scale demonstrations of evolution in action. The term *ecological genetics* is derived from its focus on trait variation with a demonstrable genetic foundation (genetics) and its orientation to studying trait variation in nature (ecological).

Three types of investigations dominate Ford's *Ecological Genetics.* First, there are demonstrations of evolution in action; the saga of industrial melanism in the moth *Biston betularia* is the best-known example. Second, there are investigations of geographic variation in trait values, such as the shell colors of snails, that document how changes in ecological factors from one location to another change the selection pressures on organisms and cause different trait values to be favored in different locations. These investigations demonstrated adaptive explanations for geographic variation in characters, complementing demonstrations of adaptive change in time. In some cases, these studies also uncovered the signature of random events as in the so-called area effects on snail shell patterns. Third, there are studies of how different selection pressures can balance one another and preserve trait variation within a single population. These studies represented a particularly fertile area of ecological genetics. Innumerable theoretical papers on the population genetics of balancing selection grew from the initial discoveries that alternate genes would continue to segregate despite being acted upon by strong natural selection.

Ecological Genetics was enormously influential. Its compilation of case studies eliminated any doubt that natural selection was an ongoing force and could be studied in real time. The studies of geographic variation conducted by Ford and his colleagues complemented the work of others, such as Göte Turesson and Jens Clausen with plants (see the alphabetical entry "Jens Clausen" in this volume), F. B. Sumner with mice, and Sergeĭ Chetverikov and Theodosius Dobzhansky with flies (see the alphabetical entries "Sergeĭ Sergeevich Chetverikov" and "Theodosius Dobzhansky" in this volume), who were making similar discoveries, often with characters less amenable to genetic analysis. The waning of the book's influence began with the discovery of biochemical variation and continued through the emergence of a sophisticated approach to studying continuously varying traits that are more representative of the stuff of evolution than discrete polymorphisms in visible characters. Nonetheless, the questions about selection and evolution presented in the book, and the conviction advanced that ecological and genetic studies are necessary to understand adaptive evolution, remain bedrock influences on modern evolutionary biology.

BIBLIOGRAPHY

Ford, E. B. 1975. *Ecological Genetics*. 4th ed. London: Chapman and Hall.   —*J.T.*

Eigen, Manfred (b. 1927)

Manfred Eigen is a German biophysicist and a former director of the Max Planck Institute for Biophysical Chemistry in Göttingen. In 1967, together with Ronald George Wreyford Norrish and George Porter, he won the Nobel Prize in chemistry for the study of extremely fast chemical reactions. Beginning in the 1960s, Eigen's research interests turned to biochemical questions, and during the 1970s and 1980s he formulated an elaborate theory on the emergence of life on earth founded on mathematical calculations and experiments in viral populations.

Eigen's fundamental theoretical and philosophical postulate was that natural selection was the organizing principle responsible not only for evolution but also for the emergence of life. This claim was based on new advancements in the field of molecular biology. Starting in the 1960s, it became possible to describe evolution based on replication, mutation, and competition not only of organisms but also in mixtures of self-replicating molecules and enzymes catalyzing replication. Eigen's research along these lines greatly advanced the work of biochemist Sol Spiegelman and others in the 1970s that demonstrated the evolution in the test tube of viral RNA sequences. Based on these experiments, Eigen developed his model for the origin of life.

Eigen formulated a key new concept in origin-of-life study and virology called the "quasi-species." Following the competition for building blocks in a population of RNA sequences engaged in self-replication, a unique pattern of the whole population emerged based on the distribution of the various mutants. This resulting population, the quasi-species, reflected the consensus sequence representing an average of all the mutants according to their individual frequency. Eigen discovered that, based on the specific dynamic of the quasi-species, the mechanism of selection guaranteed a nonrandom and directed evolution toward a population capable of accurate and fast self-replication.

In addition to competition, Eigen also postulated stages of molecular cooperation leading to the emergence of a "hypercycle," an ensemble in which several quasi-species units participated, each enabling the replication of the adjacent unit. This cooperation was deemed crucial to overcome a catch-22 in the scenario and allow further evolution. The difficulty stemmed from the fact that RNA sequences in the quasi-species had to be limited to a certain length (about 100 nucleotides). Above this threshold, too many mutations would accumulate, destroying the sequences' identities. At the same time, in order to code for proteins, much longer chains were required. The hypercycle made it possible to pull together the information contents of each quasi-species without lengthening individual chains, thus enabling the synthesis of functional proteins. However, the cooperation of several quasi-species depended on their enclosure within a compartment, postulated by Eigen to form in the rich primordial "soup." Competition among the diverse compartments could then permit further evolution.

Eigen's theory was based on the optimistic assumption that prebiotic chemistry produced in the primordial soup a far-from-equilibrium, abundant supply of amino acids, nucleotides, and their polymers. These could serve as energy-rich building blocks and primitive catalysts for the synthesis of short genetic sequences and for the crucial processes of biological self-organization. This optimism as well as several of Eigen's theoretical and empirical assumptions relating to the quasi-species and the hypercycle were severely criticized by other researchers. Nonetheless, his contribution to ideas on the evolution of genetic information and to the later formulation of the concept of the RNA world is indisputable. So is his claim that the principle of natural selection operates not only on biological systems but on any physical systems fulfilling certain conditions and that, notwithstanding the role of random events, the emergence of life was a naturally lawful process.

BIBLIOGRAPHY

Eigen, M. 1992. *Steps towards Life.* Oxford: Oxford University Press.
Eigen, M., W. Gardiner, P. Schuster, and R. Winkler-Oswatitsch. 1981. The origin of genetic information. *Scientific American* 244, no. 4: 78–118.
Fry, I. 2000. *The Emergence of Life on Earth: A Historical and Scientific Overview.* New Brunswick, NJ: Rutgers University Press.   —I.F.

Endosymbiotic origin of eukaryotes

PRECEDENTS AND PREDECESSORS

Once contentious, the theory is now accepted that all eukaryotes (organisms composed of cells with nuclei and true chromosomes), including the familiar plant and animal organisms, evolved from symbiotic mergers of bacteria. Clear genetic evidence connects the plastids (green photosynthetic organelles) of algae to free-living photosynthetic bacteria called cyanobacteria. And because algae and plants have a distant common ancestor, this evidence connects the cyanobacteria to modern plants. So, too, the oxygen-using "power packs" in cells, the mitochondria, have been traced via genetics, physiology, and morphology to free-living respiring bacteria. For the last 3 billion years or more, through mutations and gene trading, one has the same kind of genetic engineering that goes on in biotechnology firms. This accumulation of DNA mutations and passing of sets of genes from one kind of bacterium to very different kinds is, and has been since ancient times, the way of the bacterial world. Yet mergers, whole-body fusions, and permanent alliances as ways of survival in the hardest of times has selected bacterial associations to form new individuals. Bacteria, by virtue of literal takeovers, have also evolved into something else. By fusion of different kinds of bodies, rather than by mutating and growing larger, bacteria evolved to become larger cells. The larger cells, including the ancestors to all animals and plants, indeed all cells with nuclei (all eukaryotes), evolved by symbiotic merger.

BACTERIAL ANCESTRY

Bacteria are ubiquitous organisms that are present in the fossil record long before any animals or plants. Bacteria, not plants, evolved photosynthesis and chemical synthesis, the processes that provide food to the rest of the biosphere. All bacteria share a common cellular structure known as prokaryotic. Prokaryotic cells, unlike eukaryotic cells, do not have an enclosed nucleus housing their DNA, but rather an open area called a nucleoid in which the DNA resides. Typically there are no centrioles, spindles, or microtubules (indeed there is no mitosis as in eukaryotes), and prokaryotic cells lack membrane-bounded enzyme sacs within which organic molecules are oxidized. Bacteria, which are all prokaryotic, share many but not all traits in common with the eukaryotes. Like the three main groups of more familiar, larger eukaryotes—animals, plants, and fungi (all yeasts, molds, and mushrooms are fungi)—bacteria are made of cells that are bounded by dynamic protein-rich lipid membranes. All cells from the tiniest bacteria to the largest egg cells use the enzymes of their complex energy and carbon metabolism to produce their own necessary outer membranes. The metabolic maintenance of the dynamic protein-rich, phosphorus-rich, lipid-rich membranes exchange component parts. Cells use foodstuffs as raw materials. Membranes encapsulate identity. The absolute necessity of continuity through space and time requires the presence of active, intact membranes surrounding all cells at all times. Membranes, and often additional coverings such as cell walls and cuticles, define the inside and the outside borders of all organisms, all of which are made of either prokaryotic or eukaryotic cells. In chemical terms, this implies that the inside of any organism is in a state of water, salt, sugar, protein, and lipid concentration very different from its outside. The environment beyond the outer membrane is ever-present and often changing, but it is the ubiquitous active membrane (whether naked or supplemented by skin, waxy surface, sheath, or other additional outer layer) that establishes and maintains identity of living beings.

ALL LIFE FROM THE EARLIEST CELLS

All life forms on earth make proteins inside their cells, including membrane proteins. Proteins, which are long chains of smaller units called amino acids, are synthesized on intracellular tiny spherical structures called ribosomes. Ribosomes look like tiny dots even at high magnification as seen with an electron microscope. Membrane biochemistry and protein synthesis are universal features of life that attest to the fact that all life evolved from common cellular ancestors.

Bacterial genes, those of humans as well as all other animals, are composed of DNA; the chemical composition of genes, like many other details of bacterial chemistry, are like those of the larger organisms. Most bacteria have at least 3,000 different genes that determine thousands of different enzyme proteins. Bacterial cells grow and reproduce by reproductive division, which is known as fission. The prokaryotic cell division process is far less complex in appearance than the "chromosome dance" alignment and separation process in eukaryotes. Known as mitosis, the cell division process in nucleated organisms always involves movement inside cells on little tracks called microtubules (see figure on page 536). The inside of a eukaryotic cell is incessantly moving, as seen through a microscope when the cell is alive. Even the largest prokaryotic cells divide by fission; they lack the microtubules and the microtubule-based movements typical of nucleated cells. Some claim that the discontinuity between bacteria and all other forms of life is the largest evolutionary gap in the living world.

Although clearly we (i.e., larger forms of life composed of nucleated cells) evolved, somehow, from bacterial predecessors, the differences between bacteria and eukaryotes are significant. The major differences between bacteria and the members of the animal, plant, and fungal kingdoms is that all larger forms of life are composed of cells that harbor double membrane–bounded nuclei. For that reason they are eukaryotes (eu=true, karyon=kernel, seed; Greek). All bacteria have DNA, but they lack both the nuclear membrane and the true chromosomes where the DNA is packaged with special proteins.

Bacteria, as a group, are far more metabolically diverse than plants, animals, and fungi. Whereas all the latter breathe oxygen, some bacteria can breathe sulfate, carbon dioxide, nitrate, or even arsenate. Bacteria, despite their small size, more ancestral lifestyle, and reputation as parasites and

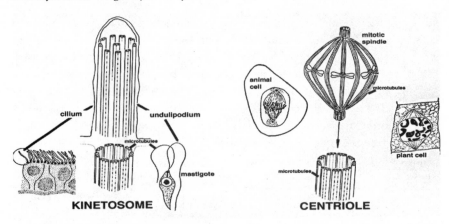

The nucleocytoskeletal organellar system of eukaryotes. The system, by hypothesis, evolved from a symbiotic merger of a motile eubacterium (a sulfide-oxidizing spirochete that became the undulipodium) with an archaebacterium (a sulfidogenic thermoacidophil codescendant with *Thermoplasma* sp.). See Margulis et al. 2005.

germs, are in fact more proficient and broadly skilled at many chemical tasks than are larger life forms.

EARLIEST NUCLEATED CELLS

How might our bacterial ancestors have evolved to form the earliest nucleated cells? The evolutionary process always involves the capacity for rampant population growth far beyond the environment to sustain it (i.e., biotic potential), the generation of inherited variation, and the "checks" on growth of the population (including its variants) by the failure of the biotic potential to be reached. The inhibition of maximal population growth by elimination of many, for whatever reason, is called natural selection. Natural selection maintains healthy reproductive populations of many types of organisms, but it does not produce them in the first place.

　Well, then, how is inherited variation generated? Although random mutation clearly exists and has been prodigiously documented, inherited variation is generated in many more different and important ways than via random mutation. Indeed, random mutation almost certainly did not produce nucleated cells from bacteria. Rather, the major mode of evolutionary variation in bacteria that led to the first nucleated cells is likely to have been via a process called symbiogenesis. This process is allied to symbiosis, which is the ecological relationship between beings of different types (differently named organisms). Symbiogenesis is simply a protracted, long-lasting physical association, usually between different species. It is an evolutionary, Darwinian process. Symbiogenesis refers to the appearance of a new trait, behavior, tissue, organ, organism, species, or other higher taxon as a result of an identifiable long-term symbiosis. Several evolution-changing symbioses have been identified with various levels of assurance. Although the earliest change is the most difficult to prove, symbiogenesis led to the eukaryotic organism itself, with swimming motility (see figure below) and the capability for mitosis; the second, now proven definitively, led to the presence of mitochondria inside a cell, thus allowing the cell to breathe oxygen. The last symbiosis, the most recent and therefore the one about whose history we are most confident, was limited to the ancestors of algae and plants. This symbiosis, which rendered our planet green, led to the presence of chloroplasts inside the cells. By acquisition of photosynthetic prokaryotes (cyanobacteria), the photosynthetic eukaryotes, algae, and plants appeared on earth.

But how did the defining feature of a eukaryote, the nucleus itself, evolve? The first nucleated microorganisms likely evolved by the merger of two different kinds of bacteria but under anoxia (i.e., environmental conditions that lacked oxygen). One type of bacterium, capable of fermentation, produced hydrogen sulfide gas from ambient elemental sulfur globules, probably of volcanic origin. This organism, called a sulfidogen because of its gas production, hypothetically acquired an entire genome of another type of bacterium by symbiogenesis. The second organism also lived on organic foodstuffs (e.g., sugars, acetate, or organic acids). But the second bacterium, which is not a sulfidogen, was a whiz at locomotion: it swam more than 100 micrometers a minute. The nucleus itself evolved only after integration, that is, after the first

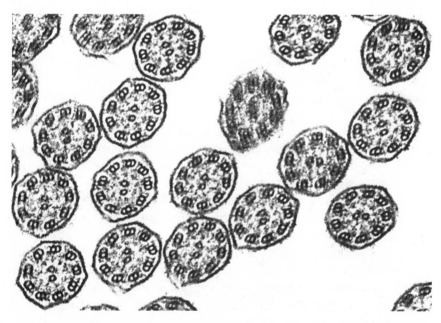

The [9(2)+2] microtubule array of all undulipodia in transverse thin section as seen with electron microscopy. The observation that this seme (multigenic trait of evolutionary importance) is conserved in sperm tails, vertebrate sensory cilia, zoospores of mastigote molds, and myriad other intrinsically motile cell projections suggests its common ancestry.

bacterium (the sulfidogenic archaebacterium) merged with the swimming bacterium to form the symbiotic ancestor of all nucleated cells.

The earliest products of the first proposed symbiogenetic merger were amitochondriate protists, cells that did not possess mitochondria. The group, called archaeprotists (Margulis and Schwartz 1998), still survives in airless muds and anoxic intestines. Many genera, descendants of these amitochondriate cells, still live today, but they must be sought in strange habitats because their relatives with mitochondria in this oxygen-rich world greatly outnumber them. Indeed, entire families such as the Trichomonadida and Calonymphida still thrive in organic-rich anoxic habitats such as the guts of termites and wood-eating roaches.

PROTOCTIST DIVERSITY: KEY TO CELL EVOLUTION

The great group of relict nucleated microorganisms, the earliest nucleated cells, and their extant descendants is called Protoctista. This group, which today is estimated to contain some 250,000 species, is considered its own kingdom in the five-kingdom classification scheme. The protoctists include all the algae, the so-called protozoa, the slime molds, and many more obscure groups. The three-domain molecular-biological classification of Carl Woese lumps plants, animals, and fungi together in a single kingdom and the protoctists are put with all nucleated organisms into Domain Eukarya. Woese separates the prokaryotes into two bacterial groups, based mainly on differences in the details of their ribosome's RNA. One is the Archeae (= archaebacteria) and the other is Eubacteria. In the scheme of nucleated cell origins outlined above, the merger was between an archaebacterium (the sulfidogenic archean, similar to today's *Thermoplasma acidophila*) and a eubacterium (similar to today's *Spirochaeta*, a genus of snakelike swimmers).

The protoctists include all photosynthetic eukaryotes that do not grow from embryos, as plants do (see figure). Familiar examples include diatoms; red, green, and brown seaweeds (including kelp); chrysophytes; phytoplankton; and all other algae. (The blue-greens are not included; they are not algae but cyanobacteria Protozoa (an obsolete term insofar as it suggests animals), ciliates, slime nets, foraminifera, chytrids, and many other even more obscure taxa are in the great kingdom of Protoctista. It is in this unruly group of eukaryotic microbes and their many multicellular descendants that so many features of nucleated organisms evolved: mitochondriate and plastidic cells, mitosis and meiotic sex, animal cell junctions (septate junctions, desmosomes), and animal-style multicellularity. Even eyes (in certain dinoflagellates), cell penetration devices (in Apicomplexans, the group to which the malarial parasite belongs), and hunting and shell-making (in the predatory and agglutinating foraminifera, respectively) evolved in these groups of protoctists.

Hypothetically, the origin of nucleated cells (eukaryosis) occurred by the middle of the Proterozoic eon. The dates for evolution of any innovation must be set by observations in the fossil record. Therefore, it is likely that the symbiogenetic evolutionary sequence described above occurred prior to the deposition in

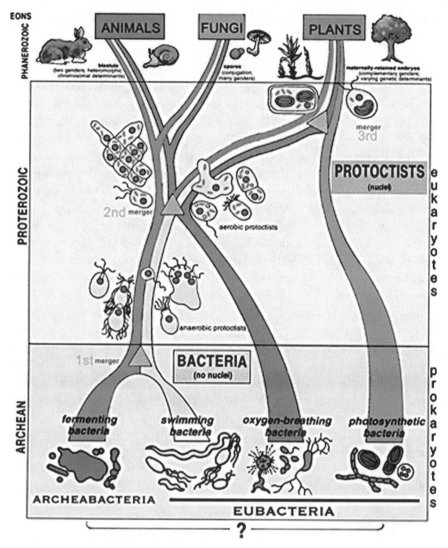

Origin of eukaryotes from specific bacterial lineages: serial endosymbiotic theory. *Left to right:* Thermoplasma-like archaebacteria merge with spirochete eubacterial to form amitochondriate protists in the lower Proterozoic eon. Intracellular motility, including phagocytosis, potentiates the ingestion but lack of digestion of oxygen-breathing bacterial ancestors of mitochondria by the mid-Proterozoic. Some of these aerobic protists ingest and retain cyanobaceria that become the ancestors to the plastids of algae and plants.

sediments of such well-preserved microfossils as *Vandalosphaeridium* (Samuelsson et al. 1999) or the spiny spheres in the Doushantou.

Published evidence documents that the first structures that appeared in the origin of the nucleated cell were not free nuclei. Rather, the nucleus began tethered in a peculiar structure, an organellar system still inside many protoctists, called the karyomastigont. The karyomastigont is an example of a

Darwinian imperfection and oddity from which a path of history may be reconstructed. The significance of the karyomastigont, an organellar system in which a nucleus is embedded, was described in the early years of the twentieth century (Janicki 1915). Minimally the karyomastigont consists of a nucleus, a nuclear connector (called a rhizoplast in early protozoological literature), and an undulipodium (the 9+0 kinetosome/centriole and its shaft, the 9+2 microtubular axoneme). (Some prefer to write this universal pattern of ninefold symmetry of the microtubules in the shaft more accurately as [9(2)+2].) When the centriole in mitosis develops into the kinetosome of the motility organelle, we are seeing the legacy of the swimmer component of the earliest eukaryotes. The structure and behavior of protoctists that lack mitochondria under anoxic conditions and today's motility symbioses in many organisms with swimming bacteria, such otherwise oddities and peculiarities, are understood in the context of their evolution. The nucleus itself is most likely a product of genetic-level integration of bacterial symbionts (archaebacterium sulfidogen plus swimming eubacterium) that began as part of the karyomastigont from which it was released.

This evolutionary scenario for the origin of eukaryotes is based on the invaluable descriptions by Harvard professor Lemuel R. Cleveland (1892–1971) and Harold Kirby Jr. (1900–1952), chairman of the Department of Zoology at the University of California at Berkeley. Cleveland provided details of cell motility (movements of the karyomastigonts and other cell parts in mitosis) and Kirby came to understand calonymphid evolution by recognizing the ability of karyomastigonts to reproduce independently of the nuclei in mitochondrial cell lineages that probably never acquired mitochondria. Preservation of the published work of these scholars (including the unprecedented 16-mm black and white films by Cleveland) would be of interest to those who wish to delve further into the details of cell evolution.

BIBLIOGRAPHY

Cleveland, L. R., and A. V. Grimstone. 1964. The fine structure of the flagellate *Mixotricha paradoxica* and its associated microorganisms. *Proceedings of the Royal Society of London, Series B* 157: 668–683.

Janicki, C. 1915. Examinations of parasitic flagellates. Part II. The genera *Devescovina, Parajoenia, Stephanonympha, Calonympha*. On the parabasal apparatus. On the constitution of the nucleus and amitosis. *Zeitschrift für Wissenschaftliche Zoologie* 112: 573–691.

Kirby, H., Jr. 1941. Devescovinid flagellates of termites. *University of California Publications in Zoology* 45, nos. 1–5.

Margulis, L., M. E. Dolan, and J. H. Whiteside. 2005. "Imperfections and oddities" in the origin of the nucleus. *Paleobiology* 31: 175–191.

Margulis, L., and D. Sagan. 1998. *What Is Sex?* New York: Simon and Schuster.

———. 2000. *What Is Life?* Berkeley: University of California Press.

Margulis, L., and K. V. Schwartz. 1998. *Five Kingdoms: An Illustrated Guide to the Phyla of Life on Earth.* 3rd ed. New York: W. H. Freeman and Company.

Samuelsson, J., P. R. Dawes, and G. Vidal. 1999. Organic-walled microfossils from the Proterozoic Thule supergroup, Northwest Greenland. *Precambrian Research* 96: 1–23.

Searcy, D. G. 1987. Phylogenetic and phenotypic relationships between the eukaryotic nucleocytoplasm and thermophilic Archaebacteria. *Annals of the New York Academy of Sciences* 503: 168–179.

Woese, C. 2002. On the evolution of cells. *Proceedings of the National Academy of Sciences USA* 99: 8742–8747.　　　　　　　　　　　　　—*L.M. and D.Sa.*

Ethology and the study of behavioral evolution

While many individuals have contributed to the development of ethology, Austrian Konrad Lorenz (1903–1989) and Dutchman Niko Tinbergen (1907–1988) are regarded as its central figures, whose efforts established ethology as a recognized subdiscipline of evolutionary biology. When Lorenz traveled to Leiden in 1936, he met Tinbergen and they became friends and collaborators. Lorenz was known for his great passion for watching animals and his visionary breadth, while Tinbergen brought a more analytical and experimental approach to the study of animal behavior.

As well as being practitioners of ethology, both Lorenz and Tinbergen wrote about its methodology. Tinbergen identified four questions that ethologists should attempt to answer about any specific behavior: How does it develop? What are its immediate causes? What is its function? How did it evolve? Lorenz stressed the importance of spending hours watching animals behaving spontaneously before formulating any hypotheses, and he professed himself shocked by the deep ignorance of animals that he thought he detected in behavioristic psychologists such as John B. Watson (1878–1958). The development of an ethogram for a species—a standardized list of observable behaviors—is the starting point for ethological research.

In 1973, Lorenz and Tinbergen shared the Nobel Prize for medicine with German zoologist Karl von Frisch (1886–1982). They were cited "for their discoveries concerning organization and elicitation of individual and social behavior patterns." Von Frisch is best known for his discovery of the dance language of honey bees. Von Frisch's student, Martin Lindauer, conducted a comparative study of the dances of several closely related bee species to reconstruct their phylogeny, similar to Lorenz's study of behavioral homologies in the *Anatidae* (waterfowl). The idea that behavior could be as reliable as morphology in reconstructing the relationships among species was significant for the acceptance of ethology as an important part of evolutionary biology.

Early ethologists were interested in the concept of instinct; indeed, Lorenz and Tinbergen met at a workshop on this topic. Lorenz is associated with the idea of a fixed action pattern—a complex behavior that can be reliably and repeatedly elicited by a specific sign stimulus. Tinbergen showed that a particular pecking behavior of herring gull chicks could be elicited by relatively simple models of the adult beak with a prominent red spot. While the general concept of a fixed action pattern applies across numerous species, particular sign stimuli and fixed action patterns are highly species-specific and can only be understood in the context of the particular developmental, adaptive, and evolutionary contexts faced by members of the species. Lorenz's concept of instinct was strongly criticized in the 1950s by American psychologist Daniel Lehrman for being insufficiently attuned to developmental processes. Tinbergen's inclusion of development among his four questions represented his attempt to accommodate Lehrman's critique. Subsequently, many years of fruitless "nature versus nurture" debates and conflicts over definitions have made current ethologists leery of the concept of instinct. Nevertheless, the study of inherited behavioral adaptations is still an important part of ethology.

Areas of particular current interest and activity within ethology include predator-prey interactions and the various systems of antipredatory alarm calls found in a wide range of species, the social dominance structures that also vary widely across different species, comparative studies of social play, the memory demands of food caching and retrieval, tool manufacture and use, and the existence of culturally transmitted traits.

Ethology can be understood by contrast to alternative approaches to the study of behavioral evolution in biology and psychology. Within biology, behavioral ecology applies population-level models of evolution to the understanding of animal behavior. Behavioral ecologists freely borrow models from economics and game theory, which represent individual organisms as rational optimizers and allow analysis of interactions between environment and behavior on reproductive fitness and the distribution of behavioral traits. Unlike traditional ethology, behavioral ecology has been less concerned with immediate causes of the behavior of individual animals or its development. Mechanisms responsible for generating optimal behavior are often simply assumed and not explained. Nevertheless, there has been significant crossover between the two subdisciplines, with many of the originators of behavioral ecology having first trained as ethologists.

Comparative psychology also takes an explicitly evolutionary approach to animal behavior. However, unlike biologists, who tend to be interested in differences among individuals and among species, psychologists are trained to look for general patterns of behavior and principles of learning, sometimes referred to as laws of learning. Where comparative psychologists do see differences between species, they often regard these as revealing steps on an evolutionary ladder to full-blown human cognition, and they are often drawn to investigating tasks to mark thresholds of cognitive sophistication. Thus, for example, there has been much interest among comparative psychologists in finding out which species can recognize themselves in mirrors, can pass a false belief attribution task, or can learn a human language. Many ethologists prefer to observe animals in natural or naturalistic habitats. Some scientists, to be found on both sides of this divide, are pluralists who believe that the integrative study of animal behavior requires both field and laboratory investigation; others are less ecumenical.

Ethology has had several important offshoots. Both Lorenz and Tinbergen extended their approaches to the study of human behavior: Lorenz in his study of aggression, and Tinbergen (with his wife Elisabeth) in studying autistic children. The willingness of ethologists to treat humans as scientific objects of study, like any other animal, also importantly foreshadowed Edward O. Wilson's extension of evolutionary models of the social behavior of insects to explain human sociality, thus giving rise to the field of sociobiology, a field that also is tightly connected to behavioral ecology.

Another important offshoot of ethology is the comparative study of neural function known as neuroethology. Specific adaptations of neural systems to particular behavioral tasks have been described in a variety of species, such as the startle response of fish, the coordinated vocalizations of frogs, the directional hearing of owls, or the unusual sensory mechanisms of star-nosed moles.

Also, "cognitive ethology" is the label coined by Donald Griffin (1915–2003) to describe the reintroduction of issues of cognition and consciousness to the evolutionary study of animal behavior. Griffin made his early reputation conducting careful physical analyses of the echolocation capabilities of bats. Next, he turned his attention to questions of animal mind and awareness. Many disagree with his approach because they find it excessively anecdotal and anthropomorphic. Nevertheless, the cognitive revolution that took place in human psychology was slow in coming to the ethology, and Griffin was instrumental in getting other scientists to take seriously the idea that questions about the evolution of mind could be addressed by studying topics ranging from cognitive maps in honeybees to intentional communication in primates.

BIBLIOGRAPHY

Bekoff, M., C. Allen, and G. M. Burghardt, eds. 2002. *The Cognitive Animal: Empirical and Theoretical Perspectives on Animal Cognition.* Cambridge, MA: MIT Press.

Burkhardt, R. 2005. *Patterns of Behavior: Konrad Lorenz, Niko Tinbergen, and the Founding of Ethology.* Chicago: University of Chicago Press.

Cheney, D. L., and R. M. Seyfarth. 2007. *Baboon Metaphysics: The Evolution of a Social Mind.* Chicago: University of Chicago Press.

Frisch, K. von. 1967. *The Dance Language and Orientation of Bees.* L. E. Chadwick, trans. Cambridge, MA: Harvard University Press.

Griffin, D. R. 2001. *Animal Minds: Beyond Cognition to Consciousness.* Chicago: University of Chicago Press.

Lehrman, D. S. 1953. A critique of Konrad Lorenz's theory of instinct. *The Quarterly Review of Biology* 28: 337–363.

Lindauer, M. 1961. *Communication among Social Bees.* Cambridge, MA: Harvard University Press.

Lorenz, K. 1981. *The Foundations of Ethology.* New York: Springer-Verlag.

Tinbergen, N. 1963. On the aims and methods of ethology. *Zeitschrift für Tierpsychologie* 20: 410–463.

Watson, J. B. 1913. Psychology as the behaviorist views it. *Psychological Review* 20: 158–177.

Wilson, E. O. 1975. *Sociobiology.* Cambridge, MA: Harvard University Press.

—C.A.

Eugenics

Eugenics is fundamentally the generalization to humans of the practices of plant and animal breeders, who propagate the best and cull the worst of their stocks. The idea that such methods might be applied to humans has a long history, reaching back at least to the ancient Greeks. Thus, in Plato's *Republic* (1945, 157–158), Socrates tells Glaucon that "anything like unregulated unions would be a profanation in a state whose citizens lead the good life," and Glaucon acknowledges that he is careful to mate only the best of his sporting dogs and game birds so as to avoid the otherwise rapid deterioration of his stock.

Apart from a few efforts to systematically shape human heredity, as in the program adopted by the utopian community at Oneida, New York, in the 1860s, eugenics remained largely conceptual until the turn of the twentieth century. It was only then that organizations expressly devoted to this aim were established in the United States, Britain, and Germany. There are multiple reasons why the need to implement eugenics on a large scale came to seem urgent around this time, but perhaps the most important was a new worry about the possibility of biological degeneration.

In his *On the Origin of Species* (1859), Charles Darwin argued that improvement in plants and animals resulted from a fierce competitive struggle, a process encapsulated in the famous expression "the survival of the fittest," coined by Herbert Spencer in 1866 and later adopted by Darwin. Although Darwin did not discuss his own species in the *Origin*, many of his contemporaries wondered about its implications for humans. In particular, they fretted over the possibility that a relaxation of the struggle meant that human progress would be slowed, halted, or even reversed. Darwin himself was deeply disturbed by this prospect, and in *The Descent of Man* (1871, 168) he publicly expressed his concerns about the counterselective effects of charity, vaccinations against smallpox, the building of asylums for the sick and insane, and other accoutrements of civilized society, noting that as a result, "the weak members of civilised societies propagate their kind. No one who has attended to the breeding of domestic animals will doubt that this must be highly injurious to the race of man."

Darwin's cousin, Francis Galton, was among the first to explore the social implications of the theory of evolution by natural selection. Galton noted in his memoirs that he had been inspired by reading the *Origin* to pursue a long-standing interest in the topics of heredity "and the possible improvement of the Human Race." In "Hereditary Talent and Character" (1865), expanded to the book *Hereditary Genius* (1869), Galton showed that men who had achieved distinction in science, literature, and the law were more likely than members of the public at large to have had male relatives who were themselves eminent. In a raft of later works, Galton extended this research and developed what he saw as its corollaries. Galton thought that his inquiries proved beyond doubt that all human abilities and traits of character and temperament were transmitted from parent to child in the hereditary material (the nature of which was then unknown) and thus were largely fixed at birth. He also assumed that people varied greatly in their hereditary endowments, and that those with the fewest talents and the worst characters were reproducing at a shockingly rapid rate. It followed that if something were not done to prevent it, civilization would collapse. The solution was for humans to take charge of their own evolution, a process he termed *eugenics* in 1883.

The idea that human differences were innate and immutable was nothing new—nor were debates about its validity. In "Civilisation" (1835), the philosopher John Stuart Mill argued that the study of history teaches that human nature has taken an infinite variety of forms and must thus be highly pliable. In *Principles of Political Economy* (1848, 319), he famously asserted: "Of all the vulgar modes of escaping from the consideration of the social and moral influences on the human mind, the most vulgar is that of attributing the diversities of conduct and character to inherent natural differences." But Mill was swimming against the tide. Thus Galton's originality lay not in the claim that nature trumps nurture, but in the effort to prove through statistical studies of inheritance that the claim was true. One of those convinced was Darwin, who wrote in his autobiography (1958, 43) that he was "inclined to agree with Francis Galton in believing that education and environment produce only a small effect on the mind of any one, and that most of our qualities are innate." What gave Galton's work its significance was the climate of anxiety in which it appeared: there was widespread fear among middle-class people that the process of selection had been halted in civilized societies, and that the mentally and morally worst were now swamping the best.

But while Galton sounded an alarm, he had few practical proposals to deal with the crisis, and these encouraged the gifted to have more children, rather than discouraging the stupid and reckless from having any. The emphasis on what Galton called "positive" rather than "negative" eugenics did not reflect any squeamishness on his part about the latter. In Galton's day, those middle-class people most attracted to eugenic principles also generally adhered to the prevailing Whig ideology, according to which the functions of the state should be kept to a minimum. Proposals to expand the reach of the government into the intimate sphere of family life would thus have to overcome an antistatist worldview. That is a central reason why eugenics only became a matter for legislation in the twentieth century. Another reason was the increasing acceptance of August Weismann's "hard" theory of heredity, which seemed to bolster Galton's view that the hereditary material was impervious to the environment. A third reason was apparent evidence of a decline in population quality. In Britain, the large number of recruits deemed unfit for service in the Boer War (1899–1902) generated much alarm. In the United States, the low scores achieved by recent immigrants from Southern and Eastern Europe on IQ tests administered by the army during World War I seemed to confirm fears that the country was being inundated by newcomers who were feebleminded as well as fecund.

In Britain, the land of its birth, eugenics remained mostly a matter of propaganda. However, in North America, Scandinavia, Germany, Eastern Europe, and Japan, among other places, laws were passed permitting the compulsory sterilization of the feebleminded, criminals, sexual deviates, and others deemed unfit to procreate. In the 1927 case of *Buck v. Bell,* the U.S. Supreme Court upheld the constitutionality of compulsory sterilization in a decision signed by Chief Justice Oliver Wendell Holmes, a political progressive. He famously wrote: "It is better for all the world if, instead of waiting to execute degenerate offspring for crime, or to let them starve for their imbecility, society can prevent those who are manifestly unfit from continuing their kind." In the United States, eugenic arguments were also invoked in support of restricting immigration from Southern and Eastern Europe, arguments that played a role (although how large a one is debated) in passage of the national-origins quota system adopted in 1924. But eugenic zeal reached its apex (or nadir) in Germany, where atrocities against mental patients, Jews, and others were justified by the ostensible need for a program of "racial hygiene."

As the case of Holmes suggests, support for eugenics once appealed to diverse constituencies and was invoked in support of disparate agendas: the defense of pacifism as well as war, "free love" as well as traditional marriage, and access to birth control information and devices as well as their repression. In its heyday in the 1920s and 1930s, eugenics garnered support not just from social and political conservatives but also from Fabian and even Marxian socialists and reformers of many stripes. In time, eugenics came to be equated with its worst instantiations, and above all, with Nazism. As a result, the word is today usually reserved for practices that the speaker or writer detests. Thus critics of prenatal diagnosis deem it eugenics while its advocates strenuously resist the label. And legislation barring first-cousin marriages is not regarded as eugenics, notwithstanding its compulsory character. Selective breeding is still practiced, and in some forms it is widely approved, but one legacy of eugenics' association with atrocities is a distorted discourse in which we shrink from calling these practices by their rightful name.

BIBLIOGRAPHY

Broberg, G., and N. Roll-Hansen, eds. 2005. *Eugenics and the Welfare State: Sterilization Policy in Denmark, Sweden, Norway, and Finland.* Rev. paperback ed. East Lansing: Michigan State University Press.

Buck v. Bell. 1927. 274 U.S. 200.

Carlson, E. A. 2001. *The Unfit: A History of a Bad Idea.* Cold Spring Harbor, NY: Cold Spring Harbor Laboratory Press.

Darwin, C. [1859] 1964. *On the Origin of Species.* Introduction by E. Mayr. Cambridge, MA: Harvard University Press.

———. [1871] 1981. *The Descent of Man, and Selection in Relation to Sex.* Introduction by J. T. Bonner and R. M. May. Princeton, NJ: Princeton University Press.

———. 1958. *The Autobiography of Charles Darwin, 1809–1882: With Original Omissions Restored.* N. Barlow, ed. New York: Harcourt Brace.

Galton, F. 1865. Hereditary talent and character. *Macmillan's Magazine* 12: 157–166, 318–327.

———. 1869. *Hereditary Genius: An Inquiry into Its Laws and Consequences.* London: Macmillan.

———. 1908. *Memories of My Life.* London: Methuen.

Gillham, N. W. 2001. *A Life of Sir Francis Galton: From African Exploration to the Birth of Eugenics.* Oxford: Oxford University Press.

Kevles, D. J. 1995. *In the Name of Eugenics: Genetics and the Uses of Human Heredity.* Reprint ed. Cambridge, MA: Harvard University Press.

Kline, W. 2001. *Building a Better Race: Gender, Sexuality, and Eugenics from the Turn of the Century to the Baby Boom.* Berkeley: University of California Press.

Mill, J. S. [1835] 1977. Civilisation. In J. M. Robson, ed., *Collected Works of John Stuart Mill.* Vol. 18, *Essays on Politics and Society,* Pt. I, 117–148. Toronto: University of Toronto Press.

———. [1848] 1965. *Principles of Political Economy, with Some of Their Applications to Social Philosophy.* In J. M. Robson, ed., *Collected Works of John Stuart Mill.* Vols. 2–3. Toronto: University of Toronto Press.

Paul, D. B. 2003. Darwin, social Darwinism, and eugenics. In J. Hodge and G. Radick, eds., *The Cambridge Companion to Darwin,* 214–239. Cambridge: Cambridge University Press.

Plato. 1945. *The Republic of Plato.* F. M. Cornford, trans. London: Oxford University Press.

Turda, M., and P. J. Weindling, eds. 2007. *"Blood and Homeland": Eugenics and Racial Nationalism in Central and Southern Europe, 1900–1940.* Budapest: Central European University Press.   —D.B.P.

Evolution: The Modern Synthesis (Julian Huxley)

Julian Huxley had exceptional skills as a commercial writer. He knew how to collate vast amounts of information and how to extract essential points. *Evolution: The Modern Synthesis* (1942) is a perfect example. Huxley decided to create an encyclopedia of recent developments in evolutionary studies. Holding his survey together were a small number of carefully chosen threads. Cherished among these was natural selection. Although he made other points, stressing selection's importance was Huxley's central aim. This book served, especially in Britain, as a training manual for the "synthetic" theory of evolution.

The encyclopedia approach in *Evolution* moved systematically. Huxley devoted several chapters to genetic phenomena, especially those that create and maintain variation. He then shifted to describe causes of variation at the genome level, such as polyploidy. With a background in developmental biology, Huxley appreciated the importance of these phenomena. Shifting from his focus on organisms and ecological phenomena, Huxley next considered information about local populations, varieties, and subspecies. Three themes structured his survey: variation, processes of divergence, and processes of isolation. He stressed the importance of polytypic species and geographical variation. Clines were his special interest, a phenomena he had helped popularize a decade before. These sections of *Evolution* read simply as literature surveys, with Huxley trying to compile information as is done in a textbook. The chief virtue of these sections is their breadth. Huxley made extensive use of knowledge gleaned from his contacts, who spanned both Europe and the English-speaking world.

Evolution was no mere survey. Huxley placed natural selection and adaptation solidly at the center of his study. Selection dominates all evolutionary processes, he insisted; it was the explanation of first resort. To be sure, Huxley identified a wide range of other biological processes (e.g., drift, isolation, hybridization, and changing mutation rates). He frequently gave them key roles in certain confined topics. Nevertheless, selection remained his most potent evolutionary mechanism. For Huxley, adapting to a wide range of selection pressures was life's most frequent pattern.

Huxley considered himself a philosopher as much as a biologist. The final chapters of *Evolution* consider evolutionary trends and progress. Huxley presented several specific (and objective, he thought) criteria to measure progress: greater control over the environment, greater independence from the environment, and greater efficiency. The evolutionary advantages of these qualities seemed obvious to Huxley. Natural selection drove toward these qualities. This connection was key. Huxley presented his approach as an improvement over his predecessors, who required supernatural drivers and guiding forces. His scheme did not. Overall, humans served as Huxley's pinnacle of progress, with our intelligence (consisting of true speech, conceptual thought, and complex emotions) capping evolution's course. Huxley held out for further progress, too, suggesting humans had a moral imperative to fuel that process. *Evolution* ends with Huxley considering ways humans might act either to hasten or to steer their own evolutionary future.

For several generations of biologists, *Evolution* provided an important historical narrative for evolutionary studies. Huxley presented Charles Darwin as a genius who discovered in natural selection one of life's central principles. This genius was eclipsed in the years 1880–1920, supposedly owing to Darwin's failure to understand inheritance. The rise of genetics solved the problem. Its recent growth, and the development of other specialities since, vastly improved our understanding of particular pieces to evolution's puzzle. This narrative made Huxley's role seem natural. The time was ripe, he said, for someone to collate the many separate endeavors into a comprehensive, synthetic whole. *Evolution* was Huxley's attempt at that synthesis. Huxley's overall message was that Darwinism had returned in improved form, a phoenix rising from turn-of-the-century ashes.

Evolution delivered no brilliant new insight. Instead, it provided a competent survey of recent developments in evolutionary studies. Within that research community, Huxley promoted natural selection as an organizing theme. Later generations followed the same path and pointed to *Evolution* for its enthusiastic support in that direction.

Huxley published a second edition of *Evolution* in 1963. Colleagues published a third edition, with a second new introduction, in 1974. In each, the main chapters remained essentially unchanged.

BIBLIOGRAPHY

Huxley, J. [1942, 1963] 1974. *Evolution: The Modern Synthesis.* 3rd ed. London: Allen and Unwin. —*J.C.*

Evolutionary computing and genetic algorithms

Evolutionary computing embodies a collection of techniques that use ideas from biological evolution to solve computational problems (Mitchell and Taylor 1999). Many such problems require searching through a huge space of possibilities for complex solutions. These solutions are usually difficult for human programmers to devise.

Biological evolution is an appealing source of inspiration for addressing difficult problems. Evolution is, in effect, a method for searching among an enormous number of possibilities for "solutions" that allow organisms to survive and to reproduce in their environments.

The origins of evolutionary computing can be traced to Alan Turing's (1950) suggestion of an "obvious connection between machine learning and evolution." This observation excited people's imagination toward evolving solutions to problems using biological evolution as a metaphor. In the late 1950s, several groups independently demonstrated the possibilities of this approach.

In the late 1960s, John Holland, Ingo Rechenberg, and Lawrence Fogel independently introduced fundamental evolutionary computing approaches: genetic algorithms (Holland 1992), evolution strategies (Rechenberg 1973), and evolutionary programming (Fogel 1998), respectively. The general term of such approaches is *evolutionary algorithms.* The most widely used form of evolutionary algorithms are genetic algorithms.

The simplest version of a genetic algorithm consists of a population of encoded solutions to a given problem, a fitness function that assigns a numerical value to each individual, and a set of genetic operators to be applied to the population to create a new population. These typically include selection, crossover, and mutation. Starting with a randomly generated population, a genetic algorithm iteratively transforms the population until some stop criteria is met.

Evolutionary computing is finding use in a variety of commercial and scientific applications, including integrated circuit design, factory scheduling, robot behavior, morphology design, computer animation, image processing, supersonic jet design, financial market prediction, and drug design.

An interesting variant of evolutionary computing, termed *genetic programming*, was introduced by John R. Koza (1992). Genetic programming may be viewed as a genetic algorithm that evolves not just parameters, but rather truly executable computer programs. This method has been successfully demonstrated in a variety of applications, including optimal foraging.

Recent trends in evolutionary computing include the automated design and manufacture of physical robots. For example, Lipson and Pollack (2000) have combined simulation of behavior and morphology to automatically evolve and construct actual robots. Further work by Murata and Yamaguchi (2005) employs genetic algorithms to self-configure robots that can assemble and repair themselves in simple situations.

The parallels between biological evolution, searching through a space of genotypes, and computer evolution, searching through a space of computer programs or other data structures, have crystallized to create the scientific enterprise of evolutionary computing. This field has proven useful in a variety of commercial optimization problems, showing promise for many complex scientific and engineering problems.

BIBLIOGRAPHY

Fogel, D. B. 1998. *Evolutionary Computation: The Fossil Record.* New York: IEEE Press.
Holland, J. H. 1992. *Adaptation in Natural and Artificial Systems: An Introductory Analysis with Applications to Biology, Control and Artificial Intelligence.* Cambridge, MA: MIT Press.
Koza, J. R. 1992. *Genetic Programming: On the Programming of Computers by Means of Natural Selection.* Cambridge, MA: MIT Press.
Lipson, R., and J. B. Pollack. 2000. Automatic design and manufacture of robotic life forms. *Nature* 406: 974–978.
Mitchell, M., and C. E. Taylor 1999. Evolutionary computation: An overview. *Annual Review of Ecology and Systematics* 30: 593–616.
Murata, T., and M. Yamaguchi. 2005. Neighboring crossover to improve GA-based Q-learning for multi-legged robot control. Paper presented at the Proceedings of the 2005 Conference on Genetic Evolutionary Computation, Washington, DC, June 25–29.
Rechenberg, I. 1973. *Evolutionsstrategie: Optimierung Technischer Systeme und Prinzipien der Biologischen Evolution.* Stuttgart: Frommann-Holzboog.
Turing, A. 1950. Computing machinery and intelligence. *Mind* 59: 433–460.

—E.E.V. and C.E.T.

Evolutionary progress

Modern organisms seem clearly more advanced than ancient ones. Compare an elephant with a modern bacterium. (The comparison is apt in that certain modern bacteria are probably very similar to the first fossil organisms known, from rocks 3.5 billion years old.) The notion of an ordering among organisms dates at least to Aristotle, who arranged living forms on a linear scale based on degree of perfection, also called the *scala naturae* or great chain of being. For more than 2,000 years after Aristotle, the chain was understood to be static, meaning that organisms and their rankings did not change. But in the early nineteenth century Jean-Baptiste de Lamarck added an evolutionary component, so that organisms moved up the chain as they evolved. A half century later the Darwinian view challenged the ordering itself. Darwin saw evolution as a process of branching and divergence rather than linear ascent. But Darwin nevertheless recognized progress, writing famously in *On the Origin of Species* of "that vague yet ill-defined sentiment, felt by many paleontologists, that organisation on the whole has progressed" (Darwin 1859, 363).

In evolutionary discourse since Darwin, progress has been acknowledged by many evolutionists, but the idea has also come to be seen as troublesome, both philosophically and scientifically. It is widely recognized that intuitions on evolutionary progress have been contaminated by cultural influences. We have a tendency to read into the history of life the social and technological progress we think we see in human affairs. Also, progress is now recognized as having two independent components that are frequently conflated, a value-neutral claim that long-term directional change has occurred and a value claim that the trend has been for the better. Among contemporary evolutionists the existence of a long-term trend is widely accepted. Most would agree that something is increasing. But what, exactly? And increasing in what sense? Is it progress? If so, progress in what sense? These issues have not been resolved, but the modern discourse has enabled us at least to formulate them fairly clearly.

Discussions of evolutionary progress can be found in the works of biologists, paleontologists, philosophers, and historians, including Thomas Huxley (1893), Herbert Spencer (1857), Julian Huxley (1942), Thomas Goudge (1961), Ledyard Stebbins (1969), Ernst Mayr (1988), George Gaylord Simpson (1949), Stephen Jay Gould (1988), Francisco Ayala (1974), Leigh Van Valen (1973), Geerat Vermeij (1987), Robert Richards (1992), Michael Ruse (1996), and others. Interestingly, among the scientists most of the discussion has been relegated to popular works. Ruse argues that the notion of progress has remained central in evolutionary studies, often its proof the chief motivating factor behind the work that people do, but that in the second half of the twentieth century the increasing professionalization of the field forced discussion of this value-laden subject out of the technical literature.

PROGRESS IN WHAT SENSE?

There are two ways in which a trend in some feature of organisms might have value. The feature could be valuable to us, or it could be valued by the evolutionary process, so to speak, in the sense of being preserved. From a scientific standpoint the first alternative is problematic. In science our personal or cultural (religious, social) values should not be relevant or at least not central. Indeed, they are a distraction. For example, if the actual defining directionality in evolution were increasing energy usage, an infatuation with some feature we value, such as intelligence, might cause us to overlook it. Of course, we might be interested in the evolutionary sources of our values rather than trends generally. For example, it might be that intelligence is favored in evolution by natural selection, that we are the product of that trend, and that we value intelligence on account of its advantages for us. But here too the evaluative component of our interest is beside the point, except as original motivation. Thus it is hard to see how progress, if it is defined as a trend in features valuable to us, could be a scientifically useful notion.

The second alternative is to interpret valuable to mean that which is preserved, or, because we are interested in trends, that which increases over time. Progress then becomes a purely scientific term, but now doubts arise about the aptness of the word. It might be apt if the evolutionary process preserves features that strike us as positive, like intelligence, fitness, or complexity. But suppose we discover that the main long-term trend is an increase in a feature about which we are ambivalent, like energy usage, or a feature that seems downright bad, like fragility (leading to an increasing probability of extinction). It would sound odd to call either of these progress.

I think that these problems are fatal, and we should abandon talk of evolutionary progress and instead devote our energies to the study of trends. But the notion of progress has great cultural momentum and seems likely to persist, at least outside the technical literature, for some time to come. In what follows, I accede to the follies of my time and use the word *progress,* but only in the second sense, to mean that which is preserved. For convenience, I will also use the phrase *degree of advancement,* as though the variables involved in progress were either known or measurable. (As will be seen, at present they are not.)

A TREND IN WHAT SENSE?

Discussion of progress has been muddled, even recently, by confusion about how trends in groups work. In current trend theory it is not enough to say that some variable increases. To be clear, one must specify which of several group statistics is increasing—mean, maximum, or minimum—because trends in the different statistics have different interpretations. A trend in the mean is an increase in the average degree of advancement among all species in existence at a given time. In the figure, A shows a hypothetical long-term trend. Life begins with a single species, a single lineage, with some low level of advancement. As time passes, new species arise, and some species become extinct, with change occurring in the origin of every new species. The mean at any given time slice is the average level of advancement of all species in existence at that time. A trend is an increase in that average, that is, movement of the average to the right. A claim that a trend has occurred, without any qualifier, usually means an increase in the mean.

A trend in the maximum is a rise in the degree of advancement achieved by the most advanced species at a given time, in other words, a rise in the highest level of advancement achieved. (In the figure, B is the same as A, but annotated to show the trend in the maximum, the dotted line on the right.) Maxima are of special interest because humans are generally thought to represent the most progressive species in existence and, if so, would represent the maximum in a time slice that included the present. We would be the last species on the right at the very top of the graph. Finally, a trend in the minimum is an increase in the degree of advancement of the least advanced species (the dotted line on the left in figure B). Minima are of special interest because apparently the lowest level of advancement has not changed. Bacteria have existed, and indeed have dominated the biosphere, throughout the history of life. (In other words, it is clear that figures A and B are not a good representation of the history of life.)

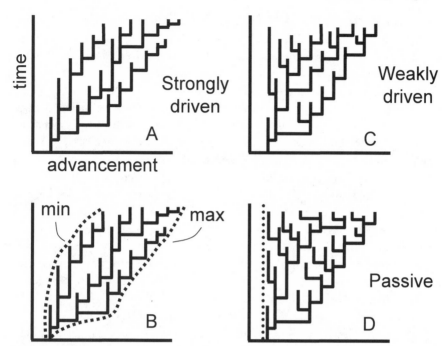

(A) A hypothetical long-term trend in advancement, that is, progress. Life begins with a single species, with some low level of advancement (the single vertical black line segment at the bottom left). As time passes, new species arise, and some species become extinct, with change occurring in the origin of every new species. Here all changes are increases, so the trend is said to be strongly driven. (B) Same as A, with dotted lines added to show the trends in the maximum and minimum. (C) A weakly driven trend in which increases predominate, but a number of decreases also occur. (D) A passive trend in which increases and decreases are equally frequent, but a lower limit on change (the vertical dotted line) forces lineages to diffuse to the right. See text for discussion.

Discussion of progress has not been concerned with the existence of a trend, perhaps because a trend is taken for granted (despite our inability to measure it). Rather, it has been about the pattern of change, the underlying mechanism. In figure A, natural selection strongly favors advancement, so that all changes are increases. The mean, maximum, and minimum all increase. Figure C is the same except that the net selective advantage is weak, meaning that selection sometimes favors decreases. (The classic examples are the evolution of certain reduced forms, like parasites.) The result is that less advanced species persist or are replaced by decreases from above, and the minimum stays roughly the same. The trends in A and C are said to be *driven*, or more precisely, A is strongly driven, whereas C is weakly driven. Figure D is different in that there is no overall selective advantage to increase. Increases and decreases occur equally often, so the only tendency is to spread symmetrically, both up and down. However, there is a boundary, a lower limit on degree of advancement, that blocks the spread of the group on the left. This boundary can be thought of as the lowest level of advancement consistent with being alive. The result is that the mean and maximum increase while the minimum stays the same (as in the weakly driven case). The trend is said to be *passive*.

The current controversy over mechanism began in 1996 with a book on progress by Stephen Jay Gould in which he raised the passive mechanism to the level of plausibility. At issue is the role of selection in progress, that is, whether the (presumed) trend in the history of life is weakly driven by selection or merely passive. (A strongly driven mechanism is ruled out because the true minimum is thought to have remained stable: bacteria persist.) Gould has been widely misinterpreted as claiming that evolution is random, but the implication of the passive mechanism is not randomness. Within each evolving lineage selection does operate, and organisms are adapting. Rather, the implication is that evolving lineages as a group are governed by many different selective forces, complexly configured, so that selection favors increases and decreases equally often. The process can be understood as purely deterministic, like the diffusion of a gas.

Unfortunately, as will be seen, there are hardly any data available, so from a scientific perspective the debate about mechanism is taking place in a near vacuum. Still, theory has produced some insights. For example, it is now clear that in the debate over mechanism two common observations are irrelevant. The first is the persistence of nonprogressive species, the stable minimum. This observation is not helpful because the minimum is predicted to remain unchanged in both cases, passive and weakly driven. Likewise, the rise in the maximum and the current existence of some highly advanced species, like humans, are not helpful. A rising maximum is expected in all cases.

A TREND IN WHAT?

Many candidates for a long-term trend have been proposed: (1) fitness; (2) complexity; (3) ability to sense, control, or respond to the environment; (4) body size; (5) intelligence; (6) versatility or evolvability, meaning the ability to change in evolution; (7) the reverse of evolvability, also called degree of entrenchment; (8) energy intensiveness or rate of energy usage; and (9) depth and sophistication of mechanisms of inheritance. For the most part advocates of all these candidates have taken a case-building approach, with the result that the literature offers a number of clever and sometimes-powerful arguments in favor of each. Unfortunately, some of the most interesting—like fitness, complexity, energy intensiveness, and ability to control the environment—are difficult to operationalize. Thus there are hardly any data on the scale of life as a whole, for any trend statistic, for any of these candidate variables.

For some of these variables, many would want to count a long-term trend, if one could be documented, as progress. One variable is fitness, understood as ability to survive and reproduce. It might seem that Darwin's principle of natural selection virtually guarantees a trend in fitness. Later organisms should be more fit than earlier ones, on the whole, having beaten them, in Darwin's terms, "in the race for life" (Darwin 1859, 363). But as Darwin also knew, natural selection produces adaptation to local environments, and on geological timescales these change dramatically. Thus selection might produce just constant change with no net improvement, no net increase in fitness. On the other hand, on long timescales organisms with general adaptations, suiting them to a wider variety of environments, could be favored over those more narrowly adapted. Are there general adaptations that offer advantages on timescales of hundreds of millions or billions of years? There are some biological mechanisms that may have persisted that long, such as certain cellular transport mechanisms and parts of certain metabolic pathways. But we do not know how to identify general adaptations in any rigorous way, or how to test whether they have become more prevalent. All that is clear now is that both alternatives, a long-term trend and its absence, are possible in principle, and therefore this is an empirical issue.

The most widely recognized candidate for a long-term trend is complexity. As a possible basis for progress it is really a stealth candidate because a direct connection with progress is rarely acknowledged. But the way complexity is used in evolutionary studies suggests that it functions as a kind of code word for progress, superficially value free and therefore scientific sounding but still subtly connoting advancement. The absence of a widely known technical definition makes this usage problematic. Nobody knows what complexity is, so one can say anything at all about it. The literature of recent decades does contain a technical definition, or rather several definitions for each of the several senses of complexity. These include complexity in the sense of number of part types, number of physiological or behavioral processes, and number of steps in development (see figure on page 556). However, little empirical work has been done, and with one exception, no trend that spans the history of life has yet been documented. Nor do we know very much about the mechanisms involved in change in these variables, passive versus driven. (On a smaller scale there is some evidence within the animals, over the past 500 million years, that there has been a trend in number of part types and that it has been passive.)

The exception is complexity in the sense of nestedness. Organisms are nested, to some extent, like Chinese boxes, with multiple levels of parts within wholes. Over the history of life, organisms with ever greater numbers of levels have evolved. The first living things 3.5 billion years ago were bacteria. About 2 billion years ago some bacteria joined together in a symbiotic relationship to produce the first eukaryotic cells, similar to modern protozoans such as *Amoeba*. They were superficially similar to bacterial cells, but because they evolved as colonies of bacteria, they occupied the next level up. Then about 600 million years ago clones of eukaryotic cells joined to produce the first multicellular individual (probably a kind of algae), the ancestor of all modern multicellulars, from mushrooms to magnolias to muskrats. Finally, about 480 million years ago clones of multicellular individuals joined to form the first colonial animal, a coral-like animal called a bryozoan. Modern colonial organisms include bryozoans, corals,

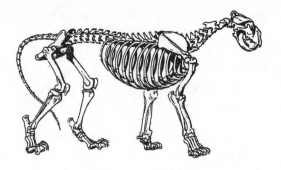

A classic example of why definitions of progress in terms of moves to complexity are problematic. The upper skeleton is of a lion and the lower skeleton is of a whale. By any measurement the backbone of the whale is the simpler, and yet adaptively the whale's backbone is what is needed for a marine mammal, just as the lion's backbone is what is needed for a terrestrial predator. Whales descended from land tetrapods (four-legged animals) so their backbones mark a selection-driven move from complexity to simplicity. (From McShea 1991, 315.)

---

the social insects, and many vertebrates, from fish to humans. We do not know if even higher levels—colonies of colonies—have been reached. There are no clear-cut cases. Humans seem to associate at many levels, but only weakly at the higher ones. Interestingly, if humans have not truly reached the colony-of-colonies level, and if no other organism has either, this raises a novel possibility: the only variable in which we can document a trend, complexity in the sense of nestedness, in fact does show a trend, but that trend might have ended, perhaps 480 million years ago. Perhaps the age of progress is over.

BIBLIOGRAPHY

Ayala, F. J. 1974. The concept of biological progress. In F. J. Ayala and T. Dobzhansky, eds., *Studies in the Philosophy of Biology*, 339–354. London: Macmillan.

Goudge, T. A. 1961. *The Ascent of Life*. Toronto: University of Toronto Press.

Gould, S. J. 1988. On replacing the idea of progress with an operational notion of directionality. In M. H. Nitecki, ed., *Evolutionary Progress*, 319–338. Chicago: University of Chicago Press.

———. 1996. *Full House: The Spread of Excellence from Plato to Darwin*. New York: Harmony Books.

Huxley, J. S. 1942. *Evolution: The Modern Synthesis*. London: Allen and Unwin.

Huxley, T. H. 1893. *Evolution and Ethics*. London: Macmillan.

Mayr, E. 1988. *Toward a New Philosophy of Biology: Observations of an Evolutionist.* Cambridge, MA: Belknap Press of Harvard University Press.

McShea, D. W. 1991. Complexity and evolution: What everybody knows. *Biology and Philosophy* 6, no. 3: 303–325.

———. 1998. Possible largest-scale trends in organismal evolution: Eight "live hypotheses." *Annual Review of Ecology and Systematics* 29: 293–318.

Richards, R. J. 1992. *The Meaning of Evolution: The Morphological Construction and Ideological Reconstruction of Darwin's Theory.* Chicago: University of Chicago Press.

Ruse, M. 1996. *Monad to Man: The Concept of Progress in Evolutionary Biology.* Cambridge, MA: Harvard University Press.

Simpson, G. G. 1949. *The Meaning of Evolution.* New Haven, CT: Yale University Press.

Spencer, H. 1857. Progress: Its law and cause. *Westminster Review* 67: 244–267.

Stebbins, G. L. 1969. *The Basis of Progressive Evolution.* Chapel Hill: University of North Carolina Press.

Van Valen, L. 1973. A new evolutionary law. *Evolutionary Theory* 1: 1–30.

Vermeij, G. J. 1987. *Evolution and Escalation.* Princeton, NJ: Princeton University Press. —D.W.M.

Evolutionary psychology

Some behavioral scientists define evolutionary psychology as simply the study of human behavior and psychology from an evolutionary perspective. In this broad sense, evolutionary psychology is a field of inquiry. Fields of inquiry are defined not by specific theories about the phenomena they study, but by the kinds of questions they pose about them. Fields of inquiry thus differ from paradigms, which are defined by the specific theories and methodologies with which they answer the questions that define a field of inquiry. As a field of inquiry, evolutionary psychology began with Charles Darwin's publication of *The Descent of Man* in 1871 and *The Expression of the Emotions in Man and Animals* in 1872. It was not until a century later that clearly articulated paradigms began to emerge in evolutionary psychology, each with distinct theories about, and specific methods for studying, the evolution of human behavior and mentality.

The first of these paradigms was human sociobiology, which emerged in the 1970s. The core idea of sociobiology was that behavior has evolved under natural and sexual selection just as organic form has. For example, females of many species choose mates based on the quality of male courtship displays. If male courtship displays vary in quality, and that variation is heritable, sexual selection will tailor male courtship behavior to female preference. In this way, sociobiologists argued, a form of behavior can be an adaptation, whose presence in a population is to be explained by the principles of evolutionary theory. Accordingly, human sociobiology sought to offer evolutionary explanations of how humans are behaviorally adapted to social life with one another—to explain human behavioral adaptations for dominance hierarchies, for manifesting and dealing with aggression, and for mating. Indeed, the central theoretical problem of human sociobiology was to explain the evolution of altruism, the performance of acts that benefit others at a fitness cost to the actor.

In the field of evolutionary psychology today, human sociobiology has been superseded by several alternative paradigms, two of which are particularly notable. One of these confusingly goes by the name "evolutionary psychology." To avoid confusion, and to distinguish the field of inquiry from the paradigm, I will refer to this paradigm as "Evolutionary Psychology" (capitalized).

Evolutionary Psychology is a marriage of cognitive psychology with evolutionary biology. In contrast with sociobiology, Evolutionary Psychologists argue that, when a behavior has evolved under selection, it is not the behavior that is an adaptation, but rather the psychological mechanism that causes that behavior. The goal of Evolutionary Psychology is thus to discover and describe the information-processing structure of our psychological adaptations. Because adaptation is a slow process, Evolutionary Psychologists believe that our psychological adaptations evolved during the Pleistocene (1.8 million to 10,000 years ago) to solve the problems of survival and reproduction faced by our hunter-gatherer ancestors. Moreover, Evolutionary Psychologists argue that a distinct psychological mechanism evolved for each distinct problem faced by our hunter-gatherer ancestors. They believe that the human mind contains hundreds or thousands of these mechanisms, collectively constituting a universal human nature. Because the environments inhabited by humans have changed dramatically and rapidly since the Pleistocene, however, Evolutionary Psychologists argue that our "Stone Age" psychological mechanisms often produce maladaptive behavior in the modern environments for which they are not designed. Thus, Evolutionary Psychologists claim, in order to discover the evolved design of the mind, we must "reverse engineer" the mind from the vantage of our evolutionary past, figuring out the problems our ancestors faced and then hypothesizing the psychological mechanisms that evolved to solve them. Evolutionary Psychologists then conduct standard psychological experiments to determine whether people exhibit the behavior or preferences that such mechanisms would produce.

To illustrate, consider a fundamental asymmetry between the sexes. A woman's lifetime reproductive output is limited by the number of pregnancies she can carry to term, and her maximum reproductive output can be achieved with a single mate. In contrast, a man's lifetime reproductive output is limited only by the number of eggs he can fertilize, and thus his maximum reproductive output can be achieved only with multiple mates. Among ancestral humans, then, having multiple mates would have entailed significant reproductive benefits for males, but not for females. Evolutionary Psychologists argue that this would have created selection pressure for men to evolve a greater desire for sexual variety than women, making men more interested in polygamous mating than women. Evolutionary Psychologists have tested for this sex difference in desire, and they have found that, on average, men want nearly five times more sexual partners over the course of a lifetime than women. Evolutionary Psychologists conclude that male desire for sexual variety is a psychological adaptation, which motivates men to seek more sex partners than women.

The other notable paradigm that grew out of human sociobiology is human behavioral ecology. Behavioral ecology is the study of how animal behavior is adaptively responsive to conditions in animals' physical and social environments. The fundamental premise of behavioral ecology is that selection has designed animals to be maximally efficient in performing a variety of tasks that are essential to survival and reproduction, such as foraging for food, capturing prey, eluding predators, or wooing mates. Accordingly, behavioral ecologists view animals as capable of flexibly altering their behavior in response to environmental conditions in order to maximize their chances of survival and reproductive success. Behavioral ecologists test these assumptions about animal behavior with optimality models and evolutionary game theoretic models. These models specify a particular task that is essential to survival or reproduction, postulate a range of behavioral strategies that animals can pursue in performing that task, and assign fitness costs and benefits to each of the strategies. This enables behavioral ecologists to calculate which behavioral strategy optimizes the average ratio of benefits to costs to the animals they are studying. Behavioral ecologists then predict that the studied animals will pursue that optimal strategy, testing their prediction against observed behavior. Human behavioral ecology is simply the application of these techniques to the study of human behavior.

To illustrate, consider polyandry, a marital system in which one woman has more than one husband and which is practiced in four of the 849 societies in the ethnographic record. At first glance, polyandry appears maladaptive for males in light of the evolutionary logic underlying Evolutionary Psychology's claim about an evolved sex difference in desire for polygamy. But human behavioral ecologists have discovered that the co-husbands in polyandrous marriages are typically brothers. This helps to offset the costs of polyandry to the co-husbands, because their resources are pooled to rear only offspring to which they are all genetically related. Polyandry also typically occurs among brothers who have inherited farmland that is too small to be divided into parcels that could each sustain a family. Finally, where polyandrous marriages occur, alternative sources of income are not available to the brothers; cultivating the family farm is the only viable means of subsistence. Under these ecological conditions, human behavioral ecologists argue, polyandrous marriage is actually an adaptive choice for the brothers who enter it, rather than a maladaptive by-product of evolved desires encountering conditions for which they were not designed. Under these conditions, the brothers achieve greater reproductive success by maintaining joint possession of the farm, working it together, marrying one woman, and rearing their joint offspring than they would by marrying monogamously.

Human behavioral ecology differs from Evolutionary Psychology in several key respects. First, whereas Evolutionary Psychology strives to discover psychological adaptations to Pleistocene environments, human behavioral ecology studies how human behavior is adaptive to environmental conditions.

Second, Evolutionary Psychology believes that human behavior is often mal-adaptive in ecological conditions that do not resemble those of our hunter-gatherer ancestors, but human behavioral ecology believes humans behave adaptively in a wide range of environments. Third, while Evolutionary Psychology tries to reverse engineer the evolved nature of the human mind from the vantage of our Pleistocene past, human behavioral ecology applies evolutionary principles to observable human behavior. Finally, whereas Evolutionary Psychology strives to discover a uniform and universal human nature, human behavioral ecology strives to determine how environmental differences between individuals affect behavioral differences between them.

BIBLIOGRAPHY

Barrett, L., R. Dunbar, and J. Lycett. 2002. *Human Evolutionary Psychology.* Princeton, NJ: Princeton University Press.

Buller, D. J. 2005. *Adapting Minds: Evolutionary Psychology and the Persistent Quest for Human Nature.* Cambridge, MA: MIT Press.

Buss, D. M. 2007. *Evolutionary Psychology: The New Science of the Mind.* 3rd ed. Boston: Allyn and Bacon.

Laland, K. N., and G. R. Brown. 2002. *Sense and Nonsense: Evolutionary Perspectives on Human Behaviour.* Oxford: Oxford University Press.

Wilson, E. O. 1978. *On Human Nature.* Cambridge, MA: Harvard University Press.

—D.J.B.

"Evolution in Mendelian Populations" (Sewall Wright)

The classic paper "Evolution in Mendelian Populations" by American population geneticist Sewall Wright was published in *Genetics* in 1931. In it, and in "The Roles of Mutation, Inbreeding, Crossbreeding and Selection in Evolution," a shorter paper given the following year at an international genetics congress, in which he introduced his famous metaphor of an "adaptive landscape" (see figure), Wright presented his "shifting balance theory of evolution." His fellow population geneticist across the Atlantic, R. A. Fisher (1930), had just given his neo-Darwinian account of evolutionary change, one that saw such change coming as the result of mutations into a large population, and selection then eliminating all but the best. Wright had worked for a number of years at the U.S. Department of Agriculture. From studies of the breeding of cattle, he was convinced that the best way to get change is by isolating small groups, working within them to get the best of all possible characteristics, and then finally reintroducing the highest quality to the whole herd (Provine 1986).

Wright's shifting balance theory followed the same pattern: isolation of groups, improvement within the group, and then mingling with the whole. In his congress paper, Wright represented the process as something occurring on a landscape, with the best (the fittest) being at the tops of peaks and the worst down in the valleys. He saw evolution as being a matter of moving from one peak to another higher peak. The challenge was how to get down the side of the first peak so that one could start climbing up the other. Because selection

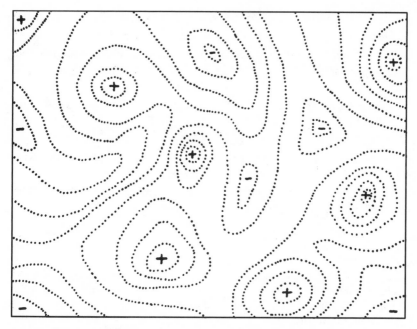

Adaptive landscapes. This picture, showing how the fitter organisms climb to the tops of hills of adaptive fitness, illustrates perhaps the most famous evolutionary metaphor of the twentieth century (from Wright 1932). It is possible that Wright came to the metaphor from reading an article by J. B. S. Haldane, "A Mathematical Theory of Artificial and Natural Selection, Part VIII. Metastable Populations," *Proceedings of the Cambridge Philosophical Society* 28 (1931), 137–142, which contains a pictorial representation of peaks of fitness, drawn by the geneticist C. H. Waddington. Wright's biographer William Provine points out that the landscape is condensing a huge amount of information about the effects of different genes and that consequently, ultimately, the landscape may not be entirely coherent. See Provine 1986.

would be working against this, Wright introduced the controversial notion of "genetic drift," also at times known as the "Sewall Wright effect." He argued that mathematics shows that, in small groups, the chance effects of breeding may well counter the effects of selection, so new features can be produced without (or even despite) natural selection. These new features may themselves be of great value, especially in the whole, and can lead to further evolutionary change.

This is a very non-Darwinian theory, although later versions, particularly under the influence of Theodosius Dobzhansky, gave a much larger role to natural selection. (Dobzhansky's masterwork, *Genetics and the Origin of Species*, first published in 1937, gave strong emphasis to Wright. Later editions gave more credence to natural selection, thanks to studies Dobzhansky was doing on chromosomal variations of fruit flies in nature. See Ruse 1999.) In part the influence on Wright came from Henri Bergson, who had complained that natural selection cannot yield innovations. But a far greater influence was Herbert Spencer (particularly as passed on by Wright's teacher, L. J. Henderson). Populations are seen in a kind of balance, homogeneous; when something disrupts the balance, causing fragmentation and heterogeneity, the whole strives to return to homogeneity, at a higher level. This is a classic case of Spencerian dynamic equilibrium.

In recent years, the shifting balance theory has been very severely criticized, although most evolutionists still think that there is some role for genetic drift (see Coyne, Barton and Turelli 1997). At the molecular level, beneath the effects of natural selection, drift is thought to be very important (Kimura 1983) and is the basis of the major way of calculating significant dates in evolution (through "genetic clocks"). (See the main essay "Molecular Evolution" by Francisco J. Ayala in this volume.)

BIBLIOGRAPHY

Bergson, H. 1911. *Creative Evolution*. New York: Holt.
Coyne, J. A., N. H. Barton, and M. Turelli. 1997. Perspective: A critique of Sewall Wright's shifting balance theory of evolution. *Evolution* 51, no. 3: 643–671.
Dobzhansky, T. [1937] 1951. *Genetics and the Origin of Species*. 3rd ed. New York: Columbia University Press.
Fisher, R. A. 1930. *The Genetical Theory of Natural Selection*. Oxford: Oxford University Press.
Henderson, L. J. 1917. *The Order of Nature*. Cambridge, MA: Harvard University Press.
Kimura, M. 1983. *The Neutral Theory of Molecular Evolution*. Cambridge: Cambridge University Press.
Provine, W. B. 1986. *Sewall Wright and Evolutionary Biology*. Chicago: University of Chicago Press.
Ruse, M. 1996. *Monad to Man: The Concept of Progress in Evolutionary Biology*. Cambridge, MA: Harvard University Press.
———. 1999. *Mystery of Mysteries: Is Evolution a Social Construction?* Cambridge, MA: Harvard University Press.
Spencer, H. 1862. *First Principles*. London: Williams and Norgate.
Wright, S. 1931. Evolution in Mendelian populations. *Genetics* 16: 97–159.
———. 1932. The roles of mutation, inbreeding, crossbreeding and selection in evolution. *Proceedings of the Sixth International Congress of Genetics* 1: 356–366.   —M.R.

Evolution of language

Although many animals communicate using vocalizations, only humans use complex language, based on a discrete number of arbitrary bits of sounds and gestures, to comprehend and produce a seemingly infinite number of novel messages. Listening to speech and reading are sensory aspects of language, while talking, writing, and body language are motor aspects. In the 1960s, Charles Hockett listed 15 universal design features that distinguish human language from communication systems in other animals. Hockett viewed three features as particularly important for human language: (1) displacement, or being able to communicate about matters that are remote in time and space; (2) productivity, which is the ability to communicate an open-ended variety of novel messages; and (3) duality, in which a small number of meaningless phonological elements may be combined to create an infinite set of meaningful utterances. Two other interesting design features are reflexiveness, in which one can use language to communicate about language, and prevarication, whereby language can be used to deliver false messages. Although language-trained apes such as the well-known bonobo, Kanzi, show subtle signs of utilizing some of Hockett's design features in their signed or computer-facilitated communications with people, their communications remain extremely simple despite decades of intense schooling. By the time a human child is four years old, she is likely to fasten a beady eye on her parents and ask, "Where did I come from?" The highly trained four-year-old chimpanzee, on the other hand, is likely to sign a spontaneous communication on the order of "cookie, cookie, hurry." Apes, in fact, never use their language skills to ask questions. Further, their larynxes are incapable of producing the full range of human speech sounds. How and when, then, did hominins evolve language during the 5 to 7 million years since they diverged from the ancestors of their closest genetic cousins, the chimpanzees?

Scholars have been fascinated by this question since at least Charles Darwin's time. Nonetheless, in 1866 the influential Linguistic Society of Paris famously banned all communications dealing with the origin and evolution of language because of its speculative nature, which is revealed in the amusing names that have been given to various hypotheses. The "bow-wow" hypothesis, for example, suggests that language began as imitations of natural sounds such as "moo," "buzz," and "crash"; the "yo-heave-ho" hypothesis attributes the emergence of language to rhythmic chants and grunts associated with physical labor; and the "pooh-pooh" hypothesis suggests language originated from emotional cries or exclamations. Although interesting, these hypotheses do not lend themselves to scientific testing.

Notions about when and how language first arose are as controversial today as they were a century ago. Some researchers think that language is millions of years old; others believe it originated a mere 40,000 years ago. Among the latter, some archeologists think that language could not have arisen before the "cultural explosion" of the Upper Paleolithic in Europe that was characterized by cave paintings, portable art, living sites, deliberate burials, personal adornments, hunting and fishing technology, and signs of trade. Upper Paleolithic toolmakers used and engraved ivory and bone in addition to stone, and their tools were better made than earlier ones and frequently more specialized for particular tasks. Although most anthropologists would agree that Upper Paleolithic innovations indicate that language had arisen by 40,000 years ago, they disagree about when it initially emerged. Some archeologists believe that the prior material record is too simple to justify an assumption of linguistic behaviors. Others think that too much emphasis has been placed on the Western European Upper Paleolithic; they suggest, instead, that numerous potential signs of language including barbed spear points, evidence of trade networks, and the use of red pigment (ochre) occurred much earlier (perhaps as long ago as 130,000 years) in the African record of the Middle Stone Age. Rather than seeing language as having emerged suddenly and relatively recently, they believe it evolved more gradually over much longer periods of time.

Other scientists hypothesize that language originated still earlier. Skeletal biologists who study the evolution of the parts of the skull and neck that facilitate speech believe that the larynx and throat had evolved to their modern forms by 150,000 years ago, thus allowing the rapid vocalizations that characterize modern languages. They also acknowledge that speech need not entail the full range of contemporary vocalizations and that rudimentary language may have existed at least a million years ago.

Paleoneurologists, who study the evolution of the brain, believe that language emerged even earlier, around or before 2 million years ago. They study casts of the insides of fossil skulls (endocasts) that reproduce details from the outside part of the brain (or cerebral cortex) that were imprinted on the walls of the braincase when hominins (or other animals) were alive. They also chart the increase in cranial capacity (as a surrogate for brain size) that occurred during the course of hominin evolution. Although speaking depends on activity in various parts of the brain, Broca's speech area on the left side of the frontal lobes is especially important for coordinating the intricate movements of the tongue, mouth, and larynx that are necessary for speech. Broca's area is associated with a specific pattern of convolutions that does not appear in the brains of our closest genetic cousins, the great apes. This part of the brain is sometimes reproduced on endocasts. Hints revealed on endocasts from fossil hominins suggest that the frontal lobe speech area may already have begun developing by 1.9 million years ago. Studies of stone tools from the same time also suggest that hominins were in the process of becoming right-handed, which apes as a population are not. It is important to note, however, that brains had not yet enlarged to anywhere near their modern sizes, and that the evidence from endocasts regarding Broca's area is difficult to interpret. The suggestion of rudimentary language at this early point in time should therefore be viewed as interesting but speculative.

Language and population-level right-handedness both depend heavily on the left side of the brain. Nonhuman primates have brains that are also asymmetrical (or lateralized), to some degree, but they do not begin to approach human levels of neurological asymmetry. In addition to language functions, the left hemispheres of humans are differentially involved in skilled movements, such as those of the right hand, as well as analytical, time-sequencing processes. The right hemisphere, on the other hand, excels at more global pursuits, such as visuospatial and mental imaging, recognizing faces, appreciating humor, and hearing and producing music. It also controls the left hand. Although it is the left side of the brain that speaks, it is the musical right hemisphere that provides the tone of voice, or prosody, which colors our utterances. In this sense, language depends on the right as well as the left hemisphere.

Some scholars, however, think that prosody had nothing to do with the emergence of language and, instead, developed as a kind of auditory frill that eventually invaded language. They also assert that musical skills were not a target of natural selection in the same way that linguistic abilities were, and explicitly reject the notion that language could have evolved from primatelike call systems, although they acknowledge that tone of voice likely did. This "top-down" school of thought therefore focuses on understanding how humans evolved the ability to link symbolic referential sounds into a potentially infinite number of rule-governed meaningful sentences. An opposing "continuity hypothesis," on the other hand, posits that language emerged as a result of incremental evolution from the primate call systems of our early ancestors and sees its roots reflected in the combinations of prosodic calls that various wild primates have in their repertoires. Some continuitists (e.g., John Newman) extend the discussion even further into the past by focusing on contact calls between mother and infant mammals.

No discussion of language evolution would be complete without mentioning the well-known gestural-origins hypothesis, which recognizes and builds on the deep psychological links between thought, speech, and gesture. Language, including mother-infant interactions, encompasses vocal, visual, tactile, and gestural components. Humans do not just "talk" with our voices; we also use flailing hands, facial expressions, body posture, and (sometimes) physical contact with people or objects to convey meaning. Nonhuman primates also combine such gestures with emotional vocalizations, and occasionally olfactory signals, in their blend of multimodal communication. The difference is that, in addition to emotional calls and intonations, the human primate produces vocalizations that have the linguistic properties outlined above. Advocates of gestural origins think that the natural selection that led eventually to language initially targeted gestural communications, be they defined narrowly (e.g., manual gestures) or more broadly (articulatory gestures that produce visual, acoustic, vocal, or tactile signals).

My own ideas are in keeping with the gestural hypothesis, writ broadly. I believe that the roots of language are extremely ancient and that prosody represents a kind of signature that is left over from events that preceded the emergence of the first language. Although the body language that human and chimpanzee mothers use with their infants is virtually identical, only human mothers deliver a constant stream of baby talk, which is very prosodic. Such "motherese," or musical speech, is universal among humans, and its acoustic features help babies around the world learn the structures of their native languages. Chimpanzee mothers direct few, if any, special vocalizations toward their infants, who remain physically attached to their mothers by clinging, unaided, to their hairy chests. Because of developmental effects related to the evolution of walking upright, human infants have lost this ability, although they do retain a grasping reflex. Early hominin mothers must have carried their helpless infants in their arms, at least until baby slings were invented. Under these circumstances, soothing maternal vocalizations would have been adaptive for reassuring nearby infants that were, of necessity, periodically removed from their mothers' embraces. Tone of voice may thus have evolved as a kind of disembodied extension of mothers' cradling arms. Taking a cue from how modern infants begin to acquire language, it is reasonable to surmise that natural selection sculpted incremental changes in mother-infant prosodic, tactile, visual, and gestural interactions, which facilitated the emergence of linguistic features and, eventually (perhaps far down the line), the first language. Needless to say, like the other ideas about language origins, this "putting the baby down" hypothesis is controversial.

Because of the wide disparity in current conjectures regarding the evolution of language, one might be tempted to conclude that research on this subject has not improved much since the days of the bow-wow, yo-heave-ho, and pooh-pooh hypotheses. This is not true, however. Advances in medical imaging technology and the neurosciences are currently permitting elucidation of the neurological substrates of language (e.g., the role of mirror neurons); the innate nature of language is becoming clearer from fascinating research on deaf individuals; and genes that have a role in the development of normal spoken language (e.g., FOXP2) are beginning to be identified. Language is what sets us apart from the other primates, and today we have more sophisticated ideas about how it evolved than we did a hundred years ago. Because of the information explosion that is currently under way, we should soon know more.

BIBLIOGRAPHY

Armstrong, D. F., W. C. Stokoe, and S. E. Wilcox. 1995. *Gesture and the Nature of Language.* Cambridge: Cambridge University Press.

Falk, D. 2004a. *Braindance: New Discoveries about Human Origins and Brain Evolution.* Revised and expanded ed. Gainesville: University Press of Florida.

———. 2004b. Prelinguistic evolution in early hominins: Whence motherese? *Behavioral and Brain Sciences* 27: 491–541.

———. 2009. *Finding Our Tongues: Mothers, Infants and the Origins of Language.* New York: Basic Books.

Hockett, C. 1966. The problem of universals in language. In J. Greenberg, ed., *Universals of Language,* 1–29. Cambridge, MA: MIT Press.

Holden, C. 1998. No last word on language origins. *Science* 282: 1455–1458.

Kenneally, C. 2007. *The First Word: The Search for the Origins of Language.* New York: Viking.

Laitman, J. T., J. S. Reidenberg, and P. J. Gannon. 1992. Fossil skulls and hominid vocal tracts: New approaches to charting the evolution of human speech. In J. Wind et al., eds., *Language Origin: A Multidisciplinary Approach,* 395–407. Dordrecht, The Netherlands: Kluwer Academic Publishers.

Masataka, N. 2003. *The Onset of Language.* Cambridge: Cambridge University Press.

McBrearty, S., and A. S. Brooks 2000. The revolution that wasn't: A new interpretation of the origin of modern human behavior. *Journal of Human Evolution* 39: 453–563.

Newman, J. D. 2004. Motherese by any other name: Mother-infant communication in non-hominin mammals. *Behavioral and Brain Sciences* 27: 519–520.

Pinker, S. 1994. *The Language Instinct.* New York: William Morrow and Company.

Savage-Rumbaugh, S., S. G. Shanker, and T. J. Taylor. 2001. *Apes, Language, and the Human Mind.* New York: Oxford University Press.   —D.F.

Evolution of life histories

Why do some plants, such as redwood trees, begin to reproduce only many years after germinating and live for hundreds of years, whereas others, such as the annuals in your garden, mature young and live only one season? Why do some animals, such as humans, produce one offspring at a time and rear comparatively few throughout life, whereas others, such as the ocean-dwelling krill, produce thousands of young, most of which cannot survive? What are the consequences of these differences for the ecology and the evolution of these organisms? These, and a variety of related questions, are in the realm of life-history evolution.

An organism's life history is its pattern of survival and reproduction, plus the key stages and transitions in the life cycle that influence its survival and reproduction. Life-history traits include growth and development before reproduction; age and size at sexual maturity; the number, size, and sex of offspring; and the age of death. Species, and even populations within species, often vary substantially in life-history characteristics in predictable and intriguing ways. The goal of life-history studies is to understand this variation in the patterns of growth and reproduction, the ecological conditions that generate this variation, and the ecological and evolutionary consequences of this variation. The study of life histories, whether empirical or theoretical, requires integrating concepts of genetics and development with ecology and with an understanding of the organism's life cycle.

Actuarial studies by insurance companies are an example of life-history analysis in humans. Insurance actuaries use life tables to describe the patterns of human mortality at specific ages to assess how age influences the risk of death. Ecological life table analysis is based on age-specific birth and death rates, from which scientists can calculate a variety of useful variables such as the net reproductive rate (the total number of offspring produced), future expected reproduction (called reproductive value), and emergent characteristics such as the relationship of reproduction to mortality. These variables are directly related to fitness, a function of total reproduction (the net reproductive rate) and the schedule of reproduction; fitness is higher for individuals that produce relatively more offspring more quickly. Because of the direct link between life-history traits and fitness, natural selection shapes the diversity of life-history strategies observed in nature. These variables are also central to how ecologists predict population fluctuations, a topic of importance to the management of harvested or protected wild species.

Modern life-history analysis traces to David Lack in the 1940s and Lamont Cole in the 1950s. David Lack (1947) was a population biologist and ornithologist who infused Darwinian thinking into the study of life histories. His most influential early life-history hypothesis was that the number of eggs that birds lay in a nesting attempt (their "clutch size") should be that which maximizes the total offspring fledged from that clutch. This has become known as the "Lack Clutch Size." To maximize the production of fledglings, parents must balance the number of hatchlings with their ability to rear them to independence. If parents attempt to rear more offspring than they can support, some or all of the young may die. Likewise, parents producing fewer offspring than they can rear will fledge fewer than they could, and will likewise have lower reproductive success than if they had produced a larger clutch.

Experimental studies on many species of birds, in which clutches are artificially enlarged, have found that clutch sizes produced by birds are generally smaller than the Lack Clutch Size (e.g., VanderWerf 1992). This discrepancy between the Lack and the observed clutch sizes stimulated a great diversity of experimental and theoretical study that in turn generated many adaptive hypotheses, mainly derived from optimization analyses (e.g., "optimality models"; Orzack and Sober 1994) that became commonplace by the 1970s and 1980s. For example, models predict and experimental studies have shown that current year reproductive effort affects future reproductive success of parents; that is, there is a trade-off between current and future reproductive effort (see reviews in Godfray et al. 1991 and Roff 1992). When this and other trade-offs (such as that between clutch size and egg size; Smith and Fretwell 1974) are incorporated into models of clutch size, we can explain much of the observed deviation between real clutch sizes and Lack's predicted clutch sizes.

Following on Lack's introduction of Darwinian thinking into the study of life-history variation, Lamont Cole (1954) introduced a mathematical framework to the study of life-history ecology by demonstrating how changes in demographic variables influence the rate of increase of a population (Roff 1992). Most famously, he raised a simple mathematical conundrum known as Cole's paradox: an organism that lives forever, producing n offspring per year, leaves no more descendants, all else being equal, than one that lives for just one year (an annual organism), but produces $n+1$ offspring. The reason is that only one offspring is required to replace the parent in our annual organism, after which both the immortal and annual organism produce the same number of offspring. Given the impossibility of immortality, why do not all organisms evolve toward the shortest, simplest life history? Many biologists, including William D. Hamilton, George Williams, and Eric Charnov (Charnov and Schaffer 1973), have tackled this conundrum, and its solution was seminal to the formalization of Darwinian thinking in modern ecology. The solution to Cole's paradox requires recognition that each life-history decision causes a shift in the probability of accruing certain benefits, and certain costs, in terms of survival and reproduction. For example, if reproductive success increases with age (e.g., due to benefits of parental experience) and the risk of mortality between reproductive periods is low, then selection may favor reproduction for multiple seasons. In contrast, when the risk of mortality is high, selection may favor allocating more resources to current reproduction.

Allocation trade-offs such as these constrain how life histories evolve. Resources, including time, energy, and essential nutrients, must be divided among growth, maintenance, and reproduction. Resources allocated to one function cannot be allocated to other functions. Understanding these trade-offs, and how behavioral or developmental decisions at one stage of the life cycle influence survival and reproduction at other stages (and total fitness), is the central question in the study of life-history evolution. The Darwinian perspectives of Lack and Cole brought a new way of thinking into the study of life histories that has led to myriad insights and a field of inquiry that is very active today.

Modern life-history analyses aim to understand how evolutionary history, the environment, development, and inheritance interact to create and constrain life-history variation. An important recent insight is that life histories can evolve rapidly when environments change (e.g., the experimental introduction of predators, commercial exploitation by humans, and expansion onto new host species by parasites of plants and animals) (Carroll et al. 2007). Selective predation or harvesting of large individuals, as in fisheries, has led to the evolution of substantially reduced age and size at maturity over tens of generations (Hutchings and Fraser 2008); this happens because selection favors reproduction before animals are eaten or harvested. Thus, while shared evolutionary history tends to create similarities among related organisms, genetic variation for life-history traits is abundant and allows the rapid evolution of life-history strategies. Such ongoing adaptation allows evolutionary biologists to weigh the relative influences of phylogeny, genetic variation, developmental systems, and the environment in shaping life histories, so that cause-and-effect relationships become clearer.

The Darwinian perspective on life histories can bring insights to medicine and other fields of biology, including public health concerns such as senescence and life span in aging human societies. The evolution of life span is influenced by an organism's reproductive value, the measure of future expected reproduction. In most organisms once breeding starts reproductive value declines with age. Thus mortality late in life affects total reproduction much less than does mortality early in life, so traits affecting mortality late in life are under weaker selection than are traits affecting mortality early in life. It is this decline in the intensity of selection with age that allows the evolution of senescence. Coupled with the selective advantage of short generation times, traits that favor successful reproduction early in life will be favored even at a cost to ultimate longevity (an example of antagonistic pleiotropy); indeed, genes have been identified that have these effects. Traits with negative effects only on the old can evolve in populations through genetic drift (i.e., the accumulation of deleterious mutations).

More powerful mathematical tools, new technologies (e.g., genomics), and more insightful experimentation are permitting biologists to better understand life-history variation within and among species and the role of such variation in generating new species. The resulting insights are increasingly being translated into the applied sciences where biologists are studying how life-history evolution influences the conservation of biotic resources, the yield and the sustainability of agriculture, and how the evolution of ontogeny, reproduction, and disease relates to human health.

BIBLIOGRAPHY

Carroll, S. P., A. P. Hendry, D. N. Reznick, and C. W. Fox. 2007. Evolution on ecological time-scales. *Functional Ecology* 21: 387–393.

Charnov, E. L., and W. M. Schaffer. 1973. Life-history consequences of natural selection: Cole's result revisited. *American Naturalist* 107: 791–793.

Cole, L. C. 1954. The population consequences of life history phenomena. *Quarterly Review of Biology* 29: 103–137.

Godfray, H. C. J., L. Partridge, and P. H. Harvey. 1991. Clutch size. *Annual Review of Ecology and Systematics* 22: 409–429.

Hutchings, J. A., and D. J. Fraser. 2008. The nature of fisheries- and farming-induced evolution. *Molecular Ecology* 17: 294–313.

Lack, D. 1947. The significance of clutch size. *Ibis* 89: 309–352.

Orzack, S. H., and E. Sober. 1994. Optimality models and the test of adaptation. *American Naturalist* 143: 361–380.

Roff, D. A. 1992. *The Evolution of Life Histories: Theory and Analysis.* New York: Chapman and Hall.

Smith, C. C., and S. D. Fretwell. 1974. The optimal balance between the size and number of offspring. *American Naturalist* 108: 499–506.

VanderWerf, E. 1992. Lack's clutch size hypothesis: An examination of the evidence using meta-analysis. *Ecology* 73: 1699–1705.   —*S.P.C. and C.W.F.*

Evolution of sex

> Here, you see, it takes all the running you can do, to keep in the same place.
>
> —The Red Queen, in Lewis Carroll's
> *Through the Looking Glass* (1872)

Of the many meanings of the word *sex*, the most fundamental refers to the production of new genomes by the recombination of preexisting genomes. Many bacteria and a few viruses are capable of occasional, limited sex, but the most thoroughly sexual creatures are the eukaryotes, which include animals, plants, and other organisms with nuclei in their cells. Humans are fairly typical eukaryotes in being obligately sexual: the only way to make a new human is to split your genome in half and to get another half from some other human who has done the same. Despite the central role that sex plays in organismal life cycles, biologists do not understand what its function is.

The first biologist to suggest a plausible hypothesis for the advantage of sexuality was August Weismann (1889). He proposed that sex increased the variability within a population and thus increased the effectiveness of natural selection. This is an unusual sort of advantage; typically, natural selection promotes traits that confer an advantage to the individual that bears the trait. Weismann, however, saw sex as a feature of a population that conferred an advantage to that population. Weismann's explanation was influential and went largely unchallenged until the 1960s, when George C. Williams objected to its reliance on group selection and insisted that sex should have a short-term, individual-level advantage like any other trait. On the contrary, sex seemed to have an enormous short-term, individual-level disadvantage. A clonal female transmits all of her genes to all of her offspring. A sexual female transmits only half as many genes to the next generation, and wastes half of her parental investment propagating the genes of some unrelated male. Thus sex carries an enormous (twofold) cost. The contrast between this huge theoretical disadvantage of sex and its prevalence in the real world has been called "the paradox of sex." In the 1970s, recognition of this paradox spurred several leading evolutionary theorists to write hand-wringing books highlighting the importance and difficulty of this "queen of problems in evolutionary biology" (Bell 1982). Though it was easy to imagine possible benefits of sexuality, it was hard to see how sexual individuals could be more than twice as fit, on average, as asexual individuals, which is what seems to be required to overcome the twofold cost of sex. Theorists found that sexual individuals can have this magnitude of advantage over asexual individuals only in a world that changes radically from one generation to the next. Specifically, sex can outcompete asexuality only if there is some phenomenon that (1) kills a large proportion of each population, or has an equivalently dire effect on reproduction; (2) is sensitive to variation between genomes, killing individuals with certain genotypes and sparing those with other genotypes; and (3) changes radically from one generation to the next in terms of which genotypes are killed and which are spared. Further, to provide a general explanation for sex, the phenomenon would have to be ubiquitous in the living world, affecting most multicellular organisms. What could this phenomenon be? In recent decades, two major candidates have been recognized: infectious disease (Hamilton 2001) and harmful mutations (Kondrashov 2001).

The idea that infectious diseases, caused by parasites and pathogens, could make sex advantageous is often known as the Red Queen hypothesis. Of all the aspects of an organism's environment, its parasites and pathogens best fit the theorists' description of the phenomenon that gives an advantage to sex. Because parasites and pathogens have large populations and reproduce quickly compared to their hosts, they can evolve to attack particular host genotypes. One striking prediction of the hypothesis is that strong fluctuations should occur in the frequencies of alleles affecting immunity; alleles that are common in one generation should become rare in the next. The Red Queen hypothesis has been successful in explaining the evolutionary dynamics of hosts and parasites in a few well-studied species of snails and water fleas, but scientists do not yet know how broadly applicable it will turn out to be.

The other phenomenon with the potential to make sex necessary is harmful mutation. With high mutation rates, sex can be beneficial because some recombinant offspring will carry fewer harmful alleles than either parent. This is only true if rates of harmful mutation are remarkably high, with every individual carrying on average more than one new harmful mutation. To test the mutation theory directly, scientists need to estimate how many new harmful mutations appear in each individual. This is a difficult quantity to measure, but estimates are now available for a few organisms. Humans seem to have a high enough rate of harmful mutation to make sex useful, but the rate seems to be too low in worms and flies. An important detail is that sex only helps to increase average fitness if harmful alleles interact synergistically, that is, if they become proportionately more harmful as more of them accumulate in a genome. A weakness of the mutation theory is that it does not seem to account for ecological patterns in the distribution of asexual lineages as well as the Red Queen hypothesis does.

When a genome replicates, it tends to deteriorate. Mutations occur and they tend to be harmful. The environment also tends to get worse from year to year: your enemies get better at attacking you, especially parasites and pathogens with short generation times that evolve rapidly to adapt to your genotype. These two processes tend to make maintaining a static genome both difficult (due to mutation) as well as a bad idea (due to parasites). Sex seems to be an adaptation to deal with one or another of these problems, or perhaps to deal with both of them at once. There has been a recent trend toward combining the Red Queen and mutational models and recognizing that sex has multiple advantages. Even George Williams, whose quest for an individual-level advantage of sex launched the modern study of its evolution, has conceded that its species-level advantages might be crucial after all (Williams 1992). Asexual species seem to be much more prone to extinction than sexual species. Hence, most surviving species are sexual, and many, including mammals, seem to have features that prevent them from ever becoming asexual. Although the problem of sex has recently been reduced from a mystery to a mere controversy, scientists still need a good deal more information to understand exactly how sex helps individual organisms to have more successful offspring and how it prevents populations from going extinct.

BIBLIOGRAPHY

Bell, G. 1982. *The Masterpiece of Nature*. Berkeley: University of California Press.

Hamilton, W. D. 2001. *Narrow Roads of Gene Land*. Vol. 2, *The Evolution of Sex*. Oxford: Oxford University Press.

Judson, O. 2002. *Dr. Tatiana's Sex Advice to All Creation*. New York: Metropolitan Books.

Kondrashov, A. S. 2001. Sex and U. *Trends in Genetics* 17: 75–77.

Ridley, M. 1994. *The Red Queen: Sex and the Evolution of Human Nature*. New York: Macmillan.

Weismann, A. 1889. *Essays upon Heredity and Kindred Biological Problems*. Oxford: Clarendon Press.

Williams, G. C. 1992. *Natural Selection: Domains, Levels, and Challenges*. Oxford: Oxford University Press.   —B.B.N.

Ewens, Warren J. (b. 1937)

Although they may seem esoteric at first glance, many of the statistical aspects of population genetics, such as the precise distribution of fitness among alternative forms of a gene, play vital roles in determining how genetic changes unfold in evolution. Few scientists have uncovered these vital roles more thoroughly or carefully than Warren J. Ewens, whose work is the foundation of several areas of research in evolutionary population genetics.

Trained in statistical genetics at the Australian National University, Ewens received his doctoral degree in 1964 under the supervision of P. A. P. Moran, who was a student of Sir Ronald Fisher. Ewens is known primarily for his work on the sampling distribution of neutral alleles, which are alleles that have no effects on the fitness of their carriers. The neutral allele theory developed by Motoo Kimura and Tomoko Ohta (see the alphabetical entries "Motoo Kimura" and "Tomoko Ohta" in this volume) wedded the study of biochemical and molecular polymorphisms to patterns of variation among species. One of the critical tests of this theory compared the observed statistical distribution of allele numbers and frequencies at a single gene to the distributions expected under neutrality. In a series of papers in the early 1970s, Ewens derived these expected distributions under a variety of assumptions about the nature of mutations. The Ewens Sampling Distribution, as it came to be called, not only allowed the neutral theory to be evaluated against empirical data, but also provided the foundation on which several additional theoretical advances were built. These included various refinements of the neutral theory and the development of methods for using the distribution of alleles among a set of neighboring populations to estimate rates of migration and gene flow.

Three of Ewens's other contributions stand out. First, he showed how interpreting the genetic load, the difference between the fitness of the best allele available in a population and the average fitness of all other alleles, depended on the precise statistical distribution of fitness. The concept of the genetic load lays beneath several controversies in evolutionary genetics, from the so-called Haldane's dilemma (how can an evolving population bear the cost of so many unfit genotypes) to the likelihood that molecular variation is indeed neutral. Ewens's critical study of the genetic load changed the way scientists thought about these issues. Second, Ewens derived precise expressions for the effective population size, which is a measure of how many individuals are really contributing to the next generation. This work clarified several areas of confusion, especially surrounding the effect of catastrophic decreases in population size, a question of interest in conservation biology. Third, he and his colleagues derived new methods for detecting associations between genes and human diseases, methods that solve long-standing statistical problems based on the nonrandom samples that emerge from medical records.

Ewens has a legendary reputation as a superb lecturer and classroom instructor. He has inspired generations of students throughout his career in Australia, primarily at LaTrobe University, and in the United States at the University of Pennsylvania. He received the E. J. Pitman Medal from the Statistical Society of Australia in 1996 and, in 1999, he was elected as a fellow of the Royal Society.

BIBLIOGRAPHY

Ewens, W. J. 2004. *Mathematical Population Genetics.* 2nd ed. New York: Springer.
Ewens, W. J., and G. R. Grant. 2001. *Statistical Methods in Bioinformatics.* New
    York: Springer.   —*J.T.*

Exobiology

With the advent of space exploration, two fields of research—origin-of-life
studies and the search for extraterrestrial life—were melded into a single new
discipline. In 1959, Nobel Prize–winning microbiologist Joshua Lederberg
coined the term *exobiology* for this new interdisciplinary field, offering the
rationale that one needed to know the conditions under which life's origin
had been possible on earth in order to know what kind of chemical conditions to look for on other planets. Furthermore, he argued that sterilization
of spacecraft was crucial before they landed on other planets, lest contamination of those worlds by earth microbes or even organic molecules from earth
render it impossible forever to know with certainty what kind of native organics or organisms had been there originally. At first largely an American
science because the National Aeronautics and Space Administration (NASA)
was the single largest source of funding, work in these areas had the potential
to be politicized by the Cold War, as did all the science related to space exploration. The Soviet Union was highly secretive, for example, about what
sterilization standards, if any, had been applied to its interplanetary missions
targeted for impact or soft landing on the moon, Venus, and Mars. Similarly,
the United States was highly secretive about how its expertise in sterilization
was related to classified germ warfare research. Gradually, however, international conferences sponsored by the Committee for Space Research
(COSPAR) developed international spacecraft sterilization standards, eventually incorporated into the 1967 Outer Space Treaty. International origin-of-
life conferences and a wide range of NASA-sponsored meetings on the origin
of life, organic molecules in space, Precambrian paleofossils, and other topics
also gradually consolidated an international community of scientists working
on pieces of the puzzle as disparate as biochemistry, geochemistry, planetary
geology, and radio astronomy. In 1968 the first exobiology journal began
publication (now titled *Origins of Life and Evolution of the Biosphere*). In
1972 exobiologists founded their first professional society, the International
Society for Study of the Origins of Life (ISSOL). Only with the technology of
space exploration, these scientists foresaw, would it become possible to develop a truly universal science of biology—to understand (or test) what universal laws existed for biochemistry and evolutionary history.

From the beginning, exobiology pursued academic laboratory research as
well as the development of specific instruments to fly on spacecraft, such as
the two U.S. Viking Mars landers. In 1976 these craft found no clear evidence
for life at the Martian surface but provided enough data to suggest that if life
did exist on Mars, it would probably only be found deep enough below the
surface to be shielded from solar ultraviolet rays. In addition, the development

"THERE ARE CERTAIN FEATURES IN WHICH THEY ARE LIKELY TO RESEMBLE US. AND
AS LIKELY AS NOT THEY WILL BE COVERED WITH FEATHERS OR FUR. IT IS
NO LESS REASONABLE TO SUPPOSE, INSTEAD OF A HAND, A
GROUP OF TENTACLES OR PROBOSCIS LIKE ORGANS"

Beings from outer space have long been the stuff of fantasy and nightmare. This
drawing, by William R. Leigh, appeared in H. G. Wells, "The Things That Live
on Mars," *Cosmopolitan Magazine*, March 1908, 335–343. It appeared just
at the time when there was much popular discussion about whether the "canals"
on Mars were the products of real intelligences. Note how often such pictures
assume that evolution is progressive and noncontingent, with humanlike forms
the inevitable outcome of the process, anywhere and everywhere.

of radio astronomy techniques to search for signals from intelligent extrater-restrials (SETI) was part of exobiology from the outset, although Congress cut off NASA funding for SETI work in 1992. By the 1980s and 1990s, new astronomical techniques were developed to search for planets of stars beyond the solar system, with the first such planet being detected in 1995. Labora-tory work has pursued numerous goals, including study of the organic com-pounds found in meteorites, returned moon rocks, and (via spacecraft) comets, and study of ancient rocks for fossilized microbes and/or organic molecules that might be associated with once-living organisms. This work has led to pushing back to 3 billion years or more the oldest known fossil mi-crobes from earth, narrowing the time window for the process of chemical evolution to have occurred. However, it has simultaneously revealed that earth is being bombarded by organic molecules formed in huge amounts in interstellar space, so that the original precursors of life on earth need not all have been synthesized here from purely inorganic starting materials. Exobiol-ogy research has also spun off numerous other important discoveries: Lynn Margulis's theory, now universally accepted, that eukaryotic cells arose via formation of serial symbiotic colonies of prokaryotic cells; Carl Woese's dis-covery of the Archaea, a group of bacteria as different from other bacteria as they are from eukaryotes; James Lovelock's Gaia hypothesis, which claims that all life on earth acts as a coordinated homeostatic system maintaining conditions within the limits life can tolerate; the theory that a huge asteroid impact at the end of the Cretaceous period was responsible for the mass ex-tinction that included the dinosaurs; and the "nuclear winter" hypothesis, predicting that a similar mass extinction could follow a nuclear war of any significant size, and spurring on much of global climate modeling science. The broadening of the research agenda resulted in the exobiology field being renamed astrobiology in 1996.

Following the Soviet biochemist Aleksandr Oparin, origin-of-life research assumed in the 1950s that earth's primitive atmosphere must have been chemically reducing, that is, it contained no free oxygen, carbon mostly as methane, and nitrogen as ammonia. The 1953 Urey-Miller experiment showed that spontaneous synthesis of amino acids and other relevant bio-molecules could occur rapidly under such conditions, even assuming random chemistry. By the late 1970s, however, exobiology research suggested earth's early atmosphere was more likely to have been chemically neutral (or per-haps even slightly oxidizing), under which conditions the synthesis of organic molecules is drastically reduced. Alternative pathways of chemical evolution are under investigation, including synthesis of organic polymers using clays as templates and catalysts; high-temperature, high-pressure chemistry such as might occur at undersea hydrothermal vents, where many researchers now think life may have first originated; and combining meteorite-derived organ-ics to see what kinds of lifelike protocells will be produced. It is now widely believed that the chemistry that led from inorganic precursors to the first biopolymers occurs nonrandomly, more rapidly, and spontaneously under a wider range of conditions than was believed in the 1950s and 1960s. Biased in the direction of biologically relevant molecules, life probably originated fairly quickly but "neither by chance nor design," as science historian Iris Fry states (2000, 109). Research has also focused on the RNA world, a period early in life's history when RNA existed before proteins but was capable of acting as both an information-carrying molecule and a catalyst of metabolism.

BIBLIOGRAPHY

Dick, S., and J. E. Strick. 2004. *The Living Universe: NASA and the Development of Astrobiology.* New Brunswick, NJ: Rutgers University Press.

Fry, I. 2000. *The Emergence of Life on the Earth: A Historical and Scientific Overview.* New Brunswick, NJ: Rutgers University Press.

Graham, L. 1987. *Science, Philosophy and Human Behavior in the Soviet Union.* New York: Columbia University Press.

Hazen, R. 2005. *Genesis: The Scientific Quest for Life's Origins.* New York: Joseph Henry Press.

Lederberg, J. 1960. Exobiology: Approaches to life beyond the earth. *Science* 132: 393–400.

Lovelock, J. 1979. *Gaia: A New Look at Life on Earth.* London: Oxford University Press.

Margulis, L. 1970. *Origin of Eukaryotic Cells.* New Haven, CT: Yale University Press.

Miller, S. 1953. A production of amino acids under possible primitive earth conditions. *Science* 117: 528–529.

Oparin, A. 1938. *The Origin of Life.* New York: Macmillan.

Woese, C. 1987. Bacterial evolution. *Microbiological Reviews* 51: 221–271.

Wolfe, A. 2002. Germs in space: American life scientists, space policy, and public imagination, 1958–1964. *Isis* 93: 183–205. —*J.E.S.*

F

Felsenstein, Joseph (b. 1942)

Although all scientists aspire to change how their colleagues think about an issue, only a few manage to do so. A select few change their colleagues' thinking on more than one issue. Joseph Felsenstein has changed the thinking of evolutionary biologists on at least three critical issues.

A student of population geneticist Richard Lewontin, Felsenstein wrote his dissertation on methods for inferring phylogenies, receiving his doctoral degree in 1968 from the University of Chicago. After postdoctoral work at Edinburgh with Alan Robertson, he devoted the early years of his career to population genetics; his dissertation research did not appear in scientific journals until 1973. After this refractory period, he returned to the study of phylogenetic methods and has made seminal contributions to phylogenetics and population genetics, along the way promoting a synthesis of the two.

Felsenstein pioneered a statistical approach to phylogenetic inference (see the alphabetical entry "Phylogenetics" in this volume). He recognized that, when inferring a phylogeny, not only would the lengths of the branches be subject to sampling error, but so would the actual shape of the entire tree. Sampling error introduces uncertainty about the identity of close relatives and recent common ancestors, which in turn creates uncertainty about the order of appearance of major features in groups like birds or fish. Neither traditional methods nor the emerging cladistic methods confronted this problem satisfactorily. Felsenstein introduced the statistical technique of maximum likelihood to the study of phylogenies. The initial reluctance of evolutionary biologists to embrace the technique was overcome when Felsenstein showed that cladistic methods would point to the wrong phylogeny whenever different branches of an evolutionary tree experienced sufficiently different rates of character evolution. This result has become a textbook example, and maximum likelihood and related methods have become the standard approaches for understanding evolutionary relationships.

Biologists are now following Felsenstein's lead on two other topics that emerged from his synthesis of phylogenetics and population genetics. In the late 1970s, several evolutionary biologists pointed out that traditional comparisons of character associations among species could be misinterpreted because such comparisons did not recognize the phylogenetic relationships among the species being compared. Felsenstein developed a powerful method for integrating phylogenetic affinities into comparative analyses. His method, called independent contrasts, and its descendants have become standard tools for comparative biology. In the early 1990s, Felsenstein pointed out that populations of the same species are part of a complex pedigree of ancestor-descendant relationships and that, through this pedigree, one could estimate long-term rates of genetic exchange among populations and how genetically large the populations have been. Methods developed by Felsenstein and his colleagues have been applied to several problems in evolutionary biology, and some of these methods have become important tools in conservation biology.

In addition to these accomplishments, Felsenstein wrote a series of papers that explored the evolution of genetic linkage, which is the statistical association among alleles at different genes. He examined the circumstances under which tighter and looser linkage (stronger and weaker statistical associations) evolve and explored how close linkage affects a variety of evolutionary processes. Linkage among genes is a vital element in explanations for the evolution of sexual reproduction (see the alphabetical entry "Evolution of sex" in this volume) and it plays a crucial role in speciation (see the main essay "The Pattern and Process of Speciation" by Margaret B. Ptacek and Shala J. Hankinson in this volume).

In 1992 Felsenstein was elected to the American Academy of Arts and Sciences. In 1993 he was the second recipient of the Sewall Wright Award of the American Society of Naturalists, and in 1999 he was elected to the U.S. National Academy of Sciences.

BIBLIOGRAPHY

Felsenstein, J. 2004. *Inferring Phylogenies*. Sunderland, MA: Sinauer Associates.
Slatkin, M. 1995. The Sewall Wright Award: Joseph Felsenstein. *American Naturalist* 145: ii–v.

—*J.T.*

Fisher, Ronald Aylmer (1890–1962)

Sir Ronald Aylmer Fisher was born February 17, 1890, in London, England. Fisher and his twin brother, who died in infancy, were the youngest of eight children born to Katie Heath and George Fisher. Showing his mathematical gifts early, Fisher would change the face of both statistics and biology. Because of extreme myopia, his early training was unconventional. Indeed, some of his most influential early tutoring was done without pencil or paper, giving Fisher a remarkable ability to do complex mathematics in his head.

Fisher matriculated at Gonville and Caius College, Cambridge, in 1909. He graduated in 1912 with high honors in mathematics, and he would stay on for another year to study physics and statistical mechanics. Fisher's interests were broad, including astronomy, mathematics, physics, and biology. In 1914 he became a statistician at the Mercantile and General Investment Company in London. Contrary to his desires, Fisher was kept out of World War I because of his poor vision. Between 1915 and 1919 he taught mathematics and physics at a variety of public schools. In 1917 he married Ruth Eileen Gatton Guinness, with whom he would parent eight children. One of Fisher's six daughters, Joan Fisher Box, authored the most well-known biography of him in 1978.

At Rothamsted Experimental Station, between 1919 and 1933, Fisher made substantial contributions to statistics and genetics. In statistics, he introduced the concept of likelihood (Fisher 1921). The likelihood of a parameter is proportional to the probability of the data and it gives a function that usually has a single maximum value, that is, the maximum likelihood. And the next year Fisher introduced a new conception of statistics, the aim of which was the reduction of data and the problematic of which was the specification of the kind of population from which the data came, estimation, and distribution. Soon thereafter, Fisher published *Statistical Methods for Research Workers* (1925), in which he articulated methods for the design and evaluation of experiments. The 1925 edition was the first of many. In genetics, Fisher's first paper (in 1922) adumbrated a mathematical synthesis of Darwinian natural selection with the recently rediscovered laws of Mendelian heredity. That, and other work (e.g., on the evolution of dominance), would culminate in *The Genetical Theory of Natural Selection* (1930), one of the principal texts, along with those of J. B. S. Haldane (1892–1964) and Sewall Wright (1889–1988), completing the reconciliation of Darwinism and Mendelism and establishing the field of theoretical population genetics and its application to eugenics. Remarkably, *The Genetical Theory of Natural Selection* was dictated by Fisher to his wife, Ruth, during the evenings. The book was revised and reissued in 1958. In 1929 Fisher was elected to the Royal Society.

Fisher left Rothamsted to take the Galton Chair of Eugenics at University College, London, in 1933. Karl Pearson (1857–1936) was retiring from the Chair, which was then divided into two, so that both Fisher and Pearson's son, Egon, could be appointed. Fisher controlled the genetics section; Pearson controlled the statistics section. (Fisher had turned down Pearson's offer of chief statistician at Galton Laboratories for the position at Rothamsted in 1919.) When he took the Galton professorship at University College, he was thoroughly mired in controversy over the foundations of statistics with Karl Pearson and his followers. Fisher's revolutionary work in statistics came at around the time that Pearson's own work was showing weaknesses; unfriendly competition catalyzed the controversy, and Pearson would take ill feelings toward Fisher, who had his own ill feelings toward Pearson, to his grave. The controversy in statistics with Pearson would not be the only controversy in which Fisher was involved. Beginning in 1929, Fisher and American physiological geneticist Sewall Wright would engage in an animus-filled controversy over the genetic basis of evolutionary change that would last until Fisher's death in 1962.

In 1935 Fisher's *The Design of Experiments* appeared and, like *Statistical Methods for Research Workers,* would be expanded and reissued many times. These two works, and Fisher's *Statistical Tables for Biological, Agri-*

cultural, and Medical Research (with Frank Yates, 1938) revolutionized agricultural research. Fisher's work in statistics was revolutionary at the field's conceptual foundations. Moreover, Fisher's work in genetics, high-lighted mainly by *The Genetical Theory of Natural Selection,* would, with good company in Haldane and Wright, revolutionize biology. Fisher's ac-complishments did not go unrecognized. He was awarded the medal of the Royal Society in 1938, the Darwin medal in 1948, and the Copley medal in 1955. In 1943 Fisher was appointed Balfour Professor of Genetics at Cam-bridge University; in 1957 he became president of Gonville and Caius Col-lege, Cambridge; and in 1952 he was created Knight Bachelor by Queen Elizabeth II. Numerous other honors were bestowed upon him.

After a luminous career as statistician and biologist, Fisher died at the age of 72 on July 29, 1962, in Adelaide, Australia, where he had retired in 1959. Fisher has been described as charming and warm among friends, but with a volatile temper. Fisher's temper, combined with an unwavering commitment to his views, drove him to heated controversies with other scientists. Fisher's writings have been described as difficult; much of what he contributed was more effectively conveyed by others more capable of simplifying his presenta-tion and his genius.

BIBLIOGRAPHY

Box, J. F. 1978. *R. A. Fisher: The Life of a Scientist.* New York: Wiley.
Fisher, R. A. 1921. On the "probable error" of a coefficient of correlation deduced from a small sample. *Merton* 1: 3–32.
———. 1922. On the dominance ratio. *Proceedings of the Royal Society of Edinburgh* 42: 321–341.
———. 1925. *Statistical Methods for Research Workers.* London: Oliver and Boyd.
———. 1930 [1958]. *The Genetical Theory of Natural Selection.* 2nd ed. New York: Dover Publications.
———. 1935. *The Design of Experiments.* London: Oliver and Boyd.
Fisher, R. A., and F. Yates. 1938. *Statistical Tables for Biological, Agricultural, and Medical Research.* London: Oliver and Boyd.   —R.A.S.

Fishes

The word *fishes* is about as slippery as the organisms to which it refers are to hold. Most people generally get a rather similar mental image when the term *fish* is used, but this rarely includes thinking of eels or even sharks, although these too are fishes. Defining a fish as "a water-dwelling, backboned animal moving by either (or both) lateral body undulations or oscillations of paired and/or unpaired fan-like body appendages called fins and which obtains oxy-gen (and releases numerous dissolved wastes) via specialized structures called gills" goes a long way toward including all organisms that are scientifically considered fishes (Helfman et al. 1997; Pough et al. 2005). So diverse are the living types of fishes, however, that this definition is not clear cut. Inclusion of known fossil forms further stretches beyond accuracy almost any attempt at a comprehensive definition. Two reasons account for the near impossibility of clearly defining what seems at first glance to be a straightforward noun. First, a number of distinct evolutionary lineages, albeit evidentially derived from a common ancestor, survive today having enormously diverged in almost all respects through the influence of natural selection. Second, fishes (the plural applies to more than one species, the singular to individuals of a single species no matter how many of them) are the most biodiverse of living backboned animals (vertebrates). Extant (living) fishes number no less than 27,000 named species, comprising about 50% of all extant vertebrates; they are adapted to inhabit the nearly 71% of the earth's surface covered by oceans and seas, as well as the additional 1 to 2% of the planet's surface occupied by lakes, rivers, and other freshwater and estuarine environments. Fishes are found everywhere there is surface water, and often in subterranean waters as well; they vary in morphological, physiological, and behavioral characteristics from evolutionary adaptations to the widely variable conditions of the earth's waters.

Four very distinctive fish lineages (clades), plus another usually, if reluctantly, considered "fishlike," are extant today. A dozen, perhaps considerably more, equally distinctive lineages are known only from fossils, some of which extend back at least 480 million years to bony fragments from the Ordovician, and possibly even from impressions in rocks from the early Cambrian, another 40 million years earlier. The living lineages are the fishlike lancets (Cephalochordates); the near vertebrate but backboneless hagfishes (Myxiniforms), with 50 or more living species; the jawless and truly vertebrate lampreys (Petromyzontiforms), with about 40 living species; the cartilaginous sharks, skates, and rays (Chondrichthyes), with at least 880 living species; and the overwhelming majority of extant species, the bony fishes (Osteichthyes), with varying estimates of the number of living species somewhere around 27,000. The living hagfishes and lampreys share a common primitive trait with many fossil lineages in that they lack jaws and paired fins. They feed by pinching or grasping with horny teeth located around their oral opening. They use unpaired ribbonlike fins in the midlines of their eel-like scaleless bodies to control movement. The Chondrichthyes and Osteichthyes not only have well-developed jaws, usually rimmed by harder-than-bone mineralized teeth, but also paired fins that permit enormous three-dimensional control of movement. Detailed study of fossils of both groups shows similar trends through time in the adaptations of the two clades to the natural selection demands of water as a medium in which to move, to feed, and to avoid predation. Fins become more flexible and able to assume varying angles of attack relative to the flow of dense viscous water over the fishes' bodies. Jaws and associated structures become loosened from the rest of the head skeleton so that they may protrude toward food items, and the oral-branchial cavity becomes increasingly flexible so that it may be expanded to create suction that draws water, and the food it contains, into the mouth. As fins and hydrodynamic body shape provide for increased agility, heavy external armor of scales, spines, and head bones is thinned or lost altogether, making near neutral buoyancy relatively easier to achieve. Although demonstrating similar responses of the vertebrate body to the natural selection demands of living in water, the separate evolutionary clades of the Chondrichthyes and Osteichthyes have often used very different mechanisms to achieve a common adaptation. For example, to overcome the effect of gravity on their denserthan-water bodies, chondrichthyes have evolved large, oil-rich livers that provide buoyancy. Osteichthyes have modified an auxiliary respiratory outpocketing of the esophagus into a gas-filled buoyancy bladder. Such differences demonstrate the random nature of adaptation based on chance genome variation.

The extant Chondrichthyes are flesh eaters and generally large organisms. A typical shark is about 2 meters long. Evidence from analysis of the shark genome indicates that they have evolved at a much slower rate than have advanced Osteichthyes. This may account for the relatively small number of species in existence today. Living Osteichthyes are highly variable in diet and size, but they are generally smaller than sharks, with some being less than a centimeter in adult maximum length. Their dietary breadth and small size allow them to thrive in many more habitats than any other clade of fishes, especially in freshwater environments. Forty-one percent of all fish species live in freshwater habitats, which are highly divided and isolated one from another, while 58% of living species occur in the sea, which (in spite of its enormous extent) is a more continuous habitat than the epicontinental freshwaters. Indeed some freshwater bony fishes show astoundingly high rates of evolution. The cichlids of South America, Africa, and Southern India are especially speciose, with literally hundreds of species having evolved in situ in each of the several geologically young African Rift Valley lakes. Because they have extensive parental care of eggs and young, cichlid offspring can become behaviorally imprinted on their parents and show a preference for individuals similar to their parents in choosing their own mates (assortative mating), resulting in rapid genetic isolation and selection leading to new species.

The random pattern of evolution without direction or clear selective values is also abundantly demonstrated by fishes. Almost all Osteichthyes belong to the vast group known as Teleosts, including eels, herrings, salmons, minnows, catfishes, lantern and other deep-sea fishes, cods, anglerfishes, basses, perches, almost all coral reef fishes, and so on. There are 25,000 or more species. Teleosts first appeared about 200 million years ago in the mid-Triassic. Their unique uniting feature is a structurally rather minor shift in the internal support of a portion of the tail (caudal) fin that does not appear to give them any clear adaptive advantage over other contemporary fishes. Nevertheless, this one lineage has been the genesis of most fishes living today, illustrating the random and chance nature of evolution.

BIBLIOGRAPHY

FishBase: A Global Information System on Fishes. http://www.fishbase.org/home .htm.

Helfman, G. S., B. B. Collette, and D. E. Facey. 1997. *The Diversity of Fishes.* Malden, MA: Blackwell Science.

Maisey, J. G. 1996. *Discovering Fossil Fishes.* New York: Henry Holt and Company.

Pough, F. H., C. M. Janis, and J. B. Heiser. 2005. *Vertebrate Life.* 7th ed. Parts I and II, 1–156. Upper Saddle River, NJ: Prentice Hall.   —*J.B.H.*

Fitch, Walter M. (b. 1929)

Philosopher Isaiah Berlin (1953) characterized intellectual endeavor as being conducted by either hedgehogs or foxes. Hedgehogs spend their lives focusing intently on a single coherent issue, while foxes consider numerous, at times even unrelated, problems. For the most part, this dichotomy holds up well for scientific inquiry, but there are notable exceptions who upset the rule. Walter M. Fitch, who managed to make lasting contributions to the field of molecular evolution, is one such exception. In particular, Fitch's career has been distinguished by groundbreaking advances in methodology, the kind of effort that requires the digging ability of a hedgehog, as well as fundamental conceptual advances across a range of topics, achievements that necessitate the mental agility of a fox.

Trained as a biochemist, Fitch has spent the better part of his career at the University of Wisconsin and later at the University of California at Irvine. When he began his comparative research on protein sequences, the notion of molecular evolution was little more than a fantasy; he is as much responsible as anyone for the emergence and development of this nascent research program into a major, legitimate, and distinct discipline at the nexus of evolutionary and molecular biology. Together with Masatoshi Nei, another luminary of molecular evolution, Fitch founded the Society for Molecular Biology and Evolution and its journal, *Molecular Biology and Evolution*, which remains the premier journal in the field and is one of the highest-ranking journals in the evolutionary biology field. The contributions that Fitch has made are numerous, so a mere sample of what Fitch has given to molecular evolution is presented by focusing on his work in three distinct areas: methods, concepts, and applications.

A crowning achievement of molecular evolution is the use of phylogenetic methods to reconstruct the evolutionary histories of genes, proteins, and, by extension, species. When related protein sequences began to accumulate in the late 1950s and early 1960s, there were no methods available to analyze them from an evolutionary standpoint. Numerical taxonomy methods had only just been formalized into algorithms, but these approaches could not be readily applied to sequence data. Working with Emanuel Margoliash, Fitch published a seminal paper in 1967 that reported the first distance-based method for phylogenetic reconstruction of protein sequences. Distances were computed as the proportion of amino acid differences between all pairs of sequences under consideration. The Fitch-Margoliash method amounted to a clever implementation of the least squares statistical principle, whereby a phylogeny was reconstructed such that the distances along the branches of the tree showed the least difference with the pairwise protein sequence distances. A critical feature of the method was a weighting scheme that allowed for the fact that greater distances were likely to be relatively less reliable. While the algorithmic implementation of the Fitch-Margoliash method has faded from use, distance methods are currently the most widely used approach for reconstructing phylogenies, and the general criterion of weighted least squares has remained viable. The utility and reliability of the Fitch-Margoliash method is underscored by the eukaryotic phylogeny reported in their paper, based on cytochrome c sequences, which is remarkably accurate given what we know now.

One of the most important concepts in gene evolution is the distinction between so-called orthologs and paralogs. Fitch coined these terms in 1970 and used them to distinguish between genes that share a common ancestor due to speciation events (orthologs) and genes that diverged via gene duplication (paralogs). This distinction is relevant not only for basic molecular evolution studies but also for the understanding of protein function. Indeed, by the late 1990s, the ortholog-paralog dichotomy became an indispensable conceptual tool for investigators trying to annotate the products of completely sequenced genomes. This is because orthologs tend to encode proteins with the same, or very similar, functions, while paralogs often encode proteins that have diverged in function. Information transfer based only on naïve sequence comparisons can obscure these facts and lead to misannotations based on the conflation of orthologous and paralogous sequences. More nuanced sequence comparison approaches that embrace Fitch's articulation of the evolutionary process by discriminating between orthologous and paralogous relationships (2000) allow for more reliable and useful annotations.

Evolutionary biology is an explicitly historical science, and molecular evolutionists are presented with the unique challenge of inferring the nature of past events based solely on contemporaneous data, namely gene and protein sequences of extant organisms. Despite the lack of access to historical data, such as the fossils utilized by paleontologists or the ancient texts deciphered by historians, molecular evolutionists have successfully developed an arsenal of tools that provide a window to the past. Evolutionary science is far less adept, however, about making predictions concerning the future course of events. In an audacious series of studies, Fitch and his colleagues attempted to defy this trend by actually predicting future evolutionary events. Their effort did not represent a mere academic exercise. The object of study was the influenza virus, and the idea was to try and predict the course of future flu epidemics. Such a prediction would provide a basis for selecting among influenza strains to use in the production of a vaccine. To do this, Fitch relied on a phylogenetic approach comparing the relationship among flu strain genomic sequences to the dates when they were collected. A clear trend manifested itself with the influenza phylogeny showing a successful trunk lineage that represented the evolutionary history of the influenza strains that dominated epidemics from one year to the next. Focusing on individual sites in the sequence, Fitch and his colleagues (Fitch et al. 1997; Bush et al. 1999) were able to identify specific variable positions in the influenza sequence that could be used to help predict those strains that will cause future epidemics. This bold departure from the elucidation of past events, the status quo for almost all of evolutionary biology, to the prediction of future evolutionary trajectories stands as a testament to forward thinking and prescience that has marked Walter Fitch as a truly exceptional molecular evolutionist.

BIBLIOGRAPHY

Berlin, I. 1953. *The Hedgehog and the Fox*. London: Weidenfeld and Nicolson.
Bush, R. M., C. A. Bender, K. Subbarao, N. J. Cox, and W. M. Fitch. 1999. Predicting the evolution of human influenza A. *Science* 286: 1921–1925.
Fitch, W. M. 1970. Distinguishing homologous from analogous proteins. *Systematic Zoology* 19: 99–113.
———. 1971a. Rate of change of concomitantly variable codons. *Journal of Molecular Evolution* 1: 84–96.
———. 1971b. Toward defining the course of evolution: Minimum change for a specific tree topology. *Systematic Zoology* 20: 406–416.
———. 2000. Homology: A personal view on some of the problems. *Trends in Genetics* 16: 227–231.
Fitch, W. M., R. M. Bush, C. A. Bender, and N. J. Cox. 1997. Long term trends in the evolution of H(3) HA1 human influenza type A. *Proceedings of the National Academy of Sciences USA* 94: 7712–7718.
Fitch, W. M., and E. Margoliash. 1967. Construction of phylogenetic trees. *Science* 155: 279–284.
Fitch, W. M., and E. Markowitz. 1970. An improved method for determining codon variability in a gene and its application to the rate of fixation of mutations in evolution. *Biochemical Genetics* 4: 579–593.   —I.K.J.

Ford, Edmund Brisco (1901–1988)

E. B. Ford, usually called Henry by his colleagues and friends, is best known as the chief proponent of the research program known as ecological genetics. Intended to throw light on the role of genetic diversity in the adaptation of organisms to their environments, ecological genetics begins with observations of variation in natural populations. Breeding studies then determine the manner of inheritance of the different variations. Comparative analyses of different habitats and correlations of environmental and genetic variables lead to hypotheses about the role of variation in evolutionary adaptation. Finally, experimental manipulation of environmental variables results in the confirmation or rejection of the hypotheses. Ford's research philosophy is set out in his seminal book, *Ecological Genetics*.

Ford spent his entire scientific career at Oxford University. He matriculated as an undergraduate in 1920, receiving a BA in zoology in 1924 and a BSc in 1927. He was appointed Departmental Demonstrator in Zoology (1927), then University Demonstrator (1929) and Lecturer at University College (1933). In 1939, he became Reader in Genetics, the first explicit appointment in that subject in Oxford. He established the Genetic Laboratories in 1951 and received an appointment as Professor of Ecological Genetics in

1963, a post he held until his retirement in 1969. Ford was elected a Fellow of All Souls College in 1958. He was a Fellow of the Royal Society and received its Darwin Medal in 1954.

Ford came to the study of genetics and evolution through his lifelong interest in the Lepidoptera. Before attending Oxford, he began a study with his father of a population of *Melitaea (Euphydras) aurinia,* the Marsh Fritillary. Their personal observations embraced the years from 1917 to 1935, and locality records going back to 1881 make this an unusually lengthy study. In the paper published with his father in 1930, Ford attributed the unusual variability of the population during periods of abundance to the relaxation of selection, which allowed suboptimal phenotypes to survive. Throughout his life he maintained his interest in moths and butterflies, publishing two volumes on these animals in Collins's New Naturalist series. His *Butterflies* was a bestseller for the series.

Ford's earliest work at Oxford was a collaborative study with Julian Huxley of a gene that affected the rate of darkening of the eye color in *Gammarus chevreuxi.* They showed that the rate depends not only on the Mendelian gene, but also on the size of the eye, the temperature, and the rate of bodily growth. Although not the first to attribute evolutionary importance to rate genes, the *Gammarus* studies provided an important early example.

Ford is perhaps best known for his work on genetic polymorphism. He was the first to offer a critical definition of the phenomenon in a paper prepared for Julian Huxley's *The New Systematics* (1940), refined as follows in Ford's *Ecological Genetics* (1964, 84). "Genetic polymorphism is the occurrence together in the same locality of two or more discontinuous forms of a species in such proportions that the rarest of them cannot be maintained by recurrent mutation." He distinguished between transient polymorphism, in which one morph is in the process of replacing another, and stable or balanced polymorphism, maintained by stabilizing selection. He noted that genetic polymorphism often involved large phenotypic effects, close linkage among loci contributing to those effects, and strong selection, leading to what he called "supergenes." He interpreted the ABO blood groups, Rhesus phenotypes, Batesian mimicry in butterflies, and heterostyly in plants as examples of supergenes, a position borne out by subsequent research.

Although not given to mathematical modeling, Ford collaborated with R. A. Fisher in the 1947 study of a color polymorphism in *Panaxia dominula,* the Scarlet Tiger moth. Since heterozygotes at this locus are distinguishable, Ford was able to track gene frequency and population size in a long series of yearly censuses. Fisher's analysis established that the changes in gene frequency were too great to be brought about by genetic drift alone, a point of contention at that time. Although challenged by Sewall Wright on the basis of effective population sizes, subsequent work on this colony has established that the original analysis was correct.

Ford was also interested in the effects of polygenes on variation in natural populations. For many years he studied variation in the patterns of spots on the hindwings of *Maniola jurtina,* the Meadow Brown butterfly. The patterns were remarkably constant over wide areas of Europe, but they showed great changes over slight distances in southwest England and the Isles of Scilly. Spot patterns on the large islands tended to be similar, while those from smaller islands differed radically. Ford attributed the constancy of pattern over time to stabilizing selection and the difference between large and small islands to the effect of averaging over diverse habitats on the large islands.

In addition to his genetic work, Ford made contributions in several other fields. He published on the chemistry of butterfly pigments, on the archaeology of southwest England, and on the contents of churches in the Oxford district.

Perhaps Ford's greatest impact on the study of variation in natural populations has come about through his interactions with his students and colleagues. He influenced the work of Philip M. Sheppard, Cyril A. Clarke, A. J. Cain, H. B. D. Kettlewell, Laurence M. Cook, Lincoln P. Brower, John R. G. Turner, Ronald A. Fisher, and many others. Ford was perhaps too inclined to cite only his friends and colleagues, but, as Bryan Clarke (1995, 162) makes clear in his *Royal Society Biographical Memoir,* "still *Ecological Genetics . . .* is necessary fare for serious workers in the field, still it is the clearest and least compromising statement of the 'selectionist' view, and still it is a joy to read."

BIBLIOGRAPHY

Clarke, B. C. 1995. Edmund Brisco Ford. *Biographical Memoirs of Fellows of the Royal Society of London* 41: 147–168.

Creed, E. R., E. B. Ford, and K. G. McWhirter. 1964. Evolutionary studies on *Maniola jurtina:* The Isles of Scilly, 1958–59. *Heredity* 19: 471–488.

Fisher, R. A., and E. B. Ford. 1947. The spread of a gene in natural conditions in a colony of the moth *Panaxia dominula* L. *Heredity* 1: 143–174.

Ford, E. B. 1945. *Butterflies.* London: Collins.

———. 1964 [1975]. *Ecological Genetics.* 4th ed. London: Chapman and Hall.

—*J.M.*

Form and Function (Edward S. Russell)

Edward S. Russell's *Form and Function* is a uniquely valuable resource in the history of animal morphology. It is a lively exploration of the most important morphological theorists from the nineteenth through the early twentieth centuries. How could a report on long-dead morphologists be lively? The liveliness comes from the fact that Russell organizes the theorists around a central debate that persisted throughout the nineteenth century and persists even today. The debate is in the title. *Form and Function* could just as well have been titled *Form versus Function.* To put the matter simply, which came first, form or function? Why do bodies of animals have the forms they do? Do the forms exist because they allow functions to be served? If so, functionalism is favored and form is a by-product of function. Or are organismic forms or structures a product of autonomous form-generating processes? If the latter, then the functions that body parts serve are the by-products, not the causes of form. Russell referred to the pro-form position as "pure morphology." The pro-function position took various forms during the nineteenth century. Prior to Charles Darwin, functionalists were teleologists, most of whom were natural theologians. After Darwin, most functionalists were believers in the importance of natural selection or Lamarckism. The contrast was illustrated in the 1830 debate between Georges Cuvier and Étienne Geoffroy Saint-Hilaire (Coleman 1964; Appel 1987; Ruse 2003). Cuvier claimed that the forms of animal bodies and body parts were dictated by the functions that had to be served, their so-called conditions of existence. Geoffroy claimed that animals shared abstract body plans, and he insisted on the unity of type. Darwin reported on this debate in *On the Origin of Species,* but came down on the side of function. Natural selection—a process that shapes form to suit function—was the cause of evolutionary change.

> It is generally acknowledged that all organic beings have been formed on two great laws: Unity of Type, and the Conditions of Existence. By unity of type is meant that fundamental agreement in structure, which we see in organic beings of the same class, and which is quite independent of their habits of life. On my theory, unity of type is explained by unity of descent. The expression of conditions of existence, so often insisted on by the illustrious Cuvier, is fully embraced by the principle of natural selection. For natural selection acts by either now adapting the varying parts of each being to its organic and inorganic conditions of life; or by having adapted them during long-past periods of time: the adaptations being aided in some cases by use and disuse, being slightly affected by the direct action of the external conditions of life, and being in all cases subjected to the several laws of growth. Hence, in fact, the law of the Conditions of Existence is the higher law; as it includes, through the inheritance of former adaptations, that of Unity of Type. (Darwin 1859, 206)

Later-nineteenth-century evolutionists swung back toward form again. They doubted the power of natural selection. Instead, they believed that rules of formal, or sometimes embryological, construction were responsible for the bodies of animals (see Ruse 1996 and Richards 2008).

Form and Function contains a number of useful translations of original texts and reproductions of illustrations from hard-to-find morphological treatises. (Copies of Russell's redrawings of morphological plates from the nineteenth century have often been passed off as reproductions of the originals by later historians and biologists.) Ironically, the book received little attention at its publication in 1916. It began to be taken more seriously by morphologists and some evolutionary theorists around 1970. The reason, a second irony, was that the study of morphology went through a dry spell during the early twentieth century. The modern synthetic theory of evolution (see the main essay "The History of Evolutionary Thought" by Michael Ruse in this volume), which was inaugurated in the 1930s and well established by 1950, saw little importance in traditional morphological studies. However, a growing minority of evolutionary theorists, some of them critics of the synthetic theory, were reconsidering the importance of morphology during the 1970s. One of these, George V. Lauder, a new assistant professor at the University of Chicago in 1981, learned that the University of Chicago Press was reprinting historically important scientific texts. He nominated *Form and Function* and wrote a useful introduction. Lauder discussed why morphology had so little relevance to the synthetic theory of evolution, but he reported that morphologists were once again beginning to consider the nineteenth-century topics of structural patterns and principles of organismic design. So Russell's 1916 book was reprinted because of the rebirth of interest in morphology around 1980. Lauder and other twentieth-century morphologists drew insights from Russell's history that were relevant to contemporary debates in evolutionary theory (Lauder 1996; Amundson and Lauder 1994). Russell himself favored functionalism over pure morphology, but his book presented the pure morphologists in a sympathetic light. Recognition of the virtues of his book was not limited to morphologists, however. Ernst Mayr, architect of the synthetic theory of evolution and a functionalist to the core, commented on *Form and Function:* "There is probably no other branch of biology for which we have as superb a history as for morphology: Russell (1916) is unsurpassed to this day, remarkable for its fresh analysis of the primary sources" (Mayr 1982, 879n15).

BIBLIOGRAPHY

Amundson, R., and G. V. Lauder. 1994. Function without purpose: The uses of causal role function in evolutionary biology. *Biology and Philosophy* 9: 443–469.

Appel, T. A. 1987. *The Cuvier-Geoffroy Debate: French Biology in the Decades before Darwin.* New York: Oxford University Press.

Coleman, W. 1964. *Georges Cuvier, Zoologist: A Study in the History of Evolution Theory.* Cambridge, MA: Harvard University Press.

Darwin, C. 1859. *On the Origin of Species.* London: John Murray.

Lauder, G. V. 1996. The argument from design. In M. R. Rose and G. V. Lauder, eds., *Adaptation,* 55–91. San Diego, CA: Academic Press.

Mayr, E. 1982. *The Growth of Biological Thought.* Cambridge, MA: Harvard University Press.

Richards, R. J. 2008. *The Tragic Sense of Life: Ernst Haeckel and the Struggle over Evolutionary Thought.* Chicago: University of Chicago Press.

Ruse, M. 1996. *Monad to Man: The Concept of Progress in Evolutionary Biology.* Cambridge, MA: Harvard University Press.

———. 2003. *Darwin and Design: Does Evolution Have a Purpose?* Cambridge, MA: Harvard University Press.

Russell, E. S. 1916 [1982]. *Form and Function.* London: John Murray. Reprint, Chicago: University of Chicago Press.   —R.A.

Frisch, Karl Ritter von (1886–1982)

Karl Ritter von Frisch was born in Vienna to a family distinguished on his father's side by three generations of physicians and on his mother's side by a number of noted academics. A lover of animals as a youth, he went on to have a distinguished career as an experimental physiologist. The teachers who influenced him most were his uncle, Sigmund Exner, professor of physiology at the University of Vienna, and Richard Hertwig, professor of zoology at the University of Munich. Both impressed upon him the critical importance of careful experimentation as a complement to his talents as a thoughtful observer of living animals. Frisch responded by becoming a master at devising simple experiments to test the behavior of animals under natural conditions.

When Frisch was a young researcher, his early work on the ability of minnows to adapt their coloration to dark or light backgrounds brought him into conflict with the claims of Carl von Hess, a prominent Munich ophthalmologist. Studying how various animals respond to light, Hess had concluded that fish and invertebrates respond to the brightness of colors but are otherwise color-blind. Frisch, both as a naturalist and as someone who had raised and kept many different species of fish, knew that some fish are capable of changing color and that certain species are especially colorful at mating time. This suggested to him that fish have a sense of color, a hypothesis that he proceeded to test and confirm experimentally. Hess, however, rejected Frisch's findings, and the debate that ensued between them lasted more than a decade. Early in the debate Frisch proceeded to address the question of color sense in honeybees. Thus began a whole series of researches that were to make Frisch world-famous.

Frisch's first studies of honeybees examined the insects' sense of color and shape. He next studied the bees' sense of smell. In the course of his experiments he discovered that scout bees are somehow able to communicate to their fellow workers in the hive the existence of a desirable food source. In 1920 he published his first reports on what he called the "language" of the bees. He understood the bees' "language" to involve two distinctive kinds of dances, a round dance and a waggle dance, performed by bees returning to the hive. Initially he supposed that the round dance signified a nectar source and the waggle dance signified a pollen source. It was not until the fall of 1944 that he discovered that the different dances are related to the distance of the food source from the hive and not to the nature of the food. The following spring he arrived at the remarkable conclusion that the waggle dance, performed by bees returning from food sources more than 100 meters from the hive, serves to communicate both the distance and the direction of the food source. In his further studies of honeybees he demonstrated that they perceive polarized light and in addition have an "internal clock" that enables them on long flights to adjust to the sun's changing place in the sky.

Frisch spent much of his career at the University of Munich, where he succeeded Richard Hertwig in 1925 as professor of zoology and director of the

Zoological Institute. In the early 1930s, with help from the Bavarian government and the Rockefeller Foundation, Frisch made the Zoological Institute at Munich the most advanced scientific center of its kind in Europe. He managed to retain his position throughout the Third Reich, though not without some difficulties from the Nazi authorities. After Allied bombing destroyed much of the institute in 1944, he continued his studies in Austria, first at his summer home in Brunnwinkl and then as professor of zoology at the University of Graz (1946–1950). He returned to the University of Munich in 1950, where he remained until his retirement in 1958. Among his many honors was the 1973 Nobel Prize for Physiology or Medicine, which he shared with Konrad Lorenz and Niko Tinbergen.

BIBLIOGRAPHY

Burkhardt, R. W., Jr. 1990. Frisch, Karl Ritter von. In *Dictionary of Scientific Biography,* supplement 2, 17: 312–320.
Frisch, K. von. 1967. *A Biologist Remembers.* Lisbeth Gombrich, trans. Oxford: Pergamon Press.   —R.W.B.

G

The Galápagos Archipelago

The Galápagos Archipelago is a group of 14 small islands (the largest about 70 miles long) located on the equator about 500 miles off the coast of Ecuador, to which they belong. They are volcanic in origin (some are still slightly active) and of recent geological age. They were formed as the continental plate drifted easterly. Harsh, with cinder and ash underground, the climate is hot and there is little or no fresh water. First discovered by Westerners in 1532, they were initially called the *Encantadas,* or the Enchanted Isles. The American novelist Herman Melville, who visited them on a whaling trip, made a sour allusion to this name, describing them as "evilly enchanted ground" (1987, 129), adding: "Man and wolf alike disown them" (127). For evolutionists they are far from evil and they are certainly enchanted.

The islands soon took on the name of their most famous denizens, Galápagos being Spanish for tortoise. Not just any kind of tortoise, but giant monsters that crowded the isles until their numbers were reduced to near extinction, thanks to the killing and taking by mariners who found that the brutes provided excellent fare for long sea voyages. The islands were visited sporadically by sailors, chiefly buccaneers, and were given English names, first in the seventeenth century after prominent personages (like King James II) and then in the eighteenth century after worthies of that time. Today, they carry Spanish names given by Ecuador.

The Galápagos Archipelago was visited in 1835 by HMS *Beagle* on its trip around the globe, carrying Charles Darwin as the ship's naturalist. Darwin was fascinated by the wildlife, especially the tortoises and the many small birds that live on the islands. As was his custom, he made large collections of both flora and fauna, although as yet he did not truly realize the importance and significance of keeping separate the individuals from different islands. What did strike Darwin at once was the seeming relationship between the organisms on the Galápagos and the South American mainland, and also the fact that the island dwellers generally had slight but distinctive differences. None of this seemed to make sense from a creationist viewpoint of organic origins. Why would the organisms be similar to the mainland organisms, and why South America rather than Africa? And why the differences between the islands?

His curiosity piqued, Darwin mulled over the biogeographical distributions of the island specimens, concentrating especially on several species of mockingbird that he knew came from different islands and that seemed different physically. He thought also about the tortoises that the archipelago's governor had described as distinctive to their respective islands. As he tried to put in order the many small finches that he had collected, he saw that they too revealed differences on different islands. Could these all be mere varieties or were they something more, in fact and in theory?

When Darwin returned to England, he gave his specimens to specialists to catalog. At this point Darwin felt obliged to slip over to an evolutionary position. The birds particularly were undoubtedly different species, and this could be explained naturally only on the supposition that original founders had come to the archipelago and changed and evolved as they and their successors moved from island to island. A crucial part of this thinking was that evolutionary change involves division of populations and subsequent speciation. Darwin therefore always thought of evolution's history as being fundamentally one of splitting and diverging. Its pattern was coral-like (his first metaphor) or a tree of life (the metaphor that he made famous in *On the Origin of Species*). Here he differed from earlier evolutionists who thought of evolution as essentially a move up a single line, with variations along minor side branches.

Although the Galápagos took somewhat of a secondary role as his thinking matured, Darwin always used island biogeography as a key support of evolutionism, frequently acknowledging the importance of his visit to the Galápagos. Probably in part because of this significance, the great Swiss-American ichthyologist Louis Agassiz, the most important post-*Origin* scientist who never accepted any form of evolutionism, visited the Galápagos late in life in 1872. Perhaps unsurprisingly, he felt that his special creationist vision of life's history was vindicated by the Galápagos and its inhabitants. Agassiz thought there was insufficient time to produce so varied a display on islands geologically so young (see Larson 2001, especially chapter 4). However, no one really followed in his path. Soon the Galápagos Archipelago was established as a prize case for evolution, although, in line with the general post-*Origin* discrediting of natural selection, most would have agreed with Agassiz that some other mechanism was needed to produce the animals and plants as they are found today.

Such was the state of affairs through the 1930s, when Darwinian selection was melded with Mendelian genetics and modern evolutionary theory. Neo-Darwinism, or the synthetic theory of evolution, was born. However, there was no an immediate rush to bring the Galápagos under the net of natural selection. There was a general feeling, especially among Americans, that small differences between groups are rarely adaptive. The population geneticist Sewall Wright (1931, 1932) gave theoretical backing to this prejudice by arguing that many differences are due to so-called genetic drift, the randomness that is brought by the vagaries of breeding. The differences between Galápagos organisms were often put down to just such a cause. No one denied that selection was important overall, but not in the kinds of details revealed by the islands.

The key figure in changing this opinion was an English schoolteacher,

David Lack, who through the encouragement of Julian Huxley went to the Galápagos just before World War II to study the birds. He then began a massive collection of the birds—finches particularly—that was owned by the American Museum of Natural History in New York. Writing up his findings, Lack at first put all significant change down to drift. Reflecting during the war years on his findings, Lack swung entirely to a selective explanation. He was no doubt much influenced by the German-born, American-residing taxonomist Ernst Mayr. Lack roomed with Mayr and his family in New York, precisely at the time that Mayr was finishing his pro-selection contribution to the synthesis *Systematics and the Origin of Species* (1942). Lack argued in *Darwin's Finches* (1947) that all differences between the species are adaptive, brought on by struggle between groups for resources. Although the title of Lack's book has now given the popular name to the finches and has publicly linked Darwin with the island inhabitants, it also had the somewhat unfortunate effect of making people think that it was the finches exclusively that made Darwin an evolutionist. At least, if not more important, were the mockingbirds (see Sulloway 1982).

After Lack, the Galápagos birds were one of the prime pieces of evidence in the neo-Darwinian synthesis. However, controversy continued over the exact nature of the selective force that had brought on differences. Lack, following British thinking, had made competition between groups the key factor. In America, in line with general thinking in that country about the nature of organisms, there was more enthusiasm for a position that made the ecology of the islands more significant. This view of selection and its effects was explored and strongly defended by California biologist Robert Bowman (1961), who made a long visit to the islands in 1952–1953. He found that the existence of alternative food supplies was the key factor in selective change.

The British were unimpressed, and thus matters remained at an impasse until the 1970s. The husband-and-wife team of Peter and Rosemary Grant, together with a succession of coworkers and students, began what has since become one of the classic long-term studies of natural selection (Grant 1986, 1991; B. R. Grant and P. R. Grant 1989; P. R. Grant and B. R. Grant 1995, 2007; Weiner 1994). Working on the islet Daphne Major (see figure), they ringed and watched and recorded every bird, noting the number of fluctuations and physical changes from one year to the next, especially noting the effects brought on by drought and by record rainfall. They found that selection works constantly, and that changes can occur rapidly from one year to the next. Moreover, there seems to be truth in the claim that food supplies and availabilities are a significant factor in selective change, but not excluding competition between different groups.

Although the Galápagos has been battered by human occupancy, from the early seamen who took the tortoises, through the American forces in World War II who built an airbase there, to the present Ecuadorian fishermen, there is now realization by scientists and governmental authorities of what an important natural phenomenon the Galápagos Archipelago represents. Serious efforts are being made to protect and conserve the islands and their inhabitants.

The islet Daphne Major in the Galápagos Archipelago, where Peter and Rosemary Grant did their decades-long study of Darwin's finches. The volcanic origins of the islet, a hill with the top blown off and a crater formed within, are clearly visible. It has been possible to work out relationships between the denizens of the islands from the sequential order in which the islands were formed, by volcanic activity, as the earth's plates slipped around its surface.

Now they are one of the most popular of destinations for tourists. Darwin's visit to the island, rather than Melville's, has proven to have the longer-lasting effect.

BIBLIOGRAPHY

Bowman, R. 1961. *Morphological Differentiation and Adaptation in the Galápagos Finches.* Berkeley: University of California Press.

Grant, B. R., and P. R. Grant. 1989. *Evolutionary Dynamics of a Natural Population: The Large Cactus Finch of the Galápagos.* Chicago: University of Chicago Press.

Grant, P. R. 1986. *Ecology and Evolution of Darwin's Finches.* Princeton, NJ: Princeton University Press.

———. 1991. Natural selection and Darwin's finches. *Scientific American,* October: 82–87.

Grant, P. R., and B. R. Grant. 1995. Predicting microevolutionary responses to directional selection on heritable variation. *Evolution* 49: 241–251.

———. 2007. *How and Why Species Multiply: The Radiation of Darwin's Finches.* Princeton, NJ: Princeton University Press.

Lack, D. 1947. *Darwin's Finches.* Cambridge: Cambridge University Press.

Larson, E. J. 2001. *Evolution's Workshop: God and Science on the Galápagos Islands.* New York: Basic Books.

Mayr, E. 1942. *Systematics and the Origin of Species.* New York: Columbia University Press.

Melville, H. 1987. The Encantadas, or Enchanted Isles. In *Piazza Tales and Other Prose Pieces, 1839–1860: The Northwestern-Newberry Edition of the Writings of Herman Melville*. Vol. 9. H. Hayford, A. A. Macdougall, and G. T. Tanselle, eds. Evanston, IL: Northwestern University Press.

Sulloway, F. J. 1982. Darwin and his finches: The evolution of a legend. *Journal of the History of Biology* 15: 1–53.

Weiner, J. 1994. *The Beak of the Finch: A Story of Evolution in Our Time*. New York: Knopf.

Wright, S. 1931. Evolution in Mendelian populations. *Genetics* 16: 97–159.

———. 1932. The roles of mutation, inbreeding, crossbreeding and selection in evolution. *Proceedings of the Sixth Annual Congress of Genetics* 1: 356–366.

—M.R.

Galton, Francis (1822–1911)

Francis Galton was born near Birmingham, England. He originally intended to pursue a career in medicine, but he switched to Cambridge University to study mathematics, influenced by his cousin Charles Darwin (1809–1882). Darwin and Galton were grandsons by different marriages of Erasmus Darwin (1731–1802), a physician, scientist, poet, and inventor. Charles Darwin would be a crucial influence on Galton throughout much of his career. In fact, it was Galton's reading of *On the Origin of Species* (1859) that gave him the idea of improving the human race through selective breeding, for which he coined the term *eugenics*.

Several years after graduating from Cambridge with an ordinary degree, Galton organized and financed an African expedition that resulted in the first European exploration of northern Namibia. He made careful measurements of latitudes, longitudes, and altitudes, reflecting his lifelong interest in the application of numerical and quantitative methods to whatever happened to interest him. He published his results in the *Journal of the Royal Geographical Society* in 1852 and was awarded a gold medal by the society the same year largely because of his precise, quantitative work. *Tropical South Africa* (1853), Galton's book about his journey, was a success, but *The Art of Travel* (1855), a guidebook for amateur and professional alike who ventured into the bush, proved a triumph. It went through many editions, growing bigger each time. In 2001 Phoenix Press reissued the fifth edition, which had first been published in 1872. Galton was an active member of the Royal Geographical Society for many years. His comments at society meetings frequently dealt with quantitative matters (e.g., latitudes, boiling point thermometer readings). He also became interested in meteorology and discovered the anticyclone, a weather feature characteristic of a high-pressure system.

The part of Galton's career for which he is probably best remembered began after he had read Darwin's *On the Origin of Species*. Because domestic animals and cultivated plants were the products of selective breeding, it seemed reasonable to Galton that the human race might be improved similarly. The key was to demonstrate that desirable human traits such as intellectual ability were inherited. He attempted to do just this in a two-part article entitled "Hereditary Talent and Character," published in a popular periodical called *MacMillan's*

Magazine (1865), and in the book that followed, *Hereditary Genius* (1869). In both the article and the book Galton tried to demonstrate that what he referred to as "talent and character" were inherited. The book traced pedigrees of judges, statesmen, and others. Galton's basic assumption was that if he picked an eminent judge, for example, that judge's closest male relatives (e.g., father and son) had a greater chance of being distinguished than those further removed (e.g., grandfather, grandson). Women were not included, probably reflecting both Victorian prejudice and male dominance in Victorian society. Galton concluded that his results supported his hypothesis. Although others pointed out that environment (e.g., the father's good position might ensure a plumb job for his son) could also underlie the correlation, Galton attempted to dismiss this contention.

Familiar with the bell curve by the time he wrote *Hereditary Genius*, Galton calculated a hypothetical normal distribution for the estimated 15 million males in the United Kingdom according to their natural abilities. He also was anxious to collect real data that he could analyze statistically, particularly with regard to inheritance. Because this was not easy to do with human beings, he decided to use sweet peas as a model because they were easy to grow and supposedly self-fertilizing. In his experiments he found that seed diameter was normally distributed, but the seed diameter from progeny of large-seeded and small-seeded plants drifted toward the mean of the population as a whole. He called this "regression to the mean," a statistical property that has been repeatedly demonstrated since (e.g., in the case of different classes of mutual funds, such as ones specializing in growth versus international stocks). Galton also discovered that when he graphed the diameters of parental seeds versus progeny seeds, the points fell on a straight line. He had drawn the first regression line and from it he calculated the first regression coefficient. Later, Galton obtained the human data he so desired when he organized an anthropometric laboratory at the 1884 International Health Exhibition held in South Kensington, London. The laboratory reopened in the Science Galleries of the South Kensington Museum after the exhibition closed. Collecting quantitative data from both parents and children, he was now able to demonstrate regression to the mean for measurable human characteristics (e.g., height). When he plotted forearm length against height he discovered another important statistical concept, correlation (i.e., tall men have long forearms). He calculated the first correlation coefficient, myriads of which have been computed since. Galton also became interested in fingerprints and their classification; his work was central to the development of fingerprinting as a forensic technique.

Galton collected many of his important findings in *Natural Inheritance* (1889). He also acquired his first protégées, including Karl Pearson. A fine mathematician, Pearson went far beyond Galton in formulating statistical theory. Meanwhile, Galton was promoting eugenics and the idea was gaining traction in the prevailing Social Darwinist environment at the end of the nineteenth century. Positive eugenics encouraged the selective reproduction of the supposedly fit, while negative eugenics aimed to eliminate the reproduction of those deemed unfit. Unfortunately, negative eugenics prevailed.

In the United States, passage of state eugenic sterilization laws resulted in the involuntary sterilization of thousands of individuals regarded to be mentally deficient or "feebleminded." Eugenic developments in the United States were followed with interest in Europe and elsewhere, particularly in Germany. When the Nazis came to power they passed an involuntary sterilization law that resulted in the sterilization of hundreds of thousands of individuals. After World War II, eugenic sterilization gradually ended in the United States and Europe. While eugenics is Galton's most well-known legacy, he also must be credited with important achievements in fields as diverse as statistics, fingerprinting, meteorology, and exploration.

BIBLIOGRAPHY

Brookes, M. 2004. *Extreme Measures: The Dark Visions and Bright Ideas of Francis Galton.* New York: Bloomsbury Publishing.

Bulmer, M. 2003. *Francis Galton: Pioneer of Heredity and Biometry.* Baltimore: Johns Hopkins University Press.

Forrest, D.W. 1974. *Francis Galton: The Life and Work of a Victorian Genius.* New York: Taplinger Publishing Co.

Galton, F. 1852. Recent expedition into the interior of South-Western Africa. *Journal of the Royal Geographical Society* 22: 140–163.

———. 1853. *Tropical South Africa.* London: John Murray.

———. 1855. *The Art of Travel: Or, Shifts and Contrivances Available in Wild Countries.* London: John Murray.

———. 1865. Hereditary talent and character. *Macmillan's Magazine* 12: 157–166, 318–327.

———. 1869. *Hereditary Genius.* London: Macmillan.

———. 1889. *Natural Inheritance.* London: Macmillan.

———. 2001. *The Art of Travel, or, Shifts and Contrivances Available in Wild Countries.* London: Phoenix Press.

Gillham, N. W. 2001. *A Life of Sir Francis Galton: From African Exploration to the Birth of Eugenics.* New York: Oxford University Press.

Kevles, D. 1995. *In the Name of Eugenics.* Cambridge, MA: Harvard University Press.

Kühl, S. 1994. *The Nazi Connection.* New York: Oxford University Press.

Paul, D. B. 1995. *Controlling Human Heredity, 1865 to the Present.* Atlantic Highlands, NJ: Humanities Press.

Pearson, K. 1914–1930. *The Life, Letters and Labours of Francis Galton.* 3 vols. Cambridge: Cambridge University Press.

Reilly, P. R. 1991. *The Surgical Solution: A History of Involuntary Sterilization in the United States.* Baltimore: Johns Hopkins University Press.   —N.W.G.

Game theory

The study of animal behavior thrives on paradoxes, baffling inconsistencies between intuition and evidence that engage attention and stimulate further investigation. Efforts to resolve such paradoxes rely increasingly on analytical tools called games.

A game in this sense is a mathematical model of strategic interaction, arising when the outcome of an individual's actions depends on actions taken by others. Thus a game has three key components. First, there are at least two interacting individuals, called players. In an evolutionary game, the set of players is an ecotype, that is, a population of animals in a given ecological environment (for example, a population of spiders in a grassland habitat and a riparian population of the same species would form two different ecotypes). Second, each player has a set of feasible strategies that is constrained by the information structure of the interaction (for example, animals can condition their behavior on whether they are owners or intruders only if they are aware of such roles). Third, the pattern of interaction must be well defined and accompanied by a formula for how each player's reward from the interaction depends on its strategy and on those of the other players. In evolutionary games, the rewards are measured in terms of expected future reproductive success.

For a game to be useful, it must be possible to identify a strategy or strategies from among those feasible as the "solution" for a given purpose. In an evolutionary game, this is the behavior expected to evolve by natural selection. If a behavior is fixed in a real population, then it must at least be true that every feasible alternative behavior would yield a lower reward, otherwise the alternative behavior would have spread into the population. Thus the relevant solution concept—introduced by John Maynard Smith (1982)—is that of an evolutionarily stable strategy (ESS), a population strategy that yields a higher reward than any feasible mutant strategy (see also Dawkins 1976).

Scientists strive to unravel a paradox by establishing conditions for an ESS to exist in a model population with assumed components, and by analyzing its properties when it does exist. If the ESS in the model population fails to match observed behavior in the real population, then one or more of the assumptions are modified—about ecotype, information structure and strategy set, or pattern of interaction and reward. The ESS is recalculated as often as necessary. In other words, if a paradox of animal behavior exists, then we have wrongly guessed which game best models how a real population interacts. To resolve this paradox, we must guess again—if necessary, repeatedly—until eventually we guess correctly.

Games are valuable because they allow scientists to explore the logic of a verbal argument rigorously, assuming biologically realistic ecotypes, and to determine when it is true and when it is false. Game theory often demonstrates what is difficult to intuit. For example, game theory has shown that victory by stronger animals need not imply that strength is being assessed in contests, and has yielded a test for such an ESS, namely, whether energy reserves of losers correlate positively with contest durations. For further details of this example and others, see Mesterton-Gibbons and Adams (1998).

BIBLIOGRAPHY

Dawkins, R. 1976. *The Selfish Gene*. Oxford: Oxford University Press.
Maynard Smith, J. 1982. *Evolution and the Theory of Games*. Cambridge: Cambridge University Press.
Mesterton-Gibbons, M., and E. S. Adams. 1998. Animal contests as evolutionary games. *American Scientist* 86: 334–341.   —M.M.-G.

The Genetical Theory of Natural Selection (R. A. Fisher)

Sir R. A. Fisher's *The Genetical Theory of Natural Selection*, published in 1930, is celebrated as the first major reconciliation of Darwinian natural selection and Mendelian heredity. The 12-chapter manuscript, created by Mrs. Fisher from Fisher's dictation, was the culmination of Fisher's thought on the problems of genetics and natural selection begun in the late 1910s.

The first seven chapters of *Genetical Theory* set out Fisher's synthesis of Darwin's mechanism of natural selection and Mendelian genetics. Fisher considered the first two chapters, on the nature of inheritance and the "fundamental theorem of natural selection," the most important of the book. Indeed, these two chapters accomplish the key piece of the reconciliation.

Fisher's first chapter considers implications of a synthesis of natural selection with, alternatively, blending and Mendelian inheritance. He demonstrates that on the Mendelian theory, natural selection may be the main cause of a population's variability. The demonstration importantly resolved a persistent problem for Darwin's theory of descent with modification, one that had led biologists to abandon natural selection as an evolutionary cause. Darwin's acceptance of blending inheritance required him to imagine causes controlling mutation because of enormous mutation rates demanded by the blending theory. Because Mendelian heredity did not demand such enormous mutation rates, Fisher was able to eliminate these controlling causes and revive natural selection as an important evolutionary cause.

Fisher's second chapter develops, mathematically, his genetical theory of natural selection. Three key elements may be distilled from Fisher's "heavy" mathematics. The first is a measure of average population fitness: Fisher's "Malthusian parameter," the reproductive value of all genotypes at all stages of their life histories. The second is a measure of variation in fitness, which Fisher partitions into genetic and environmental components. The third is a measure of the rate of increase in fitness, that is, the change in fitness due to natural selection. For Fisher, "the rate of increase of fitness of any species is equal to the [additive] genetic variance in fitness" (p. 35). This last element is Fisher's "fundamental theorem of natural selection" and is the centerpiece of his natural selection theory. Under this rubric, Fisher then offers a geometrical proof that cumulative evolution is primarily the result of low pressures of natural selection on mutations of small effect.

By and large, the middle chapters of *Genetical Theory* are explorations of cases, such as dominance, sexual selection, and mimicry, to support the preceding theoretical work. Nevertheless, in the fourth and fifth chapters, Fisher expands his theoretical discussion to more general issues concerning the causes of genetic variation, including random genetic drift.

The last five chapters of *Genetical Theory* explore natural selection in human populations, particularly social selection in human fertility. Fisher's central observation is that the development of economies in human societies structures the birth rate so that it is inverted with respect to social class. In the final chapter, Fisher offers strategies for countering this effect. Despite

Fisher's espousal of this eugenics thesis in this part of the book, he intends the discussion to be taken as an inseparable extension of the preceding part.

The Genetical Theory of Natural Selection is a point of departure in evolutionary thought, responsible in part for the origination of theoretical population genetics and what is commonly called the modern synthetic theory of evolution. *Genetical Theory* was followed by Sewall Wright's and J. B. S. Haldane's major works in 1931 and 1932, respectively. Fisher's views on the role of natural selection in evolution are widely accepted today.

BIBLIOGRAPHY

Fisher, R. A. 1930 [1958]. *The Genetical Theory of Natural Selection.* 2nd ed. New York: Dover Publications. Released in a variorum edition by Oxford University Press in 1999, edited by J. H. Bennett.
Haldane, J. B. S. 1932. *The Causes of Evolution.* London: Longmans.
Wright, S. 1931. Evolution in Mendelian populations. *Genetics* 16: 97–159.

—R.A.S.

The Genetic Basis of Evolutionary Change (Richard Lewontin)

In 1974, Richard Lewontin, a professor at Harvard University, published an important assessment of the state of evolutionary genetics. Lewontin's 1974 book, *The Genetic Basis of Evolutionary Change,* considered the tremendous changes within evolutionary genetics wrought by the introduction of techniques from molecular biology and the subsequent development of the neutral theory of molecular evolution. Lewontin's book is noteworthy for its analysis of the nature of evolutionary genetics and the dynamics of scientific controversy and controversy resolution.

Lewontin and his colleague, J. L. Hubby, had touched off a molecular revolution in 1966 when they introduced the biochemical technique of electrophoresis as a means of resolving genetic differences at the molecular level. Lewontin believed that this new electrophoretic analysis of genetic variability resolved the earlier dispute between the classical and balance positions over variability and selection. The high levels of variability detected with electrophoresis seemed to support the balance position and the importance of balanced polymorphisms in evolution. The classical position's belief was that most mutations were deleterious and most were selection purifying. They struggled to explain electrophoretic variability until Motoo Kimura and other biologists began to argue that most of the detected polymorphism was in fact neutral, neither selected for nor against. In *The Genetic Basis of Evolutionary Change,* Lewontin emphasized the continuity between the earlier classical-balance controversy and the then-raging neutralist-selectionist controversy. For him, the neutralist position was really just the classical position brought in line with newer molecular results. While this interpretation of the controversy makes a good case for continuity in evolutionary genetics, it neglects the growing contribution of biochemical research to the development of molecular evolution.

Having argued that one controversy has been transformed into another, Lewontin advances an explanation for why controversy in evolutionary genetics is inevitable. The connection and persistence of these controversies was a result of a deeper, irresolvable ideological conflict. Advocates of the classical and neutralist positions held a conservative attitude toward change and variability; their fondness for stability led them to see selection as a means of accepting or rejecting variations. Advocates of the balance position, however, embraced change and variability; they favored the idea that variation could be actively maintained in a population by balancing selection. These deeply held convictions regarding the value of change explained for Lewontin why his electrophoretic data did not resolve the classical-balance controversy, but inspired its transformation.

The Genetic Basis of Evolutionary Change should not be understood as reducing evolutionary genetics to conflicting ideologies. Lewontin presents a careful analysis of the challenges of producing a theory of genetic change that captures how genetic processes interact and develop over time. According to Lewontin, information about molecular level changes was finally providing a body of data that would allow theories of evolutionary genetics to be tested and refined. In light of new molecular data, Lewontin claimed that the entire relationship between theory and fact would have to be reconsidered. In addition, Lewontin argued that theories of individual genes would have to be replaced with models that considered genomes and their complex interactions. Indeed, *The Genetic Basis of Evolutionary Change* ends with the charge that the real problems of evolutionary genetics can only be addressed by understanding the interactive context of alleles.

BIBLIOGRAPHY

Dietrich, M. R. 1994. The origins of the neutral theory of molecular evolution. *Journal of the History of Biology* 27: 21–59.

Hubby, J. L., and R. C. Lewontin. 1966. A molecular approach to the study of genic heterozygosity in natural populations I. The number of alleles at different loci in *Drosophila pseudoobscura. Genetics* 54: 577–594.

Kimura, M. 1983. *The Neutral Theory of Molecular Evolution.* Cambridge: Cambridge University Press.

Lewontin, R. C. 1974. *The Genetic Basis of Evolutionary Change.* New York: Columbia University Press.

Lewontin, R. C., and J. L. Hubby. 1966. A molecular approach to the study of genic heterozygosity in natural populations. II. Amount of variation and degree of heterozygosity in natural populations of *Drosophila pseudoobscura. Genetics* 54: 595–609.   —M.R.D.

Genetics and the Origin of Species
(Theodosius Dobzhansky)

With great charisma and a passion for the subject, Theodosius Dobzhansky became a rallying point for the evolutionary synthesis. *Genetics and the Origin of Species* (1937) was his principal contribution to that movement, and it became an essential text in evolutionary biology. On one level, *Genetics* is a splendid summary of evolutionary genetics in the 1930s, both zoological and botanical. It brought together diverse information and created an easy-to-understand model for evolution leading to new species. It also presented an unexpected level of sophistication and confidence to biologists suspicious of evolutionary studies. Like no other contribution to the subject, *Genetics* gave a burst of energy to speciation studies as a research program.

Much of *Genetics* surveyed knowledge about the genetic structure of populations. Recent developments in technique had given geneticists sensitive tools for monitoring new mutations, changes in gene frequency, and changes in chromosome structure. Dobzhansky presented considerable data on the extent of this variation in both laboratory and wild populations. He also described experimental systems that monitored changes to the genetic structure of populations over time and over changing environmental conditions. In part, Dobzhansky's survey celebrated the hard-earned empirical side of population genetics. It also aimed to present the subject as rigorous, with the evolutionary scientists in full experimental control of their data.

Dobzhansky defines evolution in narrow terms: a change in the frequency of alleles within a population. This is purposively operational. *Genetics* offers many examples of such a change. It also presents a detailed discussion of mechanisms driving those changes. This discussion, along with the later discussion of isolation and divergence, make for the two theoretical spines in *Genetics*. Dobzhansky argued natural selection drives adaptation and is a principal agent in shifting the genetic structure of populations. Selection uses the raw material of genome variation resulting from random mutation as well as from the many other genetic and chromosomal processes at work within organisms.

Although potent, selection was not Dobzhansky's only evolutionary mechanism. He was heavily influenced by Sewall Wright's "shifting balance" theory of population genetics and Wright's notion of adaptive landscapes (see Wright 1931, 1932). Wright's ideas also emphasized population size and structure, migration, mutation pressure, and chance. Much of a whole chapter in *Genetics* is devoted to describing Wright's approach. In fact, many evolutionary biologists learned about Wright's models from reading *Genetics*. For Dobzhansky, shifting balance and adaptive landscapes were essential tools for explaining the genetics of evolution.

Speciation involves more than simply change in allele frequencies. For Dobzhansky, it required isolation and divergence. This is the second theoretical spine in *Genetics*. Dobzhansky examined both processes in detail leading to a definition of species as reproductively isolated and physiologically incapable of interbreeding with other groups. Once isolated, divergence resulted from both directional and random processes. Dobzhansky argued that in the continuum of life, species were the only "natural" units. Their unique genetic configuration, isolated and following a peculiar destiny, was central to that thinking.

How populations became isolated was a subject of considerable study in the 1930s. Many disciplines contributed to that subject, and many "isolating mechanisms" were proposed (e.g., geographic, ecological, behavioral, mechanical, physiological). Anything that prevented genetic exchange counted in that discussion. *Genetics* surveys prominent examples, stressing the importance of geographic and physiological mechanisms. Sterility in hybrids between two species served as a special case study in this discussion. A whole chapter in *Genetics* is devoted to this topic, illustrating how isolating processes close off the tap of gene flow.

Following Wright's shifting balance theory, Dobzhansky explained divergence largely in terms of selection and drift, with circumstances influencing the particular balance of forces. Importantly, Dobzhansky is not a knee-jerk adaptationist. Even when it played a determinant role in shaping organisms, he thought selection was many layered. It frequently shifted direction and intensity, and it frequently pressed populations in contradictory directions. Dobzhansky presented experimental scenarios in which such contradictory pressures had been distinguished. Dobzhansky did not require divergence for speciation. Famously, he named *Drosophila miranda* as a new species even though it was externally identical to *D. pseudoobscura*. In this case, individuals failed to breed so the absence of fertile offspring was sufficient for naming a new species.

In the wider frame, Dobzhansky largely restricted his focus in *Genetics* to evolution within species and to the formation of species, in his strict sense of reproductively isolated communities. To explain the origin of higher categories (e.g., new genera, families, and classes), he saw no reason to invoke additional mechanisms, suggesting macroevolution merely extended the processes of microevolution.

Key to *Genetics*'s impact was Dobzhansky's concern for objectivity, rigor, and empirical studies. His book was a bold assertion of confidence in the period's approach to evolutionary studies. *Genetics* asked narrowly defined questions and built carefully on a thick evidence base. Dobzhansky combined this with close ties to mathematical models and rigorous experimental tests. In part, Dobzhansky wrote as a representative of a new wave of experimental biologists attacking evolutionary topics. He also wrote with an eye toward skeptics from other disciplines anxious to avoid unsavory complications. Sympathetic colleagues pointed to *Genetics* as a sign of a new era in evolutionary studies. The product of Dobzhansky's 1936 Jesup lectures at Columbia University, *Genetics* revived the Columbia Biological Series, which would become an outlet for other key books in the making of modern evolutionary biology, that theory usually referred to as the evolutionary synthesis (Cain 1993).

Dobzhansky published a second edition of *Genetics and the Origin of Species* in 1941 and a third in 1951. His *Genetics of the Evolutionary Process* (1970) was intended as *Genetics*'s fourth edition. Each new edition included some significant changes to Dobzhansky's views. In particular, finding significant systematic changes of chromosomal variations in the wild, Dobzhansky grew to see a much stronger role for selection in the evolutionary process. Using one of his powerful metaphors, Stephen Jay Gould (1983), no great lover of universal adaptationism, referred to this as the "hardening" of the synthesis, although others (for example, Ruse 1999) see it as a natural development of a scientific theory in the face of new evidence.

BIBLIOGRAPHY

Cain, J. A. 1993. Common problems and cooperative solutions: Organizational activity in evolutionary studies 1936–1947. *Isis* 84: 1–25.

Dobzhansky, T. 1937. *Genetics and the Origin of Species*. New York: Columbia University Press. (Second edition 1941, third edition 1951.)

———. 1970. *Genetics of the Evolutionary Process*. New York: Columbia University Press.

Gould, S. J. 1983. The hardening of the synthesis. In M. Grene, ed., *Dimensions of Darwininsm*. Cambridge: Cambridge University Press.

Ruse, M. 1999. *Mystery of Mysteries: Is Evolution a Social Construction?* Cambridge, MA: Harvard University Press.

Wright, S. 1931. Evolution in Mendelian populations. *Genetics* 16: 97–159.

———. 1932. The roles of mutation, inbreeding, crossbreeding and selection in evolution. *Proceedings of the Sixth International Congress of Genetics* 1: 356–366.

—J.C.

Geoffroy Saint-Hilaire, Étienne (1772–1844)

Étienne Geoffroy, known as Saint-Hilaire, was born in Étampes on April 15, 1772. Sent to Paris to study for holy orders, he entered medical studies in order to fulfill his scientific ambitions. Placed under the patronage of the crystallographer René-Juste Haüy, whom he saved from almost certain death in September 1792, in June 1793 Geoffroy was appointed professor of zoology at the newly created Muséum national d'histoire naturelle, not without some belated opposition. In 1795 he invited Georges Cuvier to share his own apartment in Paris; they would write five innovative joint essays. Unlike Cuvier, who stayed home to better his career, in 1798 Geoffroy accepted the invitation to join the expedition to Egypt organized by General Napoléon Bonaparte. His work on the fauna of the Nile region was noted in France, as were his privately communicated speculations on life and electricity, increasingly suspicious in the conservative climate leading to the Napoleonic Empire. Back to Paris in 1802, shortly to leave again for Spain and Portugal, he was badly lagging behind Cuvier. He was made a member of the Institut only in 1807 and was appointed professor at the newly created science faculty in Paris in 1809. The bulk of his major work was published after 1815, and during the 1820s Geoffroy led the scientific opposition to the conservative governments in power, of which Cuvier was a supporter and at times a representative. During the 1830s his persistent attacks on his deceased former friend, his mystical overtones, and his philosophical pretensions contributed to his isolation within the Parisian scientific community, though his reputation was considerable throughout Europe.

Geoffroy's *Philosophie anatomique* (1818), his eulogies of Lamarck, and his repeated lip service to transformist theories made him a hero of radical or peripheral naturalists who accused Cuvier of prostituting his science to political conservatism. Deeply influenced by Haüy's crystallography and by Buffon's hints concerning the unity of animal types, Geoffroy initially conceived vertebrates as being composed of the same set of organs and parts, at times almost invisible in some families, fully developed in others. His celebrated studies on the vestigial presence in reptiles of all the parts that compose mammalian auditory organs elicited the admiration of Goethe and Hegel and proved, in his eyes, the theory of analogues, or, in the terminology established by Richard Owen, homology. He believed that structural or mechanical constraints during development might have been responsible for the different "balancing" of various sets of parts in different families and even classes. Different ratios of expression of an organ or parts induced different animal forms, thus producing different functions. During the 1820s and 1830s he toyed with the idea that changing environmental conditions might mechanically affect the embryo, thus producing modified organ dispositions, modified functions, and modified animals. His growing theoretical boldness made Geoffroy extend his theory of the unity of type to all animal forms. His claim that arthropods were built on the same plan as vertebrates gave Cuvier the opportunity to engage his former friend in a heated, famous dispute on the unity of type that started at the Académie des sciences in 1830. Although historians have repeatedly asserted that victory went to Cuvier, many in Europe sided with Geoffroy and with philosophical anatomy.

BIBLIOGRAPHY

Appel, Toby A. 1987. *The Cuvier-Geoffroy Debate: French Biology in the Decades before Darwin.* Oxford: Oxford University Press.

Corsi, P. 1988. *The Age of Lamarck: Evolutionary Theories in France, 1790–1830.* Berkeley: University of California Press.

Geoffroy Saint-Hilaire, É. 1818. *Philosophie anatomique.* Paris: Mequignon-Marvis.

Le Guyader, H. 2004. *Etienne Geoffroy Saint-Hilaire, 1772–1844: A Visionary Naturalist.* Chicago: University of Chicago Press.   —P.C.

Gillespie, John H. (b. 1944)

John Gillespie has been the most prominent champion of the argument that patterns of molecular evolution are explained better by the action of natural selection in variable environments than by the so-called neutral theory. His work has included the development of complex mathematical models that describe natural selection under varying conditions and the comparison of various patterns of molecular variation within and among species to the predictions of those, and other, models.

A student of Kenichi Kojima, Gillespie received his doctoral degree in 1970 from the University of Texas at Austin. He approached the question of molecular variation by considering relatively simple models of how natural selection might operate when some forms of a protein were favored under some environmental conditions and others under different conditions, with conditions changing from one generation to the next. In this early work, Gillespie began exploring the conditions under which varying environments created balancing selection, a form of selection in which fitness differences among alleles that controlled protein variation are canceled out over the long term by the fluctuations in environmental conditions. At first glance, this cancellation of fitness differences would seem to devolve into the neutral theory, which postulates that the majority of differences among protein variants produce no differences or only slight differences in the fitness of individuals that carry them. Although in some cases models of variable selection predict the same patterns as do neutral models, many of the long-term patterns of gene substitution are different. In particular, models of variable selection make different predictions about the existence of a molecular clock, which is the apparently regular pattern with which one form of a protein replaces another within a group, such as mammals or birds, over long periods. Gillespie's work showed that whether a clock emerges depends upon the precise distribution of fitness differences and the timescale over which one measures substitutions. These results stand in stark contrast to many aspects of widely accepted neutral models, calling those models into question.

Besides offering the most comprehensive alternative to the neutral theory, Gillespie is responsible for a number of other insightful contributions. He has developed statistical approaches to testing hypotheses about evolutionary rates for both molecular data and data on quantitative characteristics, such as body size or morphological traits. Gillespie's explorations of fitness discounting—how an individual might compromise performance in one condition with the need to perform in many, variable conditions—led to a number of advances by others in the area of life history evolution in a variable environment. His studies of the evolution of mutation and migration rates in variable environments changed biologists' views of how evolvable those rates might be.

The highly sophisticated mathematics that Gillespie employed has restricted his readership among general evolutionary biologists, many of whom may know him more for his delightful introductory textbook on population genetics. However, his work on protein evolution is the major alternative to neutral theory and, as such, occupies a prominent position in the discipline.

Gillespie, J. H. 1991. *The Causes of Molecular Evolution.* New York: Oxford University Press.
———. 2004. *Population Genetics: A Concise Guide.* 2nd ed. Baltimore: Johns Hopkins University Press.

—J.T.

Goethe, Johann Wolfgang von (1749–1832)

Johann Wolfgang von Goethe, well known for his literary masterpieces, was born in 1749 in Frankfurt am Main, during the time of the Holy Roman Empire. He completed *Die Leiden des jungen Werther* (The Sorrows of Young Werther) in 1774. *Faust I*, Goethe's luminous, rich treatment of Europe's struggle with secularization, was first published in 1808. *Faust II*, which recasts and broadens the episodes of *Faust I*, was completed a year before Goethe's death in Weimar in 1832.

Goethe's reputation as a poet should not, however, eclipse his insight and prominence as a natural scientist. Indeed, Goethe believed that science was the product of verse, and that art and the study of nature were inseparable. Goethe's approach to morphology is a fine case in point. For Goethe, the skeleton and the parts of the organic body were to be looked upon as a whole, despite the anatomist's proclivity for dissection. Physiological knowledge of the parts is dependent upon the (artist's) vision of the form of the whole. In 1784, Goethe wrote an essay on the intermaxillary bone and formulated a vertebral theory of the skull, maintaining that the bones of the skull could be derived from the vertebrae. The initial form, according to Goethe, could be recognized despite the considerable change that had occurred.

Goethe's approach to morphology does not emphasize *causa finalis* at the expense of *causa formalis*. The productive approach is not to maintain that a bull has been given horns to butt, but to ask how he might have developed the horns that he, in fact, so uses. Goethe's remarks, and his rejection of the teleological approach, to some extent are precursory to those who criticized the adaptationist's program—the business of understanding every biological trait in terms of selective pressures. Nevertheless, Goethe is more sanguine about locating real explanatory elements in organic nature than, say, Immanuel Kant, who famously denied that there would ever be a Newton of organic nature who would explain even a single blade of grass without an appeal to teleological language. Kant is, perhaps, correct, if the appropriate explanatory elements are to be understood in terms of *causa efficiens*. Goethe, however, distances himself from this kind of view by maintaining that while the nature of an organism is not explained entirely in mechanical terms, there are nevertheless explanatory forms that can be apprehended on the basis of human experience.

The kind of thing Goethe has in mind can be illustrated by his thinking on the organization of plants. The leaf, according to Goethe, is a basic organ. The modification thereof issues in the various parts of a single plant, as well as different plants. He seems to have thought that by systematically observing various plants as the differing instantiations of a developmental principle, a paradigm of such a principle, one could understand the possible permutations of the leaf: "All their shapes are alike, yet none the same as the next and the lot of them suggests a secret law, a sacred puzzle. Could I only, sweet lady, convey to you in a word the happy solution." The poet's occupation is frequently thought to express concisely in verse that which is difficult to articulate. So, too, the biologist in Goethe seems to seek to express the elusive forms that underwrite the experiences of nature.

BIBLIOGRAPHY

Kant, I. 2000. *Critique of the Power of Judgment*. P. Guyer, ed. Cambridge: Cambridge University Press.
Richards, R. J. 2002. *The Romantic Conception of Life*. Chicago: University of Chicago Press.
Seamon, D., and A. Zajonc, eds. 1998. *Goethe's Way of Science: A Phenomenology of Nature*. Albany: State University of New York Press.
Steuer, D. 2002. In defense of experience: Goethe's natural investigations and scientific culture. In L. Sharpe, ed., *Cambridge Companion to Goethe*, 160–168. Cambridge: Cambridge University Press.   —*J.K.*

Goldschmidt, Richard Benedict (1878–1958)

Richard Goldschmidt was one of the most controversial geneticists of the twentieth century. Known as a skilled experimentalist and a creative theorist, his ideas about the nature of the gene and the role of systemic mutations (or large rearrangements of chromosomes) in evolution have been considered heretical, while at the same time they helped to crystallize the so-called modern synthesis, whose participants were, at the very least, united in their opposition to Goldschmidt. More recently, Stephen Jay Gould (1982, in his introduction to the reprint of Goldschmidt's *The Material Basis of Evolution*) and others (e.g., Dietrich 2003) have reevaluated Goldschmidt and argued that his ideas deserve at least as much consideration as the prevailing dogma of adaptationism.

Richard Goldschmidt was born in 1878 in Frankfurt am Main into a prominent German Jewish family and grew up in a well-to-do milieu. He began his studies of medicine in Heidelberg, attending lectures of both Otto Bütschli and Carl Gegenbaur. However, his real interest was zoology, and in 1898 he moved to Munich to study with Richard Hertwig. His first academic appointment was as Hertwig's assistant in Munich. As part of his teaching duties he organized the laboratory in comparative anatomy, which in part explains his vast knowledge of animal morphology, anatomy, and cell biology. In his own research he focused on problems of genetics and cytology, especially on the problem of sex determination. Goldschmidt worked with the Gypsy moth *Lymantria,* whose geographical variants, when mated to each other, produced an assortment of so-called intermediate intersexes. To account for these intermediate forms, Goldschmidt proposed a dynamic and quantitative model of genetics that was based on the underlying physiological action of gene products, hence the name physiological genetics. The basic idea of physiological genetics was that different variants of gene products should have different physiological properties, such as reaction times, which can account for the different observable phenotypes.

Because of his work in genetics and physiology, Goldschmidt was appointed as one of the founding directors of the new Kaiser Wilhelm Institute for Biology in Berlin in 1913. Soon his interests also included the genetic differences between geographic races of *Lymantria*. These differences between races only represented microevolutionary events. To account for larger evolutionary transformations, Goldschmidt postulated the existence of macromutations. His ideas on this subject were also motivated by his study of phenocopies and homeotic mutations, both representing large coordinated phenotypic changes. Goldschmidt's notion of "hopeful monsters," which he used to describe these novel phenotypes, is one of the most unfortunate metaphors in the history of biology and was soon ridiculed by his critics. Ernst Mayr's *Systematics and the Origin of Species* (1942) was written in major part in order to refute Goldschmidt. However, as we have learned more recently in the context of evolutionary developmental biology, large phenotypic changes are often the result of evolutionary changes in the regulatory networks of genes, a concept that is, at least in spirit, very close to Goldschmidt's earlier ideas. Goldschmidt died in 1958 in Berkeley, where he had found refuge after his emigration from Nazi Germany.

BIBLIOGRAPHY

Dietrich, M. R. 2003. Richard Goldschmidt: Hopeful monsters and other "heresies." *Nature Reviews Genetics* 4: 68–74.
Goldschmidt, R. [1940] 1982. *The Material Basis of Evolution.* Introduction by S. J. Gould. New Haven, CT: Yale University Press.
———. 1958. *Theoretical Genetics.* Seattle: University of Washington Press.
———. 1960. *In and Out of the Ivory Tower.* Seattle: University of Washington Press.
Mayr, E. 1942. *Systematics and the Origin of Species.* New York: Columbia University Press.   —*M.D.L.*

Gould, Stephen Jay (1941–2002)

Stephen Jay Gould was born and raised in New York City, in a family and cultural setting that nurtured his interest in natural history, his instinctive iconoclasm, and his left-leaning politics. He attended Antioch College, and at Columbia University he earned his PhD under the direction of paleontologist Norman D. Newell. He also became associated with his lifelong friend and collaborator, Niles Eldredge. Harvard called, and Gould joined the faculty on which he would serve throughout his career.

In his early work, Gould focused on changes in shape that occur in the growth of individuals and in the evolution of animal species and higher taxa. Using quantitative methods of allometry, he showed how changes in proportion of individual characters, in relation to body size, reflect shifts in the rates and timing of their development (see Gould 1966). This proved to be a powerful tool for analysis of adaptive function and patterns of evolution in extinct organisms. Moreover, these shifts in timing, known as heterochrony, provide a potential mechanism for rapid evolutionary change, triggered by modest changes in regulatory genes. In *Ontogeny and Phylogeny* (1977b), Gould assessed the history of the study of heterochrony and showed how it was related to an emerging field that would soon become known as evolutionary developmental biology, now nicknamed evo-devo.

Studies of trilobites and land snails provided Eldredge and Gould with evidence showing that species persist for long periods of time with little if any directional change. However, evolution can occur rapidly in locally isolated populations, giving rise to new species that may seem from the fossil record to have appeared instantaneously. Eldredge and Gould (1972) advocated a model of "punctuated equilibrium" as an alternative to gradualism, that was implicit in most conventional accounts of evolution by natural selection (see also Gould and Eldredge 1977, 1993). This hypothesis prompted a lively controversy that was never fully resolved; both patterns of change can be shown to occur. Many paleontologists tended to favor punctuated equilibrium, whereas most geneticists held to gradualism, both groups swayed by the nature of their data and the timescales over which it is observed. (Stebbins and Ayala 1981 provides geneticists' response to punctuated equilibrium. A historical overview of the controversy can be found in Sepkoski and Ruse 2008.)

Gould always emphasized the role of historical circumstance in evolution. In "The Spandrels of San Marco and the Panglossian Paradigm: A Critique of the Adaptationist Programme" (1979), coauthored with Richard Lewontin, Gould challenged the notion that every aspect of form and behavior of living organisms can be explained directly in terms of natural selection. Natural barriers or constraints of ancestral design and pathways of development, characters with no immediate adaptive function that emerge as by-products of selection for other features, and above all chance must be taken into account. (Selzer 1993 analyzes this paper from a literary perspective.) Relatedly, Gould was a lifelong enthusiast for the ideas of the Scottish morphologist D'Arcy Wentworth Thompson, who always argued that much animal form could be explained as the result of the laws of physics and chemistry without reference to natural selection (see Gould 1971).

Gould was a vigorous opponent of the concept of evolutionary "progress." To him, it implied a predetermined direction that has no basis in principle or on the evidence of life's evolutionary history. Change is unpredictable. Many factors influence the fates of individual species, lead to diversification of major groups of organisms, or cause mass extinctions that have periodically cleared the stage, making way for the evolution of new casts of characters from a few survivors. Gould (1988) is among the several papers that tackle progress. *Full House: The Spread of Excellence from Plato to Darwin* (1996) is full-length discussion directed at the general reader as much as at the professional evolutionist.

Building on the concept of species selection elaborated by Steven Stanley (1975, 1979), Gould argued that patterns of evolution in clades (groups of related species that share a common ancestor) arise from a process of selection different in kind from that which occurs in evolution at the level of populations and speciation. If this macroevolution involves more than the sum of

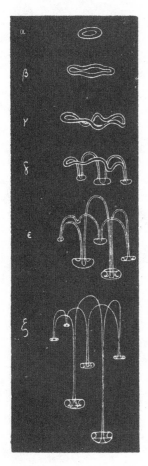
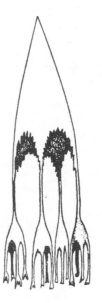

A   B

A particular hero of Stephen Jay Gould was the Scottish morphologist D'Arcy Wentworth Thompson. Illustrations like these, from Thompson's *On Growth and Form*, showing how (apparently purely as a function of physical forces) jellyfish (B) have the same shape as falling drops of ink in water, or oil in paraffin (A), went far to convince Gould that natural selection is not the sole or necessarily even the prime causal force in evolution. Such phenomena have been christened "order for free" by the theoretical biologist Stuart Kauffman (1995, 185). There is some ambiguity in Thompson and in later writers as to whether the claim is that phenomena like the shape of the jellyfish are neither very adaptive nor nonadaptive (and hence selection is not involved) or whether they are adaptive but that selection was not needed to form them. See Gould 1971.

evolutionary change within each species, it is thereby decoupled from the direct effects of natural selection, acting on populations. This led Gould to propose an expanded evolutionary theory, in which selection based on distinct criteria operates at successive levels in the taxonomic hierarchy. Species selection is a focus of current research and experimental tests. The idea that selection operates among clades has yet to be translated into testable hypotheses.

Because he challenged prevailing views, some rightly saw Gould as a critic of neo-Darwinian orthodoxy. Nonetheless, he was an ardent admirer of Charles Darwin, seeing his own work as an expansion of Darwin's enterprise and by no means a refutation of it. Gould's earlier work is most accessible to those seeking an introduction to his ideas; his final magnum opus, *The Structure of Evolutionary Theory* (2002), is a rich but unwieldy tome, more readily quarried than read in its entirety.

In addition to his original research, Gould published numerous essays, most notably in the magazine *Natural History*. These were collected in a series of best-selling books, starting with *Ever since Darwin* (1977a). Other works addressed measures of human brain size and intelligence (Gould 1981), the bizarre extinct animals of Cambrian seas (Gould 1989), and the relation between science and religion (Gould 1999). He testified as an expert witness in the Arkansas court case where "creation science" was shown to be a religious doctrine. Gould never hesitated to join an intellectual fight. As a result, Gould became the premier public face of evolution in the United States during the last quarter of the twentieth century.

BIBLIOGRAPHY

Allmon, W. D., P. H. Kelley, and R. M. Ross, eds. 2008. *Stephen Jay Gould: Reflections on His View of Life.* New York: Oxford University Press.

Eldredge, N., and S. J. Gould. 1972. Punctuated equilibria: An alternative to phyletic gradualism. In T. J. M. Schopf, ed., *Models in Paleobiology*, 82–115. San Francisco: Freeman, Cooper and Co.

Gould, S. J. 1966. Allometry and size in ontogeny and phylogeny. *Biological Reviews of the Cambridge Philosophical Society* 41: 587–640.

———. 1971. D'Arcy Thompson and the science of form. *New Literary History* 2: 229–258.

———. 1977a. *Ever since Darwin.* New York: Norton.

———. 1977b. *Ontogeny and Phylogeny.* Cambridge, MA: Harvard University Press.

———. 1981. *The Mismeasure of Man.* New York: Norton.

———. 1988. On replacing the idea of progress with an operational notion of directionality. In M. H. Nitecki, ed., *Evolutionary Progress*, 319–338. Chicago: University of Chicago Press.

———. 1989. *Wonderful Life: The Burgess Shale and the Nature of History.* New York: Norton.

———. 1996. *Full House: The Spread of Excellence from Plato to Darwin.* New York: Paragon.

———. 1999. *Rocks of Ages: Science and Religion in the Fullness of Life.* New York: Ballantine.

———. 2002. *The Structure of Evolutionary Theory.* Cambridge, MA: Harvard University Press.

Gould, S. J., and N. Eldredge. 1977. Punctuated equilibria: The tempo and mode of evolution reconsidered. *Paleobiology* 3: 115–151.

———. 1993. Punctuated equilibrium comes of age. *Nature* 366: 223–227.

Gould, S. J., and R. C. Lewontin. 1979. The spandrels of San Marco and the Panglossian paradigm: A critique of the adaptationist programme. *Proceedings of the Royal Society of London, Series B* 205: 581–598.

Selzer, J., ed. 1993. *Understanding Scientific Prose.* Madison: University of Wisconsin Press.

Sepkoski, D., and M. Ruse, eds. 2008. *The Paleobiological Revolution.* Chicago: University of Chicago Press.

Stanley, S. M. 1975. A theory of evolution above the species level. *Proceedings of the National Academy of Sciences* 72: 646–650.

———. 1979. *Macroevolution, Pattern and Process.* San Francisco: W. H. Freeman.

Stebbins, G. L., and F. J. Ayala. 1981. Is a new evolutionary synthesis necessary? *Science* 213: 967–971.   —*R.D.K.T.*

Grant, B. Rosemary (b. 1936), and Grant, Peter R. (b. 1936)

Rosemary and Peter Grant are renowned for their long-term studies of Darwin's finches in the Galápagos Islands. They and their collaborators have performed one of the most comprehensive studies of any group of species, describing the feeding ecology, breeding biology, and numerical dynamics of populations; documenting natural selection on morphological traits and patterns of mate choice; discovering hybridization between species; and reconstructing the phylogenetic relationships among them. Their work stands as the signal case study of evolutionary biology.

The Grants received undergraduate degrees in 1960, Rosemary (B.R.G.) with honors from Edinburgh and Peter (P.R.G.) with honors from Cambridge. They met and married while attending the University of British Columbia. P.R.G. took his doctorate in 1964 from the University of British Columbia; B.R.G. took hers in 1985 from Uppsala University. After a postdoctoral year at Yale, P.R.G. held faculty positions at McGill University, the University of Michigan, and Princeton University while B.R.G. held research positions at these universities. P.R.G.'s early work focused on the ecology of island bird populations, including patterns of distribution across archipelagoes and morphological variation among species and among populations on different islands. B.R.G.'s early work on Darwin's finches examined variation among individuals in bill morphology and song characteristics, leading to her early papers on selection on bill morphology and the possibilities for sympatric speciation. The Grants' commitment to studying Darwin's finches took them to the Galápagos for six months of every year (since 1973), despite the challenges of logistics, limited diet, and the spartan lifestyle of living in tents.

The Grants have received a host of honors, individually and together, including the Darwin Medal of the Royal Society of London, the Balzan Prize, and the Darwin-Wallace Award of the Linnean Society of London. Both Grants were elected to the American Academy of Arts and Sciences in 1997; they were both elected as Fellows of the Royal Society of London (B.R.G. in 2007, P.R.G. in 2003) and foreign members of the U.S. National Academy of Sciences (B.R.G. in 2008, P.R.G. in 2007).

BIBLIOGRAPHY

Grant, B. R., and P. R. Grant. 1989. *Evolutionary Dynamics of a Natural Population: The Large Cactus Finch of the Galápagos.* Chicago: University of Chicago Press.

Grant, P. R. 1986. *Ecology and Evolution of Darwin's Finches.* Princeton, NJ: Princeton University Press.

Grant, P. R., and B. R. Grant. 2007. *How and Why Species Multiply: The Radiation of Darwin's Finches.* Princeton, NJ: Princeton University Press.   —*J.T.*

Grant, Robert Edmond (1793–1874)

Robert Edmond Grant, an Edinburgh-born sponge expert, introduced comparative anatomy to Britain in the 1820s and 1830s. He incorporated the studies of Jean-Baptiste Lamarck and Étienne Geoffroy Saint-Hilaire, accepting the "transformism," or evolution, accompanying their works. Grant also inducted the young Charles Darwin into Lamarck's views. Darwin, bored by medicine at Edinburgh University in 1825–1827, studied North Sea invertebrates with Grant, who had become an MD in 1814. As an old man writing his *Autobiography,* Darwin recalled Grant's admiration for Lamarck.

Grant championed Henri de Blainville's continuous chain of life (rather than Georges Cuvier's discrete *embranchements*) and Geoffroy's belief that the same organs occurred throughout the chain—hence Grant's claim in 1825 to have found the homology of the vertebrate's pancreas in mollusks. Studying sponges in 1825–1828, Grant coined the group's name, Porifera, and he mooted a simple, defenseless, freshwater form as the "parent" of the armored marine species. Darwin examined the larvae of the local bryozoan *Flustra* under Grant. Grant believed these larvae were related to the "ova" of algae and pointed back to the spontaneously generated "monads" lying at the junction of the plant and animal kingdoms. At Edinburgh Darwin was already broaching issues surrounding the birth of individuals and species, even if he rejected Grant's monadism upon beginning his own evolutionary speculations.

As the first professor of comparative anatomy at London University in 1827, Grant alerted the metropolitan elite to the Lamarckian threat. He joined the Geological Society council in 1832, just as Charles Lyell was publishing his antievolutionary volume *Principles of Geology.* Grant's fossil lectures promoted the "direct generation" of successive species, one giving birth to another. And his students were examined on the conditions responsible for "originating and effacing the temporary organic film on our planet" (Zoology Examination Papers, 1857–58, p. 6, in "Grant on Zoological Subjects," College Collection DG 76, University College London).

Grant took the Fullerian chair of physiology at the Royal Institution in 1837, but that marked his apogee. He had lost his council seat at the Zoological Society in 1835, when the zoo's aristocratic patrons gave his conservative rival, Richard Owen, preferential treatment. At the Geological Society, Owen countered the fossil implications of Grant's transmutationism. Grant lost resources and his finances teetered. The 1840s found him living in a slum, his shabby swallow-tail coat as unfashionable as his lectures. A rising young Thomas Henry Huxley in 1852 merely laughed at Grant's "eccentricity."

At the end, Grant envisaged planetary cooling as the motor to power the evolution of warm-blooded mammals and birds from dinosaurs and pterosaurs. In

Tabular View of the Primary Divisions of the Animal Kingdom (1861) he praised Darwin's *On the Origin of Species* while reminding his old pupil of their Edinburgh days together. But the *Tabular View*'s countless evolutionary trees sprouting from spontaneously generated monads showed how deeply the two men differed. A seminal figure in the radical 1830s, Grant outlived his age, and few mourned his passing in 1874.

BIBLIOGRAPHY

Desmond, A. 1984a. Robert E. Grant: The social predicament of a pre-Darwinian transmutationist. *Journal of the History of Biology* 17: 189–223.

———. 1984b. Robert E. Grant's later views on organic development: The Swiney Lectures on "palaeozoology," 1853–1857. *Archives of Natural History* 11: 395–413.

———. 1989. *The Politics of Evolution: Morphology, Medicine, and Reform in Radical London.* Chicago: University of Chicago Press.

Sloan, P. R. 1985. Darwin's invertebrate program, 1826–1836: Preconditions for transformism. In D. Kohn, ed., *The Darwinian Heritage*, 71–120. Princeton, NJ: Princeton University Press.   —A.D.

Gray, Asa (1810–1888)

Asa Gray was one of the leading botanists of mid-nineteenth-century America. He became one of Charles Darwin's most active supporters, using arguments derived from the geographic distribution of plants to defend the theory of evolution. A deeply religious man, he also wrote extensively on the theological implications of Darwinism, attempting to show how evolution by natural selection could be reconciled with the belief that the world was designed by a wise and benevolent God.

Gray was born in Sauquoit, New York, on November 18, 1810. He trained in medicine and practiced for a few years, but abandoned this for part-time teaching so that he would have more time for botany. He was appointed Fisher Professor of Natural History at Harvard in 1842, where it was understood that he would concentrate exclusively on botany. He visited Europe to consult with other botanists, but his own work focused mainly on North American plants. He worked on a *Flora of North America* with John Torrey (1838–1843) and eventually produced a highly successful *Manual of the Botany of the Northern United States* (1848). He retired in 1873 and died on January 30, 1888.

Gray's work on the classification and geographic distribution of North American plants made him suspicious of the "creationist" position being advocated by Louis Agassiz. Charles Darwin confided in Gray before the publication of *On the Origin of Species* in 1859, and an abstract of Darwin's letter to Gray, dated September 5, 1857, formed part of the joint paper by Darwin and A. R. Wallace in 1858 (reprinted in Darwin and Wallace 1958, 264–267). Gray subsequently defended the theory of evolution against attacks from Agassiz at the American Academy of Arts and Sciences.

Several of his papers, supporting evolutionism on the basis of botanical geography, were collected in his *Darwiniana* (1876). Gray saw how the dispersal of plants from an original location, coupled with adaptation of populations to the local environments found in their new homes, could account for the observed distribution of species. The similarity between the plants of eastern North America and Japan could be explained by noting that in earlier geological times, these plants had been widespread across Asia and North America but had been driven southward during the ice ages, with only the two isolated fragments of the earlier population now remaining.

As a staunch Presbyterian, Gray was anxious to show that Darwin's theory did not destroy the Argument from Design. In a series of articles (reprinted in *Darwiniana*) he began by arguing that any process by which species could be adapted to their environment was compatible with the belief that the evolutionary process had been established by a wise and benevolent Creator. Later on, however, he conceded that the wastefulness and cruelty of natural selection generated difficulties for this approach, and he suggested that variation is led along beneficial lines by the Creator. Darwin objected to this view because it left Gray supporting the position of theistic evolutionism, in which the direction of development is determined by supernatural rather than natural means.

BIBLIOGRAPHY.

Darwin, C. R., and A. R. Wallace. 1958. *Evolution by Natural Selection.* Cambridge: Cambridge University Press.

Dupree, A. H. 1959. *Asa Gray.* Cambridge, MA: Harvard University Press.

Gray, A. 1848. *Manual of the Botany of the Northern United States.* Boston: J. Munroe.

———. 1876. *Darwiniana: Essays and Reviews Pertaining to Darwinism.* New York: Appleton.

Torrey, J., and A. Gray. 1838–1843. *A Flora of North America.* 2 vols. New York: Wiley and Putnam.   —*P.J.B.*

Group selection

Charles Darwin identified a fundamental problem with social life and its potential solution in the following famous passage from *The Descent of Man* (Darwin 1871, 166): "It must not be forgotten that although a high standard of morality gives but a slight or no advantage to each individual man and his children over other men of the same tribe, yet that an increase in the number of well-endowed men and advancement in the standard of morality will certainly give an immense advantage to one tribe over another."

This problem and its potential solution became the basis of a theoretical framework in evolutionary biology called multilevel selection (Sober and Wilson 1998). Because natural selection is based on relative fitness, a trait that benefits the group as a whole does not change in frequency within the group. A trait that benefits others at the expense of the self actually decreases in frequency within the group and will ultimately become extinct in the absence of other evolutionary forces. This describes the first part of Darwin's passage quoted above. Fortunately, groups of individuals who benefit each other will contribute more to the total gene pool than more selfish groups, as described in the second part of Darwin's passage. Traits that are "for the good of the group" can evolve by natural selection, but only by a process of selection among groups in a larger population and often in opposition to selection among individuals within single groups.

It is impossible to describe the concept of group selection without also describing its turbulent history. Many biologists during the first half of the twentieth century did not share Darwin's insight; they assumed that adaptations straightforwardly evolve all levels of the biological hierarchy, from genes to ecosystems. This position became known as naïve group selection and was widely criticized in the 1960s, especially by George C. Williams in his book *Adaptation and Natural Selection* (1966), which became a modern classic. Williams affirmed the importance of multilevel selection as a theoretical framework, agreeing with Darwin that group-level adaptations can evolve only by a process of between-group selection. Williams then made the empirical claim that between-group selection is almost invariably weak compared to within-group selection. It was this empirical claim that turned multilevel selection theory into "the theory of individual selection," which became the consensus for the rest of the twentieth century. Subsequent evolutionary theories of social behavior, such as kin selection, reciprocal altruism, game theory, and selfish gene theory, were developed explicitly as alternatives to group selection.

The 1960s consensus was based upon three arguments: (1) between-group selection is theoretically implausible, requiring a delicate combination of parameter values to prevail against within-group selection; (2) there is no convincing empirical evidence for group selection; and (3) the theories proposed as alternatives do not invoke multilevel selection in their own right. All three arguments began to be questioned, even as early as the 1970s, although it was difficult to argue in favor of a topic that for many had become taboo. Modern theoretical models of group selection are far more plausible than their predecessors. Williams himself acknowledged the importance of group selection for traits such as sex ratio and disease virulence (Williams 1992; Williams and Nesse 1991). All evolutionary theories of social behavior must assume the existence of multiple groups to be biologically realistic. When these groups are identified, cooperative and altruistic traits are selectively disadvantageous within the groups and require between-group selection to evolve, as always envisioned by multilevel selection theory (Sober and Wilson 1998).

A major event in evolutionary theory occurred in the 1970s with the discovery that the single organisms of today were the groups of past ages (Margulis 1970; Maynard Smith and Szathmary 1995). Evolution proceeds not only by small mutational change, but also by single-species groups and multi-species symbiotic associations becoming so integrated that they become higher-level organisms in their own right. These major transitions of life, as they are called, have occurred repeatedly and possibly include the origin of life itself as groups of cooperative molecular reactions. Despite multilevel selection theory's turbulent past, it is the accepted framework for studying major transitions. The balance between levels of selection is not static but can itself evolve. A major transition occurs when selection within groups is suppressed, enabling between-group selection to become the primary evolutionary force.

The paradigm of major transitions did not appear until the 1970s and was not generalized until the 1990s, but already it seems likely to explain the evolution of social insect colonies (Wilson and Hölldobler 2005) and humans as the first ultrasocial primate species (Boehm 1999). The traits associated with human morality appear designed to suppress selection within groups, similar to genetic mechanisms such as chromosomes and the rules of meiosis, enabling selection among groups to become the primary evolutionary force. Cultural processes can cause groups to become very different in their phenotypic properties, even when they are genetically similar. Multilevel selection has become the accepted framework for the study of human cultural evolution, in exactly the same way as the study of major transitions (Richerson and Boyd 2004).

The revival of group selection might seem extraordinary until we remember that Williams (1966) accepted multilevel selection as a theoretical framework, rejecting group selection on the basis of an empirical claim that it is always weak compared to within-group selection. When Williams changed his mind about the theoretical and empirical evidence, he himself reverted back to multilevel selection. For female-biased sex ratios (Williams 1992, 49), he stated, "I think it desirable . . . to realize that selection in female-biased Mendelian populations favors males, and that it is only the selection among such groups that can favor the female bias." For the evolution of virulence in disease organisms (Williams and Nesse 1991, 8), he stated, "The evolutionary outcome will depend on relative strengths of within-host and between-host competition in pathogen evolution."

Unfortunately, the field as a whole has not yet achieved a new consensus. Books and articles written from a multilevel perspective appear alongside other books and articles that continue to treat group selection as a rejected concept. Textbooks have been particularly slow to reflect the many developments that have taken place since the 1960s. The turbulent history of this important subject is reviewed by Borrello (2005) in an article appropriately titled "The Rise, Fall, and Resurrection of Group Selection."

BIBLIOGRAPHY

Boehm, C. 1999. *Hierarchy in the Forest: Egalitarianism and the Evolution of Human Altruism.* Cambridge, MA: Harvard University Press.

Borrello, M. E. 2005. The rise, fall, and resurrection of group selection. *Endeavor* 29: 43–47.

Darwin, C. 1871. *The Descent of Man, and Selection in Relation to Sex.* New York: D. Appleton and Co.

Margulis, L. 1970. *Origin of Eukaryotic Cells.* New Haven, CT: Yale University Press.

Maynard Smith, J., and E. Szathmary. 1995. *The Major Transitions of Life.* New York: W. H. Freeman.

Richerson, P. J., and R. Boyd. 2004. *Not by Genes Alone: How Culture Transformed Human Evolution*. Chicago: University of Chicago Press.

Sober, E., and D. S. Wilson. 1998. *Unto Others: The Evolution and Psychology of Unselfish Behavior*. Cambridge, MA: Harvard University Press.

Williams, G. C. 1966. *Adaptation and Natural Selection: A Critique of Some Current Evolutionary Thought*. Princeton, NJ: Princeton University Press.

———. 1992. *Natural Selection: Domains, Levels and Challenges*. Oxford: Oxford University Press.

Williams, G. C., and R. M. Nesse. 1991. The dawn of Darwinian medicine. *Quarterly Review of Biology* 66: 1–22.

Wilson, E. O., and B. Hölldobler. 2005. Eusociality: Origin and consequences. *Proceedings of the National Academy of Sciences* 102: 13367–13371.

—*D.S.W.*

H

Haeckel, Ernst (1834–1919)

Ernst Heinrich Philipp August Haeckel was Charles Darwin's foremost champion at the turn of the twentieth century. More people prior to World War I learned of evolutionary theory through his voluminous publications than through any other source. His *Natürliche Schöpfungsgeschichte* (Natural history of creation, 1868) went through 12 increasingly augmented German editions (1868–1920) and was translated into the major European languages. Erik Nordenskiöld, in the first decades of the twentieth century, judged it "the chief source of the world's knowledge of Darwinism" (1936, 515). The crumbling detritus of this synthetic work can still be found scattered along the shelves of most used-book stores. *Die Welträthsel* (The world puzzles, 1899), which placed evolutionary ideas in a broader philosophical and social context, sold over 40,000 copies in the first year of its publication and well over 15 times that number during the next quarter century—and this only in the German editions. (By contrast, during the three decades between 1859 and 1890, Darwin's *On the Origin of Species* sold only some 39,000 copies in six English editions.) By 1912, *Die Welträthsel* had been translated, according to Haeckel's own meticulous tabulations, into 24 languages, including Armenian, Chinese, Hebrew, Sanskrit, and Esperanto. The young Mohandas Gandhi had requested permission to render it into Gujarati; he believed it the scientific antidote to the deadly religious wars plaguing India. Haeckel achieved many other popular successes, and he produced more than 20 large technical monographs on various aspects of systematic biology and evolutionary history. His studies of radiolarians, medusae, sponges, and siphonophores remain standard references today. These works not only informed the public, but they drew to Haeckel's small university in Jena the largest share of Europe's great biologists of the next generation, among whom were Richard Hertwig (1850–1937) and his brother Oscar Hertwig (1849–1922), Anton Dohrn (1840–1909), Hermann Fol (1845–1892), Eduard Strasburger (1844–1912), W. O. Kovalevsky (1842–1883), Nikolai Miklucho-Maclay (1846–1888), Arnold Lang (1855–1914), Richard Semon (1859–1918), Wilhelm Roux (1850–1924), and Hans Driesch (1867–1941). Haeckel's influence extended far into succeeding generations of biologists, many of whom recalled reading his popular works as young students.

Haeckel received his medical degree from Würzburg in 1858, after which he planned to do his habilitation with Johannes Müller (1801–1858) at

Berlin. Müller's suicide led him to turn to Carl Gegenbaur (1826–1903) at Jena, who became his adviser. During his research work in southern Italy and Sicily, he fell in with a group of German artists, among whom was the poet Hermann Allmers (1821–1902), who became a lifelong friend. Haeckel, a gifted painter (see figure on page 624), thought of giving up biological research for the life of a Bohemian; only his betrothal to Anna Sethe (1835–1864), his first cousin, kept him focused on establishing a professional career. With a small tract by Müller as his inspiration, Haeckel concentrated his research on the little-known group of radiolaria, creatures about the size of a pinhead that secrete exoskeletons of unusual geometries. While completing his habilitation back in Berlin, he read Darwin's *Origin* in Georg Heinrich Bronn's (1800–1862) German translation and immediately became a convert. His research finally yielded, in 1862, *Die Radiolarien*—a magnificent two-volume folio having extraordinarily beautiful plates based on his own illustrations. The book won the admiration of Darwin, who received the volumes by way of introduction to this new disciple. Haeckel's research had the added benefit of allowing him to marry Anna and to begin his life as extraordinarius professor in the medical school at Jena.

Haeckel's brilliant beginning turned dark in 1864 when his wife of eighteen months suddenly died. He suffered a nervous collapse, and during his recovery he wrote his parents that he could no longer accept their religious creed. Rather, he would put his faith in something more reliable, namely, the Darwinian promise of progressive transformation. He then developed that conviction in considerable detail in a large two-volume, theoretical application of Darwinian ideas to all areas of biology, including human evolution. His *Generelle Morphologie der Organismen* (General morphology of organisms, 1866) laid down the fundamental conceptions that he would cultivate for the rest of his career. He made central an idea that he found intimated in Darwin but more carefully worked out by Fritz Müller (1821–1897) in his book *Für Darwin* (1864), namely, the principle of recapitulation—the proposition that the embryo of a given species would pass through the same morphological stages as the phylum had in its evolutionary descent. Haeckel's *Generelle Morphologie* formulated several new perspectives, outfitting them with neologisms that gave his treatise a formidable cast: phylum, ontogeny, ecology, and a host of other terms that had a shorter life span. He also introduced tree diagrams to illustrate the descent of species and to suggest their morphological and temporal distance from one another. The book concluded by advancing a Goethean monism as the appropriate metaphysical position for the naturalist: God and nature, mind and body were to be regarded as expressions of the same underlying *Urstuff*. Darwin and Thomas Henry Huxley (1825–1895) initially sought to have an abridged version of the *Generelle Morphologie*, shorn of its polemical barbs, translated into English.

In order to seek a wider audience for his theoretical treatise, Haeckel delivered a series of popular lectures in 1868 summarizing his Darwinian morphology. The series was published the same year under the title *Natürliche Schöpfungsgeschichte*, and it achieved immediate notoriety. In an initial

Ernst Haeckel was a brilliant artist and always filled his books with his own drawings. This one, revealingly labeled the "Apotheosis of Evolutionary Thought" (from the supplement to Haeckel's *Wanderbilder*, 1905), is an interesting reflection on his own private life—he was grieving the death of a young woman, to whom he was not married but with whom he had just had a passionate (spiritual but also physical) affair—as well as the changing times. Sigmund Freud's *Three Essays on the Theory of Sexuality* (greatly influenced by Haeckel's thinking on recapitulation) was published in the same year this picture appeared.

review, Ludwig Rütimeyer (1825–1895), an embryologist at Basel, charged Haeckel with fraud. He observed that in illustrating the principle of recapitulation—or the biogenetic law, as it became known—Haeckel had represented very young embryos of a dog, chicken, and turtle as morphologically identical. Rütimeyer maintained, however, that Haeckel had made the case by using the same woodcut three times. In the next edition of the book (1870), Haeckel used only one illustration of a vertebrate embryo at a very early stage and said it might as well be the embryo of a dog, chicken, or turtle because you cannot tell the difference. The damage, however, was done, and the charge of fraud would haunt Haeckel for the rest of his days.

Despite the controversy, *Natürliche Schöpfungsgeschichte* made a powerful impact on its readers, especially on the topic of human evolution. He represented nine species of human beings along a tree of evolutionary development, with the Papuans and Hottentots at the lowest branches, closer to roots in the *Urmensch,* or ape-man, and with the Caucasian branch at the highest level, carrying at the top reaches the Mediterraneans, Germans, Jews, and Arabs. Although Haeckel shared many of the racial views common to nineteenth-century Europeans, he was decidedly not anti-Semitic, an attitude which one of his disaffected students held against him. He argued, following his friend, the linguist August Schleicher (1821–1868), that grades of human mental ability expressed grades of language complexity and that the European and Semitic languages helped create a correspondingly complex mind—a general thesis that Darwin adopted in *The Descent of Man.*

In 1867, after visiting Darwin and other British scientists in England, Haeckel traveled to the Canary Islands with three research associates. He performed the kinds of experiments on developing siphonophore embryos that would garner fame for Wilhelm Roux and Hans Driesch some 20 years later. He also began work on a systematic analysis of calcareous sponges that would yield a three-volume study, *Die Kalkschwämme* (The calcareous sponges, 1872). In this work, Haeckel attempted to provide what Bronn maintained was necessary to show the viability of Darwin's theory, namely, empirical proof that species descent was more than a theoretical possibility. Haeckel also argued, employing the biogenetic law, that in ancient times an organism, having the structure of a primitive sponge (and the form taken by metazoans in gastrulation), plied the ancient seas. This became his gastraea theory.

Because of his various investigations of marine invertebrates, Haeckel received the commission in the late 1870s to describe systematically several classes of organisms dredged up by HMS *Challenger.* Over a 10-year period, he composed several large volumes on medusae, calcareous sponges, siphonophores, and radiolaria—with more pages produced than by any other author in the series of *Challenger* reports. The commission indicated the high regard of the scientific community for his work in marine biology. That regard was also expressed by the many honorary degrees and awards he received during his lifetime.

Resentment by the biologically and religiously orthodox continued to build against Haeckel throughout the 1870s, and it has not abated in to this day. In 1874, the Swiss embryologist Wilhelm His (1831–1904) published *Unsere Körperform und das physiologische Problem ihrer Entstehung* (Our bodily form and the physiological problem of its origin), which repeated the earlier charges of fraud against Haeckel and instituted new ones. Among other claims, His asserted that Haeckel had represented the human embryo with an exaggeratedly long tail—a controversy that became known as the *Schwanzfrage*. In 1877, Rudolf Virchow (1821–1902) rejected his onetime student's efforts to have evolutionary theory taught in the lower schools in Germany. Virchow charged that evolutionary thought abetted socialists and communists, a claim that Huxley thought quite scurrilous because of its inflammatory character in Bismarck's Germany—although, in fact, many Marxists (e.g., August Bebel, 1840–1913) did find Darwinism congenial. At the turn of the century, religious opponents of Haeckel's *Welträthsel* and of his newly established Monist League renewed the claims of falsehood. These many charges had their foundation in Haeckel's acknowledged slip in 1868, but thereafter they gained force mostly from intellectual recalcitrance and religious dogmatism. More recently, Daniel Gasman (1971) and Stephen Jay Gould (1977) argued that Haeckel's biology supported Nazi racism, although they conveniently ignored Haeckel's philo-Semitism, an attitude quite unusual for the period. Michael Richardson reexamined Haeckel's illustrations of embryos, and he too suggested Haeckel's malfeasance. Richardson compared Haeckel's illustrations with photographs of embryos, and easily showed the deviations. However, Haeckel had adapted illustrations from then-contemporary sources. He showed that when you lined up depictions rendered by experts, the similarity of evolutionarily related types at earlier stages of embryogenesis became manifest—a phenomenon acknowledged by today's embryologists. Creationists and Intelligent Design theorists have cited the older German literature and Richardson's photographs to indict not only Haeckel but all of evolutionary theory.

BIBLIOGRAPHY

Darwin, C. 1860. *Über die Entstehung der Arten im Thier- und Pflanzen-Reich durch natürliche Züchtung oder, Erhaltung der vervollkommneten Rassen im Kampfe um's Daseyn.* H. Bronn, trans. Stuttgart: Schweizerbart'sche Verlagshandlung.

Di Gregorio, M. 2005. *From Here to Eternity: Ernst Haeckel and Scientific Faith.* Göttingen: Vandenhoeck and Ruprecht.

Driesch, H. 1891. Entwicklungsmechanische Studien. *Zeitschrift für wissenschaftliche Zoologie* 53: 160–184.

Gasman, D. 1971. *The Scientific Origins of National Socialism.* New York: Science History Publications.

Gould, S. J. 1977. *Ontogeny and Phylogeny.* Cambridge, MA: Harvard University Press.

Haeckel, E. 1862. *Die Radiolarien (Rhizopoda Radiaria): Eine Monographie.* 2 vols. Berlin: Georg Reimer.

———. 1866. *Generelle Morphologie der Organismen.* 2 vols. Berlin: Georg Reimer.

————. 1868. *Die Natürliche Schöpfungsgeschichte*. Berlin: Georg Reimer.

————. 1872. *Die Kalkschwämme*. 3 vols. Berlin: Georg Reimer.

————. 1899. *Die Welträthsel*. Bonn: Emil Strauss.

His, W. 1874. *Unsere Körperform und das physiologische Problem ihrer Entstehung*. Leipzig: Vogel.

Krausse, E. 1984. *Ernst Haeckel*. Leipzig: Teubner.

Müller, F. 1864. *Für Darwin*. Leipzig: Engelmann.

Nordenskiöld, E. [1920–1924] 1936. *The History of Biology*. New York: Tudor Publishing.

Pennisi, E. 1997. Haeckel's embryos: Fraud rediscovered. *Science* 277: 1435.

Richards, R. J. 2005. The aesthetic and morphological foundations of Ernst Haeckel's evolutionary project. In M. Kemperink and P. Dassen, eds., *The Many Faces of Evolution in Europe, 1860–1914*. Amsterdam: Peeters.

————. 2008. *The Tragic Sense of Life: Ernst Haeckel and the Struggle over Evolutionary Thought*. Chicago: University of Chicago Press.

Roux, W. 1888. Beiträge zur Entwickelungsmechanik des Embryo. *Archiv für pathologische Anatomie und Physiologie und für klinische Medicin* 94: 113–153, 246–291.   —R.J.R.

Haldane, J. B. S. (1892–1964)

John Burdon Sanderson Haldane, known familiarly as J. B. S., was born on November 5, 1892, in Edinburgh, Scotland, to Louisa Kathleen Haldane and the physician John Scott Haldane. J. B. S.'s younger sister was Naomi Haldane (later Mitchison). The Haldane family is descended from Scottish aristocrats. J. B. S. Haldane is known principally as one of the three architects of the modern synthetic theory of evolution, along with R. A. Fisher (1890–1962) and Sewall Wright (1889–1988). Haldane also made original contributions to biochemistry. In addition, he was a popular science, political, and fiction writer.

Haldane was educated at the Dragon School, Eton College, and New College, Oxford, in England. He entered New College in 1911 by way of a mathematics scholarship won at Eton. A year later, Haldane won first-class honors in mathematics and subsequently, in 1914, first-class honors in classics and philosophy. Haldane did not possess a science degree, yet his interest in science developed at an early age, mainly due to the influence of his father and sister. In 1901, when Haldane was eight, he and his father attended a lecture where the biologist A. D. Darbishire (1879–1915) discussed the newly rediscovered principles of Mendelism. Along with his sister, who was breeding guinea pigs, Haldane did experiments looking for patterns of Mendelian heredity in 1908. By 1912, Haldane believed he had evidence of linkage (i.e., the inheritance of two alleles together), and in 1915, before entering World War I as a member of the Scottish Black Watch, his view was published in the *Journal of Genetics*.

In 1919, Haldane became a fellow of New College, Oxford. He worked diligently on linkage, publishing several papers. In 1922 he moved to Cambridge University, where he would stay until 1932. At Cambridge, Haldane became, with Fisher and Wright, a key figure in the synthesis of Darwinian natural selection with the principles of Mendelian heredity. Indeed, Haldane would play no small role in ushering in theoretical population genetics. In 1924, Haldane published the first of a series of 10 mathematical papers exploring the reconciliation of Darwinism with Mendelism; the last in the series was published in 1934, just two years after publication of his book *The Causes of Evolution* (1932).

Causes was the capstone text not only of Haldane's series of papers on evolution; it was the capstone text of the origins of theoretical population genetics. Haldane's problematic was aligned with his contemporaries; he wanted to understand the nature and significance of natural selection in Mendelian populations. Indeed, the book is known today primarily for its critical comparison of Fisher's, Wright's, and Haldane's own mathematical explorations of this issue. But the book is much more than that. Haldane accomplishes a synthesis not merely of Darwinism and Mendelian heredity, but also, under the rubric of population genetics, of chromosomal mechanics, cytology, and biochemistry.

During his tenure at Cambridge, Haldane's work was by no means restricted to evolution. In 1923, Haldane published his famous *Daedalus, or Science and the Future,* a short popular book on science in its social context. The book is said to have influenced Haldane's friend Aldous Huxley's *Brave New World* (1932). Haldane further contributed Haldane's principle, the idea that sheer size often defines what bodily equipment animals require, in his 1927 essay "On Being the Right Size," in the collection *Possible Worlds and Other Essays.* In 1929, Haldane's speculations on the origins of life were prescient, coinciding with those of Aleksandr Oparin (1894–1980). In 1930, Haldane's *Enzymes* laid down the foundations of enzyme kinetics. Despite his frenetic professional pace, Haldane wed Charlotte Franken Burghes in 1924. This caused some controversy because Charlotte was married at the time (to Jack Burghes) and had to divorce before marrying Haldane. J. B. S. later divorced Charlotte and married biologist Helen Spurway in 1945.

Haldane left Cambridge for University College, London, and became Chair of Biometry in 1937. Also in 1937, Haldane published his paper, "The Effects of Variation on Fitness," arguing that the decline of the average fitness of a population due to mutation of the optimal allele was proportional to the mutation rate but not to the induced decrease in fitness. Haldane's conclusion would later be called the "principle of the mutation load." At this time, Haldane became a member of the Communist Party and, in fact, wrote a number of political papers on Marxism for *The Daily Worker.* In 1950, Haldane broke with the party due largely to Lysenkoism and to the crimes of Soviet leader Joseph Stalin. This was a period in the Soviet Union during which science was politicized and many defenders of Darwin were sent to the gulags.

Haldane was awarded the Darwin Medal of the Royal Society in 1953, and he was awarded an honorary LLD from the University of Paris in 1956. Haldane left Cambridge in 1957 and emigrated to India and ultimately became an Indian citizen. The move was a protest of British policy during the Suez Crisis. In India, Haldane was a research professor at the Indian Statistical Institute. In his famous paper, "The Cost of Natural Selection" (1957), Haldane argued that there is a cap on the rate of adaptive evolution because of a substitution load, that is, because a substitution (by mutation or change in environment) could not start more frequently than every 300 generations. Haldane's claim was later called Haldane's dilemma.

Haldane was awarded the Darwin-Wallace Medal of the Royal Society in 1958, and in 1961 received the Kimber Medal of the U.S. National Academy of Sciences. After a prodigious career, Haldane died of colon cancer in Bhubaneswar, Orissa, India, on December 1, 1964. Serious illness, however, did not dampen Haldane's spirit. In his poem, "Cancer Is a Funny Thing," he wrote, "I know that cancer often kills / But so do cars and sleeping pills / And it can hurt one till one sweats / So can bad teeth and unpaid debts / A spot of laughter, I am sure / Often accelerates one's cure / So let us patients do our bit / To help the surgeons make us fit" (Clark 1968). Haldane has been described as larger than life due to his service in the Black Watch, his famous wit, and his political views. Haldane was, to be sure, a biologist and then some.

BIBLIOGRAPHY

Clark, R. W. 1968. *JBS: The Life and Work of J. B. S. Haldane*. London: Coward-McCann.

Haldane, J. B. S. 1923. *Daedalus, or Science and the Future*. London: Kegan Paul, Trench, and Trubner.

———. 1924. A mathematical theory of natural and artificial selection, I. *Transactions of the Cambridge Philosophical Society* 23: 19–41.

———. 1927. *Possible Worlds and Other Essays*. London: Chatto and Windus.

———. 1929. The origin of life. *Rationalist Annual*: 3.

———. 1930. *Enzymes*. London: Longmans.

———. 1932. *The Causes of Evolution*. London: Longmans.

———. 1934. A mathematical theory of natural and artificial selection, X. *Genetics* 19: 412–429.

———. 1937. The effects of variation on fitness. *American Naturalist* 71: 337–349.

———. 1957. The cost of natural selection. *Journal of Genetics* 55: 511–524.

Haldane, J. B. S., A. D. Sprunt, and N. M. Haldane. 1915. Reduplication in mice. *Journal of Genetics* 5: 133–135. —R.A.S.

Hamilton, W. D. (1936–2000)

William Donald Hamilton was one of the major evolutionary biologists of the twentieth century. Many of his contemporaries compared him to Charles Darwin, partly because both men shared a deep love of natural history and partly because both attempted to solve some of biology's biggest mysteries. Perhaps the largest of these mysteries is the question of altruism, performing acts that benefit others at a cost to the altruist. Superficially, evolutionary theory suggests that mutants exhibiting more "selfish" behavior should be able to wipe out altruism, because by definition the altruists have reduced fitness while the selfish individuals prosper at their expense. Yet altruism abounds in nature. African kingfishers feed and protect young that are not their own;

honeybees commit suicide when they sting intruders, and even more notably from an evolutionary perspective, worker bees are sterile, sacrificing their own reproduction to help the hive flourish.

Hamilton's insight for solving the puzzle was deceptively simple: self-sacrificial behavior can evolve and persist if it benefits relatives of the altruist because they share the altruist's genes. Thus aiding a sibling, cousin, or other relation can still be favored by selection, even if it means forgoing the production of the altruist's own offspring. Hamilton expressed this in an equation that has come to be called Hamilton's rule: altruism is favored when $b/c > 1/r$, where b is the benefit, c is the cost of the act, and r is the degree of relatedness between the donor and the recipient of the altruism. One should therefore be more likely to see the most sacrifice for those most closely related, such as siblings, and less sacrifice for those less closely related, such as third cousins. The process is called kin selection, and the idea has been used with great success to explain a variety of processes in animals, including the cooperative birds and suicidal bees mentioned earlier, although it does not always provide a complete answer. The bees, it turns out, are through a genetic peculiarity of their insect group very closely related to their hive mates, more so than they would be to their own offspring, and hence helping their mother (the queen) to raise their sisters and brothers, who will become future reproductives, will be favored under many circumstances.

In addition to his work on altruism, Hamilton wrote about the evolution of animal groups in "Geometry for the Selfish Herd" (1971), about sex ratios, about aging and senescence, and about the origins and function of cooperation and fighting. Much of his work is marked by a willingness to entertain new ideas and view old information with a fresh eye, and he was a rather solitary scientist throughout most of his life. The idea of kin selection went relatively unappreciated for several years after its initial publication in 1964, but eventually it became one of the cornerstones of modern evolutionary biology. His ideas almost always became key components of their field, whether immediately or after a lag while other scientists caught up to him.

Hamilton eventually became intrigued with one of the greatest problems of evolution, the existence of sexual reproduction. This led to an interest in parasites and their effects on host ecology, behavior, and evolution that lasted for the rest of his career. Sex is a conundrum because, just as the altruist should lose in a race with a selfish competitor, an asexual species (one that simply reproduces itself by making genetically identical copies) should outcompete a sexual species that requires two individuals to produce a single offspring. Certainly some organisms, including some crustaceans, lizards, insects, and plants, do reproduce asexually. Most, however, pay the price of sex, with all of its consequences. As Hamilton asked in a 1975 book review, "Why all this silly rigmarole of sex? Why this gavotte of chromosomes? Why all these useless males, this striving and wasteful bloodshed, these grotesque horns, colors, . . . and why, in the end, novels . . . about love?" (p. 175).

Hamilton, along with a few other scientists, suggested that selection must favor the shuffling of genotypes between generations and hence sex. But the selective pressures must fluctuate in either space or time, or else asexual reproduction still wins in the long term. For a variety of reasons best explained in detail using population genetic models, parasites are ideally suited to provide such fluctuating selection because they continually evolve new means of attack as their hosts evolve defenses against them. Hamilton outlined how the continual barrage of insults by pathogens large and small could keep sexual reproduction advantageous in long-lived multicellular organisms.

He then turned his attention to sexual selection and mate choice, suggesting that there, too, pressure from parasites serves a crucial role in the evolution of showy sexual ornaments like the tail of the peacock. If only healthy males resistant to parasites can produce these ornaments, females can use the flashy traits as indicators of genes that will help make their offspring resistant as well. Parasites are of particular significance because they provide continued pressure to maintain genetic variability; just as with sex, unless hosts change their defenses, they will be overrun by the rapid evolution of their diseases. Females thus use the ornaments to find resistant mates, even though the specific genes conferring this resistance change over time. This idea has been tested in numerous animal species. While parts of his theory remain debatable, parasites do seem to play a significant role in mate choice and the development of some sexual signals.

Natural history, particularly of insects, was always of keen interest to Hamilton, and his theory was always grounded in a sound knowledge of how organisms behaved in the real world. He died in 2000 at the rather early age of 63, following complications of malaria contracted on an expedition to the Congo to follow up on a controversial theory that AIDS arose from faulty polio vaccine. He won many of biology's prizes and awards, including the Kyoto Prize, the Crafoord Prize (Sweden's Nobel Prize equivalent for non-medical biology), the Distinguished Animal Behaviorist Award from the Animal Behavior Society, and the Sewall Wright Award from the American Society of Naturalists. Because his work influenced so many fields, many disciplines, from entomology to parasitology, claim him as one of their own, further testimony to the broad reach of evolution in science.

BIBLIOGRAPHY

Hamilton, W. D. 1971. Geometry for the selfish herd. *Journal of Theoretical Biology* 31: 295–311.

———. 1975. Gamblers since life began: Barnacles, aphids, elms. *Quarterly Review of Biology* 50: 175–180.

———. 1980. Sex versus non-sex versus parasite. *Oikos* 35: 282–290.

———. 1995a. *Narrow Roads of Gene Land*. Vol. 1, *Evolution of Social Behavior*. New York: W. H. Freeman.

———. 1995b. *Narrow Roads of Gene Land*. Vol. 2, *Evolution of Sex*. Oxford: Oxford University Press.

———. 2005. *Narrow Roads of Gene Land: The Collected Papers of W. D. Hamilton*. Vol. 3, *Last Words*. Mark Ridley, ed. Oxford: Oxford University Press.

—M.Z.

Hennig, Willi (1913–1976)

In competition with Carl Linnaeus, Willi Hennig is perhaps the most influential systematist of all time. After Charles Darwin's presentation of the theory of evolution, many biologists, including Darwin himself, expressed the idea that organisms should be classified based on their evolutionary relationships—their phylogeny. However, it was Hennig who first presented a complete theory for phylogeny reconstruction and developed the principles for phylogeny-based classification.

As a teenager, Hennig was an ardent amateur zoologist. He collected beetles and butterflies, started an herbarium, and worked as a volunteer in the Dresden Museum. He took an interest in the science of systematics and, in a high school essay written at the age of 18, he clearly identified the reconstruction of evolutionary relationships as the main task of the discipline. Biologists at the time lacked a clear understanding of how organismal relationships could be reliably reconstructed, but Hennig discussed the merits of different types of evidence of phylogenetic affinity and presented original ideas that foreshadowed his later work. However, it would take him almost two decades to develop a rigorous theory of phylogenetic inference.

Hennig studied zoology, botany, and geology at the University of Leipzig, where he defended his thesis on the copulatory apparatus of Diptera (flies, mosquitoes, and midges) in 1936, at the age of 23. In 1937, he joined the German Entomological Institute, where he spent most of his early professional career. Although comparative morphology and taxonomy of different groups of Diptera remained the major focus of his work, Hennig was a prolific writer, publishing more than 9,000 pages and thousands of drawings during his lifetime. In addition to the Diptera monographs, his writings included outstanding theoretical and empirical contributions as well as textbooks spanning the entire field of zoological systematics.

Hennig was conscripted as an infantryman when World War II broke out in 1939. After recovering from a serious injury he suffered in Russia in 1942, he was ordered to work with malaria control. Shortly before the end of the war he was sent to Greece and northern Italy, where, in British captivity, he continued with malaria work until the fall of 1945. It was during this period that he wrote the draft of his most important work, *Grundzüge einer Theorie der phylogenetischen Systematik* (*Principles of a Theory of Phylogenetic Systematics*, 1950).

Hennig finished the manuscript after the war, and the book appeared in 1950 in a small edition and without an index because of the shortage of paper at the time. The text was widely discussed among German-reading zoologists, including Professor Lars Brundin at the Swedish Museum of Natural History in Stockholm, who became a powerful proponent of the new ideas. However, Hennig's theories were not introduced to the English scientific literature until a revised version of the book was printed in the United States in 1966; several related papers appeared in English-language journals around the same time.

Long before Hennig, it was generally assumed that the evolutionary relatedness of two species was correlated with their overall similarity in morphology, behavior, and other traits. It was also widely recognized that independently evolved similarity, commonly called convergence, was false evidence of genealogical relationship. But Hennig realized that many of the similarities inherited from a common ancestor, called homologies, were useless as grouping criteria; only the shared derived traits (synapomorphies) indicate true phylogenetic relationship. Because a synapomorphy originated in a common ancestor shared by only some organisms, it groups those organisms into an evolutionary unit (monophyletic group or clade). The shared primitive traits (plesiomorphies) do not define clades. For instance, in a sample of dog, lizard, chimp, and man, the dog and the lizard cannot be grouped on the ground that they share a tail because this trait was inherited from the common ancestor of all four species; it is a plesiomorphy. The lack of a tail, however, is a shared derived feature that can be used to group the chimp and man.

To Hennig, basing biological classification on phylogeny meant that all named taxa must be monophyletic groups. This idea was strongly opposed at the time by researchers like Ernst Mayr (1969), who insisted that classification must be a compromise between evolutionary relationship and overall similarity ("evolutionary systematics"), and by Robert Sokal and Peter Sneath (1963), who argued that grouping by overall similarity was preferable because phylogenetic relationships were so difficult to reconstruct ("phenetics"). Hennig's principles are almost universally accepted now, even though it will take decades before a consistent Hennigian classification is universally adopted because of the resilience of old group concepts—such as "invertebrates," "gymnosperms," and "protozoans"—and the lack of knowledge about phylogeny. The rules for naming clades are still much debated, especially the Linnaean idea of recognizing formal ranks (genus, family, order, etc.). Hennig proposed that ranks be tied to age; each rank would be reserved for groups of a specific date of origin. Accurate dating of clades seemed completely unrealistic at the time, but the refinement of phylogenetic methods and the growing amount of molecular data may yet lead to a reconsideration of this idea.

Hennig discussed at length the principles by which convergence could be sorted from homology and plesiomorphy from apomorphy. His techniques included detailed analysis of similarity, the sequence of appearance of characters in the fossil record or in ontogeny (the development from egg to adult), and the geographic distribution of characters. Conflicts were to be manually resolved using checking and rechecking, carefully weighting the arguments against each other. A basic idea was that homology should be assumed in the absence of evidence to the contrary, commonly referred to as Hennig's auxiliary principle. Hennig never discussed quantitative techniques for resolving conflicts, such as parsimony and compatibility analysis, but his texts clearly point toward them, and some early parsimony papers attributed the basic idea to Hennig. However, the mathematics of phylogeny reconstruction was independently discovered at about the same time by other systematists and molecular biologists unaware of Hennig's work.

The debate following the presentation of phylogenetic systematics was plagued by outrageous controversy (Hull 1988). Some of Hennig's enemies labeled him "cladist" because of his emphasis on recognizing monophyletic groups (clades), and Hennig's more zealous followers quickly adopted the term. After Hennig's death, several prominent cladists argued that the basis of classification should not be evolutionary relationship but the most parsimonious arrangement of the available data (Ridley 1986). Others (phylogenetic systematists) maintained that parsimony was only one of several tools that could be used to find the evolutionary relationships on which proper classification should rest. Debates inspired by Hennig's revolutionary book still continue today, more than 50 years after its original publication.

BIBLIOGRAPHY

Anonymous. 2005. Willi Hennig. Das Biologie Wiki, http://www.biologie.de/biowiki/Willi_Hennig.

Dupuis, C. 1984. Willi Hennig's impact on taxonomic thought. *Annual Review of Ecology and Systematics* 15: 1–24.

Felsenstein, J. 2003. A digression on history and philosophy. In *Inferring Phylogenies*, chap. 10. Sunderland, MA: Sinauer Associates.

Hennig, W. 1950. *Grundzüge einer Theorie der phylogenetischen Systematik*. Berlin: Deutscher Zentralverlag.

———. 1966. *Phylogenetic Systematics*. Urbana: University of Illinois Press.

Hull, D. L. 1988. *Science as a Process*. Chicago: University of Chicago Press.

Mayr, E. 1969. *Principles of Systematic Zoology*. New York: McGraw-Hill.

Ridley, M. 1986. *Evolution and Classification: The Reformation of Cladism*. New York: Longman.

Schlee, D. 1978. Willi Hennig 1913–1976: Eine biographische Skizze. *Entomologica Germanica* 4: 377–391.

Sokal, R. R., and P. H. A. Sneath. 1963. *Principles of Numerical Taxonomy*. San Francisco: W. H. Freeman. —F.R.

Hobbit *(Homo floresiensis)*

"Hobbit" is the nickname (after J. R. R. Tolkien's 1937 novel, *The Hobbit*) for a tiny skeleton (cataloged as LB1) that represents a new human species *(Homo floresiensis)*, which lived from 95,000 to 12,000 years ago. This remarkable find was announced in 2004 by Michael Morwood and other scientists from Australia and Indonesia. LB1 was excavated from 18,000-year-old deposits in Liang Bua, a cave on the Indonesian island of Flores. The nearly complete skeleton stands a little over a meter tall, and is believed to represent an adult woman. The skeleton and fragments of eleven other individuals show that *Homo floresiensis* manifested a mixture of features found in very early relatives of humans (e.g., "Lucy," dated to over 3 million years ago), early *Homo*, more recent *Homo erectus*, and living people. LB1's combination of features of the brain, teeth, shoulder, wrist, pelvis, legs, and feet has not been found in any other hominins, living or extinct. *Homo floresiensis* had very thick bones, relatively short legs, long feet, a short big toe, a tiny braincase, and an ape-sized brain-to-body ratio.

To date, the remains of twelve *Homo floresiensis* individuals have been found in association with faunal remains of dwarfed elephant-like *Stegodon*, giant Komodo dragons, rats, and bats. Charred bones and clusters of reddened, fire-cracked rocks were also found in Liang Bua. High densities of stone tools were excavated from the cave, including cores, anvils, retouched tools, and flaking debris. Because of these associated finds, the discoverers of *Homo floresiensis* suggest that the species made tools, hunted, used fire, and butchered and cooked meat. Although the suggestion of higher cognitive abilities in such a small-brained species remains controversial, it is supported by an analysis of the impressions of the brain that were imprinted inside LB1's braincase. A cast of the braincase (or endocast; see figure) reveals that the brain was highly convoluted and uniquely organized (wired), with a number of advanced features. For example, the tips of the frontal lobes were expanded in a region that is known to be important for monitoring internal thought and for planning ahead. These and other details of LB1's endocast contradict the suggestion by a few researchers that it may be from a modern human afflicted with a pathology known as microcephaly.

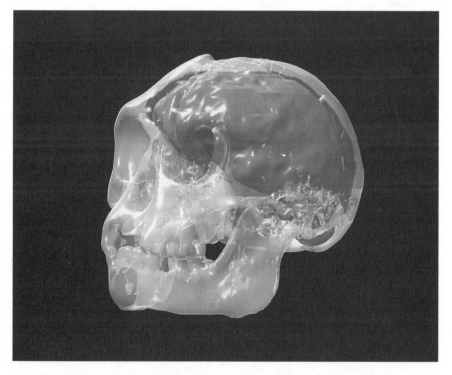

Virtual skull of LB1 ("Hobbit") that was reconstructed using three-dimensional computer tomography. The skull has been flood-filled to reveal a virtual endocast that reproduces the shape and other details of "Hobbit's" brain.

The discovery of *Homo floresiensis* was extremely surprising because it showed that *Homo sapiens* was not the only hominin around 18,000 years ago (and before), as had been widely believed. Scientists are mixed on their interpretations about where this new species came from. Some think that it is a descendant from bigger-bodied *Homo erectus* ancestors that became dwarfed after migrating to the island of Flores. This hypothesis is in keeping with the well-known island rule, whereby large animals become smaller (like the above-mentioned *Stegodon*) and small animals become larger (e.g., the giant Komodo dragons) on islands where resources are limited. Other workers believe that *Homo floresiensis* may have been descended from a small ancestor, such as those in Lucy's genus *(Australopithecus)*, and that its ancestors simply remained small after arriving on the island of Flores. Time will tell.

BIBLIOGRAPHY

Brown, P., T. Sutikna, M. J. Morwood, R. P. Soejono, Jatikmo, E. W. Saptomo, and R. A. Due. 2004. A new small-bodied hominin from the Late Pleistocene of Flores, Indonesia. *Nature* 431: 1055–1061.
Falk, D., C. Hildebolt, K. Smith, M. J. Morwood, T. Sutikna, P. Brown, Jatmiko, E. W. Saptomo, B. Brunsden, and F. Piror. 2005. The brain of LB1, *Homo floresiensis*. *Science* 308: 242–245.
Morwood, M. J., P. Brown, Jatikmo, T. Sutikna, E. W. Saptomo, K. E. Westaway, R. A. Due, R. G. Roberts, T. Maeda, S. Wasisto, and T. Djubiantono. 2005. Further evidence for small-bodied hominins from the Late Pleistocene of Flores, Indonesia. *Nature* 437: 1012–1017.   —D.F.

Homology

Homology is one of the central concepts in biology. It refers to the similarity between morphological characters, genes, and other biological traits, such as behaviors, among different species. Some of the best-known examples of homology are (1) the wings of birds, the arms of humans, and the forelimbs of other tetrapods and (2) the bony elements in the gill arches of sharks and the three middle ear bones (malleus, incus, and stapes) in humans. In both cases, the similarity between characters is established independently of their size, shape, and function. This fact is reflected in Richard Owen's definition of a homologue as "the same organ in different animals under every variety of form and function," to be distinguished from an analogue as "a part or organ in one animal which has the same function as another part or organ in a different animal" (Owen 1843, 379, 374).

What is already clear from Owen's definition is that in order to establish the existence of homology, we need to have a set of criteria that allow us to determine the sameness between different characters. In the past, comparative anatomy and embryology provided the tools for homology assessments; today we also use molecular and sequence data. The morphological homology criteria include (1) position—homology is more likely if two characters are found in a similar position within two organisms; (2) structure—homology is more likely if two characters share many individual features; (3) transition—homology between dissimilar characters can be established if they are linked through a series of intermediate forms; and (4) several congruence criteria that capture the distribution of similar characters in closely related species. The molecular homology criteria capture the similarity between molecules, such as DNA sequences, based on patterns of common descent. Here the number and the position of nucleotide substitution are often used as an approximation for the closeness of two sequences. Because of the high frequency of gene duplications in the course of evolution, a further distinction has to be made between orthologous genes, which diverge as a result of independent evolution after a speciation event, and paralogous genes, which diverge within a species after a gene duplication event.

The concept of homology dates back to pre-Darwinian and pre-evolutionary times. Aristotle found similarity between organisms to be something noticeable, and even before Owen formally defined the concepts of homology and analogy, the ideas captured by these concepts were at the heart of the famous debate between Georges Cuvier and Étienne Geoffroy Saint-Hilaire in the early 1830s (see Appel 1987). While Cuvier held that the diversity of life is best understood in the context of distinct higher groups (so-called *embrachements*) and within those groups based on the functional needs of organisms, Geoffroy argued for the existence of an underlying common organizational plan of all animals, which he called universal analogy or unity of type. Owen recognized that identifying corresponding parts in different specimens and arranging different forms according to a nested hierarchy of similarities provides a rational system for ordering the diversity of life. He also introduced the notion of the archetype as an abstract representation of the underlying similarities among a group of organisms.

In the context of the Darwinian revolution of the second half of the nineteenth century, the homology concept was historicized. The explanation for the observed patterns of similarity between organisms was now to be found in their common evolutionary history, and homologies became the main tool for the reconstruction of these phylogenetic relationships. Along with morphological criteria, embryological evidence was also increasingly employed to ascertain homology relationships between different structures. This practice was based on the assumption that similar structures should also develop from similar embryonic anlagen. However, this program of evolutionary morphology would soon run into difficulties. First, if all structures develop from a single cell, it was not clear which embryonic stages should be used to establish sameness between two adult structures. Second, regenerating tissues, such as amphibian lenses, often develop from a different source of cells. The conclusion that the original and the regenerated lens are not homologous was clearly problematic. Another application of the homology concept was more fruitful. The recognition that behavioral patterns can be treated the same way as morphological structures and that one can therefore establish homologies also between behaviors was the foundation of the science of ethology.

Today, the homology concept is again at the center of several theoretical discussions within biology. This has resulted in a variety of different proposals and definitions of homology that can be grouped, according to the research questions they address, into (1) historical homology concepts and (2) biological homology concepts. Historical homology concepts are prominent in phylogeny reconstruction. The most widely used approach, cladistics or phylogenetic systematics, characterizes groups of species (clades) by the presence of shared derived characters (synapomorphies). These synapomorphies are historical homologues—their sameness is a consequence of their phylogenetic (ancestor/descendant) relationships. The resulting phylogenetic system represents a nested hierarchy of groups, each characterized by a unique set of shared derived characters or homologues.

The biological homology concept, on the other hand, also emphasizes, in addition to the common phylogenetic history, the developmental origin of homologous structures. The similarities of phenotypic structures are seen as a consequence of a shared developmental program. As Günter Wagner defined it: "Structures from two individuals or from the same individual are homologous if they share a set of developmental constraints, caused by locally acting self-regulatory mechanisms of . . . differentiation" (Wagner 1989, 62). The biological homology concept thus not only captures the distribution of homologues among groups of species, but also intends to explain the mechanistic causes for the observed similarities, which are to be found in the details of the developmental system. The biological homology concept is therefore at the core of evolutionary developmental biology, the recent synthesis of evolutionary and developmental biology.

BIBLIOGRAPHY

Appel, T. 1987. *The Cuvier-Geoffroy Debate: French Biology in the Decades before Darwin*. Oxford: Oxford University Press.

Bock, G. R., and G. Cardew, eds. 1999. *Homology*. Novartis Foundation Symposium 222. Chichester, U.K.: Wiley.

Hall, B. K., ed. 1994. *Homology: The Hierarchical Basis of Comparative Biology*. San Diego: Academic Press.

Laubichler, M. D. 2000. Homology in development and the development of the homology concept. *American Zoologist* 40: 777–788.

Owen, R. 1843. *Lectures on the Comparative Anatomy and Physiology of the Invertebrate Animals, Delivered at the Royal College of Surgeons*. London: Longman, Brown, Green, and Longmans.

Wagner, G. P. 1989. The biological homology concept. *Annual Review of Ecology and Systematics* 20: 51–69.   —M.D.L.

Hooker, Joseph Dalton (1817–1911)

Joseph Dalton Hooker was a leading botanist of Victorian Britain and one of Charles Darwin's staunchest supporters. He was one of the first scientists to be informed of Darwin's theory and played a leading role in helping Darwin to formulate his ideas, especially in the application of evolution theory to biogeography.

Hooker was born on June 30, 1817, the son of Sir William Jackson Hooker, who was director of the Royal Botanical Gardens at Kew. He obtained an MD at Glasgow in 1839 and later that year joined HMS *Erebus* on a four-year exploratory voyage to New Zealand, Tasmania, and the Antarctic. He subsequently produced a six-volume survey of the botany of these regions. In 1847, he traveled to Sikkim, Nepal, and the Eastern Himalayas, sending back exotic plants such as rhododendrons, which soon became favorites in European gardens. He published *Flora Indica* in 1855, the year in which he became assistant to his father at Kew. In 1865, he succeeded his father as director, continuing Kew's role as a leading center for the economic botany of the British Empire. Hooker died in 1911.

In 1844, Darwin informed Hooker of his ideas on evolution and in 1847 sent him the "Essay" in which his theory was first described in detail. The two argued for many years over the theory's implications for the geographic distribution of plants. Hooker believed that dispersal had taken place over ancient land connections between the existing continents, which had subsequently been submerged by geological activity. He was particularly active in defending the view that there had once been a great continent in the southern hemisphere, of which Antarctica, Australia, and New Zealand are but scattered remnants. Darwin argued instead for the dispersal of seeds across the oceans by flotation, winds, and birds. Hooker was at first skeptical of evolutionism, but eventually he became one of the earliest converts to the theory.

Along with Charles Lyell, Hooker advised Darwin on the course of action he should take after receiving Alfred Russel Wallace's paper in 1858. Hooker's introductory essay to his *Flora Tasmaniae* of 1860 was one of the first major public endorsements of the theory. He spoke after Thomas Henry Huxley in the famous debate with Bishop Samuel Wilberforce at the Oxford meeting of the British Association in 1860, and some observers thought his contribution was more persuasive.

Darwin and Hooker continued to debate biogeography in the following decades. In 1878, Hooker used his presidential address to the Royal Society of London to defend his theory of an ancient continent in the southern hemisphere. At the same time, he endorsed the view favored by Darwin and many others that the main source of more highly evolved species was the northern regions of Eurasia and North America. Later in his life, he began to back away from the theory of sunken land bridges, under pressure from Darwin, Wallace, and other supporters espousing the permanence of the continents and oceans.

BIBLIOGRAPHY

Alan, M. 1967. *The Hookers of Kew*. London: Michael Joseph.

Hooker, J. D. 1845. *The Cryptogamic Botany of the Antarctic Voyage of H.M. Discovery Ships* Erebus *and* Terror *in the Years 1839–43*. 6 vols. London: Reeve Bros.

———. 1855. *Flora Indica: Being a Systematic Account of the Plants of British India*. London: W. Pamplin.

———. 1860. *Flora Tasmaniae*. London: L. Reeve.
—*P.J.B.*

Host parasite evolution

Parasites can exert natural selection on host populations, and vice versa, resulting in changes in allele frequencies over time and hence evolution. Parasites and hosts can also exert reciprocal, similtaneous natural selection on each other, resulting in allele frequency changes in both partners and hence coevolution.

There are several active research programs in the area of parasite evolution. One of the most active areas is on the evolution of parasite virulence. The research program is mainly engaged in answering the question: Why do some parasites make their hosts very sick, while other parasites have little effect on their hosts? The conventional wisdom in this area prior to the 1980s, especially in the medical community, was that parasites should evolve to be increasingly benign so as not to harm the host population. But this kind of "good-of-the-species" reasoning was rejected in the 1980s by theoretical biologists. They showed that parasites should in fact be under natural selection to maximize their own transmission to the next host, without regard to the effects on the host population. In some cases, transmission is maximized by rapid replication within the host, leading to high rates of host mortality (and/or morbidity). Such an outcome might be expected, for example, in parasites that are easily moved between hosts by vectors, such as mosquitoes (e.g., malaria). Hence, even if the host is incapacitated by disease, the infectious propagules can be moved between hosts by the vectors (or, similarly, by water). In general, if disease transmission does not require a healthy, mobile host, selection may favor parasites that reproduce rapidly at the host's expense.

Studies on the evolution of parasite virulence have important implications for human health. For example, medical practices that increase the ease of disease spread not only increase the number of infected hosts, but they can also result in selection for more virulent strains. Similarly, parasite strains that are most resistant to antibiotics will be favored by selection. If, for example, antibiotics are admistered in low doses that are not sufficient to kill the entire parasite population within each host, then the antibiotic resistant strains will increase in frequency. The evolution of antibiotic resistance is a major problem in human health and in the health of domesticated animals; it represents a classic case of rapid evolutionary change.

Parasites are also under natural selection to evade detection and elimination by their hosts. Conversely, hosts are under natural selection to find and kill their parasites. Mutations that increase the infection success of parasites will be favored by natural selection, while mutations that increase detection success of parasites by hosts will be similarly favored. This can lead to two types of coevolutionary interactions. In one type, called an "arms race," beneficial mutations in the parasite population sweep to fixation, meaning that they replace the alternative alleles in the population, followed by similar succcesssful sweeps of mutations for defense in the host population.

In another general type of coevolutionary interaction, beneficial alleles do not go to fixation. Here host alleles interact with alleles at other host loci, and their combination yields a multilocus genotype that defines "self." The host then indentifies parasites that do not match its "self" genotype. This sets up natural selection for parasite genotypes that can infect the most common host genotypes for self. However, once these successful parasite genotypes become common, they impose selection against the most common host genotypes, driving them down in frequency. Then a new host genotype becomes the most common genotype, and the selection on parasites changes to a different multilocus parasite genotype. In theory, this sets up a continous oscillation in both host and parasite genotypes, which would function to maintain genetic diversity in both the host and the parasite populations. The theory is now known as the Red Queen hypothesis, because it would seem as though both host and parasite genotypes are running as fast as they can to stay in the same place. (The name comes from the famous remark by the Red Queen to Alice in Lewis Carroll's *Through the Looking Glass:* "Now, *here*, you see, it takes all the running *you* can do, to keep in the same place.") The Red Queen hypothesis currently stands as one of the best-supported theories for the maintenance of genetic diversity at disease-resistance loci, which are known to be highly polymorphic. It also stands as one of the best-supported theories that attempt to explain why asexual production of all-female offspring does not more commonly replace sexual reproduction in natural populations. The basic idea here is that parasites would impose selection against asexual clones as they become locally common, and thus prevent the elimination of the sexual population in the short term.

Parasites have also been implicated as a selective force in the timing of reproduction in some animals. The idea is that animals would be under parasite-mediated natural selection to reproduce at a smaller size and younger age in populations that are subject to a high risk of infection by virulent parasites than populations that are at relatively low risk of infection. The reason is that delaying reproduction in high-risk populations might result in the production of fewer offspring, if, in fact, the host becomes infected before maturing.

Finally, parasites have also been implicated in the manipulation of host behavior in order to enhance the transmission of their infective propagules to new hosts. For example, several studies have now shown that parasites induce behaviors in their hosts that increase the likelihood that the host will be eaten by predators. But this is only known to occur in parasites for which the predator is also another host in the parasite's life cycle; and it is known that the parasite survives and establishes itself in the new host following consumption. Thus it would appear that some parasites have responded to natural selection to enhance their rates of transmission by altering the behavior of one of the hosts in their multihost life cycles.

BIBLIOGRAPHY

Bell, G. 1982. *The Masterpiece of Nature: The Evolution and Genetics of Sexuality.* Berkeley: University of California Press.
Ewald, P. W. 1994. *Evolution of Infectious Disease.* Oxford: Oxford University Press.

Garrett, L. 1994. *The Coming Plague: Newly Emerging Diseases in a World Out of Balance*. New York: Farrar, Straus and Giroux.

Neese, R. M., and G. C. Williams. 1994. *Why We Get Sick*. New York: Random House.

Zimmer, C. 2000. *Parasite Rex: Inside the Bizarre World of Nature's Most Dangerous Creatures*. New York: Free Press.   —C.L.

Hrdy, Sarah Blaffer (b. 1946)

Sarah Blaffer Hrdy is one of America's most eminent anthropologists and a member of the National Academy of Sciences. An anthropologist and primatologist by training, she works within the broad framework of sociobiological theory, to which she has made significant contributions. Hrdy is most well known for calling attention to the various evolutionary strategies used by female primates, which has brought a radical shift away from primatologists' traditional focus on males as the prime social actors. More recently, Hrdy has reevaluated the role of mothers and their relationship to their children. Although she relies heavily on the "comparative method" central to sociobiology, when she writes about sex differences Hrdy is an equal-opportunity critic of both feminists for ignoring biology and of evolutionary psychologists for ignoring social history.

Hrdy was born in 1946 to a wealthy Southern family. She graduated from Radcliffe with a BA in 1969. It was in a Harvard anthropology course that she met her future husband, Daniel Hrdy (a Czech name), whom she married in 1972; he later became a medical doctor. While rearing their three children, she became increasingly focused on the need to update Attachment Theory with a view to the evolution of early hominids as cooperative breeders. (According to this theory, developed by John Bowlby, early attachment to caregivers is crucial for infants' normal social and emotional development.)

Hrdy entered graduate school in anthropology in 1970, about the same time that Edward O. Wilson was beginning to write *Sociobiology: The New Synthesis;* she completed her PhD in 1975, the year it was published. Irven DeVore's first female student, she was heavily influenced by Wilson and by Robert Trivers. All three supervised her fieldwork on the Hanuman langurs of India, published as *The Langurs of Abu: Female and Male Strategies of Reproduction* in 1977. She had initially undertaken her research on infanticide by adult males in langurs, believing it was a pathological response to crowding. During her first field season she had to discard that hypothesis. Infanticidal behavior was highly selective: males targeted infants born to females they had not mated with. Recognizing that females were far from passive, Hrdy began to study female counterstrategies. For example, by mating with multiple males before they usurped control of the troop, females strove to forestall infanticide later on.

In her next book, *The Woman That Never Evolved* (1981), Hrdy provocatively kept females center stage and incorporated the primate origins of patriarchy in her analysis. Meanwhile, the work on infanticide remained controversial, inspiring the first international conference on the topic, followed by *Infanticide: Comparative and Evolutionary Perspectives* (1984), coedited with Glenn Hausfater. By then, Hrdy had become increasingly interested in dilemmas confronting working mothers throughout history and deep into prehistory. She set out to examine across species and cultures the options that mothers have employed, from seeking the assistance of "allomothers" to neglect and abandonment. (Allomothers are members of a group who assist a mother in rearing her infant.) When *Mother Nature* appeared in 1999, *Library Journal* named it one of the best books of the year. Increasingly drawn to the question of infant needs, in 2005 she coedited a Dahlem Workshop report on attachment and bonding.

Hrdy's books have been translated into several languages. Although her ideas of infanticide were long regarded with suspicion within the anthropology community, in 2002, at the annual meeting of the American Anthropological Association, *Mother Nature* was awarded the Howell Prize for Outstanding Contribution in Biological Anthropology.

Hrdy held the position of Professor of Anthropology at the University of California at Davis from 1984 to her early retirement in 1996. As Professor Emerita, she continues doing research, writing books, and serving on the editorial boards of *Evolutionary Anthropology* and *Human Nature*. Hrdy and her husband live in California, where they are involved in habitat restoration and sustainable agriculture.

BIBLIOGRAPHY

Carter, C. S., L. Ahnert, K. E. Grossmann, S. B. Hrdy, M. E. Lamb, S. W. Porges, and N. Sachser, eds. 2006. *Attachment and Bonding*. Cambridge, MA: MIT Press.

Hausfater, G., and S. Hrdy, eds. [1984] 2008. *Infanticide: Comparative and Evolutionary Perspectives*. New Brunswick, NJ: Aldine Transaction.

Hrdy, Sarah Blaffer. 1977 [1980]. *The Langurs of Abu: Female and Male Strategies of Reproduction*. Cambridge, MA: Harvard University Press. (1980, paperback edition with new preface.)

———. 1981 [1999]. *The Woman That Never Evolved*. Cambridge, MA: Harvard University Press. (1999, paperback edition with new preface.)

———. 1999. *Mother Nature: A History of Mothers, Infants and Natural Selection*. New York: Pantheon.

Web page. http://www.citrona.com/sarahbhrdy.htm.   —U.S.

Hutton, James (1726–1797)

James Hutton, a Scottish-born doctor and gentleman farmer, is considered the father of geology in Britain. He lived much of his later life in Edinburgh and was a major figure in the so-called Scottish Enlightenment. A contemporary of David Hume and Adam Smith, he championed both the Plutonist theory of geology as well as the methodology of uniformitarianism. As a Plutonist, he argued—against the so-called Neptunists, who thought that the world had been formed by precipitation from water—that the center of the earth is very hot and that rocks are formed by molten lava pushing up to the surface, where it solidifies. At the same time, as a uniformitarian, he argued that the way to understand geology is by assuming that causes today are the same as those of yesterday, and that given enough time, geological formations can be created naturally.

Hutton published these ideas in a massive two-volume work titled *Theory of the Earth* (1795), a treatise whose length was exceeded only by its obscurity. Fortunately a follower, John Playfair, published a more readily accessible version in 1802, and Hutton's geological position became recognized and established. However, it was opposed by the more popular theorizing of Georges Cuvier (1813), who supposed that there are many convulsions in earth history, and it was not until 1830 that Hutton's uniformitarianism found a major champion against the Frenchman's catastrophism. This came in the three-volume *Principles of Geology* by Charles Lyell, probably the work that was the greatest influence on the young Charles Darwin.

From the viewpoint of evolutionary thinking, Hutton's importance lies in the way in which he extended earth history. He famously spoke of the world as showing "no vestige of a beginning, no prospect of an end" (1788, 304), and to Playfair claimed that on looking at strata, "the mind seemed to grow giddy by looking so far into the abyss of time." (Playfair 1805) Obviously, this kind of thinking was going against biblical history read literally. Although Hutton was accused of atheism, more probably he was (like many late-eighteenth-century thinkers) more of a deist, one who sees the glory of God in his unbroken laws.

Although historically insignificant in the sense that this belief he held was not an influence on later thinkers, it is probable that Hutton shared with others of his day (notably Erasmus Darwin, the grandfather of Charles) an inclination to some kind of transformism in the organic world. Indeed, there are passages in his writing that sound almost like an anticipation of natural selection. He wrote in his *Principles of Knowledge* that "if an organised body is not in the situation and circumstances best adapted to its sustenance and propagation, then, in conceiving an indefinite variety among the individuals of that species, we must be assured, that, on the one hand, those which depart most from the best adapted constitution, will be the most liable to perish, while, on the other hand, those organised bodies, which most approach to the best constitution for the present circumstances, will be best adapted to continue, in preserving themselves and multiplying the individuals of their race" (Hutton 1794, 3).

Clearly much influenced by the work that he and others were doing on animal and plant breeding in the advancement of better agriculture, Hutton noted that dogs survive thanks to "swiftness of foot and quickness of sight" and hence "the most defective in respect of those necessary qualities, would be the most subject to perish, and that those who employed them in greatest perfection . . . would be those who would remain, to preserve themselves, and to continue the race" (Hutton 1794). However, one should not read too much into such passages, because they were almost commonplace in the first half of the nineteenth century. It took Charles Darwin in *On the Origin of Species* to turn such ideas into a full theory of evolution.

BIBLIOGRAPHY

Cuvier, G. 1813. *Essay on the Theory of the Earth.* R. Kerr, trans. Edinburgh: W. Blackwood.

Hutton, J. 1788. Theory of the earth; Or an investigation of the laws observable in the composition, dissolution, and restoration of land upon the globe. *Transactions of the Royal Society of Edinburgh* 1, no. 2: 209–304.

———. 1794. *An Investigation of the Principles of Knowledge and of the Progress of Reason, from Sense to Science and Philosophy.* Edinburgh: A. Strahan and T. Cadell.

———. 1795. *Theory of the Earth, with Proofs and Illustrations.* Edinburgh: William Creech.

Lyell, C. 1830–1833. *Principles of Geology: Being an Attempt to Explain the Former Changes in the Earth's Surface by Reference to Causes Now in Operation.* 3 vols. London: John Murray.

Playfair, J. 1802. *Illustrations of the Huttonian Theory of the Earth.* Edinburgh: William Creech.

———. 1805. Biographical account of the late Dr. James Hutton F. R. S. Edin. *Transactions of the Royal Society of Edinburgh* 5, no. 3: 39–99.

Repcheck, J. 2003. *The Man Who Found Time: James Hutton and the Discovery of the Earth's Antiquity.* London: Simon and Schuster.

Rudwick, M. J. S. 2005. *Bursting the Limits of Time.* Chicago: University of Chicago Press. —*M.R.*

Huxley, Julian S. (1887–1975)

Julian Sorell Huxley was a biologist, a popular writer, and a member of Britain's intelligentsia. In evolutionary studies, he promoted synthesis during the 1930s and 1940s. He contributed to the revival of selection theory with the publication of *Evolution: The Modern Synthesis* (1942) and he concentrated attention onto Charles Darwin as the subject's theoretical center. Huxley promoted evolutionary taxonomy, which linked classification to the study of speciation, while also seeking objective methods for taxonomic work. Huxley's contributions of new knowledge were far less important than his infectious enthusiasm and encouragement, as well as his ability to combine scattered concepts or ideas into general principles and meaningful visions.

Born in London, Huxley attended Eton College (1900), then Balliol College, Oxford (1906), where he received a first-class degree in natural science (zoology). He spent the next year at the Naples Zoological Station. His first book, *The Individual in the Animal Kingdom* (1912), was an attempt to give a scientific account of evolution inspired by the recently published *Evolution Creatice* by the French philosopher Henri Bergson (Huxley always retained a fondness for the vitalism of Bergson while realizing that it could never be true science). In the course of the book, using as an example the competition at the time between the British and German navies, Huxley introduced the idea of a biological "arms races" between competing lines of organisms, an idea that has been much endorsed by such thinkers as Richard Dawkins (1986). In 1910, Huxley received his first faculty appointment, in zoology at Oxford. While in Oxford he undertook his first major research success, a study of courtship rituals and sexual selection in the great crested grebe. He also developed skill in experimental embryology.

Huxley's career included many employers. In 1912, he accepted an offer from the Rice Institute in Houston to develop a new department of biology, which proved difficult. When World War I erupted, he returned to England. During the war, Huxley served mainly in army intelligence. In 1919, he returned to Oxford's zoology department and concentrated his attention on developmental biology and physiological genetics. In 1920, he reported on the ability of hormones to provoke metamorphosis in axolotls. This received considerable press attention and served as a catalyst for his lifelong involvement in the media. Huxley also investigated the regulation of rates of gene expression and problems of relative growth, such as allometry. In 1925, he moved to King's College London, accepting the position of professor in zoology and laboratory director. Two years later, Huxley reduced his position to an honorary one and left King's to pursue a career as a freelance writer. His contributions to his science are summarized in *Problems of Relative Growth* (Huxley 1932) and *The Elements of Experimental Embryology* (Huxley and de Beer 1934). These earned Huxley election to the Royal Society in 1938.

Huxley's most important freelance project was a collaboration with H. G. and G. P. Wells to survey biology and the history of evolution, titled *The Science of Life* (1929). During the 1930s, Huxley contributed to newspapers, magazines, radio, and film. He took on book and editorial projects, as well as speaking tours (see for example Huxley 1934). Seemingly everything involving public science in Britain in the 1930s included Huxley in some way. He is best known for his role on the BBC radio program "The Brains Trust" during World War II. Huxley's ability to access the media proved essential later in his life.

In 1935, Huxley returned to institutional employment, becoming secretary of the Zoological Society of London. This entailed managing the society and its zoological gardens in Regent's Park and Whipsnade. While secretary, Huxley kept up his public contributions. He turned his research interests to systematics and evolution. Huxley was particularly interested in evolution leading to adaptation and to the formation of new species. Polytypic species and geographical varieties (subspecies) were of interest as they suggested "evolution in action," a phrase Huxley often repeated. He coined the term *cline* for characters that show gradual change over a geographic gradient. He also encouraged research into natural and sexual selection. During this period, Huxley organized *The New Systematics* (1940) and wrote *Evolution: The Modern Synthesis* (1942).

The outbreak of World War II in Europe radically changed Huxley's commitments. He wrote extensively against fascism and against Nazi racist biology, and he stressed the importance of internationalism and democracy (Huxley 1941, 1943). After 1939, Huxley campaigned vigorously for American intervention in Europe. With so much of his attention elsewhere, some members of the Zoological Society of London felt Huxley had abandoned the society. They forced his resignation in 1942.

After several contributions to the war effort and to planning postwar reconstruction, in 1946 Huxley became first director-general of the United Nations Educational, Scientific, and Cultural Organisation (UNESCO). This role suited Huxley perfectly, although he frequently failed to manage the political aspects of the directorship. He stayed only two years in office. Huxley gave UNESCO an agenda based on scientific humanism, extending a broad vision he had advocated for many years, that humanity can rely only upon itself for its future evolution (Huxley 1948). Our destiny is in our own hands; we have to take responsibility for ensuring progress. Huxley campaigned for family planning programs, international conservation of nature, and universities in developing countries. He also used science to leverage political discussions, especially regarding race and inequality. If nothing else, Huxley tried to embed a vision of progress that was based on deliberate human action.

After leaving UNESCO in 1948, Huxley lived as a freelance writer and scientist-celebrity. He wrote extensively (including influential work against Lysenkoism), traveled, and consulted for government and many other groups. Progress and cultural development were two main themes of his ongoing campaigns (Huxley 1954a, 1954b, 1959, 1964). While he revised some previous work and encouraged many biologists, he did not undertake original research again. Huxley died in 1975.

As a scientist, Huxley had specialist expertise in several subjects, especially behavior, development, and evolutionary biology. Throughout his career, however, Huxley normally presented himself as a "general" biologist. This focused attention on biological processes and principles, and gave priority to theory and explanation. While his career shifted from one place to another and his interests shifted from specialty to specialty, Huxley always linked his work back to the spine of "general" biology. The particulars of individual topics became less important; the underlying general principles were what gave coherence and meaning to particulars. Without them, biology was simply fact gathering.

Huxley's preference for general principles is best illustrated in work he intended for nonspecialist audiences. In books like *Ants* (1930) or *At the Zoo* (1936), and in his film *Private Life of the Gannets* (1934), he used particular examples to illustrate general biological phenomena. Gannets, for instance, provided a case for the study of development, parental care, adaptations, mating behavior, flight, ecology, and so on. This approach helped Huxley succeed admirably in projects requiring the encyclopedic collection of information and the extraction of key themes. *Evolution: The Modern Synthesis* is a superb example. This reported no original research of Huxley's making. Instead, the book's accomplishment—and a landmark one—was Huxley's coordination of information into an overall vision about evolution and the myriad forms of selection in nature. It helped shape evolutionary studies in Britain for a generation.

In his politics, Huxley was a progressive and a humanist (Huxley 1927). He believed humanity was responsible for shaping its own future, and he turned to science and rational planning to solve social problems. The Tennessee Valley Authority's electrification project, modernist architecture, and urban planning were among his favorite examples. He fervently supported Population and Economic Planning, a social policy group. Huxley was not a eugenicist in its common meaning (i.e., actively removing undesirables), but he supported positive incentives to improve desirable qualities as well as birth control for population management. At the same time, he lobbied for improvements to public health and poor living conditions. Huxley argued strongly against the use of race as a scientific or political concept and bitterly opposed fascism. He supported boosting aid to developing countries, especially aid toward improving infrastructure, education, and agriculture. Much of his agenda in UNESCO supported these aims.

In 1919, Huxley married Marie Juliette Baillot (1895–1994). They had two sons. Like his grandfather Thomas Henry Huxley, Julian Huxley's life was punctuated by periods of severe manic depression, which sometimes debilitated him or led to major transitions in his life (such as leaving academia). He wrote detailed and informative memoirs and is now the subject of increasing interest because of the wide scope of his thinking (Waters and van Helden 1992; Ruse 1996).

BIBLIOGRAPHY

Dawkins, R. 1986. *The Blind Watchmaker*. New York: Norton.

Huxley, J. S. 1912. *The Individual in the Animal Kingdom*. Cambridge: Cambridge University Press.

———. 1927. *Religion without Revelation*. London: Ernest Benn.

———. 1930. *Ants*. London: Jonathan Cape.

———. 1931. *What Dare I Think? The Challenge of Modern Science to Human Action and Belief*. New York: Harper and Brothers.

———. 1932. *Problems of Relative Growth*. London: Methuen.

———. 1934. *If I Were Dictator*. New York: Harper and Brothers.

———. 1936. *At the Zoo*. London: Allen and Unwin.

———. 1940. *The New Systematics*. Oxford: Clarendon Press.

———. 1941. *Man Stands Alone*. New York: Harper.

———. 1942. *Evolution: The Modern Synthesis*. London: Allen and Unwin.

———. 1943. *TVA: Adventure in Planning*. London: Scientific Book Club.

———. 1948. *UNESCO: Its Purpose and Its Philosophy*. Washington, DC: Public Affairs Press.

———. 1954a. The evolutionary process. In J. Huxley, A. C. Hardy, and E. B. Ford, eds., *Evolution as a Process*, 1–23. London: Allen and Unwin.

———. 1954b. Scientific humanism, evolution, and human destiny. Unpublished lecture given in Los Angeles, October 16. HP 70.9, Huxley Papers, Rice University, Houston.

———. 1959. Introduction. In P. Teilhard de Chardin, *The Phenomenon of Man*, 11–28. London: Collins.

———. 1964. *Essays of a Humanist*. London: Chatto and Windus.

———. 1970. *Memories*. London: Allen and Unwin.

———. 1973. *Memories II*. London: Allen and Unwin.

Huxley, J. S., and G. de Beer. 1934. *The Elements of Experimental Embryology.* Cambridge: Cambridge University Press.

Ruse, M. 1996. *Monad to Man: The Concept of Progress in Evolutionary Biology.* Cambridge, MA: Harvard University Press.

Waters, C. K., and A. van Helden, eds. 1992. *Julian Huxley: Biologist and Statesman of Science.* Houston: Rice University Press.

Wells, H. G., J. S. Huxley, and G. P. Wells. 1929. *The Science of Life.* London: Amalgamated Press.   —*J.C.*

Huxley, Thomas Henry (1825–1895)

Thomas Henry Huxley was an invertebrate anatomist, paleontologist, biology educator, and popularizer of evolution. Huxley's nickname, "Darwin's Bulldog," conceals more than it reveals: he was a publicist for the emerging scientific profession and used Darwinism's naturalistic ideology to undercut rival pulpits.

Trained as a surgeon at an anatomy school, Sydenham College (1841–1842), and at Charing Cross Hospital (1842–1845), Huxley sailed on HMS *Rattlesnake* to Australia's Great Barrier Reef during the years 1846–1850. His papers on the siphonophores (similar to the Portuguese man-of-war), which established that all coelenterates comprised two foundation membranes (ectoderm and endoderm), won him the Royal Society's Royal Medal in 1852.

The jobless tyro denounced Robert Chambers's transmutationist *Vestiges of the Natural History of Creation* (1854) for its mannikin-like divine laws pulling life upward. Charles Darwin's naturalism, however, was appealing. Called to Down House in 1856 to air his objections to evolution, Huxley came away converted. Now the lecturer in natural history at London's School of Mines, Huxley immediately caricatured supernatural "Creation" in class. Publicly he pushed farther than Darwin: where Richard Owen in 1857 distanced humans from apes taxonomically, Huxley in 1858 drew parallels between human and gorilla anatomy. Lecturing to working men in 1859, Huxley said outright that humans had animal ancestors—the "vilest and beastliest paradox ever vented," fumed the evangelical *Witness* on hearing Huxley in Edinburgh. But Huxley was playing to new audiences; priestcraft-pricking assertions of human evolution were familiar in street prints. A self-professed "plebeian," he conjured an image of noble-future-from-lowborn-beginnings that meant nothing to the Anglican aristocracy, but the freethinking mechanics lapped it up.

The publication of Darwin's *On the Origin of Species* (1859) enabled the new science professional to usurp old theological questions: Whence mankind and what basis morality? Against the conservative clergy, Huxley deployed what he called Darwin's "Whitworth gun in the armoury of liberalism." His Manichean portrayal of science versus theology made news. Even today he is recalled for upbraiding Bishop Samuel Wilberforce at the British Association for the Advancement of Science meeting in 1860. Wilberforce had played on Victorian sensibilities about the sanctity of womanhood by inquiring whether the apes were on Huxley's grandfather's or grandmother's side. In preferring an ape ancestor over a bishop who prostituted his talents, Huxley, the Puritan, oozed moral sobriety to shame the worldly Wilberforce.

Such confrontations only exacerbated the Victorians' "crisis of faith." In 1861, Huxley denied Owen's claim that apes lack the human's third cerebral lobe, whose cavity has a posterior horn with a projecting "hippocampus minor." Charles Kingsley in *The Water Babies* laughed about "hippopotamus majors," and the affray brought human evolution squarely to the public's attention. Huxley made his newly acquired *Natural History Review* an anti-Owen organ and set the trend with his article "On the Zoological Relations of Man with the Lower Animals" (1861). Huxley's working-class talks, coupled with a lecture on fossil humans (focusing on the recently unearthed Neanderthal Man), formed the cornerstone of his *Evidence as to Man's Place in Nature* (1863).

Huxley studied Devonian fishes with lobe fins and similar-looking amphibians, called labyrinthodonts, from the strata surrounding coal seams. Rather than draw evolutionary connections, he emphasized that labyrinthodonts had persisted unaltered for immense periods. Initially, Huxley believed that the major groups had evolved before Silurian times and had changed little since. Only when Darwin's German supporter, Ernst Haeckel, published on phylogeny, or racial ancestry, did Huxley reconsider. In 1867, Huxley reclassified birds genealogically, and in 1868, he famously suggested that the small dinosaur *Compsognathus* could have been ancestral to birds (and may have borne feathers). He dramatized these relationships and promoted O. C. Marsh's horse fossils from Nebraska on his American tour of 1876. One illustration in *American Addresses* (1877), showing the increasing leg and teeth size from the small three-toed Eocene *Orohippus* to the tiptoeing *Equus,* was cited as the definitive fossil evidence of evolution for generations. Later he described the evolution of the secondary palate in crocodiles.

Huxley's attitude toward Darwin's mechanism remained equivocal. Where Darwin envisaged variation occurring in minuscule steps, Huxley in 1859 pictured an inner force producing new species at once. His entry on "Evolution" in the *Encyclopaedia Britannica* (1878) mooted natural selection only once, and Huxley's academic work bypassed it completely. Huxley reserved his discussion for popular talks: here natural selection was likened to Napoleon's troops retreating from Moscow and crossing the Beresina River, "everyone heeding only himself," the fittest alone reaching the far bank. Mostly, Darwin's Malthusian mechanics served Huxley in combating socialism or debunking creationism. Also, he avoided evolution in class because he was training science schoolmasters after the Education Act (1870). Establishing his first proper laboratory in South Kensington in 1871, he fashioned a practical "biology" based on the common plant and animal "Types." This influenced school curricula until the mid-twentieth century. But inculcating a discrete "Type" system left little room for evolutionary discussions. Opening Birmingham's new Science College, Huxley contrasted the recycling of the classics at the Anglican universities with science's experimental approach. The undervalued scientific researcher, he contended, would be better qualified to lead industrial society. This delighted the steel magnates who financed him. As Dissenters (outside the Anglican Church), they were excluded from taking Oxford and Cambridge degrees and saw Huxley's "new Reformation" legitimizing their growing civic authority.

Huxley made the word *evolution* fashionable around 1870. For him it had a naturalistic meaning, whereas for the Catholic St. George Mivart and the spiritualist Alfred Russel Wallace it conveyed the fiat of a higher power. Hence, Huxley coined the word *agnosis* in 1869. Biologists, as cultural leaders, had to be "agnostic" about supernatural issues. They had to approach science with clean hands to be trusted. Huxley helped mold the image of the scientist as the neutral professional, and of evolution as demystified science. He spent his last years arguing his antimiraculous case in theological controversies, shaping a century of evolution-creation debate.

Although evolution was fine for demolishing Anglican doors closed to reform, Darwinism's weak-to-the-wall upshot perturbed Huxley. A brutal selection had raised mankind, but it could not explain our care of the weak. In "Evolution and Ethics" (1893) he portrayed ethical man renouncing nature's "moral indifference." Ethics were formed by evolution, yet mankind's "nature within nature" somehow worked against the primitive selective forces.

Ultimately, the supremacy of science, naturalism, and agnosticism were what concerned Huxley as a populist. This was evident as his works were gathered. Whereas his *Scientific Memoirs* (five volumes, 1898–1902) were descriptive, his nine volumes of essays (1893–1894) were scintillating prose works showing the Victorian age accommodating itself to evolution. Reading them, H. L. Mencken called Huxley "perhaps the greatest virtuoso of plain English who has ever lived."

BIBLIOGRAPHY

Barr, A. P., ed. 1997. *Thomas Henry Huxley's Place in Science and Letters: Centenary Essays*. Athens: University of Georgia Press.

Desmond, A. 1997. *Huxley: From Devil's Disciple to Evolution's High Priest*. Reading, MA: Addison-Wesley.

Di Gregorio, M. 1984. *T. H. Huxley's Place in Natural Science*. New Haven, CT: Yale University Press.

Lyons, S. L. 1999. *Thomas Henry Huxley: The Evolution of a Scientist*. Amherst, NY: Prometheus Books.

Paradis, J. G., and G. C. Williams, eds. 1989. *Evolution and Ethics*. Princeton, NJ: Princeton University Press.

White, P. 2003. *Thomas Huxley: Making the "Man of Science."* Cambridge: Cambridge University Press.

—A.D.

Industrial melanism

Industrial melanism refers to the evolution of dark body colors in animal species that live in habitats blackened by industrial soot. Of approximately 100 reported examples, most are insect species, and of these the vast majority are moths. The most thoroughly documented example is the peppered moth, *Biston betularia*.

The common name describes the appearance (or phenotype) of the typical adult, a pale moth "peppered" with black-and-white scales (see figure). Melanic phenotypes, named carbonaria, are essentially solid black. Intermediates, or insularia, also occur but receive relatively little attention in general summaries. Extensive genetic analysis confirms that the phenotypes are determined by multiple alleles at a single gene locus.

The melanic phenotype was first discovered near Manchester, England, in 1848. By 1895, about 98% of the specimens near Manchester were melanic; this once rare phenotype had spread across the industrial regions of Great Britain. With only one generation per year, the nearly complete reversal in phenotype frequency was astonishingly rapid. Well away from industrial regions, melanic specimens remained rare.

British lepidopterist J. W. Tutt advanced the following Darwinian explanation in 1896: Peppered moths are active at night and rest during daylight hours, hiding on the surfaces of trees. Most insectivorous birds hunt by day and locate prey primarily by vision. Moths that are well concealed against backgrounds upon which they stay motionless are more likely to escape detection by birds than are conspicuous moths. Tutt proposed that typical peppered moths in undisturbed environments gain protection from visual predators by their resemblance to lichens on tree bark. In manufacturing regions where lichens have been destroyed by pollution and the surfaces of trees have been blackened by soot, the typical phenotypes lose their camouflage and fall victim to bird predators, whereas the melanic phenotypes in such polluted habitats escape detection long enough to pass their genes on to progeny.

Tutt's ideas were not tested until the 1950s when Bernard Kettlewell (1973) initiated a series of experiments to determine whether or not birds actually ate melanic and typical peppered moths selectively. His experiments included three main components: (1) quantitative rankings of conspicuousness (to the human eye) of typical and melanic phenotypes placed on various

Pale and dark (melanic) forms of the peppered moth, *Biston betularia.*

backgrounds; (2) direct observations of predation by birds on moths placed onto tree trunks; and (3) recapture rates of marked moths released onto trees in polluted and unpolluted woodlands. The design and execution of Kettlewell's pioneering experiments have been subjected to intense scrutiny over the years. Criticisms about his methods (including the densities of the moths used, the positions on trees where the moths were placed, the time of day the moths were released, and the use of laboratory-reared and wild-caught moths) inspired nine additional experiments by other researchers. The most recent was a seven-year study by Michael Majerus, completed in 2007. Specifically, Majerus released living moths into arenas at dusk so they could determine their own hiding places on trees before dawn. He adjusted the phenotypic proportions to match local frequencies and kept the densities low. Through binoculars he made direct observations of nine different bird species eating peppered moths. By comparing the survival rates of the moth phenotypes used in these predation experiments, he calculated selection coefficients that virtually matched the selection coefficients derived from the rate of decline of the melanic phenotype in the local population over the course of his study. The weight of evidence now supporting Tutt's hypothesis that birds eat peppered moth phenotypes selectively is substantial and remains uncontradicted by experiment.

Because the reflectance of light from the surface of tree bark is strongly negatively correlated with atmospheric levels of suspended particles, the testable prediction is that melanic phenotypes should be more common in sooty, polluted regions than they are in unpolluted regions. This prediction was clearly met by the frequency of melanic peppered moths distributed among 30,000 specimens from populations surveyed at 83 locations across Britain between 1952 and 1970.

The geographic distribution of melanism in peppered moths is more strongly correlated with atmospheric levels of sulfur dioxide (SO_2) than with smoke. SO_2, as a gas, is more widely dispersed from point sources than particulate matter that tends to settle locally as soot. Although selection may operate locally, gene flow from the migration of individuals among populations has contributed to more gradual clines in melanism in peppered moths, a relatively mobile species, than in more sedentary moth species (e.g., *Gonodontis* [= *Odontoptera*] *bidentata*) in which frequency distributions of melanic phenotypes are more sharply subdivided over their ranges. The power of gene flow to obscure genetic adaptation to local conditions resulting from selection was little appreciated by early researchers, who entertained various forms of nonvisual selection in attempts to account for apparent anomalies in clines. Molecular techniques using noncoding DNA sequences have now been developed for peppered moths by Ilik Saccheri and his associates (Daly et al. 2004). Their analysis of 12 microsatellite loci from populations across a 125-kilometer transect from north Wales to Leeds, England, indicates high levels of dispersal and gene flow. This and future work should help quantify the role of migration in the evolution of melanism in this species.

In 1956 Britain initiated the Clean Air Acts to establish so-called smokeless zones in heavily polluted regions. Following significant reductions in atmospheric pollution, melanic peppered moths have declined in frequency as the typical form recovered. Indeed, the decline in melanism in the latter half of the twentieth century is far better documented than the increase in melanism in the latter half of the nineteenth century. An uninterrupted annual record 18 kilometers west of Liverpool begun in 1959 by Cyril Clarke (Clarke et al. 1985) includes nearly 19,000 specimens showing a drop in the frequency of melanics from a high of 93% to below 5% by 2003 (Grant 2005). The most recent national survey, taken in 1996 (Grant et al. 1998) shows marked declines in melanism have occurred everywhere in Britain. The predicted correlation between changes in the levels of pollution and in the frequencies of melanic phenotypes in peppered moth populations has been firmly established.

Bruce Grant and Denis Owen (Grant et al. 1996) have documented parallel evolutionary changes in both directions in the North American peppered moth, *Biston betularia cognataria*. Melanic phenotypes in the subspecies result from alleles at the same gene locus (Grant 2004), but the rise and spread of melanism in American populations started about 50 years later than in Britain. Early museum collections do not include any melanic specimens prior to 1929 in the vicinity of Detroit, Michigan, but by 1959 the frequency of melanics had reached 90%. Clean air legislation was then inaugurated in the United States in 1963. Records for southeastern Michigan show that atmospheric SO_2 and suspended particles subsequently declined significantly. By 1994 the frequency of melanic peppered moths had fallen to 20%, dropping to 5% by 2001 (Grant and Wiseman 2002). The significance of a sharp decline in melanism in American peppered moths coincident with reductions in atmospheric pollution is complemented by comparisons of SO_2 concentrations from southwestern Virginia, where melanism never exceeded 5% over the same time interval. The Michigan location recorded higher concentrations of SO_2 than the Virginia location in 23 of the 25 years that measurements were taken at both places.

American and British peppered moth populations are now approaching monomorphism for their respective typical phenotypes, correlating with reduced levels of atmospheric pollution on both sides of the Atlantic. While correlations alone do not establish causal relationships, common correlations suggest common causes.

Evolution is defined at the operational level as a change in gene (allele) frequency in a population over time. The changes we have observed (and continue to observe even now) in peppered moth populations meet that definition precisely. Of the four primary evolutionary forces—mutation, genetic drift, gene flow, and selection—only strong selection can account for such rapid, directional changes. Whether or not differential bird predation of moth phenotypes is the sole, proximal agent of selection, the crucial evidence for natural selection in the evolution of melanism is based on the genetic changes that have occurred in populations. This evidence is indisputable and wholly independent of experiments aimed at identifying particular components of fitness (e.g., camouflage from predators and physiological differences among genotypes).

In all areas of active research, participants routinely challenge assumptions and interpretations. While we continue to investigate the relative importance of gene flow, lichens, predators, potential nonvisual selection, and the behavior and ecology of this moth species, industrial melanism remains an outstanding example of natural selection based on the most massive data set on record documenting long-term allele frequency changes in natural populations.

BIBLIOGRAPHY

Clarke, C. A., G. S. Mani, and G. Wynne. 1985. Evolution in reverse: Clean air and the peppered moth. *Biological Journal of the Linnean Society* 26: 189–199.

Cook, L. M. 2003. The rise and fall of the *carbonaria* form of the peppered moth. *Quarterly Review of Biology* 78: 399–417.

Daly, D., K. Waltham, J. Mulley, P. C. Watts, A. Rosin, S. J. Kemp, and I. J. Saccheri. 2004. Trinucleotide microsatellite loci for the peppered moth *(Biston betularia). Molecular Ecology Notes* [doi:10.1111/j.1471-8286.2004.00607l.x].

Grant, B. S. 1999. Fine tuning the peppered moth paradigm. *Evolution* 53: 980–984.

———. 2004. Allelic melanism in American and British peppered moths. *Journal of Heredity* 95: 97–102.

———. 2005. Industrial melanism. *Encyclopedia of Life Sciences,* www.els.net [doi:10.1038/npg.els.0004150].

Grant, B. S., A. D. Cook, D. F. Owen, and C. A. Clarke. 1998. Geographic and temporal variation in the incidence of melanism in peppered moth populations in America and Britain. *Journal of Heredity* 89: 465–471.

Grant, B. S., D. F. Owen, and C. A. Clarke. 1996. Parallel rise and fall of melanic peppered moths in America and Britain. *Journal of Heredity* 87: 351–357.

Grant, B. S., and L. L. Wiseman. 2002. Recent history of melanism in American peppered moths. *Journal of Heredity* 93: 86–90.

Kettlewell, H. B. D. 1973. *The Evolution of Melanism: The Study of Recurring Necessity*. Oxford: Clarendon Press.

Majerus, M. E. N. 1998. *Melanism: Evolution in Action*. Oxford: Oxford University Press.

———. 2007. The peppered moth: The proof of Darwinian evolution. http://www .gen.cam.ac.uk/Research/Majerus/SwedenPepperedmoth2007.ppt; http://www .gen.cam.ac.uk/Research/Majerus/Swedentalk220807.pdf.

Tutt, J. W. 1896. *British Moths*. London: George Routledge and Sons.

—B.S.G.

Insects

It is difficult to argue that evolution has produced any single group more spectacular than insects. Their numbers alone are staggering: over 820,000 species have been described, and estimates for the actual number of species exceed 2 million. Major groups of insects are still being discovered. In 2002, for example, an entirely new order of mantidlike insects was described, and hundreds of new species are described from the tropics every year.

The diversity of insects is no less staggering. Insects occupy every habitable region on earth. They have invaded every habitat except the open sea, and they have radiated into nearly every conceivable feeding niche. Many species build shelters and nests, ranging from the casings of sand and debris constructed by caddis fly larvae to the intricate mound and tunnel systems of tropical termite colonies. While the majority of insects lead solitary lives, there is a spectrum of social organization that includes family aggregations in caterpillars, small family groups of wood roaches, colonial nesting in wasps, and highly refined, complicated societies of siblings that form bee, wasp, ant, and termite colonies. The ants alone include species that are scavengers, predators, and gardeners; some ant species tend aphids for their sugary secretions and others capture ants of other species and enslave them as workers in their own colonies.

Insects provide some of nature's most spectacular phenomena and display some of nature's most sophisticated features. The biblical plague of locusts reflects a spectacular natural occurrence, the eruption and organized movement of millions of individuals of a species named *Schistocerca gregaria*. Periodical cicadas emerge in the forests of eastern North America by the millions on 13- and 17-year cycles, only to mate, reproduce, and die within weeks. Monarch butterflies make seasonal migrations across hundreds of miles between North America and Mexico. Many insects appear to defy biological conventions: there are species that consist only of females, and many species can produce many individuals, from dozens to hundreds, from a single egg. Insects have sophisticated sensory systems, seeing in wavelengths that humans cannot, such as the ultraviolet, and hearing at frequencies that humans cannot. Insects are not easy to poison; they have the capacity to assimilate a variety of toxic molecules, sometimes breaking them into harmless components and sometimes sequestering them as defensive compounds that deter predators.

There are striking patterns in the organization and history of insect diversity. To appreciate these patterns, it is important first to place insects in context. Insects, along with myriapods (centipedes and millipedes), are one of the four major groups that comprise the arthropod phylum. The arthropods are the segmented animals, and the other major arthropod groups, often called subphyla, are the trilobites, the chelicerates (spiders and their relatives), and the crustaceans (a diverse group that includes crabs, lobsters, and many others). The insects are a distinct class, the Hexapoda (six legged), within the subphylum they occupy with the myriapods. While each of the arthropod subphyla has been quite successful in number of species, the proliferation and variety of hexapods are approached only by the crustaceans.

The characteristic features of insects include having their segmented bodies organized into three major divisions (head, thorax, abdomen); three pairs of legs emerging from the central division (the thorax); an exoskeleton (a hardened outer shell that is not always so hard, as in caterpillars); a ventral nerve cord; and a variety of distinct morphological and physiological features. Insects have no lungs, breathing instead through sphericles, which are tubes in the body that open at the body surface. Like all arthropods, insects grow by molting, shedding their external skin and exoskeleton periodically and expanding into a new skin and exoskeleton.

One of the striking patterns in the diversity of insects is that nearly all of the 31 extant orders are old; individuals that represent most of these orders can be found in very old fossil deposits. This is quite different from nearly every other major animal group, in which many orders are known only from fossils or in which orders range widely in age. The earliest insect fossil is a springtail that is estimated to be 396 million years old. By the end of the Permian, about 250 million years ago, most of the known orders had appeared and fossils of them are widespread from that point onward. The only major orders that had not appeared by the Permian are the ectoparasites of birds and mammals, which appeared only after the appearance of birds and mammals in the fossil record. To calibrate these observations, remember that the coelacanths, *Latimeria spp.*, are called "living fossils" because there are recognizable members of this group preserved as 380 million-year-old fossils. By this standard, we might call most insect orders living fossils.

The only insect order that has gone extinct was a group called the Palaeodictyoptera, whose members resembled dragonflies in their appearance. Unlike dragonflies, these insects were herbivorous, emerging in the mid-Carboniferous (about 330 million years ago), not much later than many other insects, and disappearing at the end of the Permian. The striking aspect of the Palaeodictyoptera's disappearance is that they appear to have been enormously successful, comprising about 50% of the known species of insects from those periods. Researchers have no idea why so successful a group disappeared so suddenly.

While nearly all of the orders and many of the families are very old, there have been four periods when the diversity of species within families increased dramatically. In the first of these periods, the early Carboniferous (about 380 million years ago), winged insects appeared and began to multiply. In the second, toward the end of the Permian, the earliest holometabolous insects appeared, and this group began diversifying soon afterward. Holometabolous insects have a complex life cycle of egg, larva, pupa, and adult; the larva and adult are dramatically different in appearance and lifestyle, and they represent almost two distinct organisms. Holometabolous insects include the familiar groups of flies, butterflies, beetles, ants, and lacewings. In contrast, the insects we call hemimetabolous, which appear in the fossil record before holometabolous insects, have a simpler life cycle in which the organism that emerges from the egg, called a nymph, resembles a simpler and smaller version of the adult. Common hemimetablous insects include grasshoppers, crickets, genuine bugs like stinkbugs, and roaches. Holometabolous insects have been enormously successful, representing about 90% of existing insect species.

The third period of insect expansion occurred in the Cretaceous, from 130 to 65 million years ago, when about half of the modern families appeared and insect species diversity increased dramatically. This expansion began immediately after the initial radiation of flowering plants. The fourth period of expansion, which some scholars consider as a continuation of the Cretaceous radiations, was in the Tertiary, the period that began 63 million years ago. The continued radiation of insects with phytophagous habits, meaning that they feed on plants in one way or another, was certainly a continuation of the Cretaceous expansion. As witness to this, about 95% of the recent Lepidopterans (butterflies, moths, and skippers), nearly all of which are phytophagous, emerged in the Tertiary. But the Tertiary radiation also included the emergence of the decidedly nonphytophagous ants, bees, wasps, and higher termites with their sophisticated social systems as well as the orders of lice, which are ectoparasites of birds and mammals.

Two ties bind the evolutionary histories of insects and flowering plants to each other. First, the overwhelming majority of flowering plants are pollinated by insects, and insect pollination service has promoted diversification and speciation. Insects have been eating pollen for quite some time. Pollen is recognizable in the guts of some fossil insects in the Permian, over 250 million years ago. The striking change that coincided with the emergence of flowering plants was the appearance of specialized features of the head and mouth that facilitated not the eating of pollen but the transport of pollen as the insects fed on nectar. These first appear in some flies in the early Cretaceous. By the latter half of the Cretaceous, 90 million years ago, most specialized pollination systems known today had evolved and are clearly seen in the fossil record. And should one doubt that insect pollination has contributed to the success of flowering plants, it is worth remembering that groups of plants pollinated by animals have diversified at twice the rate of those groups that are pollinated by wind or water.

Second, phytophagous insect diversity has followed the diversification of flowering plants. To be sure, many insects have been feeding on plants for most of their existence. Fossil plant tissue shows evidence of being pierced by mites and collembolans since the late Devonian (about 400 million years ago), and the earliest entirely herbivorous order of insects was the Palaeodictyopterida, which appeared about 330 million years ago. About 83% of fossil plant leaves from the Permian (which ended 250 million years ago) show damage from herbivores. However, it was in the Cretaceous and Tertiary when the major radiations of phytophagous species occurred, following closely the major radiations of flowering plants.

This joint radiation reflects a coevolutionary process through which plants evolve defenses against insect feeding, insects evolve features to overcome those defenses, and specialized associations of insect and plant groups emerge. The evidence for this process has come from experimental ecological studies. Insects act as agents of natural selection on plant defenses; individual plants that have defensive features, such as denser hairs that deter herbivores or higher concentrations of toxic chemicals in their tissues, are able to survive, grow larger, and leave more offspring. Conversely, plant defenses act as agents of selection on insect traits; individuals that can overcome defensive structures or detoxify toxic compounds leave more offspring. As a plant species refines its defenses through continued selection and evolution, only specific groups of insects continue to overcome those defenses through evolution. As an insect species refines a specialized ability to exploit a species or genus of plants, it appears to lose its ability to exploit others that employ different defensive compounds. As a result, populations of plants that produce new types of effective defenses diverge from their sibling populations, often being free for a while of insect pests but finding themselves battling with different insects than their sibling populations. Hybrids between these plant populations fare more poorly than either parent and the process of speciation has begun, driven by a continual and unforgiving arms race.

Through this process, phytophagous insects and plants have driven each other reciprocally toward mutual specializations and diversification. The end result is that most plants are exploited by specific groups of insects and most phytophagous insects are specialized on particular plants. While there are associations at broad levels—for example, larvae of individual butterfly families are found only on a few plant families—there are many associations at finer levels. Monarch butterflies feed only on milkweeds (genus *Asclepias*) and pipevine swallowtails feed only on pipevines and snakeroots (genus *Aristolochia*); in each case the plants harbor highly toxic compounds that deter nearly all other insects.

The study of insects has contributed enormously to the development of evolutionary biology. The mimicry of noxious butterflies by palatable ones was regarded as one of the great challenges for Charles Darwin's hypothesis of natural selection, and studies of mimicry occupied a central position in the canon of early experimental work on natural selection. The domestication of the fruit fly (or, more properly, vinegar fly), *Drosophila melanogaster,* by Thomas Hunt Morgan in the early twentieth century produced a model organism that has been used around the world for experimental studies of adaptation and speciation. The challenge of explaining how the sophisticated social structure of ants, wasps, and bees evolved, with its characteristic of sterile fe-male workers raising their sisters, inspired W. D. Hamilton to develop a formal theory of kin selection. Kin selection now underlies much of our understanding of sociality and many patterns of behavioral evolution throughout the animal world. Studies of individual variations in moth and butterfly wing patterns by E. B. Ford and his colleagues were one of the foundations of ecological gene-tics, the experimental study of natural selection and evolutionary change out-side of the laboratory.

More recently, many different insect groups have been used as model sys-tems for evolutionary studies. The flour beetles, *Tribolium spp.*, have been used in pathbreaking studies of group selection. Studies of water striders have illuminated the conflict between the sexes over mating, and how that conflict has shaped the physiology and behavior of the sexes, better than any other system studied to date. The damselfly species that segregate by type of lake (with or without fish) have been used to illuminate how rapidly a group can diversify in novel habitats and how adapting to a novel ecological cir-cumstance often comes with the cost of losing the ability to persist in the original conditions.

BIBLIOGRAPHY

Berrenbaum, M. R. 1995. *Bugs in the System: Insects and Their Impact on Human Affairs*. Reading, MA: Addison-Wesley.
Eisner, T. 2003. *For Love of Insects*. Cambridge, MA: Belknap Press of Harvard University Press.
Grimaldi, D., and M. S. Engel. 2005. *Evolution of the Insects*. New York: Cambridge University Press.   —J.T.

J

Jablonski, David (b. 1953)

David Jablonski is among the most prominent of a generation of paleontologists who, having been trained in the midst of the renaissance in paleobiology that took place in the late 1970s and early 1980s, have extended the study of large-scale patterns and processes of evolution and extinction begun by pioneering paleontologists such as Stephen Jay Gould, David M. Raup, and Steven Stanley. Jablonski's particular area of study has been the dynamics of extinctions and their resulting impact on patterns in evolutionary history. He has approached this subject primarily through statistical, quantitative analysis, which uses computer modeling and simulation to process large quantities of data drawn from the fossil record. In addition to important analyses of extinctions in the geologic past, Jablonski's work has made contributions to our understanding of the modern dynamics of extinction and to the current global biodiversity crisis.

One of Jablonski's primary accomplishments has been to challenge or disprove a number of theoretical assumptions about evolution and extinction dynamics by using a variety of quantitative tools not available to earlier paleontologists. His recent work, for example, has shown that Cope's rule—the theory that evolution tends to favor larger body sizes—is based on sampling error. Scientists tend to study larger, more "exciting" organisms, which produces the misleading impression that large organisms are more common (and thus more evolutionarily successful) than small ones. In a study of the originations and extinctions of 190 lineages of mollusks over many millions of years, Jablonski argued that, on the contrary, body size confers no particular advantage. In a similar study, he showed that estimates of a major increase in biodiversity over the past 50 to 100 million years are not an exaggeration. For a number of years, scientists believed that the recent fossil record shows greater marine animal diversity because the recent record is better preserved; this sampling artifact is known as "the pull of the recent." By studying the fossil ancestry of hundreds of genera of living bivalves, Jablonski led a group of scientists who showed that the recent fossil record is remarkably accurate— between 87% and 95% over the past 50 million years. This work helps confirm empirically what paleontologists have suspected for some time: that a major explosion in biodiversity followed the Palaeozoic era, roughly 250 million years ago.

Perhaps Jablonski's most significant research involved studying the dynamics and patterns of extinctions. In this area, Jablonski has continued to test evolutionary hypotheses with quantitative analysis. One question he has addressed is whether survivorship during mass extinctions is predictable or random, and whether the evolutionary trends that follow major extinctions display a uniform pattern. A related study asked whether the survivors of major extinction events (for example, the mass extinction that killed the dinosaurs 65 million years ago) are in fact successful evolutionary competitors afterward. He found a number of possible routes genera may follow after such an event, ranging from thriving expansion and diversity to stagnation and eventual extinction. He concluded that while there were some predictable factors involved, the paths appear to vary based on regional environmental differences, an observation that supports the hypothesis of randomness.

Surprisingly, Jablonski confirmed there was often little correlation between extinction survival and postextinction evolutionary success. In many cases, formerly struggling genera thrived, while dominant organisms were wiped out. A third alternative, called "Dead Clade Walking," describes genera that were not immediately killed off but were doomed to eventual decline, suggesting that secondary "ripples" of extinction often follow millions of years after mass extinction events. Overall, these studies and other work have established Jablonski at the forefront of efforts to use new methodologies (and often the latest technology) to answer many of the questions about the accuracy and usefulness of the fossil record that have vexed paleontologists since Charles Darwin's day. In general, Jablonski's work supports the conviction that the fossil record is reliable and thus can be used to infer the relative importance of different evolutionary mechanisms patterning the history of life on earth. By better understanding the dynamics of extinction, Jablonski has also highlighted the ways extinctions, which are normally seen as negative processes, contribute to the overall dynamics of evolution.

BIBLIOGRAPHY

Jablonski, D. 1999. The future of the fossil record. *Science* 284: 2114–2116.
———. 2001. Lessons from the past: Evolutionary impacts of mass extinctions. *Proceedings of the National Academy of Sciences USA* 98: 5393–5398.
———. 2004. Extinction: Past and present. *Nature* 427: 589.   —D.Se.

Jenkin, Henry Charles Fleeming (1833–1885)

Henry Charles Fleeming (pronounced Fleming) Jenkin made major contributions to physics and electrical engineering in mid-nineteenth-century England. Remembered by evolutionists as the author of a review of Charles Darwin's *On the Origin of Species,* he argued that natural selection would not work if heredity was a process that blended parental characteristics. This review is often supposed to have shown that the move to a particulate model of heredity, such as Mendelian genetics, was essential if selection was to become a plausible theory. In fact, although Darwin was impressed by Jenkin's review, other selectionists realized that if variation was seen as a property of populations, not individuals, selection was plausible whatever the mechanism of heredity.

Jenkin was born in Dungerness, Kent, England, on March 25, 1833. His family moved to Italy and Jenkin received an MA from the University of Genoa in 1851. He returned to England and spent 10 years working on the design and manufacture of submarine telegraph cables, after which he set himself up as a consulting engineer in London. He took out numerous patents, some with his friend William Thomson (later Lord Kelvin). In the 1860s, Jenkin wrote reports for the British Association for the Advancement of Science's Committee on Electrical Standards, as a consequence of which the ohm was accepted as an absolute standard of resistance.

Jenkin had wide scientific interests, and the *North British Review* asked him to comment on Darwin's theory from a mathematical viewpoint. In his review, published in 1867, Jenkin endorsed Thomson's argument that the earth must have cooled down too rapidly for natural selection to have produced the amount of progressive evolution observed in the history of life. He also repeated the traditional view that within each species, variation is restricted within definite limits. More originally, he analyzed how natural selection could accumulate favorable variations. Arguing that each such variation was a distinct saltation, or sport of nature, he explored what would happen to the new character as the individual reproduced, on the assumption that heredity blends the characteristics of the male and female parents. He noted that in artificial selection, such new characters could be concentrated within a small population and could thus have significant effect. This had happened in the case of the short-legged Ancon sheep, where the farmer had bred only from the single ram born with the new characteristic. In a wild population, however, the favored individuals would breed with unchanged members of the population, and the characteristic would be halved in each generation until its effect was too small to count.

Darwin had never thought of variations as large-scale sports, but he did think of them as individual character changes. He was thus impressed by Jenkin's review, modifying later editions of his book to emphasize that other processes were at work. Other Darwinists, including Alfred Russel Wallace, pointed out, however, that if one thought of variation as a range of character differences within a population, Jenkin's argument was ineffective because favored individuals are plentiful rather than rare.

BIBLIOGRAPHY

Bowler, P. 2003. *Evolution: The History of an Idea*. 3rd ed. Berkeley: University of California Press.

Darwin, C. [1859] 1872. *On the Origin of Species*. 6th ed. London: John Murray.

Hull, D. L., ed. 1973. *Darwin and His Critics*. Cambridge, MA: Harvard University Press.

Jenkin, F. 1867. The origin of species. *North British Review* 46: 277–318.

Ruse, M. 1999. *The Darwinian Revolution: Science Red in Tooth and Claw*. 2nd ed. Chicago: University of Chicago Press.

Vorzimmer, P. J. 1970. *Charles Darwin: The Years of Controversy*. Philadelphia: Temple University Press. —*P.J.B.*

K

Kettlewell, H. B. D. (1907–1979)

Henry Bernard David Kettlewell was a founding member of a loosely associated group often referred to as E. B. Ford's Oxford School of Ecological Genetics, which included such luminaries as A. J. Cain, Cyril Clarke, and Philip Sheppard. While there was never any formal structure, the members were marked by a common approach to the study of genetic variation in natural populations, and, in particular, by a commitment to the pervasive role of natural selection as the most important agent of evolutionary change. Kettlewell was a gifted naturalist who left medical practice at the age of 41 to pursue his boyhood love of entomology. He began working full time as a senior researcher in E. B. Ford's newly founded subunit of genetics at Oxford in 1951.

Kettlewell is best known for his research on the phenomenon of industrial melanism, which refers to a rapid increase in the frequency of rare dark forms in many moth species in the vicinity of manufacturing centers as a consequence of the first large-scale air pollution associated with the Industrial Revolution in England and Continental Europe. The first and most famous example of this trend occurred in the peppered moth, *Biston betularia*. The dark form in this species was first discovered near Manchester in 1848. Within the space of only 50 years it became so common that the pale form was locally extinct. (Tutt 1891 is the classic nineteenth-century discussion of the topic.)

Kettlewell became famous for a series of field experiments conducted in the early 1950s documenting the reason why the dark form was becoming more common. In polluted areas, dark coloration protected the moths, resting on soot-darkened tree trunks, against bird predation. In an elegant field experiment conducted in the summer of 1953, Kettlewell released large numbers of marked pale and dark moths into a polluted wood near Birmingham and attempted to capture as many as possible over the next several nights. The logic behind his investigation was that, all things being equal, the recapture rates for the two forms should be the same. If, on the other hand, the dark form was at an advantage, owing to its relative inconspicuousness, one would expect the recapture rate for it to be higher because more of the dark form survived during the interval between release and recapture. And this is precisely what Kettlewell found. Kettlewell conducted a companion experiment in an unpolluted wood in Dorset in the summer of 1955 where he found the reverse. In this wood he was able to recapture more of the pale form than the dark. Kettlewell supplemented these experiments with a film record documenting that when birds were given a choice, they passed over inconspicuous moths in favor of the moth that was more conspicuous in its background resting site. (See Kettlewell 1955 and 1956 for the original reports and Kettlewell 1973 for a summation and overall discussion.)

While textbooks often portray Kettlewell's investigations as unproblematic, researchers (including Kettlewell himself) recognize that the initial investigations involved numerous assumptions that are open to question (e.g., the resting site of moths). This has led those critical of evolutionary theory to claim that the phenomenon is no longer an example of natural selection, and further that there is a conspiracy by textbook writers to hide this fact. (Hooper 2002 is a sensationalistic account of the work by a journalist and Wells 2000 is a severe, religiously motivated critique by a biologist who is a follower of the Reverend Sun Myung Moon, founder of the Unification Church.) At least eight studies directly address perceived problems in Kettlewell's initial investigation. The studies have independently confirmed that bird predation is the most important factor in the spread of the dark form. Perhaps the most convincing evidence that it is an example of natural selection, however, comes from the predicted rapid decline of the dark form since the introduction of "Clean Air" legislation in the United Kingdom and elsewhere. (Majerus 1998 gives a full overview of modern research and Rudge 2002, 2005 address the claims of fraud.)

BIBLIOGRAPHY

Hooper, J. 2002. *Of Moths and Men: The Untold Story of Science and the Peppered Moth*. New York: Norton.

Kettlewell, H. B. D. 1955. Selection experiments on industrial melanism in the lepidoptera. *Heredity* 9: 323–342.

———. 1956. Further selection experiments on industrial melanism in the lepidoptera. *Heredity* 10: 287–301.

———. 1973. *The Evolution of Melanism*. Oxford: Clarendon.

Majerus, M. E. N. 1998. *Melanism: Evolution in Action*. Oxford: Oxford University Press.

Rudge, D. W. 2002. Cryptic designs on the peppered moth. *International Journal of Tropical Biology and Conservation (Revista de Biología Tropical)* 50, no. 1: 1–7.

———. 2005. Did Kettlewell commit fraud? Re-examining the evidence. *Public Understanding of Science* 14, no. 3: 249–268.

Tutt, J. W. 1891. *Melanism and Melanochroism in British Lepidoptera*. London: Swan Sonnenschein.

Wells, J. 2000. *Icons of Evolution: Science or Myth?* Washington, DC: Regnery.

—D.W.R.

Kimura, Motoo (1924–1994)

Motoo Kimura is best known for the idea that most evolution at the DNA level occurs by random processes, rather than by Darwinian natural selection. Prior to this, evolutionists had attributed essentially all of evolution to natural selection. Kimura argued that for most evolutionary changes, chance is more important.

Within a cell, a characteristic number of chromosomes can be found (for example, 46 in humans). Each chromosome is a long, tightly coiled thread of

DNA. DNA consists of a sequence of four kinds of bases, adenine, cytosine, guanine, and thymine, abbreviated A, C, G, and T. A gene is a stretch of a few thousand bases that is responsible for a specific function; the sequence of bases in that region determines the function. Most of the rest of the DNA has no known function and is often referred to as "junk DNA." A mutation is a permanent change in the DNA; for example, a C may change to a T. If the change occurs within a gene, it will often have some visible effect on the individual inheriting it. Changes in the "junk" usually have no observable effect. If a particular gene or other sequence of DNA changes its frequency in the population by random processes, this is called random genetic drift.

In 1968 Kimura put forth the idea—very daring at the time—that the great bulk of changes in the DNA are purely random, the result of mutation and random drift. Kimura argued that whether the new mutation increases in the population and ultimately displaces its predecessor DNA is usually not the result of natural selection but, much more often, simply chance. This is the essence of Kimura's idea. This might be called the mutation-random drift theory of evolution; Kimura called it the neutral theory. The same idea was discovered independently by Jack King and Thomas Jukes (1969), but because most of the subsequent developments are due to Kimura, the theory is ordinarily attributed to him.

It often happens in the history of science that mathematics, developed for its own sake, turns out later to have practical uses. It turned out that a substantial body of theory had already been developed by Kimura himself. He was fortunate that the neutral theory found its application, not decades or centuries later, but within his own lifetime. Starting when he was a graduate student, Kimura solved a number of difficult problems involving random processes, such as the probability that mutation will ultimately increase in the population and become the prevailing form, the number of generations it would take for such a process to occur, and the number of individuals that would be affected by the mutation during this process. Kimura worked out the answers, whether the mutation is neutral, beneficial, or deleterious. Success for the last is extremely rare, but not negligible in geological times.

It turns out that the rate of evolutionary substitution of one neutral base for another is simply equal to the mutation rate. Because mutation rates are reasonably well known, this permits a prediction of evolutionary rates that can be checked against the historical record. One of the most important consequences of this discovery is that it makes possible a "molecular clock." If we know the mutation rate, we can determine the rate of molecular evolution. For studies of evolution, the timescale is determined by a molecular clock, which depends on Kimura's neutral theory.

It is now generally accepted that the great bulk of DNA changes in mammals and other vertebrates is neutral, following Kimura's theory. The theory is still controversial, however, regarding changes in functional regions of DNA. It is clear that in some cases, chance predominates; in others, it is natural selection. The answers will have to come from case-by-case analysis.

Motoo Kimura was born in Okazaki, Japan, on November 13, 1924. He showed an early aptitude for mathematics, but his deepest interest was botany. He greatly enjoyed a microscope given to him by his father. He never lost his interest in flowers, and in his later life he was a successful orchid breeder. He was in high school during the early days of World War II, and in 1944 was admitted to the Kyoto Imperial University. He was not involved in military service, but wartime life in Japan was hard for everyone. After the war, he joined the laboratory of Japan's premiere geneticist, Hitoshi Kihara. Kihara recognized Kimura's talents and permitted him to spend his time studying the literature of mathematical population genetics.

With the help of American geneticists working for the Atomic Bomb Casualty Commission, Kimura was able to obtain a scholarship to study in the United States. After a year at Iowa State University, he transferred to the University of Wisconsin and became a student of James F. Crow. He had the opportunity for daily contact with Sewall Wright, one of the founders of mathematical population genetics. After receiving his doctorate in 1956, he returned to Japan and for the rest of his life was employed by the National Institute of Genetics in Mishima, Japan. He continued his collaboration with Crow, and each took every opportunity to spend time with the other. Together they authored a number of papers and a textbook. Kimura died of amyotrophic lateral sclerosis on November 13, 1994, on his 70th birthday.

Although Kimura is best known for the neutral theory and the attendant mathematics, he was also known for applying diffusion models to other problems. These involved the solution of partial differential equations, at which he was particularly adept. He considered the effects of random fluctuations in the intensity of natural selection, which can lead to processes that mimic random genetic drift. He developed a model of migration between colonies, in which migrants move to one of the neighboring colonies. Altogether, he wrote six books and over 650 scientific papers. At the time of his death he was Japan's best-known geneticist and he had been the recipient of many international honors.

BIBLIOGRAPHY

Crow, J. F. 1997. Motoo Kimura. *Biographical Memoirs of Fellows of the Royal Society of London* 43: 255–265.

Kimura, M. 1983. *The Neutral Theory of Molecular Evolution.* Cambridge: Cambridge University Press.

King, J. L., and T. H. Jukes. 1969. Non-Darwinian evolution: Random fixation of selectively neutral mutations. *Science* 164: 788–798.

Takahata, N., ed. 1994. *Population Genetics, Molecular Evolution and the Neutral Theory: Selected Papers.* Chicago: University of Chicago Press.   —J.F.C.

L

Lack, David Lambert (1910–1973)

David Lambert Lack was a British ornithologist and ecologist famous for his studies of Galápagos finches, one of the key examples of natural selection in the wild. Born in London, Lack began bird-watching while in boarding school in the Kent and Norfolk countryside. He was addicted, winning school prizes for natural history three consecutive years. He entered Magdalene College, Cambridge, in 1929, but like Charles Darwin, Lack found little inspiration in his formal studies. Instead, he took to the field with members of the Cambridge Ornithological Club. Lack also began publishing systematic observations and conclusions on bird behavior. By graduation in 1933, he had written a dozen papers, and the club sponsored him to write *The Birds of Cambridgeshire* (1934).

Lack surprised friends and family by not pursuing an academic career. Believing that fieldwork was unpopular in Cambridge faculties, Lack instead accepted a job as biology master at Dartington Hall, a progressive school in Devon. Lack was encouraged to use the field as a teaching space. He had students trap and band robins in the nearby woods, and he encouraged them to make observations on territory and behavior. These student projects became the basis for a four-year study of robins.

Superficially, Lack's *The Life of the Robin* (1943) was a typical life history. However, Lack's writing hid a revolution in method. Empirically meticulous, he accepted an observation as real "only where the action concerned has been seen on at least six occasions" (Lack 1943, 6). He also used the new technique of banding birds. This permitted long-term studies of known individuals, and it allowed him to ask complex questions about life history and behavior. Lack believed that such methods made ethology more rigorous and scientific. He later added to his studies of robins with *Robin Redbreast* (1950), a collection of related literature, music, and folk culture. He repeated this approach to a single bird with *Swifts in a Tower* (1956).

Julian Huxley became a mentor and promoter. Under his encouragement, Lack visited the Galápagos Islands (from December 1938 to April 1939) to study finches. As had many before him, Lack first supposed that the absence of predators and competitors for food meant that natural selection was extremely weak on the islands. This was reinforced by American colleagues with whom Lack consulted on the way home to England. At the California Academy of Sciences and the American Museum of Natural History, Lack was shown how his observations fit well into the "genetic drift" models proposed in the 1930s by Sewall Wright and fashionable in America. A technical monograph (Lack 1945, written in 1940) detailed his conclusions drawn from this influence.

Lack reversed his views shortly after returning to England and to Huxley's influence. "Unexpectedly, a reconsideration of the original material led to a marked change in viewpoint regarding competition between species and the beak differences between the finches" (Lack 1947, vi). In *Darwin's Finches* and many popular articles thereafter, Lack made two points about natural selection in Galápagos finches.

First, he connected finch beaks to diet, arguing that they were well adapted to specific feeding styles: large seeds, small seeds, insects, and fruit (see figure). Natural selection within each species reinforced specialization and created highly adapted forms. Second, Lack focused on cases where several species occupied identical ecologies in the same location. In these cases, he said, natural selection was especially intense: "Since the chance of their being equally well adapted is negligible, one of them should eliminate the other completely" (Lack 1947, 62). On small islands, Lack found only one species of ground finch, suggesting that competition led one species to dominate and others to local extinction. On islands where two or more species were found, Lack realized that these species were never found with exactly the same ecologies. Natural selection was causing species to highlight their differences, thereby reducing competition. This caused divergence, as different species

Figure 1. Examples of Darwin's finches. Note the very different beak shapes, adapted to different feeding stuffs—big strong beaks for hard nuts and the like, fine beaks for small seeds and the like.

adapted to different habitats. Changing diets was the usual avenue for this divergence, Lack suggested. In this interpretation, beaks took on special importance; they offered a spectacular, nuanced example of natural selection in the wild. Inspired by this success, Lack made similar studies of passerine birds (1944), and of the cormorant and the shag (1945).

Darwin's Finches made Lack famous. He had transformed a case against natural selection into a premier example for it. His research came at a time when evolutionists were calling for more studies of "natural" rather than "artificial" (laboratory) examples of selection. It did the same for ecology's principle of competitive exclusion, made theoretically important by G. F. Gause in the 1930s. "Lack's finches," as many later called them, also boosted English efforts to develop evolutionary ecology, championed by Oxford's E. B. Ford and reinforced by Arthur Cain and Philip Sheppard's studies of the land snail, *Cepaea,* and Bernard Kettlewell's work on the peppered moth, *Biston.* Although the finch case was criticized in the 1950s, especially by Robert Bowman and Daniel Simberloff, for overstating the extent and intensity of competition, it never lost its allure. By the 1970s, long-term studies under way by Peter and Rosemary Grant confirmed and extended Lack's conclusions.

In 1945, Lack became director of the Edward Grey Institute of Field Ornithology, part of the Department of Zoological Field Studies at Oxford. The institute had two goals. First, it was a center for exchange of ornithological data and oversaw national census work in collaboration with the British Trust for Ornithology. This put Lack in a central position in the ornithological community. Second, the institute sponsored original research in field ornithology. Lack focused this research on population ecology and the regulation of population numbers (Lack 1954). He argued that only three factors controlled population sizes: disease, food availability, and parasites or predators. He focused the institute's research on precisely how this regulation took place. Lack's theories were severely criticized, especially by V. C. Wynne-Edwards and other "group selectionists." He replied with *Population Studies in Birds* (1966). His final book, *Ecological Isolation in Birds* (1971), emphasized ecological isolation as the driving force of diversity.

Not a strong administrator, Lack was more comfortable in the field than in the director's chair. He was criticized for focusing the institute's attention too narrowly. Lack's strengths as an ornithologist, however, far exceeded his weaknesses as an administrator. He was inspiring and mentored many students.

Lack converted to Christianity after World War II. He was confirmed in the Anglican Church in 1951, and he remained devout throughout his life. In *Evolutionary Theory and Christian Belief* (1957), Lack argued that science and evolution could help to explain many unresolved themes in religion. Chance processes, such as natural selection, could be creative. Death was natural, not evil. Evolution did not speak to the origin of life, only to its subsequent history. The Bible and nature did not contradict: "The true significance of the first chapter of Genesis is to assert that God made the universe and all in it, that He saw that it was good, and that He placed man in a special relationship to Himself" (Lack 1957, 34). Lack also sought to better understand areas of genuine conflict. For instance, many aspects of man's moral being seemed to him to be unexplained by theories about the moral evolution of animals.

Lack's "My Life as an Amateur Ornithologist" (1973) is an autobiography to 1945, written when he became director of the Edward Grey Institute. For him this marked an important transition from "amateur" to "professional." W. H. Thorpe (1974) provides the kind of knowledgeable memorial only a close friend could produce. Lack was awarded the Godman-Salvin Gold Medal of the British Ornithologists Union in 1958 and the Darwin Medal of the Royal Society in 1972.

BIBLIOGRAPHY

Grant, P., and R. Grant. 1986. *Ecology and Evolution of Darwin's Finches.* Princeton, NJ: Princeton University Press.

Lack, D. 1934. *The Birds of Cambridgeshire.* Cambridge: Cambridgeshire Bird Club.

——. 1943. *The Life of the Robin.* London: Witherby.

——. 1944. Ecological aspects of species-formation in passerine birds. *Ibis* 86: 260–286.

——. 1945. The Galápagos finches (Geospizinae): A study of variation. *Occasional Papers of the California Academy of Sciences* 21: 1–159.

——. 1947. *Darwin's Finches: An Essay on the General Biological Theory of Evolution.* Cambridge: Cambridge University Press.

——. 1950. *Robin Redbreast.* Oxford: Oxford University Press.

——. 1954. *The Natural Regulation of Animal Numbers.* Oxford: Clarendon Press.

——. 1956. *Swifts in a Tower.* London: Methuen.

——. 1957. *Evolutionary Theory and Christian Belief: The Unresolved Conflict.* London: Methuen.

——. 1966. *Population Studies of Birds.* Oxford: Clarendon Press.

——. 1971. *Ecological Isolation in Birds.* London: Blackwell.

——. 1973. My life as an amateur ornithologist [with appreciations by six colleagues]. *Ibis* 115: 421–431.

Perrins, C. M. 2004. Lack, David Lambert (1910–1973). In H. C. G. Matthew and B. Harrison, eds., *Oxford Dictionary of National Biography.* Oxford: Oxford University Press.

Thorpe, W. H. 1974. David Lambert Lack, 1910–1973. *Biographical Memoirs of Fellows of the Royal Society of London* 20: 271–293.

Weiner, J. 1994. *The Beak of the Finch: A Story of Evolution in Our Time.* New York: Alfred Knopf.   —J.C.

Lamarck, Jean-Baptiste (1744–1829)

Jean-Baptiste Lamarck was the first biologist to set forth a broad, comprehensive theory of organic evolution. Others before him had raised the possibility that new organic forms had appeared in the course of the earth's history, but Lamarck was the first to develop at length the idea that all the different forms of life on earth had been produced successively, beginning with the very simplest forms and proceeding gradually to the most complex.

Lamarck initially made his scientific reputation as a botanist, interacting in the 1770s with the circle of botanists at the Jardin du Roi (the King's Garden) in Paris. His career changed dramatically in 1793 when the Jardin du Roi was reconstituted as the National Museum of Natural History. He was named one of the professors of the new institution, but not to a chair of botany. He was given instead the professorship of "insects, worms, and microscopic animals." He soon characterized this part of the animal kingdom as the "animals without vertebrae." Although his expertise in this area was limited initially to what he knew from having been an avid collector of shells, he became over the next two decades the world's leading authority on invertebrate zoology.

Classifying the invertebrates and developing a course of instruction about them were not Lamarck's only concerns in the 1790s. From the late 1770s onward, he had speculated broadly about what he took to be the basic principles of physics, chemistry, and meteorology. In the 1790s, he also began to think about the history of the earth. This new geological interest was directly related to his studies of shells and to the emerging problem of explaining the similarities and differences between living and fossil shell types. While many of his contemporaries were inclined to see his broad speculative efforts as examples of unproductive "system-building," he complained of critics who dealt only with "small facts," who failed to address important issues, and who (he believed) were afraid that their reputations would be hurt by his ideas.

Lamarck first set forth his revolutionary notion of species change in an introductory lecture he delivered to the students in his invertebrate zoology class in 1800. By 1802, he was arguing that all the different forms of life on earth, from the simplest to the most complex, had been successively produced by natural causes operating over immense periods of time. His most famous expositions of his ideas appeared in 1809 in his *Philosophie zoologique* (Zoological Philosophy), and then in 1815 in the introduction to his great, seven-volume work on invertebrate zoology, *Histoire naturelle des animaux sans vertèbres* (Natural History of the Invertebrates).

Lamarck sought to account for more than change at the species level. His evolutionary theorizing addressed in addition the nature and origin of life, the successive production of the different forms of animal organization, and the different faculties to which these forms of organization give rise. He maintained that the very simplest forms of life had been, and were continuing to be, generated spontaneously. These forms then became successively diversified, he claimed, as the result of two very different sorts of causes: (1) "the power of life" or "power that tends unceasingly to make organization more complex" (Lamarck 1815–1822 1: 134) and (2) the modifying influence of particular circumstances. He explained the first of these as a function of the natural action of fluids moving through living tissues. It was responsible, he said, for the orderly scale of progression of the animal classes. The second of these causes, in contrast, was responsible for the lateral ramifications from the main scale. It involved the part of Lamarck's theory that has since become most famous: the idea of the inheritance of acquired characteristics.

Lamarck proposed that animals were induced to develop new habits when they were confronted with changing or new environments. Changes in habits inevitably led to certain organs being used more and other organs being used less, thereby strengthening the former and weakening the latter. Differences thus acquired were passed on to the next generation, albeit too gradually to be apparent on a human timescale. Over great periods of time, however, this resulted in significant organic change. Lamarck was confident that the distinctive features of such diverse creatures as giraffes, moles, storks, and snakes could be understood as the cumulative result of the inheritance of characters acquired because of long-maintained habits. Although the basic idea of "the inheritance of acquired characters" was not novel with him, he stands out as the first scientist to argue at length that the operation of this process on a geological timescale could produce species change.

It was not until near the end of the nineteenth century, well after Charles Darwin had advanced his own theory of evolution, that the idea of the inheritance of acquired characteristics, or "use-inheritance," came to be widely regarded as the "Lamarckian" explanation of species change in contrast to Darwin's idea of evolution by natural selection. This took place even though Lamarck's ideas about use and disuse constituted just one part of his theorizing and Darwin himself strongly endorsed use-inheritance as a mechanism seconding natural selection in the evolutionary process. The first serious challenge to the idea of use-inheritance came from German biologist August Weismann in the 1880s. Although the idea persisted into the twentieth century, it became increasingly discredited when it failed to gain experimental confirmation and the empirical evidence typically cited in its support was interpreted as being amenable to other explanations.

Lamarck left his mark on science through his contributions to botanical and especially zoological systematics, his pioneering work in invertebrate paleontology, and his evolutionary theorizing. Although he is primarily remembered today for an idea that biologists no longer accept—the idea of the inheritance of acquired characteristics—he nonetheless stands out in the history of biology as the first scientist to set forth a comprehensive theory of organic evolution.

BIBLIOGRAPHY

Burkhardt, R. W., Jr. 1995. *The Spirit of System: Lamarck and Evolutionary Biology.* New ed. Cambridge, MA: Harvard University Press.

Corsi, P. 1988. *The Age of Lamarck: Evolutionary Theories in France, 1790–1830.* J. Mandelbaum, trans. Berkeley: University of California Press.

Lamarck, J.-B. 1809. *Philosophie zoologique, ou exposition des considérations relatives à l'histoire naturelle des animaux; à la diversité de leur organisation et des facultés qu'ils en obtiennent; aux causes physiques qui maintiennent en eux la vie et donnent lieu aux mouvemens qu'ils exécutent; enfin, à celles qui produisent, les unes le sentiment, et les autres l'intelligence de ceux qui en sont doués.* 2 vols. Paris: Dentu.

———. 1815–1822. *Histoire naturelle des animaux sans vertèbres, présentant les caractères généraux et particuliers de ces animaux, leur distribution, leurs classes, leurs familles, leurs genres, et la citation des principales espèces qui s'y rapportent; précédée d'une introduction offrant la détermination des caractères essentiels de l'animal, sa distinction du végétal et des autres corps naturels; enfin, l'exposition des principes fondamentaux de la zoologie.* 7 vols. Paris: Déterville.   —*R.W.B.*

Lande, Russell (b. 1951)

Russell Lande is widely recognized as the architect of modern theories of phenotypic evolution and as an innovator in conservation biology. Born in 1951 in Jackson, Mississippi, Lande spent the formative years of his youth in Shreveport, Louisiana; Roswell, New Mexico; and Seal Beach, California. The son of a biology teacher and a petroleum geologist, he explored the swamps of Louisiana and body surfed on the beaches of southern California. He played Little League and bred hooded rats as a student in grammar school, and he was a member of math and honor societies in high school.

Lande's interests in ecology were awakened when he was an undergraduate at the University of California at Irvine. A talented student in mathematics and the physical sciences, Lande looked for opportunities to take a quantitative approach in the biological sciences. That opportunity was provided in a weekly reading club organized by graduate students Ted Case and Michael Gilpin. Their influence encouraged Lande to apply to graduate school at the University of Chicago and to work with professors Richard Levins and Richard C. Lewontin, eminent practitioners of the modeling approach in ecology and evolutionary biology. Shortly after his arrival at Chicago, Levins and Lewontin moved to Harvard. Lande followed suit.

As a graduate student at Harvard, Lande focused on developing theory for the inheritance of quantitative characters and using it to model phenotypic evolution. A chance remark by Lewontin helped set him on this course. While lecturing on Mendelism, Lewontin turned from the chalkboard and quipped, "Of course all ecologically important characters are quantitative, polygenic traits." Three papers (1976a, 1976b, 1977) that are now classics in the fields of quantitative genetics and phenotypic evolution constituted his doctoral thesis.

Lande continued to develop his novel approach to phenotypic evolution as a postdoctoral student with James Crow at the University of Wisconsin at Madison. He also had regular conversations with Sewall Wright, then retired from the University of Chicago and living in Madison. When Lande showed Wright the draft of his 1979 paper, another classic, Wright said that he did not care for the approach because it did not model the frequencies of the genes that underlay the quantitative traits. A year later, Wright volunteered that he might have been hasty in his judgment.

Lande's development of quantitative genetic theory continued when he joined the faculty at the University of Chicago. During his Chicago period, Lande applied his framework to the evolution of life-history characters, sexual dimorphism and sexually selected characters, the measurement of phenotypic selection, phenotypic plasticity, and many other problems in phenotypic evolution. Each of these contributions stands out because novel theoretical formulations are combined with insightful literature reviews and applications to actual data.

Beginning in the 1980s, Lande's interests shifted to conservation biology, as he took the lead in characterizing the demographic plight of the Northern Spotted Owl by using a combination of classical, modern, and original demographic theory. During this ongoing period, Lande joined the faculty at the University of Oregon and later moved to the University of California at San Diego. Lande is currently a Royal Society Research Professor at Imperial College, London.

BIBLIOGRAPHY

Lande, R. 1976a. The maintenance of genetic variability by mutation in a polygenic character with linked loci. *Genetical Research* 26: 221–234.
———. 1976b. Natural selection and random genetic drift in phenotypic evolution. *Evolution* 30: 314–334.
———. 1977. Statistical tests for natural selection on quantitative characters. *Evolution* 31: 442–444.
———. 1979. Quantitative genetic analysis of multivariate evolution, applied to brain:body size allometry. *Evolution* 33: 402–416.   —S.J.A.

Leakey, Louis Seymour Bazett (1903–1972)

Louis (pronounced Lewis) S. B. Leakey, the son of an English missionary to Kenya in East Africa, was born at a church mission station at Kabete, northwest of Nairobi in Kikuyu country. It was the policy of the Church Missionary Society to rotate its missionaries back to England every fourth year so that they would not forget their English roots. The Leakeys were brought back in 1904 and stayed for two years, returning to Kenya in 1906. Their next English sojourn was between 1910 and 1913, at which time Louis began formal schooling. In between those times, they lived at the mission station at Kabete. There young Louis joined a peer group of Kikuyu boys and became fluent in their language. Adopted into the group, he was formally initiated as a Kikuyu. Although he often said that he thought of himself as a Kikuyu, he acted and sounded like a good British colonialist for the rest of his life.

World War I interrupted plans to send him to school in England. When the family returned there in 1919, Leakey was duly sent to a public school in Dorset. The phenomenon referred to in England as a "public" school is really a private institution. With the aid of a governess, he was instructed in mathematics, Latin, and French. Although of a longtime English family, Leakey's father had been born in France, so French was often spoken at the Leakey dinner table. While Louis could speak the language, he evidently did so with a pronounced English accent, as did many of his countrymen.

Leakey entered St. John's College, Cambridge, as a freshman in the autumn of 1922. He asked if he could offer Kikuyu as one of the two modern languages required, and the university turned to authorities in London to ask if there were qualified Kikuyu speakers in England. The answer was yes, so Cambridge duly gave Leakey the permission he had requested. Only later did they discover that he was one of only two Kikuyu speakers known, so he was then requested to train the person who became his own examiner. He subsequently received top honors in his language exams. He also received top honors in his major fields of anthropology and archaeology, giving him what Cambridge called a "double first" in 1926. This earned him a research fellowship from St. John's College for three years, renewed again for three more.

Leakey had been back to East Africa as part of a dinosaur-hunting group in the summer after his sophomore year, but with the fellowship he was able to pursue work in the area of his training and interests, namely, African prehistory. The first two East African archaeological expeditions allowed him to document the basis for an archaeological sequence of East Africa. In the autumn of 1930, he was awarded a PhD from Cambridge University on the basis of *The Stone Age Cultures of Kenya Colony* (1931). Unlike American universities, there were no courses or exams required in order to earn a doctorate. The sole requirement was an acceptable doctoral dissertation, and a published book counted.

His African archaeological work yielded human skeletal material as well as tools. Because Leakey was not trained in skeletal analysis, he sought the assistance of Sir Arthur Keith, a highly respected anatomist and the curator of the Hunterian Museum at the Royal College of Surgeons in London. Keith was to have a profound effect on Leakey's views on human evolution. Keith, despite declaring his belief in "Darwinism" and spending nearly the last 25 years in a house on the Darwin estate, was in fact profoundly anti-Darwinian (Keith 1915). He was largely the reason why many professional students of human evolution in England and America refused to accept the available evidence for human evolution.

In 1931, Leakey led his third East African archaeological expedition on a visit to the site that would be identified with Leakey's impact on the field for all subsequent time: Olduvai Gorge in what is now called Tanzania. The substance of Leakey's international reputation was grounded in his work at Olduvai for more than a third of a century. In the first year, he found quantities of stone tools dating back to the very earliest made by the ancestors of what evolved into true humans. In July 1959, while Louis was suffering from a tropical fever in camp, his wife, Mary Leakey, found the first fossil of a maker of those tools (M. Leakey 1984). It had an ape-sized brain and an enormous dentition, but it lacked the projecting canines of a true ape and it was an erect-walking biped. Clearly it was related to those members of the genus *Australopithecus,* previously known only from South Africa and including forms that did evolve into the genus *Homo.* As he did so often, Leakey gave it a new generic and specific name, *Zinjanthropus boisei.* The genus name did not stick, although informally it is still known as "Zinj." Not only was it the first East African and the most spectacular member of the early human relatives referred to as hominids, but the techniques of radioactive isotope analysis had been developed to the extent that it could be given an absolute date of over

1.5 million years. Previously, the Australopithecines known only from South Africa were assumed to be approximately half a million years old and much too recent to have played a role in human evolution. Even though Zinj was a sideline, not on the path toward true humans, its great antiquity and primitive hominid traits provided substance for Charles Darwin's (1871) suggestion that Africa was the original locale for the evolution of human form.

Leakey took full advantage of the opportunity provided by Zinj. He wrote a series of technical as well as popular descriptive articles. Even more effective was his presence on the lecture platform. A charismatic and colorful public speaker, his frequent international appearances brought him earnings that supported further fieldwork, especially at Olduvai Gorge and elsewhere in East Africa. This also encouraged others to engage in the search for early hominids and their tools elsewhere in East Africa. Hardly a year goes by now without further discoveries.

Leakey also promoted fieldwork on living great apes—chimpanzees and gorillas in Africa and orangutans in Southeast Asia—by a series of protégés who have greatly expanded what we know about the life and ways of humans' closest nonhuman relatives.

BIBLIOGRAPHY

Cole, S. 1975. *Leakey's Luck: The Life of Louis Seymour Bazett Leakey.* New York: Harcourt Brace Jovanovich.

Darwin, C. 1871. *The Descent of Man, and Selection in Relation to Sex.* London: John Murray.

Keith, A. 1915. *The Antiquity of Man.* London: Williams and Norgate.

Leakey, L. S. B. 1931. *The Stone Age Cultures of Kenya Colony.* Cambridge: Cambridge University Press.

———. 1974. *By the Evidence: Memoirs, 1932–1951.* New York: Harcourt Brace Jovanovich.

Leakey, M. 1984. *Disclosing the Past: An Autobiography.* London: Weidenfeld and Nicolson.

Morell, V. 1995. *Ancestral Passions: The Leakey Family and the Quest for Humankind's Beginnings.* New York: Simon and Schuster.   —C.L.B.

Lederberg, Joshua (1925–2008)

Joshua Lederberg was one of the leading microbiologists and geneticists of the twentieth century. His experiments explored the nature of fundamental genetic mechanisms in bacteria and thus shed light on processes that helped shape the evolution of the most abundant organisms on earth. For his contributions, he received a Nobel Prize in 1958, at the age of 33. Lederberg also advised policymakers on such matters as emerging infectious diseases and biological warfare, and he was awarded the Presidential Medal of Freedom in 2006.

Lederberg made three major contributions to understanding bacterial genetics. First, as a graduate student working with Edward Tatum at Yale University, Lederberg showed that bacteria could recombine their genes to produce new genotypes. This experiment demonstrated the existence of a sexual process in bacteria that was not previously known. Subsequent research showed that the process Lederberg discovered involves an extrachromosomal element, or plasmid, that mediates the movement of bacterial genes from one cell to another. This process, known as conjugation, also provided a powerful tool that enabled scientists to map the physical position of bacterial genes on chromosomes, in a manner logically similar to the mapping studies pioneered earlier by Thomas Hunt Morgan and colleagues with fruit flies.

Second, working with his student, Norton Zinder, at the University of Wisconsin, Lederberg discovered that viruses that infect bacteria could serve as vectors to move genetic material from one strain to another. They called this virus-mediated process transduction and it, too, played an important role in facilitating years of increasingly detailed and sophisticated studies of bacterial genetics.

Third, Lederberg and his first wife, Esther Lederberg, developed the replica-plating experiment. This elegant experiment demonstrated that bacterial mutations—in particular, those that confer resistance to viral infections—occurred prior to the bacteria being exposed to viruses. Therefore, the mutations could not have been caused by the bacteria responding in a directed manner to the selective agent. This work showed that mutations arose at random, with selection providing the directional force responsible for genetic adaptation. In other words, bacteria are subject to the same basic Darwinian mechanisms as higher organisms. While an earlier experiment by Salvador Luria and Max Delbrück (1943) had supported the same conclusion, the replica-plating experiment settled the issue even for skeptics because it provided a striking visual demonstration. That is, rare mutants that arose on a master petri dish, which went undetected absent any selection, yielded mutant descendants in the same relative locations following transfer of the population, using a velvet pad, to several replica plates where bacteria were subject to selection for resistance to viral infection. In fact, by screening cells from the corresponding location on the master plate, the Lederbergs could isolate mutations without directly exposing the bacteria to the selective conditions that favored the mutant phenotype.

At the time that Lederberg and his colleagues performed these pioneering experiments, little was known about bacterial evolution. For the next few decades, the field of microbiology mostly pursued genetic and molecular approaches to understanding cellular structure and function, paying little attention to evolution. However, the past few decades have seen increasing work on microbial evolution, which has followed several approaches and imperatives. First, the field of molecular evolution revealed the relationships of bacterial species to one another and to other organisms, including the discovery by Carl Woese of the Archaea, a group of single-celled, non-nucleated organisms that are distinct in many respects from true bacteria. The approach of molecular evolution has also been used to examine the extent of gene exchange within and among bacterial species, thus extending our understanding of the genetic mechanisms discovered by Lederberg in the laboratory to their importance in nature. Second, owing to their rapid generations and other tractable features, bacteria are now often used in experiments that seek to understand the dynamics of phenotypic and genomic evolution as well as to test particular hypotheses about evolutionary phenomena. Finally, there is growing attention to the role that evolution plays in the emergence of new infectious diseases, including the acquisition of virulence factors that allow pathogens to infect their hosts as well as the spread of antibiotic resistance. The mechanisms of genetic exchange discovered by Lederberg have been found to be very important in the emergence of many new bacterial pathogens.

BIBLIOGRAPHY

Elena, S. F., and R. E. Lenski. 2003. Evolution experiments with microorganisms: The dynamics and genetic bases of adaptation. *Nature Reviews Genetics* 4: 457–469.

Lederberg, J. 1997. Infectious disease as an evolutionary paradigm. *Emerging Infectious Diseases* 3: 417–423.

Lederberg, J., and E. M. Lederberg. 1952. Replica plating and indirect selection of bacterial mutants. *Journal of Bacteriology* 63: 399–406.

Luria, S. E., and M. Delbrück. 1943. Mutations of bacteria from virus sensitivity to virus resistance. *Genetics* 28: 491–511.

National Library of Medicine. N.d. Profiles in science. The Joshua Lederberg papers. http://profiles.nlm.nih.gov/BB/.

Tatum, E. L., and J. Lederberg. 1947. Gene recombination in the bacterium *Escherichia coli. Journal of Bacteriology* 53: 673–684.

Woese, C. R., O. Kandler, and M. L. Wheelis. 1990. Towards a natural system of organisms: Proposal for the domains Archaea, Bacteria, and Eucarya. *Proceedings of the National Academy of Sciences USA* 87: 4576–4579.

Zinder, N. D., and J. Lederberg. 1952. Genetic exchange in *Salmonella. Journal of Bacteriology* 64: 679–699.   —R.E.L.

Lessons in Comparative Anatomy (Georges Cuvier)

In his ever-busy life, the years 1795 to 1802 were particularly hectic ones for Georges Cuvier. On July 2, 1795, the professors' assembly at the Muséum national d'histoire naturelle approved his nomination as assistant to Jean-Claude Mertrud, the anatomy professor, and he started in earnest to arrange the comparative anatomy collection. He gave lectures at the museum and read scientific papers to a variety of societies (the Société Philomatique and the Natural History Society in particular) and at the Institut de France (instituted in 1795 to replace the old Academies, including the Académie français and the Académie des sciences). Cuvier took full advantage of the museum library that benefited, as did the museum itself, from important additions acquired from those European countries conquered by the French armies.

During 1798 and 1799, assisted by André-Marie-Constant Duméril (1774–1860), Cuvier worked on the notes Duméril had taken at his lectures in the previous four years. The first two volumes of the *Leçons d'anatomie comparée (Lessons in Comparative Anatomy)* appeared in March 1800; three more volumes, edited by Georges Louis Duvernoy (1777–1855), were published by 1805. Volume 1 dealt with the organs of movement; volume 2, the organs of sensation; volumes 3 and 4, digestion, circulation, respiration, and the organs for producing sounds; and volume 5, the organs of generation and secretion. Cuvier devoted his first lecture to methodological considerations on "the economy of life." Living organisms, he had declared in the preface to the work, are "machines . . . that cannot be taken apart without being destroyed" (Cuvier 1800, 1: V). Lecture 1 expanded upon this concept. Cuvier quoted Immanuel Kant to state that the raison d'être, the intimate "cause" of every single living part or organ, was in the "whole," that is, the organism in itself. Drawing upon previous botanical and anatomical work, notably Antoine-Laurent de Jussieu's concept of "correlation and subordination of parts" (1789) and Felix Vicq-d'Azyr's application of the principle to anatomy (1805), Cuvier showed that systems of organs (digestion, locomotion, nutrition, vision, etc.) are in every animal class related to each other by "laws of coexistence" (Cuvier 1800, 1: 57). Thus the presence and configuration of one organ or part legitimated cogent deductions as to the structure of the whole organism and its way of life. It was in any case the structure and functions of internal organs that constituted the key to all well-grounded divisions within the animal kingdom. Therefore the progress of comparative anatomy proved essential to the development of a natural system of classification.

Cuvier placed strong emphasis on the impossibility of spontaneous generation by insisting that all known organisms are descended from other organisms. Life, he agreed with his friend and colleague Xavier Bichat, is the property to temporarily resist the destructive action of physicochemical agents. He did not deny, however, that the progress of chemistry would in time unveil the intimate structure and properties of each component of the "animal machine." Finally, he devoted a long section to disprove the existence of a chain of beings, gradually ascending from the most simple organisms to man. Gradual variation is possible only within the same system of internal organs, and there are no known forms making the transition between each of the main anatomical and functional animal plans comparative anatomy was discovering and describing.

BIBLIOGRAPHY

Coleman, W. 1964. *Georges Cuvier, Zoologist: A Study in the History of Evolution.* Cambridge, MA: Harvard University Press.

Cuvier, G. 1800. *Leçons d'anatomie comparée.* Vols. 1–2. C. Duméril, ed. Paris: Baudouin.

———. 1805. *Leçons d'anatomie comparée.* Vols. 3–5. G. L. Duvernoy, ed. Paris: Baudouin.

Jussieu, A. L. de. 1789. *Genera plantarum.* Paris: Hérissant.

Outram, D. 1984. *George Cuvier: Vocation, Science and Authority in Post-revolutionary France.* Manchester: Manchester University Press.

Rudwick, M. 1997. *Georges Cuvier, Fossil Bones and Geological Catastrophes: New Translations and Interpretations of the Primary Texts*. Chicago: University of Chicago Press.

Vicq-d'Azyr, F. 1794 [1805]. *Oeuvres recueillies et publiées avec des notes et un discours sur sa vie et ses ouvrages par M. Jacques L. Moreau de la Sarthe*. 6 vols. Paris: Baudouin. —*P.C.*

Lewontin, Richard (b. 1929)

Richard Lewontin has proved to be every bit as influential in population genetics as his mentor, Theodosius Dobzhansky. Like Dobzhansky, Lewontin is well known for articulating a very definite perspective of the field and the most important problems to be solved—including, like Dobzhansky, an especially puzzling "paradox." Like Dobzhansky, Lewontin is well known for a highly original approach to measuring variation at the genetic level and for his own views of how best to explain that variation. Like Dobzhansky, Lewontin tweaked molecular biologists for suggesting that molecular processes were more fundamental than the evolutionary processes that actually gave rise to them. Like Dobzhansky, Lewontin has been deeply concerned with the broader moral and political significance of genetics and evolutionary biology. And, like his teacher, he has mentored scores of graduate and postdoctoral students. But these are similarities at a very high level of generality. Otherwise, these two influential population geneticists are very different!

Along with Dobzhansky, Ernst Mayr, George Simpson, and other evolutionary biologists, Lewontin greeted the fanfare surrounding the advent of molecular genetics in the 1960s with a cautionary reminder to keep in mind the

> distinction between efficient and final cause in biology. The molecular configurations of living organisms are the efficient causes of biological phenomena but not their final causes. That is, except in a trival sense, the laws of genetics are not the result of the structure of DNA, but rather DNA has been chosen by natural selection from among an immense variety of molecules precisely because it fits the requirements of an evolved genetic system. DNA is only the tactic adopted in the course of working out an evolutionary strategy. That is why some organisms can get on without it. (Lewontin 1964, 566)

But he saw considerable promise in the new field and was at the vanguard of applying molecular genetic techniques to evolutionary issues. In particular, he helped to pioneer the use of gel electrophoresis to document variation close to the genetic level—namely, at the level of amino acid differences in proteins (due to nucleotide differences in the genes that code for those proteins). In so doing, he helped establish the field of molecular population genetics. His student, Martin Kreitman, was the first population geneticist to study variation directly at the DNA nucleotide level, using more recent DNA sequencing techniques.

The findings of Lewontin and his early collaborator in this work, Jack Hubby, suggested—given the inherent biases of their methods—that nearly every gene locus is polymorphic, and that a third of all loci are heterozygous (Hubby and Lewontin 1966; Lewontin and Hubby 1966; Lewontin 1974). This might seem to be strong evidence in favor of Dobzhansky's "balance" view of evolution (i.e., that there is considerable genetic variation within natural populations, maintained largely by selection in favor of heterozygotes) and against the "classical" position of Dobzhansky's archrival, H. J. Muller (i.e., that natural selection in favor of optimal genes eliminates variation within populations). To his teacher's chagrin, Lewontin argued instead that the results were paradoxical, and not at all easy to explain.

There was just too much variation to make sense from the balance perspective, Lewontin argued, citing similar considerations proposed by Motoo Kimura and James Crow. One major problem is that half of the offspring of a heterozygote cross *(Aa×Aa)* are homozygotes *(AA, aa).* Heterozygote superiority at a particular gene locus thus entails the production of many inferior homozygotes and a corresponding decrease in the reproductive capacity of the population. The greater the number of loci at which heterozygotes are superior, the greater the damage done to the reproductive capacity of the population, until the population would be quite unable to perpetuate itself.

Moreover, Lewontin argued, no alternative to the balance view makes any better sense of documented patterns of genetic variation. It is a "paradox of variation" that there is too much to be explained in terms of natural selection, no matter whether it is a variation-maintaining sort of selection such as the balance theorists had proposed, or a variation-reducing sort such as that proposed by the classical theorists. There was also too much variation to be explained in terms of the selective insignificance or neutrality of the variation, a position that Lewontin dubbed "neo-classical."

What is/was to be done? Or as Lewontin posed the question in his influential text, *The Genetic Basis of Evolutionary Change* (1974, 267), "How can such a rich theoretical structure as population genetics fail so completely to cope with the body of fact? Are we simply missing some critical revolutionary insight that in a flash will make it all come right, as the Principle of Relativity did for the contradictory evidence on the propagation of light? Or is the problem more pervading, more deeply built into the structure of our science? I believe it is the latter."

One reason for the impasse, he argued, is that population genetic theory is not "empirically sufficient." For example, it includes parameters that cannot be measured directly, or with sufficient accuracy, to distinguish clearly between alternative causal accounts. This reflects Lewontin's more general epistemological concerns—indeed, his skepticism. He is well known—even infamous—for persistently questioning whether geneticists and evolutionary biologists can possibly know what they want to know and often claim to know. His critiques of adaptationists (including his 1979 article "The Spandrels of San Marco and the Panglossian Paradigm," coauthored with Stephen Gould) and genetic determinists are largely epistemological. Adaptationists and genetic determinists do not know enough about the inheritance of the traits they study and the evolutionary forces acting on those traits. Moreover, no one knows. This is a central motif of Lewontin's *Genetic Basis* and his subsequent reviews of the state of population genetics. The title of one such review well illustrates his discontent even with methods of his own devising: "Electrophoresis in the Development of Evolutionary Genetics: Milestone or Millstone?" (1991).

A second reason for the failure of theory to make sense of data in population genetics is that patterns of genetic variation can be due in large part to "history"—for instance, not just which environments a population experiences, but the order in which those environments are encountered (Lewontin 1967). Population genetic theory, however, is an equilibrium theory that discounts historical contingencies. Lewontin's discussion of the ahistorical character of population genetics is partly epistemological: history is ignored largely because the past is so often unrecoverable. But it is also partly political. In population genetics, as in other areas of the natural and social sciences, equilibrium theories prevailed in the twentieth century in part because of a very general "preoccupation with stability" (Lewontin 1974, 269).

For Lewontin, the way forward in science involves, in part, exposing the ideologies that help to perpetuate incorrect views of the world. For example, he argued in *Genetic Basis* that the classical/balance controversy could not be understood without taking into account the conflicting political ideals of Muller and Dobzhansky.

A common mistake of Muller and Dobzhansky, according to Lewontin, was their genetic determinist assumption that existing patterns of genetic diversity have moral and political implications. For his own part, he has claimed, "I do not believe that the ultimate issues depend on how much diversity exists" (Lewontin et al. 2001, 43). Instead of trying to decide how to organize society on the basis of genetic findings, we should use our best understanding of genetics, psychology, sociology, and so forth to determine what interventions would result in the sort of society we aspire to (Lewontin et al. 2001, 43–44; Lewontin et al. 1984). Along the same lines, Lewontin has often been concerned to rebut the charge that his skepticism concerning genetic determinism renders him a cultural determinist instead. As he quotes Karl Marx, "The materialist doctrine that men are the products of circumstances and upbringing, and that, therefore, changed men are products of other circumstances and changed upbringing, forgets that it is men that change circumstances and that the educator himself needs educating" (Lewontin et al. 1984, 267).

The one case where Lewontin has claimed political significance for his own findings in population genetics has to do with the sorts of studies that he initiated in his classic 1972 paper "The Apportionment of Human Diversity," in which he argued that the proportion of variation within so-called races is far greater (more than 10 times) than the variation between races (see also Lewontin et al. 2001, 43). Racial differences, and all the other social and political differences that have accompanied them, are thus not biologically based in the way that many had supposed. According to Lewontin, this is really more a case of showing that genetics is not as politically significant as believers in the genetic basis of racial differences had imagined.

Lewontin's epistemological and political interests merge in his dialectical materialism, which plays a cautionary, heuristic role in his work:

> Dialectical materialism is not, and never has been, a programmatic method for solving particular physical problems. Rather, dialectical analysis provides an overview and a set of warning signs against particular forms of dogmatism and narrowness of thought. It tells us, "Remember that history may leave an important trace. Remember that being and becoming are dual aspects of nature. Remember that conditions change and that the conditions necessary to the initiation of some process may be destroyed by the process itself. Remember to pay attention to real objects in time and space and not lose them in utterly idealized abstractions. Remember that qualitative effects of context and interaction may be lost when phenomena are isolated." And above all else, "Remember that all the other caveats are only reminders and warning signs whose application to different circumstances of the real world is contingent." (Levins and Lewontin 1985, 191–192)

BIBLIOGRAPHY

Gould, S. J., and R. C. Lewontin. 1979. The spandrels of San Marco and the Panglossian paradigm: A critique of the adaptationist programme. *Proceedings of the Royal Society of London, Series B* 205: 581–598.

Hubby, J. L., and R. C. Lewontin. 1966. A molecular approach to the study of genic heterozygosity in natural populations. I. The number of alleles at different loci in *Drosophila pseudoobscura*. *Genetics* 54: 577–594.

Levins, R., and R. Lewontin. 1985. The problem of Lysenkoism. In R. Levins and R. Lewontin, eds., *The Dialectical Biologist*, 163–196. Cambridge, MA: Harvard University Press.

Lewontin, R. C. 1964. A molecular messiah: The new gospel in genetics? *Science* 145: 566–567.

———. 1967. The principle of historicity in evolution. In P. S. Moorhead and M. M. Kaplan, eds., *Mathematical Challenges to the Neo-Darwinian Interpretation of Evolution*, 81–88. Philadelphia: Wistar Institute Press.

———. 1972. The apportionment of human diversity. *Evolutionary Biology* 6: 381–398.

———. 1974. *The Genetic Basis of Evolutionary Change*. New York: Columbia University Press.

———. 1991. Electrophoresis in the development of evolutionary genetics: Milestone or millstone? *Genetics* 128: 657–662.

Lewontin, R. C., and J. L. Hubby. 1966. A molecular approach to the study of genic heterozygosity in natural populations. II. Amount of variation and degree of heterozygosity in natural populations of *Drosophila pseudoobscura*. *Genetics* 54: 595–609.

Lewontin, R. C., D. Paul, J. Beatty, and C. B. Krimbas. 2001. Interview of R. C. Lewontin. In R. S. Singh, C. B. Krimbas, D. B. Paul, and J. Beatty, eds., *Thinking about Evolution: Historical, Philosophical and Political Perspectives*, 22–61. Cambridge: Cambridge University Press.

Lewontin, R. C., S. Rose, and L. J. Kamin. 1984. *Not in Our Genes: Biology, Ideology, and Human Nature*. New York: Pantheon.

Singh, R. S., and C. B. Krimba, eds. 2000. *Evolutionary Genetics from Molecules to Morphology*. Cambridge: Cambridge University Press.

Singh, R. S., and M. K. Uyenoyama, eds. 2004. *The Evolution of Population Biology*. Cambridge: Cambridge University Press.   —J.B.

Linnaeus, Carl (or Carl von Linné) (1707–1778)

Born and raised in Sweden, Carl Linnaeus was passionate about botany from his youth. By the time he went to Holland in 1735 to finish his medical degree, he possessed an impressive knowledge of the known kinds of plants and animals. There he was hired to manage and catalog the huge private museum, zoo, herbarium, and botanical garden of George Clifford, a wealthy merchant. By the time Linnaeus returned to Sweden in 1738, he had published numerous works, all in Latin, that set out the rules, framework, and catalogs of all the plant genera then known. He was a founder and the first president of the Swedish Academy of Sciences. In 1741 he was hired by the University of Uppsala, where he taught for the next 33 years. Exerting a dominating influence upon natural history (botany and zoology), he attracted students from across Europe, some of whom he sent on collecting trips overseas. He exchanged letters with distant naturalists who sent him specimens. Most important, he frequently produced new editions of his catalogs of plant and animals: *Systema Naturae* (12 editions between 1735 and 1768), *Genera Plantarum* (six editions from 1737 to 1764), and *Species Plantarum* (three editions from 1753 to 1764). These works provided naturalists of all nations with one central, well-organized register of names. Linnaeus's catalogs were the search engines of their day.

Linnaeus worked in the context of the worldwide collection of natural history specimens that accompanied the exploration, trade, and colonization carried out by European governments and entrepreneurs. His particular talents, which included unlovable traits like egotism and an obsession with making lists, equipped him to produce an expandable register of the diversity of living things. Although he expected the next generation to complete the project, biologists in later centuries, including our own, continue to encounter species not previously described. The project of making an inventory of life is still unfinished.

Linnaeus's system was strictly hierarchical, grouping similar species into a genus, similar genera into an order, orders into a class, and classes into a kingdom. Later taxonomists added families (below order) and phyla (below kingdom), as well as suborders, superorders, and the like; this expanded the hierarchy without changing its principles. Linnaeus began the convenient practice of attaching a proper name to each individual group, using a traditional one where possible (such as Aves and Pisces, Latin for bird and fish, respectively) but coining a new one if needed (such as Mammalia for the class that included whales and humans). He also introduced what is called binominal nomenclature, that is, always using two words to name a species, the generic name plus a modifier, such as *Canis lupus* (wolf), *Canis familiaris* (dog), and *Homo sapiens* (human). (Current practice is to italicize such names.) In order to fit the thousands of species into manageable volumes, he invented abbreviations, omitted illustrations, and made descriptions as short as possible. He included citations of other authors, so his works functioned as a bibliographic guide or index to the scientific literature.

Linnaeus's system for flowering plants stimulated discussion of the philosophical foundations of classification. At the start of his career he had learned about recent theories interpreting the central parts of a flower, the stamen and pistil, as organs of male and female sexual reproduction. He invented classes based on the number and arrangement of stamens and orders based on details of the pistil or the fruit. These groups were artificial, being based on only a few characteristics, but they gave a key that even inexpert botanists could use (as long as the specimen was in flower). Linnaeus looked forward to the day when these artificial groups could be replaced by more natural ones. At the level of genera and species, Linnaeus used a wider range of features, including the leaves and seeds, and allowed his general impression of the whole plant to affect his judgment. His "sexual system" was thus a confusing mixture of artificial classes and orders with natural genera and species. In his classification of animals, he used various characteristics, whatever seemed to work for each group, such as teeth for mammals, wings for insects, and fins for fishes. Many otherwise reliable historians state that Linnaeus aimed at the logical definitions of medieval "Aristotelians," but the old terms *genus* and *species* functioned very differently from Aristotle's, and the ill-founded claim that Linnaeus was an essentialist badly misrepresents both his aims and procedures.

At the start of his career Linnaeus insisted that species cannot change, but in later years he saw evidence of hybridism and mutation, which led him to think that God had created only one species per genus, with the other species arising later. Most of his followers preferred to stick with the fixity of species. Because visible variation within species is often very slight, especially when the entire range is not sampled, the assumption of fixity had the positive result of encouraging the discovery and description of previously unknown species. Linnaeus had a lively interest in the dynamics of nature, encouraging his students to study which kinds of plants various animals preferred to eat, at what date different migrating birds returned, and how insects fit into the balance of nature.

The most remarkable development toward the end of Linnaeus's career was that the structure he had imposed in his classification seemed very often to be congenial to the patterns of similarities and gaps of apparently natural groups. The old idea of a chain of being, a linear scale, did not fit, but neither did the prediction of Georges-Louis Leclerc, Comte de Buffon and Jean-Baptiste Lamarck that connections run in all directions. Instead, a pattern of clumps, sets within sets, seemed nature's own. This unexpected and gradually emerging picture was the great phenomenon left to Linnaeus's heirs. Unlike Lamarck's theory, which grew out of the chain of being, Charles Darwin's theory directly addressed the hierarchical pattern of Linnaean taxonomy by positing a high rate of extinction and a principle of divergence.

BIBLIOGRAPHY

Blunt, W. 1971. *The Compleat Naturalist: A Life of Linnaeus.* New York: Viking Press.

Frängsmyr, T., ed. 1994. *Linnaeus: The Man and His Work.* Canton, MA: Science History Publications.

Larson, J. L. 1971. *Reason and Experience: The Representation of Natural Order in the Work of Carl von Linné.* Berkeley: University of California Press.

Linnaeus, C. 1743. *Genera plantarum, eorumque characteres naturales, secundum numerum, figuram, situm, & proportionem omnium fructificationis partium. Editio secunda, nominibus plantarum Gallicis locupletata.* Paris: Michael Antonius David.

———. 1753. *Species plantarum: Exhibentes plantas rite cognitas, AD genera relatas, cum differentiis specificis, nominibus trivialibus, synonymis selectis, locis natalibus, secundum systema sexuale digestas.* 2 vols. Introduction by W. T. Stearn. London: Ray Society.

———. 1758. *Systema naturae per regna tria naturae, secundum classes, ordines, genera, species, cum characteribus, differentiis, synonymis, locis.* London: British Museum (Natural History).

Müller-Wille, S. 2001. Gardens of paradise. *Endeavour* 25: 49–54.

Project Linnaeus. Website. http://www.c18.rutgers.edu/pr/lc/proj.lin.html.

Winsor, M. P. 2006. Linnaeus's biology was not essentialist. *Annals of the Missouri Botanical Garden* 93, no. 1: 2–7.   —M.P.W.

Living fossils

Living fossils are defined informally; the term does not have a precise technical definition. Living animals and plants that have changed very little in form and behavior, compared with their remote ancestors, are said to be living fossils. Also, living relics of groups that were much more diverse far back in the geologic past may be called living fossils. Because evolutionary change tends to be concentrated in episodes of rapid diversification early in the expansion of new groups of organisms, these categories largely overlap.

Charles Darwin first applied the contradictory expression "living fossil" to the platypus (see figure) and the lungfish, as well as *Polypterus,* the African bichirs and their relative, the reedfish. These animals all retain primitive characteristics that show them to be more or less intermediate between typical mammals and reptiles, or major groups of living bony fishes, as Darwin noted. The platypus lays eggs, but suckles its young. It has hair and is warm blooded, but it has a less constant body temperature than typical mammals. Earlier naturalists, who saw the animal world as a fixed set of separately created "types," regarded these creatures as anomalies. To Darwin, and now to us, these living fossils are the least modified descendants of remote, extinct ancestors. Those ancestors, when they lived, were critical links in the branching evolutionary history of higher animals and plants.

The duck-billed platypus, the mammal that lays eggs. Darwin was fascinated by such odd organisms, thinking them highly significant for his purposes. "Species and groups of species, which are called aberrant, and which may fancifully be called living fossils, will aid us in forming a picture of the ancient forms of life" (Darwin 1859, 486).

Living relatives of animals long recognized as fossils, but thought to be extinct, are occasionally discovered. They are most apt to be referred to as living fossils, especially in the popularization of science. The most familiar of these is the coelacanth. The first fossil coelacanth, of Permian age (255 million years ago), was described by Louis Agassiz in 1839. Today, 70 or more fossil species representing 27 genera are known, ranging from the Late Devonian (365 million years ago) to the Late Cretaceous (80 million years ago). The living coelacanth, *Latimeria*, retains characteristic limb-like stalked fins and a joint within its relatively primitive braincase that separates the nasal organs and eyes from the ears and brain. *Latimeria* was first discovered off South Africa in 1938. Many additional specimens of the same species have since been caught off southeast Africa and adjacent to the Comoro Islands, near Madagascar. Recently, one specimen of a different but very closely related species has been found on the other side of the Indian Ocean, off the Celebes, in Indonesia.

Latimeria has not been found in the fossil record, so the evolutionary time span of the species *Latimeria chalumnae* is unknown. It differs in only very modest ways from its nearest relative, a Late Cretaceous coelacanth assigned to the genus *Macropoma*. Strictly speaking, it is the conservative characteristics of *Latimeria* that are unequivocally living fossils, and not the species or genus—a cluster of closely related species—in which they occur.

There seems to be no single explanation for the long-continued survival of small groups of closely related evolving lineages, with limited speciation and little change in form. Arthur Conan Doyle imagined dinosaurs and pterodactyls surviving in his *Lost World,* based in part on his knowledge of the inaccessible Roraima Plateau, on the border of Venezuela and Guyana. Some living fossils do occur in remote, isolated regions, where they have avoided interaction with newly evolved competitors. The platypus and its spiny, hedgehog-like relative, the echidna, live only in Australia. Sphenodontids, relatively unspecialized ancestors of modern lizards, were moderately diverse and widely distributed around the world during the early Mesozoic. Today, two species of tuatara, their only living close relatives, are restricted to small islands off New Zealand.

In contrast, the few living species of *Nautilus* and the horseshoe crabs have wide geographic distributions. *Nautilus pompilius,* the most common and variable of these cephalopods with elegant chambered shells, ranges from the Great Barrier Reef to New Guinea, the Philippines, and oceanic archipelagos as far out in the Pacific as Fiji. Five distinct species of *Nautilus* are known. Typically, these animals live at depths of 100 to 400 meters or more, on steeply sloping sea bottoms, close in against coral reefs. Populations living off isolated oceanic islands display limited variation. Some constitute distinct species, while others are regarded as local subspecies of *Nautilus pompilius.* This has prompted the recent inference that *Nautilus* is currently expanding its diversity, in a modest burst of speciation. To put this in context, note that the species and genera of ancient Paleozoic nautiloids have been assigned to over 125 families, with an extraordinary variety of shell shapes and modes of life. The prolific diversity of nautiloids was repeatedly cut back in major Paleozoic extinction events. A further sharp reduction occurred near the end of the Triassic (210 million years ago). Since then, the group has dwindled to the single genus that survives today.

The horseshoe crabs have never been very diverse, throughout their long history. Forms not dissimilar to the living animals first appeared during the Silurian (approximately 425 million years ago). Horseshoe crabs' external skeletons are not as easily preserved as mollusk shells, so their fossil record may be less complete. However, no more than eight genera can currently be inferred to have lived at any one time. Three genera are living today. One familiar species, *Limulus polyphemus,* is abundant and widely distributed, from Maine to Yucatan in the western Atlantic. In the Indo-Pacific, two species representing two different genera range from the Bay of Bengal to the Philippines, and a third species ranges from Borneo to Japan. Like *Limulus polyphemus,* this last species has a large climatic range. It is also tolerant of varying salinity and other environmental variables. In this respect, the horseshoe crabs are quite unlike *Nautilus,* which is temperature-sensitive and cannot survive in the warm surface waters of the tropics, where it lives.

Among living plants, the spore-bearing clubmosses and horsetails are relicts of groups that dominated vegetation of the late Paleozoic. Both are widely distributed geographically. The horsetails, also known as scouring rushes, are limited to about 30 species of a single genus. Typically, they are opportunistic colonists of wet slopes and riverbanks that are frequently disturbed. A similar habitat was formerly preferred by the gingko tree, which may be the oldest living species. Fossil leaves and wood up to 100 million years old are morphologically indistinguishable from the single surviving species, *Gingko biloba,* of this ancient group. This plant is so distinctive that it remains uncertain whether it is more closely related to conifers or to the tree fern–like cycads. Once widespread, the gingko survived only in the Yangtze River basin in China until its domestication. Now, its beauty and tolerance of environmental insults have brought the gingko to many city streets.

The deep ocean floor is neither as low in diversity of animal life nor as uniform and constant in its physical conditions as was once supposed. Nonetheless, it has more than its share of living fossils. The creeping mollusk, *Neopilina,* with its cap-shaped shell and serial sets of muscles and other organs, is a deep-sea relict of two families that lived in shallow water during the early Paleozoic. Protobranchs with simple gills predominate among bivalves at abyssal depths; on the continental shelves they have largely been displaced by more advanced forms with varied burrowing and feeding adaptations. Among single-celled foraminifera, only those that build their skeletons from grains of preexisting sediment can live at the greatest depths, below the zone at which calcium carbonate dissolves. Forams of this sort still occur in shallow water habitats, but in Cambrian time (about 510 million years ago) only this primitive group was present. Most striking of all are the stalked echinoderms popularly known as sea lilies. Among the most prominent of all animals in shallow-water Paleozoic faunas, about 80 species of stalked crinoids survive today, worldwide, on continental slopes and at abyssal depths. These distributions, together with those of *Nautilus* and the coelacanth, corroborate evidence from the fossil record of successive displacements toward deeper water, first of descendants of the Cambrian fauna and then those of a later Paleozoic fauna, as new communities evolved in nearshore settings.

The variety of circumstances in which living fossils have survived indicates that there is no one cause of this phenomenon. There are many ways and places in which to be marginalized without being eliminated by natural selection. Living fossils do not constitute a special caste of organisms, except for the fact that they tend to be among the least specialized members of the major groups to which they belong. At any given time in earth's history, there has been a spectrum of living things: rapidly evolving groups of ephemeral species, groups of average stability and tenure, and groups that depart little from the stable adaptations of their ancestors, as long as their modes of life remain viable. Trilobites were the preeminent living fossils of Permian time, but they finally succumbed in the great extinction at the end of the Paleozoic. Their long decline from exuberant diversity in the mid-Paleozoic resulted from changes in community structure that are not yet understood in detail. Chance, outmoded adaptation, and competition all surely contributed to their ultimate demise. Today's living fossils have similar evolutionary histories and future prospects.

BIBLIOGRAPHY

Darwin, C. 1859. *On the Origin of Species*. London: John Murray.

Eldredge, N., and S. M. Stanley, eds. 1984. *Living Fossils*. New York: Springer-Verlag.

Forey, P. 1998. A home from home for coelacanths. *Nature* 395: 319–320.

Royer, D. L., L. J. Hickey, and S. L. Wing. 2003. Ecological conservatism in the "living fossil" *Gingko*. *Paleobiology* 29: 84–104.

Schopf, T. J. M. 1984. Rates of evolution and the notion of "living fossils." *Annual Review of Earth and Planetary Sciences* 12: 245–292.

Thomson, K. S. 1991. *Living Fossil: The Story of the Coelacanth*. New York: Norton.   —R.D.K.T.

Lorenz, Konrad Zacharias (1903–1989)

Konrad Lorenz, the son of a rich and distinguished Viennese orthopedic surgeon, grew up with a passion for raising animals. Later, when he became famous for his studies of animal behavior, he claimed that his mature scientific practices were essentially continuous with practices he developed in his youth as an animal lover. He insisted furthermore that by rearing and observing animals under seminatural conditions he was able to learn things about animal behavior that were not readily accessible to laboratory scientists or field naturalists.

Following his father's wishes, Lorenz enrolled in 1923 as a medical student at the Second Anatomical Institute of the University of Vienna, eventually earning his MD there in 1928. He thereupon enrolled at the university's Zoological Institute, from which he received a PhD in 1933. Meanwhile, at his private home research station in the village of Altenberg, close to Vienna, he had begun making on tame, free-flying birds the observations that would help propel him to prominence in the 1930s, first among German ornithologists and then among animal psychologists. His research provided the observational and conceptual foundations of the new science of ethology, the biological study of behavior, an enterprise in which Lorenz's work was ably supported and complemented by the contributions of Niko Tinbergen, a Dutch naturalist.

Lorenz's initial contribution to the evolutionary understanding of living things was his promotion of the idea that the methods of comparative anatomy could be applied to animal behavior patterns just as effectively as they could be applied to animal structures. He maintained that comparing the innate behavior patterns of different species was sometimes even more valuable than comparing their physical characters when it came to reconstructing phylogenies. His most important publication in this regard was a major monograph in 1941 comparing innate behavior patterns in ducks and geese. He later put his broad, comparative perspectives to use in thinking more generally about evolutionary epistemology, promoting the view that the way the human mind apprehends the world is itself the product of a long evolutionary process.

In addition to his efforts to understand animal and human behavior in evolutionary terms, Lorenz sought to make sense of the physiological causation and social function of "innate," species-specific behavior patterns. Looking at birds in particular, he insisted that they are adapted to their environments not so much by acquired knowledge as by highly differentiated instinctive behavior patterns, built up over time by evolution as a result of their survival value. For these "fixed motor patterns" to be effective, they needed to be "released" by stimuli emanating from appropriate objects in their environment. In the 1930s he constructed a theory of instinct featuring "releasers," "innate releasing mechanisms," action-specific energies, and innate, fixed motor patterns. At the same time, he called attention to the phenomenon of "imprinting." He found that young birds such as jackdaws or geese do not instinctively recognize members of their own species but instead normally acquire this information at a brief, critical period of their early development. They are "imprinted" upon the object that will subsequently serve to release certain of their instinctive behavior patterns. However, if a young greylag gosling, for example, is exposed to a human being instead of a mother greylag goose in the imprinting period, the gosling will become imprinted upon humans rather than its own species.

With his charismatic, exuberant personality, his extensive knowledge of the natural behavioral repertoires of a host of different animal species, his bold theorizing, and his skills as a popularizer, Lorenz was able to attract many recruits to the new science of ethology. Controversial in the 1960s for his popular book, *On Aggression,* and recurrently criticized for statements he made earlier as a biologist under the Third Reich, Lorenz is remembered most for attracting the attention of zoologists and psychologists alike to the importance of studying animal behavior from a biological perspective and for founding the science of ethology. His work was recognized by many honors, including the award of the 1973 Nobel Prize for Physiology or Medicine, which he shared with Karl von Frisch and Niko Tinbergen.

BIBLIOGRAPHY

Burkhardt, R. W., Jr. 2005. *Patterns of Behavior: Konrad Lorenz, Niko Tinbergen, and the Founding of Ethology.* Chicago: University of Chicago Press.

Lorenz, K. Z. 1941. Vergleichende Bewegungsstudien an Anatiden. In *Festschrift O. Heinroth.* Ergänzungsband 3. *Journal für Ornithologie* 89: 194–293.

———. 1966. *On Aggression.* M. K. Wilson, trans. New York: Harcourt, Brace and World.

———. 1970–1971. *Studies in Animal and Human Behaviour.* R. Martin, trans. 2 vols. Cambridge, MA: Harvard University Press.

Nisbett, A. 1976. *Konrad Lorenz.* New York: Harcourt Brace Jovanovich.

R.W.B.

Lucy *(Australopithecus afarensis)*

In every scientific field of endeavor there are certain discoveries and breakthroughs that are enduring and achieve somewhat of an iconic nature. Nearly all are superseded or significantly altered when new finds come to light or deeper insights are made, but the core innovation or discovery is everlasting. There is no doubt that constructs like the Big Bang, the theory of evolution, the theory of gravitation, continental drift, the laws of inheritance, and many others have been enhanced or modified from their original manifestations, but the central idea is still there and the concept continues to be known by its original iteration.

Looking back on my career as a paleoanthropologist—someone who studies human origins—I had occasion in 1974, in the remote desert region of Ethiopia known as the Afar Triangle, to make one of those enduring discoveries. She was a 3.2 million-year-old human ancestor who has become perhaps the most widely celebrated discovery of a fossil human of the twentieth century. Known by the affectionate name of Lucy (named after the Beatles song "Lucy in the Sky with Diamonds"; see figure), this 40% complete skeleton (not including the bones of the hands and feet) epitomizes our fossil ancestry. Lucy has become a touchstone for comparing and evaluating all new fossil human finds. Even for those who have difficulty remembering her scientific name, *Australopithecus afarensis,* bringing up the name Lucy rings a bell for everyone: "Oh, yeah, that ancient skeleton from Africa!"

One of the essential questions that all humans ask is, "Where did we come from?" There are two ways to answer this question. One is to evoke a creation story of how we were conceived in some sort of supernatural manner. Such explanations that are based on belief differ from society to society, and acceptance is premised on providing some sort of special event or supernatural cause, the evidence for which is invisible and available only to those who have faith.

On the other hand, paleoanthropologists use a scientific approach to comprehend the cause, course, and timing of human evolution. We need not invoke special explanations or supernatural causes. The study of human origins is based on observable evidence that is available to everyone, regardless of their belief system.

There is a fossil record for humankind buried in the earth's sediments. Lucy comes from a period of time when our ancestors looked very apelike, with projecting faces, small brains (one quarter the size of a modern human brain), relatively long arms (left over from tree-living precursors), and curved hand and foot bones (also from our arboreal past). However, these ancestors, sometimes called "ape-men," possessed a cardinal feature that places them on the evolutionary branch of humans rather than apes: bipedalism. Bipedalism is the ability to stand and walk on two legs, a behavior that stands in powerful contrast to all other mammals, which get around on all fours. Additionally, these "ape-men," in contrast to present-day humans, were precultural, meaning they were incapable of making and using even rudimentary stone tools (they may have utilized perishable raw materials like wood as digging sticks, but these items did not fossilize).

Originally documented with the 1924 discovery in South Africa of the "Taung Child" (the skull of an immature *Australopithecus africanus*) by the late Raymond Dart, the die was cast that Africa—or as Charles Darwin referred to it, the Dark Continent—was the original homeland for humanity. In the decade after the publication of Darwin's controversial, lasting, and pro-

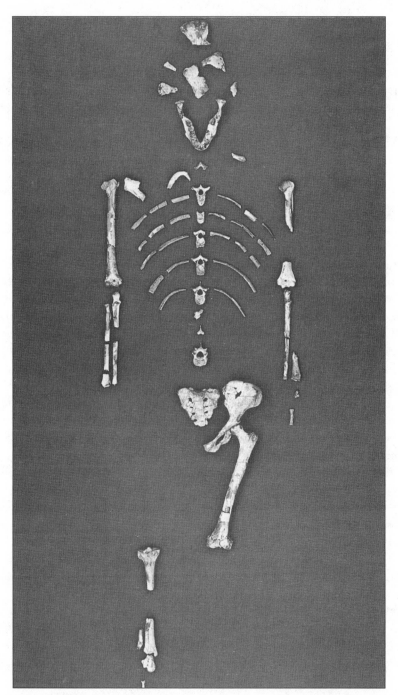

Australopithecus afarensis. This specimen, "Lucy," was discovered by Donald C. Johanson in the Afar triangle of Ethiopia in 1974.

foundly insightful volume, *On the Origin of Species* (1859), it became apparent to scholars that here was an idea so powerful that it was capable of elucidating why species look and act they way they do. Furthermore, it explained why there is a seemingly endless diversity of life on earth.

Darwin knew that living creatures look similar to each other because of common ancestry. He believed they are each uniquely distinct because the process of natural selection has crafted them to act in manners reflecting their particular modes of life. Observations by Darwin (1871) and his close friend Thomas Henry Huxley (1863) led them to make a testable prognostication. The premise was that if humans and the African apes look alike, they must have had a common ancestor, and the differences are related to how they behave (apes climb trees, humans walk on the ground). If the theory was correct, and evolution reflects common ancestry as well as descent with modification (e.g., humans lost their ability to locomote on all fours when they became bipedal), and that the African apes and humans sprang from a long-distant common beginning, the oldest traces of humanity reside in Africa. That proposition is what the Taung Child, Lucy, and thousands of other African fossils, have vindicated. Our direct ancestors first emerged in Africa.

The common ancestor to the African apes and humans still eludes researchers and rests in some unexplored and unexcavated African stratum. However, paleoanthropologists now picture Lucy's species as having lived at an important inflection point in human evolution. Prior to Lucy, the fossil record is sparse. Now, human ancestor fossils are coming to light at 4 million years *(A. anamensis)*, at 4.4 million years *(Ardipithecus ramidus)*, and perhaps even as far back as 6 to 7 million years *(Sahelanthropus tchadensis)*. And true to the predictions made by the theory of evolution, these older species are looking more and more apelike, similar to the presumed more primitive common ancestor.

Subsequent to Lucy, in more recent African sediments we see evidence of those features we think of as more humanlike. At 2.6 million years, stone tools, fashioned by some unknown human hand, are found associated with butchered remains of scavenged animals—meat had entered the human's diet. Then, at around 2 million years, a near doubling in brain size *(Homo habilis, Homo rudolfensis, Homo ergaster)* is seen in larger skulls. The modern human body form (long legs and relatively short arms) appears shortly thereafter, and the first out-of-Africa experience is documented in the Republic of Georgia at 1.8 million years *(Homo erectus)*.

Following a long period of time, during which human species migrated, evolved, diversified, and became extinct, even more modern humanlike ancestors such as *H. antecessor* and *H. heidelbergensis* (500,000 to 1 million years ago) walked the earth. The first tantalizing glimpses of art, engraved ochre, and the specialized and highly insightful manufacture of bone tools appear in Africa some 70,000 to 100,000 years ago, two to three times earlier than in Europe. Now research has confirmed the occurrence of anatomically modern humans *(Homo sapiens,* supposedly "wise man") at close to 200,000 years ago in Africa (Ethiopia).

Humans, capable of reading and writing, entered Europe around 40,000 years ago. Sometimes known as Cro-Magnons, these human ancestors had originally evolved in Africa, and their cultural and intellectual capacities led to the demise of such species as Neandertals (30,000 years ago) and ultimately to the peopling of the entire globe.

We have come from humble beginnings. Each and every feature that defines us as human—bipedalism, tool making, enlarged brains, reduced tooth size, bodies of modern proportions, perhaps even language—first made their appearance in Africa. Yes, Lucy has left a legacy that unites us all as human beings with a common beginning in Africa, and it is my hope that this enlarged, enlightened, and profound insight into the question of "Where did we come from?" will unite all humankind into a fruitful and long-lived common future.

BIBLIOGRAPHY

Darwin, C. 1859. *On the Origin of Species*. London: John Murray.
———. 1871. *The Descent of Man, and Selection in Relation to Sex*. London: John Murray.
Huxley, T. H. 1863. *Evidence as to Man's Place in Nature*. London: Williams and Norgate.
Johanson, D. C. 2004. Lucy, thirty years later: An expanded view of *Australopithecus afarensis*. *Journal of Anthropological Research* 60, no. 4: 465–486.
Johanson, D. C., and M. A. Edey. 1981. *Lucy: The Beginnings of Humankind*. New York: Simon and Schuster.
Johanson, D. C., and B. Edgar. 2005. *From Lucy to Language*. New York: Simon and Schuster.
Kimbel, W. H., Y. Rak, and D. C. Johanson. 2004. *The Skull of* Australopithecus afarensis. New York: Oxford University Press.
Tattersall, I., and J. Schwartz. 2000. *Extinct Humans*. Boulder, CO: Westview Press.

—D.C.J.

Lyell, Charles (1797–1875)

Invariably remembered, even on his gravestone, as the author of *Principles of Geology* (first edition in three volumes: 1830, 1832, and 1833), Charles Lyell was the first child of an English mother, from a landed Yorkshire (Swaledale) family, and of a Scottish father, laird of a large Angus county estate bought only 15 years earlier by the boy's grandfather, a farmer's lad who had made his fortune while serving at sea in the New World. From infancy, however, Lyell lived in southern England, always in London itself after graduating with his BA from Oxford in 1819. Extensive travels, largely in geology's service, took him often to continental Europe and four times, in the 1840s and 1850s, to North America. He worked sporadically for a few years as a lawyer in the 1820s, and briefly (1831–1833) held a professorial chair in his science at the new King's College, London. Otherwise, he depended on his private means supplemented by proceeds from writing and, in the United States especially, from lecturing, so devoting himself almost exclusively to his science.

Eventually knighted, a friend of Prince Albert, and an adviser to the government, he became a prominent national figure with a house on Harley Street, famous for its medical grandees. His final resting place is in Westminster Abbey.

Three years, from mid-1823 to mid-1826, were the most formative for the views upheld by Lyell in *Principles*. Earlier, as an undergraduate, Lyell had attended William Buckland's geology lectures and then done fieldwork with publication in mind, while remaining uncritical of Oxford-taught geology. By mid-1826, however, he was publicly arguing for a fundamental and comprehensive alternative to the teachings of the Oxford school of geology and to the teachings of Parisian Georges Cuvier, the school's primary mentor (Cuvier 1813). In making this shift, Lyell was above all integrating the Huttonian theory of the earth, as defended by John Playfair (1802), whom Cuvier had consistently opposed, with the proposals of Constant Prévost (1823, 1825), who was opposing Cuvier's own interpretations of the Tertiary formations of the Paris Basin. Lyell's engaging with Playfair in his writings and with Prévost in person were probably brokered by William Fitton, a secretary at this time, along with Lyell, of the prestigious Geological Society of London. Fitton, an Irish-born, Scottish-trained physician, was a supportive friend of the Frenchman and an admiring successor to Playfair as the main writer on geology for the *Edinburgh Review*. By late 1823, Fitton had rejected Buckland's Noachian diluvial geology in the *Edinburgh Review* (see also Fitton 1824, 1828), while it was embraced in the rival Tory, Anglican *Quarterly Review* by Edward Copleston. Copleston, a future bishop, was Oxford's poetry professor, a longtime rebuttor of Playfair's criticisms of his university, and Buckland's boyhood friend (Buckland was even rumored to have helped Copleston with the writing of his review).

The *Edinburgh Review* often defined itself by its liberal, Scottish, Whig, reforming opposition to Tory, Anglican, and Oxonian hegemonies in the nation's politics, religion, letters, and science. Determining Lyell's view on any issue in geology or beyond can start from his family's and his own alignments with the *Edinburgh Review*'s stance. When Lyell wrote in 1826–1827 on education and scientific institutions as well as on geology for the *Quarterly Review,* he did so as a conscious dissident intruder. It was as a dissident intruder too that he taught at King's College. With Copleston as an overseer, King's College was designed to be a Tory, Anglican foundation, a counterweight to the young University College, London, which had been founded by liberal Scottish Whigs and headed by Lyell's own father-in-law, a brother of an *Edinburgh Review* founder. Lyell's new reforming ambition in the mid-1820s was to refute and replace wholesale the Oxford school's neo-Cuvierian geology, just when that school had achieved dominance in Cambridge, in the metropolis, and so in the nation. Arguably, therefore, this ambition is explicable as the ambition of a young man who had come to think that starting out in the Oxford school was not a place where a liberal, Scottish, Whig lawyer like himself should ever, geologically speaking, have gone. Conversely, he now had to act in accord with what he had come to appreciate as his authentic familial and personal affiliations.

That reform, as expounded in its fullest extent in *Principles,* was (1) to conform geology to the *vera causa* evidential-explanatory ideal (effects are to be explained by causes known to exist from independent evidence) exemplified by Newtonian celestial mechanics; (2) to give the science a systemic theory—a new version of the acosmogonical Huttonian theory (itself now conformable, potentially at least, to that ideal)—that was theoretical but avoiding divisive, inconclusive, cosmogonical speculation; and (3) to free the science from Mosaic biblical associations. These ends, as Lyell saw them, could only be met if geology presumed that the past changes recorded in the rocks were all produced by causes still working in the same circumstances and with the same intensities and so producing, in the present and on into the future, the same sorts and sizes of effects. It was indeed a presumption, Lyell insisted, entitled to a priori preference over its denial and so wrongly rejected by most geologists because they had prematurely decided that they already knew enough of present causes to judge them inadequate for those past effects. By reforming geology by giving it these principles, Lyell could achieve his prime ambition for the science and for himself: to save geology from its current low repute and even ridicule by giving it sound foundations and elevated status comparable to the foundations and status of celestial mechanics, taken to be the highest science of all. As the son and grandson of arriviste Scottish gentry, Lyell worried intensely about the foundations and status of the science that he had made his avowed avocation and even, since 1827, his prospective profession in place of the law. Notorious for his snobbery, Lyell's elitist anxiety was at once social and cognitive.

The ultimate status of mankind—our filiation, dignity, and fate—raised transcendent anxieties that, Lyell insisted, only religion can address. A deist (a believer in God but not in the Bible), he held to a Divine creation, design, and government not only for nature but also for man, whose immortal, immaterial soul comes under God's moral law in this life and the next. He saw Jean-Baptiste Lamarck's (1809) system of zoology, extensively discussed and rejected in the early editions of *Principles,* as directly threatening these commitments about man. Accordingly, he rejected any reading of the fossil record as a progression, from low life to high, so as to discredit any scheme such as Lamarck's that has humans produced from animal ancestors. Lyell came then to join Cuvier, Lamarck's other opponent, in replacing any unified progressive necessitating plan with contingent, fitting adaptations in the history of life on earth. Charles Darwin's theorizing retained this favoring of the many adaptive contingencies over the one necessitating plan. While recognizing such contrasts with Lamarck, Lyell always read this new Darwinian theorizing as raising, if extended to man, the same old intolerable threat to theology and morality.

Childless and destined not to inherit the family estate, Lyell, although hugely influential over several decades of geological science, never founded any school of comprehensively consensual disciples. In his geology, Darwin was very much Lyell's most loyal protégé, even after developing his own dissident biology. Lyell's neo-Huttonian system of the earth, criticized from the start over its claims for a constant, undeclining thermal economy for the planet, was more widely thought implausible from the 1850s on after William Thomson (Lord Kelvin) and others raised their novel thermodynamical objections to these claims (Burchfield 1975). The intricate complexity of the young Charles Darwin and young Alfred Russel Wallace's agreements and disagreements with Lyell, and with Lyell's agreements and disagreements with Lamarck, put the proximate and ultimate debts of evolutionary biology to Lyell beyond quick-and-easy summary generalization. Those debts are more rather than less deep because of that very complexity.

BIBLIOGRAPHY

Burchfield, J. D. 1975. *Lord Kelvin and the Age of the Earth*. New York: Science History Publications.

Cuvier, G. 1813. *Essay on the Theory of the Earth*. R. Kerr, trans. Edinburgh: W. Blackwood.

Fitton, W. H. 1824. Inquiries respecting the geological relations of the beds between the Chalk and the Purbeck Limestone in the South-east of England. *Annals of Philosophy* 8: 365–383.

———. 1828. Address delivered on the anniversary, February 1828. *Proceedings of the Geological Society of London* 6: 50–62.

Lamarck, J.-B. 1809. *Philosophie zoologique*. Paris: Dentu.

Lyell, C. 1830–1833. *Principles of Geology: Being an Attempt to Explain the Former Changes in the Earth's Surface by Reference to Causes Now in Operation*. 3 vols. London: John Murray.

Lyell, K. M., ed. 1881 [1970]. *Life, Letters and Journals of Sir Charles Lyell, Bart.* 2 vols. London: John Murray. Republished in facsimile, Farnborough, U.K.: Gregg International Publishers.

Playfair, J. 1802. *Illustrations of the Huttonian Theory of the Earth*. Edinburgh: William Creech.

Prévost, C. 1823. De l'importance de l'étude des corps organisés vivants pour la géologie positive. *Memoirs de Societé D'Histoire Naturelle de Paris* 1: 259–268.

———. 1825. De la formation des terrains des environs de Paris. *Bulletin de la Societé Philomatique de Paris:* 74–77, 88–90.

Rudwick, M. J. S. 2008. *Worlds before Adam: The Reconstruction of Geohistory in the Age of Reform*. Chicago: University of Chicago Press.

Wilson, L. G. 1972. *Charles Lyell: The Years to 1841*. New Haven, CT: Yale University Press.

———. 1998. *Lyell in America: Transatlantic Geology, 1841–1853*. Baltimore: Johns Hopkins University Press.   —*J.H.*

Lynch, Michael (b. 1951)

When considering the impact a scientist has had, one must take into account the breadth and depth of the intellectual contributions. By those criteria, very few of his peers have had a greater impact on evolutionary biology than Michael Lynch. Further, his career has been marked by that most valued of academic achievements: "steady progress."

Lynch began his career as an ecologist, studying the community structure of zooplankton as a student of Joseph Shapiro at the University of Minnesota. In a series of papers stemming from his dissertation research, he articulated a comprehensive model of life-history evolution in Cladocera, incorporating disparate elements including predation, competition, and energetics, and culminating in the first of many influential synthetic works (Lynch 1980). His observation that a trade-off between present and future reproductive success need not be necessary was a challenge to the conventional wisdom of life-history theory. The underlying philosophy of always seeking the simplest explanation has been the hallmark of his career and has put him at the center of controversy almost uninterruptedly for the past quarter century.

The 1970s and 1980s witnessed two important developments in evolutionary biology: the emergence of evolutionary quantitative genetics and the maturation of the field of molecular evolution. By the mid-1980s, Lynch had reinvented himself as a quantitative geneticist and over the next decade established himself as a leader in the field. Along with Russell Lande, Lynch was instrumental in codifying a neutral theory of evolutionary quantitative genetics (e.g., Lynch and Hill 1986). The textbook *Genetics and Analysis of Quantitiative Traits*, coauthored with Bruce Walsh (1998), is the industry standard for the field of quantitative genetics.

Evolutionary quantitative genetics historically focused on traits under stabilizing or positive directional selection. Along with Tomoko Ohta and Alex Kondrashov, Lynch was among the first to recognize the importance of slightly deleterious mutations to the evolutionary process. Along with collaborators Wilfried Gabriel, Reinhard Burger, and John Conery, Lynch began a systematic study of the effects of deleterious mutations on the probability of population extinction (Lynch et al. 1993, 1995). The resulting synthesis established the "mutational meltdown" in the lexicon, and, more important, established the relevance of the relationship between fitness and population size in the mind-set of biologists outside the realm of theoretical population genetics.

In recent years, in collaboration with Conery and developmental geneticist Allan Force, Lynch has forged a remarkable synthesis of several seemingly disparate features of genome evolution, including duplicate genes, introns, and transposable elements (Lynch and Conery 2001, 2003a, 2003b; Force et al. 1999; Lynch and Force 2000a, 2000b; Lynch et al. 2001). Operating from first principles of population genetics, Lynch showed that population size and slightly deleterious mutations may interact in such a way as to produce the genomic features that distinguish prokaryotes from eukaryotes and single-celled from multicellular organisms. True to form, Lynch provided a simple, controversial explanation in lieu of a complicated, controversial one. His recent book *The Origins of Genome Architecture* (2007) provides a highly readable synthesis of this work. It is as ambitious in scope and message ("Nothing in evolution makes sense except in light of population genetics") as Motoo Kimura's classic *The Neutral Theory of Molecular Evolution* (1983).

Lynch has also made many important contributions on topics as diverse as the phylogenetic comparative method, the evolution of parthenogenesis, the meaning of inbreeding depression, and statistical population genetics.

BIBLIOGRAPHY

Force, A., M. Lynch, F. B. Pickett, A. Amores, Y.-L. Yan, and J. Postlethwait. 1999. The preservation of duplicate genes by complementary degenerative mutations. *Genetics* 151: 1531–1545.

Kimura, M. 1983. *The Neutral Theory of Molecular Evolution*. Cambridge: Cambridge University Press.

Lynch, M. 1980. The evolution of cladoceran life histories. *Quarterly Review of Biology* 55: 23–42.

———. 2007. *The Origins of Genome Architecture*. Sunderland, MA: Sinauer Associates.

Lynch, M., R. Bürger, D. Butcher, and W. Gabriel. 1993. The mutational meltdown in asexual populations. *Journal of Heredity* 84: 339–344.

Lynch, M., and J. S. Conery. 2000. The evolutionary fate and consequences of duplicate genes. *Science* 290: 1151–1154.

———. 2001. Gene duplication and evolution: Response to Long and Thornton and Zhang et al. *Science* 293: 1551.

———. 2003a. The evolutionary demography of duplicate genes. In A. Meyer and Y. Van de Peer, eds., *Genome Evolution*, 35–44. Dordrecht, The Netherlands: Kluwer Academic Publishers.

———. 2003b. The origins of genome complexity. *Science* 302: 1401–1404.

Lynch, M., J. Conery, and R. Bürger. 1995. Mutation accumulation and the extinction of small populations. *American Naturalist* 146: 489–518.

Lynch, M., and A. Force. 2000a. Gene duplication and the origin of interspecific genomic incompatibility. *American Naturalist* 156: 590–605.

———. 2000b. The probability of duplicate gene preservation by subfunctionalization. *Genetics* 154: 459–473.

Lynch, M., and W. G. Hill. 1986. Phenotypic evolution by neutral mutation. *Evolution* 40: 915–935.

Lynch, M., M. O'Hely, B. Walsh, and A. Force. 2001. The probability of preservation of a newly arisen gene duplicate. *Genetics* 159: 1789–1804.

Lynch, M., and B. Walsh. 1998. *Genetics and Analysis of Quantitiative Traits*. Sunderland, MA: Sinauer Associates.   —*C.B.*

M

Malthus, Thomas Robert (1766–1834)

Thomas Robert Malthus, known as Bob to his intimates, was an English economist. He was an ordained clergyman in the Anglican Church and for many years taught political economy at Haileybury College, a school that produced civil servants destined to work in India. As an individual he was warm and friendly and much liked; he was an effective speaker despite a cleft lip.

His claim to fame is a book that first appeared in 1798, *An Essay on the Principle of Population as It Affects the Future Improvement of Society, with Remarks on the Speculations of Mr. Godwin, M. Condorcet and Other Writers*. Malthus was dismayed at what he felt was the false optimism of many writers and therefore tried to show that grandiose plans for human improvement are impossible. He did this through a famous inequality, drawing on an earlier argument of Benjamin Franklin and arguing that although human population numbers tend to increase geometrically (1, 2, 4, 8, . . .), food supplies can at maximum go up arithmetically (1, 2, 3, 4, . . .). Hence there are inevitable "struggles for existence," and plans for state aid only exacerbate the problem (Malthus 1798, 14).

Malthus's chief intent was natural theological. He was concerned to see how God had arranged that humans do something here on earth rather than idle time away aimlessly. The Malthusian inequality is the answer. However, his work was taken up with enthusiasm by social reformers, who used his calculations as the basis for draconian reforms of the Poor Laws, where workhouses were made so unpleasant that the poor would do anything to stay out of them. One way of avoiding such degradation would have been through what Malthus termed *prudential restraint* (from breeding), a subclause that Malthus added to later, much-enlarged editions of his book in response to criticism that he seemed to have implied that God left no option but bloody interhuman strife.

It was the sixth edition of Malthus's book that Charles Darwin read at the end of September 1838. Generalizing to the animal and plant worlds and realizing that there could be no prudential restraint in those areas, Darwin argued that there will be a universal struggle for existence, and that this is the motive force behind a natural form of selection. Twenty years later, when Alfred Russel Wallace was independently formulating an evolutionary hypothesis with selection at its center, he too remembered the Malthusian inequality and made it a crucial element of his thinking. Like Malthus, the evolutionists stressed that struggle might be silent and nonobvious, where organisms battle for life in hard conditions. However, Darwin and Wallace differed in their readings of the struggle, with the former always inclined to think that it is between individuals and the latter prepared to allow that sometimes it occurs between groups.

BIBLIOGRAPHY

Malthus, T. R. 1798. *An Essay on the Principle of Population.* London: Printed for J. Johnson, in St. Paul's Church-Yard. Reprint, New York: Macmillan, 1966.
———. [1826] 1914. *An Essay on the Principle of Population.* 6th ed. London: Everyman.
Young, R. M. 1985. *Darwin's Metaphor: Nature's Place in Victorian Culture.* Cambridge: Cambridge University Press.   —M.R.

Mammals

We commonly speak of mammals as being descended from reptiles, but although modern reptiles belong to one of the major divisions of amniotes, the sauropsids (which also include birds and dinosaurs), mammals and their fossil ancestors belong to the other major division, the synapsids. Nonmammalian synapsids are often called mammal-like reptiles, a rather misleading name because these animals are not closely related to those modern forms that we call reptiles.

Although the synapsid lineage has evolved many unique features, such as the milk and fur typical of mammals, synapsids have also retained a number of more primitive features that have been lost or modified in sauropsids. Mammals retain a glandular skin and did not evolve the thicker scales with contributions from beta-keratin seen in sauropsids (as well as the more general alpha-keratin seen in all amniotes); thus they have been able to evolve mammary glands, sweat glands, and glands that produce various odiferous secretions. Unlike sauropsids, which have the derived ability to excrete uric acid and thus produce a concentrated type of semisolid waste, synapsids excrete a relatively dilute urine, always stored in a urinary bladder (which is lost in many sauropsids, including birds). This enables mammals to use urine for social interactions such as scent-marking. The typically mammalian use of the penis for urination in males, as well as for sperm transmission, is also only possible with this type of urinary setup. Differences in the internal anatomy of synapsids and sauropsids also suggest that various features evolved convergently within the two groups; in particular, endothermy (warm-bloodedness) clearly evolved convergently in birds and mammals.

Synapsids and sauropsids both appeared during the Late Carboniferous, around 300 million years ago. At this time both lineages looked rather lizard-like, but synapsids had a characteristic hole in the skull behind the eye socket, which allowed for jaw-muscle expansion and attachment. (Most sauropsids later evolved an analogous condition in the skull, but with two holes on each side, known as the diapsid condition.) Synapsids were initially the dominant large land vertebrates, especially prominent during the Permian (299–251 million years ago). The earliest forms were the pelycosaurs, which spanned about the body-size range seen in domestic dogs today and included carnivores, herbivores, and fish eaters. Some pelycosaurs had distinct "sails" on their backs, which were probably used as environmental heat exchangers. Pelycosaurs were mainly found in geographic areas that were in tropical or subtropical zones at that time, such as North America.

By the Late Permian the pelycosaurs were largely extinct, replaced by their descendants, the therapsids. Therapsids had moved their legs more underneath their body, with a stance that was now more upright than the sprawling, lizardlike posture of pelycosaurs, enabling them to run and breathe at the same time; this is suggestive of the evolution of a higher metabolic rate. Later therapsids (convergently among different lineages) evolved various features suggestive of increasing levels of metabolic rate, such as increased volume in the skull for jaw muscles and more complexly differentiated teeth, indicative of increased amounts of food processing. The therapsids included small insectivores and medium to large-sized carnivores (some as big as lions) and herbivores (some as big as bison, with hornlike knobs on their heads that might have been used in interspecific combat). But in contrast with modern mammals or even with contemporaneous sauropsids, there were no gliders, swimmers, or bipedal runners. Later in the Permian therapsids became common in South Africa, which had cold winters at this time. This changing geographic distribution also suggests an increased ability for thermoregulation.

After the devastating end-Permian extinctions two main lineages of therapsids survived: the dicynodonts, pig- to cow-sized herbivores, and the cynodonts, initially small and insectivorous. Cynodonts later evolved larger herbivorous and carnivorous forms, although none was much bigger than a Labrador dog. Cynodonts were ancestral to mammals and had numerous anatomical features suggestive of a high metabolic rate, which suggests some degree of mammal-like endothermy. More complex teeth and evidence of an increased volume of jaw muscles show that they were processing more food. An increased rate of lung ventilation is shown by evidence of a muscular diaphragm (reduction of the posterior ribs) and by evidence of turbinate (scroll-like) bones in the nasal cavity. Turbinates act both to warm incoming air and to reclaim water from expired air, because water loss from the lungs can be a significant problem with a high rate of ventilation.

In the later Triassic therapsids started to face competition from the radiation of larger sauropsids, including the thecodonts (ancestors of dinosaurs). Global changes in vegetation, such as the replacement of the more archaic seed ferns by the conifers, may have influenced these evolutionary trends. Another issue that relates to this replacement may be that atmospheric oxygen levels were apparently lower in the Triassic than in the Permian, and it can be inferred from the biology of the living relatives of thecodonts (crocodiles and birds) that superior lung function may have provided them with a competitive edge. Additionally, the eggs of birds and crocodiles are more highly resistant to desiccation than those of other amniotes; this feature may have given the thecodonts a competitive edge in the apparently drier conditions of the Triassic. Late Triassic cynodonts became progressively smaller, perhaps downsizing to escape competition. At the very end of the Triassic (around 200 million years ago) only four lineages of synapsids remained. Two rat-sized lineages of cynodonts persisted into the Early Jurassic, along with the first true mammals. Only mammals prospered past this time, although a relict cynodont (and also a relict dicynodont) are now known from the Early Cretaceous (~140 million years ago). The earliest mammals of the Jurassic were shrew sized, with teeth indicative of a shrewlike insectivorous diet. The major large land vertebrates were now the dinosaurs, a condition that continued throughout the rest of the Mesozoic.

The earliest mammals had relatively larger brains than cynodonts, as well as modifications of the backbone to allow for flexion up and down. In later mammals, with a more mobile shoulder, this feature allowed for bounding locomotion, but initially it may have been important to allow mammals to lie on their side to suckle their young. The teeth of mammals now interlocked precisely, indicating actual chewing (mastication) of the food, and formed only two sets (milk teeth and permanent teeth, as in humans). The evolution of essentially a single, nonreplacing adult dentition was probably important for precisely occluding teeth because continually replacing teeth could not be maintained in precise alignment. This type of tooth replacement is probably also indicative of the evolution of lactation because the eruption of dentition could be delayed until the young needed to eat solid food. The evolution of suckling required the evolution of lips that could form a seal and muscular cheeks; these features were probably precursors of the complex series of facial muscles, seen only in mammals, that humans now use for speech and facial expressions. Like the modern-day monotremes (platypuses and echidnas), the most primitive of the modern mammals, these early mammals must have been egg laying, would have lacked nipples and an external ear, and probably had a dense fur coat.

During the Jurassic several different lineages of early mammals evolved and diversified, mainly insectivorous or carnivorous in their diet. Few were bigger than a rat, and they were mainly extinct by the Cretaceous. Our knowledge of Mesozoic mammals has greatly increased in the past few years, and we can now add digging, swimming, gliding, and somewhat larger (opossum-sized) carnivorous creatures to the diversity of known forms. In the Late Jurassic (~180 million years ago) a distinct omnivorous/herbivorous lineage appeared, the multituberculates (so called because of their complex cheek teeth). The multituberculates were a highly successful lineage of rodentlike mammals that survived into the Cenozoic (until around 40 million years ago), when they may have been eclipsed by the evolution of true rodents. During the later part of the Early Cretaceous, a new type of mammal appeared, the therians, including the first true marsupials and placentals. (Interestingly, the first monotremes also date from this time.)

Therians possessed new, more complex types of cheek teeth (tribosphenic molars) that allowed their owners to process a broader spectrum of available dietary items. Therians also had a more hingelike ankle joint and a shoulder blade that could now swing freely, contributing to the stride length and making possible the characteristic bounding gait of many modern mammals. Both modern marsupials and placentals are also viviparous (giving birth to babies rather than laying eggs), although they do it in somewhat different ways, and it is not clear if viviparity was present in their common ancestor or if they evolved this condition independently. Cretaceous mammals were diverse, similar in numbers of species to the Cretaceous dinosaurs, but were still small, none bigger than the size of a small dog. Although the Cenozoic era is commonly called the age of mammals, it in fact represents only a third of mammalian evolutionary history, when mammals radiated out into larger body sizes to fill the types of large land vertebrate niches previously held by the dinosaurs. The evolution of Cenozoic mammals is best understood in the context of the changing patterns of the earth's continents and the subsequent changes in global climate. Although the supercontinent Pangaea started to break up during the Mesozoic, the southern continents were still united in Gondwana at the start of the Cenozoic, and the northern continental blocks had not yet reached their final positions. Movement, fragmentation, and coalescence of continental plates created changes in ocean currents and episodes of mountain building that resulted in global climatic changes. To a first approximation the world changed during the Cenozoic from initially globally warm, tropical-like conditions to a world that became increasingly colder at the higher latitudes and increasingly drier throughout.

These changing global climatic conditions consequently influenced the types of mammals that evolved. For example, the ecological niches now filled by animals such as polar bears and camels were not in existence until around 5 million years ago, with the development of an Arctic ice cap and the desertification of tropical regions. Additionally, the isolation of different seed groups of mammals on different continental blocks (although this mainly occurred via migration rather than the mammals being carried on the separating continents) resulted in a great amount of convergent evolution of mammalian types. The best known of these are the examples between Australian marsupials and placental mammals elsewhere (e.g., placental wolf and marsupial "wolf," or thylacine), but many other examples exist between different groups of placentals. For example, the African golden mole is more closely related to elephants than it is to the true moles of the Northern Hemisphere.

Of the various mammalian ecomorphological types (particular body types adapted for specific ecological roles), two types that have repeatedly evolved at different times and places in the past are missing from today's world. One is the saber-toothed predator. This type of mammal evolved twice within true cats, within a now-extinct family of "false saber-tooth" carnivores, and in South American marsupials. (Australian marsupials produced a similar type of heavily built, muscular predator, the marsupial lion, which had evolved from herbivorous forms and so lacked the distinctive canines.) A second type is the large (bison-sized) clawed herbivore that could rise on its hind legs to pull down branches. Mammals like this evolved among the horse-related chalicotheres in the Northern Hemisphere and Africa, among native ungulates and giant ground sloths in South America, and among wombat-related forms in Australia.

A brief history of Cenozoic mammalian distribution is as follows. At the start of the Cenozoic, Africa was isolated from the other continents and evolved its own specialized fauna (e.g., elephants); primates were also known early on in Africa, but other northern immigrants (such as carnivores and ungulates, or hoofed mammals) did not reach there until the early Miocene, around 20 million years ago. The monotremes were the original Australian inhabitants; marsupials (immigrating from South America via a then-ice-free Antarctica) were not known until the early Eocene, around 55 million years ago. South America was isolated until the middle Pliocene (2.5 million years ago), when the Isthmus of Panama formed. Its original fauna consisted of edentates (sloths, armadillos, and anteaters), marsupials (opossums and borhyaenids, now-extinct predatory forms), and native ungulates (now all extinct). Around 40 million years ago it gained its stocks of primates and caviomorph (related to guinea pigs) rodents, most likely from animals rafting over from Africa. About 2.5 million years ago many forms migrated from North America, and now over half the South American fauna is of recent northern origin (e.g., cats, foxes, tapirs, deer, and mouselike rodents). North America and Eurasia have long had fairly broad faunal connections, but today their mammal faunas are more similar than in many past times. For example, North American deer and bison are recent (within the past 3 million years) Eurasian immigrants.

During the early Cenozoic (the Paleocene and early Eocene epochs, from 65 to 50 million years ago) the world was in general warm and equable and, with the browsing pressure of the herbivorous dinosaurs removed, was largely covered in forests; tropical-like forests extended even within the confines of the Arctic Circle. The types of mammals present were largely arboreal insect and fruit eaters, with a few terrestrial omnivores and herbivores. Small predators included early members of the modern order Carnivora, but larger predators were not diverse and belonged to groups now extinct. Most of these mammals have been termed *archaic* types, meaning that they belonged to lineages that did not persist until the present day (indeed, most went extinct before the end of the Paleocene). The start of the Eocene saw the first appearance of primate and modern orders of ungulates (with the first rodents appearing in the latest Eocene) and also the highly specialized bats and whales. The first known fossil bats are already clearly specialized fliers, and we have little information on their ancestry. However, in the past decade a spectacular series of early fossil whales has been collected that clearly shows the transition from a land animal to a secondarily aquatic one.

Around 50 million years ago temperatures in the higher latitudes started to fall, resulting in increased seasonality and winter frosts. By the start of the Oligocene (33 million years ago) the tropical forests were confined to the equatorial region, and deciduous woodland spread across much of the Northern Hemisphere. The archaic mammals were now all extinct, primates were confined to the tropics, and more modern types of large predators (dogs and cats) made an appearance. Rodents and rabbits (or equivalent types of small herbivores) were common, and the ungulates were larger and mostly leaf eaters rather than fruit eaters. In South America large rodents (like today's capybara) took over some of the small ungulate niches. A warming trend during the early Miocene, around 20 million years ago, followed by subsequent cooling and drying during the later Miocene (12–5 million years ago), brought with it the spread of savanna-type habitats (treed grasslands) at higher latitudes. Long-legged ungulates, with teeth modified for eating grass, became prevalent in North America (horses), Eurasia (antelope), and South America (extinct native forms). However, the equivalent types of marsupial, the larger kangaroos, were not in evidence until the Pliocene in Australia. Further cooling and drying in the late Cenozoic (from around 8 million years ago) resulted in the transformation of these productive savanna habitats into less productive treeless prairie, with the loss of large mammal diversity. Plio-Pleistocene (from 5 million years ago) cooling resulted in the ice ages (largely experienced by the Northern Hemisphere), and specialized mammals (such as musk ox and reindeer today and woolly rhino and mammoths in the recent past) inhabited the new high-latitude habitats such as tundra. However, ice-age global cooling and drying resulted in the development of extensive savanna habitats in tropical Africa and Asia, which today preserve the type of large-mammal diversity that was formerly common worldwide.

BIBLIOGRAPHY

Agustí, J., and M. Antón. 2002. *Mammals, Sabertooths, and Hominids: 65 Million Years of Mammalian Evolution in Europe.* New York: Columbia University Press.

Benton, M. J. 2004. *Vertebrate Palaeontology.* 3rd ed. Chaps. 5 and 10. Malden, MA: Blackwell Publishing.

Cifelli, R. L. 2001. Early mammalian radiations. *Journal of Paleontology* 75: 1214–1226.

Janis, C. M. 2001. Victors by default: The mammalian succession. In S. J. Gould, ed., *The Book of Life*, 2nd ed., 166–217. New York: W. W. Norton and Co.

Kemp, T. S. 2005. *The Origin and Evolution of Mammals.* Oxford: Oxford University Press.

Kielan-Jaworowska, Z., R. L. Cifelli, and Z.-X. Luo. 2004. *Mammals from the Age of Dinosaurs: Origins, Evolution, and Structure.* New York: Columbia University Press.

Pough, F. H., C. M. Janis, and J. B. Heiser. 2008. *Vertebrate Life.* 8th ed. Upper Saddle River, NJ: Benjamin Cummings.

Thewissen, J. G. M., and S. Bajpai. 2001. Whales as a poster child for macroevolution. *BioScience* 51: 1037–1049.

Turner, A., and M. Antón. 2004. *Evolving Eden: An Illustrated Guide to the Evolution of the African Large-Mammal Fauna.* New York: Columbia University Press.

Vaughan, T. A., J. A. Ryan, and N. Czaplewski. 2000. *Mammalogy.* 4th ed. Philadelphia: Saunders Publishing.

—C.J.

Man's Place in Nature (Thomas Henry Huxley)

Thomas Henry Huxley's *Man's Place in Nature* (1863) (the full title is *Evidence as to Man's Place in Nature,* but it is always known by the shorter title, which appeared on the spine) was the first English book after Darwin's *On the Origin of Species* to relate humans and apes anatomically and in evolutionary terms (the *Origin* had avoided the subject). And it did so in punchy prose for general consumption.

Its inception lay in Huxley's dispute with the antitransmutationist Richard Owen. Owen, an expert on apes (including the gorilla, discovered in the 1840s), placed mankind in a new subclass, Archencephala (ruling brain), in 1858. He insisted that humans possess unique cerebral hemispheres, which divide into a third lobe, whose lateral ventricle contains a protruding "hippocampus minor."

Huxley, a biologist at London's School of Mines, hated Owen's hauteur and providentialism and responded in 1858 by stating that "there is very little greater interval *as animals* between the *Gorilla* & the *Man* than exists between the *Gorilla* & the [baboon]" (lecture 10, March 16, 1858, Royal Institution Series, Principles of Biology, January 19–March 23, 1858, Huxley Papers, Imperial College, London, 36.100). This sentiment stands at the heart of *Man's Place in Nature.* Criticizing "Theology & Parsondom" in his talks to artisans in 1859, Huxley added that humans and animals "*must* have proceeded from one another in the way of progressive modification" (Huxley 1859). It was, he told a friend, "as respectable to be modified monkey as modified dirt" (letter to Frederick Dyster, January 30, 1859, Huxley Papers, Imperial College, London, 15.106). Before the *Origin* appeared, Huxley had polarized the issue of human evolution versus creation.

Owen reasserted his claim at the meeting of Oxford's British Association for the Advancement of Science in 1860. From the floor came Huxley's pointed contradiction, and from his pen a substantiating essay "On the Zoological Relations of Man with the Lower Animals" (1861), which showed that ape brains possess every "unique" human feature. The dispute, which exploded publicly in the *Athenaeum* in 1861, alerted respectable society to human evolution.

The pauper presses had long relished an aristocracy-humbling bestial origin, and Huxley was tapping this groundswell for support. He portrayed evolution as self-betterment: it granted dignity to humble origins. Huxley's artisan talks, running through 1862, formed the second chapter of *Man's Place in Nature* (there were only three chapters). Serendipity provided the third. Sir Charles Lyell, who was writing his *Antiquity of Man,* asked Huxley's views on Neanderthal Man (found in 1856). Huxley's ensuing Royal Institution lecture "On Fossil Remains of Man" (Huxley 1858–1862) would finish *Man's Place in Nature,* and he garnered accounts of living apes for an opening chapter. If *Man's Place in Nature* was less assertive on evolution than his lectures, it would still prick Victorian vanities, not least by suggesting that humans might go back far into geological time.

An initial printing of 1,000 copies hit the bookstalls in February 1863. Darwin had never "read anything grander" than the closing of the middle

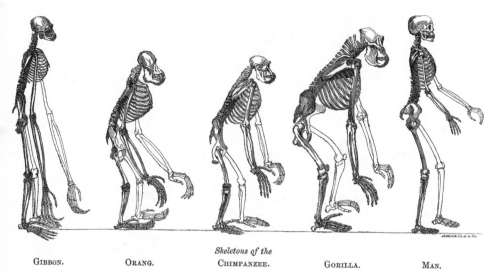

Skeletons of the
GIBBON.   ORANG.   CHIMPANZEE.   GORILLA.   MAN.

When it came to evolution, Thomas Henry Huxley had very different motives from Charles Darwin. The latter wanted to provide an overall scientific theory or paradigm; the former wanted material to fight the Christians. There is little wonder that Huxley at once picked up on that about which Darwin was so reticent, namely human evolution. This frontispiece of *Man's Place in Nature* is designed to show beyond doubt that we humans are from the same stock as the great apes. Most people agreed, although to this day Christians usually argue that our souls are inserted miraculously.

chapter (letter to Huxley, February 26, 1863, Darwin 1999, 11: 180). But the religious press was stunned, and many noted the lack of any recognition of mind and speech. Brisk sales necessitated an immediate 1,000-copy reprint. The only bar to its falling into grubbier hands was the price: six shillings, expensive for a 159-page tome. Even so, the Catholic St George Mivart was grieved to see translations at Italian railway stations alongside *"obscenities,"* showing that the street-level fascination stretched across Europe (unpublished letter from Mivart to Charles Darwin, April 25, 1870, Darwin Collection, University Library, Cambridge, Folder 171; emphasis in original).

Lyell's *Antiquity of Man* was published within days of *Man's Place in Nature*. The first made man "a hundred thousand years" old, said one reviewer, scooping the modern worldview, and the second gave him "a hundred thousand apes for his ancestors" (Anonymous 1863).

BIBLIOGRAPHY

Anonymous. 1863. *Evidence as to Man's Place in Nature* [review]. *Athenaeum*, February 28, 287–288.

Darwin, C. 1999. *The Correspondence of Charles Darwin*. Vol. 11. Cambridge: Cambridge University Press.

Desmond, A. 1997. *Huxley: From Devil's Disciple to Evolution's High Priest.* Reading, MA: Addison-Wesley.

Gross, C. G. 1993. Hippocampus minor and man's place in nature: A case study in the social construction of neuroanatomy. *Hippocampus* 3: 403–415.

Huxley, T. H. 1858–1862. On fossil remains of man. *Proceedings of the Royal Institution of Great Britain:* iii, 420–422.

———. 1859. Science and religion. *The Builder,* January 15.

———. 1861. On the zoological relations of man with the lower animals. *The Natural History Review:* 67–84.

———. 1863. *Evidence as to Man's Place in Nature.* London: Williams and Norgate.

Lyell, C. 1863. *The Antiquity of Man.* London: John Murray.

Lyons, S. L. 1997. Convincing men they are monkeys. In A. P. Barr, ed., *Thomas Henry Huxley's Place in Science and Letters: Centenary Essays,* 95–118. Athens: University of Georgia Press.

Owen, R. 1858. On the characters, principles of division, and primary groups of the class mammalian. *Journal of the Proceedings of the Linnaean Society (Zoology)* 2: 1–37.

Wilson, L. G. 1996. The gorilla and the question of human origins: The brain controversy. *Journal of the History of Medicine and Allied Sciences* 51: 184–207.   —A.D.

Margulis, Lynn (b. 1939)

Lynn Margulis, a longtime professor of geosciences at the University of Massachusetts, Amherst, is properly celebrated for her strong advocacy of the origin of eukaryote cells (cells with a nucleus) from more primitive prokaryote cells (cells without a nucleus). Some version of this theory goes back to the nineteenth century, when it was pushed by the Russian lichen specialist Konstantin Sergivich Merezhkovsky (1855–1921), and it was advocated in the 1920s by the Swedish American Ivan Wallen. But it was Margulis who took up the idea, now known as the *endosymbiotic theory,* offered microscopic evidence, and pushed it (in her book *Origin of Eukaryotic Cells,* 1970) until it became orthodoxy.

Margulis claimed that certain functioning parts of the cell (organelles) are in fact former prokaryotes that have been taken up and now function for the benefit of the whole eukaryotic cell. Included here are mitochondria and chloroplasts, respectively, the power plants of the cell and (in plants) the organelles that perform photosynthesis. Definitive proof of Margulis's position came in the 1980s when it was discovered that the DNA of these organelles is different from that of the cell's nucleus but bears significant similarities to that of various prokaryotes, for instance, in being circular, having the same size, and being able to perform the same functions as the appropriate organelles or functions similar to those of the organelles.

Margulis's thinking is deeply influenced by her philosophy of life, which sees harmony and mutualism in nature rather than fighting and antagonism. Although some have argued that the endosymbiotic theory can best be seen in terms of capture and slavery—the organelles by the main cell—for her it is rather a matter of different life forms coming together in a successful attempt to work as a unified whole. She sees evolution less in terms of a ruthless Darwinian struggle than as a process based on the sideways transfer of genetic material from one organism to another, carried via microorganisms like bacteria. This view has been received with some skepticism, although recent work on the human genome strongly suggests that genetic material is transferred sideways in this fashion. Since Darwinism works on genetic variation, however obtained, it is probably better to think of this sort of phenomenon as a complement to Darwinian evolution than as an opposing view.

In recent years, also guided by her philosophy of harmony and mutualism, Margulis (together with Dorion Sagan, her son from her marriage to Carl Sagan) has been a strong advocate of the Gaia hypothesis, which sees the earth as an organism and as self-regulating, at least until humans disrupted it so significantly that it could no longer function properly. This advocacy has expectedly heaped criticism on Margulis's head—many conventional biologists equate the Gaia hypothesis with pre-Christian nature worship—and equally expectedly she is more than prepared to weather the storm, convinced of the rightness of her cause.

BIBLIOGRAPHY

Margulis, L. 1970. *Origin of Eukaryotic Cells: Evidence and Research Implications for a Theory of the Origin and Evolution of Microbial, Plant, and Animal Cells on the Precambrian Earth.* New Haven, CT: Yale University Press.

———. 2000. *Symbiotic Planet: A New Look at Evolution.* New York: Basic Books.

Margulis, L., and D. Sagan. 1997. *Slanted Truths: Essays on Gaia, Symbiosis and Evolution.* New York: Copernicus Books.   —M.R.

Marsh, Othniel Charles (1831–1899)

Othniel Charles Marsh was the most influential paleontologist in late-nineteenth-century America. He exploited the wealth of his uncle, George Peabody, and his connections with the U.S. Geological Survey to explore the rich fossil beds of the West. He was an enthusiastic supporter of Darwinism and used many of his fossil discoveries to provide evidence in support of evolution.

Marsh was born in Lockport, New York, in 1831. He was trained at the Sheffield Scientific School of Yale University and after three years of study in Europe returned to become professor of paleontology at Yale in 1866. His uncle financed the Peabody Museum at Yale and provided funds for the many expeditions that Marsh sent to the West. Although Marsh was a reserved man, he was politically astute and became a powerful figure in the National Academy of Sciences (he served as president from 1883 to 1895). He was paleontologist of the U.S. Geological Survey from 1882 to 1892 and exploited this connection to boost his own privately financed expeditions. He engaged in a fierce feud with Edward Drinker Cope over access to western fossils and over the naming and description of species (see also the alphabetical entry on Cope in this volume). He died in 1899.

Marsh's teams discovered many new species of fossil mammals and reptiles.

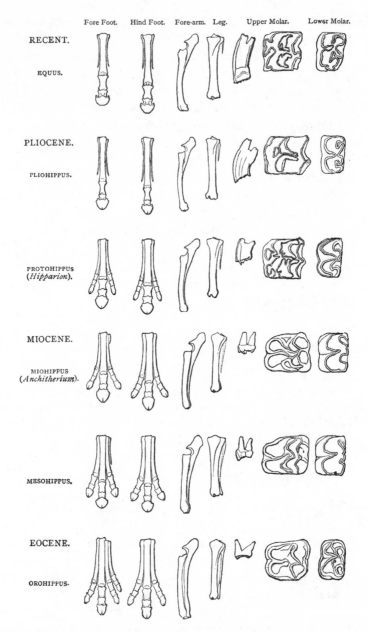

On a lecture tour to the United States, Thomas Henry Huxley was bowled over by Othniel Marsh's collection of horse fossils. To make the case for evolution in a public lecture, he used this reconstruction of horse evolution (something we now know to be much more branch-like), predicting that soon a five-toed ancestor would be discovered. To his delight, soon thereafter a specimen with rudimentary fifth toes was discovered. (Marsh called it *Eohippus*, "dawn horse." It was then discovered that specimens had been found earlier, although not identified as horses. The earlier name, *Hyracotherium*, takes precedence, although obviously Marsh's name has stuck.)

His collection of horse fossils (see figure) generated what appeared to be a continuous sequence of evolution from the small four-toed Eocene *Eohippus* to the modern horse, a sequence that Thomas Henry Huxley called "demonstrative evidence of evolution" (Huxley 1888, 90). The fossils seemed to indicate a trend of increased specialization by adaptation to running on the open plains, although the true story of the horse family turned out to be far more complex. Marsh also discovered toothed birds from the Cretaceous, which he described in a monograph, *Odontornithes* (1880). Like the better-known *Archaeopteryx* from Germany, these were hailed as important confirmation of the view that birds had evolved from reptiles. The head of one of Marsh's birds, *Ichthyornis,* was subsequently found to belong to a marine reptile. Marsh discovered and named a large number of dinosaurs and also studied the evolution of a group of giant early mammals he called Dinocerata. In his monograph on this group (1886) he traced the origin of the Ungulata back to a hypothetical Cretaceous ancestor. Like Huxley, he believed that the origin of orders and classes lay much further back in geological time than the earliest known fossils.

Marsh was an enthusiastic supporter of Darwinism, although he did not concern himself with the details of how evolution worked. He believed that natural selection would gradually favor the increasing size of the brain because intelligence was an important adaptive advantage. In his 1877 address *The Introduction and Succession of Vertebrate Life in America* he argued that paleontology provided clear proof of this adaptive trend toward increasing brain size in the course of evolution. Marsh's law of brain growth is a typical product of nineteenth-century progressionist evolutionism.

BIBLIOGRAPHY

Huxley, T. H. 1888. *American Addresses, with a Lecture on the Study of Biology.* New York: Appleton.
Marsh, O. C. 1877. *Introduction and Succession of Vertebrate Life in America: An Address Delivered before the American Association for the Advancement of Science at Nashville, Tenn., August 30, 1877.* [New Haven, CT: Tuttle, Morehouse & Taylor.]
———. 1880. *Odontornithes: A Monograph on the Extinct Toothed Birds of North America.* Washington, DC: Report of the Geological Exploration of the Fortieth Parallel, vol. 7.
———. 1886. *Dinocerata: A Monograph of an Extinct Order of Gigantic Mammals.* Washington, DC: Monographs of the U.S. Geological Survey, vol. 10.

—*P.J.B.*

Mass extinctions

Mass extinctions are global calamities of brief duration in which large numbers of species from diverse habitats are destroyed. They are thus distinct from more regional extinction events that often occur because of normal geological processes. For example, species that are restricted to a marine embayment or a lake may go extinct if their habitat is gradually infilled with sediment. In contrast, the global nature of mass extinctions requires some global-scale catastrophe, often involving the near shutdown of all photosynthetic activity. Such disasters require major changes of the earth's climate and ocean circulation system. The most famous of these global disasters is the giant meteorite impact that is thought to have terminated the reign of the dinosaurs 65 million years ago. This is known as the K-T event because it marks the boundary between the Cretaceous period, which is spelled with a K in German, and the following Tertiary. However, this is not the only mass extinction of the fossil record. Plots of extinction rates through time (see figure) reveal great variability, with five major peaks clearly seen and perhaps twice this number of smaller extinction events. The major peaks are known as the big 5 mass extinctions, with the K-T event being the most recent. However, by far the biggest mass extinction of all time was around 250 million years at the boundary between the Permian and Triassic periods. This P-Tr or end-Permian event saw the loss of more than 90% of all species. This catastrophe was as devastating on land as it was in the seas and oceans and, uniquely, was also a severe crisis for terrestrial plants. In other extinction events plants generally fared much better than animals.

CAUSES OF MASS EXTINCTIONS

The idea that mass extinctions have punctuated the history of life has been around for a long time. The reason that most mass-extinction events occur at major boundaries of geological time, such as between the Cretaceous and Tertiary, is that they mark major discontinuities in the history of life and thus are useful markers for the subdivision of geological time. These times were obvious even when such intervals were being defined in the early nineteenth century. However, the intensive study of the causes of mass extinctions is a much more recent science that can be dated to the publication of a study by Luis Alvarez, his son Walter, and two colleagues in 1980 (Alvarez et al. 1980). This group measured the concentrations of the exceedingly rare trace metal iridium in sediments that straddled the Cretaceous-Tertiary transition. For the most part this metal occurs in tiny amounts in the earth's crust, but the Alvarezes and colleagues found a dramatic enrichment precisely at the level of the mass extinction. They suggested that this iridium came from a chondritic meteorite (one of the most common types of meteorite) and got into the sediment from fallout in the aftermath of a giant impact. The iridium anomaly has now been detected in dozens of locations around the world, and this, together with the discovery of the impact crater in the Yucatán Peninsula of Mexico, has ensured that their hypothesis is now widely accepted. The crater is located near the village of Chicxulub and is buried beneath younger sediments. It may be as much as 180 km in diameter, making it one of the largest impact craters known on earth.

In the original 1980 article the link between the impact and the resultant mass extinction was thought to have come about from the global darkness that would follow such a large impact. The energy of such an event would pulverize both the meteorite and target rocks into dust and fine droplets of

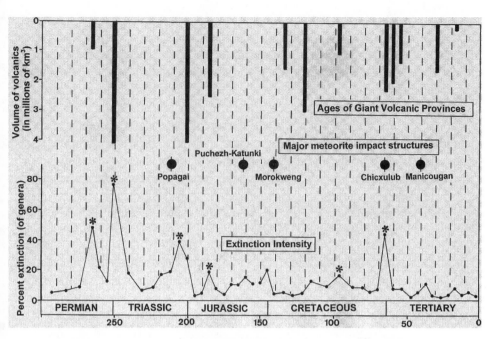

Comparison of extinction rates with the timing of large volcanic provinces and large meteorite craters over the past 300 million years. Note that all the recognized extinction events (marked with an asterisk) coincide with the eruption of large volcanic provinces; however, only the Chicxulub crater coincides with a mass-extinction event.

liquid and blast them into the outermost reaches of the atmosphere. There they would stay suspended in the upper atmosphere for several months and perhaps even years and thus block most sunlight. The consequences would be devastating for the photosynthesizers that form the base of the food chain in most ecosystems. The resultant cascade of extinctions would pass on up through the food chain to the higher consumers, such as *Tyrannosaurus rex,* which would die of starvation, not to mention the cold caused by the lack of sunlight. It appears that the magnificent dinosaur dynasty met a miserable end. This original kill scenario still holds sway with scientists, but with the addition of various other nasty effects. For example, the Chicxulub impact site is very unusual in that much calcium sulfate is present in the target strata. This would have been vaporized on impact and then rained back as a con- centrated type of acid rain (sulfuric acid). From the point of view of dinosaurs, this corner of Mexico was a particularly unfortunate place for a meteorite to hit.

The 1980s and 1990s saw much debate on the connection of the meteorite impact with extinction, and many scientists considered the K-T scenario ap- plicable to all extinction events. Some workers even saw a periodicity to ex- tinctions in which the K-T event was just one of several events that were spaced every 26 million years. A regular periodicity requires a regular expla- nation that is unlikely to come from random earthbound processes. Thus various astronomical scenarios were proposed whereby the earth was subjected to devastating meteorite and/or comet showers every 26 million years. However, subsequent study of the fossil record and improved age dating of the rocks has shown that several of the 26-million-year extinction events do not exist (for example, there was no extinction event 26 million years after the death of the dinosaurs). None of the extinction events that remain show a regular spacing. For example, there was a mass extinction 200 million years ago, known as the end-Triassic event, followed by a smaller-magnitude extinction event 20 million years later in the early Jurassic. The next extinction events did not occur until the Cretaceous, over 80 million years later, and these were all very minor affairs compared with the K-T event.

Attempts to link other extinction events with meteorite impacts have so far proved rather unconvincing, mainly because of a lack of evidence. For example, the Manicougan Crater of northern Ontario, Canada, is roughly 70% the size of the Chicxulub crater. It is one of the largest craters known but seems to have had little or no effect on extinction rates. The crater was formed 220 million years ago at a time marked by very low extinction rates. However, it is worth noting that a giant impact crater dated to the boundary of the Permian and the Triassic has recently been reported from the northwest Australian continental shelf, thus linking in time a big crater with the biggest mass extinction. However, the claims for this crater have been treated with skepticism by the general scientific community, which regards the structure as more likely to be of volcanic origin, and it remains to be seen whether these claims gain acceptance.

If the link between impacts and extinctions is generally weak, the same cannot be said for giant volcanic provinces. The surface of the earth is marked by several such provinces that are composed of layer upon layer of basalt lava flows that typically exceed 1 million cubic kilometers in volume. These erupted from long fissures or cracks, and individual flows often exceeded 1,000 cubic kilometers in volume, a figure that is two or three orders of magnitude larger than those seen during eruptions in historical times. Within the past 10 years improved age dating of these volcanic provinces has revealed two remarkable facts. First, most of the provinces appear to have erupted in less than 1 million years; a geological blink of an eye. Thus the environmental consequences of such eruptions would have been concentrated into a short interval. Second, many of the volcanic provinces have been found to precisely coincide with extinction events. Indeed, it now appears that every extinction event in the past 300 million years, including the mass extinctions, coincides with the eruption of a giant volcanic province. This includes the K-T event, when the Deccan Trap province of India was formed. However, not every giant volcanic province coincides with an extinction event. In other words, there are more volcanic provinces than there are extinctions. Obviously, it would be nice and neat if every province coincided with a mass extinction, but the history of the earth is definitely not neat.

How can volcanism cause extinction? Much research has focused on the climatic effects of the two major volcanic gases associated with the eruption of basalt lavas: carbon dioxide and sulfur dioxide. As is well known, carbon dioxide is a greenhouse gas, and the injection of large volumes into the atmosphere will cause global warming. In contrast, sulfur dioxide forms aerosols (by reaction with water vapor) that block sunlight and cause cooling. Thus the most noticeable effect of modern large volcanic eruptions, such as the eruption of Mount St. Helens in Washington State in 1980, is cooling of the climate. However, such effects are exceedingly short lived because the aerosols are rained out of the atmosphere in a year or so. Geological evidence shows that the eruption of giant volcanic provinces is nearly always associated with global warming, suggesting that the eruption of carbon dioxide may be the more important climatic effect of volcanism. However, it is one thing to raise global temperature by a few degrees and entirely another to wipe out nearly all life on earth. Therefore, many current extinction scenarios see the eruption of flood basalt provinces as just a trigger in a complex chain of events that leads to a runaway greenhouse scenario.

CONSEQUENCES OF MASS EXTINCTIONS

If the cause of mass extinctions is still actively debated, the consequences of mass extinctions are abundantly clear. They have long been recognized as important turning points in the evolution of life. An intriguing aspect of these events is that groups that were insignificant before the extinction rise to dominance in the aftermath. The quintessential example of this is the success of the mammals in the aftermath of the K-T extinction. Mammals had a long history before this mass extinction, but they spent this time literally and metaphorically living in the shadow of the dinosaurs. Only with the demise of the dinosaurs were mammals able to evolve into the larger sizes and diverse body forms we see today. Somewhat paradoxically, the initial success of the dinosaurs may also owe something to a mass-extinction event. Dinosaurs first appeared in the Triassic around 230 million years ago, but for the first 30 million years of their history they were generally only a relatively small component of terrestrial land animal communities, which included many other reptile-like groups. Only after the end-Triassic mass extinction 200 million years ago did dinosaurs quickly rise to total dominance of terrestrial communities.

In effect mass extinctions provide opportunities for the survivors to radiate into the vacated niches left behind by the extinct species. However, the pace of this evolutionary radiation is highly inconstant and varies in an interesting way from community to community. For example, there is a distinct difference in the rates of recovery of creatures that live in the water column of the oceans compared with those that live on the seafloor. Ammonoids were a highly successful group, distantly related to squid, that had attractive spiral shells. They suffered numerous extinction events but always recovered rapidly within a million years or so (except for the K-T extinction, which finally wiped them out). In contrast, clams that live on the seafloor often took tens of millions of years to recover from extinction events. The end-Permian extinction event also shows an interesting dichotomy in the recovery rates of plants and terrestrial vertebrates. Plant communities only began to radiate and return to normal around 10 million years after the extinction event, whereas the terrestrial vertebrates were diversifying rapidly within 1 million years of the extinction. In fact, the recovery interval after the end-Permian extinction was exceptionally long. Preextinction diversity levels were not achieved again until nearly 100 million years after the mass extinction. Indeed, for many seafloor communities the recovery did not even begin until 8 to 10 million years after the crisis.

The extreme length of time it has taken for earth's ecosystems to recover should serve as a warning for the current man-made mass-extinction event. Geological history reveals that the earth's ecosystems are able to recover from even the most severe environmental catastrophes, but the timetable of recovery is spread over millions of years, a time span that is immeasurably long from a human perspective. If the time taken for communities to recover from modern degradation is of a similar duration, then it is clear that it is best to stop the damage from happening in the first place.

BIBLIOGRAPHY

Alvarez, L. W., W. Alvarez, F. Asaro, and H. V. Michel. 1980. Extraterrestrial cause for the Cretaceous-Tertiary extinction: Experimental results and theoretical application. *Science* 208: 1095–1108.

Becker, L., R. J. Poreda, A. R. Basu, K. O. Pope, T. M. Harrison, C. Nicholson, and R. Iasky. 2004. Bedout: A possible end-Permian impact crater offshore of northwestern Australia. *Science* 304: 1469–1476.

Benton, M. J. 2003. *When Life Nearly Died: The Greatest Mass Extinction of All Time*. London: Thames and Hudson.

Courtillot, V. 1999. *Evolutionary Catastrophes: The Science of Mass Extinction*. Cambridge: Cambridge University Press.

Hallam, A. 2004. *Catastrophes and Lesser Calamities: The Causes of Mass Extinctions*. Oxford: Oxford University Press.

Hallam, A., and P. B. Wignall. 1997. *Mass Extinctions and Their Aftermath*. Oxford: Oxford University Press.

Looy, C. V., W. A. Brugman, D. L. Dilcher, and H. Visscher. 1999. The delayed resurgence of equatorial forests after the Permian-Triassic ecologic crisis. *Proceedings of the National Academy of Sciences USA* 96: 13857–13862.

Wignall, P. B. 1992. The day the world nearly died. *New Scientist,* January 25, 51–55.

———. 2001. Large igneous provinces and mass extinctions. *Earth-Science Reviews* 53: 1–33.   —P.B.W.

Maynard Smith, John (1920–2004)

John Maynard Smith was the leading British evolutionary biologist of the second half of the twentieth century. He developed a keen interest in natural history as a child, largely without any help from adults. As a schoolboy at Eton College he became acquainted with the writings of J. B. S. Haldane, largely because Haldane's outspoken Marxism was anathema to his teachers, whose job was to train the next generation of the British ruling class. After studying engineering at Cambridge University and doing war work in aircraft factories, he studied zoology at University College London (UCL), where Haldane was the Weldon Professor of Biometry. He became a postgraduate student of Haldane's and took up an appointment in the zoology department at UCL without taking his PhD. He left UCL in 1965 to become the first dean of the School of Biological Sciences at the new University of Sussex, where he remained for the rest of his life, although he was dean during only a small part of this time. As an administrator he was successful in developing a school with considerable research strengths, especially in experimental psychology, neurobiology, and population biology. For many years his own group at Sussex was a mecca for visitors and postdoctoral scholars from around the world.

Haldane exerted a lifelong influence on Maynard Smith. They were both renowned for the clarity of their lectures and writings, as well as the breadth of their knowledge and interests. Like Haldane, Maynard Smith was a Communist for many years, but he left the party after the Hungarian uprising of 1956 while remaining broadly left-wing in his political views. They were both active and successful communicators of science to the general public, in Maynard Smith's case through television as well as writing. But while Haldane was irascible and violent, Maynard Smith was kindly and beloved by his colleagues. He was a highly entertaining conversationalist and was exceptionally good at interacting with fellow scientists, irrespective of their seniority. He was remarkably open minded but never fooled by nonsense or pretentiousness.

He started his biological career in the 1950s by working on the genetics of *Drosophila subobscura,* which Haldane's group had developed as a tool for evolutionary studies. Apart from some basic work on genetic mapping, he made two notable contributions through this experimental work. He studied the effects of inbreeding on male mating behavior and reproductive success in *D. subobscura,* which caused him to recognize the significance of sexual selection by female choice of mates. This topic was largely ignored by most early twentieth-century evolutionary biologists, apart from R.A. Fisher in *The Genetical Theory of Natural Selection* (1930). He even anticipated the currently fashionable good-genes theory of the evolution of female mate choice. Maynard Smith also used *D. subobscura* as a model system for the study of aging. He provided an ingenious demonstration of the survival cost of reproduction and obtained evidence against the somatic mutation theory of aging. *Drosophila* is now a major tool for the biology of aging. Both of these lines of research were years ahead of their time and probably did not gain as much recognition as they deserved. He also did some pioneering work on the developmental genetics of *Drosophila,* stimulated by Curt Stern's concept of a prepattern and Alan Turing's theories of pattern formation by reaction-diffusion processes.

After moving to Sussex, Maynard Smith devoted his time to theoretical work, especially in evolution, and gave up doing experiments. This was partly because Haldane, by whom he always felt overshadowed, had died and partly because of the burden of building up a new department. His theoretical work was characterized by the use of rather simple mathematical methods that concealed a good deal of ingenuity in thinking up approaches to biologically significant problems. This sometimes met with disapproval from applied mathematicians, who often prefer to use complex methods to solve insignificant problems. Maynard Smith contributed importantly to early work on molecular variation and evolution, using the then newly proposed neutral theory as the basis for several important publications. The most influential of these is with his Sussex colleague John Haigh on genetic hitchhiking (Maynard Smith and Haigh 1974). This showed quantitatively how the spread of an advantageous mutation reduces variation at linked neutral loci, by dragging a chromosomal segment along with it. This concept has become very important in relation to the avalanche of data on natural variability at the level of DNA sequences, since a valley of reduced variability can be interpreted as the signal of a recent fixation of an advantageous mutation.

Another field to which he made important contributions was the evolution of sex and genetic systems, starting in the late 1960s. His work helped free this field from its domination by the rather woolly group-selectionist ideas of Cyril Darlington and George Ledyard Stebbins, which had had a severely negative effect on its development. In particular he emphasized the problem of the cost of sex, whereby an otherwise selectively neutral asexual variant arising in a sexual population experiences a transmission advantage and rapidly spreads to fixation. Maynard Smith was the first person to appreciate the difficulty this poses for explaining the prevalence of sexual reproduction among eukaryotes. He made numerous population-genetic models of processes that could provide an advantage to sex or increased recombination, and he summed up the state of the field in his 1978 book *The Evolution of Sex,* which is still the best survey of this subject as a whole. The study of the evolution of sex and breeding systems is now a flourishing discipline within evolutionary biology, in no small part because of Maynard Smith's own work and his encouragement of others.

Maynard Smith's most influential single contribution was the development of the concept of the *evolutionarily stable strategy* (ESS), initially in collaboration with the late George Price. This flowed out of his long-standing interest in animal behavior and his desire to understand why animal conflicts usually do not end in serious fighting. It uses the principle that, for a trait value to represent an equilibrium under natural selection, a necessary condition is that all possible deviant trait values are selectively disadvantageous when introduced at a low frequency into a population where most individuals have the specified trait value. Determining an ESS provides a powerful means of predicting the outcome of selection in cases where frequency-dependent fitnesses are generated by the biological context, such as sex ratios or social behavior. Although this approach had been used before for dealing with sex ratios, notably by R. A. Fisher and W. D. Hamilton, Maynard Smith developed ESS theory explicitly and applied it to many previously intractable evolutionary problems. It is one of the most scientifically fruitful applications of games theory. A huge theoretical and empirical literature has since developed, which applies ESS methods to many different biological problems. These methods are a mainstay of much of the theory that underlies behavioral ecology. Maynard Smith reviewed his contributions in his 1982 book *Evolution and the Theory of Games,* which is characteristically brief but lucid and informative.

After formal retirement in 1985, Maynard Smith started to collaborate with Brian Spratt's microbial genetics group, then at Sussex, on the analysis of data on molecular variation and evolution in bacteria. An important result of this work was evidence for much more exchange of genetic information among bacterial cells in nature than previously believed. It also generated new methods for examining the effects of infrequent recombinational exchange among members of bacterial populations on patterns of variation and evolution at the DNA-sequence level (Maynard Smith et al. 2000). Maynard Smith achieved the remarkable feat of becoming a leader in a new field well after establishing his reputation as an elder statesman of evolutionary biology.

Maynard Smith also made many important contributions to general evolutionary questions, including such topics as group selection versus kin selection (the latter term was invented by him), sympatric speciation, punctuated equilibrium, and the evolutionary role of developmental constraints. He also wrote a series of excellent textbooks on various aspects of theoretical biology, culminating in his 1989 book *Evolutionary Genetics.* Late in life he and Eörs Szathmáry developed a set of speculative ideas about the major transitions in evolution, from the evolution of life itself and the evolution of cells to the evolution of language. Maynard Smith suffered increasingly from the effects of mesothelioma in the last two years of his life but continued with his research until his death in April 2004. His last public lecture was a brief but characteristically stimulating talk on his bacterial work, at the December 2003 meeting of the UK Population Genetics Group in Sussex.

BIBLIOGRAPHY

Fisher, R.A. 1930. *The Genetical Theory of Natural Selection.* Oxford: Oxford University Press.

Maynard Smith, J. 1958a. The effects of temperature and of egg-laying on the longevity of *Drosophila subobscura. Journal of Genetics* 35: 832–842.

———. 1958b. Sexual selection. In S. A. Barnett, ed., *A Century of Darwin,* 230–244. London: Heinemann.

———. 1978. *The Evolution of Sex.* Cambridge: Cambridge University Press.

———. 1982. *Evolution and the Theory of Games.* Cambridge: Cambridge University Press.

———. 1989. *Evolutionary Genetics.* Oxford: Oxford University Press.

Maynard Smith, J., E. J. Feil, and N. H. Smith. 2000. Population structure and evolutionary dynamics of pathogenic bacteria. *BioEssays* 22: 1115–1122.

Maynard Smith, J., and J. Haigh. 1974. The hitch-hiking effect of a favourable gene. *Genetical Research* 23: 23–35.

Maynard Smith, J., and E. Szathmáry. 1995. *The Major Transitions in Evolution.* Oxford: Oxford University Press.

—B.C.

Mayr, Ernst Walter (1904–2005)

Ernst Walter Mayr was a German-born American ornithologist who played a central role in evolutionary theory and systematics during the twentieth century. He was a strong advocate of two ideas: the theory of speciation by geographic isolation and the Biological Species Concept. Mayr also made important contributions to biogeography, systematics, and the study of whole organisms during a period when biology as a profession became reductionist.

Mayr was born into a professional German family. His father died in 1917, and the family experienced considerable financial trouble during the 1920s. Mayr's interest in natural history started with his father. He began university in 1923, studying medicine at the University of Greifswald. Despite high marks, Mayr switched to zoology in 1925. He already had developed ties to professional zoologists, of which the most important were with Erwin Stresemann at the Museum of Natural History at the University of Berlin. Stresemann became Mayr's mentor. Under his guidance Mayr completed a PhD in 1926 on the European distribution and biogeography of the serin finch, *Serinus canaria serinus*. Afterward Stresemann hired Mayr as an assistant.

From 1928 to 1930 Mayr explored New Guinea and the Solomon Islands, collecting for several major museums. After his return Mayr sought a permanent museum position, but several promising offers in Europe failed to materialize. Rather than remain in Berlin, Mayr accepted temporary work at the American Museum of Natural History (AMNH) in New York. He began in January 1931. In 1932 Walter Rothschild sold his ornithological collection to the museum: 280,000 bird specimens, among other things. Mayr was given a permanent position to curate the new materials. This kept him in New York. He never undertook major fieldwork again.

When the National Socialists rose to power in Germany, Mayr chose not to return home. In 1935 he married Margarete (Gretel) Simon (1912–1990), and they had two daughters. Gretel played an important role in Mayr's career. She served as secretary and assistant and also looked after students and visiting colleagues. In 1950 both were naturalized as U.S. citizens.

Mayr continued at the AMNH until 1953, when he accepted an Agassiz Professorship at the Museum of Comparative Zoology (MCZ) at Harvard University. He served as director of the MCZ from 1961 to 1970. For Mayr the Harvard appointment was a symbolic change in status. He was now a full professor, like his mentor. Mayr became an emeritus professor at Harvard in 1975 and remained in Cambridge thereafter, with the MCZ as his base.

As an ornithologist Mayr specialized in birds of the southwest Pacific, about which he wrote many technical studies and several field guides. From the 1930s Mayr was active in the Linnaean Society of New York and contributed to supervising many doctoral students. He also made considerable contributions to several editions of James Peters's *Check-list of Birds of the World*. After Peters died in 1952, Mayr supervised the *Check-list*'s next edition. He also was active in many professional societies, such as the American Ornithologists' Union.

Mayr's contributions to evolutionary studies began in the late 1930s when he focused on the process of speciation and made contributions to general theory. *Systematics and the Origin of Species* (1942) received praise internationally and became a cornerstone in the synthetic theory of evolution. Mayr was part of a small group of biologists who thought that recent developments in population genetics and biogeography made possible a rigorous study of evolutionary processes. During the 1940s he concentrated on this as his specialty, working closely with Theodosius Dobzhansky, among others.

In approach Mayr emphasized geographic isolation and polytypic species. He passionately advocated the theory of allopatric speciation, which required that a population be physically isolated before divergence into a new species could occur. In this sense geographically distinct populations were species in the making. On the same lines Mayr championed the biological species concept. This placed a premium on interbreeding and gene flow in defining species. The biological species concept focused attention on isolating mechanisms and barriers to gene flow. Although Mayr thought that many isolating mechanisms existed as special cases, he considered geographic isolation the most important mechanism in nature. His book *Animal Species and Evolution* (1963) presented his views at their most developed. In addition to his empirical and intellectual contributions, Mayr played a crucial role in shaping evolutionary studies into a community. Largely on his own, Mayr organized the Society for the Study of Evolution (1946) and launched the journal *Evolution* (1947). Each required a great deal of effort, but they provided conversational spaces for evolutionary studies and helped develop a sense of common purpose and expectation.

Mayr's impact on systematics involved several levels. As a theoretician Mayr promoted evolutionary systematics together with George Simpson and Arthur Cain, among others. This had special relevance for taxonomic groups near the species level: subspecies, species, and species groups. The basic principle of this approach was to use classification to express evolutionary relationships and to use an understanding of evolutionary processes to make taxonomic decisions. For instance, the presence of geographic barriers was deemed sufficient to draw boundaries in classifications. Likewise, characters thought to be adaptations to local conditions were highlighted as the features that made a group distinct. This approach had its critics. In the 1960s numerical taxonomy and cladistics were important rivals.

Professionally Mayr worked to increase the status of systematics within biology. On the one hand, this required reform within the discipline. Mayr struggled to convince colleagues that they had a responsibility to keep up-to-date in biological subjects like genetics, ethology, and physiology. He also struggled to draw them into contributing to general biological subjects, including evolution. On the other hand, Mayr became politically active within biology in the late 1940s. While he was at the MCZ, he lobbied funding bodies and other groups so they would not ignore systematic biology when

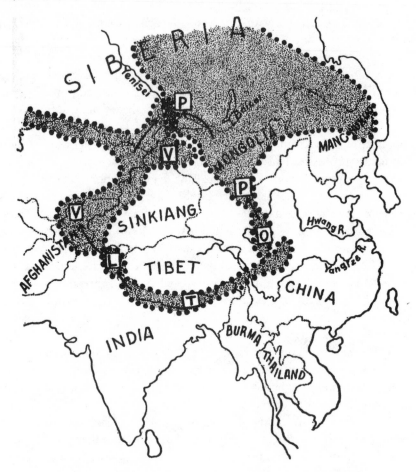

In his *Systematics and the Origin of Species* (1942), Ernst Mayr made much of such phenomena as "rings of races," where members of touching subspecies can interbreed but when the end populations circle back and touch, there is reproductive isolation. Here we see species in the making. If the middle populations (of the warbler *Phylloscopus trochiloides*) were eliminated, we would have two separate species. The subspecies in the ring are V=*viridanus;* L=*ludlowi;* T=*trochiloides;* O=*obscuratus;* and P=*plumbeitarus.* The cross-hatched area, in the district between the western Sayan Mountains and the Yenisei River, is where *viridanus* and *plumbeitarsus* overlap.

supporting the field. His vision for systematics is best presented in *Methods and Principles of Systematic Zoology* (1953), which became a standard training manual.

In the second half of his life, Mayr wrote extensively on the history and philosophy of biology, especially on themes related to natural history and evolution. His book *The Growth of Biological Thought* (1982) was a significant contribution to this topic. Darwin became an important subject for Mayr, as did many other key evolutionists of the past 100 years (e.g., Karl Jordan). Mayr's contributions as a historian centered on two themes. First, he gave credit to those who developed the ideas he himself defended. Second, he promoted the work of naturalist traditions in the history of biology, placing them on equal footing with other traditions in the life sciences, such as medicine and experimental biology.

As a philosopher Mayr is chiefly responsible for promoting the distinction between proximate and ultimate causes. By investigating the operation and interaction of structural elements, he said, experimental biology focused on proximate causes. They asked "how" questions, such as "How does this work?" In contrast, evolutionary biology sought ultimate causes, in other words, answers to "why" questions. As did his historical writing, the distinction between proximate and ultimate causes arose during conflicts in the 1950s and 1960s between evolutionary and molecular biologists.

BIBLIOGRAPHY

Bock, W. 1994. Ernst Mayr, naturalist: His contributions to systematics and evolution. *Biology and Philosophy* 9: 267–327.

———. 2004. Ernst Mayr at 100: A life inside and outside of ornithology. *The Auk* 121: 637–651.

Mayr, E. 1942. *Systematics and the Origin of Species*. New York: Columbia University Press.

———. 1953. *Methods and Principles of Systematic Zoology*. New York: McGraw-Hill.

———. 1963. *Animal Species and Evolution*. Cambridge, MA: Belknap Press.

———. 1982. *The Growth of Biological Thought*. Cambridge, MA: Belknap Press.

Stresemann, E. 1975. *Ornithology: From Aristotle to the Present*. Cambridge, MA: Harvard University Press.   —J.C.

Meme

The term *meme* was coined by Richard Dawkins in his 1976 book *The Selfish Gene* to describe units of cultural evolution (as opposed to genes, which are units of biological evolution). In the intervening years the term has been popularized, as evidenced by its inclusion in *The Oxford English Dictionary*, the spawning of a dedicated (and now-defunct) journal, and several monographs. Memes can be such things as songs, tools, styles of dress, religious beliefs, hairstyles, aphorisms, and myths. Dawkins originally employed memes to serve as an instance of "Universal Darwinism"—the thought that any system with discrete units that exhibit differential fitness values (variation), selection, and replication (heredity) will evolve by Darwinian means. Like organisms, ideas appear to exhibit these characteristics. We are continually inundated with ideas. Thankfully, most of this information is discarded (otherwise we would have memory overload). Thus ideas are in competition with each other for survival within our minds; the fit ones reside in our minds and get replicated, while the unfit ones get discarded. Replication occurs through imitation, be it through the written word, demonstration, or discussions at the dinner table.

It is important not to confuse memes, which are ideas, with the referents of the ideas. For example, the meme *arrowhead sharpening* is the *idea* of a manner of producing an arrowhead, not the arrowhead itself. This distinction preserves the genotype/phenotype dichotomy within biology, or, in Dawkins's terminology, the difference between replicators and vehicles. Replicators are the entities that copy themselves and code for structures that interact in the world (some theorists prefer the term *interactor* to *vehicle*). This raises several points that distinguish memetic evolution from competing conceptions of cultural evolution. Cultural evolution is often seen as simply improving the fitness of the group or group members. For example, fishhooks and arrowheads evolved to convey a fitness advantage to the groups and individuals who used such artifacts. Memes, however, can be quite successful without conveying a fitness advantage to their hosts, even to the point of being deleterious to their vectors. Dawkins famously claims that religion is a particularly successful but toxic meme. Memes are fit when they have high fidelity and high fecundity unto themselves. According to proponents, memes hold implications for our conception of culture and human minds.

The success of the *meme* meme, however, belies its success as a term of art in the academy. Critics have identified several problems with memes, not the least of which is that it is simply recasting an old idea in new packaging. Perhaps it is not the memes themselves that are explanatory but the underlying psychological mechanisms responsible for the selection and replication of certain memes that are useful for explaining cultural evolution. Furthermore, it is unclear exactly what a meme is. It is easy to come up with examples of memes, but a clear and accepted definition remains elusive. For example, is a meme a part of a Beethoven symphony or the entire symphony? Until an operational definition is found, the fruitfulness of this concept remains questionable. Finally, some have argued that the lack of fidelity of memes, as well as their seemingly nondigital nature, suggests a Lamarckian rather than a Darwinian evolutionary process. The true impact of memes for understanding cultural evolution remains largely undecided.

BIBLIOGRAPHY

Blackmore, S. 1999. *The Meme Machine*. New York: Oxford University Press.
Dawkins, R. 1976. *The Selfish Gene*. Oxford: Oxford University Press.
Dennett, D. 1995. *Darwin's Dangerous Idea*. London: Penguin.   —J.Z.

Mendel, Gregor (1822–1884)

Gregor Johann Mendel has been called "the priest who held the key to evolution" (Eiseley 1959, 205), but in his own day few biologists had ever heard of him, Charles Darwin included. His family was poor, but he sought support in a monastery where fortune shone upon him by giving him the means to carry out his remarkable and ambitious experiments.

Mendel, the firstborn of Anton Mendel and Rosine Mendel (née Schwirtlich), was raised in German-speaking Silesia in the little village of Heinzendorf. At 11 he attended schools far from home, and at 18 he entered the Philosophical Institute in Olmütz, completing his courses there in 1843.

The strain of study plus tutoring to provide his own funds had meanwhile caused bouts of severe depression and extended time back home to recover. This experience lay behind his decision to apply to the Augustinian Monastery of Saint Thomas in Brünn. He knew, he explained, that he needed to be freed "from the bitter struggle for existence" that had been his lot hitherto (Iltis 1954).

At the monastery Mendel proved unable to cope emotionally with the often-sad tasks of a priest, but his abbot, the understanding Cyril Napp, found him work as a substitute teacher in Znaim. Teacher certification, however, required an examination. This he took twice but through stress failed both times. After the first attempt the abbot sent Mendel to the University of Vienna to obtain adequate foundations in the sciences. There he assisted in the practical demonstrations for the physics courses and was exposed to the most recent work in plant cytology.

Mendel's teacher, the botanist Franz Unger, was a victim of attacks in Vienna from Catholic quarters for his advocacy of species transmutation during Mendel's time in Vienna (1851–1853) and subsequently. Mendel must have known of this and of Unger's transplant experiments in which characters acquired by plants in novel habitats were lost on returning to their natural locales. Later Mendel repeated such experiments and found in them no support for the inheritance of acquired characters, thus ruling out Lamarckian transmutations. Species multiplication by hybridization, however, was a possibility mentioned by Unger, for he knew the work of the German botanists Joseph Kölreuter and Carl von Gärtner, who had hybridized plants, seeking to establish whether species remain constant or can change through hybridization.

Having returned from Vienna in 1853, Mendel began the following year to test varieties of the edible pea for constancy of type in preparation for the hybridization experiments that occupied him between 1856 and 1862. His aim, he explained, was to "determine the number of different forms in which hybrid progeny appear, permit classification of these forms in each generation with certainty, and ascertain their numerical interrelationships." To undertake this program he chose varietal differences advertised by nonblending character differences that could be clearly distinguished. Aware of the need to achieve statistically significant results, he grew large populations. The task was "far-reaching," but he warned that it was the only way to solve the important question: "the developmental history [*Entwicklungsgeschichte*] of organic forms" (Stern and Sherwood 1966, 2). If some hybrids bred true, they could form the basis for the development of new species, but if they returned or "reverted" to the originating forms, no novelty would result.

Evidently this was the question he set out to solve in 1854 when he began testing his research material. Charles Darwin's theory was at that time unpublished. Not until 1862 did Mendel read the German translation of *On the Origin of Species*. Mendel's most frequent marginalia therein refer to passages presenting and opposing the doctrine of special creation. Here Darwin was attacking a target that Mendel, it appears, did not accept either.

The context of Mendel's famous experiments was thus not Darwin's theory of evolution. It was instead the debate among German-speaking biologists, several of whom, like Unger, had espoused elements of the speculative teaching of *Naturphilosophie,* principally the notion of the successive development of higher and higher forms of life over an extended period of time. One of the Augustinians in Brünn, Matouš Klácel, also supported a developmental view of species origins, but unlike his friend Mendel, Klácel was a speculative philosopher who never carried out a rigorous series of experiments. Mendel was drawn chiefly to the experimental literature of the German hybridists. His chief resource was Carl von Gärtner, who, in his book *Experiments and Observations of Hybridization in the Plant Kingdom* (*Versuche und Beobachtungen über die Bastarderzeugung im Pflanzenreich,* 1849), found that fertile hybrids yielded progeny that tended sooner or later to revert to one or the other of the originating species. Gärtner listed only a few hybrids that appeared to remain constant in their offspring, and it is significant that Mendel reproduced many of these hybrids and found that they did revert. Mendel went on to choose hawkweeds *(Hieracium)* because of the multitude of forms found in the wild. Although the botanist Carl Nägeli encouraged him in this direction, it was not he who persuaded Mendel to choose this difficult genus, as is often asserted. Mendel reasoned that polymorphic genera might owe their multitudes of species to hybridization, but of a different kind from that he had found in his classic experiments with the edible pea *(Pisum).*

Mendel explained the results of his *Pisum* experiments by suggesting that there must be a process of segregation (see figure) of the differing potentials

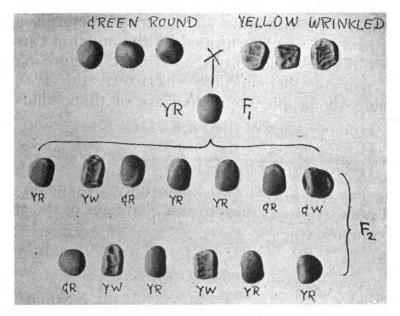

William Bateson was Mendel's mouthpiece in Britain in the early years of the twentieth century. This diagram shows his representation of Mendel's first law, the law of segregation. (From Bateson 1909.)

present in the hybrid plant when the germ cells are formed such that germ cells are either of one kind or the other, not mixed. Then if in fertilization all combinations of pollen cells and egg cells are possible, populations of progeny of the two kinds will be yielded that will approximate to the statistical ratios that he found in his experiments. Where unions are between pollen and egg cells of the same kind, both originating forms are recovered pure. Thus the old idea of once tainted, always tainted was wrong. The assumption that a new variant will simply be diluted by reproduction with the normal form was also wrong. New combinations of characters can thus persist. This was directly relevant to the problem of the *swamping effect* of outbreeding that Darwin had found daunting for his theory. Mendel's concept of germinal segregation is thus seen in retrospect as the key that Darwin lacked.

Also from segregation flowed the possibility of combining hereditary traits in all sorts of combinations and demonstrating that the traits behave independently of one another. Mendel identified the source of hereditary variation with this process of *recombination*. Therefore, he objected to Darwin's claim that domestic animals and cultivated plants are more variable than the corresponding wild species because of the effect of "changed conditions of life." Instead, he opted for the greater opportunities for cross-pollinations.

The acceptance of Mendel's work in 1900 came along with an uncritical adherence to nonblending characters, thus promoting skepticism toward the creative role of natural selection. If new characters are produced in one nonblending step, there is no space for natural selection to accumulate slight differences little by little. This materially delayed the integration of Mendelian genetics into Darwinian theory.

BIBLIOGRAPHY

Bateson, W. 1909. *Mendel's Principles of Heredity*. Cambridge: Cambridge University Press.

Eiseley, L. 1959. *Darwin's Century: Evolution and the Men Who Discovered It*. London: Victor Gollancz.

Gärtner, C. von. 1849. *Versuche und Beobachtungen über die Bastarderzeugung im Pflanzenreich*. Stuttgart: K. F. Herring.

Iltis, Mrs. H. 1954. Gregor Mendel's autobiography. *Journal of Heredity* 45: 231–234. Reprinted in R. Olby, *Origins of Mendelism*, 2nd ed., 196–198. Chicago: University of Chicago Press, 1985.

Olby, R. 1985. *Origins of Mendelism*. 2nd ed. Chicago: University of Chicago Press.

Orel, V. 1996. *Gregor Mendel: The First Geneticist*. Oxford: Oxford University Press.

Stern, C., and E. Sherwood, eds. 1966. *The Origin of Genetics: A Mendel Source Book*. San Francisco: W. H. Freeman.

Weiling, F. 1991. Historical study: Johann Gregor Mendel, 1822–1884. *Journal of Medical Genetics* 40: 1–25.

www.Mendelweb.org. Includes the English text of Mendel's 1865 paper on *Pisum* in English and German and "Essays and Commentary," which contains, among others, the following essays: Olby, R. 1997. Mendel, Mendelism and genetics; Paul, D. B., and B. A. Kimmelman. 1988. Mendel in America: Theory and practice, 1900–1919; and Sapp, J. 1990. The nine lives of Gregor Mendel.

—R.O.

Michener, Charles D. (b. 1918)

Charles Michener, entomologist at the University of Kansas, is the world's leading expert on the classification, morphology, natural history, and evolution of bees, especially the diverse solitary and primitively social bees whose biology contrasts with that of the familiar honeybee *(Apis mellifera)*. Michener has used his broad knowledge of these insects to test ideas about the evolution of insect societies and to help modernize the science of taxonomy. He and his many students have traveled extensively in the United States, Africa, Asia, Europe, and Latin America to collect and observe bees in their natural habitats. As a result of this work the bees have become a model taxon for comparative studies of evolution.

Michener's research is a premier example of how evolutionary biology uses comparative studies of behavior and morphology to understand the origin of new traits and the diversification of species. Because of Michener's work and his encouragement of young bee researchers, thousands of species of bees can be identified with certainty. This satisfies the first requirement of comparative research: ability to identify the organisms. In addition, Michener's work treats the phylogenetic relationships of different groups of bees, summarizes knowledge of their biology, and explains how to rear and observe them in the laboratory.

Michener's more than 400 publications, including two major books on bees, are a veritable encyclopedia of information that has been exploited by Michener and others to trace the evolution of such phenomena as parasitism and the transition from solitary to social in group-living insects. His taxonomic insight has made him a leader in the theory and practice of systematics, the study of classification and relationships among groups of organisms. At the age of 81, Michener published his magnum opus *The Bees of the World* (2000), a 913-page tome that treats 1,200 genera and subgenera and more than 16,000 species in a unified classification and summarizes much of the current information on this important group, insects that are essential to ecological and human communities as pollinators of flowering plants, including many crops.

Michener's interest in bees began when as a boy he began to collect insects near his home in southern California. Encouraged by his mother, an amateur naturalist, and by P. H. Timberlake, an authority on bee taxonomy, he published his first essay on bees at the age of 16. His contagious enthusiasm for bees has inspired a large number of students and colleagues and has put the University of Kansas on the map as a center for bee research and scientific leadership. Michener also wrote important essays on the taxonomy of butterflies, although these are now less known than his work on bees, and he was at the center of controversy during debates over numerical taxonomy in the 1950s and 1960s. Along with Robert Sokal, whose views were somewhat more radical, Michener advocated the use of objective, quantitative data in classifications *(numerical taxonomy)* as an antidote to some of the overly intuitive and less scientific procedures of the time. But he also believed that quantitative methods "must not be allowed to separate the systematist from the organisms themselves for only by knowing their behavior, ecological relationships, the functions of their various structures, etc., can he understand and interpret taxonomic data" (Michener 1963, 170). In an account of the numerical-taxonomy wars, Hull (1988, 118) described Michener as "a quiet, gentle man . . . who does not enjoy participating in raucous polemics . . . [but] in scientific combat [he is] also highly effective, some might say deadly."

Michener's cordial, unassuming manner has made him an effective statesman of science, especially in countries like Japan, Mexico, and Brazil, where he has conducted fieldwork and promoted entomological research; and among entomologists in museums and laboratories throughout the world. For many years Michener was the only Kansas member of the National Academy of Sciences, one of only two Kansans to have joined the academy in its 184-year history when he was elected in 1965. In 1967 he served as president of the Society for the Study of Evolution. Michener's career shows the value, for hypothesis testing and for the understanding of biological diversity, of deep and broad knowledge of a particular group of organisms.

BIBLIOGRAPHY

Hull, D. L. 1988. *Science as a Process.* Chicago: University of Chicago Press.
Michener, C. D. 1963. Some future developments in taxonomy. *Systematic Zoology*
   12: 151–172.
———. 1974. *The Social Bees.* Cambridge, MA: Harvard University Press.
———. 2000. *The Bees of the World.* Baltimore: Johns Hopkins University Press.
West-Eberhard, M. J. 2001. The importance of taxon-centered research in biology.
   In M. J. Ryan, ed., *Anuran Communication,* 3–7. Washington, DC:
   Smithsonian Institution Press.   —M.J.W.-E.

Miller, Stanley L. (1930–2007)

Stanley L. Miller spent most of his career as professor of chemistry at the University of California, San Diego. In the early 1950s he performed a series of breakthrough experiments that signaled the beginning of the systematic research of prebiotic chemistry and the emergence of life. From then on, together with many colleagues and students, Miller investigated the possible steps and chemical compounds involved in the emergence of life. As an acknowledgment of his eminent contribution to the establishment of the origin-of-life field, in 1973 he was the first scientist working in this area to be elected to the National Academy of Sciences. In recognition of his achievements and his role in the creation of the International Society for the Study of the Origins of Life (ISSOL), the society established after Miller's death the Stanley L. Miller Award to be presented to young outstanding origin-of-life scientists.

In 1952–1953, working as a graduate student in the laboratory of chemist and Nobel laureate Harold Urey at the University of Chicago, Miller attempted to simulate the conditions conductive to life on the primordial earth. In these experiments he was testing for the first time the Oparin-Haldane hypothesis on the emergence of life formulated in the 1930s (see the alphabetical entry "Aleksandr I. Oparin" in this volume). Miller built a glass apparatus that contained in separate flasks connected by tubes a mixture of gases mimicking the early hydrogen-rich (reducing) atmosphere and water representing the ocean. As an energy source he employed electric discharges that stood for lightning. After running the experiment for a week, Miller analyzed the aqueous solution—the "ocean"—for the dissolved products of the interacting atmosphere's constituents. He found abundant synthesis of organic compounds, including amino acids, the building blocks of proteins. Significantly, the results were not statistically random: Only a small number of biologically relevant organic compounds were produced in high yields in the experiment out of the vast number of potentially relevant organic substances. Glycine and alanine, the most common amino acids in proteins, were the major products. In work performed by Miller and other researchers in the 1970s, it was found that the same amino acids and in the same relative quantities were obtained when the content of meteorites that reached earth was analyzed. Since meteorites are considered relics of the formation of the solar planets, their organic content may indicate that chemical processes similar to the Miller reactions were common in the solar system.

Similar experiments by other chemists demonstrated the synthesis under prebiotic conditions of several constituents of nucleic acids, for example, the nitrogen base adenine. More recently the existence of a primordial hydrogen-rich atmosphere that enabled organic synthesis has been questioned, and many researchers postulate instead that organics were imported to earth from space by comets, meteorites, and dust particles. Miller nevertheless believed that conditions for organic synthesis on earth might have existed. Using a more potent high-energy source, Miller succeeded in demonstrating, beginning in the 1980s, the synthesis of amino acids and nucleic acid bases in a gas mixture that was less reducing than his 1953 "atmosphere."

Miller was a promoter of the genetic conception of the origin of life, according to which living systems were first and foremost self-replicating molecular entities. Throughout his career the most fundamental question that preoccupied him was the nature of the first genetic material that could have replicated, mutated, and evolved to form a primitive living system. Miller shared the conviction of many scientists in the field that a watershed in the emergence of life was an RNA world, a chemical world in which RNA molecules functioned both as genetic material and as catalysts. The idea that RNA molecules could have fulfilled this dual role very early in the evolution of life was suggested independently in the 1960s by Leslie Orgel, Francis Crick, and Carl Woese. The more radical claim that RNA was the sole player in the very emergence of life became popular in the 1980s after the surprising discovery in present-day cells of ribozymes. These are RNA molecules that in addition to their genetic role also act as enzymes, thus carrying out the function that until then was known to be accomplished only by proteins.

On the basis of numerous experiments, however, Miller, as well as his colleague Leslie Orgel and other researchers, became convinced that life could not have begun with the RNA world. They realized that the prebiotic synthesis of an RNA sequence that could replicate itself was highly improbable.

Consequently, Miller's research from the 1990s until his death was devoted to the search for the constituents of a pre-RNA world that were easier to synthesize and that could have evolved into the constituents of the RNA world. Several such candidates were suggested by Miller and his group.

It is important to note that from its inception Miller's research on the origin of life was tightly linked to the wider context of the search for life on other planets. After the founding of the National Aeronautics and Space Administration (NASA) in 1958, the young Miller was among the main contributors to the establishment of the field of exobiology, later to be called astrobiology. Several of Miller's later experiments lent credence to the hypothesis that the potential for prebiotic chemistry exists beyond earth, perhaps under conditions prevalent on one of Jupiter's moons, Europa, and possibly on other Jovian moons.

In order to discredit the study of the emergence of life, creationists often rely on the empirical debates among origin-of-life scientists over specific prebiotic scenarios and on the many theoretical and empirical difficulties still faced by researchers. Miller, who strongly denounced creationists for promoting ignorance about science, never doubted that a chemical solution to the problem of the emergence of life will eventually be found.

BIBLIOGRAPHY

Fry, I. 2008. Miller, Stanley Lloyd. In N. Koertge, ed., *New Dictionary of Scientific Biography*, vol. 5, 154–158. Detroit: Charles Scribner's Sons.

Miller, S. L. 1953. A production of amino acids under possible primitive earth conditions. *Science* 117: 528–529.   —I.F.

Mimicry

Mimicry occurs when the features of one species resemble those of another and, through that resemblance, confer some advantage on the mimic. Mimicry is a striking phenomenon in nature; its study has played an important role in evolutionary biology from the earliest arguments about natural selection, and several challenging puzzles about the evolution of complex mimicry systems remain to be solved.

Mimicry is rampant in nature (see Wickler 1968; Randall 2005). In its simplest form, mimicry occurs when a particular feature of one organism resembles a very different organism. The most familiar examples are cases of aggressive mimicry, where a predator lures its prey with a part of its own body that resembles the food of the prey. The fleshy red protuberance on the tongue of an alligator snapping turtle mimics a bloodworm and lures bloodworm-eating fish into the turtle's mouth. The lures of anglerfish and batfish are other well-known examples. In these cases, the mimetic feature confers a clear benefit to its owner and the evolution of the feature creates no disadvantage for the organism being imitated.

The more complicated form of mimicry occurs when one organism resembles a similar one, as in one butterfly resembling another, but with the twist that the resemblance offers a benefit to the mimic but a potential disadvantage to the model (the organism being imitated). This happens when model and mimic have very different characteristics or ecological roles and the success of the mimic threatens the model. For example, on coral reefs in the Pacific, small fish called "cleaner wrasses" find food by picking crustacean ectoparasites from the scales of larger fish, which visit specific spots on the reef where the wrasses occur. There is another species of small fish, a blenny, that mimics the shape and coloration of the cleaner wrasse but, rather than pick parasites from the larger fish, takes small bits from their tail fins. The blenny's mimicry of the wrasse allows it to close in on its target when it might otherwise be eaten itself by the larger reef fish. But one might expect the annoyance of a reef fish at having a piece removed from its tail will provoke less tolerance of the model, the cleaner wrasse, and imperil the wrasse's ability to gain the easy meal. To understand the persistence of this mimicry system, we must understand the reciprocal effects of model and mimic on each other's fitness and the conditions under which those effects are balanced.

The most challenging complex mimicry systems are the hundreds of cases in which dangerous or toxic animals are imitated by harmless or palatable ones. In each case the harmless or palatable species gains its advantage through the ability of other animals to learn the association between a visual or olfactory cue with the dangerous animal and then associate that cue with the harmless mimic and leave it alone.

The most well-studied cases are those involving palatable butterflies that resemble unpalatable, toxic ones. Henry Bates (1862) brought this phenomenon to the attention of biologists when he described how certain Amazonian butterflies in the family Pieridae bore a striking resemblance to unrelated butterflies in the family Heliconiidae. Bates had noticed that the heliconiid butterflies flew languidly yet were rarely attacked by birds, while the pierids flew more furtively and elusively, as if avoiding exposure to birds. The heliconiids are, in fact, toxic and unpalatable, and the pierids are nontoxic and palatable. Bates described this system as a dramatic demonstration of the power of natural selection, pointing out the advantage gained by an individual of a palatable species whose features resembled those of a noxious species that would be avoided by predators.

Inspired by Bates's observations and suggestions, evolutionary biologists from Charles Darwin onward plunged into studying the nature and evolution of this type of mimicry. With greater scrutiny came the appreciation that mimicry is a complicated phenomenon; as a result, the study of mimicry came to occupy a central position in evolutionary biology as a model system for understanding complicated forms of natural selection that could mold complex patterns from simple genetic components.

Complex mimetic systems can be classified into three types. In Batesian mimicry a harmless or palatable species resembles a dangerous or unpalatable, toxic one. It is not difficult to understand how the predator learns to avoid a toxic butterfly; a predator that eats one becomes ill and vomits, after which taste aversion leads the predator to avoid prey that resemble the offending food item. It is also not difficult to see Bates's argument that a mutant of a palatable species that resembles an unpalatable one will have higher fitness through a reduced risk of predation. If a better resemblance confers a lower risk of predation, natural selection will refine that resemblance until mimic and model are indistinguishable.

Despite the simplicity of the basic arguments, Batesian mimicry raises several important questions for which we have only partial answers. First, Batesian mimics must remain rarer than their models lest predators encounter more mimics than models and associate the shared pattern with palatability rather than the reverse. The precise relationship expected among the abundances of model and mimic is unclear, especially when a single species serves as the model for several mimetic species in the same location (e.g., the pipevine swallowtail of the southeastern United States and its mimics).

Second, if Batesian mimics must resemble their models closely, then the genes that control pattern must be inherited together so that the genetic instructions for the pattern are not scattered during sexual reproduction (see the alphabetical entry "Evolution of sex" in this volume). Research has shown that those genes are inherited as a block, a process facilitated in butterflies by the fact that there is no crossing-over in females. The initial studies of inheritance of mimetic patterns suggested that they were controlled by single genes of large effect, a finding used by some scientists to argue that such genes were obviously the important raw material for evolution. This contrasted with the Darwinian view that the important raw material for evolution was the cumulative impact of many genes with small individual effects. This dispute was clarified only when detailed studies of inheritance revealed the concerted inheritance of the several genes that controlled the basic mimetic patterns and the effects of other genes that modified the expression of these controlling genes (see Nijhout 2003 for a technical discussion). Although new research has revealed the location of these genes in some species (Clark et al. 2008), the evolutionary pathway that brought those genes into a closely linked complex that is inherited as a block remains to be described.

The description of such an evolutionary pathway is even more complicated when different populations of the same mimetic species imitate very different-looking models in different geographic locations. This is an especially striking pattern within the heliconiid butterflies, and new research into this pattern is starting to reveal these pathways (Kronforst and Gilbert 2008).

Third, the study of Batesian mimicry led quickly to the question of how unpalatability and toxicity could evolve, along with a specific visual signal that a predator would remember. If a mutant that is unpalatable and toxic must die so that the predator can learn to avoid other individuals with the same appearance and initial taste, it is unclear how the mutant gene can spread via the classic Darwinian selection among individuals. Fisher suggested a kin-selection argument based on the supposedly gregarious nature of many toxic butterflies, but recent research on phylogenetics (Beltran et al. 2007 and others cited therein) has indicated that gregariousness evolved after toxicity and warning coloration in many of these species, and so the kin-selection argument cannot be correct. Other hypotheses have been suggested, but the problem remains unsolved.

The second type of complex mimicry is called Müllerian mimicry, which describes the resemblance of two or more unpalatable or dangerous species to each other. This pattern was also described by Bates, who observed that many tropical butterfly species formed mimicry rings in which a group of species exhibited a common mimetic pattern. Further, Bates observed that such patterns changed in wholesale fashion from one geographic area to another. Several species within the genus *Heliconius* in the American tropics offer one of the best-documented cases of a mimicry ring. Müllerian mimicry and mimicry rings pose a number of perplexing questions. The advantage of a mutual resemblance of unpalatable species is clear at first; if a certain number of individuals must die to teach predators to avoid a pattern, then if that number is divided between two species, the risk of predation is lower for individuals of each species, and a mutual resemblance will evolve. Of course, the problem of how toxicity evolves in the first place still exists because this argument begins once two species are each toxic. And although many explanations have been suggested, it remains unclear why mimicry rings change their patterns abruptly every few hundred miles.

The third type of complex mimicry is the most puzzling. In these cases all the species involved are toxic or dangerous but they vary substantially in that noxiousness. This makes the rings difficult to describe as either purely Batesian complexes or purely Müllerian rings. The most spectacular of these cases might be the mimicry group that includes the deadly coral snakes of tropical America and a series of mildly venomous species that share the brightly banded color pattern. The difficulty in explaining this pattern is that a bite from the extremely deadly coral snake kills most would-be predators, and dead predators cannot learn to avoid what killed them.

Two explanations have been offered for this type of mimicry ring. In the first explanation, called Mertensian mimicry, the mildly noxious species is in fact the model from which predators learn and the deadly species is the mimic, the resemblance being favored by the fact that attack rates on the deadly animal decrease as predators learn from the less deadly encounters. The second explanation, called quasi-Batesian mimicry, reverses model and mimic and postulates that the mildly noxious species mimics the highly noxious one. This hypothesis was developed for butterfly rings in which species varied in abundance, as well as toxicity, the mildly toxic species being the rarer. The relative abundances resemble the arguments about model and mimic abundances that emerge from purely Batesian mimicry, hence the name *quasi-Batesian*. The mechanisms posited for quasi-Batesian mimicry involve very specific arguments about how animals learn, including arguments about whether learning is permanent and how long learning lasts without reinforcement (see Mallet and Joron 1999 for a deeper discussion). These opposing explanations remain to be resolved (Mallet 2001).

The study of mimicry epitomizes evolutionary biology as a discipline. Simple mimicry is readily observed by even a casual natural historian and begs the question of its origin and maintenance. Complex mimicry systems inspire deeper questions, which have provoked nearly a century and a half of sophisticated research, from Bates's original observations to modern work in genetics, development, ecology, neurobiology, and mathematical and computational modeling. While the discovery of mimicry is an old story, understanding its evolution is very much a modern one that continues to unfold.

BIBLIOGRAPHY

Bates, H. W. 1862. Contributions to an insect fauna of the Amazon Valley. *Transactions of the Linnaean Society of London* 23: 495–566.

Beltran, M., C. D. Jiggins, A. V. Z. Brower, E. Bermingham, and J. Mallet. 2007. Do pollen feeding, pupal-mating and larval gregariousness have a single origin in *Heliconius* butterflies? Inferences from multilocus DNA sequence data. *Biological Journal of the Linnean Society* 92: 221–239.

Clark, R., S. M. Brown, S. C. Collins, C. D. Jiggins, D. G. Heckel, and A. P. Vogler. 2008. Colour pattern specification in the Mocker swallowtail *Papilio dardanus*: The transcription factor invected is a candidate for the mimicry locus H. *Proceedings of the Royal Society of London, Series B* 275: 1181–1188.

Kronforst, M. R., and L. E. Gilbert. 2008. The population genetics of mimetic diversity in *Heliconius* butterflies. *Proceedings of the Royal Society of London, Series B* 275: 493–500.

Mallet, J. 2001. Causes and consequences of a lack of coevolution in Müllerian mimicry. *Evolutionary Ecology* 13: 777–806.

Mallet, J., and M. Joron. 1999. Evolution of diversity in warning color and mimicry: Polymorphisms, shifting balance, and speciation. *Annual Review of Ecology and Systematics* 30: 201–233.

Nijhout, H. F. 2003. Polymorphic mimicry in *Papilio dardanus*: Mosaic dominance, big effects, and origins. *Evolution and Development* 5: 579–592.

Randall, J. E. 2005. A review of mimicry in marine fishes. *Zoological Studies* 44: 299–328.

Wickler, W. 1968. *Mimicry*. London: Weidenfeld and Nicolson.   —*J.T.*

Modern reptiles

Reptiles constitute one of the most morphologically and ecologically diverse groups of tetrapod (four-legged) vertebrates. The roughly 16,000 species of living reptiles range in body size from tiny lizards and snakes that weigh less than 1 gram to saltwater crocodiles that range up to nearly 7 meters in length and over 1,100 kilograms in weight. The living reptiles include species that are live-bearing and egg-laying, ectothermic (cold-blooded) and endothermic (warm-blooded), marine and terrestrial, herbivorous and predaceous—in short, most of the ecological, physiological, and morphological variation found in tetrapods. Some of this variation can be extremely labile evolutionarily— examples abound of pairs of closely related lizard species where one is live-bearing and the other egg-laying, for example. One of the most important controversies in reptilian evolutionary biology is exactly what to include in this group, and in particular, the inclusion of birds within reptiles (see the alphabetical entry "Birds" in this volume). A phylogenetic definition of reptiles that most experts would agree on is "the most recent common ancestor of a turtle, a snake, a lizard, and a crocodile, and all species derived from that ancestor." If that definition is used, the living reptiles include lizards, snakes, amphisbaenians (a group of limbless, lizardlike animals), rhynchocephalians (the tuataras), crocodilians, birds, and turtles. It also includes a great variety of extinct lineages, including the marine ichthyosaurs and plesiosaurs, the flying pterosaurs, and the ever-popular ornithischian (bird-hipped) and saurischian (lizard-hipped) dinosaurs (see the alphabetical entry "Dinosaurs: The model system for evolution" in this volume).

Phylogenetically the living reptiles are often split into three primary lineages. The lepidosaurs include the lizards, snakes, amphisbaenians, and tuataras and are the numerically dominant group of nonavian reptiles. Archosaurs include crocodilians and birds, as well as a vast diversity of extinct species, including the traditional dinosaurs. Finally, the testudines (sometimes also known as the chelonians) include the turtles and tortoises, a small group of morphologically unique vertebrates. Among reptile biologists one of the key recent research efforts has focused on how these major groups are related to each other, and what these relationships tell us about the history of diversification of tetrapod vertebrates.

Turtles are often considered the most primitive group of reptiles on the basis of their shared condition of the lack of an opening (or apse) in the cheek region of the skull (the anapsid condition). Mammals and their relatives have a single opening (and are classified as synapsids), and all other reptiles have a pair of openings in the cheek region (diapsids). On the basis of their extremely distinctive morphology and the notion that the anapsid condition is the simplest and therefore the most primitive, turtles have often been considered the first lineage to split evolutionarily from the other reptiles. However, recent molecular and morphological research indicates that turtles, while certainly bizarre and specialized, may well be a form of modified diapsid that has secondarily lost the openings characteristic of the group. Exactly where turtles fall within the diapsids remains controversial; some analyses favor a closer affinity to the lepidosaurs, while others favor an archosaurian affinity.

Although the relationship of turtles to other tetrapods is controversial, recognizing a turtle is not. Both the shell (a fused, modified rib cage) and the placement of the shoulder girdle inside the rib cage (as opposed to outside the ribs in all other tetrapods) are unique features in the history of vertebrate evolution. For the past 220 million years turtles have maintained these unique features and have been a consistent feature of life on earth. The living turtles now contain about 325 species that are widely distributed across the temperate and tropical regions of the world. The living turtles are often divided into two major groups. The pleurodires, or side-necked turtles (about 75 species), have a horizontally retractable neck joint, while the cryptodires (about 250 species) have vertically oriented neck joints that enable them to retract their head straight back into the shell. Pleurodires dominate the southern continents (Africa, South America, and Australia), while cryptodires are found throughout Eurasia and North America.

The lepidosaurs are the most diverse and numerous of the living reptiles. They include two groups: the morphologically primitive tuataras (a pair of lizardlike species restricted to several offshore islands in New Zealand) and the ~7,000 species of lizards, snakes, and their relatives (which are collectively called the squamate reptiles). Squamates are most diverse in the world's tropical regions, although they occur on all continental and most island land masses except polar and extreme high-elevation regions. One of the most ubiquitous and persistent evolutionary themes in squamate reptiles is the repeated evolutionary loss of both the digits and limbs; by one estimate complete limblessness has evolved a minimum of 62 independent times in the group. The most famous example is the snakes, or serpents (see figure). With nearly 3,000 living species, they are by far the most successful lineage of limbless squamates, but many other examples abound, particularly among the skinks. Other important innovations in the squamates are the evolution of complex venom systems, which reach their greatest level of morphological complexity and precision in the vipers and rattlesnakes. Here the highly modified skull and teeth allow injection of precisely modulated amounts of venom directly into prey with a fang like a hypodermic needle that is attached to the maxillary bone; when not in use, the fangs can be folded back into the mouth.

Finally, the living archosaurs consist of the birds (not considered further here) and the 22 living species of alligators, crocodiles, and their relatives.

The pine snake, *Pituophis melanoleucus,* a member of the colubrid family, inhabits the pine forests of the southern United States and the Pine Barrens of New Jersey. Pine snakes burrow in sandy soil and are often found in association with gopher tortoise colonies. Snakes are the most successful lineage of limbless reptiles, with nearly 3,000 species, and exhibit a remarkable range of adaptive features including, in many species (but not pine snakes), a complex system for injecting venom into prey and attackers.

Uniformly persecuted for their skins and meat, by 1970 all crocodilian species on earth were considered threatened or endangered by the International Union for the Conservation of Nature (www.iucn.org), an international organization that monitors biodiversity declines. However, by 1995 conservation efforts around the world had reduced this by about half, and as of this writing many species appear to be recovering to healthy population sizes. Reflecting their relatively close phylogenetic relationship to birds, crocodilians are both the most vocal and provide the greatest levels of parental care of any group of nonavian reptiles.

One of the classic questions in reptilian evolutionary biology is "Why did the dinosaurs go extinct?" Although it is true that many lineages of dinosaurs did go extinct roughly 65 million years ago at the end of the Cretaceous period (and that extinction was triggered by a meteorite collision with the earth), it is important to remember that phylogenetically, birds actually fall within the dinosaurs and are therefore a kind of dinosaur. Thus in a very real sense the answer to this question is that the dinosaurs did not go extinct. That pigeon that you saw in the park today was a perfectly good dinosaur; so was the chicken that you ate for dinner.

BIBLIOGRAPHY

Ernst, C. H., and R. W. Barbour. 1989. *Turtles of the World.* Washington, DC: Smithsonian Institution Press.

Greene, H. W. 1997. *Snakes: The Evolution of Mystery in Nature.* Berkeley: University of California Press.

Lee, M. S. Y., T. W. Reeder, J. B. Slowinski, and R. Lawson. 2004. Resolving reptile relationships: Molecular and morphological markers. In J. Cracraft and M. J. Donoghue, eds., *Assembling the Tree of Life,* 451–467. New York: Oxford University Press.

Pianka, E. R., and L. J. Vitt. 2003. *Lizards: Windows to the Evolution of Diversity.* Berkeley: University of California Press.

Pough, F. H., R. M. Andrews, J. E. Cadle, M. L. Crump, A. H. Savitsky, and K. D. Wells. 2004. *Herpetology.* 3rd ed. Upper Saddle River, NJ: Pearson Prentice Hall.   —H.B.S.

Modes of Speciation (Michael J. D. White)

A common misconception among students of science is that theories must be correct to produce progress in their disciplines. For example, students misunderstand why they learn the theory of the inheritance of acquired characters of Jean-Baptiste Lamarck, remembering him only as "that guy who was wrong about heredity." However, that is a shortsighted view: Lamarck presented a comprehensive theory of evolution at a time when the processes were poorly understood. Even if his contribution was imprecise in detail, it positively influenced our understanding of evolution today.

With that in mind, it is a pleasure to comment on cytologist Michael J. D. White's classic 1978 book *Modes of Speciation.* This book was the most comprehensive treatment of speciation until the appearance of Coyne and Orr's 2004 text. The most famous aspect of White's book is his emphasis on the importance of chromosomal rearrangements in speciation. White noted that ~99% of closely related animals differ by chromosomal inversions, translocations, fusions, or other rearrangements, and that the rate at which these rearrangements arise is high by evolutionary standards (1/500 individuals has a newly arisen rearrangement). He also noted that most rearrangements diminish fertility when they are heterozygous because heterozygosity induces irregularities in meiosis. Thus when diverging populations come to differ in arrangements, their F1 hybrids are sterile, and speciation has occurred.

Alas, White's view of *mechanical chromosomal speciation* fell into disfavor in the next 15 years. Theoretical arguments showed that extremely stringent conditions were necessary for the fertility reduction to be strong (hence causing speciation) while also not eliminating the new arrangement when it first arises (the individual who first bears this mutation will necessarily be a heterozygote). Empirical work showed that rearrangement heterozygotes are sometimes not as sterile as initially believed. Instead, studies now suggest different ways in which rearrangements contribute to speciation (e.g., effects on recombination) rather than the direct fertility reductions discussed by White. His emphasis on the importance of rearrangements was correct, but his explanation for how they caused speciation is not widely accepted today. However, there are exceptions: one elegant study of *Saccharomyces* yeast showed a direct effect of a translocation on hybrid fitness (Delneri et al. 2003).

White tackled several other contentious issues. He reviewed genetic data regarding species differences and concluded that there was no evidence supporting Ernst Mayr's influential view of a "genetic revolution" that occurs during speciation, during which a large fraction of the genome changes because of new selective pressures after colonization. White also strongly favored interpretations of sympatric speciation in many insect groups, such as Rhagoletis fruit flies, still studied today in this context. Contrary to many other authors of his time (and ours), he criticized "ethological speciationists" who strongly advocate sexual discrimination and other behaviors as being of paramount importance to speciation. Instead, he suggested that hybrid problems may be more common early in speciation. Although his view is not widely accepted today, Coyne and Orr (2004) also note the overemphasis placed by some authors on sexual discrimination when the relative importance of effects from such discrimination versus hybrid problems in speciation has not been definitively demonstrated.

White's text is a richly detailed review of the literature on speciation. The discussions of empirical studies of speciation are thorough and eloquent. White was an outstanding cytologist, but like many biologists, he became convinced that the phenomena he studied were of paramount importance in speciation. Although some of his interpretations are unpopular today, no one can doubt that this text was influential in motivating several outstanding studies of speciation. Some of his controversial views may yet be vindicated; many continue to be active areas of study. This quote from his text still applies: "And, at the present time, it would seem that sweeping generalizations about methods of speciation should be avoided as far as possible (it may be more justifiable to generalize in 20 years time)" (White 1978, 324).

BIBLIOGRAPHY

Coyne, J. A., and H. A. Orr. 2004. *Speciation*. Sunderland, MA: Sinauer Associates.
Delneri, D., I. Colson, S. Grammenoudi, I. N. Roberts, E. J. Louis, and S. G. Oliver. 2003. Engineering evolution to study speciation in yeasts. *Nature* 422: 68–72.
Ortíz-Barrientos, D., J. Reiland, J. Hey, and M. A. F. Noor. 2002. Recombination and the divergence of hybridizing species. *Genetica* 116, nos. 2–3: 167–178.
White, M. J. D. 1978. *Modes of Speciation*. San Francisco: W. H. Freeman.

—M.A.F.N.

Monad to Man: The Concept of Progress in Evolutionary Biology (Michael Ruse)

Evolutionary ideas became popular in Western civilization just when concepts of human progress were flourishing. The Industrial Revolution and the growth of science itself led educated people to expect limitless progress in human knowledge and in human society. Many commentators have suggested that ideological beliefs about human progress were a motivating factor in beliefs about evolution and especially in beliefs about the progressiveness of evolution. Michael Ruse set out to investigate whether personal beliefs of scientists about progressiveness in human society and in evolution were correlated.

Monad to Man is the result of this study. Ruse concludes that scientists' beliefs about the progressiveness of evolution were indeed correlated with their beliefs about social progress. This provides a challenge to the objectivity of the scientific belief in evolutionary progress. Judgments of progressiveness cannot be demonstrated on straightforward empirical grounds; they are normative assessments of what counts as "forward" change. Could the metaphorical inference from social progress to evolutionary progress be taken seriously by scientists? To answer this question Ruse asks three questions about a large number of historical and contemporary evolutionists. First, do a scientist's ideas about biological progress correspond to what one would expect if the scientist were influenced by ideas about social progress? Second, was that scientist conscious of the relation between social progress and biological progress? Third, to what extent did the biological views outstrip the evidence? The larger the gap between evidence and theory, the more likely it is that the scientist was carried across that gap by his or her social biases.

Ruse's answers are intriguing. First, almost all nineteenth-century evolutionists and most twentieth-century evolutionists were progressionists with respect both to society and to evolution. But second, they seem to have recognized that this created a bias. The recognition is revealed in differences between their treatment of evolutionary topics and their nonevolutionary research. Evolution was a topic for public lectures and popular writings, but it was usually not a topic for hardcore scientific research. Thomas Henry Huxley, Darwin's strongest supporter, was a perfect example of this tendency. He wrote dozens of popular articles on evolution but largely confined his scientific publications to papers on comparative morphology.

The alternative to treating evolution as a popular (but not a research) topic was to find a way to study evolution that did not assume progression. A framework for such a study finally appeared with the synthetic theory of evolution in the 1930s and 1940s, but many of the architects of that synthesis were themselves progressionists. Only in recent years has the association between progressionism and evolutionism faded. This is partly because the framework for nonprogressionist evolutionary studies has advanced, and partly because the belief in social progressionism has faded.

The book's value is not solely in its conclusions about progressionism. It combines a sweeping history of the science of evolution with intricate details about individual scientists' researches, prejudices, and personal lives. It can also sensitize readers to subtle progressionist tendencies in their own thought.

BIBLIOGRAPHY

Ruse, Michael. 1996. *Monad to Man: The Concept of Progress in Evolutionary Biology*. Cambridge, MA: Harvard University Press.   —R.A.

Morgan, Thomas Hunt (1866–1945)

Originally trained as an embryologist, Thomas Hunt Morgan became one of the principal architects of the Mendelian chromosome theory of heredity in the period 1910–1935. He was the first geneticist to win the Nobel Prize in Physiology or Medicine (1933) for the work he and his laboratory group carried out demonstrating that Mendelian genes were physical parts of chromosomes, the rod-shaped bodies found in the nucleus of all eukaryotic cells. This work had enormous influence not only in the rapidly developing field of genetics but also in embryology and especially evolutionary biology. The work of Morgan and his group resolved the question whether the hereditary factors postulated in the work of Gregor Mendel (1822–1884) in the nineteenth century were merely hypothetical constructs that were convenient for explaining certain patterns of heredity or represented real, material components of the organism that were transmitted as discrete bodies from parent to offspring. This was not an unimportant debate since critics of the new genetics maintained that it was of limited value because the basic units, genes, had no physical reality.

T. H. Morgan was born in Lexington, Kentucky, in 1866, one year after the end of the U.S. Civil War and the same year in which Mendel published his essay on hybridization in peas *(Pisum)*. Morgan came from a distinguished family that included his uncle John Hunt Morgan (1825–1864), Confederate general and leader of Morgan's Raiders during the Civil War, and Francis Scott Key (1779–1843), author of "The Star-Spangled Banner." Educated at the State University of Kentucky (now the University of Kentucky), he entered Johns Hopkins University in 1886 as a student of morphology under William Keith Brooks (1848–1908). He received his PhD in 1891 with a thesis on the evolutionary relationship of the sea spiders (Pycnogonida) to the Crustacea and Arachnida. From a detailed analysis of early embryology he concluded that the group was most closely related to the Arachnida, but that the two groups must have diverged from a common ancestor in the distant past. After a summer (1891) and then a year (1894–1895) at the Naples marine laboratory (Stazione Zoologica) working closely with the embryologist Hans Driesch (1867–1941), Morgan gave up the largely descriptive and speculative science of morphology and became a committed experimentalist. Experimentation, he came to believe, provided a more rigorous and analytical way to approach biological research than the speculative and untestable phylogenetic hypotheses of the morphologists. Moreover, the questions experimentalists such as Driesch were asking concerned important biological processes occurring in the present (for example, the mechanism of embryonic differentiation) rather than historical (phylogenetic) relationships. He collaborated with Driesch on a series of experiments on the development of ctenophores (popularly called comb jellies), marine invertebrates that have a superficial resemblance to jellyfish. The excitement of being able to analyze complex physiological or developmental processes directly by manipulating the organism or its environment convinced Morgan this was the direction in which modern biology needed to proceed.

Morgan held three teaching positions during his career: at Bryn Mawr College (1891–1904), Columbia University (1904–1928), and the California Institute of Technology (1928–1945), where he founded the Division of Biology. Throughout all these years he spent summers at the Marine Biological Laboratory in Woods Hole, Massachusetts, which he and other biologists considered the American Naples. During his Bryn Mawr period Morgan worked mostly on problems of experimental embryology, using organisms such as frogs, earthworms, and sea urchins. It was only after arriving at Columbia University, starting around 1907, that he became familiar with the fruit fly, *Drosophila melanogaster*. The work for which he and his laboratory group at Columbia subsequently (after 1910) became most famous, establishing the Mendelian chromosome theory of heredity, was initiated and carried out during the Columbia years (see figure). Although he transported his "fly lab" to Pasadena in 1928, his administrative duties took him away from day-to-day fly work, and when he did return to the laboratory, problems of embryonic development resurfaced as his main interest.

Two major aspects of Morgan's work relate to the development of evolutionary theory in the twentieth century. The first was his early opposition to Darwin's theory of natural selection (1903–1915) as a mechanism for how evolution, especially adaptation, could occur. The second was his skepticism about both the Mendelian and chromosome theories of heredity.

Morgan's initial objections to Darwin's mechanism of natural selection were voiced in his book *Evolution and Adaptation* (1903), in which he argued that selection could never produce wholly new species by acting on slight individual differences (referred to at the time as continuous or fluctuating variations), as Darwin had claimed. Because such variations were small, even if they conferred some advantage, they would tend to get "swamped out" in successive generations. In addition, Morgan argued that fluctuating variations occurred only rarely and by chance, so that selection, acting on such variations, could

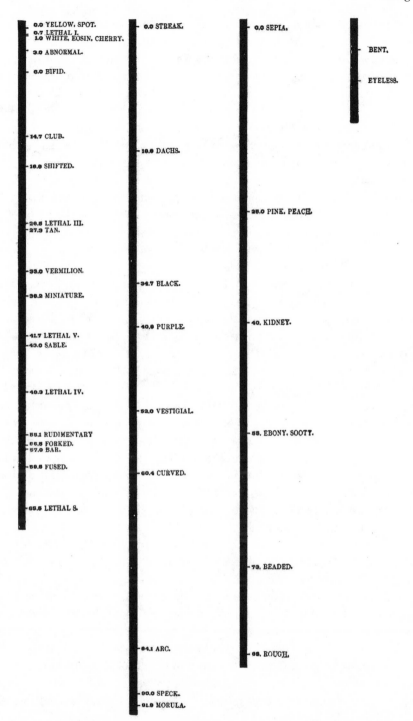

Thomas Hunt Morgan and his young associates (A. H. Sturtevant, H. J. Muller, and Calvin B. Bridges) mapped the order of genes on the four (pairs of) chromosomes of *Drosophila*. This is the frontispiece of their book, *The Mechanism of Mendelian Heredity* (1915).

never produce the fine-tuned adaptations seen throughout the living world. Between 1903 and 1910 Morgan advocated Hugo de Vries's (1848–1935) mutation theory, which claimed that new species arose in one step by large-scale mutations that could make an offspring a different species from its parents. This theory, for which there appeared to be sound experimental evidence in de Vries's model organism, the evening primrose *(Oenothera lamarckiana),* was especially appealing to Morgan and many of his contemporaries not only because it circumvented the objections to Darwinian theory outlined earlier, but also because it was claimed to be an *experimental* approach to evolution. De Vries was able to hybridize different varieties and species of *Oenothera* and produce the supposedly new, mutated species at will. Although de Vries's mutation theory was later abandoned because it was found to be the result of a variety of chromosomal anomalies peculiar to *Oenothera* (and thus did not produce new species), it served for many at the time, like Morgan, as a viable alternative to Darwinian natural selection.

During the same period (1903–1910) Morgan was also skeptical of the Mendelian theory as an explanation of the results of hybridization and the chromosome theory as an explanation of such phenomena as sex determination. To Morgan the Mendelian theory, like numerous other particulate theories of heredity in the late nineteenth century (those of Darwin, Ernst Haeckel, August Weismann, or Carl Nägeli, to name a few), was couched in terms of hypothetical "factors" or particles and was therefore completely speculative. Aside from the fact that the existence of these supposed factors could not be confirmed, Morgan's objections to Mendelian theory were threefold. First, if, as Mendelians seemed to claim, sex was a hereditary trait, the Mendelian theory could not account for the 1:1 sex ratio found in most animals (and plants). Second, Morgan thought that Mendelians treated phenotypic traits (an organism's appearance) as discrete, sharply differentiated categories when in reality most (for example, animal coat color) ranged over a wide spectrum. Morgan also noted that Mendelians seemed to invent as many pairs of factors as they needed to explain any given hereditary condition; for example, if one pair of factors would not suffice to explain a particular pattern of heredity, two would be invoked (as in cases of epistasis, where one pair of factors affects the expression of another). He also questioned the Mendelian assumption of segregation, the separation of the two factors for any trait during germ-cell formation so that each ended up in a different gamete.

With regard to the chromosome theory, until 1910 Morgan claimed that there was no evidence that chromosomes had anything to do with heredity, including the claim by his colleague E. B. Wilson (1856–1938) and former student Nettie M. Stevens (1861–1912) that the so-called accessory chromosomes (what we today refer to as X and Y) are responsible for determination of sex in a wide variety of animals. Although chromosomes were clearly visible in microscopes at the time, and much descriptive work had been done detailing their movements in mitosis (the process of distributing the chromosomes during cell division) and meiosis (the process of separating homologous chromosomes during reduction division in gamete formation), the significance of the chromosomes was still unclear. Finally, as an embryologist Morgan had a strong aversion to both Mendelian and chromosome theories because they seemed like reversions to the long-rejected theory of preformation, which had claimed that the process of embryonic development was merely the unfolding of miniature embryonic parts into adult characters. Referring complex and dynamic biological processes back to invisible particles in the germ cells, Morgan claimed, might seem at first glance to be explanatory but in fact ultimately explained nothing.

Ironically Morgan began to change his mind on all three theories between the spring of 1910 and the winter of 1911. The major impetus came first from his own studies of *Drosophila*. Morgan had begun breeding the fruit fly in his Columbia laboratory around 1908 to see if he could produce mutations of de Vriesian type in animals. He had gotten nowhere when one day he observed among his cultures of normally red-eyed flies one that had white eyes. It was a male, and he bred it to a red-eyed female; the offspring were all red-eyed. But when he bred these offspring with each other, he obtained a ratio of roughly one white-eyed fly for every three red-eyed flies. This was basically the expected Mendelian ratio for a cross between two parents heterozygous for a recessive trait (in this case white was said to be recessive to red, which was dominant to white). More important, however, Morgan noted that all the white-eyed flies were males. In his first published paper on eye-color inheritance (1910), Morgan assumed that the factor for white eye must somehow segregate with the factor for maleness in the formation of gametes. But this explanation was not wholly satisfactory since there was no apparent biological basis for assuming such selective segregation unless, as he soon came to see, the factors for determining eye color resided on the accessory chromosomes (in particular the X). Since females had two X chromosomes, even if one of them had the new mutant factor for white, it would be masked by the red factor on the homologous, or partner chromosome. However, in males, with only one X chromosome, the mutant factor would not be masked and thus would show up as the white-eyed phenotype. By 1911 Morgan was convinced that Mendel's theory could be given a concrete material basis if it were assumed that factors, or what Wilhelm Johannsen had in 1909 labeled *genes*, occupied a specific physical region, or locus, of the chromosome.

Subsequently the idea of genes as arranged linearly along the chromosome was given further support by observations on what William Bateson (1861–1926) and others had identified as linkage, the inheritance of two or more traits together. Although Bateson himself did not interpret this phenomenon in terms of the chromosome theory, others, including Morgan and his students, did. Indeed, Morgan used the idea of linkage to devise a way of actually mapping a gene's position on the chromosome. On occasion, Morgan noted, linkages appeared to be broken and the genes recombined, as if the two homologous chromosomes on which a group of genes were located had actually exchanged parts:

A   B   Male parental chromosome

a   b   Female parental chromosome

↓

A   b

      Offspring chromosomes

a   B

Morgan supported this idea from work by a Belgian cytologist, F. A. Janssens (1863–1924), who had described in a 1909 paper what he called *chiasmatypie,* or the intertwining of the two homologous members of a chromosome pair during meiosis, and what appeared to be their subsequent breaking and rejoining. Morgan and his student A. H. Sturtevant (1891–1971) reasoned that the frequency with which recombination could be observed in offspring was a function of the distance apart of any two genes on the chromosome, and this provided a way to estimate at least the relative positions of genes with respect to each other on the same chromosome. This mapping procedure soon became for Morgan and his laboratory group—including, in addition to Sturtevant, Calvin B. Bridges (1889–1938) and H. J. Muller (1890–1967)—the basis of an elaborate research program that was extended by other workers to hundreds of organisms (corn, the mouse, and the rat) over the next 30 years. It was for establishing the validity of the chromosome theory of heredity that Morgan was awarded the Nobel Prize in 1933.

Morgan's own turnabout with regard to genetics and the chromosome theory led him to see that Mendelian variations—the difference between red and white eyes or between wild-type and beaded wings—might be exactly the sorts of variations on which natural selection could act. In 1916, for a series of lectures at Princeton (published as *A Critique of the Theory of Evolution*), Morgan noted that the many variations he had observed in *Drosophila* were discrete, stable, and inherited. Whereas earlier he had argued that small individual or fluctuating variations would always tend to be swamped, the discrete Mendelian variations would not. These could be accumulated indefinitely and thus could lead eventually to the formation of a new species. In his later writings on evolution (1925 and 1932) he amplified these views, clearly having come to accept the basic outlines of Darwin's theory. It was for others, however, with more mathematical leanings to use the new genetics to formulate the mature form of Darwinian theory that came to be known as the evolutionary synthesis.

BIBLIOGRAPHY

Allen, G. 1968. Thomas Hunt Morgan and the problem of natural selection. *Journal of the History of Biology* 1: 113–139.

———. 1978. *Thomas Hunt Morgan: The Man and His Science*. Princeton, NJ: Princeton University Press.

———. 1980. The evolutionary synthesis: Morgan and natural selection revisited. In E. Mayr and W. Provine, eds., *The Evolutionary Synthesis: Perspectives on the Unification of Biology,* 356–382. Cambridge, MA: Harvard University Press.

Maienschein, J. 1991. Thomas Hunt Morgan. Chapter 8 of *Transforming Traditions in American Biology, 1890–1915*. Baltimore: Johns Hopkins University Press.

Morgan, T. H. 1903. *Evolution and Adaptation*. New York: Macmillan.

———. 1910. Sex limited inheritance in *Drosophila*. *Science* 32: 120–122.

———. 1916. *A Critique of the Theory of Evolution*. Princeton, NJ: Princeton University Press.

Morgan, T. H., A. H. Sturtevant, H. J. Muller, and C. B. Bridges. 1915. *The Mechanism of Mendelian Heredity*. New York: Holt.

Shine, I., and S. Wrobel. 1977. *Thomas Hunt Morgan, Pioneer of Genetics*. Lexington: University of Kentucky Press.   —G.E.A.

Muller, Hermann Joseph (1890–1967)

When H. J. Muller was about 10 years old, he went with his father to see an exhibit on evolution at the American Museum of Natural History. He remembered being impressed by the small size of fossil horses and how their size and anatomy changed over millions of years. Although Muller is best known as a geneticist and won his Nobel Prize (in 1946) for work proving that X-rays induce gene mutations, he considered himself an evolutionist in his professional and personal life. Muller was third-generation American and attended Columbia University on a scholarship. There he took courses with two great American biologists, Edmund Beecher Wilson and Thomas Hunt Morgan. Wilson was a cytologist whose writings helped convince a younger generation of biologists who took his classes that the cell was deeply involved in heredity, development, and evolution. Morgan, skeptical at first, eventually proved Wilson's theories by using the tiny fruit fly, *Drosophila melanogaster*. In his senior year Muller met two sophomores with whom he founded a biology club. Alfred Henry Sturtevant and Calvin Blackman Bridges enjoyed sharing with Muller their ideas on the new field of genetics that was creating such a stir among their teachers. These five individuals were the key figures in putting together the field of classical genetics and making fruit flies universally known among scientists.

Muller received his PhD for working out some complexities that arose when Morgan and Sturtevant discovered crossing over. Muller's work led him to a study of what he called gene-character relations. It took him about 10 years to publish these results because the experiments were tedious and extended over many generations. For Muller it was a crucial demonstration of how Darwinian natural selection worked through the selection of modifier genes that could either intensify or diminish a chief gene's major effect on an organ system or trait. Muller used genes for wing shape that were highly variable, and he isolated and mapped the modifier genes that caused the variations. Muller, like Darwin, recognized that evolution is a gradual process and that new organ systems do not arise suddenly but are built up gradually.

Muller worked first at Rice University in Texas and then at the University of Texas in Austin. While at these institutions he worked on the problems of mutation and the gene. He restricted the term *mutation* to a change in the individual gene and argued that the gene was the basis of life because all components of the cell and the organism were products of genes, as shown by the abnormalities produced in them when the normal genes were altered by mutation. Muller also measured the first spontaneous mutation rates at a time when they were so rare that he said that finding a new mutation was like finding a dollar bill on the street. By 1926 Muller had worked out the tools for measuring mutation rates by designing fruit-fly stocks in which he could detect a class called recessive lethals. If one of these were induced on a normal X chromosome in a sperm, the resulting daughter could be bred, and among her sons one category of her two X chromosomes in her immature reproductive cells would produce fertilizations that killed half her sons. The other half of her sons would be killed by the chromosome Muller designed. This chromosome, called ClB (containing a suppressor of crossing over, C; a lethal, l; and a visible dominant eye shape called Bar eyes, B) allowed Muller to examine thousands of vials of control and X-rayed descendants for vials containing only daughters and no sons. He found these in abundance with the doses he used (about 4,000 roentgens). Those doses are 10 times greater than the dose that would kill half of all humans exposed to it. It is not that fruit flies are hardier than humans. Many years later Muller showed that X-rays break chromosomes, as well as induce gene mutations. Broken chromosomes in dividing cells lead to cell death or, if these cells are sperm, to aborted embryos. Independently Barbara McClintock showed the same cycle of breaking leading to cell abnormalities in corn (maize) and called this the breakage-fusion-bridge cycle. Muller immediately recognized that this cycle in fruit flies and maize must be the basis of radiation sickness, especially when that occurred among tens of thousands of atomic-bomb victims in Hiroshima and Nagasaki. Muller received his Nobel Prize at a time when the atomic age was being ushered in, and he quickly became an advocate for radiation protection in industry, medicine, and the military applications of atomic weapons.

Muller made many contributions to the study of evolution. He was the first to propose an interpretation of why polyploidy (having three or more sets of chromosomes instead of two) is rare in animals and common in plants. In animals sex is usually determined by sex chromosomes, which are rare in plants. This makes it impossible for triploid animals to produce fertile males. He demonstrated that the Y chromosome (not normally found in females) gradually loses its genes and shrinks in size (over millions of years of evolution). Muller also analyzed hybrids of the related species *D. melanogaster* and *D. simulans* and succeeded in getting one of the tiny fourth chromosomes of *simulans* into an otherwise all-*melanogaster* fly. What impaired these flies was not some vague physiological incompatibility but specific gene differences between the two that he was able to identify.

In the 1950s Muller turned his attention to human evolution and worked out equations to follow the fate of gene mutations in a population. He estimated that in the absence of natural selection the load of mutations present in a population would double in about four centuries. He also estimated that this doubling of the load of mutations was the equivalent of subjecting humans to a radiation dose of about 150 roentgens. Muller suggested that eventually society would have to practice some form of voluntary selective breeding to reduce this mutational load, or humanity would suffer a disproportionate amount of time and cost to patch up its infirmities.

BIBLIOGRAPHY

Carlson, E. A. 1983. *Genes, Radiation, and Society: The Life and Work of H. J. Muller*. Ithaca, NY: Cornell University Press.

Muller, H. J. 1968. *Studies in Genetics: The Selected Papers of H. J. Muller*. Bloomington: Indiana University Press. [Muller selected excerpts from most of his papers, and this book contains a complete bibliography of his works.]

—E.A.C.

N

Natural history

Natural history is the careful observation of nature, from a study of how soil changes from one region to another to how female butterflies search for a suitable plant upon which to lay their eggs. The practitioners of natural history are termed *naturalists,* and well before scientists called themselves geologists or botanists, individuals who studied the natural world were called naturalists. A true naturalist cultivates expertise in areas from the course of rivers to the habitats of plants.

Natural history was the foundation for the discovery of evolution. Naturalists like Jean-Baptiste Lamarck and Erasmus Darwin observed the habits of animals and noted the match between habits and particular values of traits, like the oft-cited example of giraffes exploiting their long necks to forage in trees. Naturalists noted the relationships between soil types and the occurrence of individual species, and naturalists who traveled widely recorded the distributions of plants and animals in space, noting carefully where similar species replaced one another geographically. Charles Darwin's appreciation of natural history led him to observe how much the tortoises of the Galápagos differed in the shape of their shells from one island to another and how the beak sizes and shapes of the finches of the archipelago varied with their food habits. When Darwin coupled these observations with his appreciation of the young age of the archipelago, the idea of evolution was planted firmly in his mind.

Although the impetus for the discovery of evolution grew out of natural history, the empirical study of natural selection was driven by it. The British naturalist W. F. R. Weldon's observations of marine animals, coupled with his appreciation of the changing nature of their habitat, led him to begin what might well be called the first attempt to document the action of natural selection (see the alphabetical entry "W. F. R. Weldon" in this volume). Naturalists who had made careful studies of individual variation in a character like snail color and banding, mindful of Darwin's hypothesis of natural selection, began to study the fates of individuals with different features and thus gave rise to the quantitative study of selection. The thousands of studies of natural selection that have been published over the decades are the lineal descendants of the work of these dutiful naturalists.

Although a glance through the professional journals of evolutionary biology may suggest that natural history is a lost pursuit, it continues to thrive even as it appears superficially much changed. Naturalists document animals and plants with digital images rather than paintings and drawings and map the distribution of plants and animals with geographic information systems rather than surveyors' instruments. Although the tools used to describe natural features have changed, the need to do so and the value in doing so remain. In particular, as climate changes and human activity alter landscapes, naturalists are playing critical roles in documenting changes in the distributions of organisms in response to these alterations and through their documentation are providing the raw material for studying what may be a new era of adaptation to a rapidly changing earth.

BIBLIOGRAPHY

Bartram, W., and F. Harper. 1998. *The Travels of William Bartram: Naturalist's Edition.* Athens: University of Georgia Press.

Hölldobler, B., and E. O. Wilson. 1994. *Journey to the Ants: A Story of Scientific Exploration.* Cambridge, MA: Belknap Press of Harvard University Press.

Wilson, E. O. 1994. *Naturalist.* Washington, DC: Island Press.   —J.T.

Natural Selection and Heredity (Philip M. Sheppard)

Philip Sheppard's *Natural Selection and Heredity,* which appeared in 1958, just before the centenary of Charles Darwin's *On the Origin of Species,* was a major contribution to the effort by British evolutionists to persuade people, not least their fellow biologists, that the theory of evolution through natural selection was now a fully functioning area of research, a "paradigm," in the language of Thomas Kuhn (1962). This was still not a generally accepted fact, and indeed the review of the book in *Heredity* was amazingly hostile, saying that although the writing was clear, "considered as a whole, however, the book is disappointing." It might do for "advanced University students," but "in surveying modern views on the theory of natural selection, however, it is much less satisfactory" (Rees 1958, 523). Obviously, even back then, not everyone agreed, because the book was to go through four editions.

The work exemplifies beautifully the basic structure of what in America was called the synthetic theory of evolution and in England was called neo-Darwinism. First, the core of the theory is presented. This is natural selection underpinned by Mendelian genetics. Recognizing that even more than 50 years after Mendel's laws were rediscovered, the basic ideas of genetics were not widely known, Sheppard therefore gives a short primer on the essentials. Those interested in the diffusion of ideas will note that the first edition, published five years after the groundbreaking work of James Watson and Francis Crick, makes no mention of DNA. By the fourth edition (1975), however, not only is DNA introduced comfortably and without comment, but there is also a whole new chapter on protein evolution, drawing on the work of people like Richard Lewontin who had discovered significant molecular differences in organisms, differences unremarked by classical studies and techniques.

With his foundation now given, Sheppard goes on to discuss a number of important issues in neo-ontological evolutionary studies. Particular attention is paid to polymorphism—the ways in which natural selection can at the same time hold different forms in a population. Much of this naturally draws on Sheppard's own work, particularly the studies of snails that he had done earlier in the decade, with fellow Darwinian A. J. Cain, in the Oxford countryside. Sheppard had spent a year in the laboratory of Theodosius Dobzhansky in America, and it is perhaps because of this that he is not entirely dismissive of genetic drift as a causal factor in evolution, although the strong impression given is that it is natural selection that really counts in determining the genetic composition of populations. By the 1975 edition Sheppard is even less keen on drift, although he does admit that it might matter in protein evolution.

Other topics that get major discussion are the evolution of dominance, protective coloration, and mimicry. Here we surely see something of a nationalistic bias, for these were all topics of major interest to Ronald Fisher in his *Genetical Theory of Natural Selection* (1930) and were mentioned little in the leading works of American evolutionists. Dobzhansky's *Genetics and the Origin of Species* (third edition, 1951) has very little on mimicry, and even that is gone by the time of the de facto fourth edition, *Genetics of the Evolutionary Process* (1970). Lewontin, a Marxist convert, has somewhat contemptuously remarked that the English interest in butterflies, the chief source of information about mimicry, reflects the dilettantish, upper-class nature of English evolutionary biology (see Lewontin 1974). With good reason, A. J. Cain (of very humble origins) reacted vehemently against this. But for those interested in the cultural factors that influence science, if the emphasis is put only on particular problems, Sheppard's very attractively written little book is highly suggestive. It is very much written by an Englishman who loves his countryside and the denizens thereof.

BIBLIOGRAPHY

Dobzhansky, T. 1951. *Genetics and the Origin of Species*. 3rd ed. New York: Columbia University Press.

Fisher, R. A. 1930. *The Genetical Theory of Natural Selection*. Oxford: Oxford University Press.

Kuhn, T. 1962. *The Structure of Scientific Revolutions*. Chicago: University of Chicago Press.

Lewontin, R. C. 1974. *The Genetic Basis of Evolutionary Change*. New York: Columbia University Press.

Rees, H. 1958. Review of *Natural Selection and Heredity*. *Heredity* 12: 522–523.

Sheppard, P. M. 1958. *Natural Selection and Heredity*. London: Hutchinson.

—M.R.

Natural theology

Theology deals with our knowledge of the nature and existence of God. Revealed theology is the area that treats of our understanding of God through faith and authority, be this authority the church or the Bible. That Jesus died on the cross for our sins is a revealed theological claim. Natural theology is the area that treats of our understanding of God through reason. That God exists necessarily from his very definition (the ontological argument) is a natural theological claim.

In the Christian religion natural theology takes a back seat to revealed theology. However, especially given the Greek influence on Christianity, it has always had an important role. Catholics particularly make much of natural theology, and famously Saint Thomas Aquinas gave a series of proofs of the existence of God, as well as discussions of how we might reasonably understand his nature. Protestants have regarded reason in religion with more suspicion but have generally allowed some place for natural theology. The Anglican Church particularly, seeking a middle way between the extreme authority of the Catholics and the extreme biblicalism of many Protestants, has traditionally given natural theology a significant place.

Evolutionary theory, especially Darwinism, impinges on natural theology in several respects. One obvious point of tension occurs over the existence of miracles. Many natural theologians have argued that the existence of miracles—breakings of the laws of nature—points to the active intervention of the deity. Water could not have been turned into wine had God not done so. However, any scientific theory, in particular an evolutionary theory, presupposes that the world works according to law, and the success of such a theory supports the ubiquity of law and hence the nonexistence of miracles.

In the face of this objection, the believer has two options, both of which have been taken. On the one hand, it can be argued that there is a difference between the order of nature and the order of grace. Normally the world does proceed according to law, but sometimes God feels the need to intervene. Evolutionary theory does not prove that miracles are impossible, only that they will be rare, a point accepted by the believer. On the other hand, it can be argued that miracles do not necessarily imply a breaking of law. It is more a matter of meaning. Jesus's resurrection, for instance, did not involve a physical rise from the dead but the feeling in the hearts of his disciples that he still lived. Evolutionary theory has nothing to say on this.

Another place where evolutionary theory and natural theology seem to come into contact is the problem of evil. A classic argument against the existence of God is the existence of physical and moral evils, such as earthquakes and Auschwitz. No all-powerful, all-good god could have allowed these to happen. Darwinism seemingly confirms the existence of physical evil, for it posits a ubiquitous struggle for existence, involving pain and suffering always throughout the living world.

One response might come from an equally classic counter to the problem of evil, namely, that God cannot do the impossible. Perhaps the only world that God could have created was one that includes pain and suffering. If God decided not to let burning cause pain, then he would have had to allow that people would never learn to fear fire. It is better to have pain with fire than fire without pain. Perhaps the only way in which God could have created the living world was through a mechanism like selection, and hence he cannot be blamed for its bad consequences. (Kitcher 2007 makes this criticism, that Darwinism exacerbates the problem of evil, and Ruse 2001 argues that it may not be quite such a problem for the Christian.)

The major place where evolutionary theory, Darwinism in particular, comes into touch with natural theology is the question of design (Ruse 2003). From Plato on, it was recognized that the living world is not random but shows complexity, an integrated and functioning complexity, for example, the hand and the eye and the nose. How could this be, given that normally blind laws lead to randomness? Only through the intervention of a designer, comes the reply. This argument was picked up by Christians, Saint Thomas particularly, and made the basis of the most convincing of the arguments for the existence of God: the living world is as if designed; this is because it is designed, and the designer is God.

Philosophers, most particularly David Hume, showed that in major respects, even if there is a designer, it cannot be much like the Christian god. Apart from anything else, why stop at one designer? However, Hume had to admit that it does seem as though there is something at work. Using a version of what is known as the argument to the best explanation, Hume agreed that blind law does not lead to designlike phenomena, and given that the organic world is designlike, there must be something or someone behind it all. With this possibility, subsequent natural theologians, notably Archdeacon William Paley at the beginning of the nineteenth century, went on arguing for the existence of God on the basis of the designlike nature of the living world.

Darwin's mechanism of natural selection offers an alternative to the god hypothesis. It is agreed that the living world is designlike, but it is argued that this is because of natural selection rather than God. No longer is the best explanation the Christian god. Darwin has shown that the world is indeed the product of blind law, but blind law of a special kind, namely, that which produces adaptive functioning by a process of competition and attrition.

Note that Darwin does not here disprove the existence of God. In fact, he himself was a believer in a deity when he wrote *On the Origin of Species,* even though later in life (mainly because of the problem of evil) he became an agnostic. Rather, Darwin shows that the designlike nature of the world does not necessitate a god. In the words of today's popular science writer Richard Dawkins, "Darwin made it possible to be an intellectually fulfilled atheist" (1986, 6).

Does this now mean that the believer should have no theological interest in the living world? Many of today's Christians deny this strongly. Even though one can no longer have a natural theology—a proof of God—one can have a theology of nature. One believes in God on faith (revealed theology), but then one's appreciation of God's power and majesty is fleshed out by one's encounter with the living world. The beauty of a rose, the charm of a chaffinch, and the exuberance of a faun show that the Creator was great and loving despite all the hatred and evil in the world.

It should be mentioned that there is today an active group, the so-called Intelligent Design Theorists, who argue that selection cannot explain the design-like nature of the living world, that it shows "irreducible complexity," and that hence we must make explanatory reference to a designing intelligence. Although members of this group deny that this intelligence must necessarily be identified with the Christian god, in fact this is the direction taken by most of them. As might be expected, the evolutionary community regards this version of "creationism lite" with some disdain, arguing on general grounds that one should never invoke miracles in science and on specific grounds that they are just not needed in evolutionary studies. Natural selection can do all that is required. (See Dembski and Ruse 2004 for a collection, edited by two of the leading controversialists, presenting both sides to the problem.)

In conclusion, it must be stressed that, even though natural selection does rather trump traditional natural theology, the connection is of more than just historical interest. Darwinians agree fully with natural theologians that the designlike nature of the living world is *the* crucial thing that needs explaining. Moreover, like natural theologians, Darwinians use the metaphorical language of design unreservedly. George Williams, in his seminal *Adaptation and Natural Selection,* writes: "Whenever I believe that an effect is produced as the function of an adaptation perfected by natural selection to serve that function, I will use terms appropriate to human artifice and conscious design. The designation of something as the means or mechanism for a certain goal or function or purpose will imply that the machinery involved was fashioned by selection for the goal attributed to it" (1966, 9). A beautiful and historically redolent example is furnished by the trilobite eye. The eye was above all the organ on which the natural theologians seized as something that had to have a designer—Paley (1802) made much of it. The eye could not have come about by pure chance. Darwinians agree. The trilobites, those long-gone marine invertebrates (they appeared more than 500 million years ago and lasted over 250 million years) had compound eyes like flies (see figures on page 760). If one cuts through a lens vertically, one finds that it is complex, with two parts, separated by a curved line (see the figure on page 761). Why is this? What design principle does it serve? Researchers on the problem realized that in the seventeenth century, René Descartes and Christiaan Huygens had solved the problem. It is the kind of lens you build if you want to avoid chromatic aberration (an indistinct focus coming because the different component colors of white light have different refractive indices). Nature beat the scientists by hundreds of millions of years (Clarkson and Levi-Setti 1975)!

Critics of universal Darwinism, like D'Arcy Wentworth Thompson (1917) and Stephen Jay Gould (1997) generally do not want to claim that this is a wrongheaded solution; but they warn that one swallow does not make a summer. Natural theology leads you to expect ubiquitous adaptation, and this is precisely what should be proven not assumed as part of the unseen baggage of a now-outmoded theological metaphor.

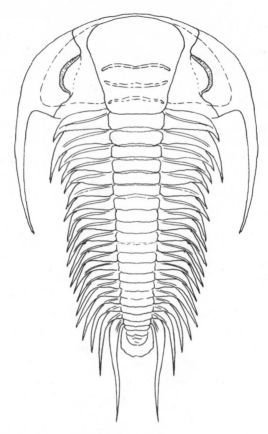

Trilobite (*Paradoxides gracilis*, Middle Cambrian specimen from Bohemia).

Trilobite eye (*Reedops sternbergi*, Devonian trilobite from Bohemia).

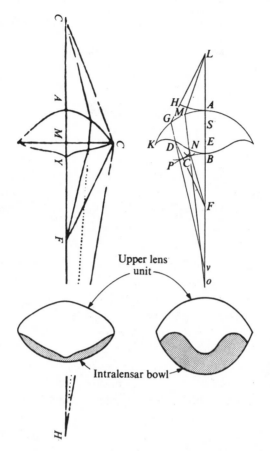

Top: René Descartes *(left)* and Christiaan Huygens *(right)* showing how to avoid chromatic aberration. *Bottom:* cross-sections of trilobite eyes. (From Clarkson and Levi-Setti 1975.)

BIBLIOGRAPHY

Clarkson, E. N. K., and R. Levi-Setti. 1975. Trilobite eyes and the optics of Descartes and Huygens. *Nature* 254: 663–667.

Dawkins, R. 1986. *The Blind Watchmaker*. New York: Norton.

———. 1995. *A River Out of Eden*. New York: Basic Books.

Dembski, W. A., and M. Ruse, eds. 2004. *Debating Design: Darwin to DNA*. Cambridge: Cambridge University Press.

Gould, S. J. 1997. The Darwinian fundamentalists. *New York Review of Books* 44, no. 10: 34–37.

Kitcher, P. 2007. *Living with Darwin: Evolution, Design, and the Future of Faith*. New York: Oxford University Press.

Paley, W. [1802] 1819. *Natural Theology* (*Collected Works*, vol. 4). London: Rivington.

Ruse, M. 2001. *Can a Darwinian Be a Christian? The Relationship between Science and Religion*. Cambridge: Cambridge University Press.

———. 2003. *Darwin and Design: Does Evolution Have a Purpose?* Cambridge, MA: Harvard University Press.

Thompson, D. W. 1917. *On Growth and Form*. Cambridge: Cambridge University
    Press.
Williams, G. C. 1966. *Adaptation and Natural Selection*. Princeton, NJ: Princeton
    University Press.   —M.R.

The Nature of Selection (Elliott Sober)

The Nature of Selection by Elliott Sober is an important book that addresses a wide range of issues in the philosophy of biology. It is one of the few examples of philosophical work that has attracted the attention of biologists (most notably Ernst Mayr). The book includes discussions of the tautology problem, units of selection, adaptationism, chance and probability in evolutionary theory, the evolution of altruism, causation, and foundational issues concerning scientific explanation in the biological sciences. Here we can discuss only a few highlights.

One of the most important distinctions that Sober makes in this book is that between *selection of* and *selection for*. In his simple example we consider a toy that contains small green beads and large white beads. When the toy is shaken, the small beads pass through a set of small holes and are thereby selected. These beads have two properties: they are both small and green. But they are selected for the property of being small; the fact that green beads are selected is merely accidental. So we have selection *of* small green beads, but selection *for* only the property of being small.

Sober brings this distinction to bear upon the problem of the units of selection. Using a discussion in G. C. Williams's *Adaptation and Natural Selection* (1966), he distinguishes between individual and group-level adaptation in the following way. We suppose that a herd of deer consists of fast individuals who are able to escape predators. As a result, two things happen. The first is that individual deer survive and reproduce. The second is that the herd survives and reproduces. But because the survival of the herd is merely a by-product of the individuals' survival, this is a case in which there is *selection of* fast herds. This selection of the fast herd is in contrast to the *selection for* fast individual deer. In this way Sober argues that the debate about group selectionism hinges upon the question whether there can be *selection for* groups. This discussion suggests that the problem of the units of selection is an empirical question, which is not to be settled a priori.

A theme that runs through this book, as well as the rest of Sober's work, is that philosophical analysis can clarify biological concepts in such a way as to inform empirical investigation. On his view there are substantive and distinctly philosophical questions that confront the biological sciences, but those philosophical issues bear directly upon how biologists go about their work.

BIBLIOGRAPHY

Mayr, E. 2004. 80 years of watching the evolutionary scenery. *Science* 305: 46–47.
Sober, E. 1984. *The Nature of Selection: Evolutionary Theory in Philosophical
    Focus*. Cambridge, MA: MIT Press.
Williams, G. C. 1966. *Adaptation and Natural Selection*. Princeton, NJ: Princeton
    University Press.   —Z.E.

Neandertals

Pronounced Nay-*an*-der-tahl and now spelled Neandertal, the name first appeared as Neanderthal when the famous fossil was discovered in the summer of 1856. Neander is the Greek rendition of Neumann (new man). The Greek form was adopted by the family of a seventeenth-century German hymn composer, Joachim Neumann. Thal was German for "valley" before the spelling change of 1901, when it became Tal. Neandertal then was Newman's valley near Düsseldorf, now in the industrial heart of the western part of Germany, and the skeleton was found by marble quarry workers who were clearing out a cave called the Feldhofer grotto. The workmen thought that the skeleton might have belonged to a prehistoric cave bear, so they were not very careful in their treatment, and the more fragile and smaller parts of the face and teeth were not recovered. Subsequently other representatives of the population to which the original Neandertal individual belonged were found in Belgium, Croatia, France, Italy, and Spain, as well as farther east in Iraq and Israel. In fact, another specimen had been discovered below the north face of the Rock of Gibraltar in 1848, but it remained unstudied in the collections of the Royal College of Surgeons in London until the first decade of the twentieth century. Those individuals have been called Neandertals ever since.

Neandertal skeletal material was found in the Belgian province of Namur at the commune of Spy (pronounced Spee) in 1886 in a stratified archaeological context with stone tools and animal bones. The tools were recognized to be of the same kind as those found at the southwestern French site of Le Moustier, dug in 1863. In honor of that type site, the tools associated with the Neandertals are always referred to as Mousterian. The use of radioactive elements to establish absolute dating of prehistoric material was not possible until after World War II, although it was realized that the Mousterian was older than the Upper Paleolithic traditions associated with modern-appearing humans such as Cro-Magnon, found in 1868. Much earlier than the Neandertals was the material found by a Dutch army physician in Java in 1891 and 1892 that he called *Pithecanthropus* (we now call it *Homo) erectus*. All of this material was meticulously studied and put into an evolutionary perspective during the first decade of the twentieth century by the Strassburg anatomist Gustav Schwalbe (1844–1916). He did not deal with the actual mechanisms whereby change was produced but presented a straightforward picture of continuity and change through time. This was just about the last time that the human fossil record was treated in an evolutionary perspective.

In the summer of 1908 two brothers who had recently been ordained as priests were pursuing archaeological excavation near the village of La Chapelle-aux-Saints in Corrèze in southwestern France. On August 3 they uncovered a complete and well-preserved Neandertal burial. They communicated their discovery to Abbé Henri Breuil, a professor at the Collège de France in Paris. Abbé Breuil was already well known in archaeology, and he recommended that the brothers put their skeleton into the hands of his friend

Marcellin Boule (1861–1942), a paleontologist who was a professor at the Muséum d'Histoire Naturelle in Paris. Boule's monograph describing La Chapelle-aux-Saints came out in three installments in 1911, 1912, and 1913 and set the tone for the study of human "evolution" for the remainder of the twentieth century. There was nothing even faintly evolutionary about it. Gustav Schwalbe wrote one of the reviews of Boule's monograph, noting the anatomical errors Boule had made in his reconstruction but, curiously enough, accepting the patently antievolutionary cast of the interpretation. The very next year World War I broke out, and one of the casualties of that conflict was the evolutionary orientation in the German treatment of the human fossil record.

Before the war both the German and the English literature treated Neandertals as an earlier stage in human evolution that had become transformed through time into anatomically modern form. Under Boule's influence the Neandertals were perceived in caricature and regarded as having been replaced by an invasion of modern-appearing humans who came from "the East." There is no archaeological or anatomical support for this interpretation, although it is the view now held by most students of human "evolution." Various estimates of Neandertal stature show males ranging from 164 to 175 centimeters (5'5" to 5'9") and females ranging from 152 to 158 centimeters (5'0" to 5'1"). Male body weight has been estimated to run from 90 to 100 kilograms (198 to 220 pounds). Cranial capacity was larger than the modern average but perfectly in proportion to their greater body weight. Clearly the Neandertals were characterized by a degree of skeletal robustness and muscularity well beyond the more recent human average. With intellectual and linguistic capabilities probably at modern levels, the Neandertals qualify as an archaic form of *Homo sapiens*.

The conversion of Neandertal into modern human form was accomplished by a reduction in those manifestations of robustness as a result of technological innovations that allowed them to gain subsistence with less expenditure of effort. Analysis of trace elements shows that their intake of animal protein was approximately the same as that which characterizes the wolf. The use of traps and snares, however, greatly reduced the amount of effort needed to bring food home. Most mutations reduce the trait that they control, and when selection that maintains robustness is relaxed, its manifestations will show a reduction. The cooking of food will reduce the intensity of selection that maintains tooth size. During the last glaciation Neandertals were using earth ovens to thaw food that froze, and a reduction in tooth size can be shown from 130,000 years ago in a straight line of 1% every 2,000 years until the end of the Pleistocene around 10,000 years ago.

Preliminary mtDNA comparisons show that the Neandertals differ more from the condition in living humans than the average difference between populations of the living. The distinction between Neandertals and living humans, however, is not as great as the mtDNA differences between populations of chimpanzees or even the differences within populations of chimpanzees.

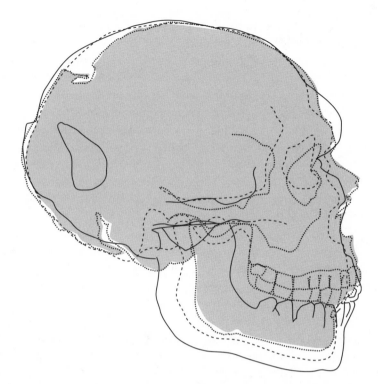

This drawing shows how Neandertal craniofacial form was converted into the modern state by a reduction of the dentition and its supporting facial architecture while the contours of the brain case remained effectively unchanged. The Neandertal is the La Ferrassie 1 male of southwestern France, of approximately 70,000 years ago, superimposed over an early Upper Paleolithic male from the Czech Republic, Předmostí 3, of 25,000 to 29,000 years ago, and a recent Faeroe Islands male.

BIBLIOGRAPHY

Brace, C. L. 1995a. Bio-cultural interaction and the mechanism of mosaic evolution in the emergence of "modern" morphology. *American Anthropologist* 97, no. 4: 711–721.

———. 1995b. *The Stages of Human Evolution.* 5th ed. Englewood Cliffs, NJ: Prentice Hall.

———. 2000. *Evolution in an Anthropological View.* Walnut Creek, CA: AltaMira Press.

Morin, E. 2004. Late Pleistocene population interaction in western Europe: An evolutionary approach to the origins of modern humans. PhD diss., University of Michigan, Ann Arbor.

Trinkaus, E., and P. Shipman. 1992. *The Neandertals: Changing the Image of Mankind.* New York: Alfred A. Knopf.   —C.L.B.

Nei, Masatoshi (b. 1931)

Masatoshi Nei has been Evan Pugh Professor at Pennsylvania State University since 1990. He obtained his PhD from Kyoto University in 1959. He was assistant professor, Kyoto University, 1958–1962; geneticist, National Institute of Radiological Sciences, Chiba, 1962–1969; associate to full professor, Brown University, 1969–1972; and professor, University of Texas, Houston, 1972–1990.

In evolution there is a great effort to infer the evolutionary relationships among organisms. This branch of biology is commonly known as systematics. In the 1960s evolutionists started to use biochemical and molecular data, mainly protein electrophoretic data, to study systematics; electrophoresis is a technique used to measure the mobility of a protein on a gel. In 1972 Nei developed a method for estimating the genetic distances between populations from electrophoretic data. This method soon became the standard method in the study of evolutionary relationships among populations or closely related species, and the measure obtained by use of the method is known as Nei's genetic distance. The reason for its popularity is that the method gives an estimate of the number of amino acid differences between two proteins and thus the degree of protein-sequence divergence between the two populations from which the two proteins were taken. Therefore, this measure is easy to understand. The method also has better statistical properties than previous methods. In 1979 Nei developed a method for estimating the number of nucleotide substitutions per site between two DNA sequences from restriction-enzyme data. This was another milestone in the study of genetic distance. Later he developed several excellent methods for estimating the number of nucleotide substitutions per site between two DNA sequences from DNA-sequence data. In 1987 he and N. Saitou developed the neighbor-joining method for reconstructing evolutionary (phylogenetic) trees. This has become one of the most popular methods in phylogenetic reconstruction since the late 1980s. Now almost every essay on phylogenetic study uses this method. In summary, Nei has been a prominent leader in the study of molecular systematics (phylogenetics) since the 1970s. His methods have helped clarify the evolutionary relationships among organisms and genes. In addition, he has made many seminal contributions to the study of molecular evolution, such as the neutral theory of molecular evolution (devised by Motoo Kimura and based on the claim that much molecular change is beneath the force of natural selection and thus genetic drift is a major factor), as well as to our understanding of the evolution of major histocompatibility genes. For his contributions, Nei was elected to the U.S. National Academy of Sciences in 1997.

BIBLIOGRAPHY

Kimura, M. 1983. *The Neutral Theory of Molecular Evolution*. Cambridge: Cambridge University Press.

Nei, M. 1972. Genetic distance between populations. *American Naturalist* 106: 283–292.

Nei, M., and W.-H. Li. 1979. Mathematical model for studying genetic variation in terms of restriction endonucleases. *Proceedings of the National Academy of Sciences USA* 76: 5269–5273.

Saitou, N., and M. Nei. 1987. The neighbor-joining method: A new method for reconstructing phylogenetic trees. *Molecular Biology and Evolution* 4: 406–425.

—*W.-H.L.*

O

Ohta, Tomoko (b. 1933)

Tomoko Ohta has been a leading figure in developing a theoretical foundation for understanding molecular evolution. In her work she has combined a deep understanding of patterns of molecular variation within and among species with creative mathematical modeling to approach some of the most fundamental questions about molecular evolution.

A student of Kenichi Kojima at North Carolina State University, Ohta returned to Japan after completing her doctoral degree in 1967 and began a famous collaboration with Motoo Kimura (see the alphabetical entry "Motoo Kimura" in this volume). Their early essays developed models for the dynamics of neutral alleles (i.e., alleles that do not confer differences in fitness on their carriers). In 1971 she and Kimura published two landmark essays that showed how the neutral theory could connect mutation rates, levels of protein polymorphism in natural populations, and long-term rates of amino acid substitutions. These essays presented a complete picture of the neutral theory and demonstrated its potential power as an explanation for seemingly disparate phenomena. This work provided a theoretical foundation for several lines of research, from using protein variation to estimate population parameters such as the effective population size (a measure of how many individuals actually contribute to the next generation) to the use of molecular variation in phylogenetic studies.

In exploring the neutral theory further, Ohta was among the first to point out that natural populations harbored more rare alleles than neutral theory predicted. To account for this pattern, she postulated that some molecular variants would be slightly deleterious and others moderately deleterious. Her hypothesis led to different statistical distributions of fitness effects that were aligned better with observed data on polymorphism and rates of evolution.

Ohta also did pioneering work on evolution in multigene families like the hemoglobins. These families are sets of tandemly repeated copies of genes and are major parts of the genome. Ohta recognized the importance of these families in evolution and developed most of the early theory that predicted how genetic drift would promote divergence in the DNA sequences of these genes among copies within a single individual, among individuals in a population, and among populations. In 1984 Tomoko Ohta was elected a foreign member of the American Academy of Arts and Sciences. In 2002 she was elected a foreign associate of the U.S. National Academy of Sciences and was recognized by the emperor of Japan as a Person of Cultural Merit.

BIBLIOGRAPHY

Kimura, M., and T. Ohta. 1971. *Theoretical Aspects of Population Genetics.* Princeton, NJ: Princeton University Press.

Ohta, T. 1980. *Evolution and Variation of Multigene Families.* Berlin: Springer-Verlag.   —J.T.

On Growth and Form (D'Arcy Wentworth Thompson)

On Growth and Form, first published in 1917, is still in print as an abbreviated version in paperback, with an introduction by Stephen Jay Gould. This in itself marks the book as something remarkable. It is hard to think of other biological texts of a similar vintage that have a comparable enduring scientific relevance; apart from volumes reprinted primarily for their historical interest, one thinks of Darwin's works and not much else. The book is and has always been seen as the distinctly idiosyncratic opus of a highly eccentric biologist, D'Arcy Wentworth Thompson (professor of zoology at St. Andrews University in Scotland, 1917–1948), who is effectively remembered solely on the strength of this book. Why has it survived?

The book presents the case for an account of biological structures based on mechanical and physical principles. It aspires especially to a mathematical methodology; chemistry hardly features at all. Using an astonishing series of examples across the whole range of phylogeny and of scales of magnitude, it shows how various types of biological "form" can potentially be explained by various types of physical forces. Topics covered include scaling, ratios of surface to volume, growth curves, chromosome patterns, cell structure and cell division (explained in terms of surface tension), multicellular arrays (as in honeycombs), shells, horns, flower patterns (analyzed in terms of Fibonacci series), zebra striping, and shaping of eggs and bone structure (explained as adaptations to imposed mechanical stresses). The closing chapter is undoubtedly the best known and the section most directly related to evolution. It deals with Thompson's "transform" method—a depiction of differences in morphology of related species (e.g., fish) or whole structures, such as skulls, in the form of coordinated geometric deformations (familiar nowadays in the guise of computerized morphing) (see figure). The nature and value of this "method" is discussed elsewhere (Horder 2005); it is actually no more than a descriptive device and is not based on, or an indication of, any developmental or evolutionary mechanisms as such. Among present-day approaches to evolution, the closest comparisons are morphometrics and especially allometry, which, through J. S. Huxley, was the most direct and explicit development that arose out of Thompson's book.

The survival of this archetypically classic book is all the more puzzling because it is distinctly dated in both style—it is written in scholastic and even literary mode—and content. It dates from a time when genetics was in its infancy, Darwinian evolution was actively being questioned in favor of Lamarckism, and there were many gaps in understanding of basic cell biology. More surprising still, Thompson adopted positions on these themes that have

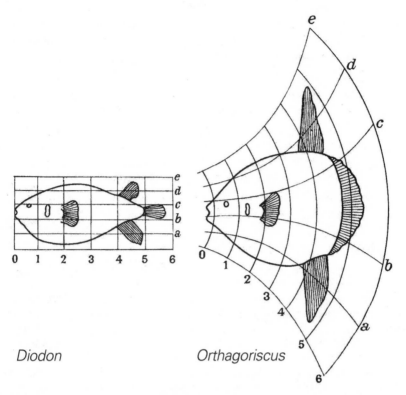

Diodon   *Orthagoriscus*

Left: outline of the porcupine fish *(Diodon)*. *Right:* outline of a sunfish *(Orthagoriscus mola)*. Note that Thompson was working intuitively, without any knowledge of the mathematics involved. Today, with computers, we can replicate his findings and have a full understanding of the moves needed to go from the fish form on the left to the fish form on the right. (From Thompson 1948, 1064.)

proved to be "wrong" (see Horder 2005). His position regarding phylogenesis was saltationist. The book is in every way a product of its time. Thompson's real mission was—as a counter to the vacuous vitalistic explanations still prevalent—to show that biological phenomena were explicable along the lines of the already scientifically highly developed discipline and methods of physics.

How is it possible that this book is as influential now as it was 90 years ago? One factor might well be the illustrations. The book is richly illustrated with some instantly striking and memorable pictures that in themselves sum up Thompson's message; they can be, and have often been, readily reproduced. Probably more significant is the way in which a steady succession of key biologists were inspired by the work. Across several generations (Richard Goldschmidt, Julian Huxley, Evelyn Hutchinson, Alan Turing, J. B. S. Haldane, and Peter Medawar to Stephen Gould and John Bonner) the book was a symbol and ideal of a rigorous scientific approach to many of the major problems of biology. The book appealed all the more because of its audacious sweep and range. This list includes the names of some of the most influential biologists of the twentieth century, many of whom made Thompson widely known through their popular writings. The book might appear to be of most direct relevance to developmental biologists, but they have rarely taken it up; it has appealed mainly to "holists" (morphologists and especially paleontologists) or on methodological grounds (biomathematicians). In an age dominated by molecular biology, genetics, and population thinking—when morphology is less and less taught in universities—this book stands today as a reminder of ultimate evolutionary problems and explanatory challenges posed by whole organisms. The affection with which it has been held through the generations is indicated by the fact that D'Arcy Wentworth Thompson is the dedicatee of Huxley's *Problems of Relative Growth* (1932) and Gould's *Ontogeny and Phylogeny* (1977).

BIBLIOGRAPHY

Gould, S. J. 1977. *Ontogeny and Phylogeny*. Cambridge, MA: Belknap Press of Harvard University Press.

Horder, T. J. 2005. Chapter 64. In I. Gratton-Guinness, ed., *Landmark Writings in Western Mathematics: Case Studies, 1640–1940, 823–832*. Amsterdam: Elsevier.

Huxley, J. S. 1932. *Problems of Relative Growth*. London: Methuen.

Keller, E. F. 2002. *Making Sense of Life: Explaining Biological Development with Models, Metaphors, and Machines*. Cambridge, MA: Harvard University Press.

Thompson, D. W. [1917] 1948. *On Growth and Form*. 2nd ed. Cambridge: Cambridge University Press.

———. 1992. *On Growth and Form*. Abridged version. Cambridge: Cambridge University Press.　　　　　　　　　　　　　　　　　　—T.J.H.

On the Origin of Species (Charles Darwin)

Charles Darwin's major work, *On the Origin of Species by Means of Natural Selection, or the Preservation of Favoured Races in the Struggle for Life*, was published in November 1859. There were 1,250 copies, and it was sold out on the first day (more precisely, the publisher sold all of the copies to booksellers at his biannual sale). Darwin started immediately to prepare a second edition, and in all there were six editions, of which the last appeared in 1872.

The work is divided into two parts. In the first part Darwin presented his mechanism of natural selection and offered reasons to think it plausible. He did this first through analogy with artificial selection, pointing out the successes of animal and plant breeders. He then went on to show (in an argument modeled on that of Thomas Robert Malthus) how a struggle for existence is a consequence of the growth of populations to a level greater than available supplies of food and space can support, and then how, given naturally occurring variation, the struggle leads to differential survival and reproduction, which Darwin called natural selection.

In the second part of the *Origin*, Darwin applied his mechanism across the spectrum of biological areas of inquiry—instinct, biogeographic distributions, paleontology, anatomy, classification, embryology, and others—showing how natural selection provides explanations in all these areas and how, conversely, these areas confirm the truth and power of selection. This is not so much a circular argument as one that is reinforcing. The argument is known technically as a *consilience of inductions*. It was promoted by the English philosopher of science William Whewell, and it was from this source that Darwin learned of its importance and power.

There were major changes in the successive editions of the *Origin*. After Darwin published, very few in the scientific community continued to deny the truth of evolution. However, Darwin's theory of natural selection was subjected to stringent scientific analysis, and two objections seemed very serious. The first was that Darwin had no adequate theory of heredity—what we today would call a theory of genetics—and without such a theory it was not obvious that a mechanism like selection could have a lasting effect. The second objection, based on physicists' calculations about such things as the heating power of the sun, was that the earth was far too young for so leisurely a mechanism as selection to have taken effect.

As we now know, the first objection was well taken, and it was not until Mendel's theory of heredity was rediscovered that this objection could be countered. The second objection was not well taken, because the heating effect of radioactive decay (and the consequent much older age of the earth) was not then known. However, Darwin felt obliged to speak to both problems, and to counter them he started to rely more and more on supplementary mechanisms to selection, most particularly so-called Lamarckism, the inheritance of acquired characters.

Unfortunately, the end product of the various changes was no longer an elegant extended essay but a lengthy and cumbersome tract. It is for this reason that today's readers generally prefer the first edition of the *Origin*.

BIBLIOGRAPHY

Darwin, C. 1859. *On the Origin of Species by Means of Natural Selection, or the Preservation of Favoured Races in the Struggle for Life*. London: John Murray.

———. 1959. *The Origin of Species by Charles Darwin: A Variorum Text*. M. Peckham, ed. Philadelphia: University of Pennsylvania Press.

Richards, R. J., and M. Ruse, eds. 2008. *The Cambridge Companion to the "Origin of Species."* Cambridge: Cambridge University Press.   —M.R.

Ontogeny and Phylogeny (Stephen Jay Gould)

In *Ontogeny and Phylogeny* a detailed analysis of the relationship between growth of the individual organism and its ancestry is tightly integrated with a historical study of recapitulation and other patterns of correspondence between stages in development and those of evolutionary history.

The analogy between individual development and the "great chain of being" was familiar to Aristotle and other classical authors. Gould shows how this idea was reshaped in the context of nineteenth-century German idealism to interpret studies of the developing embryo. Karl Ernst von Baer saw increasing divergence from a common initial body plan as vertebrates develop distinct adult forms. Ernst Haeckel accepted Darwin's evidence for evolution but not the adequacy of his mechanism. He inferred that structural innovation involves the addition of new terminal stages to ontogeny. The resulting recapitulation of abbreviated ancestral forms became codified as the biogenetic law, from which ancestral relationships could be inferred and lost forms predicted, thereby overcoming deficiencies of the fossil record.

In the absence of a satisfactory theory of individual variation and heredity, Haeckel's ideas gained wide currency in the post-Darwinian era. Haeckel insisted that terminal addition and recapitulation were the mechanism, the efficient cause of evolutionary change. Gould shows that this theory depended on inheritance of acquired characters and an unacknowledged vitalism that were characteristic of most evolutionary thought at that time.

It is often supposed that recapitulation fell from grace because evidence accumulated that evolutionary novelties can appear at any stage in the life cycle. In fact, such insertions had been accommodated as permissible exceptions to the biogenetic law, even by Haeckel himself. Gould argues that it was work in experimental embryology and the emergence of Mendelian genetics that made the theory untenable. Natural selection can and does act on all stages of the life cycle. Thereafter, bereft of mechanism and predictive power, the patterns of relationship between stages in development and those of evolutionary history were subjected to a succession of paralyzing classifications. These inhibited further work on what had come to be called heterochrony.

In *Ontogeny and Phylogeny* Gould made three major contributions that revived heterochrony as a vigorous field of inquiry. First, he shifted the focus of study from mapping patterns to the analysis of developmental mechanisms that can be studied experimentally in living organisms and inferred from analogous patterns in fossils of extinct forms. Second, he established a simple model in which the timing of overall development and the onset of reproductive maturity can be advanced or retarded in real terms, against clock time. This shows that the two fundamental patterns of heterochrony can each arise in two ways, with radically different demographic and ecological consequences for the organisms involved, as shown in the following table.

Somatic development	Reproductive maturation	Shift in pattern
Acceleration	No change in timing	*Recapitulation* (by acceleration)
No change in timing	*Acceleration*	*Paedomorphosis* (by truncation) = *Progenesis*
Retardation	No change in timing	*Paedomorphosis* (by retardation) = *Neoteny*
No change in timing	*Retardation*	*Recapitulation* (by prolongation)

Third, using a clock model, he showed that the categories of heterochrony are end members in a continuum of potential evolutionary change that can be analyzed quantitatively, using methods of allometry to study evolving patterns and experiments in developmental biology to elucidate the processes involved. In short, he laid the groundwork for a calculus of heterochrony, later developed more formally in conjunction with a group of colleagues.

The most critical of the distinctions Gould elucidated, in terms of its evolutionary implications, is that between two processes that give rise to paedomorphic, juvenilized descendants. Progenesis occurs where selection acts more strongly in favor of early reproduction than it does on morphology, giving rise to novel combinations of characters as a result of the shift in timing of development. This process involves a shake-up in morphology, loss of adult specialization, and population growth accelerated by the short life cycle. It has a unique potential to generate rapid change and the sudden emergence of new higher taxa. Gould cited well-known examples from earlier literature; many others have since been documented. Neoteny, on the other hand, extends the time allotted to early stages of development, when growth is most rapid. It is implicated in rapid size increase, notably including that of hominid brains, increased socialization, and other forms of specialization.

Gould inferred that heterochrony is under the control of regulatory genes, which had been conceived but not yet documented when he wrote *Ontogeny and Phylogeny*. His prediction that renewed interest in heterochrony would stimulate productive interaction between molecular genetics and evolutionary biology has been amply borne out. But the influence of Gould's book has not been limited to this field. Its broad scope and significance are such that it has been cited in articles published in over 500 scientific journals.

BIBLIOGRAPHY

Alberch, P., S. J. Gould, G. F. Oster, and D. B. Wake. 1979. Size and shape in ontogeny and phylogeny. *Paleobiology* 5: 296–317.
Gould, S. J. 1977. *Ontogeny and Phylogeny*. Cambridge, MA: Harvard University Press.
McNamara, K. J. 1997. *Shapes of Time: The Evolution of Growth and Development*. Baltimore: Johns Hopkins University Press.   —R.D.K.T.

Oparin, Aleksandr I. (1894–1980)

Aleksandr I. Oparin, a leading Russian biochemist, was among the most influential pioneers of the study of the emergence of life. In a booklet published in the Soviet Union in 1924, and later in a more detailed book, *Origin of Life* (1936), Oparin presented his breakthrough theory on the origin of life on earth. Oparin's ideas inspired many researchers worldwide and were instrumental in establishing in the 1950s a scientific field devoted to the subject.

Realizing the enormous complexity of the living cell, revealed by the rising field of biochemistry, Oparin rejected both simplistic mechanistic scenarios of the origin of life from matter and vitalistic claims of an unbridgeable gap between matter and life. Instead, relying on new astronomical and geological theories and data, he formulated a naturalistic evolutionary scenario that postulated specific physicochemical conditions, conducive to life, on the primordial earth. Oparin first suggested the synthesis of organic compounds in the reducing, that is, hydrogen-rich primordial atmosphere on earth, spurred by various sources of energy. These compounds were dissolved in the first seas to form an organic "soup" and later underwent chemical evolution toward more complex monomers and polymers. Similar ideas were independently formulated in a 1929 article by the British geneticist J. B. S. Haldane. Although Haldane's impact on the study of the origin of life was comparatively minor, the combination of tenets relating to the reducing primordial atmosphere, the soup, and chemical evolution was later named the Oparin-Haldane hypothesis. Some historians of science raised the intriguing suggestion that Oparin's and Haldane's Marxist affiliations (as well as similar inclinations of other pioneering researchers in the field) contributed to their origin-of-life conception.

Reflecting the then-contemporary discovery of enzymes and metabolic cycles and proclaiming the interdependent, homeostatic nature of biological organization as the hallmark of life, Oparin suggested in the 1930s a detailed origin-of-life scenario, backed by experimental work. He postulated the separation of microscopic enclosed structures made of organic polymers from the broth by a process of colloidal coagulation. He proposed that these structures grew by absorption of organic matter, demonstrated a primitive metabolism, and divided into "offspring." Through inaccurate inheritance of the original organization and primitive natural selection among competing offspring, these structures, Oparin believed, could evolve into more complex entities.

Even in the 1950s and later, when the nature of the genetic material was discovered and molecular biology was on the rise, Oparin and his group, and several supporting researchers, still insisted on the primacy of "metabolism" over "genetics" in the origin of life. They suggested a later evolution of a primitive genetic apparatus within an evolved complex metabolic entity. This view, upheld today by quite a few origin-of-life scientists, reflects the philosophical conception of life as an organized, interactive whole rather than as a "naked gene." Supporters of this view also point to the enormous difficulties involved in the prebiotic synthesis of a replicating molecule. At the same time genetically inclined critics reject the notion that a metabolic entity that lacks genetic material can participate in the necessary processes of evolution through natural selection.

This controversy notwithstanding, Oparin is recognized by both supporters and critics for his contribution to origin-of-life research, as embodied in the pioneering Oparin-Haldane hypothesis. Although various elements of this hypothesis have recently been rejected or called into question, its philosophical message—the idea of the natural, gradual evolution of life from matter—remains the fundamental framework for any research on the emergence of life.

BIBLIOGRAPHY

Fry, I. 2000. *The Emergence of Life on Earth: A Historical and Scientific Overview.* New Brunswick, NJ: Rutgers University Press.

Haldane, J. B. S. 1967. The origin of life. In J. D. Bernal, *The Origin of Life*, 242–249 (Appendix II). London: Weidenfeld and Nicolson. (Originally published in *The Rationalist Annual*, 3–10, in 1929.)

Kamminga, H. 1988. Historical perspectives: The problem of the origin of life in the context of developments in biology. *Origins of Life and Evolution of the Biosphere* 18: 1–11.

Oparin, A. I. [1936] 1953. *Origin of Life.* S Morgulis, trans. New York: Dover Publications.

———. 1967. The origin of life. A. Synge, trans. In J. D. Bernal, *The Origin of Life*, 199–234 (Appendix I). London: Weidenfeld and Nicolson. (Originally published in Moscow in 1924.) —*I.F.*

Organismic evolution and radiation before the Cambrian

When we think of evolution, we think of the development of life over the last ~550 million years, the Phanerozoic eon of geological time—that part of earth history that records the well-known progression from spore-producing to seed-producing to flowering plants and from animals without backbones to fish, land-dwelling vertebrates, and then birds and mammals. Yet rocks of this age are like the tip of an enormous iceberg, for they record only a brief late chapter of a very much longer evolutionary story, the history of Precambrian life, that extends some seven times further into the geological past, to ~3.5 billion years ago (see figure on page 776).

The Phanerozoic world was much like that today, one populated by animals and plants, many relatively large, most specialized for particular settings, and all fairly familiar, even those such as trilobites and dinosaurs that are now long extinct. And the rules of Phanerozoic evolution, the rise and then the ultimate demise of myriad species, are known to all: *speciation, specialization, extinction.* But the world of the Precambrian was vastly different. Instead of large multicellular organisms, the Precambrian was dominated by diverse widespread microbes and single-celled algae, virtually all of which are too small to be seen without use of a high-powered microscope. Among the most successful of these were cyanobacteria, microbial jacks-of-all-trades that were able to thrive in remarkably varied settings. Most notably, many of these early organisms evolved little or not at all over billions of years and have never gone extinct. Unlike the history of Phanerozoic life, evolution during the Precambrian followed a different set of rules: *speciation, generalization, long-term survival.*

EARLY PRECAMBRIAN LIFE

The most obvious and easily detected fossil evidence of Precambrian microbes is the typically mound-shaped stromatolites that they build (see figure at top of page 777), finely layered, more or less cabbagelike structures (page 777, bottom) that are common worldwide in Precambrian limestones and are known from rocks as old as ~3.5 billion years (Hofmann et al. 1999). Although most

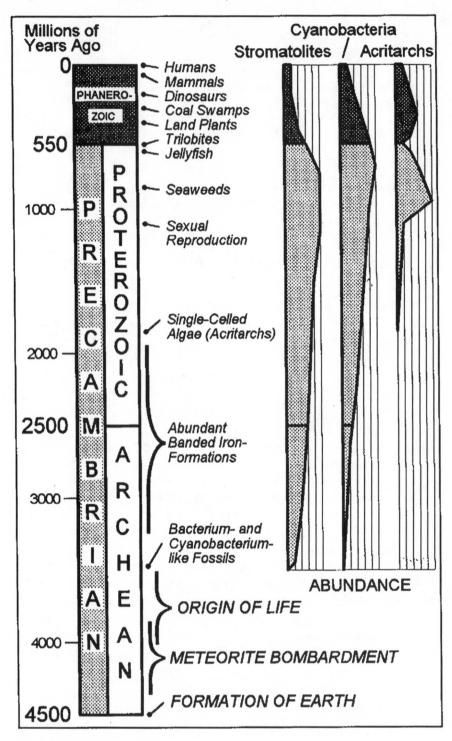

Overview of the history of life.

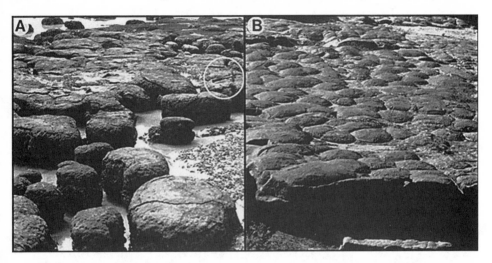

Calcareous stromatolites forming today at Shark Bay, Western Australia (A), closely resemble fossil limestone stromatolites (B) of the ~2.3 billion-year-old Transvaal Dolomite at Boetsap, Cape Province, South Africa. The geological hammer circled in part (A) shows the scale of both photos.

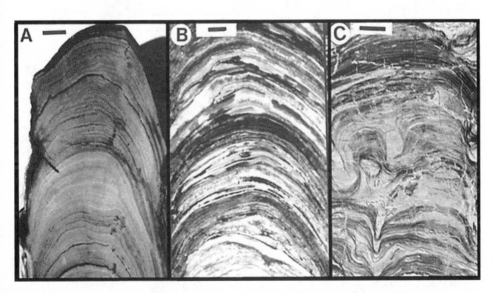

Comparison of vertically sliced stromatolites of various ages shows that microbe-built stromatolitic mat-like layers changed little over geological time. Thin wavy mats ~1.3 billion years old (A, from the Belt Supergroup of Montana, United States) are essentially indistinguishable from those ~2 billion years old (B, from the Albanel Formation of Quebec, Canada), and both closely resemble mats dating from 3.35 billion years ago (C, from the Fig Tree Group of the eastern Transvaal, South Africa). (Bars for scale represent 1 cm.)

such structures are composed of the mineral calcite (which, as it crystallizes to form limestone, destroys the delicate organic-walled microbes that build the stromatolitic layers), some become petrified by fine-grained quartz that can preserve such tiny microorganisms in exquisite detail (see figures here and on page 779). Such cellular microbial fossils are known from hundreds of Precambrian deposits worldwide, the oldest petrified in rocks nearly 3.5 billion years in age (see figure on page 780; Schopf 1993). Remarkably, many such microbes, particularly ancient cyanobacteria (see figure on page 781), appear to have changed little or not at all over literally billions of years (Schopf 1994).

Two lessons about early evolution stand out. First, life emerged quite early in the history of the planet. The earth, the moon, and all other bodies of the

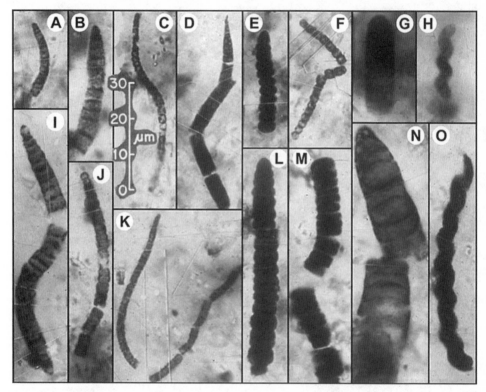

Filamentous cyanobacteria three-dimensionally petrified in stromatolitic cherts of the ~850 million-year-old Bitter Springs Formation of central Australia. Many of the fossils shown are morphologically indistinguishable from modern cyanobacteria of the taxonomic family Oscillatoriaceae. For example, the spiral filaments shown in parts (H) and (O) closely resemble the living oscillatoriacean *Spirulina* (C on page 781); and the fossil shown in part (M), composed of distinct disc-shaped medial cells and capped by a rounded terminal cell, is identical to *Oscillatoria*, the oscillatoriacean used to illustrate cyanobacteria in first-year college biology courses worldwide. (A, B, I, and J) *Cephalophytarion;* (C and K) *Cyanonema;* (D) *Caudiculophycus;* (E and L) *Filiconstrictosus;* (F) *Veteronostocale;* (G) *Palaeolyngbya;* (H and O) *Heliconema;* (M) *Oscillatoriopsis;* (N) *Obconicophycus.*

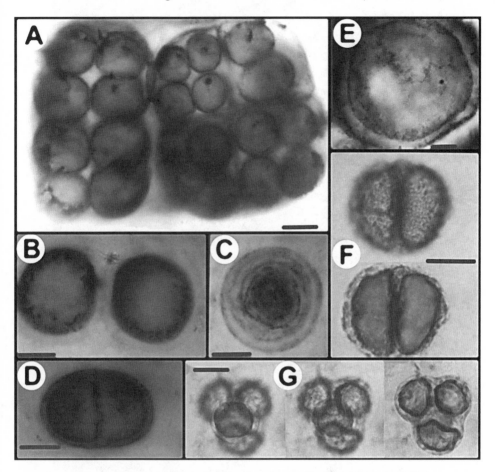

Spheroidal single and colonial cyanobacterial unicells three-dimensionally petrified in stromatolitic cherts of the ~1 billion-year-old Sukhaya Tunguska Formation of Siberia, Russia (A–D); the ~770 million-year-old Skillogalee Dolomite of South Australia (E); and the ~850 million-year-old Bitter Springs Formation of Northern Territory, Australia (F, G). Many of the fossils shown are morphologically indistinguishable from modern coccoidal cyanobacteria. For example, the sheath-enclosed unicells shown in parts (F) and (D) are essentially identical to the living chroococcacean *Gloeocapsa* (cf. E on page 781). (A) *Eoentophysalis*; (B) unnamed chroococcaceans; (C) *Gloeodiniopsis*; (D) *Gloeocapsa*-like cell pair; (E) unnamed ensheathed unicell; (F) *Eozygion*; (G) *Eotetrahedrion*. (Bars for scale represent 10 μm.)

solar system formed ~4.5 billion years ago. Yet the earth, because of early impacts by ocean-vaporizing meteorites, seems to have become continuously habitable only ~3.9 billion years ago (see figure on page 776). Firm evidence for life ~3.5 billion years ago—in the form of stromatolites, cellular fossils, and carbon isotopic evidence of microbial photosynthesis—means that life had become a flourishing success within a scant 400 million years after the environment had become clement. Evidently life's origin occurred more rapidly and was perhaps appreciably easier than has generally been imagined.

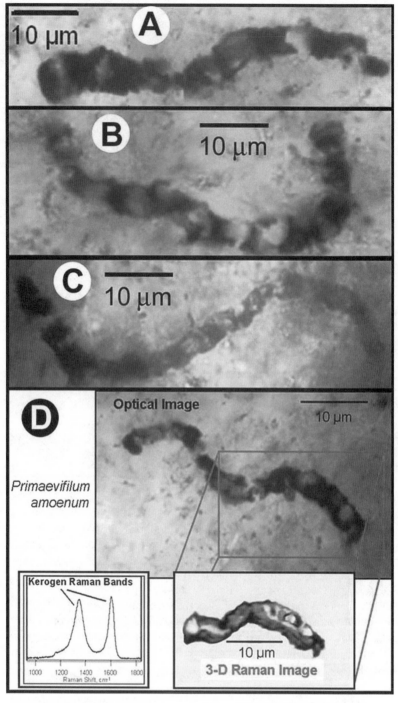

Four specimens of the cellular filamentous fossil microbe *Primaevifilum amoenum*, among the oldest fossils now known, petrified in the ~3.465 billion-year-old Apex chert of northwestern Western Australia. In part (D) an optical photomicrograph of a specimen is shown together with its Raman spectrum and its three-dimensional Raman image (of the area denoted by the rectangle in the optical photo), chemical data that by direct analysis establish that such especially ancient fossils, like younger petrified fossils (see figures on pages 778, 779, 781, and 783), are composed of carbonaceous organic matter.

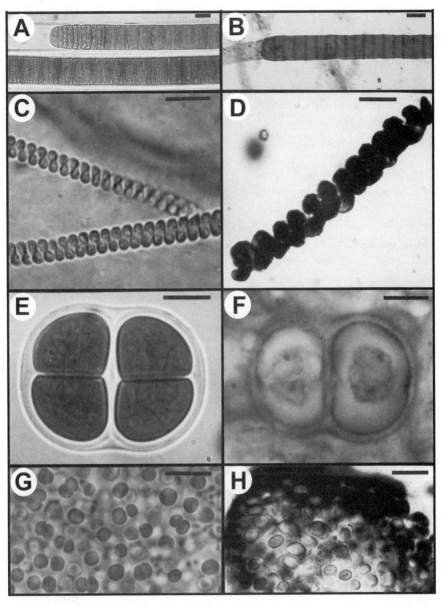

Living stromatolite-building cyanobacteria from Baja, Mexico (A, C, E, and G), and Precambrian fossil look-alikes preserved as flattened organic compressions in shaley sediments (B, D) or three-dimensionally petrified in stromatolitic cherts (F, H). (A) *Lyngbya*, compared with (B) *Paleolyngbya* from the ~950 million-year-old Lakhanda Formation of Siberia, Russia; (C) *Spirulina*, compared with (D) *Heliconema* from the ~850 million-year-old Miroedikha Formation of Siberia, Russia; (E) *Gloeocapsa*, compared with (F) *Gloeodiniopsis* from the ~1.55 billion-year-old Satka Formation of Bashkiria, Russia; (G) *Entophysalis*, compared with (H) *Eoentophysalis* from the ~2.1 billion-year-old Belcher Supergroup of Hudson Bay, Canada. (Bars for scale represent 10 μm.)

Nevertheless, direct evidence of the emergence of life is unknown and may be unknowable because geological processes have destroyed virtually all rocks more than 3.5 billion years of age. The known fossil record extends almost as far into the geological past as the surviving record of rocks.

Second, once life had gained a foothold and the major players had emerged, numerous microbes, most notably cyanobacteria, evolved hardly at all over an enormous span of geological time. This, of course, flies in the face of the standard picture of the evolutionary process, the Phanerozoic-centered view that species originate, become successful, and then go extinct as they become supplanted by better-adapted forms of life. Virtually all higher organisms fit the rule. Why are microbes so different? The answer is that evolution itself evolved; evolution has had a two-stage history. During the first stage, beginning with life's origin, perhaps as early as ~4 billion years ago, the world was dominated by nonsexual microbes, early-evolved ecological generalists adept at coping with diverse environmental settings. Eventually, however, as the second stage of evolution came to the fore, these primitive microbes came to be outcompeted by more advanced forms of life specialized for particular environments. Although the early-evolved microbes thus lost out, they never went extinct, chiefly because the history of past successes encoded in their genes enabled these jack-of-all-trades generalists to survive and then thrive in settings (hot springs, ice fields, desert crusts, hypersaline lagoons) far too harsh for their later-evolved competitors.

LATER PRECAMBRIAN LIFE

In the evolution of evolution, oxygen and sex were key. Today's world is dominated by life that needs oxygen to survive, but this was not always true. Indeed, until ~2.3 billion years ago (Bekker et al. 2004) life got along quite well without an oxygen-rich environment, simply because oxygen was scarce and microbes of the time used more primitive anaerobic metabolic means to keep their cells alive. Although the fossil record shows that cyanobacteria, microbes that give off oxygen by photosynthesis, were present far earlier (Schopf 1999), the oxygen they produced was quickly sponged from the environment by its chemical combination with iron dissolved in seawater to lay down thick beds of rocks rich in iron oxide and known as banded iron formations (see figure on page 776). By about 2.3 billion years ago the iron had been used up; the earth had rusted, and oxygen began to build up in the atmosphere.

Life responded. Advanced organisms are all oxygen dependent, using this highly reactive (and, to primitive earlier-evolved microbes, highly toxic) gas to oxidize or "burn" glucose, life's universal fuel. Also, unlike microbes, all advanced organisms have cell nuclei (and are therefore classed as eukaryotes, in contrast with prokaryotes, their more primitive microbial ancestors). Within only a few hundred million years after the onset of earth's oxygen-rich environment, large balloon-shaped algal unicells (acritarchs) first appeared, fossils of which can be hundreds of microns across, much larger than nonnucleated prokaryotes (see figure on page 783). Large-celled phytoplank-

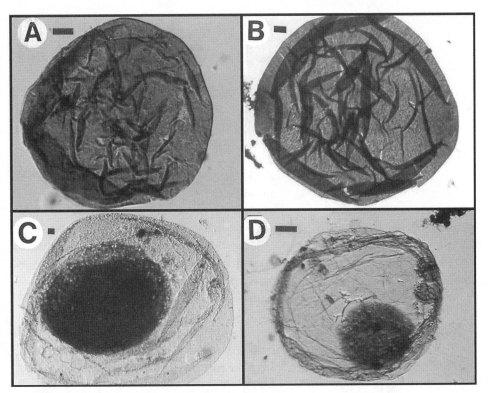

Large-celled eukaryotic acritarchs preserved as flattened compressions in carbonaceous black shales of the 950 million-year-old Lakhanda Formation of Siberia, Russia (A); the ~800 million-year-old Akberdin Formation of Bashkiria, Russia (B); and the ~850 million-year-old Miroedikha Formation (C) and ~580 million-year-old Redkino Formation (D), both of Siberia. (A and B) *Kildinella*; (C) *Pterospermopsimorpha*; (D) *Leiosphaeridia*. (Bars for scale represent 10 μm.)

ton that lived in offshore waters and, therefore, were mainly preserved by compression in shales and siltstones rather than by being petrified in stromatolites are among the earliest eukaryotes known. Importantly, however, the fossil record shows that they, like their prokaryotic ancestors, evolved little over an enormous period of time, from ~1.8 to ~1 billion years ago.

All evidence suggests that in cellular structure these early-evolved single-celled algae were much like later-evolved Phanerozoic life. Yet in comparison with later life they evolved at an incredibly sluggish pace (Schopf 1992). For this there is an easy explanation, namely, that these early acritarchs probably reproduced entirely by cloning (mitosis, body-cell division), and therefore, except for mutations, each new cell was an exact copy of its parent. By reproducing in this way, they followed the cardinal rule of life: avoid change, never evolve at all. All organisms, each in their own ways, are well adapted to their local environment; otherwise they would not be as successful. Life's credo might well be "If it ain't broke, don't fix it." In fact, of course, biomolecules do get "broken" by mutations, but this is a problem life solved early. Living systems have many biochemical repair mechanisms, molecular processes that correct changes caused by mutation and return damaged molecules to their original state. The lack of evolutionary change among such cloning acritarchs, despite their eukaryotic organization, thus makes sense. Evolution is a result of small changes that slip through the system unfixed and that over time can ultimately add up. We see the results of evolution in the fossil record only because of the vastness, the absolute enormity, of geological time.

Why and when did evolution finally speed up? Though no one knows for certain, the best bet for the "why" is sex. In and of itself this advanced reproductive process—based on meiosis (sex-cell division, an evolutionary derivative of mitosis) and syngamy (the fusion of gametes, sperm and egg, to produce a single-celled zygote that has a full complement of gene-carrying chromosomes)—produces new genetic combinations in each and every offspring. The advent of sexual reproduction would have markedly speeded the development of new species that because of their novel genetics could compete and eventually dominate in habitats previously owned by their nonsexual prokaryotic ancestors, a development well evidenced by rRNA phylogenetic trees based on molecular biology.

As for the "when" of the speedup, fossils of unicellular eukaryotes record an abrupt major increase in evolutionary rates ~1 billion years ago (Schopf 1992). Most lineages of higher, multicellular organisms date from the so-called Cambrian explosion of life, beginning ~550 million years ago. To evolve to this level, evolution must have speeded up in the half a billion years between 1 billion years ago and the beginning of the Cambrian. And evolution did indeed accelerate: phytoplankton gave rise to multicellular seaweeds by ~850 million years ago (Butterfield et al. 1990); and primitive protozoans, present as early as ~950 million years ago (Bloeser et al. 1977; Porter et al. 2003), had by ~600 million years ago led to development of soft-bodied multicelled animals (see top figure on page 785). But during this period life also suffered its earliest episode of major extinction when all the largest-celled acritarchs died out (Schopf 1992). Soon thereafter animals developed shelly protective armor (see bottom figure on page 785), an advance that marks the beginning of the Cambrian period and thus of the Phanerozoic eon, and animal diversity markedly and rapidly expanded as many new major types first appeared.

THE CAMBRIAN EXPLOSION

Although the evolutionary impact of the great radiation of life during the Cambrian period has been recognized since the time of Darwin, its cause remains a puzzle. At present two major schools of thought compete. Physical scientists (geologists, paleontologists) look to major changes in earth's environment (the worldwide glaciation of a "snowball earth," perhaps coupled with an abrupt rise in oxygen abundance). In contrast, life scientists (biologists, molecular biologists) favor mechanisms intrinsic to life itself (the onset of the processes that regulate organismic development, such as the biochemical networking of genes that govern cell differentiation) that enabled life to

Three members of the latest Precambrian Ediacaran Fauna of soft-bodied animals, preserved as imprints in sandy sedimentary rocks of the ~560 million-year-old Pound Quartzite of South Australia. (A) *Tribrachidium,* a triradially symmetrical fossil possibly related to corals; (B) *Dickinsonia,* an ovate evidently wormlike animal; (C) *Spriggina,* a segmented animal having a distinctive thickened head shield *(upper right).* (Bars for scale represent 1 cm.)

Small skeletal fossils *(Chancelloria)* of the type that mark the beginning of the Phanerozoic Eon of geological time, dissolved out of the Early Cambrian (~540 million-year-old) Ajax Limestone of South Australia. The intricately sculptured fragments are thought to have fit together to make up the body armor of a sluglike animal. (Bar for scale represents 1 mm.)

evolve to greater complexity. With new evidence and new ideas being added at a rapid clip, this is one of the most volatile and controversial areas of evolutionary biology (Cameron 2004). A firm answer to this hotly debated longstanding problem may soon be at hand, very likely based on a merging of the evidence of molecular biology and the known fossil record (Valentine 2004).

BIBLIOGRAPHY

Bekker, A., H. D. Holland, P.-L. Wang, D. Rumble III, H. J. Stein, J. L. Hannah, L. L. Coetzee, and N. J. Beukes. 2004. Dating the rise of atmospheric oxygen. *Nature* 427: 117–120.

Bloeser, B., J. W. Schopf, R. J. Horodyski, and W. J. Breed. 1977. Chitinozoans from the Late Precambrian Chuar Group of the Grand Canyon. *Science* 195: 676–679.

Butterfield, N. J., A. H. Knoll, and K. Swett. 1990. A bangiophyte red alga from the Proterozoic of arctic Canada. *Science* 250: 104–107.

Cameron, R. A. 2004. Hunting for origins. *Science* 305: 613–614.

Hofmann, H. J., K. Grey, A. H. Hickman, and R. I. Thorpe. 1999. Origin of 3.45 Ga coniform stromatolites in Warrawoona Group, Western Australia. *Geological Society of America Bulletin* 111: 1256–1262.

Porter, S. M., R. Meisterfeld, and A. H. Knoll. 2003. Vase-shaped microfossils from the Neoproterozoic Chuar Group, Grand Canyon: A classification guided by modern testate amoebae. *Journal of Paleontology* 77: 409–429.

Schopf, J. W. 1992. Evolution of the Proterozoic biosphere: Benchmarks, tempo, and mode. In J. W. Schopf and C. Klein, eds., *The Proterozoic Biosphere: A Multidisciplinary Study,* 583–600. New York: Cambridge University Press.

———. 1993. Microfossils of the Early Archean Apex chert: New evidence of the antiquity of life. *Science* 260: 640–646.

———. 1994. Disparate rates, differing fates: The rules of evolution changed from the Precambrian to the Phanerozoic. *Proceedings of the National Academy of Sciences USA* 91: 6735–6742.

———. 1999. *Cradle of Life: The Discovery of Earth's Earliest Fossils.* Princeton, NJ: Princeton University Press.

Valentine, J. W. 2004. *On the Origin of Phyla.* Chicago: University of Chicago Press.
—*J.W.S.*

Orgel, Leslie E. (1927–2007)

Leslie Eleazer Orgel, a leading origin-of-life scientist, spent most of his career at the Salk Institute for Biological Studies in San Diego, heading the Chemical Evolution Laboratory. Born in England, he earned his undergraduate degree and doctorate in chemistry from Oxford University. Later, at Cambridge University he helped develop a theory of chemical bonding in metals, ligand field theory. In 1964 Orgel moved to the Salk Institute, where he devoted many years of research to elucidating the chemical processes that led to the emergence of life on the primordial earth. In addition to many prizes and awards, Orgel was elected a fellow of the Royal Society in 1962 and in the United States, in 1990, a member of the National Academy of Sciences.

In the 1960s Orgel suggested that very early in the evolution of life molecules of RNA could have functioned both as genetic material and as catalysts.

This idea, later called the RNA-world hypothesis, was independently proposed also by the codiscoverer of DNA structure Francis Crick and the American microbiologist Carl R. Woese. The RNA-world theory gained credence in the early 1980s when ribozymes, RNA molecules that showed catalytic activities, were discovered in present-day cells.

The claim that RNA molecules were the sole players in the origin of life, serving as precursors to DNA and proteins, could potentially solve a fundamental chicken-and-egg puzzle: how could the genetic molecule, DNA, evolve without protein enzymes, and how could proteins evolve without DNA? With the idea in mind of RNA as a sole player with no need for proteins, Orgel's work in the 1960s and 1970s demonstrated the synthesis of complementary chains on templates of short sequences of RNA without catalysis by protein enzymes. However, further research made Orgel realize that a complete self-replication—copying the copies of the original sequences—was impossible without an enzyme.

Although Orgel was instrumental in advancing the RNA-world hypothesis and believed that the RNA world did play a crucial role in the early evolution of life, he became one of the most severe critics of the application of this theory to the very origin of life. On the basis of numerous experiments he came to realize the improbability of the prebiotic synthesis of an RNA sequence capable of catalyzing its own replication—the central requirement of the RNA-world concept. Orgel's outstanding chemical expertise, combined with his candid and critical evaluation of the strengths and difficulties of prebiotic chemistry, led him to an open-minded search after much simpler genetic materials that could have been easily synthesized primordially. Among the candidates for a pre-RNA world, Orgel studied several molecules different from but still similar in structure to RNA. He also searched for replicating polymers unrelated to nucleic acids, for example, positively and negatively charged amino acids that would pair with each other.

Throughout his career Orgel was closely associated with the National Aeronautics and Space Administration (NASA) and the search for life on other planets. In the early 1970s he was part of the team that designed the gas chromatograph and mass spectrometer sent to Mars on board one of the Viking landers. Upon landing on Mars in 1976, the miniaturized instrument checked samples of Martian soil for traces of organic molecules and found none. Because of existing data relevant to Mars and other solar-system planets, this result was surprising and was interpreted by most scientists (though not all of them) as indicating an active destruction of organics on Mars by solar ultraviolet radiation. Together with his friend and Salk Institute colleague Francis Crick, Orgel raised in 1973 a highly speculative hypothesis about the possibility of *directed panspermia* (*panspermia* is Greek for mixture of seeds or seeds everywhere). Crick and Orgel wondered whether life on earth was seeded intentionally by a highly developed extraterrestrial civilization that sent an unmanned spaceship loaded with different species of bacteria to many planets in the Milky Way.

Orgel was the first to admit that the idea of directed panspermia lacked any empirical evidence, and he certainly believed that the emergence of life on earth was more likely, but he thought that the idea should be entertained by origin-of-life researchers, at least on sleepless nights. This bold speculation was one indication among many of Orgel's honest assessment of the enormous difficulties faced by the origin-of-life field. He conceded on many occasions that our understanding of the problem is still greatly lacking. At the same time he was convinced that only persistent research could reveal the detailed path from a lifeless planet to the first organized living systems.

BIBLIOGRAPHY

Orgel, L. E. 1994. The origin of life on the earth. *Scientific American,* October: 53–61.
———. 1998. The origin of life—A review of facts and speculations. *Trends in Biochemical Sciences* 23: 491–495.   —I.F.

Origin of vertebrates

The origin of vertebrates, more than of any other animal group, has been the subject of vigorous historical debate, and this continues to be the case. However, a renaissance in integrative organismic biology has reinvigorated debate, sparked by fundamental new insights into the origin of vertebrate characteristics from within the field of developmental genetics. In turn, this has stimulated anatomists and paleontologists to test the hypotheses that have emerged from within this field. Much of this work has revealed how cherished assumptions are unfounded and has provided new perspectives on the origin of the major animal group to which we ourselves belong.

The starting point of the modern perspective on the origin of vertebrates is the new-head hypothesis. In essence it observes that although the differences between vertebrates and their nearest invertebrate relatives are vast, virtually all these characters owe their origin to just a few embryological novelties and are mainly associated with the head. In the main, these are the neural crest and neurogenic placodes (specialized regions of embryonic ectoderm), both of which form specialized sensory organs, while neural crest cells also differentiate into skeletal, connective, and muscle tissue. The hypothesis was also framed within an evolutionary scenario, drawing upon evidence from comparative anatomy, developmental biology, and paleontology and arguing that the establishment of the vertebrate body plan was associated with a trend of increasingly active food acquisition, from passive filter feeding to predation. The new-head hypothesis has proved extremely influential, to the extent that most of the research during the ensuing decades has been aimed at testing its assumptions and postulates in fields as diverse as molecular developmental genetics, molecular phylogenetics, comparative anatomy, and paleontology.

MOLECULAR DEVELOPMENTAL GENETICS

Molecular developmental genetics is concerned with the role of gene expression and gene interaction in relation to embryology. It did not even exist as a coherent discipline when the new-head hypothesis was formulated, but it is from within developmental genetics that the greatest challenges to the hypothesis have emerged.

First, the hypothesis that vertebrates possessed a literal new head, that essentially they are invertebrates with a head added to the front, was tested. This was achieved by examining in lancelets (presumed the closest invertebrate relative of the vertebrates) the patterns of expression of genes in the head of vertebrates. These genes, known as *Hox* genes, are found in all animals and are expressed very early in development, directing and regulating the expression of other genes. Thus they have a fundamental role in organizing the early embryo, especially its head-tail axis. Although they are present in all animals, vertebrates have the most *Hox* genes of all, and some are expressed within the head-tail axis of the embryo only at the level of the hindbrain. The results of the experiments revealed that a portion of the front end of the lancelet body is equivalent to the vertebrate hindbrain, and thus the new head, though a vertebrate novelty, was a development of an invertebrate chassis rather than a wholly new domain.

The discoveries in this new field are too many to recount here, but before moving on we must make mention of the discovery of a possible nascent neural crest and epidermal placodes in the tunicates, invertebrate chordate relatives of the vertebrates. Although many invertebrate groups possess migratory cell populations, those of tunicates exhibit the same patterns of gene expression and some of the fates of neural crest cells. Tunicates also possess rudimentary sense organs that develop from ectodermal thickenings, like neurogenic placodes, and exhibit a common suite of genes during their development. Thus although the repertoire of embryological derivatives is far smaller, it appears that tunicates possess neural crest cells and epidermal placodes in all but name. The question remains, however, why tunicates should possess these fundamental embryological novelties when lancelets, closer relatives of vertebrates, do not.

COMPARATIVE GENOMICS

Comparisons of vertebrates and invertebrates at the anatomical and embryological levels clearly mark out vertebrates as in some essence more complex than their spineless relatives. This theme has now been extended to the level of the genome, the library of an organism's genetic constitution. It was discovered that vertebrates possess a greater number of *Hox* genes than any other animal, but that quality is not particularly significant because the precise number varies within invertebrates. *Hox* genes are sometimes organized in a cluster, meaning that they are located within the same physical domain on a chromosome, and all animals possess just a single *Hox* cluster except vertebrates, which possess at least two *Hox* clusters, each equivalent to the one found in invertebrates. This simple feature might suggest only that the origin of vertebrates coincided with a duplication of the *Hox* cluster—no mean observation given the fundamental role played by *Hox* genes during embryology. However, further comparative surveys of vertebrate and invertebrate genomes have revealed that the same phenomenon is encountered in countless other genes and gene families. This has led to the conclusion that the origin of vertebrates coincided either with wholesale gene duplication, possibly at the chromosome level, or the prevailing hypothesis, that the entire genome was duplicated. Under either scenario this would have provided a vast number of novel functionally redundant genes that were available for co-option to new functions, possibly explaining the origin of vertebrates. In fact, it has been argued since the early 1970s that only genome-scale gene duplication could provide for so many fundamental novelties evolving in concert, such as at the origin of vertebrates and the origin of jawed vertebrates. We now know that similar events or episodes have punctuated vertebrate evolution, since genome duplication is also implicated in the origin of jawed vertebrates and the origin of teleost fishes—events in which vertebrate anatomy has been adapted at a scale comparable with that of the origin of vertebrates, at least on the evidence of living vertebrates.

PALEONTOLOGY

The excitement created by developmental genetics has led directly to renewed interest in other disciplines, both in testing new hypotheses on the origin of vertebrates and in exploring their implications. This has been especially true of paleontology, where attempts have been made to chase the origin of vertebrates and vertebrate-specific characters ever deeper into geological history. This has been achieved on a number of fronts, for example, the reassignment of long-known but problematic fossil groups, such as the conodonts, as well as new discoveries resulting from systematic and serendipitous searching of the rock record. There is now compelling evidence for vertebrates extending to the Lower Cambrian (520 million years ago) on the basis of a well-preserved soft-bodied vertebrate, *Myllokunmingia,* from the Chengjiang fossil deposit of South China. Skeletonizing vertebrates were present by the Late Cambrian (501 million years ago) at the very latest, as evidenced by the presence of conodonts. What is more, all these early fossils are from rocks deposited under marine conditions, providing compelling evidence for the marine origin of vertebrates.

Paleontology provides not only a timescale for vertebrate origin but also insight into the sequence in which vertebrate-specific characters were acquired, constraining whether or not vertebrate characters evolved in concert or in a staggered fashion over a protracted episode. Unfortunately, there are precious few remains of unequivocal primitive vertebrates in the fossil record, an artifact of their lack of mineralized tissues. Nevertheless, the fossil record of vertebrates more advanced than the most primitive living vertebrates is sufficient to reveal that many characters generally deemed vertebrate specific were actually acquired much later, among now-extinct lineages intermediate between the living lampreys and living jawed vertebrates. In fact, our knowledge of these extinct intermediates is sufficient to exclude the possibility of a major anatomical event to coincide with the genome duplication that can be inferred to have occurred.

Knowledge of the evolutionary relationships of both fossil and living primitive vertebrates has also provided a test of the new-head scenario that food acquisition drove early vertebrate evolution. Undoubtedly the living groups lancelets (filter-feeding invertebrates), lampreys (parasitic jawless vertebrates), and jawed vertebrates (primitively predators) bear this out, but rigorous functional analyses of extinct jawless vertebrates have demonstrated that this progressive pattern is a gross oversimplification of actual events. Some extinct jawless vertebrate groups like the conodonts were undoubtedly sophisticated feeders (scavengers/predators), but most other groups appear to have been deposit feeders, making their living by slurping mud.

Although the fossil record of the most primitive vertebrates has been downplayed here, there has been no shortage of potential candidates identified, including additional fossils from the Lower Cambrian Chengjiang deposit *(Haikouella, Yunnanozoon)* and the Middle Cambrian Burgess Shale *(Pikaia, Nektocaris)*. Note should also be made of the spectrum of bizarre echinoderm-like organisms referred to as the calcichordates, carpoids, or stylophorans. The interpretation of these remains has long been extremely controversial, but some specialists have perceived that among these organisms lie almost every possible kind of intermediate between the deuterostome phyla and the chordate subphyla: tunicates, lancelets, and vertebrates. Changed understanding in interpretation of the interrelationships of the deuterostome phyla (echinoderms, hemichordates, chordates) has, however, shown many of the most important characters to be primitive features of deuterostomes, not derived characteristics of tunicates, lancelets, and vertebrates.

MOLECULAR PHYLOGENETICS

Traditional perceptions of the interrelationships of the deuterostome phyla recognize hemichordates as the closest relatives of the chordates. However, phylogenetic analysis of molecular data has demonstrated that echinoderms and hemichordates are each others' closest relatives, constituting the group Ambulacraria, to the exclusion of chordates.

Historically the relationships of the chordate subphyla have fluctuated, and either tunicates or lancelets have been considered the closest relatives of vertebrates. The modern view is that lancelets are more closely allied to the vertebrates, but this has begun to be questioned with the discovery that tunicates appear to possess evolutionary equivalents of the neurogenic placodes of vertebrates, as well as cells with migratory properties and even some of the fates seen in neural crest cells.

Have these characteristics evolved in parallel among vertebrates and tunicates, or were they lost in amphioxus? This has been partly resolved by analyses of large molecular data sets that have unequivocally identified tunicates as the closest living relatives of vertebrates. More germane are the implications of these discoveries for the view that the origins of the neural crest and neurogenic placodes were the key innovations that underpinned vertebrate evolution. Clearly they were present and had evolved some of their embryological fates among invertebrate chordates well before the origin of vertebrates. Nevertheless, the divergence of vertebrates clearly coincides with an unfolding explosion in the embryological potential of the neural crest and neurogenic placodes.

CONCLUSIONS

Our understanding of the origin of vertebrates has been reinvigorated in recent years, largely as a result of a more holistic approach to understanding the developmental and evolutionary establishment of the major animal body plans. Together these data provide for a more protracted episode of embryological evolution underpinning the vertebrate origin of vertebrates rather than an explosive radiation resulting from one or two key innovations. Many of the features deemed characteristically vertebrate are now recognized to have a prehistory among invertebrates, including the neural crest and neurogenic placodes. Thus in many of its details the new-head hypothesis was wrong, but these are just details. Despite discoveries of events of genome duplication, discoveries of fossil vertebrates tens of millions of years older than the oldest previously known, and a fundamental overhaul in our understanding of the invertebrate relatives of the vertebrates, our perceptions on the origin of vertebrates remain rooted in the neural crest, neurogenic placodes, and the development of a new head.

BIBLIOGRAPHY

Benton, M. J. 2005. *Vertebrate Palaeontology.* Oxford: Blackwell.

Carroll, S. B., J. K. Grenier, and S. D. Weatherbee. 2001. *From DNA to Diversity: Molecular Genetics and the Evolution of Animal Design.* Malden, MA: Blackwell Science.

Delsuc, F., H. Brinkmann, D. Chourrout, and H. Philippe. 2006. Tunicates and not cephalochordates are the closest living relatives of vertebrates. *Nature* 439: 965–968.

Donoghue, P. C. J., A. Graham, and R. N. Kelsh. 2008. The origin and evolution of the neural crest. *BioEssays* 30: 530–541.

Donoghue, P. C. J., and M. A. Purnell. 2005. Genome duplication, extinction and vertebrate evolution. *Trends in Ecology and Evolution* 20: 312–319.

Janvier, P. 1996. *Early Vertebrates.* Oxford: Oxford University Press.

Mazet, F., and S. M. Shimeld. 2005. Molecular evidence from ascidians for the evolutionary origin of vertebrate cranial sensory placodes. *Journal of Experimental Zoology, Part B: Molecular and Developmental Evolution* 304B: 340–346.

Northcutt, R. G. 2005. The new head hypothesis revisited. *Journal of Experimental Zoology, Part B: Molecular and Developmental Evolution* 304B: 274–297.

Purnell, M. A. 2002. Feeding in extinct jawless heterostracan fishes and testing scenarios of early vertebrate evolution. *Proceedings of the Royal Society of London, Series B* 269: 83–88.

Swalla, B. J., and A. B. Smith. 2008. Deciphering deuterostome phylogeny: Molecular, morphological and palaeontological perspectives. *Philosophical Transactions of the Royal Society B: Biological Sciences* 363: 1557–1568.

—*P.D.*

Osborn, Henry Fairfield (1857–1935)

Henry Fairfield Osborn was one of the most influential paleontologists in early twentieth-century America. He was independently wealthy and adopted a patrician attitude toward his colleagues, based on his own position in high society. He was active in promoting the discovery of fossils both in the American West and in central Asia, although in later life he seldom went into the field himself. He described many new fossil species and studied the origin and development of many important groups of organisms. Although he was a convinced evolutionist, he rejected Darwinism and moved toward a theory based on parallel evolution driven by orthogenetic trends. At the same time, however, he was aware that at certain crucial points in the history of life evolution had led to the development of new forms with immense potential, and he popularized the concept of *adaptive radiation*.

Osborn was born in Fairfield, Connecticut, on August 8, 1857. His father was the president of the Illinois Central Railroad. He studied at the College of New Jersey (later Princeton University), where he met William Berryman Scott. The two became interested in paleontology and undertook fossil-hunting expeditions in the West in 1877 and 1878. On the latter occasion they met Edward Drinker Cope and both came under the spell of his non-Darwinian evolutionism based on the idea of parallel trends governing the development of each major group. After studying in Britain with T. H. Huxley and Francis Maitland Balfour, Osborn taught at Princeton until 1891, when he moved to Columbia University in New York, where he founded the Department of Biology. He also built up links with the American Museum of Natural History, where he created the Department of Vertebrate Paleontology. Osborn died in 1935.

Osborn organized many expeditions throughout his career and was active in describing the fossils they discovered. He worked on the origin of mammals, accepting that they evolved from the mammal-like reptiles of South Africa. He thought that the monotremes (platypuses and echidnas) were a separate development from a different reptilian stock. He continued the work of Cope on the evolution of mammalian teeth. His *Age of Mammals in Europe, Asia and North America* (1910) was a major overview of mammalian evolution. He published monographs on the giant extinct titanotheres (see figure on page 794) and on the fossil Proboscidea (elephants). His *Men of the Old Stone Age* (1916) was an influential survey of human paleontology.

Initially Osborn favored Cope's neo-Lamarckism, and he always retained Cope's vision that evolution was composed of multiple parallel trends. In his later writings, though, he favored an orthogenetic explanation of the trends, invoking mysterious forces within the organism that drove variation in predetermined directions. At the same time, however, he was active in exploring the correlation between major evolutionary innovations and the more dramatic changes that have transformed the earth's climate at key points in geological time. Osborn's vision of evolution was thus one of rare outbursts of innovative adaptive radiation within newly formed groups, followed by long periods of steady parallel development among the branches thus established.

The titanothere was a large mammal that roamed North America about 30 million years ago. Henry Fairfield Osborn was fascinated by it, in part because being big he felt it worthy of the attention of a paleontologist as distinguished as he, and in part because of the bizarre forms that their noses took, each species odder than the one before. Today we are inclined to think that the noses are the product of sexual selection, but Osborn saw them as evidence of a kind of momentum, "orthogenesis," in evolution, which takes organisms to adaptive heights and then over from the useful to the dysfunctional and absurd. See Osborn 1929b.

BIBLIOGRAPHY

Bowler, P. J. 1996. *Life's Splendid Drama.* Chicago: University of Chicago Press.
Gregory, W. K. 1937. Biographical memoir of Henry Fairfield Osborn 1857–1935. *National Academy of Sciences of the United States of America Biographical Memoirs:* 51–119.
Osborn, H. F. 1910. *The Age of Mammals in Europe, Asia and North America.* New York: The Macmillan Company.
———. 1916. *Men of the Old Stone Age: Their Environment, Life, and Art.* London: George Bell.
———. 1929a. *From the Greeks to Darwin: The Development of the Evolution Idea through Twenty-Four Centuries.* New York: Scribner's.
———. 1929b. *The Titanotheres of Ancient Wyoming, Dakota and Nebraska.* Washington, DC: U.S. Government Printing Office.
Ranger, R. 1991. *An Agenda for Antiquity: Henry Fairfield Osborn and Vertebrate Paleontology at the American Museum of Natural History, 1890–1935.* Tuscaloosa: University of Alabama Press.
Ruse, M. 1996. *Monad to Man: The Concept of Progress in Evolutionary Biology.* Cambridge, MA: Harvard University Press.   —*P.J.B.*

Owen, Richard (1804–1892)

Richard Owen was an English comparative anatomist and paleontologist who founded the British Museum (Natural History). Born in Lancaster, Owen was the second of six children and the younger of two sons. In 1824 he matriculated at Edinburgh University but stayed only about half a year. In 1827, at the age of 22, Owen started on his lifelong career as a museum curator, being made assistant curator to William Clift (1775–1849) at the Hunterian Museum of the Royal College of Surgeons. In 1842 Owen succeeded Clift as conservator of the Hunterian Museum. From the outset Owen wished to turn the Hunterian Museum into a national museum of natural history. This proved impossible, and in 1856 he resigned from the College of Surgeons to become the first superintendent of the natural history collections at the British Museum, in which position Owen agitated for a separate natural history museum. The fulfilment of his ambition came in 1881 when the British Museum (Natural History) in South Kensington opened its doors to the public.

During the early part of his career, being crucially dependent on the patronage of Oxbridge Anglicans who were followers of William Paley, Owen perfected the functionalist approach in comparative anatomy. Examples of this work are his *Memoir on the Pearly Nautilus* (1832), as well as his work of a decade later on the extinct moas from New Zealand. Traditionally Owen has been portrayed as an archcreationist, opposed to any sort of evolution, whose anti-Darwinian machinations were the main obstacle in the fight to establish the truth of organic evolution. This portrayal is inaccurate: over a period of some four decades, from the mid-1840s to the mid-1880s, Owen explicitly and repeatedly expressed—in articles, monographs, a textbook, and letters—his belief in a natural origin of species. Owen's change from creation to species transformation went hand in hand with his shift of approach from functionalist to transcendental morphology, which latter approach he perfected in *On the Archetype and Homologies of the Vertebrate Skeleton* (1848).

To demonstrate the inadequacy of the functionalist method, Owen cited the development of the skull because it had been the skull with which transcendental anatomy, in the work of Goethe, Carl Gustav Carus, Lorenz Oken, and others, had been most publicly associated. The skull of the human fetus at the time of birth consists of some 28 separate pieces that ultimately unite into an unyielding whole. The functional purpose of the loose, disassembled nature of the fetal skull was believed to be that it facilitated childbirth by making possible a change of shape of the head. However, such a function exists only in placental mammals; in other vertebrates, such as marsupial mammals and birds, no supple and adjustable cranium is needed to make parturition safer, and yet in those animals, too, the skull bones are uncoalesced at the time of birth. A more comprehensive view had to be taken of what governs the fact that in all vertebrates, as a rule, the skull is composed of the same number of pieces, arranged in the same general way.

Owen then borrowed and further developed the vertebrate theory of the skull, that is, the notion that the skull, like the rest of the skeleton, should be understood as metamorphosed vertebrae. The basic building block of the skeleton was the vertebra, and in principle the skeleton was a series of ideal vertebrae, most closely approximated by the relatively simple skeleton of fishes. Owen called this fishlike concatenation of virtually undifferentiated ideal vertebrae the vertebrate archetype (see figure). The meaning of an organ did not derive from a specific function but from its place in the architectural makeup of the whole organism, that is, from its homological relations to the archetype. Skull parts are uncoalesced in all vertebrate fetuses because the loose bones are homologically related to separate cephalic vertebrae.

With his notion of archetypes, Owen shifted the evidence for the existence of a supreme designer from concrete adaptations to an abstract plan—from special to general teleology. Divine contrivance was to be recognized not so much in the characteristics of individual species but in their common ground plan. In other words, species had originated by natural means, not by miraculous creation. To this belief Owen gave cautious, some would say cryptic, expression in a variety of publications, for example, in *On the Nature of Limbs* (1849). Strong criticism from among his Anglican Paleyite patrons turned Owen into a closet evolutionist, but he enormously resented it when later he was portrayed by Charles Darwin (1809–1882), with whom Owen had enjoyed a good working relationship, as a creationist.

The question remained, however, what these "natural means" were. With respect to this issue Owen and Darwin parted company. To Owen natural selection could only explain the extinction of species, not their origin. He believed in an orthogenetic-saltational process of organic unfolding, driven by an inherent tendency to chance, not by external forces such as natural selection. The phenomenon of metagenesis provided Owen with a visualizing aid for this kind of evolution. In his booklet *On Parthenogenesis, or the Successive Production of Procreating Individuals from a Single Ovum* (1849) he described the phenomenon of alternating generations in the reproductive cycles of, for example, aphids, jellyfish, or flukeworms. Characteristic of such metagenetic cycles is that the individual generations can differ in form from each other as much as different species or even genera, families, and orders do. One could imagine that under particular circumstances the cycle would be broken, and the separate stages would go on reproducing. In this way wholly new genera or even orders might originate.

Owen's involvement in the question of origins was, by and large, incidental to his museum work. He was above all a museum man whose main ambition was not solving the problem of the origin of species but establishing a separate national museum of natural history. The issue of evolution was subordinate to that of Owen's museum-building agenda. To a significant extent he made his scientific work, his choice of subject matter, and his theoretical approach serve the politics of institutionalization. The traditional preoccupation with the differences between Owen and Darwin has hindered a balanced appreciation of Owen's considerable accomplishments.

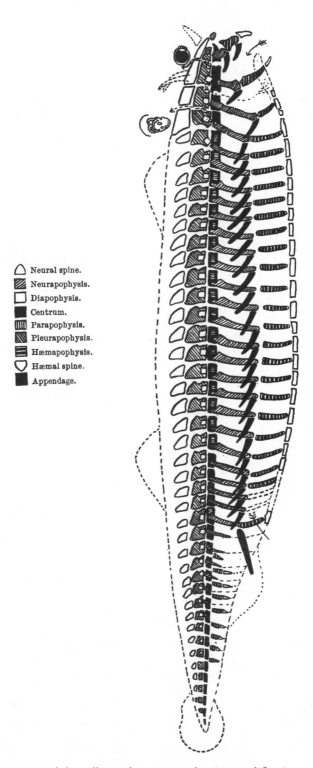

Neural spine.
Neurapophysis.
Diapophysis.
Centrum.
Parapophysis.
Pleurapophysis.
Hæmapophysis.
Hæmal spine.
Appendage.

Richard Owen argued that all vertebrates are adaptive modifications of a basic archetype. Evolutionists like Charles Darwin assumed that the archetype represented an ancestor. Owen probably did too, but he kept open the option that it was more an ideal representation. (From Owen 1849a.)

BIBLIOGRAPHY

Desmond, A. 1989. *The Politics of Evolution: Morphology, Medicine, and Reform in Radical London*. Chicago: University of Chicago Press.

Owen, R. 1832. *Memoir on the Pearly Nautilus (Nautilus Pompilius, Linn.) with Illustrations of Its External and Internal Structure*. London: W. Wood.

———. 1848. *On the Archetype and Homologies of the Vertebrate Skeleton*. London: Richard and John E. Taylor.

———. 1849a. *On the Nature of Limbs*. London: John Van Voorst.

———. 1849b. *On Parthenogenesis, or the Successive Production of Procreating Individuals from a Single Ovum*. London: John Van Voorst.

———. 1894. *The Life of Richard Owen*. 2 vols. London: John Murray.

———. 2007. *On the Nature of Limbs: A Discourse*. New ed. R. Amundson, ed. Chicago: University of Chicago Press.

Richards, E. A question of property rights: Richard Owen's evolutionism reassessed. *British Journal for the History of Science* 20: 129–171.

Rupke, N. A. 1994. *Richard Owen*. New Haven, CT: Yale University Press. Revised ed., Chicago: University of Chicago Press, 2009.

Sloan, P. R., ed. 1992. *Richard Owen: The Hunterian Lectures in Comparative Anatomy, May–June, 1837*. Chicago: University of Chicago Press.

—N.A.R.

P

Paley, William (1743–1805)

Archdeacon William Paley was one of the clearest and most forceful defenders of the middle-of-the-road kind of Christianity epitomized by the Church of England. He was not a very original thinker, but he had an eye for an example, a sense of purpose, and an overall skill at marshaling an argument that is possessed by only the very best textbook writers. He tried hard to keep emotion out of his works, but in public life he was a dedicated opponent of the slave trade and in private life was completely committed to the wonder and joy of God's creation, even when at the end of his life he suffered painfully from cancer of the bladder.

Two of Paley's books, both standard reading for an undergraduate at Cambridge in the years when Charles Darwin was a student, concern us here. First, there is his work on revealed religion, that part of religion that deals with faith and commitment. *Evidences of Christianity* (1794) has David Hume's skeptical arguments about miracles in its sights. It makes the case that it is reasonable to be a Christian, given the testimony of Jesus's disciples and the martyrdom that they were prepared to undergo rather than deny his divinity. The claim is that miracles require something out of the ordinary, an intervention by a creator, and the word of Jesus's followers guarantees the veracity of their reports.

As a student Darwin accepted this argument completely. However, on the *Beagle* voyage he came under the influence of Charles Lyell's *Principles of Geology* (1830–1833), which argued that no miracles are needed to explain the wonders of the world—the mountains, the seas, the lakes, and more. Just regular causes operating over much time are enough. Darwin was convinced by this, and as his belief in miracles in this sphere declined, so did his belief in miracles generally. Given that his was a Christianity that depended crucially on the existence of miracles, this meant that his faith faded away. Darwin never became an atheist. Toward the end of his life he became an agnostic. But in the 1830s he joined Lyell in a form of deism—God as unmoved mover—and since for this God laws are the greatest mark of his magnificence, Darwin was on the way to evolution.

Paley's other important pertinent work was his *Natural Theology* (1802), a work on the intellectual reasonableness of Christianity, with its celebrated opening about finding a watch in a desert and inferring a watchmaker. Thus Paley argued for the existence of God, claiming that because the eye is like a telescope, and since the telescope has a telescope maker, so the eye must have an eye maker, namely, God. Again Hume is the spur, for the celebrated Scottish philosopher had claimed that the world gives no evidence of design of a deity like the Christian god.

At first Darwin absorbed Paley's thinking completely, and when he came to doubt the Christian god, he continued to think that Paley was absolutely right that the organic world shows designlike features. In searching for a mechanism of evolutionary change, Darwin therefore sought a mechanism that could explain such features, adaptations. This mechanism he found in natural selection. The Darwinian mechanism of the *Origin of Species* does not simply push for change; it pushes for change in a particular direction, namely, adaptive advantage, like the beak of the finch.

This is the position still taken by Darwinians today. Richard Dawkins, in *The Blind Watchmaker* (1986), argues that Paley's solution is absolutely wrong. There is no need of a Christian god to explain adaptation. But he agrees that Paley's question is the right one. The most important question for the evolutionary biologist is just how one explains adaptive complexity. Which, referring back to Paley and his *Natural Theology*, all goes to demonstrate the lasting importance of a really good textbook.

BIBLIOGRAPHY

Darwin, C. 1859. *On the Origin of Species*. London: John Murray.
Dawkins, R. 1986. *The Blind Watchmaker*. New York: Norton.
Lyell, C. 1830–1833. *Principles of Geology: Being an Attempt to Explain the Former Changes in the Earth's Surface by Reference to Causes Now in Operation*. 3 vols. London: John Murray.
Paley, W. [1794] 1819. *Evidences of Christianity (Collected Works*, vol. 3). London: Rivington.
Paley, W. [1802] 1819. *Natural Theology (Collected Works*, vol. 4). London: Rivington.
Ruse, M. 2003. *Darwin and Design: Does Evolution Have a Purpose?* Cambridge, MA: Harvard University Press.   —M.R.

Pearson, Karl (1857–1936)

Karl Pearson was an important British statistician, a major enthusiast for eugenics (the belief that one can and should direct the course of human evolution, particularly by restrictions on breeding), a widely read philosopher of science, and an important influence on the history of modern evolutionary theory.

Pearson was a brilliant student, graduating third wrangler (the ordered list in the mathematics honors course) from Cambridge, but at the same time he immersed himself in many other areas of interest, including German literature to such an extent that shortly thereafter he was offered a post in the Cambridge languages department. He trained for the law (in the footsteps of his father) but never practiced, soon becoming a professor of mathematics at University College, London, his academic home for his whole career. He lectured widely to nonacademic audiences on such topics as Marx, German religious thought, and social issues, such as the relative roles of men and women in society. He wrote a classic philosophical work in the positivist tradition, *The Grammar of Science* (1892, second edition 1900), in which he argued that ideas exist in the mind, and this is our source of knowledge. Hence all is relative to mind—"the field of science is much more consciousness than an external world" (1900, 52), and "law in the scientific sense is thus essentially a product of the human mind" (1900, 36). It is no surprise that the young Albert Einstein devoured this volume.

Pearson's work in the field of statistics will forever keep him in the halls of scientific heroes, for he contributed to developing the ideas of linear regression (estimating one variable given the values of another variable) and correlation (estimating the connection between two variables), as well as the famous chi-square test (estimating the chances that variables follow a specific frequency, for instance, that observed numbers of males and females support the idea that the numbers of the sexes are evenly balanced). Later in his career Pearson was embroiled in a ferocious controversy with the no-less-eminent evolutionist and statistician Ronald A. Fisher over technical questions in statistics. In major part this was a clash between two dominant personalities, but it was also a difference between rival aims—Pearson using large samples to infer correlations and Fisher using small samples to ferret out causes.

It is hard today to feel sympathy for one as committed as was Pearson to eugenics. "It is a false view of human solidarity, a weak humanitarianism, not a true humanism, which regrets that a capable and stalwart race of white men should replace a dark-skinned tribe which can neither utilize its land for the full benefit of mankind, nor contribute its quota to the common stock of human knowledge" (Pearson 1900, 369). Let us simply observe that in the early days of genetics, many thought that eugenics offered a relatively quick and easy method of improving humankind. We can also note that much of Pearson's seminal work in statistics stemmed from his eugenic aims, an overall program he inherited from a major influence (in his thinking about statistics) and general mentor, Francis Galton (the cousin of Charles Darwin), and a program shared by his rival Fisher.

Pearson's major contributions to evolutionary thinking came from his interaction with his colleague at University College, W. F. R. (Raphael) Weldon, with whom in 1901 Pearson founded the journal *Biometrika*. Weldon was the most important observational and experimental evolutionary biologist (with the possible exception of Henry Walter Bates) between the publication of Darwin's *On the Origin of Species* in 1859 and the neo-Darwinians (or synthetic theorists) who began sustained empirical work on evolutionary problems in the 1930s. Weldon was convinced that natural selection is a major factor in evolutionary change. To support this view he turned to large-scale projects, measuring features of organisms that he hypothesized might show the effects of selection. Some of his most important work was done on dimensions of crabs found in the Bay of Naples. He followed this work with like studies of crabs off the coast of Devon in England and then turned to experiments to see if his findings could be replicated under controlled conditions. (In particular, Weldon believed that mouth sizes were a function of relative needs to feed and to avoid the clogging effects of silt.)

Weldon turned to the mathematically gifted Pearson for help in reducing his findings to mathematics, which could then be interpreted and translated back into empirical claims. In one important finding Weldon believed that his Naples crabs were not one single population but two overlapping populations (see figure) and showed that selection was dividing the populations according to different adaptive needs and advantages. It was Pearson who was able to show that Weldon's findings were not random or meaningless but formed two overlapping normal curves, thus suggesting that it is indeed true that selection is a major distorting influence—distorting in the sense of driving different individuals to different adaptive solutions.

Unfortunately, the significance of this work was soon overshadowed by a major controversy that pitted Weldon and Pearson against Weldon's former student William Bateson. Weldon and Pearson, the leaders of the faction known as biometricians, were committed (as ardent Darwinians) to the belief that the causes of variation must be such as to produce very slight changes in the members of populations. Whatever the causal factors might be, the differences produced must be virtually continuous; otherwise, natural selection could not operate (or worse, became unnecessary). At the beginning of the twentieth century, Bateson, who was searching for the causes of variation as a project in its own right (that is, without regard for the significance of selection), seized on the rediscovery of the ideas of the Moravian monk Gregor Mendel. Henceforth Bateson was the leader of the British Mendelians, who were opposed to the biometricians in the belief that the significant changes in evolution came through one-step, newly appearing variations, what today we would call mutation, or rather macromutation. (This is a form of evolutionism called saltationism, from the Latin *saltus,* "jump." In one form or another it has been endorsed by evolutionists from Thomas Henry Huxley to Stephen Jay Gould and is generally characterized by a belittling of the importance of selection and of adaptation, the chief reason for invoking selection over other mechanisms of change.)

It was not long before more temperate heads started to realize that in major respects the biometrician-Mendelian debate was founded on the false belief that they were incompatible. As soon as one sees that mutations can have minor as well as major effects, selection can come into play, and there is thus no need to choose between the two sides, and indeed much reason to synthesize insights from both camps. Unfortunately, before this really happened, Weldon (who took the dispute very personally) died suddenly, right in the middle of an exhaustive study of the British stud book (which records features of racehorses and thus offers promising material for the study of the nature of heredity). Pearson was devastated and, blaming Bateson, continued the feud long after it had exhausted its usefulness or its very rationale.

Living to a ripe old age, after his retirement Pearson had the somewhat dubious pleasure of seeing his professorship divided into two, one part of which was given to his son, who continued his work, and the other to his archrival Fisher, who in his seminal *The Genetical Theory of Natural Selection* (1930)

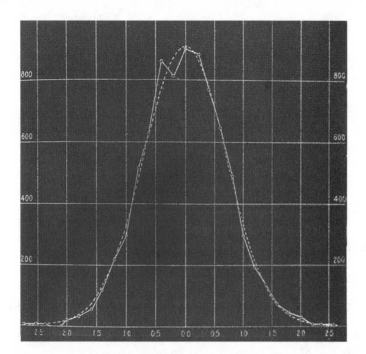

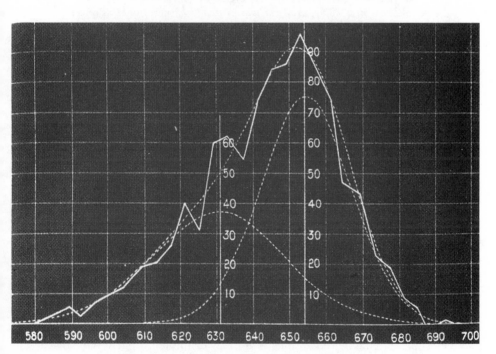

Karl Pearson's close friend Raphael Weldon produced these graphs of crab frontal breadths, the first *(top)* from specimens caught in Plymouth Sound (England) and the second *(bottom)* from specimens caught in the Bay of Naples (Italy). The first shows a typical normal curve. The second, as Weldon discovered empirically, can be shown as the combination of two normal curves. Weldon turned to Pearson to give a mathematical analysis, showing that at Naples there are two populations, breaking apart under different selective pressures. (From Weldon 1893.)

offered the definitive blending of Darwinian natural selection with Mendelian genetics.

BIBLIOGRAPHY

Fisher, R. A. 1930. *The Genetical Theory of Natural Selection.* Oxford: Oxford University Press.
MacKenzie, D. A. 1981. *Statistics in Britain 1865–1930: The Social Construction of Scientific Knowledge.* Edinburgh: Edinburgh University Press.
Pearson, E. S. 1936. Karl Pearson: An appreciation of some aspects of his life and work. Part 1. *Biometrika* 28: 193–257.
———. 1937–1938. Karl Pearson: An appreciation of some aspects of his life and work. Part 2. *Biometrika* 29: 161–248.
Pearson, K. 1892. *The Grammar of Science.* London: Walter Scott.
———. 1894. Contributions to the mathematical theory of evolution. *Philosophical Transactions A* 185: 71–110.
———. 1900. *The Grammar of Science.* 2nd ed. London: Black.
Provine, W. B. 1971. *The Origins of Theoretical Population Genetics.* Chicago: University of Chicago Press.
Ruse, M. 1996. *Monad to Man: The Concept of Progress in Evolutionary Biology.* Cambridge, MA: Harvard University Press.
———, ed. 2009. *Philosophy after Darwin: A Reader.* Princeton, NJ: Princeton University Press.
Weldon, W. F. R. 1893. On certain correlated variations in *Carcinus moenas. Proceedings of the Royal Society of London* 54: 318–329.
———. 1895. Attempt to measure the deathrate due to the selective destruction of Carcinus moenas with respect to a particular dimension. *Proceedings of the Royal Society of London* 57: 360–379.
———. 1898. Presidential address to the Zoological Section of the British Association. *Transactions of the British Association* (Bristol), 887–902. London: John Murray.   —M.R.

Phenotypic plasticity

Phenotypic plasticity is the ability of an individual to express different features under different environmental conditions. For example, flowering plants produce broader leaves when they grow in shady conditions, and fish are smaller when they develop in crowded conditions. Some of these changes reflect unavoidable consequences of adverse conditions, like the stunting of fish when they are crowded. Others reflect general effects of factors like temperature on development, as when tadpoles grow larger before metamorphosing when they experience cooler conditions. But many examples of phenotypic plasticity are actually adaptations to a variable environment; they reflect the evolution of a developmental system that produces different traits under different conditions because no single trait is best suited for all conditions.

Adaptive plasticity is one of the most remarkable products of Darwinian evolution. For adaptive plasticity to evolve, there must be an ecological situation in which no one value of a trait works well in all conditions. In addition, the developmental machinery to build different trait values must be integrated with a sensory system that detects reliable cues about the prevailing environmental condition so that suitable trait values are expressed in a timely manner. Given these requirements, it seems incredible to argue that such features could evolve through Darwinian evolution.

Yet adaptive plasticity is ubiquitous. Some of the most familiar examples are traits or features whose expression is reversible; that is, trait values change back and forth within an individual's lifetime. For example, tadpoles change their foraging patterns in response to the presence or absence of predators. Tadpoles that forage blithely in front of predators fare poorly, as do tadpoles that forage only furtively when there is no risk of predation. Many examples of plasticity are not reversible. Water fleas in ponds develop spines and a thicker carapace in response to the presence of a predatory fly larva in the water; once developed, the carapace is not altered appreciably even if the predators disappear. A thicker carapace and its attendant spines reduce the ability of the predatory flies to capture and eat water fleas. However, energy invested in building a thicker carapace is diverted from other needs, so individuals who form a thicker carapace in the absence of appreciable predation risk have lower fitness than individuals who forgo that thicker carapace.

Adaptive plasticity has two characteristics that distinguish it from other types of environmental responses in development. First, it is specific. That is, the development of certain traits responds to specific cues; traits in a species that respond to one environmental agent do not respond to a different one, and the same features in different species do not respond to the same agent. For example, damselfly species that coexist with fish behave differently in the presence of fish than in their absence, but species that do not coexist with fish fail to respond to their presence and are more likely to be eaten. Wild parsnip populations with a history of heavy attack from herbivores respond to leaf damage by secreting compounds toxic to insect herbivores into leaf tissue to deter further herbivory; populations without a history of heavy herbivore attacks do not respond to leaf damage in this manner.

Second, adaptive plasticity is precise; a trait with adaptive plasticity responds only to a particular range of variation in an environmental factor, and the same trait in different species may respond to a different range of variation in that same factor. Insect diapause is a classic example. Diapause is a state of arrested development in which the animal can survive long periods of challenging conditions like low temperatures or drought by lying dormant. It is a distinct physiological state that is expressed in response to particular cues, and the animal exits diapause when conditions change. Different insect species enter diapause in different stages, some in the egg, others as larvae or pupae. Populations of the same insect species at different latitudes enter winter diapause in response to different combinations of temperature and day length.

Adaptive plasticity evolves because it gives individuals with the capacity to adjust their development to the prevailing conditions the ability to outperform, in the long run, individuals who express the same trait values or features regardless of environmental conditions. The subtlety in studying plasticity is the phrase "in the long run." In any single circumstance the individual with the capacity to adjust its development to express the most suitable feature will perform just as well as the individual who expresses the same feature constitutively (that is, expression is the same regardless of condition), but it will outperform all the individuals who express unsuitable features constitutively. Individuals with the capacity to adjust development have high fitness in all conditions, whereas individuals with constitutive development patterns for the same set of features have high fitness in some conditions but low fitness in most conditions. In the long run, over many generations or many locations, individuals with the capacity to adjust development have the highest *average* fitness.

Of course, a given individual does not live in many generations or at many locations at once, so in this case the fitness benefits are associated with the collection of genes that control the plastic developmental system. This point gives phenotypic plasticity its intellectual richness. There is an extensive body of theory on how natural selection on individual animals can drive evolutionary changes in developmental systems that involve dozens or even hundreds of genes. This is not a trivial problem; a successful individual does not pass its entire genome to its offspring. The identification of networks of genes that control development and the ability to measure the expression levels of those genes have opened the door to testing this theory in ways that were not possible in the past.

The study of plasticity has a long history in evolutionary biology and has attracted the attention of some of its leading figures (see the alphabetical entries "Ivan Ivanovich Schmalhausen" and "Conrad Hal Waddington" in this volume). There is a growing body of research on the ecological consequences of adaptive plasticity, particularly in systems of induced defenses like that of the wild parsnip, and the many facets of phenotypic plasticity continue to inspire research in this area, as well as in evolutionary biology.

BIBLIOGRAPHY

DeWitt, T. J., and S. M. Scheiner, eds. 2003. *Phenotypic Plasticity: Functional and Conceptual Approaches*. New York: Oxford University Press.
Miner, B. G., S. E. Sultan, S. G. Morgan, D. K. Padilla, and R. A. Relyea. 2005. Ecological consequences of phenotypic plasticity. *Trends in Ecology and Evolution* 20: 685–692.
Pigliucci, M. 2001. *Phenotypic Plasticity: Beyond Nature and Nurture*. Baltimore: Johns Hopkins University Press.
West-Eberhard, M. J. 2003. *Developmental Plasticity and Evolution*. New York: Oxford University Press.   —*J.T.*

Phylogenetics

Phylogenetics is the tracing of genealogical relationships among species and the use of those relationships to classify species and groups of species into a hierarchical tree of life. It is part of the practice of systematics, the other branch of which is taxonomy (the distinguishing of species from one another and the diagnosis of species boundaries). Phylogenetics is one of evolutionary biology's central specialties, perhaps *the* central specialty, and there is no other specialty that has been as controversial in its theory and practice.

Modern phylogenetics is predicated on the principle that the classification of organisms must reflect a hierarchy of evolutionary relationships. Those relationships emerge from evolutionary history. A single species may, in time, split into two daughter species, each of which may split in their turn. As some species divide and others do not, and as some go extinct without descendants while others spawn many descendants, a genealogy emerges that resembles the branching pattern of a tree, with some major branches producing many smaller branches and others not.

Several conventions guide the use of the phylogenetic tree in classification. The groups on adjacent branches of a genealogical tree must be each other's closest relatives because they share an immediately recent common ancestor. If the branches are at the tips of the tree, each group is an individual species; if the adjacent branches are lower, each group is a collection of species found on the higher branches that emerge from each lower branch. All the species along a major branch of the genealogy constitute a group, which might be a genus, family, or order, depending on the position of the major branch, and the hierarchical labeling of the various branches produces a hierarchical classification of all the species on the tree. This classification thus carries information about the relationships of groups at all levels to one another and is called a natural classification.

The traditional Linnaean classifications created a hierarchy based on general similarities but made no presumptions about the explicit evolutionary relationships among groups of species. For example, the Linnaean system recognizes five classes of existing reptiles—turtles, alligators, tuataras, lizards, and snakes—but does not specify their relationships to one another or their relationships to groups known only from fossil specimens. This is no surprise, given that the system was devised long before Charles Darwin's *On the Origin of Species* was published. A phylogenetic classification would clarify which of those groups are most closely related, which are closer to the base of the tree, representing older lineages, and what are the relationships of modern reptiles to birds and mammals, as well as dinosaurs and other extinct forms (see the alphabetical entries "Birds," "Dinosaurs: The model system for evolution," and "Modern reptiles" in this volume).

Although employing evolutionary relationships for classification was itself once a controversial idea (see the alphabetical entries "Ernst Walter Mayr" and "Willi Hennig" in this volume), the modern controversies revolve around the theory and practice of estimating genealogical relationships. Phylogenetic inference is extremely challenging because it attempts to reconstruct the past by analyzing features of the present. More specifically, it uses existing characteristics of species within an evolutionary model to reconstruct the most likely paths through which those species diverged from a temporal sequence of common ancestors. The heart of the challenge is that one has data only from existing species, which occupy the tips of the branches on the tree, and from these one is attempting to find a series of nested evolutionary relationships and reconstruct the branching patterns of the entire tree.

The fundamental premise for reconstructing the phylogenetic tree is that evolutionary relationships are revealed by synapomorphies, which are states of homologous characters (see the alphabetical entry "Homology" in this volume) shared by species or groups of species through inheritance from a common ancestor in which the character states evolved. The possession of these shared, derived character states defines a group, and the emergence of these character states defines where the overall genealogy divides into two branches. For example, in all the fishes we call elasmobranchs (chimaeras, sharks, and rays), males have claspers on their pelvic fins; this character state defines the elasmobranchs and distinguishes them from other so-called chondrichthyian fish, which lack claspers (so their character state is "absence of claspers"). The chondrichthyian fish are themselves defined by characteristic mineral deposits in their skeletons, and the presence of claspers defines a branch point where the elasmobranchs separate from the other chondrichthyian fish. As this description implies, more restricted synapomorphies define subgroups and subsequent branch points. Within the elasmobranchs the location of gill arches (bones that support the gills) and the articulation of the dorsal fin distinguish and define the chimaeras and separate them from the sharks and rays (see the alphabetical entry "Fishes" in this volume for more about these relationships). Synapomorphies are thus nested one within another, and the hierarchical classification defined by the synapomorphies emerges from the hierarchy of branching events in the genealogical pattern.

The controversies in phylogenetic inference arise from different convictions about how to reconstruct a tree that is consistent with these nested synapomorphies. First, identifying nested synapomorphies is not sufficient for reconstructing evolutionary lineages; although synapomorphies define groups of species at various levels of branching, they do not reveal the location of the branches on the tree. Specifically, they do not reveal which branches are lower and which are higher. To make this decision, which is called rooting the tree, the synapomorphies must be compared with the character states of the group that would be considered the closest relative to the entire group being analyzed. That group is called the *outgroup,* and the root is placed at the point defined by the synapomorphies that distinguish the ingroup from the outgroup. This establishes the initial direction of evolutionary change within the ingroup and points to the direction in which subsequent branches emerge. Obviously the need for the outgroup extends the analysis to an additional level, and different choices of outgroup will produce different points for the root of the tree and lead to different conclusions.

Second, it is often quite difficult to identify individual synapomorphies, particularly at higher levels on a tree. This is particularly true if the characters that are being used in the study are labile and likely to change frequently. When characters are very labile, two species may share the same character state not because they each inherited it from a common ancestor but because the character evolved to that same state independently in each species. This condition, called homoplasy, can hide genuine relationships and suggest incorrect ones. Homoplasy can be particularly troublesome when DNA sequences are being used in a phylogenetic analysis; a given site in the molecule can have only one of four states (the bases adenine, guanine, cytosine, or thymine), and so homoplasy is quite likely at any single site if mutation rates are comparatively high. In many cases where characters are labile, groups can be identified only through combinations of character states. But identifying groups in this way moves a phylogenetic analysis closer to a more traditional statistical grouping of species based on overall similarity, whether similarity is quantified through DNA sequences or a statistical measure of morphological similarity. In these situations a robust phylogenetic analysis is possible only if one postulates a model for how the character states change in the course of evolution.

Herein lies one source of controversy because there are many models of character change from which to choose, and different models can lead to different reconstructions of evolutionary history. In cases like those that employ base-pair changes in DNA sequences, the models are based on properties of the mutational process, and using different properties can lead to different models of character change. In cases like those that employ morphological characters such as the shapes of bones in vertebrates or the number of bundle scars for plants, the models are based on plausible assumptions about the steps involved in the transformation of one character state into another; different assumptions may lead to different sequences of synapomorphies and different phylogenetic conclusions.

When many characters are being studied in many species, there will rarely, if ever, be a single obvious phylogenetic tree consistent with the similarities and differences among species. The challenge in practical phylogenetic inference is distinguishing the many possible trees and selecting the one that represents the best choice for the data. To understand the challenge, consider that a study of three species, A, B, and C, must distinguish among three possible trees or, as they are denoted in practice, topologies: A and B may be closest relatives, A and C may be closest relatives, or B and C may be closest relatives. This is equivalent to considering the three species as each occupying a branch radiating from a central point but placing the root of the tree along one branch. If A and B are closest relatives, then the root of the tree must be somewhere on the branch that leads from the central point to species C. For four species, A, B, C, and D, there are three unrooted, bifurcating trees, corresponding to close pairings of A and B versus C and D, A and C versus B and D, and A and D versus B and C. In each case there are five branches; one connects each pair of close relatives to the other, and within each pair there is one branch from each member to a point that represents the common ancestor. The root of this tree could be placed along any of the five branches, leading to 15 possible topologies for only four species. This logic can be generalized into a mathematical formula that reveals the enormous number of possible topologies for even a small number of species (e.g., seven species could be described by any of 10,395 possible topologies).

The greatest source of controversy in phylogenetic analysis has been over how to select the best choice of topology for the data in hand. One possibility is to use parsimony analysis, which is based on the premise that the best choice is the topology with the fewest total number of evolutionary changes in the character states. Another possibility is to use maximum likelihood, the premise of which is that the best choice is the topology that maximizes the probability of the data in hand, given a particular model of evolutionary character change. The most parsimonious topology need not be the one with the highest likelihood and vice versa, and this realization has led to spirited and sometimes-bitter debates between the proponents of each method. Moreover, for more than a modest number of species, it is impossible to enumerate all the possible topologies to identify either the most parsimonious or the one with the highest likelihood. In these cases computationally intensive methods, which exploit various mathematical properties of topologies, are employed to estimate the best choice.

Phylogenetic analyses have proliferated over the past 25 years with the advent of widespread computational power. There is now a major effort around the world to describe the phylogenetic relationships among all living organisms and develop a genuine tree of life (see www.tolweb.org). These efforts have changed our understanding of relationships in many groups of organisms, from revealing the relationships between dinosaurs and birds (see the alphabetical entry "Birds" in this volume) to changing our entire view of major groups like the Crustaceans (see the alphabetical entry "Crustaceans" in this volume). In addition, research on phylogenetic methods has proved fertile ground for collaborations among biologists, statisticians, and computer scientists and has often led to innovative projects in other areas (see www .morphbank.net).

Phylogenetic analyses have changed the face of evolutionary biology by illuminating our understanding of many other topics. Combining powerful phylogenetic analyses with data on the distributions of species has led to new insights in biogeography, and examining the phylogenetic distribution of adaptive features has led to greater understanding of the evolution of complex traits (see the main essay "Adaptation" by Joseph Travis and David Reznick in this volume). Phylogenetic methods have been employed to study the origins of local epidemics, identify strains of influenza to be targeted in annual vaccine development, and investigate the origin of novel pathogens (see the main essay "Evolutionary Biology of Disease and Darwinian Medicine" by Michael F. Antolin in this volume).

BIBLIOGRAPHY

Avise, J. C. 2004. *Molecular Methods, Natural History, and Evolution.* 2nd ed. Sunderland, MA: Sinauer Associates.
Felsenstein, J. 2004. *Inferring Phylogenies.* Sunderland, MA: Sinauer Associates.
Kitching, I. J., P. L. Forey, C. J. Humphries, and D. M. Williams. 1998. *Cladistics: The Theory and Practice of Parsimony Analysis.* 2nd ed. Oxford: Oxford University Press.
Pough, F. H., C. M. Janis, and J. B. Heiser. 2002. *Vertebrate Life.* 6th ed. Upper Saddle River, NJ: Prentice Hall.   —*J.T.*

Piltdown Man

Piltdown is the name given to what had been assumed to be an early form of human brought to public attention in 1912. A partial skull and an incomplete mandible plus the bones of fossil mammals were uncovered by a local lawyer (country solicitor in English terminology), Charles Dawson (1864–1916) of Uckfield, Sussex (see figure). The discovery was made rather casually in a gravel pit at Barkham Manor on Piltdown Common not far from Lewes in eastern Sussex, southern England. Dawson had been a collector of prehistoric material for Arthur (later Sir Arthur) Smith Woodward (1864–1944), keeper of geology at the British Museum of Natural History in London. For a good 40 years Piltdown was considered a genuine, if puzzling, human fossil.

Starting in 1953, however, closer scrutiny and testing demonstrated that Piltdown was in fact a deliberate fabrication. Right from the beginning competent anatomists had stated that the skull and mandible came from different species. The skull was clearly human, but the mandible was obviously that of an anthropoid ape. Some guessed that it was a female chimpanzee, and

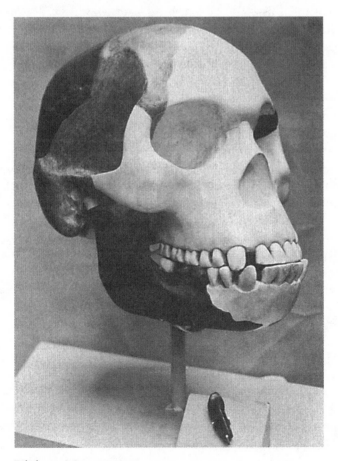

Piltdown Man.

others suggested, correctly, that it was a female orang. At first, even after it was evident that Piltdown was an artificial juxtaposition of unrelated species, there still was no realization that its antiquity was also fraudulent. Eventually it was shown that the appearance of great age and fossilization was the result of staining with a recipe of oxides of iron and manganese, as well as chromium. Piltdown can be regarded as a candidate for the most spectacular scientific fraud of the twentieth century.

Oddly enough, suspicion of fraudulence never fell on Dawson, although he had pursued a career of deception for 35 years. Among his some half-dozen acts of fakery, he was known to have been involved in concocting representatives of supposed early artifacts, fabricating a map of an early iron forge, and plagiarizing a historical account of Hastings Castle, Sussex, from an earlier and unpublished document. He was also known to have dipped specimens he had collected for Smith Woodward in bichromate of potash in order to "harden" them. He made no secret of the fact that he had treated the Piltdown cranial fragments with ferric ammonium sulfate (iron alum) and then had stained them with potassium chromate.

Smith Woodward visited the initial Piltdown site but made no effort to see that a careful excavation and study were made. His principal expertise was in the study of fossil fish, and he and his department at the British Museum were widely regarded as being geologically incompetent. He did bring in Arthur (later Sir Arthur) Keith (1866–1955), a specialist in human anatomy, from the Hunterian Museum at the Royal College of Surgeons in London to assist in the anatomical assessment and reconstruction of the Piltdown material. Although Keith was said by some to have been the perpetrator of the fraud, he never visited the site and had no access to the source of the faunal material that had been planted there along with the human skull and ape jaw. Keith had visited France the year before the Piltdown find was announced and had had a kind of conversion experience there. His earlier outlook had been to arrange the available human fossils according to the times of their existence and declare that this illustrated the course of human evolution. In his 1911 visit to France he abandoned that view for the French stance that modern human form had existed in remote antiquity somewhere in "the East" and perhaps Africa. Subsequently it had come to Europe as invaders and extinguished the "primitive" residents, the Neandertals. This view was clearly at odds with evolutionary biology, but then a Darwinian outlook was never accepted in French biology.

One of the other players in the scene was the French Jesuit priest Pierre Teilhard de Chardin (1881–1955). Teilhard had visited Dawson in 1908 while he was staying at a Jesuit retreat house in Hastings, Sussex. It was at that retreat that he read the 1907 volume *L'évolution créatrice (Creative Evolution)* by Henri Bergson, a somewhat dotty, romantic, and very nonmechanistic treatise. Insofar as French science accepts the idea of evolution at all, it tends to be a Bergsonian and not a Darwinian version. In 1913 Teilhard returned to Ore Place in Hastings for a full year. In August he visited Piltdown, where Smith Woodward and Dawson were washing and sieving the gravel. They suggested that Teilhard sift through the results, and it was while doing so that he found an apelike canine tooth. Some commentators on the Piltdown scenario have suggested that Teilhard may have been the perpetrator of the fraud.

In 1914 Dawson and Smith Woodward found a Piltdown artifact made from a piece of fossilized elephant bone shaped into the form of a cricket bat. The completely humorless Smith Woodward accepted it as a prehistoric artifact even though it was clear that it had been shaped by a metal cutting tool. Then in 1915 Dawson wrote to Smith Woodward that across the River Ouse at Netherhall Farm he had found pieces of a human skull and a molar tooth along with a rhinoceros molar. These were then called Piltdown II. As it happened, Dawson held the position of steward at both Barhkam Manor and Netherhall Farm. Piltdown II was not announced until 1917, partly because of the larger concern for the issues of World War I and partly because Dawson died in 1916.

Even after Piltdown Man's fraudulent nature was documented beyond any possible doubt just after the midpoint of the twentieth century, it was more than another 40 years before the identity of the forger was conclusively demonstrated. In the mid-1970s contractors cleared loft space in the abandoned southwest tower of the Natural History Museum before maintenance work on the roof. There they found a trunk with the letters MACH, clearly the initials of Martin A. C. Hinton (1883–1961), who was deputy keeper of zoology and a fossil rodent expert at the time of the Piltdown fabrication. Some time went by before a recent fossil rodent expert looked in the trunk, and there at the bottom, beneath hundreds of vials of rodent dissections, was a collection of carved and stained elephant and hippopotamus teeth and various bones that looked like the Piltdown collection. Hinton had been well known for his elaborate practical jokes and resented the handling of his salary in the museum by the humorless and incompetent Smith Woodward. But it was not until the mid-1990s that Hinton's role was completely understood. Certainly only an insider at the Natural History Museum would have had access to the fossils planted at the Piltdown sites since those fossils represented part of the museum collections assembled as a result of years of museum-sponsored excavations in various parts of the world. It seems plausible that Dawson was brought in on the plot, but there is no way to be absolutely sure of that now.

BIBLIOGRAPHY

Bergson, H. 1907. *L'évolution créatrice*. Paris: Alcan.

Gee, H. 1996. Box of bones "clinches" identity of Piltdown palaeontology hoaxer. *Nature* 381: 261–262.

Millar, R. 1972. *The Piltdown Men*. New York: Ballantine Books.

Spencer, F. 1990. *Piltdown: A Scientific Forgery*. New York: Oxford University Press.

Thomson, K. S. 1991. Piltdown Man: The great English mystery story. *American Scientist* 79, no. 3: 194–201.

Weiner, J. S. 1955. *The Piltdown Forgery*. London: Oxford University Press.

—C.L.B.

Pleistocene extinctions

The history of life on earth is punctuated by episodes of rapid, widespread extinctions in one or more groups of organisms (see the main essay "Paleontology and the History of Life" by Michael Benton and the alphabetical entry "Mass extinctions" in this volume). One of the most notorious of these episodes is the loss of the so-called Pleistocene megafauna. Between about 50,000 years ago and 10,000 years ago, on all the continents except Africa, most of the large (over 40 kg mass) mammals and birds and some of the large reptiles went extinct. This notoriety springs from the facts that it was among the first wholesale extinctions to be recognized, it included some of the most spectacular mammals ever described, and human hunting appears to have played a major role in causing it.

The mammalian fauna of 50,000 years ago included a high number of very large species. Australia contained echidnas of 20 to 30 kg in mass, kangaroos over 200 kg in mass (the largest extant kangaroos are about 90 kg), a variety of giant birds and enormous monitor lizards, and a genus of large grazing marsupials named *Diprotodon* that were up to 2,000 kg in mass. North America harbored giant ground sloths, mammoths, and mastodons, and the Eurasian fauna included saber-tooths, woolly rhinos, cave bears, and the giant Irish elk. By 10,000 years ago over 90 genera of large mammals were extinct. Although there was some regional heterogeneity, on all the continents save Africa there was nearly total extinction of species whose individuals were larger than 300 kg in mass. In general, extinction rates were inversely proportional to body size; for example, in North America the extinction rate for species larger than 1,000 kg was 100%, that for species between 32 and 1,000 kg was about 50%, and that for species between 10 and 32 kg was only 20%. Deeper analysis suggests that extinction rates were inversely proportional to reproductive rate, which is itself inversely proportional to body size, meaning that the Pleistocene extinctions were very selective, eliminating species with slower reproductive rates. In fact, no other episode of extinctions associated with a glacial-interglacial transition had extinction rates as selective as these.

This striking episode of extinction was recognized as early as the nineteenth century, and its timing, coincident with the waning of the last glacial period and the start of the present interglacial period, suggested that climate change and its associated effects were responsible. Pulses of extinctions are associated with periods of rapid climate change, especially the transition from glacial to interglacial periods, throughout the history of life. Many features of the environment besides temperature regime changed in this transition; seasonality increased as summers became significantly warmer, the growing season was shortened, habitats became increasingly fragmented, and the geographic distribution of plants and plant associations changed dramatically, especially at the higher latitudes. The initial hypotheses about climate change drew upon all these effects; that the Pleistocene extinction rates were much higher than those seen at other glacial-interglacial transitions was ascribed to the possibility that this last transition was more rapid and more extreme than any of the others.

Although the environmental effects of a glacial-interglacial transition surely played a role in the demise of the Pleistocene megafauna, they do not appear sufficient to explain the magnitude of the extinctions nor their selectivity. For one thing, oxygen isotope studies of marine benthic deposits, which reflect sea surface temperatures, do not support the argument that the magnitude and rapidity of this glacial-interglacial transition were unusual when compared with other transitions. Second, the extinctions did not discriminate between grazers (animals that ate grasses) and browsers (animals that ate woody vegetation), and the habitat changes associated with this transition should have put grazers at a distinct disadvantage as woody vegetation increased in abundance. Third, examinations of pollen profiles from long-buried sediments indicate that although there were rapid changes in the composition of plant species nearly everywhere across the globe, regions with higher extinction rates did not experience greater or more rapid changes than regions of lower extinction rates.

A suggestive line of evidence is the fact that the megafauna did not disappear at this time on New Zealand and many oceanic islands, but extinction rates rose dramatically on these islands when humans arrived. Within a short period after the arrival of humans, all the giant moas of New Zealand and a host of other species on the oceanic islands were gone. Deposits of enormous bone piles with clear evidence of butchering and remains found near sites of large firepits point clearly to the effect of human hunting in causing these extinctions.

Of course, it may seem incredible to accept that early humans who hunted with stone weapons could drive so many large-bodied species to extinction. It is clear that humans hunted the Pleistocene megafauna; there is ample evidence of kill sites (large deposits of bones), and stone points have been found in what would have been the body cavities of many skeletal deposits. A small industry in mathematical modeling has suggested that the hypothesis of human overkill is plausible under some assumptions. After all, extinction is a certainty if death rates exceed birth rates long enough, and all that human hunting needed to accomplish was to add enough to the natural death rates to raise them above birth rates. The highly selective nature of these extinctions with respect to reproductive rates is consistent with this hypothesis. On the other hand, although extinctions in North America may have accelerated after the arrival of humans in significant numbers (10,000–20,000 years ago), humans had been present in Europe for tens of thousands of years before the evidence of extensive extinctions and present in Africa, where the extinction rates were lowest, for much longer.

It may be that no single hypothesis is sufficient to account for the magnitude, rate, and distribution of extinctions and that different combinations of effects came into play in different regions of the world. Nonetheless, there is no discounting the effect of humans in contributing to one of the most striking extinction episodes in the last 65 million years of the history of life. For many scientists this may be a harbinger of what is to come in our near future as global climate change occurs in conjunction with the substantial direct and indirect effects of humans on the natural world.

BIBLIOGRAPHY

Kemp, T. S. 2005. *The Origin and Evolution of Mammals.* Oxford: Oxford University Press.

Koch, P. L., and A. D. Barnosky. 2006. Late Quaternary extinctions: State of the debate. *Annual Review of Ecology and Systematics* 37: 215–250.

Martin, P. S. 2005. *Twilight of the Mammoths: Ice Age Extinctions and the Rewilding of America.* Berkeley: University of California Press.   —*J.T.*

Principles of Geology (Charles Lyell)

Charles Lyell published the first edition of the three volumes of *Principles of Geology* in 1830, 1832, and 1833, respectively. Often called the most influential of all geological books, it is also arguably among the dozen most influential texts ever on the living world. The subtitle identified it as "an attempt to explain the former changes of the earth's surface, by reference to causes now in operation." In this attempt Lyell explicitly assumed that ever since the oldest fossil-bearing rocks then known had been laid down (rocks of the Carboniferous groups), the same causes had been at work on the planet's surface, had worked, moreover, in the same general circumstances and with the same average intensity, and so had produced the same sorts and sizes of effects throughout this vast prehuman past, on into the present, and so into an indefinitely prolonged future. There was, then, in Lyell's knowing echo of the eighteenth-century Scottish theorist of the earth, James Hutton, no sign in any rocks of the earth's beginning or of its end.

The treatise has three parts. First, the attempt to ascribe past changes to present causes is defended; second, the resources for the attempt are assembled by inquiring what causes are now operating in the physical and organic worlds; third, the attempt is carried out, most extensively for the changes recorded in the Tertiary formations of Europe and also, more briefly, for the secondary and primary rocks.

After the fifth edition of 1837 this third part was made a separate book initially titled *Elements of Geology.* Lyell was preparing a twelfth edition of *Principles* when he died in 1875. The main changes in the treatise's teachings were required by his embracing in the 1840s the new doctrine of a glacial epoch or ice age and by his partial, not very enthusiastic acceptance in the 1860s of Darwin and Wallace's new theory of the origin of species by natural selection.

The early, most influential editions taught three main theses. First, the igneous and aqueous agencies have always been in a providentially designed, stable, antagonistic balance, the igneous unleveling and the aqueous leveling the earth's surface, so that there is continued destruction of old land here and formation of new there, with the overall proportion of land to sea staying roughly constant. Second, global climate changes are entirely due to reversible changes in the distribution of land and sea relative to the poles and equator;

there has been a steady global cooling due to this cause throughout the time since the Carboniferous rocks were formed, a cooling that must eventually reverse itself in the future (see figure). Third, ever since that remote time the planet has been fit somewhere for even the highest type of organisms: mammals; and because adaptational requirements alone determine what types of life live where and when, mammals must have lived then, even though their fossils have not been found yet. So species of all the known main animal and

The frontispiece of the first volume of Charles Lyell's *Principles of Geology*. To avoid having to accept a gradually cooling earth (on the basis of the palm tree fossils found around Paris, thus suggesting a warmer time in the past), Lyell proposed his theory of climate. The picture shows a ruined Roman temple on the Italian coast. Lyell and other geologists ascribed the dark bands on the columns to borings by marine mollusks when the columns were long immersed upright in the sea. For Lyell, this and other evidence showed that the land here had subsided and then later risen. Over the whole earth such local changes could lead, he held, to changes in the distribution of land and sea relative to the poles and the equator, and so—land being a better absorber of radiant solar heat than water—could cause the average temperature of the planet to change. The steady planetary cooling since the Carboniferous rocks were formed, cooling indicated by fossil finds, was due to this causation, he taught— causation that will reverse itself and its effects in the long run of the future.

plant types have been coming in, by creation, and going out, by extinction, throughout the past, are doing so in the present, and will continue to do so in the future, although at a rate—perhaps one species a year—that makes it unlikely that any species origins have yet been witnessed by competent naturalists. Each species only lives and dies once. As a separate creation it starts at a single providentially chosen spot, as a first pair or lone hermaphrodite, and then spreads its range, increases its numbers, varies slightly to fit new conditions but not enough to change or transmute into another species, and is eventually extinguished when it loses out in competitive imbalances caused by climatic and other environmental disturbances. These adaptational and ecological determinations of the times and places of species births, lives, and deaths are, then, together with the continual coming and going of avenues (e.g., land bridges) and barriers (e.g., mountain ranges) to migrations, the sole determinants of the different representations at different places and periods of groups of species such as genera, families, and classes.

The first of these three main theses was the least novel, being, as Lyell himself insisted, a modification of Huttonian theory. However, the combination of the second and third was a totally unprecedented proposal; no one had expounded anything remotely like it before; and all three theses together allowed for an entirely new integration of biogeography with geology. Those few of Lyell's contemporaries who committed themselves to any comprehensive account of the earth's physical and organic histories mostly favored, in Britain especially, a completely different synthesis wherein the earth had cooled and calmed irreversibly ever since a molten fluid beginning and had hence become progressively fitter for higher and higher types of life, which had accordingly been created sporadically at special times in the earth's past in a progressive succession consummated by man's appearance as the crown of creation, since when no further species origins had occurred.

Some younger geologists and naturalists in the next generation, with a yen for grand synthetic theorizing about the earth and the history and geography of life upon it, saw themselves as challenged, above all, to adjudicate between these two conflicting comprehensive proposals made by Lyell and his opponents. No one took up this challenge more sustainedly than Charles Darwin and Alfred Russel Wallace. In their independent but strikingly parallel agreements and disagreements with Lyell and his opponents, eventuating in their independent but remarkably similar theories about a tree of life and natural selection, they were both arguably more influenced by his *Principles* than by everything else they read put together.

Lyell's *Principles* could be such a decisive influence on them for two reasons. First, his novel agenda for geology had led him to write, as no one had before, nearly 200 pages all about species: what they are and how they come and go in their successive, continual births, lives, and deaths. It is therefore thanks to Lyell that Darwin's 1859 book is called *On the Origin of Species* and that Wallace's 1855 paper is titled with a phrase from *Principles:* "On the Law Which Has Regulated the Introduction of New Species." Second, Lyell's species chapters open with a dozen pages on what he called "Lamarck's system." Lyell's exposition of this system is structured quite differently from Lamarck's original; so Lyell's version includes what Lamarck's does not: a theory of common ancestry, and therefore ramifying descents, for all species. Rejecting the gradual transmutationism of this system, Lyell presents his own hypothesis of specially created fixed species as an alternative. Lyell's entire geographic-geological treatment of species and the confrontation of his version of Lamarck with a special creationist hypothesis did more than any other text to define the decisive issues about species as Darwin and Wallace engaged them.

Darwin, having embraced all the main teachings of *Principles* by 1834 (halfway through his voyage on HMS *Beagle*), first disagreed in early 1835 with Lyell's theory of the causes of species extinctions (later, on reading Malthus in 1838, he would return to Lyell's view); then, most likely in mid-1836, came to favor species transmutations, and finally in mid-1837 constructed his own first modified successor to Lamarck's whole system as expounded by Lyell. Some eight years later Wallace, having already read and embraced *Principles,* was converted to species transmutation in 1845 by reading Robert Chambers's anonymously published *Vestiges of the Natural History of Creation*. But Wallace came to insist on taking species transmutation versus special creation of fixed species as an issue separable from other issues raised by *Vestiges* or by Lyell's version of Lamarck, especially spontaneous generation and necessary progress. Before Darwin and Wallace came independently to very similar theories about the causes of species origins, what the two men converged on was common descent—with irregularly arboriform (treelike) species transmutations—as a lawful generalization about species origins that was more consistent with biogeographic evidence, most especially, and very generally more explanatory than was Lyell's providentialist-adaptationist special creationism. This convergence on this disagreement with Lyell presupposed Darwin's and Wallace's independent acceptance of Lyell's integration of geology and biogeography (summarized at the opening of Wallace's 1855 essay) and the decisive corollary of this integration: that the representation, the distribution in time and space, of supraspecific groups, genera, families, orders, and so on, is determined by whatever determines the timing and placing of species origins and by the continual changes to avenues and barriers to migrations. Where the two young Lyellians disagreed with their mentor was in independently agreeing that adaptational requirements alone do not determine that law, for although species are exquisitely adapted, common ancestries and so hereditary descents together with adaptations do that determining. When the two went on, again independently, to converge on what Darwin called natural selection as the main cause of adaptive species formations, both were knowingly following the precedent set by Lyell's ecological, geographic, and populational theory of the causes of species extinctions. This last convergence could only occur as it did because it had been preceded by years of other independent, convergent engagements, agreements and disagreements, with *Principles*.

BIBLIOGRAPHY

Herbert, S. 2005. *Charles Darwin, Geologist.* Ithaca, NY: Cornell University Press.

Hodge, M. J. S. 1991. *Origins and Species: A Study of the Historical Sources of Darwinism and the Contexts of Some Other Accounts of Organic Diversity from Plato and Aristotle On.* New York: Garland Publishers.

———. 2009. *Darwin Studies: A Theorist and His Theories in Their Contexts.* Aldershot, UK: Ashgate Publishing.

Hodge, M. J. S., and G. M. Radick, eds. 2003. *The Cambridge Companion to Darwin.* Cambridge: Cambridge University Press.

Laudan, R. 1987. *From Mineralogy to Geology: The Foundations of a Science, 1650–1830.* Chicago: University of Chicago Press.

Lyell, C. 1830–1833. *Principles of Geology: Being an Attempt to Explain the Former Changes in the Earth's Surface by Reference to Causes Now in Operation.* 3 vols. London: John Murray. Facsimile reprint, Chicago: University of Chicago Press, 1990.

Rudwick, M. J. S. 2008. *Worlds before Adam: The Reconstruction of Geohistory in the Age of Reform.* Chicago: University of Chicago Press.   —*J.H.*

R

Race

The concept of race is one of the most controversial in all evolutionary biology. It has been used to justify the slavery of Africans in the New World, to rationalize the murder of millions of Jews, Slavs, and Gypsies by the Nazis, and as a reason for one group to oppress another throughout human history to the present day. From its use in plant and animal biology to recognize genetically based population variation within a species, the concept of race has become a sociocultural classification of human diversity with highly charged interpretations and connotations. The biological foundations and applications of existing racial diversity are more complicated and nuanced.

Classifications of humans can be found in antiquity; the story of Noah and his sons was one traditional way of explaining human diversity (the descendants of Canaan were supposed to be cursed and black because his father Ham had disrespectfully stared at Noah when he was naked and drunk). The modern tradition in the use of the term *race* starts with the German physician Johann Friedrich Blumenbach (1752–1840), who divided the human species into five groups: Caucasian (white), Mongolian (yellow), Malayan (brown), Negroid (black), and American (red). This was not very different from the use of the concept of race in plant and animal biology in the nineteenth and early to mid-twentieth centuries, in which races were designated as subgroups of a single species, recognized by a few visible differences in coloration or form.

In the nineteenth century the evolutionary explanations of racial diversity were explored by many scholars, notably by Charles Darwin in his *The Descent of Man, and Selection in Relation to Sex* (1871). Although as a good Englishman he had little doubt of the superiority of his own group, Darwin certainly did not think that humans are divided into several species. Unlike most of his contemporaries, he did not think that the more stringent demands of living in colder climates had led to the intellectual superiority of the white humans. Darwin was inclined to ascribe racial differences to the effects of sexual selection. He thought that different isolated groups would develop different, culturally based notions of beauty, which would lead to different features being established as the norm in each group.

Darwin also explored the arguments offered at the time to rationalize the colonization and subjugation by Europeans of other groups. Although Darwin believed that European civilization was superior to those of other groups, he was not, contrary to some misreadings of his writing, an advocate of genocide.

He suggested that a major factor in any triumph by European colonists was the asymmetrical spread of disease, with Western infections being especially dangerous for native peoples without a history of exposure. It is also worth noting that Darwin deplored the ill effects of native peoples' taking up unattractive European habits, particularly the excessive consumption of alcohol.

Racial thinking, especially a more prejudicial version of racial thinking than any in which Darwin indulged, continued into the twentieth century, reaching its abhorrent pinnacle in the policies of the National Socialists. Ironically, the experience of African Americans in the war against the Nazis sparked the start of the struggle for civil rights for those same African Americans. With time and the gradual success of that struggle, an increasingly large segment of American society became increasingly uncomfortable with classifications based on racial differences. Race became, in many intellectual quarters, a so-called social construction, meaning something with cultural rather than biological status. Nonetheless, debates, often fierce, continue to erupt over the interpretation of differences among groups in measures like IQ scores or the preponderance of certain conditions and diseases. In many areas of the world, oppression and genocide continue as a direct result of longstanding rivalries and bitterness between groups of people with different historical ancestries living in proximity.

To appreciate what biological foundation there might be for a concept of race, one must return to its use in plant and animal biology. Even there racial distinctions were drawn on the basis of one or a very few characters and were used to pose interesting questions about the significance of those differences. As it turns out, in humans, as in other animals, the distinctions that have been used traditionally to define races represent variation in a tiny proportion of the genes. The effects of those genes are visible to all, however, and, following Darwin, influence characters that once probably shaped, via culture, "attractiveness" for each group. In that light it is no wonder that so few genetic distinctions have been so broadly overinterpreted.

This is not to assert that there are no meaningful genetic distinctions among certain human groups. The distribution of genes for lactose tolerance (the ability to digest the sugar in dairy products) varies among groups and reflects the history of reliance upon dairy products for food, especially protein, particularly in cold climates. But this is not a racial distinction; two different tribes of African herders also tolerate lactose, but through a mutation distinct from the one that allows lactose consumption in northern European whites. The original distribution of genes that control sickle-cell syndrome reflected the history of exposure to malaria in one region of Africa; this is not a racial distinction either, inasmuch as the sickling trait appears in some groups of African ancestry but not in others, not even in others with a history of exposure to malaria.

Darwinian approaches to medicine (see the main essay "Evolutionary Biology of Disease and Darwinian Medicine" by Michael F. Antolin in this volume) have also uncovered differences among groups in apparent predispositions to particular diseases. In some cases differential vulnerability is associated with different forms of a gene in different groups or genetic markers (gene differences that are probably not involved with the disease but are located near genes that probably are) that differ between groups. In some of these cases, "groups" correspond to the traditional racial classifications, but in others they correspond to groups with different ethnic heritages within a traditional race.

When specific genes cannot be identified, it is very difficult to ascribe differences between groups, racial or ethnic, to genetic bases because it is often the case, especially when one considers traditional racial groups, that many additional factors are confounded with group identity. The controversy over interpreting IQ scores is a profound example. Even if we put aside what IQ scores measure, differences in average IQ scores among groups have many sources. Educational history and socioeconomic stratum, among other variables, affect these scores, and if individuals from different groups have different histories or occupy different socioeconomic strata, it is difficult, if not impossible, on the basis of those data alone to ascribe the differences to any single cause.

A common counterargument is that twin studies have shown that variation in IQ scores can have a genetic foundation. Although this is the case, this level of variation does not imply that variation among groups is genetically based. If there are genes whose effects influence IQ scores, they are likely to behave like the preponderance of genetic variation in humans, variable within groups but distributed similarly among groups.

Today probably most evolutionary biologists would agree that to some extent the notion of *race* in humans is primarily cultural—if you label someone "black," even though one parent is white and the other black, it is hard to think that it is other than a cultural act. Of course, cultural differences can lead and have led to genetic distinctions, as surveys of small, isolated groups confirm time and again. For example, tradition had it that the Kalash, a somewhat isolated group in northern Pakistan, were descended from European stock. Genetic analyses have shown that they are not like their neighbors and bear instead an overall genetic resemblance to groups far to their west. But these distinctions are not meaningful ones in that they do not represent adaptations to local conditions or distinctions that influence any measure of behavior, intelligence, fertility, or vitality.

Given the dreadful history of racial thinking and the intellectual poverty of its foundation, it would seem past time to consign this thinking to the dustbin of history. But consigning racial thinking to the dustbin does not mean pretending that there are no genetic distinctions among groups of humans. Those distinctions are clear, whether they are observable in facial features or in the DNA sequences of mutations that affect susceptibility to disease. The important points are that they represent a fraction of human variation and that they occur as much among groups within traditional races as among traditional racial categories. As migration, mixing, and mating continue (and these continue at an accelerated rate in the modern world), the distinctions that have caused so much ill will represent an ever smaller fraction of human genetic variation. Ultimately, we are forced to confront the fact that the prejudices behind racial thinking are culturally founded and that they also ought to be consigned to the dustbin of history.

BIBLIOGRAPHY

Cavalli-Sforza, L. L., and F. Cavalli-Sforza. 1995. *The Great Human Diasporas: The History of Diversity and Evolution.* Reading, MA: Addison-Wesley.

Cavalli-Sforza, L. L., P. Menozzi, and A. Piazza. 1994. *The History and Geography of Human Genes.* Princeton, NJ: Princeton University Press.

Darwin, C. 1871. *The Descent of Man, and Selection in Relation to Sex.* London: John Murray.

Richerson, P. J., and R. Boyd. 2005. *Not by Genes Alone: How Culture Transformed Human Evolution.* Chicago: University of Chicago Press.

Rosenberg, N. A., J. K. Pritchard, J. L. Weber, H. M. Cann, K. K. Kidd, L. A. Zhivotovsky, and M. Feldman. 2002. Genetic structure of human populations. *Science* 298: 2381–2385.

Stearns, S. C., ed. 1999. *Evolution in Health and Disease.* Oxford: Oxford University Press.   —M.R. and J.T.

Raup, David M. (b. 1933)

David Raup was a central figure in the renaissance of paleontology's contribution to evolutionary theory that took place in the 1970s. His work has primarily focused on applying statistical analysis and computer simulation to the fossil record, and therefore he has pioneered techniques that depart from the traditional paleontological orientation toward descriptive studies of taxonomy, morphology, and stratigraphy. Arguably Raup's most important contribution to evolutionary theory has been to draw attention to the possibility that the broad patterns of species origination (speciation) and extinction are largely governed by stochastic or random processes. The implications of this suggestion are far reaching and have contributed generally to a challenge to the ubiquity of neo-Darwinian processes as the generator of evolutionary novelty.

Raup's original training was in fairly traditional paleontology, and by his own account he began his career fully steeped in the neo-Darwinian tenets of the modern synthesis. After receiving his PhD from Harvard in 1957 (where he studied under Bernhard Kummel and Ernst Mayr), he moved to Johns Hopkins University and eventually to the Universities of Rochester and Chicago. By the mid-1960s he had come to recognize the potential computers had to help paleontologists model complex biological processes involving many variables. Initially Raup focused on computer simulation of mollusk shells, which allowed him to explore geometric constraints on morphology by examining both real and hypothetical shell morphologies. Although this work fell squarely in the traditional study of systematics, it led to applications that opened new horizons for mathematically literate paleontologists.

The turning point came in 1971 when Raup participated in a symposium at the Geological Society of America's annual conference titled "Models in Paleontology." It was during this meeting that Niles Eldredge and Stephen Jay Gould's theory of punctuated equilibrium was unveiled, and at the instigation of Thomas J. M. Schopf, Raup and Gould embarked on a lifelong collaboration. The next year Raup, Gould, Schopf, and the population biologist Daniel Simberloff met at the Woods Hole Oceanographic Institution, where they set the bold task of finding new ways of looking at the history of life. Challenging the widely held assumption that the fossil record was a poorly preserved and incomplete document, Raup conceived the key insight that the patterns in the fossil record might make more sense if they were understood to be in part the result of random rather than strictly deterministic processes. In order to test this idea, Raup wrote a computer program that would produce simulated evolutionary diagrams, where the likelihood of a particular lineage producing a new species or going extinct was set as a randomly generated variable (see figure on page 826). The results, which were published in 1973, were striking: Raup and his group argued that randomly generated phylogenetic trees were virtually identical to the patterns found in the actual fossil record. Although this did not prove that evolution was necessarily a random process, it did suggest that there was no reason to assume that traditional causal factors, such as selective fitness, always determined the course of evolution.

In the years that followed, Raup continued to develop models of macroevolutionary patterns, and while the original stochastic work was discussed and revised, it had a major impact on the way paleontologists approached the fossil record. By the early 1980s Raup's interest settled on the problem of extinction, which he determined had been dramatically overlooked since *On the Origin of Species*. In collaboration with his colleague J. John Sepkoski Jr., Raup began examining Sepkoski's massive database of the marine fossil record. He and Sepkoski subjected this data to statistical analysis and determined that major extinction events—mass extinctions where up to 95% of existing taxa were wiped out in a short period of time—appeared to occur at regular 26-million-year intervals in the history of life. The resulting publication, "Periodicity of Extinctions in the Geologic Past" (1984), attracted significant attention both in scientific circles and from the popular media. In part this attention derived from the independent discovery by Walter and Luis Alvarez of evidence for a massive extraterrestrial impact at the boundary between the Cretaceous and Tertiary periods—the exact moment when the dinosaurs went extinct. Raup and Sepkoski's analysis placed a major extinction event at the K-T boundary, and this sparked an intense renewal of debates concerning the causes of major extinctions. The periodicity hypothesis also, however, raised the controversial possibility that mass extinctions, which have had a major role in constraining the path of evolution throughout life's history, may be determined by factors that have a nonbiological and even extraterrestrial origin.

Raup's contribution to evolutionary theory is significant and rests primarily on drawing attention to the possibility that nonselective factors may play an important role in the history of life, and to the notion that extinction, as the opposite of evolution, must receive significant attention. Equally important, however, was Raup's role in pioneering mathematical and computer-aided simulation of macroevolution, which has changed the face of paleontological research. The techniques that Raup literally invented from scratch in the 1960s and 1970s are now routinely taught to beginning graduate students in paleontology, and Raup's primary legacy may be as the person who brought paleontology into the computer age.

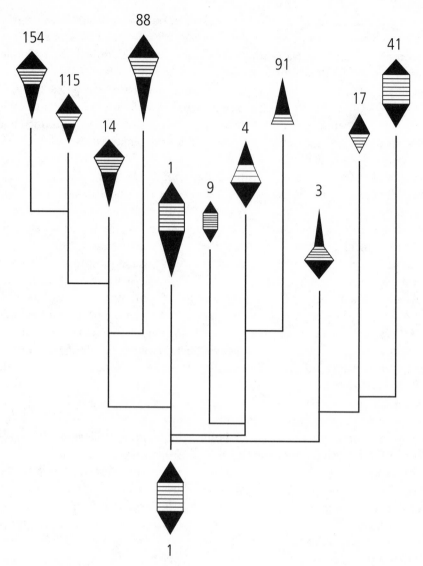

Anyone looking at this picture would think it a perfectly normal phylogentic representation of a group that speciated for fairly normal adaptation-based reasons. In fact, it is of an imaginary group of "triloboids" and was generated by random factors, thus throwing into question the argumentation of many evolutionary claims. (From Raup and Gould 1974, 319.)

BIBLIOGRAPHY

Raup, D. M. [1986] 1999. *The Nemesis Affair: A Story of the Death of Dinosaurs and the Ways of Science.* Rev. 2nd ed. New York: W. W. Norton.
———. 1991. *Extinction: Bad Genes or Bad Luck?* New York: W. W. Norton.
Raup, D. M., and S. J. Gould. 1974. Stochastic simulation and evolution of morphology—towards a nomothetic paleontology. *Systematic Zoology* 23: 305–322.

Raup, D. M., S. J. Gould, T. J. M. Schopf, and D. S Simberloff. 1973. Stochastic
    models of phylogeny and the evolution of diversity. *Journal of Geology* 81, no.
    5: 525–542.
Raup, D. M., and J. J. Sepkoski Jr. 1984. Periodicity of extinctions in the geologic
    past. *Proceedings of the National Academy of Sciences USA* 81: 801–805.

—D.Se.

Ray, John (1627–1705)

John Ray was the leading naturalist in late-seventeenth-century Britain. He
made important contributions to the development of the modern system of
describing and classifying species. He was interested in fossils and the possibility that the earth's structure has changed since the Creation. He is also remembered for his writings on natural theology, which tried to show that
nature was designed by a wise and benevolent god.

Ray was born in Black Notley, Essex, on November 29, 1627. His father
was the village blacksmith. He was educated at Braintree grammar school
and went up to Trinity College, Cambridge, in 1644. He became a fellow of
Trinity in 1649 but left in 1662. Legend holds that this was because he refused to accept the anti-Puritan edict imposed by Charles II, but he had already arranged to begin working privately with a wealthy patron, Francis
Willughby, with whom he traveled on the Continent from 1663 to 1666.
When Willughby died in 1672, he left Ray enough funding for him to retire
to Black Notley. He died in 1705.

Ray originally studied botany and published a number of important works
on the classification of plants, culminating with his *Historia Plantarum*
(1686–1704). He helped rid natural history of the old emblematic approach in
which the role of a species in myth and heraldry was as important as its physical character. He emphasized the distinction between monocotyledons and dicotyledons but insisted that as many characters as possible should be used to
work out the position of a species in the system. He recognized that many so-
called species were merely local varieties formed by natural processes, only the
species themselves being original products of creation. Ray subsequently extended his classification work to the animal kingdom.

Ray was interested in fossils and in the natural processes that affected the
earth's surface, although he was deeply troubled by the threat his geological
studies posed for the Genesis story of creation and the deluge. He published
his *Miscellaneous Discourses Concerning the Dissolution and Changes of the
World* in 1692. Noah's flood might explain how fossil-bearing rocks were
now exposed on dry land, but Ray was aware that natural processes still affected the earth's surface. He realized that some fossils represented species no
longer alive, and he eventually became so concerned about the theological
implications of extinction that he doubted that these fossils really were the
remains of living creatures.

Ray always believed that species were designed and created by God. He
attacked Descartes' materialistic biology and the theory of spontaneous

generation. His *Wisdom of God Manifested in the Works of the Creation* (1691) became one of the seminal works of British natural theology. Here Ray argued that the structure of living species provided evidence that they were designed by a wise and benevolent god. This included the human species, the eye and the hand being very obvious signs of design, but Ray held that every species was adapted both in its physical form and in its instincts to a particular environment.

BIBLIOGRAPHY

Raven, C. E. 1942. *John Ray, Naturalist: His Life and Work*. Cambridge: Cambridge University Press.

Ray, J. 1686–1704. *Historia Plantarum*. 3 vols. London: H. Fairthorne.

———. 1691. *The Wisdom of God Manifested in the Works of the Creation*. London: Samuel Smith.

———. 1692. *Miscellaneous Discourses Concerning the Dissolution and Changes of the World*. London: Samuel Smith.   —P.J.B.

Recapitulation

Recapitulation theory has its roots in the early-nineteenth-century German world picture of nature philosophy or *Naturphilosophie*. Endorsed by poets like Johann Wolfgang von Goethe, philosophers like Friedrich Schelling, and scientists like Lorenz Oken, *Naturphilosophie* saw deep bonds between different parts of the world that were revealed by underlying patterns of form. The similarities (homologies) between the skeletal forelimbs of vertebrates and the six-sided symmetry of the snowflake were favorite examples. *Naturphilosophen* (some of whom were evolutionists and some of whom were not) saw parallels between the development of the individual and the development of life on earth, both of which supposedly showed progress to a better end and were somehow driven by a kind of world force or momentum. The fullest exposition of this kind of thinking came from the Swiss-American ichthyologist Louis Agassiz, who declared that life exhibits a threefold parallel in which the range of organisms alive today forms the third string: "One may consider it as henceforth proved that the embryo of the fish during its numerous families, and the type of fish in its planetary history, exhibit analogous phases through which one may follow the same creative thought like a guiding thread in the study of the connection between organized beings" (Agassiz 1885, 1: 369–370; see figure).

Agassiz was ever an opponent of evolutionary ideas, but after the publication of *On the Origin of Species* in 1859, expectedly it was not long before the parallelism was taken up and given a firmly evolutionary interpretation. This occurred in the major work *Generelle Morphologie der Organismen* (1866) by the German zoologist Ernst Haeckel. Famously he argued that the development of the individual (ontogeny) parallels the development of the race or the species (phylogeny). He expressed this in his so-called biogenetic law: ontogeny recapitulates phylogeny. This kind of thinking was taken up with enthusiasm by biologists of the day, and indeed it is no exaggeration to say that thanks

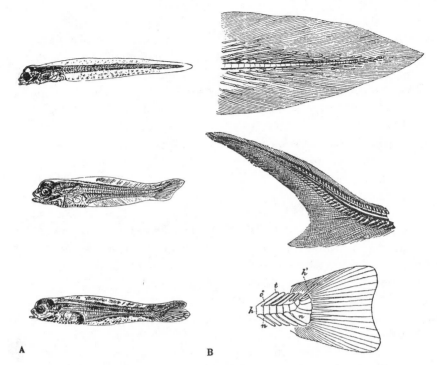

Louis Agassiz's favorite case of recapitulation (Gould 1977, 67). On the left (A), from top to bottom, we see the development of the teleost flatfish *Pleuonectes*. The tail starts diphycercal (the vertebrae go right to the tip), then becomes heterocercal (upper lobe larger then lower), and ends homocercal (lobes equal). On the right (B), from top to bottom, we have the tails of the adults of three living fish, from "primitive" (first to appear in the fossil record) to "advanced" (last to appear in the fossil record): *Protopterus* (lungfish) with a diphycercal tail, sturgeon with a heterocercal tail, and salmon with a homocercal tail.

particularly to the fact that people were much more interested in tracing life's history than in exploring ecological and evolutionary causal factors for change, much of the evolutionary work of the late nineteenth century owed more to Haeckel's biogenetic law than it did to Darwin's natural selection.

This thinking also spread into the wider cultural domain. It was, for instance, significant in theories of child development, where the American psychologist and educationalist G. Stanley Hall argued that children go through set stages of development, and it is pointless and counterproductive to try to force them beyond their level. Such ideas, particularly given that they were firmly progressionist, with the best at the end, also fit nicely with then-prevalent thinking about primitive and advanced civilizations and how the former were stages through which the latter had now advanced.

A matter of some interest and controversy is where precisely Charles Darwin fit with respect to recapitulation. He certainly saw more advanced organisms (and he truly did think them more advanced) in some sense revealing their origins through their embryology. There is some debate about whether he got this insight from people directly influenced by *Naturphilosophie,* or whether he was more inclined to a modified version due to the Estonian embryologist Karl Ernst von Baer. Although Baer was never a full evolutionist, he argued that all organisms (within major groups) come from (in the sense "are modeled on and reveal themselves through time from") some basic form but then go their different ways. Hence an organism early in development will share similarities with the embryos of past and present forms but will not necessarily show similarities with (as the biogenetic law claims) adult forms of past organisms. (However, it complicates matters that if past forms did not develop much in ontogeny, their adult forms would be like their embryonic forms and hence like the embryonic forms of today.) It is probable that Darwin gathered ideas from all sides, although he was always strongly opposed to the necessity of progress, the idea that was central to *Naturphilosophie.* For Darwin, progress was something that came through the successes of selection. (Richards 1992, 2003, 2008 argues strongly that Darwin owed much to the *Naturphilosophen.* Ruse 1996 disagrees strongly, thinking that Baer was the chief influence and that generally Darwin was very wary of German metaphysical thinking.)

By the end of the nineteenth century criticisms of the biogenetic law were rising. Too often it was clear that it did not hold, and that its employment led only to contradictions and confusions. Increasingly, young biologists turned from phylogeny tracing to more profitable experimental fields like cytology and (later) genetics. By the time of the evolutionary synthesis in the 1930s, recapitulation was completely discredited, and this is probably a major reason why the synthesis did not really include embryology. It was thought irrelevant. (There were other reasons, primarily that the early population geneticists found it easiest to build models that ignored development and went straight from genotype to phenotype.)

In the past three decades there has been an explosion of interest in embryology and its connections to evolution. Spurred by findings in molecular biology, evolutionary development (or evo-devo) is now one of the hottest areas in the Darwinian synthesis. With this, particularly thanks to the urging of the late Stephen Jay Gould, has come renewed interest in the sorts of things that excited Haeckel and those in his tradition. No one has gone back to the old recapitulatory ideas—although this does not stop creationists from consistent misrepresentation of the situation with claims that all evolutionists are firmly committed to the biogenetic law—but there is increased appreciation that it is not stupid to seek parallels between ontogeny and phylogeny.

There is certainly some evidence for some such parallels. The backbone, common to all vertebrates, does appear relatively early in development. Conversely, the human cerebrum (responsible for motor coordination) appears toward the end of development. In a like vein, good evidence that the ancestors of the whales once walked on land is the fact that in whale development they show leg extremities that then shrink into nonfunction. To a Darwinian all this is hardly mysterious. Evolution is opportunistic and builds on what is already there, often by adding on to the end. But at times selection is going to remove periods of development entirely, especially if they are not needed and are costly to the organism (just as animals in caves lose functioning eyes). So although recapitulation in its traditional form will never return, the sorts of issues with which it dealt are once again biologically respectable.

BIBLIOGRAPHY

Agassiz, E. C., ed. 1885. *Louis Agassiz: His Life and Correspondence.* 2 vols. Boston: Houghton Mifflin.

Agassiz, L. 1859. *Essay on Classification.* London: Longman, Brown, Green, Longmans, and Roberts.

Baer, K. E. von. [1828–1837] 1853. Über Enwickelungsgeschichte der Thiere (Fragments related to Philosophical Zoology: Selected from the works of K. E. von Baer). In *Scientific Memoirs.* A. Henfry and T. H. Huxley, ed. and trans. London: Taylor and Francis.

Carroll, S. B. 2005. *Endless Forms Most Beautiful: The New Science of Evo Devo and the Making of the Animal Kingdom.* New York: Norton.

Darwin, C. 1859. *On the Origin of Species.* London: John Murray.

Gould, S. J. 1977. *Ontogeny and Phylogeny.* Cambridge, MA: Belknap Press of Harvard University Press.

Haeckel, E. 1866. *Generelle Morphologie der Organismen.* 2 vols. Berlin: Georg Reimer.

Hall, G. S. 1906. *Youth: Its Education, Regimen, and Hygiene.* New York: Appleton.

Knoll, A., and S. B. Carroll. 1999. Early animal evolution: Emerging views from comparative biology and geology. *Science* 284: 2129–2137.

Richards, R. J. 1992. *The Meaning of Evolution: The Morphological Construction and Ideological Reconstruction of Darwin's Theory.* Chicago: University of Chicago Press.

———. 2003. *The Romantic Conception of Life: Science and Philosophy in the Age of Goethe.* Chicago: University of Chicago Press.

———. 2008. *The Tragic Sense of Life: Ernst Haeckel and the Struggle over Evolutionary Thought.* Chicago: University of Chicago Press.

Ruse, M. 1996. *Monad to Man: The Concept of Progress in Evolutionary Biology.* Cambridge, MA: Harvard University Press.

Russell, E. S. 1916. *Form and Function: A Contribution to the History of Animal Morphology.* London: John Murray.   —M.R.

Rensch, Bernhard (1900–1990)

Bernhard Rensch was one of the leading German evolutionary biologists of the twentieth century. His contributions to the theories of biogeography, speciation, and macroevolution, which he first synthesized in relative isolation during World War II, have been recognized as the German equivalent of the Anglo-American modern synthesis. After the war Rensch's extensive contacts with leading evolutionary biologists, such as Ernst Mayr, Theodosius Dobzhansky, Julian Huxley, and J. B. S. Haldane, contributed to the spread of the modern synthesis in Germany.

Bernhard Rensch was born on January 21, 1900, in Thale, Germany. He had an early interest in natural history, as well as an outstanding artistic talent. After World War I he studied biology, chemistry, and philosophy in

Halle, where he took his PhD in 1922 with the geneticist Valentin Haecker, himself a student of August Weismann. His teacher in philosophy was Theodor Ziehen, who worked primarily in epistemology and the relationship between physics, physiology, and psychology. Rensch's interest in evolution soon led him to the Museum of Natural History in Berlin, where he was put in charge of the mollusk collections. His second interest was in ornithology. He was in the right place at the right time because Berlin was one of the centers for ornithology during the 1920s, and Rensch was in close contact with both Erwin Stresemann and the young Ernst Mayr.

In 1927, as part of his museum duties Rensch organized a zoological-anthropological expedition to the Sunda Islands. His experiences in the tropics confirmed his earlier ideas about the distribution and evolution of species and contributed to his interest in macroevolutionary processes. His first contribution was the idea of a *Rassenkreis* (literally a circle of races), a concept that unites subspecies and closely related species, which geographically replace each other, into larger complexes. Related to the notion of *Rassenkreis* was Rensch's idea that geographic races are the precursors of new species. Studying the distribution of species, Rensch could confirm several evolutionary rules, such as Bergmann's rule, correlating latitude with body size in animals (the higher the latitude, the bigger the animals), or Cope's rule, stating that lineages tend to increase in size over evolutionary times. Initially he tried to explain these phenomena of parallel evolution (e.g., animals in colder climates tend to be larger) in neo-Lamarckian terms as caused by direct influence of the environment on inherited characters. However, he soon realized that all these phenomena can be explained by the action of mutation, selection, and the principles of developmental physiology, such as pleiotropic gene action or allometry. Rensch developed his own version of an evolutionary synthesis at the end of World War II. His book *Neuere Probleme der Abstammungslehre* was published in 1947 and was translated into English as *Evolution above the Species Level* in 1959. When it was possible for Rensch to get access to the Anglo-American literature on evolution, the so-called modern synthesis, he and everybody else was struck by the kind of intellectual parallelism between these two syntheses, and today Rensch is always included as one of the founders of the modern synthesis.

Later in his career Rensch developed his own system of natural philosophy, which was an outgrowth of his lifelong work in evolutionary biology. His theory had two main aspects, a theory of universal determinism that included psychological, cognitive, and social phenomena, and his theory of panpsychic identism *(Identismus)*. The latter aimed to explain the relationship between mind and matter as the result of developmental and evolutionary processes, which for Rensch explained all aspects of human and animal existence.

BIBLIOGRAPHY

Heberer, G., ed. 1943. *Die Evolution der Organismen.* Jena: Gustav Fischer.
Mayr, E., and W. B. Provine, eds. 1980. *The Evolutionary Synthesis: Perspectives on the Unification of Biology.* Cambridge, MA: Harvard University Press.

Rensch, B. 1947. *Neuere Probleme der Abstammungslehre*. Stuttgart: Enke.
———. 1959. *Evolution above the Species Level*. New York: Columbia University Press.   —*M.D.L.*

Robertson, Alan (1920–1989)

Alan Robertson was born on February 21, 1920, in Preston, England, and died in Edinburgh on April 25, 1989. His early education was at the Liverpool Institute, followed by a BA in chemistry in 1941 from Gonville and Caius College, Cambridge University. Robertson's postgraduate research in physical chemistry at Cambridge resulted in several publications, but he did not complete his PhD because of the outbreak of World War II. During the war he worked in operational research with C. H. Waddington. He subsequently joined Waddington in the Agricultural Research Council Animal Breeding and Genetics Research Organization (ABGRO) (subsequently the ARC Unit of Animal Genetics) in Edinburgh, where he remained for the rest of his career. He received a DSc from the University of Edinburgh in 1951 for his work on animal genetics and was appointed an honorary professor in 1967.

Robertson's seminal research contributions were in three major areas: theoretical and applied animal breeding; the genetics of quantitative traits; and population genetics and evolution (Hill and Mackay 1989). Robertson demonstrated the theoretical benefits of progeny testing in dairy cattle breeding (Robertson and Rendel 1950) and suggested using the contemporary-comparison method of progeny testing to implement it on a national level. His theoretical work explored the effect of population structure on limits to selection response (Robertson 1960), as well as optimal experimental designs for estimating quantitative genetic parameters, including correlated responses to artificial selection. Robertson also used *Drosophila* as a model system to evaluate the validity of existing quantitative genetic theory, to determine how the theory needed to be extended to cope with discrepancies between observed and predicted results, and to investigate the nature of quantitative genetic variation. He and his colleagues showed that theory predicted short-term responses to artificial selection quite well (Clayton et al. 1957b), but that long-term responses (Clayton and Robertson 1957) and responses in correlated traits (Clayton et al. 1957a) were difficult to predict on the basis of parameters estimated from the base population. He showed that the latter predictions require knowledge of the frequencies and distributions of effects of segregating alleles that affect quantitative traits, including effects on reproductive fitness; as well as knowledge about rates and effects of new spontaneous mutations that affect quantitative traits. The theory led to one of the first experiments to estimate spontaneous mutation rates for quantitative traits and many experiments to estimate the strength of natural selection acting on quantitative traits. Robertson was convinced that future progress in quantitative genetics depended on identifying the genetic loci responsible for the observed variation. He and his colleagues were responsible for one of the first studies to identify quantitative trait loci (QTLs) by association with molecular markers in outbred populations and for a large-scale effort to identify QTLs in *Drosophila* using a combination of artificial selection, introgression, and progeny testing (Shrimpton and Robertson 1988). In this regard Robertson is truly the father of modern molecular quantitative genetics. His many accomplishments were honored by appointment to the Order of the British Empire in 1965; election as a fellow of the Royal Society of London in 1966 and of Edinburgh in 1964; and election as a foreign associate of the National Academy of Sciences of the United States of America in 1979, a foreign honorary member of the Genetics Society of Japan, and a member of the Spanish Real Academia de Ciencias Veterinarias.

BIBLIOGRAPHY

Clayton, G. A., G. R. Knight, and A. Robertson. 1957a. An experimental check on quantitative genetical theory. III. Correlated responses. *Journal of Genetics* 55: 171–180.

Clayton, G. A., J. A. Morris, and A. Robertson. 1957b. An experimental check on quantitative genetical theory. I. Short-term responses to selection. *Journal of Genetics* 55: 131–151.

Clayton, G. A., and A. Robertson. 1957. An experimental check on quantitative genetical theory. II. The long-term effects of selection. *Journal of Genetics* 55: 152–170.

Hill, W. G., and T. F. C. Mackay, eds. 1989. *Evolution and Animal Breeding: Reviews on Molecular and Quantitative Approaches in Honour of Alan Robertson.* Wallingford, U.K.: C.A.B. International.

Robertson, A. 1960. A theory of limits in artificial selection. *Proceedings of the Royal Society of London, Series B* 153: 234–249.

Robertson, A., and J. M. Rendel. 1950. The use of progeny testing with artificial insemination in dairy cattle. *Journal of Genetics* 50: 21–31.

Shrimpton, A. E., and A. Robertson. 1988. The isolation of polygenic factors controlling bristle score in *Drosophila melanogaster*. 2. Distribution of third chromosome bristle effects within chromosome sections. *Genetics* 118: 445–459.

—T.M.

Romer, Alfred Sherwood (1894–1973)

Alfred Sherwood Romer was an American vertebrate paleontologist and an expert on early reptiles. He produced several highly successful textbooks and served as director of Harvard's Museum of Comparative Zoology (1946–1961).

Romer was born in New England, and this culture defined his identity. After a difficult childhood Romer entered Amherst University in 1914, studying German literature and history. During the Great War he volunteered for the American Field Service and served in the U.S. Air Service in Europe before returning to the United States in 1919. He entered Columbia University to study vertebrate paleontology and earned his doctorate for work on the evolution of locomotion in reptiles. Afterward he taught zoology at Bellevue Medical College. In 1923 Romer moved to the University of Chicago to succeed Samuel Williston as professor of vertebrate paleontology.

In Chicago Romer continued his interest in early (Permian) reptiles and amphibians, working closely with his preparator, Paul Miller, and making extensive collections of the Permian "red beds" in Texas. In Chicago Romer also wrote general textbooks that expertly summarized his field. His *Man and the Vertebrates* (1933a) presented comparative vertebrate anatomy and evolution to general audiences. It also included brief hints of his social philosophy. For specialists, *Vertebrate Paleontology* (1933b, with later editions in 1945 and 1966) covered the same scientific ground, omitting the social commentary. Both became standard textbooks in universities.

In Chicago paleontology was considered a geological science. Romer considered himself a biologist. By 1934 he found himself frustrated with criticism of his approach; he then left Chicago for a professorship at Harvard University and the Museum of Comparative Zoology (MCZ). He flourished in this new setting. During his directorship of the MCZ Romer struggled to modernize. He sought better funding, tried to revive the program of collecting, and recruited to improve the calibre of the museum's staff and students. By his retirement in 1961 the MCZ had recovered its first-class reputation. Afterward Romer returned to research and exploration. His death in November 1973 was unexpected and came during his most active research period. Nearly half his bibliography was published after his 65th birthday.

Romer's wife, Ruth Hibbard Romer, was central to his career. They married in 1924 and raised three children. Ruth accompanied her husband on his expeditions, not only collecting at his side but also administering his research and social life. He relied heavily on her for maintaining his hectic schedule.

Two themes flowed through Romer's research. One was comparative anatomy, especially the functional implications of changing structure. Romer studied extinct animals as a biologist studies living organisms. For instance, he tried to reconstruct changing patterns of muscles as the vertebrate skeleton evolved. He also was interested in the evolutionary history of cartilage, bone, and the nervous system. A second theme concerned major transitions in evolutionary history: fish to amphibians, amphibians to reptiles, and reptiles to mammals. Romer was a confirmed neo-Darwinian; however, he preferred to study patterns of evolutionary change rather than to theorize about evolutionary processes. He defended this increasingly conservative approach within his discipline by arguing that theories regarding evolutionary mechanisms were worthless without basic structural knowledge of the forms concerned. Despite a disinterest in theory generally, Romer was one of the first vertebrate paleontologists to defend the idea of drifting continents.

Romer worked hard to increase the organization and prestige of his field within both biology and geology. For example, in 1940 he helped organize the Society of Vertebrate Paleontology. In this society Romer demanded informality and equality. He imposed his own set of "Quaker rules" on the society, allowing an opportunity for all who attended its meetings to discuss any subject they wished.

BIBLIOGRAPHY

Colbert, E. 1982. Alfred Sherwood Romer (1894–1973). *Biographical Memoirs of the National Academy of Sciences* 53: 265–294.
Romer, A. 1933a. *Man and the Vertebrates*. Chicago: University of Chicago Press.
————. 1933b. *Vertebrate Paleontology*. Chicago: University of Chicago Press.

—*J.C.*

S

Schmalhausen, Ivan Ivanovich (1884–1963)

Ivan Ivanovich Schmalhausen was one of the leading Russian evolutionary biologists of the twentieth century. Trained in the tradition of evolutionary morphology under his mentor Alexej Severtzov (1866–1936), Schmalhausen later developed his own version of evolutionary theory that merged morphological and embryological concepts with selectionist ideas and genetics. Specifically, he introduced the idea of stabilizing selection to account for the relative constancy of phenotypic forms across a range of different environmental conditions. Stabilizing selection, as defined by Schmalhausen, leads to canalization of phenotypes as a consequence of developmental regulation that buffers organisms from external and internal disturbances, thus minimizing the phenotypic effects of genetic and environmental variation. Today Schmalhausen's ideas are still widely discussed, and their historical significance as part of a Russian equivalent of the Anglo-American modern synthesis is also increasingly recognized.

Ivan Schmalhausen was born in Kiev (Ukraine, then part of Russia) in 1884 into a family of German immigrants to Russia. His father, Ivan Fedorovich Schmalhausen, was a professor of botany in Kiev and one of the founders of Russian paleobotany. Already during his school years Ivan was focusing mostly on the sciences, an emphasis he continued from 1901 onward at the University of Kiev, studying evolutionary morphology under the zoologist Alexej Severtzov. Schmalhausen got his first appointment as an assistant to Severtzov when the latter moved to Moscow, where he was offered the chair in comparative anatomy. In 1914, while researching his PhD thesis, Schmalhausen worked briefly at the Stazione Zoologica di Napoli, where he learned additional experimental techniques that would further expand his morphological studies. After the turmoil of World War I and the Russian Revolution Schmalhausen briefly held the chair in zoology and comparative anatomy in Voronezh before he returned to Kiev as professor of zoology in 1921. In 1936 he was invited to become the chair of the department of Darwinism at Moscow University and to head the Moscow Institute of Evolutionary Morphology, founded by his teacher Severtzov. Despite his prominent position—he was awarded the Order of the Red Banner in 1945—Schmalhausen became a target of the politically motivated campaign for neo-Lamarckism headed by Trofim Lysenko. Like many other geneticists and evolutionary biologists, Schmalhausen lost his positions and opportunities for research. After Stalin's death in 1953 he was gradually rehabilitated, and in 1955 he became the head of the embryology laboratory at the Zoological Institute in Leningrad, where he worked on his major monograph on the origin of terrestrial vertebrates, as well as several other projects, especially the incorporation of cybernetic ideas into evolutionary theory. Schmalhausen was also one of the recipients of the Darwin Medal of the Leopoldina in Halle, one of the oldest academies in Europe. He was honored in 1959, four years before his death in 1963.

Schmalhausen's theories stress the organism as the focal unit of evolution. They incorporate elements of evolutionary morphology, such as a concern for the integration of the organism as a whole, the relationship between ontogenetic and phylogenetic patterns and processes, and an emphasis on the correlations between different structures and ontogenetic stages. Because of his interest in questions of ontogeny and phylogeny Schmalhausen is often seen as a forerunner of present-day evolutionary developmental biology (evo-devo). Schmalhausen was also influenced by the Russian traditions of biogeography and biogeochemistry. This influence is reflected in his use of the term *biogeocenosis,* rather than ecosystem, as the elementary structural unit of the biosphere. He preferred this term because it emphasizes the integration of geographic, geological, chemical, and biotic factors and their interactions. It is within this conceptual framework of biogeocenosis and developmental evolution that Schmalhausen developed his most famous idea, stabilizing selection. Schmalhausen distinguished between dynamic and stabilizing selection. The former leads to continuous adaptations to ever-changing environmental (or biogeocenotic) conditions, while the latter maintains complex adaptations by canalizing the developmental and physiological processes of organisms. The problem that Schmalhausen addressed with this distinction is the following: if one assumes that natural selection will only act on existing variation within populations and that these populations live in an ever-changing environment, then it is difficult to understand how complex adaptations could emerge in the first place or be maintained thereafter in light of constantly shifting fitness values of any given variant. However, if one assumes that natural selection also acts on the developmental and physiological systems of an organism, those systems that cause the integration of the organism as a whole, then one can see how there would be a selective advantage to those variants of these systems that better shield the organism from external and internal disturbances. The integrity and autonomy of the organism will thus be maintained in the face of environmental fluctuations. As a consequence of this form of stabilizing selection, these developmental systems will become more canalized, that is, more closely integrated, which in turn will constrain the future expressions of phenotypic variation. As a result of stabilizing selection we thus see an apparent directionality in the patterns of phylogeny. However, because these patterns are the result of selection, no additional evolutionary factors, such as directed evolution, are needed.

Schmalhausen's concept of stabilizing selection emphasizes the importance of regulatory processes in development, as well as within biogeocenosis. It is exactly this emphasis on regulatory processes that connects Schmalhausen to present-day evolutionary developmental biology and the concept of regulatory evolution.

BIBLIOGRAPHY

Adams, M. B. 1982. Severtsov and Schmalhausen: Russian morphology and the evolutionary synthesis. In E. Mayr and W. B. Provine, eds., *The Evolutionary Synthesis: Perspectives on the Unification of Biology,* 193–228. Cambridge, MA: Harvard University Press.

Levit, G. S., U. Hossfeld, and L. Olsson. 2006. From the "modern synthesis" to cybernetics: Ivan Ivanovich Schmalhausen (1884–1963) and his research program for a synthesis of evolutionary and developmental biology. *Journal of Experimental Zoology, Part B: Molecular and Developmental Evolution* 306B: 89–106.

Schmalhausen, I. I. 1949. *Factors of Evolution: The Theory of Stabilizing Selection.* I. Dordick, trans. Philadelphia: Blakiston Co. Reprint, Chicago: University of Chicago Press, 1986.

———. 1968. *The Origin of Terrestrial Vertebrates.* L. Kelso, trans. New York: Academic Press.   —*M.D.L.*

Science as a Process (David Hull)

David Hull, emeritus professor at Northwestern University, is one of the major philosophers of biology writing today. He established whole areas of inquiry through his several influential books and many articles. His *Darwin and His Critics* (1973) and *Philosophy of Biological Science* (1974) focused attention on such questions as the nature of species, the units of selection, and the character of explanation in biological science, especially evolutionary science. He argued tirelessly for the conception of species as individuals, an idea that made it possible to think of selection at multiple levels, from genes through organisms, demes, and species. He introduced in a clear way the distinction between *replicators* and *interactors* in selection theory: replicators (e.g., genes) preserve structure through reproduction, while interactors (e.g., organisms) are differentially affected by their environments and serve as vehicles for replicators. Hull brought to his considerations not only the intellectual acumen of a remarkable philosopher but the sensitivies of a biologist with deep experience, especially in systematics.

Hull's major philosophical work, *Science as a Process: An Evolutionary Account of the Social and Conceptual Development of Science* (1988), laid the ground for a successful middle way between bitterly opposed philosophical camps. On the one side were the social constructionists, who suggested that scientific ideas had no purchase on an independent nature but merely reflected social power in the communities where they were produced; on the other side were the logical empiricists, who assumed that the well-honed techniques of logical articulation and empirical investigation delivered reliable, knowledge about the structure of nature. In *Science as a Process* Hull attempted to catch scientists at work, when their social positions, institutional affiliations, and personal antagonisms gave shape to the ideas they advanced through the usual instrumentalities of science. Hull did not doubt that knowledge of the world was achievable, but he was cannily aware of the evolution of that knowledge against a variegated environment of social relationships and natural determinates.

Science as a Process is a three-story book. On the ground floor lies a tale of contest between distinct approaches to systematics that developed from the 1960s through the 1980s. During this period traditionally Darwinian efforts to construct phylogenetic lineages—Haeckelian trees trimmed of unwarranted metaphysical assumptions—were challenged both by the new numerical taxonomists, or pheneticists, as they became known, and by the rival group of cladistic phylogeneticists. The numerical taxonomists, who developed their ideas initially at the University of Kansas, maintained that traits of organisms should be given equal weight and numerically assessed through artfully designed computer algorithms. Clusters of similarity would be the mode of grouping organisms together into taxonomic categories. The cladists, who gathered at the American Museum of Natural History in New York, were inspired by the ideas of an obscure East German entomologist, Willi Hennig. They contended that phylogenetic relationships (of a very truncated sort) could be established by arranging organisms into dichotomous taxa on the basis of shared sets of traits derived from a common ancestor. This mode of analysis would produce nested sister-group pairings that could be depicted in cladograms, a now-common graphic device. Neither the technique of phenograms (diagrams showing physical or morphological resemblance) nor that of cladograms represented the developmental history of species, and they were, for that reason, regarded as deficient by the more traditional evolutionary systematists, such as G. G. Simpson and Ernst Mayr. The pheneticists and cladists returned the compliment by charging traditional systematists with hyperbolic speculation. Hull himself admitted his preference for considerations that included a causal account of proposed systematic relationships.

The second story of Hull's book consists of a history of the general fortunes of evolutionary theory through the period during which the systematists conducted their battles. In this history Hull portrayed the main features of evolutionary thought when individuals such as Richard Lewontin, Stephen Jay Gould, Richard Dawkins, and Edward O. Wilson began reshaping the contours of Darwinian theory.

In addition to his philosophical activities, Hull served as an editor of the journal *Systematic Zoology*, in the pages of which battles between the warring factions of systematists took place. This allowed him to follow the struggles deep into the trenches of scientific activity, for example, in the refereeing of scientific articles. On the basis of this kind of experience, he constructed a third story to *Science as a Process* in which he proposed an evolutionary theory of the development of scientific ideas. It was a powerful framework that employed the traditional Darwinian mechanisms of cooperation and competitive struggle to explain the development not of species but of knowledge. He also applied to revelatory effect his own neologisms of interactors (e.g., articles, books, lectures, and brains of scientists) and replicators (e.g., ideas, theories, problems, and approaches) to give a compelling account of real advances in the sciences. In offering his theory of scientific growth, Hull became one of the leaders of the new group of evolutionary epistemologists.

BIBLIOGRAPHY

Hull, D. 1973. *Darwin and His Critics: The Reception of Darwin's Theory of Evolution by the Scientific Community.* Cambridge, MA: Harvard University Press.
———. 1974. *Philosophy of Biological Science.* Englewood Cliffs, NJ: Prentice-Hall.
———. 1988. *Science as a Process: An Evolutionary Account of the Social and Conceptual Development of Science.* Chicago: University of Chicago Press.

—R.J.R.

Scientific creationism

The term *scientific creationism,* and its synonym, *creation science,* first came into use in the early 1970s, when advocates of what had been called flood geology sanitized their Bible-based views to get them into American classrooms. In 1970 a creationist group in southern California formed the Creation-Science Research Center (CSRC). That fall Henry M. Morris, who had given up an engineering professorship at Virginia Tech to join the center, offered a course titled Scientific Creationism. By the spring of 1972 Morris and other creationist leaders were urging followers to adopt "scientific creationism" as the label of choice for their antievolutionary agenda. That year Morris split with colleagues at the CSRC and set up the Institute for Creation Research, which became the hub of scientific creationism.

A 1981 Arkansas law that mandated "balanced treatment" in teaching creation and evolution defined creation science as "(1) Sudden creation of the universe, energy, and life from nothing; (2) The insufficiency of mutation and natural selection in bringing about development of all living kinds from a single organism; (3) Changes only within fixed limits of originally created kinds of plants and animals; (4) Separate ancestry for man and apes; (5) Explanation of the earth's geology by catastrophism, including the occurrence of a worldwide flood; and (6) A relatively recent inception of the earth and living kinds" (Act 590 of 1981, General Assembly, State of Arkansas, section 4). What most distinguished creation science from other alternatives to evolution was the fifth article, espousing catastrophism. Scientific creationists identified Noah's flood as "the real crux of the conflict between the evolutionist and creationist cosmologies" (Morris et al. 1974, 252).

During the first half of the twentieth century Christian fundamentalists had split over three interpretations of Genesis 1: the *gap theory,* which held that the first chapter of Genesis described two creations, the first "in the beginning," at some unspecified time in the distant past, and the second about 6,000 years ago, when God created Adam and Eve in the Garden of Eden; the *day-age theory,* which equated the "days" of Genesis 1 with vast geological ages; and the *flood-geology theory,* which allowed for no life on earth before the Edenic creation and assigned most of the fossil-bearing rocks to the catastrophic work of the biblical Deluge. Until the 1960s the vast majority of Christian antievolutionists who left a record of their views on Genesis embraced either the gap or day-age schemes, which permitted the acceptance of the fossil evidence for an old earth. The young-earth flood geologists, in contrast, compressed virtually the entire geological column into the year of the Flood.

Flood geology was the invention of George McCready Price (1870–1963), a scientifically self-taught Seventh-day Adventist teacher from eastern Canada (see figure). As an Adventist Price accepted the authority of the prophetess Ellen G. White, who during one of her trancelike visions claimed to have witnessed the Creation, which occurred in a literal week. She also taught that Noah's flood had sculpted the surface of the earth, burying the plants and animals found in the fossil record, and that the Christian Sabbath should be celebrated on Saturday rather than Sunday, as a memorial of a six-day creation. Shortly after the turn of the century Price dedicated his life to a scientific defense of White's version of earth history, dubbed the "science of creationism" by one of Price's disciples (Harold Clark in *Back to Creationism,* 1929, 126–139). Although Price was widely lauded in antievolutionist circles, few non-Adventist creationists before midcentury adopted his peculiar interpretation of Genesis 1.

For flood geologists, the size of Noah's ark became a major concern. Given its limited capacity, it could not have housed representatives of all the animal species that zoologists had identified. Thus in order to avoid making the ark "more crowded than a sardine can," flood geologists typically pulled back from defending the special creation of *species* to insisting on the divine origin of the Edenic *kinds* (letter from D. J. Whitney to B. C. Nelson, June 18, 1928, in Nelson Papers, property of Paul Nelson, quoted in Numbers 2006, 127). The need to have species evolve rapidly from the kinds saved on the ark (which came to be known as microevolution) pushed the young-earth creationists to grasp "every bit of *modification during descent* that can reasonably be asserted" (letter from D. J. Whitney to B. C. Nelson, June 3, 1928, in Nelson Papers, property of Paul Nelson, quoted in Numbers 2006, 127).

The popularity of flood geology rose dramatically with the publication in 1961 of *The Genesis Flood* by Morris and John C. Whitcomb Jr. Unlike Price, who at times allowed for the presence of a lifeless earth before Eden, Morris and Whitcomb limited the age of the entire universe to no more than 10,000 years and argued that some physical laws, such as the second law of thermodynamics, did not exist until Adam and Eve sinned. In 1963 Morris joined nine other scientifically trained creationists to form the Creation Research Society, an organization committed to the propagation of flood geology.

The high tide of scientific creationism came in the early 1980s when Arkansas and Louisiana passed laws that mandated the teaching of creation science whenever evolution was taught. In 1987, however, the U.S. Supreme Court

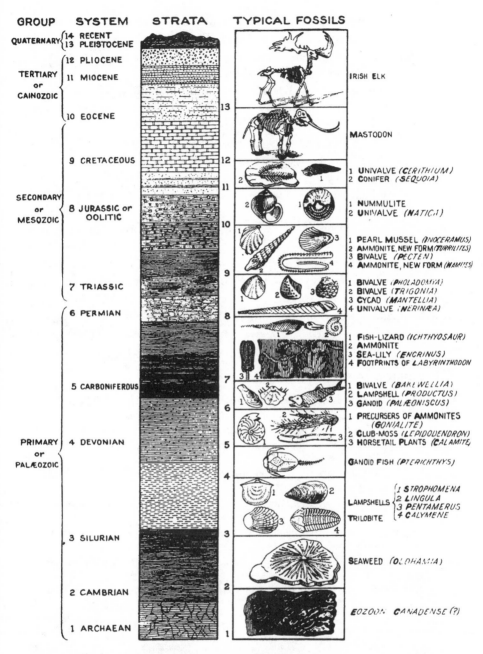

GROUP	SYSTEM	STRATA	TYPICAL FOSSILS

QUATERNARY {14 RECENT / 13 PLEISTOCENE}

TERTIARY or CAINOZOIC {12 PLIOCENE / 11 MIOCENE / 10 EOCENE}

SECONDARY or MESOZOIC {9 CRETACEOUS / 8 JURASSIC or OOLITIC / 7 TRIASSIC}

6 PERMIAN

PRIMARY or PALÆOZOIC {5 CARBONIFEROUS / 4 DEVONIAN / 3 SILURIAN / 2 CAMBRIAN / 1 ARCHAEAN}

IRISH ELK

MASTODON

1 UNIVALVE *(CERITHIUM)*
2 CONIFER *(SEQUOIA)*

1 NUMMULITE
2 UNIVALVE *(NATICA)*

1 PEARL MUSSEL *(INOCERAMUS)*
2 AMMONITE, NEW FORM *(TURRILITES)*
3 BIVALVE *(PECTEN)*
4 AMMONITE, NEW FORM *(NAMITES)*

1 BIVALVE *(PHOLADOMYA)*
2 BIVALVE *(TRIGONIA)*
3 CYCAD *(MANTELLIA)*
4 UNIVALVE *(NERINÆA)*

1 FISH-LIZARD *(ICHTHYOSAUR)*
2 AMMONITE
3 SEA-LILY *(ENCRINUS)*
4 FOOTPRINTS OF *LABYRINTHODON*

1 BIVALVE *(BAKEWELLIA)*
2 LAMPSHELL *(PRODUCTUS)*
3 GANOID *(PALÆONISCUS)*

1 PRECURSERS OF AMMONITES *(GONIALITE)*
2 CLUB-MOSS *(LEPIDODENDRON)*
3 HORSETAIL PLANTS *(CALAMITE)*

GANOID FISH *(PTERICHTHYS)*

LAMPSHELLS {1 STROPHOMENA / 2 LINGULA / 3 PENTAMERUS / 4 CALYMENE}
TRILOBITE

SEAWEED *(OLDHAMIA)*

EOZOON CANADENSE (?)

Superficially this picture, by George McCready Price, looks very much like the picture of life's history given just after the *Origin* by Richard Owen (see Figure 14 in the main essay "The History of Evolutionary Thought" by Michael Ruse in this volume). The Darwinians and by then almost certainly Owen himself took the column to be evidence of evolution, something that had taken place over millions of years. To the contrary, Price saw this as evidence of Noah's flood, with the more nimble organisms climbing to the tops of hills before they were drowned. (From Price 1925.)

ruled that such laws violated the First Amendment to the Constitution, which requires the separation of church and state. Despite this setback creation scientists flourished to the point that they virtually co-opted the term *creationism* for the formerly marginal ideas of Price.

When Intelligent Design (ID) Theory appeared in the 1990s as a more ecumenical option for antievolutionists, evolutionist critics attempted to discredit it as a creationist alias or, more colorfully, "the same old creationist bullshit dressed up in new clothes" (Webb 1996, 5). Creation scientists themselves, though fond of the design argument, faulted ID advocates for their "lack of reliance on the literal statements of Scripture and the construction of alternative models of origin, which involve long period of years" (Haas 1997, 1).

Public-opinion polls in the early twenty-first century, though failing to distinguish young- from old-earth creationists, showed that 45% of Americans believed that "God created human beings pretty much in their present form at one time within the last 10,000 years or so." Fifty-seven percent said that they accepted or "leaned toward" creationism (Numbers 2006, 1). Contrary to widespread expectations, scientific creationism had begun to attract followings outside the United States, even among non-Christians. A century and a half of elite evolutionism had left many unbelievers.

BIBLIOGRAPHY

Clark, H. W. 1929. *Back to Creationism*. Angwin, CA: Pacific Union College Press.
Coleman, S., and L. Carlin, eds. 2004. *The Cultures of Creationism: Antievolutionism in English-Speaking Countries*. Aldershot, U.K.: Ashgate.
Haas, J. W., Jr. 1997. On intelligent design, irreducible complexity, and theistic science. *Perspectives on Science and Christian Faith* 49 (March): 1.
Larson, E. J. 2003. *Trial and Error: The American Controversy over Creation and Evolution*. 3rd ed. New York: Oxford University Press.
Morris, H. M., et al. 1974. *Scientific Creationism*. San Diego, CA: Creation-Life Publishers.
Numbers, R. L. 1998. *Darwinism Comes to America*. Cambridge, MA: Harvard University Press.
———. 2006. *The Creationists: From Scientific Creationism to Intelligent Design*. Expanded ed. Cambridge, MA: Harvard University Press.
Price, G. M. 1925. *The Predicament of Evolution*. Nashville, TN: Southern Publishing Association.
Toumey, C. P. 1994. *God's Own Scientists: Creationists in a Secular World*. New Brunswick, NJ: Rutgers University Press.
Webb, D. K. 1996. Letter to the editor. *Origins and Design* 17 (Spring): 5.
Whitcomb, J. C., Jr., and H. M. Morris. 1961. *The Genesis Flood: The Biblical Record and Its Scientific Implications*. Philadelphia: Presbyterian and Reformed Publishing Company.   —R.L.N.

Scopes trial

In 1925, in Dayton, Tennessee, after the passage of a state law that banned the practice of teaching evolution, a young high-school teacher, John Thomas Scopes, was put on trial for violating the prohibition. He was prosecuted by three-time presidential candidate William Jennings Bryan and defended by the noted lawyer Clarence Darrow. Matters descended to near-farcical level when, denied the opportunity to present his own expert witnesses, Darrow put Bryan on the stand and examined him on the veracity of the Bible. In the end Scopes was found guilty and fined $100. On appeal this conviction was overturned by the state's appeal court on the technicality that the jury, not the judge, should have fixed the penalty, and since the state decided not to take the decision further to a yet higher court, this brought legal matters to an end.

The background to this highly publicized event—passions inflamed by the incendiary writings of *Baltimore Sun* reporter H. L. Mencken—lies in the nineteenth century. After the American Revolution, Protestant Christianity moved quickly to fill the gaps left by the collapse of the older British laws and customs. This was a Christianity that relied increasingly on the Bible for guidance about moral and social customs and was inclined to take literally anything to be found in the Good Book. The Civil War saw a divide in the nation theologically, as well as socially and politically. If anything, the Bible

The whole dispute over evolution and the threat it posed to biblical literalism offered so many opportunities to cartoonists of all persuasions that they must have concluded collectively that there really is a benevolent God! (By E. J. Pace; from Bryan 1924.)

taken literally supports slavery. This was a major reason why, in the North after the conflict, increasingly the accounts in the Bible, especially the Genesis stories of Creation, were interpreted in a metaphorical or allegorical fashion. In the South, however, and in the West as people moved out across the Great Plains, the Bible, if anything, took on added importance, and taking it literally was less a matter of theology and more of regional pride. Inflamed by the teaching of charismatic evangelists, notably Dwight L. Moody—"I look upon this world as a wrecked vessel. God has given me a lifeboat and said to me 'Moody, save all you can'" (Daniels 1877, 432)—literalists took the Bible as a major defense against the "modernism" of the North.

Evolution was one of the great temptations, along with the theater and Sunday newspapers, and opposition to it was increasingly seen as a major obligation of the true Christian thinker. Matters came to a head after World War I, after the demon drink had been vanquished by the constitutional amendment that prohibited alcohol. Now energies were released to fight theories about natural origins of organisms, especially humans. Paradoxically, what happened in Tennessee in 1925 was a function of modernism and its success. Forward-looking thinkers have always stressed the importance of education, and one result of their efforts was that high-school enrollment in the United States shot from 200,000 in 1890 to over 2 million in 1920. Tennessee was one state that felt this effect, and, moreover, it felt the effect of the urge to upgrade the quality of teaching by the use of good textbooks. The students in Dayton were exposed to *Civil Biology* by George Hunter (1914), a work that unambiguously discussed and endorsed evolution.

This was religiously offensive to the people of the South, although the cynic will note that virtually no one who participated in the trial of Scopes was driven entirely by disinterested motives. The citizens of Dayton were keen to have a trial because it would put them on the map, which it certainly did, if not entirely in ways that they had anticipated. Bryan was ever a man looking for a cause and, in leading the troops against the forces of modernism, felt that he had found it. Darrow likewise was not averse to publicity. The year before he had successfully saved the child killers Nathan Leopold and Richard Loeb from execution, arguing that neither had genuine free will, and they were thus compelled to do their crime. This was part of his anti-Christian agnosticism, and appearing for Scopes fed his vanity. Even the American Civil Liberties Union (ACLU) was happy to get involved as a way of publicizing its intent and need of funds.

Thirty years later the Scopes trial was revived vividly in a lightly fictionalized play (and later movie), *Inherit the Wind*. However, this was in the 1950s, the time of the Cold War and of the repression of dissent in America. The trial was used as a vehicle for these concerns rather than as an opportunity to give an exact rendering of historical events. Although it is a great show, many of the original characters, especially those on the prosecuting side, are presented in a caricatured form. The Bryan figure, Matthew Harrison Brady, is simply made into a bigoted buffoon who eats himself to death shortly after the trial. In fact, general opinion was that Bryan acquitted himself well on the witness stand. He was certainly no blind fundamentalist, as the literalists were now called. He admitted that he himself was quite prepared to accept that the six days of creation were six very long periods of time.

Moreover, although evolutionists rather congratulated themselves that in Dayton they had shown how crude and simplistic the literalist Christians of the South were, in many respects it was these literalists who had the last laugh. School-textbook publishers always put profits ahead of principles, and knowing how controversial the teaching of evolution had now become, they gutted their wares of any discussion whatsoever about the topic. This state of affairs persisted through the 1950s until, after the Russian success with Sputnik, American science education generally was upgraded (mainly through government-sponsored new texts), and this upgrading included evolution. Naturally this aggravated literalists, and so the nation was set for another round of quarreling over science and religion, which we still live with today.

BIBLIOGRAPHY

Anonymous. 1925. *The World's Most Famous Court Trial: Tennessee Evolution Case. A Word-for-Word Report of the Famous Court Test of the Tennessee Anti-Evolution Act, at Dayton, July 10 to July 21, 1925, Including Speeches and Arguments of the Attorneys, Testimony of Noted Scientists, and Bryan's Last Speech*. Cincinnati: National Book Company.

Bryan, W. J. 1924. *Seven Questions in Dispute: Shall Christianity Remain Christian?* New York: Fleming H. Revell.

Daniels, W. H. 1877. *Moody: His Words, Works, and Workers*. New York: Nelson and Phillips.

Hunter, G. 1914. *A Civic Biology: Presented in Problems*. New York: American Book Company.

Larson, E. J. 1997. *Summer for the Gods: The Scopes Trial and America's Continuing Debate over Science and Religion*. New York: Basic Books.

Ruse, M. 2005. *The Evolution-Creation Struggle*. Cambridge, MA: Harvard University Press.   —M.R.

Seed plants

The evolution of seed plants, those plants with protein-coated pollen grains and seeds, is a remarkable chronicle of diversification, specialization, and, in the case of flowering plants, rapid evolutionary radiation into a wide variety of growth forms that inhabit every region of the globe, including the seas.

Seed plants include two major groups, gymnosperms and angiosperms. The gymnosperms comprise all plants whose seeds emerge naked, that is, not enclosed in a protective structure such as a fruit. The gymnosperms include about 900 species of trees (conifers, cycads, and the ginkgo) or shrubs (the genus *Ephedra*, from which the cold medicine ephedrine is derived, and a few other groups). These are the surviving members of what was once the dominant group of land plants. The conifers are the most abundant and diverse of the remaining gymnosperms and have a nearly worldwide distribution. The other gymnosperms are more restricted in their natural distribution and, as a result, less familiar. The gymnosperms have separate male and female reproductive organs, usually on the same individual (called monoecy), although some species have individuals that are either completely male or female (called dioecy). Pollen is usually transported to stigmas by wind.

The angiosperms, or flowering plants, include all plants whose seeds are enclosed in a protective structure, like a fruit, formed from the tissues of the female ovary (called the carpel). There are over 350,000 existing species of angiosperms, and they display a dazzling diversity of growth forms, including tiny, floating aquatic plants like duckweed, familiar herbs like lilies and irises, shrubs like blueberries, giant trees such as oaks, vines, and even a wide variety of parasites of other plants like Indian pipes. Flowering plants have nearly every breeding system imaginable; the majority of species have perfect flowers (male and female organs in the same flower), but there are quite a few species with monoecy, some with dioecy, and other species in which some individuals have perfect flowers and others have flowers of only one gender. Although most species are cross-pollinated, some are self-fertilizing, and others use a mixture of normal cross-pollination and either self-fertilization or clonal reproduction. The majority of flowering plants are pollinated by insects, but a significant number of species are pollinated by other animals, especially birds and bats, and many species disperse their pollen by wind (especially grasses and sedges). Some aquatic plants are water pollinated, and several are pollinated by rodents.

Angiosperms and some of the gymnosperms have a remarkable reproductive system that involves a double fertilization. Pollen spores carry two nuclei, each of which carries a haploid complement of genes (a complement with half the genes necessary for the whole organism). One nucleus fertilizes the female ovule, which has a haploid complement of its own, to form the embryo. The other nucleus fuses with other cells associated with the ovule to form an organ called the endosperm, which helps provide nutrition to the developing embryo. This process is well described in the major angiosperm groups but remains poorly understood in some angiosperms and most gymnosperms in which it has been observed. Double fertilization is a dramatic reproductive innovation, and understanding how it occurs, and whether it occurs in the same fashion in gymnosperms and angiosperms (not to mention whether it occurs in all species), would provide critical clues to the relationships between angiosperms and gymnosperms and help us understand the early evolutionary innovations that allowed angiosperms to become so successful.

The first seed plants appear in the fossil record toward the end of the Devonian, approximately 360 million years ago, and, like many other groups, their diversification accelerated over time. Seed plants began to diversify significantly in the Permian, about 270 million years ago, and by about 250 million years ago gymnosperms had supplanted ferns and other land plants as the most common plant type. Gymnosperms maintained that dominance for over 150 million years. Angiosperms made their first appearance early in the Cretaceous, about 130 million years ago, and began their spectacular radiation soon thereafter. The radiation of angiosperms accelerated dramatically at the end of the Cretaceous; by about 50 million years ago angiosperms were diversifying rapidly and dramatically.

The accelerated diversification of angiosperms is remarkable because it is tied to the diversification of insects. More specifically, the radiation of the angiosperms coincided with the radiation of insects that fed on nectar or pollen and those that fed on living plant tissue (called phytophagous insects). Indeed, the coevolution of flowering plants and these insects is one of the most spectacular chapters in the history of life on earth. But that chapter has two different stories because the coevolution of flowers and pollinators is largely a story of mutualism between plant and animal, whereas the coevolution of plant form and chemistry with phytophagous insects is a tale of antagonism and arms races.

Insect pollination brings obvious benefits to each participant. The insects feed on the nectar secreted by the flower and obtain cheap energy from its high sugar content. In the process the animal is smeared with pollen grains that stick to its exoskeleton. As the insect moves from flower to flower to seek more nectar, pollen grains from the previous flower may rub off and contact the stigma, and the insect may pick up new pollen grains. The benefit to the plant is the efficient transport of pollen from one flower to another. The earliest hypotheses about the evolution of flowers postulated that natural selection favored shapes, colors, and flowering patterns that attracted more insects with pollen, providing a fitness benefit to the individual plant by increasing its seed production. In recent years biologists have paid more attention to whether the flower is selected strongly to attract insects that disperse more pollen and thereby increase an individual's fitness through its male function. Although the idea of selection for greater male reproductive success in flowers may seem odd at first reading, research has shown that selection on male function is an important force in shaping a number of flower characteristics.

There are many striking variations on this basic pattern. In milkweeds and orchids the pollen is transported in whole sacs called pollinia, and fertilization is a complex process that requires proper placement of the pollinia near the female parts of the flower. In some carnivorous pitcher plants pollination requires the insect to walk into and through parts of the flower, almost like a maze, so that pollen is placed properly on the stigma. These and similar variations occur when only certain species of insects serve to pollinate the flowers, a result of a long evolution of mutualism. In these cases the insects are especially drawn to those specific plants, and the flowers are modified in ways that facilitate movements in and around the flowers by those insects. The most spectacular variations are found in certain figs and yuccas in which the mutualisms between plant and insect are obligate; in the figs the fig wasp completes its entire life cycle in the flower and fruit, obtaining food and habitat in exchange for its services.

Many features of flowers represent adaptations for attracting insects. The most familiar are the sweetly scented nectars that attract butterflies, beetles, and many other insects. Some flowers produce quite different scents; the flowers of the star anise have a notably foul smell and attract flies that breed in decaying animal bodies. Certain groups of tropical orchids have complex scents whose components have been mixed and matched such that different species of bees are attracted preferentially to different species of orchid. Many flowers have features visible only in ultraviolet light; insects see in the ultraviolet range, and the so-called nectar guides and other markings, which humans cannot see, attract insects and direct them to particular parts of the flower. The colors of flowers also serve to attract insects of different types; however, recent research has suggested that in some habitats flower color serves as a thermal trap and creates microclimates in the flower that facilitate pollen germination and fertilization.

The saga of phytophagous insects and plants is quite different. A long arms race between insects that evolved to exploit plant tissue for food and plants that evolved to discourage this exploitation has produced a variety of interesting features of plants. Perhaps the most remarkable of these features is the array of chemicals found in leaves and shoots that are distasteful or toxic and that inhibit herbivory by many insects. From alkaloids to terpenes, these chemicals protect the plant from many would-be diners (see figure). These are some of the very chemicals that make certain plants useful for humans; caffeine and nicotine ought to discourage human as well as insect consumption, but at low doses they have effects that attract humans to their consumption. Certain compounds that impede cell division, like taxol, have been converted to powerful medicines to fight particular diseases.

For the plants, however, protection is not perfect, and even plants with some of the most toxic chemical constituents suffer predators. For example, milkweeds are filled with cardiac glycosides that discourage most insects, but the leaves are eaten by larvae of monarch butterflies, the roots by larvae of milkweed beetles, and the seeds by milkweed bugs. Each of these species feeds only on milkweed; each has evolved to sequester the toxin in specialized cells. Not only does this permit them to consume milkweed parts, but the presence of the toxin in their bodies makes them distasteful to many of their potential predators and provides protection against them.

Milkweeds and other plants with very few, specialized herbivores represent one extreme of the continuum of plant-insect relationships. The more typical situation is that an individual plant species is exploited by many different types of insects, but an individual insect species is found on only a very few species of plants. This appears to be the result of the insect's increasing focus on one or a few closely related species and the trade-off between the features needed to overcome one sophisticated plant defense system and those needed for the systems of other, less closely related plants. The result is a pattern in which certain groups of insects are found only on certain groups of plants. For example, the heliconiid butterflies are found only on passionflowers.

Seed plants and their features have played prominent roles in evolutionary biology. Darwin himself studied the contrived reproductive biology of orchids, and Mendel's discovery of particulate inheritance (discrete genes) emerged from his studies of peas. Plants were used in some of the most influential early experiments on selection, in particular the work of Hugo de

Jack-in-the-pulpit, *Arisaema triphyllum,* is an herbaceous perennial found along streams, riverbanks, and moist bottomlands throughout eastern North America. It is a member of the Araceae, the family that includes canna lillies and other popular ornamental houseplants. The flowers are borne on a cylindrical structure called a spathe (the upright "jack"), which is enclosed by the cup-shaped envelope with the overhanging flap (the "pulpit"). An individual plant produces either male or female flowers and changes its gender back and forth over the years depending on its size and recent history of reproduction. The leaves and underground tuber (called a "corm") contain calcium oxate crystals and alkaloid compounds that make the plant toxic to mammals and most insects. Seed plants have evolved an enormous diversity of flowering systems and an even greater diversity of defensive features to deter herbivory; the evolution of herbivorous insects is intertwined with the evolution of plant defenses.

Vries. Some of the earliest studies of local adaptation, which showed that different populations of the same species were adapted to different local conditions, were done with plants, in particular the common garden experiments of Göte Turesson and the famous reciprocal transplant experiments by Jens Clausen and colleagues (see the alphabetical entry "Jens Clausen" in this volume). The work of Anthony Bradshaw and his students on metal tolerance in certain grasses produced case studies of rapid evolution in response to anthropogenic environmental changes, and the studies of Janis Antonovics and students on anther smut disease in campion helped convince evolutionary biologists and ecologists alike of the enormous role that disease plays in natural populations (see the alphabetical entry "Janis Antonovics" in this volume).

Molecular biology has facilitated enormous advances in our understanding of flowering plants, has revealed considerable surprises, and has opened wholly new questions. The phylogenetic relationships among seed plants have become clearer; these relationships point to a small shrub from the island of New Caledonia, *Amborella trichopoda,* the water lilies, and a small group of diminutive, mosslike aquatic plants as the remaining representatives of the earliest angiosperm families. The emerging picture of the relationships among the flowering plants has raised new questions about the distribution of double fertilization in angiosperms and gymnosperms, in particular whether this complicated reproductive trait might have evolved multiple times. Several genes that control flower development have been discovered, and the control of the nested components of flowers—sepals, petals, anthers, and stigmata—has been uncovered. This discovery has contributed to the emerging picture of how seemingly simple sets of a few genes can combine to produce an enormous variety of organismic features, indicating that what may seem at first like vast differences between organisms may in fact be easily generated by the effects of selection on a system of interacting genes.

The novelty of discovery sometimes resurrects old ideas, and the study of seed plants offers an entertaining example. Mutations in three of the genes that control flower development cause the plant that carries those mutations to produce a whorl of leaflike structures instead of flowers. This suggests that flowers may represent the product of selection's co-opting the genes that control the pattern of leaf development for a novel purpose. This idea was advocated, albeit in different terms, by no less a figure than Johann Wolfgang von Goethe. Sometimes the old masters knew best.

BIBLIOGRAPHY

Judd, W. S., C. S. Campbell, E. A. Kellogg, P. F. Stevens, and M. J. Donoghue. 2002. *Plant Systematics: A Phylogenetic Approach.* Sunderland, MA: Sinauer Associates.

Nicklas, K. J. 1997. *The Evolutionary Biology of Plants.* Chicago: University of Chicago Press.

Soltis, D. E., P. S. Soltis, P. K. Endress, and M. W. Chase. 2005. *Phylogeny and Evolution of Angiosperms.* Sunderland, MA: Sinauer Associates.

Willis, K. J., and J. C. McElwain. 2002. *The Evolution of Plants.* New York: Oxford University Press.   —*J.T.*

The Selfish Gene (Richard Dawkins)

The Selfish Gene (1976) is a highly influential and popular book in which Richard Dawkins articulates a "gene's-eye" perspective on evolution. It has become one of the standard statements of genic selectionism. The book is also notable for its speculative discussion of cultural evolution and the corresponding notion of *memetic evolution.*

According to Dawkins, the real actors in evolutionary processes are not individuals, families, or groups but genes. In the arms race of evolution the genes that have been most successful have been the ones that could best protect themselves and be reproduced into the next generation. So just as the protein coating that surrounds a virus has the function of protecting the genetic information stored within, the body, brain, and behavior of an animal are to be understood as serving exactly the same function. As Dawkins puts it, animals (including human beings) are "lumbering robots" (p. 21) whose sole function is to protect the genes and ensure that the genes are reproduced.

Much of the book is devoted to an extended argument that the behavior of organisms can be understood as serving the best interests of genes. Familiar phenomena such as favoritism toward members of one's own family, incest avoidance, and parental care are easily explained by this gene's-eye perspective. (Critics, such as Sober and Wilson 1997, argue that many of these phenomena are better explained by other mechanisms, group selection in particular.) Dawkins also argues that such phenomena as free riding, reciprocal altruism, and maintenance of viable population size are also easy to explain if we adopt the gene's-eye perspective. A notable feature of the book, for which it deserves much praise, is that many of these arguments are supported by lucid but informal game-theoretic considerations. In particular, those discussions offer an accessible introduction to John Maynard Smith's theory of evolutionarily stable strategies (ESSs).

Having forcefully argued for the usefulness and explanatory power of the gene's-eye perspective, Dawkins goes on to apply a similar metaphor to cultural evolution. According to Dawkins, we may understand cultural evolution as taking place within a sort of gene pool of ideas. In his metaphor ideas—called memes—compete with each other to replicate themselves inside the brains of human beings. Memes are to be understood as replicators with the same relevant properties as those possessed by genes. For example, each meme will have a fitness that is its propensity to spread through a population of intelligent individuals. Similarly, mutation among memes will occur when an error is made in copying a meme from the brain of one individual to the brain of another.

In this way memetic evolution is to be understood as obeying the same dynamics as genetic evolution. It is perhaps ironic that this theory of memetic evolution has spread so successfully. Indeed, much subsequent work on cultural evolution implicitly or explicitly assumes that cultural evolution and biological evolution obey the same dynamics in exactly the way that Dawkins advocates. In the concluding passages of *The Selfish Gene,* Dawkins reveals some of the ambiguity about human beings that runs through his book. On the one hand, he clearly wants to see us as part of the general selfish-gene picture. On the other hand, with the introduction of memes, he suggests that perhaps we can escape our biology. "We are built as gene machines and cultured as meme machines, but we have the power to turn against our creators. We, alone on earth, can rebel against the tyranny of the selfish replicators" (p. 215).

BIBLIOGRAPHY

Blackmore, S. 2000. *The Meme Machine*. Oxford: Oxford University Press.
Dawkins, R. 1976. *The Selfish Gene*. Oxford: Oxford University Press. (Thirtieth anniversary edition with a new preface, 2006.)
Sober, E., and D. S. Wilson. 1997. *Unto Others: The Evolution of Altruism*. Cambridge, MA: Harvard University Press.
Sterelny, K. 2007. *Dawkins vs. Gould: Survival of the Fittest*. London: Totem Books.
Sterelny, K., and P. Kitcher. 1988. The return of the gene. *Journal of Philosophy 85*, no. 7: 339–362.   —Z.E.

Sepkoski, J. John "Jack," Jr. (1948–1999)

Jack Sepkoski was fortunate to enter the field of paleontology at a transitional moment in the discipline's history. When he began graduate study at Harvard in 1970, the field had not yet seen some of the breakthrough theoretical ideas of the 1970s (such as Niles Eldredge and Stephen Jay Gould's theory of punctuated equilibria or Steven Stanley's work on species selection), but paleontologists were already in the process of directing their work toward evolutionary theory. Sepkoski's education was therefore a mixture of old and new: from Bernhard Kummel he learned traditional global biostratigraphy (the understanding of fossil deposits in the context of their place and time), while Gould taught him broader, macroevolutionary thinking. He also began to refine an interest in using computers to quantify and model trends in the fossil record, which involved using complex multivariate statistics. Along with his colleague David M. Raup, Sepkoski was one of the first paleontologists to attempt serious statistical analysis of the fossil record, and his work is partially responsible for the emergence of an important branch of paleontology known as quantitative paleobiology.

Sepkoski's major contribution to evolutionary theory was to develop and test methodologies and resources for determining patterns in the fossil record. The foundation for this work came from his career-long project to compile and update a database listing all known fossil families and genera of marine animals. This list, which was begun in the mid-1970s and currently stands at more than 3,500 families and 37,000 genera, tracks the first and last appearances of taxa in the fossil record and provides the best and most accurate estimate of the originations and extinctions of marine organisms available to scientists. Drawing on this resource, Sepkoski performed a series of pioneering analyses of the record of marine diversity during the Phanerozoic eon (the period of time from about 540 million years ago to the present, during which most life on earth evolved). His interest in this study was to compare the rates of diversification and extinction in the actual record with mathematical models for population growth derived from biological theories of island biogeography. Sepkoski found that these matched: rates of origination and extinction of taxa in the fossil record mirror rates of birth and death for populations of organisms competing for resources on islands. This finding suggests that taxonomic groups (families, genera, and, by implication, species) are subject to the same population pressures as individuals, and that the historical pattern of clade diversity is directly analogous to biological population growth. Beyond this conclusion's intrinsic importance, this work also indirectly supported the theory of species selection developed concurrently by Gould and Stanley, among other paleontologists, which treated taxonomic groups (like species) as analogous to individual organisms.

From his careful analysis of fossil data, Sepkoski was able to extrapolate several major theoretical articulations of the nature of diversity in the history of life. The most important is known as his three-faunas model: in a series of essays published between 1976 and 1984, Sepkoski argued that the pattern of macroevolution during the Phanerozoic could be represented by three overlapping sigmoidal (S-shaped) curves. These curves represent three distinct periods of evolution, equilibrium, and decline for a characteristic set of organisms, followed by the rise of a new faunal group. Although this model will probably be Sepkoski's most far-reaching and lasting accomplishment, he gained considerable notoriety for participating in a widely publicized debate concerning mass extinctions. In the early 1980s he and Raup published an analysis of the fossil record that showed distinct pulses of mass extinction at fairly regular intervals during the post-Paleozoic era (250 million years ago to the present). The fact that these pulses happened roughly every 26 million years was particularly significant: their regularity suggested a recurring phenomenon, quite possibly of abiotic origin. Although Sepkoski declined to speculate about the source of these seemingly periodic events, others certainly did, and a number of highly speculative candidates were put forward, ranging from terrestrial causes (regular periods of volcanism or climatic upheaval) to extraterrestrial factors (cosmic radiation, earth's cyclical motion through the galactic plane) and science fiction (the so-called Nemesis hypothesis, which imagined that an orbiting "death star" passed close enough to the solar system every 26 million years to send comets from the nearby Oort Cloud hurtling toward the earth). Despite often being associated with these theories, Sepkoski himself considered his participation in these debates accidental and noted that his work on periodicity occupied a relatively minor place in his overall scientific output. Despite his untimely death in 1999 at the age of 50, Sepkoski contributed a number of important approaches to mathematical analysis of biodiversity and paleoecology that have helped shape the current study of macroevolution.

BIBLIOGRAPHY

Ruse, M. 1999. *Mystery of Mysteries: Is Evolution a Social Construction?* Cambridge, MA: Harvard University Press.

Sepkoski, J. J., Jr. 1984. A kinetic model of Phanerozoic taxonomic diversity. III. Post-Paleozoic families and mass extinctions. *Paleobiology* 10: 246–267.

———. 1994. What I did with my research career; or, How research on biodiversity yielded data on extinction. In W. Glen, ed., *The Mass Extinction Debates: How Science Works in a Crisis,* 132–144. Stanford, CA: Stanford University Press.

—D.Se.

Sheppard, Philip M. (1921–1976)

Philip Sheppard was the brightest of the young men gathered around E. B. Ford in the school of ecological genetics founded at Oxford University in the years after World War II. Coming off three years as a prisoner of war, Sheppard rapidly made his mark, although it is an indication of the state of biological education that he did not get a first-class degree because he spent too much of his time thinking about genetics and not enough on the traditional subjects of morphology and physiology. Sheppard was swept up in the work that Ford and Ronald Fisher were doing on the moth *Panaxia* and then began to make his own important way in the selection studies he did with fellow Oxonian A. J. Cain. They looked at the snail *Cepaea nemoralis* and showed the falsity of earlier claims that the varied markings of the snail—banded, unbanded, pink, yellow, brown, and so forth—had no adaptive significance. To the contrary, they were able to prove that the markings play a major role in protecting the snails from predators, chiefly thrushes, and that the markings vary according to the background—beechwood floor, ditches by hedgerows, meadows, and the like.

Showing that by now neo-Darwinian studies were becoming very much a joint Anglo-American endeavor, Sheppard spent a year (1954–1955) working on *Drosophila* in the laboratory of Theodosius Dobzhansky at Columbia University in New York. But although this clearly strengthened Sheppard's knowledge of and interests in underlying genetic factors in evolution, studies of organisms in nature were always close to Sheppard's heart, and these continued and increased when he moved to Liverpool University in 1956. Together with Cyril Clarke, Sheppard conducted lengthy and massive studies on butterflies, uncovering the nature and causes of mimicry. This is perhaps the most venerable topic in Darwinian studies, going back to the 1860s and the pathbreaking work of Henry Walter Bates, but had long been a topic of controversy, primarily because mimicry often seems to be the function of one gene, and so it is hard to see how natural selection could be the significant cause. In response to such a mutationist view, pushed by Reginald Punnett, in his *Genetical Theory of Natural Selection* (1930) Fisher argued that many genes can be involved in making for mimicry that can then be packaged as one or a few supergenes. It was the work of Sheppard and Clarke that showed this to be true.

Although the 1950s were exciting times for Darwinian evolutionary studies, in other respects things were difficult. The newly formed molecular biology was conquering the life sciences, taking grants and posts and students. Led by Ford, the evolutionists fought back (Ruse 1996). Major funding at Liverpool was given by the Nuffield Foundation (sponsored by Lord Nuffield, England's leading car industrialist, the equivalent of Henry Ford in America). The foundation's mandate was medical research, and Sheppard and Clarke convinced it that studies of fast-breeding organisms like butterflies could serve as models for human diseases with genetic underpinnings. This led to major and in respects very successful work on such topics as blood groups and associated illnesses and, specifically with regard to the Rhesus blood-group system, on the problems associated with babies whose blood is incompatible with that of their mothers.

Sheppard died at the age of 55 of acute leukemia, recognized as one of the most important evolutionists of the mid-twentieth century, not least for his influential and often-revised text *Natural Selection and Heredity* (first edition, 1958).

BIBLIOGRAPHY

Bates, H. W. [1863] 1892. *The Naturalist on the River Amazons*. London: John Murray.

Cain, A. J., and P. M. Sheppard. 1950. Selection in the polymorphic land snail *Cepaea nemoralis* (L). *Heredity* 4: 275–294.

———. 1952. The effects of natural selection on body colour in the land snail *Cepaea nemoralis*. *Heredity* 6: 217–231.

———. 1954. Natural selection in Cepaea. *Genetics* 39: 89–116.

Clarke, C. A., R. B. McConnell, and P. M. Sheppard. 1960. ABO blood groups and secretor character in rheumatic carditis. *British Medical Journal* 1, no. 5165: 21–23.

Clarke, C. A., D. A. Price Evans, R. B. McConnell, and P. M. Sheppard. 1959. Secretion of blood group antigens and peptic ulcer. *British Medical Journal* 1, no. 5122: 603–607.

Clarke, C. A., and P. M. Sheppard. 1955. A preliminary report on the genetics of the Machaon group of swallowtailed butterflies. *Evolution* 9: 182–201.

———. 1959a. The genetics of some mimetic forms of *Papilio dardanus*, Brown, and *Papilio glaucus*, Linn. *Journal of Genetics* 56: 236–260.

———. 1959b. The genetics of *Papilio dardanus*, Brown, I Race *cenea* from South Africa. *Genetics* 44: 1347–1358.

———. 1960a. The genetics of *Papilio dardanus*, Brown, II Races *dardanus, polytrophus, meseres,* and *tibullus*. *Genetics* 45: 439–457.

———. 1960b. The evolution of dominance under disruptive selection. *Heredity* 14: 73–87.

———. 1960c. The evolution of mimicry in the butterfly *Papilio dardanus*. *Heredity* 14: 163–173.

———. 1960d. Supergenes and mimicry. *Heredity* 14: 175–185.

———. 1962. The genetics of the mimetic butterfly, *Papilio glaucus*. *Ecology* 43: 159–161.

Fisher, R. A. 1930. *The Genetical Theory of Natural Selection*. Oxford: Oxford University Press.

Punnett, R. C. 1915. *Mimicry in Butterflies*. Cambridge: Cambridge University Press.

Ruse, M. 1996. *Monad to Man: The Concept of Progress in Evolutionary Biology.*
    Cambridge, MA: Harvard University Press.
Sheppard, P. M. 1958. *Natural Selection and Heredity.* London: Hutchinson.

—M.R.

Simpson, George Gaylord (1902–1984)

George Gaylord Simpson was an American paleontologist and an expert on mammal evolution. He made profound contributions to evolutionary theory, systematics, and historical biogeography. He built lasting links between paleontology and biology.

Simpson was raised in Colorado and attended the University of Colorado in Boulder from 1918. He began in journalism and literature but discovered a passion for paleontology during his third year. To develop this interest Simpson transferred to Yale University in 1922, where he earned a PhB (1923) and a PhD (1926). Simpson's research interests first focused on mammals that lived during the age of dinosaurs. After Yale Simpson spent a year in Britain (1926–1927) to study collections of the same mammals held in European museums.

Simpson joined the American Museum of Natural History in New York in 1927 and stayed more than 30 years. Simpson replaced his mentor, William Diller Matthew, whom Simpson held in high esteem. The American Museum owned one of the world's largest collections of vertebrate fossils. Simpson flourished during the 1930s and 1940s despite the Great Depression. Patrons gave him funding for several expeditions. In North America Simpson collected fossils in Florida and Montana, as well as across the entire Southwest. Abroad he led expeditions to Venezuela, Brazil, and Patagonia. Simpson also nearly reached Mongolia, but Soviet officials refused him permission for political reasons. Later in his life Simpson had a reputation for being an armchair naturalist. This is unfair. He loved to travel and spent a significant amount of time away from his home institutions. He was a proud member of the Explorers Club, and he always enjoyed the adventure of travel.

In 1942 Simpson volunteered for military service and served in military intelligence in North Africa, Sicily, and Italy. Fluent in key languages, including French, German, Spanish, and Italian, Simpson was ideal for such work. In the service he learned Arabic. Simpson's service allowed him to escape a crisis for paleontology at the American Museum. Its director, Albert Parr, had recently reorganized the museum. In this reorganization the paleontology department was dissolved. It was seen as antiquated and unsuitable for a modern biological institution. Parr was an oceanographer and wanted more experimental biology. On Simpson's return in 1944, cooler heads prevailed. Simpson withdrew his resignation and was appointed head of the reorganizing department.

In 1956 Simpson suffered an accident while on an expedition in Brazil. This left him disabled. He never undertook expedition work again. Friends also noticed psychological changes. Indeed, none of his writing after the accident matched his earlier brilliance. While he was incapacitated, Simpson felt abandoned by colleagues at the American Museum. The resulting rifts led to his resignation in 1959 and a move to the Museum of Comparative Zoology (MCZ) at Harvard University. Although his new colleagues included some of the most important evolutionary biologists of the day (notably Alfred Romer, Ernst Mayr, Bryan Patterson, Arthur Crompton, and E. O. Wilson), Simpson largely worked alone. He despised teaching and had few students. Simpson formally retired in 1982 and died in 1984.

Simpson married twice. His first marriage, to Lydia Frances Pedroja (1899–1988), produced four children, but the marriage never blossomed. For more than a decade the couple fought bitterly over custody of their children. Divorce came in 1938. During this time Simpson developed a relationship with Anna (Anne) Miriam Roe (1904–1991), a childhood friend. Roe was a psychologist who trained at Columbia University and eventually became an expert on career choice. They married immediately on Simpson's divorce. They had no children together, though Roe became a much-loved mother to Simpson's children. Simpson and Roe collaborated on several projects in their careers, including a book on using statistics in zoology and a conference on the value of evolutionary perspectives in the study of behavior.

In his contributions to science Simpson was a restless genius. He had a near-photographic memory and intense analytical abilities. His bibliography included more than 750 items, although many of these fall into several core categories: systematics, historical biogeography, and evolution.

In systematics Simpson produced both general theory and specific revisions designed to better organize mammal groups, and he named many new species. This work culminated in 1945 with a new classification of all mammals, fossil and living. This was universally described as a masterpiece for its breadth and depth of knowledge. Simpson included a short introduction that prescribed methods for good taxonomy. First, he insisted on the importance of multidisciplinary knowledge. Every discipline in biology, he said, offered something for the taxonomist. Second, he argued that classification must reflect evolutionary history. This second idea put Simpson into one of several competing camps in taxonomy. Others argued that his approach made taxonomic names and arrangements too dependent on ever-changing ideas about evolutionary relationships. They wanted names and arrangements to be stable and independent of evolutionary facts. Later Simpson disputed the same points with newer approaches to taxonomy, such as phenetics and cladism.

Simpson trained at a time when paleontologists understood evolution to include three kinds of phenomena: change in form, change in distribution or range, and change in ecosystems. The second of these phenomena formed the focus of historical biogeography. Following his mentors, Simpson undertook many studies to trace changing distributions of mammals over geological time. He concentrated his attention on interchanges between North and South America. Not only did he seek the specific data of these interchanges, but he also sought explanations. What made mammals able or poor travelers? What pushed a group to move across continents? Simpson developed new methods for measuring biogeographic changes. He also proposed a series of theoretical ideas for testing when explanations were needed. On similar lines, beginning in the 1930s, Simpson severely criticized the theory of continental drift. He thought that the fossil record for mammals was best explained by theories of migration across land bridges. He kept this view until the 1960s, when evidence for seafloor spreading convinced him that continents were mobile.

Simpson is best known for his studies of evolutionary processes. *Tempo and Mode in Evolution* (1944), revised in 1953 as *Major Features of Evolution,* gained instant recognition for its novelty and scope. *Horses* (1951) applied his theories to a specific case (see figure). In their overall approach these books brought together genetics, biogeography, ecology, and mathematical modeling of populations to produce a general understanding of evolutionary patterns and processes. These books became essential reading for several generations of evolutionary biologists. Some, however, criticized him for excessive speculation. For paleontology Simpson introduced some of the latest developments in genetics and statistics. He wanted to reshape the way his

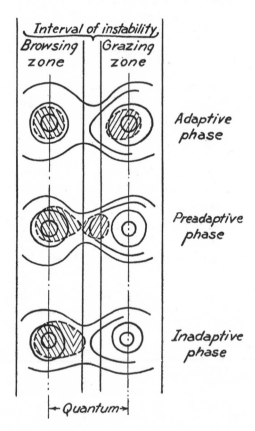

G. G. Simpson adapted Wright's adaptive landscape metaphor, reading it in terms of groups of long-gone organisms (rather than genes) to show how horses evolved from browsers to grazers (after which point, not shown here, the browsers went extinct). (From Simpson 1944, 208.)

colleagues thought about fossils, and he rejected several explanations for evolutionary change popular in his discipline. To other biologists Simpson wanted to show the importance of paleontology for identifying new evolutionary phenomena and for testing theoretical concepts. Simpson worked in tandem with other biologists who were seeking to modify biological theory in light of recent innovations in their fields. These included Theodosius Dobzhansky, Ernst Mayr, and Julian Huxley, among others. Together this group dubbed their movement a *modern synthesis,* a phrase that continues to be applied, although Simpson preferred *post-Darwinism.*

On similar lines is Simpson's 1949 book *The Meaning of Evolution.* On the surface this is a general summary of the synthetic approach. Added to this are discussions of evolutionary progress and the scientific basis for ethics. This book clearly shows the effects of Simpson's personal experience of war and its aftermath. It offers a biologist's search for meaning in life after a bloody conflict and after embracing humanism over theism. What, Simpson asks, can be learned about life's meaning by studying life's history? This book was widely discussed and translated into many languages.

In the 1970s Simpson was bitterly attacked from within evolutionary biology. Advocates of punctuated equilibrium, including Stephen Jay Gould and Niles Eldredge, presented Simpson as part of an entrenched paradigm who misled paleontologists into thinking that most evolutionary change was gradual rather than rapid. Sour relations between Simpson's defenders and detractors polarized paleontology. Simpson dismissed such attacks as efforts by ambitious youngsters eager for fame and fortune. He argued that they had simply taken a few of his ideas to an extreme and then made a smokescreen to hide their lack of evidence. Disputes about Simpson's theories continued well after his death. The acrimony of these disputes colors the way Simpson is remembered.

BIBLIOGRAPHY

Cain, J. 1992. Building a temporal biology: Simpson's program for paleontology during an American expansion of biology. *Earth Sciences History* 11: 30–36.
Laporte, L. 2000. *George Gaylord Simpson: Paleontologist and Evolutionist.* New York: Columbia University Press.
Simpson, G. G. 1944. *Tempo and Mode in Evolution.* New York: Columbia University Press.
———. 1945. The principles of classification and a classification of mammals. *Bulletin of the American Museum of Natural History* 85: i–xvi, 1–350.
———. 1949. *The Meaning of Evolution: A Study of the History of Life and of Its Significance for Man.* New Haven, CT: Yale University Press.
———. 1951. *Horses: The Story of the Horse Family in the Modern World and Through Sixty Million Years of History.* New York: Oxford University Press.
———. 1953. *Major Features of Evolution.* New York: Columbia University Press.
———. 1978. *Concession to the Improbable: An Unconventional Autobiography.* New Haven, CT: Yale University Press.
Simpson, G. G., and A. Roe. 1939. *Quantitative Zoology.* New York: McGraw-Hill.
Whittington, H. B. 1986. George Gaylord Simpson. *Biographical Memoirs of Fellows of the Royal Society of London* 32: 527–539.

—J.C.

Social Darwinism

Social Darwinism is conventionally defined as the application of Darwin's theory of natural selection to the domain of human affairs. The term is used to describe social, racial, and moral theories, as well as economic and political ones. Most commonly it refers to theories that advocate a gladiatorial struggle for existence, in which the losers deserve to lose and survival becomes proof of merit. This definition, however, does not correspond to historical reality and requires further clarification. Social Darwinism is a complex concept. Its significance changed over time and across geographic borders, acquiring remarkably different meanings.

It is important to note that social Darwinism is not the immediate offspring of Darwin's biological theories. Although Darwin sometimes expressed views that seem concordant with positions held by some social Darwinists (his description of racial struggles as part of social evolution in *The Descent of Man* is one such example), his main interest did not lie in social or political theorizing. Furthermore, attempts to draw parallels between human evolution and the evolution of society as part of an effort to justify certain political doctrines preceded Darwin's *On the Origin of Species* (1859). Herbert Spencer, for instance, who coined the expression "survival of the fittest," published essays before 1859 in which he claimed that development (i.e., evolution) was the universal law of progress. He used this idea to argue in favor of free economic competition between individuals and to support the ideology of laissez-faire liberalism. This has led some scholars to declare that social Darwinism would be more accurately named social Spencerism. Spencer's biology, though, was mainly inspired by Jean-Baptiste Lamarck's transformism.

Whether attributed to Darwin or to Spencer, so-called social Darwinism clearly played a major role in the shaping of disciplines dedicated to the study of humankind: anthropology, ethnology, and sociology. In order to be considered scientific, these social sciences drew on biological notions. Human races, cultures, and peoples were compared with species and were considered subject to the same natural laws of struggle, adaptation, and survival. They were often arranged on a kind of evolutionary scale that indicated their degree of progress. The works of American sociologist William Graham Sumner, British anthropologist E. B. Tylor, and French racial theorists Gustave Le Bon and Georges Vacher de Lapouge are a few examples of this type of approach. Francis Galton's eugenics, which he termed *the science of human betterment,* can also be classified under the label of social Darwinism. These theories and others similar to them had a profound influence on political and economic thought. They served to rationalize and justify social inequalities and cutthroat economic competition. American magnates Andrew Carnegie and John D. Rockefeller Jr., for instance, claimed that personal enrichment and business monopolies were simply manifestations of the principle of survival of the fittest. Social Darwinism was also used to justify racial superiority and discrimination and to fuel imperialism and colonial conquests (see figure). With the growth of national-

Evolution after Darwin succeeded in large part because it so readily lent itself to people's prejudices. Almost inevitably, cartoonists portrayed the Irish as ape-like. This cartoon, from the popular magazine *Puck* (1882), compounds the insult by having as a caption "King of A-Shanty," thus bringing in attitudes about the Africans (Ashantees) as well.

ism, it was harnessed to warrant the use of military power in conflicts between nations or to lend support to fascist and Nazi ideologies.

These gladiatorial interpretations of evolutionary biology do not apply automatically, or in full, to all "social Darwinists." Spencer, for example, opposed militarism, which he regarded as a form of state intervention, contrary to laissez-faire politics. More important, many thinkers used Darwin's biology and Spencer's philosophy to provide scientific legitimacy to doctrines that were the exact opposite of free-market liberalism and individualism, namely, socialism, collectivism, and Marxism. Marx himself drew a close connection between Darwin's theory and his own doctrine and declared in 1861, in a letter to Ferdinand Lasalle, "Darwin's work is most important and suits my purpose in that it provides a basis in the natural science for the historical class struggle" (Marx 1985, 246–247). Later some socialists, like the

Italian Enrico Ferri, insisted on the intimate connection between Darwinism, Spencerism, and Marx's scientific socialism. Others, like anarchists Emile Gautier and Prince Peter Kropotkin, argued that the most important law in evolutionary biology was not the struggle for existence but mutual aid. In the United States, socialist Henry George assigned an important role to environmental conditions in determining the course of progress and argued for equality and fair competition. George's views and the campaign he led for land nationalization won the enthusiastic support of Alfred Russel Wallace, codiscoverer of the theory of natural selection. Toward the turn of the twentieth century, J.-L. de Lanessan's revised version of Darwinism as the doctrine of "association in the struggle" became the predominant ideology of the French Third Republic, known as Solidarism, and the basis of its welfare regime. Evolutionary biology was also one of the main building blocks in Emile Durkheim's sociology, at least in its early formulation.

Clearly, evolutionary theory could be interpreted to support very different ideologies, depending on which of its elements was considered the most important. If struggle and survival of the fittest were viewed as the main operating factors, evolution could be used to support right-wing doctrines. But when adaptation to the environment and mechanisms of social behavior such as altruism were viewed as the determining factors, evolution could be interpreted to support left-wing doctrines.

Criticism of social Darwinism appeared as soon as the theories themselves did. A fierce attack on evolutionary ethics came from one of the greatest exponents of Darwinism, biologist Thomas Henry Huxley. He objected that one cannot legitimately derive a moral code from nature's workings, a move that British philosopher G. E. Moore later termed the *naturalistic fallacy*. Yet social Darwinism is by no means a dead or outdated enterprise. Progress made in genetic studies and molecular biology led American entomologist E. O. Wilson to declare in 1975 the birth of a new synthesis, sociobiology. It was presented as an explanation of all aspects of social behavior through biology. In economics, Nobel laureate Friedrich A. von Hayek attempted to show the correlation between evolutionary biology and free-market capitalism. On the other end of the political spectrum, Peter Singer's book *A Darwinian Left* (2000) is evidence that attempts to link socialist ideas with Darwinism have also not gone out of vogue. Biology today is certainly very different from Darwin's time and much more advanced, but its implications for understanding the ways in which human beings interact with one another are still very much a controversial issue.

BIBLIOGRAPHY

Bannister, R. C. 1979. *Social Darwinism: Science and Myth in Anglo-American Social Thought*. Philadelphia: Temple University Press.

Caudill, E. 1997. *Darwinian Myths*. Knoxville: University of Tennessee Press.

Clark, L. L. 1984. *Social Darwinism in France*. University: University of Alabama Press.

Dickens, P. 2000. *Social Darwinism*. Buckingham, U.K.: Open University Press.

Jones, G. 1980. *Social Darwinism and English Thought*. Brighton, U.K.: Harvester Press.

Lanessan, J.-L. de. 1881. *De la lutte pour l'existence et l'association pour la lutte: Etude sur la doctrine de Darwin.* Paris: Octave Doin.

Marx, K. 1985. *Karl Marx, Frederick Engels: Collected Works.* Vol. 41. R. Dixon et al., trans. New York: International Publishers.

Pittenger, M. 1993. *American Socialists and Evolutionary Thought, 1870–1920.* Madison: University of Wisconsin Press.

Singer P. 2000. *A Darwinian Left: Politics, Evolution and Coopertaion.* New Haven, CT: Yale University Press.   —*N.B.*

Sociobiology: The New Synthesis (Edward O. Wilson)

Sociobiology: The New Synthesis, by the Harvard entomologist Edward O. Wilson, was published in 1975. It was the center work of a trilogy that began with *The Insect Societies* (1971) and ended with the first of Wilson's Pulitzer Prize–winning books, *On Human Nature* (1978). In *Sociobiology* Wilson gives an overall picture of the growing field of animal social behavior, considered from a Darwinian perspective, that is, from a perspective that takes natural selection to be the key causal component behind the behavior of animals in social situations.

The first part of the book deals with mechanisms: the workings of population biology, with special emphasis on what Wilson declared to be the key notion of *altruism.* Why and how is it that animals, forged in the struggle for existence, do not always fight and compete in a brutal way? Why do they sometimes show cooperation, most obviously in the hymenoptera (the ants, the bees, and the wasps)? To answer these questions Wilson introduces some new theoretical models, especially those of William Hamilton that focus on what had become known as *kin selection,* where animals help in the reproduction of others when these others share the same genes as the helper. There is discussion also of other issues, particularly modes of animal communication; most significantly, he highlights the importance of pheromones (chemicals used carry information).

The second half of the book surveys social behavior in organisms from the slime molds through the social insects and next the vertebrates to the mammals and thence on to the so-called higher apes. The final chapter is on our own species, *Homo sapiens.* Wilson sees a paradox in animal sociality. He feels that the most integrated social beings are the slime molds, and then we get a relaxing of social behavior as we move up the animal chain. Yet humans seem somehow to have reversed this, because we are the most social beings that have been seen on this planet. Clearly this is in major part a function of the special adaptations that we have, for instance, language ability.

To explain humans Wilson introduces what he calls the autocatalysis model. This supposes that there is a certain threshold level that, once breached, allows human evolution to proceed much more rapidly because each innovation feeds into the whole and (often through feedback) speeds up the rate of change. The move to bipedalism coincidentally freed up the hands, which led to tool manufacture, which led to mental evolution, which led to increased cooperation in hunting, which led to a change of diet (more meat and hence the potential for bigger brains), and to other features and abilities that we associate with modern humans.

When it was published, Wilson's work was highly controversial. Critics from the social sciences felt that he was insensitive to the nuances of human culture, and critics from the Left accused him of simply promulgating a conservative, patriarchal view of humankind, dressed up to look like science. Particularly critical was a group that went under the name of Science for the People. Opinion divides over whether the chief effect of the criticisms was to destroy Wilson's position or to increase his book sales. Today, more than three decades later, *Sociobiology* is rightly seen as one of the landmarks of evolutionary biology, taking its place alongside the great works of the earlier part of the twentieth century. The fact that Wilson discusses humans is considered no anomaly, but something that has been done by almost every great evolutionist since Charles Darwin (and before). In major respects, more interesting than the mere fact that humans are discussed is that they are used firmly to support Wilson's belief that they are the apotheosis of evolution, the supreme point of the progress that the development of life has shown through the ages. Thus Wilson opened the way for his later, more openly philosophical works, like *On Human Nature* and *The Creation* (2006), where he has proposed explicitly that what he calls the "myth" of evolution (Wilson 1978, 192) replace the outmoded story of Christianity. For Wilson, his science has always been part of a bigger metaphysical picture.

BIBLIOGRAPHY

Ruse, M. 1996. *Monad to Man: The Concept of Progress in Evolutionary Biology.* Cambridge, MA: Harvard University Press.

Segerstrale, U. 2000. *Defenders of the Truth: The Battle for Science in the Sociobiology Debate and Beyond.* New York: Oxford University Press.

Wilson, E. O. 1971. *The Insect Societies.* Cambridge, MA: Harvard University Press.

———. 1975. *Sociobiology: The New Synthesis.* Cambridge, MA: Belknap Press of Harvard University Press.

———. 1978. *On Human Nature.* Cambridge, MA: Harvard University Press.

———. 2006. *The Creation: A Meeting of Science and Religion.* New York: Norton.

—M.R.

Spencer, Herbert (1820–1903)

Herbert Spencer was a British philosopher and one of the most influential thinkers of the Victorian age. A man of eclectic interests, Spencer wrote on such diverse topics as physics, biology, psychology, education, music, aesthetics, and religion. He is, however, mostly known for his attempt to construct a comprehensive evolutionary theory of progress in nature and society.

Spencer received little formal education. He was mainly taught at home by his father and his uncle and started working at age 16 as a civil engineer for a British railway company. Although his talent and mechanical inventions promised success in this field, he decided to quit the profession in his mid-twenties and turn to journalism and political writing. Earlier Spencer showed how passionately he cared about politics with his first important publication, a series of letters titled "The Proper Sphere of Government," published in the *Nonconformist* in 1842. They encapsulate the key elements of the great philosophical system he later developed: belief in a universal law of progress that applies to both inorganic and organic matter, and the conviction that there is a biosocial continuity that links nature, human beings, and societies by means of growing complexity. "The Proper Sphere of Government" also provided an outline of Spencer's political credo. He defended laissez-faire liberalism and opposed any form of government intervention that would hinder the free course of nature other than by maintaining order and administering justice. This position remained practically unaltered thereafter and directed most of Spencer's scientific work.

In accordance with his political tenets, Spencer got a job for a short period (1848–1853) as subeditor for the London-based financial weekly the *Economist*. Apart from a few travels to the European continent, Egypt, and the United States, Spencer spent most of his life in London. During the time of his engagement with the *Economist,* Spencer met George Eliot, with whom he had a very intimate relationship, though not as intimate as Eliot would have hoped. This was the closest Spencer ever came to marrying. He also met Thomas Henry Huxley, who later became his close friend and introduced him to the famous X club.

The work for the *Economist* left Spencer enough free time to produce his first book, *Social Statics* (1851), in which he endeavored to set down the principles of a purely scientific system of morality. He followed the book with a number of articles, notably "The Development Hypothesis," which argued in favour of Lamarck's theory of evolution, and "A Theory of Population," both published in 1852. In the latter, which was greatly inspired by Thomas Robert Malthus, Spencer coined the phrase "survival of the fittest." He thus came very close to enunciating a theory similar to Charles Darwin and Alfred Russel Wallace's natural selection. But even though Spencer declared his great satisfaction with Darwin's *On the Origin of Species,* he placed greater emphasis on the law of inheritance of acquired characteristics than Darwin did. This Lamarckian law, with its underlying assumption of an almost mechanically causal relationship between adaptation and environment, was more in tune with Spencer's deterministic reductionism. Indeed, Spencer did not allow random mutations the same evolutionary power as Darwin and regarded natural selection as only a secondary mechanism of elimination. He defined evolution as the passage from an incoherent homogeneity to a coherent heterogeneity, a process that signified for him lengthy, cumulative, and necessary progress. Darwin did not have quite the same confidence in progressive development. The two evolutionists also differed considerably in the nature and objectives of their enterprises. Darwin was first and foremost a biologist. Spencer's ultimate purpose was to provide a wide-ranging evolutionary theory that would account for progress in all domains and give scientific basis to his political and ethical individualism. As a result he focused on the study of society, where his contribution was most important.

Spencer envisioned society as a natural organism, analogous to a living body and subject to the same evolutionary laws, with one important caveat. While in biological organisms the emergence of a nervous system and the development of a brain, which functions as a central regulating organ of the body, are the signs of a highly evolved animal, it is not so for the social organism. It does not possess a "social sensorium" that exists apart from its units (Spencer 1892, 461). Society exists only for the benefit of its members, whose individual consciousnesses cannot be reduced to some collective consciousness. The presence of a central coercive authority in the social organism is thus the sign of a low phase of evolution, a transitory state that Spencer termed the *militant* type. As societies grow in dimension, the division of labor becomes more important and leads to a predominance of the industrial-economic factor. Finally, there is no longer need for a central government because industrial-type societies can auto-regulate themselves through systems of production and distribution.

Spencer's most important publication, to which he dedicated most of his life as an independent writer, is the voluminous *A System of Synthetic Philosophy,* published between 1862 and 1893. It comprises 10 volumes and gives a systematic account of his views on metaphysics, biology, psychology, sociology, and ethics. The *System of Synthetic Philosophy* was a highly praised synthesis of knowledge, and in the 1870s and 1880s Spencer's reputation rivaled that of Darwin, but his glory did not last long. August Weismann's attacks on the Lamarckian principle of the inheritance of acquired characteristics shook the very foundations on which Spencer's evolutionary edifice was built. His efforts to defend his theory elicited little support, while neo-Darwinian theories, based on the principle of natural selection, slowly became the predominant paradigm in the field. By the first decade of the twentieth century it therefore seemed that Spencer had been proved wrong, and there was no reason to further investigate his system.

It would be a mistake, however, to overlook the profound impact of Spencer's ideas. A little over a century ago he was one of the most influential writers in Britain, as well as abroad. His books were translated into most European languages and had a very large circulation in the United States, where his fame was particularly great. Indian nationalists also held his theories in high esteem, and in Japan he was studied more than any other foreign writer. Within Spencer's lifetime some 1 million copies of his books were sold, and he counted among his admirers people from many different fields of study. Few people in the history of modern Western thought can claim to have had equal success. Even at present, though the man himself may be discredited, his ideas still resonate in contemporary free-market thinking. They therefore merit serious attention, and indeed today there is a revival of interest in Spencer.

BIBLIOGRAPHY

Francis, M. 2007. *Herbert Spencer and the Invention of Modern Life.* Ithaca, NY: Cornell University Press.

Jones, G., and R. A. Peel. 2004. *Herbert Spencer: The Intellectual Legacy.* London: Galton Institute.

Peel, J. D. Y. 1971. *Herbert Spencer: The Evolution of a Sociologist.* New York: Basic Books.

Perrin, R. G. 1993. *Herbert Spencer: A Primary and Secondary Bibliography.* New York: Garland Publishing.

Spencer, H. 1843. *The Proper Sphere of Government.* London: W. Brittain.

———. 1851. *Social Statics: Or the Conditions Essential to Human Happiness Specified and the First of Them Developed.* London: Chapman.

———. 1852a. The development hypothesis. *Leader,* March 20, 280–281. Reprinted in H. Spencer, *Essays Scientific, Political and Speculative,* vol. 1, 1–7. London: Williams & Norgate, 1891.

———. 1852b. A theory of population, deduced from the general law of animal fertility. *Westminster Review,* April: 468–501.

———. 1892. *The Principles of Sociology.* Vol. 1. New York: D. Appleton & Co.

Taylor, M. W. 1992. *Man versus the State: Herbert Spencer and Late Victorian Individualism.* New York: Oxford University Press.

———. 2007. *The Philosophy of Herbert Spencer.* New York: Continuum.

Wiltshire, D. 1978. *The Social and Political Thought of Herbert Spencer.* Oxford: Oxford University Press.   —N.B.

Stanley, Steven M. (b. 1941)

Steven M. Stanley is a member of an influential group of paleontologists whose work is uniting paleontology and neontology (or biology) by demonstrating that paleontological data (like the fossil record) have genuine significance for understanding the mechanisms by which evolution happens. Stanley's primary theoretical contributions have been in the area of macroevolution, or the formation of new species and evolutionary novelties. In particular he is the originator of the theory of species selection, which holds that species are legitimate units of selection and on a macroevolutionary level act analogously to individual organisms in microevolution. An important implication of this view is that the history of life has not been the smooth, continual progression of small changes, a prediction derived from Darwin's theory of natural selection. Rather, Stanley has been an outspoken supporter of punctuationism and hierarchy theory—the notion that life's history has been broken by significant discontinuities at intervals and on various levels—which are most famously associated with Niles Eldredge and Stephen Jay Gould's theory of punctuated equilibria.

Beginning in the mid-1970s, Stanley argued that evolutionary biologists had ignored the important lessons that could be learned from the fossil record by focusing solely on the gradual accumulation of small changes in individual organisms. In a groundbreaking essay published in 1975 he challenged the neo-Darwinian view that natural selection accounts for all evolutionary patterns and concluded that this reductionist conflation of genetics with evolution is misguided. Slow, gradual change does not account for the majority of important evolutionary changes observable in the fossil record, which appear to be much more sudden and abrupt than was expected under natural selection. He reasoned that just as individuals under natural selection are favored for both the ability to survive to reproductive maturity and the ability to produce many offspring, species are favored for lengthy existence (which allows greater chance for speciation) and the tendency to produce many daughter species. In this analogy speciation is the analogue to birth, and extinction to death. A striking consequence of this view is that the course of evolution has been influenced more by random factors than scientists had formerly believed. Since Stanley's model proposed that most evolutionary change takes place at the species level, and since speciation (occurring most often by geographic isolation) is directed by fairly arbitrary environmental factors, patterns in the history of life are comparable with a statistical random walk, in which long-term trends are largely absent. The notion of progress as an evolutionary metaphor (which has had great currency since before Darwin's day) is largely absent from this view.

Stanley's work is notable in addition for its explicit attempt to promote cooperation between paleontologists and biologists. His 1968 doctoral thesis at Yale examined functional morphology in extant lineages of bivalves, reflecting his hope to bring biological methods and data to bear on paleontology. Subsequently in his career he reversed this impulse and began to agitate for greater awareness of paleontological findings in the work of biologists. Along with Gould, Eldredge, David M. Raup, Thomas J. M. Schopf, and others, he was responsible for the paleontological renaissance in the 1970s and 1980s that saw the establishment of paleobiology as a thriving discipline. He is also the author of two foundational textbooks: with Raup he coauthored *Principles of Paleontology* (1971), which became the authoritative text for undergraduate paleontology courses (and was not substantially revised for more than 30 years), and in 1979 he published the treatise *Macroevolution: Pattern and Process*, which has been described as one of the most significant books in paleontology since G. G. Simpson's classic *Tempo and Mode in Evolution*.

BIBLIOGRAPHY

Raup, D. M., and S. M. Stanley. 1971. *Principles of Paleontology.* San Francisco: W. H. Freeman and Co.

Stanley, S. M. 1975. A theory of evolution above the species level. *Proceedings of the National Academy of Sciences USA* 72, no. 2: 646–650.

———. 1979. *Macroevolution: Pattern and Process.* San Francisco: W. H. Freeman and Co.

———. 1992. *Exploring Earth and Life through Time.* New York: W. H. Freeman and Co.

———. 1996. *Children of the Ice Age: How a Global Catastrophe Allowed Humans to Evolve.* New York: Harmony Press.   —D.Se.

Stebbins, George Ledyard (1906–2000)

George Ledyard Stebbins is regarded as one of the major figures in twentieth-century evolutionary biology. His scientific career spanned most of the century and included three primary areas of research: botany, genetics, and evolution. He is especially well known for his masterful synthesis of these three areas in his 1950 book *Variation and Evolution in Plants*. More than any other, this book formed the conceptual backbone of the new science of plant evolutionary biology and guided an entire generation of plant evolutionists. Because this book also reconciled plant evolution with animal evolution and because it complemented books like Theodosius Dobzhansky's 1937 *Genetics and the Origin of Species*, Stebbins is generally regarded as the botanical architect of the evolutionary synthesis, ranking alongside not only Dobzhansky but also Ernst Mayr, G. G. Simpson, and Julian Huxley. Much of his original research delved into the genetics of the evolutionary process in plant species like the weed *Crepis*. He is best known for his work with Berkeley geneticist E. B. Babcock, which led to the idea of the polyploidy complex, a complex of reproductive forms centering on sexual diploids surrounded by polyploids. He is also known for being the master of the synthetic review article in critical areas of interest to plant evolutionists like polyploidy.

George Ledyard Stebbins was born in Lawrence, New York, on January 6, 1906, and was the third child of a wealthy financier and real-estate developer and a New York socialite. He spent his summers in the resort town of Seal Harbor, Maine, where his father had served as a developer. It was thanks to the efforts of his parents and to some of the distinguished residents, many of whom were scientists and naturalists, that Stebbins developed an early interest in natural history and in the flora of Mt. Desert Island. He attended Harvard University, a renowned center of botanical instruction in the United States in the 1920s. Although his primary training had been in plant systematics (floristics in particular) with taxonomists like M. L. Fernald in the Gray Herbarium, Stebbins moved into newer areas like cytogenetics and studied with E. C. Jeffrey, a well-known plant morphologist, and then with Karl Sax, who had been hired as a geneticist at Harvard's Arnold Arboretum. By the time of his graduation Stebbins was using the latest cytological tools to determine phylogenetic relationships in members of the Compositae (the sunflower family). His first academic position at Colgate University allowed him to continue his research into cytogenetics in species like the peony, or *Paeonia*. In 1935 he accepted a position as junior geneticist to E. B. Babcock at the University of California at Berkeley to work on the phylogenetic relationships in Old World and New World species of *Crepis*. Babcock had found *Crepis* a promising organism for understanding the genetics of the evolutionary process and had undertaken an ambitious effort with a number of workers to study it (Smocovitis 1988).

Stebbins early distinguished himself in genetic and systematics research and in 1939 became an assistant professor of genetics. During the mid- to late 1930s Stebbins became increasingly interested in central problems of concern to evolutionists. He met and interacted with Theodosius Dobzhansky, who was a frequent visitor to the Bay area, and he was an active participant in the Biosystematists Circle, a group of systematists who were supporting the adoption of tools from cytology, genetics, and ecology to solve problems in systematics. Stebbins began to pursue plant evolution full-time after 1939 when he began to teach Berkeley's course in evolution, taught out of the agriculture school. His own research turned to solving problems in plant evolution, and so Stebbins followed closely the work of the Carnegie Institution of Washington at Stanford University's team of Jens Clausen, David Keck, and William Hiesey, as well as the work of Edgar Anderson and Carl Epling, all botanists concerned with understanding evolutionary processes. Unlike most of these contemporaries, however, Stebbins was equally versed in animal evolution and followed the work in species like *Drosophila pseudoobscura* by Dobzhansky and his collaborators.

His grand synthesis of plant evolution began as part of an invitation to give the Jesup Lectures at Columbia University in 1947 (Dobzhansky had been instrumental in the invitation). Reworking his lectures for his evolution course, Stebbins completed a book-length manuscript that was published in 1950 as *Variation and Evolution in Plants*. The book was an instant success, was widely read by evolutionists, and became the authoritative source for understanding evolution in the plant world. In addition to organizing the new field that became known as plant evolutionary biology and serving as a kind of textbook to younger workers, the book also made Stebbins the single best-recognized authority on plant evolution at that time.

Stebbins continued his evolutionary researches into the 1950s when he took a new position as chair of the Genetics Department at the University of California in Davis, California. He continued to write a series of important synthetic books on plant evolution, including *Chromosomal Evolution in Higher Plants* in 1971 and *Flowering Plants: Evolution above the Species Level* in 1974. The latter set forth a series of controversial theories concerning the origin of the angiosperms and was considered critical reading for graduate students and younger scholars. He also edited an important collection exploring phenomena like adaptive radiation and island biogeography with Herbert Baker in 1965 titled *The Genetics of Colonizing Species*. Stebbins also wrote general books on evolution like *The Basis of Progressive Evolution* in 1969 and authored a widely used university textbook that went through three editions titled *Processes of Organic Evolution* in the late 1960s. In 1977 he coauthored yet another comprehensive textbook on evolution with Theodosius Dobzhansky, Francisco Ayala, and James Valentine, titled simply *Evolution*. His teaching efforts were not limited only to graduate and undergraduate instruction but also extended to developing an American high-school curriculum rich in the biological sciences that argued that evolution was the central unifying principle. He was an especially active member of the Biological Sciences and Curriculum Study's efforts to reform high-school biology education in the United States and actively fought against efforts by so-called scientific creationists to undermine the scientific legitimacy of evolutionary biology.

In the 1960s Stebbins increasingly became active in conservation efforts, especially to preserve California's native flora. He became a popular figure and teacher and able spokesperson for evolution. By the late 1970s Stebbins had emerged as one of the more visible evolutionary biologists.

He continued pursuing his researches into evolution, but in the 1960s he became more heavily involved in newer studies to solve important questions in biology dealing with the transformation of "gene to character," or understanding more precisely the mechanisms by which genes affected developmental patterns so as to lead to the origin of novel but adaptive traits and characters. He retired from the University of California at Davis in 1973 but was still active in writing articles and books on evolution until the time of his death in 2000.

BIBLIOGRAPHY

Crawford, D. J., and V. B. Smocovitis, eds. 2004. *The Scientific Papers of G. Ledyard Stebbins, Jr. (1929–2000).* Regnum Vegetabile, vol. 142. Ruggell, Liechtenstein: A. R. G. Gantner Verlag.

Dobzhansky, T. 1937. *Genetics and the Origin of Species.* New York: Columbia University Press.

Dobzhansky, T., F. J. Ayala, G. L. Stebbins, and J. W. Valentine. 1977. *Evolution.* San Francisco: W. H. Freeman.

Smocovitis, V. B. 1988. Botany and the evolutionary synthesis: The life and work of G. Ledyard Stebbins, Jr. Unpublished PhD dissertation, Cornell University.

———. 1997. G. Ledyard Stebbins Jr. and the evolutionary synthesis (1924–1950). *American Journal of Botany* 84: 1625–1637.

Smocovitis, V. B., and F. J. Ayala. 2004. G. Ledyard Stebbins (1906–2000). *Biographical Memoirs of the National Academy of Sciences* 85: 3–24.

Solbrig, O. 1979. George Ledyard Stebbins. In O. Solbrig, S. Jain, G. B. Johnson, and P. Raven, eds., *Topics in Plant Population Biology,* 1–17. New York: Columbia University Press.

Stebbins, G. L. 1950. *Variation and Evolution in Plants.* New York: Columbia University Press.

———. 1966. *Processes of Organic Evolution.* Englewood Cliffs, NJ: Prentice-Hall.

———. 1969. *The Basis of Progressive Evolution.* Chapel Hill: University of North Carolina Press.

———. 1971. *Chromosomal Evolution in Higher Plants.* Reading, MA: Addison-Wesley.

———. 1974. *Flowering Plants: Evolution above the Species Level.* Cambridge, MA: Belknap Press of Harvard University Press.

Stebbins, G. L., and H. G. Baker, eds. 1965. *The Genetics of Colonizing Species.* New York: Academic Press.

—V.B.S.

Superorganism

A superorganism is an entity that functions as an organism but is made of parts that are or can properly be considered organisms in their own right. The idea of the superorganism long antedates the birth of evolutionary ideas and, as applied to human social systems, can indeed be found in the thinking of the Greek philosophers. In his *Republic* Plato likened the ideal state to the individual soul, arguing that each has three parts: the thinking part, in the state represented by the rulers or guardians; the spirited part, in the state represented by the auxiliaries (soldiers and civil servants); and the appetitive part, in the state represented by the artisans and shopkeepers. This kind of analogy lent itself well to different philosophies, and in the nineteenth century, with the coming of evolution, there was much enthusiasm for seeing the state as an evolving entity, akin to an organism. In particular, the English evolutionist Herbert Spencer argued strongly that we see in the state methods of organization identical to methods of organization in the individual, notably the way in which both entities function best when they exhibit a strong division of labor, with a part of the body doing that and only that for which it is best suited. Today perhaps the social organic analogy is best seen in the theorizing of those who believe that units of culture are akin to genes. Richard Dawkins proposed such a theory in *The Selfish Gene* (1976), arguing that culture is divided into *memes* analogous to genes, and he and the philosopher Daniel Dennett argue that society can be analyzed as a kind of memetic genotype with strong similarities to the individual biological genotype.

The superorganism concept was hinted at (if not by that or any name) by Charles Darwin in *On the Origin of Species*. He was strongly committed to the view that natural selection works for and only for the individual. He could not see that adaptation could ever be for the benefit of the group at the expense of the individual organism. Anticipating thinkers today, Darwin made it clear in correspondence with natural selection's codiscoverer, Alfred Russel Wallace, that essentially he thought that if selection works for the group, it is always open to cheating by the individual. A cheater benefits from its own efforts, as well as those of the group, whereas a noncheater loses because of the cheater. However, individual selection failed to explain the role of sterile workers in the social insects, specifically the hymenoptera (ants, bees, and wasps). The problem is that it seems that individuals in these species give all to the group at the expense of their own labors. Darwin finally decided that the integrated groups formed by the hymenoptera, paradigmatically the honeybee hive, can be regarded as organisms in their own right. Thus selection can work on the whole, with sterile workers being regarded as part of this larger unit, just as individual body parts are integrated into a functioning whole organism.

In the 1960s and 1970s, with the coming of sociobiology, a discipline that applies biology to the study of social behavior, and thanks particularly to powerful models that showed how apparently altruistic behavior can be explained in terms of advantage to the individual (selfish-gene theory), Darwin's kind of thinking was dismissed by biologists. The idea of the superorganism was regarded as beyond the realm of the helpful, a sentiment reinforced by the idea's unfortunate political connotations—such organic thinking was a favorite of the National Socialists—and by its association with woolly metaphysics. The Gaia hypothesis of James Lovelock, in which the earth is considered an organism and its denizens (including humans) are considered its parts, was taken as an exemplar of an attractive but ultimately sterile way of understanding the problems of the environment. Recently, however, as a function of a general revival of interest in processes by which selection can work for the benefit of the group, the idea of the integrated functioning whole—the superorganism—has been enjoying a similar revival. In particular, the leading student of the social insects, Edward O. Wilson, now argues that in trying to understand many of the more complex ant societies, it is unwise and unhelpful to regard the individuals as functioning organisms in their own right. It is much more profitable to think of these individuals as parts of the whole, with natural selection acting on the colony or nest rather than on the individuals. This is an empirical hypothesis, not indebted to ideology or metaphysics, and no doubt will generate much interest, theoretical and experimental, as evolutionists continue to struggle with some of the most challenging and interesting phenomena that the living world has to show.

BIBLIOGRAPHY

Dawkins, R. 1976. *The Selfish Gene*. Oxford: Oxford University Press.
Hölldobler, B., and E. O. Wilson. 2008. *The Superorganism*. New York: Norton.
Ruse, M. 1980. Charles Darwin and group selection. *Annals of Science* 37: 615–630.   —*M.R. and J.T.*

Systema Naturae (Carl Linnaeus)

The Swedish botanist Carl Linnaeus (also known as Carl von Linné) is justly credited with the hierarchical system of biological taxonomy and nomenclature that has dominated taxonomy since its development in the 13 editions of his *Systema Naturae*. He is also often, and misleadingly, associated with an essentialist species concept committed to the fixity of species.

In 1735 Linnaeus was in the Netherlands studying for a degree in medicine. With the encouragement and help of Jan Frederik Gronovius, he published the first edition of *Systema Naturae*, consisting of a title page, 11 pages of "observations" and taxonomic tables, and two one-page leaflets. The full title, *Systema Naturae sive Regna Tria Naturae Systematice Proposita per Classes, Ordines, Genera & Species*, reveals the framework of his taxonomic system: three kingdoms of nature—mineral, plant, and animal—each divided into classes, orders, genera, and species. Most notably, in his taxonomic tables he classifies humans *(Homo)* into the order Anthropomorpha along with other primates *(Simia)*, while recognizing the subspecific taxa *Americanus, Asiaticus, Africanus,* and *Europeanus*.

In the first observation of "Observations on the Three Kingdoms of Nature," Linnaeus argues for species fixity on the grounds that every offspring closely resembles its parent. He also argues that the members of a species multiply and increase over time, and by tracing this process backward, we can see that eventually there must be some single original pair or individual for each species created by God. In subsequent observations Linnaeus claims that God's plan is to be discovered in the observation of nature, and that it is the duty of man to uncover this hidden plan so that he might better admire God's work. He also argues that by studying nature, we can provide a better foundation for building, commerce, food supply, and medicine to better achieve a healthier state.

The following editions of the *Systema Naturae* grew from the modest first edition to the 3,000-page 13th edition of 1770. This growth was largely due to the addition of new taxa to the taxonomic tables, but there were important changes as well. In the 10th edition of 1758 Linnaeus introduced his system of binomial nomenclature, where each species is identified by the now-familiar genus-species name. But most significantly, Linnaeus also revised his conception of species. In the 1740s Linnaeus was confronted with evidence that natural hybridization had occurred among plants, producing differences of kind. By 1755, he concluded that hybridization could form new species of plants (von Linné and Daldberg 1755). Consequently he abandoned his commitment to the fixity of species, observing that perhaps God had created an individual or mating pair for each genus and that new species were produced by intergeneric crosses. Significantly, he left open the question whether there are limits to this process. Linnaeus later took this idea even further, speculating in the 13th edition of the *Systema Naturae* that the original individuals or mating pairs might represent orders, rather than genera, and that even new genera could be created through hybridization.

BIBLIOGRAPHY

Frängsmyr, T., ed. 1983. *Linnaeus, the Man and His Work*. Berkeley: University of California Press.

Larson, J. L. 1968. The species concept of Linnaeus. *Isis* 59, no. 3: 291–299.

Linnaeus, C. 1964. *Systema Naturae: Facsimile of the First Edition*. M. J. S. Engel-Ledeboer and H. Engel, eds. Nieuwkoop, Netherlands: B. de Graaf.

Linné, C. von, and N. E. Daldberg. 1755. *Dissertatio botanica metamorphoses plantarum sistens*. Holmiæ: e Typographia regia.   —R.A.R.

Systematics and the Origin of Species (Ernst Mayr)

Ernst Mayr played a central role in constructing and promoting the synthetic theory of evolution, developed in the 1930s and 1940s. Later he was its chief advocate. *Systematics and the Origin of Species* was his first major contribution to that theory. Its key feature is a defense of the theory of speciation by geographic isolation.

The *polytypic species* is this book's core idea. These are species that show distinct varieties or subspecies. Mayr concentrated his research on geographic varieties. In part he did this because he was fascinated by the processes that cause populations to vary from one place to the next. Also, as a museum curator responsible for naming and cataloging specimens, Mayr struggled daily with decisions about how much variation he should attribute to varieties, subspecies, and species. Where should he draw the lines between these units? Mayr was not alone in working on this topic. In the 1930s many taxonomists struggled with the same question. Part of *Systematics* is Mayr's attempt to describe the problem and to show his taxonomist colleagues what he thought should be done about it. His own approach to polytypic species was rooted in his training under Erwin Stresemann and Bernhard Rensch in 1920s Berlin. In a sense *Systematics* adapted those German ideas for Mayr's American audience. In a sense, also, *Systematics* continued a German quarrel, between Mayr and his mentors and the geneticist Richard Goldschmidt. Although Mayr had respectful relations with Goldschmidt himself (who was now an American-residing German-Jewish refugee), he was strongly opposed to Goldschmidt's belief that evolution was essentially saltationary (going by jumps, caused by macromutations), as Goldschmidt claimed in his 1940 book, *The Material Basis of Evolution*. A major theme of *Systematics* was that evolutionary change is gradual, and many examples were chosen to show this (see the figure in the alphabetical entry "Ernst Walter Mayr" in this volume).

Mayr considered the formation of distinct local varieties as the first step that led to new species. His theory of speciation required local populations to become geographically isolated. Mayr saw no other way for a whole population to accumulate distinctive features without that physical separation. If it remained in contact with the parent species, he argued, genetic material would still flow around the species. This would dilute any regional distinctiveness. Once a population was physically separated, natural selection would adapt it to local circumstances. Those adaptations could produce distinctiveness. Alternatively, once a population was physically separated, nonadaptive changes could accumulate from either genetic drift or the founder effect. (These other scenarios were derived from Sewall Wright's shifting-balance theory of evolution, which Mayr advocated in his book.) Regardless of how distinctive features arose, Mayr argued, they sometimes resulted in reproductively isolating the local population. If that occurred, flow of genetic material was no longer possible even if contact with the parent species was reestablished. If that flow stopped, true speciation had occurred.

In *Systematics* Mayr defended reproductive isolation as a simple test for species rank. This became the core of the Biological Species Concept. Though he did not invent this concept, Mayr was its principal advocate. *Systematics* was a major defense of this idea. Mayr argued that the biological species concept made evolutionary sense, and it produced a real unit—he called it a *natural* one—for classification. At the same time Mayr recognized the limitations of this test. It was difficult to study in the field. It made sense only for a relatively small number of organisms. Regardless, Mayr insisted on this test because it offered a criterion independent of the individual judgment and personal biases of taxonomists. The ability to test such criteria gave Mayr the sense that his approach was objective, something he thought essential for scientific work.

In addition to promoting certain theories and practices, *Systematics* also had an agenda about participation in evolutionary studies. This agenda was double sided. On the one hand, Mayr promoted his own profession—museum systematics and biogeography—within evolutionary studies. He complained that too much emphasis had been placed on abstract problems in genetics. Because evolution occurred in ecological space and time, he explained, its study required the knowledge of whole organisms plus the biogeographic skills of systematists and zoologists. Notably, Mayr tied his book closely to Theodosius Dobzhansky's *Genetics and the Origin of Species* (1937) as equal partners in that conversation. On the other hand, Mayr needed his fellow "museum men" to consider evolutionary studies and other biological problems as worthy topics for study. In *Systematics* Mayr hoped to provide models that illustrated how that study could be done.

Despite its success, Mayr published only one edition of *Systematics*. His 1963 book *Animal Species and Evolution* was its sequel.

BIBLIOGRAPHY

Dobzhansky, T. 1937. *Genetics and the Origin of Species.* New York: Columbia University Press.
Goldschmidt, R. 1940. *The Material Basis of Evolution.* New Haven, CT: Yale University Press.
Mayr, E. 1940. Speciation phenomena in birds. *American Naturalist* 74: 249–278.
———. 1942. *Systematics and the Origin of Species.* New York: Columbia University Press.
———. 1948. The bearing of the new systematics on genetical problems: The nature of species. *Advances in Genetics* 2: 205–237.
———. 1963. *Animal Species and Evolution.* Cambridge, MA: Belknap Press of Harvard University Press.   —*J.C.*

T
___

Taphonomy

Taphonomy, literally the study of graves or embedding of organic remains (from the Greek *taphos*), can be characterized more generally as the study of processes of preservation and how they affect information in the fossil record. Taphonomists use lab and field experiments, pattern analysis in the fossil record, and computer simulations and other modeling methods to understand the processes behind the accumulation and destruction of the fossil record and to estimate the resolving power, completeness, and fidelity of biological information captured therein. Even for organisms with durable hard parts, taphonomic processes can potentially distort the abundance and diversity of species and morphotypes and the time, location, and habitat of first and last occurrences of species and evolutionary lineages.

Several research themes are active in taphonomy today.

1. Using the state of preservation of fossils to *reconstruct ancient environmental conditions*. Soft-tissue preservation requires extraordinary conditions—in most cases low oxygen or rapid burial to exclude scavengers and retard decomposition, and early postburial mineral replacement to stabilize cellular and tissue-grade features permanently. In contrast, mineralized skeletal elements can be highly durable in postmortem accumulation and time averaging in many settings, and they provide the great bulk of the fossil record. Patterns of damage in all remains can reveal the conditions of accumulation, providing insights into the local environment beyond those routinely extracted from the sedimentary rock that hosts the fossils or by the use of isotopic and other chemical indicators. The inherent durability of a taxon, whether mineralized or not, large-bodied or small, robust or flimsy, thus determines its likelihood of fossilization, and the state of preservation of such taxa gives insights into the local environment of accumulation (which in many cases is the original life environment).

2. Quantifying how faithfully fossil materials *capture the original composition* of fossil biotas. Important issues relevant to microevolutionary analysis include the following: (a) Whether time averaging of multiple generations within a single bed increases the observed morphological variability of a taxon because short-term fluctuations are pooled into a single sample, or instead reduces observed variability because of the preferential destruction of small or fragile morphs when burial is delayed. For some important groups, such as marine mollusks and terrestrial mammals, time averaging appears

(quite surprisingly) to have little net effect. (b) The impact of low preservation potential on observed evolutionary duration: are taxa with skeletons of low preservation potential more likely to be singletons (known from single occurrences owing to low frequency of fossilization), or are they instead more likely to have artificially lengthened geologic ranges because poor quality of preservation leads to specimens being assigned to already-known taxa (lumping)? The former probably holds when one compares soft-bodied with biomineralized groups, but the latter is more likely among mineralized groups. (c) What impact have long-term trends—such as the northward plate-tectonic migration of continents during the Phanerozoic, the evolutionary intensification of predation and sediment stirring by organisms, and broad fluctuations in global climate and the amount of sedimentary record preserved—had on preservational rates and bias? A great deal of research attention is currently focused on determining the taphonomic comparability of younger and older parts of the fossil record.

3. *Evaluating the temporal resolution* (acuity or time value) of individual beds and the *temporal completeness* (versus gappiness) of the stacks of sedimentary rocks that host fossils. That is, how many generations of organisms are represented by a single bed or collecting horizon, and what proportion of total elapsed time is archived in sedimentary successions and is thus available to sample? High temporal acuity per fossil sample is required to evaluate morphologic or genetic variability at a population level (as opposed to the species level); acuity is basically determined by the rate of sediment accumulation (fast rates promote high acuity), which varies among depositional settings. The close spacing of samples over time—regardless of whether individual samples have high or low acuity and regardless of whether sample spacing is determined by the completeness of the sedimentary record or the effort expended by the paleontologist—determines our ability to identify fluctuations in average values and variability over time. It also determines the confidence with which geologic-range end points can be known and thus our confidence in the evolutionary durations of taxa and their relative timing (e.g., did taxon A first appear before or after the extinction of taxon B?). A variety of probabilistic and analytic methods have been developed over the past few decades to evaluate these preservational aspects, which are particularly critical to evolutionary analysis.

4. *Analyzing major macroevolutionary events and trends* in the history of life. The fossil record has long been used as a unique source of information on large-scale patterns of biological diversification and extinction, complementing phylogenetic and molecular (DNA) analysis of modern-day biotas. Our confidence in the fossil record as a source of such insights has been greatly increased by taphonomic evaluation of critical intervals. For example, the failure to find metazoan (multicellular-animal) fossils in Precambrian deposits that are preservational counterparts to the Cambrian-age deposits that host rich, well-known metazoan fossils, despite intense sampling, greatly increases our confidence that these phyla truly arose in the Cambrian. Sim-

ilarly, the absence of (many) Cretaceous taxa in subsequent Tertiary-age deposits, despite the presence in the Tertiary of taxa that are preservational counterparts (in terms of body size, skeletal composition, or environment of preservation), is compelling evidence of a genuine mass extinction across the Cretaceous-Tertiary boundary of many animal groups (including the dinosaurs and many reef-building and other important marine lineages). Even records with poor temporal acuity can contribute meaningfully to the recognition of such large-scale evolutionary patterns, and a wide array of increasingly sophisticated sampling and analytic methods have been developed to permit data from scattered and relatively low-resolution records to be synthesized into rigorous pictures of evolutionary dynamics over geologic timescales.

BIBLIOGRAPHY

Behrensmeyer, A. K., S. M. Kidwell, and R. Gastaldo. 2000. Taphonomy and paleobiology. In D. H. Erwin and S. L. Wing, eds., *Deep Time: Paleobiology's Perspective,* 103–147. Lawrence, KS: Allen Press.
Kidwell, S. M., and S. M. Holland. 2002. Quality of the fossil record: Implications for evolutionary biology. *Annual Review of Ecology and Systematics* 33: 561–588.
Martin, R. E. 1999. *Taphonomy: A Process Approach.* Cambridge: Cambridge University Press.   —*S.M.K.*

Teilhard de Chardin, Pierre (1881–1955)

Pierre Teilhard de Chardin was a French-born Jesuit priest who became one of the leading paleoanthropologists (students of human evolution) in the first half of the twentieth century. Long based in China, he was involved in the discovery of Peking Man, now recognized as *Homo erectus* and dating from just under half a million years ago. (The actual finds were lost in World War II; now only casts exist.) He was also, as a student, present in England when some of the Piltdown forgeries were uncovered. Stephen Jay Gould, the paleontologist and popular science writer, accused Teilhard of the hoax, but this is almost certainly not the case.

Deeply moved by his experiences in World War I, when he served bravely as a stretcher bearer, Teilhard sought to reinvigorate Christianity by infusing it with evolutionary ideas. The result, published in his masterwork *The Phenomenon of Man* (*Le phénomène humain,* 1955), saw life as an upward climb, progressing through various stages, notably the geosphere (the physical world) and the biosphere (the living world) until it reached the noösphere (the world of human consciousness or ideas). This noösphere itself he read as part of evolution's progressive route upward until it reached its apotheosis, the Omega Point, which in some way Teilhard identified with Jesus Christ.

None of this sounds very Darwinian, and it is not. Teilhard was much influenced by the vitalist work *L'évolution créatrice* (1907) by fellow Frenchman Henri Bergson, and he saw this as supporting a version of orthogenesis, an evolutionary belief that there is a kind of world force pushing organisms up the ladder of change until they reach the highest point, namely, humankind. Teilhard garnered praise from leading evolutionary biologists during his lifetime. In Britain, Julian Huxley was enthusiastic about Teilhard's ideas; in America Theodosius Dobzhansky, a Russian-born population geneticist, felt the same way. These two were, respectively, the presidents of the British and the American Teilhard societies, although Huxley was always an atheist and Dobzhansky a lifelong member of the Russian Orthodox Church (for details, see Ruse 1996). Today there are many theologians, notably the Catholic theologian John Haught of Georgetown University and the Lutheran theologian Philip Hefner of the Lutheran School of Theology in Chicago, who find Teilhard's work deeply inspiring. The great evolutionary paleontologist George G. Simpson was a good friend of Teilhard. He admired him as a scientist and loved him as a friend, although Simpson's Calvinist childhood came through in disapproval of Teilhard's somewhat French attitude to the vows of poverty, chastity, and obedience. The mistress especially stuck in Simpson's craw. "I do not myself consider poverty, chastity, and obedience to be virtues, but anyone who voluntarily took what he recognized as sacred vows to observe them and then egregiously and consciously broke all three vows throughout his life without ever renouncing them—such a man must be considered a hypocrite" (letter to D. A. Hooijer, July 11, 1972, Simpson Papers, American Philosophical Society, Philadelphia; more references in Ruse 1996).

Equaling the enthusiasm for his ideas, Teilhard has had strong critics. From the first, the Catholic Church was suspicious of his thinking. It was believed that because of his evolutionism he denied the historical authenticity of Adam and Eve and thereby denied original sin. It was also thought that he downgraded the significance of the crucifixion of Jesus by making the world system an ongoing evolving process, rather than as having culminated in one key event. He was forbidden to publish his ideas in his lifetime, and even in the 1980s the Roman Catholic Church reaffirmed its opposition to his thinking. Fortunately, Teilhard had close nonclerical friends who saw that his work appeared publicly shortly after his death.

Others, more in the social realm, worry that Teilhard took too little notice of the ills of his day—nuclear weapons and so forth. They find unconvincing the idea that all is in a happy upward rise. They feel that he should have been more sensitive to the problems in the way of progress. But none of these criticisms equaled the hostility that was shown by some scientists. Nobel Prize–winning Peter Medawar wrote a scathing review of the English version of the *Phenomenon of Man*, although his wrath was directed as much toward Julian Huxley (who wrote the introduction) as toward Teilhard. Without denying the merits of Medawar's critique, however, it should be noted that Teilhard was in some respects his own worst enemy. Criticized by his own church, he insisted with perhaps too much vigor that his writings should be considered in the realm of science rather than theology (and hence by nature nonheretical). Because they were clearly not purely scientific, people like Medawar could be critical.

Science or nonscience, Darwinian or not, right or wrong, Teilhard's vision is today generally recognized as the most important post-*Origin* attempt to build a world picture incorporating both Christianity and evolution. Increasingly, even those who can see the major problems are coming to recognize this. The task is not to refute his vision but to improve it.

BIBLIOGRAPHY

Bergson, H. 1907. *L'évolution créatrice*. Paris: Alcan.
Dobzhansky, T. 1967. *The Biology of Ultimate Concern*. New York: New American Library.
Gould, S. J. 1980. The Piltdown conspiracy. *Natural History* 89 (August): 8–28.
Haught, J. F. 2000. *God after Darwin: A Theology of Evolution*. Boulder, CO: Westview Press.
Hefner, P. 1993. *The Human Factor: Evolution, Culture, and Religion*. Minneapolis, MN: Fortress Press.
Huxley, J. S. 1959. Introduction to P. Teilhard de Chardin, *The Phenomenon of Man*, 11–28. B. Wall, trans. London: Collins.
Medawar, P. 1961. Review of *The Phenomenon of Man*. Mind 70: 99–106. Reprinted in P. Medawar, ed., *The Art of the Soluble*. London: Methuen and Co., 1967.
Ruse, M. 1996. *Monad to Man: The Concept of Progress in Evolutionary Biology*. Cambridge, MA: Harvard University Press.
———. 2005. *The Evolution-Creation Struggle*. Cambridge, MA: Harvard University Press.
Teilhard de Chardin, P. 1955. *Le phénomène humain*. Paris: Editions de Seuil.
———. 1959. *The Phenomenon of Man*. B. Wall, trans. London: Collins.

—M.R.

Tempo and Mode in Evolution
(George Gaylord Simpson)

A paleontologist who specialized in mammals, George Gaylord Simpson wrote *Tempo and Mode in Evolution* between 1938 and 1942. Before this his research focused on key moments in the evolutionary history of mammals. While he was writing *Tempo and Mode,* Simpson also was writing a technical classification of mammals, both living and extinct. Simpson enjoyed tackling big questions.

Two such questions were the starting point for *Tempo and Mode:* (1) what, in fact, is the rate of evolution in nature, and (2) what patterns recur in evolutionary histories? Simpson identified three evolutionary rates: exceptionally slow, normal, and exceptionally rapid. In absolute terms, he said, these rates vary from group to group, but as relative rates they are distinctly different phenomena. Simpson also identified three recurring patterns: speciation (gradual differentiation of a group into species), phyletic evolution (an accumulating trend or direction), and quantum evolution (a rapid shift in fundamental qualities).

Tempo and Mode combined Simpson's description of these phenomena with a critical study of their underlying causes. For this he drew from recent developments in genetics, mathematical population genetics, and ecology. He was strongly influenced by Sewall Wright's shifting-balance theory, which he learned from Theodosius Dobzhansky. With this theory as part of his larger understanding of evolution, Simpson put great emphasis on natural selection, taking place in adaptive landscapes, for giving direction to evolution. Population size also became vitally important, especially when Simpson explained quantum evolution and exceptionally rapid rates of change. Historians frequently suggest that Simpson merely applied these theory components. That is wrong. For Simpson the fossil record provided the ultimate empirical test for a theory. It also provided raw material for new theoretical ideas, for example, his own work on rates and patterns. Simpson critically tested the theories he considered, sometimes rejecting them as unsupported by facts (e.g., generation length or changing mutation rates) and sometimes using them to make novel predictions for study (e.g., relating small population size to rapid evolution). Overall he thought of his work as synthesizing knowledge, not merely consuming someone else's ideas.

What was *Tempo and Mode*'s lasting impact? The answer splits into its impact on three audiences. *Tempo and Mode* is most remembered for integrating paleontology and population genetics within the framework of the evolutionary synthesis. In the 1930s evolutionary studies were dominated by genetics and experimental botany. Researchers in these areas saw Simpson's book as an application of their techniques that extended them into new territory and generally confirmed their approach. Another audience, paleontologists, celebrated *Tempo and Mode* for demonstrating to those other researchers that fossils were the heart of evolutionary studies and for reminding them that only paleontologists could speak to certain aspects of the subject. Simpson himself wanted to begin several conversations. He wanted paleontologists to appreciate how much more information could be extracted from the fossil record than had previously been the case. He also wanted them to use new statistical and biological techniques and to focus their attention on studying processes (e.g., evolutionary, ecological, and migratory) as much as on studying fossil objects. At the same time Simpson wanted biologists to appreciate modern paleontology as a rigorous, modern science.

Later in his life Simpson frequently distanced himself from *Tempo and Mode*, referring colleagues to his *Major Features of Evolution* (1953). He considered his 1944 book rushed and a bit too speculative and said that his later book reflected his construction of his views in more considered fashion.

BIBLIOGRAPHY

Laporte, L. 2000. *George Gaylord Simpson: Paleontologist and Evolutionist.* New York: Columbia University Press.

Simpson, G. G. 1944. *Tempo and Mode in Evolution.* New York: Columbia University Press.

———. 1953. *Major Features of Evolution.* New York: Columbia University Press.

Simpson, G. G., and A. Roe. 1939. *Quantitative Zoology.* New York: McGraw-Hill.

—J.C.

Timofeeff-Ressovsky, Nikolai Vladimirovich (1900–1981)

Nikolai Vladimirovich Timofeeff-Ressovsky was a Russian biologist who developed important ideas in the study of genetic mutations, evolutionary genetics, and the effects of radiation both on people and on ecosystems. He witnessed firsthand important political regimes, including the Russian Revolution, the rise and fall of Nazi Germany, and Stalin's political purges.

Timofeeff was born to a family of minor nobility but poor circumstances in Kaluga province, near Moscow. An admirer of Peter Kropotkin, he first sided with the anarchists' Green Army during the Russian Revolution in 1917 and then joined the Red Army in 1919. He served in numerous campaigns, including the Crimean and Polish fronts.

Before the revolution Timofeeff studied biology at Moscow University and actively participated in the city's intellectual circles. After the war he returned. In 1922 he began studying under Sergei Chetverikov, who heavily influenced his basic thinking about biology. From his mentor Timofeeff learned the importance of genetic variation within populations and mastered techniques for uncovering the variation that might be hidden in recessive genes. He also learned to think about evolution as a population-level phenomenon and to concentrate his attention on small changes in gene frequency, or microevolution.

While he was studying under Chetverikov, Timofeeff also began work in the prestigious Institute of Experimental Biology, run by N. K. Kol'tsov. Here Timofeeff obtained rigorous training in comparative anatomy, systematics, and morphology. The institute was becoming a world-class center for experimental and theoretical genetics. Timofeeff developed special interests in phenogenetics (what now is described as gene expression), the causes and effects of gene mutations, and the chemical structure of genes. Among his colleagues at the Institute were some of the Soviet Union's leading figures in the field, including S. S. Chetverikov, A. S. Serebrovsky, S. M. Gershenson, and N. P. Dubinin. In this group Timofeeff stood out as one of the most talented despite the fact that he still held no formal university degree.

The timing of his arrival at the institute was fortuitous. In the same year H. J. Muller arrived from California, carrying news of Thomas Hunt Morgan's chromosome theory of heredity and stocks of the fruit fly, *Drosophila*. This had become the organism of choice for Western genetics. In a masterful combination Timofeeff brought the two traditions of his training together with opportunities to be found in Muller's fly stocks. He quickly earned a reputation as an expert experimenter. For instance, he isolated genes that produced different effects under different developmental contexts. This confirmed the theoretician's notion of pleiotrophy.

In 1926 Timofeeff was surprised by an offer from Berlin to organize a laboratory of genetics. At that time he was relatively unknown outside a small circle of Russian biologists and still had no university degree. The offer came personally from Oskar Vogt, director of the Institute of Brain Research. Vogt had been impressed by Timofeeff's skill as an experimenter. When Timofeeff arrived, it was to new facilities and the title of head of his department.

Timofeeff remained in Germany until 1945. These were the most productive and creative years of his career. In the study of gene function, for instance, he developed new ways to study gene expression and regulation. He investigated the relation between gene structure and function. He developed an influential theory to explain how X-rays cause mutations. This became known as the hit theory, developed with Karl Zimmer and Max Delbrück. The theory compared X-rays with bombs striking targets and exploding; different targets produced different effects. Their approach inspired Erwin Schrödinger's famous book *What Is Life?* (1944). Timofeeff based his thinking on his own demonstrations of the linear relation between radiation dose and the number of mutations produced.

In population genetics Timofeeff extended Chetverikov's work on genetic variability, paying special attention to the reserves of recessive variation found in wild populations. Studying changes to gene frequencies gave him quantitative measures of microevolution.

The political situation in Germany and the Soviet Union presented Timofeeff with both opportunities and risks. Genetics received considerable support in Germany after the National Socialists came to power in 1933. But Timofeeff was ordered to return to the Soviet Union several times in the 1930s. Each time he refused. Friends privately warned him that arrest was likely in the increasingly paranoid world of Stalin's purges. Furthermore, Western genetics was being denounced aggressively by Trofim Lysenko and his followers. Because Timofeeff was universally identified as one of Russian's leading advocates of that approach, he knew that he faced considerable danger if he returned. Friends in the United States encouraged him to emigrate from Germany and secured for him the directorship of a research center in the United States. To their surprise Timofeeff declined, citing rumors of poor financial support. He did not want to be a refugee. As an incentive to remain in Germany, Timofeeff was given virtual autonomy as the director of a reorganized laboratory in the Institute of Genetics and Biophysics of the Kaiser Wilhelm Society.

After Hitler broke his pact with Stalin in 1941, Timofeeff simply could not return to Russia. His genetics research continued during World War II. He also began a collaboration with scientists at the Auer Society, a chemical firm involved in war production, including uranium refinement for German atomic projects. For it Timofeeff undertook research on the effects of exposure and on ways to improve radiation protection. This was not weapons-related research and was most likely aimed at the civilian atomic power industry.

Timofeeff's role in Nazi and military activities during the war is much debated. He did not join the Nazi Party, and he refused German citizenship, remaining a strong Russian nationalist (although he was no Communist). His son was involved with the anti-Nazi resistance, but he was arrested and died in the Mauthausen concentration camp in 1944. Timofeeff used his political connections and professional authority to protect Jewish and Russian refugees, as well as some prisoners and foreigners drafted into forced labor. At the same time Timofeeff contributed to numerous programs in Nazi science, such as discussions on how to identify carriers of recessive genetic disorders as part of race hygiene programs. Although he was expert in this area, he did not actively campaign for negative eugenics.

Timofeeff remained in Berlin through 1945. He was present when the Soviet army captured Berlin. He cooperated in the occupation. Soviet officials allowed some of his research to continue, and Timofeeff seriously discussed the possibility of moving his laboratory back to Russia. His experience in the area of radiation biology and genetics was crucial in this negotiation. Because the Soviet military was expanding its atomic programs, Timofeeff had valuable expertise.

In September 1945, however, Timofeeff was arrested by the Soviet secret police, who were following a different agenda. He was sentenced to 10 years for failing to return during the war, and he was sent into the gulag prison system. During his imprisonment Timofeeff nearly died of starvation and lost much of his vision. After more than a year in prison he was retrieved by military sponsors to work in atomic research. By the spring of 1947, after release and hospitalization, Timofeeff was sent to a closed military research center near Sverdlovsk, in the Ural Mountains, and asked to organize a laboratory for radiation biology. This included studying the effects of radiation in medicine and genetics. It also included ecosystems ecology, or radiation biogeocenology. This latter research studied how radioactive isotopes move through or accumulate in biological systems, for example, in organisms, food chains, and whole ecosystems. This assignment carried some privileges. He was reunited with his wife and surviving son. They had known nothing about Timofeeff's status since his incarceration two years earlier. Because he worked under military secrecy, Timofeeff was one of the few Soviet geneticists sympathetic to Western theories to remain free from Lysenko's purges.

The secret work continued until 1955, when Timofeeff received an amnesty and restrictions on his life were partly lifted. However, he was forced to remain in the Urals. Making the best of his situation, Timofeeff organized a biophysics laboratory in Sverdlovsk and created a series of summer schools to train biologists in genetics, ecology, and radiation biology. As an expert in the field he played an important role in assessing the contamination and long-term consequences of the 1957 Kyshtym nuclear accident.

In 1964 Timofeeff was allowed to move to Obninsk, near Moscow, where he organized another laboratory, now for the Academy of Medical Sciences. Less protected now by his military connections, Timofeeff was increasingly attacked by Lysenkoists. They accused him of Nazi collaboration and experimenting on Soviet prisoners of war. He was blocked from election to the Soviet Academy of Sciences. In 1969 Timofeeff was forced to retire. Heavy pressure from his supporters led to an appointment as a scientific consultant for the Institute of Medico-Biological Problems, directed by O. V. Gasenko. He remained at the institute, studying space medicine and various genetics problems, until his death in 1981.

Throughout his life Timofeeff depended heavily on his wife of 50 years, Helena Alexandrovna Fidler (1898–1973). An excellent experimenter herself, Helena collaborated in her husband's research, coauthored many of his essays, and fully participated in the intellectual life of the scientific community surrounding her husband. She also managed his social life and served as his professional secretary.

BIBLIOGRAPHY

Medvedev, Z. 1982. Nicolai Wladimorovich Timofeeff-Ressovsky. *Genetics* 100: 1–5.
Paul, D., and C. Krimbas. 1992. Nikolai V. Timofeeff-Ressovsky. *Scientific American* 2: 86–92.
Schrödinger, E. 1944. *What Is Life?* Cambridge: Cambridge University Press.
Timofeeff-Ressovsky, N. W. 1940. Mutations and geographical variation. In J. Huxley, ed., *The New Systematics*, 73–136. Oxford: Oxford University Press.

—J.C.

Tinbergen, Niko (1907–1988)

Niko Tinbergen was the third of five children of a Dutch schoolteacher. As a youth he was attracted to the strong, youth-oriented tradition of nature study that flourished in Holland in the early decades of the twentieth century. He took great pleasure observing animals in their natural habitats. In contrast, he was not attracted to academic zoology, which seemed to him to deal too exclusively with museum- or laboratory-based examinations of dead specimens. After concluding that it might be possible to pursue a career as an academic biologist with an orientation toward field studies, he enrolled as a zoology student at Leiden University in 1926. He earned his PhD there in 1932 with a field study of insect orientation behavior, modeled in part after the work of Karl Ritter von Frisch (see the alphabetical entry "Karl Ritter von Frisch" in this volume). Upon graduation he married and spent 14 months doing fieldwork in Greenland. He then returned to Leiden University and took up a position as an instructor in the Department of Zoology. There he developed a program of teaching and research that included field as well as laboratory studies of animal behavior. The fieldwork focused on insects and birds (especially the herring gull), while the laboratory work dealt for the most part with the behavior of a single species of fish, the three-spined stickleback.

Critical for Tinbergen's further career was his encounter and subsequent friendship with the Austrian naturalist Konrad Lorenz (see the alphabetical entry "Konrad Zacharias Lorenz" in this volume). Having first met Lorenz at a symposium on instinct held at Leiden in 1936, Tinbergen traveled to Lorenz's home near Vienna in the spring of 1937, where he worked with Lorenz for three and a half months. Their talents proved to be complementary. Tinbergen recognized that Lorenz was laying the conceptual foundations of a new science of animal behavior. Lorenz in turn appreciated that Tinbergen's experimental and analytical talents were an invaluable complement to his own largely intuitive theorizing. They were separated by the war but renewed their friendship when they met again in 1948. Although both continued to be leaders of ethology's development over the next two decades, Tinbergen's work was especially significant for the discipline's ongoing growth. His contributions included founding in 1948 the journal *Behaviour;* moving in 1949 to Oxford, where he established an influential program of animal behavior studies; and publishing in 1951 the first systematic overview of ethology, *The Study of Instinct.* He also was a successful popularizer of ethological studies, most notably in his book *Curious Naturalists* (1958) and in the film *Signals for Survival* (Hugh Falkus, director, 1968).

Tinbergen worked hard in the 1950s and 1960s to see that ethological studies grew in a coordinated fashion. Often promoting ethology as "the biological study of behavior" (e.g., Tinbergen 1963, 411), he insisted that for the discipline to thrive it needed to address simultaneously the questions of physiological causation, individual development, evolutionary history, and function. As for his own research, he found himself in the 1950s and 1960s turning with particular success to questions of behavioral function. The comparative and experimental studies that he and his students conducted on the survival value of specific behavior patterns represented a significant contribution to modern behavioral ecology. In recognition of his contributions to the study of animal behavior, Tinbergen was awarded the 1973 Nobel Prize for Physiology or Medicine jointly with Konrad Lorenz and Karl Ritter von Frisch.

BIBLIOGRAPHY

Burkhardt, R. W., Jr. 2005. *Patterns of Behavior: Konrad Lorenz, Niko Tinbergen, and the Founding of Ethology.* Chicago: University of Chicago Press.
Kruuk, H. 2003. *Niko's Nature: A Life of Niko Tinbergen and His Science of Animal Behaviour.* Oxford: Oxford University Press.
Tinbergen, N. 1951. *The Study of Instinct.* Oxford: Oxford University Press.
———. 1958. *Curious Naturalists.* London: Country Life.
———. 1963. On aims and methods of ethology. *Zeitschrift für Tierpsychologie* 20: 410–433.   —R.W.B.

Trivers, Robert (b. 1943)

Robert Trivers is best known for five early essays (1971–1976) that transformed the study of sex differences, sex ratios, and social behavior. He has also contributed to a number of other areas, including the evolution of selfish genetic elements and genomic imprinting, the causes and consequences of fluctuating asymmetry, and the behavioral ecology of *Anolis* lizards.

Trivers was educated at Harvard University, where he took a BA in history (1965) and a PhD in biology (1972). His doctoral adviser was Ernest E. Williams, and he was strongly influenced by several other mentors, including William Drury, Ernst Mayr, Edward O. Wilson, and Irven DeVore. Trivers has subsequently held faculty positions at Harvard, the University of California at Santa Cruz, and Rutgers University, where since 1994 he has been professor of anthropology and biological science.

Trivers's early essays apply gene's-eye reasoning, as pioneered by George C. Williams (1966) and William D. Hamilton (1964a, 1964b, 1967) in the 1960s, to specific problems that had not been addressed so directly or fully. The 1971 essay on reciprocal altruism asks how cooperation can evolve between unrelated individuals and identifies some evolutionary consequences to be expected (e.g., ongoing selection for improved abilities to detect cheating and for improved abilities to cheat, possibly involving tactics such as self-deception) (Trivers 1971). The 1972 essay on sexual selection identifies differences in parental investment in offspring as a key parameter that explains much about the evolution of sex differences in morphology and behavior, including the sporadic occurrence of sex-reversed species (Trivers 1972). The 1973 essay on sex-ratio adjustment in relation to parental condition argues that one sex (typically males) may often benefit more than the other from above-average parental investment, and that parents are therefore expected to bias the sexes of their offspring in relation to their present abilities to invest (Trivers and Willard 1973). The 1974 essay on parent-offspring conflict argues (from Hamilton's theory of inclusive fitness) that offspring should often attempt to extract more investment than their parents would prefer to give, and that weaning conflict therefore represents a true divergence of interests rather than a mere breakdown of communication (Trivers 1974). The 1976 essay on sex allocation in social insects tests this parent-offspring conflict model in a system where unusual coefficients of relationship and patterns of investment generate predictions that differ greatly depending on whether mothers (queens) or daughters (workers) control the sex ratio (Trivers and Hare 1976). Each of these essays has stimulated hundreds of refinements, extensions, and empirical tests and has affected thinking in a substantial area of research.

In 1985 Trivers published a textbook, *Social Evolution,* which gives a comprehensive, accessible, and richly illustrated account of the problems, data, and theories of behavioral ecology and sociobiology. In 2002 he published a book of selected essays, *Natural Selection and Social Theory,* which includes informative and highly entertaining accounts of the intellectual and personal backgrounds to each of his five early essays and six subsequent works, with brief postscripts commenting on developments since each essay first appeared. Trivers's most recent book (with Austin Burt) is *Genes in Conflict: The Biology of Selfish Genetic Elements* (2006), which synthesizes theoretical and empirical work on the full range of selfish genetic phenomena, including transposable elements, gene conversion and homing, genomic imprinting, and several varieties of nonrandom segregation, such as genome exclusion.

BIBLIOGRAPHY

Burt, A., and R. Trivers. 2006. *Genes in Conflict: The Biology of Selfish Genetic Elements.* Cambridge, MA: Harvard University Press.

Hamilton, W. D. 1964a. The genetical evolution of social behaviour I. *Journal of Theoretical Biology* 7: 1–16.

———. 1964b. The genetical evolution of social behaviour II. *Journal of Theoretical Biology* 7: 17–52.

———. 1967. Extraordinary sex ratios. *Science* 156: 477–488.

Trivers, R. 1985. *Social Evolution*. Menlo Park, CA: Benjamin/Cummings.

————. 2002. *Natural Selection and Social Theory: Selected Papers of Robert L. Trivers*. Oxford: Oxford University Press.

Trivers, R. L. 1971. The evolution of reciprocal altruism. *Quarterly Review of Biology* 46: 35–57.

————. 1972. Parental investment and sexual selection. In B. Campbell, ed., *Sexual Selection and the Descent of Man*, 136–179. Chicago: Aldine-Atherton.

————. 1974. Parent-offspring conflict. *American Zoologist* 14: 249–264.

Trivers, R. L., and H. Hare. 1976. Haplodiploidy and the evolution of the social insects. *Science* 191: 250–263.

Trivers, R. L., and D. E. Willard. 1973. Natural selection of parental ability to vary the sex ratio of offspring. *Science* 179: 90–92.

Williams, G. C. 1966. *Adaptation and Natural Selection*. Princeton, NJ: Princeton University Press.   —*J.S.*

V

Variation and Evolution in Plants
(George Ledyard Stebbins)

Variation and Evolution in Plants, published in 1950, is the masterwork of the geneticist and botanist George Ledyard Stebbins. It is the fourth and final work of a quartet of books that established the synthetic theory of evolution, the American version of the theory that emerged in the 1930s from the blending of Darwinian selection and Mendelian genetics. Stebbins's book, the only one of the four explicitly restricted to one group of organisms, came later than the others—Theodosius Dobzhansky's *Genetics and the Origin of Species* in 1937, Ernst Mayr's *Systematics and the Origin of Species* in 1942, and George Gaylord Simpson's *Tempo and Mode in Evolution* in 1944. The intention had been that the botanist Edgar Anderson would contribute to the project, but when he failed to do so, Dobzhansky explicitly and intentionally recruited Stebbins, getting him to give the Jesup Lectures at Columbia in 1946 and all the way urging him to complete and publish the work.

Given its genesis, it is not surprising that *Variation and Evolution in Plants* reflects particularly the evolutionary vision of Dobzhansky, namely, that Sewall Wright's shifting-balance theory of evolution is the correct conceptual foundation for evolutionary studies, a foundation on which one must now build using empirical data. In Stebbins's book not only is there a general foundation from Wright, but there is also careful discussion of the details of the shifting-balance theory. For example, Stebbins offered a detailed and generally sympathetic coverage of Wright's central claim that much evolution is due not to the direct effects of natural selection but to genetic drift, where random factors of breeding in small populations are a key process. However, although Stebbins thought drift an important notion, based on his own observations and perhaps influenced by a general skepticism about drift that was growing in the late 1940s, he gave no unambiguous and enthusiastic endorsement of the notion. He wrote a book in the school of the synthetic theorists, not a slavish endorsement of the thinking of others.

Indeed, despite its place in the general picture of the synthetic theory, Stebbins's book (longer and more detailed in many respects than the others) is in some respects the odd man out, for he is much concerned to show and discuss aspects of the plant world not generally represented (or significant) in the animal world. The fact that many plants are asexual, for instance, gets much discussion. So also does the fact that much plant change revolves around hybridization, where two virtually isolated groups meet and start to exchange genes. Major coverage is given to the peculiar plant processes (polyploidy) where chromosome sets are doubled or combined entire, and where new species (in the sense of being reproductively isolated from their parents) are formed in one generation.

There is also discussion of evolutionary trends in plants. Although in his other writings Stebbins makes it very clear that (like the other synthetic theorists) he is deeply committed to a progressionist view of the evolutionary world, like the other theorists he is very careful not to introduce explicit value-laden discussions about higher and lower into *Variation and Evolution and Evolution in Plants*. This work was intended to complete the paradigm of modern evolutionary biology as a functioning and respectable branch of science, and in this aim it succeeded admirably.

BIBLIOGRAPHY

Dobzhansky, T. 1937. *Genetics and the Origin of Species*. New York: Columbia University Press.
Mayr, E. 1942. *Systematics and the Origin of Species*. New York: Columbia University Press.
Ruse, M. 1996. *Monad to Man: The Concept of Progress in Evolutionary Biology*. Cambridge, MA: Harvard University Press.
Simpson, G. G. 1944. *Tempo and Mode in Evolution*. New York: Columbia University Press.
Smocovitis, V. B. 1988. Botany and the evolutionary synthesis: The life and work of G. Ledyard Stebbins, Jr. Unpublished PhD dissertation, Cornell University.
Stebbins, G. L. 1950. *Variation and Evolution in Plants*. New York: Columbia University Press.   —M.R.

Vestiges of the Natural History of Creation (Robert Chambers)

Vestiges of the Natural History of Creation, first published in 1844 (a supplement, *Explanations,* appeared in 1845), was authored anonymously by the Scottish publisher Robert Chambers. In this work Chambers argued that all organisms, living and dead, are the end results of a long, natural process of evolutionary change. Although he made much use of empirical information scrounged from a variety of sources, some scientifically respectable and some not, Chambers's main intent was to provide an overall world picture, as much philosophical as anything else. Here he differed from Charles Darwin in *On the Origin of Species,* where the intent was to put evolutionary studies on a firm scientific basis.

Chambers opened *Vestiges* by referring favorably to the so-called nebular hypothesis, which supposes that the universe itself evolved, with planets and other bodies condensing out of gases. Next he tackled the question of the origin of life. Particular critical scorn was reserved for Chambers's suggestion that the frost-fern patterns left on windowpanes in cold weather were indicative of a link between the living and the material worlds. Chambers moved next to a greater area of strength, the fossil record. In his *Principles of Geology* (1830–1833), Charles Lyell had introduced Jean-Baptiste Lamarck's theory of evolution, completely misunderstanding it as an answer to the upward-rising fossil record. As it happens, Lamarck said nothing on this subject, but thanks to Lyell this is how it was interpreted, and Chambers was nothing if not thorough in his detailed discussion of the progression from fish to land (plants) and from mammals to the higher animals. Showing, however, how far he was from Darwin, who was then working privately on his theory, or indeed from the concerns of any professional biologist, Chambers exhibited virtually no interest in questions of adaptation. Causally, he was more interested in providing a sort of quasi-German embryological theory that claimed that organisms normally go through different life forms and then get born. But if for some reason gestation is prolonged, then the organisms go on developing in the womb until finally they appear as a higher form.

The real force of the proposed causal process was that it provided something that Chambers saw as the absolute crux of the evolutionary chain of life, namely, that it is progressive. It goes from the simplest to the most complex, namely, our own species (with the possibility of something even greater later). Chambers was explicit that he saw this progress in biology as the counterpart of progress in society, something to which he was committed both as a very successful businessman and as the author of many shorter pieces for his own publications—pieces that hymned the virtues of effort and thrift and pointed the way to future success. It was hardly any surprise that those who shared his philosophy tended to like his book, and those who put real change in the hands of God, Providence, loathed *Vestiges* with a passion.

Adam Sedgwick, professor of geology at the University of Cambridge and sometime teacher of Charles Darwin, speculated that *Vestiges* must have been written by a woman and then decided that no member of the fair sex could have penned such filth: "the ascent up the hill of science is rugged and thorny, and ill suited for the drapery of the petticoat" (Sedgwick 1845, 4). Scottish scientist David Brewster, biographer of Newton, declared that it showed the general degeneracy into which Britain had sunk. "It would augur ill for the rising generation, if the mothers of England were infected with the errors of Phrenology: it would augur worse were they tainted with Materialism." The problem, Brewster gloomily concluded, was with the slackness of modern education. "Prophetic of infidel times, and indicating the unsoundness of our general education, 'The Vestiges . . .' has started into public favour with a fair chance of poisoning the fountains of science, and of sapping the foundations of religion" (Brewster 1844, 503). William Whewell, another of Darwin's mentors, collected passages from earlier writings designed to show the falsity of evolutionary thinking. In the first edition of this collection, *Indications of the Creator* (1845), he did not even mention *Vestiges* by name.

To the contrary, the poet Alfred Tennyson, read *Vestiges* (or more probably a detailed review), and decided that it offered hope that perhaps the death of his dear friend Arthur Hallam had not been in vain and that the nihilism of

Lyellian geology, showing extinction and apparently without hope of progress, could be conquered by the uplifting message of *Vestiges* (Ross 1973).

First the despair:

> Are God and Nature then at strife,
>   That Nature lends such evil dreams?
>   So careful of the type she seems,
> So careless of the single life; . . .
>   (Tennyson 1850, stanza LIV, 78)

> "So careful of the type?" but no.
>   From scaped cliff and quarried stone
>   She cries "a thousand types are gone:
> I care for nothing, all shall go."
>   (Tennyson 1850, stanza LV, 80)

Given nature "red in tooth and claw"—this famous phrase has its source here (stanza LV)—nothing seems to make any sense. But then came the hope. Perhaps Hallam was an advanced form born before his time.

> A soul shall strike from out the vast
> And strike his being into bounds,

> And moved thro' life of lower phase,
>   Result in man, be born and think,
>   And act and love, a closer link
> Betwixt us and the crowning race. . . .
>   (Tennyson 1850, Conclusion, 209)

> Whereof the man, that with me trod
>   This planet, was a noble type
>   Appearing ere the times were ripe,
> That friend of mine who lives in God.
>   (Tennyson 1850, Conclusion, 210)

Undoubtedly, something like this—the poem was loved by everyone, from the Queen on down—prepared the way for Darwin. By 1859, the date of the publication of the *Origin,* evolution was simply no longer that shocking an idea. And if this were not enough, there is good evidence that the young Alfred Russel Wallace read *Vestiges,* became an instant convert to evolution, and devoted the next decade to finding a cause. In 1858 he was successful: he discovered natural selection and sent his essay to Charles Darwin, and thus spurred, Darwin wrote the *Origin* in fifteen months and evolution was well and truly launched.

BIBLIOGRAPHY

Brewster, D. 1844. Vestiges. *North British Review* 3: 470–515.
Chambers, R. 1844. *Vestiges of the Natural History of Creation*. London: Churchill.
———. 1845. *Explanations: A Sequel to "Vestiges of the Natural History of Creation."* London: Churchill.

Darwin, C. 1859. *On the Origin of Species*. London: John Murray.

Lamarck, J.-B. 1809. *Philosophie zoologique*. Paris: Dentu.

Lyell, C. 1830–1833. *Principles of Geology: Being an Attempt to Explain the Former Changes in the Earth's Surface by Reference to Causes Now in Operation*. 3 vols. London: John Murray.

McKinney, H. L. 1972. *Wallace and Natural Selection*. New Haven, CT: Yale University Press.

Ross, R. H. 1973. *Alfred, Lord Tennyson: "In Memoriam": An Authoritative Text, Backgrounds and Sources of Criticism*. New York: Norton.

Ruse, M. 1996. *Monad to Man: The Concept of Progress in Evolutionary Biology*. Cambridge, MA: Harvard University Press.

———. 1999. *The Darwinian Revolution: Science Red in Tooth and Claw*. 2nd ed. Chicago: University of Chicago Press.

Secord, J. A. 2000. *Victorian Sensation: The Extraordinary Publication, Reception, and Secret Authorship of "Vestiges of the Natural History of Creation."* Chicago: University of Chicago Press.

Sedgwick, A. 1845. Vestiges. *Edinburgh Review* 82: 1–85.

Tennyson, A. 1850. *In Memoriam*. London: Edward Moxon.

Wallace, A. R. 1905. *My Life: A Record of Events and Opinions*. London: Chapman and Hall.

Whewell, W. 1845. *Indications of the Creator*. London: Parker.   —M.R.

Vries, Hugo de (1848–1935)

Today Hugo de Vries is best known as one of the rediscovers of Mendel's laws. His paper, "Das Spaltungsgesetz der Bastarde," together with similar studies by Carl Correns and Erich von Tschermak-Sysenegg, all published in 1900, mark the beginning of the scientific discipline of modern genetics. But in the course of his long and distinguished career, de Vries also made significant contributions to evolutionary biology, especially with his theory of intercellular pangenesis (1889) and, most importantly, with his mutation theory (1901–1903), which set the stage for the debate about the nature of evolutionary change as either gradual and continuous or as discontinuous and proceeding in "jumps" or mutations.

Hugo de Vries was born in Haarlem, The Netherlands, in 1848 into a prominent Dutch family. Showing an early fascination with botany, his interest soon shifted to plant physiology and toward an experimental approach to the study of living systems. His dissertation at the University of Leiden focused on the role of temperature on the life processes in plants. In 1870 de Vries moved to Germany to learn more about experimental plant physiology, working for several years in the Würzburg laboratory of Julius Sachs, a leading plant physiologist. He returned to The Netherlands in 1878 when he was offered an appointment at the newly founded University of Amsterdam. During the 1880s his research shifted to questions of variation and inheritance. In his theory of intercellular pangenesis he updated Charles Darwin's theory of inheritance and combined it with recent insights in cell biology. De Vries argued that the nucleus contains all hereditary factors in the form of independent particles, while only a subset of those will be active in the cytoplasm of individual cells, thus providing a mechanistic explanation of both heredity and differentiation. In honor of Darwin, he called these particles pangenes.

Further experimental studies demonstrated that these factors followed predictable patterns of inheritance. After a colleague sent him a reprint of Mendel's 1866 paper containing the latter's now-famous experiments with peas, de Vries realized that the laws of inheritance are of a more general nature. This insight marks the rediscovery of Mendel and the beginning of modern genetics. However, de Vries did not participate in the establishment of genetics because he was more interested in accounting for evolutionary transformations than merely tracing the patterns of inheritance of specific variants. He had first noticed the appearance of new variants in *Oenothera lamarckiana,* which he termed mutations, in 1886. After years of experiments and observation and a thorough review of existing literature, de Vries summarized his ideas in the two volumes of his mutation theory, postulating that species are characterized by a general tendency to produce new mutants, that these mutants represent discrete new variants, and that mutations can be either beneficial or detrimental. De Vries's ideas were widely discussed and led to a renewed interest into the nature of variation and evolutionary change. De Vries retired in 1918 and continued his studies on the nature of mutations until his death in 1935.

BIBLIOGRAPHY

Correns, C. 1900. G. Mendel's "Regel über das Verhalten der Nachkommenschaft der Rassenbastarde." *Berichte der Deutschen Botanischen Gesellschaft* 18: 156–168.

Mendel, G. 1866. Versuche über Pflanzen-Hybriden. *Verhandlungen des naturfoschenden Vereins in Brünn* 3: 3–47.

Tschermak-Sysenegg, E. von. 1900. Über künstliche Kreuzung bei *Pisum sativium. Berichte der Deutschen Botanischen Gesellschaft* 18: 232–249.

Veer, P. de. 1969. *Leven en Werk van Hugo de Vries.* Groningen: Wolters-Noordhoff.

Vries, H. de. 1889. *Intracellulare Pangenesis.* Jena: Gustav Fischer.

———. 1900. Das Spaltungsgesetz der Bastarde: Vorläufige Mitteilung. *Berichte der Deutschen Botanischen Gesellschaft* 18: 83–90.

———. 1901–1903. *Die Mutationstheorie: Versuche und Beobachtungen über die Entstehung von Arten im Pflanzenreich.* 2 vols. Leipzig: Veit and Comp.

—*M.D.L.*

W

Waddington, Conrad Hal (1905–1975)

Conrad Hal Waddington was a pioneer in linking developmental and evolutionary biology. He explored the central role of development in evolution and argued that natural selection molded developmental processes themselves in different ways under different environmental circumstances.

Waddington came to the evolution of development after showing that several dozen genes interacted to guide development of normal wings in the fruit fly, *Drosophila*. This result inspired his concept of the epigenetic (literally, beyond genetics) landscape, the set of possible developmental pathways for a structure inherent in the network of interacting genes that produce it. Waddington suggested that natural selection would fashion this network to insulate normal development from the disruptive effects of external perturbations like unfavorable temperatures or internal perturbations like mutations. This insulation, called canalization (see figure), was a characteristic of normal development that Waddington saw as an essential adaptive property of organisms.

Waddington is known more widely for describing a striking process called genetic assimilation that, at first glance, suggests a kind of neo-Lamarckism. Waddington's original experiments showed that heat shock during a critical phase of development precluded formation of a small vein in the fly wing in some individuals. Waddington exerted artificial selection for greater expression in response to heat shock; that is, he subjected all individuals to heat shock but selected for breeding only individuals that lacked the vein. As expected, after many generations more and more individuals responded to heat shock by not forming the vein. But the unexpected result was that many individuals in the selected population failed to form the vein even when they were not exposed to heat shock; even more surprising, their offspring failed to form the vein. Thus a response that was facultative at the start became constitutive; somehow the genes had "assimilated" the response to heat shock through selection of individuals that lacked the vein. Waddington argued that this process represented a disruption of canalization that could be favored by natural selection because it allowed a facultative feature important for survival in a novel environment to be expressed constitutively.

Although Waddington's explanation for genetic assimilation has been superseded by simpler ones, his view of developmental systems as integral elements of evolution has been embraced universally. A great deal of research has focused on when and how a developmental system evolves sensitivity to

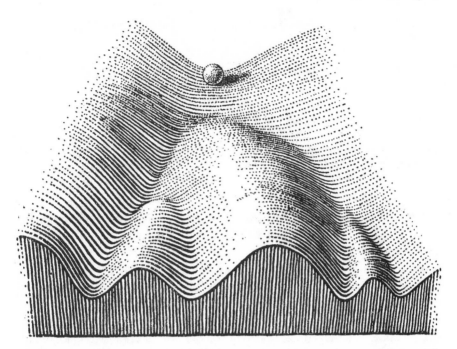

Conrad Hal Waddington was particularly interested in visual representations of scientific ideas. This figure, taken from *The Strategy of the Genes* (1957), shows how his idea of canalization works. Within certain limits, the organism will develop normally despite disturbances (in the picture, the ball being pushed up the side of the valley). But there are choices, and too great a disturbance will have lasting effects. There are obvious hints here of Wright's adaptive landscapes, and as noted elsewhere in discussion of them, it may be that Wright was influenced in his choice of metaphor by Waddington.

and not insulation from environmental influences (see the alphabetical entry "Phenotypic plasticity" in this volume). The evolution of developmental systems themselves has become an important topic in evolutionary biology (see the main essay "Evolution and Development" by Gregory A. Wray in this volume), and epigenetics and developmental networks are two of the most rapidly growing and exciting areas of modern biology.

Waddington left an enduring legacy beyond his research. His textbooks in embryology and developmental evolution guided generations of students. He also helped build the Institute of Animal Genetics at the University of Edinburgh into one of the world's leading institutions. Waddington spent nearly his entire career on the faculty at Edinburgh and was elected a fellow of the Royal Society of Edinburgh in 1948 and a fellow of the American Academy of Arts and Sciences in 1959.

BIBLIOGRAPHY

Waddington, C. H. 1957. *The Strategy of the Genes: A Discussion of Some Aspects of Theoretical Biology.* London: Allen & Unwin.

————. 1975. *The Evolution of an Evolutionist.* Ithaca, NY: Cornell University Press.

West-Eberhard, M. J. 2003. *Developmental Plasticity and Evolution.* New York: Oxford University Press.   —*J.T.*

Wade, Michael John (b. 1949)

Michael John Wade is an American evolutionary geneticist who used flour beetles of the genus *Tribolium,* as well as theoretical models, to investigate group (or interdemic) selection, kin selection, Sewall Wright's "shifting balance" theory, sexual selection, espistasis, maternal genetic effects, and speciation.

The oldest of eight children, Wade studied biology and mathematics at Boston College, graduating in 1971. Interested in combining mathematics and biology, Wade went to the University of Chicago's now-defunct Theoretical Biology program for graduate work. Coadvised by the population geneticist Montgomery Slatkin and the ecologist Thomas Park, Wade conducted the first experimental study of group selection by selecting flour beetles based not on individual characteristics of the beetles, but on aspects of the populations from which they arose. A strong response to selection was evident after only three generations of selection: populations in lines selected for high population size and those selected for low population size evolved substantial differences, showing that interdemic selection could be a powerful evolutionary force, at least under certain conditions.

From 1975 to 1998 Wade held a position at the University of Chicago, first in the Biology Department and culminating as chair of Ecology and Evolution (1991–1998). He moved to his present position at the University of Indiana in 1998. In 1980 Wade conducted the first experimental test of kin selection, observing that cannibalism in flour beetles evolved along different evolutionary trajectories depending on patterns of genetic relatedness determined by the mating system. Wade and his student Charles Goodnight used laboratory metapopulations to test the efficacy of Wright's "shifting balance" theory, a model of evolution that involves random genetic drift resulting in peak shifts in local subdivided populations and the spread of new, adaptive gene combinations across the larger population via migration and selection. They observed the largest response at intermediate levels of drift and interdemic selection.

Wade has made several important theoretical contributions. He was the first to show that allele frequency change under kin selection could be partitioned into separate components of change within and among kin groups. In sexual selection, he showed that there was a necessary relationship between the variance in reproductive success in males and that in females, explaining why selection acting on males was often many times greater than that acting on females. He showed that genes with maternal effects evolve much more readily in haplo-diploids than in diplo-diploids and, with his former student, Tim Linksvayer, argued that maternal effects theory provides a better account of the evolution of eusociality in the Hymenoptera than the haplo-diploidy hypothesis. His theoretical studies of gene interactions have shown how epistasis within populations can accelerate the adaptive divergence between isolated populations, contributing to speciation.

BIBLIOGRAPHY

Demuth, J. P., and M. J. Wade. 2006. Experimental methods for measuring gene interactions. *Annual Review of Ecology, Evolution and Systematics* 37: 289–316.
Wade, M. J. 1977. An experimental study of group selection. *Evolution* 31: 134–153.
Wade, M. J., and C. J. Goodnight. 1991. Wright's shifting balance study: An experimental study. *Science* 253: 1015–1018.
———. 1998. Perspective: The theories of Fisher and Wright in the context of metapopulations: When nature does many small experiments. *Evolution* 52: 1537–1553.　　　　　　　　　　　　　　　　　　　　　　　　　　—*N.A.J.*

Wake, David B. (b. 1936)

David Wake is one of America's leading evolutionary biologists. His numerous research contributions span many fields, including functional and comparative morphology, behavior, developmental biology and ontogeny, ecology, biogeography, population genetics, molecular evolution, taxonomy, systematics, and phylogeny. These studies have been motivated by a central and overarching interest in evolutionary patterns and the processes that produce them. Although the goal has been to identify and explore mechanisms of evolutionary diversification that are broadly applicable to all organisms, Wake's empirical work has focused almost exclusively on a single evolutionary lineage, the lungless salamanders of the family Plethodontidae. Largely through the efforts of Wake and his numerous students and collaborators over the nearly five decades that spanned the last half of the twentieth century and continue into the twenty-first, plethodontid salamanders are one of the most comprehensively investigated and fully documented instances of adaptive radiation in the history of evolutionary biology.

David Burton Wake was born on June 8, 1936, in Webster, South Dakota. Much of the Upper Midwest of North America had been extensively settled by Scandinavian immigrants in the late nineteenth and early twentieth centuries, and both of Wake's parents were of Norwegian descent. At the age of 17 Wake relocated with his immediate family to Tacoma, Washington, where he completed precollegiate education. As was typical of many children of Scandinavian immigrants living in the Pacific Northwest at the time, Wake enrolled at Pacific Lutheran College (now Pacific Lutheran University), where he completed his undergraduate degree in biology in 1958. From there he advanced to graduate school in biology at the University of Southern California (USC) in Los Angeles, where he worked under the supervision of Jay M. Savage for both master's (1960) and doctoral (1964) degrees. Savage, a leading tropical ecologist and biogeographer, was at that time launching a broad research effort on amphibians and reptiles of Central and South America, and he promoted Wake's interests in plethodontid salamanders, which have their principal species diversity in the neotropics. Savage's close association with the Los Angeles County Museum of Natural History also provided Wake with valuable experience with research collections in herpetology and facilitated his early studies in taxonomy and systematics.

Wake left USC in 1964 to assume his first full-time academic position, at the University of Chicago, where he remained for five years. In 1969 he returned to California to join the Zoology Department faculty at the University of California, Berkeley, and become curator of herpetology at Berkeley's Museum of Vertebrate Zoology (MVZ). This began a long and productive association with the MVZ, which Wake served as director for 27 years until 1998. Under Wake's direction the MVZ became one of the world's leading centers for research and teaching in evolutionary biology.

In purest terms Wake's research addresses a fundamental question in evolutionary biology: why are there so many different kinds of organisms? His work has been influential because it exemplifies a comprehensive approach to analysis of evolutionary diversification that evaluates the roles of both extrinsic and intrinsic factors and simultaneously considers a wide range of biological attributes, from molecules to organisms and their environments. The study of evolution since the modern synthesis has until recently focused largely on extrinsic factors as determinants of evolutionary change, such as the role of the external environment in mediating natural selection for adaptations. Wake's work has highlighted the need for complementary studies that evaluate the potential role of intrinsic factors, such as lineage-specific developmental traits or conserved anatomical features, which may both facilitate and constrain diversification in significant ways. A full understanding of the evolution of any taxon is likely to require such a holistic approach, which incorporates all relevant information.

Wake is the author of more than 250 peer-reviewed publications, as well as popular articles and books. He has received numerous honors and awards, including election to membership in the U.S. National Academy of Sciences in 1998.

BIBLIOGRAPHY

Wake, D. B. 1997. Incipient species formation in salamanders of the *Ensatina* complex. *Proceedings of the National Academy of Sciences USA* 94: 7761–7767.

Wake, D. B., and A. Larson. 1987. Multidimensional analysis of an evolving lineage. *Science* 238: 42–48.

Wake, D. B., and J. F. Lynch. 1976. The distribution, ecology, and evolutionary history of plethodontid salamanders in tropical America. *Science Bulletin, Natural History Museum of Los Angeles County* 25: 1–65.   —*J.H.*

Wallace, Alfred Russel (1823–1913)

Alfred Russel Wallace was one of the most brilliant theoretical and field biologists of the nineteenth century. His fame in the history of evolution rests primarily upon his discovery, independently of Charles Darwin, of the theory of evolution by natural selection. Wallace was a meticulous field observer, a prolific generator of ideas on a broad range of issues ranging from evolutionary biology to social and political concerns, and a theoretician whose work laid some of the main foundations for the scientific study of modern zoology and botany.

By 1844 the 20-year-old Wallace had collected and analyzed an extensive array of local British plants. In 1844 he met Henry Walter Bates. An accomplished entomologist, Bates encouraged Wallace to move beyond the bounds of botany to a more general study of natural history. During the next few years Wallace read Charles Lyell's three-volume *Principles of Geology* (1830–1833), Robert Chambers's (then-anonymous) *Vestiges of the Natural History of Creation* (1844), William Lawrence's *Lectures on Comparative Anatomy, Physiology, Zoology, and the Natural History of Man* (1819), and Charles Darwin's *Voyage of the Beagle* (1839)—works that dealt, either explicitly or implicitly, with evolutionary speculations, the origin of species, the geographic distribution of animals and plants, and the difference between species and varieties. When Bates and Wallace met again in the summer of 1847, the two made the fateful decision to journey to the tropics. Wallace and Bates left England on April 26, 1848, destined for Pará (now Belém), Brazil. Thus began the first of the two tropical journeys that were to transform Wallace's life and the emerging science of evolutionary biology. After a four-year exploration of the Amazon basin of South America (1848–1852), Wallace returned to London. He published *A Narrative of Travels on the Amazon and Rio Negro* (1853), which established his reputation as the foremost analyst of the geographic distribution of animals and plants (biogeography). The recognition that the distribution of closely allied species was often marked by surprisingly precise and abrupt barriers was the most important scientific achievement of his Amazonian travels. However, although Wallace had been committed to some form of general evolutionary theory since 1845, he was not yet prepared to posit an explicit evolutionary mechanism in *A Narrative of Travels*.

Convinced that another voyage of exploration was the most certain means of providing the data required for the theoretical elucidation of what was termed *the species problem* (i.e., how new species originate from preexisting ones), Wallace decided on an expedition to the Malay Archipelago. He arrived in Singapore on April 24, 1854, to begin eight years of intensive travels in the islands of Java, Borneo, Celebes, New Guinea, and Bali and in many smaller islands in the archipelago. From 1854 to 1862 Wallace covered nearly 14,000 miles and collected the vast amount of 125,000 (primarily faunal) specimens. He encountered animals, birds, and insects in bewildering variety and abundance, many of which had not been previously seen by Europeans. Wallace also observed and lived with the diverse human inhabitants of those regions. Wallace's observations on the geographic distribution of species led him to new insights about evolutionary history. He proposed that the species in the western half of the Malay Archipelago were overwhelmingly Indian in origin, whereas those in the eastern half were predominantly of Australian origin. This was a bold synthesis of evolutionary theory and copious field observation. The faunal discontinuity that separates the Indian from the Australasian segments of the archipelago is called Wallace's line in his honor. Wallace later generalized these concepts to elaborate a global paradigm for identifying the earth's fundamental biogeographic regions in his magisterial *Geographical Distribution of Animals* (1876). Details of Wallace's original biogeographic regions—and the precise location of the boundary first suggested by Wallace's line—underwent revision as more abundant data became available during the twentieth century. Theories of continental drift have also required a reinterpretation of certain of Wallace's nineteenth-century premises. Nonetheless, his biogeographic synthesis stands as a major development in evolutionary biology and continues to influence contemporary biogeographic studies.

It is, of course, Wallace's elucidation of the mechanism of evolution that constitutes his greatest scientific legacy from the Malay travels. In 1855 he published the famous essay "On the Law Which Has Regulated the Introduction of New Species." In this essay Wallace laid the foundations for what would shortly become his explicit statement of the evolutionary theory. Wallace constructed a powerful argument in support of the thesis that new species arise naturally from closely related, preexisting species, but he suggested no mechanism for such change. From 1855 to 1858 Wallace sent to England several articles on the flora and fauna of the islands he visited and dealt explicitly with the theoretical implications of the 1855 law. Finally, in February 1858—recalling passages from Thomas Malthus's *Essay on the Principle of Population* (1798), with its vivid depiction of the competitive struggles for survival among human populations—the principle of natural selection emerged in Wallace's mind as the key mechanism of evolutionary change. For Wallace evolution was a two-step process: first, the appearance of variations (later called mutations) in individual members of a species, and second, the sorting out of this variation by natural selection. Wallace argued that the existence of *heritable* variations within a species, coupled with the production of more offspring than could possibly survive (given the constraints of the environment, such as food supply), constituted the conditions under which favorable variations tended to be preserved and injurious variations eliminated. Over long periods of time—that is, over many generations—and under the continued selective influence of the environment (the so-called struggle for existence), a group of organisms would have eventually accumulated many new favorable variations. They would then differ sufficiently from their ancestors to constitute a new taxonomic status: thus the "origin of species." Wallace wrote out a draft of his complete theory in the now-classic essay "On the Tendency of Varieties to Depart Indefinitely from the Original Type" (1858), and mailed it to Darwin in England. A copy of Wallace's essay, along with extracts from an unpublished manuscript on natural selection written by Darwin in 1844, were presented together at the historic meeting of the Linnean Society (London) on July 1, 1858. This meeting, a year before the publication of Darwin's *On the Origin of Species* (1859), ensured that both Wallace and Darwin received recognition and joint priority for their momentous achievement.

On his return to England in 1862 Wallace spent the remainder of his long life elucidating the implications of evolutionary theory for a vast array of subjects ranging from biogeography, sexual selection, the phenomenon of organic mimicry (by which one animal species evolves to so closely resemble another animal or even plant species as to be mistaken for it by predators), taxonomy, physical geography and geology, and anthropology. It was Wallace's theories and writings in the last domain, human evolution, that elicited the greatest controversies of his career. Although he remained an ardent selectionist in his overall analysis of evolutionary processes, Wallace regarded natural selection as inadequate to account completely for the origin and development of certain human characteristics, notably consciousness and the moral sense. Instead, he suggested that other agencies of a nonmaterial nature had been, and continued to be, instrumental in the origin and future evolution of the human species.

By emphasizing that culture—including certain aspects of theism and of political and social ideologies—was not merely compatible with the evolutionary process but essential for comprehending the full significance of human evolution, Wallace echoed the views of many of his Victorian contemporaries. Prominent evolutionary scientists, including Asa Gray, Joseph LeConte, and St. George Mivart, among others, shared Wallace's conviction that adherence to a scientifically rigorous evolutionary biology did not preclude the recognition that certain aspects of evolutionary theory, notably teleology (the apparent purposiveness of many evolutionary processes and adaptations), called for explanations that transcended dogmatic mechanistic reductionism. Wallace typified a major strand of late Victorian thought that sought to make of evolutionary biology a crucial and scientifically verifiable base upon which to build a comprehensive theory of humans in nature and of humans and nature.

Wallace is one of the towering figures in the history and development of evolutionary biology precisely because his conception of the evolutionary process integrated so diverse a range of subjects. The issues Wallace posed continue to resonate in contemporary debates on the scope, mechanism, and, ultimately, significance of evolution in both scientific and cultural domains. At the start of the twenty-first century, when the advocates of so-called creation science present dangerous obstacles to a fuller understanding of the role evolutionary biology will and must play in modern society, Wallace's evolutionary theory and worldview provide a cogent basis for a judicious integration of science and the broader culture.

BIBLIOGRAPHY

The Alfred Russel Wallace Page. http://www.wku.edu/~smithch/index1.htm. Maintained at Western Kentucky University, Bowling Green, Kentucky.

Berry, A., ed. 2002. *Infinite Tropics: An Alfred Russel Wallace Anthology.* New York: Verso.

Camerini, J. R., ed. 2002. *The Alfred Russel Wallace Reader: A Selection of Writings from the Field.* Baltimore: Johns Hopkins University Press.

Daws, G., and M. Fujita. 1999. *Archipelago: The Islands of Indonesia, from the Nineteenth-Century Discoveries of Alfred Russel Wallace to the Fate of Forests and Reefs in the Twenty-first Century.* Berkeley: University of California Press.

Fichman, M. 2004. *An Elusive Victorian: The Evolution of Alfred Russel Wallace.* Chicago: University of Chicago Press.

Knapp, S. 1999. *Footsteps in the Forest: Alfred Russel Wallace in the Amazon.* London: Natural History Museum.

Marchant, J., ed. [1916] 1975. *Alfred Russel Wallace: Letters and Reminiscences.* New York: Arno Press.

Raby, P. 2001. *Alfred Russel Walllace: A Life.* Princeton, NJ: Princeton University Press.

Smith, C. H., ed. 1991. *Alfred Russel Wallace: An Anthology of His Shorter Writings.* Oxford: Oxford University Press.

Wallace, A. R. 1853. *A Narrative of Travels on the Amazon and Rio Negro.* London: Reeve & Co.

———. 1855. On the law which has regulated the introduction of new species. *Annals and Magazine of Natural History* 16 (2nd series): 184–196.

———. 1858. On the tendency of varieties to depart indefinitely from the original type. *Journal of the Proceedings of the Linnean Society, Zoology* 3, no. 9: 53–62.

———. 1876. *The Geographical Distribution of Animals.* 2 vols. London: Macmillan & Co.   —M.F.

Watson, James (b. 1928), and Crick, Francis (1916–2004)

In 1953 the American James Watson and the Englishman Francis Crick, working at Cambridge University, discovered the structure of the deoxyribonucleic acid (DNA) molecule, long suspected to be the carrier of the genetic information within the cell. They found that it was a pair of long molecules, twined around each other in a double helix (see figure). Very excitingly, the DNA molecule consists of four different kinds of smaller molecules: adenine (A), thymine (T), guanine (G), and cytosine (C). The opposing DNA molecules are linked together, with one of each kind of smaller molecule always specifically paired on the other DNA molecule with another of the four molecules (A with T, and G with C). Watson and Crick immediately hypothesized that the order of the smaller molecules carries the information of heredity, and so it proved later in the decade when the so-called genetic code was cracked. It was found that information from the DNA is used to make the building blocks of the cell, as well as the enzymes used to drive processes. In 1962 Watson, Crick, and a University of London researcher, Maurice Wilkins, jointly received the Nobel Prize for Medicine or Physiology for their determination of the structure of DNA.

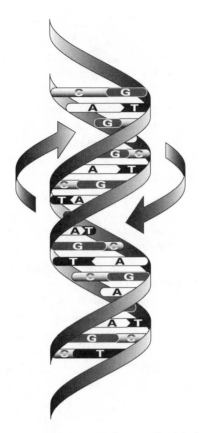

The first public representation of the now-famous double helix. (From Watson and Crick 1953.)

Initially, relationships between molecular biologists and evolutionists were very tense, with the former regarding the latter as mere "stamp collectors" and the latter regarding the former as "reductionists" and worse (Watson's autobiographical *The Double Helix* [1968] uses this kind of demeaning language). The systematist Ernst Mayr was one who turned to philosophy to prove that evolutionary theorizing has its own special principles of understanding, and that hence molecular biology can yield little or no true understanding about whole organisms and their histories. (Mayr's comments at a conference on the philosophy of biology in 1969 use inflammatory language quite equal to that of Watson. Edward O. Wilson, a junior faculty member in the same biology department as Watson in the 1950s, describes Watson as "the most unpleasant human being I had ever met" [Wilson 1994, 219]. Wilson is detailed in his discussion of the tension between molecular biologists and the more traditional organismic biologists.)

By the 1960s, however, things changed rapidly. It was found that molecular techniques could throw light on evolutionary problems, like genetic variation, that were impossible to crack in conventional ways (this was a major insight of Richard Lewontin, fully described in his *Genetic Basis of Evolutionary Change,* 1974). This has continued to the present, and evolutionists have been among the keenest to do genetic fingerprinting to determine heredity and other matters of interest (see, for instance, Davies 1992 on the hedge sparrows, the dunnocks).

At the same time molecular biology has given rise to one of the more important new hypotheses about evolutionary change, namely, the Japanese hypothesis that at the molecular level much variation escapes the forces of selection and so drifts from one peak to another. (The chief author of the "neutral theory" was Moto Kimura [1983]. With good reason, most Darwin evolutionists do not find it at all threatening because the processes occur at a level below that at which natural selection operates. See Godfrey-Smith 2001 for an insightful discussion of the relationships between selectionist and nonselectionist theories of evolution.) Although in some respects this is still a controversial idea, it has proved very fruitful when it is employed in a molecular clock, determining the dates since major evolutionary events occurred (see the main essay "Molecular Evolution" by Francisco J. Ayala in this volume). This was particularly significant in persuading paleoanthropologists (students of human evolutionary history) that the human-ape break occurred much more recently (about 5 million years ago) than the fossil record apparently suggested (see the main essay "Human Evolution" by Henry M. McHenry in this volume). Most fruitful of all, perhaps, has been the way in which molecular studies have brought development (embryology) right back into the evolutionary synthesis. The evolutionists of the 1930s tended to treat organisms rather like black boxes, jumping straight from the genes (genotypes) to the physical features (phenotypes). Now, molecular biology is opening up the ways in which the information at the genetic level is translated into physical features, and a whole new synthesis of molecular studies and developmental studies has been created, making discoveries quite undreamed of even a couple of decades ago. Boosted by studies of the complete genomes of different organisms, including humans (e.g., the Human Genome Project), evolutionary development, so-called evo-devo, is among the hottest areas of study today and promises much for the future.

We have come full circle. Molecular biology, which back in 1953 seemed the nemesis of evolutionary biology, is today in many respects its greatest supporter—both handmaiden and guide to new areas of study and discovery.

BIBLIOGRAPHY

Carroll, S. B. 2005. *Endless Forms Most Beautiful: The New Science of Evo Devo and the Making of the Animal Kingdom.* New York: Norton.

Carroll, S. B., J. K. Grenier, and S. D. Weatherbee. 2001. *From DNA to Diversity: Molecular Genetics and the Evolution of Animal Design.* Oxford: Blackwell.

Davies, N. B. 1992. *Dunnock Behaviour and Social Evolution.* Oxford: Oxford University Press.

Godfrey-Smith, P. 2001. Three kinds of adaptationism. In S. H. Orzack and E. Sober, eds., *Adaptationism and Optimality,* 335–357. Cambridge: Cambridge University Press.

Kimura, M. 1983. *The Neutral Theory of Molecular Evolution.* Cambridge: Cambridge University Press.

Lewontin, R. C. 1974. *The Genetic Basis of Evolutionary Change.* New York: Columbia University Press.

Mayr, E. 1969. Commentary. *Journal of the History of Biology* 2, no. 1: 123–128.

Watson, J. 1968. *The Double Helix.* New York: Signet Books.

Watson, J. D., and F. H. C. Crick. 1953. Molecular structure of nucleic acids. *Nature* 171: 737.

Wilson, E. O. 1994. *Naturalist.* Washington, DC: Island Books/Shearwater Books.

—M.R.

Weismann, August Friedrich Leopold (1834–1914)

August Weismann was one of the most influential zoologists and evolutionary biologists of the late nineteenth and early twentieth centuries. He is still remembered today for his theory of the continuity of the germplasm and its distinction from the somatoplasm, as well as for his staunch opposition to the view that acquired characters can be inherited. Weismann was especially concerned with the relationship of evolutionary, hereditary, and developmental processes. This emphasis makes him also a forerunner of the present-day synthesis of evolutionary developmental biology (evo-devo).

August Weismann was born in 1834 in Frankfurt am Main into a family of academics and artists. Already an ardent butterfly collector and talented musician in his youth, Weismann studied medicine in Göttingen as a practical means to make a living while pursuing his interests in science. After reading Darwin's *On the Origin of Species* in 1861, Weismann, next to Ernst Haeckel, soon became one of Darwin's strongest advocates in Germany. After moving to Freiburg in 1863, Weismann devoted himself full-time to zoology, focusing primarily on comparative embryology and cell biology (especially of insects). Severe problems with his eyesight forced him to interrupt his empirical studies for an extended period during the 1860s and 1870s. During that time Weismann devoted himself to a theoretical analysis of Darwin's ideas, emphasizing the role of heredity and developmental processes for evolution. Weismann was a strict selectionist; he rejected Darwin's idea of pangenesis and any other notion of the inheritance of acquired characters. On the basis of recent findings in cytology, especially Oscar Hertwig's observation in 1875 of the fusion of nuclei during fertilization, and of related observations of intracellular events during cell division, Weismann began to develop his own views both on the continuity of the germplasm (1885) and on the more general processes of development and heredity. Postulating a system of subcellular units (including determinants, ids, and other units that would make up the chromosomes) as carriers of heredity, Weismann also attempted to explain the differentiation and determination of particular cells during development as a consequence of a selective distribution and activation of these units during subsequent cell divisions. Such alterations of the hereditary material during development, however, required that the germplasm, those cells that contain the full complement of hereditary factors, would be set aside early during development, because Weismann's own experiments had demonstrated that somatic changes are not passed on to the next generation.

Weismann spent his whole career as a zoologist at the University of Freiburg. He was the first to hold the chair in zoology (from 1874 on) and oversaw the expansion of the zoological institute and its facilities. During his tenure Freiburg became one of the leading institutions of zoological research. Weismann also attracted some of the brightest students from all over the world, and many of his pupils later held prestigious positions. After retiring in 1912, Weismann prepared the final (third) edition of his *Vorträge über Descendenztheorie* (published in 1913). He died in 1914, deeply disappointed by the outbreak of World War I.

BIBLIOGRAPHY

Gaupp, E. 1917. *August Weismann, sein Leben und sein Werk*. Jena: Gustav Fischer.
Weismann, A. 1892. *Das Keimplasma: Eine Theorie der Vererbung*. Jena: Gustav
    Fischer.
———. 1893. *The Germ-Plasm: A Theory of Heredity*. New York: Charles
    Scribner's Sons.
———. 1902. *Vorträge über Descendenztheorie*. Jena: Gustav Fischer.   —M.D.L.

Weldon, W. F. R. (1860–1906)

W. F. R. Weldon (he was known as Raphael Weldon to intimates and family) was professor of comparative anatomy and zoology at University College, London, and then later Linacre Professor of Zoology at Oxford. His honors included being a fellow of the Royal Society and president of the British Association for the Advancement of Science. He was a leading member of the group of evolutionists known as the "biometricians," who tried to analyze continuous variation shaped by natural selection, as opposed to those taking up the new theories of heredity, the "Mendelians." A vigorous controversialist, Weldon carried on a vitriolic feud with his former student (and leading British Mendelian) William Bateson. It is thought that the strain of this fight was a contributory factor to his early death.

Charles Darwin and Alfred Russel Wallace were not the first to propose evolution as a process or the evolution of new species; Darwin cited 34 antecedents in the "Historical Sketch" at the beginning of *On the Origin of Species*. The more original aspect of their contribution was to champion natural selection as the mechanism that causes evolution. W. F. R. Weldon developed the most prominent research program of the nineteenth century that characterized the process of natural selection. He argued that such studies must be statistical in nature in that they must characterize how traits varied among individuals within a population, the extent to which this trait variation was transmitted from parent to offspring, and the extent to which trait variation influenced survival. Finally, he showed that if these three conditions prevailed, the average value of a trait in a population would change over time. He collaborated with Karl Pearson, who developed statistics for characterizing variation within populations and how this variation changed within and among generations (a major contribution was Pearson 1894). One of his prominent study systems was the shell morphology of crabs *(Carcinus moenas)* found near the laboratory of the Marine Biological Station on Plymouth Sound. He showed that the pattern of variation of frontal breadth (see figure) was well described by a normal distribution, which, as Francis Galton had previously shown, characterizes the pattern of variation in many traits in many species. He also showed that this distribution was wider in juvenile than in adult crabs, which suggested that individuals that had the widest and narrowest frontal widths were less likely to survive to maturity, or that they had been selected against. He argued that this form of selection, which we now call *stabilizing selection,* would be the most common process seen in nature. He felt that this was true because selection that caused changes in the average trait of a population, which we now call *directional selection,* would occur rapidly in response to a change in the environment and hence would only be seen if one were lucky enough to be watching a population when such a change occurred. He had such a stroke of luck when he observed that the frontal breadth of the crabs became narrower after Plymouth Sound was isolated from the ocean by an artificial breakwater and then filled with the fine clay sediment that was discharged by rivers that emptied into the sound. He showed experimentally that individuals with broader shells were less likely to survive exposure to suspended sediments. He hypothesized, but never

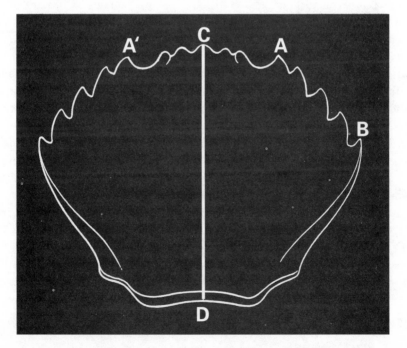

The crab whose statistics Raphael Weldon collected and Karl Pearson analyzed. The crucial measurement was the "frontal breadth," from A to A'. (From Weldon 1895.)

showed, that a narrower frontal breadth would prevent the accumulation of sediment in the gill chambers. Weldon and Pearson combined their empirical and statistical skills to create biometry, or the statistical study of variation and evolution in natural populations. Many other investigators adopted their approach. Our current understanding of evolution by natural selection indicates that all of Weldon's general conclusions were correct. Furthermore, his general approach to the study of natural selection is similar to that of modern investigators.

Given his pioneering accomplishments, one might expect that Weldon's name would appear prominently in every general biology textbook or general treatment of natural selection; however, he is all but invisible in modern literature, except in some historical reviews (where his importance is increasingly being recognized—see especially Ruse 1996). Furthermore, if we were to evaluate where the modern approach to the study of natural selection came from, the trail would lead back to the 1940s, then diffuse into stray tracks that go back only 10 or 20 years earlier. What happened? Obviously his early death was a significant contributing factor, since it left the field open to Bateson and the Mendelians. However, there were scientific factors contributing to Weldon's fall from sight. The study of evolution between 1880 and 1920 was dominated by debates over the mechanism of inheritance, since evolution demands that traits that characterize survivors be transmitted to the next generation. It was Weldon's misfortune to champion a mechanism of inheritance that was wrong. He, Pearson, and their fellow biometricians argued that heritable variation was continuous, as seen in a statistical distribution of traits in a natural population. They also failed to distinguish between variation that was caused by genes versus variation caused by the environment and, further, often seemed to imply that traits in parents that were caused by the environment could be transmitted to offspring. Finally, they argued that the transmission of traits from father and mother to offspring was like the mixing of paint. If the father was a different color from the mother, then these differences would be inextricably blended together in the offspring. The rediscovery of Mendel's laws of inheritance instead showed that inheritance is particulate, meaning that traits are faithfully transmitted from parent to offspring and the traits from each parent remain discrete, so they can emerge unchanged in future generations. Weldon's remarkable accomplishments in the study of natural selection thus died as innocent bystanders of his loss in the battle over inheritance. Pearson was more fortunate since some of his statistical innovations, such as the Pearson product-moment correlation, are still with us. The modern approach to the study of natural selection was reborn, beginning in the 1930s, after the different approaches to the study of inheritance were resolved.

BIBLIOGRAPHY

Darwin, C. 1959. *The Origin of Species by Charles Darwin: A Variorum Text.* M. Peckham, ed. Philadelphia: University of Pennsylvania Press.

Pearson, K. 1894. Contributions to the mathematical theory of evolution. *Philosophical Transactions A* 185: 71–110.

———. 1906. Walter Frank Raphael Weldon: 1860–1906. *Biometrika* 5: 1–51.

Provine, W. B. 1971. *The Origins of Theoretical Population Genetics.* Chicago: University of Chicago Press.

Ruse, M. 1996. *Monad to Man: The Concept of Progress in Evolutionary Biology.* Cambridge, MA: Harvard University Press.

Weldon, W. F. R. 1890. The variations occurring in certain decapod crustacea. I. Crangon vulgaris. *Proceedings of the Royal Society of London* 47: 445–453.

———. 1892. Certain correlated variations in Crangon vulgaris. *Proceedings of the Royal Society of London* 51: 2–21.

———. 1893. On certain correlated variations in *Carcinus moenas. Proceedings of the Royal Society of London* 54: 318–329.

———. 1895. Attempt to measure the deathrate due to the selective destruction of Carcinus moenas with respect to a particular dimension. *Proceedings of the Royal Society of London* 57: 360–379.

———. 1898. Presidential address to the Zoological Section of the British Association. *Transactions of the British Association* (Bristol), 887–902. London: John Murray.   —D.N.R.

Whewell, William (1794–1866)

William Whewell (pronounced Hule) was a major figure in the development and understanding of science in the first half of the nineteenth century in Britain. Although best known as a historian and philosopher of science, Whewell's writings ranged broadly over a huge number of other fields, in the sciences and beyond—mechanics, crystalography, mineralogy, political economy, geology, theology, educational reform, moral philosophy, international law, and architecture, to name but some (Ruse 1991). Much given to coining new terms, it was Whewell who invented the English word *scientist.* His terminological innovations are emblematic of the great influence exerted by him on the science of the age, through his writings and through his positions including president of the British Association for the Advancement of Science, fellow of the Royal Society, president of the Geological Society, and longtime master of Trinity College, Cambridge. As required of a fellow of Trinity College, Whewell was an ordained minister in the Anglican Church (Douglas 1881; Garland 1980; Todhunter 1876). Adamantly opposed to evolutionary ideas all of his life, Whewell was nevertheless one of the key players in the story. Whether, looking back a century and a half later, he would be appalled at the role he played inadvertently, or whether he would be somewhat complacently satisfied that his ideas and influence were truly significant, is a conundrum that is left for the reader—and for Whewell and his God. No doubt Whewell would happily have helped God to make the right decision.

Whewell is significant to the story of evolution in two different ways. First, he epitomized the vision of the organic world that was predicated on the Argument from Design. Charles Darwin, who knew Whewell from his undergraduate days at Cambridge and who was much in Whewell's company after

the *Beagle* voyage (in other words, just when he was becoming an evolutionist and working toward natural selection), recognized that he needed to speak
to this and offer a full counterargument. This in major part was the aim of
On the Origin of Species. Second, an irony that might not have been appreciated fully in the master's lodge at Trinity, Whewell's philosophical thinking
was a great influence on Darwin and provided a major methodological tool
in the presentation of the theory of evolution through natural selection.

Whewell was one of the contributors to the series of "Bridgewater Treatises" funded by bequest of the Earl of Bridgewater in order to show how
various disciplines gave evidence for the intelligent design of the universe
(Gillespie 1950). In his volume *Astronomy and General Physics Considered
with Reference to Natural Theology* (1833), Whewell argued that the lawful
structure of the physical world implies that there must be a designer-creator
behind everything. As he put it: "It will be our business to show that the laws
which really prevail in nature are, by their *form,* that is, by the nature of the
connexion which they establish among the quantities and properties which
they regulate, remarkably adapted to the office which is assigned to them;
and thus offer evidence of selection, design, and goodness, in the power by
which they were established" (p. 9). Later Darwin rather cheekily used a
quotation from this work opposite the title page of the *Origin:* "But with regard to the material world, we can at least go so far as this—we can perceive
that events are brought about not by insulated interpositions of Divine
power, exerted in each particular case, but by the establishment of general
laws" (Whewell 1833, 356).

For Whewell, this invocation of the Creator made otiose if not contradictory any appeal to natural causes when trying to explain origins. God did it
and the complex functioning of the world simply could not have appeared
through natural law. This kind of thinking comes through even more strongly
later in the decade, in Whewell's *History of the Inductive Sciences* (1837).
Now he was able to turn his attention from planets to organisms, arguing
that in studying organic structures it is necessary to make reference to the end
or goal for which they were designed. He thus followed Immanuel Kant in
endorsing teleological explanations in biology (German ideas were very popular in Cambridge in the 1830s). Now even more powerful was the case
against natural origins. Blind law leads to randomness. The organic world is
not random. Hence it was not caused by blind law.

When Robert Chambers's evolutionary work *Vestiges of the Natural History of Creation* appeared in 1844, Whewell responded by publishing *Indications of the Creator* (1845, second edition 1846), a compilation of extracts
from his earlier works that showed evidence of design in the natural world. In
the first edition, so contemptuous was Whewell of such naturalistic hypotheses that he did not even mention *Vestiges* by name. Note, however, that for
Whewell, as for others in his set, it was natural theology that set the barriers
against evolution, not crude literal readings of Genesis. Also note that although in respects Whewell's thinking seems to be very much a forerunner of
the thinking of today's so-called Intelligent Design Theorists, unlike them

Whewell was quite certain that an appeal to a designing god put the origins question quite outside the realm of science. Fossil organisms "were, like our own animal and vegetable contemporaries, profoundly adapted to the condition in which they were placed, we have ample reason to believe; but when we inquire whence they came into this our world, geology is silent. The mystery of creation is not within the range of her legitimate territory; she says nothing, but she points upwards" (Whewell 1837, 3: 588).

By the 1850s, unfortunately, some parts of Whewell's careful system were coming apart. By this time, under pressure from the theorizing of his friend, the comparative anatomist Richard Owen (1848, 1849), Whewell was having to rework his thinking. Owen argued that individual vertebrate animals could be seen as modified instantiations of patterns or archetype forms that existed in the Divine Mind. This view allowed Owen to explain what he had christened "homologies"—the similarities of vertebrate structures from humans to horses, from moles to mamoths. The parts were used in different ways, but, according to Owen, they were united by virtue of being a consequence of variations on the vertebrate archetype. Similarly, structures without apparent purpose, such as male nipples, did not contradict the claim that all creation was designed; rather, they were the result of the application of general archetypes. Whewell accepted this kind of reasoning, despite the fact that it weakened the appeal to universal utilitarian design. Indeed, in a controversial work appearing supposedly anonymously in 1853, *Of the Plurality of Worlds,* Whewell even extended Owen-type thinking, suggesting that this kind of archetype-based patterning is something that holds throughout the universe and not just here on earth: "Order is the first and universal Law of the heavenly work" (p. 243). But, although Whewell was still holding firm to the idea that design precludes natural causes for origins and hence evolution is impossible, he was dramatically weakening the claim that there is such strong evidence of direct design throughout the organic world that any solution suggesting natural causes is almost contradictory. There are those who claim that the difficulties that someone like Whewell was experiencing at this point were as important factors in the eventual triumph of evolutionary thought as is the positive case made by the evolutionists (see Ruse 2001, 2003).

Darwin not only knew and talked to Whewell, he also read several of the key works very carefully. Darwin's private evolution notebooks make reference to Whewell's Bridgewater Treatise and to the *History,* suggesting that Darwin's nonsupernatural explanation of adaptation was framed as a response, not only to earlier natural theologians like William Paley (1802), but also to Whewell (see Ruse 2003). There are about twenty references to Whewell's writings in Darwin's notebooks (for details, see Barrett et al. 1987). There are also extensive marginalia in the *History,* generally not complementary to Whewell (Di Gregorio and Gill 1990). For instance, against a favorable discussion of Kant's (1790) claim that teleology is a necessary component of thinking in biology, Darwin wrote "all this reasoning is vitiated; when we look at animals, on my view" (Whewell 1837, 3: 470).

For all of his opposition to Whewell, Darwin agreed that the designlike nature of the world needs explaining. (In this, they would unite in arguing against someone like the late Stephen Jay Gould, who thought that designlike effects are overvalued. See Gould and Lewontin 1979.) As significantly, in constructing his alternative explanation, despite his opposition to the theology, Darwin was influenced by a very important aspect of Whewell's philosophy of science, namely, his notion of consilience (Ruse 1975, 1979). In his *Philosophy of the Inductive Sciences* (1840) Whewell explained that what really counts in science, what really convinces you that you have the truth—that you have what Newton had called a *vera causa*—is when different areas of inquiry come together under the same hypothesis. "Accordingly the cases in which inductions from classes of facts altogether different have thus *jumped together*, belong only to the best established theories which the history of science contains. And as I shall have occasion to refer to this peculiar feature in their evidence, I will take the liberty of describing it by a particular phrase; and will term it the *Consilience of Inductions*" (2: 230). (See Kavalovski 1974 for a general discussion of Whewell's views on causation. See also Snyder 2006 for details of Whewell's later debates about causation, especially that with John Stuart Mill.) The consilience par excellence for Whewell was that of Newton, who brought together under one law—that of gravitational attraction—the facts of satellite motion, of planetary motion, and of terrestrial moving bodies. In Newton's terminology (adopted by Whewell), the inverse-square attractive force was a *vera causa*, a genuinely existing active force in nature.

Although Darwin read the *History* very carefully (even twice), it is improbable that he read the *Philosophy*. But, apart from the personal contact with Whewell in the years immediately succeeding the *Beagle* voyage, Darwin knew all about the idea of consilience from a very detailed review of Whewell's ideas by the philosopher and astronomer John F. W. Herschel (1841). It is therefore not surprising that, as required by Whewell's criterion of consilience, Darwin's theory of evolution by natural selection explains not just many facts but many different kinds of facts: facts of instinct, geology and paleontology, biogeographic distributions, systematics, comparative anatomy, and embryology. Moreover, Darwin's theory provides the causal unification of consilience by showing that all these different kinds of phenomena share a common cause, namely, the alteration of original organisms by gradual modification through the mechanism of natural selection. Expectedly, therefore, having done what his mentors had demanded, Darwin made explicit his belief that a consilience offers justification enough. Faced with criticism, in the second edition of the *Origin* (1860) he added: "I cannot believe that a false theory would explain, as it seems to me that the theory of natural selection does explain, the several large classes of facts above specified" (Darwin 1959, 748). Then, in the sixth edition (1872), he added further: "It has recently been objected that this is an unsafe method of arguing; but it is a method used in judging of the common events of life, and has often been used by the greatest natural philosophers. The undulatory theory of light has thus been arrived at; and the belief in the revolution of the earth on its own axis was until lately supported by hardly any direct evidence" (Darwin 1959, 748).

As Darwin explained in a letter of February 9, 1860, to botanist C. F. J. Bunbury:

> With respect to Nat. Selection not being a "vera causa"; it seems to me fair in Philosophy to invent *any* hypothesis & if it explains many phenomena it comes in time to be admitted as real. In your sense the undulatory theory of the *hypothetical* ether (the undulations themselves being not recognised) is not a vera causa in accounting for all the phenomena of Light. Natural selection seems to me in so far in itself not be quite hypothetical, in as much if there be variability & a struggle for life, I cannot see how it can fail to come into play to some extent. (Darwin 1993, 76)

When the *Origin* appeared, Darwin sent a copy to Whewell, as he did to all of his old teachers. Whewell responded with a polite note:

> My dear Mr Darwin
>
> I have to thank you for a copy of your book on the "Origin of Species." You will easily believe that it has interested me very much, and probably you will not be surprized to be told that I cannot, yet at least, become a convert to your doctrines. But there is so much of thought and of fact in what you have written that it is not to be contradicted without careful selection of the ground and manner of the dissent, which I have not now time for. I must therefore content myself with thanking you for your kindness
>
> believe me | Yours very truly | W Whewell
>
> C. Darwin Esqe
>
> (Letter of January 2, 1860, in Darwin 1993, 6)

Predictably, Whewell did not embrace evolutionary theory. In a preface to a new edition of the Bridgewater Treatise, *Astronomy and General Physics* (1864), continuing his tradition of not mentioning vile evolutionary tracts by name, he continued to assert, as he had written in *History of the Inductive Sciences,* that "*species have a real existence in nature,* and a transmutation from one to another does not exist" (Whewell 1837, 3: 576). According to an anecdote of Thomas Henry Huxley, Whewell as master of his college refused to allow the *Origin* on the library shelves!

Whewell's continued reluctance to accept evolutionary ideas is often described as a result of his theological prejudices. This is true, but one must understand the meaning and context of Whewell's theology. In a letter to his friend, the physicist James David Forbes, Whewell wrote in 1864:

> I have myself taken no share in the discussions on the antiquity of man; but I will not conceal from you that the course of speculation on this point has somewhat troubled me. I cannot see without some regrets the clear definite line, which used to mark the commencement of the human

period of the earth's history, made obscure and doubtful . . . It is true that a reconciliation of the scientific with the religious views is still possible, but it is not so clear and striking as it once was. But it is still a weakness to regret this; and no doubt another generation will find some way of looking at the matter which will satisfy religious men. I should be glad to see my way to this view, and am hoping to do so soon. (Todhunter 1876, 2: 435–436)

Whewell had always believed in the dictum that truth cannot be opposed to truth. Especially, as noted earlier, his was not an opposition to evolution based on a crude biblical literalism. As always, natural theology was a (if not *the*) major factor. Darwin may not have been mentioned by name in the new preface to the Bridgewater Treatise. This did not stop Whewell from quoting and then strongly disagreeing with passages in the *Origin* where Darwin tried to show that producing the eye did not necessarily have to suppose a direct intervention of the Creator. By then, however, time had moved on and a new generation of religious thinkers were trying to show why evolution proves design, rather than why design refutes evolution.

BIBLIOGRAPHY

Barrett, P. H., P. J. Gautrey, S. Herbert, D. Kohn, and S. Smith, eds. 1987. *Charles Darwin's Notebooks, 1836–1844*. Ithaca, NY: Cornell University Press.

Chambers, R. 1844. *Vestiges of the Natural History of Creation*. London: Churchill.

Darwin, C. 1871. *The Descent of Man, and Selection in Relation to Sex*. London: John Murray.

———. 1959. *The Origin of Species by Charles Darwin: A Variorum Text*. M. Peckham, ed. Philadelphia: University of Pennsylvania Press.

———. 1993. *The Correspondence of Charles Darwin*. Vol. 8. Cambridge: Cambridge University Press.

Di Gregorio, M., and N. W. Gill, eds. 1990. *Charles Darwin's Marginalia*. Vol. 1. New York: Garland.

Douglas, S., ed. 1881. *The Life and Selections from the Correspondence of William Whewell, D.D.* 2 vols. London: Kegan Paul.

Fisch, M. 1991. *William Whewell, Philosopher of Science*. Oxford: Oxford University Press.

Fisch, M., and S. Schaffer, eds. 1991. *William Whewell: A Composite Portrait*. Oxford: Oxford University Press.

Garland, M. M. 1980. *Cambridge before Darwin: The Ideal of a Liberal Education, 1800–1860*. Cambridge: Cambridge University Press.

Gillespie, C. C. 1950. *Genesis and Geology*. Cambridge, MA: Harvard University Press.

Gould, S. J., and R. C. Lewontin. 1979. The spandrels of San Marco and the Panglossian paradigm: A critique of the adaptationist programme. *Proceedings of the Royal Society of London, Series B* 205: 581–598.

Herschel, J. F. W. 1841. Review of Whewell's History and Philosophy. *Quarterly Review* 135: 177–238.

Kant, I. [1790] 1928. *The Critique of Teleological Judgement*. J. C. Meredith, trans. Oxford: Oxford University Press.

Kavalowski, V. 1974. The "vera causa" principle: An historico-philosophical study of a metatheoretical concept from Newton through Darwin. Unpublished PhD thesis, University of Chicago.

Owen, R. 1848. *On the Archetype and Homologies of the Vertebrate Skeleton*. London: Voorst.

———. 1849. *On the Nature of Limbs*. London: Voorst.

Paley, W. [1802] 1819. *Natural Theology* (*Collected Works*, vol. 4). London: Rivington.

Ruse, M. 1975. Darwin's debt to philosophy: An examination of the influence of the philosophical ideas of John F. W. Herschel and William Whewell on the development of Charles Darwin's theory of evolution. *Studies in History and Philosophy of Science* 6: 159–181.

———. 1979. *The Darwinian Revolution: Science Red in Tooth and Claw*. Chicago: University of Chicago Press.

———. 1991. William Whewell: Omniscientist. In M. Fisch and S. Shaffer, eds., *William Whewell: A Composite Portrait*, 87–116. Oxford: Oxford University Press.

———. 2001. Introduction. W. Whewell. In *Of the Plurality of Worlds: A Facsimile of the First Edition of 1853: Plus Previously Unpublished Material Excised by the Author Just Before the Book Went to Press: And Whewell's Dialogue Rebutting His Critics, Reprinted from the Second Edition*, 1–31. Chicago: University of Chicago Press.

———. 2003. *Darwin and Design: Does Evolution Have a Purpose?* Cambridge, MA: Harvard University Press.

Snyder, L. J. 2006. *Reforming Philosophy: A Victorian Debate on Science and Society*. Chicago: University of Chicago Press.

Todhunter, I. 1876. *William Whewell, D.D., Master of Trinity College, Cambridge: An Account of His Writings with Selections from His Literary and Scientific Correspondence*. 2 vols. London: Macmillan.

Whewell, W. 1833. *Astronomy and General Physics Considered with Reference to Natural Theology* (Bridgewater Treatise, 3). London: William Pickering. (7th ed., 1864.)

———. 1837. *The History of the Inductive Sciences, from the Earliest to the Present Time*. 3 vols. London: John W. Parker.

———. 1840. *The Philosophy of the Inductive Sciences*. 2 vols. London: John W. Parker.

———. 1845. *Indications of the Creator*. London: John W. Parker. (2nd ed., 1846.)

———. 1853. *Of the Plurality of Worlds*. London: John W. Parker.   —M.R.

Wilberforce, Samuel (1805–1873)

Samuel Wilberforce was the third son of William Wilberforce, a leader in the successful fight against owning and selling slaves in the British Empire. Samuel Wilberforce was ordained in the Church of England and, unlike his Low Church father, became a leader of the High Church forces in the country. Several of his relatives, including his brother-in-law Archdeacon (later Cardinal) Henry Manning, crossed over to Rome, but he was never tempted. Indeed, John Henry Newman, himself to cross over and later become a cardinal, leader of the Oxford Movement, put distance between himself and Wilberforce because it was clear that Wilberforce was first and foremost a man ambitious for social and ecclesiastical success in the national church.

Wilberforce was created bishop of Oxford in 1845, and it was in this position that he became a figure of history. In 1860 the British Association for the Advancement of Science—an annual jamboree that attracted both professionals and laypeople—held its meetings in Oxford. The year after the publication of Charles Darwin's *On the Origin of Species,* the topic of evolution was on the minds of all. Richard Owen, the anatomist and hated rival of the Darwinians, put the bishop up to debating with Darwin's bulldog, Thomas Henry Huxley. The encounter became legendary, with the bishop—so well known for his smooth debating style that he was nicknamed "Soapy Sam"—supposedly asking Huxley if he was descended from monkeys on his father's side or his mother's side, and Huxley responding that he had rather be descended from a monkey than from a bishop of the Church of England.

Legends serve a purpose, but they are not necessarily true. It is improbable that things were quite this sharp. Contemporary accounts suggest that all had a jolly good time and probably went off together afterward for good food and drink. Certainly the review that Wilberforce wrote of the *Origin* in the (conservative) *Quarterly Review* was critical but deeply respectful. But for the Darwinians the story served a wonderful purpose because the young and very middle-class Huxley took on the upper-class pillar of the establishment

Left: Bishop William Wilberforce (1869). *Right:* Thomas Henry Huxley (1871). These two cartoons, by the artist "Ape," Carlo Pellegrini, appeared in the weekly magazine *Vanity Fair.* Some suggest that the nickname "Soapy Sam" comes from Wilberforce's hand-washing habit, as shown in the cartoon.

and purportedly ground him into atoms, rather like David and Goliath or Falstaff and the robbers. There are well-known "Ape" cartoons from *Vanity Fair* of Huxley and Wilberforce, the former in Victorian business clothes and the latter in the Elizabethan dress of a bishop, epitomizing the clash between the forward-looking attitudes of science and the backward-looking attitudes of religion.

Wilberforce died after he fell off his horse. Huxley's unkind quip was that for the first time in his life Wilberforce's brain came into touch with the real world, and it killed him.

BIBLIOGRAPHY

Desmond, A. 1994. *Huxley, the Devil's Disciple.* London: Michael Joseph.
Lucas, J. R. 1979. Wilberforce and Huxley: A legendary encounter. *Historical Journal* 22: 313–330.
Newsome, D. 1966. *The Parting of Friends: The Wilberforces and Henry Manning.* London: John Murray.
Wilberforce, S. 1860. Review of "Origin of Species." *Quarterly Review* 108: 225–264.   —M.R.

Williams, George C. (b. 1926)

George Williams is one of the most influential evolutionary biologists of modern times, casting a long intellectual shadow across most of the discipline. Trained in ichthyology (he received his doctoral degree in 1955 from the University of California, Los Angeles), he published numerous essays on the ecology and evolutionary biology of fishes. But his towering reputation among evolutionary biologists rests primarily on his conceptual analyses and syntheses. His critical explorations of topics ranging from sex to senescence have changed how biologists think, and his analyses of the process of natural selection created the paradigm that continues to dominate the field.

In his remarkable book *Adaptation and Natural Selection* (1966; see the alphabetical entry in this volume) Williams made the compelling argument that, as Darwin postulated, selection was the major driving force of evolutionary change and diversification. In doing so, he tackled what were at that time several controversial aspects of natural selection, including the notions that adaptive evolution engenders progress and that increased complexity is a direct by-product of the Darwinian process. The most significant analysis in this book, and one to which Williams returned in later years, is the critical examination of the level at which selection acts, offering clear distinctions among selection at the levels of the gene, the individual, and the population. These arguments were particularly important and timely. On the one hand, William Hamilton had recently published his argument that kin selection at the level of the gene could be a powerful force in social evolution. On the other hand, ecologists were engaged in fierce debates about the relationship between Darwinian selection and the self-regulation of populations. These debates were inspiring hypotheses about whether and how selection could act at the level of the population or species (group selection; see the alphabetical entry in this volume). Williams's forceful arguments against the prevailing formulations of group selection influenced generations of evolutionary biologists and played a role in delaying the general acceptance of more sophisticated views of group selection. Indeed, teachers of evolution continue to echo Williams each time they remind students that adaptations have not emerged for the benefit of the population or species. (The philosopher Elliott Sober's *The Nature of Selection* [1984] is a sympathetic if sometimes critical examination of Williams's thinking.)

Natural selection in its myriad manifestations has been the focus of Williams's work throughout his career. In his book *Sex and Evolution* (1975) Williams focused on the question of how and why sexual reproduction evolved and became so prevalent. Previous arguments about the evolution of sexual reproduction had been focused on a limited range of population-genetic issues. Williams reexamined what others had suggested and broadened the topic to include the ecological and social considerations that surround sexual reproduction. In doing so, he brought the topic to wider attention and greater prominence and inspired several research programs dedicated to testing the ideas he had raised. Whether sexual reproduction in higher organisms has become more burden than advantage, as Williams suggested, remains under empirical scrutiny.

Williams is also responsible for one of the dominant hypotheses for the existence of senescence, the so-called antagonistic pleiotropy argument. When Williams offered his hypothesis, the prevailing idea about senescence was that it was the result of deleterious genes whose expression was delayed until later ages. This idea was based on the argument that natural selection would be more effective at purging deleterious genes that were expressed early in life because those genes would severely compromise fitness, especially if they were expressed during an individual's reproductive prime. Genes with deleterious effects expressed late in life would accumulate because they would have little effect on fitness. Williams suggested that senescence could be caused by genes that were expressed throughout an individual's life but had beneficial effects early and deleterious effects late. In this hypothesis the early beneficial effects outweighed the later deleterious effects, and so selection would not eliminate those genes. Although the distinction between these hypotheses may seem subtle, it addresses the important question of how often and how much adaptation requires compromises between genes that affect the same individual and between the multiple effects of individual genes. Indeed, the issue of compromise, especially between genetic effects at different ages, has been argued by Williams to lie beneath the larger problem of age-specific trait expression and the individual's entire life history (including the so-called cost of reproduction). Evolutionary biologists and experimental gerontologists have been devoting the past 40 years to discriminating between these hypotheses for senescence and to testing the prevalence of genetic compromises in life history in a host of contexts.

In Williams's later years he applied the principles of natural selection and adaptation to the study of disease. With Randolph Nesse he developed a series of arguments that have come to be known generally as Darwinian medicine (see the main essay "Evolutionary Biology of Disease and Darwinian Medicine" by Michael F. Antolin in this volume). In a Darwinian approach to medicine the scientist—or the physician—examines the condition in light of evolutionary history to understand the condition's ultimate origin. This is a logical extension of Williams's arguments about senescence. From maternal-fetal interactions in pregnancy to the genetic susceptibility to particular diseases, a variety of medical conditions have been illuminated by Darwinian analysis. Darwinian medicine remains a young subject, and its implications have yet to penetrate fully into clinical practice.

Williams's contributions to evolutionary biology have been recognized by numerous awards. He received the Eminent Ecologist Award from the Ecological Society of America and is an elected member of the American Academy of Arts and Sciences and the U.S. National Academy of Sciences. In 1999 he shared the Crafoord Prize with John Maynard Smith and Ernst Mayr for their pioneering and lasting contributions to evolutionary biology.

BIBLIOGRAPHY

Nesse, R. M., and G. C. Williams. 1994. *Why We Get Sick: The New Science of Darwinian Medicine.* New York: Vintage Books.
Sober, E. 1984. *The Nature of Selection.* Cambridge, MA: MIT Press.
Williams, G. C. 1966. *Adaptation and Natural Selection.* Princeton, NJ: Princeton University Press.
———. 1975. *Sex and Evolution.* Princeton, NJ: Princeton University Press.
———. 1992. *Natural Selection: Domains, Levels, and Challenges.* New York: Oxford University Press.
———. 1997. *The Pony Fish's Glow and Other Clues to Plan and Purpose in Nature.* New York: Basic Books.   —J.T.

Wilson, Edward O. (b. 1929)

Edward O. Wilson is today's leading evolutionary biologist. Born in the American South, he was an undergraduate at the University of Alabama and then shortly thereafter moved to Harvard University, where he spent the rest of his academic career, moving from graduate student to fellow to junior faculty member and eventually retiring as one of the most distinguished university professors. He has won two Pulitzer Prizes, as well as many other domestic and foreign awards, and is a member of the National Academy of Sciences and the recipient of many honorary doctoral degrees. He has recorded his life experiences in his autobiography, *Naturalist* (1994). Wilson greatly admires the militaristic culture of the South, but it is clear that the escape into nature was very satisfying for a boy and youth whose own talents lay in the direction of observation and interpretation.

His work has been wide ranging but began and has always been deeply rooted in his studies of the social insects, most particularly ants. His early work was on classification and biogeography and was based both on museum study and on fieldwork, particularly in the Far East. A major paper, cowritten with a fellow graduate student on the subspecies concept (Wilson and Brown 1953), earned the lasting ire of Ernst Mayr, the leading taxonomist and major figure in the Harvard Museum of Comparative Zoology. Increasingly in the 1960s Wilson turned to more technical issues in biogeography and with the theoretician the late Robert MacArthur was the author of a major hypothesis about the ways in which islands achieve species equilibrium, with numbers of incoming species equaling outgoing (generally through extinction) species (Macarthur and Wilson 1967). With his student Dan Simberloff, Wilson conducted classic experiments that tried to show the empirical truth of this work. One of the best-known involved an islet off the coast of Florida where Wilson and Simberloff destroyed all the animal life and then recorded the rate at which the islet was repopulated.

Even as he was completing this work, Wilson was moving into issues of animal communication and soon made pathbreaking discoveries about how insects use chemicals (pheromones) to transfer information to each other. This set the scene for the 1970s, when Wilson produced a major trilogy: *The Insect Societies* (1971), dealing with the nature of sociality among the insects; *Sociobiology: The New Synthesis* (1975), dealing with the evolutionary biology behind the sociality of all animals; and *On Human Nature* (1978), which treats exclusively of the evolutionary basis of human sociality, and for which Wilson won the first of his Pulitzer Prizes.

Sociobiology: The New Synthesis was Wilson's magnum opus and is now recognized as one of the major works in the history of evolutionary theorizing. It presented a huge review of all that was then known about the mechanisms by which natural selection can produce social behavior, followed by a detailed review from the lowest forms to the highest of how such behavior actually can be found in action. Nevertheless, because the book ends with a chapter that applies the ideas to our own species, at the time of publication the book was highly controversial. Many—especially social scientists who felt threatened for the autonomy of their own studies, and Marxist biologists who hated any noneconomic analysis of behavior—launched bitter critiques of the book. It was argued that the book was methodologically flawed, that it was racist and sexist, and that it was no more than Western capitalist ideology dressed up as objective knowledge.

Brushing off criticism, Wilson did more work in the 1980s on human sociality, especially with a young Canadian physicist, Charles Lumsden. At the same time, Wilson continued to work in basic science, thinking hard about the conceptual issues in the field of social behavior. This led to a coauthored work with the theoretician George Oster (1978) on so-called optimality models. Increasingly, however, Wilson turned back to his studies of ants. With then–fellow Harvard professor Bert Hölldobler, Wilson produced massive surveys of the ant world, culminating in the coauthored book *The Ants* (1990), for which Wilson won the second of his Pulitzer Prizes. In the 1990s Wilson continued his work on ants with a major taxonomic study of the group.

More recently, Hölldobler and Wilson have been thinking about the foundations of sociobiology, and in a new work on the "superorganism" (2008) argue for a much more holistic approach to the topic, one which takes into full account the way that social societies (especially in the insects) get integrated into wholes, which selection treats as individuals rather than as separate organisms in a colony. This was very much the position of Charles Darwin in the *Origin*. Disinterested observers sense the irony that in the 1970s Wilson's critics accused him of undue "reductionism," treating everything as understandable in terms of parts rather than wholes.

Always interested in broader issues, increasingly in more recent years Wilson has turned his attention to more philosophical issues. He has written on ethics with the philosopher Michael Ruse, and in a work with the provocative title *Consilience* (1998) has articulated his own vision of the connections between different branches of knowledge. An ardent conservationist— Wilson has made major efforts in the campaign to preserve the Brazilian rain forests—he has written extensively on biodiversity and the problems facing us today (see especially *The Diversity of Life* [1992] and *The Future of Life* [2002]). Wilson has been arguing for a kind of evolutionarily inspired humanism, in which the supreme moral commandment must be that which best ensures the future of the human species. An ardent proponent of what he calls biophilia, Wilson argues that humans have evolved in symbiotic relationship with nature, and unless they preserve the world's flora and fauna, it will be both physical and spiritual death for the human species. This we must do all we can to prevent.

Although Wilson has never been one who argues that conventional religion adds to the world's woes, and although in some respects he shows considerable sympathy for the religious mind—in part, undoubtedly, because of his own early childhood commitments to evangelical religion—he has long argued that today we must underpin our moral and religious yearnings with a more naturalistic base. It was this, rather than some neoconservative ideology, that motivated Wilson in his earlier writings on humankind (especially in *On Human Nature*) and it continues to this day, especially in a heartfelt plea (*The Creation*, 2006) to people of religion to join in the work of conservation. Ultimately, Wilson finds his naturalistic base in the upward rise of the history of life from simple to complex, from worm to human, and, in line with evolutionary thinkers of the past (notably Herbert Spencer in the nineteenth century and Julian Huxley in the twentieth century), he claims that this progress in itself is justification of all of his desired moral norms.

Needless to say, claims like these do not find universal favor with biologists or with philosophers. Many invoke David Hume's distinction between matters of fact and matters of value, arguing that even though one may believe in progress, this is something one reads into the evolutionary picture rather than something one reads from the picture. Wilson is not convinced, and whether he is right or wrong, he certainly shows the courage of his convictions. He cannot believe that beings like humans, the end product of a long, unguided process of change, do not show the labors of that change and should not be understood in terms of that change. This is true of ants and, argues Wilson, must also be true of *Homo sapiens*. (In a traditional way, Ruse 1985 and Kitcher 2003 are very critical of Wilson's thinking. More recently, although not all appreciate Wilson's progress-based approach, many are starting to move toward a naturalistic stance on ethics and for this Wilson deserves massive credit. See Ruse 1998, 2008, 2009.)

BIBLIOGRAPHY

Hölldobler, B., and E. O. Wilson. 1990. *The Ants.* Cambridge, MA: Harvard University Press.

———. 1994. *Journey to the Ants: A Story of Scientific Exploration.* Cambridge, MA: Harvard University Press.

———. 2008. *The Superorganism: The Beauty, Elegance, and Strangeness of Insect Societies.* New York: Norton.

Kitcher, P. 2003. *In Mendel's Mirror.* New York: Oxford University Press.

Lumsden, C. J., and E. O. Wilson. 1981. *Genes, Mind, and Culture.* Cambridge, MA: Harvard University Press.

———. 1983. *Promethean Fire: Reflections on the Origin of Mind.* Cambridge, MA: Harvard University Press.

MacArthur, R. H., and E. O. Wilson. 1967. *The Theory of Island Biogeography.* Princeton, NJ: Princeton University Press.

Oster, G., and E. O. Wilson. 1978. *Caste and Ecology in the Social Insects.* Princeton, NJ: Princeton University Press.

Ruse, M. 1985. *Sociobiology: Sense or Nonsense?* 2nd ed. Dordrecht: Reidel.

———. 1996. *Monad to Man: The Concept of Progress in Evolutionary Biology.* Cambridge, MA: Harvard University Press.

———. 1998. *Taking Darwin Seriously: A Naturalistic Approach to Philosophy.* 2nd ed. Buffalo, NY: Prometheus Books.

———. 2008. *Charles Darwin.* Oxford: Blackwell.

———, ed. 2009. *Philosophy after Darwin.* Princeton, NJ: Princeton University Press.

Ruse, M., and E. O. Wilson. 1985. The evolution of morality. *New Scientist* 1478: 108–128.

———. 1986. Moral philosophy as applied science. *Philosophy* 61: 173–192.

Segerstrale, U. 2000. *Defenders of the Truth: The Battle for Science in the Sociobiology Debate and Beyond.* New York: Oxford University Press.

Simberloff, D., and E. O. Wilson. 1969. Experimental zoogeography of islands: The colonization of empty islands. *Ecology* 50: 278–296.

Wilson, E. O. 1971. *The Insect Societies.* Cambridge, MA: Harvard University Press.

———. 1975. *Sociobiology: The New Synthesis.* Cambridge, MA: Harvard University Press.

———. 1978. *On Human Nature.* Cambridge, MA: Cambridge University Press.

———. 1984. *Biophilia.* Cambridge, MA: Harvard University Press.

———. 1992. *The Diversity of Life.* Cambridge, MA: Harvard University Press.

———. 1994. *Naturalist.* Washington, DC: Island Books/Shearwater Books.

———. 1998. *Consilience: The Unity of Knowledge.* New York: Vintage Books.

———. 2002. *The Future of Life.* New York: Vintage Books.

———. 2006. *The Creation: A Meeting of Science and Religion.* New York: Norton.

Wilson, E. O., and W. L. Brown Jr. 1953. The subspecies concept and its taxonomic application. *Systematic Zoology* 2: 97–111.

Wright, R. 1987. *Three Scientists and Their Gods.* New York: Times Books.

—M.R.

Wright, Sewall (1889–1988)

Sewall Green Wright, one of the architects of the modern synthetic theory of evolution with R. A. Fisher (1890–1962) and J. B. S. Haldane (1892–1964), was born on December 21, 1889, to Philip Green Wright and Elizabeth Quincy Sewall Wright. He had two younger brothers, Quincy and Theodore. The Wright family can be traced back through sixteenth-century England to the ninth-century reign of Charlemagne. Sewall Wright was born in Melrose, Massachusetts, but grew up in Galesburg, Illinois, where his father was on the faculty of Lombard College. Wright and his brothers were gifted children, reading at an early age. Indeed, in school Wright's abilities caused him torment by his classmates. In 1906 Wright graduated from Galesburg High School and matriculated at Lombard College to study mathematics and surveying; his father had encouraged him to study his own loves, poetry and music.

Wright's interests in biology took center stage during his last year, 1911, at Lombard through the mentoring of Wilhelmine Enteman Key (who was one of the first women to earn a PhD at the University of Chicago). That summer Key sent Wright to Cold Spring Harbor Laboratory on Long Island, New York, where Wright learned from world-class biologists, including Key's former mentor, Charles Davenport (1866–1944). Wright found his experience at Cold Spring Harbor rewarding, and he returned during the summer of 1912. In the fall of 1911 Wright entered the graduate program in biology at the University of Illinois, Urbana-Champaign.

By the spring of 1912 Wright completed his master's thesis on the anatomy of the trematode *Microphallus opacus*. Wright completed the thesis in short order because of a chance meeting after a lecture by the geneticist William Ernest Castle (1867–1962), of Harvard University's Bussey Institute, another student of Davenport. Castle lectured on his selection experiments on hooded rats and on mammalian genetics. Wright was fascinated and approached Castle about working with him. Castle was sufficiently impressed by Wright that Wright quickly finished his master's thesis and enrolled at Harvard University in the fall of 1912.

At Harvard Wright engaged in original experimental research in physiological genetics. Wright's research was directed by Castle, but he was also considerably influenced by the geneticist Edward Murray East (1879–1938). At Harvard and the Bussey Institute Wright worked closely with Castle on his hooded-rat selection experiments and on the genetics of small mammals. By 1915 Wright completed his doctoral thesis on coat-color inheritance in guinea pigs. Wright's research demonstrated the existence of multiple loci and alleles that affected coat color in the animal; it further set out the hypothesis that enzyme pathways and pigment precursors provided the physiological basis for observed patterns of inheritance of coat coloration. Around the time Wright was finishing his thesis, he accepted a position as senior animal husbandman at the U.S. Department of Agriculture (USDA) in Beltsville, Maryland. Wright's work as a physiological geneticist permeated his work as a biologist more generally and, in particular, his work in evolutionary theory.

Wright was with the USDA for 10 years, from 1915 to 1925. He controlled a large-scale inbreeding experiment on guinea pigs begun at the USDA in 1906. During his tenure at the USDA Wright published widely on physiological genetics, but two main publications in 1921 stand out. The first, "Correlation and Causation" (Wright 1921a), described Wright's invention of the method of path analysis, that is, the method for estimating the magnitude and significance of hypothesized causal connections between sets of variables. The other was a series of essays published under the heading "Systems of Mating" (Wright 1921b). Here Wright used path analysis to explore the numerous features of mating systems, including inbreeding, assortative mating, and the effects of selection. Wright subsequently analyzed the history of inbreeding in American shorthorn cattle using path analysis. Also in 1921 Wright married Louise Williams, a member of the faculty of biology at Smith College. The Wrights parented three children, Elizabeth Rose, Richard, and Robert.

In 1926 Wright joined the faculty of biology at the University of Chicago. It was here that Wright's contribution as an architect of the synthesis of Darwinism and Mendelism was completed. Wright's most famous essay, "Evolution in Mendelian Populations," was published in 1931. Wright demonstrated the mathematical unification of Darwinian natural selection and the principles of Mendelian heredity, and he communicated this synthesis in the form of his famous "shifting balance" theory of evolution. Wright used his famous adaptive-landscape diagram to communicate his theory a year later (Wright 1932). Using his work as a physiological geneticist, Wright emphasized genetic interaction in evolution by way of the following assumption: Because the field of gene combinations in a population is vast, genes adaptive in one combination may not be adaptive in another; consequently, the field of joint gene frequencies graded for adaptive value will produce a hilly landscape. Populations thus face the problem of shifting from one peak to another. Wright solved the problem of peak shifts with his three-phase shifting-balance process. In the first phase, subdivided populations would be subject to genetic drift, pulling them into a "valley." In the second phase, those populations would be dragged up the next peak by intrademe selection, or within-group selection. In the third phase, the global population would find the highest peak by interdeme selection, or population-structured selection.

Wright's mathematical analysis of the reconciliation between Darwinism and Mendelism agreed with those of the other two synthesis architects, R. A. Fisher and J. B. S. Haldane. However, Fisher and Wright in particular disagreed about the extent to which gene interaction was important in describing the genetic basis of evolutionary change. For Fisher, who deemphasized genetic interaction, natural selection acting on mutations of small effect was sufficient accounting for most cumulative evolutionary change. Indeed, Wright and Fisher ultimately became mired in a controversy over their alternative understandings of the evolutionary process that lasted until Fisher's death in 1962.

In 1955 Wright was subjected to mandatory retirement from the University of Chicago. He became Leon J. Cole Professor of Genetics at the University of Wisconsin, Madison, where he remained for the rest of his long career. Wright was prolific throughout. Perhaps there is no better testament to this than the publication between 1968 and 1978 of his four-volume magnum opus, *Evolution and the Genetics of Populations*. These volumes were the culmination of his work in evolutionary theory. Yet Wright's last essay, "Surfaces of Selective Value Revisited," was published the year he died (1988). This essay was a mostly favorable reaction to his biography, *Sewall Wright and Evolutionary Biology*, published by William B. Provine in 1986. Wright died at the age of 99 on March 3, 1988, due to complications from a broken pelvis after a slip on an icy sidewalk during one of his usual walks.

Sewall Wright is remembered as a towering figure among American evolutionary biologists. His adaptive-landscape diagram permeates evolutionary thought, and his statistical theory of inbreeding is standard in evolutionary genetics. Wright was the recipient of numerous awards, including the Weldon Medal of the Royal Society of London in 1947, the National Medal of Science in 1966, and the Medal of the Royal Society of London in 1980. Wright the man was described by his friends and associates as shy but warm, and unflinching when discussion turned to his interests.

BIBLIOGRAPHY

Provine, W. B. 1986. *Sewall Wright and Evolutionary Biology*. Chicago: University of Chicago Press.

Wright, S. 1921a. Correlation and causation. *Journal of Agricultural Research* 20: 557–585.

———. 1921b. Systems of mating. *Genetics* 6: 111–178.

———. 1931. Evolution in Mendelian populations. *Genetics* 3: 97–159.

———. 1932. The roles of mutation, inbreeding, crossbreeding and selection in evolution. *Proceedings of the Sixth Annual Congress of Genetics* 1: 356–366.

———. 1968–1978. *Evolution and the Genetics of Populations*. 4 vols. Chicago: University of Chicago Press.

———. 1988. Surfaces of selective value revisited. *American Naturalist* 131: 115–123.

—R.A.S.

Y

Yablokov, Alexei V. (b. 1933)

Alexei Yablokov has made substantial contributions to biogeography and conservation through his innovative use of easily measured, discretely varying external characters such as pigmentation patterns. Although biologists have always taken advantage of such variations—recall that Mendel's experiments were focused on round or wrinkled peas and yellow or green peas—Yablokov demonstrated how these characters could be used in a variety of contexts, from elucidating the family structure of dolphin groups to tracing the long-term effects of environmental pollution. Yablokov called these discrete variations in color or morphology *phenes,* and his book *Phenetics* is a synthetic treatment of how the study of phenes can be applied to almost every topical area of evolutionary biology.

Yablokov, a protégé of the population geneticist Nikolai Timofeeff-Ressovsky, received his doctoral degree in 1965 at the Novosibirsk State University. His early work in population genetics led him to the phenetic approach as a practical means of studying microevolutionary processes in natural populations and connecting them to broader problems in biogeography and conservation. These connections are described in *Phenetics,* as well as in two other books in Russian (coauthored with Timofeeff-Ressovsky and colleagues). Yablokov has also been a leader in the study of fluctuating asymmetry, which is a randomly asymmetric pattern that can reflect a variety of developmental and genetic influences, and he was one of the first scientists to apply fluctuating asymmetry to conservation issues by showing that pollution was increasing the level of asymmetry in skulls of gray seals. He has also published extensively on the ecology and genetics of whales and seals and pioneered the use of color patterns for individual identification and the tracing of families and pedigrees. In his later years he has become one of Russia's leading conservationists; he served in the Russian parliament, was science adviser to Boris Yeltsin, and started a number of conservation organizations.

Yablokov's work has been enormously influential in Russia and Europe, particularly where the use of more expensive, more direct molecular genetic methods has been impractical. His phenetic approach has two parallels. First, the use of phenes to study natural selection recalls the ecological genetics of E. B. Ford and his students and colleagues. Indeed, many of the examples offered in *Phenetics* can also be found in Ford's final edition of *Ecological Genetics.* Second, the use of phenes to study biogeography, which Yablokov called phenogeography, has its parallel in the use of biochemical and molecular markers for understanding how dispersal patterns and historical events have shaped species ranges.

Unfortunately, Yablokov's reputation in the West is limited because little of his work has been translated from the Russian. The magnitude of Yablokov's contributions emerges more clearly when one appreciates the context in which he worked. He was a real geneticist doing real genetics in the Soviet Union during the Lysenko era, he did not have the advantages that Western scientists have in equipment and resources, and he championed environmental issues in a place and time when it was unpopular and even dangerous to do so.

BIBLIOGRAPHY

Yablokov, A. V. 1986. *Phenetics: Evolution, Population, Trait.* M. J. Hall, trans. New York: Columbia University Press.

—*D.Si. and J.T.*

Z

Zoological Philosophy (Jean-Baptiste Lamarck)

Zoological Philosophy (Philosophie zoologique) was the title of Jean-Baptiste Lamarck's most famous exposition of his evolutionary views (see also the alphabetical entry "Jean-Baptiste Lamarck" in this volume). Published in 1809, the book attacked the idea of species constancy. Yet this was by no means Lamarck's only goal in this work. He sought more broadly to explain the diversity of animal organization and the particular faculties animals derived from the physical structures they possessed. In the first part of the work he addressed questions relating to natural history, with special attention to matters of animal classification, the nature of species, and the influence of the environment on animal structures. In the second and third parts of the work he went on to treat the physical causes of life and motion in the simpler animals and how it was that as animals became more complex, they gained the faculty of feeling and then finally became capable of intelligent action. This work reflected Lamarck's image of himself as more than a classifier and cataloger of nature productions. He defined his own role as that of "naturalist-philosopher."

Lamarck was forthright in rejecting the creationist notion that all species had existed from the beginning exactly as one found them in the present. He argued to the contrary that "Nature has produced all the species of animals in succession, beginning with the most imperfect or simplest, and ending her work with the most perfect, so as to create a gradually increasing complexity in their organization; these animals have spread at large throughout all the habitable regions of the globe, and every species has derived from its environment the habits that we find in it and the structural modifications which observation shows us" (*Zoological Philosophy*, 126).

After interpreting the animal scale as the temporal order in which the different levels of organization had been successively produced, Lamarck went on to explain how these different levels of organization gave rise to particular faculties (see figure). Whereas earlier writers had argued whether animals were intelligent, or whether there was such a thing as instinct, Lamarck offered a naturalistic explanation in which the different faculties of irritability, then instinct, and finally intelligence emerged as animals of increasing organic complexity were successively produced. In his account the least perfect animals were stirred to action only as the result of excitations that came from sources external to them. They were incapable of experiencing sensations

TABLEAU
Servant à montrer l'origine des différens animaux.

Vers.   Infusoires.
                                         Polypes.
                                         Radiaires.

                                         Insectes.
                                         Arachnides.
Annelides.   Crustacés.
Cirrhipèdes.
Mollusques.

Poissons.
Reptiles.

Oiseaux.

Monotrèmes.

M. Amphibies.

M. Cétacés.

M. Ongulés.
M. Onguiculés.

The tree of life as given by Lamarck in his *Philosophie zoologique* (1809). Atypically, it is to be read from top to bottom. Care should be taken with its interpretation. First, for Lamarck the branching (brought on by the inheritance of acquired characteristics) was all rather incidental to the main progressivist upthrust of the evolutionary process. Second, it should not be read as with Darwin (and others like Haeckel) as a tree of shared descent. Lamarck thought that new life forms are constantly appearing and then starting up the chain of being. Hence, if (say) lions were to go extinct today it would be possible for a younger line to produce them.

---

and likewise incapable of voluntary action. It was only when one came to the higher invertebrates that one found animals with a nervous system sufficient to give them the faculty of feeling. At this point in the scale the source of excitation of animal actions became internalized, and a special "internal feeling" became the basis for instinctive action. However, insofar as these animals lacked cerebral hemispheres (the organ of intelligence, in Lamarck's view), they remained incapable of thinking and thus of any acts of will. Only among the vertebrates, Lamarck claimed, and even then primarily among the birds and the mammals, were intelligent action and the faculty of willing to be found, and even these animals acted mostly by instinct.

Lamarck's contemporaries were inclined to treat his *Zoological Philosophy* as idle speculation. The work failed to win them over to the idea of organic evolution. Today the work stands out as a classic of evolutionary thought, although modern biologists do not endorse the specific mechanisms that Lamarck called upon to explain evolutionary change.

BIBLIOGRAPHY

Burkhardt, R. W., Jr. 1995. *The Spirit of System: Lamarck and Evolutionary Biology.* New ed. Cambridge, MA: Harvard University Press.

Lamarck, J.-B. 1809. *Philosophie zoologique, ou exposition des considérations relatives à l'histoire naturelle des animaux; à la diversité de leur organisation et des facultés qu'ils en obtiennent; aux causes physiques qui maintiennent en eux la vie et donnent lieu aux mouvemens qu'ils exécutent; enfin, à celles qui produisent, les unes le sentiment, et les autres l'intelligence de ceux qui en sont doués.* 2 vols. Paris: Dentu.

———. 1984. *Zoological Philosophy.* Hugh Elliot, trans. Chicago: University of Chicago Press. —*R.W.B.*

Contributors

Colin Allen (C.A.) is a professor in the Department of History and Philosophy of Science and in the Cognitive Science Program at Indiana University, Bloomington. He works on many issues related to the evolution of mind and cognition in non-human animals, and has coauthored and coedited books that include *Species of Mind* (1997), *The Evolution of Mind* (1998), and *The Cognitive Animal* (2002).

Garland E. Allen (G.E.A.) is professor of biology at Washington University in St. Louis, where he teaches biology and history of science courses. He is the author of numerous articles on the history of genetics and eugenics, a biography of Thomas Hunt Morgan, and a history of the life sciences in the twentieth century. He is currently working on a history of genetics in the twentieth century.

Ron Amundson (R.A.) is professor of philosophy at the University of Hawaii at Hilo. He is author of *The Changing Role of the Embryo in Evolutionary Thought* (2005). His research focuses on the relations between evolutionary and developmental biology, and on bioethical understandings (and misunderstandings) of disability.

Michael F. Antolin is professor of biology at Colorado State University, where he teaches population genetics and evolutionary biology to students ranging from entering freshmen to postgraduates. His research focuses on effects of spatial variation on evolutionary ecology of organisms including aphids, parasitoid wasps, flowering plants, prairie dogs, and plague. The crossroads between science, religion, and society offer a point of deep fascination and regular distraction.

Stevan J. Arnold (S.J.A.) is professor of zoology and curator of amphibians and reptiles at Oregon State University. He is best known for his work on techniques for measuring selection, comparing inheritance matrices, and characterizing mating systems. His other interests include sexual selection and its connection to behavioral isolation, pheromone evolution, and testing models of phenotypic evolution.

Francisco J. Ayala is University Professor and Donald Bren Professor of Biological Sciences at the University of California, Irvine. He has published over 900 articles and is author or editor of 31 books. A member of the U.S. National Academy of Sciences and the American Philosophical Society, Ayala received the 2001 U.S. National Medal of Science. The *New York Times* named him the "Renaissance Man of Evolutionary Biology."

Jeffrey L. Bada is a professor of marine chemistry at the Scripps Institution of Oceanography, University of California at San Diego. His primary research deals with the origin of life on earth and elsewhere. He is also involved with the search for life beyond earth and is the chief scientist for the Urey Organic and Oxidant Detector instrument on the 2013 European Space Agency ExoMars mission.

Charles Baer (C.B.) is an assistant professor of zoology at the University of Florida. He is interested in nematodes (roundworms) as well as general issues in evolutionary biology, including life-history strategies and the evolution of mutation rates.

Ana Barahona (A.B.) is professor of history and philosophy of science at the National University of Mexico. A pioneer in the historical and philosophical studies of science in Mexico since 1980, she founded the area of Social Studies of Science and Technology in the Faculty of Sciences. She has published work on the history of science in Mexico, genetics, and evolutionary biology, as well as science education. She is the president-elect of the International Society for the History, Philosophy and Social Studies of Biology.

John Beatty (J.B.) teaches history and philosophy of science, and social and political philosophy, in the Department of Philosophy at the University of British Columbia in Vancouver. His research focuses on the theoretical foundations, methodology, and sociopolitical dimensions of genetics and evolutionary biology.

Naomi Beck (N.B.) received her PhD from the University of Paris 1 (Panthéon-Sorbonne) and is currently employed as assistant professor at the University of Chicago. She wrote on Spencer's influence in France and Italy and now studies the role of evolutionary ideas in Hayek's free-market theories.

Michael Benton is professor of vertebrate palaeontology and former head of the Department of Earth Sciences at the University of Bristol. He is the author of many papers and books on vertebrate palaeontology and evolution, including *Introduction to Palaeobiology and the Fossil Record* (2008).

Peter J. Bowler (P.J.B.) is professor of the history of science at Queen's University, Belfast, Northern Ireland. He is the author of several books on the history of evolution theory, the most recent of which is *Monkey Trials and Gorilla Sermons* (2007).

C. Loring Brace (C.L.B.) is professor of anthropology emeritus at the University of Michigan and curator of biological anthropology emeritus at the University of Michigan Museum of Anthropology. He is the author of many papers and several books on human evolution, including *"Race" Is a Four-Letter Word* (2005), which shows the absence of a biological basis for the concept of race.

David J. Buller (D.J.B.) is Presidential Research Professor in the Department of Philosophy at Northern Illinois University. He is the author of *Adapting Minds: Evolutionary Psychology and the Persistent Quest for Human Nature* (2005) and the editor of *Function, Selection, and Design* (1999).

Richard W. Burkhardt Jr. (R.W.B.) is professor of history emeritus at the University of Illinois, Urbana–Champaign. His writings related to the history of evolutionary biology include *The Spirit of System: Lamarck and Evolutionary Biology* (1977, 1995) and *Patterns of Behavior: Konrad Lorenz, Niko Tinbergen, and the Founding of Ethology* (2005).

Joe Cain (J.C.) is senior lecturer in history and philosophy of biology at University College London. He is an expert on the history of evolutionary studies, palaeontology, and natural history in Britain and the United States.

Elof Axel Carlson (E.A.C.) is distinguished teaching professor emeritus, Department of Biochemistry and Cell Biology, Stony Brook University. His interests include the history of genetics and eugenics and the relation of science to society. He is the author or editor of 12 books, including, most recently, *Neither Gods Nor Beasts: How Science Is Changing Who We Think We Are* (2008).

Scott P. Carroll (S.P.C.) is affiliated with the Department of Entomology and Center for Population Biology at the University of California, Davis, and the School of Integrative Biology at the University of Queensland. He is coeditor of *Conservation Biology: Evolution in Action* (2008).

Brian Charlesworth (B.C.) is professor and head of the Institute of Evolutionary Biology at the University of Edinburgh. His current research interests are in population genetics, molecular evolution, and genome evolution. He has published over 200 research papers and two books.

Deborah Charlesworth is professor of plant population genetics at the University of Edinburgh. She works of the evolution of plant mating systems and sex chromosomes, and is coauthor of *Evolution: A Very Short Introduction* (2003).

Simon Conway Morris (S.C.M.) is professor of evolutionary palaeobiology at the University of Cambridge and a Fellow of St John's College. He has twin interests in the Burgess Shale and the Cambrian "explosion" (summarized in *The Crucible of Creation*, 1998) and inevitabilities in evolution based on the ubiquity of convergence (as in *Life's Solution*, 2003). He is active in the science and religion debates and lectures and broadcasts regularly on these areas.

Pietro Corsi (P.C.) is professor of the history of science at Oxford University. He is the author of books on Baden Powell and Jean-Baptiste Lamarck.

James F. Crow (J.F.C.) is professor emeritus of genetics at the University of Wisconsin–Madison. He has divided his time among teaching, administration, and research in population genetics. In addition to journal articles, he has authored three books, including (with M. Kimura) *An Introduction to Population Genetics Theory* (1970).

Daniel Dennett (D.D.) is University Professor, Austin B. Fletcher Professor of Philosophy, and codirector of the Center for Cognitive Studies at Tufts University. He is the author of *Darwin's Dangerous Idea* (1995), *Freedom*

Evolves (2003), *Breaking the Spell: Religion as a Natural Phenomenon* (2006), and many articles on mind, evolution, and science.

Adrian Desmond (A.D.) is Honorary Research Fellow in the Biology Department at University College London. His many books on Victorian evolution include *Archetypes and Ancestors* (1982), *The Politics of Evolution* (1989), and *Huxley* (1994–1997), as well as two books coauthored with Jim Moore, *Darwin* (1991) and *Darwin's Sacred Cause* (2009).

Michael R. Dietrich (M.R.D.) is an associate professor in the Department of Biological Sciences at Dartmouth College. He has written a number of articles on the history of evolutionary genetics and molecular evolution.

Philip Donoghue (P.D.) is a reader in geology at the University of Bristol, United Kingdom. His main research interests are the emergence of evolutionary novelties and the major evolutionary transitions.

Gregory M. Erickson (G.M.E.) is an associate professor in the Department of Biological Science at Florida State University. His research focuses on the life-history evolution and feeding biomechanics in nonavian and basal avian dinosaurs.

Zachary Ernst (Z.E.) is an associate professor in the Department of Philosophy at the University of Missouri–Columbia. In addition to the philosophy of biology, he writes on other subjects including game theory and logic.

Dean Falk (D.F.) is the Hale G. Smith Professor of Anthropology at Florida State University. She has authored many articles and a number of books about various aspects of human evolution.

Alan Feduccia (A.F.) is S. K. Heninger Professor Emeritus and former chairman of the Department of Biology at the University of North Carolina, Chapel Hill. He is the author of numerous papers on evolutionary biology, and his books include *The Age of Birds* (1980) and *The Origin and Evolution of Birds* (1996, 1999).

Martin Fichman (M.F.) is a professor of the history of science at York University. He is the author of many articles and four books, most recently *An Elusive Victorian: The Evolution of Alfred Russel Wallace* (2004). In addition to the history of evolutionary thought, his research interests include science studies and the cultural history of Victorian science.

Charles W. Fox (C.W.F.) is a professor in the Department of Entomology at the University of Kentucky. He studies evolutionary ecology, primarily of insects. He is coeditor of multiple books in evolutionary biology, including *Evolutionary Ecology: Concepts and Case Studies* (2001), *Evolutionary Genetics: Concepts and Case Studies* (2006), and *Conservation Biology: Evolution in Action* (2008). He is also executive editor of the journal *Functional Ecology*.

Iris Fry (I.F.) teaches the history and philosophy of biology in the Department of Humanities and Arts at the Technion-Israel Institute of Technology. She is the author of *The Origin of Life: Mystery or Scientific Problem?* (1997) and *The Emergence of Life on Earth: A Historical and Scientific Overview* (2000).

Jean Gayon (J.G.) is professor of philosophy at University Paris 1-Panthéon Sorbonne and a member of the Institute for History and Philosophy of Science and Technique. Author of *Darwinism's Struggle for Survival* (1998), his work is in history and philosophy of biology, with special emphasis on genetics and biometry, general philosophy of science, and social and ethical problems raised by the life sciences.

Nicholas W. Gillham (N.W.G.) is James B. Duke Professor Emeritus in the Department of Biology at Duke University. His research has been focused on understanding how nuclear and organelle genomes interact in controlling the biogenesis of chloroplasts and mitochondria. His books include *Organelle Heredity* (1978) and *Organelle Genes and Genomes* (1994).

Brian Goodwin is a professor of biology teaching Holistic Science at Schumacher College, a centre of transformative learning for sustainable living. His books explore creative emergence in biological and social contexts.

Bruce S. Grant (B.S.G.) is professor of biology emeritus at the College of William and Mary, where he taught genetics and evolution for 33 years until his retirement in 2001. He has done research with *Drosophila* species and parasitoid wasps, but since 1983 his research has focused on the evolution of melanism in peppered moths, a field in which he remains active.

James Hanken (J.H.) is Alexander Agassiz Professor of Zoology and director of the Museum of Comparative Zoology at Harvard University. His research focuses on the evolutionary morphology, development, and systematics of vertebrates. He has authored numerous professional and popular articles, and with Brian Hall coedited *The Skull* (1993). Beginning in 2008, he chairs the Steering Committee of the Encyclopedia of Life.

Shala J. Hankison is an assistant professor of animal behavior in the Department of Zoology at Ohio Wesleyan University. She is interested in the role of sexual selection and behavior in shaping evolution.

John Barrett Heiser (J.B.H.) has been associated with Cornell University and its ichthyology, aquatic, and marine biology programs since his graduate school days. He has studied fishes and their habitats on every continent and ocean (plus many seas) from a variety of platforms and land forms. For 15 years he was director of the Shoals Marine Laboratory on Appledore Island, Maine, where he continues to teach between bouts of leading eco-educational tours to far-flung corners of the earth.

Jonathan Hodge (J.H.) is a senior fellow in history and philosophy of science at the University of Leeds. He has written on theories of origins and species from ancient to present times. Two volumes of his papers are being published in 2008 by Ashgate Publishing. A revised edition of his *Cambridge Companion to Darwin*, coedited with Gregory Radick, will appear in 2009.

Brian Hollis (B.H.) is a graduate student in the Department of Biological Science at Florida State University. His research interests include the study of sexual selection and, broadly, the levels of selection debate.

Tim J. Horder (T.J.H.) is a senior research fellow at Jesus College in the University of Oxford. He edited *A History of Embryology* (1986) and has published on both historical themes and on his experimental research in the embryology and evolution of pattern formation.

David Houle (D.H.) is a professor the Department of Biological Science at Florida State University. His research interests are in evolutionary genetics, especially the role of genetic variation in shaping the evolutionary process.

David L. Hull (D.L.H.) is emeritus professor in the Department of Philosophy at Northwestern University. He has authored several books and numerous papers on the philosophy of biology, evolutionary theory, biological systematics, and the socioiology of science.

William Irons (W.I.) is a professor of anthropology at Northwestern University. He is interested in the effects of evolutionary forces on human nature and is co-editor of *Adaptation and Human Behavior* (2000).

Christine Janis (C.J.) is a professor of ecology and evolutionary biology at Brown University. She has published numerous papers on mammalian evolution, is the senior editor of *Evolution of Tertiary Mammals of North America* (2 vols., 1998, 2008), and contributes to the popular textbook *Vertebrate Life*.

Donald C. Johanson (D.C.J.) is director of the Institute of the Human Origins, where he holds the Virginia M. Ullman Chair, and professor in the School of Human Evolution and Social Change at Arizona State University. He has coauthored several books, including *Lucy: The Beginnings of Humankind* (1981) and *From Lucy to Language* (2005).

Norman A. Johnson (N.A.J.) is an adjunct research faculty member at the University of Massachusetts. His research interests have focused on the genetics and evolution of reproductive isolation, genetic interactions, the evolution of development, and selfish genetic elements. He is the author of *Darwinian Detectives: Revealing the Natural History of Genes and Genomes* (2007).

I. King Jordan (I.K.J.) is an associate professor in the School of Biology at the Georgia Institute of Technology. His laboratory takes a computational approach to the study of genome structure, function, and evolution. The influence of transposable element sequences on gene regulation is one particularly active area of investigation in the lab.

Kenneth Y. Kaneshiro (K.Y.K.) is director of the Center for Conservation Research & Training in the Pacific Biosciences Research Center at the University of Hawaii. His research focus has been on the role of sexual selection in the evolution and speciation of the Hawaiian *Drosophila*. He has published numerous articles on the topic.

Susan M. Kidwell (S.M.K.) is the William Rainey Harper Professor in the Department of Geophysical Sciences and Committee on Evolutionary Biology at the University of Chicago. Her research focuses on the completeness of the fossil record and on the application of skeletal remains to conservation biology.

William C. Kimler (W.C.K.) is Alumni Distinguished Undergraduate Professor at North Carolina State University, where he is as associate professor of history and a core faculty member in the Program on Science, Technology, and Society and in the W. M. Keck Center for Behavioral Biology. His research interests in the history of evolutionary ideas lie in nineteenth- and twentieth-century natural history, ecology, and animal behavior.

Jeremy Kirby (J.K.) is assistant professor of philosophy at Albion College. He has authored works that treat such themes as the history of science and philosophy, including a book on Aristotle's conception of change.

Manfred D. Laubichler (M.D.L.) is a professor of theoretical biology and history of biology in the School of Life Sciences at Arizona State University. He is coeditor of *From Embryology to Evo-Devo* (with Jane Maienschein, 2007) and *Modeling Biology* (with Gerd Müller, 2007) and serves as associate editor of *Biological Theory* and the *Journal of Experimental Zoology*.

Antonio Lazcano is professor of origins of life at the Facultad de Ciencias of the National University of Mexico. He is author of *Origen de la Vida,* which has sold over 750,000 copies. He is current president of the International Society for the Study of the Origins of Life.

James G. Lennox (J.G.L.) is professor of history and philosophy of science at the University of Pittsburgh. He has written extensively on Aristotle and Darwin, including *Aristotle: On the Parts of Animals I–IV* (2001), a translation with commentary of Aristotle's seminal work, and *Aristotle's Philosophy of Biology* (2001).

Richard E. Lenski (R.E.L.) is the John A. Hannah Distinguished Professor of Microbial Ecology at Michigan State University and a member of the National Academy of Sciences. His research focuses on the dynamics of phenotypic and genomic evolution in experimental populations of bacteria.

Wen-Hsiung Li (W.-H.L.) is James Watson Professor in the Department of Ecology and Evolution at the University of Chicago. His research interests include the evolution of gene regulation, gene duplication and its evolutionary consequences, and the development of statistical methods and computational tools for genomic data.

Curt Lively (C.L.) is a professor of biology at Indiana University. He has worked on predator-prey, plant-pollinator, and host-parasite interactions. He has also contributed to theoretical and empirical studies on the evolutionary stability of sexual reproduction.

David N. Livingstone is professor of geography and intellectual history at Queen's University, Belfast, Northern Ireland. He is the author of many books and articles on science and religion, most recently *Adam's Ancestors: Race, Religion, and the Politics of Human Origins* (2008).

Trudy Mackay (T.M.) is a professor of genetics at North Carolina State University. Her research focuses on the molecular genetic basis of variation for quantitative traits.

Jane Maienschein is Regents' Professor, President's Professor, and Parents Association Professor at Arizona State University, where she directs the Center for Biology and Society. She is the author and editor of more than a dozen books and many articles, including *Defining Biology* (1986) and *Whose View of Life: Embryos, Cloning, and Stem Cells* (2003).

Lynn Margulis (L.M.), Distinguished University Professor, Department of Geosciences, University of Massachusetts, Amherst, received the Presidential Medal of Science from President William J. Clinton in 1999. A member of the National Academy of Sciences she investigates and publishes on evolution (by symbiogenesis) and the Gaia hypothesis (of J. E. Lovelock).

Henry M. McHenry is a professor of anthropology at the University of California, Davis. He is the author of numerous studies of human fossil remains that interpret the evolutionary relationships among extinct species and the evolution of bipedal walking, brain-size increase, sexual dimorphism, and diet.

Daniel W. McShea (D.W.M.) is a paleobiologist in the Biology Department at Duke University. His research has focused the apparent trend in complexity at the largest scale, in all of life, over its entire history. He is coauthor of a recent textbook, *Philosophy of Biology: A Contemporary Introduction* (2008).

Paolo Menozzi (P.M.) is professor of ecology and chairman of the Department of Environmental Sciences at the University of Parma, Italy. He worked for many years on human population genetics and is coauthor of *The History and Geography of Human Genes* (1994). His current research focuses on genetic and ecological aspects of the evolutionary biology of organisms as different as zooplankton and forest trees.

Mike Mesterton-Gibbons (M.M.-G.) is a professor of mathematics at Florida State University, where he directs research on complex games among humans and other animals. He is the author of two books on mathematical modeling and numerous articles on evolutionary game theory.

James Murray (J.M.) is professor of biology emeritus at the University of Virginia and sometime director of the Mountain Lake Biological Station. He is the author of *Genetic Diversity and Natural Selection* (1972) and many papers on the evolution and population biology of polymorphic land snails.

Mohamed A. F. Noor (M.A.F.N.) is a professor of biology at Duke University. He is broadly interested in various themes in evolutionary genetics and genomics, but his principal focus has been on the genetics underlying and evolutionary forces driving changes leading to species formation, focusing on *Drosophila* species as model systems. One recurrent theme in his research has been the role of restricting recombination in facilitating species persistence.

Benjamin B. Normark (B.B.N.) is an associate professor of plant, soil, and insect sciences and a member of the graduate program in organismic and evolutionary biology at the University of Massachusetts, Amherst. He studies the systematics of armored scale insects and the evolution of unusual genetic systems such as haplodiploidy and parthenogenesis.

Ronald L. Numbers (R.L.N.) is Hilldale Professor of the History of Science and Medicine at the University of Wisconsin–Madison. Among his many publications are *Darwinism Comes to America* (1998) and *The Creationists* (2006). A past president of both the History of Science Society and the American Society of Church History, he is currently president of the International Union of History and Philosophy of Science.

Robert Olby (R.O.) is a historian of biology and research professor in the Department of the History and Philosophy of Science at the University of Pittsburgh. His work has been principally on the history of genetics and molecular biology. He has recently completed a biography of the late Dr. Francis Crick.

Kevin Padian (K.P.) is professor of integrative biology and curator in the Museum of Paleontology, University of California, Berkeley, and president of the National Center for Science Education.

Diane B. Paul (D.B.P.) is professor emerita of political science at the University of Massachusetts, Boston, and research associate at the Museum of Comparative Zoology at Harvard University. She has published widely on historical and policy issues in genetics. Her books include *Controlling Human Heredity: 1865 to the Present* (1995, 1998) and *The Politics of Heredity: Essays on Eugenics, Biomedicine, and the Nature-Nurture Debate* (1998).

Patricia M. Princehouse (P.M.P.) teaches evolutionary biology and history and philosophy of science at Case Western Reserve University. She received the National Center for Science Education's "Friend of Darwin" award for her activities in the Intelligent Design controversy and helped advise the legal team in *Kitzmiller v. Dover School Board*. She is the author of many articles on evolution and is currently working on a book, *Stephen Jay Gould's View of Life*.

Margaret B. Ptacek is a professor in the Department of Biological Sciences at Clemson University. Her interests lie at the intersection of behavioral ecology, population genetics, and speciation. Specifically, she is interested in processes that control genetic divergence among populations and the contributions of these processes to local adaptation and speciation.

Tom Ray (T.R.) is professor of zoology and adjunct professor of computer science at the University of Oklahoma. He has led research programs in tropical biology and rain forest conservation in Costa Rica and pioneered research on evolution in the digital medium. His current research uses genomic data to study the evolution of gene families and to explore new approaches for understanding the evolution and development of the human mind.

David N. Reznick (D.N.R.) is a professor of biology at the University of California, Riverside. He studies the process of natural selection and the evolution of complex traits. His primary study organisms are fish in the family Poeciliidae. He has published over 100 research articles and reviews in academic journals and edited volumes and is currently writing *A Reader's Guide to the Origin of Species*.

Richard A. Richards (R.A.R.) is associate professor of philosophy at the University of Alabama. He writes on Darwin and philosophical topics in biological systematics.

Robert J. Richards (R.J.R.) is the Morris Fishbein Professor of History of Science at University of Chicago and professor in the departments of history, philosophy, and psychology. Most recently, he is the author of *The Tragic Sense of Life: Ernst Haeckel and the Struggle over Evolutionary Thought* (2008).

Fredrik Ronquist (F.R.) is professor and head of the Department of Entomology at the Swedish Museum of Natural History. He is interested in computational phylogenetics, particularly the development of parsimony methods for the study of coevolution and biogeography and, more recently, Bayesian methods for phylogenetic inference.

Daniel I. Rubenstein is the Class of 1877 Professor of Zoology, chair of the Department of Ecology and Evolutionary Biology, and director of the Program in African Studies at Princeton University. He studies the social behavior of equids, the movement ecology of zebras, and human-wildlife interactions in Africa.

David W. Rudge (D.W.R.) is an associate professor in the Department of Biological Sciences at Western Michigan University with a secondary appointment in the Mallinson Institute for Science Education. He is the author of several articles on H. B. D. Kettlewell's classic investigations of natural selection in the peppered moth, which he has examined from historical, philosophical, and science education perspectives.

Nicolaas A. Rupke (N.A.R.) is a professor of the history of science and director of the Institute for the History of Science at Göttingen University. Among his publications is a monograph on Richard Owen (1994; new edition, 2008).

Michael Ruse (M.R.) is the Lucyle T. Werkmeister Professor of Philosophy and director of the Program in the History and Philosophy of Science at Florida State University. He is the author of many books on evolutionary themes, including *The Darwinian Revolution: Science Red in Tooth and Claw* (1979), *Monad to Man: The Concept of Progress in Evolutionary Biology* (1996), and *Can a Darwinian Be a Christian?* (2001).

Dorion Sagan (D.Sa.), partner of *Sciencewriters*, has authored or coauthored many articles, poems, and over 15 books translated into 11 languages. These include *Notes from the Holocene* (2007), *Up from Dragons: The Evolution of Human Intelligence* (with John Skoyles, 2002), and *Into the Cool: Energy Flow, Thermodynamics and Life* (with Eric D. Schneider, 2005). He studied art, literature, and philosophy, mainly at the University of Massachusetts, Amherst, where he graduated with a BA in history.

James William Schopf (J.W.S.) is distinguished professor of paleobiology in the Department of Earth and Space Sciences and Director of the IGPP Center for the Study of Evolution and the Origin of Life at the University of California, Los Angeles. A member of the National Academy of Sciences and a prize-winning author, his interests focus on the early (Precambrian) history of life.

Eugenie C. Scott is executive director of the National Center for Science Education (NCSE) in Oakland, California. A not-for-profit membership organization, NCSE promotes evolution education. Scott, a physical anthropologist, is the author of *Evolution vs. Creationism: An Introduction* (2004, 2008) and coeditor (with Glenn Branch) of *Not in Our Classrooms: Why Intelligent Design Is Wrong for Our Schools* (2006).

Jon Seger (J.S.) is a professor of biology at the University of Utah. He currently works mainly on the evolutionary genetics of whale lice (Amphipoda, Cyamidae), whose simple population histories and ecologies allow them to be interpreted as unusually well defined evolutionary experiments.

Ullica Segerstrale (U.S.) is a professor of sociology and the director of the Camras Scholars Program at the Illinois Institute of Technology, Chicago. She has written on issues at the intersection of science, values, and human nature. Her books include *Defenders of the Truth: The Sociobiology Debate* (2000) and the forthcoming *Nature's Oracle: A Life of W. D. Hamilton*.

David Sepkoski (D.Se.) is an assistant professor of history at the University of North Carolina, Wilmington. He specializes in the history of paleontology and evolutionary theory. He is editor, with Michael Ruse, of *The Paleobiological Revolution: Essays on the History of Recent Paleontology* (2009) and he is writing a book about the history of paleobiology.

H. Bradley Shaffer (H.B.S.) is a professor at the University of California, Davis. He works on the evolutionary biology, ecology, and conservation biology of amphibians and reptiles. Recent research projects include comparative phylogeography of amphibians and reptiles in California and the central United States, systematics of freshwater turtles and tortoises both globally and in eastern Australia, and conservation genetics of endangered California amphibians and reptiles.

Daniel Simberloff (D.Si.) is the Nancy Gore Hunger Professor of Environmental Studies and director of the Institute for Biological Invasions at the University of Tennessee. He is an evolutionary ecologist and conservation biologist who studies a range of topics that includes the ecology of invasive species and the susceptibility of communities and ecosystems to invasion by exotics.

Robert A. Skipper Jr. (R.A.S.) is associate professor of philosophy and affiliate of the Center for Environmental Studies at the University of Cincinnati. His work focuses mainly on the conceptual and historical foundations of evolutionary genetics, particularly the dynamics of controversies within the field.

Vassiliki Betty Smocovitis (V.B.S.) is professor of the history of science in the Departments of Zoology and History at the University of Florida. She is also affiliated with the Botany department. She is the author of *Unifying Biology: The Evolutionary Synthesis and Evolutionary Biology* (1996).

Trisha Spears (T.S.) has held research and instructional positions in the Department of Biological Science at Florida State University since 1989. Her research interest is in the area of crustacean molecular systematics and evolution, whereby DNA evidence is used to investigate evolutionary relationships among crustaceans, as well as the genetic variation among populations of crustacean species. She is also a past president of the Crustacean Society, an international society of carcinologists.

Kim Sterelny is a professor of philosophy at the Australian National University and the Victoria University of Wellington. He has written extensively on the intersection of philosophy and the life sciences, with a particular interest in the relationship between microevolution and macroevolution, and the evolution of human cognition and behavior.

James E. Strick (J.E.S.) is associate professor in the Department of Earth and Environment and chair of the Program in Science, Technology and Society at Franklin and Marshall College in Lancaster, Pennsylvania. Originally trained in microbiology, and later in history of science, he has published extensively on the history of ideas and experiments about the origin of life, including *Sparks of Life: Darwinism and the Victorian Debates over Spontaneous Generation* (2000) and, with Steven Dick, *The Living Universe: NASA and the Development of Astrobiology* (2004).

Charles E. Taylor (C.E.T.) is professor of ecology and evolutionary biology at the University of California, Los Angeles. He directs two research programs, one in artificial life and the evolution of self-organization and emergent behaviors and one on population genomics of *Anopheles gambiae* and the factors that determine local variation in transmission rates of malaria, using computer models and empirical studies.

Roger D. K. Thomas (R.D.K.T) is the John Williamson Nevin Professor of Geosciences at Franklin and Marshall College in Lancaster, Pennsylvania. He is a paleontologist, with interests in functional and constructional morphology, the design of animal skeletons, evolutionary theory, and the history of science. His assessment of Stephen Jay Gould's contributions to functional morphology can be found in the collection *Stephen Jay Gould: Reflections on His View of Life* (2009).

Joseph Travis (J.T.) is the Robert O. Lawton Distinguished Professor of Biological Science and dean of the College of Arts & Sciences at Florida State University. His research in evolutionary biology follows the traditional paradigm of ecological genetics, in particular examining the causes and consequences of population variation in the life histories and mating systems of livebearing fishes.

Edgar E. Vallejo (E.E.V.) is an associate professor of computer science at Instituto Tecnologico y de Estudios Superiores de Monterrey, Campus Estado de Mexico. He is the author of many papers on evolutionary computing themes, including genetic programming, evolutionary robotics, and simulations on the origin and evolution of language.

David B. Wake (D.B.W.) is a professor of integrative biology and curator in the Museum of Vertebrate Zoology at the University of California, Berkeley. An evolutionary biologist, he has published many research papers on amphibians, especially salamanders, using morphological, developmental, genetic, phylogenetic, ecological, and systematic data. The plight of amphibians, many of which are at high risk of extinction, is a current focus.

Mary Jane West-Eberhard (M.J.W.-E.) is senior scientist at the Smithsonian Tropical Research Institute. She has written extensively on the behavior and evolution of social insects, sexual selection, and speciation-related divergence and is the author of *Developmental Plasticity and Evolution* (2003).

Paul B. Wignall (P.B.W.) is a professor of paleoenvironments and the director of the Institute of Geological Sciences at the University of Leeds, United Kingdom. He is the author of many papers on the theme of mass extinctions and their causes, especially the end-Permian event.

David Sloan Wilson (D.S.W.) is professor of biology and anthropology at Binghamton University (State University of New York). He applies evolutionary theory to all aspects of humanity in addition to the rest of life. He is author of numerous books, including *Evolution for Everyone: How Darwin's Theory Can Change the Way We Think about Our Lives* (2007). He is known for championing the theory of multilevel selection, which explains how adaptations can evolve at all levels of the biological hierarchy, from genes to ecosystems.

Mary P. Winsor (M.P.W.) is emerita professor in the Institute for the History and Philosophy of Science and Technology at the University of Toronto. Her book *Starfish, Jellyfish, and the Order of Life* (1976) describes the classification of radially symmetrical animals in the first half of the nineteenth century. She traces the history of Harvard's Museum of Comparative Zoology in her book *Reading the Shape of Nature* (1991).

Gregory A. Wray is a professor of biology and director of the Center for Evolutionary Genomics at Duke University. His research interests include the evolution of development, genomes, and gene networks.

Jason Zinser (J.Z.) is a visiting assistant professor of philosophy at the University of North Florida. His dissertation centered on theoretical morphology and the concept of contingency in evolutionary theory.

Marlene Zuk (M.Z.) is a professor of biology and the associate vice provost for faculty equity and diversity at the University of California, Riverside. She studies sexual selection and host-parasite evolution in a variety of organisms, and did her PhD with W. D. Hamilton at the University of Michigan.

Illustration Credits

Page 9: By permission of English Heritage.

Page 10: By permission of estate of John Chancellor.

Page 12, bottom: MS.DAR 121, p. 36, drawn by Charles Darwin, and reproduced by kind permission of the Syndics of Cambridge University Library.

Page 26: Photo by Joseph Cain; used with permission.

Page 33: From *Dobzhansky's Genetics of Natural Populations*, ed. R. C. Lewontin et al. (New York: Columbia University Press, 1981); by permission of the publisher and the estate of Theodosius Dobzhansky.

Page 34: By permission of *Journal of Genetics*.

Page 36: By permission of Wiley-Blackwell.

Page 40, bottom: By permission of Macmillan Publishers Ltd.: Nature Publishing Group.

Page 41: Photo by Emily Ruse; used with permission.

Page 55: Figure by Jeffrey L. Bada and Antonio Lazcano.

Page 98: By permission of Alan Cheetham.

Page 118: By permission of Zhe-Xi Luo.

Page 122: Photos by Doug Schemske; used with permission.

Page 190: By permission of Macmillan Publishers Ltd.: Nature Publishing Group.

Page 202: By permission of *American Naturalist*.

Page 220: Images by FlyBase; used with permission.

Page 413: Photo by Joseph Travis.

Page 417: We have attempted in good faith, without success, to trace the copyright on this picture. If copyright is claimed, it will be acknowledged in subsequent editions of this work.

Page 421: By permission of the Berlin Museum.

Page 458: By permission of Jean Gayon.

Page 461: By permission of the Smithsonian Institution.

Page 492: By permission of the author, Gary C. B. Poore, and Museum Victoria.

Page 520: Redrawn with permission from Hayden-McNeil Publishing, Inc.

Page 527: Photo by Jarmo Halopainen; used with permission.

Page 556: By permission of Springer Science + Business Media.

Page 596: Photo by Edward G. R. Ruse; used with permission.

Page 635: By permission of Kirk Smith.

Page 653: Photo by Bruce Grant; used with permission.

Page 695: By permission of Cleveland Museum of Natural History.

Page 726: By permission of Museum of Comparative Zoology, Harvard University.

Page 741: Photo by Joseph Travis.

Page 760, bottom: By permission of Riccardo Levi-Setti.

Page 761: By permission of Macmillan Publishers Ltd.: Nature Publishing Group.

Pages 777, 778, 779, 780, 781, 783, 785: Photos by author, W. Schopf.

Page 826: By permission of Society of Systematic Biologists.

Page 851: Photo by Joseph Travis.

Pages 860 and 899: We have attempted in good faith, without success, to trace the copyright on these figures. If copyright is claimed, it will be acknowledged in subsequent editions of this work.

Page 907: By permission of Macmillan Publishers Ltd.: Nature Publishing Group.

We would like to thank Martin Young of Tallahassee, our artist, for working so diligently on the illustrations, drawing and redrawing many and making sure that all were of a quality that could be reproduced.

Index

selection, 17, 18*f*, 19, 21*f*; Hennig and, 632–634; Lamarck and, 672–673; Linnaeus and, 686–688, 875–876; Mendel and, 732–733. *See also* Phylogenetics

Clausen, Jens, **483–486**, 532, 852, 872

Clements, Frederic, 484

Cleveland, Lemuel R., 540

Clift, William, 795

Climate change, 415, 816–817, 817*f*; Pleistocene extinctions and, 814–816. *See also* Conservation

Climbing Mount Improbable (Dawkins), 327

Cline, 646

Clubmosses, 690–691

Cluster analysis, molecular evolution study and, 141–142

Clutton-Brock, T. H., 37, 245

Coat color, of rats, 469, 470*f*

Coelacanth, 689

Cognitive ethology, 543

Cohesion Species Concept, 179, 180*t*

Cole, B. J., 307–308

Cole, Lamont, 567, 568, 569

Cole's paradox, 568–569

Colonial living, in history of life theories, 83

Comets. *See* Exobiology

Comparative genomics, 789–790

Compatibilist naturalists, 321

Competitive exclusion, 671

Complexity, evolutionary progress and, 555–556, 556*f*

Complex specified information (CSI), 378, 379

Complex traits, adaptation controversies, 115–116; floral evolution in monkey flowers, 121–124, 122*f*; gill arches, ear ossicles, and jaws, 116–121, 118–119*f*

Compsognathgus, 519

Conan Doyle, Arthur, 689–690

Conery, John, 701

Conflict, social behavior and, 239–240

Conklin, Edwin, 36*f*

Consanguinuity, Inbreeding, and Genetic Drift (Cavalli-Sforza), 472

Consciousness, in history of life theories, 82

Conservation: Wilson and, 925–926; Yablokov and, 930–931. *See also* Climate change

Consilience (E. O. Wilson), 925

Consilience of inductions, 771, 916

Consilience of Inductions (Whewell), 916

Continental drift, theory of, 860, 904

Continuity hypothesis, of language evolution, 565

Contrived dualism, 384, 395n1

Convergence, 633

Convergent adaptation, 109

Conway Morris, Simon, 84, 462*f*

Cook, Laurence M., 588

Cooper, Clarence, 391, 394

Cope, Edward Drinker, 462*f*, **486–487**, 487*f*, 519, 713, 793

Cope's rule, 661, 832

Copleston, Edward, 698

Cornell University, 351, 352

"Correlation and Causation" (S. Wright), 928

Correns, Carl, 896

"Cost of Natural Selection, The" (Haldane), 629

Couder, Y., 301–305

Covington, Syms, 348

Coyne, George V., 356

Coyne, J. A., 529, 743

Crabs, 801–802, 803*f*, 911–912, 911*f*

Cracraft, Joel, 179

Creation, The (E. O. Wilson), 866, 925

Creationism: Arkansas Creation Trial, 428–429, 614; attacks on Darwin, 514–515; Dobzhansky and, 522; Gray and, 617. *See also* Antievolutionism; Scientific creationism

Creation science. *See* Scientific creationism

Crick, Francis, 734, 787, **906–909**

Critique of the Theory of Evolution, A (T. H. Morgan), 750

Crohn's disease, 440

Croizat, Leon, 450

Crompton, Arthur, 859

Crook, J. H., 238–239

Crow, James Franklin, **488–490**, 668, 675

Crustaceans, **490–499**; distinguishing features, 490; diversity, 492*f*, 499; evolutionary research, 494–499; social behavior, 491, 493

Cultural evolution, 727–728. *See also* Social behavior

Cultural renewal, antievolutionism and, 382–384